Contributions to Algebraic Geometry

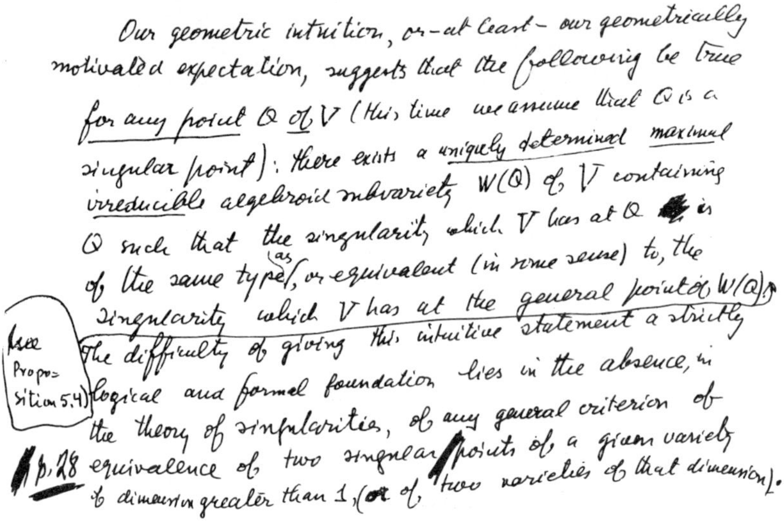
Our geometric intuition, or—at least—our geometrically motivated expectation, suggests that the following be true for any point Q of V (this time we assume that Q is a singular point): there exists a uniquely determined maximal irreducible algebroid subvariety W(Q) of V containing Q such that the singularity which V has at Q is of the same type as, or equivalent (in some sense) to, the singularity which V has at the general point of W(Q). (see Proposition 5.4) The difficulty of giving this intuitive statement a strictly logical and formal foundation lies in the absence, in the theory of singularities, of any general criterion of equivalence of two singular points of a given variety of dimension greater than 1, (or of two varieties of that dimension). p. 28

A section of Zariski's manuscript for his paper "Foundations of a general theory of equisingularity on r-dimensional algebroid. . ." The text appears on page 453.

CONTRIBUTIONS TO ALGEBRAIC GEOMETRY

In Honor of Oscar Zariski

Edited by Michael Artin and David Mumford
with the cooperation of the Editors of the
American Journal of Mathematics

The Johns Hopkins University Press
Baltimore and London

Originally published as Volume 101, Numbers 1 and 2 of the *American Journal of Mathematics,* 1979.

Manufactured in the United States of America

The Johns Hopkins University Press, Baltimore, Maryland 21218
The Johns Hopkins University Press Ltd., London

Library of Congress Catalog Card Number 79-88169
ISBN 0-8018-2307-2

DEDICATION

This book is dedicated to Oscar Zariski with affection and respect by the editors and contributors. We are very pleased to be able to include Professor Zariski's latest work with these papers, and we join together to wish our good friend a very happy 80th birthday.

The contributors would like to express their sincere appreciation to the editors of the *American Journal of Mathematics* for making this book possible.

Dear Oscar,
Happy Birthday
from all of us.

Affectionately,

Photograph Courtesy of Nan White

Oscar Zariski

CONTENTS

THE HÖLDER CONTINUITY OF QUASICONFORMAL DEFORMATIONS

By Lars V. Ahlfors*

A. Deformations of R^n.

A1. We denote the Jacobian matrix of a mapping $f\colon R^n \to R^n$ by Df and its elements by $Df_{ij} = D_j f_i$. The differential operator S defined by

$$Sf_{ij} = \frac{1}{2}(Df_{ij} + Df_{ji}) - \frac{1}{n}\delta_{ij} \sum_k Df_{kk} \tag{1}$$

arises naturally by linearization of the matrix $MF = (\det DF)^{-2/n}\, {}^t Df \cdot Df$ which measures the deviation of a homeomorphism F from conformality.

Apart from regularity conditions f is called a quasiconformal (q.c.) deformation of R^n if the square norm of Sf, defined by $\|Sf\|^2 = \Sigma_{ij}(Sf_{ij})^2$ is bounded. The relation between q.c. deformations and q.c. homeomorphisms is still obscure, but the theory of q.c. deformations is interesting in its own right.

A2. The matrix Sf is symmetric with zero trace; the linear space of $n \times n$ matrices with this property will be denoted by SM_n. The operator S has a formal adjoint S^* which maps functions with values in SM_n on vector-valued functions. For any sufficiently regular $\phi\colon R^n \to SM_n$ we define $S^*\phi$ to be the vector-valued function with components

$$(S^*\phi)_i = \sum_j D_j \phi_{ij}.$$

It follows from Stokes' formula that

$$\int \sum_{ij} Sf_{ij}\phi_{ij}\, dx = -\int \sum_i f_i (S^*\phi)_i\, dx \tag{2}$$

---

*Supported by NSF Grant MCS 77-07782.

if either f or ϕ has compact support. If ϕ is a test-function (C^∞ with compact support), then (2) defines Sf as a distribution. The integrals are over the whole space R^n, and dx refers to Lebesgue measure.

In what follows we shall assume that f is continuous and Sf locally integrable and such that (2) is satisfied for all test-functions ϕ. This determines Sf uniquely, and if $f \in C^1$ it is given by (1). We do not need to assume that the individual derivatives Df_{ij} are locally integrable.

A3. In terms of the above agreement f will be called a q.c. deformation of R^n as soon as $\|Sf\| \in L^\infty$, or $Sf \in L^\infty$ in simpler notation. In this paper we also consider a slightly more general condition, namely $Sf \in L^p$ with $p > n$. We shall show that this is sufficient to imply that f satisfies a Hölder condition. When $Sf \in L^p$ we say that f is L^p-q.c. and we denote the L^p-norm of the square norm $\|Sf\|$ by $\|Sf\|_p$.

THEOREM A. *If f is L^p-q.c. with $p > n$, then f satisfies a Hölder condition of the form*

$$|f(x) - f(y)| \le M|x - y|^{1-n/p} \tag{3}$$

for all x, y in a given compact set. If f has compact support, then (3) holds for all x, y and $M = C_{n,p}\|Sf\|_p$ where $C_{n,p}$ depends only on n and p.

For $p = \infty$ the inequality is replaced by

$$|f(x) - f(y)| \le M|x - y|(1 + \log^+ 1/|x - y|) \tag{4}$$

in the general case, and if the support of f is contained in $\{|x| < R\}$ by

$$|f(x) - f(y)| \le M|x - y| \log 3R/|x - y| \tag{4'}$$

with $M = C_{n,p}\|Sf\|_\infty$.

A4. It is sufficient to prove the theorem for the case of compact support. To see this, let λ be a scalar C^∞ function, and replace ϕ by $\lambda\phi$ in (2). One obtains

$$\int \sum_{ij} (\lambda Sf_{ij} + f_i D_j \lambda)\phi_{ij}\, dx = -\int \sum_i \lambda f_i (S^*\phi)_i\, dx,$$

which shows that $S(\lambda f)_{ij} = \lambda Sf_{ij} + \frac{1}{2}(f_i D_j\lambda + f_j D_i\lambda) - (1/n)\,\delta_{ij}$

$\Sigma_k f_k D_k \lambda$. Choose λ with compact support and identically 1 on the given compact set. Then λf has compact support and $S(\lambda f) \in L^p$. Thus the general theorem follows from the compact case.

It is also enough to prove the theorem for $f \in C^\infty$. This is shown by very familiar use of mollifiers.

A5. We assume now that f is C^∞ with compact support. Recall that the SM_n-valued functions $\gamma_{..,k}$, $k = 1, \ldots, n$, with elements

$$\gamma_{ij,k}(x) = |x|^{-n}(\delta_{ik}x_j + \delta_{jk}x_i - \delta_{ij}x_k) + (n-2)|x|^{-n-2}x_i x_j x_k$$

satisfy $S^*\gamma_{..,k}(x) = 0$ for $x \neq 0$. As a consequence Stokes' theorem yields

$$\int_{|x-y|>r} \sum_{ij} Sf(x)_{ij}\gamma_{ij,k}(x-y)\,dx$$

$$= -\int_{|x-y|=r} \sum_{ij} f_i(x)(x_j - y_j)r^{-1}\gamma_{ij,k}(x-y)\,d\sigma(x)$$

where the integral on the right is with respect to $(n-1)$-dimensional Lebesgue measure. In the limit for $r \to 0$ $f_i(x)$ can be replaced by $f_i(y)$, and by homogeneity the remaining integral equals

$$\int_{|x|=1} \sum_j x_j\gamma_{ij,k}(x)\,d\sigma(x) = \delta_{ik}c_n$$

with $c_n = [2(n-1)/n]\omega_n$ (ω_n is the measure of S^{n-1}). This gives the formula

$$f_k(y) = -\frac{1}{c_n}\int \sum_{ij} Sf(x)_{ij}\gamma_{ij,k}(x-y)\,dx. \tag{5}$$

We apply (5) to the difference $f(y) - f(0)$ and obtain by Hölder's inequality

$$|f_k(y) - f_k(0)| \leq \frac{1}{c_n}\|Sf\|_p \|\gamma_{..,k}(x-y) - \gamma_{..,k}(x)\|_q \tag{6}$$

where $p^{-1} + q^{-1} = 1$. It is to be observed that the integral

$$\int \|\gamma_{\ldots,k}(x - y) - \gamma_{\ldots,k}(x)\|^q \, dx \tag{7}$$

converges. In fact the integrand has the order $|x|^{-(n-1)q}$ and $|x - y|^{-(n-1)q}$ at the finite singular points and $|x|^{-nq}$ at ∞. The assumption $p > n$ implies $(n - 1)q < n$ while $nq > n$ for any finite p; this is enough to guarantee convergence.

If we replace x by $|y|x$ in (7) the integral becomes, in view of the homogeneous nature of the integrand,

$$|y|^{n-(n-1)q} \int \left\|\gamma_{\ldots,k}\left(x - \frac{y}{|y|}\right) - \gamma_{\ldots,k}(x)\right\|^q dx.$$

Here the integral is a constant, and after extracting the qth root the factor in front is $|y|^{1-n/p}$. In other words, (6) is of the form (3) with M a fixed multiple of $\|Sf\|_p$. To obtain (3) for arbitrary x and y one applies (6) to $f(y - x)$.

Suppose now that $p = \infty$ and that $f(x) = 0$ for $|x| > R$. Then (6) can be replaced by

$$|f_k(y) - f_k(0)| \leq \frac{1}{c_n} \|Sf\|_\infty \int_{|x|<R} \|\gamma_{\ldots,k}(x - y) - \gamma_{\ldots,k}(x)\| \, dx.$$

Homogenization replaces the integral by

$$|y| \int_{|x|<R/|y|} \left\|\gamma_{\ldots,k}\left(x - \frac{y}{|y|}\right) - \gamma_{\ldots,k}(x)\right\| dx.$$

Here the integrand is of the order $|x|^{-n}$ at ∞, and the whole is hence majorized by a multiple of $|y| \log R/|y|$ for large R, and hence for instance by a multiple of $|y| \log 3R/2|y|$ for all R and $|y| < R$. Because the result has to be applied to $f(y - x)$ it is necessary to replace R by $2R$, and (4′) is proved.

B. Deformations of the Unit Ball.

B1. The definition of q.c. deformations adopted in A2 and A3 extends without significant change to functions f defined on an arbitrary open set $\Omega \subset R^n$. The only difference is that one requires (2) to hold only for test-functions ϕ with compact support in Ω. A deformation is L^p-q.c. if $\|Sf\|$ has finite norm in $L^p(\Omega)$. Theorem A remains valid as it stands. Indeed, if f has compact support it can be extended to R^n by setting $f = 0$ outside of Ω, and even if it does not have compact support we can apply the local aspect of the theorem to λf for a suitable choice of λ. Because this application is so obvious we have preferred to formulate Theorem A only for R^n.

B2. A more important question is whether $Sf \in L^p(\Omega)$ implies an inequality of the form (3) or (4) with fixed M for all $x, y \in \Omega$ even when f does not have compact support. It would then follow that f has a continuous extension to the closure of Ω, and it would also make sense to ask whether f can be extended to an L^p-q.c. deformation of R^n.

We shall study this question only in the special case where Ω is the unit ball $B = B^n$, and even in that case we shall consider only deformations *of* B rather than *on* B. In order to explain this distinction, let f be a q.c. deformation of R^n, and define $F(x, t)$ for real t as the solution of the differential equation $\partial F(x, t)/\partial t = f(F(x, t))$ with initial value $F(x, 0) = x$. It is known ([5], [9]) that the solutions are unique and that the $F(x, t)$ form a one parameter group of q.c. homeomorphisms. Moreover, as would be expected, B remains fixed under this group if and only if the vectors $f(x)$ with $|x| = 1$ are tangent to S^{n-1}. It is therefore natural to say that a vector-valued function f, defined and continuous on the closed unit ball, is a deformation *of* B only if $\sum_i x_i f_i(x) = 0$ (or $x \cdot f = 0$) for all $x \in S^{n-1}$.

B3. As formulated above the definition requires f to be continuous on the closure of B, whereas we are interested in conditions that imply this continuity. Our procedure will therefore be to assume that f is given and continuous on B, to define a standard extension f^* to the exterior B^* of B, and then impose conditions which guarantee the existence of a continuous function on R^n which agrees with f on B and with f^* on B^*.

The standard extension is by symmetrization. The symmetric point of x with respect to S^{n-1} is $x^* = x/|x|^2$. The Jacobian matrix Dx^* has the elements $Dx^*_{ij} = |x|^{-2}(\delta_{ij} - 2x_i x_j/|x|^2)$. The author finds it convenient to use a special notation $Q(x)$ for the matrix with elements $x_i x_j/|x|^2$. In this notation $Dx^* = |x|^{-2}(1_n - 2Q(x))$ where 1_n is the unit matrix. It is immediate that $Q(x)^2 = Q(x)$ and $(1_n - 2Q(x))^2 = 1_n$, and hence $1_n - 2Q(x) \in O(n)$.

The group of Möbius transformations is generated by all similarities together with the inversion $Jx = x^*$. Given a vector-valued function f one defines its "pullback" A^*f with respect to a Möbius transformation A by $A^*f(x) = (DA)^{-1}f(Ax)$. Similarly, if ϕ has values in SM_n we define $A^*\phi(x) = (DA)^{-1}\phi(Ax)DA$. Observe that $A^*\phi(x)$ has again values in SM_n because DA is a scalar multiple of an orthogonal matrix. It is a less obvious but important fact that the conformality also implies $SA^*f = A^*Sf$ [4].

The symmetric extension of f to B^* is defined by $f^*(x) = J^*f(x)$. This means that $f^*(x) = |x|^2(1_n - 2Q(x))f(x^*)$. From $SJ^* = J^*S$ it follows that $Sf^*(x) = (1_n - 2Q(x))Sf(x^*)(1 - 2Q(x))$, and in particular $\|Sf^*(x)\| = \|Sf(x^*)\|$. In case f remains continuous and satisfies the orthogonality relation $x \cdot f = 0$ on S^{n-1}, then $Q(x)f(x) = 0$ and consequently $f^*(x) = f(x)$ for $|x| = 1$.

B4. After these preliminaries we formulate and prove

THEOREM B. Suppose that f is L^p-q.c. in B^n, $n < p \leq \infty$. In order that f and f^* have the same continuous boundary values on S^{n-1} it is necessary and sufficient that

(i) $\lim x \cdot f(x) = 0$
(ii) $\lim (1 - |x|)|f(x)| = 0$

as $|x| \to 1$.

When this is so, then f satisfies

$$\begin{aligned} |f(x) - f(y)| &\leq M|x - y|^{1-n/p} && \text{if } p < \infty \\ |f(x) - f(y)| &\leq M|x - y|(1 + \log 2/|x - y|) && \text{if } p = \infty \end{aligned} \tag{8}$$

for all $x, y \in B^n$ and $M = C_{n,p}(\|Sf\|_p + \max|f(x)|)$.

The necessity of (i) and (ii) is obvious. To prove that these conditions are sufficient, let $\lambda \in C^\infty$ be identically 1 for $|x| \leq R$, $R > 1$, and identically 0 for $|x| \geq 2R$, say. Define f_λ to be λf in B and λf^* in B^*. We shall show that the representation (5) is applicable to f_λ in spite of the possible break of continuity on S^{n-1}.

We consider $y \in B$ and choose r so that $|y| < r < 1$ and $r^{-1} < R$. Stokes' theorem is applicable, if necessary with use of a mollifier, to f_λ and $\gamma_{..,k}(x - y)$ on the union of $\{|x| < r\}$ and $\{|x| > r^{-1}\}$. It leads to an expression for $f_k(y)$ as a volume integral, as in (5), together with boundary integrals over $S^{n-1}(r)$ and $S^{n-1}(r^{-1})$. Except for a common numerical factor the boundary integrals are

$$\int_{|x|=r} \sum_{ij} f_i(x)x_j r^{-1}\gamma_{ij,k}(x-y)\,d\sigma$$

and

$$\int_{|x|=1/r} \sum_{ij} f_i^*(x)x_j r\gamma_{ij,k}(x-y)\,d\sigma.$$

By a change of variable both integrals can be extended over S^{n-1}, and their difference becomes

$$\int_{S^{n-1}} \sum_{ij} [r^{n-1}f_i(rx)\gamma_{ij,k}(rx-y) - r^{1-n}f_i^*(r^{-1}x)\gamma_{ij,k}(r^{-1}x-y)]x_j\,d\sigma.$$

By definition $f^*(r^{-1}x) = r^2(1_n - 2Q(x))f(rx)$ and thus $f_i^*(r^{-1}x) = r^2f_i(rx) - 2r^2(x\cdot f(rx))x_i$. With this substitution the integral can be written as

$$\int_{S^{n-1}} \sum_{ij} f_i(rx)x_j[r^{n-1}\gamma_{ij,k}(rx-y) - r^{n+1}\gamma_{ij,k}(r^{-1}x-y)]\,d\sigma$$

$$+\, 2r^{n+1}\int_{S^{n-1}} (x\cdot f(rx)) \sum_{ij} x_ix_j\gamma_{ij,k}(r^{-1}x-y)\,d\sigma.$$

Under the conditions (i) and (ii) both terms tend to zero when $r \to 1$. We can therefore conclude that

$$f_k(y) = -\frac{1}{c_n}\int \sum Sf_\lambda(x)_{ij}\gamma_{ij,k}(x-y)\,dx.$$

As in A4 this implies (8) with $M = C\|Sf_\lambda\|_p$ if $p < \infty$ and with some M if $p = \infty$. In any case, f is continuous across S^{n-1}, and we have proved the first part of the theorem.

It remains to estimate $\|Sf_\lambda\|_p$. We know that $Sf_\lambda = Sf$ in B and $(Sf_\lambda)_{ij} = \lambda Sf_{ij}^* + \frac{1}{2}(f_i^*D_j\lambda + f_j^*D_i\lambda) - (1/n)\delta_{ij}\sum_k f_k^*D_k\lambda$ in B^*.

Here λ and its partial derivatives are bounded, and $|f^*(x)| = |x|^2|f(x^*)|$, $\|Sf^*(x)\| = \|Sf(x^*)\|$. It is therefore immediate that $\|Sf_\lambda\|_\infty \leq C(\|Sf\|_\infty + \max|f|)$, and the second inequality (8) is thus a consequence of (4′). The first inequality follows similarly when we observe that

$$\int \|Sf^*\|^p\, dx = \int \|Sf\|^p |x|^{-2n}\, dx,$$

the integral on the left being over $R < |x| < 2R$, the one on the right over $1/2R < |x| < 1/R$.

B5. Because a constant can be subtracted at will the maximum of $|f|$ can and perhaps should be replaced by $\max|f(x) - f(y)|$. The question arises whether this term can be absorbed in $\|Sf\|_p$. The answer is in the negative, for there are even functions f with $Sf = 0$ that satisfy (i)-(ii) and have arbitrarily large oscillation. These functions are in fact of the form $f_i = (1 - |x|^2)a_i + \sum_j b_{ij}x_j$ with $b_{ij} = -b_{ji}$, and there is no limitation of the coefficients.

Theorems A and B are a minor but useful detail in the theory of q.c. deformations. In part they have been used without complete proof in some of the references.

HARVARD UNIVERSITY

REFERENCES

[1] Lars V. Ahlfors, "Kleinsche Gruppen in der Ebene und im Raume. Festband zum 70 Geburtstag von Prof. Rolf Nevanlinna," Springer Verlag, Berlin, New York, 1966, pp. 7-15.

[2] ———, "Hyperbolic Motions," *Nagoya Math. J.* (1967), pp. 163-166.

[3] ———, "Conditions for quasiconformal deformations in several variables," Contributions to Analysis. A collection of papers dedicated to Lipman Bers, pp. 19-25, Academic Press, New York, London, 1974.

[4] ———, "Invariant Operators and Integral Representations in Hyperbolic Space," *Math Scand.* **36** (1975), pp. 27-43.

[5] ———, "Quasiconformal deformations and mappings in R^n; *Journal d'Analyse Mathématique* **30** (1976), pp. 74-97.

[6] ———, "On a class of quasiconformal mappings," *Sitz. Ber. d. Österr. Akad. d. Wiss., Math. Nat. Kl.*, II, **185.** Bd, Heft 1-3 (1976), pp. 5-10.

[7] ———, "A somewhat new approach to quasiconformal mappings in R^n." *Lecture Notes in Mathematics* **599** (1976), pp. 1-6.
[8] ———, "A singular integral equation connected with quasiconformal mappings in space." To appear in *Enseignments Mathématiques.*
[9] H. M. Reimann, "Ordinary differential equations and quasiconformal mappings," *Inventiones Mathematicae* **33** (1976), pp. 247-270.

COMPACTIFYING THE PICARD SCHEME II

By Allen B. Altman & Steven L. Kleiman*

Compactified jacobians of integral, projective curves have been constructed and studied for over twenty years by many authors using various methods. (A brief summary of the work is contained in [CP].) In [B*l*] we announced a theory modeled on Grothendieck's theory of the Picard scheme. In [CP] we presented the details, except for rigidification theory. Rigidification theory is done here. Our main purpose, however, is to construct a natural compactified Picard scheme for a flat, locally projective, finitely presentable family X/S of integral varieties of arbitrary dimension.

In Section 1, we recall from [CP] some basic definitions and properties, including those of the functor of simple sheaves $Spl_{X/S}$ and its subfunctors, $Pic_{X/S}$, $Pic_{X/S}^{-}$ and $Pic_{X/S}^{=}$, the functors of invertible sheaves, pseudo-invertible sheaves and torsion-free, rank-1 sheaves. We also do rigidification theory for these functors, generalizing Grothendieck's theory outlined in [FGA, p. 232-05, Rem. 2.5]. The basic result, reformulated in (1.6), is this: whenever X/S admits a section g, then a T-point of the *fpqc*-sheaf associated to $Spl_{X/S}$ which is "invertible along g" is representable by a T-simple sheaf on X_T. It follows, see (3.4, iii), that if the smooth locus of X/S admits a section g, then the minimal compactified Picard scheme parametrizes a universal sheaf (or Poincare sheaf). Moreover, by [CP, 5.17] the Abel map is then a fibration by projective spaces of varying dimension; more precisely, it is the structure map of a $\mathbf{P}(H)$ for a certain locally finitely presentable sheaf H.

In [CP, 7.9] we showed how a further variation of Grothendieck's construction of the Picard scheme yields an algebraic space $\mathrm{Pic}_{X/S}^{=}$ which compactifies the Picard scheme. It represents the etale sheaf associated to

---

*Both authors wish to thank the mathematics department of the California Institute of Technology for warm hospitality during the summers of 1977 and 1978. The second author is also pleased to thank the Universidad Simon Bolivar for round-trip airfare between Boston and Caracas and to thank the National Science Foundation (of the U.S.) for some financial support under grant no. 7701964-MCS.

$Pic_{X/S}^{=}$. Below we prove that $\text{Pic}_{X/S}^{=}$ is, in fact, a scheme and that the minimal compactified Picard scheme, which is defined as the closure of $\text{Pic}_{X/S}$ in $\text{Pic}_{X/S}^{=}$, is a disjoint union of schemes projective locally over S.

The construction of $\text{Pic}_{X/S}^{=}$ in [CP, 7.9] and that below differ notably at the point where it is necessary to form a quotient by a flat and finitely presentable, but not proper, equivalence relation. In [CP] we simply applied Artin's theorem. Below we adapt the method Mumford used in [CS, Lectures 19–21] to form a quotient to construct the Picard scheme of a smooth, projective surface over an algebraically closed field.

The heart of Mumford's method is a remarkable theory of "λ-independent" 0-cycles. In Section 2 we present the modifications necessary to generalize the theory suitably for a singular, projective variety X of arbitrary dimension. A main result (stated more precisely in (2.8), (2.10) and (2.11)) is this: Let U be a dense open set of X. Let θ be a polynomial and r an integer. Let N be a large integer. (The minimal possible N depends only on the Hilbert polynomial of X and on θ.) Set $\nu = rN - 1$. Then there exists a ν-tuple $\bar{x} = (x_1, \ldots, x_\nu)$ of closed points of U with the following property. Let I be a torsion-free, rank-1 sheaf that (1) has Hilbert polynomial θ, (2) is generated by its global sections, (3) satisfies the relation $r = h^0(X, I)$ and (4) is invertible on U. Let x^* be an arbitrary functional on $H^0(X, I)$ and let x_i^* denote the functionals of evaluation at x_i. Then the rN functionals $x^*, x_1^*, \ldots, x_\nu^*$ may be arranged to form N sets of bases of $H^0(X, I)^*$. This result implies that a key map,

$$u(\bar{x}): O_X^{\oplus \nu!} \to I \otimes_k M(\bar{x}),$$

which is defined in (2.3), is surjective. Here $M(\bar{x})$ is a certain canonical 1-dimensional vector space.

The final construction of $Pic_{X/S}^{=}$ is done in Section 3. The matter is local on S, so we may assume X/S is projective. Fix a relatively very ample sheaf $O_X(1)$. Then $\text{Pic}_{X/S}^{=}$ breaks up as a disjoint union of subschemes $\text{Pic}_{X/S}^{=\theta}$, where θ ranges over the set of Hilbert polynomials. It suffices to construct each $Pic_{X/S}^{=\theta}$. We show that each is, in fact, proper.

Fix θ. For any m, tensoring by $O_X(m)$ defines an automorphism of $\text{Pic}_{X/S}^{=}$. Hence, to construct $\text{Pic}_{X/S}^{=\theta}$ we may replace $\theta = \theta(n)$ by $\theta(n + m)$. Taking $m \gg 0$ we may assume that on each geometric fiber $X(s)$ each torsion-free, rank-1 sheaf I with θ for Hilbert polynomial is generated by its global sections and has trivial higher cohomology groups. Set $r = \theta(0)$, fix $N \geq 1$ and set $\nu = rN - 1$. Then I is isomorphic to a quotient of

$O_{X(s)}{}^{\oplus\nu!}$. So consider the open subscheme Q of $Quot(O_{X/S}{}^{\oplus\nu!})$ parametrizing the various such quotients, and consider the natural surjection, $c:Q \to \mathrm{Pic}_{X/S}{}^{=\theta}$. Then c is flat because of the vanishing of the first cohomology group of each I, but c is not proper. Nevertheless, we construct $\mathrm{Pic}_{X/S}{}^{=\theta}$ by forming the quotient.

We form the quotient locally as follows. Fix (1) an open, retrocompact subset S_0 of S, (2) a finite, flat and finitely presentable map $p:R \to S_0$ and (3) ν quasi-sections $y_i:R \to X$ of X/S. Let A denote the open subscheme of $\mathrm{Pic}_{X/S}{}^{=\theta}$ parametrizing the sheaves I on the fibers $X(s)$ such that (1) I is invertible at the ν points $y_i(R)(s)$ of $X(s)$ and (2) a certain map $u:O_{X(s)}{}^{\oplus\nu!} \to I$, which is determined up to scalar multiple by the y_i, is surjective. Sending I to u defines a section of the restriction $c^{-1}A \to A$, and this section (defined on the level of functors) is used to form A. The various A's cover $\mathrm{Pic}_{X/S}{}^{=\theta}$ if N is sufficiently large by virtue of the theory of λ-independent 0-cycles. So patching together the A's yields $\mathrm{Pic}_{X/S}{}^{=\theta}$.

Some important questions remain. First, is the scheme $Pic_{X/S}{}^{=\theta}$ projective over S, not just proper? Indeed, is it strongly projective in the sense of [CP, 2.1] as $\mathrm{Pic}_{X/S}{}^{-\theta}$ was shown to be in [CP, 6.6] in the case that the fibers $X(s)$ are Cohen-Macaulay?

Second, consider the set C^θ of points of $\mathrm{Pic}_{X/S}{}^{=\theta}$ parametrizing those sheaves that are invertible on the smooth locus. We prove in (3.2, ii) that C^θ is a closed set and that locally over S it is contained in a quasi-projective, open subscheme of $\mathrm{Pic}_{X/S}{}^{=\theta}$ (in fact, in one of the A's). Does C^θ have a natural subscheme structure? In other words, is the functor of torsion-free, rank-1 sheaves invertible along the smooth locus representable by a closed subscheme?

Third, the minimal compactification of $\mathrm{Pic}_{X/S}{}^{\theta}$, which is its closure in $\mathrm{Pic}_{X/S}{}^{=\theta}$, is obviously contained in C^θ. Hence it is projective locally over S. Is it, if not $\mathrm{Pic}_{X/S}{}^{=\theta}$, globally projective, or better, strongly projective? Is there a good intrinsic description of its points?

Fourth, when is the closure of $\mathrm{Pic}_{X/S}{}^{\theta}$ equal to $\mathrm{Pic}_{X/S}{}^{=\theta}$? In [AIK] we proved that equality holds for a family of curves with embedding dimension 2 at each singular point. Rego [Rego] has offered another proof of this result and a proof of its converse. What can be said in higher dimension?

1. Rigidification.

(1.1) **(Simple sheaves).** Let $f:X \to S$ be a quasi-compact, quasi-separated map of schemes. An O_X-Module I is called *S-simple* [CP, 5.1] if

I is locally finitely presentable and S-flat and if for each S-scheme T the canonical map σ is an isomorphism:

$$\sigma: O_T \xrightarrow{\sim} (f_T)_* Hom_{X_T}(I_T, I_T).$$

Define a functor [CP, 5.5],

$$Spl_{X/S},$$

whose T-points are the equivalence classes of T-simple O_{X_T}-Modules I, where I and J are considered equivalent if there exists some invertible O_T-Module L and some isomorphism,

$$I \otimes_S L \simeq J.$$

It is proved [CP, 5.6, ii] that the canonical maps between the associated sheaves for the indicated topologies are monomorphisms,

$$Spl_{(X/S)} \hookrightarrow Spl_{(X/S)(Zar)} \hookrightarrow Spl_{(X/S)(et)} \hookrightarrow Spl_{(X/S)(fppf)} \hookrightarrow Spl_{(X/S)(fpqc)}.$$

(It is also proved [CP, 5.6, iii, and 7.4] that each geometric point of $Spl_{(X/S)(fppf)}$ can be represented by a simple sheaf on the corresponding geometric fiber of X/S and that if X/S is locally projective and finitely presentable, then $Spl_{(X/S)(et)}$ is representable by a quasi-separated algebraic space locally finitely presentable over S. There are examples [CP, 7.5 and 7.7] in which the algebraic space is not separated and in which it is not a scheme.)

(1.2) **(g-rigidification).** Let $f: X \to S$ be a map of schemes, and fix a section $g: S \to X$ of f. Let T be an S-scheme. An O_{X_T}-Module I will be called *g-rigidifiable* if I is invertible along $g(S)$ and if there exists an isomorphism,

$$\phi: O_T \xrightarrow{\sim} (g_T)^* I.$$

For any g-rigidifiable sheaf I, such an isomorphism ϕ will be called a *g-rigidification* and such a pair (I, ϕ) will be called a *g-rigidified sheaf.* By a map from (I, ϕ) to another g-rigidified sheaf (J, ψ) will be meant a map from I to J such that the corresponding triangle,

$$\begin{array}{ccc} g^*I & \longrightarrow & g^*J \\ {\scriptstyle\phi}\nwarrow & & \nearrow{\scriptstyle\psi} \\ & O_T & \end{array}$$

is commutative. It is obvious that the various such pairs (I, ϕ) form a category; it will be denoted by

$$\mathcal{R}_g(T).$$

The $\mathcal{R}_g(T)$, as T varies, obviously form a fibered category; it will be denoted by $\mathcal{R}_g$.

(1.3) Proposition. Let $f{:}X \to S$ be a map of schemes, and let $g{:}S \to X$ be a section of f. Let $p{:}R \to T$ be an *fpqc* map of S-schemes.

(i) p is effectively $\mathcal{R}_g$-descendible.

(ii) Let (I, ϕ) be a g_R-rigidified sheaf such that I is R-simple. Suppose that the two pullbacks of (I, ϕ) over $R \times_T R$ are isomorphic. Then (I, ϕ) is isomorphic to the pullback of a g_T-rigidified sheaf (J, ψ) such that J is T-simple.

(iii) Let P be an S-scheme, and let (I, ϕ) be a g_P-rigidified sheaf such that I is P-simple. Then every automorphism of (I, ϕ) is the identity.

Proof. Assertion (i) holds because an *fpqc* map is effectively descendible for the fibered category of quasi-coherent sheaves by standard descent theory [SGA 1, VIII, 1.1]. Assertion (ii) holds by (i) because the cocycle condition holds automatically by (iii).

To prove (iii), consider the following commutative triangle:

$$\begin{array}{ccc} Hom_{X_P}(I, I) & \xrightarrow{\;c\;} & Hom_P(g_P^*I, g_P^*I) \\ {\scriptstyle a}\nwarrow & & \nearrow{\scriptstyle b} \\ & \Gamma(P, O_P) & \end{array}$$

Here a is bijective because I is P-simple, and b is bijective because I is

g-rigidifiable. Hence c is bijective. Assertion (iii) is now immediate from the definition of a map in $\mathcal{R}_g(P)$.

(1.4) **(The functors of g-rigidified simple sheaves).** Let $f:X \to S$ be a quasi-compact, quasi-separated morphism of schemes, and fix a section g of f.

Define a functor,

$$R_g Spl,$$

whose T-points are the isomorphism classes of g_T-rigidified sheaves (I, ϕ) such that I is T-simple.

Define a subfunctor of $Spl_{X/S}$,

$$R_g Spl_{X/S},$$

whose T-points are each representable by a T-simple sheaf I which is invertible along $g_T(T)$. Note that to determine whether or not a given T-point lies in $R_g Spl_{X/S}(T)$ the particular choice of I is immaterial.

The subfunctor $R_g Spl_{X/S}$ of $Spl_{X/S}$ is *open* and *retrocompact.* Indeed, for any subset V of X, a flat sheaf I is invertible at every point of a fiber $V(s)$ if and only if $I(s)$ is invertible at every point of $V(s)$. Hence the property of being invertible along g may be checked on the fibers. On the other hand, the set U of points of S such that I is invertible at $g(s)$ is clearly open and retrocompact in S. So, a map $T \to S$ factors through U if and only if I_T is invertible along g_T. Thus $R_g Spl_{X/S}$ is a retrocompact, open subfunctor of $Spl_{X/S}$.

(1.5) **(The enlarged Picard functors).** Let $f:X \to S$ be a flat, proper, finitely presentable morphism of schemes with integral geometric fibers.

An O_X-Module I is called a *relatively torsion-free, rank*-1 (resp. *pseudo-invertible*) sheaf on X/S [CP 5.9] if it is locally finitely presentable and S-flat and if every fiber $I(s)$ is a torsion-free, rank-1 (resp. and Cohen-Macaulay) $O_{X(s)}$-Module for every geometric point s of S, or equivalently, for every point s of S. Such O_X-Modules are S-simple [CP, 5.10].

A series of subfunctors of $Spl_{X/S}$,

$$Pic_{X/S},\ Pic_{X/S}^{-} \subset Pic_{X/S}^{=} \subset Spl_{X/S},$$

is defined as follows [CP, 5.11]: The T-points of $Pic_{X/S}$ (resp. $Pic_{X/S}{}^{-}$, resp. $Pic_{X/S}{}^{=}$) are those T-points of $Spl_{X/S}$ representable by invertible sheaves (resp. relatively pseudo-invertible sheaves, resp. relatively torsion-free, rank-1 sheaves) on X_T/T. Note that $Pic_{X/S}$ is a subfunctor of $Pic_{X/S}{}^{-}$ if and only if $X(s)$ is Cohen-Macaulay for every s in S.

These subfunctors of $Spl_{X/S}$ are retrocompact, open subfunctors because for any locally finitely presentable, T-flat sheaf I on X_T the points t of T for which $I(t)$ is invertible (resp. pseudo-invertible, resp. torsion-free, rank-1) form a retrocompact, open subset of T [CP, 5.12]. (The hypothesis in [CP, 5.12] that all the fibers of f have the same dimension is easily eliminated as f is flat; with more effort it can be eliminated even when f is not flat by using the fact that the set of s in S where $\mathrm{Supp}(I(s))$ is equal to $X(s)$ is constructible [EGA IV_3, 9.5.3] and the fact that the dimension of the fibers of a map which is locally of finite type is upper semi-continuous on the source [EGA IV_3, 13.1.3].)

Let $P_{X/S}$ be any retrocompact, open subfunctor of $Spl_{X/S}$. It follows formally that the associated Zariski sheaf $P_{(X/S)(Zar)}$ is a retrocompact, open subfunctor of $Spl_{(X/S)(Zar)}$, that the associated etale sheaf $P_{(X/S)(et)}$ is an open subfunctor of $Spl_{(X/S)(et)}$, etc. The case of the etale topology is an important one. It is worked out in [CP, 5.13, ii, b)] for $Pic_{X/S}{}^{=}$ for an f with fibers of the same dimension. The other cases can be done similarly.

Assume f admits a section g. Let $P_{X/S}$ be an arbitrary subfunctor of $Spl_{X/S}$. Set

$$R_gP_{X/S} = R_gSpl_{X/S} \cap P_{X/S}.$$

For example, we obviously have

$$R_gPic_{X/S} = Pic_{X/S}.$$

Similarly, if g factors through the smooth locus of X/S, then we have

$$R_gPic_{X/S}{}^{-} = Pic_{X/S}{}^{-},$$

because a Cohen-Macaulay, torsion-free module over a regular local ring is free.

Sheaving $R_gP_{X/S}$ for the various topologies gives a series of new functors. On the other hand, we define a functor R_gP as the inverse image of $P_{X/S}$ under the canonical map from R_gSpl to $R_gSpl_{X/S}$. For example, we obtain the following functors in this way:

$$R_gPic,\ R_gPic^- \subset R_gPic^= \subset R_gSpl.$$

Note that the T-points of R_gP are those g_T-rigidified sheaves (I, ϕ) such that I represents a T-point of $P_{X/S}$.

Finally, note that a relatively torsion-free, rank-1 sheaf I on X_T/T represents a T-point of $R_gSpl_{(X/S)}$ if and only if $g_T{}^*I$ is invertible. Indeed, a torsion-free, rank-1 sheaf is invertible at a point if and only if it is generated by a single element at that point, because every such element is regular.

(1.6) THEOREM. Let $f{:}X \to S$ be a quasi-compact, quasi-separated map of schemes, and let $g{:}S \to X$ be a section of f. Let $P_{X/S}$ be a subfunctor of $Spl_{X/S}$ with the following property: Let I represent a T-point of $Spl_{X/S}$ and let $p{:}T' \to T$ be an *fpqc* covering such that the pullback of I by p represents a T'-point of $P_{X/S}$; then I represents a T-point of $P_{(X/S)}$. (For example, $Pic_{X/S}$, $Pic_{X/S}{}^-$ and $Pic_{X/S}{}^=$ clearly have this property.)

Then there is a series of canonical isomorphisms,

$$[R_gP \xrightarrow{\sim} R_gP_{X/S} \xrightarrow{\sim} R_gP_{(X/S)(Zar)} \xrightarrow{\sim} R_gP_{(X/S)(et)}$$

$$\xrightarrow{\sim} R_gP_{(X/S)(fppf)} \xrightarrow{\sim} R_gP_{(X/S)(fpqc)}$$

Proof. The first map is defined by sending a pair (I, ϕ) to the class of I. The remaining maps are the natural ones from the presheaf to its associated sheaf for the indicated topology; they are injective by (1.1).

To prove that the first map is injective, take two pairs (I, ϕ) and (J, ψ) defining the same element of $R_gP_{X/S}(T)$. Then there exist an invertible sheaf M on T and an isomorphism,

$$u{:}I \otimes f_T{}^*M \xrightarrow{\sim} J.$$

Consider the composition of isomorphisms,

$$v{:}M \xrightarrow{\sim} O_S \otimes M \xrightarrow{\sim} g_T{}^*\mathrm{I} \otimes M \xrightarrow{\sim} g_T{}^*J \xrightarrow{\sim} O_T;$$

here the first map is defined by $m \mapsto 1 \otimes m$, the second is $\phi \otimes id$, the third is equal to $g_T{}^*u$ (note that $g_T{}^*f_T{}^*M = M$ because $fg = id$), and the last map is ψ^{-1}. It is evident that the map

$$u \circ (id \otimes f_T{}^*v^{-1}){:}I \otimes f_T{}^*O_T \xrightarrow{\sim} J$$

induces an isomorphism from (I, ϕ) to (J, ψ). Thus the first map is injective, too.

It remains to prove that the map from R_gP to $R_gP_{(X/S)(fpqc)}$ is surjective. To prove this, fix a T-point ξ of the latter functor, and represent ξ by taking a suitable *fpqc* cover $R \to T$ and a suitable sheaf I on X_R. Set $M = g_R{}^*I$. Then M is invertible. Set $J = I \otimes f_R{}^*M^{-1}$. Hence the T-point ξ is also represented by J, and there is a canonical isomorphism, $\psi : O_R \xrightarrow{\sim} g_R{}^*J$.

Consider the pair (J, ψ). It represents an R-point of R_gSpl mapping to the R-point $\xi | R$ of $R_gSpl_{(X/S)(fpqc)}$. So the two pullbacks of (J, ψ) to $R \times_T R$ map to the two pullbacks of ξ to $R \times_T R$. However, the latter two are equal because ξ is a T-point. Hence the former two are equal by the injectivity of the map from R_gSpl to $R_gSpl_{(X/S)(fpqc)}$, which already has been proved. Therefore, by (1.3, ii) the pair (J, ψ) descends to a pair (K, θ) on X_T. It is evident that (K, θ) defines a T-point of R_gSpl that maps to ξ. Finally, (K, θ) defines a T-point of R_gP by the hypothesis on $P_{X/S}$.

2. The Key Map.

(2.1) **(An alternating map).** Let A be a ring (commutative with 1). Let F be an A-module. Fix integers n, m with $n \geq m \geq 1$. Define a map

$$a : \Lambda^n F \to \Lambda^m F \otimes \Lambda^{n-m} F$$

as follows. Let G be the subgroup of the symmetric group S_n consisting of those permutations that permute the first m letters and the last $(n - m)$ letters among themselves. Let C be a set of representatives of the cosets of S_n modulo G. Then set

$$a(f_1 \wedge \cdots \wedge f_n) = \sum_{\sigma \in C} \operatorname{sgn}(\sigma) f_{\sigma 1} \wedge \cdots \wedge f_{\sigma m} \otimes f_{\sigma(m+1)} \wedge \cdots \wedge f_{\sigma n}.$$

Note that the map a is well-defined, for the right hand side is alternating in $f_1, \ldots, f_n$. Note also that a is independent of the particular choice of the representatives comprising C.

(2.2) Lemma. Let A be a ring. Let K, F and J be A-modules.

(i) Consider an exact sequence,

$$0 \to K \xrightarrow{v} F \xrightarrow{w} J \to 0.$$

Fix integers n, m with $n \geq m \geq 1$. Assume that J is generated by m elements. Then the following two maps have the same image:

$$(\Lambda^m w \otimes \Lambda^{n-m} F) \circ a : \Lambda^n F \to \Lambda^m J \otimes \Lambda^{n-m} F,$$

$$(\Lambda^m J \otimes \Lambda^{n-m} v) : \Lambda^m J \otimes \Lambda^{n-m} K \to \Lambda^m J \otimes \Lambda^{n-m} F.$$

(ii) There is a natural surjective map,

$$b : \Lambda^n (J \oplus K) \to \Lambda^m J \otimes \Lambda^{n-m} K.$$

If J and K are locally free with ranks m and $n - m$, then b is an isomorphism.

Proof. (i) Let $j_1, \ldots, j_m \in F$ be elements whose images generate J. Then, clearly, any element of F is equal to a linear combination of the j's plus an element of $v(K)$. Hence $\Lambda^n F$ is generated by the elements of the form $f_1 \wedge \cdots \wedge f_n$ where $f_1, \ldots, f_p$ are p different j's (so $p \leq m$) and $f_{p+1}, \ldots, f_n$ are elements of $v(K)$. It is evident that the image of any such $f_1 \wedge \cdots \wedge f_n$ under the first map is equal to 0 if $p < m$ and to $\pm j_1 \wedge \cdots \wedge j_p \otimes f_{p+1} \wedge \cdots \wedge f_n$ if $p = m$.

(ii) The first assertion follows from (i) by taking $F = J \oplus K$. Indeed, since v is split, $\Lambda^m J \otimes \Lambda^{n-m} v$ is split, so injective.

The second assertion follows immediately from the first, because a surjective map between locally free modules with the same (finite) rank is an isomorphism.

(2.3) **(The key map $u(\bar{x})$).** Let $f : X \to S$ be a map of schemes. Let I be an O_X-Module, set

$$E = f_* I,$$

and let $\sigma : f^* E \to I$ denote the canonical map. Fix integers r and N, and set

$$\nu = (rN) - 1.$$

Fix ν sections of f

$$x_i : S \to X \qquad (i = 1, \ldots, \nu),$$

and set $\bar{x} = (x_1, \ldots, x_\nu)$. For each i, set

$$v_i = (x_i)^* \sigma : E \to I_i = (x_i)^* I.$$

For $j = 1, \ldots, N - 1$, set

$$w_j = (v_{(j-1)r+1}, \ldots, v_{rj}) : E \to J_j = \oplus^{rj}{}_{i=(j-1)r+1} I_i.$$

and (for $j = N$) set

$$w_N = (v_{(N-1)r+1}, \ldots, v_\nu) : E \to J_N = \oplus^{\nu}{}_{i=(N-1)r+1} I_i.$$

Form the composition,

$$w_n' : \Lambda^r E \to \Lambda^{r-1} E \otimes E \to \Lambda^{r-1} J_N \otimes E,$$

in which the first map is the sheaved version of the map a of (2.1) and the second map is $\Lambda^{r-1} w_N \otimes E$. Define

$$w' : (\Lambda^r E)^{\otimes N} \to (\Lambda^r J_1) \otimes \cdots \otimes (\Lambda^r J_{N-1}) \otimes (\Lambda^{r-1} J_N) \otimes E$$

$$w' = (\Lambda^r w_1) \otimes \cdots \otimes (\Lambda^r w_{N-1}) \otimes w_N'.$$

By (2.2, ii) there are surjections,

$$\Lambda^r J_j \to \otimes^{rj}{}_{i=(j-1)r+1} I_i$$

$$\Lambda^{r-1} J_N \to \otimes^{\nu}{}_{i=(N-1)r+1} I_i.$$

Combining these surjections with w' yields a natural map,

$$w(\bar{x}) : (\Lambda^r E)^{\otimes N} \to (\otimes^{\nu}{}_{i=1} I_i) \otimes E.$$

For each permutation ρ of $\{1, \ldots, \nu\}$, set

$$\bar{x}_\rho = (x_{\rho 1}, \ldots, x_{\rho \nu}).$$

The ν! various maps $w(\bar{x}_\rho)$ have canonically isomorphic images. So they define a natural map,

$$u'(\bar{x}):[(\Lambda^r E)^{\otimes N}]^{\oplus \nu!} \to (\otimes^\nu{}_{i=1} I_i) \otimes E.$$

Assume that E is locally free with rank r. Then $\Lambda^r E$ is invertible. Set

$$M(\bar{x}) = (\otimes^\nu{}_{i=1} I_i) \otimes [(\Lambda^r E)^{\otimes N}]^{-1}.$$

Then $u'(\bar{x})$ yields a natural map,

$$u^\ell(\bar{x}):O_S{}^{\oplus \nu!} \to E \otimes M(\bar{x}).$$

Finally, the pullback $f^*u^\ell(\bar{x})$ and the canonical map $\sigma:f^*E \to I$ yield the key map,

$$u(\bar{x}):O_X{}^{\oplus \nu!} \to I \otimes_S M(\bar{x}).$$

(2.4) Lemma. Let $f:X \to S$ be a map of schemes. Let I be an O_X-Module such that f_*I is locally free with a finite rank, say r. Let N be an integer, and set $\nu = rN - 1$. Let $\bar{x} = (x_1, \ldots, x_\nu)$ be a ν-tuple of sections of f. Then the sheaf $M(\bar{x})$ and the map $u(\bar{x})$ of (2.3) have the following two properties:

(i) Let L be an invertible sheaf on S. Then replacing I by $I \otimes_S L$ replaces $M(\bar{x})$ by $M(\bar{x}) \otimes L^{-1}$ and replaces $u(\bar{x})$ by a map canonically ismorphic to it.

(ii) Let T be an S-scheme, and assume that the formation of f_*I commutes with the base-change to T. Then the formation of $M(\bar{x})$ and the formation of $u(\bar{x})$ also commute with the base-change.

Proof. The proof is straightforward.

(2.5) **(Degree).** Let k be a field. Let X be a projective k-scheme equipped with an ample sheaf $O_X(1)$. Let F be a nonzero coherent sheaf on X, and let m denote the dimension of the support of F.

By the *degree* of F we mean the (leading) coefficient a_M in the Hilbert polynomial

$$\chi(F(n)) = \sum_{i=0}^{M} a_i \binom{n+i}{i},$$

where M is the dimension of the support of F. The degree of F will be denoted by deg (F). Note that deg (F) is a (strictly) positive integer. By the degree of a closed subscheme Y of X we mean the degree of the structure sheaf O_Y; that is,

$$\deg(Y) = \deg(O_Y).$$

For example, take $X = \mathbf{P}^M{}_k$ and $Y{:}H = 0$. Then

$$\deg(Y) = \deg(H).$$

This equation results immediately from the exactness of the sequence,

$$0 \to O_X(-d) \xrightarrow{\times H} O_X \to O_Y \to 0, \qquad d = \deg(H),$$

from the additivity of the Hilbert polynomial, and from the following easily-proved relation (the full relation is found in [SGA 6, XIII, 2.10, p. 630], for instance):

$$\binom{n-d+M}{M} = \binom{n+M}{M} - d\binom{n+M-1}{M-1} + \cdots.$$

Returning to the general case, we have a formula

$$\deg(F) = \Sigma_Y \ell_Y(F) \deg(Y), \tag{2.5.1}$$

where Y ranges over the M-dimensional, reduced and irreducible components of the support of F and $\ell_Y(F)$ denotes the length of the stalk of F at the generic point of Y. This formula is easily established using the additivity of the Hilbert polynomial, the additivity of length, and the lemma of devissage [EGA III_1, 3.1.2].

Let $\pi{:}X \to P$ be a finite map of projective k-schemes with ample sheaves such that

$$O_X(1) = \pi^* O_P(1).$$

Then we have

$$\deg(\pi_* F) = \deg(F). \tag{2.5.2}$$

Indeed, the Leray spectral sequence degenerates yielding

$$H^q(P, (\pi_* F)(n)) = H^q(X, F(n)).$$

(2.6) (**λ-independent 0-cycles**). Let k be a field. Let X be an M-dimensional projective k-scheme equipped with an ample sheaf. Let

$$\alpha = \sum_{i=1}^{\nu} x_i$$

be a 0-cycle on X with each x_i a k-point of X. Let λ be an integer. Then α is said to be λ-*independent* if for each coherent sheaf F on X whose support Supp (F) has codimension 1 the following inequality holds:

$$\#(\alpha . \operatorname{Supp}(F)) \leq \lambda \deg (F)^M,$$

where the left hand side denotes the number of indices i such that x_i lies in Supp (F).

In view of Formula (2.5.1), α is λ-*independent* if (and only if) for each closed integral subscheme Y of X with codimension 1 the following analogous inequality holds:

$$\#(\alpha . Y) \leq \lambda \deg (Y)^M.$$

(2.7) PROPOSITION. Let k be an infinite field. Let U be a dense open subset of $\mathbf{P}^M{}_k$. Then for every integer $\nu > 0$ there exists a 2^M-independent 0-cycle $\alpha = \Sigma^\nu{}_{i=1} x_i$ with each x_i a k-point of U.

Proof. Reason as in [CS, pp. 139–40] *mutatis mutandis.*

(2.8) PROPOSITION. Let k be an algebraically closed field. Let X be an M-dimensional projective k-scheme equipped with a very ample sheaf $O_X(1)$. Set $d = \deg (X)$. Let U be a dense open subset of X. Then for every integer $\nu > 0$ there exists a $(2^M + 1)\, d$-independent 0-cycle $\alpha = \Sigma^\nu{}_{i=1} x_i$ with each x_i a k-point of U.

Proof. Embed X in a projective space via $O_X(1)$, and k being infinite, take a linear space L of codimension $(M - 1)$ not meeting X. Consider the projection from X with center L,

$$\pi : X \to P = \mathbf{P}^M{}_k.$$

Then π is a finite, surjective map, and we have

$$\pi^* O_P(1) = O_X(1).$$

Since π is proper, $\pi_* O_X$ is coherent on P. So it is free on an open subset V, say with rank e. Since π is surjective, $e \neq 0$. It is now evident from Formula (2.5.1) and the definition of degree that the following two equations hold:

$$\deg(\pi_* O_X) = e \deg(O_P) = e.$$

Therefore Formula (2.5.2) yields $d = e$.

Given ν, write

$$\nu = \mu d + \rho \qquad \text{with } 0 \leq \rho \leq d - 1.$$

By (2.7) there exists a 2^M-independent 0-cycle $\beta = \Sigma_{j=1}{}^{\mu}\, y_j$ with y_j a k-point of V. Let $x_1, \ldots, x_\eta \in X$ be the points lying over the y_j, but with x_i repeated $\dim_k(O_{x_i} \otimes k(\pi x_i))$ times. The x_i are k-points because k is algebraically closed. Since $\pi_* O_X$ is free with rank d on ρ, it is easy to prove $\eta = \mu d$.

We may assume $\pi^{-1} V \subset U$ because π is proper and U is dense in X. Then $x_1, \ldots, x_{\mu d}$ are k-points of U. Choose any ρ additional k-points of U, say $x_{\mu d+1}, \ldots, x_\nu$, and set $\alpha = \Sigma x_i$.

Let F be a coherent sheaf on X whose support has codimension 1. Obviously,

$$\#(\alpha . \operatorname{Supp} F) \leq (d - 1) + \#((\Sigma^{\mu d}{}_{i=1}\, x_i) . \operatorname{Supp} F).$$

So, since there are d x_i's over each y_j, obviously

$$\#(\alpha . \operatorname{Supp} F) \leq (d - 1) + d\#(\beta . \operatorname{Supp} \pi_* F).$$

Since π is finite and surjective, Supp $\pi_* F$ also has codimension 1. Since β is 2^M-independent, therefore

$$\#(\alpha . \operatorname{Supp} F) \leq (d - 1) + 2^M d \deg(\pi_* F)^M.$$

Finally, Formula (2.5.2) yields the desired inequality,

$$\#(\alpha\,.\,\mathrm{Supp}\,F) \leq (2^M + 1)d\ \deg(F)^M.$$

(2.9) DEFINITION [CS, bottom of p. 141]. Let k be a field, and let $\alpha = \Sigma^{\nu}{}_{i=1}\,x_i$ be a 0-cycle on $\mathbf{P}^{r-1}{}_k$ with each x_i a k-point. Then α is said to be *strongly stable* if for each hyperplane H the following inequality is satisfied:

$$\#(\alpha\,.\,H) \leq \nu/r.$$

(2.10) PROPOSITION. Let k be a field. Let X be an integral, M-dimensional, projective k-scheme equipped with a very ample sheaf $O_X(1)$. Let I be a torsion-free, rank-1 sheaf on X, and let $\delta(I)$ denote the (leading) coefficient a_{M-1} of the following polynomial in n,

$$\chi(I(n)) - \chi(O_X(n)) = \Sigma_{i=1}^{M-1}\,a_i\binom{n+i}{i}.$$

Let U be a dense open set of X on which I is invertible. Let ν and λ be integers, and let $\alpha = \Sigma^{\nu}{}_{i=1}\,x_i$ be a λ-independent 0-cycle with each x_i a k-point of U. Assume that $I|U$ is generated by $H^0(X, I)$ and consider the map,

$$g\colon U \to \mathbf{P}^{r-1}{}_k \qquad \text{with } r = h^0(X, I),$$

defined by $I|U$ and the restriction of a basis of $H^0(X, I)$. If the inequality

$$\lambda r\delta(I)^M \leq \nu \tag{2.10.1}$$

is satisfied, then $g_*\alpha = \Sigma^{\nu}{}_{i=1}\,gx_i$ is strongly stable.

Proof. Let H be a hyperplane of $\mathbf{P}^{r-1}{}_k$. Then the inverse image $g^{-1}H$ is equal to the scheme (divisor) of zeros on U of a global section e of I. Let D denote the closure of $g^{-1}H$ in X.

Since each $x_i \in U$, obviously

$$\#(g_*\alpha\,.\,H) = \#(\alpha\,.\,D).$$

Obviously D has codimension 1 in X. So, since α is λ-independent, by definition

$$\#(\alpha . D) \leq \lambda \deg (D)^M.$$

Consider the following exact sequence, which defines F:

$$0 \to O_X \xrightarrow{e} I \to F \to 0.$$

(It is exact on the left because X is integral, and e is regular on U.) It and additivity of the Hilbert polynomial obviously yield

$$\deg (F) = \delta(I).$$

Moreover, the natural surjection $F \to F \otimes O_D$ and the additivity of the Hilbert polynomial obviously yield

$$\deg (F \otimes O_D) \leq \deg (F).$$

Since $I|U$ is invertible, clearly $F|U$ is isomorphic to $(I \otimes O_D)|U$. Since $U \cap D$ is dense in D, it follows that $F \otimes O_D$ and O_D have the same support and that their stalks are isomorphic at the generic point of each irreducible component of D. Therefore, Formula (2.5.1) yields

$$\deg (D) = \deg (F \otimes O_D).$$

Putting together the various relations above, we get

$$\#(g_* \alpha . H) \leq \lambda \delta(I)^M.$$

Therefore (2.10.1) implies that $g_* \alpha$ is strongly stable.

(2.11) Proposition. Let k be a field. Let X be a integral, M-dimensional, projective k-scheme equipped with a very ample sheaf $O_X(1)$. Set $d = \deg (X)$. Let I be a torsion-free, rank-1 sheaf on X that is generated by its global sections. Set $r = h^0(X, I)$ and, as in (2.10), let $\delta(I)$ denote the leading coefficient of the polynomial $\chi(I(n)) - \chi(O_X(n))$. Let U be a dense open set of X on which I is invertible. Let N and λ be integers satisfying the strict inequality,

$$N > \lambda \delta(I)^M.$$

Set $\nu = rN - 1$. Let $\bar{x} = (x_1, \ldots, x_\nu)$ be a ν-tuple of k-points of U such that the 0-cycle $\alpha = \Sigma x_i$ is λ-independent. Then the map

$$u(\bar{x}):O_X{}^{\oplus\nu!} \to I \otimes_k M(\bar{x})$$

of (2.3) (where $M(\bar{x})$ is a certain canonical 1-dimensional k-vector space) is surjective.

Proof. Set $E = H^0(X, I)$. By hypothesis, E generates I and $I|U$ is invertible; let $g:U \to \mathbf{P}(E)$ denote the corresponding map. Then (2.10) implies that $g_*\alpha$ is strongly stable. Fix a k-point $x \in \mathbf{P}(E)$. Then by [CS, Cor., p. 143] there is a decomposition,

$$g_*\alpha = \Sigma_{j=1}^{N-1} \beta_j + \beta_N{}^*, \tag{2.11.1}$$

where $\beta_1, \ldots, \beta_{N-1}$ are 0-cycles of r independent k-points and $\beta_N{}^*$ is a 0-cycle of $(r - 1)$ independent k-points spanning a hyperplane not containing x. For convenience, temporarily reorder the x_i so that we have

$$\beta_j = \Sigma^{rj}{}_{i=(j-1)r+1}\, x_i \qquad \text{for } j < N,$$

$$\beta_N{}^* = \Sigma^{\nu}{}_{i=(N-1)r+1}\, x_i\,.$$

For each point x_i in $\beta_N{}^*$, there is an element e_i of E which is nonzero at x_i but is zero at each of the other points in $\beta_N{}^*$, because otherwise the points of $\beta_N{}^*$ would not be independent. Therefore, the map $w_N:E \to J_N$ of (2.3) is surjective. So ker (w_N) is 1-dimensional. Obviously, ker (w_N) consists of those elements e of E vanishing on $\beta_N{}^*$. So, since x is independent of $\beta_N{}^*$ every nonzero element of ker (w_N) does not vanish at x. Identifying $\Lambda^r E$ with k and $\Lambda^{r-1} J_N \otimes E$ with E, and applying (2.2, i) with $n = r$ and $m = r - 1$, we conclude that the map $w_N{}':k \to E$ of (2.3) carries 1 onto an element of E not vanishing at x.

Similarly, for $j < N$, since the points of β_j are independent, the map $w_j:E \to J_j$ is surjective; so w_j is an isomorphism because both J_j and E are r-dimensional. Identifying $(\Lambda^r E)^{\otimes N}$ with k and identifying $(\otimes^{\nu}{}_{i=1} I_i) \otimes E$ with E, we conclude that the map $w(\bar{x}):k \to E$ of (2.2) carries 1 onto an element of E not vanishing at x.

Identify $E \otimes M(\bar{x})$ with E. Then, what we have shown is that for each k-point x of $\mathbf{P}(E)$ one component of the map $u^g(\bar{x}):k^{\oplus\nu!} \to E$ carries 1 onto an element of E not vanishing at x. Hence $u^g(\bar{x})$ is surjective; otherwise, its image would lie in a subspace of E of codimension 1, say F, and F would define a k-point x of $\mathbf{P}(E)$ at which every element of the image would vanish. Therefore, $u(\bar{x})$ also is surjective.

3. Representability

(3.1) THEOREM. Let $f:X \to S$ be a flat, locally projective, finitely presentable map of schemes, with integral geometric fibers. Then $Pic_{(X/S)(et)}{}^{=}$ is representable by a scheme $Pic_{X/S}{}^{=}$, whose connected components are proper over S.

Proof. The assertions are local on S, so we may assume that f is projective and that S is connected. Fix a relatively very ample sheaf $O_X(1)$. Then each fiber of O_X has the same Hilbert polynomial because f is flat.

Let θ be a polynomial, and consider the functor of [CP, 5.7]

$$Spl_{(X/S)(et)}{}^{\theta}$$

whose T-points are those T-points of $Spl_{(X/S)(et)}$ representable by means of an etale covering map $R \to T$ by an R-simple sheaf on X_R whose fibers each have θ for Hilbert polynomial. By [CP, 5.7] the $Spl_{(X/S)(et)}{}^{\theta}$ are open and closed subfunctors, whose union as θ varies is all of $Spl_{(X/S)(et)}$. Clearly, for each subsheaf P of $Spl_{(X/S)(et)}$, there is induced a corresponding covering of P by open and closed subfunctors P^{θ}.

It is now evident that the assertion follows from (3.2, i) below.

(3.2) THEOREM. Let $f:X \to S$ be a flat, projective, finite presentable map of schemes, with integral geometric fibers. Fix a very ample sheaf $O_X(1)$. Fix a polynomial θ.

(i) Then $Pic_{(X/S)(et)}{}^{=\theta}$ is representable by a proper S-scheme $Pic_{X/S}{}^{=\theta}$.

(ii) Let C^{θ} denote the subset of $Pic_{X/S}{}^{=\theta}$ parametrizing those torsion-free, rank-1 sheaves with Hilbert polynomial θ on the fibers of X/S that are invertible on the smooth locus. Then C^{θ} is closed, and locally over S it is contained in a quasi-projective, open subscheme of $Pic_{X/S}{}^{=\theta}$.

Proof. The proof consists of 13 steps.

Step I. We may assume, obviously, that f is surjective and S is connected.

Step II. Let $\mathcal{G}$ denote the family of classes of torsion-free, rank-1 sheaves with Hilbert polynomial θ on the fibers of X/S. Then we may assume that $\mathcal{I}$ is a 0-regular family, and that there exists an integer μ such that $Hom(\mathcal{I}, O_X)$ is a μ-regular family.

Proof. Since S is connected (Step I) and f is flat, the fibers of O_X all have the same Hilbert polynomial, say ψ. So, by [CP, 3.4, iii] with $J = O_X$, there is an m (depending only on θ and ψ) such that $\mathcal{I}$ is m-regular. Set $\eta(n) = \theta(n + m)$. Clearly, tensoring by $O_X(m)$ induces an isomorphism from $Pic_{(X/S)(et)}{}^{=\theta}$ onto $Pic_{(X/S)(et)}{}^{=\eta}$, and it carries C^θ onto C^η (assuming representability). Hence θ may be replaced by η. Doing so replaces $\mathcal{I}$ by a 0-regular family. Finally, a suitable μ exists by [CP, 6.1].

Step III. Set $r = \theta(0)$. Fix $N \geq 1$. Set $\nu = rN - 1$. Let S_0 be an open, retrocompact subset of S, let $p:R \to S_0$ be a finite, flat, finitely presentable map, and let

$$\bar{y} = (y_1, \ldots, y_\nu)$$

be a ν-tuple of quasi-sections $y_i:R \to X$ (that is, $f \circ y_i = p$). Define a subfunctor

$$A = A(\bar{y})$$

of $Pic_{X/S}{}^{=\theta}$ as follows. For each S-scheme T, let $A(T)$ be empty if T does not lie over $S_0 = p(R)$, that is, the structure map $T \to S$ does not factor through S_0, and if T does lie over S_0, then let $A(T)$ consist of those T-points of $Pic_{X/S}{}^{=\theta}$ representable by a relatively torsion-free, rank-1 sheaf I on X_T/T satisfying the following two conditions: (1) For $i = 1, \ldots, \nu$, let x_i denote the section of $f_{P \times T}$,

$$x_i = (y_i, id_{R \times T}):R \times T \to X_{R \times T},$$

and let U denote the open set of $X_{R \times T}$ on which $I_{R \times T}$ is invertible. Then the image of x_i lies in U. (2) The key map of (2.3),

$$u(\bar{x}):(O_{X \times R \times T})^{\oplus \nu!} \to I \otimes_T M(\bar{x}), \tag{3.2.1}$$

is surjective.

Note that $(f_T)_* I$ is locally free with rank r (and its formulation commutes with base-change) by base-change theory and Step II; hence $u(\bar{x})$ is defined.

Note that the formation of A commutes with base-change; that is, there is a canonical isomorphism,

$$A(\bar{y}) \times_S S' = A(\bar{y} \times_S S'),$$

for any S-scheme S'.

Step IV. The subfunctor A of Step III is a retrocompact open subfunctor of $Pic_{X/S}{}^{=\theta}$.

Proof. Using the notation of Step III, consider the open subscheme V of $R \times T$ defined by

$$V = \bigcap_{i=1}^{\nu} (x_i)^{-1} U \subset R \times T.$$

Let U' denote the open set of points of $X \times R \times T$ at which $u(\bar{x})$ is surjective. Set

$$V' = V \cap [R \times T - (f_{R \times T})(X \times R \times T - U')].$$

Since U' is open and f is proper, the term in the brackets is open. Hence V' is open. Set

$$W = T - (p_T)(R \times T - V').$$

Since V' is open in $R \times T$ and p is proper, W is open in $T \times_S S_0$, so in T. It is easy to see that W is retrocompact in T by using the method of [EGA IV$_3$, §§8, 11]; indeed, the assertion is local on S, it is compatible with base-change, and it is obviously true when S is noetherian.

Let τ be a geometric point of T. It is evident that τ is a geometric point of W if and only if, for every geometric point ρ of $R \times_S T$ lying over τ, (a) $x_i(\rho)$ lies in $U(\rho)$, which (because I is T-flat) is equal to the open set of $X(\rho)$ on which $I(\rho)$ is invertible and (b) the map $u(\bar{x}(\rho))$ is surjective (because by (2.4, ii) the formation of u commutes with base change.) It follows that a map $T' \to T$ factors through W if and only if (a) the image of each $(x_i)_{T'}$ lies in the open set $U_{T'}$ of $X \times_S R \times_S T \times_T T'$ on which the pullback of I is invertible and (b) the map $u(\bar{x}_{T'})$ is surjective. Thus W represents $A \times T$ (product over $Pic_{X/S}{}^{=\theta}$). Therefore A is a retrocompact open subfunctor.

Step V. Let ν be the integer of Step III, and set $F = O_X{}^{\oplus\nu!}$. Let Q denote the retrocompact, open subscheme of $Quot_{(F/X/S)}{}^{\theta}$ parametrizing those quotients that are torsion-free and rank-1; Q exists by [CP, 5.12, ii, a], and Q is quasi-projective and finitely presentable because $Quot_{(F/X/S)}{}^{\theta}$ is (see [CP, 2.6] for example). Finally, let

$$c:Q \to Pic_{X/S}{}^{=\theta}$$

denote the map of functors sending, for each S-scheme T, a quotient I of F_T to its class in $Pic_{X/S}{}^{=\theta}$.

Note that c is surjective on geometric points. Indeed, each geometric point π of $Pic_{X/S}{}^{=\theta}$ is, by definition, represented by a torsion-free, rank-1 sheaf I on a geometric fiber of $X(s)$. By Step II, I is 0-regular, so generated by its global sections ([CS, Prop., p. 99] or [SGA 6, XIII, 1.3, iii, p. 616]). Hence I is isomorphic to a quotient of $O_{X(s)}{}^{\oplus\nu!}$. This quotient corresponds to a geometric point q of Q. Obviously, $c(q) = \pi$.

Step VI. The functor A of Step III is representable by a quasi-projective, finitely presentable S-scheme if the map $p:R \to S$ is an open embedding.

Proof. Let B denote the inverse image of A under the map c of Step V. Then B is a retrocompact, open subfunctor of Q by virtue of Step IV. Hence B is represented by a retrocompact, open subscheme of Q, call it B. Since Q is quasi-projective and finitely presentable by Step V, B is also.

Consider the restriction of c,

$$b = c|B:B \to A.$$

Define a section $s:A \to B$ of b as follows. Given a T-point α of A, represent it by a relatively torsion-free, rank-1 sheaf I on X_T. Then, by definition of A, T lies over S_0 and the key map (3.2.1) is surjective. Since, by hypothesis, $p:R \to S$ is an open embedding, this surjection is isomorphic to a surjection,

$$O_{X\times T}{}^{\oplus\nu!} \to I \otimes_T M(\bar{x}). \tag{3.2.2}$$

The latter defines a T-point β of B. Define s by $s(\alpha) = \beta$.

The T-point β of B depends only on the T-point α of A and not on the choice of representing sheaf I, because a second choice has the form $I \otimes_T$

L, for some invertible sheaf L on T, and by (2.4, i), M is replaced by $M \otimes (L^{-1})$ and the surjection (3.2.2) is replaced by another one equal to it. Thus s is well-defined.

Since the image of each x_i lies in the set U where $I_{R\times T}$ is invertible, $M(\bar{x})$ is invertible. So I and $I \otimes_T M(\bar{x})$ represent the same T-point of $Pic_{X/S}{}^{=}$, namely, α. Hence $b(\beta) = \alpha$. Therefore s is a section of b.

Consider the following diagram of functors:

$$\begin{array}{ccc} A & \xrightarrow{\;s\;} & B \\ {\scriptstyle s}\uparrow & & \downarrow{\scriptstyle (1,\, s\circ b)} \\ B & \xleftarrow{\;\Delta\;} & B \times_S B \end{array}$$

It is clearly cartesian. Therefore, the functor A is represented by the corresponding fibered product,

$$A = B \times_{(B\times B)} B.$$

Since B is quasi-projective and finitely presentable, A is also.

Step VII. The sheaf A_{fppf} associated to the functor A of Step III is representable by a quasi-projective, finitely presentable S-scheme A.

Proof. Clearly, the map $p{:}R \to S$ may be assumed surjective. Then p is finite and locally free.

Suppose the product $A \times_S R$ is representable by a quasi-projective, finitely presentable R-scheme A'. Then the canonical descent data on the functor $A \times_S R$ define descent data on the scheme A'. The latter is effective and the descended S-scheme, A, is quasi-projective by standard descent theory [SGA 1, VIII, 7.7, p. 224], and A is finitely presentable by [EGA IV_2, 2.7.1, p. 29]. Since p is an *fppf* map, it follows that A represents A_{fppf}.

As noted in Step III, the product $A \times_S R$ is equal to the subfunctor $A(\bar{y} \times_S R)$ of $Pic_{X\times R/R}{}^{=}$. Assume the inclusion,

$$A(\bar{y} \times_S R) \subseteq A(\bar{y}, 1), \tag{3.2.3}$$

where $(\bar{y}, 1)$ denotes the ν-tuple of sections $(y_i, 1)$ of $X \times R/R$. By Step IV, both $A(\bar{y} \times_S R)$ and $A(\bar{y}, 1)$ are retrocompact, open subfunctors of

$Pic_{X\times R/R}^{=}$. Hence $A(\bar{y} \times_S R)$ is clearly a retrocompact, open subfunctor of $A(\bar{y}, 1)$. Therefore, since $A(\bar{y}, 1)$ is representable by a quasi-projective, finitely presentable R-scheme by Step VI, so is $A(\bar{y} \times_S R)$.

Finally, to establish the inclusion (3.2.3), take a T-point ξ of $Pic_{X\times R/R}^{=}$ and represent it by a sheaf I on X_T. Let U denote the open subset of X_T on which I is invertible. Consider the diagram,

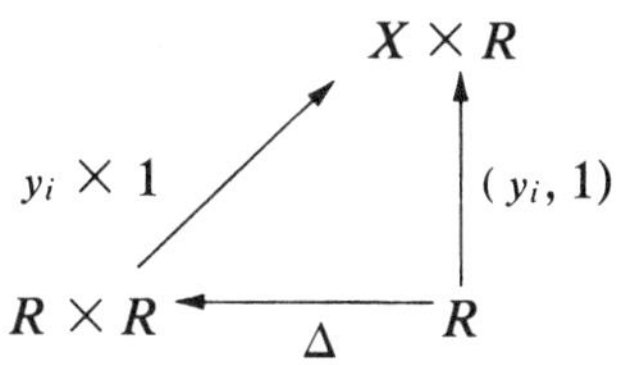

in which Δ denotes the diagonal map. Then, (1) the image of $(y_i \times 1)_T$ obviously contains the image of $(y_i, 1)_T$, hence if the former lies in U, so does the latter, and (2) if the map $u(\bar{x} \times 1)$ is surjective, so is the map $u(\bar{x}, 1)$, because the latter is equal to $\Delta^{-1}u(\bar{x} \times 1)$ since the formation of u commutes with base-change by (2.4, ii). Thus, if the T-point ξ lies in $A(\bar{y} \times_S R)$, then it also lies in $A(\bar{y}, 1)$.

Step VIII. LEMMA. Let $g:Y \to S$ be a flat, projective, finitely presentable map of schemes, and fix a point $s \in S$. Assume that the fiber $Y(s)$ has pure dimension n. Let W be an open subset of Y such that $W(s)$ is dense in $Y(s)$. Then there exist an open subset S_0 of S, a finite, flat, finitely presentable map $p:R \to S_0$, and an S-map $y:R \to W$, where R is regarded as S-scheme via p.

Proof. (inspired by part of the proof of [Raynaud, Prop. 2.2.3, Cas a), p. 32]). Let S_0 be an affine, open neighborhood of s in S. By induction on j, choose n hypersurfaces H_j of $Y|S_0$ such that H_{j+1} contains no associated point of the fiber $(H_1 \cap \cdots \cap H_j)(s)$ and no generic (or maximal) point of the fiber $([H_1 \cap \cdots \cap H_j] - W)(s)$. Note that the latter fiber has dimension at most $n - 1 - j$; hence it is empty for $j = n$.

Set $R = H_1 \cap \cdots \cap H_n$. Then, as just noted, $(R - W)(s)$ is empty. Since Y is projective, so proper, therefore $(R - W)(t)$ is empty for t in a neighborhood of s. So, by shrinking S_0, we may assume $R \subset W$. Now, R is proper and finitely presentable because it is a closed and finitely presentable subscheme of Y. Moreover, $R(s)$ is 0-dimensional by construction; hence $R(t)$ is 0-dimensional for t in a neighborhood of s. It follows that by shrinking S_0 we may assume that R is a finite S_0-scheme. Finally, R is flat over a neighborhood of s because Y is flat and R is defined by a sequence which is regular on $Y(s)$.

Step IX. The various functors A of Step III cover $Pic_{X/S}^{=\theta}$ if $N \gg 0$. (The minimal size of N depends only on $\chi(O_{X(s)}(n))$ and θ.)

Proof. Clearly we may assume that S is affine and, by the method of [EGA IV_3, §8, §11], of finite type over the integers.

Consider the map $c:Q \to Pic_{X/S}^{=\theta}$ of Step V. Since c is surjective on geometric points, it suffices to show that the various inverse images $c^{-1}A$ cover Q. Each $c^{-1}A$ is represented by an open subscheme, say B, of Q by Step VI. If the union $\cup B$ were not all of Q, then the closed set $Q - \cup B$ would contain a closed point by [EGA O_I, 2.1.2, p. 49] because Q is quasicompact, or by the Hilbert Nullstellensatz because Q is of finite type over **Z**. Therefore, a closed point q of Q being given, it suffices to construct an A whose B contains q.

The point q of Q represents a sheaf, I say, on the $k(q)$-scheme $X(q)$. Let U denote the open set on which I is invertible. Let $u_i: U^{\times\nu} \to U$ denote the ith projection, let z_i denote the section,

$$z_i = (u_i, 1): U^{\times\nu} \to X(q) \times U^{\times\nu},$$

and set $\bar{z} = (z_1, \ldots, z_\nu)$. Consider the subset V of $U^{\times\nu}$ of points x such that the map $u(\bar{z})$ is surjective all along the fiber of $X(q) \times U^{\times\nu}$ over x. Since $X(q)$ is complete, V is open. By (2.8) and (2.11), for $N \gg 0$ independent of q, a geometric point of V exists. Now, $X(q)$ is geometrically irreducible, hence $U^{\times\nu}$ is also. Therefore V is open and dense in $U^{\times\nu}$ and so also in $X(q)^{\times\nu}$.

Let $s \in S$ denote the image of $q \in Q$. Then s is a closed point of S and the field extension $k(q)/k(s)$ is finite by the Hilbert Nullstellensatz (in the form of [EGA I, 6.46]) because S and Q are of finite type over **Z** and q is a closed point of Q. Hence the canonical map,

$$\pi: X(q)^{\times\nu} \to X(s)^{\times\nu},$$

is finite. Therefore the image,

$$z = \pi(X(q)^{\times\nu} - V),$$

is a closed subset of $X(s)^{\times\nu}$, and z has smaller dimension than $X(s)^{\times\nu}$.

Since s is closed, $X(s)^{\times\nu}$ is closed in $X^{\times\nu}$. Consequently, $W = (X^{\times\nu} - Z)$ is open in $X^{\times\nu}$, and $W(s)$ is dense in $X(s)^{\times\nu}$.

There exist, by Step VIII, an open subset S_0 of S, a finite, flat, finitely presentable map $p:R \to S_0$, and an S-map

$$\bar{y} = (y_1, \ldots, y_\nu):R \to W \subset X^{\times\nu}.$$

It now follows in a straightforward way from the definitions that I defines a $k(q)$-point of $A(\bar{y})$. Therefore, the open subset B of Q representing $c^{-1}A(\bar{y})$ contains the point q.

Step X. There exist a projective, finitely presentable S-scheme Z and a Z-flat sheaf I on X_Z such that the family $\mathcal{G}$ of Step II is equal to the family of classes of fibers of I.

Proof. By Step II there is a μ such that $Hom(\mathcal{G}, O_X)$ is a μ-regular family. Set $F = O_X(\mu)$ and let Z denote the open and closed subscheme of $Quot_{(F/X/S)}$ parametrizing those quotients whose fiber over any geometric point s of S has the polynomial in n,

$$\chi(O_{X(s)}(n)) - \theta(n),$$

as its Hilbert polynomial. Then Z is a projective, finitely presentable S-scheme by [CP, 2.6]. Let I denote the canonical subsheaf of F_Z. Then I is flat and its fibers are, clearly, representatives of classes in $\mathcal{G}$.

Let s be a geometric point of S, and let J be a sheaf on the fiber $X(s)$ representing a class of $\mathcal{G}$. Then

$$H = Hom_{X(s)}(J, F(s))$$

is 0-regular by the choice of F, so H is generated by its global sections. Since J is rank-1, H is nonzero at the generic point of $X(s)$. Hence there exists a nonzero map $v:J \to F(s)$. Since J is torsion-free, v is injective. Therefore, v defines a geometric point, z say, of Z such that $I(z)$ is equal to J. Thus $\mathcal{G}$ is equal to the family of classes of fibers of I.

Step XI. Assertion (3.2, i) holds.

Proof. By Steps IV and IX, the various A form an open covering of $Pic_{X/S}{}^{=\theta}$. Hence, by general Grothendieck topology, the associated sheaves A_{fppf} form an open covering of $Pic_{(X/S)(fppf)}{}^{=\theta}$. By Step VII, each

A_{fppf} is representable by a finitely presentable S-scheme. Therefore, $Pic_{(X/S)(fppf)}{}^{=\theta}$ is representable by a locally finitely presentable S-scheme, say

$$P = Pic_{X/S}{}^{=\theta}.$$

It follows from Step X that there exist a projective S-scheme Z and a surjective S-map $Z \to P$. Consequently, P is proper and finitely presentable.

It remains to prove that the canonical map of functors,

$$\iota : Pic_{(X/S)(et)}{}^{=\theta} \to Pic_{(X/S)(fppf)}{}^{=\theta},$$

is an isomorphism. First, consider the map $c^{\#}: Q \to P$ associated to the map c of Step V. We now prove that $c^{\#}$ is smooth. Indeed, represent the canonical P-point of $Pic_{(X/S)(fppf)}{}^{=\theta}$ by an *fppf* covering map $g: T \to P$ and a sheaf I on X_P. Then by [CP, proof of 7.4] there exist a locally free sheaf H on T, an open subscheme V of $\mathbf{P}(H)$ and a cartesian diagram,

$$\begin{array}{ccc} V & \longrightarrow & Q \\ \downarrow & \square & \downarrow{\scriptstyle c^{\#}} \\ T & \xrightarrow{g} & P . \end{array}$$

Since the map $V \to T$ is smooth and the map g is an *fppf* covering, $c^{\#}$ is smooth by [EGA IV_4, 17.7.4].

Consider the following natural factorization of $c^{\#}$:

$$c^{\#}: Q \to Pic_{(X/S)(et)}{}^{=\theta} \to P,$$

in which the second map is canonically isomorphic to ι. Since $c^{\#}$ is smooth, it admits an etale quasi-section by [EGA IV_4, 17.16.3, ii]. Hence $c^{\#}$ is an epimorphism of etale sheaves. Therefore ι is an epimorphism. Now, ι is also a monomorphism by [CP, 5.6, ii]. Thus ι is an isomorphism.

There is another proof that ι is an isomorphism. It is based on the theory of rigidification. It does not require representing $Pic_{(X/S)(fppf)}{}^{=\theta}$ although it does involve functors like the A's. Briefly, it runs as follows. For each etale quasi-section y of f, define an open subfunctor P_y of $Pic_{X/S}{}^{=\theta}$ by the condition that a sheaf I on X_T be invertible along the image of y_T. If follows from (1.6) that the restriction of ι is an isomorphism from $(P_y)_{et}$ onto $(P_y)_{fppf}$. Moreover, the various P_y cover $Pic_{X/S}{}^{=\theta}$; this fact can be established by reasoning as in Step IX, using in place of (2.8) and (2.11) the fact that every torsion-free, rank-1 sheaf on an integral scheme is invertible on a dense open set and using in place of Step VIII the fact that every open subset of the smooth locus of f lying surjectively over S admits an etale quasi-section by [EGA IV$_4$, 17.16.3, ii]. It now follows formally that ι is an isomorphism.

Step XII. LEMMA. Let $g:Y \to S$ be a smooth, finitely presentable map of schemes with integral geometric fibers. Let I be a relatively torsion-free, rank-1 sheaf on Y/S. Then the following subset D of S is closed:

$$D = \{s \in S | I(s) \text{ is invertible}\}.$$

Proof. Clearly the assertion is local on S and compatible with limits as in [EGA IV$_3$, Section 8]. Hence we may assume that S is affine and noetherian.

The set D is locally constructible by [EGA IV$_3$, 9.4.7], so constructible by [EGA I, 2.3.11] since S is noetherian. Hence, by [EGA I, 2.4.5], D will be proved closed if it is proved closed under specialization.

Let t be a point of D, and let s be a specialization of t in S. By [EGA I, 5.5.3], since S is noetherian, there is a discrete valuation ring R and a map from Spec (R) into S sending the generic point to t and the special point to s. Obviously we may replace S by Spec (R) and so assume that S is the spectrum of a discrete valuation ring.

Since Y is quasi-compact, it is covered by finitely many affine, open subsets, Y_i say. Obviously, $I(s)$ is invertible if (and only if) each $(I | Y_i)(s)$ is invertible. Hence we may assume that Y is affine.

Since $I(s)$ is torsion-free and rank-1, it is invertible at some point, y say, of $Y(s)$. Since I is flat, I is invertible at y. Now, consider the dual sheaf I^*. Then I^* also is invertible at y. So, since Y is affine, there exists a global section u of I^* generating I^* at y. View u as a map from I into O_Y. Then u is an isomorphism at y.

Let Z denote the closed subscheme of Y defined by the Ideal $u(I)$ and consider the corresponding exact sequence,

$$I \xrightarrow{u} O_Y \to O_Z \to 0.$$

Since u is an isomorphism at y, so is its fiber $u(s)$. Since $I(s)$ is torsion-free and $Y(s)$ is integral, therefore $u(s)$ is injective. Hence, by [EGA IV_3, 11.3.7], since Y is flat, u is injective and Z is flat on an open subset of Y containing $Y(s)$, that is, on all of Y.

Since t was chosen in D, the fiber $I(t)$ is invertible. Hence $Z(t)$ is a divisor, so a divisorial cycle. Since Z is flat, it is equal to the closure of $Z(t)$ by [EGA IV_2, 2.8.5]. Hence Z is a divisorial cycle. Since Y is smooth over S and S is regular, Y is regular, so factorial. Therefore, Z is a divisor. Hence I is invertible. So $I(s)$ is invertible. Thus $s \in D$.

Step XIII. Assertion (3.2, ii) holds.

Proof. It follows from Step X that there exist a projective S-scheme Z and a sheaf I on X_Z defining a surjective map,

$$h:Z \to P = Pic_{X/S}{}^{=\theta}.$$

Let D denote the set of points z of Z such that $I(z)$ is invertible on the smooth locus of $X(z)$. Then D is closed in Z by Step XII applied with the smooth locus of X_Z/Z for Y. Clearly $h(D)$ is equal to the set C^θ of the assertion. Therefore C^θ is closed.

Let $P_i:X^{\times\nu} \to X$ denote the ith projection, let x_i denote the section

$$x_i = (p_i, 1)_Z:X^{\times\nu} \times Z \to X \times X^{\times\nu} \times Z,$$

and set $\bar{x} = (x_1, \ldots, x_\nu)$. Consider the subset V of $X^{\times\nu} \times Z$ of points y such that the map $u(\bar{x})$ is surjective all along the fiber of $X \times X^{\times\nu} \times Z$ over y. Since $f:X \to S$ is proper, V is open. Let p denote the projection from $X^{\times\nu} \times Z$ onto $X^{\times\nu}$ and consider the set

$$W = X^{\times\nu} - p(X^{\times\nu} \times D - (X^{\times\nu} \times D) \cap V).$$

Since Z is S-proper, p is proper. Hence, since D is closed and V is open, W is open.

Fix $s \in S$. Let W_0 denote the smooth locus of W/S. It follows from (2.8) and (2.11) and from the definition of D that if $N \gg 0$ then the fiber $W_0(s)$ has a geometric point. Hence, since $X(s)$ is geometrically irreducible, $W_0(s)$ is dense in $X(s)^{\times\nu}$. Therefore, by Step VIII, there exist an open subset S_0 of S, a finite, flat, finitely presentable map $R \to S_0$, and an

S-map

$$\bar{y} = (y_1, \ldots, y_\nu): R \to W_0 \subset X^{\times\nu}.$$

By Steps IV and VII, the sheaf $A(\bar{y})_{fppf}$ is representable by a quasi-projective, open subscheme A of P. It is evident from the definitions that A contains the restriction $C^\theta | S_0$.

(3.3) THEOREM. Let $f{:}X \to S$ be a flat, projective, finitely presentable map of schemes, with integral geometric fibers. Fix a very ample sheaf $O_X(1)$. Then, for each polynomial θ, the etale sheaf $Pic_{(X/S)(et)}{}^\theta$ (resp. $Pic_{(X/S)(et)}{}^{-\theta}$) is representable by an open subscheme $Pic_{X/S}{}^\theta$ (resp. $Pic_{X/S}{}^{-\theta}$) of $Pic_{X/S}{}^{=\theta}$, whose closure is projective locally over S. Moreover, as θ varies, the disjoint union of the schemes $Pic_{X/S}{}^\theta$ (resp. $Pic_{X/S}{}^{-\theta}$) represents the etale sheaf $Pic_{(X/S)(et)}$ (resp. $Pic_{(X/S)(et)}{}^{-}$).

Proof. In view of the preliminaries in (1.5) and the proof of (3.1), the assertion is an immediate consequence of (3.2).

(3.4) THEOREM. Let $f{:}X \to S$ be a flat, projective, finitely presentable map of schemes, with integral geometric fibers. Assume that f admits a section g.

(i) Then the following g-rigidified, enlarged Picard functors are equal:

$$R_gPic^{=},\ R_gPic_{X/S}{}^{=},\ R_gPic_{(X/S)(et)}{}^{=},\ R_gPic_{(X/S)(fppf)}{}^{=}.$$

These functors are representable by an open subscheme,

$$P = R_gPic^{=},$$

of the scheme $Pic_{X/S}{}^{=}$. Further, P parametrizes a unique universal pair (I, ϕ), in which I is a relatively torsion-free, rank-1 sheaf on X_P/P and $\phi{:}O_P \to g_P{}^*I$ is an isomorphism.

(ii) Then the following Picard functors are equal:

$$R_gPic,\ Pic_{X/S},\ Pic_{(X/S)(et)},\ Pic_{(X/S)(fppf)}.$$

These functors are representable by an open subscheme $Pic_{X/S}$ of P. Further $Pic_{X/S}$ parametrizes a universal invertible sheaf of X/S, which is determined up to tensor product with the pullback of an invertible sheaf on $Pic_{X/S}$.

(iii) Assume that g factors through the smooth locus of f. Then the following enlarged Picard functors are equal:

$$R_g Pic^-, \; Pic_{X/S}{}^-, \; Pic_{(X/S)(et)}{}^-, \; Pic_{(X/S)(fppf)}{}^-.$$

These functors are representable by an open subscheme $Pic_{X/S}{}^-$ of P. The closures of $Pic_{X/S}{}^-$ and of $Pic_{X/S}$ in $Pic_{X/S}{}^=$ are both contained in P. Further both closures and $Pic_{X/S}{}^-$ itself parametrize universal sheaves, each of which is determined up to tensor product with the pullback of an invertible sheaf.

Proof. In (1.5) the several functors are defined and their preliminary properties discussed. By (1.6) the asserted equalities of functors hold. The scheme $Pic_{X/S}{}^=$ exists by (3.1), and the asserted representabilites follow formally. The existence and uniqueness of a universal pair (I, ϕ) follows immediately from the definition of the functor $R_g Pic^=$ and its representability by P. The closure of $Pic_{X/S}$ and $Pic_{X/S}{}^-$ in $Pic_{X/S}{}^=$ are obviously contained in the (disjoint) union over θ of the closed sets C^θ of (3.2, ii), and this union in turn is obviously contained in P if g factors through the smooth locus of f. Finally, the asserted existence and uniqueness of the four universal sheaves follows immediately from the definition of the functor $R_g Pic_{X/S}{}^=$, from its representability by P, and from the fact that P contains $Pic_{X/S}$ and the fact that, if g factors through the smooth locus of f, then P also contains $Pic_{X/S}{}^-$, its closure and that of $Pic_{X/S}$; the identity map of P and the three other inclusion maps into P correspond to the four universal sheaves.

UNIVERSIDAD SIMON BOLIVAR
MASSACHUSETTS INSTITUTE OF TECHNOLOGY

REFERENCES

[Bl] Altman, A., and Kleiman, S., "Compactifying the jacobian," *Bull. A.M.S.*, **82,** 6, Nov. 1976, pp. 947–9.

[CP] Altman, A., and Kleiman, S., "Compactifying the Picard Scheme," to appear in *Advances in Mathematics.*

[SGA 1] Grothendieck, A., "Revetments Etales et Groupe Fondamental." *Lecture Notes in Math.*, **224**, Springer (1971).

[EGA O_I, I] Grothendieck, A., and Dieudonne, J., *Elements de Geometrie Algebrique I,* Springer (1971).

[EGA III—IV_4] Grothendieck, A., and Dieudonne, J., *Elements de Geometrie Algebrique,* Publ. Math. I.H.E.S., Nos. **11, 24, 28, 32** (1961, 1965, 1966, 1967).

[SGA 6] Kleiman, S., "Les Theoremes de Finitude pour le Foncteur de Picard," Theorie des Intersections et Theoreme de Riemann-Roch, Berthelot, P. *et al., Lecture Notes in Math.*, **225,** Springer (1971).

[CS] Mumford, D., "Lectures on Curves on an Algebraic Surface," *Annals of Math. Studies* No. **59**, Princeton University Press (1966).

[Raynaud] Raynaud, M., "Specialisation du foncteur de Picard," *Publ. Math. I.H.E.S.*, No. **38** (1970), pp. 27-76.

[Rego] Rego, C., "The Compactified Jacobian," revised and expanded version, Tata Institute Preprint, Feb. 1978.

ON SECOND ORDER LINEAR DIFFERENTIAL EQUATIONS WITH ALGEBRAIC SOLUTIONS

By F. Baldassarri *and* B. Dwork

Introduction. We consider second order linear differential operators

$$L = D^2 + B \cdot D + C,$$

$D = d/dx$, B, $C \in \mathbf{C}(x)$. The singular points of L consist of the poles of B and C and possibly the point at infinity.

H. A. Schwarz [15] determined all such operators with three singular points whose kernel consists of algebraic functions. His method was to show that if B and C lie in $\mathbf{R}(x)$ then the monodromy group can be calculated from the group generated by the reflections relative to three circles which meet at angles determined by the exponent differences of L. He used this to show that the solutions of L are all algebraic if and only if these angles coincide with the angles of a spherical triangle whose vertices are fixed points of three rotations any pair of which generates a finite rotation group.

The Schwarz list of 15 reduced curvilinear triangles contains a basic sublist of exponent differences

				group type	order $= M$
I	$\frac{1}{2}$	$\frac{1}{2}$	$\frac{1}{n}$	dihedral	$2n$
II	$\frac{1}{2}$	$\frac{1}{3}$	$\frac{1}{3}$	tetrahedral	12
IV	$\frac{1}{2}$	$\frac{1}{3}$	$\frac{1}{4}$	octahedral	24
VI	$\frac{1}{2}$	$\frac{1}{3}$	$\frac{1}{5}$	icosohedral	60

the first case corresponding to an infinite series of dihedral groups.

Klein asserted [10, pp. 132–135] that he had determined all differential equations of second order with only algebraic solutions. This widely accepted assertion is quite misleading. One may pose two questions

(0.1) Given $n + 1$ points $\{a_1, \ldots, a_n, \infty\}$, determine all second order linear differential equations with only algebraic solutions and with these points as singularities.

(0.2) Given a second order linear differential equation L, determine in a finite number of steps whether the solutions of L are all algebraic.

The method of Klein is better adapted to problem (0.2) than to problem (0.1). His method (cf. Theorem 3.4 below) is based on the fact that (excluding the cyclic case) L has only algebraic solutions if and only if it is a weak pullback by a rational map of an element in the basic Schwarz list. To decide whether L as given by (3.3) is the pullback of L_0 is equivalent to the determination of the existence of a rational solution x of the non-linear third order differential equation (3.7.3). With an a-priori bound for the degree of x this can be reduced to a purely algebraic problem. Klein illustrated this program [12] by exhibiting cases XII, XIV, XV of Schwarz as pullbacks of VI (the other cases on the full Schwarz list had been previously checked off in that way by Brioschi [2]). Klein did not for example show by his method that the Schwarz list is exhaustive, a reasonable step in illustrating a credible program for problem (0.1).

It is our understanding that Problem 0.1 is still open for $n \geq 3$ (i.e. four or more singular points). In the present article we carry out the Klein program for Problem (0.2). The new ingredients of our work are

(0.3) A general formula (Section 1) for the degree of the pullback mapping.

(0.4) A decision procedure (Section 6) for the existence of an algebraic solution of a *linear* first order differential equation over a Riemann Surface. (Correction: This procedure is not new. See R. H. Hisch, Bull. Amer. Math. Soc. 76 (1970), pp. 605–608).

The degree calculation is needed to carry out Klein's method for checking whether L is a pullback of II, IV or VI but the method breaks down if we must test whether the group of L is one of the two infinite series. The cyclic case is treated in Section 4 but the main point in the dihedral case is the reduction (Section 5) to problem (0.4) mentioned above.

In a subsequent article we shall explain how Klein's program may be extended to second order differential equations defined over a Riemann Surface.

In Section 8 we indicate connections with Grothendiecke's conjecture and the work of Katz. We note that the conjecture is still open for four or more singular points (unless the differential equation "comes" from algebraic geometry). Even if verified it would not respond to Problem (0.2).

We note that Forsyth [3, p. 184] discussed Problem (0.2) along the lines of Klein. His treatment cannot be taken seriously.

We regard the present article as expository. Indeed a superior decision procedure (over the sphere) has been given by L. Fuchs (cf. items 19, 20, 21, 22, 23, 25, Vol. II of his collected works). This may be explained quite easily.

For each integer m let L_m be the linear differential equation of order $m + 1$ satisfied by all homogeneous forms of degree m in solutions y_1, y_2 of the second order linear differential equation L given by (4.1). If L has only algebraic solutions then (cf. Theorem 3.8 below) L_{12} must have at least one solution which is the radical of a rational function. If L_{12} has such a solution but L_2 does not then (Fuchs p. 45 loc cit) all solutions of L are algebraic. If L_2 has such a solution, ϕ, but L does not then ϕ^2 is a rational function and $\phi = y_1 y_2$, a product of independent solutions of L. Putting $\tau = y_2/y_1$ we obtain

$$\tau' = \frac{w}{\phi}\tau \tag{0.5}$$

where $w = y_1 y_2' - y_2 y_1'$ is a constant which may be calculated from

$$-w^2 = \phi^2\left[\left(\frac{\phi'}{\phi}\right)^2 + 2\left(\frac{\phi'}{\phi}\right)' - 4Q\right] \tag{0.6}$$

In this case Problem (0.2) is reduced to determining whether τ as given by (0.5) is algebraic, a problem evidently untreated by Fuchs. Our Section 6 serves then to fill in this gap in the treatment of Fuchs. If

L has itself a solution which is the radical of a rational function then the solutions of L are all algebraic only if L satisfies the criteria for the cyclic case (cf. Section 4 below). Finally we note that the question of whether L_{12} has a solution, v, which is the radical of a rational function is elementary since v must be of the form

$$\prod_{i=1}^{n} (x - a_i)^{\alpha_i} g(x)$$

where g is a polynomial whose degree as well as the values of the α_1 are determined up to a finite set of possibilities by the exponents of L at its singularities $\{a_1, \ldots, a_n, \infty\}$.

1. **Pullbacks on Riemann Surfaces.** Let L be a second order differential equation with coefficients in an algebraic function field K of characteristic zero and algebraically closed field of constants k. Hence choosing $x \in K$, $x \notin k$, we may write

$$L = D^2 + AD + B \tag{1.1}$$

where A, $B \in K$ and $D = d/dx$. Letting Γ be the Riemann surface corresponding to K, we view L as being *equivalent* to L_1 also defined over K if a ratio of solutions (at some point of Γ) of L is also a ratio of solutions of L_1. Thus in particular we identify L with

$$L_1 = D^2 + B_1 \tag{1.2}$$

where

$$B_1 = B - \frac{1}{2} A' - \frac{1}{4} A^2 \qquad \left(A' = \frac{d}{dx} A\right),$$

since

$$L_1 = \frac{1}{\theta} \circ L \circ \theta \quad \text{where} \quad \theta'/\theta = -\frac{1}{2} A$$

(notice that θ does not lie in K but is defined locally on Γ).

We shall assume that at each point, P, of Γ, L has two independent solutions of the form

$$y_1 = t^{\alpha_i}(1 + b_{i1}t + b_{i2}t^2 + \cdots) \qquad i = 1, 2 \tag{1.3}$$

where t is a local uniformizing parameter, $\alpha_i \in k$ and the $b_{i,j}$ also lie in k. Because of our definition, the pair (α_1, α_2) is not well defined but their difference is well defined up to sign. For a fixed model we have $(\alpha_1, \alpha_2) = (0, 1)$ for almost all points and so if we choose any archimedean valuation of k and setting $\gamma(P) = |\alpha_1 - \alpha_2|$ for each point P of Γ, we see that $\gamma(P) - 1 = 0$ for almost all P. Let S be any finite subset of Γ and put

$$\Delta(S, L) = \sum_{P \in S} (\gamma(P) - 1). \tag{1.4}$$

For S large enough we obtain a limiting value $\Delta(L)$ independent of S, but depending on the valuation of k chosen for the construction.

Now let K_0 be an algebraic extension of K and let π denote the mapping $\Gamma_0 \to \Gamma$ of the corresponding Riemann surface. Let L_0 be the pullback of L by π (i.e. if τ is ratio of two solutions of L then $\tau \circ \pi$ is ratio of two solutions of L_0). Using the same valuation of k we may define $\Delta(L_0)$.

LEMMA 1.5 *Let g (resp: g_0) be the genus of Γ (resp: Γ_0), let M be the mapping degree, then*

$$M[\Delta(L) - 2(g - 1)] = \Delta(L_0) - 2(g_0 - 1). \tag{1.5.1}$$

Proof. For $P \in \Gamma$, let P_0 be a point of Γ_0 above P. Let $e(P_0)$ be the relative ramification. If α_1, α_2 are exponents at P of a model of L, then $\alpha_i \cdot e(P_0)$ are exponents at P_0 of L_0. Thus

$$\gamma(P_0) = \gamma(P) \cdot e(P_0)$$

which shows that

$$\sum_{P_0 | P} \gamma(P_0) = M \cdot \gamma(P). \tag{1.6}$$

If S is any finite subset of Γ and S_0 is the set of all points of Γ above S, then

$$\Delta(S_0, L_0) + \text{card } S_0 = M(\Delta(S, L) + \text{card } S). \tag{1.7}$$

The Hurwitz genus formula states that

$$2(g_0 - 1) - 2M(g - 1) = M \text{ card } S - \text{card } S_0 \tag{1.8}$$

provided S contains all points of Γ which ramify in Γ_0. The lemma now follows by taking S, so large that in (1.7) we may replace $\Delta(S_0, L_0)$ (resp $\Delta(S, L)$) by $\Delta(L_0)$ (resp: $\Delta(L)$) and using (1.8) to eliminate the terms involving cardinality.

In the application below $g_0 = g_1 = 0$ and the exponent differences are all rational. Besides, coordinate functions t, x will be fixed on Γ_0, Γ (that is $K_0 = k(t)$, $K = k(x)$) and by the *weak pullback* of L to (Γ_0, t) we will denote the model of the pullback of L to Γ_0, which is in the form (1.2) with respect to the coordinate function t.

The point in the lemma is that L, L_0, Γ and Γ_0 may be given without the relation π between Γ and Γ_0 being known. In that situation for given set S of Γ there is no way to determine the lifting, S_0, to Γ_0. We can determine the set T of singularities of L and T_0 of singularities of L_0 but T_0 need not coincide with $\pi^{-1}(T)$, the problem being that ramification points of π may produce singular points of L_0 and again if $\gamma(P) = 1/e$ and P_0 covers P with ramification e then P_0 is not a singularity of L_0 since the exponent difference at P_0 would be 1. The situation would be different if we viewed L as determining the solution space rather than ratios of solutions.

2. **Finite homography groups.** We review (following Halphen [6]) the theory of such groups. Let G be a finite group of homographies

$$t \to \frac{at + b}{ct + d}$$

where a, b, c, d lie in $\mathbf{C}$ and $ad - cb \neq 0$. Thus G may be viewed as a finite set of automorphisms of the function field $H = \mathbf{C}(t)$. By Lüroth's theorem, the fixed field H^G is generated by a single element y and so is of genus zero. Let $P_1, \ldots, P_s$ be the points of H^G which ramify in H, let M be the order of G and let e_i be the relative ramification of H over H^G at one point of H above P_i. It then follows that there are M/e_i places of H above P_i and each place has ramification e_i, which shows that if we use the Hurwitz genus formula we have

$$0 = 2(1 - M) + \sum_{i=1}^{s} (e_i - 1)M/e_i.$$

and so

$$\sum_{i=1}^{s} 1/e_i = s - 2 + \frac{2}{M}. \tag{2.1}$$

Since each $e_i \geq 2$, we obtain

$$s/2 \geq \Sigma\, 1/e_i > s - 2$$

which shows that $s = 2, 3$.

If $s = 2$, we use $e_i \leq M$ to conclude from (2.1) that

$$e_1 = e_2 = M$$

which shows that the Galois group G coincides with the ramification group at P_1 and hence G is cyclic.

If $s = 3$ then (2.1) becomes

$$\frac{1}{e_1} + \frac{1}{e_2} + \frac{1}{e_3} = 1 + \frac{2}{M} \tag{2.2}$$

a diophantine equation with only 4 types of solutions as given by the table in the Introduction. (The point is that we cannot have $e_i \geq 3$ for all i and so can assume $e_1 = 2$. Furthermore e_2, e_3 cannot both exceed 3. Thus $e_2 = 2$ or 3.)

Let G be a finite group of homographies. Then for each $\theta \in G$ we may choose a matrix

$$A_\theta = \begin{pmatrix} a_\theta & b_\theta \\ c_\theta & d_\theta \end{pmatrix} \in SL(2, \mathbf{C}) \tag{2.3}$$

such that

$$\theta(t) = \frac{a_\theta t + b_\theta}{c_\theta t + d_\theta}. \tag{2.4}$$

Of course A_θ is only determined mod ± 1 and hence if $\theta, \varphi \in G$, we have

$$A_\theta A_\varphi = h(\theta, \varphi) A_{\theta\varphi} \tag{2.5}$$

for suitable $h(\theta, \varphi) \in \{\pm 1\}$. It is clear that h is a 2-cocycle for G with coefficients in $\{\pm 1\}$ with trivial action of G on $\{\pm 1\}$. (The example of the 4-group generated by $x \to \pm 1/x$ shows that h need not be cohomologically trivial). For later use we define a mapping k of $G \times G$ into ± 1 and a mapping k_0 of G into ± 1 by the formulas

$$k(\phi, \theta) A_{\phi^{-1}\theta} = A_\phi{}^{-1} A_0 \tag{2.6.1}$$

$$k_0(\phi) = \prod_{\theta \in G} k(\phi, \theta) \tag{2.6.2}$$

For $u \in \mathbf{C}$, we define

$$F_u(t) = \prod_{\theta \in G} (t - \theta u) \tag{2.7}$$

if θu is finite for all θ. Whenever θu is infinite we replace the factor $t - \theta u$ by $-(a_\theta u + b_\theta)$. Thus F_u is a polynomial of degree $M - \sigma_u$ where σ_u is the number of $\theta \in G$ such that $\theta u = \infty$. We note that

$$\pm F_{\varphi u}(t) = F_u(t)/(c_\varphi u + d_\varphi)^{\sigma_u}$$

for all ϕ in G, the sign being given by $\prod h(\theta\phi^{-1}, \phi)$, the product being over all θ such that $\theta u = \infty$.

If we put $t = t_1/t_2$, we obtain

$$\Delta_u t_2{}^M F_u(t) = \prod_{\theta \in G} [t_1(c_\theta u + d_\theta) - t_2(a_\theta u + b_\theta)] \tag{2.8}$$

where $M =$ order of G

$$\Delta_u = \prod (c_\theta u + d_\theta)$$

the product being over all $\theta \in G$ such that $\theta u \neq \infty$. If $\varphi \in G$ then putting

$$\varphi_1(t_1, t_2) = a_\varphi t_1 + b_\varphi t_2$$

$$\varphi_2(t_1, t_2) = c_\varphi t_1 + d_\varphi t_2$$

we obtain $\varphi(t) = \varphi_1(t_1, t_2)/\varphi_2(t_1, t_2)$ and so

$$\begin{aligned}\Delta_u \varphi_2(t_1, t_2)^M F_u(\varphi(t)) &= \prod_{\theta \in G} [\varphi_1(t_1, t_2) \cdot (c_\theta u + d_\theta) \\ &\qquad - \varphi_2(t_1, t_2)(a_\theta u + b_\theta)] \\ &= \prod_{\theta \in G} [t_1(c_{\varphi^{-1}\theta} u + d_{\varphi^{-1}\theta}) \\ &\qquad - t_2(a_{\varphi^{-1}\theta} u + b_{\varphi^{-1}\theta})] k(\phi, \theta) \\ &= \Delta_u t_2{}^M F_u(t) k_0(\phi).\end{aligned}$$

This shows that

$$F_u(\varphi(t)) = \frac{k_0(\phi)}{(c_\varphi t + d_\varphi)^M} F_u(t). \tag{2.9}$$

The key point is that the ratio between $F_u(\varphi(t))$ and $F_u(t)$ is independent of u. Thus if v lies in another orbit, we see that F_u/F_v is an absolute invariant under G, and so lies in H^G, the subfield of $H = \mathbf{C}(t)$ fixed by G.

Putting $x = F_u(t)/F_v(t)$ we see that $H \supset H^G \supset \mathbf{C}(x)$ and hence

$$\deg H/\mathbf{C}(x) \geq M$$

while the equation

$$0 = F_u(t) - xF_v(t)$$

is a polynomial in t of degree M, which shows that

$$\deg \mathbf{C}(t)/\mathbf{C}(x) \leq M.$$

This shows that

$$H^G = \mathbf{C}(x).$$

LEMMA 2.10. *If u, v, w are points of* $\mathbf{C}$ *then the polynomials F_u, F_v, F_w are linearly dependent over* $\mathbf{C}$.

Proof. We may assume u, v, w to lie in distinct G-orbits and deg $F_w = M$. Then H^G is generated over $\mathbf{C}$ by F_u/F_w and also by F_u/F_v. This shows that F_u/F_w and F_u/F_v are related by a homography. Hence there exist a, b, c, d in $\mathbf{C}$ such that

$$\frac{F_u}{F_w} = \frac{aF_u + bF_v}{cF_u + dF_v}$$

Since F_u and F_w are relatively prime, we conclude that F_w divides $cF_u + dF_v$ but deg $F_w = M \geq$ deg $(cF_u + dF_v)$ which shows that up to a non-zero constant factor,

$$F_w = cF_u + dF_v.$$

This completes the proof of the lemma.

THEOREM 2.11. *If G is a finite non-cyclic group of homographies of order M then there exist polynomials P_1, P_2, P_3 of degree M/e_i $(i = 1, 2, 3)$ (or possibly $M/e_i - 1$ for one i) such that*

$$\frac{1}{e_1} + \frac{1}{e_2} + \frac{1}{e_3} = 1 + \frac{2}{M} \tag{2.2}$$

$$P_1^{e_1} + P_2^{e_2} + P_3^{e_3} = 0, \qquad (P_i, P_j) = 1 \quad \text{for} \quad i \neq j \tag{2.11.1}$$

$$P_i(\theta t) = P_i(t)/(c_\theta t + d_\theta)^{M/e_i} \qquad \forall \theta \in G. \tag{2.11.2}$$

Proof. As before let $H = \mathbf{C}(t)$, H^G the fixed field under G. The extension H/H^G ramifies at 3 valuations of H^G. If a_i is such a place of H^G, let P_i be the polynomial whose roots are the finite extensions of a_i to H. This means that the degree of P_i is M/e_i unless the infinite place of H lies above a_i. It is clear that $P_i^{e_i}$ is an invariant constructed from an orbit and after adjusting constant factors, equation (2.11.1) follows from Lemma 2.10, equation (2.11.2) follows from (2.9) and (2.2) has been previously established.

To formulate the converse of the above theorem we consider solutions of the equation

$$Q_1^{e_1} + Q_2^{e_2} + Q_3^{e_3} = 0 \tag{2.12}$$

in $\mathbf{C}[t]$ such that Q_1, Q_2, Q_3 are relatively prime in pairs and such that if (say)

$$\deg Q_3^{e_3} < \deg Q_1^{e_1} = \deg Q_2^{e_2} = M \tag{2.13}$$

then

$$\deg Q_3^{e_3} \equiv M \bmod e_3.$$

LEMMA 2.14.

$$\frac{1}{e_1} + \frac{1}{e_2} + \frac{1}{e_3} \geq 1 + \frac{2}{M} \tag{2.14.1}$$

and equality holds if and only if the equation

$$-x = \frac{Q_1^{e_1}}{Q_3^{e_3}}(t) \tag{2.15}$$

defines an extension $\mathbf{C}(t)$ *of* $\mathbf{C}(x)$ *ramified only at 0, 1,* ∞ *with ramification* e_1, e_2, e_3 *at each point of t-sphere above the indicated points of the x-sphere. If equality holds then there exists a constant c such that*

$$Q_2^{e_2-1} = c(e_1 Q_1' Q_3 - e_3 Q_3' Q_1) \tag{2.15.1}$$

Proof. We assume that either $\deg Q_i^{ei} = M$ for $i = 1, 2, 3$ or equation 2.13 holds. We put $(w_1, w_2, w_3) = (0, 0, 0)$ in the first case and $(0, 0, M\text{-}\deg Q_3^{e_3})/e_3$ in the second case.

We put $\sigma_\infty = 0$, in the first case and $\sigma_\infty = w_3 e_3 - 1$ in the second case. In either case it represents the contribution of the point $t = \infty$ to the Hurwitz genus formula for the genus of $\mathbf{C}(t)$ as an extension of $\mathbf{C}(x)$. This formula may be written

$$0 = 1 - M + \frac{1}{2}\sum_{i=1}^{3} N_i(e_i - 1) + \frac{1}{2}\sigma_\infty + E \tag{2.16}$$

where $N_i = \deg Q_i$ and $E \geq 0$ is a correction term introduced to allow for the possibility that

(a) Ramification may occur at points other than $x = 0, 1, \infty$
(b) the zeros of Q_i may be non-simple.

Since

$$M = (N_i + w_i)e_i,$$

we deduce from (2.16)

$$M\left(\frac{1}{e_1} + \frac{1}{e_2} + \frac{1}{e_3} - 1\right) = 2E + \begin{cases} 2 & \text{if } w_3 = 0 \\ 1 + w_3 & w_3 \neq 0. \end{cases} \tag{2.17}$$

This proves the inequality which appears in the statement of the lemma. Equality holds if and only if both $E = 0$ and $w_3 = 0, 1$. This completes the proof of the lemma.

Now assume equality holds. Let

$$H = e_1Q_1'Q_3 - e_3Q_3'Q_1$$

$$K = e_2Q_2'Q_3 - e_3Q_3'Q_2$$

By dividing equation (2.12) by $Q_3^{e_3}$ and differentiating

$$Q_1^{e_1-1}H + Q_2^{e_2-1}K = 0.$$

Since Q_1 and Q_2 are relatively prime, $Q_2^{e_2-1}$ divides H. Since Q_1 and Q_3 are relatively prime, H cannot be zero. Equation (2.15.1) now follow by comparison of degrees of two sides.

PROPOSITION 2.18. *Let e_1, e_2, e_3 be given positive integers. Let $\mathbf{C}(t)$ be algebraic over $\mathbf{C}(x)$ ramified at $x = a_1, a_2, a_3$ (and possibly* *elsewhere). Suppose that at each place of* $\mathbf{C}(t)$ *above* a_i *the ramification index is divisible by* e_i.

Let z be a multivalued function on the x-sphere having only a_1, a_2, a_3 as critical points, suppose each branch of z is locally algebraic and that each branch at a_i lies in $\mathbf{C}(((x - a_i)^{1/e_i}))$. Then $z \in \mathbf{C}(t)$.

Proof. The function z is everywhere uniform on the t-sphere and hence is rational function of t.

Definition. A solution of (2.12) is called primitive if equality holds in (2.14.1).

LEMMA 2.19. *If* (Q_1, Q_2, Q_3) *is a primitive solution of* (2.12) *and* (P_1, P_2, P_3) *is an arbitrary one, then there exist polynomials* f, g *such that*

$$P_i = Q_i\left(\frac{f}{g}\right) \cdot g^{M/e_i} \qquad i = 1, 2, 3.$$

Proof. Define z, t algebraic over $\mathbf{Z}(x)$ by setting

$$\begin{cases} -x = \dfrac{Q_1^{e_1}}{Q_3^{e_3}}(z) \\ \\ -x = \dfrac{P_1^{e_1}}{P_3^{e_3}}(t) \end{cases} \tag{2.19.1}$$

We apply proposition 2.18 and lemma 2.14 and conclude that

$$z \in \mathbf{C}(t), \qquad \text{i.e.} \quad z = f/g,$$

where f, $g \in \mathbf{C}[t]$, $(f, g) = 1$. Equation (2.19.1) gives

$$\frac{P_1^{e_1}}{P_3^{e_3}} = \frac{Q_1^{e_1}(f/g)}{Q_3^{e_3}(f/g)} = \frac{[g^{N_1}Q_1(f/g)]^{e_1}}{[g^{N_3}Q_3(f/g)]^{e_3}\ g^{e_3 w_3}}$$

$$\frac{P_2^{e_2}}{P_3^{e_3}} = x - 1 = \frac{Q_2^{e_2}(f/g)}{Q_3^{e_3}(f/g)} = \frac{[g^{N_2}Q_2(f/g)]^{e_2}}{[g^{N_3}Q_3(f/g)]^{e_3}\ g^{e_3 w_3}}$$

(with no loss in generality we take 3 as in the proof of Lemma 2.14).

By hypothesis $1 = (Q_1, Q_2)$. From this we deduce that $g^{N_1}Q_1(f/g)$ and $g^{N_2}Q_2(f/g)$ are relatively prime. Again $(P_1, P_2) = 1$ by hypothesis and the lemma follows from our relations.

COROLLARY 2.20. *If (P_1, P_2, P_3) and (Q_1, Q_2, Q_3) are both primitive solutions then the rational function f/g represents a homography.*

Proof. In this situation, equation (2.19.1) shows that z and t are rational functions of each other, hence z is homographic image of t. We can now formulate a converse of Theorem 2.11.

THEOREM 2.21. *If (Q_1, Q_2, Q_3) is a primitive solution of* (2.12) *then the equation*

$$-x = Q_1^{e_1}(t)/Q_3^{e_3}(t) \tag{2.21.1}$$

defines a galois extension $\mathbf{C}(t)$ of $\mathbf{C}(x)$ of degree M.

Proof. Consider

$$h(Y) = xQ_3^{e_3}(Y) + Q_1^{e_1}(Y)$$

as polynomial in Y with coefficients in $\mathbf{C}(x)$. Clearly t is a root and by the corollary if t_1 is another root then t_1 is homographic image of t. Thus h splits in $\mathbf{C}(t)$. It is easy to check that h is irreducible in $\mathbf{C}(x)[Y]$ and so deg $\mathbf{C}(t)/\mathbf{C}(x) = m$ as asserted.

Note: The extension $\mathbf{C}(t)/\mathbf{C}(x)$ is ramified only at 0, 1, ∞. This completes the demonstration that the finite subgroups of the group of homographies are uniquely determined up to inner automorphisms by the invariants e_1, e_2, e_3.

The existence of the indicated groups is demonstrated by writing down a primitive solution in each case.

For future use we note that these groups correspond to the case in which $s = 3$ in equation (2.1) and hence to the *non-cyclic* finite groups of homographies.

1. Dihedral group of order $2n$

$$(X_1^n + X_2^n)^2 - (X_1^n - X_2^n)^2 = 4(X_1X_2)^n$$

2. Tetrahedral

$$12\sqrt{-3}f^2 = \phi_1{}^3 - \phi_2{}^3$$
$$f = X_1X_2(X_1{}^4 - X_2{}^4)$$
$$\phi_1 = X_1{}^4 + 2\sqrt{-3}\,X_1{}^2X_2{}^2 + X_2{}^4$$
$$\phi_2 = X_1{}^4 - 2\sqrt{-3}\,X_1{}^2X_2{}^2 + X_2{}^4$$

3. Octahedral

$$W^3 - K^2 = 108f^4$$
$$f = X_1X_2(X_1{}^4 - X_2{}^4)$$
$$W = X_1{}^8 + 14X_1{}^4X_2{}^4 + X_2{}^8$$
$$K = X_1{}^{12} - 33X_1{}^8X_2{}^4 - 33X_1{}^4X_2{}^8 + X_2{}^{12}$$

4. Icosahedral

$$1728f^5 = T^2 + H^3$$
$$f = X_1X_2(X_1{}^{10} + 11X_1{}^5X_2{}^5 - X_2{}^{10})$$
$$H = -(X_1{}^{20} + X_2{}^{20}) + 228(X_1{}^{15}X_2{}^5 - X_1{}^5X_2{}^{15}) - 494X_1{}^{10}X_2{}^{10}$$
$$T = X_1{}^{30} + X_2{}^{30} + 522(X_1{}^{25}X_2{}^5 - X_1{}^5X_2{}^{25}) - 10005(X_1{}^{20}X_2{}^{10} - X_1{}^{10}X_2{}^{20})$$

3. **Pullbacks on the gauss sphere.** We consider hypergeometric differential equations in normalized form: For λ, μ, ν elements of $\mathbf{C}$, let

$$L_{\lambda,\mu,\nu} = D^2 + \frac{A}{x^2} + \frac{B}{(x-1)^2} + \frac{C}{x(x-1)} \qquad (3.1)$$

with

$$D = \frac{d}{dx}$$

$$4A = 1 - \lambda^2$$

$$4B = 1 - \mu^2$$

$$4C = \lambda^2 + \mu^2 - \nu^2 - 1.$$

This is the unique second order differential equation with rational coefficients, singular points only at 0, 1, ∞, with constant wronskian and with exponent differences λ, μ, ν at 0, 1, ∞ respectively.

If (Q_1, Q_2, Q_3) is a primitive solution of (2.12) then by setting

$$-x = Q_1^{e_1}(t)/Q_3^{e_2}(t) \tag{3.2}$$

we obtain an extension $K_0 = \mathbf{C}(t)$ of $\mathbf{C}(x) = K$ such that the pullback of $L_{\lambda,\mu,\nu} = L$ in the sense of Section 1 has $e_1\lambda$, $e_2\mu$, $e_3\nu$ as exponent differences. In particular if

$$(\lambda, \mu, \nu) = \left(\frac{1}{e_1}, \frac{1}{e_2}, \frac{1}{e_3}\right)$$

then the pullback has only unity as exponent difference, and no logarithmic singularities and is Fuchsian. Hence the pullback has no singularities and hence is d^2/dt^2.

The function t is a ratio of solutions of this equation and so this algebraic function of x is a ratio of solutions of $L_{1/e_1,1/e_2,1/e_3}$. We summarize:

THEOREM 3.2.1. *If* $\sum_{i=1}^{3} (1/e_i) > 1$ *then all solutions of* $L_{1/e_1,1/e_2,1/e_3}$ *are algebraic and* t *given by* (3.2) *is a ratio of solutions and the group of* $\mathbf{C}(t, x)/\mathbf{C}(x)$ *is dihedral, tetrahedral, octahedral, icosahedral depending on the values of* (e_1, e_2, e_3).

Let

$$L = \frac{d}{dt^2} + Q(t) \tag{3.3}$$

be a *normalized* second order linear differential operator defined over $\mathbf{C}(t)$ whose solutions are *locally algebraic.* Let τ be a ratio of solutions and let G be the group of homographies corresponding to the action of monodromy upon τ. As is well known the finiteness of G is equivalent to the condition that all solutions of L be algebraic functions. We will refer to G as the *projectivized* monodromy group of L.

THEOREM 3.4. (Klein) *If G is finite but not cyclic then L is the pullback by a rational map of one of the hypergeometric operators $L_{1/e_1,1/e_2,1/e_3}$ with $\sum_{i=1}^3 (1/e_i) > 1$. Conversely such a pullback has a finite projectivized monodromy group.*

Proof. The converse follows directly from Theorem 3.2.1.

If G is finite non-cyclic then by Theorem 2.11 there exists a primitive solution of (2.12) for suitable e_1, e_2, e_3 such that $Q_1^{e_1}/Q_3^{e_3}$ is invariant under G. Thus

$$-\xi(t) = \frac{Q_1^{e_1}}{Q_3^{e_3}}(\tau(t)) \tag{3.5}$$

is invariant under analytic continuation on the t-sphere, and hence is a rational function of t. Comparing this with equation (3.2) we see that $\tau(t)$ is a ratio of solutions of $L_{1/e_1,1/e_2,1/e_3}$ if we replace x by $\xi(t)$. This completes the proof of the theorem.

Suppose now that the projectivized group of L is finite and not cyclic. Hence according to the theorem L is a pullback of some $L_{1/e_1,1/e_2,1/e_3}$ by a rational map. This map need not be unique. To see this let $L_0 = d^2/dt^2$ so that L_0 is pullback of $L_{1/e_1,1/e_2,1/e_3}$ $(\Sigma\ (1/e_i) > 1)$ by means of equation (3.2). If we put

$$t' = \sigma t$$

where σ is an arbitrary homography and put

$$-X_1 = \frac{Q_1^{e_1}}{Q_3^{e_3}}(t') \tag{3.6}$$

then L_0 is again pullback of $L_{1/e_1,1/e_2,1/e_3}$ by X_1. If X_1 were to equal X for all σ, we could conclude that X is a constant.

The point in this example is that L_0 has the trivial projectivized group while $L_{1/e_1,1/e_2,1/e_3}$ has a non-trivial group. In general if L is a pullback of $L_{1/e_1,1/e_2,1/e_2}$ then the group of L is a subgroup of that of $L_{1/e_1,1/e_2,1/e_3}$.

THEOREM 3.7. (Klein [12]). *If L and $L_{1/e_1,1/e_2,1/e_3}$ have the same group and if $e_1 \neq e_2$ then the pullback mapping is unique.*

Proof. By hypothesis there exists $x \in \mathbf{C}(t)$ such that τ, a ratio of solutions of L, satisfies

$$-x = \frac{Q_1{}^{e_1}}{Q_3{}^{e_3}}(\tau). \tag{3.7.1}$$

Field theoretically we have the lattice shown.

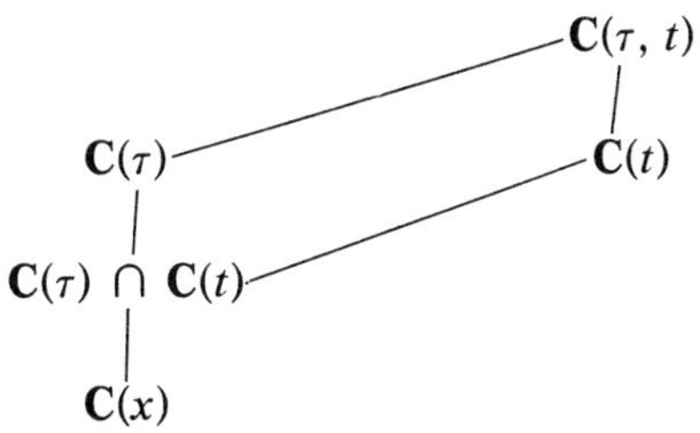

The projectivized group G of L coincides with the galois group of $\mathbf{C}(\tau, t)/\mathbf{C}(t)$ and hence by the usual identification with the group of $\mathbf{C}(\tau)/(\mathbf{C}(\tau) \cap \mathbf{C}(t))$. On the other hand the group of $L_{1/e_1,1/e_2,1/e_3}$ coincides with the group of $\mathbf{C}(\tau)/\mathbf{C}(x)$. The hypothesis of equality of groups is equivalent to the assertion that

$$C(\tau) \cap \mathbf{C}(t) = \mathbf{C}(x). \tag{3.7.2}$$

This condition fixes x up to homography. By an elementary calculation, L as given by (3.3) is the pullback of

$$L_0 = \frac{d^2}{dx^2} + q(x)$$

under the mapping $x = x(t)$ if

$$Q(t) = q(x) \cdot x'^2 + \frac{1}{2}\left(\frac{x''}{x'}\right)' - \frac{1}{4}\left(\frac{x''}{x'}\right)^2, \quad \left(x' = \frac{dx}{dt}\right) \cdot \tag{3.7.3}$$

a formula which may also be written in terms of the Schwarzian derivative. The key point is that

$$\frac{1}{2}\left(\frac{x''}{x'}\right)' - \frac{1}{4}\left(\frac{x''}{x'}\right)^2 = \frac{1}{2}\{x, t\} \tag{3.7.4}$$

remains invariant under homographies

$$x \mapsto \frac{ax+b}{cx+d}, \qquad ad - bc = 1$$

Since x is fixed up to homography, equation 3.7.3 implies that if $y = y(t)$ is also a pullback mapping then

$$q(x)x'^2 = q(y)y'^2 \tag{3.7.5}$$

$$y = \frac{ax+b}{cx+d} \tag{3.7.6}$$

and q is given by (3.1) with $(\lambda, \mu, \nu) = (1/e_1, 1/e_2, 1/e_3)$. Note that A, B, C in (3.1) are each *distinct* from zero. This shows that

$$\frac{A}{x^2} + \frac{B}{(x-1)^2} + \frac{C}{x(x-1)}$$

$$= \left[\frac{A}{y^2} + \frac{B}{(y-1)^2} + \frac{C}{y(y-1)}\right]\frac{1}{(cx+d)^4} \tag{3.7.7}$$

We consider four cases. Since a and c cannot simultaneously be zero we exhaust all possibilities.

Case 1 c, a, a-c all distinct from zero.

The right side of (3.7.7) lies in $x^{-4}\,\mathbf{C}[[x^{-1}]]$. The corresponding Laurent series for the right side is

$$(A + B + C)x^{-2} + (2B + C)x^{-3} \bmod x^{-4}$$

This shows that

$$A + B + C = 0$$
$$2B + C = 0$$

and hence $A = B$ contrary to hypothesis.

Case 2 $a = 0, c \neq 0$.

Here we may let $c = 1$, $b = -1$. The right side of 3.7.7 is

$$\frac{A}{(x + d)^2} + \frac{B}{(x + 1 + d)^2(x + d)^2} + \frac{C}{(x + 1 + d)(x + d)^2}$$

Comparing poles we see that as sets $\{0, 1\} = \{-d, -d - 1\}$. Hence $d = -1$ and now multiplying both sides of (3.7.7) by x^2 and setting $x = 0$ we obtain $A = B$ contrary to hypothesis.

Case 3 $c = 0, a \neq 0$.

We may set $a = 1 = d$. Proceeding as in the previous case we obtain $\{0, 1\} = \{-b, 1 - b\}$ as sets. Hence $b = 0$ and therefore $y = x$ as asserted.

Case 4 $a = c \neq 0$.

Here we may let $a = c = 1$, $d = b + 1$. Proceeding as in Case 2, $\{0, 1\} = \{-b, -b - 1\}$ as sets. Hence $b = -1$ and now multiplying both sides of (3.7.7) by $(x - 1)^2$ and setting $x = 1$, we obtain $A = B$ contrary to hypothesis.

THEOREM 3.8 (Klein [12]) *In the notation of equation 3.7.1, let*

$$\tau = y_1/y_2$$

with y_1, y_2 *solutions of L, and let*

$$w = y_2y_1' - y_1y_2',$$

(prime denotes differentiation with respect to t). Let $Q_3(y_1, y_2)$ denote the form of degree M/e_3,

$$Q_3(y_1, y_2) = y_2^{M/e_3} Q_3(\tau)$$

Then

$$Q_3(y_1, y_2) = C\left[\frac{w}{x'} x^{1-(1/e_1)}(1-x)^{1-(1/e_2)}\right]^{M/2e_3} \tag{3.8.1}$$

for some constant C.

Proof. Taking the derivative of both sides of equation (3.7.1) and applying (2.15.1) gives

$$x' = C\frac{w}{y_2^2}\frac{Q_1^{e_1-1}Q_2^{e_2-1}}{Q_3^{e_3+1}}$$

We multiply by $y_2^2 Q_3^{2e_3/M}/x'$. This gives

$$Q_3(y_1, y_2)^{2e_3/M} = \frac{Cw}{x'}\left(\frac{Q_1^{e_1}}{Q_3^{e_3}}\right)^{1-(1/e_1)}\left(\frac{Q_2^{e_2}}{Q_3^{e_3}}\right)^{1-(1/e_2)}$$

Equation (3.8.1) now follows from 2.15, 2.12.

4. **Cyclic Case.** Let

$$L = D^2 - Q, \qquad D = \frac{d}{dt} \tag{4.1}$$

be a given 2nd order differential operator with $Q \in \mathbf{C}(t)$. We again assume that all solutions of L are locally algebraic. Our object is to decide in a finite number of steps whether the projectivized group is cyclic. While there are of course an infinite set of cyclic groups, in the present situation there is an a-priori bound for the order of the group. The group G is generated in any case by the transformations corresponding to local monodromy around singular points. In general this gives no upper bound for the order of G but if G is cyclic then its order is bounded by the least common multiple of the denominators of the exponent differences. This will not be used in the following.

Let τ be a ratio of solutions of L. If G is cyclic then replacing τ by some homographic image if necessary, we may suppose that the elements of G are of the form

$$\tau \mapsto w\tau \tag{4.2}$$

where w runs through the mth roots of unity in $\mathbf{C}$. Hence τ^m is invariant under the monodromy group and thus an element of $\mathbf{C}(t)$, say

$$\tau^m = \xi \in \mathbf{C}(t). \tag{4.3}$$

Then by a classical calculation two independent solutions of L are given by

$$\begin{aligned} u &= \sqrt[2]{1/(\xi^{1/m})}, \\ v &= u\xi^{1/m}. \end{aligned} \tag{4.4}$$

This shows that u'/u and v'/v lie in $\mathbf{C}(t)$.

THEOREM 4.4. *Let L as given by* (4.1) *have solutions which are locally algebraic. Then the projectivized group of L is cyclic if and only if the Riccati equation*

$$y' + y^2 = Q \tag{4.5}$$

has two distinct rational solutions.

Proof. The necessity of this condition has been explained. Conversely if y is a rational solution of (4.5) then y can have only simple poles since by hypothesis the singularities of L are all regular which means that the poles of Q are at most of order two. If we write

$$Q = \sum_{i=1}^{n} \frac{A_i}{(t-\alpha_i)^2} + \frac{B_i}{(t-\alpha_i)} \tag{4.6}$$

then we must have

$$\sum_{i=1}^{n} B_i = 0 \tag{4.7.1}$$

$$-4A_i = 1 - \lambda_i^2 \tag{4.7.2}$$

where λ_i is the exponent difference at α_i ($i = 1, \ldots, n$)

$$1 - \lambda_\infty^2 = -4 \sum_{i=1}^{n} (A_i + \alpha_i B_i). \tag{4.7.3}$$

We may write a rational solution of (4.5) in the form

$$\eta = \sum_{i=1}^{n} \frac{c_i}{t - \alpha_i} + \sum_{j=1}^{N} \frac{1}{t - \theta_j} \tag{4.8}$$

where $\theta_1, \ldots, \theta_N$ are all distinct and are distinct from the singularities of L. Indeed equation (4.5) implies that

$$L = (D + \eta)(D - \eta) \tag{4.9}$$

and so each exponent of $D - \eta$ must be an exponent of L which shows that

$$c_i = \frac{1 \pm \lambda_i}{2} \qquad i = 1, \ldots, n \tag{4.10}$$

and these numbers are rational by hypothesis. Now corresponding to the rational solution of (4.5) given by (4.8), there is an algebraic solution of L given by

$$y = \prod_{j=1}^{N} (t - \theta_j) \cdot \prod_{i=1}^{n} (t - \alpha_i)^{c_i} \tag{4.11}$$

If equation (4.5) has two distinct solutions, then the ratio, τ, of the corresponding solutions of L would be the ratio of two functions of the form given by 4.11 and hence $\tau^m \in \mathbf{C}(t)$ for suitable integer m. This completes the proof of the Theorem.

COROLLARY 4.12. *Each rational solution of* (4.5) *is of the form of equation* (4.8) *where* $(c_1, \ldots, c_n)$ *satisfies* (4.10) *and*

$$N + \sum_{i=1}^{n} c_i + \frac{-1 \pm \lambda_\infty}{2} = 0.$$

Proof. We need only check the estimate for N. As in the proof of the theorem, the exponent of $D - \eta$ at infinity must also be an exponent of L at infinity. The corollary follows directly.

What is the set of all rational solutions of the Riccati equation?

THEOREM 4.13. *If the Riccati equation* (4.5) *has more than two solutions in* $\mathbf{C}(t)$ *then the projectivized group of* L *is trivial,* (*i.e. all the exponent differences are integers*).

Proof. Let η_1, η_2, η_3 be solutions of equation (4.5) and let u_i be solution of L, such that

$$\eta_i = u_i'/u_i \qquad (i = 1, 2, 3)$$

Then

$$\eta_i - \eta_j = w_{i,j}/u_i u_j$$

where $w_{i,j}$ is the associated wronskian of L. Let $\tau = u_2/u_1$, so

$$\tau = \frac{1}{u_1 u_2}\Big/\frac{1}{u_2 u_3} = \text{const.} \cdot \eta_1 - \eta_3/\eta_2 - \eta_3.$$

Thus τ lies in $\mathbf{C}(\eta_1, \eta_2, \eta_3)$. This completes the proof.

5. **Dihedral case. Part I.** Let L be as in equation (4.1). We now wish to determine whether the projectivized group is dihedral.

THEOREM 5.1. *If the projectivized group of* L *is dihedral then the Riccati equation* (4.5) *has a solution in a quadratic extension of* $\mathbf{C}(t)$.

Proof. The conventional hypergeometric differential operator

$$x(1 - x)D^2 + (\gamma - (\alpha + \beta + 1)x)D - \alpha\beta \tag{5.1.1}$$

has exponent differences $1 - \gamma, \gamma - \alpha - \beta, \alpha - \beta$ at $0, 1, \infty$ and no other singularities. Putting

$$\alpha = \frac{1 - \mu}{2}, \qquad \beta = -\frac{\mu}{2}, \qquad \gamma = \frac{1}{2}$$

we obtain a differential equation with exponent differences ½, μ, ½ which by explicit calculations (going back to Gauss) has $(1 + \sqrt{x})^\mu$ and $(1 - \sqrt{x})^\mu$ as solutions. If we replace x by $x(t)$ then the normalized pull-back is satisfied by $(1 \pm \sqrt{x})^\mu/\sqrt{w}$ where $w = (1 - x)^{\mu-1}(\sqrt{x})'$. From this the Riccati equation has solution

$$\eta = \frac{\mu}{2}\frac{x'}{x(1-x)}\sqrt{x} - \frac{1}{4}\frac{R'}{R} \tag{5.2}$$

where

$$\sqrt{R} = \frac{\mu}{2}\frac{x'}{x(1-x)}\sqrt{x}.$$

Now by Theorem 3.2, if L has a dihedral group of order $2m$ then L is the pullback of $L_{1/2,1/m,1/2}$ by a rational map $x = x(t)$. This completes the proof of the theorem.

We now study the existence of solutions of equations (4.5) in quadratic extensions.

LEMMA 5.3 (Fuchs). *Let L_2 be the third order linear differential equation satisfied by all binary quadratic forms in y_1, y_2, a pair of independent solutions of L. Suppose that the Ricatti equation* (4.5) *has no solution in* $\mathbf{C}(t)$. *Then equation* (4.5) *has a solution in a quadratic extension field of* $\mathbf{C}(t)$ *if and only if L_2 has a solution whose square lies in* $\mathbf{C}(t)$.

Proof. If equation (4.5) has a solution η_1 in a quadratic extension field, then we may write

$$\eta_1 = \gamma + \sqrt{R} \tag{5.3.1}$$

with γ and R in $\mathbf{C}(t)$. By hypothesis $\sqrt{R} \notin \mathbf{C}(t)$ and hence $\eta_2 = \gamma - \sqrt{R}$ is a distinct solution of (4.5). We may choose y_i solution of L $(i = 1, 2)$ such that

$$\eta_i = y_1'/y_i \tag{5.3.2}$$

and then

$$2\sqrt{R} = \eta_1 - \eta_2 = w/y_1y_2 \tag{5.3.3}$$

where w is the wronskian of L. Since w is a constant, we conclude that $1/\sqrt{R}$ is a constant multiple of $y_1 y_2$ and hence satisfies L_2. This proves the assertion is one direction.

Conversely if $z \in \mathbf{C}(t)$ and $\sqrt{z}$ is a solution of L_2 then $\sqrt{z}$ is a quadratic form in solutions of L. Since such a quadratic form may be factored there are two possibilities:

$$\sqrt{z} = u_1{}^2 \tag{5.3.4}$$

$$\sqrt{z} = u_1 u_2 \tag{5.3.5}$$

where u_1, u_2 are independent solutions of L. The first case is ruled out by the hypothesis that equation 4.5 has no solution $\frac{1}{4}\, z'/z \in \mathbf{C}(t)$. We now put $\eta_i = u_i'/u_i$ $(i = 1, 2)$ and calculate from (5.3.5),

$$\eta_1 + \eta_2 = \frac{1}{2}\frac{z'}{z}, \tag{5.3.6}$$

while precisely as 5.3.3

$$\eta_1 - \eta_2 = w/u_1 u_2 = w/\sqrt{z} \tag{5.3.3$'$}$$

where again w is a determination of the wronskian of L. Equations (5.3.3′), (5.3.6) show that (4.5) has a solution in a quadratic extension of $\mathbf{C}(t)$. This completes the proof of the lemma.

We now briefly discuss the existence of a solution z of L_2 whose square lies in $\mathbf{C}(t)$. Such a solution must be of the form

$$t = \prod_{i=1}^{n} (t - \alpha_i)^{k_i} \circ g \tag{5.4}$$

where $\alpha_1, \ldots, \alpha_n$ are the finite singularities of L, g is a polynomial different from zero at each α_i. Each k_i must be a half integer with

$$k_i = 1 \pm \lambda_i, 1 \tag{5.5}$$

while

$$\Sigma\, k_i + \deg g = 1, 1 \pm \lambda_\infty . \tag{5.6}$$

This gives a finite set of possibilities for the k_i and the degree of g and the determination of g is reduced to a problem in linear algebra.

6. **Dihedral Case. Part II.** Let L be a 2nd order differential operator as in equation (4.1). In Section 5 we showed that if L has the dihedral group then L decomposes into linear factors in a quadratic extension of $\mathbf{C}(t)$. Such decompositions were studied in that section. To determine whether the group is indeed dihedral we must now be prepared to answer the following question: Given η algebraic over $\mathbf{C}(t)$ does the equation

$$y' = y\eta \tag{6.1}$$

have a solution algebraic over $\mathbf{C}(t)$? (For our application we may assume η is in a quadratic extension of $\mathbf{C}(t)$). This is equivalent to the condition that there exist $m \in \mathbf{N}$ such that $m\eta$ is logarithmic derivative of element of $\mathbf{C}(t, \eta)$. The difficulty is that of finding an a-priori bound for m.

In the classical theory of hypergeometric functions this problem did not arise, as, for $n = 2$, the field $\mathbf{C}(t, \eta)$ is either $\mathbf{C}(\sqrt{t})$, $\mathbf{C}(\sqrt{t-1})$ or $\mathbf{C}(t, \sqrt{t(t-1)})$ and so in all cases of genus zero and hence equation (6.1) may be analyzed without difficulty.

We now consider a generalization of the above problem.

Let ω be a differential with at most simple poles and with rational residues at each pole on a curve C defined over a field K of characteristic zero. To determine in a finite number of steps whether $m\omega$ is a logarithmic differential for some (unknown) $m \in \mathbf{Z}$.

By solving this question we solve the problem of Section 5. As general references for this discussion see [13, 16]. At each point P of C we write

$$\omega = n_p \frac{dt_p}{t_p} + \omega_p. \tag{6.2}$$

where t_p is a local parameter at P, ω_p is regular at t_p and n_p is a rational number which we may assume to be in $\mathbf{Z}$. Let

$$\mathfrak{L} = \Sigma\, n_p P \tag{6.3}$$

a divisor of C of degree zero.

There are two steps

I. Find an integer m such that either $m\mathfrak{L}$ is principal or $\mathfrak{L}$ is of infinite order in the Jacobian of C.

II. Decide whether $m\mathfrak{L}$ is principal and if it is find θ in the function field of C such that

$$(\theta) = m\mathfrak{L}$$

Suppose I and II carried out, then put

$$\omega_1 = \frac{d\theta}{\theta} - m\omega. \tag{6.4}$$

We note that ω_1 is a differential of the first kind. If $\omega_1 = 0$ then the problem has an affirmative solution. If $\omega_1 \neq 0$ then $N\omega_1$ is never logarithmic regardless of $N \in \mathbf{N}$ and hence the problem has a negative solution.

I. We choose a non-singular model for C in $\mathbf{P}^3$. Let K be a field of definition of both C and of the divisor $\mathfrak{L}$.

Case 1. K is an algebraic number field.

Let $\mathfrak{p}_1$, $\mathfrak{p}_2$ be primes of K extending distinct rational primes p_1, p_2 such that the reductions $\overline{C_1}$, $\overline{C_2}$ of C are non-singular. Let G_j $(j = 1, 2)$ be the group of points on the Jacobian $J(\overline{C_j})$ of the reduced curve $\overline{C_j}$ which are rational over the residue class field $\overline{K_j}$ and which are of order prime to p_j. Let e_j be the exponent of G_j. Thus $(e_j, p_j) = 1$ and

$$e_j G_j = 0. \tag{6.5}$$

LEMMA 6.6. *If the image of $\mathfrak{L}$ in $J(C)$ is of finite order then*

$$e_1 e_2 \mathfrak{L} \sim 0.$$

Proof. If $\mathfrak{L}$ is of finite order then there exists $p_1{}^i m$, $(p_1, m) = 1$ such that

$$p_1{}^i m\mathfrak{L} \sim 0. \tag{6.6.1}$$

Hence $p_1{}^i\mathfrak{L}$ is of order prime to p_1. But the mapping

$$J(C)(K) \to J(\overline{C}_1)(\overline{K}_1)$$

of the group of K-rational points of $J(C)$ into the group of $\overline{K}_1$-rational points of $J(\overline{C}_1)$, is injective on the subgroup of elements of order prime to p_1. The image of $p_1{}^i\mathfrak{L}$ is rational over $\overline{K}_1$ and of order prime to p_1, hence is annihilated by e_1. Thus

$$e_1 p_1{}^i\mathfrak{L} \sim 0 \tag{6.6.2}$$

likewise

$$e_2 p_2{}^j\mathfrak{L} \sim 0$$

for some unknown j. We choose $a, b \in \mathbf{Z}$ such that

$$ap_1{}^i + bp_2{}^j = 1$$

and conclude that

$$e_1e_2\mathfrak{L} = e_1e_2(ap_1{}^i\mathfrak{L} + bp_2{}^j\mathfrak{L}) \sim 0.$$

This completes the proof of the lemma.

Case 2. K is finitely generated over $\mathbf{Q}$.

We choose a specialization $K \to K'$ such that K' is an algebraic number field and $C \to C'$, a non-singular curve. Since

$$J(C)(K) \underset{\text{tors}}{\longrightarrow} J(C')(K')_{\text{tors}}$$

is an injection, we may repeat case 1. This completes the treatment of problem I.

II. We choose a plane curve model for C with only ordinary singular points $P_1, \ldots, P_n$ the multiplicity of P_j being s_j. Let $D = \sum_{j=1}^n s_jP_j$. Put

$$L = L_0 - L_\infty$$

L_0 and L_∞ being positive divisors.

Construct a curve B having intersection with C containing $D + L_0$. Write

$$C \cap B = D + L_0 + X,$$

where X is a positive divisor. Let $l =$ degree B. Then

$$L_0 \sim L_\infty$$

if and only if there exists a curve B' of degree l such that

$$B' \cap C = L_\infty + D + X.$$

This construction involves only linear algebra and the calculation of X. If $L_0 \sim L_\infty$ then $L_0 - L_\infty = (B/B')$. The solution of part II involves the application of this procedure to $m\mathcal{L}$.

7. **Decision Procedure.** Let L be a second order differential operator defined over $\mathbf{C}(t)$. We may suppose L is given by (3.3). Our object is to give a decision procedure for determining whether all the solutions are algebraic.

We may assume that all singularities of L are regular, and that all exponents are rational numbers. Our object is to determine whether the projective monodromy group G of L is one of the five finite types. We apply the procedure of Section 4 to determine whether G is cyclic. If it is we are done. Conversely if we exclude the possibility that G is cyclic, we then use the procedure of Section 5, Section 6 to determine whether G is dihedral. Having decided that G is neither cyclic nor dihedral we consider successively the three remaining finite possibilities, i.e. G may be the tetrahedral, octahedral or icosahedral group. The reason for proceeding in this sequence is we wish to use the uniqueness theorem of Klein (3.7).

We now consider whether L is the pullback by $x = x(t)$ of $L_{1/e_1,1/e_2,1/e_3}$ (cf. equation 3.1) where

$$\left(\frac{1}{e_1},\frac{1}{e_2},\frac{1}{e_3}\right)=\left(\frac{1}{2},\frac{1}{3},\frac{1}{3}\right)$$

$$\left(\frac{1}{2},\frac{1}{3},\frac{1}{4}\right)$$

$$\left(\frac{1}{2},\frac{1}{3},\frac{1}{5}\right)$$

and where we may assume that we have already excluded previous elements in this list of three possibilities.

We are to decide whether there exists a rational function x which satisfies (3.7.3). Here $q(x)$ is given explicitly by (3.1), i.e.

$$4q(x)=\frac{1-\dfrac{1}{e_1{}^2}}{x^2}+\frac{1-\dfrac{1}{e_2{}^2}}{(x-1)^2}+\frac{\dfrac{1}{e_1{}^2}+\dfrac{1}{e_2{}^2}-\dfrac{1}{e_3{}^2}-1}{x(x-1)}$$

Let $a_1, \ldots, a_n, \infty$ be the singularities of L. Let the exponent differences be $\lambda_1, \lambda_2, \ldots, \lambda_n, \lambda_\infty$. We know that these numbers are rational and we take them to be *positive.* We define

$$\Delta(L)=(\lambda_1+\cdots+\lambda_n+\lambda_\infty)-(n-1)$$

Thus

$$\Delta(L_{1/e_1,1/e_2,1/e_3})=\frac{1}{e_1}+\frac{1}{e_2}+\frac{1}{e_3}-1.$$

It follows from Section 1 that if L is a pullback of $L_{1/e_1,1/e_2,1/e_3}$ by $x = x(t)$ then the mapping degree (i.e. the maximum of the degrees of numerator and denominator) must be

$$M \doteq \Delta(L)/\Delta(L_0).$$

We normalize by insisting that the denominator is monic and we interpret equation 3.7.3 as a set of algebraic relations among the at most $2M + 1$ unknown coefficients of the numerator and denominator of x (we must consider $M + 1$ possibilities as the exact degree of the denominator is not fixed). We thus arrive at $M + 1$ algebraic sets which have in all, at most one rational point. By elimination theory we should not only be able to decide whether a rational solution of 3.7.3 exists but we should indeed be able to determine the solution itself.

8. **Conjecture of Grothendieck.** Let L again be a second order differential operator but for simplicity we suppose that the coefficients lie in $\mathbf{Q}(t)$. For almost all p we may consider L_p, the reduction of L modulo p on $\mathbf{F}_p(t)$. Clearly L_p acts as linear operator on $\mathbf{F}_p(t)$ as vector space over $\mathbf{F}_p(t^p)$. Let K_p be the dimension of the kernel of L_p viewed as vector space over $\mathbf{F}_p(t^p)$.

As a special case of a general conjecture of Grothendieck, it is conjectured:

(8.1) If $K_p = 2$ for almost all p then all the solutions of L are algebraic functions.

The converse of this statement is known.

The general conjecture has been verified by Katz [8] for the case in which L is a suitable direct factor of the Fuchs-Picard equation. In particular it is known to be valid in the case of second order operators with 3 singular points. The conjecture is completely open in the case of Heun's differential equation [17, p. 576] (second order, four singular points), the point being that in the case of 3 singular points, the integral formula of Euler reveals the cohomological nature of the gauss hypergeometric functions but no integral formula is known for the solutions of the Heun differential equation.

For the case of 3 singular points there is a simple algorithm for the calculation of K_p. For 4 or more singular points no such algorithm is known. Thus even if Grothendieck's conjecture were confirmed, it is not clear that it would give a response to problem (0.2).

It may be useful to explain briefly how one could use Katz's result to verify Schwarz's list. We know from Ihara that the differential equation for $F(a, b; c; x)$ has $K_p = 2$ if and only if the minimal representative mod p of c lies between the corresponding representatives of a and of b. Using the criterion of Grothendieck it is easy to check that each equation in Schwarz's list has only algebraic solutions.

To show that Schwarz's list is exhaustive one must start with the fact that the list consists of *classes* of exponent differences. Two sets $(\lambda_1, \lambda_2, \lambda_3)$ and $(\lambda_1', \lambda_2', \lambda_3')$ of exponent differences are said to be equivalent [14, p. 119] if

$$\lambda_1' = \epsilon_1\lambda_i + \mu_i$$

where $\epsilon_i = \pm 1$, each $\mu_i \in \mathbf{Z}$ and $\mu_1 + \mu_2 + \mu_3 \equiv 0 \bmod 2$. The central point is that if the sets of exponent differences are equivalent then the monodromy groups are isomorphic. This can be shown either by Riemann's method of calculating the monodromy group or by Gauss's relations among contiguous hypergeometric functions [5, Section 34].

Consequently we may assume that $(\lambda_1, \lambda_2, \lambda_3)$ form a reduced set of exponent differences, i.e.

$$i \geq \lambda_1 > 0 \tag{8.2}$$

$$1 \geq \lambda_1 + \lambda_j \quad \text{for} \quad i \neq j. \tag{8.3}$$

We follow Schwarz's theory of curvilinear triangles to show that if all solutions are algebraic then

$$\sum_{i=1}^{3} \lambda_i > 1. \tag{8.4}$$

We now use Section 4 to show that the cyclic case can only occur with exponent differences $(1/n, 1, 1/n)$, and the method of Section 5 to show that the dihedral monodromy group can occur only if the exponent differences are $1/2$, $1/2$, $1/m$. This leaves only the tetrahedral, octahedral and icosahedral groups. The elements of these groups have orders 2, 3, 4, 5, and hence the λ_i have these integers as denominators. This condition together with (8.2) shows that there are only a finite number of possibilities; use of (8.3) and (8.4) further reduces the list. Furthermore orders 4 and 5 do not occur simultaneously. By these considerations we arrive at the Schwarz list together with the additional candidates:

Exponent differences	Spherical excess
$\frac{1}{2}, \frac{2}{5}, \frac{2}{5}$	$\frac{3}{10}$
$\frac{2}{3}, \frac{1}{3}, \frac{1}{4}$	$\frac{1}{4}$
$\frac{2}{5}, \frac{2}{5}, \frac{1}{3}$	$\frac{2}{15}$
$\frac{2}{3}, \frac{2}{5}, \frac{1}{5}$	$\frac{4}{15}$
$\frac{3}{5}, \frac{2}{5}, \frac{1}{5}$	$\frac{1}{5}$
$\frac{3}{5}, \frac{2}{5}, \frac{2}{5}$	$\frac{2}{5}$
$\frac{3}{5}, \frac{1}{3}, \frac{1}{3}$	$\frac{4}{15}$

These are excluded by means of Grothendieck's criterion, i.e. by the converse of (8.1).

It is not clear that this mixture of methods of Schwarz and Katz is any improvement over the original method of Schwarz. We note that for the case of 4 singular points there is no theory of reduced set of exponent differences. In the case of 4 singular points, the monodromy group depends not only on the exponential differences and singular points but also upon an additional parameter. One does not know how to infer isomorphism of monodromy groups from relations concerning the parameters [1, p. 311-329].

REFERENCES

1. L. Bieberbach, *Theorie der gewöhnliclen differential gleichungen,* Berlin 1953.
2. M. Brioschi, "La théorie des formes dans l'integration des équations différentielles linéaires du second ordre" *Math. Ann.* **11** (1877), pp. 401-411.
3. A. Forsyth, *Theory of Differential Equations,* Vol. IV, Cambridge, 1900-1902.
4. E. Goursat, "Sur l'équation différentielle linéaire qui admet pour intégrale la série hypergéométrique," *Ann. Sc. de l'Ecole Norm. Sup.* **X** (1881) supplément, pp. 3-142.
5. ______, *Leçons sur les séries hypergéométriques,* Paris 1936, 1938.
6. G. H. Halphen, "Sur la reduction des équ. diff. lin. ..." *Oeuvres,* Vol. **3**, Paris 1921, pp. 1-260.
7. E. L. Ince, *Ordinary Differential Equations,* London 1927.
8. N. Katz, "Algebraic solutions of differential equations," *Inv. Math.* **18** (1972), pp. 1-118.
9. ______, "Deligne's work on Hilbert's 21st Problem" *AMS Symposia Pure Math.* **XXVIII** (1976), pp. 537-538.
10. F. Klein, "Lectures on the Icosahedron," London 1913.
11. ______, "Ueber lineare Differentialgleichungen," *Math. Ann.* **11** (1877), pp. 115-118.
12. ______, "Ueber lineare Diff." *Math. Ann.* **12** (1878), pp. 167-179.
13. S. Lang, *Abelian Varieties,* New York 1959.
14. E. Poole, *Linear Differential Equations,* Oxford 1936.
15. H. A. Schwarz, "Ueber diejenigem Fälle, in welchen die Gaussische hypergeometrische Reihe eine algebraische Function ihres vierten Elementes darstellt.," *Journal f.d. reine und angew. Math.* **75**, pp. 292-335.
16. R. J. Walker, *Algebraic Curves,* Princeton 1950.
17. Whittaker and Watson, *Course of Modern Analysis,* 4th Edition, Cambridge 1927.

ON A THEOREM OF BOGOMOLOV ON CHERN CLASSES OF STABLE BUNDLES

By D. GIESEKER*

Let k be an algebraically closed field. Let X be a normal projective variety of dimension n defined over k and let H be a very ample divisor on X. Let E be a coherent torsion free sheaf on X. The rank of E, denoted rk E, is the rank of E at a generic point of X. We can find divisors $H_1, \ldots, H_{n-1}$ linearly equivalent to H so that the scheme theoretic intersection $H_1 \cap \cdots \cap H_{n-1} = C$ is a nonsingular curve on X, so that X is non-singular at each point of C, and so that E is locally free at each point of C. E_C will denote the restriction of E, so E_C is locally free on C. We define

$$\mu(E) = \frac{\deg E_C}{\operatorname{rk} E}.$$

One checks that $\mu(E)$ depends only on H and E.

Definition 0.1. E is weakly H semi-stable if whenever Q is a coherent torsion free quotient of E, then

$$\mu(E) \leq \mu(Q).$$

The aim of this note is to present proofs of the following two Theorems:

THEOREM 0.2 (char $k = 0$). If dim $X = 1$, and E is semi-stable, then $S^n(E)$ is semi-stable.

* Partially supported by N.S.F.

THEOREM 0.3 (char $k = 0$). If $\dim X = 2$, and E is weakly semi-stable bundle with respect to H, then

$$c_1^2(E) \leq \frac{2r}{r-1} \cdot c_2(E),$$

where $r = \operatorname{rk} E$.

Theorem 0.2 was deduced by Hartshorne when $k = \mathbf{C}$ using the relation between stability and unitary representations discovered by Narashiman and Seshadri. However, our proof is purely algebraic. Theorem 2 is due to Bogomolov. He deduces a somewhat stronger result by studying the sections of the symmetric powers of E and using geometric invariant theory. Our proof is to define X and E over a ring $R \subseteq k$ of finite type over $\mathbf{Z}$, to reduce E mod maximal ideals m of R, and to study sections of Frobenius powers $E_m{}^{(p)}$ of E, where $p =$ char R/m.

Mumford has shown that Kodaira's vanishing theorem for surfaces is a consequence of Bogomolov's result. We also give a proof of this result. Szpiro has found a proof of Kodaira vanishing in this case by reduction mod p.

We note Theorem 0.2 fails if char $k \neq 0$ [3].

We will abbreviate weakly H semi-stable to semi-stable when no confusion is likely to arise.

1.

LEMMA 1.1. If E is weakly semi-stable on X with respect to H, and if $f: Y \to X$ is a finite separable map with Y normal, then $f^*(E)$ is weakly semi-stable with respect to $f^{-1}(H)$.

Proof. Let F be a subsheaf of $f^*(E)$ so that $\mu(F) > \mu(E)$. Let $g: Y' \to Y$ be a finite map with Y' normal so that $K(Y')$ is Galois over $K(X)$, where $K(Y')$ is the function field of Y'. Let G be the Galois group of $K(Y')/K(X)$, and let $h = g \circ f$. Then

$$\mu(g^*(F)) > \mu(h^*(E)),$$

where μ is computed with respect to $h^*(H)$. Thus replacing Y' by Y, we may assume that Y is Galois over X, and $Y/G = X$. We may assume that if F' is any subsheaf of $f^*(E)$, then $\mu(F') \leq \mu(F)$. This implies F is semi-stable. Now there is a non-zero map

$$\varphi: \prod_{\sigma \in G} (F^{\sigma}) \to f^*(E),$$

where $F^{\sigma} = \sigma^*(F)$, $\sigma: Y \to Y$. Now $\prod_{\sigma \in G} F^{\sigma}$ is semi-stable, so if $F' = \operatorname{Im} \varphi$, then

$$\mu(F') \geq \mu(F) > \mu(f^*(E)).$$

However, F' is invariant under G and so descends to a sub-sheaf F'' of E. Then

$$\mu(F'') > \mu(E).$$

Thus our lemma is established.

THEOREM 1.2 (char $k = 0$). Let X be a non-singular projective curve and E be a semi-stable bundle. Let $\mathcal{O}(1)$ denote the tautological bundle on $\mathbf{P}(E)$. Then the following hold:

i) $S^n(E)$ is semi-stable.
ii) If $\deg E \geq 0$, $\mathcal{O}(1)$ is pseudo-ample.
iii) If $\deg E > 0$, E is ample.

See Hartshorne [5] for a transcendental proof of Theorem 1.2.

Proof. We prove ii) first. Recall that $\mathcal{O}(1)$ is pseudo-ample if $\mathcal{O}(1)$ has non-negative degree on each curve in $\mathbf{P}(E)$. So supposing $\mathcal{O}(1)$ is not pseudo-ample, there is a curve C' on $\mathbf{P}(E)$ so that $\mathcal{O}(1)$ has negative degree on C'. Let $\tilde{C}'$ be the normalization of C', let $p: \tilde{C}' \to X$ be the natural map, and L the pull back of $\mathcal{O}(1)$ to $\tilde{C}'$. Now L is a quotient of $p^*(E)$, and $\deg L < 0$. This contradicts the fact that $p^*(E)$ is semi-stable (Lemma 1.1). Thus $\mathcal{O}(1)$ is pseudo-ample.

Next we prove iii. By definition, E is ample if $\mathcal{O}(1)$ is ample. Suppose $\deg E > 0$. Let $f: Y \to C$ be a finite map of degree larger than $\operatorname{rk} E$ with Y a smooth curve, and let P be a point on Y. Then $\deg (f^*(E) \otimes \mathcal{O}(-P)) > 0$, and $f^*(E) \otimes \mathcal{O}(-P)$ is semi-stable. If L is the tautological bundle on $\mathbf{P}(f^*(E))$, and F is the divisor on $\mathbf{P}(f^*(E))$ consisting of the fiber over P, then $L(-F)$ is pseudo-ample, since $L(-F)$ is the tautological bundle for $f^*(E)(-P)$. If V is any subvariety of $\mathbf{P}(f^*(E))$ of codimension k, then $(L^k.V) > 0$. So L is ample by Nakai's criterion. Thus $f^*(E)$ is ample, and hence E is ample.

Now suppose $S^n(E)$ is not semi-stable. Let Q be a quotient of $S^n(E)$ so that $\mu(Q) < \mu(S^n(E))$. Choose a finite map $f: Y \to X$ of degree d so that

$$\frac{n}{d} < \mu(S^n(E)) - \mu(Q).$$

Then we can find an integer k so that

$$\mu(f^*(Q)) \leq nk < \mu(S^n(f^*(E))).$$

Let L be a line bundle of degree $-k$ on Y. Then $Q' = f^*(Q) \otimes L^{\otimes n}$ is a quotient $S^n(E')$, where $E' = f^*(E) \otimes L$. Further, E' is semi-stable,

$$\mu(E') = \frac{1}{n}(\mu(S^n(f^*(E))) - nk) > 0$$

and $\mu(Q') \leq 0$. Now E' is ample, and hence $S^n(E')$ is ample [6]. Let $\mathcal{O}(\xi)$ be the tautological bundle on $\mathbf{P}(S^n(E'))$, and let $q = \text{rk } Q'$. Then $\mathbf{P}(Q')$ is contained in $\mathbf{P}(S^n(E'))$, and

$$\deg Q' = \xi^q.\mathbf{P}(Q').$$

But ξ is ample since $S^n(E')$ is ample. Thus

$$\deg Q' > 0.$$

This contradicts $\mu(Q') \leq 0$. So $S^n(E)$ is semi-stable.

PROPOSITION 1.3 (char $k = 0$). Let U be a variety and $X \xrightarrow{\pi} U$ a smooth projective map so that all the fibers are curves. Let E be a bundle on X. Then there is an open $V \subseteq U$ so that one of the two following conditions holds:

i) For each closed point $P \in V$, E_P is semi-stable on X_P, where $X_P = \pi^{-1}(P)$.

ii) There is a sub-bundle F of E over $\pi^{-1}(V)$ so that for each $P \in V$, F_P is semi-stable on X_P and $\mu(F_P) > \mu(E_P)$.

Proof. We can suppose $U = \text{Spec } R$. Let K be the quotient field of R, $\eta = \text{Spec } K$, $\bar{\eta} = \text{Spec } \bar{K}$. Suppose $E_{\bar{\eta}}$ on $X_{\bar{\eta}} = X \times_{\eta} \bar{\eta}$ is semi-stable. Then Maruyama [7] has shown that (i) holds, regardless of the characteristic of k.

Now suppose $E_{\bar{\eta}}$ is not semi-stable. A sub-bundle F of $E_{\bar{\eta}}$ is called maximally destabilizing if whenever F' is any sub-bundle of $E_{\bar{\eta}}$, then $\mu(F) \geq \mu(F')$ and if $\mu(F) = \mu(F')$, then $\text{rk } F \geq \text{rk } F'$. Note that there is a unique maximally destabilizing sub-bundle F, and that F is semi-stable. The existence of F is clear. Suppose F and F' are maximally destabilizing. Both F and F' are semi-stable so $F \oplus F'$ is semi-stable with $\mu(F \oplus F') = \mu(F) = \mu(F')$. Let Q be the image of $F \oplus F'$ in $E_{\bar{\eta}}$. Then $\mu(Q) \geq \mu(F)$, so $\mu(Q) = \mu(F)$. Thus $Q = F$, so $F = F'$.

Now the Galois group Gal $(\bar{K}, K)$ acts on $E_{\bar{\eta}}$, and if F is maximally destabilizing, then F^{σ} is maximally destabilizing. Thus F is invariant under Gal $(\bar{K}, K)$. So there is a sub-bundle F' of E_{η} on X_{η} so that $F_{\bar{\eta}}' = F$. Now F' is defined over some open sub-set V' of U. Further, by applying the first part of this Proposition, we see that for some open $V \subseteq V'$, F_P is semi-stable for $P \in V$.

2.

In this section, X_k will be a smooth projective surface defined over an algebraically closed field k of characteristic zero. Let E_k be a locally free sheaf of rank r on X_k. There is a subring R of k of finite type over $\mathbf{Z}$, a scheme $X_R \xrightarrow{\pi} \text{Spec } R$ so that π is projective and smooth, and a locally free sheaf E_R on X_R so that $X_R \times_R k = X_k$ and $E_R \times_R k = E_k$. For each maximal ideal $m \subseteq R$, let $k_m = R/m$ and $p_m = \text{char } R/m$. Note $p_m > 0$. Denote $X \times_R k_m$ by X_m, and $E \times_R k_m$ by E_m. If $\{U_\alpha\}$ is an open cover so that E_m is given by transition matrices $M_{\alpha\beta} \in GL(r, \mathcal{O})$ $(U_\alpha \cap U_\beta)$ let $F_m(E_m)$ denote the bundle obtained from the transition matrices $M_{\alpha\beta}{}^{(p_m)}$, where if

$$M_{\alpha\beta} = (a_{ij})$$

then

$$M_{\alpha\beta}{}^{(p_m)} = (\alpha_{ij}{}^{p_m}).$$

Sometimes we denote $F_m(E_m)$ by $E_m{}^{(p_m)}$.

LEMMA 2.1. Suppose that if $G \subseteq E_k$ is locally free, then $\mu(G) \leq 0$. Let L_R be a line bundle on X_R. Given $\epsilon > 0$, there is an open subscheme

U of Spec R so that if m is any closed point in U, then

$$h^\circ(X_m, F_m(E_m) \otimes L_m) \leq \epsilon p_m^2.$$

Proof. We work by induction on the rank of E. We may assume L_R is ample on X_R. A linear pencil of hyperplanes $\{H_t^1\}$ in $\mathbf{P}_k^n \supseteq X_k$ is called good if the base locus of $\{H_t^1\}$ intersects X_k transversally. We let $H_t = H_t^1 \cap X_k$. We can blow up X to obtain $\tilde{X}_k$,

$$\begin{array}{ccc} \tilde{X}_k & \xrightarrow{\pi_1} & X_k \\ \downarrow{\scriptstyle \pi_2} & & \\ \mathbf{P}_k^1 & & \end{array}$$

so that $H_t = \pi_1(\pi_2^{-1}(t))$ and so that π_2 is smooth over an open $V = \mathbf{P}^1 - \{z_1, \ldots, z_k\}$.

Assume first that $E_k | H_t$ is not semi-stable on H_t for generic $t \in \mathbf{P}_k^1$. We can find a locally free subsheaf $\tilde{G}_k \subseteq \tilde{E}_k = \pi_1^*(E_k)$ so that $\tilde{Q}_k = \tilde{E}_k/\tilde{G}_k$ is locally free except for a finite number of points, and so that for generic $t \in \mathbf{P}_k^1$, $\tilde{Q}_k | \pi_1^{-1}(t) = (\tilde{Q}_k)_t$ is semi-stable and has negative degree by Proposition 1.3. By Theorem 1.2 iii, $(\tilde{Q}_k)_t^V$ is ample. For generic t and some l, $(S^l(\tilde{Q}^V) \otimes \tilde{L}^V)_t$ is ample. Note $\tilde{X}_k$ and $\tilde{Q}_k$ are defined over some ring of finite type $R' \supseteq R$. We may assume $R' = R$. Then for generic $t \in \mathbf{P}_R^1$, $(S^l(\tilde{Q}^V) \otimes \tilde{L}^V)_t$ is ample. So for a generic maximal ideal $m \in$ Spec R, and generic $t \in \mathbf{P}_{k_m}^1$, $(S^l(\tilde{Q}^V) \otimes \tilde{L}^V)_t$ is ample. Hence $F_m(S^l(\tilde{Q}^V) \otimes \tilde{L}^V)_t$ is ample [6]. So

$$(S^l(F_m(\tilde{Q}^V) \otimes \tilde{L}^V))_t = F_m(S^l(\tilde{Q}^V \otimes L^V)) \otimes (\tilde{L})_t^{\otimes p_m - l}$$

is ample. Thus $H^\circ((S^l(F_m(\tilde{Q}) \otimes \tilde{L})_t) = 0$ for generic $t \in \mathbf{P}_{k_m}^1$. (Note: $(S^l(E_m^V))^V = S^l(E_m)$ if $l < p_m$.). Thus $H^\circ((F_m(\tilde{Q}) \otimes \tilde{L})_t) = 0$. Since this holds for generic $t \in \mathbf{P}_{k_m}^1$, $H^\circ(\tilde{X}_m, F_m(\tilde{Q}) \otimes \tilde{L}_m) = 0$. So $H^\circ(X_m, F_m(E_m) \otimes L_m) = H^\circ(X_m, F_m(G) \otimes L_m)$. Since by induction the Lemma holds for G, we have established the Lemma when E_t is not semi-stable.

We now assume E/H_t is semi-stable whenever we have a good linear pencil. In particular, we can find a smooth curve C_k on V_k so that E_{C_k} is semi-stable on C_k and $C_k^2 > (4 \text{ rk } E/\epsilon)$. Let M_k be a line bundle of degree 1 on C_k. By enlarging R, we may assume that C and M are defined over R. Now $E_{C_k}{}^{\vee} \otimes \mathcal{O}_{C_k}(P)$ is ample by Theorem 1.2. Hence for a generic maximal ideal $m \in \operatorname{Spec} R$,

$$F_m(E_m{}^{\vee}) \otimes (\mathcal{O}(p_m P) \otimes \mathcal{O}_C(dC) \otimes L)_m$$

is ample if $d \geq 0$. Hence

$$F_m(E_m{}^{\vee}) \otimes (\mathcal{O}_C(dC) \otimes L)_m$$

is ample if $dC^2 \geq p_m + \deg_C(L)$. Thus

$$\begin{aligned} h^\circ(F_m(E_m) \otimes (\mathcal{O}_C(-dC) \otimes L)_m) &\leq \operatorname{rk} E(p_m + \deg_C L) && \text{if } d \geq 0 \\ &\leq 0 && \text{if } d \geq \frac{\epsilon}{2 \operatorname{rk} E} \cdot p_m \end{aligned}$$

The first inequality follows from the fact that

$$H^\circ(F_m(E_m) \otimes \mathcal{O}(-p_m P) \otimes L^{\vee}) = 0.$$

Thus

$$\sum_{d=0}^{\infty} h^\circ(F_m(E) \otimes \mathcal{O}_C(-dC)_m \otimes L_m) \leq \epsilon p_m{}^2.$$

Thus

$$h^\circ(X_m, F_m(E_m) \otimes L_m) \leq \epsilon p_m{}^2.$$

THEOREM 3.2 (Bogomolov) (char $k = 0$). Suppose E_k is a semi-stable bundle of rank r on X_k. Then

$$c_1{}^2(E) \leq \frac{2r}{r-1} c_2(E).$$

Proof. We define

$$\Phi(E) = c_1{}^2(E) - \frac{2r}{r-1} c_2(E).$$

Then $\Phi(E) = \Phi(E \otimes L)$. We claim we can assume $\Lambda^r E = \mathcal{O}_X$. There is a non-singular surface Y and a finite map $\pi: Y \to X$ and a line bundle M on Y so that $M^{\otimes r} \cong \pi^*(\Lambda^r E)$ (Lemma 2.1 [1]). Let $E' = \pi^*(E) \otimes M^\vee$. Then E' is semi-stable by Lemma 1.1, and

$$\Phi(E') = (\deg \pi)\Phi(E).$$

Replacing E by E', we may assume $\Lambda^r E \cong \mathcal{O}_X$. Suppose $c_2(E) < 0$. We can define X and E over a subring R of k of finite type over $\mathbf{Z}$ so that if m is a maximal ideal of R, then

$$h^\circ(X_m, F_m(E_m)) \leq \tfrac{1}{4} p_m{}^2$$

and

$$h^2(X_m, F_m(E_m)) = h^\circ(X_m, F_m(E)^\vee \otimes \Omega_{X_m/k_m}{}^2) \leq \tfrac{1}{4} p_m{}^2.$$

Thus by Riemann-Roch

$$r\chi(\mathcal{O}_{X_m}) - p_m{}^2 c_2(E_m) = \chi(F_m(E)) \leq \tfrac{1}{2} p_m{}^2.$$

(Note that $c_2(F_m(E_m)) = p_m{}^2 c_2(E_k)$ by [4]. Since $c_2(E_k) \in \mathbf{Z}$, we have a contradiction if $c_2(E) < 0$. So $c_2(E) \geq 0$.

Theorem (Kodaira) (char $k = 0$). If L is ample on X_k, then $H^1(X, L^\vee) = 0$.

Proof. By using Lemma 2.1 of [1], we may assume there is a finite $\pi: Y \to X$ so that $M^{\otimes 2} = \pi^*(L)$. Now L is a direct factor of $\pi_*(\pi^*(L))$ by the trace map, so $H^1(X, L) \neq 0$ implies $H^1(Y, \pi^*(L)) \neq 0$. By replacing X_k by Y, we may assume $L_k = M_k{}^{\otimes 2}$. Now any $\alpha \in H^1(X_k, L_k)$ determines an extension

$$0 \to M_k{}^\vee \to E_k \to M_k \to 0. \qquad *$$

We may assume the extension is defined over a ring R of finite type over $\mathbf{Z}$. Now $H^1(X_m, (L_m)^{\otimes p_m}) = 0$, so the extension

$$0 \to F_m(M_m)^\vee \to F_m(E_m) \to F_m(M_m) \to 0$$

splits. So $H^\circ(F_m(E)) = H^\circ(M_m{}^{\otimes pm})$. Thus E is not semi-stable by Lemma 2.1. Let L be a subsheaf of E with $\mu(L) > \mu(E)$ and E/L torsion-free. Then

$$H^\circ(F_m(L_m)) \to H^\circ(F_m(M_m))$$

is an isomorphism. Since $F_m(M_m) = M_m{}^{\otimes pm}$ is generated by global sections, the map from $L_m{}^{\otimes pm}$ to $M_m{}^{\otimes pm}$ is an isomorphism. Thus the map from L_m to M_m is an isomorphism. So the map from L_k to M_k is an isomorphism. So the sequence $*$ splits, and $H^1(L^V) = 0$.

UNIVERSITY OF CALIFORNIA, LOS ANGELES

REFERENCES

[1] S. Bloch-D. Gieseker, "The positivity of the Chern classes of an ample vector bundle," *Inv. Math.* **12** (1971), pp. 112-117.
[2] D. Gieseker, "On the moduli of vector bundles on an algebraic surface," *Ann. of Math.*, **106**, No. 1, 1977, pp. 45-60.
[3] ———, "Stable vector bundles and the Frobenius morphism," *Ann. E.N.S.* **6** (1973), pp. 95-101.
[4] ———, "*p*-ample bundles and their Chern classes," *Nagoya Math. J.*, **43** (1971), pp. 91-116.
[5] R. Hartshorne, "Ample vector bundles on curves," *Nagoya Math. J.*, **43** (1971), pp. 73-90.
[6] ———, "Ample vector bundles," *Publ. IHES*, **29** (1966), pp. 319-350.
[7] Maruyama, M. "Openness of a family of torsion free sheaves," *J. Math. Kyoto U.*, **16**, 1976.

A CURVATURE FORMULA

By Andrew M. Gleason

Let A be a smooth curve of finite length $L(A)$. For each n let P_n be the longest polygon of not more than n sides that is inscribed in A. Then $L(A) - L(P_n) \to 0$, and the question naturally arises, how fast?

For a circular arc of length L and radius R one easily finds that

$$\lim_{n\to\infty} n^2(L(A) - L(P_n)) = L^3/24R^2.$$

Considering that any smooth arc of finite length can be well approximated by a finite sequence of circular arcs, it seems natural to conjecture that

$$\lim n^2(L(A) - L(P_n))$$

always exists and is some sort of integral of the curvature. In this paper we shall prove the following theorem.

Theorem. Let A be a curve in euclidean space (possibly of infinite dimension) of class C^2 and length L. For each positive integer n, let P_n be the longest polygon of n edges properly inscribed in A. Then

$$\lim n^2(L - L(P_n)) = \frac{1}{24}\left(\int_A \kappa^{2/3}\, ds\right)^3$$

This result has been stated by Fejes Toth [3, p. 43] for convex curves in the plane.

Smooth will mean of class C^2. Since we intend that a curve may cross, or even retrace, itself, it must be considered in terms of parametrizations. Technically, a curve is an equivalence class of parametrizations. Then a smooth curve is represented by a C^2-function $f:[a, b] \to H$, where H is an inner product space and $f'(s) \neq 0$ for all s. The dimension of H is unimportant and may even be infinite. In proofs we shall always assume that the dimension is at least two because otherwise everything is trivial. Since there is always an equivalent arc-length parametrization, we shall assume that we start with one. Then $\|f'(s)\| = 1$ for all s, and $\|f''(s)\| = \kappa(s)$ is the curvature. A subcurve of A is defined by f restricted to a subinterval of $[a, b]$.

A polygon of n sides inscribed in the curve A is a polygon having consecutive vertices $f(s_0), f(s_1), \ldots, f(s_n)$, where $s_0 = a \leq s_1 \leq \cdots \leq s_n = b$. It is convenient to allow some of the s's to coincide. The length of a curve is the supremum of the lengths of all inscribed polygons of whatever number of sides, but since we are dealing with an arc-length parametrization, $L(A)$ is simply $b - a$. Let $C(A)$ be the length of the chord of A, that is the segment connecting the endpoints $f(a)$ and $f(b)$.

The proof of the theorem is given in a series of lemmas directed towards proving opposite inequalities.

LEMMA 1. Suppose A is a smooth curve of length L having curvature everywhere at most κ_0, where $\kappa_0 L < 2\pi$. Then the chord of A is at least as long as that of a circular arc of length L and curvature κ_0, with equality only if A is itself such a circular arc.

We omit the proof of this lemma since A. Schur has proved a stronger result. See [2, p. 36].

LEMMA 2. Let φ be a non-negative continuous function defined on $[a, b]$ (where $a < b$). For any positive integer n it is possible to choose $s_1, s_2, \ldots, s_{n-1}$ so that $s_0 = a < s_1 < \cdots < s_{n-1} < s_n = b$ and the numbers

$$J_i = (s_i - s_{i-1}) \max \{\varphi(t) : s_{i-1} \leq t \leq s_i\}$$

are all equal. If such a choice is made for each n, then

$$\lim_{n \to \infty} nJ_i = \int_a^b \varphi(s)\, ds.$$

Proof. If φ is everywhere zero, the result is trivial, so we shall assume that φ is somewhere positive.

Parametrize the $(n - 1)$-simplex σ as usual by n-tuples $(u_1, u_2, \ldots, u_n)$ where $u_i \geq 0$ and $\Sigma\, u_i = 1$. Let $(u_1, u_2, \ldots, u_n)$ correspond to the partition of $[a, b]$ given by $s_i = a + (b - a)(u_1 + u_2 + \cdots + u_i)$ and define a map ψ of σ into itself by

$$\psi(u_1, u_2, \ldots, u_n) = (w_1, w_2, \ldots, w_n)$$

where $w_i = v_i / \sum v_i$ and

$$v_i = u_i \max\{\varphi(t) : s_{i-1} \le t \le s_i\}.$$

Since the maximum value of a continuous function over an interval (which may degenerate to a point) depends continuously on the end-points of that interval, ψ is a continuous function. Note that $\sum v_i$ is a Riemann upper sum for $\int_a^b \varphi(s)\, ds$ and is, therefore, positive under our assumption on φ.

From the defining relations for ψ it follows that $w_i = 0$ whenever $u_i = 0$; hence, ψ maps every face of σ into itself. It is well-known that this implies that ψ is surjective. Hence in particular there is a point $(u_1, u_2, \ldots, u_n)$ such that

$$\psi(u_1, u_2, \ldots, u_n) = \left(\frac{1}{n}, \frac{1}{n}, \ldots, \frac{1}{n}\right).$$

The partition of $[a, b]$ corresponding to this point satisfies the requirements.

Let $\epsilon > 0$ be given. Let $E = \{s : \varphi(s) < \epsilon\}$ and choose δ so that $|s - s'| < \delta \Rightarrow |\varphi(s) - \varphi(s')| < \epsilon$. Let M be an upper bound for φ. Then $nJ_i = J_1 + J_2 + \cdots + J_n \le M(b - a)$. Now suppose $n > M(b - a)/\delta\epsilon$. Then for any i, either $[s_{i-1}, s_i] \subseteq E$ or $\max\{\varphi(t) : s_{i-1} \le t \le s_i\} \ge \epsilon$, in which case $\epsilon(s_i - s_{i-1}) \le J_i \le M(b - a)/n$ whence $s_i - s_{i-1} < \delta$. In either case the oscillation of φ over $[s_{i-1}, s_i]$ is at most ϵ; hence $nJ_i = \sum J_i$, which is a Riemann sum for $\int \varphi\, ds$, differs from $\int \varphi\, ds$ by less than $\epsilon(b - a)$. It follows that $nJ_i \to \int \varphi\, ds$. □

We remark in passing that, provided φ is not everywhere zero, the partition points $s_i, \ldots, s_{n-1}$ are uniquely determined; in fact, if p is any point in the interior of the simplex there is just one point x with $\psi(x) = p$.

We can now prove one half of the main result.

LEMMA 3. Let A be a smooth curve and let P_n be the longest polygon of n sides inscribed in A. Then

$$\limsup_{n \to \infty} n^2(L(A) - L(P_n)) \le \frac{1}{24}\left(\int_A \kappa^{2/3}\, ds\right)^3.$$

Remark. The existence of a longest polygon of n sides follows from compactness; it need not be unique, but that doesn't matter.

Proof. Let $I = \int_A \kappa^{2/3}\, ds$. For each n we shall inscribe in A a polygon Q_n of n sides (not necessarily the longest such polygon) and prove that

$$\limsup n^2(L(A) - L(Q_n)) \leq \frac{1}{24} I^3. \tag{1}$$

The claimed inequality then follows *a fortiori.*

Fix an integer n, and let $s_1, s_2, \ldots, s_{n-1}$ be chosen so that $s_0 = a < s_1 < s_2 < \cdots < s_n = b$ and all of the numbers

$$J_i = (s_i - s_{i-1})(\max\{\kappa(t): s_{i-1} \leq t \leq s_i\})^{2/3}$$

are equal (see Lemma 2). Then $J_i = K_n/n$, where K_n is the Riemann upper sum for $\int \kappa^{2/3}\, ds$ corresponding to the partition of $[a, b]$ determined by the s's.

Let Q_n be the polygon with successive vertices at $f(s_0), f(s_1), \ldots, f(s_n)$. Let A_i be the subcurve of A corresponding to the parameter interval $[s_{i-1}, s_i]$. If n is large enough, then for each $i = 1, 2, \ldots, n$, the total curvature of A_i will be less than 2π. This follows because the total curvature of A_i is at most $J_i\kappa_0^{1/3}$, where κ_0 is the globally maximum value of the curvature, while $J_i = K_n/n \leq \kappa_0^{2/3}L(A)/n$. Assuming that n is sufficiently large, lemma 1 tells us that

$$C(A_i) \geq \frac{2}{\kappa_i} \sin \frac{1}{2} \kappa_i L(A_i),$$

where κ_i is the maximum value of the curvature on A_i. (Note that κ_i is positive unless the whole curve A is straight, in which case the lemma is trivial.) For any positive x, $\sin x > x - x^3/6$. Taking $x = \kappa_i L(A_i)/2$, we find

$$L(A_i) - C(A_i) < \frac{1}{24} \kappa_i^2 L(A_i)^3 = \frac{1}{24} J_i^3 = \frac{1}{24n^3} K_n^3.$$

Adding these inequalities, we have

$$L(A) - L(Q_n) < \frac{1}{24n^2} K_n^3.$$

Since $\lim K_n = I$, this yields (1). □

LEMMA 4. Consider a compact smooth curve A and subcurves B of A. Then

$$|L(B) - C(B) - \frac{1}{24}\kappa_B{}^2 L(B)^3| = o(L(B)^3)$$

as $L(B) \to 0$, where κ_B is the curvature of A at any point of B.

Remark. This formula has been used by Haantjes [4] as the definition of curvature in abstract metric spaces. In the terminology of Blumenthal, the lemma says that the Haantjes curvature of a compact smooth curve exists uniformly. See [1, p. 76].

Proof. Let $f:[a, b] \to H$ be an arc-length parametrization of A. Then f'' is a continuous function on $[a, b]$. Let ω be a modulus of continuity function for f''; that is, ω is a symmetric continuous function such that $\omega(0) = 0$, ω is increasing on $[0, \infty)$, and $\|f''(t) - f''(t')\| < \omega(t - t')$ for all $t, t' \in [a, b]$.

Let B be a subcurve of A corresponding to the parameter interval $[s_0, s_1]$, and let e_1 and e_2 be, respectively, the unit tangent and unit principal normal to A at s_0. (If $\kappa(s_0) = 0$, let e_2 be any unit vector orthogonal to e_1.) Then $f'(s_0) = e_1$ and $f''(s_0) = \kappa(s_0)e_2$. To simplify the notation we assume, without loss of generality, that $s_0 = 0$, and write s for s_1 and κ_0 for $\kappa(0) = \kappa(s_0)$. In what follows we consider $t \in [0, s]$. The numbers M_1 to M_5 are constants which depend only on A, not on the subcurve B.

Integrating f'' twice we see that

$$f'(t) = e_1 + \kappa_0 t e_2 + u(t)$$

and

$$f(t) - f(0) = te_1 + \tfrac{1}{2}\kappa_0 t^2 e_2 + v(t)$$

where $\|u(t)\| \leq t\omega(t)$, $\|v(t)\| \leq \tfrac{1}{2}t^2\omega(t)$, and $v(t) = \int_0^t u(r)\,dr$.

Because $f'(t)$ is always a unit vector, we have

$$1 = \|f'(t)\|^2 = 1 + \kappa_0{}^2 t^2 + 2(e_1, u(t)) + \alpha(t),$$

where $|\alpha(t)| \leq M_1 t^2 \omega(t)$. Therefore,

$$\kappa_0^2 t^2 + 2(e_1, u(t)) = -\alpha(t).$$

Integrating, we find

$$\left| \tfrac{1}{3} \kappa_0^2 s^3 + 2(e_1, v(s)) \right| \leq \int_0^s |\alpha(t)|\, dt \leq \tfrac{1}{3} M_1 s^3 \omega(s).$$

Also

$$\| f(s) - f(0) \|^2 = s^2 + \tfrac{1}{4} \kappa_0^2 s^4 + 2s(e_1, v(s)) + \beta(s),$$

where $|\beta(s)| < M_2 s^4 \omega(s)$. Combining this with (1), we obtain

$$\| f(s) - f(0) \|^2 = s^2 \left(1 - \frac{1}{12} \kappa_0^2 s^2 + \gamma(s) \right),$$

where $|\gamma(s)| < M_3 s^2 \omega(s)$. Hence

$$C(B) = \| f(s) - f(0) \| = s \left(1 - \frac{1}{24} \kappa_0^2 s^2 \right) + \delta(s),$$

where $|\delta(s)| < M_4 s^3 \omega(s)$. If $\kappa_B = \kappa(s')$ with $s' \in [0, s]$, then $|\kappa_B - \kappa_0| \leq \| f''(s') - f''(0) \| \leq \omega(s') \leq \omega(s)$. Finally, recalling that $L(B) = s$, we have

$$\left| L(B) - C(B) - \frac{1}{24} \kappa_B^2 L(B)^3 \right| \leq M_5 L(B)^3 \omega(L(B)),$$

and the lemma is proved. □

LEMMA 5. Let A be a compact smooth curve. For any $\epsilon > 0$ there is an integer N such that

$$n^2(L(A) - L(P)) > \frac{1}{24} \left(\int_A \kappa^{2/3}\, ds \right)^3 - \epsilon$$

for all polygons P inscribed in A having n sides with $n > N$.

Proof. We will prove the following statement from which the lemma follows immediately.

For any $\delta > 0$ there is an integer r such that

$$(n + r)^{2/3}(L(A) - L(P))^{1/3} > 24^{-1/3} \int_A \kappa^{2/3}\, ds - \delta L(A),$$

for all polygons P having n sides.

Let $\delta > 0$ be given. Since the function $x \mapsto x^{1/3}$ is uniformly continuous, the preceding lemma shows that there is an $\eta > 0$ such that $L(B) < \eta$ implies

$$|(L(B) - C(B))^{1/3} - 24^{-1/3}\kappa_B{}^{2/3}L(B)| < \delta L(B).$$

Choose r so that $L(A)/r < \eta$. To any polygon P of n sides adjoin not more than r vertices so that the resulting polygon Q with vertices corresponding to $s_0 = a,\ s_1, \ldots, s_t = b$, satisfies $s_i - s_{i-1} < \eta$. Choose points $s_i' \in [s_{i-1}, s_i]$ so that

$$\int_{s_{i-1}}^{s_i} \kappa^{2/3}\, ds = \kappa(s_i')^{2/3}(s_i - s_{i-1}).$$

Let A_i be the subcurve corresponding to $[s_{i-1}, s_i]$. Then

$$24^{-1/3}\kappa(s_i')^{2/3}(s_i - s_{i-1}) < (L(A_i) - C(A_i))^{1/3} + \delta L(A_i).$$

Summing on i, we obtain

$$24^{-1/3} \int_A \kappa^{2/3}\, ds < \sum_{i=1}^{t} (L(A_i) - C(A_i))^{1/3} + \delta L(A).$$

Now Holder's inequality with the exponents 3/2 and 3 gives

$$\sum_{i=1}^{t} 1^{2/3}(L(A_i) - C(A_i))^{1/3} \le t^{2/3}(\Sigma\, (L(A_i) - C(A_i)))^{1/3}$$

Hence we have

$$24^{-1/3} \int_A \kappa^{2/3}\, ds < t^{2/3}(L(A) - L(Q))^{1/3} + \delta L(A).$$

Since $t \leq n + r$ and $L(P) \leq L(Q)$, we obtain

$$(n + r)^{2/3}(L(A) - L(P))^{1/3} > 24^{-1/3} \int_A \kappa^{2/3}\, ds - \delta L(A),$$

as claimed. □

It is clear that the theorem follows from Lemmas 3 and 5.

HARVARD UNIVERSITY

REFERENCES

1. L. M. Blumenthal, *Theory and applications of distance geometry,* Oxford, 1953.
2. S. S. Chern, ed., *Studies in global geometry and analysis,* Math. Assn. of Amer., 1967.
3. L. Fejes Toth, *Lagerungen in der Ebene auf Kugel und im Raum,* Springer, 1953.
4. J. Haantjes, "Distance geometry. Curvature in abstract metric spaces," *Nederl. Akad. Wetensch., Proc.* **50**, pp. 496–508; *Indagationes Math.* **9**, pp. 302–314 (1947).

A THEOREM CONCERNING THE DIFFERENTIAL EQUATIONS SATISFIED BY NORMAL FUNCTIONS ASSOCIATED TO ALGEBRAIC CYCLES

By PHILLIP A. GRIFFITHS*

In this paper we shall give a general discussion of normal functions culminating in a result characterizing those normal functions arising from algebraic cycles as being solutions to a family of ordinary differential equations parametrized by the hypersurfaces of large degree passing through the given cycle or through one homologous to it. In Section 1 the theorem will be informally discussed for the case of curves on a surface where a minimum of technical machinery is necessary. Along the way we give proofs of the main classical results in the theory of normal functions and find some new information on the Hodge bundles arising from a Lefschetz pencil of curves. Then in Section 2 we turn to higher dimensions. Following a discussion of the definition and basic properties of normal functions we analyze the Hodge bundles arising from the cotangent spaces to the intermediate Jacobians in a Lefschetz pencil (c.f. (2.13c) and (2.14c)), and then shall reprove the result (c.f. (2.9) for the statement) characterizing the fundamental classes of normal functions by their Hodge type. Finally, after some general observations on Picard-Fuchs equations we formulate and prove our main result Theorem 2.2, and then conclude the paper with some observations concerning the problem of constructing algebraic cycles.

Unless otherwise specified, homology will be with **Z**-coefficients and cohomology with **C**-coefficients. Hopefully the other notations and terminology are standard.

1. Discussion of the Theorem for Curves on a Surface.

a) Let S be a smooth algebraic surface and $L \to S$ a very ample line bundle, which we shall think of as the hyperplane bundle for the embedding $S \hookrightarrow \mathbf{P}^N$ induced by the complete linear system $|L|$. Recall

* Research partially supported by NSF Grant MCS-78-07348.

that a Lefschetz pencil is a pencil of curves $|C_t|$ selected from $|L|$ such that the generic curve C_t is smooth and the singular curves $C_{t_1}, \ldots, C_{t_N}$ are irreducible with one ordinary double point not in the base locus $B = C_0 \cdot C_\infty$ of the pencil. A general pencil is obtained by selecting a line in the dual projective space $\mathbf{P}^{N*}$ parametrizing the curves in $|L|$, and the pencil is Lefschetz when this line meets the hypersurface $S^* \subset \mathbf{P}^{N*}$ dual to S transversely at smooth points. It is classical how one understands the topology of S in terms of the monodrony on a general $H_1(C_t)$ where t varies over $\mathbf{P}^1 - \{t_1, \ldots, t_N\}$, and we shall review this below as needed. At the moment we shall discuss the Hodge bundle associated to the pencil.

Over the $\mathbf{P}^1$ parametrizing the curves in the pencil there is a vector bundle $E \to \mathbf{P}^1$, the Hodge bundle, whose fibre over $t \in \mathbf{P}^1 - \{t_1, \ldots, t_N\}$ is the vector space $H^0(\Omega_{C_t}{}^1)$ of holomorphic differentials on the smooth curve C_t. Each singular curve C_{t_α} has a normalization $\tilde{C}_{t_\alpha}$, a smooth curve of genus one less than that of a generic C_t and having on it two marked points which when identified give back the singular curve C_{t_α}. The fibre E_{t_α} then consists of the meromorphic differentials on $\tilde{C}_{t_\alpha}$ having at most logarithmic singularities at the marked points.

Alternatively, we may describe the sections of the Hodge bundle by using residues. Given an irreducible curve $C \in |L|$ defined by a section $s \in H^0(\mathcal{O}_S(L))$, there is the Poincaré residue sequence

$$0 \longrightarrow \Omega_S{}^2 \xrightarrow{\ s\ } \Omega_S{}^2(L) \xrightarrow{\ \mathrm{Res}\ } \omega_C \longrightarrow 0. \tag{1.1}$$

When C is smooth, $\omega_C = \Omega_C{}^1$ is the usual sheaf of holomorphic differentials; when C has a double point ω_C exactly corresponds to meromorphic 1-forms on $\tilde{C}$ having at most logarithmic singularities over the double point. For all t, then, the fibre of the Hodge bundle is

$$E_t = H^0(\omega_{C_t}).$$

Denoting by $\eta \in H^1(\Omega_S{}^1)$ the Chern class of $L \to S$ and using $h^1(\Omega_S{}^2(L)) = 0$, the exact cohomology sequence of (1.1) gives a commutative diagram

$$\begin{array}{ccccccccc} 0 \to H^0(\Omega_S{}^2) & \to & H^0(\Omega_S{}^2(L)) & \to & H^0(\omega_C) & \to & H^1(\Omega_S{}^2) & \to & 0. \\ & & & & \big\uparrow & \nearrow_{\eta} & & & \\ & & & & H^0(\Omega_S{}^1) & & & & \end{array} \tag{1.2}$$

By the Hard Lefschetz theorem, η induces an isomorphism, and the image of

$$H^0(\Omega_S{}^1) \to H^0(\omega_{C_t})$$

may be thought of as the fixed part of the variation of Hodge structure defined by the curves C_t; this image therefore defines a trivial summand of the Hodge bundle $E \to \mathbf{P}^1$.

To analyze the variable part of E we choose sections s_0 and $s_\infty \in H^0(\mathcal{O}_S(L))$ defining C_0 and C_∞ (assumed smooth) and set $s_t = s_0 + ts_\infty$. For $\omega \in H^0(\Omega_S{}^2(L))$ and $t' = 1/t$, since

$$\operatorname{Res}\left(\frac{\omega}{s_0 + ts_\infty}\right) = t' \operatorname{Res}\left(\frac{\omega}{t's_0 + s_\infty}\right)$$

we deduce that the exact sequence in (1.2) globalizes over $\mathbf{P}^1$ to give

$$0 \to H^0(\Omega_S{}^2) \to H^0(\Omega_S{}^2(L)) \otimes \mathcal{O}(1) \to E \to H^1(\Omega_S{}^2) \to 0. \tag{1.3}$$

Here the vector spaces $H^q(\Omega_S{}^2)$ and $H^0(\Omega_S{}^2(L))$ are thought of as trivial bundles over $\mathbf{P}^1$, $\mathcal{O}(1)$ has the usual meaning, and we are retaining E to denote the corresponding sheaf $\mathcal{O}_{\mathbf{P}^1}(E)$. With the usual characters

$$\begin{cases} g = h^{1,0}(C_t) \\ q = h^{1,0}(S) \\ p_g = h^{2,0}(S), \end{cases}$$

we infer from (1.3) that the Grothendieck decomposition of E is given by given by

$$E \cong \mathcal{O}^{(q)} \oplus \left(\bigoplus_{i=1}^{g-q} \mathcal{O}(k_i + 1)\right), \qquad k_i \geq 0, \tag{1.4}$$

so that the variable part $E_v = \oplus\ \mathcal{O}(k_i + 1)$ of the Hodge bundle is positive. By tensoring (1.3) with $\mathcal{O}(-2)$ and taking cohomology, using $H^1(\mathbf{P}^1, \mathcal{O}(-2)) \cong \mathbf{C}$ we deduce the isomorphism

$$H^0(\mathbf{P}^1, E \otimes \mathcal{O}(-2)) \cong H^0(\Omega_S{}^2), \tag{1.5}$$

a fact which may be seen directly as follows: Each holomorphic 2-form ω on S may be uniquely written

$$\omega = \varphi(t) \wedge dt,$$

where easy local considerations show that $\varphi(t) \in H^0(\omega_{C_t})$. Since dt has a pole of order 2 at $t = \infty$ we infer that $\varphi(t) \in H^0(\mathbf{P}^1, E \otimes \mathcal{O}(-2))$ vanishes to 2nd order there.

We remark that the exact nature of the numbers k_i in (1.4) has to do with the linear series $|K - \mu L|$ for $\mu = 0, 1, \ldots$. For example, from

$$p_g = \Sigma\, k_i$$

we see that the canonical series is empty exactly when all $k_i = 0$. Similarly, if $|K - L|$ is empty then the k_i are all either zero or one, and so (1.4) becomes

$$E \cong \mathcal{O}^{(q)} \oplus \mathcal{O}(1)^{(g-q-p_g)} \oplus \mathcal{O}(2)^{p_g},$$

a situation which may always be achieved by choosing L sufficiently ample.

b) To discuss how the topology of S is described by the Lefschetz pencil it is useful to take two points of view. One is intuitive as in the Borel tract [6] or in the beautiful exposition [9], and the other is by applying Morse theory and the Leray spectral sequence as in Andreotti-Frankel [1].

In either case some local analysis around the critical values is necessary, and for this the essential points are the following: In the complex t-plane mark a non-critical reference point t_0 and draw non-intersecting paths γ_α from t_0 to t_α. Then there is for each α a vanishing cycle $\delta_\alpha \in H_1(C_{t_0})$ such that as t moves along γ_α the displaced cycle $\delta_\alpha(t)$ on C_t shrinks to the double point as $t \to t_\alpha$. The locus of $\delta_\alpha(t)$ along this path then gives a cone $\Delta_\alpha \in H_2(S, C_{t_0})$ with $\partial\Delta_\alpha = \delta_\alpha$. If $\gamma \in H_1(C_{t_0})$ is any other cycle and we displace γ along a closed loop turning once around t_α, the change in γ is measured by the Picard-Lefschetz formula

$$\gamma \to \gamma + (\gamma, \delta_\alpha)\, \delta_\alpha. \tag{1.6}$$

To give what intuitive remarks about topology we shall require we recall that the primitive cohomology $H_{\text{prim}}^2(S, \mathbf{Q})$ is represented via Poincaré duality (and ignoring torsion) by 2-cycles $\Gamma \in H_2(S)$ such that the intersection cycle $\Gamma \cdot C_{t_0} = 0$ in $H_0(C_{t_0})$. For t in the punctured t-plane $\mathbf{P}^1 - \{t_1, \ldots, t_N\}$ we locally write

$$\Gamma \cdot C_t = \partial \gamma_t$$

where $\gamma_t \in H_1(C_t, C_t \cap \Gamma_t)$ is a path on C_t which maps to $\Gamma \cdot C_t$ in $H_0(C_t \cap \Gamma_t)$. An obvious extension of (1.6) to the relative case gives that as t turns once around t_α,

$$\gamma_t \to \gamma_t + \lambda_\alpha \delta_\alpha$$

where $\lambda_\alpha = (\gamma_t, \delta_\alpha)$ is an integer. Since γ_t is invariant when t transverses a sufficiently large circle (we assume that $t = \infty$ is not a critical value)

$$\sum_\alpha \lambda_\alpha \delta_\alpha = 0 \quad \text{in} \quad H_1(C_{t_0}). \tag{1.7}$$

By virtue of this relation, the relative 2-cycle

$$\sum_\alpha \lambda_\alpha \Delta_\alpha \in H_2(S, C_{t_0}) \tag{1.8}$$

is in the image of the mapping

$$H_2(S) \to H_2(S, C_{t_0}),$$

and an easy argument shows that Γ is represented by the cycle (1.8). Summarizing, the primitive homology is given by cycles (1.8) satisfying (1.7) where the coefficients λ_α are determined by considering the variation of a chain γ_t with $\partial \gamma_t = \Gamma \cdot C_t$.

We now describe the same situation in the Andreotti-Frankel terminology. By blowing up S at the base locus $B = C_{t_0} \cdot C_\infty$ of the pencil we obtain the smooth surface $X = \bigcup_{t \in \mathbf{P}^1} C_t$, which then admits an obvious fibering $f: X \to \mathbf{P}^1$. The Leray spectral sequence for the constant sheaf $\mathbf{Z}$ degenerates at E_2 (Andreotti-Frankel, loc. cit.), and the primitive part of $H^2(S, \mathbf{Q}) \subset H^2(X, \mathbf{Q})$ is easily seen to lie in the term $E_2^{1,1} = H^1(\mathbf{P}^1, R_f^1 \mathbf{Z})$.

Now then to describe $H^1(\mathbf{P}^1, R_f^1 \mathbf{Z})$ we choose the covering $\{U_0, U_1,$

$\ldots, U_N\}$ of $\mathbf{P}^1$ where $U_0 = \mathbf{P}^1 - \{t_1, \ldots, t_N\}$ and U_α is an ϵ-disc around t_α. For this covering the 1-cocycles are just

$$\prod_\alpha H^0(U_\alpha{}^*, R_f{}^1\,\mathbf{Z}),$$

which we may think of as being given by an invariant 1-cycle in each punctured disc $U_\alpha{}^*$. For a primitive class given by (1.8) the corresponding Čech class is $\Pi_\alpha\ \{\lambda_\alpha \cdot \delta_\alpha\}$. Similarly, a class in $H_{\text{prim}}{}^0(B)$ is represented by a zero-cycle D of degree zero supported on the base locus. On each curve C_t we may write $D = \partial\gamma_t$ for some 1-chain γ_t whose variation around $t = t_\alpha$ leads again to a Čech cocycle in $\Pi_\alpha\ H^0(U_\alpha{}^*, R_f{}^1\,\mathbf{Z})$. These two subspaces of $H^0(\mathbf{P}^1, R_f{}^1\,\mathbf{Z}) \subset H^2(X, \mathbf{Z})$ are orthogonal under the cup product, and again modulo torsion there is a decomposition

$$H^1(\mathbf{P}^1, R_f{}^1\,\mathbf{Z}) \cong H_{\text{prim}}{}^0(B, \mathbf{Q}) \oplus H_{\text{prim}}{}^2(S, \mathbf{Q}). \tag{1.9}$$

c) Having described the Hodge bundle and homology in our pencil $|C_t|$ we now turn to the family of Jacobians. In general, if $C \in |L|$ has only ordinary double points the (generalized) Jacobian variety $J(C)$ is defined to be

$$H^0(\omega_C)^*/H_1(C^*, \mathbf{Z})$$

where $C^* = C\text{-}\{\text{double points}\}$, and where $H_1(C^*, \mathbf{Z})$ is embedded as a discrete subgroup of $H^0(\omega_C)^*$ by integration. By elementary duality theory (or directly)

$$J(C) \cong H^1(\mathcal{O}_C)/H^1(C, \mathbf{Z}).$$

Returning to our pencil, it is easily verified that

$$J = \bigcup_{t \in \mathbf{P}^1} J(C_t)$$

forms naturally a complex manifold (c.f. Jambois [5]) such that $J \to \mathbf{P}^1$ is an analytic fibre space of complex Lie groups. The cross sections of this fibre space are by definition the group of *normal functions*.

The origin of the word "normal functions" may be explained this way. Given $\omega \in H^0(\Omega_C^2(L))$, as explained above in Section 1a) we obtain the holomorphic section

$$\omega(t) = \operatorname{Res}\left(\frac{\omega}{s_0 + ts_\infty}\right) \in H^0(\omega_{C_t}) = E_t$$

of the Hodge bundle. Since $J(C_t) = E_t^*/(\text{periods})$, we may view a normal function ν as a multi-valued holomorphic linear function on E_t for $t \in \mathbf{P}^1 - \{t_1, \ldots, t_N\}$ which has a suitable growth property (noted below) as $t \to t_\alpha$. Suppose we let

$$\nu(t) = \langle \nu, \omega(t) \rangle$$

denote this function. Analytic continuation of $\nu(t)$ around $t = t_\alpha$ adds on a period

$$\pi_\alpha(t) = \lambda_\alpha \int_{\delta_\alpha} \omega(t) \tag{1.10}$$

where $\delta_\alpha \in H_1(C_t, \mathbf{Z})$ is the vanishing cycle. In this way the sections of $J \to \mathbf{P}^1$ may be represented by suitable multi-valued analytic functions whose local monodromy has the form (1.10).

Since analytic continuation around a sufficiently large circle leaves $\nu(t)$ invariant, and since ω was arbitrary, we infer that

$$\sum_\alpha \lambda_\alpha \delta_\alpha = 0 \quad \text{in} \quad H_1(C_{t_0}, \mathbf{Z}).$$

Thus, by the discussion centered around (1.7) and (1.8), to each normal function ν we may assign the primitive cycle $\sum_\alpha \lambda_\alpha \Delta_\alpha \in H_{\text{prim}}{}^2(S, \mathbf{Q})$; this is called the *fundamental class* η_ν of the normal function.

Normal functions and their fundamental classes may also be described using the fibering $f: X \to \mathbf{P}^1$ of the blown up surface $X = \bigcup_{t \in \mathbf{P}^1} C_t$. The basic observations are that

$$R_f{}^1\mathcal{O} \cong E^* \tag{1.11}$$

is the sheaf of Lie algebras associated to the analytic fibre space $J \to \mathbf{P}^1$, and that the sheaf of discrete subgroups is

$$R_f{}^1\mathbf{Z} \subset R_f{}^1\mathcal{O}.$$

In other words, denoting still by J the sheaf of holomorphic sections of the family of Jacobians, we have the exact sheaf sequence

$$0 \to R_f{}^1\mathbf{Z} \to R_f{}^1\mathcal{O} \to J \to 0. \tag{1.12}$$

To compute the cohomology sequence we refer to (1.4) to conclude that

$$H^0(\mathbf{P}^1, R_f{}^1\mathcal{O}) \cong H^1(S, \mathcal{O}).$$

It follows that, with the non-standard notation $J(S) = H^1(S, \mathcal{O})/H^1(S, \mathbf{Z})$ for the Picard variety of S, the fixed part of the family $\{J(C_t)\}$ is $J(S)$ and the exact cohomology sequence of (1.12) leads to a commutative diagram

$$\begin{array}{ccccccccc} 0 \to J(S) \to H^0(\mathbf{P}^1, J) & \xrightarrow{\delta} & H^1(\mathbf{P}^1, R_f{}^1\mathbf{Z}) & \to & H^1(\mathbf{P}^1, R_f{}^1\mathcal{O}) & \to \cdots \\ & & \Big\downarrow \alpha & & \Big\| \beta & \\ & & H_{\text{prim}}{}^2(S, \mathbf{Q}) & \to & H^2(\mathcal{O}_S). & \end{array} \tag{1.13}$$

Here the vertical map α is given by the Leray spectral sequence according to (1.9) and the composition $\alpha\delta(\nu) = \eta_\nu$ gives the fundamental class of ν, as may be verified using the Čech covering described above. As for the vertical map β, from (1.5) and (1.11)

$$\begin{aligned} H^1(\mathbf{P}^1, R_f{}^1\mathcal{O}) &\cong H^1(\mathbf{P}^1, E^*) \\ &\cong H^0(\mathbf{P}^1, E \otimes \mathcal{O}(-2))^* \\ &\cong H^0(\Omega_S{}^2)^* \\ &\cong H^2(\mathcal{O}_S), \end{aligned}$$

where we have used Serre duality twice. This allows us to derive the basic result, due to Lefschetz, characterizing the fundamental classes of normal functions:

A class $\eta \in H_{\text{prim}}^2(S, \mathbf{Z})$ is the fundamental class of a normal function if, and only if, η is of Hodge type (1, 1). (1.14)

To prove this result we need only show that the mapping $H_{\text{prim}}^2(S, \mathbf{Q}) \to H^{0,2}(S)$ in (1.13) is that given by the Hodge decomposition, and then the theorem follows by exactness of the cohomology sequence. This verification is "just" a commutativity check.

Actually, it is quite instructive to give the classical global argument, for which the basic formula is the following: If ν is a normal function with fundamental class $\Sigma_\alpha \lambda_\alpha \Delta_\alpha$, and if $\omega(t) = \text{Res}\,(\omega/s_0 - ts_\infty)$ is a holomorphic section of the Hodge bundle, then we claim that

$$\nu(t) = \sum_{\alpha=1}^{N} \frac{\lambda_\alpha}{2\pi\sqrt{-1}} \int_{t_0}^{t_\alpha} \frac{\pi_\alpha(s)\,ds}{t - s} \tag{1.15}$$

where $\pi_\alpha(s)$ is the period (1.10) of $\omega(s)$ over the vanishing cycle δ_α.

Proof. On the complex t-sphere minus the cuts γ_α from t_0 to t_α we may view the normal function $\nu(t)$ as being a single-valued holomorphic function having the following properties:

i) near the critical point t_α the difference

$$\nu(t) - \frac{\lambda_\alpha}{2\pi\sqrt{-1}}\,\pi_\alpha(t)\log(t - t_\alpha)$$

is holomorphic and single-valued (thus $\nu(t)$ has the jump $\lambda_\alpha\pi_\alpha(t)$ across the cut γ_α), and

ii) $\nu(t)$ vanishes to the same order $k + 1$ ($k \geq 0$) at $t = \infty$ as the section $\omega(t)$ of $E \to \mathbf{P}^1$.

On the other hand, suppose we consider the potential-theoretic integral on the right hand side of (1.15). This integral has the jump $\lambda_\alpha\pi_\alpha(t)$ across γ_α, and the difference

$$\frac{\lambda_\alpha}{2\pi\sqrt{-1}} \int_{t_0}^{t_\alpha} \frac{\pi_\alpha(s)\,ds}{s - t} - \frac{\lambda_\alpha}{2\pi\sqrt{-1}}\,\pi_\alpha(t)\log(t - t_\alpha)$$

is single-valued and holomorphic at $t = t_\alpha$. Moreover, using (1.7) which implies that $\Sigma_\alpha \lambda_\alpha\pi_\alpha(t) \equiv 0$, the integral is single-valued outside a large circle. To examine its behavior near infinity we set $t' = 1/t$ and write it as

$$t'\left(\sum_{\alpha=1}^{N} \frac{\lambda_\alpha}{2\pi\sqrt{-1}} \int_{t_0}^{t_\alpha} \frac{\pi_\alpha(s)\, ds}{t's - 1}\right). \tag{1.16}$$

It follows that the difference of $\nu(t)$ and the integral is single-valued and holomorphic on the t-sphere and vanishes at $t = \infty$, which then establishes (1.15).

Now (1.14) is an immediate consequence. Using (1.16) we have near $t = \infty$

$$\nu(t) = -\sum_{l=0}^{\infty} \left(\sum_{\alpha=1}^{N} \frac{\lambda_\alpha}{2\pi\sqrt{-1}} \int_{t_0}^{t_\alpha} s^l \pi_\alpha(s)\, ds\right) t'^{l+1}. \tag{1.17}$$

If $\omega(t)$ vanishes to order $k + 1$ at infinity the same must be true of $\nu(t)$, and therefore if $k \geq 1$

$$0 = \sum_{\alpha=1}^{N} \frac{\lambda_\alpha}{2\pi\sqrt{-1}} \int_{t_0}^{t_\alpha} \pi_\alpha(s)\, ds = \int_\Gamma \varphi \tag{1.18}$$

where $\Gamma = \Sigma_a \lambda_\alpha \Delta_\alpha$ is the fundamental class of ν and $\varphi = \omega(t) \wedge dt \in H^0(\Omega_S{}^2)$ (c.f. (1.15)). This exactly says that Γ has Hodge type (1, 1).

Conversely, if $\Gamma = \Sigma_\alpha \lambda_\alpha \Delta_\alpha$ is any primitive cycle on S so that (1.7) is satisfied, then we may *define* a section of $J \rightarrow \mathbf{P}^1$ outside of $t = \infty$ by the formula (1.15). The conditions that this section extend across $t = \infty$ are exactly that Γ be orthogonal to $H^0(\Omega_S{}^2)$, and this establishes (1.14).

d) We now take up the interplay between normal functions and the algebraic curves on S. If Z is a primitive algebraic 1-cycle (=divisor in this case) on S, then each intersection

$$Z_t = Z \cdot C_t = \sum_i p_i(t) - q_i(t)$$

is a divisor of degree zero on the curve C_t, and so represents a point $\nu_Z(t) \in J(C_t)$. Thinking of normal functions as multivalued sections of the Hodge bundle, the definition of $\nu_Z(t)$ is given by the usual abelian sum

$$\nu_Z(t) = \sum_i \int_{q_i(t)}^{p_i(t)} \omega(t) \tag{1.19}$$

where $\omega(t)$ is any local holomorphic section of $E \to \mathbf{P}^1$. This formula makes sense provided that Z does not pass through any of the double points of the singular C_{t_α}, a circumstance which may always be achieved by moving Z in its linear equivalence class and noting that by Abel's theorem the abelian sum (1.19) remains invariant.

We denote by ν_Z the normal function associated to the cycle Z. The basic topological property of the map {divisors} → {normal functions} is

The fundamental class of ν_Z is equal to the fundamental class of Z. (1.20)

Indeed, write $Z_t = \partial\gamma_t$ for a 1-chain γ_t so that (1.19) becomes

$$\nu_Z(t) = \int_{\gamma_t} \omega(t),$$

and then as t turns around t_α the analytic continuation

$$\nu_Z(t) \to \nu_Z(t) + \lambda_\alpha \int_{\delta_\alpha} \omega(t)$$

is exactly reflected by the transformation

$$\gamma_t \to \gamma_t + \lambda_\alpha\, \delta_\alpha .$$

Consequently, both fundamental classes are represented by the cycle $\sum_\alpha \lambda_\alpha \Delta_\alpha$.

The main existence theorem, which is due to Poincaré, states:

Every normal function is of the form ν_Z for some algebraic cycle. (1.21)

When combined with (1.14) and (1.20) we obtain Lefschetz's original proof of his famous (1, 1) theorem.

The existence theorem (1.21) amounts to the Jacobi inversion theorem with dependence on parameters. Namely, select one of the base points p of the pencil $|C_t|$ as a base point (different use of the word) on each Riemann surface C_t. If for generic t the point $\nu(t) \in J(C_t)$ is not in the translate of the theta divisor by $-gp$, then as ω varies over $H^0(\mathbf{P}^1, E)$ the equations

$$\nu(t) = \sum_i \int_p^{p_i(t)} \omega(t) \tag{1.22}$$

will have a unique solution

$$D_t = \sum_{i=1}^{g} p_i(t).$$

This divisor obviously varies holomorphically with t, and at points where either $t = t_\alpha$ is a critical value or else $\nu(t) \in J(C_t)$ becomes a special divisor easy arguments show that D_t remains uniquely determined. In this way we have traced out an algebraic curve $D = \cup D_t$ on S, which provides the essential step in the proof of (1.21) in this case.

In the situation where the equations (1.22) do not determine a unique D_t for a generic t, we will have dim $|D_t| = r > 0$. Then by imposing r independent base point conditions at p we generically determine a unique D_t, and the argument proceeds as before.

In concluding this discussion we note that if we vary the normal function by the image of the fixed part

$$J(S) \to H^0(\mathbf{P}^1, J)$$

in (1.13), we may construct a family of linearly *in*equivalent curves parametrized by the Picard variety of S. This was the first complete proof of this existence theorem and provided the original motivation for the introduction of normal functions.

e) Now, as indicated in the introduction and will be explained more fully below, the formal aspects of the above discussion carry over to higher dimensions with intermediate Jacobians replacing Jacobians. In particular, the analogue of Lefschetz's characterization (1.14) of the fundamental classes goes through (c.f. Section 2b)), but the fundamental existence theorem (1.21) does not generalize due to the failure of the inversion theorem for intermediate Jacobians. This suggests that we re-examine the normal functions for curves on a surface having in mind the following general philosophy: In higher dimensions a particular Hodge structure will in general not arise from a geometric situation, but a non-trivial global variation of Hodge structure should come from algebraic geometry. Now we are not able to reprove the existence theorem without using Jacobi inversion, but we are able to show by suitable differentiation that, in a certain sense, knowing the normal function ν_Z allows us to determine the equations of Z.

Our discussion of this will be facilitated by introducing some classical notation. We imagine S as being generically projected into $\mathbf{P}^3$ having there an affine equation

$$f(x, y, z) = 0$$

where f is a polynomial of degree d, and where the Lefschetz pencil is given by the plane sections $y = t$. If D is the double curve of S, then it is well known that by residues

$$\Omega_S{}^2 \cong \mathcal{O}_S(d - 4)(-D)$$

and consequently $\Omega_S{}^2(L) \cong \mathcal{O}_S(d - 3)(-D)$. Sections in $H^0(\Omega_S{}^2(L))$ are then given by polynomials $p(x, y, z)$ of degree $d - 3$ passing through the double curve (adjoint conditions), and the corresponding section of the Hodge bundle is

$$\omega(t) = \operatorname{Res}\left(\operatorname{Res}\left(\frac{p(x, y, z)\, dx \wedge dy \wedge dy}{(y - t) f(x, y, z)}\right)\right)$$

$$= \frac{p(x, t, z)\, dx}{\dfrac{\partial f}{\partial z}(x, t, z)}. \tag{1.23}$$

Recall that a normal function is given by a multi-valued holomorphic function $\nu(t)$ on $\mathbf{P}^1 - \{t_1, \ldots, t_N\}$ which has certain prescribed behavior at the critical values $t = t_\alpha$ and at $t = \infty$, and whose indeterminacy is given by period

$$\pi(t) = \int_\delta \omega(t), \qquad \delta \in H_1(C_t, \mathbf{Z}). \tag{1.24}$$

Now it is well-known that these periods satisfy a linear differential equation (Picard-Fuchs equation)

$$P(t, d/dt)\pi(t) = \sum_{k=0}^{l} r_k(t) \frac{d^k \pi(t)}{(dt)^k} = 0 \tag{1.25}$$

with rational functions $r_k(t)$ as coefficients, and it is suggested that we apply the Picard-Fuchs operator $P(t, d/dt)$ to the normal function $\nu(t)$ to obtain a single-valued *rational* function $r(t)$ which should then reflect properties of ν.

More precisely, we denote by D the Gauss-Manin connection for the cohomology bundle $\mathbf{H} = \bigcup_{t \neq t_\alpha} H_{DR}{}^1(C_t)$ over $\mathbf{P}^1 - \{t_1, \ldots, t_N\}$. Its characteristic property is that we can differentiate under the integral

$$\frac{d}{dt}\int_\delta \omega(t) = \int_\delta \frac{D\omega(t)}{dt}. \tag{1.26}$$

In fact in our case we may take

$$\begin{aligned} \frac{D\omega(t)}{dt} &= \frac{\partial \omega(t)}{\partial t} \\ &= \frac{\partial}{\partial t}\left(\frac{p(x, t, z)\, dx}{\dfrac{\partial f}{\partial z}(x, t, z)} \right) \end{aligned}$$

to be the result of formally differentiating $\omega(t)$. What is important here is that $D\omega(t)/dt$ is a rational 1-form of the 2nd kind on C_t whose cohomology class in $H_{DR}{}^1(C_t)$ is well defined. If the genus of C_t is g then the classes

$$\omega(t), \frac{D\omega(t)}{dt}, \ldots, \frac{D^{2g}\omega(t)}{(dt)^{2g}}$$

will be dependent in $H_{DR}{}^1(C_t)$, and so if they generically span a subspace of dimension l there will be a minimal linear relation on C_t

$$\sum_{k=0}^{l} r_k(t) \frac{D^k \omega(t)}{(dt)^k} = d\varphi(t), \tag{1.27}$$

where φ is the restriction of a rational function on S and the $r_k(t)$ are rational functions of t. It is clear that by using (1.26) the relation (1.27) implies the differential equation (1.25).

In order to simplify the following discussion we will assume that the irregularity $q(S) = 0$ (the general case will be taken up in Section 2 below). Then there is no fixed part to the family of Jacobians $J(C_t)$, and according to Lefschetz [6] the global monodromy group acting on $H^1(C_{t_0})$ is irreducible (in general, it is irreducible on the variable part of the cohomology). It follows that the order l of the equation (1.25) is $2g$, since the classes $\{\omega, D\omega(t)/dt, D^2\omega(t)/dt^2, \cdots\}$ span an invariant subspace of $H_{DR}{}^1(C_t)$ which must then be all of this group. The general solution to (1.25) is then of the form

$$\sum_{j=1}^{2g} c_j \int_{\delta_j} \omega(t), \qquad c_j \in \mathbf{C}, \tag{1.28}$$

when $\delta_1, \ldots, \delta_{2g}$ give a basis for $H_1(C_t)$.

Now to each normal function $\nu(t)$ we associate the rational function $P(t, d/dt)\nu(t)$. A basic result is Manin's theorem [7]: *The kernel of the mapping*

$$\nu(t) \to P(t, d/dt)\nu(t) \tag{1.29}$$

consists exactly of the torsion elements in the group of normal functions.

Proof. If the right-hand side of (1.29) is zero then $\nu(t)$ has the local form (1.28). By analytic continuation around the critical value $t = t_\alpha$ we infer that

$$\lambda_\alpha \int_{\delta_\alpha} \omega(t) = \sum_j c_j(\delta_j, \delta_\alpha) \int_{\delta_j} \omega(t), \tag{1.30}$$

where the λ_α are the coefficients in the fundamental class $\sum \lambda_\alpha \Delta_\alpha$ of ν.

By differentiation we deduce that (1.30) holds for any class in $H^1(C_t)$; i.e.

$$\sum_j c_j(\delta_j, \delta_\alpha) = \lambda_\alpha \in \mathbf{Z}$$

for all α. Since the global monodromy group is irreducible, the $2g \times N$ matrix $(\delta_j, \delta_\alpha)$ has maximal rank $2g$. It follows that the δ_j are *rational* numbers, and hence that some multiple $m\nu = 0$. Q.E.D.

f) By virtue of (1.29) we may ask if additional information may be obtained by further differentiation, and our main result (Theorem 1.36) will be that this is indeed the case.

As a preliminary to stating this theorem we need to observe that for the pair consisting of a smooth manifold N and submanifold S the relative de Rham cohomology $H^*(N, S)$ may be computed from the complex of differential forms φ on N such that $\varphi \mid S \equiv 0$. In case N is a smooth algebraic variety and S an algebraic subvariety we may similarly define the algebraic de Rham hypercohomology and extend Grothendieck's algebraic de Rham theorem to this situation.

Returning to our consideration of curves on a surface, for an arbitrary integer $k > 0$ we let $E \in |kL|$ and assume for the moment that E does not have multiple components, so that a $E_t = E \cdot C_t$ is smooth for t in a Zariski open set $U \subset \mathbf{P}^1$. Given a section $\omega(t) \in H^0(\omega_{C_t})$ of the Hodge bundle as above, we may consider the relative class $\omega(t) \in H_{DR}{}^1(C_t, E_t)$. Denoting by D_E the Gauss-Manin connection for the relative cohomology bundle $\bigcup_{t \in U} H_{DR}{}^1(C_t, E_t)$, exactly as in the preceding discussion there will be a Picard-Fuchs operator

$$P(t, d/dt, E) = \sum_{k=0}^{l} r_k(t, E) \frac{d^k}{(dt)^k} \qquad (1.31)$$

corresponding to a minimal relation

$$\sum_{k=0}^{l} r_k(t, E) \frac{D_E{}^k \omega(t)}{(dt)^k} = 0$$

in $H_{DR}{}^1(C_t, E_t)$. It is also the case that the coefficients $r_k(t, E)$ are rational functions of $t \in \mathbf{P}^1$ and of $E \in |kL|$, at least provided we normalize by requiring $r_l(t, E) = 1$.

We will now show: *For generic E and $\omega(t)$, the equation* (1.31) *has the maximal possible rank*

$$l = 2g + kd - 1 \tag{1.32}$$

where $d = c_1{}^2(L)$ is the degree of L.

Proof. We consider the exact homology sequence

$$0 \to H_1(C_t) \to H_1(C_t, E_t) \to H_0(E_t) \to \mathbf{Z} \to 0.$$

The kernel of $H_0(E_t) \to \mathbf{Z}$ corresponds to zero cycles of degree zero supported on E_t, and so the global monodromy group operating here is irreducible provided that the curve E is irreducible (it is essentially the Galois group of the branched covering $E \to \mathbf{P}^1$). Consequently, the global monodromy group acting on $H_1(C_t, E_t)$ preserves the subspace $H_1(C_t)$ and acts irreducibly on this subspace and on the quotient space. If the span of $\{\omega(t), D_E\omega(t)/dt, \cdots\}$ fails to be all of $H_{DR}{}^1(C_t, E_t)$, then writing

$$E_t = \sum_{i=1}^{kd} p_i(t)$$

and recalling that $p \in C_t$ denotes a basepoint, we will have in a small t-disc

$$0 = \sum_i c_i \int_p^{p_i(t)} \omega(t) + \int_\gamma \omega(t) \tag{1.33}$$

where γ is a constant linear combination of cycles in $H_1(C_t)$. By analytic continuation of this equation around a branch point of $E \to \mathbf{P}^1$ where, e.g., p_1 and p_2 come together and interchange, we deduce that

$$(c_1 - c_2) \int_{p_1(t)}^{p_2(t)} \omega(t) = 0.$$

Since $\omega(t)$ is assumed generic this implies that $c_1 = c_2$. Continuing in this way we infer from the irreducibility of E that all c_i are equal, and so (1.33) becomes

$$0 = c\left(\sum_i \int_q^{p_i(t)} \omega(t)\right) + \int_\gamma \omega(t).$$

By analytic continuation of this equation around the critical points $t = t_\alpha$ we deduce that $(\gamma, \delta_\alpha) = 0$ for all vanishing cycles δ_α, and hence $\gamma = 0$. Q.E.D.

We remark that, just as for ordinary polynomials, the operator $P(t, d/dt, E)$ may become reducible for special $E \in |kL|$. What is important for our purposes is the observation that if under a specialization $E \to E_0$ the coefficients $r_k(t, E)$ specialize to rational functions $r_k(t, E_0)$ which are not identically infinite, then the solutions to the O.D.E. $P(t, d/dt, E_0)\pi(t) = 0$ are obtained by specializing solutions of $P(t, d/dt, E)\pi(t) = 0$, and hence are constant linear combinations of periods of $\omega(t)$ over cycles in $H_1(C_t, E_{0_t})$.

Now suppose that Z is a primitive algebraic cycle on S with corresponding normal function ν_Z. If we want, by a linear equivalence we may think of Z as the difference of two smooth curves. Then for any $E_0 \in |kL|$ passing through Z and for which $P(t, d/dt, E_0)$ exists as a specialization—which will be the case for generic $E_0 \in |kL|$ containing Z provided that k is sufficiently large—the normal function ν_Z satisfies the relative Picard-Fuchs equation

$$P(t, d/dt, E_0)\nu_Z(t) = 0. \tag{1.34}$$

Conversely, suppose that a given normal function $\nu(t)$ satisfies (1.34) for some E_0. Then by our remark about the solutions to (1.34)

$$\nu(t) = \int_\gamma \omega(t),$$

where γ is a constant linear combination of cycles in $H_1(C_t, E_{0_t})$. By considering the monodromy around the critical values $t = t_\alpha$ we deduce that

$$(\gamma, \delta_\alpha) = \lambda_\alpha \in \mathbf{Z}$$

for all α, and then $\sum_\alpha \lambda_\alpha \Delta_\alpha \in H_2(S, C_{t_0})$ gives the fundamental class

η_ν of the normal function, which is consequently in the image of

$$H_2(E_0, \mathbf{Q}) \to H_2(S, \mathbf{Q}) \tag{1.35}$$

since the locus of the 1-chain $\gamma(t)$ in the slit plane will give a 3-chain Γ with $\partial\Gamma \equiv \Sigma_\alpha \lambda_\alpha \Delta_\alpha$ modulo E_0.

Summarizing, we have established the following:

THEOREM. *For a given normal function ν and each divisor $E \in |kL|$ we consider the equation*

$$P(t, d/dt, E)\nu(t) = 0. \tag{1.36}$$

Thinking of E as variable, those divisors for which $P(t, d/dt, E)$ is defined and (1.36) *is satisfied are exactly the divisors for which the fundamental class η_ν is in the image of the mapping* (1.35).

2. Discussion of the Theorem in Higher Dimensions.

a) We want to extend the result just given to higher dimensional varieties. For reasons stemming from the Lefschetz hyperplane theorem and Hard Lefschetz Theorem the crucial case is the middle dimension[1]—i.e., we consider primitive algebraic n-cycles Z on a smooth projective variety M of complex dimension $2n$. As in the case $n = 1$ of curves on a surface we assume given a very ample line bundle $L \to M$ and Lefschetz pencil $|V_t|$ selected from the complete linear system $|L|$. The critical values $t = t_1, \ldots, t_N$ correspond to V_{t_α} having acquired one ordinary double point. As $t \to t_\alpha$ there will be a vanishing cycle $\delta_\alpha \in H_{2n-1}(V_t)$, and the Picard-Lefschetz transformation on $H_{2n-1}(V_t)$ as t turns around t_α is

$$\gamma \to \gamma \pm (\gamma, \delta_\alpha)\,\delta_\alpha. \tag{2.1}$$

Marking a path from a fixed point t_0 to t_α the locus of the vanishing cycles δ_α along this path traces out the relative cycle $\Delta_\alpha \in H_{2m}(M, V_0)$. Ignoring questions of torsion the primitive part of $H^{2n}(M, \mathbf{Q})$ is exactly

[1] c.f. the discussion in [6] reducing the Lefschetz (1, 1) theorem in general to the case of curves on a surface.

represented by relative cycles $\sum_\alpha \lambda_\alpha \Delta_\alpha$ satisfying $\sum_\alpha \lambda_\alpha \delta_\alpha = 0$ in $H_{2n-1}(V_0)$. In fact the topological discussion in the case $n = 1$ (c.f. Section 1b)) goes over pretty much verbatim, the only real differences occurring between odd and even dimensions. Thus a primitive algebraic n-cycle Z on M satisfies either of the equivalent conditions: $Z \cdot V_t = 0$ in $H_{2n-2}(V_t; \mathbf{Q})$; or, in cohomology, the fundamental class η_Z maps to zero under the restriction mapping $H^{2n}(M, \mathbf{Q}) \to H^{2n}(V_t, \mathbf{Q})$. Given such a primitive algebraic n-cycle Z, its corresponding relative cycle $\sum_\alpha \lambda_\alpha \Delta_\alpha$ is obtained by writing

$$Z_t = Z \cdot V_t = \partial \gamma_t$$

when $\gamma_t \in H_{2n-1}(V_t, Z_t)$. As t turns around the critical value t_α, by an obvious extension of (2.1) we have

$$\gamma_t \to \gamma_t \pm \lambda_\alpha \delta_\alpha .$$

Then it is proved as before that these λ_α are the coefficients in the relative cycle corresponding to Z.

Before taking up normal functions we must discuss a little Hodge theory. For a smooth projective variety V we denote by $H^m(V)$ the complex cohomology and recall the Hodge decomposition

$$\begin{cases} H^m(V) = \bigoplus_{p+q=m} H^{p,q}(V), \\ H^{p,q}(V) = \overline{H^{q,p}(V)}. \end{cases}$$

It will be convenient to consider also the Hodge filtration defined by

$$F^k H^m(V) = \bigoplus_{p \geq k} H^{p,m-p}(V).$$

We may think of $H^m(V)$ as either de Rham cohomology or Čech cohomology for the constant sheaf $\mathbf{C}$. In each of these cases the Hodge filtration has a useful description which we shall now recall. Denoting by $A^{p,q}(V)$ the smooth (p, q) forms on V, by $F^k A^m(V) = \bigoplus_{p \geq k} A^{p,m-p}(V)$ the Hodge filtration on forms, by $Z(F^k A^m(V))$ the d-closed forms, and finally by $\hat{\Omega}_V{}^k$ the sheaf of closed holomorphic k-forms on V we have the following (c.f. [3]):

There are natural isomorphisms

$$\begin{cases} Z(F^kA^m(V))/d(F^kA^{m-1}(V)) \cong H^{m-k}(V, \hat{\Omega}_V{}^k) \\ \qquad\qquad\qquad\qquad\qquad\quad \cong F^kH^m(V) \end{cases} \tag{2.2}$$

Now suppose that V has dimension $2n - 1$. Setting

$$\begin{cases} H^+(V) = F^nH^{2n-1}(V), \qquad \text{and} \\ H^-(V) = H^{2n-1}(V)/H^+(V) \\ \qquad\quad\;\; \cong H^+(V)^* \end{cases}$$

the intermediate Jacobian[2] is defined to be

$$\begin{aligned} J(V) &= H^-(V)/H^{2n-1}(V, \mathbf{Z}) \\ &\cong H^+(V)^*/H_{2n-1}(V, \mathbf{Z}) \end{aligned}$$

where the second isomorphism results from duality. Thus, a point $\nu \in J(V)$ is given by a linear function on $H^+(V)$ taken modulo those linear functions arising by integration over cycles, which are called periods. Each algebraic $(n - 1)$-cycle $Z \subset V$ which is homologous to zero defines a point ν_Z in $J(V)$ by the linear function

$$\omega \to \int_\gamma \omega \tag{2.3}$$

where γ is a real $(2n - 1)$ chain with $\partial\gamma = Z$ and

$$\omega \in Z(F^nA^{2n-1}(V))/d(F^nA^{2n-2}(V)) \cong H^+(V)$$

using (2.2). It is straightforward to verify that (2.3) is independent modulo periods of the choice of γ and of the choice of form representing a class in $H^+(V)$, and so represents a well-defined point in $J(V)$. When $n = 1$ we just have the familiar mapping from divisors of degree zero to the Jacobian variety of a curve.

---

[2]We shall only be concerned with the intermediate Jacobian in the middle dimension.

Some properties of the mapping (2.3) which are more or less immediate from the definition are these:

i) ν_Z varies holomorphically with Z;
ii) consequently, ν maps rationally equivalent cycles to the same point, and we are free to utilize the moving lemma to put cycles in general position; and
iii) if the Hodge decomposition is skew to the rational cohomology in the sense that

$$H^{2n-1}(V, \mathbf{Q}) \cap (H^{n,n-1}(V) \oplus H^{n-1,n}(V)) = (0), \qquad (2.4)$$

then ν maps algebraically equivalent cycles to the same point.

The next properties have to do with a holomorphic family $\{V_t\}_{t\in B}$ of smooth V_t's. We denote by $\mathbf{H} \to B$ the cohomology bundle with Gauss-Manin connection D; thus

$$\mathbf{H} = \bigcup_{t\in B} H^{2n-1}(V_t)$$

and the equation $D\gamma_t = 0$ defines the local condition that cycles $\gamma_t \in H^{2n-1}(V_t)$ be displaced from a fixed cycle γ_{t_0} by diffeomorphisms $V_{t_0} \xrightarrow{\sim} V_t$. We recall that the

$$F^k\mathbf{H} = \bigcup_{t\in B} F^kH^{2n-1}(V_t)$$

give holomorphic sub-bundles and $D(F^k\mathbf{H}) \subset \Omega_B{}^1(F^{k-1}\mathbf{H})$. The relevant properties of (2.3) with dependence on parameters are these:

iv) the intermediate Jacobians $J(V_t)$ vary holomorphically with t, and $J = \bigcup_{t\in B} J(V_t)$ forms in a natural way an analytic fibre space of complex tori over B;
v) if $Z_t \subset V_t$ is an algebraic $(n - 1)$-cycle which varies holomorphically with t and which is homologous to zero, then $\nu_{Z_t} \in J(V_t)$ varies holomorphically with t; and finally
vi) the section ν_Z of the fibre space $J \to B$ defined by $\nu_Z(t) = \nu_{Z_t}$ as in v) just above is *quasi-horizontal* in the sense that

$$D\nu_Z = 0 \qquad (2.5)$$

where

$$D : \mathcal{O}(J) \to \Omega^1(F^{n+1}\mathbf{H})$$

is the mapping induced from the Gauss-Manin connection (c.f. [3]).

Our last properties have to do with the behavior of $J(V_t)$ as t approaches a critical value $t = t_\alpha$ in the Lefschetz pencil $|V_t|$ on M, and they are:

vii) there is naturally defined a generalized intermediate Jacobian $J(V_{t_\alpha})$ (whose precise properties will be recalled as needed below) such that

$$J = \bigcup_{t \in \mathbf{P}^1} J(V_t)$$

forms naturally a fibre space of complex Lie groups over $\mathbf{P}^1$ extending the previously defined family over $\mathbf{P}^1 - \{t_1, \ldots, t_N\}$; and

(viii) given a primitive algebraic n-cycle Z on M with $Z_t = Z \cdot V_t$, the cross-section $\nu_Z(t) \in J(V_t)$ defined for $t \neq t_1, \ldots, t_N$ by v) above extends across the critical values to give a quasi-horizontal section of $J \to \mathbf{P}^1$.

Definition. The *normal functions* associated to the Lefschetz pencil $|V_t|$ are the quasi-horizontal holomorphic cross-sections of $J \to \mathbf{P}^1$.

Intuitively we may view a normal function ν as follows: First we define the *Hodge bundle*

$$E = \bigcup_{t \in \mathbf{P}^1} T_0(J(V_t))^*$$

to be the bundle of dual spaces to the Lie algebras of the $J(V_t)$; note that for t not a critical value the fibre

$$E_t \cong F^n H^{2n-1}(V_t);$$

the remaining fibres E_{t_α} will be identified below. If $\omega(t)$ is any meromorphic section of the Hodge bundle $E \to \mathbf{P}^1$, then considering ν as a multi-valued section of E^* we obtain a multi-valued meromorphic function

$$\langle \nu, \omega(t) \rangle \tag{2.6}$$

on $\mathbf{P}^1 - \{t_1, \ldots, t_N\}$. Under analytic continuation around $t = t_\alpha$ this function changes by adding on a period $\lambda_\alpha \int_{\delta_\alpha} \omega(t)$ where δ_α is the vanishing cycle. Assuming that $t = \infty$ is not a critical value, as t describes a circle $|t| = r$ of sufficiently large radius the function (2.6) analytically continues back to itself, and consequently

$$\sum_{\alpha=1}^{N} \lambda_\alpha \int_{\delta_\alpha} \omega(t) = 0$$

for all $\omega(t)$. It follows that

$$\sum_{\alpha=1}^{N} \lambda_\alpha \, \delta_\alpha = 0 \qquad \text{in } H_{2n-1}(V_{t_0}),$$

and therefore $\Sigma_\alpha \lambda_\alpha \Delta_\alpha \in H_{2n}(M, V_{t_0})$ defines a primitive class $\eta_\nu \in H_{\text{prim}}{}^{2n}(M, \mathbf{Q})$, which we shall call the *fundamental class* of the normal function ν.

In more precise terms, if we let

$$X = \bigcup_{t \in \mathbf{P}^1} V_t$$

be obtained by blowing up the base locus $B = V_{t_0} \cdot V_\infty$ of the pencil, then $X \xrightarrow{f} \mathbf{P}^1$ is an analytic fibre space and it is a consequence of Lefschetz theory that (c.f. [1] and (1.9))

$$H^1(\mathbf{P}^1, R_f{}^{2n-1}\mathbf{Z}) \otimes \mathbf{Q} \cong H_{\text{prim}}{}^{2n}(M, \mathbf{Q}) \oplus H_{\text{prim}}{}^{2n-2}(B, \mathbf{Q}). \tag{2.7}$$

On the other hand there is an exact sheaf sequence

$$0 \to R_f{}^{2n-1}\mathbf{Z} \to E^* \to J \to 0,$$

of which a piece of the cohomology sequence is

$$H^0(\mathbf{P}^1, J) \xrightarrow{\delta} H^1(\mathbf{P}^1, R_f{}^{2n-1}\mathbf{Z}) \to H^1(\mathbf{P}^1, E^*). \tag{2.8}$$

The normal function ν gives a section in $H^0(\mathbf{P}^1, J)$, and the fundamental class η_ν is just the projection of $\delta(\nu)$ in $H_{\text{prim}}{}^{2n}(M, \mathbf{Q})$ using (2.7). That these descriptions agree is accomplished by computing $H^1(\mathbf{P}^1, R_f{}^{2n-1}\mathbf{Z})$ from the covering $\{U_0, U_1, \ldots, U_N\}$ of $\mathbf{P}^1$ where $U_0 = \mathbf{P}^1 - \{t_1, \ldots, t_N\}$ and U_α is a small disc around t_α, just as was done above (1.9).

Of course if the normal function ν arises from a primitive algebraic cycle Z on M then their fundamental classes are the same; this follows from the above discussion. In particular

$$\eta_\nu \in H^{n,n}(M) \cap H_{\text{prim}}{}^{2n}(M, \mathbf{Q}).$$

Conversely, it was proved by Zucker [10] that:

A cohomology class $\eta \in H_{\text{prim}}{}^{2n}(M, \mathbf{Q})$ is the fundamental class of a normal function if and only if it has Hodge type (n, n). (2.9)

This result had been previously established by Spencer Bloch and the author for sufficiently ample Lefschetz pencils, and we shall give this argument in the next section as a preliminary to discussing Picard-Fuchs equations.

The upshot is that we have a complete generalization of the Lefschetz part of the classical theory of normal functions, but of course what is missing in Poincaré's existence theorem. Now the existence theorem would undoubtedly follow from the Jacobi inversion theorem for the $J(V_t)$'s (c.f. [11]), but it is well known that for $n > 1$ this result is false. In fact, assuming for simplicity of notation that $H^{2n-1}(M) = 0$ (e.g., take $M \subset \mathbf{P}^{2n+1}$ to be a smooth hypersurface), by taking $|V_t|$ from a sufficiently ample linear system we may assume that $h^{2n-1,0}(V_{t_0}) \neq 0$. Recalling that the global monodromy action of $\pi_1(\mathbf{P}^1 - \{t_1, \ldots, t_N\})$ on $H^{2n-1}(V_{t_0}, \mathbf{Q})$ is irreducible, it follows that (2.4) is satisfied for generic t, since otherwise the left-hand side would give an invariant subspace. Therefore the group of invertible points on a generic $J(V_t)$ is a countable subgroup, one which will however be infinite provided that there is a primitive algebraic n-cycle Z on M with $\eta_Z \neq 0$ in $H^{2n}(M, \mathbf{Q})$. In a sense the basic difficulty here is that for higher weight not all Hodge structures come from geometry, even in the broadest motiv-theoretic sense. However, since a non-trivial variation of Hodge structure is in general supposed to come from geometry, and in fact since it seems that the family of determinantal varieties constructed from the differential of the variation of Hodge structure obtained by looking at the linear maps of different ranks must be non-trivial, what is suggested is that we investigate further the infinitesimal properties of normal functions.

b) Before taking up Picard-Fuchs equations it will be useful to analyze the Hodge bundle $E \to \mathbf{P}^1$. We recall (c.f. Schmid [8]) that the general fibre is $E_t = H^+(V_t)$, and those sections of E defined in a punctured disc $0 < |t - t_\alpha| < \epsilon$ around the critical value t_α and which extend are defined by the growth condition

$$\left\| \int_\gamma \omega(t) \right\| = 0(-\log |t - t_\alpha|). \tag{2.10}$$

In order to isolate the essential points we shall consider the case where $\dim M = 4$ and the line bundle $L \to M$ is chosen sufficiently ample to insure that

$$H^p(M, \Omega_M{}^q(kL)) = 0 \quad \text{for} \quad p, k > 0. \tag{2.11}$$

Along the way we shall then simply state the general conclusions which may be drawn from similar arguments (for details, c.f. Zucker's paper [10]).

To begin, by the Lefschetz hyperplane theorem we may consider $H^3(M)$ as a subgroup of $H^3(V_t)$ $(t \neq t_1, \ldots, t_N)$, and in fact we may write

$$H^3(V_t) = H^3(M) \oplus H_{\mathrm{var}}{}^3(V_t)\ ^3 \tag{2.12}$$

where $H^3(M)$ represents the fixed part of the variation of Hodge structure defined by the $\{F^kH^3(V_t)\}$, and where $H_{\mathrm{var}}{}^3(V_t)$ is the orthogonal complement of this fixed part relative to the cup product on $H^3(V_t)$. We shall represent $H_{\mathrm{var}}{}^3(V_t)$ by residues of meromorphic forms on M having poles along V_t and where the filtration by order of pole corresponds to the Hodge filtration, and then it will be easy to see how this representation varies with t.

Suppose then that $V \in |L|$ is a smooth divisor, denote by $\Omega_M{}^p(kV)$ the sheaf on M of meromorphic p-forms having a pole of order $\leq k$

---

[3] This decomposition is defined on rational cohomology.

along V and by $\hat{\Omega}_M{}^p(kV)$ the subsheaf of closed forms. We claim then that there are a pair of exact sequences[4]

$$0 \to \Omega_M{}^4 \to \Omega_M{}^4(V) \xrightarrow{\text{Res}} \Omega_V{}^3 \to 0 \tag{2.13}$$

$$\begin{cases} 0 \to \hat{\Omega}_M{}^3(V) \to \Omega_M{}^3(V) \xrightarrow{d} \Omega_M{}^4(2V) \to 0 \\ 0 \to \hat{\Omega}_M{}^3 \to \hat{\Omega}_M{}^3(V) \xrightarrow{\text{Res}} \hat{\Omega}_V{}^2 \to 0 \end{cases} \tag{2.14}$$

(recall that $\Omega^4 = \hat{\Omega}^4$ since dim $M = 4$), where in both cases "Res" is the Poincaré residue operator.

Proof. In local coordinates x, y_1, y_2, y_3 where V is given by $x = 0$, a section of $\Omega_M{}^4(2V)$ is

$$\psi = \frac{f(x, y)\, dx \wedge dy_1 \wedge dy_2 \wedge dy_3}{x^2}.$$

Setting

$$\varphi = -\frac{f(x, y)\, dy_1 \wedge dy_2 \wedge dy_3}{x} - \frac{g(x, y)\, dx \wedge dy_2 \wedge dy_3}{x}$$

we will have $d\varphi = \psi$ provided that $\partial f/\partial x = \partial g/\partial y_1$, and this implies the surjectivity of d in (2.14). The Poincaré residues are defined by showing that a section η of $\hat{\Omega}_M{}^p(V)$ has the local form

$$\eta = \sigma \wedge \frac{dx}{x} + \tau$$

where σ and τ are holomorphic forms, and then setting Res $\eta = \sigma | V$. It is well-known that this procedure is well-defined and takes closed forms into closed forms.[5]

[4] When dim $V = 2n - 1$ there are n such sequences.

[5] What is actually going on here is that $\Omega_M{}^k(V)$ are the closed forms in the log complex $\Omega_M{}^p(\log V)$, and the Poincare residue Res: $\Omega_M{}^p(\log V) \to \Omega_V{}^{p-1}$ is defined and commutes with d.

Using 2.2 and (2.11) the exact cohomology sequences of (2.13) and (2.14) give commutative diagrams

$$\left\{\begin{array}{ccccccccc} 0 \to & H^{4,0}(M) & \to & H^0(\Omega_M{}^4(V)) & \xrightarrow{\text{Res}} & H^{3,0}(V) & \to & H^{4,1}(M) & \to 0 \\ & & & & & \uparrow & \overset{\eta}{\nearrow} & & \\ & & & & & H^{3,0}(M) & & & \end{array}\right. \tag{2.13a}$$

$$\left\{\begin{array}{ccccccccc} & & H^0(\Omega_M{}^3(V)) & \to & H^0(\Omega_M{}^4(2V)) & \to & H^1(\hat{\Omega}_M{}^3(V)) & \to 0 & \\ H^0(\hat{\Omega}_V{}^2) & \to & H^1(\hat{\Omega}_M{}^3) & \to & H^1(\hat{\Omega}_M{}^3(V)) & \xrightarrow{\text{Res}} & H^1(\hat{\Omega}_V{}^2) & \to & H^2(\hat{\Omega}_M{}^3) \\ \uparrow & \overset{\eta}{\nearrow} & & & & & \uparrow & \overset{\eta}{\nearrow} & \\ H^0(\Omega_M{}^2) & & & & & & H^1(\Omega_M{}^2) & & \end{array}\right. \tag{2.14a}$$

where $\eta \in H^1(\hat{\Omega}_M{}^1)$ represents $c_1(L)$. By the Hard Lefschetz Theorem the η in (2.13a) and the η on the right in (2.14a) are isomorphisms, so that by (2.2) and (2.12)

$$\begin{cases} \text{Res}\, H^0(\Omega_M{}^4(V)) = F^3 H_{\text{var}}{}^3(V) \\ \text{Res}\, H^1(\hat{\Omega}_M{}^3(V)) = F^2 H_{\text{var}}{}^3(V). \end{cases}$$

Combining these identifications with the definition of primitive classes we obtain from (2.13a) and (2.14a) the sequences

$$0 \to F^4 H_{\text{prim}}{}^4(M) \to H^0(\Omega_M{}^4(L)) \to F^3 H_{\text{var}}{}^3(V) \to 0 \tag{2.13b}$$

$$\left\{\begin{array}{l} 0 \to F^3 H_{\text{prim}}{}^4(M) \to H^1(\hat{\Omega}_M{}^3(V)) \to F^2 H_{\text{var}}{}^3(V) \to 0 \\ \qquad\qquad\qquad\qquad\quad \| \\ \qquad\qquad H^0(\Omega_M{}^4(2L))/dH^0(\Omega_M{}^3(L)) \end{array}\right. \tag{2.14b}$$

We now think of V as varying in the pencil $|V_t|$. The Hodge bundle decomposes

$$E = E_f \oplus E_v$$

into a fixed and variable part. By (2.12)

$$E_f = F^2H^3(M),$$

while for $t \neq t_1, \ldots, t_N$

$$\begin{aligned}(E_v)_t &= F^2H_{\text{var}}{}^3(V_t) \\ &\cong F^3H_{\text{prim}}{}^4(M) \setminus (H^0(\Omega_M{}^4(2L))/dH^0(\Omega_M{}^3(L)))\end{aligned} \tag{2.15}$$

by (2.14b). In fact, the isomorphism (2.15) also holds for the critical values $t = t_\alpha$: If $\omega \in H^0(\Omega_M{}^4(2L))$ gives a form with a double pole along V_{t_α}, then the formula

$$\lim_{t \to t_\alpha} \int_{\delta_\alpha(t)} \operatorname{Res} \omega = C \overset{te}{=} \omega(p_\alpha) \tag{2.16}$$

is valid and defines the point residue of ω (c.f. [4]), where here p_α denotes the double point of V_{t_α}. If $\tilde{V}_{t_\alpha} \to V_{t_\alpha}$ is the canonical desingularization, then this point residue is the obstruction to ω inducing a class $\operatorname{Res} \omega \in F^2H^3(\tilde{V}_{t_\alpha})$. In general $\operatorname{Res} \omega$ gives a differential of the 3rd kind on $\tilde{V}_{t_\alpha}$, and its residues there are just $\pm C \overset{te}{=} \omega(p_\alpha)$. The situation is in fact completely analogous to curves acquiring double points.

Concerning the growth estimate (2.10), suppose that M has dimension $2n$ and $\omega \in H^0(\Omega_M{}^{2n}(kL))$. Then according to [10]

$$\int_\gamma \operatorname{Res} \omega = O(-|t|^{n-k} \log |t|),$$

so that for $k \leq n - 2$ we may write

$$\lim_{t \to t_\alpha} F^kH^{2n-1}(V_t) = F^kH^{2n-1}(\tilde{V}_{t_\alpha}),$$

while for $k = n - 1$ this is only true on that part of $F^kH^{2n-1}(V_t)$ having zero residue in the above sense. In particular,

$$\begin{cases} h^{2n-1-k,k}(\tilde{V}_{t_\alpha}) = h^{2n-1-k,k}(V_t), & k \leq n-2 \\ h^{n,n-1}(\tilde{V}_{t_\alpha}) = h^{n,n-1}(V_t) - 1. & \end{cases}$$

In all these cases the canonical extensions of the Hodge bundles is given by taking residues of forms in $H^0(\Omega_M{}^{2n}(kL))$.

Now using the above notation E_v for the variable part of the Hodge bundle and letting $E_v{}^{3,0} \subset E_v$ be the part corresponding to $F^3H_{\text{var}}{}^3(V_t)$, the sequences (2.13b) and (2.14b) globalize to

$$0 \to F^4H_{\text{prim}}{}^4(M) \to H^0(\Omega_M{}^4(L)) \otimes \mathcal{O}(2) \to E_v{}^{3,0} \to 0 \tag{2.13c}$$

$$\begin{cases} 0 \to F^3H_{\text{prim}}{}^4(M) \to \mathfrak{F} \to E_v \to 0 \\ 0 \to H^0(\Omega_M{}^3(L)) \otimes \mathcal{O}(1) \to H^0(\Omega_M{}^4(2L)) \otimes \mathcal{O}(2) \to \mathfrak{F} \to 0 \end{cases} \tag{2.14c}$$

For example, if $\omega \in H^0(\Omega_M{}^4(2L)$ and if s_0 and $s_\infty \in H^0(M, L)$ are sections defining V_0 and V_∞ respectively, then $s_0 + ts_\infty$ defines V_t and $\omega/(s_0 + ts_\infty)^2 \in H^0(\Omega_M{}^4(2V_t))$. Setting $t = 1/t'$

$$\operatorname{Res}\left(\frac{\omega}{(s_0 + ts_\infty)^2}\right) = t'^2 \operatorname{Res}\left(\frac{\omega}{(t's_0 + s_\infty)^2}\right),$$

and this defines the map $H^0(\Omega_M{}^4(2L)) \otimes \mathcal{O}_{\mathbf{P}^1}(2) \to E_v$. We shall analyze the above two sequences individually.

For the first we tensor with $\mathcal{O}_{\mathbf{P}^1}(-1)$ and $\mathcal{O}_{\mathbf{P}^1}(-2)$ and take cohomology sequences to obtain

$$\begin{cases} H^0(\Omega_M{}^4(L)) \cong H^0(\mathbf{P}^1, E_v \otimes \mathcal{O}(-1)) \\ H^0(\Omega_M{}^4) \cong H^0(\mathbf{P}^1, E_v \otimes \mathcal{O}(-2)). \end{cases}$$

The second isomorphism means that the top degree holomorphic differentials on M are uniquely of the form

$$\varphi = \omega(t) \wedge dt$$

where $\omega(t) \in H^{3,0}(V_t)$ vanishes to 2nd order at $t = \infty$, generalizing a phenomenon we noted previously for curves on a surface.

Coming to (2.14c) we again tensor with $\mathcal{O}_{\mathbf{P}^1}(-2)$ and take cohomology to obtain

$$\begin{cases} 0 \to H^0(\mathfrak{F}(-2)) \to H^0(\mathbf{P}^1, E \otimes \mathcal{O}(-2)) \to F^3H_{\text{prim}}{}^4(M) \to 0 \\ H^0(\mathfrak{F}(-2)) \cong H^0(\Omega_M{}^4(2L)). \end{cases}$$

The forms in $H^0(\Omega_M{}^4(2L))$ give differentials in $H^0(\Omega_M{}^4(V_0 + V_\infty))$ by the map $\omega \to \omega/s_0s_\infty$. Taking double residues we obtain a holomorphic 2-form on the base locus B. Consequently the above sequence may be rewritten as

$$0 \to F^2H_{\text{var}}{}^2(B) \to H^0(\mathbf{P}^1, E \otimes \mathcal{O}(-2)) \to F^3H_{\text{prim}}{}^4(M) \to 0. \quad (2.15)$$

We are now ready to sketch the proof of (2.9). Referring to (2.8) and (2.7) and using that $H^0(\mathbf{P}^1, E \otimes \mathcal{O}(-2))$ is dual to $H^1(\mathbf{P}^1, E^*)$ we obtain from these together with (2.15) the diagram

$$\begin{array}{l} H^0(\mathbf{P}^1, J) \to H_{\text{prim}}{}^4(M, \mathbf{Q}) \oplus H_{\text{prim}}{}^2(B, \mathbf{Q}) \\ \qquad\qquad 0 \\ \qquad\qquad \downarrow \\ \qquad\qquad F^3H_{\text{prim}}{}^4(M)^* \cong H_{\text{prim}}{}^{1,2}(M) \oplus H_{\text{prim}}{}^{0,4}(M) \\ \qquad\qquad \downarrow \\ \quad\longrightarrow H^1(\mathbf{P}^1, E^*) \\ \qquad\qquad \downarrow \\ \to \qquad F^2H^2(B)^* \cong H^{0,2}(B) \end{array} \quad (2.16)$$

What must be verified now is that the "obvious" maps in (2.16) render the diagram commutative. The obvious maps are just the projections

$$\begin{cases} H_{\text{prim}}{}^4(M, \mathbf{Q}) \to H_{\text{prim}}{}^{1,3}(M) \oplus H_{\text{prim}}{}^{0,4}(M) \\ H_{\text{prim}}{}^2(B, \mathbf{Q}) \to H^{0,2}(B) \end{cases}$$

in the Hodge decompositions of cohomology. The proof of this is the same as the case $n = 1$ given above. Once this is done, ignoring questions of torsion we will have found a section $\nu \in H^0(\mathbf{P}^1, J)$ whose fundamental class is a given element of $H_{\text{prim}}{}^4(M, \mathbf{Q}) \cap H^{2,2}(M)$. The differential

$$\begin{array}{c} D\nu \in H^0(\mathbf{P}^1, \Omega_{\mathbf{P}^1}{}^1(E^{3,0*})) \\ \| \\ H^0(\mathbf{P}^1, E^{3,0*} \otimes \mathcal{O}(-2)) \\ \| \\ 0 \end{array}$$

by (2.13c). Thus the quasi-horizontality condition is automatically satisfied for any $\nu \in H^0(\mathbf{P}^1, J)$, and this completes the proof of (2.9).

c) The discussion in Section 1d) and e) concerning Picard-Fuchs equations and their relative counterparts now carries directly over. In fact we can be quite explicit. Given $\omega \in H^0(\Omega_M{}^4(L))$ we have a holomorphically varying section

$$\omega(t) = \mathrm{Res}\left(\frac{\omega}{s_0 + ts_\infty}\right) \in H^3(V_t) = \mathbf{H}_t. \tag{2.17}$$

The Gauss-Manin connection is then given by

$$\frac{D\omega(t)}{dt} = \mathrm{Res}\left(\frac{-s_\infty \omega}{(s_0 + ts_\infty)^2}\right).$$

Since the global monodromy action on the variable part of $H^3(V_{t_0})$ is irreducible, we infer that for generic t

$$\mathrm{span}\left\{\omega(t), \frac{D\omega(t)}{dt}, \frac{D^2\omega(t)}{(dt)^2}, \cdots\right\} = H_{\mathrm{var}}{}^3(V_t).$$

Recalling that $\dim H_{\mathrm{var}}{}^3(V_{t_0}) = 2g$ is even, there will be a unique normalized O.D.E.

$$\sum_{k=0}^{2g} r_k(t) \frac{d^k\pi(t)}{(dt)^k} = 0, \qquad r_{2g}(t) = 1, \tag{2.18}$$

satisfied by the periods $\pi(t) = \int_\gamma \omega(t)$ of $\omega(t)$. We denote by $P(t, d/dt)$ the differential operator on the left in (2.18). If ν is any normal function viewed as a multi-valued holomorphic section of E^*, then setting $\nu(t) =$

$\langle \nu, \omega(t) \rangle$

$$P(t, d/dt)\nu(t) = r(t) \tag{2.19}$$

is a *rational* function of t. Indeed, it is single-valued and has no essential singularity at the critical values. As before the kernel of the map

$$\nu \to P(t, d/dt)\nu$$

consists exactly of the torsion elements in $H^0(\mathbf{P}^1, J)/J(M)$, where $J(M)$ is the intermediate Jacobian of M viewed as the fixed part of the family $\{J(V_t)\}$.

We may estimate the degree of the rational function $r(t)$ as follows: The singularities of $P(t, d/dt)$ occur at the critical values $t = t_\alpha$ together possibly with the points where the $D^k\omega(t)/(dt)^k$ fail to span $H_{\text{var}}{}^3(V_t)$. The latter are, however, only apparent singularities of $\nu(t)$, while the former are regular singular points. Since around t_α the function $\nu(t)$ has the form

$$\nu(t) = (\text{holomorphic function}) + (\text{period}),$$

it follows that $r(t)$ has a pole of order $\leq 2g$ at each t_α. Thus we obtain the bound:

$$\deg r(t) \leq h_{\text{var}}{}^1(V_{t_0}) \cdot N \tag{2.20}$$

where N is the number of critical values in the pencil $|V_t|$. Geometrically, N represents the degree of the hypersurface M^* dual to M embedded in projective space by the complete linear system $|L|$.

If now $E \in |kL|$ is such that a generic $E_t = E \cap V_t$ is smooth (thus E may have isolated singularities), and if $\omega(t)$ is given by (2.17), then $\omega(t) \in H^{3,0}(V_t)$ restricts to zero as a form on V_t and so defines a relative class in $H_{DR}{}^3(V_t, E_t)$. Letting D_E denote the Gauss-Manin connection for the relative cohomology bundle $\cup H_{DR}{}^3(V_t, E_t)$ there will be a minimal relation

$$P(t, d/dt, E)\omega(t) = \sum_{k=0}^{l} r_k(t, E) \frac{D_E{}^k\omega(t)}{(dt)^k} = 0 \tag{2.21}$$

in $H_{DR}{}^3(V_t, E_t)$. As usual we normalize by taking $r_l(t, E) = 1$.

THEOREM. *For a given normal function ν and each divisor $E \in |kL|$ we consider the equation*

$$P(t, d/dt, E)\nu(t) = 0. \tag{2.22}$$

Thinking of E as variable, those E_0 for which the operator $P(t, d/dt, E_0)$ is obtained by specialization from a generic E[6] *and for which (2.22) is satisfied are exactly the divisors for which the fundamental class η_ν is in the image of*

$$H_{2n}(E_0, \mathbf{Q}) \to H_{2n}(M, \mathbf{Q}). \tag{2.23}$$

The proof of this result is, at this juncture, pretty much verbatim that for the special case of curves on a surface discussed in Section 1e) above. Rather than belabor the details it seems more interesting to offer some observations on the theorem vis á vis the problem of constructing algebraic cycles.

According to (2.9), given a class $\eta \in H_{\text{prim}}{}^{2n}(M, \mathbf{Q}) \cap H^{n,n}(M)$ there will be a normal function ν whose fundamental class is η. We then consider the mapping

$$E \to P(t, d/dt, E)\nu(t), \tag{2.24}$$

and view it as a rational mapping

$$|kL| \to \mathbf{P}^{m(k)}, \tag{2.25}$$

where $\mathbf{P}^{m(k)}$ is obtained by adding the hyperplane at infinity to the vector space of rational functions of some fixed degree $d(k)$.[7] We then claim that:

The Hodge conjecture is true if, and only if, for k sufficiently large the image of the rational mapping (2.24) passes through the origin in $\mathbf{P}^{m(k)}$. (2.26)

Proof. If Hodge is true, then $\eta = \eta_Z$ is the fundamental class of some algebraic cycle Z and (2.22) will be satisfied for those $E_0 \in |kL|$

---

[6] We will see below that this is not a serious qualifier.

which pass through Z. (Here, there is some problem about the specialization question and this will be discussed below.)

Conversely, we assume by induction that Hodge is true for smooth varieties of dimensions less that dim M. By the various Lefschetz theorems we then reduce to primitive cohomology in the middle dimension (c.f. Section 2a) above). Given $\eta \in H_{\text{prim}}{}^{2n}(M, \mathbf{Q}) \cap H^{n,n}(M)$ coming from a normal function ν we infer that η is in the image of (2.23) for some divisor E_0. Now E_0 must be singular (c.f. the following discussion), but by applying mixed Hodge theory [2] we may conclude that η ultimately comes from (n, n) classes on smooth varieties of dimension less than dim M, and then by the induction assumption η will be algebraic. Q.E.D.

Now of course an obvious possibility would be to show that, for k sufficiently large, the rational mapping (2.25) is surjective. This could be done by counting dimensions and estimating the rank of the Jacobian matrix at a generic E. In fact the latter is not too difficult; a generic fibre is given by all E for which

$$P(t, d/dt, E)\nu(t) = r(t, E) = r(t)$$

is a fixed rational function. Differentiating this equation with respect to E we obtain a homogeneous equation satisfied by $\nu(t)$, and the number of these may be estimated. It is also possible to estimate $m(k)$ by using the regularity theorem to estimate the degree of the rational function $r(t, E)$ in a manner similar to (2.20). When carried out my crude count showed that the dimensions on both sides of (2.25) grow like

$$C \overset{tc}{=} k^n + (\text{lower order terms}),$$

but the coefficient of k^n appears to be larger for the right-hand side of (2.25). This is probably correct since one does not expect to get off so easily.

It is also quite instructive to think about the divisors E_0 for which (2.22) might be satisfied. The basic observation is

[7] If we write $P(t, d/dt, E)$ in the form $\Sigma_k\, [p_k(t, E)/q_k(t, E)][d^k/(dt)^k]$ where the p_k and q_k are polynomials, then as E specializes to E_0 it may happen that some $q_k(t, E_0) \equiv 0$. In this case under the mapping (2.24) E_0 will go to the hyperplane at infinity.

If $\eta \neq 0$ is in the image of the mapping (2.23), then E_0 must be singular. (2.27)

Indeed, the Lefschetz decomposition into primitive homology implies that the primitive cycles are exactly those which are *not* supported on a smooth divisor in the linear system $|kL|$.

Now, given this it none-the-less seems reasonable to hope that for sufficiently large k we may find an E_0 with only ordinary double points which supports η.[8] This raises the question of just how allowing E_0 to have double points might increase the rank of $H_{2n}(E_0)$, and this happens as follows: Let $\Delta = p_1 + \cdots + p_\delta$ be the zero-cycle of double points on E_0 and denote by $\mathcal{I}_\Delta$ the ideal sheaf of Δ. We picture E_0 as the limit of smooth divisors E_t as $t \to 0$. Then to each p_k there corresponds a vanishing cycle $\delta_k(t) \in H_{2n-1}(E_t)$ which shrinks to the point p_k as $t \to 0$. Suppose now that some linear combination

$$\Sigma \lambda_k \delta_k(t) = 0 \quad \text{in} \quad H_{2n-1}(E_t). \tag{2.28}$$

Writing $\Sigma_k \lambda_k \delta_k(t) = \partial\gamma(t)$ for a $2n$-chain $\gamma(t)$ and letting $t \to 0$ we obtain a cycle $\gamma = \gamma(0)$ in $H_{2n}(E_0)$. Conversely, any new cycle in $H_{2n}(E_0)$ arises in this way.

To put the condition (2.28) in algebro-geometric form we recall from Section 2b) above that, if $\tilde{E} \to E_0$ denotes the canonical desingularization, we have

$$h^{2n-1-k,k}(\tilde{E}) = h^{2n-1-k,k}(E), \qquad k \leq n-2,$$

while the expected value for $h^{n,n-1}(\tilde{E})$ is given by

$$h^{n,n-1}(\tilde{E}) = h^{n,n-1}(E) - \deg \Delta. \tag{2.29}$$

Now, assuming for simplicity of notation that L has been replaced by kL so as to have $E_0 \in |L|$, the postulated equation (2.29) will be correct exactly when the zero-cycle Δ imposes independent conditions on the linear system $|K + nL|$. This follows from our discussion of point

[8] This may be proved by writing Z as a difference of effective smooth subvarieties in the (rational equivalence ring) $\otimes$ **Q**. Many of the following observations were known to Spencer Block, and also to Herb Clemens in connection with his work on double solids.

residues centered around (2.16). The correct formula for the right-hand side of (2.29) is

$$h^{n,n-1}(\tilde{E}) = h^{n,n-1}(E) - \deg \Delta + h^1(\mathcal{I}_\Delta(K + nL)), \tag{2.30}$$

where the superabundance $h^1(\mathcal{I}_\Delta(K + nL))$ exactly measures the failure of the points of Δ to impose independent conditions on $|K + nL|$.[9]

All of this is very clear for curves on a surface. If M is a surface and $E_0 \in |L|$ is an *irreducible* curve with only double points, then on the one hand the linear series $|\mathcal{I}_\Delta(K + L)|$ cuts out the canonical series on E_0, while on the other hand the genus of E_0 is given by

$$\pi(E_0) = \pi(E_t) - \deg \Delta.$$

Consequently, Δ imposes independent conditions on $|K + L|$, and the only way we can have $h^1(\mathcal{I}_\Delta(K + L)) \neq 0$ is for E_0 to become reducible; in this case $\dim H_2(E_0)$ clearly goes up accordingly.

In general, we may say that the zeros of the equation (2.22) as E varies are detecting by analytic and topological means the solutions to the following algebro-geometric problem:

Among zero cycles Δ on M find those for which

$$\begin{cases} \dim |\mathcal{I}_\Delta{}^2(kL)| \geq 0 \\ \dim |\mathcal{I}_\Delta(K + nkL)| < \dim |K + nkL| - \deg \Delta. \end{cases}$$

Fortunately the Lefschetz (1, 1) theorem for curves on a surface was not formulated in this way or else it might not have been proved.

BIBLIOGRAPHY

[1] A. Andreotti and T. Frankel, "The second Lefschetz theorem on hyperplane sections," *Global Analysis* (papers dedicated to K. Kodaira), Princeton Univ. Press (1969), pp. 1-20.

[9] This observation follows from the exact cohomology sequence of $0 \to \mathcal{I}_\Delta(K + nL) \to \mathcal{O}_M(K + nl) \to \mathbf{C}_\Delta \to 0$ and $h^1(\mathcal{O}_M(K + nL)) = 0$.

[2] P. Deligne, "Théorie de Hodge, II," *Publ. Math. I.H.E.S.*, **40** (1971), pp. 5–57.

[3] P. Griffiths, "Periods of integrals on algebraic manifolds III," *Publ. Math. I.H.E.S.*, **38** (1970), pp. 125–180.

[4] P. Griffiths, "On the periods of certain rational integrals I and II," *Annals of Math.*, **90** (1969), pp. 460–541.

[5] T. Jambois, "The theorem of Torelli for singular curves," *Trans. Amer. Math. Soc.*, to appear.

[6] S. Lefschetz, *L'analysis et la géometrié algébrique*, Gauthier-Villars, Paris (1924).

[7] Ju. I Manin, "Rational points of algebraic curves over function fields," *Amer. Math. Soc. Translations*, **50** (1966), pp. 189–234.

[8] W. Schmid, "Variation of Hodge structure: The singularities of the period mapping," *Inventiones Math.*, **22** (1973), pp. 211–319.

[9] O. Zariski, Algebraic surfaces, Springer-Verlag, Berlin-Heidelberg-New York (1971).

[10] S. Zucker, "Generalized intermediate Jacobians and the theorem on normal functions," *Inventiones Math.*, **33** (1976), pp. 185–222.

[11] S. Zucker, "The Hodge conjecture for cubic fourfolds," *Compositio Math.*, **34** (1977), pp. 199–209.

MINIMAL HILBERT MODULAR SURFACES WITH $p_g = 3$ AND $K^2 = 2$

By F. HIRZEBRUCH *and* A. VAN DE VEN*

I. Introduction. In this paper we deal again with certain Hilbert modular surfaces, namely with the four surfaces $Y_T{}^\circ(p)$ ($p = 313$, 653, 677 and 773) for the symmetric Hilbert modular group.

Let p be a prime with $p \equiv 1(4)$. The non-compact Hilbert modular surface $X(p)$ can be compactified and the result desingularised to obtain an algebraic surface $Y(p)$. On this surface certain curves—all of them explicitly given—can be (successively) blown down. It is conjectured that the resulting algebraic surface $Y^\circ(p)$ is minimal for $p \geq 29$, that is, if the surface is not rational. The conjecture has been proved for many primes, in particular for all p with $29 \leq p \leq 317$ (see [2]). However, the general case still appears to be very difficult.

The surface $X(p)$ is a quotient of the product of the upper half plane with itself, and the interchange of the two factors in this product gives rise to an involution on $X(p)$. This involution can be extended to an involution T of $Y^\circ(p)$. The quotient is another smooth algebraic surface $Y^\circ(p)/T$. Once again, a number of curves can be blown down on $Y^\circ(p)/T$, and also in this case the resulting surface $Y_T{}^\circ(p)$ is conjectured to be minimal for almost all p, namely for those p, for which $Y_T{}^\circ(p)$ is a surface of general type. But ([3], 5.6): "Diese Vermutung ist bisher noch in keinem einzigen Fall bewiesen worden. Insbesondere sollte man feststellen, ob die vier Flächen $Y_T{}^\circ(p)$ mit $p_g = 3$ . . . minimal sind."

It is the main purpose of this paper to show that indeed the surfaces $Y_T{}^\circ(p)$ are minimal for $p = 313$, 653, 677 and 773.

In general, let $X_{\min}$ be the minimal model of a simply-connected surface X of general type with $c_1{}^2(X) = 2$ and $p_g(X) = 3$. A priori, $c_1{}^2(X_{\min})$ can assume any of the values 2, 3, . . . , 36 (not more, because of the inequality $c_1{}^2(Y) \leq 3c_2(Y)$ due to Y. Miyaoka [8], holding for every surface of general type; compare [11]). In our case, the only other information

*Supported by I.H.E.S., Bures-sur-Yvette, and SFB Theoretische Mathematik, Bonn.

readily available consists of the existence of many curves together with much of their intersection behaviour.

However, this information suffices to construct in the four cases under consideration a positive divisor D with $D^2 = DK = K^2$, which does not contain any exceptional curve. From the existence of such a divisor, together with the fact that the surfaces $Y_T{}^\circ(p)$ are simply-connected, the minimality follows. This will be explained at the beginning of Section II (Proposition II.1).

In Section II we shall also prove a more general statement about canonical divisors on surfaces of general type (Lemma II.6). Though this general result is not needed for the minimality of the four surfaces $Y_T{}^\circ(p)$ we consider, it seems nevertheless natural to include it here. Finally, in the last part of Section II and in Section III we describe how to obtain on $Y_T{}^\circ(p)$ ($p = 353, 653, 677, 773$) a positive divisor D with $D^2 = DK = K^2 = 2$.

This paper should be seen as an appendix to [3], which therefore will serve as a general reference. We are much indebted to Y.Miyaoka, whose remarks have been most helpful to us.

II. A criterion for minimality.

By a *surface* we shall always mean a compact, connected complex manifold of complex dimension 2. As usual, we shall denote by K_X either the *canonical divisior class* of X or one canonical divisor, and by $p_g(X)$ ($= \dim|K_X| + 1$) the *geometric genus* of X.

A *surface of general type* is by definition a surface of Kodaira dimension 2. Such surfaces are always algebraic.

A surface is called *minimal* if it does not contain any exceptional curves. Starting from any surface X a minimal surface is obtained by blowing down a finite number of times. If X is a surface of general type, then such a minimal model is again a surface $X_{\min}$ of general type, which is uniquely determined by X up to algebraic isomorphisms.

PROPOSITION II.1. *Let X be a surface with $K_X{}^2 > 0$, on which there exists a positive divisor D with $D^2 = KD = K^2$. Suppose that at least one of the following conditions is satisfied:*

(i) $H_1(X, Z) = 0$;

(ii) *there is a (positive) canonical divisor, containing D.*

Then X is a surface of general type and D a canonical divisor on X.

Furthermore, if D does not contain any exceptional curve, then X is a minimal surface.

Proof. By Enriques-Kodaira classification X is an algebraic surface, either rational or of general type. Since $K(D - K) = (D - K)^2 = 0$, we have by the Hodge algebraic index theorem (compare [7], Lemma 3) that $D - K$ is homologous to zero mod torsion. In case (i) this means that $D - K$ is homologous to zero over the integers and by the vanishing of the Picard variety it is linearly equivalent to zero, i.e. D is a canonical divisor. In case (ii) we have that $K - D$ can be represented by a non-negative divisor, which must be zero (a strictly positive divisor cannot represent a torsion class in homology), hence also in this case D is a canonical divisor. Since $p_g(X) \geq 1$ we have that X is of general type. The last statement of the proposition now follows from the fact that every (positive) canonical divisor contains every exceptional curve.

The next part of Section II is dedicated to the proof of Lemma II.6, which generalizes the preceding proposition.

Let D be a positive divisor on same surface. A *splitting* of D consists of two positive divisors D_1 and D_2 with $D = D_1 + D_2$ (it is *not* required that D_1 and D_2 have no irreducible components in common). If $D = \Sigma\, c_i C_i$, then we shall say that an irreducible component C_i is *totally contained* in D_1 if C_i appears in D_1 with multiplicity c_i (and hence not at all, i.e. with multiplicity 0, in D_2).

A positive divisor D is called *numerically connected* if for every splitting $D = D_1 + D_2$ the inequality $D_1 D_2 \geq 1$ holds.

THEOREM II.2 *Positive canonical divisors on a minimal surface of general type are numerically connected.*

For a proof of this theorem we refer to [7], Lemma 4, or [1], p. 181. Let on some surface $T = \{E_1, \ldots, E_m\}$ be a σ-tree, i.e. a connected configuration of (smooth, rational) curves $E_1, \ldots, E_m$, obtained by successive blowing-up. (The curves are not taken in any particular ordering.) Then we can speak of the canonical divisor

$$K_T = \sum_{i=1}^{m} k_i E_i$$

of the tree T, where the integers k_i are uniquely determined by the condition that

$$K_T E_i = -E_i^2 - 2$$

for $i = 1, \ldots, m$. The numbers $k_1, \ldots, k_m$ depend only on the structure of the tree, not on the surface, and they are all strictly positive.

PROPOSITION II.3. *Let $f: X \to Y$ be the blowing-down of an exceptional curve E on X. Then*

$$K_X = f^*(K_Y) + E$$

and $f^*(D)E = 0$ for any divisor D of Y.

More generally, if $f: X \to Y$ is the blowing-down of the disjoint σ-trees $T_1, \ldots, T_m$, then

$$K_X = f^*(K_Y) + \sum_{i=1}^{m} K_{T_i}$$

PROPOSITION II.4. *Let $T = \{E_1, \ldots, E_m\}$ be a σ-tree on a surface X.*

(i) *If $D = \sum e_i E_i$ with $e_i \in Z$, then $D(K_T - D) \geq 0$.*

(ii) *If in addition D is positive, and if none of its components contains an exceptional curve, then $D(K_T - D) \geq 1$.*

Proof. (i) By the definition of a σ-tree and because of Proposition II.3, there exist elements $h_1, \ldots, h_m \in H_2(X, Z)$ with $h_i h_j = -\delta_{ij}$, such that K_T is homologous to $\sum_{i=1}^m h_i$, and D is homologous to a sum $\sum_{i=1}^m a_i h_i$. Hence

$$D(K_T - D) = \sum_{i=1}^{m} a_i(a_i - 1) \geq 0.$$

(ii) Here we use a different argument. The fact that the intersection matrix $(E_i E_j)$ is negative definite implies that if E_i is not an exceptional curve, then $E_i^2 \leq -2$, and hence $K_T E_i \geq 0$. So if D is positive and if it does not contain any exceptional curve, we have $D(K_T - D) > DK_T \geq 0$.

Now let X be any surface of general type, and let $f: X \to X_{\min}$ be the projection from X into its minimal model. Then f is the blowing down of a certain number of σ-trees T_i ($i = 1, \ldots, N$). If K is any positive canonical divisor on X, then $K_{\min} = f_*(K)$ is a positive canonical divisor on $X_{\min}$, and K can be written as

$$K = f^*(K_{\min}) + \sum_{i=1}^{N} K_{T_i}$$

Let $p_i = f(T_i)$, and suppose that $p_1, \ldots, p_n$ are points of $K_{\min}$, but $p_{n+1}, \ldots, p_N$ are not. Then the underlying point set of K consists of exactly $N - n + 1$ connected components. A *trivial splitting* $K = K_1 + K_2$ will be any splitting of K, for which K_1 consists of all the curves in some of these components, taken with their total multiplicity in K, and K_2 consists of all the curves in the other components, also taken with their total multiplicity in K.

PROPOSITION II.5. *Let $K = K_1 + K_2$ be a non-trivial splitting with $K_1 \cdot K_2 \leq 0$. Then both K_1 and K_2 have at least one component which is an exceptional curve.*

Proof. It is sufficient to show that if $K_1 \cdot K_2 \leq 0$ for a splitting $K_T = K_1 + K_2$ of a σ-tree, or for a splitting $f^*(K_{\min}) + \Sigma_{i=1}{}^n K_{T_i} = K_1 + K_2$, then both K_1 and K_2 contain at least one exceptional curve (with some positive multiplicity). For the case of a tree this follows from Proposition II.4 (ii).

In the case of $f^*(K_{\min}) + \Sigma_{i=1}{}^n K_{T_i}$ we start by observing that for any splitting we have $K_1 \cdot K_2 \geq 0$. Indeed, $K_1 = f^*(K_1') + D_1$ and $K_2 = f^*(K_2') + D_2$ where $D_1 + D_2 = \Sigma_{i=1}{}^n K_{T_i}$ and where $K_1' + K_2' = K_{\min}$ is a splitting unless K_1' or K_2' is the zero divisor. Since $f^*(K_1') \cdot D_2 = f^*(K_2') \cdot D_1 = 0$, the inequality $K_1 \cdot K_2 \geq 0$ follows from Theorem II.2 and Proposition II.4 (i).

Now we can prove by induction on the total number of irreducible curves in the union of $T_1, \ldots, T_n$ the following statement: If for a splitting of $K = f^*(K_{\min}) + \Sigma_{i=1}{}^n K_{T_i}$ with $K = K_1 + K_2$ all exceptional curves are totally contained in K_1, then $K_1 \cdot K_2 \geq 1$. The induction can start because of Theorem II.2. We put $K = \Sigma_{i=1}{}^k a_i A_i$ (so some of the A_i are irreducible curves in the trees $T_1, \ldots, T_n$, and some of them are proper transforms of irreducible curves of $X_{\min}$) and assume—as we may—that A_1 is an exceptional curve. If $K_1 = a_1 A_1$, then $K_1 K_2 \geq 1$. So from this point on we exclude this case.

Let $g: X \to X'$ be the blowing down of A_1, and $g_*(K_1) = K_1'$,

$g_*(K_2) = K_2'$. Then $K_1' + K_2'$ is a splitting of the canonical divisor $g_*(K)$ of X'. We have

$$K_1 = g^*K_1' - b_1A_1, \qquad K_2 = g^*K_2' - b_2A_1,$$

where the integers b_1, b_2 are determined by $g^*K_1' \cdot A_1 = g^*K_2' \cdot A_1 = 0$. Therefore, $b_1 = K_1A_1$ and $b_2 = K_2A_1$. Because A_1 is not contained in K_2, we have $b_2 \geq 0$. By Proposition II.3

$$b_1 + b_2 + 1 = 0.$$

Thus

$$K_1K_2 = K_1'K_2' - b_1b_2 = K_1'K_2' + b_2^2 + b_2$$

Since $K_1'K_2' \geq 0$, we have $K_1K_2 \geq 1$ unless $b_2 = 0$. If $b_2 = 0$, then all exceptional curves in g_*K are totally contained in K_1', hence $K_1K_2 = K_1'K_2' \geq 1$ by induction. This concludes the proof of Proposition II.5.

Let D be a divisor on $X_{\min}$. If the blowing up from $X_{\min}$ to X starts in points outside of D only, then D is a divisor on X in a natural way. Such a divisor on X will be called an *untouched* divisor from $X_{\min}$.

LEMMA II.6. *Let X be a surface of general type and $D = \Sigma\, d_iD_i$ a positive divisor on X, such that*

(i) *no D_i is an exceptional curve;*
(ii) *there is at least one positive canonical divisor on X, containing D;*
(iii) $K_XD = D^2$.

Then D is an untouched canonical divisor from $X_{\min}$. Furthermore, $X = X_{\min}$ if and only if $D^2 = K_X^2$.

Proof. The first statement is an immediate consequence of Proposition II.5. As to the second statement, we have $D^2 = K_{\min}{}^2 = K_X{}^2$ if and only if $X = X_{\min}$.

Next we define certain types of configurations formed by curves on a surface. As is customary, we shall simply call a smooth rational curve C with $C^2 = -2$ (and therefore $KC = 0$) a (-2)-*curve*. A *chain* of (-2)-curves is an ordered set of (-2)-curves $C_1, \ldots, C_k$ with $C_iC_j = 1$ if $j = i \pm 1$, and $C_iC_j = 0$ otherwise ($i \neq j$).

Definition. A *configuration of type* I consists of a smooth elliptic curve A with $KA = 2$ (and $AA = -2$), and two disjoint chains of (-2)-curves $B_1, \ldots, B_k$ and $C_1, \ldots, C_l$, such that $AB_1 = AB_k = AC_1 = AC_l = 1$, but $AB_i = AC_i = 0$ for all other indices i.

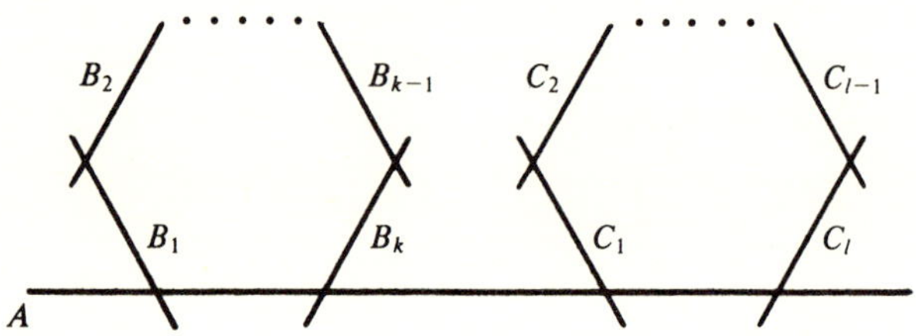

Type I

By the corresponding divisor we shall mean the divisor

$$D_I = A + \sum_{i=1}^{k} B_i + \sum_{i=1}^{l} C_i, \quad \text{for which} \quad D_I^2 = K_X D_I = 2.$$

A configuration of type II consists of two disjoint smooth rational curves A_1, A_2 with $KA_1 = KA_2 = 1$ and thus $A_1^2 = A_2^2 = -3$, and four disjoint (-2)-chains $B_1^{(i)}, \ldots, B_{k_i}^{(i)}$, $i = 1, \ldots, 4$, with $A_1 B_1^{(i)} = A_2 B_{k_i}^{(i)} = 1$, but all other $A_j B_k^{(i)} = 0$.

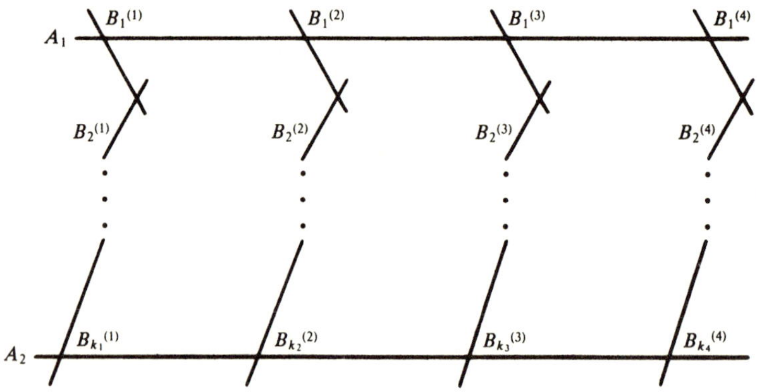

Type II

In this case the corresponding divisor is $D_{II} = A_1 + A_2 + \Sigma_{i,j} B_j^{(i)}$, which again satifies $D_{II}^2 = K_X D_{II} = 2$.

A configuration of type III consists of a smooth rational curve A with $A^2 = -3$, and a number (at least one, at most seven) of *disjoint*

(−2)-*chains*

$$B_j^{(i)}, \qquad |j| < r_i,$$

with $\Sigma\, r_i = 7$ and

$$B_j^{(i)}A = 1 \quad \text{if} \quad j = 0, \qquad \text{but } B_j^{(i)}A = 0 \quad \text{if} \quad j \neq 0.$$

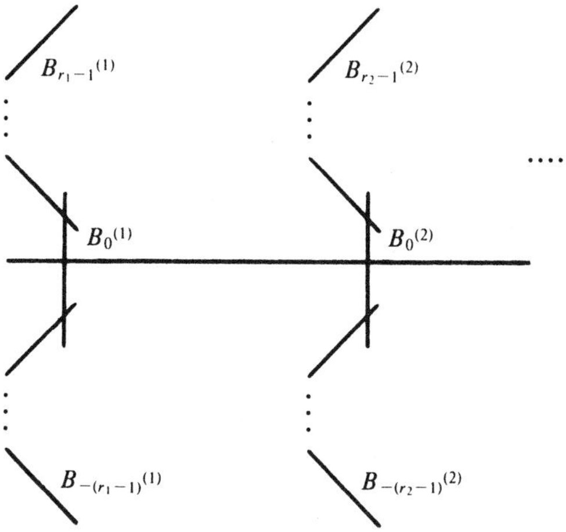

Type III

In this case the corresponding divisor is the divisor

$$D_{\text{III}} = 2A + \sum_i \sum_{|j|<r_i} (r_i - |j|)B_j^{(i)}$$

which again satisfies $D_{\text{III}}{}^2 = K_X D_{\text{III}} = 2$.

A *Moishezon surface* is by definition a minimal surface of general type X with $K_X{}^2 = 2$ and $p_g(X) = 3$. These surfaces are extreme with respect to the Noether inequalities for minimal surfaces of general type (compare [1], p. 208):

$$p_g(X) \leq \begin{cases} \tfrac{1}{2}K_X{}^2 + 2 & (K_X{}^2 \text{ even}) \\ \tfrac{1}{2}K_X{}^2 + \tfrac{3}{2} & (K_X{}^2 \text{ odd}) \end{cases}$$

They have been studied by Moishezon ([9], p. 140) and they appear as a special case in Horikawa's extensive study of all surfaces which are extreme with respect to the Noether inequalities (e.g. [5], [6]). Every Moishezon surface is known to be the minimal desingularisation of a double covering of the projective plane, ramified over a curve of degree 8.

THEOREM II.7. *Let X be a surface with* $K_X{}^2 = 2$ *and* $p_g(X) = 3$, *on which there exists at least one configuration of type* I, II *or* III. *Then X is a Moishezon surface, and the corresponding divisor is a canonical divisor.*

Proof. If $H_1(X, Z) = 0$, then the result follows immediately from Proposition II.1. Without this additional assumption we have to show (Proposition II.1 (iii)) that there is a canonical divisor containing the corresponding divisor D_{I}, D_{II} or D_{III}. This is clear for a configuration of type I or II (take a canonical divisor passing through a point of the first chain and through a point of the second chain).

In case III we take a canonical divisor K, passing through two different points of A. Since $KA = 1$, this divisor must contain A, and consequently all of the curves $B_j{}^{(i)}$.

If we have a (-2)-chain B_j with $|j| < r$ and $r \geq 2$, where we have omitted the upper index, then we have to prove that the multiplicity μ_j of B_j in K is at least $r - |j|$. We assume inductively that this is true for the smaller chain obtained by omitting B_{r-1} and $B_{-(r-1)}$ and hence have $\mu_j \geq r - 1 - |j|$ for $|j| < r$. If $\mu_j \geq r - j$ for some $j \geq 2$, then the inequality $2\mu_{j-1} \geq \mu_j + \mu_{j-2}$ implies that also $\mu_{j-1} \geq r - j + 1$. Since $\mu_{r-1} \geq 1$, we obtain successively $\mu_j \geq r - j$ for $j = r - 1, r - 2, \ldots, 1$, and in the same way $\mu_j \geq r - |j|$ for $-(r - 1) \leq j < 0$. But μ_0 must satisfy

$$2\mu_0 \geq \mu_1 + \mu_{-1} + \mu \geq 2(r - 1) + 1 = 2r - 1,$$

where μ is the multiplicity of A in K, so we also have $\mu_0 \geq r$. Finally,

$$1 = KA \geq \mu AA + \Sigma r_i = -3\mu + 7,$$

so $\mu \geq 2$. Hence D_{III} is contained in K.

Remark. The mapping of a Moishezon surface X on the projective plane is defined by $x \mapsto g(x) = (W_1(x){:}W_2(x){:}W_3(x))$ where W_1, W_2, W_3 is a base of the vector space of all holomorphic 2-forms on X. This map g is defined everywhere. It is of degree 2, the branching locus in the plane being a curve B of degree 8 without multiple components. If L is a line in the plane, then $g^*(L)$ is a positive canonical divisor on X and all positive canonical divisors are thus obtained. If we have a canonical divisor D_{I} on a Moishezon surface, then $D_{\mathrm{I}} = g^*(L)$ where L is a line intersecting B four times transversally (intersection multiplicity 1) and two times with multiplicity 2 in two singular points of B of type a_k and a_l respectively (a_n has the local equation $x^{n+1} + y^2 = 0$). In the case of type II, we have $D_{\mathrm{II}} = g^*(L)$ where L is a line intersecting B four times with multiplicity 2 in four singular points of B of type $a_{k_1}, \ldots, a_{k_4}$. Finally, for type III we have $D_{\mathrm{III}} = g^*(L)$ where L is contained in B. Then $B = L + C$ where C has degree 7 and L intersects C with multiplicities $r_1, r_2, \ldots, (\sum r_i = 7)$ in singular points of B which are of type a_{2r_i-1} and have locally the form $x^{2r_i} - y^2 = 0$ where C is given by $y - x^{r_i} = 0$ and L by $y + x^{r_i} = 0$.

We could just as well consider the configurations arising from a line L intersecting B with multiplicity 2 in one or three points of type a_k and transversally otherwise. Also other singularities of B could be used. But these configurations are not needed in the sequel.

III. The four Hilbert modular surfaces $Y_T{}^\circ(p)$ with geometric genus 3. We shall use the results of [3] very freely. The four primes in question are $p = 313, 653, 677$ and 773. In these cases the geometric genus is 3 and $c_1{}^2[Y_T{}^\circ(p)]$, the selfintersection number of a canonical divisor, equals 2. The class number $h(p)$ equals 1 for these four primes. Thus there is only one cusp in each case, and we shall list the cusp resolutions as in [3], p. 142. Using the curves introduced in [3] we shall find a configuration of type I (with $k = 6$ and $l = 3$) for $p = 313$, a configuration of type I (with $k = 3$, $l = 4$) for $p = 653$, a configuration of type III (with $r_1 = r_2 = 2$, $r_3 = r_4 = r_5 = 1$) for $p = 677$, and, finally, a configuration of type II (with $k_1 = k_2 = 4$, $k_3 = 5$, $k_4 = 2$) for $p = 773$.

The prime $p = 313$. We first note that 2, 3, 11 are quadratic residues, 5 and 7 quadratic non-residues mod 313. Consider the curves $F_{35}{}^*$, $F_{16}{}^*$, $F_{25}{}^*$, $F_{27}{}^*$. They are rational according to Tabelle 3 in [3], p. 127. As in [3] we use the notation involving the first Chern class which produces a sign change when comparing with the canonical divisor. By Tabelle 3 in [3] we have $c_1{}^*(35) = c_1{}^*(16) = c_1{}^*(25) = 0$, whereas $c_1{}^*(27) = -1$. But $F_{27}{}^*$ passes through the intersection of the curves $S_{23}{}^*$ and $S_{24}{}^*$ of the cusp resolution, because the quadratic form of discriminant 313 associated to this intersection point is [2, 19, 6] which represents 27. It follows that the value of the first Chern class of $Y_T{}^\circ(313)$ on $F_{27}{}^*$ is zero ([3], 2.2, 2.4). All the four curves $F_{35}{}^*$, $F_{16}{}^*$, $F_{25}{}^*$, $F_{27}{}^*$ are (-2)-curves.

List of cusp resolutions

	$p = 313$		$p = 653$		$p = 677$		$p = 773$	
k	b_k	N_k	b_k	N_k	b_k	N_k	b_k	N_k
0	19	1	27	1	27	1	29	1
1	2	12	2	19	3	13	2	17
2	3	11	2	23	2	37	4	11
3	2	24	3	13	2	59	2	37
4	2	33	3	11	2	79	2	61
5	2	38	5	7	2	97	2	83
6	2	39	2	31	2	113	2	103
7	2	36	2	53	2	127	2	121
8	2	29	2	73	2	139	2	137
9	2	18	2	91	2	149	2	151
10	7	3	2	107	2	157	2	163
11	5	4	2	121	2	163	2	173
12	3	8	2	133	2	167	2	181
13	2	13	2	143	2	169	2	187
14	4	6	2	151			2	191
15	3	9	2	157			2	193
16	2	18	2	161				
17	2	19	2	163				
18	3	12						
19	2	22						
20	2	26						
21	2	24						
22	2	16						
23	10	2						
24	4	6						
25	2	22						
26	2	36						
27	2	48						
28	2	58						
29	2	66						
30	2	72						
31	2	76						
32	2	78						

We also consider the curve F_{22}^*. It is elliptic and $c_1^*(22) = -3$ (see Tabelle 4 in [3], p. 128). But F_{22}^* intersects S_{25}^* transversally which is blown down when passing to $Y_T°(313)$, (see [3], 3.2). Therefore $c_1^*[F_{22}^*] = -2$ on $Y_T°(313)$.

We now give a list of four quadratic forms which determine exactly four special points on our surface $Y_T{}^\circ(313)$, (see [4]; [3], 1.2).

form	*discriminant*	
[22, 24, 35]	$-8 \cdot 313$	$h(-8) = 1$
[35, 7, 16]	$-7 \cdot 313$	$h(-7) = 1$
[22, 3, 25]	$-7 \cdot 313$	$h(-7) = 1$
[25, 14, 27]	$-8 \cdot 313$	$h(-8) = 1$

If a special point has the form $[a, b, c]$, then $F_a{}^*$, $F_c{}^*$ intersect transversally in this point. Using also the curves $S_j{}^*$ of the cusp resolution, we get the following configuration of type I.

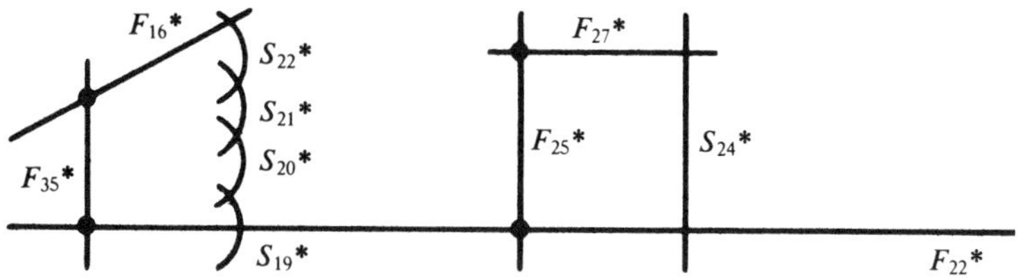

The curves $S_{19}{}^*, \ldots, S_{22}{}^*$ are (-2)-curves according to the list of the cusp resolutions. Originally, $S_{24}{}^*$ has selfintersection number -4, but $S_{25}{}^*$ and $F_6{}^*$, which intersect $S_{24}{}^*$ on $Y^\circ(313)/T$ where blown down when passing to $Y_T{}^\circ(313)$. Compare [3], 4.3 Lemma. Hence also $S_{24}{}^*$ is a (-2)-curve on $Y_T{}^\circ(313)$. By the intersection theory of modular curves [4] it follows that the four special points mentioned above and marked by a dot in the diagram are the only intersections of the curves $F_N{}^*$ considered.

The prime $p = 653$. Note that 7, 11, 13 are quadratic residues and 2, 3, 5 quadratic non-residues mod 653. Consider the curves $F_{25}{}^*$, $F_{28}{}^*$, $F_{13}{}^*$, $F_{54}{}^*$, $F_{66}{}^*$, $F_{216}{}^*$ on $Y_T{}^\circ(653)$. The first five curves mentioned are (-2)-curves ([3], in particular Tabelle 3 on p. 127). According to Tabelle 4 in [3] the curve $F_{216}{}^*$ is elliptic and $c_1{}^*(216) = -6$, but $H(4 \cdot 216 - 653) = H(211) = h(-211) = 3$ and therefore $c_1{}^*[F_{216}{}^*] = -3$ on $Y^\circ(653)/T$, ([3], 2.4 (13)). Moreover, the form [216, 35, 36] has discriminant $-43 \cdot 653$ with $h(-43) = 1$. The exceptional curve $F_{36}{}^*$ intersects $F_{216}{}^*$ on $Y^\circ(653)/T$, and $F_{216}{}^*$ is a curve on $Y_T{}^\circ(653)$ with $c_1{}^*[F_{216}] = -2$.

We now give a list of seven quadratic forms:

form	*discriminant*	
1) [216, 63, 25]	$-27 \cdot 653$	$h(-27) = 1$
2) [25, 29, 28]	$-3 \cdot 653$	$h'(-3) = 1/3$
3) [28, 81, 216]	$-27 \cdot 653$	$h(-27) = 1$
4) [216, 28, 13]	$-16 \cdot 653$	$h(-16) = 1$
5) [13, 14, 54]	$-4 \cdot 653$	$h'(-4) = 1/2$
6) [54, 43, 66]	$-19 \cdot 653$	$h(-19) = 1$
7) [66, 109, 92]	$-19 \cdot 653$	$h(-19) = 1$

The forms 1), 3), 4), 6), 7) define a special point on $Y_T{}^\circ(653)$. These points are distinct. The curves $F_{66}{}^*$ and $F_{216}{}^*$ intersect transversally in the special point belonging to the form 7), because this form represents 216 primitively. The forms 2) and 5) correspond to quotient singularities of order 3 and 2 respectively which when resolved give rise to curves $V_{25}{}^*$ and $U_{13}{}^*$ on $Y_T{}^\circ(653)$. The form 2) represents 24. Therefore, the exceptional curve $F_{24}{}^*$ on $Y^\circ(653)/T$ intersects $V_{25}{}^*$, and $V_{25}{}^*$ which has originally selfintersection number -3, becomes a (-2)-curve on $Y_T{}^\circ(653)$.

We recall that in the notation of [3]

$$V_{25}{}^* = V_{28}{}^{*(1)} = V_{24}{}^*, \qquad U_{13}{}^* = U_{54}{}^*.$$

We get the following configuration of type I

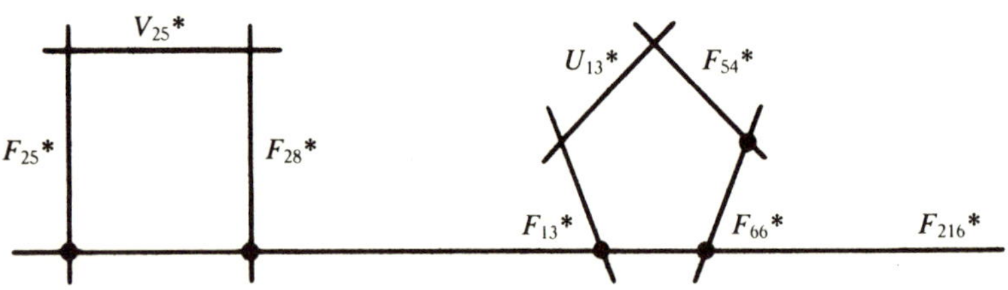

The special points belonging to the forms 1), 3), 4), 6), 7) are marked by a dot. It can be checked [4] that there are no other intersections of the curves considered.

The prime p = 677. Note that 2, 3, 5, 7, 11, 17 are quadratic non-residues mod 677. Consider the curves F_{22}^*, F_{34}^*, F_{56}^*, F_{25}^*, F_{60}^*, F_{96}^* on $Y_T{}^\circ(677)$. The first 5 curves are (-2)-curves, but $c_1^*(96) = -1$. Since $[96, 31, 36]$ has discriminant $-19 \cdot 677$ and F_{36}^* is exceptional on $Y^\circ(677)/T$, it follows that F_{96}^* is a (-2)-curve on $Y_T{}^\circ(677)$.

The quadratic form $[22, 31, 34]$ of discriminant $-3 \cdot 677$ represents primitively 22, 34, 25, 60, 96. It corresponds to a quotient singularity of order 3 and defines a curve V^* of selfintersection number -3 on $Y_T{}^\circ(677)$ with

$$V^* = V_{22}^{*(1)} = V_{34}^{*(1)} = V_{25}^* = V_{60}^{*(1)} = V_{96}^*$$

We get the following diagram of type III

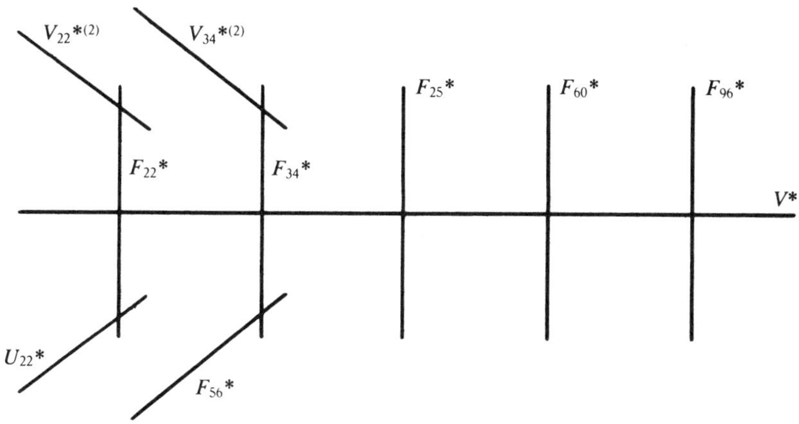

The form $[22, 9, 24]$ of discriminant $-3 \cdot 677$ belongs to the quotient singularity of order 3 which gives rise to $V_{22}^{*(2)}$. Since F_{24}^* is exceptional on $Y^\circ(677)/T$, the curve $V_{22}^{*(2)}$ is a (-2)-curve on $Y_T{}^\circ(677)$. The same holds for $V_{34}^{*(2)}$ because $[34, 3, 15]$ has discriminant $-3 \cdot 677$ and F_{15}^* is exceptional on $Y^\circ(677)/T$. The curves F_{56}^* and F_{34}^* intersect in the special point with quadratic form $[34, 13, 56]$ of discriminant $-11 \cdot 677$.

The prime $p = 773$. Note that 11, 17 are quadratic residues and 2, 3, 5, 7, 13, 23 quadratic non-residues mod 773. The curves F_{40}^*, F_{25}^*, F_{60}^*, F_{17}^*, F_{44}^*, F_{35}^*, F_{39}^* are (-2)-curves (Tabelle 3 in [3], p. 127), whereas F_{46}^*, F_{96}^* are smooth rational curves of selfintersection number -3. We have the following diagram of type II

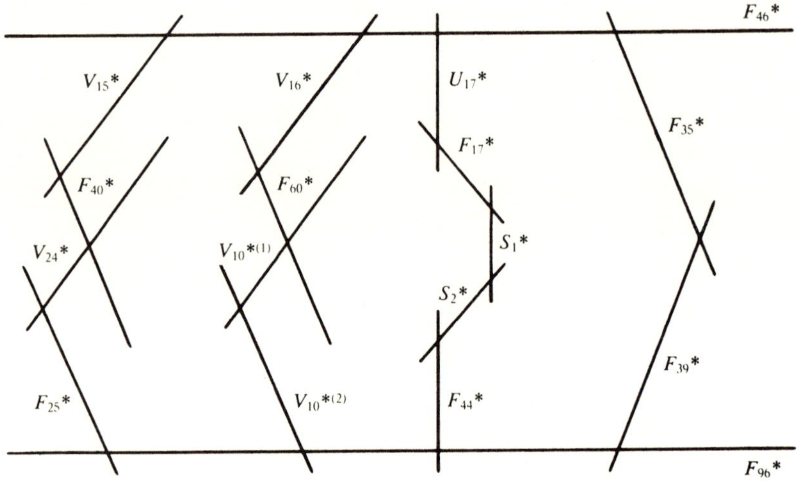

The quadratic forms which establish this diagram are the following

form	*discriminant*	
[46, 71, 40]	$-3 \cdot 773$	represents 15 primitively
[40, 41, 25]	$-3 \cdot 773$	represents 24 primitively
[25, 18, 96]	$-12 \cdot 773$	
[46, 25, 16]	$-3 \cdot 773$	represents 60 primitively
[60, 9, 10]	$-3 \cdot 773$	
[10, 39, 96]	$-3 \cdot 773$	
[46, 6, 17]	$-4 \cdot 773$	
[44, 47, 96]	$-19 \cdot 773$	
[46, 16, 35]	$-8 \cdot 773$	
[35, 7, 39]	$-7 \cdot 773$	
[39, 17, 96]	$-19 \cdot 773$	

By the list of cusp resolutions, originally S_2^* has selfintersection number -4, but F_{11}^* and S_3^* were blown down when passing to $Y_T{}^\circ(773)$.

By the preceding calculations (where details have been omitted) we have proved in virtue of Theorem II.7 the following result.

THEOREM III.1. *The modular surface for the symmetric Hilbert modular group of the field* $Q(\sqrt{p})$ (p *prime* $\equiv 1 \bmod 4$) *has geometric genus* 3 *if and only if* $p = 313, 653, 677, 773$. *The non-singular model* $Y_T{}^\circ(p)$ *for each of these surfaces is minimal. The canonical map of* $Y_T{}^\circ(p)$ *to the projective plane is well-defined and of degree* 2. *The surfaces are Moishezon surfaces. They are minimal desingularisations of double branched covers of the projective plane along curves of degree* 8.

Remarks:

1. The preceding theorem implies that (for the four primes considered) the surface $Y_T{}^\circ(p)$ admits an involution which cannot be induced from an automorphism of $H \times H$; probably this involution is the only nontrivial automorphism of $Y_T{}^\circ(p)$. It has many interesting properties, for example for $p = 773$ it maps the curve $F_{17}{}^*$ to the curve $S_2{}^*$ of the cusp resolution. A similar phenomenon is known in the theory of modular curves: The curve $\overline{H/\Gamma_0(37)}$ has genus 2 and hence is hyperelliptic, and the hyperelliptic involution carries the cusps to "finite" points (see, for example, [10]).

2. Let p be a prime $\equiv 1 \bmod 4$, and $\mathcal{O}$ the ring of integers in $Q(\sqrt{p})$. The cusp forms f for $SL_2(\mathcal{O})$ of weight 2 satisfying $f(z_1, z_2) = -f(z_2, z_1)$ are in bijective correspondence with the holomorphic sections of the canonical line bundle of $Y_T{}^\circ(p)$. Such a cusp form has a zero-divisor (f) on $X(p) = H^2/SL_2(\mathcal{O})$ which contains the curves F_1 and F_p, because these curves are pointwise fixed under the involution of $X(p)$ induced by $(z_1, z_2) \mapsto (z_2, z_1)$. The divisor (f) also contains all curves which were blown down when passing from $Y(p)$ to $Y^\circ(p)$ and from $Y^\circ(p)/T$ to $Y_T{}^\circ(p)$. For the four primes 313, 653, 677, 773 we have described a specific positive canonical divisor on $Y_T{}^\circ(p)$ which determines a cusp form f of weight 2 up to a non-zero factor. This cusp form f is "modular" in the sense that (f) consists of modular curves F_N only. Taking into account the blow-downs and the branching locus F_p of the covering $Y^\circ(p) \to Y^\circ(p)/T$ we have proved the existence of certain "modular" cusp forms f of weight 2 where (f) is the sum of the curves F_N which are listed in the following table

p	Curves F_N contained in (f). List of the numbers N
313	1, 2, 3, 4; 6, 8, 9, 11; 16, 22, 25, 27, 35; 313
653	1, 4; 6, 7, 9, 10, 11, 15, 16, 24, 36; 13, 25, 28, 54, 66, 216; 653
677	1, 4; 6, 9, 10, 14, 15, 16, 24, 36; 22, 25, 34, 56, 60, 96; 677
773	1, 4; 6, 9, 10, 11, 14, 15, 16, 24, 26, 36; 17, 25, 35, 39, 40, 44, 46, 60, 96; 773

Each curve occurs with multiplicity 1, except F_{22}, F_{34} for $p = 677$ which have multiplicity 2.

We have first listed the curves blown down in $Y(p)$, then those blown down in $Y^{\circ}(p)/T$, then the curves occurring in the positive canonical divisor of $Y_T{}^{\circ}(p)$ and finally F_p.

3. *Problems*:

a) Write down polynomials for the branch curves of degree eight belonging to our four primes.
b) Write down the Fourier series of the "modular" cusp forms (see the preceding table).

UNIVERSITY OF BONN
UNIVERSITY OF LEIDEN

REFERENCES

[1] E. Bombieri, "Canonical models of surfaces of general type," *Publ. Math. I.H.E.S.*, **42** (1973), pp. 171-219.

[2] G. van der Geer and A. Van de Ven, "On the minimality of certain Hilbert modular surfaces." In *Complex Analysis and Algebraic Geometry.* Iwanami Shoten and Cambridge University Press 1977; pp. 137-150.

[3] F. Hirzebruch, "Modulflächen und Modulkurven zur symmetrischen Hilbertschen Modulgruppe," *Ann. Scient. Ec. Norm. Sup.*, **11** (1978), pp. 101-166.

[4] F. Hirzebruch and D. Zagier, "Intersection numbers of curves on Hilbert modular surfaces and modular forms of Nebentypus," *Inventiones Math.*, **36** (1976), pp. 57-113.

[5] E. Horikawa, "Algebraic surfaces of general type with small $c_1{}^2$, I.," *Ann. Math.* **104** (1976), pp. 357-387.

[6] E. Horikawa, "Algebraic surfaces of general type with small $c_1{}^2$, II.," *Inventiones Math.* **31** (1976), pp. 43-85.

[7] K. Kodaira, "Pluricanonical systems on algebraic surfaces of general type," *J. Math. Soc. Japan*, **20** (1968), pp. 170-192.

[8] Y. Miyaoka, "On the Chern numbers of surfaces of general type," *Inventiones Math.* **42** (1977), pp. 225-237.

[9] B. Moishezon, "Surfaces of fundamental type," Chap. VI in: *Algebraic surfaces by the members of the seminar of I. R. Šafarevič, Proceedings of the Steklov Institute* **75** (1965); AMS Providence, Rhode Island, 1967.

[10] A. P. Ogg, "Über die Automorphismengruppe von $X_0(N)$," *Math. Ann.* **228** (1977), pp. 279-292.

[11] A. Van de Ven, "Some recent results on surfaces of general type," Séminaire Bourbaki 1976/77, no 500; *Lecture Notes in Mathematics* **677**, Springer-Verlag 1978, pp. 155-165.

ON THE RING OF MODULAR FORMS OF DEGREE TWO OVER Z

By JUN-ICHI IGUSA*

Introduction. We shall first explain some general problems: let $\mathfrak{S}_n$ denote the Siegel upper-half space of degree n and τ a point of $\mathfrak{S}_n$; then $\Gamma_n = Sp_{2n}(\mathbf{Z})$ acts discontinuously on $\mathfrak{S}_n$; we shall denote by $A(\Gamma_n)$ the corresponding ring of modular forms. We recall that $A(\Gamma_n)$ is a graded normal integral ring which is finitely generated over $\mathbf{C}$ and that $\mathrm{Proj}\ (A(\Gamma_n))_{\mathbf{C}}$ is the union of $\Gamma_r \backslash \mathfrak{S}_r$ for $0 \leq r \leq n$; we refer to [2] for the first systematic exposition of this theory.

We shall replace $\mathbf{C}$ by $\mathbf{Z}$ as follows: if f is an element of $A(\Gamma_n)$, then $f(\tau)$ can be expanded into a Fourier series of the following form:

$$f(\tau) = \sum_{s \geq 0} a_f(s)\mathbf{e}(\mathrm{tr}\ (s\tau)) = \sum_{s_i \geq 0} (\sum a_f(s) \cdot \prod_{i<j} q_{ij}{}^{2s_{ij}}) \prod_{i=1}^{n} q_i{}^{s_i},$$

in which $q_i = \mathbf{e}(\tau_i) = \exp\ (2\pi(-1)^{1/2}\tau_i)$, $q_{ij} = \mathbf{e}(\tau_{ij})$ and s runs over the set of half-integral positive (semi-definite) matrices of degree n. We have denoted by s_i resp. s_{ij} for $i \neq j$ the (i, i)-th resp. (i, j)-th entries of s. For any subring R of $\mathbf{C}$ we shall denote by $A_R(\Gamma_n)_k$ the R-module consisting of those f in $A(\Gamma_n)_k$ such that $a_f(s)$ is in R for every s and by $A_R(\Gamma_n)$ the sum of $A_R(\Gamma_n)_k$ for $k = 0, 1, 2, \ldots$ taken in $A(\Gamma_n)$; then $A_R(\Gamma_n)$ forms a graded integral ring over R. The problems (and the paper) concern about the graded ring $A_{\mathbf{Z}}(\Gamma_n)$ over $\mathbf{Z}$.

In the special case where $n = 1$ if we denote by ϕ_k the Eisenstein series of weight k normalized as $\phi_k(\tau) = 1 + \cdots$ and if we put $\delta = 2^{-6}3^{-3}(\phi_4{}^3 - \phi_6{}^2)$, then

$$A_{\mathbf{Z}}(\Gamma_1) = \mathbf{Z}[\phi_4, \phi_6, \delta];$$

this is a classical theorem. In the general case there exists a finite

*This work was partially supported by the National Science Foundation.

number of homogeneous elements $f_1, f_2, \ldots, f_N$ of $A_{\mathbf{Z}}(\Gamma_n)$ such that

$$A(\Gamma_n) = \mathbf{C}[f_1, f_2, \ldots, f_N];$$

cf. Eichler [3], Baily [1]. Also it is easy to see that $A_{\mathbf{Z}}(\Gamma_n)$ is normal and $A_R(\Gamma_n) = R \otimes_{\mathbf{Z}} A_{\mathbf{Z}}(\Gamma_n)$ for every subring R of $\mathbf{C}$. Moreover if we denote by $A_{\mathbf{F}_p}(\Gamma_n)_k$ the "coefficientwise reduction mod p" of $A_{\mathbf{Z}}(\Gamma_n)_k$, then $A_{\mathbf{Z}}(\Gamma_n)_k$ is a free $\mathbf{Z}$-module of rank $N_k = \dim (A(\Gamma_n)_k)$ and $A_{\mathbf{F}_p}(\Gamma_n)_k$ is an N_k-dimensional vector space over $\mathbf{F}_p$ for every p. In spite of all these the basic problem as to whether $A_{\mathbf{Z}}(\Gamma_n)$ is finitely generated over $\mathbf{Z}$ is an open question for $n \geq 2$. It is very likely that not only the answer is affirmative but also Proj $(A_{\mathbf{Z}}(\Gamma_n))$ is the union of the schemes of moduli over $\mathbf{Z}$ of r-dimensional principally polarized abelian varieties for $0 \leq r \leq n$. And we may not be too far from a solution of these problems thanks to some works by Mumford and others such as the theory of theta functions over a field of arbitrary characteristic; cf. [13], [14].

We shall explain the following related problem: let $A_{\mathbf{F}_p}(\Gamma_n)$ denote the subring of $\mathbf{F}_p[q_{ij}, q_{ij}^{-1}][[q_1, \ldots, q_n]]$ generated by $A_{\mathbf{F}_p}(\Gamma_n)_k$ for $k = 0, 1, 2, \ldots$; then we have the canonical homomorphism $\mathbf{F}_p \otimes_{\mathbf{Z}} A_{\mathbf{Z}}(\Gamma_n) \to A_{\mathbf{F}_p}(\Gamma_n)$. We can replace $A_{\mathbf{Z}}(\Gamma_n)$ by $A_R(\Gamma_n)$, where R is the ring of p-integral rational numbers, provided that we take the tensor product over R. Let ψ_k denote the normalized Eisenstein series of degree n and of weight $k > n + 1$; then at least if p is regular, the Fourier coefficients of ψ_{p-1} are all p-integral and the image of ψ_{p-1} in $A_{\mathbf{F}_p}(\Gamma_n)$ is 1; this follows from Siegel's results in [18]. Therefore if we denote by $\bar{\psi}_{p-1}$ the image of ψ_{p-1} in $\mathbf{F}_p \otimes_{\mathbf{Z}} A_{\mathbf{Z}}(\Gamma_n)$ and by $(\bar{\psi}_{p-1} - 1)$ the principal ideal generated by $\bar{\psi}_{p-1} - 1$, then we will get the following surjective homomorphism:

$$(\mathbf{F}_p \otimes_{\mathbf{Z}} A_{\mathbf{Z}}(\Gamma_n))/(\bar{\psi}_{p-1} - 1) \to A_{\mathbf{F}_p}(\Gamma_n).$$

In the case where $n = 1$ Swinnerton-Dyer showed that this is an isomorphism for every $p > 3$; cf. [19]; see also Serre [16]. In view of Maass [10] and our structure theorem in [8] this remains valid for $n = 2$ if p does not divide the $(p - 3)$rd Bernoullian number.[2] It is very likely that the above isomorphism theorem also holds for $n \geq 3$.

[2] After this work was completed, we have been informed from S. Nagaoka that some such irregular primes can really become exceptional; the smallest exceptional p is 16843.

In this paper we shall restrict ourselves to the case where $n = 2$ and settle the first-formulated problems by completely determining the structure of $A_{\mathbf{Z}}(\Gamma_2)$; we shall show, among others, that $A_{\mathbf{Z}}(\Gamma_2)$ has a minimal set of generators over $\mathbf{Z}$ consisting of fifteen homogeneous elements of weights

$$4, 6, 10, 12, 12, 16, 18, 24, 28, 30, 35, 36, 40, 42, 48.$$

It appears that such an explicit result is impossible to obtain in the general case.

We shall say a few words about our proof. In general for any graded ring A we shall denote by $P(A)$ the following graded ring:

$$P(A) = \sum_{k \geq 0} P^2(A_k);$$

in which $P^2(A_k)$ is the kernel of $A_k \otimes A_k \to A_k \wedge A_k$ for every $k \geq 0$. Let W denote the homomorphism $A(\Gamma_2) \to P(A(\Gamma_1))$ defined after Witt [21] as

$$(Wf)(\tau_1, \tau_2) = f\begin{bmatrix} \tau_1 & 0 \\ 0 & \tau_2 \end{bmatrix},$$

then W is surjective but its restriction $A_{\mathbf{Z}}(\Gamma_2) \to P(A_{\mathbf{Z}}(\Gamma_1))$ is not. In fact $A_{\mathbf{Z}}(\Gamma_2)_k \to P^2(A_{\mathbf{Z}}(\Gamma_1)_k)$ for $k \geq 1$ is surjective if and only if $k \not\equiv 0 \bmod 12$; moreover if d_n denotes the smallest positive integer such that $d_n(\delta^n \otimes \delta^n)$ is contained in $W(A_{\mathbf{Z}}(\Gamma_2)_{12n})$, then

$$d_n = 12/\text{G.C.D.}(12, n) = 12, 6, 4, 3, \ldots$$

for $n = 1, 2, 3, 4, \ldots$. The crucial point of our proof is to establish this fact in a more general form; this will be carried out in Sections 1-2. In Section 3 we shall show that Proj $(A_{\mathbf{Z}}(\Gamma_2))$ has the expected structure; its main part is the proof of the following fact: let $\mathcal{R}$ denote the

In this connection we might mention that Deligne gave an interpretation of $\bar{\psi}_{p-1}$ as a normalized Hasse invariant for $n = 1$; cf. [16]. We have learned from him that in the general case if we replace $\bar{\psi}_{p-1}$ by the determinant of a normalized Hasse-Witt matrix, then its image in $A_{\mathbf{F}_p}(\Gamma_n)$ is 1 for every $p > n + 2$.

homogeneous component of degree 0 of the ring of quotients of $A_{\mathbf{Z}}(\Gamma_2)$ by powers of the cusp form of weight 10; then the homomorphism ρ introduced in [9] gives an isomorphism of $\mathcal{R}$ to the (coordinate) ring of the scheme of moduli over $\mathbf{Z}$ of curves of genus 2 constructed in [7] by "arithmetic invariants."

Finally it is with pleasure that the author gratefully acknowledges the influence of a letter from Maass in the spring of 1977, which has later materialized as [12]; indeed without the stimulating correspondence with Maass the author would not have completed this work (which he had in mind for some time).

1. Construction of a set of generators.

1. We recall that $A(\Gamma_2)$ is generated over $\mathbf{C}$ by homogeneous elements $\psi_4, \psi_6, \chi_{10}, \chi_{12}, \chi_{35}$ each with the subscript as its weight. Subject to the additional condition that χ_{10}, χ_{12} are cusp forms, the above five elements are unique up to constant factors; and the normalization has been made as follows:

$$\psi_4(\tau) = 1 + \cdots,$$

$$\psi_6(\tau) = 1 + \cdots,$$

$$\chi_{10}(\tau) = (q_1 q_2 + \cdots)(\pi\tau_{12})^2 + \cdots,$$

$$\chi_{12}(\tau) = (q_1 q_2 + \cdots) + \cdots,$$

$$\chi_{35}(\tau) = (q_1{}^2 q_2{}^2(q_1 - q_2) + \cdots)(\pi\tau_{12}) + \cdots;$$

these are their Fourier expansions rewritten as power series in $\pi\tau_{12}$; cf. [8] for the details.

We shall try to construct a set of generators of $A_{\mathbf{Z}}(\Gamma_2)$ over $\mathbf{Z}$ out of them: in view of some results in Maass [10], [12] and of the table of Fourier coefficients in Resnikoff and Saldaña [15] we put

$$X_4 = \psi_4, \qquad X_6 = \psi_6, \qquad X_{10} = -2^2\chi_{10},$$
$$X_{12} = 2^2 3\chi_{12}, \qquad X_{35} = 2^2 i\chi_{35};$$

anticipating our later results, especially Lemma 14, we also put

$$Y_{12} = 2^{-6}3^{-3}(X_4{}^3 - X_6{}^2) + 2^4 3^2 X_{12},$$

$$X_{16} = 2^{-2}3^{-1}(X_4X_{12} - X_6X_{10}),$$

$$X_{18} = 2^{-2}3^{-1}(X_6X_{12} - X_4{}^2X_{10}),$$

$$X_{24} = 2^{-3}3^{-1}(X_{12}{}^2 - X_4X_{10}{}^2),$$

$$X_{28} = 2^{-1}3^{-1}(X_4X_{24} - X_{10}X_{18}),$$

$$X_{30} = 2^{-1}3^{-1}(X_6X_{24} - X_4X_{10}X_{16}),$$

$$X_{36} = 2^{-1}3^{-2}(X_{12}X_{24} - X_{10}{}^2X_{16}),$$

$$X_{40} = 2^{-2}(X_4X_{36} - X_{10}X_{30}),$$

$$X_{42} = 2^{-2}3^{-1}(X_{12}X_{30} - X_4X_{10}X_{28}),$$

$$X_{48} = 2^{-2}(X_{12}X_{36} - X_{24}{}^2).$$

Our first objective is to prove the following theorem:

THEOREM 1. *The above-defined fifteen homogeneous elements* $X_4, X_6, \ldots, X_{48}$ *of* $A(\Gamma_2)$ *of respective weights* 4, 6, ..., 48 *are all in* $A_Z(\Gamma_2)$, i.e., *they all have integral Fourier coefficients.*

For the time being we shall exclude X_{35} and examine the remaining fourteen elements of even weights; concerning the first two elements we have the following information:

LEMMA 1. *If we put*

$$X_4 = 1 + 2^4 3T, \qquad X_6 = 1 + 2^3 3^2 U,$$

then both T and U have integral Fourier coefficients.

Actually we have $T \equiv 0 \bmod 5$ in the sense that all Fourier coefficients of T are multiples of 5; we similarly have $U \equiv 0 \bmod 7$; cf. Maass [10].

LEMMA 2. *If* $X_{10}, X_{12}, Y_{12}, X_{16}$ *are in* $A_Z(\Gamma_2)$, *then* $X_{18}, X_{24}, X_{28}, X_{30}, X_{36}, X_{40}, X_{42}, X_{48}$ *are also in* $A_Z(\Gamma_2)$.

Proof. Since X_{12}, Y_{12} are in $A_Z(\Gamma_2)$ by assumption, we have

$$X_4{}^3 - X_6{}^2 \equiv 2^4 3^2(T - U) \equiv 0 \bmod 2^6 3^3;$$

hence $T \equiv U \bmod 2^2 3$. Since X_{10}, X_{12}, X_{16} are in $A_Z(\Gamma_2)$ by assumption, we have

$$X_4X_{12} - X_6X_{10} \equiv X_{12} - X_{10} \equiv 0 \bmod 2^2 3;$$

hence we can write $X_{12} = X_{10} + 2^2 3V$ with $V \equiv 0 \bmod 1$. Consequently we have

$$X_{18} = 2^{-2}3^{-1}(X_6X_{12} - X_4{}^2X_{10}) \equiv V \bmod 2,$$

$$X_{24} = 2^{-3}3^{-1}(X_{12}{}^2 - X_4X_{10}{}^2) \equiv X_{10}V \bmod 2.$$

For X_{28}, X_{30}, X_{36} we shall use the following expressions:

$$X_{28} = 2^{-4}3^{-2}(X_4{}^2X_{10}{}^2 + X_4X_{12}{}^2 - 2X_6X_{10}X_{12}),$$

$$X_{30} = 2^{-4}3^{-2}(X_6X_{12}{}^2 + X_4X_6X_{10}{}^2 - 2X_4{}^2X_{10}X_{12}),$$

$$X_{36} = 2^{-4}3^{-3}(X_{12}{}^3 + 2X_6X_{10}{}^3 - 3X_4X_{10}{}^2X_{12});$$

then we have

$$X_{28} \equiv X_{30} \equiv V^2 - 4X_{10}TV + 4X_{10}{}^2T^2 \bmod 2^2 3,$$

$$X_{36} \equiv X_{10}V^2 \bmod 2^2.$$

In particular X_{18}, X_{24}, X_{28}, X_{30}, X_{36} are in $A_{\mathbf{Z}}(\Gamma_2)$. We also have

$$X_4X_{36} - X_{10}X_{30} \equiv X_{10}V^2 - X_{10}V^2 = 0 \bmod 2^2,$$

$$X_{12}X_{30} - X_4X_{10}X_{28} \equiv X_{10}(X_{30} - X_{28}) \equiv 0 \bmod 2^2 3,$$

$$X_{12}X_{36} - X_{24}{}^2 \equiv X_{10}{}^2V^2 - (X_{10}V)^2 = 0 \bmod 2^2;$$

hence X_{40}, X_{42}, X_{48} are in $A_{\mathbf{Z}}(\Gamma_2)$. q.e.d.

2. We have reduced the proof of Theorem 1 to the proof of the fact that X_{10}, X_{12}, Y_{12}, X_{16}, and X_{35} are in $A_{\mathbf{Z}}(\Gamma_2)$. In order to carry out this part of the proof we shall use "theta constants"; in the following we shall briefly review some of their properties: let m', m'' denote elements of $\mathbf{Z}^n$ and put $\mathfrak{m} = (m' \; m'')$; then the theta constant $\theta_{\mathfrak{m}}$ with "characteristic" $\mathfrak{m}$ is a holomorphic function on $\mathfrak{S}_n$ defined as

$$\theta_{\mathfrak{m}}(\tau) = \sum_{p \in \mathbf{Z}^n} \mathbf{e}(\tfrac{1}{2}(p + \tfrac{1}{2} \cdot m')\tau^t(p + \tfrac{1}{2} \cdot m') + \tfrac{1}{2}(p + \tfrac{1}{2} \cdot m')^t m'').$$

The function $\theta_{\mathfrak{m}}$ is different from the constant 0 if and only if $\mathfrak{m}$ is even in the sense that the integer $m'{}^t m''$ is even; if $\mathfrak{n} = (n' \ n'')$ is another element of $\mathbf{Z}^{2n}$, then we have

$$\theta_{\mathfrak{m}+2\mathfrak{n}} = (-1)^{m'{}^t n''} \theta_{\mathfrak{m}}.$$

Therefore we have only to consider theta constants with even characteristic in which entries are 0, 1. There are $2^{n-1}(2^n + 1)$ such characteristics; they are (00), (01), (10) for $n = 1$ and (0000), (0001), (0010), (0011), (0100), (0110), (1000), (1001), (1100), (1111) for $n = 2$.

In the case where $n = 1$ we shall use ω instead of τ and write $q = \mathbf{e}(\omega)$; then by using the notation in the Introduction we have

$$\theta_{00}{}^4 = \theta_{01}{}^4 + \theta_{10}{}^4, \qquad \theta_{00}{}^8 + \theta_{01}{}^8 + \theta_{10}{}^8 = 2\phi_4,$$

$$(\theta_{00}\theta_{01}\theta_{10})^8 = 2^8\delta, \qquad \delta(\omega) = q \cdot \prod_{n=1}^{\infty} (1 - q^n)^{24}.$$

Furthermore if we put

$$F_0(r) = \sum_{p=1}^{\infty} r^{p^2}, \qquad F_1(r) = \sum_{p=1}^{\infty} r^{(p-1/2)^2},$$

in which $r = \mathbf{e}(\tfrac{1}{2} \cdot \omega)$, then we will have

$$\theta_{00}(\omega) = 1 + 2F_0(r), \qquad \theta_{01}(\omega) = 1 + 2F_0(-r), \qquad \theta_{10}(\omega) = 2F_1(r).$$

We observe that $F_0(-r)$ is well defined because $F_0(r)$ is a power series in r.

Similarly in the case where $n = 2$ if we put

$$F_0(r_1, r_2) = F_0(r_1) + F_0(r_2) + \sum_{p_1, p_2 = 1}^{\infty} A_{p_1, p_2} r_1{}^{p_1{}^2} r_2{}^{p_2{}^2},$$

$$F_1(r_1, r_2) = F_1(r_2) + \sum_{p_1,p_2=1}^{\infty} B_{p_1,p_2} r_1^{p_1^2} r_2^{(p_2-1/2)^2},$$

$$F_2(r_1, r_2) = F_1(r_2, r_1),$$

$$F_3(r_1, r_2) = \sum_{p_1,p_2=1}^{\infty} C_{p_1,p_2} r_1^{(p_1-1/2)^2} r_2^{(p_2-1/2)^2},$$

$$F_4(r_1, r_2) = \sum_{p_1,p_2=1}^{\infty} (-1)^{p_1+p_2-1} C_{p_1,p_2} r_1^{(p_1-1/2)^2} r_2^{(p_2-1/2)^2},$$

in which $r_1 = \mathbf{e}(1/2 \cdot \tau_1)$, $r_2 = \mathbf{e}(1/2 \cdot \tau_2)$, $q_{12} = \mathbf{e}(\tau_{12})$, and

$$A_{p_1,p_2} = q_{12}^{\,p_1p_2} + q_{12}^{\,-p_1p_2},$$

$$B_{p_1,p_2} = q_{12}^{\,p_1(p_2-1/2)} + q_{12}^{\,-p_1(p_2-1/2)},$$

$$C_{p_1,p_2} = q_{12}^{\,(p_1-1/2)(p_2-1/2)} + q_{12}^{\,-(p_1-1/2)(p_2-1/2)},$$

then we will have

$$\theta_{0000}(\tau) = 1 + 2F_0(r_1, r_2), \qquad \theta_{0001}(\tau) = 1 + 2F_0(r_1, -r_2),$$
$$\theta_{0010}(\tau) = 1 + 2F_0(-r_1, r_2), \qquad \theta_{0011}(\tau) = 1 + 2F_0(-r_1, -r_2),$$
$$\theta_{0100}(\tau) = 2F_1(r_1, r_2), \qquad \theta_{0110}(\tau) = 2F_1(-r_1, r_2),$$
$$\theta_{1000}(\tau) = 2F_2(r_1, r_2), \qquad \theta_{1001}(\tau) = 2F_2(r_1, -r_2),$$
$$\theta_{1100}(\tau) = 2F_3(r_1, r_2), \qquad \theta_{1111}(\tau) = 2F_4(r_1, r_2).$$

We observe that $F_0(\pm r_1, \pm r_2)$, $F_1(-r_1, r_2)$, and $F_2(r_1, -r_2)$ are well defined because $F_0(r_1, r_2)$ is a power series in r_1, r_2 and, e.g., $F_1(r_1, r_2)$ is a power series in r_1.

We have explained the homomorphism $W: A(\Gamma_2) \to P(A(\Gamma_1))$ in the Introduction. We shall denote by S the homomorphism $A(\Gamma_2) \to A(\Gamma_1)$ defined as

$$(Sf)(\omega) = \lim_{y\to\infty} f\begin{bmatrix} \omega & 0 \\ 0 & yi \end{bmatrix};$$

this is the Φ-operator introduced by Siegel in [17]. We observe that the same prescription can be applied to theta constants; and for every even characteristic $\mathfrak{m} = (m_1 m_2 m_3 m_4)$ we will have

$$S(\theta_{\mathfrak{m}}) = \begin{cases} \theta_{m_1 m_3} & m_2 = 0 \\ 0 & m_2 = 1, \end{cases}$$

$$W(\theta_{\mathfrak{m}}) = \begin{cases} \theta_{m_1 m_3} \otimes \theta_{m_2 m_4} & \mathfrak{m} \neq (1111) \\ 0 & \mathfrak{m} = (1111). \end{cases}$$

More precisely we have

$$\begin{aligned} \theta_{\mathfrak{m}}(\tau) &= \sum_{n=0}^{\infty} (2^{2n}/(2n)!)\theta_{m_1 m_3}{}^{(n)}(\tau_1)\theta_{m_2 m_4}{}^{(n)}(\tau_2)\tau_{12}{}^{2n} \\ &= \theta_{m_1 m_3}(\tau_1)\theta_{m_2 m_4}(\tau_2) + \cdots \end{aligned}$$

for $\mathfrak{m} \neq (1111)$ and

$$\begin{aligned} \theta_{1111}(\tau) &= (\pi/2i) \sum_{n=0}^{\infty} (2^{2n}/(2n+1)!)(\theta_{00}\theta_{01}\theta_{10})^{(n)}(\tau_1)(\theta_{00}\theta_{01}\theta_{10})^{(n)}(\tau_2)\tau_{12}{}^{2n+1} \\ &= (\pi/2i)(\theta_{00}\theta_{01}\theta_{10})(\tau_1)(\theta_{00}\theta_{01}\theta_{10})(\tau_2)\tau_{12} + \cdots. \end{aligned}$$

The point is that every f in $A(\Gamma_2)$ can be written as a polynomial in the theta constants and that $W(f)$, $S(f)$, and even the expansion of $f(\tau)$ into a power series in τ_{12} can easily be determined once the exact form of the polynomial is known; cf. [8].

Now in order to conveniently describe such a polynomial we use the theory of characteristics; we shall briefly review its basic terminology: a triple of even characteristics is called "syzygous" if their sum is even; otherwise it is called "azygous." A set of even characteristics is called syzygous resp. azygous if all triples in the set are syzygous resp. azygous. We shall consider sets consisting of distinct even characteristics; then in the case where $n = 2$ there are sixty syzygous triples, sixty azygous triples, fifteen syzygous quadruples, and fifteen azygous quadruples.

If we denote the ten characteristics in the order we have written earlier by 1, 2, ..., 9, 0, then the sixty azygous triples are as follows:

(1, 2, 5), (1, 2, 6), (1, 2, 9), (1, 2, 0), (1, 3, 7), (1, 3, 8),

(1, 3, 9), (1, 3, 0), (1, 4, 5), (1, 4, 6), (1, 4, 7), (1, 4, 8),

(1, 5, 8), (1, 5, 0), (1, 6, 7), (1, 6, 9), (1, 7, 0), (1, 8, 9),

(2, 3, 5), (2, 3, 6), (2, 3, 7), (2, 3, 8), (2, 4, 7), (2, 4, 8),

(2, 4, 9), (2, 4, 0), (2, 5, 7), (2, 5, 0), (2, 6, 8), (2, 6, 9),

(2, 7, 9), (2, 8, 0), (3, 4, 5), (3, 4, 6), (3, 4, 9), (3, 4, 0),

(3, 5, 7), (3, 5, 9), (3, 6, 8), (3, 6, 0), (3, 7, 0), (3, 8, 9),

(4, 5, 8), (4, 5, 9), (4, 6, 7), (4, 6, 0), (4, 7, 9), (4, 8, 0),

(5, 6, 7), (5, 6, 8), (5, 6, 9), (5, 6, 0), (5, 7, 8), (5, 9, 0),

(6, 7, 8), (6, 9, 0), (7, 8, 9), (7, 8, 0), (7, 9, 0), (8, 9, 0).

The sixty syzygous triples are the remaining triples and the fifteen syzygous quadruples are as follows:

(1, 2, 3, 4), (1, 2, 7, 8), (1, 3, 5, 6), (1, 4, 9, 0), (1, 5, 7, 9),

(1, 6, 8, 0), (2, 3, 9, 0), (2, 4, 5, 6), (2, 5, 8, 9), (2, 6, 7, 0),

(3, 4, 7, 8), (3, 5, 8, 0), (3, 6, 7, 9), (4, 5, 7, 0), (4, 6, 8, 9).

Finally the fifteen azygous quadruples are as follows:

(1, 2, 5, 0), (1, 2, 6, 9), (1, 3, 7, 0), (1, 3, 8, 9), (1, 4, 5, 8),

(1, 4, 6, 7), (2, 3, 5, 7), (2, 3, 6, 8), (2, 4, 7, 9), (2, 4, 8, 0),

(3, 4, 5, 9), (3, 4, 6, 0), (5, 6, 7, 8), (5, 6, 9, 0), (7, 8, 9, 0).

3. After the above digression we go back to the proof of Theorem 1; we shall complete the proof in four more lemmas:

LEMMA 3. *We can write*

$$X_{10} = 2^{-12} \cdot \prod_{\text{ten}} \theta_m{}^2, \qquad X_{12} = 2^{-15} \cdot \sum_{\text{fifteen}} (\theta_{m_1}\theta_{m_2} \cdots \theta_{m_6})^4,$$

in which the summation is extended over the set of fifteen complements of syzygous quadruples; and we have

$$X_{10} \equiv (F_1F_2F_3)^4 \bmod 2^2, \qquad X_{12} \equiv 0 \bmod 1.$$

Proof. The first part is not new; cf. [8]-II, [9]. We shall prove the second part: we have

$$X_{10}(\tau) = \prod_{\text{four}} (1 + 2F_0(\pm r_1, \pm r_2))^2 \cdot F_1(r_1, r_2)^2 F_1(-r_1, r_2)^2 \cdot F_2(r_1, r_2)^2 F_2(r_1, -r_2)^2 \cdot F_3(r_1, r_2)^2 F_4(r_1, r_2)^2.$$

We recall that $F_0(\pm r_1, \pm r_2)$, $F_1(\pm r_1, r_2)$, $F_2(r_1, \pm r_2)$, $F_3(r_1, r_2)$, $F_4(r_1, r_2)$ all have integral Fourier coefficients (as Fourier series in ${r_1}^{1/4}$, ${r_2}^{1/4}$, ${q_{12}}^{1/4}$). Since in any commutative ring $a \equiv b \bmod 2$ implies $a^2 \equiv b^2 \bmod 2^2$ and since $F_1(r_1, r_2) \equiv F_1(-r_1, r_2)$, $F_2(r_1, r_2) \equiv F_2(r_1, -r_2)$, $F_3(r_1, r_2) \equiv F_4(r_1, r_2) \bmod 2$, we get

$$X_{10}(\tau) \equiv F_1(r_1, r_2)^4 F_2(r_1, r_2)^4 F_3(r_1, r_2)^4 \bmod 2^2;$$

in the lemma we have dropped τ and r_1, r_2.

We observe that only the complements of (1, 5, 7, 9), (1, 6, 8, 0), (2, 5, 8, 9), (2, 6, 7, 0), (3, 5, 8, 0), (3, 6, 7, 9), (4, 5, 7, 0), (4, 6, 8, 9) have non-zero contributions to X_{12} mod 2. Since $a \equiv b \bmod 2$ implies $a^4 \equiv b^4 \bmod 2^3$, the contribution to X_{12} mod 1 coming from any one of the above eight complements is

$$2^{-3}(F_1(r_1, r_2)F_2(r_1, r_2)F_3(r_1, r_2))^4 \bmod 1;$$

this implies $X_{12} \equiv 0 \bmod 1$. q.e.d.

We can actually show that $X_{10} \equiv X_{12} \bmod 2^2$; at any rate Lemma 3 implies that X_{10}, X_{12} are in $A_{\mathbf{Z}}(\Gamma_2)$. We shall separate a similar lemma for Y_{12}, X_{16} into two lemmas:

LEMMA 4. *Put*

$$Z_{12} = 2^{-11} \cdot \sum_{\text{sixty}} (\theta_{m_1}\theta_{m_2}\theta_{m_3})^8,$$

in which the summation is extended over the set of sixty azygous triples;

then we have

$$Z_{12} = Y_{12} - 2^5 X_{12} \equiv F_1{}^8 + F_2{}^8 + F_3{}^8 \bmod 2.$$

Proof. We first observe that Z_{12} is an element of $A(\Gamma_2)_{12}$; in fact Z_{12} is $2^{-8}(azy)_2$ in the notation of [8]-II, p. 403. Since the homomorphism S maps Eisenstein series to Eisenstein series, we have

$$S(Y_{12}) = 2^{-6}3^{-3}(\phi_4{}^3 - \phi_6{}^2) = \delta;$$

we also have

$$S(Z_{12}) = 2^{-11} \cdot 2^3(\theta_{00}\theta_{01}\theta_{10})^8 = \delta.$$

Therefore $Z_{12} - Y_{12}$ is a cusp form of weight 12, hence $Z_{12} - Y_{12} = c \cdot X_{12}$ for some constant c. We compute the coefficients of $q_1 q_2 = r_1{}^2 r_2{}^2$ in the power-series expansions of $W(X_{12})(\tau_1, \tau_2)$, $W(Y_{12})(\tau_1, \tau_2)$, $W(Z_{12})(\tau_1, \tau_2)$ and we respectively get $2^2 3$, $2^5 3^2 5$, $2^5 3 \cdot 11$; hence $c = -2^5$.

We might mention our method of computation: since $X_{12} = 2^2 3 \chi_{12}$, the coefficient for X_{12} is $2^2 3$. We have

$$W(Y_{12}) = 2^{-6}3^{-3}(\phi_4{}^3 \otimes \phi_4{}^3 - \phi_6{}^2 \otimes \phi_6{}^2) + 2^4 3^2 W(X_{12}),$$

$$\phi_4(\omega) = 1 + 2^4 3 \cdot 5 \cdot q + \cdots, \qquad \phi_6(\omega) = 1 - 2^3 3^2 7 \cdot q + \cdots;$$

hence the coefficient for Y_{12} is $2^5 3^2 5$. Finally we have

$$W(Z_{12}) = 2^{-11}(\Theta \otimes (\theta_{00}\theta_{01}\theta_{10})^8 + (\theta_{00}\theta_{01}\theta_{10})^8 \otimes \Theta),$$

in which

$$\begin{aligned}\Theta &= \theta_{00}{}^{24} + \theta_{01}{}^{24} + \theta_{10}{}^{24} + 3\theta_{00}{}^{16}(\theta_{01}{}^8 + \theta_{10}{}^8) \\ &\quad + 3\theta_{00}{}^8(\theta_{01}{}^{16} + \theta_{10}{}^{16}) + 3\theta_{01}{}^8\theta_{10}{}^8(\theta_{01}{}^8 + \theta_{10}{}^8) \\ &= (\theta_{00}{}^8 + \theta_{01}{}^8 + \theta_{10}{}^8)^3 - 2 \cdot 3(\theta_{00}\theta_{01}\theta_{10})^8;\end{aligned}$$

and this is $2^3(\phi_4{}^3 - 2^6 3\delta)$. Therefore we get

$$W(Z_{12}) = (\phi_4{}^3 - 2^6 3\delta) \otimes \delta + \delta \otimes (\phi_4{}^3 - 2^6 3\delta);$$

hence the coefficient for Z_{12} is $2^5 3 \cdot 11$.

We shall prove the congruence for Z_{12}: an azygous triple contains either at most one or exactly two of 1, 2, 3, 4; if we denote by I resp. II the contributions to Z_{12} coming from the corresponding partial sums, then we obviously have $\mathrm{I} \equiv 0 \bmod 2^5$. As for II it consists of twenty-four triples: (1, 2, 5) to (1, 4, 8), (2, 3, 5) to (2, 4, 0), and (3, 4, 5) to (3, 4, 0). Since $a \equiv b \bmod 2$ implies $a^8 \equiv b^8 \bmod 2^4$, we get

$$\mathrm{II} \equiv F_1(r_1, r_2)^8 + F_2(r_1, r_2)^8 + F_3(r_1, r_2)^8 \bmod 2. \quad \text{q.e.d.}$$

In the above proof we have determined $W(Y_{12})$ and $W(Z_{12})$; those with $Z_{12} = Y_{12} - 2^5 X_{12}$ imply $W(X_{12}) = 2^2 3(\delta \otimes \delta)$; we shall later recover this fact in the proof of Lemma 12.

LEMMA 5. *We can write*

$$X_{16} = 2^{-18} \cdot \sum_{\text{fifteen}} (\theta_{\mathrm{m}_1}\theta_{\mathrm{m}_2}\theta_{\mathrm{m}_3}\theta_{\mathrm{m}_4})^8,$$

in which the summation is extended over the set of fifteen azygous quadruples; and we have

$$X_{16} \equiv (F_1F_2)^8 + (F_2F_3)^8 + (F_3F_1)^8 \bmod 2^2.$$

Proof. For a moment we shall denote by Y_{16} the expression involving theta constants; we observe that Y_{16} is an element of $A(\Gamma_2)_{16}$. In fact by the "transformation formula" of theta constants we can easily see that the fifteen terms are simply permuted under the action of Γ_2. Since $W(X_{12}) = 2^2 3(\delta \otimes \delta)$, we have

$$W(X_{16}) = 2^{-2}3^{-1}W(X_4)W(X_{12}) = \phi_4\delta \otimes \phi_4\delta;$$

we also have

$$W(Y_{16}) = 2^{-18}(\theta_{00}\theta_{01}\theta_{10})^8 \otimes (\theta_{00}\theta_{01}\theta_{10})^8 \cdot (\theta_{00}{}^8 + \theta_{01}{}^8 + \theta_{10}{}^8) \otimes (\theta_{00}{}^8 + \theta_{01}{}^8 + \theta_{10}{}^8),$$

and this is $\delta\phi_4 \otimes \delta\phi_4$. Therefore $W(X_{16} - Y_{16}) = 0$ and hence $X_{16} - Y_{16}$ is divisible by X_{10}; cf. [8], p. 199, where we used Ψ instead of W. We might mention that this divisibility property was later proved directly by Freitag [4] and Hammond [6]. At any rate we can write $X_{16} - Y_{16} = c \cdot X_6 X_{10}$ with some constant c. We compute the coefficients of $q_1 q_2$

in the Fourier expansions of X_6X_{10}, X_{16}, Y_{16} and we respectively get -2, 1, 1. In fact the table of Fourier coefficients in [8], pp. 190-191 will give

$$(X_6X_{10})(\tau) = (-2 + q_{12} + q_{12}^{-1})q_1q_2 + \cdots, \qquad X_{16}(\tau) = q_1q_2 + \cdots,$$

in which the unwritten parts consist of terms of degree at least 3 in q_1, q_2; we can also use the table in [15], p. 108. On the other hand, by using the Fourier expansions of theta constants we get $Y_{16}(\tau) = q_1q_2 + \cdots$; the contribution to the coefficient of q_1q_2 comes only from (1, 4, 5, 8), (1, 4, 6, 7), (2, 3, 5, 7), (2, 3, 6, 8). Therefore we get $-2c = 1 - 1 = 0$, hence $c = 0$.

We shall prove the congruence for $X_{16} = Y_{16}$: if we denote by I resp. II the contributions to X_{16} coming from the last three quadruples resp. the first twelve quadruples, then obviously I $\equiv 0 \bmod 2^{14}$. And by using "$a \equiv b \bmod 2$ implies $a^8 \equiv b^8 \bmod 2^4$" we get

$$\mathrm{II} \equiv \sum_{1 \leq i < j \leq 3} F_i(r_1, r_2)^8 F_j(r_1, r_2)^8 \bmod 2^2. \quad \text{q.e.d.}$$

4. We shall finally examine X_{35} to complete the proof of Theorem 1; the following lemma is more than we need:

LEMMA 6. *We have*

$$X_{35} = 2^{-37}5^{-3}\left(\prod_{\text{ten}} \theta_{\mathrm{m}}\right)\left(\sum_{\text{sixty}} \pm (\theta_{\mathrm{m}_1}\theta_{\mathrm{m}_2}\theta_{\mathrm{m}_3})^{20}\right),$$

in which the summation is extended over the set of sixty azygous triples, and

$$\begin{aligned} X_{35}^2 = {} & (-2^2X_4^2X_{16} + Y_{12}^2)X_{36} \cdot X_{10} + (-2^6X_4X_{12}^3 \\ & + 2^3Y_{12}X_{28})X_{10}^3 + (2Y_{12}X_{18} + 2^{10}X_{30})X_{10}^4 \\ & + 3 \cdot 61X_4^2X_{12} \cdot X_{10}^5 - 2 \cdot 73X_4X_6 \cdot X_{10}^6 + 2^{10}5^5X_{10}^7. \end{aligned}$$

In particular X_{35} is in $A_{\mathbf{Z}}(\Gamma_2)$.

Proof. The expression for $X_{35} = 2^2i\chi_{35}$ in terms of theta constants is really its definition; cf. [8]-II, pp. 403–404. And the formula for X_{35}^2 is a restatement of the formula in [9], pp. 849–850 expressing χ_{35}^2

as a polynomial in ψ_4, ψ_6, χ_{10}, χ_{12}; the verification is straightforward. The point is that in the expression of ${X_{35}}^2$ as a polynomial in X_4, X_6, ..., X_{36} all coefficients become integers; in particular ${X_{35}}^2$ has integral Fourier coefficients. We shall show that X_{35} itself has integral Fourier coefficients: let d denote the L.C.M. of the denominators of all Fourier coefficients of X_{35}, i.e., the smallest positive integer such that $d \cdot X_{35}$ is in $A_{\mathbf{Z}}(\Gamma_2)$; such a d exists (and it is a divisor of $2^{37}5^3$). We have only to show that $d = 1$. If $d \neq 1$, then we choose a prime factor p of d. Since ${X_{35}}^2$ is in $A_{\mathbf{Z}}(\Gamma_2)$, we get $(d \cdot X_{35})^2 = d^2{X_{35}}^2 \equiv 0 \bmod p$. Since $\mathbf{F}_p[q_{12}, {q_{12}}^{-1}][[q_1, q_2]]$ is an integral ring, we get $d \cdot X_{35} \equiv 0 \bmod p$; hence d can be replaced by $p^{-1}d$. This contradicts the minimality of d. q.e.d.

2. Proof of the completeness.

5. In Section 1 we have constructed fifteen homogeneous elements $X_4, X_6, \ldots, X_{48}$ of $A_{\mathbf{Z}}(\Gamma_2)$ each with the subscript as its weight; in this Section we shall prove the following theorem:

THEOREM 2. *The fifteen elements $X_4, X_6, \ldots, X_{48}$ form a minimal set of generators of $A_{\mathbf{Z}}(\Gamma_2)$ over $\mathbf{Z}$.*

We shall first show that they generate $A_{\mathbf{Z}}(\Gamma_2)$ over $\mathbf{Z}$, i.e., every homogeneous element f of $A_{\mathbf{Z}}(\Gamma_2)$ can be written as an isobaric polynomial in $X_4, X_6, \ldots, X_{48}$ with integral coefficients. We shall often use the following remark: if we put

$$t = q_{12} + {q_{12}}^{-1} - 2 = (2\pi i\tau_{12})^2(1 + \cdots),$$

then ${q_{12}}^n + {q_{12}}^{-n}$ can be written as a polynomial in t of degree n with integral coefficients for every $n \geq 0$.

LEMMA 7. *Let f denote a homogeneous element of $A_{\mathbf{Z}}(\Gamma_2)$ of an even weight k such that $W(f) = 0$; then f is divisible by X_{10} in $A_{\mathbf{Z}}(\Gamma_2)$,* i.e., *$f = X_{10}f'$ with f' in $A_{\mathbf{Z}}(\Gamma_2)$ of weight $k - 10$.*

Proof. We observe that $W(f) = 0$ implies that f is a cusp form; hence

$$f(\tau) = (q_1q_2) \sum_{s_1, s_2 \geq 1} \left(\sum_{s_{12}} a_f(s){q_{12}}^{2s_{12}}\right){q_1}^{s_1-1}{q_2}^{s_2-1}.$$

Since the sum of $a_f(s)q_{12}^{2s_{12}}$ is equal to

$$a_f\begin{pmatrix} s_1 & 0 \\ 0 & s_2 \end{pmatrix} + \sum_{e=2s_{12}\geq 1} a_f(s)(q_{12}^{e} + q_{12}^{-e}),$$

in which $e < 2(s_1s_2)^{1/2}$, it can be written as a polynomial in t with integral coefficients; then $W(f) = 0$ implies that this polynomial is divisible by t. Therefore we can write

$$f(\tau) = (q_1q_2t)\cdot P(q_1, q_2, t)$$

with $P(q_1, q_2, t)$ in $\mathbf{Z}[[q_1, q_2, t]]$, more precisely in $\mathbf{Z}[t][[q_1, q_2]]$; in particular we have $X_{10}(\tau) = (q_1q_2t)\cdot P_0(q_1, q_2, t)$ with $P_0(q_1, q_2, t)$ in $\mathbf{Z}[[q_1, q_2, t]]$. The point is that $P_0(0, 0, 0) = 1$ and hence $P_0(q_1, q_2, t)$ is a unit of $\mathbf{Z}[[q_1, q_2, t]]$. We know that f is divisible by X_{10} in $A(\Gamma_2)$, i.e., $f = X_{10}f'$ with f' in $A(\Gamma_2)$ necessarily of weight $k - 10$; cf. the proof of Lemma 5. By what we have shown above $f'(\tau)$ is an element of $\mathbf{Z}[[q_1, q_2, t]]$; hence f' is in $A_{\mathbf{Z}}(\Gamma_2)$. q.e.d.

LEMMA 8. *Let f denote a homogeneous element of $A_{\mathbf{Z}}(\Gamma_2)$ of an odd weight k; then f is divisible by X_{35} in $A_{\mathbf{Z}}(\Gamma_2)$,* i.e., $f = X_{35}f'$ *with f' in $A_{\mathbf{Z}}(\Gamma_2)$ of weight* $k - 35$.

Proof. Since f is necessarily a cusp form, we have

$$f(\tau) = \sum_{s>0} a_f(s)\mathbf{e}(\operatorname{tr}(s\tau)).$$

Moreover for every u in $GL_2(\mathbf{Z})$ if we put $s[u] = {}^tusu$, then we have

$$a_f(s[u]) = \det(u)^k a_f(s) = \det(u)a_f(s);$$

we also have

$$s\left[\begin{pmatrix} 0 & 1 \\ 1 & 0 \end{pmatrix}\right] = \begin{pmatrix} s_2 & s_{12} \\ s_{12} & s_1 \end{pmatrix}, \quad s\left[\begin{pmatrix} 1 & 0 \\ 0 & -1 \end{pmatrix}\right] = \begin{pmatrix} s_1 & -s_{12} \\ -s_{12} & s_2 \end{pmatrix},$$

$$\begin{pmatrix} s_1 & s_{12} \\ s_{12} & 1 \end{pmatrix}\left[\begin{pmatrix} 1 & 0 \\ -2s_{12} & 1 \end{pmatrix}\right] = \begin{pmatrix} s_1 & -s_{12} \\ -s_{12} & 1 \end{pmatrix}.$$

By putting these together we see that

$$a_f\begin{pmatrix} s_2 & s_{12} \\ s_{12} & s_1 \end{pmatrix} = a_f\begin{pmatrix} s_1 & -s_{12} \\ -s_{12} & s_2 \end{pmatrix} = -a_f(s)$$

and that $a_f(s) = 0$ if $s_1 = s_2$ or $s_{12} = 0$ or $\min(s_1, s_2) = 1$; it appears that the above observation was first made by Maass in [11], p. 282. Therefore we can write

$$f(\tau) = (q_1q_2)^2(q_1 - q_2)(q_{12} - q_{12}^{-1}) \cdot Q(q_1, q_2, t),$$

in which $Q(q_1, q_2, t)$ is equal to

$$\sum_{s_1>s_2\geq 2}\Big(\sum_{e=2s_{12}\geq 1} a_f(s)(q_{12}^{e-1} + \cdots + q_{12}^{-(e-1)})\Big)$$
$$\cdot (q_1^{s_1-3}q_2^{s_2-2} + \cdots + q_1^{s_2-2}q_2^{s_1-3});$$

actually e is further restricted as $e < 2(s_1s_2)^{1/2}$. Therefore $Q(q_1, q_2, t)$ can be considered as an element of $\mathbf{Z}[t][[q_1, q_2]]$, hence an element of $\mathbf{Z}[[q_1, q_2, t]]$. If we denote the Q for X_{35} by Q_0, then we have $Q_0(0, 0, 0) = 1$; hence $Q_0(q_1, q_2, t)$ is a unit of $\mathbf{Z}[[q_1, q_2, t]]$. The rest of the proof is entirely similar to that of Lemma 7. q.e.d.

For the sake of simplicity we put

$$x_4 = \phi_4 \otimes \phi_4, \qquad x_6 = \phi_6 \otimes \phi_6, \qquad x_{12} = \delta \otimes \delta,$$
$$y_{12} = \phi_4{}^3 \otimes \delta + \delta \otimes \phi_4{}^3;$$

then we have the following table:

$$W(X_4) = x_4, \quad W(X_6) = x_6, \quad W(X_{12}) = 2^2 3x_{12}, \quad W(Y_{12}) = y_{12},$$
$$W(X_{16}) = x_4x_{12}, \quad W(X_{18}) = x_6x_{12}, \quad W(X_{24}) = 2 \cdot 3x_{12}{}^2,$$
$$W(X_{28}) = x_4x_{12}{}^2, \quad W(X_{30}) = x_6x_{12}{}^2, \quad W(X_{36}) = 2^2x_{12}{}^3,$$
$$W(X_{40}) = x_4x_{12}{}^3, \quad W(X_{42}) = x_6x_{12}{}^3, \quad W(X_{48}) = 3x_{12}{}^4.$$

The formulas for $W(X_4)$, $W(X_6)$, $W(X_{12})$, $W(Y_{12})$, $W(X_{16})$ have already been mentioned in Section 1-3; the remaining formulas follow from the formulas for $W(X_4)$, $W(X_6)$, $W(X_{12})$.

6. We shall closely examine the restrictions of the homomorphisms $S: A(\Gamma_2) \to A(\Gamma_1)$ and $W: A(\Gamma_2) \to P(A(\Gamma_1))$ to $A_{\mathbf{Z}}(\Gamma_2)$; we shall recall the definition of P: if E is a module over a commutative ring R with 1, then we shall denote by $P^2(E)$ the submodule of $E \otimes E$ generated over R by elements of the form $e \otimes e$; if E has a free R-basis $e_1, e_2, \ldots, e_n$, then $P^2(E)$ has $e_i \otimes e_i$ for $1 \leq i \leq n$ and $e_i \otimes e_j + e_j \otimes e_i$ for $1 \leq i < j \leq n$ as a free R-basis. If $A = A_0 + A_1 + A_2 + \cdots$ is a graded ring over $R = A_0$, then $P(A)$ is the graded ring defined by $P(A)_k = P^2(A_k)$ for every $k \geq 0$:

$$P(A) = \sum_{k \geq 0} P^2(A_k).$$

The following lemma concerns about the restriction of S to $A_{\mathbf{Z}}(\Gamma_2)$:

LEMMA 9. *We have*

$$A_{\mathbf{Z}}(\Gamma_1)_{4e} = \sum_{0 \leq i \leq [e/3]} \mathbf{Z}\phi_4{}^{e-3i}\,\delta^i,$$

where the sum is direct, and $A_{\mathbf{Z}}(\Gamma_1)_{4e+6} = A_{\mathbf{Z}}(\Gamma_1)_{4e} \cdot \phi_6$ *for* $e = 0, 1, 2, \ldots$; $A_{\mathbf{Z}}(\Gamma_1)_k = 0$ *for every other k. Furthermore S maps* $A_{\mathbf{Z}}(\Gamma_2)_k$ *surjectively to* $A_{\mathbf{Z}}(\Gamma_1)_k$ *for every k.*

Proof. We know that

$$A_{\mathbf{Z}}(\Gamma_1) = \mathbf{Z}[\phi_4, \phi_6, \delta], \qquad \phi_6{}^2 = \phi_4{}^3 - 2^6 3^3 \delta$$

and that ϕ_4, ϕ_6 are algebraically independent over $\mathbf{C}$; the first part of the lemma follows from this classical structure theorem. We recall that if

$$f(\tau) = \sum_{s \geq 0} a_f(s)\mathbf{e}(\mathrm{tr}\,(s\tau))$$

is the Fourier expansion of an element f of $A(\Gamma_2)_k$, then Sf is in $A(\Gamma_1)_k$ and

$$(Sf)(\omega) = \sum_{n=0}^{\infty} a_f\begin{pmatrix} n & 0 \\ 0 & 0 \end{pmatrix} q^n$$

is its Fourier expansion. Therefore S certainly maps $A_{\mathbf{Z}}(\Gamma_2)_k$ to $A_{\mathbf{Z}}(\Gamma_1)_k$ for every k. We have seen that X_4, X_6, Y_{12} are in $A_{\mathbf{Z}}(\Gamma_2)$ and that $S(X_4) = \phi_4$, $S(X_6) = \phi_6$, $S(Y_{12}) = \delta$. Therefore S maps $A_{\mathbf{Z}}(\Gamma_2)$ surjectively to $A_{\mathbf{Z}}(\Gamma_1)$. q.e.d.

LEMMA 10. *We have*

$$P(A_{\mathbf{Z}}(\Gamma_1))_{4e} = \sum_{0\leq i\leq j\leq [e/3]} \mathbf{Z}x_4{}^{e-3j}x_{12}{}^{i}y_{12}{}^{j-i},$$

where the sum is direct, and $P(A_{\mathbf{Z}}(\Gamma))_{4e+6} = P(A_{\mathbf{Z}}(\Gamma_1))_{4e}\cdot x_6$ *for* $e = 0, 1, 2, \ldots$; $P(A_{\mathbf{Z}}(\Gamma_1))_k = 0$ *for every other* k. *Furthermore* W *maps* $A_{\mathbf{Z}}(\Gamma_2)_k$ *to* $P(A_{\mathbf{Z}}(\Gamma_1))_k$ *for every* k.

Proof. For the sake of simplicity we shall write P_k instead of $P(A_{\mathbf{Z}}(\Gamma_1))_k$. By Lemma 9 we have the direct-sum decomposition in the lemma with $y_{12}{}^n$ for $n = j - i$ replaced by $\phi_4{}^{3n} \otimes \delta^n + \delta^n \otimes \phi_4{}^{3n}$. Since their difference:

$$\sum_{0<p<n} \binom{n}{p} (\phi_4{}^{3(n-p)}\,\delta^p \otimes \phi_4{}^{3p}\,\delta^{n-p})$$

is an element of $P_{12(n-2)}\cdot x_4{}^3x_{12}$, by an induction on e we get the decomposition for P_{4e} as stated in the lemma. And $P_{4e+6} = P_{4e}\cdot x_6$, etc. also follow from Lemma 9.

We shall show that f in $A_{\mathbf{Z}}(\Gamma_2)_k$ implies $W(f)$ in P_k: if $k = 4e$, then we can write

$$W(f) = \sum_{0\leq i\leq j\leq [e/3]} c_{ij}x_4{}^{e-3j}x_{12}{}^{i}y_{12}{}^{j-i}, \qquad (*)$$

in which c_{ij} are, for a moment, complex numbers. Since this implies

$$S(f) = \sum_{0\leq j\leq [e/3]} c_{0j}\phi_4{}^{e-3j}\,\delta^j,$$

we see that c_{0j} are integers. Suppose that $c_{0j}, \ldots, c_{i_0-1,j}$ for some $i_0 \geq 1$ are all integers; we shall show that c_{i_0j} are also integers. We recall that $W(f)$, more precisely $W(f)(\tau_1, \tau_2)$, is an element of $\mathbf{Z}[[q_1, q_2]]$ and that

$$x_4{}^{e-3j}x_{12}{}^{i}y_{12}{}^{j-i} = (q_1q_2)^i((q_1 + q_2)^{j-i} + \cdots).$$

Therefore if we compare the coefficients of $q_1^{i_0}q_2^{j}$ on both sides of (*) for $j \geq i_0$ starting from $j = i_0$, then we successively get $c_{i_0,j} \equiv 0 \bmod 1$ for $j = i_0, i_0 + 1, \ldots$. This completes the induction. The proof in the case where $k = 4e + 6$ is entirely similar; and $W(f) = 0$ for every other k. q.e.d.

There is a remarkable difference between the above two lemmas: we have shown that $S(A_{\mathbf{Z}}(\Gamma_2)) = A_{\mathbf{Z}}(\Gamma_1)$ but we have not said anything about the surjectivity of W. In order to determine what $W(A_{\mathbf{Z}}(\Gamma_2))$ is, we need two preliminary lemmas:

LEMMA 11. *Let u, v denote letters and R the subring of $\mathbf{Z}[u, v]$ generated over $\mathbf{Z}$ by uv^n, $d_n v^n$, in which $d_n = 12/\text{G.C.D.}(12, n)$, for all $n \geq 0$; then*

$$u,\ uv,\ uv^2,\ uv^3,\ 12v,\ 6v^2,\ 4v^3,\ 3v^4$$

form a set of generators of R over $\mathbf{Z}$.

Proof. It is clear that d_n depends periodically on n with period 12 and that $d_n = 12, 6, 4, 3, 12, 2, 12, 3, 4, 6, 12, 1$ for $n = 1, 2, \ldots, 12$; it is also clear that

$$12v^5 = (12v)(3v^4) - (6v^2)(4v^3), \qquad 2v^6 = (6v^2)(3v^4) - (4v^3)^2,$$
$$12v^7 = (4v^3)(3v^4), \qquad 3v^8 = 11(3v^4)^2 - (6v^2)(4v^3)^2,$$
$$4v^9 = 3(12v)(3v^4)^2 - 5(4v^3)^3, \qquad 6v^{10} = (3v^4)(2v^6),$$
$$12v^{11} = (4v^3)(3v^8), \qquad v^{12} = (3v^4)(3v^8) - 2(2v^6)^2.$$

Therefore if we denote by R' the subring of R generated over $\mathbf{Z}$ by the eight elements in the lemma, then R' contains $d_n v^n$ for every $n \geq 0$. Since

$$uv^n = uv^{n-3} \cdot 4v^3 - uv^{n-4} \cdot 3v^4$$

for $n \geq 4$, by an induction on n we see that R' also contains uv^n for every n; hence $R = R'$. q.e.d.

Actually we can easily show that the eight elements in the lemma form a minimal set of generators of R over $\mathbf{Z}$.

7. If f is a homogeneous element of $A(\Gamma_2)$ of even weight, then we can rewrite its Fourier expansion as a power series in

$$t = q_{12} + q_{12}^{-1} - 2 = (2\pi i\tau_{12})^2(1 + \cdots);$$

therefore we can define not only Wf but also $W'f$ as

$$f(\tau) = (Wf)(\tau_1, \tau_2) + (W'f)(\tau_1, \tau_2)t + \cdots.$$

We observe that $(W'f)(\tau_1, \tau_2)$ is an element of $\mathbf{C}[[q_1, q_2]]$ and that $(W'f)(\tau_1, \tau_2)$ is in $\mathbf{Z}[[q_1, q_2]]$ if f is in $A_{\mathbf{Z}}(\Gamma_2)$. We shall denote by ϕ_2 the Eisenstein series of degree 1 and of weight 2; it is defined as

$$\phi_2(\omega) = (½\pi i)(\log \delta)'(\omega) = 1 - 24 \cdot \sum_{n=1}^{\infty} \sigma_1(n)q^n,$$

in which $\sigma_1(n)$ denotes the sum of all divisors of n and the prime indicates $d/d\omega$; cf. [16], [19]. In accordance with other notation we put $x_2 = \phi_2 \otimes \phi_2$.

LEMMA 12. *We have*

$$W'(X_{10}) = \delta \otimes \delta = x_{12}, \qquad W'(X_{12}) = \phi_2\delta \otimes \phi_2\delta = x_2x_{12}.$$

Proof. We shall use the expressions

$$X_{10} = 2^{-12} \cdot \prod_{\text{ten}} \theta_{\mathfrak{m}}{}^2, \qquad X_{12} = 2^{-15} \cdot \sum_{\text{fifteen}} (\theta_{\mathfrak{m}_1}\theta_{\mathfrak{m}_2} \cdots \theta_{\mathfrak{m}_6})^4$$

in Lemma 3. If $\mathfrak{m} = (m_1m_2m_3m_4)$ is an even characteristic, then we have

$$\theta_{\mathfrak{m}}(\tau) = \theta_{m_1m_3}(\tau_1)\theta_{m_2m_4}(\tau_2) - (½\pi^2)\theta_{m_1m_3}{}'(\tau_1)\theta_{m_2m_4}{}'(\tau_2)t + \cdots$$

for $\mathfrak{m} \neq (1111)$ and

$$\theta_{1111}(\tau) = (\pi/2i)(\theta_{00}\theta_{01}\theta_{10})(\tau_1)(\theta_{00}\theta_{01}\theta_{10})(\tau_2)\tau_{12} + \cdots,$$

$$\theta_{1111}{}^2(\tau) = 2^{-4}(\theta_{00}\theta_{01}\theta_{10})^2(\tau_1)(\theta_{00}\theta_{01}\theta_{10})^2(\tau_2)t + \cdots;$$

cf. Section 1-2. Therefore by using $(\theta_{00}\theta_{01}\theta_{10})^8 = 2^8\delta$ we get $W'(X_{10}) =$

$\delta \otimes \delta$. As for $W'(X_{12})$ we have only to examine those sextuples which do not contain (1111): there are six such sextuples and they are the complements of (1, 4, 9, 0), (1, 6, 8, 0), (2, 3, 9, 0), (2, 6, 7, 0), (3, 5, 8, 0), (4, 5, 7, 0). The contribution to $X_{12}(\tau)$ of the complement of (1, 4, 9, 0), which is (2, 3, 5, 6, 7, 8), is

$$2\delta(\tau_1)\delta(\tau_2)\{1 - (2/\pi^2)((\log\theta_{00})'(\tau_1)(\log\theta_{01})'(\tau_2) + \cdots + (\log\theta_{10})'(\tau_1)(\log\theta_{01})'(\tau_2))t + \cdots\},$$

in which (00)(01), $\cdots$, (10)(01) come from (2, ..., 8), i.e., from (0001), ..., (1001). We have similar contributions to $X_{12}(\tau)$ from the remaining five sextuples. Therefore by adding all contributions we get

$$X_{12}(\tau) = 12\delta(\tau_1)\delta(\tau_2) - (1/(2\pi)^2)\delta'(\tau_1)\delta'(\tau_2)t + \cdots;$$

we observe that the coefficient of t is $(\phi_2\delta)(\tau_1)(\phi_2\delta)(\tau_2)$. q.e.d.

REMARK. We can apply the same method to

$$X_{35} = 2^{-37}5^{-3}\left(\prod_{\text{ten}}\theta_{\mathrm{m}}\right)\left(\sum_{\text{sixty}} \pm (\theta_{\mathrm{m}_1}\theta_{\mathrm{m}_2}\theta_{\mathrm{m}_3})^{20}\right)$$

in Lemma 6: if we number the sixty azygous triples in the order we have written in Section 1-2 by 1, 2, ..., 60, then we take the plus-sign for 1, 4, 6, 7, 10, 11, 14, 15, 18, 20, 21, 24, 25, 29, 30, 31, 33, 36, 37, 40, 41, 43, 44, 48, 50, 51, 53, 56, 58, 59. We have

$$\prod_{\text{ten}}\theta_{\mathrm{m}}(\tau) = (\pi/2i)(\theta_{00}\theta_{01}\theta_{10})^4(\tau_1)(\theta_{00}\theta_{01}\theta_{10})^4(\tau_2)\tau_{12} + \cdots,$$

$$\sum_{\text{sixty}} \pm W(\theta_{\mathrm{m}_1}\theta_{\mathrm{m}_2}\theta_{\mathrm{m}_3})^{20} = \Theta \otimes (\theta_{00}\theta_{01}\theta_{10})^{20} - (\theta_{00}\theta_{01}\theta_{10})^{20} \otimes \Theta,$$

in which

$$\begin{aligned}\Theta &= \theta_{00}{}^{60} - \theta_{01}{}^{60} - \theta_{10}{}^{60} - 3(\theta_{00}{}^{40}\theta_{01}{}^{20} + \theta_{00}{}^{40}\theta_{10}{}^{20} \\ &\qquad - \theta_{00}{}^{20}\theta_{01}{}^{40} - \theta_{00}{}^{20}\theta_{10}{}^{40} + \theta_{01}{}^{40}\theta_{10}{}^{20} + \theta_{01}{}^{20}\theta_{10}{}^{40}) \\ &= (\theta_{00}\theta_{01}\theta_{10})^{12}(5^3(\theta_{01}{}^8 + \theta_{01}{}^4\theta_{10}{}^4 + \theta_{10}{}^8)^3 - 2\cdot 3(\theta_{00}\theta_{01}\theta_{10})^8) \\ &= (\theta_{00}\theta_{01}\theta_{10})^{12}(5^3\phi_4{}^3 - 2^9 3\delta);\end{aligned}$$

and hence $X_{35}(\tau)$ is equal to

$$2^2\delta^2(\tau_1)\delta^2(\tau_2)(\phi_4{}^3(\tau_1)\delta(\tau_2) - \delta(\tau_1)\phi_4{}^3(\tau_2))(-\pi i\tau_{12}) + \cdots$$
$$= (2^2 i)q_1{}^2 q_2{}^2(q_1 - q_2)(\pi\tau_{12}) + \cdots.$$

We might mention that this is how we obtained the first term of the expansion for $\chi_{35}(\tau)$ in [8]-II, p. 404. We might also mention that from the above calculation we get

$$W'(X_{35}{}^2) = 2^2 x_{12}{}^4(\phi_4{}^3 \otimes \delta - \delta \otimes \phi_4{}^3)^2 = 2^2 x_{12}{}^4(y_{12}{}^2 - 2^2 x_4{}^3 x_{12})$$

in agreement with a consequence of the formula for $X_{35}{}^2$ in Lemma 6.

8. We are ready to determine the graded subring $W(A_{\mathbf{Z}}(\Gamma_2))$ of $P(A_{\mathbf{Z}}(\Gamma_1))$; we shall formulate our results as follows:

LEMMA 13. *Let A denote the graded subring of $A_{\mathbf{Z}}(\Gamma_2)$ generated over $\mathbf{Z}$ by $X_4, X_6, \ldots, X_{48}$; then A and $A_{\mathbf{Z}}(\Gamma_2)$ have the same W-image. Moreover we have*

$$W(A_{\mathbf{Z}}(\Gamma_2))_{12n} = \sum_{0\le i\le j<n} \mathbf{Z}x_4{}^{3(n-j)}x_{12}{}^i y_{12}{}^{j-i} + \sum_{0\le i\le n} \mathbf{Z}d_i x_{12}{}^i y_{12}{}^{n-i} \qquad (\#)$$

for $n = 1, 2, 3, \ldots$ and $W(A_{\mathbf{Z}}(\Gamma_2))_k = P(A_{\mathbf{Z}}(\Gamma_1))_k$ for every other k.

Proof. For the sake of simplicity we shall again write P_k instead of $P(A_{\mathbf{Z}}(\Gamma_1))_k$. We recall that

$$x_4,\ x_4x_{12},\ x_4x_{12}{}^2,\ x_4x_{12}{}^3;\quad x_6,\ x_6x_{12},\ x_6x_{12}{}^2,\ x_6x_{12}{}^3;$$
$$12x_{12},\ 6x_{12}{}^2,\ 4x_{12}{}^3,\ 3x_{12}{}^4;\quad y_{12}$$

are in $W(A)$. Therefore by Lemma 11 we see that $x_4{}^{a+1}x_{12}{}^b y_{12}{}^c$, $x_4{}^a x_6 x_{12}{}^b y_{12}{}^c$ for $a, b, c \ge 0$ and $d_i x_{12}{}^i y_{12}{}^j$ for $i, j \ge 0$ are all in $W(A)$. In view of Lemma 10, therefore, we will have $W(A)_k = P_k$ if $k \not\equiv 0 \bmod 12$ and also for $k = 0$. Furthermore the right-hand side of (#) is contained in $W(A)_{12n}$ for $n = 1, 2, 3, \ldots$. Therefore we have only to show that if

$$\sum_{i+j=n} c_i x_{12}{}^i y_{12}{}^j$$

is in $W(A_{\mathbf{Z}}(\Gamma_2))$, then $c_i \equiv 0 \bmod d_i$ for $0 \leq i \leq n$; and the proof is as follows:

By assumption the above sum can be written as $W(f)$ for some f in $A_{\mathbf{Z}}(\Gamma_2)_{12n}$. We recall that every homogeneous element of $A(\Gamma_2)$ of even weight can be uniquely written as an isobaric polynomial in X_4, X_6, X_{10}, X_{12} with complex coefficients; cf. [8]. Since

$$X_6{}^2 = X_4{}^3 + 2^{10}3^5X_{12} - 2^6 3^3 Y_{12},$$

if the weight is a multiple of 4, then it can be uniquely written as

$$\sum_{i \geq 0} F_i(X_4, X_{12}, Y_{12})X_{10}{}^{2i} + \sum_{j \geq 0} G_j(X_4, X_{12}, Y_{12})X_6 X_{10}{}^{2j+1},$$

in which $F_i(X_4, X_{12}, Y_{12})$, $G_j(X_4, X_{12}, Y_{12})$ are isobaric polynomials in X_4, X_{12}, Y_{12} with complex coefficients. In particular we have

$$f = \sum_{i+j=n} (12)^{-i}c_i X_{12}{}^i Y_{12}{}^j + \sum_{a,b} c_{ab}X_4{}^{3(n-a-b)-4}X_{12}{}^a Y_{12}{}^b \cdot X_6 X_{10}$$
$$+ \text{ a sum of terms containing } X_{10}{}^2,$$

in which the second summation is restricted as $a,\ b \geq 0$ and $3(n - a - b) - 4 \geq 0$, i.e., as $n - 2 \geq a + b$ and $a,\ b \geq 0$.

We apply W' to both sides of the above identity; then by using Lemma 12 we get

$$W'(f) = \sum_{i+j=n} \{(12)^{-1}ic_i x_2 x_{12}{}^i y_{12}{}^j + jc_i W'(Y_{12})x_{12}{}^i y_{12}{}^{j-1}\}$$
$$+ \sum_{a,b} (12)^a c_{ab} x_4{}^{3(n-a-b)-4} x_6 x_{12}{}^{a+1} y_{12}{}^b.$$

Since c_i are integers by Lemma 10 and since $W'(f)(\tau_1, \tau_2)$, $W'(Y_{12})(\tau_1, \tau_2)$ are elements of $\mathbf{Z}[[q_1, q_2]]$, we get

$$\sum_{i+j=n} c_i{}' x_2 x_{12}{}^i y_{12}{}^j \equiv \sum_{a,b} c_{ab}{}' x_4{}^{3(n-a-b)-4} x_6 x_{12}{}^{a+1} y_{12}{}^b \bmod 1, \qquad (*)$$

in which $c_i{}' = (12)^{-1}ic_i$ and $c_{ab}{}' = -(12)^a c_{ab}$; and "mod 1" means that the difference evaluated at (τ_1, τ_2) is in $\mathbf{Z}[[q_1, q_2]]$. Since $c_i \equiv 0$

mod d_i and $c_i' \equiv 0 \bmod 1$ are equivalent, we have only to show that $c_i' \equiv 0 \bmod 1$ for every i.

We lexicographically order all non-negative integer pairs: $(0, 0) < (0, 1) < (0, 2) < \cdots < (1, 0) < (1, 1) < \cdots$. If $c_i' \not\equiv 0 \bmod 1$ for some i, we shall denote the smallest of such i's by i_0 and we put $j_0 = n - i_0$; then $i_0, j_0 \geq 0$. We shall denote by E the possibly empty set of all (a, b), where $n - 2 \geq a + b$ and $a, b \geq 0$, satisfying $(a + 1, b) < (i_0, j_0)$; we shall show that $c_{ab}' \equiv 0 \bmod 1$ for every (a, b) in E. If $c_{ab}' \not\equiv 0 \bmod 1$ for some (a, b) in E, we shall denote the smallest of such (a, b)'s by (a_0, b_0); and we compare the coefficients of $(q_1q_2)^{a_0+1} q_2^{b_0}$ on both sides of $(*)$, more precisely both sides of $(*)$ evaluated at (τ_1, τ_2). Then by definition and by using

$$x_2(\tau_1, \tau_2), x_4(\tau_1, \tau_2), x_6(\tau_1, \tau_2) = 1 + \cdots,$$

$$x_{12}(\tau_1, \tau_2) = q_1q_2(1 + \cdots), \qquad y_{12}(\tau_1, \tau_2) = (q_1 + q_2) + \cdots,$$

we get $c_{a_0b_0}' \equiv 0 \bmod 1$; but this is a contradiction. We observe that for no (a, b), where $n - 2 \geq a + b$ and $a, b \geq 0$ as always, we have $(a + 1, b) = (i_0, j_0)$ because otherwise we will get $n - 2 \geq i_0 + j_0 - 1 = n - 1$. In view of these we may restrict the summation in (i, j) as $i \geq i_0$, i.e., as $(i, j) \geq (i_0, j_0)$, and the summation in (a, b) as $(a + 1, b) > (i_0, j_0)$; this is a key point. We finally compare the coefficients of $(q_1q_2)^{i_0}q_2^{j_0}$ on both sides of $(*)$. Then we get $c_{i_0}' \equiv 0 \bmod 1$; but this is a contradiction. Therefore $c_i' \equiv 0 \bmod 1$ for every i. q.e.d.

We can now prove the "completeness" in Theorem 2: take any homogeneous element f of $A_{\mathbf{Z}}(\Gamma_2)$ of weight k; the problem is to show that f is in A. By Lemma 8 we may assume that k is even. By Lemma 13 we have $W(f) = W(f')$ for some f' in A_k; then by Lemma 7 we will have

$$f = f' + X_{10}f''$$

for some f'' in $A_{\mathbf{Z}}(\Gamma_2)$ of weight $k - 10$. By repeatedly applying this argument or rather by an induction on k we see that f is in A.

The above proof actually gives an algorithm to express f as a polynomial in $X_4, X_6, \ldots, X_{48}$ with integral coefficients; and it is by this algorithm that we have obtained the expression for $X_{35}{}^2$ in Lemma 6.

9. We shall show that the set of generators

$$X_4, X_6, X_{10}, X_{12}, Y_{12}, \ldots, X_{35}, \ldots, X_{48}$$

of $A_Z(\Gamma_2)$ over **Z** is minimal, i.e., none of them is redundant. We recall that X_4, X_6, X_{10}, X_{12} are algebraically independent over **C** and that X_{35} is the only element of the set with odd weight. Therefore none of X_4, X_6, X_{10}, X_{35} is redundant; in particular neither X_{10} nor X_{35} is redundant. In order to prove the irredundancy of the remaining thirteen elements, we may apply the homomorphism W: it is enough to show that none of

$$x_4, x_6, 12x_{12}, y_{12}, x_4x_{12}, x_6x_{12}, 6x_{12}^2,$$
$$x_4x_{12}^2, x_6x_{12}^2, 4x_{12}^3, x_4x_{12}^3, x_6x_{12}^3, 3x_{12}^4$$

is redundant. We observe that $W(A_Z(\Gamma_2))$ is a graded subring of $\mathbf{Z}[x_4, x_6, x_{12}, y_{12}]$ and that

$$x_6^2 = x_4^3 + (12)^6x_{12} - (12)^3y_{12}$$

is the only relation between x_4, x_6, x_{12}, y_{12}. Therefore none of x_4, x_6, $12x_{12}$, y_{12} is redundant. We shall divide the rest of the proof into two parts:

Let x denote one of the remaining six elements of weights divisible by 4 and assume that it is redundant; then we have $x = F + G$, in which F is an isobaric polynomial in the nine elements of weights divisible by 4, but not including x, and G is a similar polynomial involving x_6, x_6x_{12}, $x_6x_{12}^2$, $x_6x_{12}^3$ both with integral coefficients. Since the weight of x is a multiple of 4, the above four elements must appear altogether even times in each monomial in G. By replacing x_6^2 by the above expression, which does not affect F, we will get a relation in $\mathbf{Z}[x_4, x_{12}, y_{12}]$; we recall that x_4, x_{12}, y_{12} are algebraically independent over **C**. We can therefore apply the homomorphism $\mathbf{Z}[x_4, x_{12}, y_{12}] \to (\mathbf{Z}/12\mathbf{Z})[x_4, x_{12}, 0]$ and we get

$$x = F' + \text{a sum of terms containing } x_4^3,$$

in which F' is an isobaric polynomial in

$$x_4,\ x_4x_{12},\ 6x_{12}^2,\ x_4x_{12}^2,\ 4x_{12}^3,\ x_4x_{12}^3,\ 3x_{12}^4$$

with coefficients in $\mathbf{Z}/12\mathbf{Z}$ and F' does not contain x nor elements of weights higher than that of x. We successively take $x = x_4x_{12}, \ldots,$ $x = 3x_{12}^4$ and, in each case, we get a contradiction.

Let x denote one of x_6x_{12}, $x_6x_{12}^2$, $x_6x_{12}^3$ and assume that it is redundant; then we get a relation of the form

$$x_6x_{12}^n = \sum_{0\leq i<n} (F_i + G_i)x_6x_{12}^i$$

for some n among 1, 2, 3, in which F_i, G_i are similar to the above F, G and they are both of weight $12(n - i)$ for $0 \leq i < n$. In particular neither F_i nor G_i contains $3x_{12}^4$. We cancel x_6 from both sides and apply the homomorphism $\mathbf{Z}[x_4, x_{12}, y_{12}] \rightarrow \mathbf{F}_2[0, x_{12}, 0]$; then we get $x_{12}^n = 0$, a contradiction. This completes the proof.

We recall that a minimal set of generators of a graded ring has a certain uniqueness property; in the present case if we have another minimal set of generators of $A_{\mathbf{Z}}(\Gamma_2)$ over $\mathbf{Z}$, then the new set necessarily consists of

$$X_4' = \pm X_4; \qquad X_6' = \pm X_6; \qquad X_{10}' \equiv \pm X_{10} \bmod X_4X_6;$$

$$X_{12}' \equiv c_1X_{12} + c_{12}Y_{12}, \qquad Y_{12}' \equiv c_{21}X_{12} + c_2Y_{12} \bmod X_4^3,$$

$$c_1,\ c_{12},\ c_{21},\ c_2 \text{ integers, } c_1c_2 - c_{12}c_{21} = \pm 1;$$

$$X_{16}' \equiv \pm X_{16} \bmod (X_4^4, X_4X_{12}, X_4Y_{12}); \text{ etc.,}$$

in which, e.g., mod $(X_4^4, X_4X_{12}, X_4Y_{12})$ means that the difference is a linear combination of X_4^4, X_4X_{12}, X_4Y_{12} with integral coefficients.

3. Structure of Proj $(A_{\mathbf{Z}}(\Gamma_2))$

10. For the sake of clarity we shall briefly review some basic definitions in [5]: if R is a commutative ring with 1, then we shall denote by Spec (R) the "prime spectrum" of R; as a set it consists of all prime ideals of R. If K is a field, then we shall denote by Spec $(R)_K$ the set

of all homomorphisms $R \to K$, called "K-valued geometric points" of Spec (R); we shall only use a universal field as K. We sometimes reduce Spec (R), i.e., replace Spec (R) by Spec $(R)_{\text{red}}$ defined as the prime spectrum of R/(radical of R); as sets they are same.

If A is a graded ring: $A = A_0 + A_1 + A_2 + \cdots$, then we shall denote by Proj (A) the "homogeneous prime spectrum" of A; as a set it consists of all homogeneous prime ideals of A not containing $A_+ = A_1 + A_2 + \cdots$. For any positive integer d we put

$$A^{(d)} = A_0 + A_d + A_{2d} + \cdots;$$

then we have Proj (A) = Proj ($A^{(d)}$). We shall only use $d = 2, 4$ and, different from the usual practice, we shall keep the degrees in A for homogeneous elements of $A^{(d)}$. If f is a homogeneous element of A of positive degree, then we shall denote by $A_{(f)}$ the ring consisting of all a/f^n, in which $n \geq 0$ and a is a homogeneous element of A of degree $n \cdot \deg(f)$. If A is generated over A_0 by homogeneous elements $f_1, f_2, \ldots, f_N$ of positive degrees, then we have

$$\text{Proj}\,(A) = \bigcup_{i=1}^{N} \text{Spec}\,(A_{(f_i)});$$

in such a case Proj $(A)_{\text{red}}$ can be defined as the union of Spec $(A_{(f_i)})_{\text{red}}$. We also have the concept of "open and closed subschemes" of Proj (A); for instance if f is a homogeneous element of A of positive degree, then Spec ($A_{(f)}$) resp. Proj (A/fA) are open resp. closed subschemes of Proj (A); and by taking unions resp. intersections we get more general open resp. closed subschemes. We observe that as a point set Proj (A) becomes the disjoint union of Spec ($A_{(f)}$) and Proj (A/fA). If $\mathfrak{M}$ is a, say open, subscheme of Proj (A) and K is a universal field, then we shall denote by $\mathfrak{M}_K$ the set of all K-valued geometric points of $\mathfrak{M}$; it can be defined by representing $\mathfrak{M}$ as a union of prime spectrums.

We shall first take $A_{\mathbf{Z}}(\Gamma_2)^{(2)}$ as A and X_{10} as f; for the sake of simplicity we put

$$\mathfrak{A} = A_{\mathbf{Z}}(\Gamma_2)^{(2)}, \qquad \mathfrak{A}' = \mathfrak{A}/X_{10}\mathfrak{A};$$

then Proj ($A_{\mathbf{Z}}(\Gamma_2)$) = Proj ($\mathfrak{A}$) decomposes into Spec ($\mathfrak{A}_{(X_{10})}$) and Proj

($\mathfrak{A}'$). We shall examine the geometric meaning of Spec ($\mathfrak{A}_{(X_{10})}$) by identifying $\mathfrak{A}_{(X_{10})}$ with another ring which appeared in our earlier work.

Consider a binary sextic

$$P(u; x_0, x_1) = u_0 x_0^6 + u_1 x_0^5 x_1 + \cdots + u_6 x_1^6$$

with variable complex coefficients $u_0, u_1, \ldots, u_6$; if $(x_0, x_1) \to (y_0, y_1)$ is the obvious action of $SL_2(\mathbf{C})$ and if we rewrite the above sextic as $P(v; y_0, y_1)$, then $(u_0, u_1, \ldots, u_6) \to (v_0, v_1, \ldots, v_6)$ defines a 7-dimensional (irreducible) linear representation of $SL_2(\mathbf{C})$. The corresponding invariant polynomials are called the invariants of a binary sextic.

Consider a plane curve of genus 2 defined by

$$xy^2 + (1 + ax + bx^2)y + x^2(c + dx + x^2) = 0,$$

in which a, b, c, d are variable complex numbers; then the plane curve defined by

$$y^2 = (1 + ax + bx^2)^2 - 4x^3(c + dx + x^2)$$

is birationally equivalent to the original curve. An invariant J of a binary sextic is called arithmetic if the value of J at the right-hand side of the above equation multiplied by x_1^6 after the substitution $x = x_0/x_1$ becomes a polynomial in a, b, c, d with integral coefficients. If we decompose $P(u; x_0, x_1)$ into linear factors as

$$P(u; x_0, x_1) = u_0 \cdot \prod_{i=1}^{6} (x_0 - \xi_i x_1),$$

put $(ij) = \xi_i - \xi_j$, and define the classical invariants "A, B, C, D" as

$$A = u_0^2 \cdot \sum_{\text{fifteen}} (12)^2(34)^2(56)^2,$$

$$B = u_0^4 \cdot \sum_{\text{ten}} (12)^2(23)^2(31)^2(45)^2(56)^2(64)^2,$$

$$C = u_0^6 \cdot \sum_{\text{sixty}} (12)^2(23)^2(31)^2(45)^2(56)^2(64)^2(14)^2(25)^2(36)^2,$$

$$D = u_0^{10} \cdot \prod_{i<j} (ij)^2,$$

then not only A, B, C, D but more significantly

$$J_2 = 2^{-3}A, \qquad J_4 = 2^{-5}3^{-1}(2^2J_2{}^2 - B),$$
$$J_6 = 2^{-6}3^{-2}(2^3J_2{}^3 - 2^5 5J_2J_4 - C),$$
$$J_8 = 2^{-2}(J_2J_6 - J_4{}^2), \qquad J_{10} = 2^{-12}D$$

are all arithmetic. And we proved that

$$\mathrm{Spec}\,(\mathbf{Z}[J_2, J_4, J_6, J_8, J_{10}]_{(J_{10})})$$

is the scheme of moduli over $\mathbf{Z}$ of curves of genus 2 and, in particular, the set of its geometric points is bijective to the set of birational classes of curves of genus 2 for all characteristics; we refer to [7] for the details.

On the other hand we defined in [9] a degree-preserving injective homomorphism ρ from $A(\Gamma_2)$ to the ring of invariants of a binary sextic; we started from a curve of genus 2 defined by $y^2 = P(u; x, 1)$ and we associated to the curve a point τ of the Siegel upper-half space $\mathfrak{S}_2$ by the standard normalization of its period matrix; then for any homogeneous element f of $A(\Gamma_2)$ of weight k the prescription of ρf was

$$f(\tau) = \mu^k \cdot (\rho f)(P(u; x, 1)),$$

in which $\mu \neq 0$ depends on the definition of τ. And by normalizing μ suitably we found that

$$\rho(\psi_4) = 2^{-2}B, \qquad \rho(\psi_6) = 2^{-3}(AB - 3C),$$
$$\rho(\chi_{10}) = -2^{-14}D, \qquad \rho(\chi_{12}) = 2^{-17}3^{-1}AD;$$

we refer to op. cit. for the details.

11. We are ready to state our result on the structure of $\mathfrak{A}_{(X_{10})}$ for $\mathfrak{A} = A_{\mathbf{Z}}(\Gamma_2)^{(2)}$; it is what we have expected:

LEMMA 14. *The homomorphism ρ gives rise to the following isomorphism*:

$$\mathfrak{A}_{(X_{10})} = A_{\mathbf{Z}}(\Gamma_2)_{(X_{10})} \xrightarrow{\sim} \mathbf{Z}[J_2, J_4, J_6, J_8, J_{10}]_{(J_{10})};$$

therefore Spec $(\mathfrak{A}_{(X_{10})})$ *gives the scheme of moduli over* $\mathbf{Z}$ *of* 2-*dimensional jacobian varieties.*

Proof. By using the results we have recalled we get the following table:

$$\rho(X_4) = J_2^2 - 2^3 3 J_4, \qquad \rho(X_6) = J_2^3 - 2^2 3^2 J_2 J_4 + 2^3 3^3 J_6,$$
$$\rho(X_{10}) = J_{10}, \qquad \rho(X_{12}) = J_2 J_{10},$$
$$\rho(Y_{12}) = -J_2^2 J_8 + J_2 J_4 J_6 + 2^5 J_4 J_8 - 3^3 J_6^2 + 2^4 3^2 J_2 J_{10},$$
$$\rho(X_{16}) = (J_2 J_4 - 2 \cdot 3^2 J_6) J_{10},$$
$$\rho(X_{18}) = (J_2^2 J_4 - 2 \cdot 3 \cdot 5 J_2 J_6 + 2^6 3 J_8) J_{10},$$
$$\rho(X_{24}) = J_4 J_{10}^2, \qquad \rho(X_{28}) = (J_2 J_6 - 2^4 J_8) J_{10}^2,$$
$$\rho(X_{30}) = (J_2^2 J_6 + 2^3 J_2 J_8 - 2^2 3^2 J_4 J_6) J_{10}^2,$$
$$\rho(X_{36}) = J_6 J_{10}^3, \qquad \rho(X_{40}) = (-2 J_2 J_8 + 3 J_4 J_6) J_{10}^3,$$
$$\rho(X_{42}) = (2 J_2^2 J_8 - J_2 J_4 J_6 - 2^5 J_4 J_8) J_{10}^3, \qquad \rho(X_{48}) = J_8 J_{10}^4.$$

The point is that no non-integral coefficients appear in the ρ-images of $X_4, X_6, \ldots, X_{48}$. For the sake of simplicity we put $\mathbf{Z}[J] = \mathbf{Z}[J_2, J_4, J_6, J_8, J_{10}]$; then ρ gives an injective degree-preserving homomorphism of $\mathfrak{A}$ to $\mathbf{Z}[J]$ satisfying $\rho(X_{10}) = J_{10}$. Therefore ρ gives rise to an injective homomorphism of $\mathfrak{A}_{(X_{10})}$ to $\mathbf{Z}[J]_{(J_{10})}$. This homomorphism is surjective, hence bijective, because for any non-negative integers $e_1, e_2, \ldots, e_5$ satisfying $e_1 + 2e_2 + 3e_3 + 4e_4 = 5e_5$ we have

$$\rho(X_{12}^{e_1} X_{24}^{e_2} X_{36}^{e_3} X_{48}^{e_4} / X_{10}^{6e_5}) = J_2^{e_1} J_4^{e_2} J_6^{e_3} J_8^{e_4} / J_{10}^{e_5}.$$

And finally we use (a special case of) Torelli's theorem in Weil [20], pp. 37–38 to identify the moduli schemes over $\mathbf{Z}$ of curves of genus 2 and 2-dimensional jacobian varieties. q.e.d.

12. We shall examine the geometric meaning of the complement of Spec $(\mathfrak{A}_{(X_{10})})$, which is Proj $(\mathfrak{A}')$ for $\mathfrak{A}' = \mathfrak{A}/X_{10}\mathfrak{A} = W(\mathfrak{A})$; cf. Lemma 7. We shall first recall a well-known geometric meaning of Proj $(\mathfrak{A}'')$ for

$$\mathfrak{A}'' = A_{\mathbf{Z}}(\Gamma_1) = \mathbf{Z}[\phi_4, \phi_6, \delta];$$

we have

$$\mathfrak{A}_{(\delta)}'' = \mathbf{Z}[j], \qquad (\mathfrak{A}''/\delta\mathfrak{A}'')^{(2)} = \mathbf{Z} + 0 + \mathbf{Z} + \mathbf{Z} + \cdots,$$

in which $j = \phi_4{}^3/\delta$, and Spec ($\mathbf{Z}[j]$) gives the scheme of moduli over $\mathbf{Z}$ of elliptic curves. In this context we can regard its complement Proj ($\mathfrak{A}''/\delta\mathfrak{A}''$) = Spec ($\mathbf{Z}$) as the scheme of moduli over $\mathbf{Z}$ of cubic curves with ordinary double point. Therefore Proj ($\mathfrak{A}''$) is a completed scheme of moduli over $\mathbf{Z}$ of elliptic curves.

We shall go back to Proj ($\mathfrak{A}'$): we decompose the homomorphism $S: A(\Gamma_2) \to A(\Gamma_1)$ into $W: A(\Gamma_2) \to P(A(\Gamma_1))$ and $S': P(A(\Gamma_1)) \to A(\Gamma_1)$; then we have $S'(x_4) = \phi_4$, $S'(x_6) = \phi_6$, $S'(x_{12}) = 0$, $S'(y_{12}) = \delta$. We have shown in Section 2 that every element of $\mathfrak{A}_{4e}'$ can be uniquely written as

$$\sum_{0 \leq i \leq j \leq [e/3]} c_{ij} x_4{}^{e-3j} x_{12}{}^i y_{12}{}^{j-i}$$

with c_{ij} in $\mathbf{Z}$ subject to the condition that $c_{ij} \equiv 0 \bmod d_i$ if $3j = e$. Since its S'-image is

$$\sum_{0 \leq j \leq [e/3]} c_{0j} \phi_4{}^{e-3j} \delta^j,$$

in which ϕ_4, δ are algebraically independent over $\mathbf{C}$, the original element is in the kernel of S' if and only if $c_{0j} = 0$ for all j; then it is nilpotent modulo $x_{12}{}^{12}\mathfrak{A}'$. Since $\mathfrak{A}_{4e+6}' = \mathfrak{A}_{4e}' \cdot x_6$, we see that the kernel of $S': \mathfrak{A}' \to \mathfrak{A}''$ is the radical of $x_{12}{}^{12}\mathfrak{A}'$. Hence we get

$$\text{Proj}\,(\mathfrak{A}'/x_{12}{}^{12}\mathfrak{A}')_{\text{red}} = \text{Proj}\,(\mathfrak{A}'');$$

and we know its geometric meaning. Therefore we have only to examine the geometric meaning of Spec ($\mathfrak{A}_{(x_{12}{}^{12})}'$).

LEMMA 15. *If we put* $\mathfrak{A}^* = P(\mathfrak{A}'')$, *then we get*

$$\mathfrak{A}^* = \mathbf{Z}[x_4, x_6, x_{12}, y_{12}], \qquad \mathfrak{A}_{(x_{12})}^* = \mathbf{Z}[j \otimes j, j \otimes 1 + 1 \otimes j];$$

hence Spec ($\mathfrak{A}_{(x_{12})}^*$) *is the scheme of moduli over* $\mathbf{Z}$ *of products of elliptic curves. Furthermore the morphism* Spec ($\mathfrak{A}_{(x_{12})}^*$) $\to$ Spec ($\mathfrak{A}_{(x_{12}{}^{12})}'$)

coming from the inclusion of $\mathfrak{A}'$ *in* $\mathfrak{A}^*$ *is the normalization of* Spec $(\mathfrak{A}_{(x_{12}{}^{12})}')$ *and it is bijective.*

Proof. The first part follows from Lemma 10 and the second part follows from what we have recalled; we shall prove the last part: if we put

$$u = x_4{}^3/x_{12}, \qquad v = y_{12}/x_{12},$$

then we can write $\mathfrak{A}_{(x_{12})}{}^* = \mathbf{Z}[u, v]$ and

$$\mathfrak{A}_{(x_{12}{}^{12})}' = \mathbf{Z}[uv^n, d_n v^n]_{n \geq 0} = \mathbf{Z}[u, uv, \ldots, 12v, 6v^2, \ldots];$$

this follows from Lemma 10 and Lemma 13. Therefore $\mathfrak{A}_{(x_{12})}{}^*$ is the integral closure of $\mathfrak{A}_{(x_{12}{}^{12})}'$ within its field of quotients, i.e., the morphism Spec $(\mathfrak{A}_{(x_{12})}{}^*) \to$ Spec $(\mathfrak{A}_{(x_{12}{}^{12})}')$ is the normalization of Spec $(\mathfrak{A}_{(x_{12}{}^{12})}')$; in particular it is surjective.

We shall show that the normalization morphism is also injective, hence bijective: let $\mathfrak{p}^*$ denote a prime ideal of $\mathbf{Z}[u, v]$ and $\mathfrak{p}'$ resp. $\mathfrak{p}_0$ denote the intersections of $\mathfrak{p}^*$ and $\mathbf{Z}[uv^n, d_n v^n]_{n \geq 0}$ resp. $\mathbf{Z}$; let $\mathbf{Z}[u, v] \to (\mathbf{Z}/\mathfrak{p}_0)[u', v']$ denote a geometric point with kernel $\mathfrak{p}^*$; then

$$\mathbf{Z}[uv^n, d_n v^n]_{n \geq 0} \to (\mathbf{Z}/\mathfrak{p}_0)[u'(v')^n, d_n(v')^n]_{n \geq 0}$$

is a geometric point with kernel $\mathfrak{p}'$. We observe that $\mathfrak{p}^*$ is (uniquely) determined by $\mathfrak{p}_0$ and (u', v') and that $\mathfrak{p}'$ determines $\mathfrak{p}_0$ and

$$u', u'v', 12v', 6(v')^2, 4(v')^3, 3(v')^4.$$

Therefore if either $u' \neq 0$ or $\mathfrak{p}_0 \neq 2\mathbf{Z}, 3\mathbf{Z}$, then $\mathfrak{p}'$ determines $\mathfrak{p}^*$. Suppose that $\mathfrak{p}_0 = 2\mathbf{Z}$; then $\mathfrak{p}'$ determines $(v')^4$. But in the universal field over $\mathbf{F}_2$ certainly $(v')^4$ determines v'; hence $\mathfrak{p}'$ determines $\mathfrak{p}^*$. Suppose next that $\mathfrak{p}_0 = 3\mathbf{Z}$; then in the same way we see that $\mathfrak{p}'$ determines $\mathfrak{p}^*$. q.e.d.

We shall summarize some of our results in the following theorem:

THEOREM 3. *The set of geometric points of* Proj $(A_{\mathbf{Z}}(\Gamma_2))$ *is a disjoint union of* Proj $(A(\Gamma_2))_{\mathbf{C}}$ *and similar projective varieties for all positive characteristics; more precisely* Proj $(A_{\mathbf{Z}}(\Gamma_2))$ *consists of the scheme of moduli over* $\mathbf{Z}$ *of 2-dimensional principally polarized abelian varieties and the scheme of moduli over* **Z** *of elliptic curves completed by including cubic curves with ordinary double point.*

In other words, if we denote by $\mathfrak{M}_n$ the scheme of moduli over **Z** of n-dimensional principally polarized abelian varieties for $n = 0, 1, 2$, then $(\mathfrak{M}_n)_{\mathbf{C}} = \Gamma_n\backslash\mathfrak{S}_n$ and Proj $(A_{\mathbf{Z}}(\Gamma_2))$ becomes the union of $\mathfrak{M}_2$, $\mathfrak{M}_1$, $\mathfrak{M}_0$. We recall that a curve of genus 2 over **C** with the dihedral group of order 12 or the octahedral group as its reduced group of automorphisms degenerates mod 2, 3 or mod 2; cf. [7], p. 648. Nevertheless its jacobian variety simply becomes the product of elliptic curves, in fact the product of a supersingular elliptic curve with itself. Therefore although the coordinates in Spec $(\mathbf{Z}[J]_{(J_{10})})$ are not all integral, we may say that such a jacobian variety determines an "integral point" of $\mathfrak{M}_2$; and our last remark is that $\mathfrak{M}_2$, not Spec $(\mathbf{Z}[J]_{(J_{10})})$, is the exact analogue of Spec $(\mathbf{Z}[j])$.

THE JOHNS HOPKINS UNIVERSITY

REFERENCES

1. W. L. Baily, Jr., "*Automorphic forms with integral Fourier coefficients,*" *Lecture Notes in Math.* **155**, pp. 1-8, Springer (1970).
2. H. Cartan, *Fonctions automorphes,* Séminaire E.N.S. (1957/1958).
3. M. Eichler, "*Zur Begründung der Theorie der automorphen Funktionen in mehreren Variablen,*" *Aequationes Math.* 3 (1969), pp. 93-111.
4. E. Freitag, *Zur Theorie der Modulformen zweiten Grades,* Göttingen Nachrichten 1965, Nr. 11.
5. A. Grothendieck, *Éléments de géométrie algébrique* I, Pub. math. I.H.E.S. (1960), No. 4; II, ibid. (1961), No. 8.
6. W. F. Hammond, "*On the graded ring of Siegel modular forms of genus two.*" *Am. J. Math.* 87 (1965), pp. 502-506.
7. J. Igusa, "*Arithmetic variety of moduli for genus two.*" *Ann. Math.* **72** (1960), pp. 612-649.
8. ______, "*On Siegel modular forms of genus two.*" *Am. J. Math.* **84** (1962), pp. 175-200; II, ibid. 86 (1964), pp. 392-412.
9. ______, "*Modular forms and projective invariants.*" *Amer. J. Math.* **89** (1967), pp. 817-855.
10. H. Maass, *Die Fourierkoeffizienten der Eisensteinreihen zweiten Grades,* Dan. Vid. Selsk. 34 (1964), Nr. 7; *Über* ____, ibid. **38** (1972), Nr. 14.

11. ______, "*Konstruktion von Spitzenformen beliebigen Grades mit Hilfe von Thetareihen,*" *Math. Ann.* **226** (1977), pp. 275-284.
12. ______, "*Lineare Relationen für die Fourierkoeffizienten einiger Modulformen zweiten Grades,*" *Math. Ann.* **232** (1978), pp. 163-175.
13. D. Mumford, "*On the equations defining abelian varieties* I-III," *Inventiones Math.* **1** (1966), pp. 287-354; **3** (1967), pp. 75-135, 215-244.
14. ______, *Varieties defined by quadratic equations,* Questioni sulle varieta algebraiche, Corsi dal C.L.M.E., Edizioni Cremonese, Roma (1969).
15. H. L. Resnikoff and R. L. Saldaña, "*Some properties of Fourier coefficients of Eisenstein series of degree two,*" *Crelles J. Math.* **265** (1974), pp. 90-109.
16. J.-P. Serre, *Congruences et formes modulaires,* Séminaire Bourbaki 1972, Exposé 416.
17. C. L. Siegel, "*Einführung in die Theorie der Modulkunktionen* n-*ten Grades,*" *Math. Ann.* **116** (1939), pp. 617-657; Collected Works II (1966), pp. 97-137.
18. ______, *Uber die Fourierschen Koeffizienten der Eisensteinschen Reihen, Dan. Vid. Selsk.* 34 (1964), Nr. 6; Collected Works III (1966), pp. 443-458.
19. H. P. F. Swinnerton-Dyer, "*On l-adic representations and congruences for coefficients of modular forms,*" *Lecture Notes in Math.* **350,** pp. 1-55, Springer (1973).
20. A. Weil, *Zum Beweis des Torellischen Satzes,* Göttingen Nachrichten 1957, Nr. 2.
21. E. Witt, "*Eine Identität zwischen Modulformen zweiten Grades,*" Abh. Math. Seminar Hamburg 14 (1941), pp. 323-337.

TOWARD THE INVERSION OF ABELIAN INTEGRALS, II

By GEORGE KEMPF*

This series of papers attempts to deal with one basic problem in the theory of algebraic curves. The problem revolves around the lack of any good description of the Picard bundles on Jacobian varieties. Some more information may be found in [1 and 2]. The contents of this paper are limited to determining the dimension of the space of sections of some Picard bundles and giving their interpretation in terms of sheaves on curves attached to the basic curve.

The basic problem is essentially non-abelian in the sense of Weil [5]. The seed planted in Weil's paper has grown through the years to bear beautiful fruits. One may hope that this work will also lead to greater human understanding. We must acknowledge our debt to our predecessors for having developed such powerful general tools as we return to the garden of algebraic geometry to try to unearth a more deeply buried harvest.

§1. **Generalities on abelian varieties.** Let X and Y be a pair of dual abelian varieties. Let $\mathcal{P}$ be a Poincaré sheaf on $X \times Y$ with a given isomorphism $\mathcal{P}|_{(0,0)} \approx k$. Let $\mathfrak{M}$ be an ample invertible sheaf on X. One may ask the question, "How does $\Gamma(X, \mathfrak{M})$ vary as $\mathfrak{M}$ varies in an algebraic family?"

More precisely, what is the sheaf $\pi_{Y*}(\pi_X{}^*\mathfrak{M} \otimes_{\mathcal{O}_{X\times Y}} \mathcal{P})$ on Y? Before I can answer this question, I must begin by proving

LEMMA 1.1. a) *$\pi_{Y*}(\pi_X{}^*\mathfrak{M} \otimes_{\mathcal{O}_{X\times Y}} \mathcal{P})$ is a locally free $\mathcal{O}_Y$-module of rank equal to the dimension of the vector space $\Gamma(X, \mathfrak{M})$.*

b) *The formation of the direct image via π_Y of the above sheaf commutes with base extension of Y. In particular, for any point y of P, we have a natural*

*Partly supported by NSF Grant MCS75-05578.

isomorphism, $\pi_{Y*}(\pi_X{}^*\mathcal{M} \otimes_{\mathcal{O}_{X\times Y}} \mathcal{P}) \otimes_{\mathcal{O}_Y} k(y) \approx \Gamma(X, \mathcal{M} \otimes_{\mathcal{O}_X} \mathcal{P}|_{X\times y})$.

c) $R\pi_{Y*}{}^i(\pi_X{}^*\mathcal{M} \otimes_{\mathcal{O}_{X\times Y}} \mathcal{P})$ *is zero if* i *is positive.*

d) $H^i(Y, \pi_{Y*}(\pi_X{}^*\mathcal{M} \otimes_{\mathcal{O}_{X\times Y}} \mathcal{P}))$ *is zero unless* $i = g$. *For* $i = g$, *it is isomorphic to the one-dimensional vector space* $\mathcal{E}xt_{\mathcal{O}_X}{}^g(k(0), \mathcal{M})$.

Proof. The reasoning here is typical of Mumford's discussion in his book [4]. I will recall some of it for the benefit of the reader. For any point y of Y, $\pi_X{}^*\mathcal{M} \otimes \mathcal{P}|_{X\times y} \approx \mathcal{M} \otimes_X \mathcal{P}|_{X\times y}$ is an ample sheaf on X as $\mathcal{M}$ is ample and $\mathcal{P}|_{X\times y}$ is algebraically equivalent to the trivial sheaf $\mathcal{O}_X$. Thus, for all y, $H^i(X, \mathcal{M} \otimes_{\mathcal{O}_X} \mathcal{P}|_{X\times y})$ is zero for positive i. From this, the theory of base extension implies that b) and c) must be true. Furthermore, it implies that $\pi_{Y*}(\pi_X{}^*\mathcal{M} \otimes_{\mathcal{O}_{X\times Y}} \mathcal{P})$ is a locally free coherent sheaf on Y. The rank of this sheaf is constant as Y is connected and may be evaluated by using the isomorphism of statement b) when $y = 0$. This proves a).

It remains to prove d). By c), a Leray spectral sequence for π_Y degenerates to an isomorphism

$$H^i(X \times Y, \pi_X{}^*\mathcal{M} \otimes_{\mathcal{O}_{X\times Y}} \mathcal{P}) \approx H^i(Y, \pi_{Y*}(\pi_X{}^*\mathcal{M} \otimes \mathcal{P}))$$

for all i. We may use the Leray spectral sequence for π_X to compute the cohomology of the above sheaf on $X \times Y$. In fact,

$$R\pi_{X*}{}^i(\pi_X{}^*\mathcal{M} \otimes_{\mathcal{O}_{X\times Y}} \mathcal{P}) \approx \mathcal{M} \otimes_{\mathcal{O}_X} R\pi_{X*}{}^i\mathcal{P}$$

by the projection formula. In his book, Mumford has shown that $R\pi_{X*}{}^i\mathcal{P} \approx \mathcal{E}xt_{\mathcal{O}_X}{}^i(k(0)), \mathcal{O}_X)$, which is zero unless $i \neq g$ and is a one-dimensional sky-scraper sheaf supported by 0 if $i = g$. Thus, the other Leray spectral sequence degenerates into isomorphisms $H^i(X \times Y, \pi_Y{}^*\mathcal{M} \otimes_{\mathcal{O}_{X\times Y}} \mathcal{P}) \approx \Gamma(X, \mathcal{E}xt_{\mathcal{O}_X}{}^i(k(0), \mathcal{M}))$. The statement d) follows by composing the two isomorphisms and using the calculation of the $\mathcal{E}xt$.

Q.E.D.

We may hope to determine the sheaf $\pi_{Y*}(\pi_X{}^*\mathcal{M} \otimes \mathcal{P})$ on Y by unraveling it; i.e. pulling it back via an isogenous covering of Y. We have an obvious candidate for this isogeny which is the morphism $\phi_{\mathcal{M}}: X \to Y$ given by $\mathcal{M}$. Recall that, for any point x of X, $\phi_{\mathcal{M}}(x)$ is the point of Y such that $T_x{}^*\mathcal{M} \otimes \mathcal{M}^{-1}$ is isomorphic to $\mathcal{P}|_{X\times\phi_{\mathcal{M}}(x)}$. Better yet, there is an isomorphism of $(\pi_1, \phi_{\mathcal{M}} \circ \pi_2)^*\mathcal{P}$ with $(\pi_1 + \pi_2)^*\mathcal{M} \otimes_{\mathcal{O}_{X\times X}} \pi_1^*\mathcal{M}^{\otimes -1} \otimes_{\mathcal{O}_{X\times X}} \pi_2^*\mathcal{M}^{\otimes -1} \otimes_k \mathcal{M}|_0$. The reason that I have added the last factor $\mathcal{M}|_0$ is so that the last sheaf will have a canonical trivialization when restricted to 0×0. As $(\pi_1, \phi_{\mathcal{M}} \circ \pi_2)^*\mathcal{P}$ inherits such a trivialization from $\mathcal{P}$'s, the above isomorphism will be unique as I will always use the one which respects the trivializations.

The unraveling is successful in this case as may be seen in

LEMMA 1.2. *There is a natural isomorphism.*

$$\phi_{\mathcal{M}}^*(\pi_{Y*}(\pi_X^*\mathcal{M} \otimes_{\mathcal{O}_{X\times Y}} \mathcal{P})) \approx \Gamma(X, \mathcal{M}) \otimes_k \mathcal{M}|_0 \otimes_k \mathcal{M}^{\otimes -1},$$

of sheaves on X.

Proof. By Lemma 1.1.b, we have a base extension isomorphism

$$\mathcal{H} \equiv \phi_{\mathcal{M}}^*(\pi_{Y*}(\pi_X^*\mathcal{M} \otimes_{\mathcal{O}_{X\times Y}} \mathcal{P})) \approx \pi_{2*}(\pi_1\mathcal{M} \otimes_{\mathcal{O}_{X\times X}} (\pi_1, \phi_{\mathcal{M}}(\pi_2))^* \mathcal{P}).$$

By the previous isomorphism, we have

$$\pi_1^*\mathcal{M} \otimes_{\mathcal{O}_{X\times X}} (\pi_1, \phi_{\mathcal{M}}(\pi_2))^*\mathcal{P} \approx \pi_1^*\mathcal{M} \otimes_{\mathcal{O}_{X\times X}} (\pi_1 + \pi_2)^*\mathcal{M}$$
$$\otimes_{\mathcal{O}_{X\times X}} \pi_1^*\mathcal{M}^{\otimes -1} \otimes_{\mathcal{O}_{X\times X}} \pi_2^*\mathcal{M}^{\otimes -1} \otimes_k \mathcal{M}|_0 \approx (\pi_1 + \pi_2)^*\mathcal{M}$$
$$\otimes_{\mathcal{O}_{X\times X}} \pi_2^*\mathcal{M}^{\otimes -1} \otimes_k \mathcal{M}|_0,$$

where the last isomorphism is just a cancellation of factors. Thus,

$$\mathcal{H} \approx \pi_{2*}[(\pi_1 + \pi_2)^*\mathcal{M} \otimes_{\mathcal{O}_{X\times X}} \pi_2^*\mathcal{M}^{\otimes -1} \otimes_k \mathcal{M}|_0],$$

which is isomorphic to

$$\pi_{2*}[(\pi_1 + \pi_2)^*\mathcal{M}] \otimes_{\mathcal{O}_X} \mathcal{M}^{\otimes -1} \otimes_k \mathcal{M}|_0$$

by the projection formula. Thus, to prove the lemma, it remains to give a natural isomorphism,

$$\pi_{2*}[(\pi_1 + \pi_2)^*\mathcal{M}] \approx \Gamma(X, \mathcal{M}) \otimes_k \mathcal{O}_X.$$

This last isomorphism is given by a shearing trick of Mumford.

We have a commutative diagram,

$$\begin{array}{ccc} S{:}X \times X & \xrightarrow{\approx} & X \times X, \\ & {\scriptstyle \pi_2}\searrow \quad \swarrow{\scriptstyle \pi_2} & \\ & X & \end{array}$$

where S is the shearing homomorphism, which sends (x_1, x_2) to $(x_1 + x_2, x_2)$. Now we have an obvious isomorphism $S^*(\pi_1{}^*\mathfrak{M}) \approx (\pi_1 + \pi_2)^*\mathfrak{M}$, which induces an isomorphism $\pi_{2*}(\pi_1{}^*\mathfrak{M}) \approx \pi_{2*}[(\pi_1 + \pi_2)^*\mathfrak{M}]$ as the diagram commutes. To finish, we need only note that the Künneth formula provides an isomorphism, $\Gamma(X, \mathfrak{M}) \otimes_k \mathcal{O}_X \approx \pi_{2*}(\pi_1{}^*\mathfrak{M})$.

Q.E.D.

The problem of determining the sheaf, $\pi_{Y*}(\pi_X{}^*\mathfrak{M} \otimes \mathcal{P})$, is now reduced to finding the natural action of the kernel H of $\phi_{\mathfrak{M}}$ on $\phi_{\mathfrak{M}}{}^*(\pi_{Y*}(\pi_X{}^*\mathfrak{M} \otimes \mathcal{P}))$ in terms of the isomorphism of Lemma 1.2. The essence of this question is measuring how unnatural the isomorphism of Lemma 1.2 is with respect to translation by H on X.

Recall that H itself only operates projectively on $\mathfrak{M}$ and its sections $\Gamma(X, \mathfrak{M})$, but Mumford has constructed his Theta group G, which acts linearly on $\mathfrak{M}$. In fact, G is a central extension of H by $\mathbf{G}_m$ (i.e., we have an exact sequence of group schemes, $\{1\} \to \mathbf{G}_m \to G \to H \to \{0\}$). Then, if G acts on X by translations through H, $\mathfrak{M}$ has an G-linearization, where $\mathbf{G}_m$ acts by scalar multiplication. This induces an G-representation on the sections $\Gamma(X, \mathfrak{M})$ where $\mathbf{G}_m$ again acts by scalar multiplication.

We will next determine the H-linearization of $\phi_{\mathfrak{M}}{}^*(\pi_{Y*}(\pi_X{}^*\mathfrak{M} \otimes \mathcal{P}))$ in terms of these G-actions in

LEMMA 1.3. *Let G act on $\Gamma(X, \mathfrak{M})$, $\mathfrak{M}$ and X as above. Then,*

a) *the sheaf $\Gamma(X, \mathfrak{M}) \otimes_k \mathfrak{M}^{\otimes -1}$ on X has a G-linearization given by letting G act diagonally on the tensor product via its given action on $\Gamma(X, \mathfrak{M})$ and the action on $\mathfrak{M}^{\otimes -1}$ which is contragradient to the given action on $\mathfrak{M}$.*
b) *The induced $\mathbf{G}_m$-linearization of $\Gamma(X, \mathfrak{M}) \otimes_k \mathfrak{M}^{\otimes -1}$ is trivial. Hence, $\Gamma(X, \mathfrak{M}) \otimes_k \mathfrak{M}^{\otimes -1}$ inherits an H-linearization where H acts on X by translation.*

c) *This H-linearization of* $\Gamma(X, \mathcal{M}) \otimes_k \mathcal{M}^{\otimes -1} \otimes_k \mathcal{M}|_0$ *given in* b) $\otimes$ *the trivial action on the lines* $\mathcal{M}|_0$ *is H-equivariantly isomorphic to the canonical H-linearization of* $\phi_{\mathcal{M}}{}^*(\pi_{Y*}(\pi_X{}^*\mathcal{M} \otimes \mathcal{P}))$ *via the isomorphism of Lemma* 1.2.

Proof. The only point to a) is that the usual constructions from the representations of groups carry over to G-linearizations. For the first part of b), note that t in $\mathbf{G}_m$ acts on $\Gamma(X, \mathcal{M})$ by multiplication by t and on $\mathcal{M}^{\otimes -1}$ by t^{-1}. Hence, the diagonal action on $\Gamma(X, \mathcal{M}) \otimes_k \mathcal{M}^{\otimes -1}$ is trivial. The rest of b) follows because $\mathbf{G}_m$ is the kernel of the epimorphism $G \to H$. The only serious question is about part c).

The essential point is to understand to what extent the isomorphism,

$$(\pi_1, \phi_{\mathcal{M}} \circ \pi_2)^*\mathcal{P} \approx ((\pi_1 + \pi_2)^*\mathcal{M} \otimes \pi_1{}^*\mathcal{M}^{\otimes -1} \otimes \pi_2{}^*\mathcal{M}^{\otimes -1} \otimes_k \mathcal{M}|_0,$$

is consistent with the action of $H \approx \operatorname{Ker}(\pi_1, \phi_{\mathcal{M}} \circ \pi_2)$ on the first sheaf. Here, we are dealing with sheaves on $X \times X$, where H is acting by translation through the second factor alone. We want to understand the transcription of the H-action via the isomorphism. At this point, it is easier to work with the action of G on the above sheaf because we can state the

Claim. The action of G on $(\pi_1 + \pi_2)^*\mathcal{M} \otimes \pi_1{}^*\mathcal{M}^{\otimes -1} \otimes \pi_2{}^*\mathcal{M}^{\otimes -1} \otimes_k \mathcal{M}|_0$ is the diagonal action where G acts on $(\pi_1 + \pi_2)^*\mathcal{M}$ by $(\pi_1 + \pi_2)^*$ of the G-linearization of $\mathcal{M}$, G acts on $\pi_2{}^*\mathcal{M}^{\otimes -1}$ by $(\pi_2)^*$ of the contragradient G-linearization of $\mathcal{M}^{\otimes -1}$, and G acts trivially on the two factors $\pi_1{}^*\mathcal{M}^{\otimes -1}$ and $\mathcal{M}|_0$.

Before I prove this claim, I want to explain how it implies the lemma. We now retrace the steps of Lemma 1.2. By the functoriality of base extension, the G-linearization of $\phi_{\mathcal{M}}{}^*(\pi_{Y*}(\pi_X{}^*\mathcal{M} \otimes_{\mathcal{O}_{X \times P}} \mathcal{P})) \equiv \mathcal{H}$ may be computed as π_{2*} of the G-linearization of $\pi_1{}^*\mathcal{M} \otimes_{\mathcal{O}_{X \times X}} (\pi_1, \phi_{\mathcal{M}}(\pi_2))^*\mathcal{P}$ as π_2 is a G-equivariant morphism for the two G-actions. As G only acts on the second factor of $X \times X$, the G-linearization of the above sheaf is the diagonal action where G acts trivially on the first factor $\pi_1{}^*\mathcal{M}$. By the claim, we know how to transcribe the action of G on the second factor. As the G-action on the two cancelled factors is trivial, we may conclude that the G-linearization of $\mathcal{H}$ is transcribed via the isomorphism with $\pi_{2*}[(\pi_1 + \pi_2)^*\mathcal{M}] \otimes_{\mathcal{O}_X} \mathcal{M}^{\otimes -1} \otimes_k \mathcal{M}|_0$ which is G-linearized diagonally by π_{2*} of $(\pi_1 + \pi_2)^*$ of the G-linearization of $\mathcal{M}$, the contragradient G-linearization of $\mathcal{M}^{\otimes -1}$ and trivial action on the last factor. Here, we have the G-equivariance of the projection formula.

To finish the lemma, we need to check the G-equivariance of the shearing isomorphism.

$$\pi_{2*}[(\pi_1 + \pi_2)^*\mathfrak{M}] \approx \Gamma(X, \mathfrak{M}) \otimes_k \mathcal{O}_X.$$

The lemma will be true if this is equivariant where G acts on $\Gamma(X, \mathfrak{M})$ via its representation and trivially on $\mathcal{O}_X$. Now, the shearing morphism S is G-equivariant. Thus, we may transcribe the G-linearization of $\pi_{2*}[(\pi_1 + \pi_2)^*\mathfrak{M}]$ to $\pi_{2*}(\pi_1{}^*\mathfrak{M})$ with π_{2*} of $\pi_1{}^*$ of the G-linearization on $\mathfrak{M}$. As the Kunneth formula respects G-linearization, the transcription of this action to $\Gamma(X, \mathfrak{M}) \otimes_k \mathcal{O}_X$ is $\Gamma(X, -)$ of the G-linearization of $\mathfrak{M}$ tensored with the trivial action on $\mathcal{O}_X$. This is what we wanted.

It remains to prove the claim. This claim is exactly how Mumford constructs the Poincaré sheaf on $X \times X/H$ in his development, but I will say some words for the benefit of the unbelievers. The point is that there are very limited possibilities for H-linearizations of $(\pi_1 \times \phi_{\mathfrak{M}} \circ \pi_2)^*\mathcal{P}$. If we have the known one, then any other can be gotten from it by twisting by a character χ of H as $\Gamma(X \times X, \mathcal{O}_{X\times X}) = k$. One may detect the character χ by restricting the H-linearization to $0 \times X$. The known H-linearization restricted to $0 \times X$ must be trivial since $\mathcal{P}|_{0\times Y}$ is trivial.

So to check formally the claim, we first note that the stated G-action is trivial on $\mathbf{G}_m$. Hence, it comes from an H-action. Then, note that the two G-actions on the factors cancel out on $0 \times X$, where $\pi_1 + \pi_2 = \pi_2$. By the above remarks, this is enough. Q.E.D.

§2. **Application of the preceeding theory to Picard sheaves.** Let $\mathcal{Q}$ be a branding sheaf on $C \times P_d$, where P_d is the dth Picard variety of a smooth complete curve C of genus g. Thus, $\{\mathcal{Q}|_{C\times x} | \ x \in P_d\}$ is a universal family of invertible sheaves of degree d on C parametrized by P_d. As such, the sheaf $\mathcal{Q}$ is only determined by this last property upto isomorphism and tensorization by $\pi_{P_d}{}^*\mathfrak{N}$, where $\mathfrak{N}$ is an invertible sheaf on P_d.

A branding sheaf $\mathcal{Q}$ is called normalized if, for any or all points c of C, $\mathcal{Q}|_{c\times P_d}$ is an algebraically trivial sheaf on P_d. In part I of this paper, we have classified all normalized brandings of P_d. We will next extend this classification to all brandings.

Let $\mathcal{L}$ be an invertible sheaf of degree one on C with p being the corresponding point in P_1. Let $\mathfrak{M}$ be any invertible sheaf on the Jacobian

J of C. Fix a Poincare sheaf $\mathcal{P}$ on $J \times J$ which expresses the autoduality of the Jacobian.

Consider the invertible sheaf $\mathcal{Q}(\mathcal{L}, \mathcal{M})$ on $C \times P_d$ given by $\pi_C{}^*\mathcal{L}^{\otimes d} \otimes_{\mathcal{O}_{C\times P_d}} ((\mathrm{I}_1 - p) \times T_{dp}{}^{-1})^* (\mathcal{P} \otimes_{\mathcal{O}_{C\times J}} \pi_J{}^*\mathcal{M})$, where $I_1 : C \to P_1$ is the universal abelian integral. We have previously seen in [1] that $\mathcal{Q}(\mathcal{L}, \mathcal{O}_{P_d})$ represent uniquely each isomorphism class of normalized brandings of P_d. More generally, we have

LEMMA 2.1 *Any branding $\mathcal{Q}$ of P_d is isomorphic to $\mathcal{Q}(\mathcal{L}, \mathcal{M})$ for some $\mathcal{L}$ and $\mathcal{M}$. Furthermore, $\mathcal{Q}(\mathcal{L}, \mathcal{M})$ and $\mathcal{Q}(\mathcal{L}', \mathcal{M}')$ are isomorphic if and only if* a) *$\mathcal{M}$ and $\mathcal{M}'$ are algebraically equivalent sheaves on J and* b) *$(p' - p) + d\phi_{\mathcal{M}}(p' - p)$ is the point in J corresponding to $\mathcal{M}' \otimes \mathcal{M}^{\otimes -1}$.*

Proof. As $\mathcal{Q}(\mathcal{L}, \mathcal{O}_{P_d}) \otimes_{P_d} T_{dp}{}^{-1*}(\mathcal{M}) \approx \mathcal{Q}(\mathcal{L}, \mathcal{M})$ and $T_{dp} : J \to P_d$ is an isomorphism, we see that the first statement is true by the above remarks. Furthermore, $\mathcal{Q}(\mathcal{L}, \mathcal{M})\,|_{c\times P_d}$ is algebraically equivalent to $T_{dp}{}^{-1}\mathcal{M}$ for any point c of C. From this, it follows that a) is true when $\mathcal{Q}(\mathcal{L}, \mathcal{M}) \approx \mathcal{Q}(\mathcal{L}', \mathcal{M}')$.

Thus, it remains to prove that $\mathcal{Q}(\mathcal{L}, \mathcal{M}) \approx \mathcal{Q}(\mathcal{L}', \mathcal{M}')$ if and only if b) is true when a) is true. Let h be the point of J corresponding to $\mathcal{M}' \otimes \mathcal{M}^{-1} \equiv \mathcal{H}$ on J. Let c be a point of C. Then, $\mathcal{Q}(\mathcal{L}, \mathcal{M}) \approx \mathcal{Q}(\mathcal{L}', \mathcal{M}') \Leftrightarrow \mathcal{Q}(\mathcal{L}, \mathcal{M})_{c\times P_d} \approx \mathcal{Q}(\mathcal{L}', \mathcal{M}')_{c\times P_d} \Leftrightarrow (T_{dp}{}^{-1})^*\{\mathcal{P}|_{\{c-p\}\times J} \otimes \mathcal{M}\} \approx (T_{d(p'-p)}{}^{-1})^*\{\mathcal{P}|_{\{c-p'\}\times J} \otimes \mathcal{M}'\} \Leftrightarrow \mathcal{P}|_{\{c-p\}\times J} \otimes \mathcal{M} \approx \mathcal{P}|_{\{c-p'\}\times J} \otimes (T_{d(p'-p)}{}^{-1})^*\mathcal{M} \otimes \mathcal{H} \Leftrightarrow \mathcal{P}|_{\{p'-p\}\times J} \approx \mathcal{H} \otimes (T_{d(p'-p)}{}^{-1})^*\mathcal{M} \otimes \mathcal{M}^{\otimes -1} \Leftrightarrow p' - p = h + \phi_{\mathcal{M}}(-d(p' - p)) \Leftrightarrow h = p' - p + d\phi_{\mathcal{M}}(p' - p) \Leftrightarrow$ b) is true. Q.E.D.

Next, we will recall the definition of the Picard sheaves on P_d. If $d > 2g - 2$, the sheaf $\pi_{P_d *}\mathcal{Q} \equiv \mathcal{W}(\mathcal{Q})$ is called the Picard sheaf associated to the branding $\mathcal{Q}$. Similarly, $R^1\pi_{P_d *}\mathcal{Q}$ is $\mathcal{W}(\mathcal{Q})$ when $d < 0$. The Picard sheaves are locally free coherent sheaves on P_d. We will give a method to determine their global sections in special cases.

A branding sheaf $\mathcal{Q}$ on P_d is lazy if, for any or all points c of C, $\mathcal{Q}|_{c\times P_d}$ is an ample invertible sheaf on P_d. Clearly, $\mathcal{Q}(\mathcal{L}, \mathcal{M})$ is a lazy branding of P_d if and only if $\mathcal{M}$ is an ample sheaf on the Jacobian J.

The method, that we have in mind, uses the divisorial correspondence principles of part I and section one of this paper. The preparatory step is

LEMMA 2.2 *Let $\mathcal{Q}$ be a lazy branding of a Picard variety P_d of de-*

gree $d > 2g - 2$. Then, we have a natural isomorphism $H^i(P_d, \mathcal{W}(\mathcal{Q})) \approx H^i(C, \pi_{C*}\mathcal{Q})$ for all i.

Proof. By Lemma 3.3 of Part I, we have a natural isomorphism $H^i(C \times P_d, \mathcal{Q}) \approx H^i(P_d, \mathcal{W}(\mathcal{Q}))$. Furthermore, if we can show that (*) $R^j\pi_{C*}\mathcal{Q}$ is zero for positive j, the Leray spectral sequence for π_C will give us natural isomorphisms, $H^i(C \times P_d, \mathcal{Q}) \approx H^i(C, \pi_{C*}\mathcal{Q})$. Thus, if (*) is true, we may find the isomorphisms of the lemma by composition.

It remains to verify the truth of (*). At the same time, we will compute $\pi_{C*}\mathcal{Q}$ in detail for later use. By Lemma 2.1, we may assume that $\mathcal{Q}$ has the form $\mathcal{Q}(\mathcal{L}, \mathfrak{M})$, where $\mathfrak{M}$ is an ample sheaf on J and $\mathcal{L}$ is an invertible sheaf of degree one on C.

By Lemma 1.1, we know about the direct images via π_1 of $\mathcal{P} \otimes \pi_2{}^*\mathfrak{M}$. From that lemma, we know that $\pi_{C*}\{((I_1 - p) \times \pi_J)^*(\mathcal{P} \otimes \pi_2{}^*\mathfrak{M})\} \approx (I_1 - p)^*\{\pi_{1*}(\mathcal{P} \otimes \pi_2{}^*\mathfrak{M})\}$ and that the higher direct images via π_C of $((I_1 - p) \times \pi_J)^*(\mathcal{P} \otimes \pi_2{}^*\mathfrak{M})$ are zero. As $T_{dp}{}^{-1}$ is an isomorphism, it follows that

$$\pi_{C*}\{((I_1 - p) \times T_{dp}{}^{-1})^*(\mathcal{P} \otimes \pi_2{}^*\mathfrak{M})\} \approx (I_1 - p)^*\{\pi_{1*}(\mathcal{P} \otimes \pi_2{}^*\mathfrak{M})\}$$

is the only non-zero (higher) direct image via π_C. Thus, by the projection formula and the definition of $\mathcal{Q}(\mathcal{L}, \mathfrak{M})$, $\pi_{C*}(\mathcal{Q}(\mathcal{L}, \mathfrak{M}))$ is naturally isomorphic to $(I_1 - p)^*\{\pi_{1*}(\mathcal{P} \otimes \pi_2{}^*\mathfrak{M})\} \otimes_{\mathcal{O}_C} \mathcal{L}^{\otimes d}$ and $R^j\pi_{C*}(\mathcal{Q}(\mathcal{L}, \mathfrak{M}))$ is zero for $j > 0$. This proves (*). Q.E.D.

Before we proceed further, I want to make some definitions. The type of branding $\mathcal{Q}$ of P_d is a class $M(\mathcal{Q})$ of algebraically equivalent invertible sheaves on J. By definition, $M(\mathcal{Q})$ contains all sheaves of the form $T_x{}^*(\mathcal{Q}|_{c \times P_d})$ for any point c of C where $T_x: J \to P_d$ is translation by an arbitrary element x of P_d. Clearly, these sheaves are algebraically equivalent as C and P_d are connected.

We will also use the homomorphism $\Psi_{\mathcal{Q}}: J \to J$ where $\Psi_{\mathcal{Q}} = \phi_{\mathfrak{M}}$ for any $\mathfrak{M}$ in $M(\mathcal{Q})$. This homomorphism is well-defined because $\phi_{\mathfrak{M}}$ only depends on the algebraic equivalence class of $\mathfrak{M}$. Clearly, the type of $\mathcal{Q}(\mathcal{L}, \mathfrak{M})$ is the class of $\mathfrak{M}$ and $\Psi_{\mathcal{Q}(\mathcal{L}, \mathfrak{M})} = \phi_{\mathfrak{M}}$.

We are now in a position to state the main result of this section, which is

THEOREM 2.3 *Let* $\mathcal{Q}(\mathcal{L}, \mathcal{M})$ *be a lazy branding of* P_d *where* $d > 2g - 2$. *Then,*

a) *there is a natural isomorphism*

$$H^i(P_d, \mathcal{W}(\mathcal{Q}(\mathcal{L}, \mathcal{M}))) \approx H^i(C, \mathcal{L}^{\otimes d} \otimes_{\mathcal{O}_C} (I_1 - p)^* \pi_{1*}(\mathcal{P} \otimes \pi_2^* \mathcal{M})).$$

b) *Let* $\tilde{C} = \Psi_{\mathcal{Q}}^{-1}(I_1(C) - p)$ *and* $r \equiv (I_1 - p)^{-1} \circ \Psi_{\mathcal{Q}}: \tilde{C} \to C$. *Then,* $\tilde{C}$ *is a Galois covering of* C *with Galois group scheme* $H \equiv \ker \Psi_{\mathcal{Q}}$. *There is a natural H-isomorphism.*

$$r^*\{(I_1 - p)^*(\pi_{1*}(\mathcal{P} \otimes \pi_2^* \mathcal{M}))\} \approx \Gamma(J, \mathcal{M}) \otimes_k (\mathcal{M}|_{\tilde{C}})^{\otimes -1} \otimes_k \mathcal{M}|_0,$$

where the H-action on the latter term is the H-action on $\Gamma(J, \mathcal{M}) \otimes_k \mathcal{M}^{\otimes -1}$ *of Lemma* 1.3 *restricted to the H-equivariant closed subscheme* $\tilde{C}$ *of J.*

Proof. Part a) follows from Lemma 2.2 and the computation in its proof. For part b), have natural isomorphisms,

$$\begin{aligned} r^*\{(I_1 - p)^*(\pi_{1*}(\mathcal{P} \otimes \pi_2^* \mathcal{M}))\} &\approx (\Psi_{\mathcal{Q}}|_{\tilde{C}})^*(\pi_{1*}(\mathcal{P} \otimes \pi_2^* \mathcal{M})) \\ &\approx \Psi_{\mathcal{Q}}^*(\pi_{1*}(\mathcal{P} \otimes \pi_2^* \mathcal{M}))|_{\tilde{C}}. \end{aligned}$$

By Lemma 1.3.c, we have a H-isomorphism between

$$\Psi_{\mathcal{Q}}^*(\pi_{1*}(\mathcal{P} \otimes \pi_2^* \mathcal{M})) \quad \text{and} \quad \Gamma(J, \mathcal{M}) \otimes_k \mathcal{M}^{\otimes -1} \otimes_k \mathcal{M}|_0.$$

This is enough to complete the proof. Q.E.D.

COROLLARY 2.4. *In the situation of Theorem* 2.3,

a) *there is a natural isomorphism between* $\Gamma(P_d, \mathcal{W}(\mathcal{Q}(\mathcal{L}, \mathcal{M})))$ *and the H-invariants in* $\Gamma(\tilde{C}, \Psi_{\mathcal{Q}}|_{\tilde{C}}^* \mathcal{L}^{\otimes d} \otimes_{\mathcal{O}_{\tilde{C}}} (\mathcal{M}|_{\tilde{C}})^{\otimes -1} \otimes_k \Gamma(J, \mathcal{M}) \otimes_k \mathcal{M}|_0)$.

b) $H^j(P_d, \mathcal{W}(\mathcal{Q}(\mathcal{L}, \mathcal{M})))$ *is zero if* $j > 1$.

c) *If the degree of* $\ker \Psi_{\mathcal{Q}}$ *is prime to the characteristic of k, the statement* a) *is true when* Γ *is replaced by* H^1.

Proof. b) follows from the theorem a) because $H^j(C, \mathcal{F})$ is zero for any quasi-coherent sheaf $\mathcal{F}$ on a curve C. For a), $r^*\mathcal{F}$ has a H-linearization and $\mathfrak{F} \approx H$-invariants in $r_*r^*\mathfrak{F}$. Thus, $\Gamma(C, \mathfrak{F}) \approx H$-invariants in $\Gamma(C, r_*r^*\mathfrak{F}) \approx \Gamma(\tilde{C}, r^*\mathfrak{F})$. The statement a) is a special case.

For c), the degree of ker $\Psi_{\mathfrak{Q}} \nmid$ char (k) if and only if H is a linearly reductive (in particular, reduced) group scheme. Thus, under the assumption of c), we have a canonical H-eigenspace decomposition $r_*r^*\mathfrak{F} = \bigotimes \mathcal{G}_X$, where $\mathcal{G}_X$ are the subsheaves of χ-eigenvectors in $r_*r^*\mathfrak{F}$. By functoriality, $H^i(C, r_*r^*\mathfrak{F})$ decomposes into disjoint eigenspaces $H^i(C, \mathcal{G}_X)$ for all i. As $\mathcal{G}_1 = \mathfrak{F}$, $H^i(C, \mathfrak{F})$ is exactly the H-invariants in $H^i(C, r_*r^*\mathfrak{F})$. As r is a finite morphism, $H^i(C, r_*r^*\mathfrak{F})$ is naturally isomorphic to $H^i(\tilde{C}, r^*\mathfrak{F})$. Hence, $H^i(C, \mathfrak{F}) \approx H$-invariants in $H^i(\tilde{C}, r^*\mathfrak{F})$. The statement c) is a special case. Q.E.D.

Before we go onto the next section, we may note that the Riemann-Hurwitz formula implies the

LEMMA 2.5. *The morphism $r: \tilde{C} \to C$ has degree equal to the rank of $H \equiv \dim_k \Gamma(H, \mathcal{O}_H)$. If the degree of $\Psi_{\mathfrak{Q}}$ is prime to the characteristic of k (more generally, if H is reduced), then genus $\tilde{C} = 1 +$ (rank H) $(g - 1)$ and r is a smooth morphism.*

§3. **Standard Picard sheaves.** A branding sheaf $\mathfrak{Q}$ on $C \times P_d$ is standard of altitude n if the homomorphism $\Psi_{\mathfrak{Q}}: J \to J$ is the homomorphism $n \cdot 1_J$. This means that the class $M(\mathfrak{Q})$ contains $\mathcal{O}_J(n\theta')$, where Θ' is a translate of the theta divisor.

Thus, a standard branding sheaf $\mathfrak{Q}$ is lazy if and only if its altitude is positive. A normalized branding in the sense of part I is a standard branding of altitude zero.

One does not expect that there are non-standard brandings for a "general" curve C. I will only deal with standard brandings. In this section, we will see how to take advantage of the standardness to make the material of the last section more definite.

Let's fix an invertible sheaf $\mathfrak{L}$ of degree one on C. Let p be the corresponding point of P_1. Let n be a non-negative integer. Define $\tilde{C}_n$ to be the curve $(n \cdot 1_J)^{-1}(I_1(C) - p)$ in the Jacobian. Let $r_n: \tilde{C}_n \to C$ be the covering morphism, $(I_1 - p)^{-1} \circ (n \cdot 1_J)$.

Let $J_n = \text{Ker}\,(n \cdot 1_J)$ be the n-torsion subgroup scheme of the Jacobian. Then, r_n is a Galois covering with Galois group scheme J_n, which is a finite abelian group scheme of rank n^{2g}. If char $(k) \nmid n$, then $n \cdot 1_J: J \to J$ is a smooth finite morphism of degree n^{2g} and J_n is a reduced scheme with n^{2g} points.

We may clarify the meaning of Lemma 2.5 in our present situation.

LEMMA 3.1 *The morphism $r_n:\tilde{C} \to C$ has degree n^{2g}. If* char (k) $\nmid n$, *r_n is a smooth morphism and*

$$\text{genus}\,(\tilde{C}_n) = 1 + n^{2g}(g - 1).$$

We next recall a well-known fact [6 and 4].

PROPOSITION 3.2. *Let $\mathfrak{M}$ be an invertible sheaf on J. For any integer n, $(n \cdot 1_J)^*\mathfrak{M}$ is algebraically equivalent to $\mathfrak{M}^{\otimes n^2}$.*

Another useful fact [6] is

PROPOSITION 3.3. *Let Θ' be a divisor on J, which is contained in the translation equivalence class of the theta divisor. Then, the total intersection number $[\Theta' : I_1(C) - p] = g$.*

From these two results, we may conclude the following.

COROLLARY 3.4. *Let $\mathfrak{M}$ be an invertible sheaf on the Jacobian J, which is algebraically equivalent to $\mathcal{O}_J(m\Theta')$ for any integer m. For any non-zero integer n,*

$$\deg\,(\mathfrak{M}\,|\,_{\tilde{C}_n}) = m \cdot g \cdot n^{2g-2}.$$

Proof. By the projection formula, $\deg\,((r_n{}^*\mathfrak{M})\,|\,_{\tilde{C}_n}) =$ (degree of r_n)$\cdot \deg\,(\mathfrak{M}|_C)$, which equals $n^{2g} \cdot \deg\,(\mathfrak{M}|_C)$ by Lemma 3.1. By Proposition 3.3, we know that $\deg\,(\mathfrak{M}|_C) = m \deg\,(\mathcal{O}_J(\Theta')|_C) = m \cdot g$. Thus,

$$\deg\,((r_n{}^*\mathfrak{M})\,|\,_{\tilde{C}_n}) = m \cdot g \cdot n^{2g}.$$

On the other hand, by Proposition 3.2, we have $\deg\,((r_n{}^*\mathfrak{M})|_{\tilde{C}_n}) = \deg\,(\mathfrak{M}^{\otimes n^2}|_{\tilde{C}_n}) = n^2 \deg\,(\mathfrak{M}|_{\tilde{C}_n})$. Hence, $n^2 \deg\,(\mathfrak{M}|_{\tilde{C}_n}) = m \cdot g \cdot n^{2g}$, which gives the result. Q.E.D.

Let $\mathfrak{M} = \mathcal{O}_J(n\Theta')$ be an invertible sheaf on J, where Θ' is contained in the translation class of the theta divisor. Then, the invertible sheaf $\mathcal{Q}(\mathcal{L}, \mathfrak{M})$ on $C \times P_d$ is a standard branding of altitude n. Furthermore, we have seen in the last section that all standard brandings of altitude n have this form up to isomorphism.

Assume that the altitude n is positive. Then, $\mathcal{Q}(\mathcal{L}, \mathfrak{M})$ is lazy. We will next study the important sheaf, $(n \cdot 1_J|_{\tilde{C}_n})^*\mathcal{L}^{\otimes d} \otimes_{\mathcal{O}_{\tilde{C}_n}} (\mathfrak{M}|_{\tilde{C}_n})^{\otimes -1}$.

LEMMA 3.5. *With the above notations,*

a) *the degree of the invertible sheaf* $(n\cdot 1_J|_{\tilde{C}_n})^*\mathfrak{L}^{\otimes d}\otimes_{\mathcal{O}_{\tilde{C}_n}}(\mathfrak{M}|_{\tilde{C}_n})^{\otimes -1}$ *is* $(d\cdot n - g)\cdot n^{2g-1}$.

b) *Assume that char* $(k)\nmid n$. *The degree of this sheaf* > 2 *genus* $\tilde{C}_n - 2$ *if and only if* $n(d-2g+2) > g$.

Proof. We have $\deg(\mathfrak{M}|_{\tilde{C}_n}) = g\cdot n^{2g-1}$ by Corollary 3.4. As the sheaf $\mathfrak{L}$ on C has degree one,

$$\deg((n\cdot 1_J|_{\tilde{C}_n})^*\mathfrak{L}^{\otimes d}) = d\cdot \deg(n\cdot 1_J) = d\cdot n^{2g}.$$

Therefore, the given sheaf has degree $= dn^{2g} - gn^{2g-1} = (dn-g)n^{2g-1}$. This proves a).

For b), note that 2 genus $(\tilde{C}_n) - 2 = 2(1 + n^{2g}(g-1)) - 2 = n^{2g}(2g-2)$ using Lemma 3.1. Thus,

$$\begin{aligned}(dn-g)n^{2g-1} > n^{2g}(2g-2) &\Leftrightarrow dn - g > n(2g-2)\\ &\Leftrightarrow (d-2g+2)n > g\end{aligned}$$

This proves b). Q.E.D.

We will next recall another well-known fact.

LEMMA 3.6. *For any positive integer n,* $\dim_k \Gamma(J, \mathcal{O}_J(n\Theta')) = n^g$ *if* Θ' *is contained in the translation class of the theta divisor.*

Proof. As $n > 0$, $n\Theta'$ is an ample divisor on J. Thus, from [4], we know that

$$\dim_k \Gamma(J, \mathcal{O}_J(n\Theta')) = (\deg \phi_{\theta'})^{1/2}\cdot n^g.$$

As $\phi_{\theta'}$ has degree one by autoduality, the Lemma follows. Q.E.D.

For our lazy standard branding $\mathfrak{Q}(\mathfrak{L}, \mathfrak{M})$, we may now be more explicit about the sheaves involved in the statement of Theorem 2.3.

THEOREM 3.7. *Let* $\mathfrak{Q}(\mathfrak{L},\mathfrak{M})$ *be a standard lazy branding of* P_d *of altitude n. Then,*

a) *the coherent sheaf* $\mathfrak{L}^{\otimes d}\otimes_{\mathcal{O}_C}(I_1 - p)^*\pi_{1*}(\mathcal{P}\otimes\pi_2^*\mathfrak{M})$ *on C is locally free of rank* n^g *and has degree* $(d\cdot n - g)n^{g-1}$.

b) *The pull-back via* r_n *of this sheaf is isomorphic to* $(n\cdot 1_J|_{\tilde{C}_n})^*$

$\mathfrak{L}^{\otimes d} \otimes_{\mathcal{O}_{\tilde{C}_n}} \Gamma(J, \mathfrak{M}) \otimes_k (\mathfrak{M}|_{\tilde{C}_n})^{\otimes -1} \otimes_k \mathfrak{M}|_0$, *which is locally free of rank* n^g *and has degree* $(dn - g)n^{3g-1}$.

Assume that char $(k) \nmid n$ *and* $n(d - 2g + 2) > g$. *Then,*

c) *the first cohomology groups of both these sheaves are zero, and*
d) $\dim_k \Gamma(C, \mathfrak{L}^{\otimes d} \otimes_{\mathcal{O}_C} (I_1 - p)^* \pi_{1*}(\mathcal{P} \otimes \pi_2^* \mathfrak{M})) = n^{g-1}\{n(d - g + 1) - g\}$ *and* $\dim_k \Gamma(\tilde{C}_n, (n \cdot 1_J|_{\tilde{C}_n})^* \mathfrak{L}^{\otimes d} \otimes_{\mathcal{O}_{\tilde{C}_n}} \Gamma(J, \mathfrak{M}) \otimes_k (\mathfrak{M}|_{\tilde{C}_n})^{\otimes -1} \otimes_k \mathfrak{M}|_0) = n^{3g-1}\{n(d - g + 1) - g\}$.

Proof. The statement b) implies the truth of a), because pulling back via r_n leaves the rank unchanged and multiplies the degree by deg $r_n = n^{2g}$ (see Lemma 3.1). To prove b), note that the rank of the given sheaf on $\tilde{C}_n$ equals the $\dim_k \Gamma(J, \mathcal{O}_J(n\Theta'))$ as $\mathfrak{M} = \mathcal{O}_J(n\Theta')$. By Lemma 3.6, we know this number is n^g. The degree of this sheaf equals $\dim_k \Gamma(J, \mathcal{O}_J(n\Theta'))$ times $\deg((n \cdot 1_J|_{\tilde{C}_n})^* \mathfrak{L}^{\otimes d} \otimes_{\mathcal{O}_{\tilde{C}_n}} \mathfrak{M}|_{\tilde{C}_n})$. Thus, its degree is given by $n^g\{(d \cdot n - g) \cdot n^{2g-1}\}$ by Lemma 3.5.a). The desired results follow directly.

To prove c), we know by Lemma 3.5.b) that the degree of the invertible sheaf $(n \cdot 1_J|_{\tilde{C}_n})^* \mathfrak{L}^{\otimes d} \otimes_{\tilde{C}_n} \mathfrak{M}|_{\tilde{C}_n}$ is large enough so that its first cohomology group must vanish. Therefore, the first cohomology group of the sheaf in part b) must be zero as it is the direct sum of several copies of this invertible sheaf. By Corollary 2.4.c, this implies that the first cohomology group of $\mathfrak{L}^{\otimes d} \otimes_{\mathcal{O}_C} (I_1 - p)^* \pi_{1*}(\mathcal{P} \otimes \pi_2^* \mathfrak{M})$ must also be zero. This proves c).

Part d) follows from the other parts and the fact that $\dim \Gamma(C, \mathfrak{F}) - \dim H^1(C, \mathfrak{F}) = \deg(\mathfrak{F}) + \mathrm{rank}(\mathfrak{F})(-g + 1)$ for any coherent sheaf $\mathfrak{F}$ on a curve C of genus g. Q.E.D.

We will next recapitulate what we have proven so far about the Picard sheaves associated to standard lazy brandings.

THEOREM 3.8 Let $\mathfrak{Q}(\mathfrak{L}, \mathfrak{M})$ *be a standard lazy branding of altitude* n *of a Picard variety* P_d, *where* $d > 2g - 2$. *Then,*

a) *the rank of the Picard sheaf* $\mathfrak{W}(\mathfrak{Q}(\mathfrak{L}, \mathfrak{M}))$ *is* $d - g + 1$.
b) *we have natural isomorphisms for each* i, $H^i(P_d, \mathfrak{W}(\mathfrak{Q}(\mathfrak{L}, \mathfrak{M}))) \approx H^i(C, \mathfrak{L}^{\otimes d} \otimes_{\mathcal{O}_C} (I_1 - p)^* \pi_{1*}(\mathcal{P} \otimes \pi_2^* \mathfrak{M}))$.
c) $r_n^*(\mathfrak{L}^{\otimes d} \otimes_{\mathcal{O}_C} (I_1 - p)^* \pi_{1*}(\mathcal{P} \otimes \pi_2^* \mathfrak{M}))$ *is naturally isomorphic to the sheaf* $(n \cdot 1_J|_{\tilde{C}_n})^* \mathfrak{L}^{\otimes d} \otimes_{\mathcal{O}_{\tilde{C}_n}} \Gamma(J, \mathfrak{M}) \otimes_k (\mathfrak{M}|_{\tilde{C}_n})^{\otimes -1} \otimes_k \mathfrak{M}|_0$, *where we know the action of the Galois group scheme* J_n *of the covering* $r_n: \tilde{C}_n \to C$ *of Theorem* 2.3.b.

Assume that $n(d-2g+2) > g$ *and* $\operatorname{char}(k) \nmid n$. *Then,*

d) *the higher cohomology groups of the Picard sheaf* $\mathcal{W}(\mathcal{Q}(\mathcal{L}, \mathfrak{M}))$ *are zero.*

e) $\dim_k \Gamma(P_d, \mathcal{W}(\mathcal{Q}(\mathcal{L}, \mathfrak{M}))) = (d-g+1)n^g - gn^{g-1}$.

f) *The space of sections* $\Gamma(P_d, \mathcal{W}(\mathcal{Q}(\mathcal{L}, \mathfrak{M})))$ *is isomorphic to the invariants in* $\Gamma(J, \mathfrak{M}) \otimes_k \mathfrak{M}|_0 \otimes_k \Gamma(\tilde{C}_n, (n\cdot 1_J|_{\tilde{C}_n})^*\mathcal{L}^{\otimes d} \otimes_{\mathcal{O}\tilde{C}_n} (\mathfrak{M}|_{\tilde{C}_n})^{\otimes -1})$ *under the action of the finite group* J_n.

Proof. The statement a) is a special case of Lemma 3.1.1.b of part I of this paper. The statements b) and c) are special cases of Theorem 2.3.a) and b). The point d) follows from the isomorphism of part b) and Theorem 3.7.c. Similarly, e) follows from Theorem 3.7.d. The idea for the proof of f) already has occurred in Corollary 2.4.a. Q.E.D.

§4 **Reinterpretation of the Previous General Theory.** In the previous sections, we have interpreted the space of sections of a Picard bundle in terms of either the sections of a locally free sheaf on a curve or the invariants under a group of the sections of different locally free sheaves. In this section, we will begin to see how to describe this space more simply as the sections of an invertible sheaf on an appropriate curve.

I will first review the necessary background from the general theory of abelian varieties. To simplify the discussion, I will avoid the degenerate cases, where the relatively easy theory of Mumford [3] does not apply. One could prove more general theorems but that would require the development here of Mumford's unpublished generalizations.

Let $\mathfrak{M}$ be an ample invertible sheaf on an abelian variety X. Assume that the characteristic of k does not divide degree $\phi_{\mathfrak{M}} = \operatorname{rank} H$, where $H = \operatorname{Ker} \phi_{\mathfrak{M}}$ as before. We have the theta group, $1 \to \mathbf{G}_m \to G \xrightarrow{\rho} H \to 0$, of $\mathfrak{M}$. Let V be a G-module. By restriction, we have an induced $\mathbf{G}_m$-module structure on V. The G-module V is said to have weight m if $V \approx \oplus k\cdot t^m$ as a $\mathbf{G}_m$-module where t is the coordinate function on $\mathbf{G}_m$.

One basic fact, which is equivalent to the non-degeneracy of the theta group G (i.e. $\mathbf{G}_m$ is the center of G), is the following

THEOREM 4.1 [3]. *Let W be an non-zero irreducible G-module of weight m, where $m = \pm 1$. Then,*

a) $(\dim_k W)^2 = \operatorname{rank} H$.

b) *Any other G-module V of weight k is isomorphic as a G-module to a direct sum $\oplus$ W of several copies of W.*

In our case, we already know an example of such a G-module W. In fact, we have

THEOREM 4.2 [3]. *The space* $\Gamma(X, \mathfrak{M})$ *with its natural* G*-module structure is a non-zero irreducible* G*-module of weight* 1. *Therefore, the contragradiant* G*-module* $\Gamma(X, \mathfrak{M})^\wedge$ *is a non-zero irreducible* G*-module of weight* -1.

The next result will describe the internal structure of these irreducible representations better. Let K be a fixed maximal subgroup of H such that $K' = \rho^{-1}(K)$ is an abelian group. A character $\chi: K' \to \mathbf{G}_m$ has weight m if its restriction to $\mathbf{G}_m$ is the m-power homomorphism. With this notation, we have

THEOREM 4.3 [3]. *If* W *is an irreducible* G*-module of weight* m, *where* $m = \pm 1$, *then* V *restricted to* K' *is isomorphic to the sum of eigenspaces* $\oplus V_\chi$ *where* χ *runs through the characters of* K' *of weight* m. *Furthermore, each eigenspace* V_χ *is a one-dimensional* k*-vector space.*

These theorems have the following corollary, which will be useful.

COROLLARY 4.4. *Let* V *be a* G*-module of weight* -1. *The diagonal action of* G *on* $\Gamma(X, \mathfrak{M}) \otimes_k V$ *has weight* 0 *and induces an action of* H.

a) *The space of* K*-invariants* $(\Gamma(X, \mathfrak{M}) \otimes_k V)^K$ *is the direct sum* $\oplus (\Gamma(X, \mathfrak{M})_\chi \otimes_k V_{\chi^{-1}})$, *where* χ *runs through the characters of* K' *of weight* $+1$.

b) *Its subspace* $(\Gamma(X, \mathfrak{M}) \otimes_k V)^H$ *projects isomorphically onto any of the summands* $\Gamma(X, \mathfrak{M})_\chi \otimes_k V_{\chi^{-1}}$ *of* $(\Gamma(X, \mathfrak{M}) \otimes_k V)^K$.

Proof. By Theorem 4.1, we know that $\Gamma(X, \mathfrak{M}) \equiv W_1$ is an irreducible G-module of weight $+1$. Thus the contragradient representation W_{-1} is an irreducible G-module of weight -1. As $V \approx \oplus W_{-1}$ by Theorem 4.1, we need only prove this corollary when $V = W_{-1}$. Furthermore, we know that $(W_1 \otimes W_{-1})^G = (W_1 \otimes W_{-1})^H$ is one-dimensional, or, better yet, spanned by the tensor representing the identity of W_1, by Schur's lemma.

Now, to prove the statement a), one simply applies Theorem 4.3 to W_1 and W_{-1}. This will give the whole K-eigenspace decomposition of $W_1 \otimes W_{-1}$, which is more information than needed in a). Also, as each eigenspace $W_{1,\chi}$ is one dimensional by Theorem 4.3, the projection $(W_1 \otimes W_{-1})^H \to W_{1,\chi} \otimes W_{-1,\chi^{-1}}$ is a non-zero linear mapping between two lines. Therefore, the projection is an isomorphism and the statement b) has been proven. Q.E.D.

Let $\mathcal{P}$ be a Poincaré sheaf on $X \times Y$ as in section 1. We want to use the methods of this section to improve our picture of the sheaf $\pi_{Y*}(\pi_X{}^*\mathfrak{M} \otimes_{\mathcal{O}_{X\times Y}} \mathcal{P})$ on the dual abelian variety Y.

Let $f:S \to Y$ be any morphism of schemes and let $\mathfrak{F}$ be a quasi-coherent $\mathcal{O}_S$-module. I want to determine a natural expression for the $\Gamma(S, \mathcal{O}_S)$-module,

$$\Gamma(S, f^*(\pi_{Y*}(\pi_X{}^*\mathfrak{M} \otimes_{\mathcal{O}_{X\times Y}} \mathcal{P})) \otimes_{\mathcal{O}_S} \mathfrak{F}) \equiv D(f, \mathfrak{F}).$$

By Lemma 1.1.b), $D(f, \mathfrak{F})$ is naturally isomorphic to

$$\Gamma(S, \pi_{S*}(\pi_X{}^*\mathfrak{M} \otimes_{\mathcal{O}_{X\times S}} (1_X \times f)^*\mathcal{P}) \otimes_{\mathcal{O}_S} \mathfrak{F}).$$

Consider the Cartesian square

$$\begin{array}{ccc} \bar{f}:\bar{S} & \longrightarrow & X \\ \big\downarrow & & \big\downarrow \phi_{\mathfrak{M}} \\ f:S & \longrightarrow & Y \end{array}$$

where $\phi_{\mathfrak{M}}$ is the classifying morphism for $\mathfrak{M}$. Then the morphism $g:\bar{S} \to S$ is a étale Galois covering with Galois group $H = \mathrm{Ker}\,(\phi_{\mathfrak{M}})$. We intend to compute $D(f, \mathfrak{F})$ by the formula $D(f, \mathfrak{F}) = [D(\phi_{\mathfrak{M}} \circ \bar{f}, g^*\mathfrak{F})]^H$.

Recall that in Section 1 we actually computed the H-action on $\phi_{\mathfrak{M}}{}^*(\pi_{Y*}(\pi_X{}^*\mathfrak{M} \otimes_{\mathcal{O}_{X\times Y}} \mathcal{P}))$. The essential result is contained in Lemma 1.3.c, which identifies this H-linearized sheaf with $\Gamma(X, \mathfrak{M}) \otimes_k \mathfrak{M}^{\otimes -1} \otimes_k \mathfrak{M}|_0$ with a rather obvious H-action. We can now determine $D(\phi_{\mathfrak{M}} \circ \bar{f}, g^*\mathfrak{F})$ after we give the appropriate action. We have a natural H-action on $g^*\mathfrak{F}$. As $H = G/\mathbf{G}_m$, we get a weight-zero G-action on $g^*\mathfrak{F}$ which we may tensor with the contragradient G-action of $\mathfrak{M}^{\otimes -1}$ to get the obvious weight minus one G-action on the global sections $\Gamma(\bar{S}, \bar{f}^*\mathfrak{M}^{\otimes -1} \otimes_{\mathcal{O}_{\bar{S}}} g^*\mathfrak{F})$. With this understood, we have seen the truth of

LEMMA 4.5. *The H-module $D(\phi_{\mathfrak{M}} \circ \bar{f}, g^*\mathfrak{F})$ is naturally isomorphic to the H-module induced by the G-action on*

$$\Gamma(X, \mathfrak{M}) \otimes_k \Gamma(\bar{S}, f^*\mathfrak{M}^{\otimes -1} \otimes_{\mathcal{O}_{\bar{S}}} g^*\mathfrak{F}) \otimes_k \mathfrak{M}|_0$$

which is the tensor product of the G-action on each factor. (*Here, you should recall that G acts trivially on the line* $\mathfrak{M}|_0$).

Now, we are ready to apply our previous corollary to the situation of this lemma. The result is

COROLLARY 4.6. *Given any character* χ *of* K' *of weight one. The* $\Gamma(S, \mathcal{O}_S)$*-module* $D(f, \mathfrak{F})$ *is naturally isomorphic to*

$$\Gamma(X, \mathfrak{M})_\chi \otimes_k [\Gamma(\bar{S}, \bar{f}^*\mathfrak{M}^{\otimes -1} \otimes_{\mathcal{O}_{\bar{S}}} g^*\mathfrak{F})]_{\chi^{-1}} \otimes_k \mathfrak{M}|_0.$$

Next, we want to recall the necessary descent theory to give a more geometric interpretation of this result. The line $\Gamma(X, \mathfrak{M})_\chi$ is isomorphic to the space of sections of an invertible sheaf, say $\mathfrak{M}_\chi$, on the quotient X/K. In fact, for any open subset U of X/K, $\mathfrak{M}_\chi$ may be defined by the formula,

$$\Gamma(U, \mathfrak{M}_\chi) \equiv \Gamma(\pi^{-1}U, \mathfrak{M})_\chi,$$

where π denotes the quotient morphism $X \to X/K$.

Analogously, we may define an invertible sheaf $\mathcal{K}(f, \chi)$ on $\bar{S}/K$ by setting $\mathcal{K}(f, \chi) = (\bar{f}^*\mathfrak{M}^{\otimes -1})_{\chi^{-1}}$. In other words, $\mathcal{K}(f, \chi)$ is the pull-back of $(\mathfrak{M}_\chi)^{\otimes -1}$ via $\tilde{S}/K \to X/K$. Now, we can state what we have proven, in a definitive form.

PROPOSITION 4.7. *For any character* χ *of* K' *which has weight one, the* $\Gamma(S, \mathcal{O}_S)$*-module* $D(f, \mathfrak{F})$ *is naturally isomorphic to*

$$\Gamma(X/K, \mathfrak{M}_\chi) \otimes_k \Gamma(\bar{S}/\mathrm{K}, \mathcal{K}(f, \chi) \otimes_{\mathcal{O}_{\bar{S}/K}} h^*\mathfrak{F}) \otimes_k \mathfrak{M}|_0$$

where $h: \bar{S}/K \to S$ *is the morphism induced by* g.

Remark. This last vector space is really a $\Gamma(\bar{S}/K, \mathcal{O}_{\bar{S}/K})$-module. Also $\Gamma(X/K, \mathfrak{M}_\chi)$ is a line.

§5. **Picard sheaves revisited.** Now we are ready to apply the abstract results of the last section. First we return to the basic situation of section two.

Let $\mathcal{Q}$ be a lazy branding of the dth Picard variety P_d of our fixed curve C. In this section we will assume that the endomorphism $\psi_{\mathcal{Q}}: J \to J$ of the Jacobian J is separable. Thus its kernel H is a reduced subgroup of J.

More explicitly, let $\mathfrak{Q}$ be $\mathfrak{Q}(\mathfrak{L}, \mathfrak{M})$ where $\mathfrak{M}$ is an ample invertible sheaf on J and $\mathfrak{L}$ is an invertible sheaf on C of degree one corresponding to the point p of P_1. Let $1 \to \mathbf{G}_m \to G \xrightarrow{\rho} H_{\mathfrak{Q}}$ be the theta group of $\mathfrak{M}$. We will also fix a maximal subgroup $K_{\mathfrak{Q}}$ of $H_{\mathfrak{Q}}$ such that $K_{\mathfrak{Q}}' \equiv \rho^{-1}(H_{\mathfrak{Q}})$ is abelian.

Our endomorphism $\psi_{\mathfrak{Q}}$ may now be canonically factored as the composition $J \xrightarrow{\beta_{\mathfrak{Q}}} J/K_{\mathfrak{Q}} \xrightarrow{\alpha_{\mathfrak{Q}}} J$ where $\beta_{\mathfrak{Q}}$ is the quotient homomorphism. Let $C_{\mathfrak{Q}}$ be the curve $\alpha_{\mathfrak{Q}}^{-1}(I_1(C) - p)$. We have the abelian Galois coverings, $\tilde{C} = \psi_{\mathfrak{Q}}^{-1}(C) \xrightarrow{\delta_{\mathfrak{Q}}} C_{\mathfrak{Q}} \xrightarrow{\gamma_{\mathfrak{Q}}} C$, where the composition $\gamma_{\mathfrak{Q}} \circ \delta_{\mathfrak{Q}}$ is the morphism r of Theorem 2.3.b). Here, the coverings $\delta_{\mathfrak{Q}}$, $\gamma_{\mathfrak{Q}}$ and r have Galois groups equal to $K_{\mathfrak{Q}}$, $H_{\mathfrak{Q}}/K_{\mathfrak{Q}}$ and $H_{\mathfrak{Q}}$.

At last we have enough notation to state

THEOREM 5.1. *Given a character $\chi : K_{\mathfrak{Q}}' \to \mathbf{G}_m$ of weight one, we may construct the invertible sheaf $\mathfrak{M}_\chi$ on $J/K_{\mathfrak{Q}}$. Then, if $d > 2g - 2$, there is a natural isomorphism,*

$$\Gamma(P_d, \mathfrak{W}(\mathfrak{Q}(\mathfrak{L}, \mathfrak{M}))) \xrightarrow{\approx} \Gamma(J/K_{\mathfrak{Q}}, \mathfrak{M}_\chi) \otimes_k$$
$$\Gamma(C_{\mathfrak{Q}}, \gamma_{\mathfrak{Q}}^*\mathfrak{L}^{\otimes d} \otimes \mathfrak{M}_\chi^{\otimes -1}|_{C_{\mathfrak{Q}}}) \otimes_k \mathfrak{M}|_0 .$$

Furthermore, we have a similar isomorphism if $\Gamma(P_d, -)$ is replaced by $H^i(P_d, -)$ and $\Gamma(C, -)$ by $H^i(C, -)$.

Proof. We need only unify the ideas of Section 2 with those of Section four. In fact, one first uses the isomorphism of Lemma 2.2 together with its proof. From these facts, one may construct an isomorphism,

$$\Gamma(P_d, \mathfrak{W}(\mathfrak{Q}(\mathfrak{L}, \mathfrak{M}))) \xrightarrow{\approx} D(f, \mathfrak{F}),$$

where $\mathfrak{F} = \mathfrak{L}^{\otimes d}$ and $f = I_1 - p : C \to J$. After that, one may simply use Proposition 4.7 to compute $D(f, \mathfrak{F})$. This should yield the desired isomorphism. The statement about the higher cohomology groups may be proven similarly using the idea in the proof of Corollary 2.4.c). Q.E.D.

It only remains to specialize this last result to the case of the standard brandings of Section three. Let $\mathfrak{Q} = \mathfrak{Q}(\mathfrak{L}, \mathfrak{M})$ be a standard lazy branding on P_d of altitude n. We will assume that n is prime to the characteristic of k.

In the standard case, ψ is just multiplication by the altitude n and

$H_{\mathcal{Q}}$ is the group J_n of n-torsion points of J. Recall that, in Section three, $\tilde{C}$ was denoted by $\tilde{C}_n$. Now we will denote $K_{\mathcal{Q}}$ by K_n and $\tilde{C}_n/K_n$ by C_n. Thus we have the covering sequence,

$$\tilde{C}_n \xrightarrow{\delta_n} C_n \xrightarrow{\gamma_n} C,$$

where both of these coverings have degree n^g.

Now we may restate the last result in the standard case in

THEOREM 5.2. *Let $\mathcal{Q}(\mathcal{L}, \mathfrak{M})$ be a standard lazy branding of P_d, where $d > 2g - 2$. Let χ be a character of $\rho^{-1}(K_n)$ of weight one. Then the invertible sheaf $\mathfrak{M}_\chi$ gives a principal polarization of J/K_n and we have an isomorphism,*

$$\Gamma(P_d, \mathcal{W}(\mathcal{Q}(\mathcal{L}, \mathfrak{M}))) \xrightarrow{\approx} \Gamma(J/K_n, \mathfrak{M}_\chi) \otimes_k$$
$$\Gamma(C_n, \gamma_n{}^*\mathcal{L}^{\otimes d} \otimes \mathfrak{M}_\chi{}^{\otimes -1}|_{C_n}) \otimes_k \mathfrak{M}|_0.$$

Proof. This is a purely notational modification of Theorem 5.1.

Q.E.D.

When applying this result, one must remember that $\Gamma(J/K_n, \mathfrak{M}_\chi)$ and $\mathfrak{M}|_0$ are one-dimensional and the term $\gamma_n{}^*\mathcal{L}^{\otimes d} \otimes \mathfrak{M}_\chi{}^{\otimes -1}|_{C_n}$ is an invertible sheaf of degree $(d \cdot n - g) \cdot n^{g-1}$ on the curve C_n of genus $1 + n^g(g - 1)$.

THE JOHNS HOPKINS UNIVERSITY

REFERENCES

[1] G. Kempf, "Toward the inversion of abelian integrals I," *Ann. Math.*, to appear.
[2] ______, "Deformations of Symmetric Products," to appear.
[3] D. Mumford, "On the Equations Defining Abelian Varieties." I, *Inv. Math.* **1** (1966), pp. 287–354.
[4] ______, *Abelian Varieties*, Oxford University Press, Oxford, 1970.
[5] A. Weil, "Généralisation des fonctions abéliennes," *Jour. de Math. Pures Appl.*, **17** (1938), pp. 47–87.
[6] ______, *Courbes algébriques et variétés abéliennes*, Hermann, Paris, 1971.

RINGS WITH DISCRETE DIVISOR CLASS GROUP: THEOREM OF DANILOV-SAMUEL

By JOSEPH LIPMAN†

Introduction. In [3], Danilov has studied normal noetherian domains A for which the canonical map of divisor class groups i^*: $\mathcal{C}\ell(A) \to \mathcal{C}\ell(A[[T]])$ is bijective; such A are said to have *discrete divisor class group* (abbreviated DCG).[1] One of Danilov's main results [3, p. 374, Theorem 1] is

(D–S). *If the normal noetherian domain A is such that the localization A_p has DCG at every prime ideal p for which* $\text{depth}(A_p) = 2$, *then A itself has DCG.*[2]

This result generalizes a theorem of Samuel [10, p. 5, Theorem 3.2]. (Danilov states [3, p. 368] that Samuel's proof is incomplete.) Danilov's proof, in which certain additional mild restrictions are placed on A, uses difficult cohomological results of Grothendieck [5]. The *purpose of this note* is to give a proof of (D–S) which uses nothing deeper than a well-known theorem of Rees on the connection between depth and Ext, and moreover needs no additional hypotheses on A.

Actually we obtain somewhat more than (D–S). Danilov defines a map j^*: $\mathcal{C}\ell(A[[T]]) \to \mathcal{C}\ell(A)$ such that $j^* \circ i^* =$ identity (cf. [4, p. 109, Theorem 18.8], or else just use the equivalent definition given in Section 1 below); so *A has DCG $\Leftrightarrow$ j^* is injective.* But in fact Danilov's definition works more generally to give a map $j_{B,t}$: $\mathcal{C}\ell(B) \to \mathcal{C}\ell(B/tB)$ for any noetherian normal ring B and nonunit $t \in B$ such that B/tB is also normal. [Thus the DCG property begins to look like one of "Lefschetz type," i.e., it has to do with the comparison of something on Spec(B) to the corresponding

† Supported by NSF grant GP 29216.

[1] All rings are understood to be commutative. The definition of i^* can be found, e.g., in [4, Sect. 6 (cf. especially pp. 29–31, 35)]. A survey of the theory of rings with DCG is given in [8, Sects. 1–3].

[2] The converse "A has DCG $\Rightarrow$ A_M has DCG for every multiplicative set M in A" is an open question. (For an affirmative answer in case A is an excellent $\mathbf{Q}$-algebra, cf. [3, p. 371, Prop. 4].)

thing on the "hypersurface section" Spec(B/tB). This indicates why Danilov finds [5] so useful.] This being so, what we show is

(D–S)′. *Let B be a normal noetherian domain and t a regular element* (= nonzerodivisor) *in the Jacobson radical of B such that the ring $A = B/tB$ is a normal domain. Suppose that for every prime ideal p in A such that* depth(A_p) = 2, *there exists a normal noetherian flat B-algebra C such that*

$$C/tC \cong A_p \qquad \textit{(isomorphic as A-algebras)}$$

and such that $j_{C,t}\colon \mathcal{C}\ell(C) \to \mathcal{C}\ell(A_p)$ is injective. Then $j_{B,t}\colon \mathcal{C}\ell(B) \to \mathcal{C}\ell(A)$ is injective.[3]

(Taking $B = A[[T]]$, $t = T$, $C = A_p[[T]]$ in (D–S)′, we get (D–S).)

Using a familiar interpretation of $\mathcal{C}\ell(A)$ as being the group of isomorphism classes of reflexive rank one A-modules—an interpretation which, for lack of convenient references, we review briefly in Section 0—we reformulate the injectivity of $\mathcal{C}\ell(B) \to \mathcal{C}\ell(B/tB)$ as a simple statement about free modules. This gives us a straightforward reduction of (D–S)′ to Theorem 1 (Section 2), which is the main result. (This reduction is carried out in Section 1; Theorem 1 and its proof (Section 2) are independent—except for motivation—of anything which precedes them.)

As a corollary of Theorem 1, we obtain a generalization of a result of [5], including the following statement:

If B is a noetherian local ring of depth ≥ 4 and if t is a regular nonunit in B, then B/tB parafactorial $\Rightarrow$ B parafactorial.

0. Reflexive Modules and the Divisor Class Monoid. Let A be a *noetherian integral domain.* For any A-module E, let E^* be the A-module $\mathrm{Hom}_A(E, A)$. Write E^{**} for $(E^*)^*$. Recall that E is said to be *reflexive* if the canonical map $\alpha\colon E \to E^{**}$ is *bijective.* $\{[\alpha(x)](f) = f(x)$ for all $x \in E$, $f \in E^*.\}$

Let K be the field of fractions of A. We shall say that an A-module E has *rank one* if E is *finitely generated* over A and the K-vector space $E \otimes_A K$ is *one-dimensional.* If E has rank one, then so does E^*; and if

[3] In (D–S)′, if $j_{C,t}$ is injective then so is $j_{B_q,t}$, where q is the inverse image of p in B. So the hypothesis in (D–S)′ could be changed to: "*Suppose that $j_{B_q,t}$ is injective for all prime ideals q containing t and such that* depth(B_q) = 3."

two A-modules E_1 and E_2 both have rank one, then so does $E_1 \otimes_A E_2$. For any rank one reflexive A-module E, the canonical map $E \to E \otimes_A K$ is injective, and so E is isomorphic to an A-submodule of K.

If E_1 and E_2 are rank one A-submodules of K, then any A-homomorphism of E_1 into E_2 extends uniquely to a K-homomorphism of $E_1 \otimes_A K$ $(=K)$ into $E_2 \otimes_A K$ $(=K)$; it follows that we can identify $\mathrm{Hom}_A(E_1, E_2)$ with the A-module

$$E_2 : E_1 = \{x \in K \mid xE_1 \subseteq E_2\}.$$

The canonical map $E_1 \to E_1^{**}$ is then identified with the inclusion of E_1 into $A:(A:E_1)$. So E_1 is *reflexive* if and only if $E_1 = A:(A:E_1)$, i.e. if and only if E_1 is a *divisorial fractionary ideal* of A (cf. [2, Section 1.1, Definition 2 and Proposition 1]). And E_1 is *A-isomorphic* to E_2 if and only if there is an $x \neq 0$ in K such that $xE_1 = E_2$, i.e. E_1 and E_2 determine *equivalent divisors* [2, Section 1.2, Proposition 4].

With this in mind, it is now straightforward to see that the elements of the *divisor class monoid*—denoted $\mathcal{C}\ell(A)$—as defined in [2, Section 1.2], are in one-to-one correspondence with the *isomorphism classes of reflexive rank one A-modules.*

The *monoid structure* on $\mathcal{C}\ell(A)$ can be described as follows (verification left to the reader): For any rank one A-module E let

$$[E]_A = (\text{isomorphism class of } E^{**}) \in \mathcal{C}\ell(A).$$

(We may write $[E]$ for $[E]_A$ if no confusion results.) Multiplication in $\mathcal{C}\ell(A)$ is such that

$$[E_1][E_2] = [E_1 \otimes_A E_2]$$

for any two rank one A-modules E_1, E_2. So *multiplication is determined by* $\otimes$ (even though the tensor product of two reflexive A-modules need not itself be reflexive!). The *identity element* of $\mathcal{C}\ell(A)$ is $[A]$. One checks that $[E_1] = [E_2]$ if and only if E_1^* and E_2^* are isomorphic; in particular, $[E]$ *is the identity element of* $\mathcal{C}\ell(A) \Leftrightarrow E^*$ *is free.*

1. Reformulation of (D–S)′. Let B, t, A be as in the statement of (D–S)′. It is easily seen that $j_{B,t}\colon \mathcal{C}\ell(B) \to \mathcal{C}\ell(A)$ is such that for each rank one reflexive B-module F,

$$j_{B,t}([F]_B) = [F \otimes_B A]_A = [F/tF]_A.$$

[Since the localization B_q of B at the prime ideal $q = tB$ is a U.F.D., therefore the reflexive rank one B_q-module $F \otimes_B B_q$ is *free,* and so the *dimension* of

$$(F/tF) \otimes_A (B_q/qB_q) = (F \otimes_B B_q) \otimes_{B_q} (B_q/qB_q)$$

over the field B_q/qB_q is *one*, i.e. F/tF is a rank one A-module (not necessarily reflexive!).] In view of the characterization of the identity element of $\mathcal{C}\ell(A)$ given at the end of Section 0, we conclude that $j_{B,t}$ *is injective* $\Leftrightarrow$ *the following condition holds*:

$(*)_{B,t}$. With $A = B/tB$, if F is any reflexive rank one B-module such that $\mathrm{Hom}_A(F/tF, A)$ is a free A-module, then $\mathrm{Hom}_B(F, B)$ is a free B-module (i.e. F itself is free).

Next, we need a simple observation:

LEMMA. *Let B, t, A, p, C be as in* (D–S)′. *Let F be a reflexive rank one B-module such that* $\mathrm{Hom}_A(F/tF, A)$ *is a free A-module. Then $(F/tF) \otimes_A A_p$ is a free A_p-module.*

Proof. Since C is *flat* over B, therefore $F_C = F \otimes_B C$ is a reflexive rank one C-module [2, Section 4.2, Proposition 8]. Furthermore

$$F_C/tF_C = (F \otimes_B C) \otimes_C (C/tC) = (F \otimes_B A) \otimes_A (C/tC) = (F/tF) \otimes_A A_p$$

whence

$$\mathrm{Hom}_{A_p}(F_C/tF_C, A_p) = \mathrm{Hom}_A(F/tF, A) \otimes_A A_p,$$

and so $\mathrm{Hom}_{A_p}(F_C/tF_C, A_p)$ is a free A_p-module. Since by assumption $j_{C,t}$ is injective, therefore $(*)_{C,t}$ holds, and we conclude that F_C is free over C, whence $(F/tF) \otimes_A A_p$ $(= F_C/tF_C)$ is free over A_p $(=C/tC)$. Q.E.D.

One final remark: B, t, A being as above, if F is any finitely generated reflexive B-module then F is *torsion-free* (in other words, if $0 \neq f \in F$ and b is a regular element of B, then $bf \neq 0$); and furthermore *F/tF is a torsion-free A-module.* (Proof left to reader; the only hypothesis really needed is that $F = \mathrm{Hom}_B(G, B)$ for some B-module G.) It follows that if p is a prime ideal of depth ≤ 1 in the normal noetherian domain A, then $(F/tF) \otimes_A A_p$ is a free A_p-module.

It should now be evident—and up to this point nothing very substantial has been done—that (D–S)$'$ follows from Theorem 1 below.

2. The Basic Result.

THEOREM 1. *Let B be a noetherian ring and let t be a regular element in the Jacobson radical of B. Let F be a finitely generated B-module such that t is not a zero-divisor in F. Set $A = B/tB$, $E = F/tF$, and suppose that*

(i) $\mathrm{Hom}_A(E, A)$ *is a free A-module, and*
(ii) $E_p = E \otimes_A A_p$ *is a free A_p-module for every prime ideal p in A such that* $\mathrm{depth}(A_p) \leq 2$.

Then $\mathrm{Hom}_B(F, B)$ *is a free B-module.*[4]

Proof. Consider the exact sequence

$$0 \to \mathrm{Hom}_B(F, B) \xrightarrow{t} \mathrm{Hom}_B(F, B) \to \mathrm{Hom}_B(F, A)$$
$$\to \mathrm{Ext}^1_B(F, B) \xrightarrow{t} \mathrm{Ext}^1_B(F, B) \to \mathrm{Ext}^1_B(F, A)$$

(obtained from the short exact sequence $0 \to B \xrightarrow{t} B \to A \to 0$). If we can show that $\mathrm{Ext}^1_B(F, A) = 0$, then $\mathrm{Ext}^1_B(F, B) = t(\mathrm{Ext}^1_B(F, B))$, so by Nakayama's lemma [1, Chapter 2, Section 3.2, Proposition 4], $\mathrm{Ext}^1_B(F, B) = 0$, whence

$$\mathrm{Hom}_B(F, B)/t(\mathrm{Hom}_B(F, B)) \cong \mathrm{Hom}_B(F, A) = \mathrm{Hom}_A(E, A);$$

since $\mathrm{Hom}_A(E, A)$ is a free A-module, it follows from [1, Chapter 2, Section 3.2, Proposition 5] (applied to the ideal tB) that $\mathrm{Hom}_B(F, B)$ is free over B, as required.

Now $\mathrm{Ext}^1_B(F, A)$ *is canonically isomorphic to* $\mathrm{Ext}^1_A(E, A)$. [This can be seen—for example—as follows: Let $G_. = \cdots \to G_2 \to G_1 \to G_0 \to 0$

[4] One can check, using [7, p. 879, Proposition 2.2], that (ii) can be replaced by the following two weaker conditions:

(ii)$'$ E_p is a torsion-free A_p-module whenever $\mathrm{depth}(A_p) = 1$;
(ii)$''$ E_p is a reflexive A_p-module whenever $\mathrm{depth}(A_p) = 2$.

[(ii)$'$ is automatic if F is reflexive.]

be a projective resolution of the B-module F (so that $F = H_0(G.)$). A brief examination of the long exact homology sequence of the exact sequence of complexes

$$0 \to G. \xrightarrow{t} G. \to G./tG. \to 0$$

shows that $G./tG.$ is an A-projective resolution of the A-module E. (It should be kept in mind here that, by assumption, multiplication by t in $H_0(G.)$ is an injective map.) Hence, for all $i \geq 0$, and any A-module M

$$\begin{aligned} \mathrm{Ext}^i{}_A(E, M) &= H^i(\mathrm{Hom}_A(G./tG., M)) \\ &= H^i(\mathrm{Hom}_B(G., M)) = \mathrm{Ext}^i{}_B(F, M).] \end{aligned}$$

Let us show then that $\mathrm{Ext}^1{}_A(E, A) = 0$.

Let $\alpha: E \to E^{**}$ be the canonical map of E into its "bidual" (over A) (cf. Sect. 0), and let K, I, C be the kernel, image, and cokernel (respectively) of α. From the exact sequences

$$0 \to K \to E \to I \to 0, \qquad 0 \to I \to E^{**} \to C \to 0$$

we obtain exact seqences

$$\mathrm{Ext}^1{}_A(I, A) \to \mathrm{Ext}^1{}_A(E, A) \to \mathrm{Ext}^1{}_A(K, A),$$

$$\mathrm{Ext}^1{}_A(E^{**}, A) \to \mathrm{Ext}^1{}_A(I, A) \to \mathrm{Ext}^2{}_A(C, A).$$

Since E^* is free (by assumption), also E^{**} is free, and $\mathrm{Ext}^1{}_A(E^{**}, A) = 0$. So it will be enough to show that

$$\mathrm{Ext}^2{}_A(C, A) = \mathrm{Ext}^1{}_A(K, A) = 0.$$

Now for any prime p in A, there is a *natural identification of* $\alpha \otimes_A A_p$ *with the canonical map of* $E \otimes_A A_p$ *into its bidual* (*over* A_p). [Indeed, for any A-algebra R, we have a natural commutative diagram

$$\begin{array}{ccc} E \otimes_A R & \xrightarrow{\ \beta\ } & \mathrm{Hom}_R(\mathrm{Hom}_R(E \otimes_A R, R), R) \\ {\scriptstyle \alpha \otimes_A R}\ \big\downarrow & & \big\downarrow\ {\scriptstyle \gamma} \\ E^{**} \otimes_A R & \xrightarrow[\ \delta\ ]{} & \mathrm{Hom}_R(E^* \otimes_A R, R) \end{array}$$

where β is the natural map of $E \otimes_A R$ into its bidual (over R); γ is obtained by applying $\mathrm{Hom}_R(\cdot, R)$ to the natural map

$$\omega_E\colon \mathrm{Hom}_A(E, A) \otimes_A R \to \mathrm{Hom}_R(E \otimes_A R, R);$$

and δ is ω_{E^*}. And if R is *flat* over A, then γ and δ are isomorphisms [1, Chapter 1, Section 2.10, Proposition 11].]

Thus if $\mathrm{depth}(A_p) \leq 2$, so that $E \otimes_A A_p$ is free, then $\alpha \otimes_A A_p$ is bijective, i.e. $K_p = C_p = 0$. By a theorem of Rees ([9, p. 31, Theorem 1.3] or also [5, p. 33, Proposition 2.9]), this means that

$$\mathrm{Ext}^i(K, A) = \mathrm{Ext}^i(C, A) = 0 \qquad (i \leq 2). \qquad \text{Q.E.D.}$$

We note, in closing, the following generalization of [5, p. 133, Lemma 3.16]:

COROLLARY. *Let B, t, and A be as in Theorem* 1 [*so that* Spec (A) *can be identified with the closed subscheme $t = 0$ of Spec (B)*]. *Let X be an open subset of* Spec (B), *let $Y \subset X$ be an open subset of* Spec (A) *such that* depth $(A_p) \geq 3$ *for all prime ideals p in* Spec $(A) - Y$,[5] *and let $i\colon Y \to X$ be the inclusion map. If $\mathfrak{F}$ is any locally free coherent sheaf on X with $i^*\mathfrak{F} \cong \mathcal{O}_Y{}^n$, then $\mathfrak{F} \cong \mathcal{O}_X{}^n$. In particular, the canonical map* Pic $(X) \to$ Pic (Y) *is injective.*

Proof. There exists a finitely generated B-module F such that, if $F^\sim$ is the corresponding coherent sheaf on Spec (B), then $F^\sim|X \cong \mathfrak{F}$ [6, p. 318, Cor. (6.9.5)]. We may replace F by its bidual, i.e. we may assume that F is reflexive. (In particular, t is not a zero-divisor in F.) We shall show that $E = F/tF$ satisfies conditions (i) and (ii) of Theorem 1; then $\mathrm{Hom}_B(F, B)$ is free, so F itself, being reflexive, is free, and the Corollary results.

For (i), let $j\colon Y \to$ Spec (A) be the inclusion map. Since Y contains all primes p such that depth $(A_p) \leq 1$, we have that: for any finitely generated *free* A-module G, with associated sheaf $G^\sim$ on Spec (A), the canonical map $G^\sim \to j_*j^*G^\sim$ is an isomorphism (cf. for example [5, p. 37, Cor. 3.5]). Applying $\mathrm{Hom}_A(\cdot, A)$ to a finite presentation of E, we obtained an exact sequence

$$0 \to E^* \to G_1 \to G_2$$

[5] This means precisely that $H^0(Y, \mathcal{O}_Y) = A$ and $H^1(Y, \mathcal{O}_Y) = 0$.

with G_1 and G_2 finitely generated free A-modules; in the resulting commutative diagram (with exact rows)

$$\begin{array}{ccccccc} 0 & \longrightarrow & E^{*\sim} & \longrightarrow & G_1^{\sim} & \longrightarrow & G_2^{\sim} \\ & & \Big\downarrow{\scriptstyle\phi} & & \Big\downarrow & & \Big\downarrow \\ 0 & \longrightarrow & j_*j^*E^{*\sim} & \longrightarrow & j_*j^*G_1^{\sim} & \longrightarrow & j_*j^*G_2^{\sim} \end{array}$$

we see then that ϕ is an isomorphism. But

$$j^*E^{\sim} = i^*\mathfrak{F} \cong \mathcal{O}_{Y}^{n}$$

and hence

$$j^*E^{*\sim} \cong \mathcal{O}_{Y}^{n}0 \qquad [=j^*(A^n)^{\sim}].$$

Applying j_*, we conclude that $E^{*\sim} \cong (A^n)^{\sim}$, so E^* is free.

As for (ii), if $\text{depth}(A_p) \leq 2$, then $p \in Y$; but then E_p is just the stalk at p of $i^*\mathfrak{F}$, so E_p is free. Q.E.D.

PURDUE UNIVERSITY

REFERENCES

1. N. BOURBAKI, "Algebre Commutative," Chapters 1 and 2, Act. Sci. Ind. No. 1290, Hermann, Paris, 1961 (English translation, 1972).
2. ______, Chapter 7, Act. Sci. Ind. No. 1314, Hermann, Paris, 1965.
3. V. I. DANILOV, "Rings with a discrete group of divisor classes," *Math. USSR Sb.* **12** (1970), pp. 368–386 (*Mat. Sb.* **83**(125) (1970), pp. 372–389).
4. R. M. FOSSUM, "The Divisor Class Group of a Krull Domain," Ergeb. der Math., Bd. **74**, Springer-Verlag, Berlin and New York, 1973.
5. A. GROTHENDIECK, "Cohomologie locale des faisceaux coherents et theoremes de Lefschetz locaux et globaux," North-Holland Publ., Amsterdam, 1968.
6. A. GROTHENDIECK and J. DIEUDONNE, "Elements de geometrie algebrique," Springer-Verlag, Berlin and New York, 1971.
7. J. LIPMAN, "Free derivation modules on algebraic varieties," *Amer. J. Math.* **87** (1965), pp. 874–898.

8. ______, "Unique factorization in complete local rings," *Proc. Symp. in Pure Math.*, 29, American Math. Soc., Providence, 1975, pp. 531-546.
9. D. REES, "The grade of an ideal or module," *Proc. Cambridge Philos. Soc.* **53** (1957), pp. 28-42.
10. P. SAMUEL, "On unique factorization domains," *Illinois J. Math.* **5** (1961), pp. 1-17.

ON POLARIZED VARIETIES OF DIMENSION 3 (I)

By T. MATSUSAKA[1]

Introduction. The problem we are interested in dealing with here is to find a reasonable upper bound for $h^2(V, \mathcal{O}(rX))$ for a nondegenerate divisor X on a non-singular projective 3-fold V, which satisfies $l(rX) > 0$, so that we can claim the following: $l(rX)$ stays positive and maintains a reasonable value while a polarized (V, X) is being deformed globaly. The problem involves in part to find such a positive integer r too. Of course, a main purpose of doing this is to extend our result in [4] to polarized 3-folds over the universal domain of characteristic p.

We first attempted this in [5], essentially under the assumption that either a suitable pluricanonical system or antipluricanonical system is free from fixed curves. But unfortunately our arguments contained some incompleteness in the discussion of the first case. In this paper, we shall extend the ideas and results in [5] to the case when a suitable pluricanonical system is not empty. Since the case when all pluri-genera vanish require additional discussions, this case will be treated in a paper which will follow this.

After discussing some preliminary results, we discuss an elementary result on ruled surfaces U which states that $l(K_U + 3X) > 0$ when X is arithmetically effective and $X^{(2)} > 0$ (X may not be positive), which is not self-evident for characteristic p. Then after expounding effects of monoidal transformations, we discuss a criterion of a subvariety D of codimension 1 of a non-singular projective 3-fold being a ruled surface. These are necessitated by a possibility that a fixed component of $\Lambda(rX)$ may be a ruled surface, which requires a separate treatment. After these preparations, a choice of r is made in Section 1, Chap. II, and then we directly estimate an upper bound for $h^2(\mathcal{O}(rX))$, showing that our requirement is satisfied.

---

[1] This work was done while the author was partially supported by N.S.F.

Chapter I

1. Preliminaries. Basically, we shall follow terminology and conventions of [10]. First we shall list some of the basic notations we shall use, which do not require detailed explanation.

$\sim$	Linear equivalence of divisors.
$\Lambda(X)$	The complete linear system of divisors determined by X.
Λ_{red}	The reduced part of the linear system Λ, i.e. Λ—the fixed part.
$\lvert Y\rvert$	The support of the cycle Y.
$I(Y, Z)$	The intersection number of Y and Z.
$I(Y, Z)_D$	The intersection number of Y and Z, on the ambient variety D.
$Y^{(m)}$	The self-intersection class, with respect to numerical equivalence.
$Y_D{}^{(m)}$	The self-intersection class, with respect to numerical equivalence on D.
K_V	A canonical divisor on V.

$l(X) = \dim \Lambda(X) + 1$ $(= h^0(V, \mathcal{O}(X))$ if V is non-singular).

$g[Y]$	The proper transform of a cycle Y by a birational transformation g.

We shall use results contained in [5], Section 1 freely. In particular, we shall need the following results.

THEOREM 1. *Let U^n be a non-singular projective variety and X a non-degenerate U-divisor* (*ample in the sense of* [2]). *Let $d = X^{(n)}$, $R_n(x) = \Sigma_0{}^{n-2}\, dx^{i+1}$, $R_n'(x) = (dx + 1)(x^{n-1}/(n! + 1) + c)$ where c is a constant which satisfies $c \geq 2$. Assume that*

$$\dim \Lambda(rX) - R_n(r) - R_n'(r) - (dr^{n-1} + n - 1) > 0.$$

Then $\Lambda(rX)$ contains a linear subsystem Λ' satisfying the following two conditions: (a) *Λ' defines a non-degenerate rational map f' of U into a projective space, without contractible subvariety of codimension* 1 *on U and* $\dim f'(U) = n$; (b) *When D is a subvariety of U of codimension* 1 *on U with* $|D| \not\subset |X|$, $\dim \mathrm{Tr}_D \Lambda_{\text{red}}' \geq r^{n-1}/(n! + 1) + c - 1$.

Remark. As $\dim f'(U) = n$, Λ_{red}' is not composed of a pencil, and a generic divisor of it over a suitable defining field is of the form

$p^t D$ with an absolutely irreducible D. Then f' is of the form $\varphi \circ g$, where g is a rational map defined by the linear system Λ_1 defined over the same field, with a generic divisor D over it, and φ a rational map of $g(U)$ induced by a Frobenuis automorphism of the universal domain. The minimum sum of Λ_1, taken p^t-times, contains Λ_{red}' as a linear subsystem and a general member of it is an absolutely irreducible divisor by the theorem of Bertini. Therefore we have the following

COROLLARY. *In the above Theorem, Λ' may be chosen so that a general member of Λ_{red}' is an absolutely irreducible divisor.*

Following [14], we shall say that a divisor Y on a projective non-singular surface U is *arithmetically effective* if $I(Y, Z) \geq 0$ for all positive U-divisor Z.

LEMMA 1 (Kodaira). *Let X be a positive divisor on a projective non-singular surface U such that $X^{(2)} > 0$. If $X = Y + Z$, $Y > 0$, $Z > 0$ and $I(X, Y) \geq 0$, $I(X, Z) \geq 0$, then $I(Y, Z) > 0$.*

Kodaira's proof (cf. (3), Lemma 4) goes through without any change.

LEMMA 2 (Kodaira). *Let X be a positive divisor on a non-singular projective surface. Assume that $X^{(2)} > 0$ and that X is arithmetically effective. Let $X = \Sigma_1^m D_i$, where the D_i are positive divisors. When j is a fixed index, there is a permutation σ of $(1, \ldots, m)$ such that $\sigma(j) = 1$, with the following property: When σ changes $(D_1, \ldots, D_m)$ to $(D_1', \ldots, D_m')$, $I(\Sigma_1^{s-1} D_i', D_s) > 0$ for $2 \leq s \leq m$.*
Cf. [3], Lemma 5.

The following Lemma is not contained in (5).

LEMMA 3. *Let U be a normal projective surface and A a U-divisor such that $\Lambda(mA)$ contains a hypersurface section of U for large m. Let Z be a positive U-divisor such that $\Lambda(mZ)$ is not composed of a pencil and free from a fixed component for large m. Let u be a positive integer such that $uA^{(2)} > I(A, Z)$. Then for large integer $t \equiv 0$ (3), $\Lambda(tuA)$ contains $(t/3)Z + F$, where F is a positive U-divisor.*

Proof. Let us take $t \equiv 0$ (3) large enough so that tA, tuA are linearly equivalent to hypersurface sections of U. By our assumption, $(tuA)^{(2)} > I(tuA, tZ)$.

Let k_0 be a field of rationality for U, A, Z and B a generic divisor of $\Lambda((t/3)Z)$ over k_0. For large t, B is absolutely irreducible by the theorem of Bertini. Let k be a field, containing k_0, over which B is defined. There is a polynomial $e_0t^2 + e_1t + e_2$ with rational coefficients, such that $e_0 = (uA)^{(2)}/2$ and that $l(tuA) = e_0t^2 + e_1t + e_2$ for large t. Then, for large t, $0 < l(tuA) - (1/2)I(tuA, tZ) < l(tuA) - I(tuA, B) - 2$. Let $I(tuA, B) = s - 1$ and $x_1, \ldots, x_s$ independent generic points of B over k. Because of the above inequality, there is a divisor T in $\Lambda(tuA)$ such that $|T|$ contains the x_i. If B is not a component of T, $T \cap B$ is proper on U and it contains the x_i. Hence $I(tuA, B) \geq \deg T \cdot B \geq s > I(tuA, B)$, which is a contradiction. Therefore B is a component of T and our lemma is proved.

2. Ruled surfaces. *Ruled surfaces in this paper will mean surfaces whose function fields are pure transcendental extensions of subfields of dimension* 1. In this paragraph, we shall be interested mainly in ruled surfaces in characteristic p.

LEMMA 4. *Let W, W' be non-singular projective surfaces and g a birational morphism of W' on W such that g^{-1} is a quadratic transformation. When X' is a W'-divisor which is arithmetically effective (and satisfies $X'^{(2)} > 0$). Then $g(X')$ is arithmetically effective (and satisfies $g(X')^{(2)} > 0$).*

Proof. Let E' be the exceptional curve of the first kind on W', relative to g. We have $I(g^{-1}(A), E') = 0$, $I(g^{-1}(A), g^{-1}(B)) = I(A, B)$ for all W-divisors A, B (cf. (13)). We have $g^{-1}(g(X')) = X' + aE'$. As $I(g^{-1}(g(X')), E') = 0$ and $E'^{(2)} = -1$, it follows that $I(X', E') = a$. As X' is arithmetically effective, $a \geq 0$. As $g(X')^{(2)} = g^{-1}(g(X'))^{(2)}$, $g(X')^{(2)} = I(g^{-1}(g(X)), X' + aE') = I(g^{-1}(g(X')), X') = X'^{(2)} + aI(E', X') \geq X'^{(2)}$.

Let Y be an absolutely irreducible divisor on W. $I(g(X'), Y) = I(g^{-1}(g(X')), g^{-1}(Y)) = I(X' + aE', g^{-1}(Y)) = I(X', g^{-1}(Y)) \geq 0$. These prove our lemma.

Let $U = U_m, U_{m-1}, \ldots, U_0$ be non-singular projective surfaces and g_i a birational morphism of U_i on U_{i-1} such that g_i^{-1} is a quadratic transformation. Denote by E_i the exceptional curve of the first kind on U_i relative to g_i. Let $X = X_m$ be a U-divisor and $X_i = g_{i+1} \circ \cdots \circ g_m(X)$, which is a divisor on U_i. We shall denote by K_i a canonical divisor on U_i. Then we have the following result.

LEMMA 5. *Assume that X is arithmetically effective and $X^{(2)} > 0$.*

Then X_i is arithmetically effective and $X_i^{(2)} > 0$. $g_i^{-1}(K_{i-1}) + E_i \sim K_i$. Moreover, when we let $K_i + X_i \sim g_i^{-1}(K_{i-1} + X_{i-1}) - a_iE_i$, we get $X_i = g_i^{-1}(X_{i-1}) - (a_i + 1)E_i$. If $I(X_i, E_i) > 0$, then $a_i \geq 0$. If $I(X_i, E_i) = 0$, then $a_i = -1$.

Proof. The first assertion follows from Lemma 4. The second is well-known (cf. [13]). The third statement is an immediate consequence of the second. $I(X_i, E_i) = -(a_i + 1)I(E_i, E_i) = a_i + 1$. The last two assertions follow from this.

LEMMA 6.

$$X_i = (g_1 \circ \cdots \circ g_i)^{-1}(X_0) - (a_1 + 1)(g_2 \circ \cdots \circ g_i)^{-1}(E_1) - \cdots - (a_j + 1)(g_{j+1} \circ \cdots \circ g_i)^{-1}(E_j) - \cdots - (a_{i-1} + 1)g_i^{-1}(E_{i-1}) - (a_i + 1)E_i.$$

Proof. This follows at once from our definitions.

From now on, we shall assume *throughout this paragraph* that $U_0 = A \times B$, where A, B are complete non-singular curves. We further assume that *the genus of B is zero.* We let genus $(A) = g$.

LEMMA 7. *$X_0 \equiv u(A \times y) + v(x \times B)$, where $\equiv$ denotes algebraic equivalence and x, y arbitrary points on A, B respectively. Moreover, $u \geq a_i + 1$, $v \geq a_i + 1$ for all i if X is arithmetically effective. If X satisfies $X^{(2)} > 0$, then $2uv > \Sigma\,(a_i + 1)^2$.*

Proof. The first algebraic equivalence relation is well-known and can be deduced easily from the theory of Jacobian varieties (cf. [11]). Let $h = g_1 \circ \cdots \circ g_i$. $I(X_i, h^{-1}(A \times y)) = I(h^{-1}(X_0), h^{-1}(A \times y)) = I(X_0, A \times y) = v$ by Lemmas 5, 6.

There is a point y_0 on B such that $A \times y_0$ contains the point $h(E_i)$. Then $h^{-1}(A \times y_0) = F + E_i$, $F > 0$ and $I(X_i, h^{-1}(A \times y_0) - E_i) \geq 0$ if X is arithmetically effective by Lemma 4. As

$$\begin{aligned} I(X_i, h^{-1}(A \times y_0) - E_i) &= I(X_i, h^{-1}(A \times y_0)) + I(X_i, -E_i) \\ &= v + I(g_i^{-1}(X_{i-1}) - (a_i + 1)E_i, -E_i) \\ &= v + I(-(a_i + 1)E_i, -E_i) \\ &= v - (a_i + 1), \end{aligned}$$

we must have $v \geq a_i + 1$. A similar argument shows that $u \geq a_i + 1$.

Finally, when $X^{(2)} > 0$, Lemma 5, applied to the case when $i = m$, yields

$$X^{(2)} = (g_1 \circ \cdots \circ g_m)^{-1}(X_0)^{(2)} - \Sigma\,(a_i + 1)^2 > 0,$$

as $(g_j \circ \cdots \circ g_m)^{-1}(E_{j-1})^{(2)} = E_{j-1}{}^{(2)} = -1$ and $I(L_j, L_t) = 0$ if $L_j = (g_j \circ \cdots \circ g_m)^{-1}(E_{j-1})$ for $j \neq t$. Since $(g_1 \circ \cdots \circ g_m)^{-1} \cdot (X_0)^{(2)} = X_0{}^{(2)} = 2uv$, our last assertion is proved.

LEMMA 8. *Assume that X is arithmetically effective and* $X^{(2)} > 0$. *Then* $u, v > 0$ *and* $l(K_0 + X_0) = (u - 1)(v + g - 1)$. *Furthermore, when* $u \geq 2, g \geq 1$,

$$(u - 1)(v + g - 1) \geq uv(u - 1)/u.$$

Proof. As in the last part of the proof above, $X_0{}^{(2)} = 2uv$ and $X_0{}^{(2)} > 0$ by Lemma 4. Therefore u and v are both positive. The formula for $l(K_0 + X_0)$ is well-known. Assume that $(u - 1)(v + g - 1) < uv$. Then

$$1 > (u - 1)(v + g - 1)/uv = (1 - 1/u)(1 + (g - 1)/v) \geq 1 - 1/u.$$

Consequently, $(u - 1)(v + g - 1) \geq uv(u - 1)/u$. This finishes our proof.

COROLLARY. *Assume that X is arithmetically effective,* $X^{(2)} > 0$, $u \geq 2$ *and* $g \geq 1$. *Then*

$$l(K_0 + X_0) = (u - 1)(v + g - 1) > \Sigma\,(1/2)a_i(a_i + 1).$$

Proof. By our lemma and by Lemma 7, $(u - 1)(v + g - 1) > \Sigma\,(1/2)(a_i + 1)^2(u - 1)/u$. Then the above holds when the right hand side is limited to those i with $a_i \geq 1$. For such i, $(a_i + 1)^2(u - 1)/u \geq a_i \cdot (a_i + 1)$ as $u \geq a_i + 1$. Thus

$$(u - 1)(v + g - 1) > \Sigma\,(1/2)a_i(a_i + 1),$$

since $a_i \geq -1$ by Lemma 5. Our lemma is thereby proved.

LEMMA 9. *Assume that* $u \geq 2$ *and* $g \geq 1$. *If* $l(K_{i-1} + X_{i-1}) > (1/2)a_i(a_i + 1)$, $l(K_i + X_i) \geq l(K_{i-1} + X_{i-1}) - (1/2)a_i(a_i + 1)$.

Proof. Note that $K_i + X_i \sim g_i^{-1}(K_{i-1} + X_{i-1}) - a_i E_i$. Let $F_i = g_i^{-1}(K_{i-1} + X_{i-1})$.

We first assume that $a_i \geq 1$. We choose $F_i' \sim F_i$, $E_i' \sim E_i$ so that $F_i' \cap E_i$, $E_i' \cap E_i$ are proper intersections on U_i. For w with $1 \leq w \leq a_i$, assume that $l(F_i - (w-1)E_i) > 0$. We have an exact sequence

$$0 \to \mathcal{O}(F_i' - (w-1)E_i' - E_i) \to \mathcal{O}(F_i' - (w-1)E_i')$$
$$\to \mathcal{O}_{E_i}(-(w-1)E_i' \cdot E_i) \to 0.$$

$-(w-1)E_i' \cdot E_i$ is a divisor on E_i of degree $w - 1 \geq 0$. Since E_i is a non-singular rational curve, the corresponding exact cohomology sequence and the theorem of Riemann-Roch yield $l(F_i - (w-1)E_i) - l(F_i - wE_i) \leq w$. In other words, $l(F_i - wE_i) \geq l(F_i - (w-1)E_i) - w$. Therefore, whenever $l(K_{i-1} + X_{i-1}) > (1/2)a_i(a_i + 1)$, successive application of the above formula shows that $l(F_i - a_iE_i) \geq l(F_i) - (1/2)a_i(a_i + 1)$, since $l(F_i) = l(K_{i-1} + X_{i-1})$.

Next we assume that $a_i = 0$. Then $K_i + X_i \sim g_i^{-1}(K_{i-1} + X_{i-1})$ and $l(K_i + X_i) = l(K_{i-1} + X_{i-1})$.

Finally, assume that $a_i = -1$. We have $K_i + X_i \sim g_i^{-1}(K_{i-1} + X_{i-1}) + E_i$. Hence $l(K_i + X_i) \geq l(g_i^{-1}(K_{i-1} + X_{i-1}))$ and our lemma is true for these two cases too. As $a_i \geq -1$ by Lemma 5, our lemma is proved.

COROLLARY. *Let $g \geq 1$ and X an arithmetically effective divisor on U, satisfying $X^{(2)} > 0$. If $u \geq 2$, then $l(K + X) > 0$.*

Proof. By the Corollary of Lemma 8,

$$l(K_0 + X_0) - (1/2) \Sigma\, a_i(a_i + 1) > 0.$$

This, together with several applications of Lemma 9, furnishes a proof of our corollary.

PROPOSITION 1. *Let U be a non-singular ruled surface with a surjective birational morphism on $A \times B$, where A, B are complete non-singular curves such that genus $(B) = 0$. Let X be an arithmetically effective U-divisor with $X^{(2)} > 0$. If genus $(A) \geq 1$, $l(K_U + 2X) > 0$.*

Proof. Our birational morphism is the product of inverses of quadratic transformations (cf. [13]). By Lemma 7, u in our case satisfies $u \geq 2$. Therefore our claim follows from the corollary above.

PROPOSITION 2. *Let the notations and assumptions be as in Proposition* 1, *except that we assume genus* $(A) =$ *genus* $(B) = 0$. *Then*

$$l(K_U + 3X) > 0.$$

Proof. If $(1/2)(K_U + X)^{(2)} - (1/2)I(K_U, K_U + X) \geq 0$, $l(K_U + X) \geq 1$ by the theorem of Riemann-Roch (cf. (12)), as $p_a(U) = 0$. Therefore, we shall assume that the left hand side of the above is negative, i.e. $X^{(2)} + I(K_U, X) < 0$. As $X^{(2)} > 0$, we must have $I(K_U, X) < 0$. When that is so, $(1/2)X^{(2)} - (1/2)I(K_U, X) > 0$, which implies that $l(X) > 1$ and $\Lambda(X)$ is not empty by the theorem of Riemann-Roch.

Let $\Sigma_1^m D_i$ be a member of $\Lambda(X)$, where the D_i are subvarieties of U of codimension 1. As $X^{(2)} > 0$ and X is arithmetically effective, there is a sequence of positive divisors (changing indices, when necessary) $F_1, F_2, \ldots, F_m = \Sigma_1^m D_i$, such that $F_i = D_1 + \cdots + D_i$, $I(F_i, D_{i+1}) > 0$ (cf. Lemma 2). Let $D_i' \sim D_i$ so that $D_i' \cap D_j$ is proper on U and that every component of it is simple on D_j for all i, j. Choose K_U also so that $K_U \cap D_j$ is proper on U and its components are simple on D_j for all j. We have an exact sequence

$$0 \to \mathcal{O}\left(K_U + \sum_1^s D_i' - D_s\right) \to \mathcal{O}\left(K_U + \sum_1^s D_i'\right)$$

$$\to \mathcal{O}_{D_s}\left(K_{D_s} + \sum_1^{s-1} D_i' \cdot D_s\right) \to 0.$$

When $s \geq 2$, the first cohomology group of the last sheaf vanishes by our choice of indices. Hence

$$h^1\left(\mathcal{O}\left(K_U + \sum_1^s D_i\right)\right) \leq h^1\left(\mathcal{O}\left(K_U + \sum_1^{s-1} D_i\right)\right) \quad (s \geq 2).$$

For $s = 1$, a direct computation, based on the exact cohomology sequence derived from above, using $h^1(\mathcal{O}(K_U)) = 0$, $h^2(\mathcal{O}(K_U)) = h^1(\mathcal{O}(K_{D_1})) = 1$ (cf. (2)), shows that $h^1(\mathcal{O}(K_U + D_1)) = 0$. It follows that $h^1(\mathcal{O}(K_U + X)) = 0$. This means, under our hypothesis, that $l(K_U + X) = (1/2)(K_U + X)^{(2)} - (1/2)I(K_U, K_U + X) + 1 = 0$, i.e. $(1/2)X^{(2)} + (1/2)I(K_U, X) + 1 = 0$. Then $l(K_U + 3X) > 0$ by the Riemann-Roch inequality. Our proposition is thereby proved.

3. Monoidal transformations

1. Let V^3 be a non-singular projective 3-fold and D a subvariety of V of dimension 2. Then there is a non-singular projective variety V^* and a birational morphism g of V^* on V such that g^{-1} is a successive monoidal transformation with absolutely irreducible centers which are non-singular and that $D^* = g^{-1}[D]$ is non-singular (cf. [1]). Moreover, we may take g so that g^{-1} is everywhere biregular on $V - |D|$.

Denote by S_2 the set of points of V with two dimensional fibres and by S_1 the set of points of V with one dimensional fibres. As is well-known and easy to see, S_2 is a finite set of points and S_1 is a locally closed subset of V such that $S_1 \cup S_2$ is a closed subset of V. We start with the following observation.

Remark. Let A be a V-divisor and g' the restriction of g on D^*. *Then we can find a V-divisor $A' \sim A$, such that at each point x of $|A'| \cap D$, either g'^{-1} is biregular or the fibre above x is a finite set of points.* This is because the set of points on D with 1-dimensional fibres (with respect to g'^{-1}) is a finite set of points.

Finally, we shall denote by sing (D) the set of singular points of D.

LEMMA 10. *Let A and B be two properly intersecting V-divisors satisfying the following conditions:* (*i*) *$A \cap B \cap D$, $A \cap D$, $B \cap D$ are proper on V and every component of them are simple on D*; (*ii*) *$|A| \cap S_2$, $|B| \cap S_2$, $|A| \cap |B| \cap S_1$ are empty*; (*iii*) *$|A| \cap S_1$, $|B| \cap S_1$ are finite sets*; *when g' denotes the restriction of g on D^*, fibres of g' above points in $|A| \cap D$, $|B| \cap D$ are finite sets of points. Then we have the following results: (a) $g^{-1}(A) \cap g^{-1}(D)$ are proper on V^* and $g^{-1}(A)\cdot g^{-1}(D) = g^{-1}(A\cdot D) = g'^{-1}(A\cdot D)$; (b) $g^{-1}(A)\cdot D^* = g'^{-1}(A\cdot D) = g'^{-1}[A\cdot D]$;* (c) *$g^{-1}(A) \cap g^{-1}(B) \cap g^{-1}(D)$, $g^{-1}(A) \cap g^{-1}(B) \cap g^{-1}[D]$ are proper on V^* and $g^{-1}(A\cdot B\cdot D) = g^{-1}(A)\cdot g^{-1}(B)\cdot g^{-1}(D) = g^{-1}(A)\cdot g^{-1}(B)\cdot g^{-1}[D]$.*

Proof. Let G be the graph of g on $V^* \times V$ and T a component of $G \cap V^* \times A \cap V^* \times D$. Let T' be the projection of T on V. If $T' \not\subset S_1 \cup S_2$, T is a proper component as g^{-1} is biregular along T'. If $T' \subset S_1 \cup S_2$, it is contained in $S_1 \cup S_2 \cap |A| \cap D \subset S_1 \cap |A| \cap D$ (cf. (ii)) and dim T is at most 1 by the definition of S_1 and (iii). Therefore, $G \cap V^* \times A \cap V^* \times D$ is a proper intersection on $V^* \times V$. It follows that $g^{-1}(A) \cap g^{-1}(D)$ is proper on V^* and $g^{-1}(A\cdot D) = g^{-1}(A)\cdot g^{-1}(D)$ (cf. [10], Chap. VIII, Th. 11).

Similar arguments show that $g^{-1}(A) \cap g^{-1}(B)$ is proper on V^* and $g^{-1}(A \cdot B) = g^{-1}(A) \cdot g^{-1}(B)$. Similarly, using $|A| \cap |B| \cap S_i = \emptyset$ for all i, we see that $(g^{-1}(A) \cap g^{-1}(B)) \cap g^{-1}(D)$ is proper on V^* and $g^{-1}((A \cdot B) \cdot D) = g^{-1}(A \cdot B) \cdot g^{-1}(D)$. It follows that $g^{-1}(A \cdot B \cdot D) = g^{-1}(A) \cdot g^{-1}(B) \cdot g^{-1}(D)$.

The above analysis of $G \cap V^* \times A \cap V^* \times D$ also shows that $g^{-1}(A \cdot D) = g'^{-1}(A \cdot D) + E^*$, where E^* is a V^*-cycle such that $g(|E^*|) \subset |A| \cap S_1 \cap \operatorname{sing}(D)$. Then $|E^*| = \emptyset$ by the latter half of (iii). Thus (a) and the first half of (c) is proved.

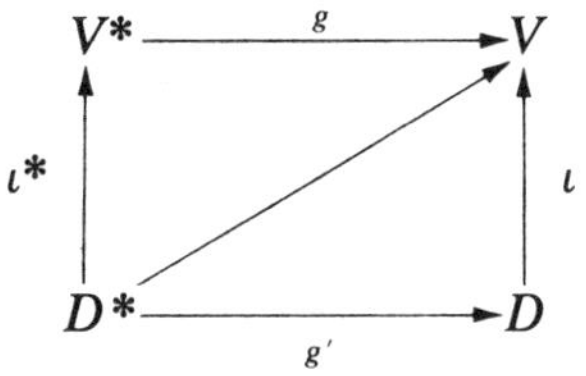

Denote by ι, ι^* respectively the injections of D, D^* into V, V^*. Then we have the commutative diagram of morphisms shown above. Let G' be the graph of $g \circ \iota^* = \iota \circ g'$ on $D^* \times V$. (ii) and (iii) show that $G' \cap D^* \times A$ is a proper intersection on $D^* \times V$. Then $(g \circ \iota^*)^{-1}(A) = \iota^{*-1}(g^{-1}(A)) = g^{-1}(A) \cdot D^*$ (cf. (10), Chap. VIII). When we apply the same arguments to $(\iota \circ g')^{-1}(A)$ on the open set $U = V^* \times V - V^* \times (\operatorname{sing}(D) \cup S_1 \cup S_2)$, we get

$$(\iota \circ g')^{-1}(A) = g'^{-1}(\iota^{-1}(A)) = g'^{-1}[A \cdot D] \text{ on } U.$$

Therefore, $(\iota \circ g')^{-1}(A) - g'^{-1}[A \cdot D]$ is a union of components T^* such that $|g'(T^*)| \subset \operatorname{sing}(D) \cap |A|$ or $\subset S_1 \cup S_2 \cap |A| \cap D = S_1 \cap |A| \cap D$ by (ii). If the former is the case, $g'(T^*)$ is a singular point of D by (i). Hence $g'(T^*)$ must be contained in $S_1 \cup S_2$. It follows that $|g'(T^*)| \subset S_1 \cap |A| \cap D$ in all cases and $g'(T^*)$ is a point of D by (iii). Moreover, by the same (iii), T^* is then a point. Thus

$$(\iota \circ g')^{-1}(A) = g'^{-1}[A \cdot D].$$

As $\iota \circ g' = g \circ \iota^*$, the first half of (b) is proved.

By the first halves of (a) and (c), $g^{-1}(B) \cdot g^{-1}(A) \cdot g^{-1}(D) = g^{-1}(B) \cdot g^{-1}(A \cdot D)$. $g^{-1}(B) \cdot g^{-1}(A \cdot D) = g^{-1}(B) \cdot g'^{-1}(A \cdot D)$ by (a). Moreover,

$g'^{-1}(A \cdot D) = g'^{-1}[A \cdot D] + F^*$, where the components of F^* are transforms of simple points of D by g'^{-1}. Hence $g'(|F^*|)$ is contained in $|A| \cap (S_1 \cup S_2 \cap D) = |A| \cap S_1 \cap D$. This, together with (iii), shows that $\dim F^* = 0$. Thus, $g'^{-1}(A \cdot D) = g'^{-1}[A \cdot D]$ and the remaining parts of (b) and (c) are proved.

2. Here we shall make some applications of the last lemma, which will be needed later.

LEMMA 11. *Let V^3 be a non-singular projective variety, D a subvariety of V of codimension 1 and X a V-divisor such that $\Lambda(X)$ has no base point. Let V^* be a non-singular projective variety and g a birational morphism of V^* on V such that g^{-1} is a successive monoidal transformation with absolutely irreducible non-singular centers, that $D^* = g^{-1}[D]$ is nonsingular and that g^{-1} is everywhere biregular on $V - |D|$. Then*

$$I(K_V + D, D, X) \geq I(K_{V^*} + D^*, D^*, g^{-1}(X)) = I(K_{D^*}, g^{-1}(X)).$$

Proof. Let a canonical divisor K_V of V, $D' \sim D$, $X' \sim X$ be such that pairs (K_V, X'), (D', X') satisfy (i), (ii) and (iii) of Lemma 10. Let the E_i^* be the contractible subvarieties of codimension 1 on V^* and I the set of indices i such that $g(E_i^*)$ is a curve. As is well-known and easy to see (for instance, generalizing a proof of [13], Prop. II.5.6), we have

$$K_{V^*} \sim g^{-1}(K_V) + \sum_{i \in I} E_i^* + \sum_{j \notin I} a_j E_j^*, \qquad a_j > 0.$$

By our definition of g,

$$g^{-1}(D') \sim g^{-1}(D) = g^{-1}[D] + \sum_{i \in I} b_i E_i^* + \sum_{j \notin I} c_j E_j^*,$$

with $b_i \cdot c_j \neq 0$ for all i, j. By Lemma 10, (c),

$$g^{-1}((K_V + D') \cdot D \cdot X') = g^{-1}(K_V + D') \cdot g^{-1}[D] \cdot g^{-1}(X'),$$

and the three factors on the right intersect properly on V^*. $g^{-1}(K_V + D') \sim K_{V^*} + D^* + \sum_{i \in I} (b_i - 1)E_i^* + E^*$, where $b_i - 1 \geq 0$ and every component of E^* contracts to points of D. It follows that $|E^*| \cap$

$|g^{-1}(X')| = \emptyset$ by our choice of X'. Therefore, $I(K_V + D, D, X) = I(g^{-1}(K_V + D), g^{-1}[D], g^{-1}(X)) = I(K_{V^*} + D^* + \Sigma_{i \in I}(b_i - 1)E_i^*, D^*, g^{-1}(X)) \geq I(K_{V^*} + D^*, D^*, g^{-1}(X))$ as $I(E_i^*, D^*, g^{-1}(X)) \geq 0$ since $\Lambda(g^{-1}(X))$ has no base point. Our lemma is thereby proved.

COROLLARY. *When A is a V-divisor,*

$$I(K_V + A + D, D, X) \geq I(K_{V^*} + g^{-1}(A) + D^*, D^*, g^{-1}(X)).$$

LEMMA 12. *Let V^3 be a non-singular projective variety, D a subvariety of V of codimension 1 and X a V-divisor such that $\Lambda(X)$ has no base point. If $I(K_V, D, X) \leq 0$, $I(D^{(2)}, X) \leq 0$ and if at least one of them is a sharp inequality, D is a ruled surface.*

Proof. Let V^* be a non-singular projective variety and g a birational morphism of V^* on V such that g^{-1} is a successive monoidal transformation with absolutely irreducible non-singular centers, that $D^* = g^{-1}[D]$ is non-singular and that g^{-1} is everywhere biregular on $V - |D|$.

By our assumption, $I(K_V + D, D, X) < 0$. Moreover, the linear equivalence class of $K_{V^*} + D^*$ induces on D^* that of K_{D^*}. It follows that $I(K_{V^*} + D^*, D^*, g^{-1}(X)) = I(K_{D^*}, g^{-1}(X)) < 0$ by Lemma 11. This shows that $\Lambda(mK_{D^*})$ does not exist for all m and D^* is a ruled surface (cf. (7)).

LEMMA 13. *Let V^3 be a non-singular projective variety, D a subvariety of V of codimension 1 and X a non-degenerate V-divisor. Assume that $\Lambda(mK_V) \neq \emptyset$ for large m. When $I(D^{(2)}, X) > 0$, then $I(K_V$, D, X$)$ ≥ 0. Moreover, we have in general $I(K_V + 3X + D, D, X) \geq 0$.*

Proof. If $I(D^{(2)}, X) > 0$, let Z be an absolutely irreducible non-singular divisor in $\Lambda(mX)$ for large m. As we may take Z so that $D' = D \cdot Z$ is absolutely irreducible, $\Lambda(m'D')$ on Z is not empty for large m' by the theorem of Riemann-Roch. As D' is absolutely irreducible, it then follows that $\Lambda(m'D')$ has no fixed component for large m'. Letting m'' to be a large positive integer, let T be a divisor in $\Lambda(m''K_V)$. We may assume, without loss of generality, that $T \cap Z$ is proper on V. We can find $D'(m')$ in $\Lambda(m'D')$ so that $(T \cap Z) \cap D'(m')$ is proper on Z. Then deg $((T \cdot Z) \cdot D'(m')) \geq 0$ and hence $I(K_V, D, X) \geq 0$.

As to the second assertion, let us consider first the case when

D is a ruled surface. Let V^* be a projective non-singular variety and g a birational morphism of V^* on V such that g^{-1} is a successive monoidal transformation with absolutely irreducible non-singular centers, that g^{-1} is everywhere biregular on $V - |D|$ and that $D^* = g^{-1}[D]$ is non-singular. By the Corollary of Lemma 11, $I(K_V + 3X + D, D, X) \geq I(K_{V^*} + g^{-1}(3X) + D^*, D^*, g^{-1}(X)) = I(K_{D^*} + 3X^*, X^*)$, where $X^* = g^{-1}(X) \cdot D^*$, taking X so that X and D intersect properly on V.

There is a non-singular projective surface D^{**} and a birational morphism h of D^{**} on D^* such that D^{**} is a transformation of $A \times B$ by a successive quadratic transformations with absolutely irreducible non-singular centers (cf. [1]), where A and B are complete non-singular curves with genus $(B) = 0$. $h^{-1}(X^*)$ is arithmetically effective and satisfies $h^{-1}(X^*)^{(2)} > 0$. It follows that $l(K_{D^{**}} + 3h^{-1}(X^*)) > 0$ by Propositions 1 and 2. Then $l(K_{D^*} + 3X^*) > 0$ and $I(K_{D^*} + 3X^*, X^*) \geq 0$ as X^* is arithmetically effective on D^*. When D is not ruled, our result follows from (7) and from the Corollary of Lemma 11. This proves our lemma.

Chapter II

1. Preliminaries. In this chapter, we shall fix the following notations and assumptions.

V^3	Non-singular projective variety of dimension 3;
X	Non-degenerate V-divisor;
d	$= X^{(3)}$;
ξ	$= I(K_V, X^{(2)})$;
$R_3(x)$	$= dx + dx^2$;
$R_3'(x)$	$= (dx + 1)(x^2/7 + 1)$;
$R_3''(x)$	$= 63(\xi + 3d)^2(2 + (\xi + 3d)x^{(e-1)/e})x^2 + 21(\xi + 3d)dx$, where e is a fixed positive integer such that $e \geq 2$;
r	A positive integer which satisfies $l(rX) > R_3(r) + R_3'(r) + R_3''(r)$; $r \geq 2$; $(\chi(V, \mathcal{O}(rX)) > R_3(r) + R_3'(r) + R_3''(r))$*);

*This condition is not actually for the purpose of this paper. But it is necessary for our final goal.

k An algebraically closed field, over which V, X and K_V are rational;

Λ A linear subsystem of $\Lambda(rX)$, defined over k, satisfying the following conditions:

(II.1.1) A non-degenerate rational map f defined by Λ is such that $\dim f(V) = 3$ and that there is no contractible subvariety of V of codimension 1 for f;

(II.1.2) When D is a subvariety of V of codimension 1, $\dim \mathrm{Tr}_D \Lambda_{\mathrm{red}} \geq r^2/7 + 1$;

(II.1.3) A generic divisor of Λ_{red} over k is absolutely irreducible;

(II.1.4) When $W + \Sigma_1{}^n D_i$ is a generic divisor of Λ over k, $W = D_0$ being a generic divisor of Λ_{red} over k, $I(\Sigma_0{}^{s-1} D_i, D_s, X) > 0$ for all s with $1 \leq s \leq n$.

Remark. The existence of Λ follows from Theorem 1, its Corollary and from Lemma 2.

Definition. Let U be a projective variety defined over k and Z a U-divisor, rational over k. Let (U^*, g) be the normalization of U over k. Then the set of $g(Z^*)$ with $Z^* \in \Lambda(g^{-1}(Z))$ will be called the *complete linear system* $\Lambda(Z)$ *defined by* Z on U. Clearly $\Lambda(Z)$ is a linear system defined by a module of functions of finite type. When Z^* is a generic divisor of $\Lambda(g^{-1}(Z))$ over k, $g(Z^*)$ is a generic divisor of $\Lambda(Z)$ over k. We shall use the similar definition for the complete linear system on U, defined by a linear system of divisors defined in terms of a module of functions of finite type.

In this chapter, we shall be interested in finding an upper bound for $h^1(V, \mathcal{O}(K_V - rX))$. Let $D_i' \sim D_i$ be k-rational such that $D_i' \cap D_j$ is proper on V and that every component is simple on D_j for all i, j. We also choose a k-rational K_V which has the similar properties as the D_i'. Let $Y_s = K_V - \Sigma_0{}^s D_i'$ and $Y_{-1} = K_V$. Then we have an exact sequence

$$0 \to \mathcal{O}(Y_{s-1} - D_s) \to \mathcal{O}(Y_{s-1}) \to \mathcal{O}_{D_s}(Y_{s-1} \cdot D_s) \to 0.$$

From this, we get the following exact sequence:

$$0 \to H^0(V, \mathcal{O}(Y_{s-1} - D_s)) \to H^0(V, \mathcal{O}(Y_{s-1})) \to H^0(\mathcal{O}_{D_s}(Y_{s-1} \cdot D_s)) \to H^1(V, \mathcal{O}(Y_{s-1} - D_s) \to H^1(V, \mathcal{O}(Y_{s-1})).$$

From this and from $Y_{s-1} - D_s \sim Y_s$, we get

$$h^1(V, \mathcal{O}(Y_s)) + h^0(V, \mathcal{O}(Y_{s-1})) - h^0(V, \mathcal{O}(Y_s)) \leq h^1(V, \mathcal{O}(Y_{s-1})) + h^0(D_s, \mathcal{O}(Y_{s-1} \cdot D_s)).$$

When we sum these up for all s, we get

$$\text{(II.1.5)} \quad h^1(V, \mathcal{O}(K_V - rX)) \leq -h^0(V, \mathcal{O}(K_V)) + h^1(V, \mathcal{O}(K_V)) + h^0(V, \mathcal{O}(K_V - rX)) + \sum_0^n h^0(D_s, \mathcal{O}(D_s \cdot Y_{s-1})).$$

Therefore, we shall be interested mainly in finding upper bounds for $h^0(D_s, \mathcal{O}(Y_{s-1} \cdot D_s))$.

2. Estimation of $h^0(D_s, \mathcal{O}(Y_{s-1} \cdot D_s))$ (I).

LEMMA 14. *Let a_s be the coefficient of D_s in the reduced expression for $W + \Sigma_1{}^n D_i$. Then we have the following results:*

(i) $a_s I(D_s, W, rX) \leq dr^3$;

(ii) *When $\Lambda(mK_V)$ is not empty for large m,*

$$I(D_s{}^{(2)}, X) \geq -I(K_V + 3X, rX, X) = -\xi r - 3dr;$$

(iii) *When $\Lambda(mK_V)$ is not empty for large m,*

$$I(r^2 X^{(2)}, D_s) \geq I(rX, W, D_s) - (\xi + 3d)a_s r^2.$$

Proof. W is a generic divisor of Λ_{red} over k and the D_i are fixed components of Λ. It follows that $I(D_i, W, rX) \geq 0$ for all i. Therefore $a_s I(D_s, W, rX) \leq \Sigma\, a_i I(D_i, W, rX) = I(rX, W, rX) \leq (rX)^3 = dr^3$.

When $\Lambda(mK_V)$ is not empty for large m, Lemma 13 shows that $I(K_V + D_i + 3X, D_i, X) \geq 0$ and that $I(K_V, D_i, X) \geq 0$ if $I(D_i{}^{(2)}, X) > 0$. Therefore, if $I(D_i{}^{(2)}, X) \leq 0$, we have $I(K_V + 3X, D_i, X) \geq 0$, which is valid also when $I(D_i{}^{(2)}, X) > 0$.

As

$$I(D_s{}^{(2)}, X) \geq -I(K_V + 3X, D_s, X)$$

$$\geq \Sigma - a_i I(K_V + 3X, D_i, X)$$
$$= -I(K_V + 3X, rX, X),$$

(ii) is proved.

We have $I(rX, rX, D_s) = I(rX, \Sigma_{i \neq s} a_i D_i, D_s) + a_s I(rX, D_s^{(2)}) \geq I(rX, W, D_s) + a_s I(rX, D_s^{(2)})$. Then (iii) follows from this and from (ii).

LEMMA 15. *Let* $\Lambda_s = \Lambda(Tr_{D_s} \Lambda_{\text{red}})$ *be the complete linear system defined by* $\text{Tr}_{D_s} \Lambda_{\text{red}}$ *on* D_s. *Then it is defined over k. A generic divisor* C_s *of* $\Lambda_{s,\text{red}}$ *over k is absolutely irreducible. Moreover, we have the following results:*

(i) *When* $I(X^{(2)}, D_s) > dr^{1/e}$ *and* $c_1 = [r^{(e-1)/e} + 1]$, $\Lambda(tc_1 rX \cdot D_s)$ *contains a divisor* $(t/3)C_s + F$, $F > 0$, *for large* $t \equiv 0\ (3)$;

(ii) *When* $I(X^{(2)}, D_s) \leq dr^{1/e}$, $I(X, D_s, W) > dr \cdot r^{1/e}$ *and* $c_2 = [(\xi + 3d)r^{(e-1)/e} + 2]$, $\Lambda(tc_2 rX \cdot D_s)$ *contains* $(t/3)C_s + F$, $F > 0$, *for large* $t \equiv 0\ (3)$, *provided that* $\Lambda(mK_V) \neq \emptyset$ *for large* m;

(iii) *When* $I(X^{(2)}, D_s) \leq dr^{1/e}$, $I(X, D_s, W) \leq dr \cdot r^{1/e}$ *and* $c_3 = [dr^{1/e} + 2]$, $\Lambda(tc_3 rX \cdot D_s)$ *contains* $(t/3)C_s + F$, $F > 0$, *for large* $t \equiv 0\ (3)$.

Proof. As Λ_{red} is defined over k, Λ_s is obviously defined over k. By (II.1.1), $\text{Tr}_{D_s} \Lambda_{\text{red}}$ is not composed of a pencil. Then $\Lambda_{s,\text{red}}$ is a complete linear system without fixed component, which is not composed of a pencil. Then C_s is absolutely irreducible by the theorem of Bertini. In the discussions which follow, we shall omit the index s, since there will be no danger of confusion by doing so.

Case (i). $c_1(rX \cdot D)_D{}^{(2)} \geq r^{(e-1)/e}(rX \cdot D)_D{}^{(2)} = r^{(e-1)/e} I((rX)^{(2)}, D) > dr^3$. It follows that $c_1(rX \cdot D)^{(2)} > I(rX, W, D) \geq I(rX \cdot D, C)_D$ by Lemma 14, (i). Hence the Case (i) follows from Lemma 3.

Case (ii). By (iii), Lemma 14, $I(r^2 X^{(2)}, D) \geq I(rX, W, D) - (\xi + 3d)ar^2$, where a is the coefficient of D in the reduced expression for $W + \Sigma_1{}^n D_i$. By (i), Lemma 14, $a \leq dr^3/I(rX, W, D) < dr^3/dr^2 \cdot r^{1/e} = r^{(e-1)/e}$. It follows that $I(r^2 X^{(2)}, D) + (\xi + 3d)r^2 \cdot r^{(e-1)/e} > I(rX, D, W)$. Moreover, $I(r^2 X^{(2)}, D) \geq r^2$. From these we conclude that

$c_2 I(r^2X^{(2)}, D) > I(rX, D, W) \geq I(rX \cdot D, C)_D$. When that is so, our result follows also from Lemma 3.

Case (iii). As $I(r^2X^{(2)}, D) \geq r^2$, $(dr^{1/e} + 1)(rX \cdot D)_D{}^{(2)} > dr^2 \cdot r^{1/e} \geq I(rX, D, W) \geq I(rX \cdot D, C)_D$. Therefore $c_3(rX \cdot D)_D{}^{(2)} > I(rX \cdot D, C)_D$. Then our result follows from Lemma 3 too.

LEMMA 16. *Let* Λ_s, C_s *be as in the previous lemma. When* $c_4 \geq 21(\xi + 3d)$, *we have* $I(K_V \cdot D_s - \Sigma_1{}^{s-1} D_i' \cdot D_s - c_4 C_s, X) < 0$, provided that $\Lambda(mK_V)$ is not empty for large m.

Proof. Let us assume that $I(K_V \cdot D_s - \Sigma_1{}^{s-1} D_i' \cdot D_s - \nu C_s, X) \geq 0$ for some positive integer ν. Let us consider the case when $I(C_s, rX) < r^2$ first.

As $I(\Sigma_1{}^{s-1} D_i, D_s, X) > 0$ by (II.1.4), our assumption leads to $I(K_V, D_s, trX) - \nu I(C_s, trX) > 0$ for all positive t. As $I(r^2X^{(2)}, D_s) \geq r^2 > I(C_s, rX)$, we may take $u = 1$ in Lemma 3, and conclude that $\Lambda(trX \cdot D_s)$ contains $(t/3)C_s + F$, $F > 0$, for large $t \equiv 0\ (3)$.

Let K be a field, containing k, over which C_s is defined. Take t large enough so that $trX \cdot D_s$ is linearly equivalent to a hypersurface section of D_s. When we denote by A a generic divisor of $\Lambda(trX \cdot D_s)$ over K, A is absolutely irreducible and we have a specialization $A \to A' = (t/3)C_s' + F'$ over K, where (C_s, F) and (C_s', F') are generic specializations of each other over k so that C_s and A' intersect properly on D_s. As C_s and A intersect properly on D_s too, $C_s \cdot A'$ is a specialization of $C_s \cdot A$ over K by the compatibility of specializations with the operation of intersection-product (cf. (9)). Moreover, $C_s \cdot A' - (t/3)C_s \cdot C_s' > 0$. It follows that $I(C_s, trX) \geq (t/3) \deg C_s \cdot C_s'$.

When we denote by h a non-degenerate rational map of D_s defined by $\Lambda_{s,\mathrm{red}}$, $\dim h(D_s) = 2$ and $\dim \Lambda_{s,\mathrm{red}} \geq r^2/7 + 1$ by (II.1.1) and (II.1.2). $\deg C_s \cdot C_s' \geq \deg h(D_s)$ and the dimension of the linear system of hyperplane sections of $h(D_s)$ is bounded above by $\deg h(D_s) + 1$ (cf. [6]). It follows that $I(C_s, trX) \geq (t/3) \deg C_s \cdot C_s' \geq (t/3)(\dim \Lambda_{s,\mathrm{red}} - 1) \geq (t/3)r^2/7$. Thus, we get $I(K_V, D_s, trX) - (t/3)\nu r^2/7 > 0$. By Lemmas 12, 13, $I(K_V + 3X, D_i, X) \geq 0$ for all i. When we combine these two, we find that $I(K_V + 3X, rX, trX) > (t/3)\nu r^2/7$.

Next, consider the case when $I(C_s, rX) \geq r^2$. Then our assumption leads to $I(K_V, D_s, rX) - \nu r^2 \geq 0$ by (II.1.4) as before. From this we get $I(K_V + 3X, rX, trX) \geq \nu r^2$ in the same way as above. Our lemma follows from these at once.

3. Estimation of $h^0(D_s, \mathcal{O}(K_V \cdot D_s - \Sigma_1^{s-1} D_i' \cdot D_s))$ (II).

In this paragraph, we shall assume that $\Lambda(mK_V) \neq \emptyset$ for large m. As in Lemmas 15, 16, we shall denote by Λ_s the complete linear system on D_s defined by $\mathrm{Tr}_{D_s} \Lambda_{\mathrm{red}}$ and by C_s a generic divisor of $\Lambda_{s,\mathrm{red}}$ over k, which is absolutely irreducible by Lemma 15.

LEMMA 17. *Assume that $H^0(D_s, \mathcal{O}(K_V \cdot D_s - \Sigma_1^{s-1} D_i' \cdot D_s)) \neq \emptyset$. Then*

$$h^0(D_s, \mathcal{O}(K_V \cdot D_s - \Sigma_1^{s-1} D_i' \cdot D_s)) < 21(\xi + 3d) \dim \mathrm{Tr}_{C_s} H^0(D_s, \mathcal{O}(K_V \cdot D_s - \Sigma_1^{s-1} D_i' \cdot D_s)).$$

Proof. For the sake of simplicity, we shall omit the index s and let $Y = K_V \cdot D - \Sigma_1^{s-1} D_i' \cdot D$. Let $\nu < 21(\xi + 3d)$ and $C^{(1)}, \ldots, C^{(\nu)}$ independent generic specializations of C over k. Let $M^{(0)} = M = H^0(D, \mathcal{O}(Y))$, and when $M^{(i-1)}$ is already defined, we define $M^{(i)}$ to be the submodule of $M^{(i-1)}$, consisting of those functions f such that the coefficient of $C^{(i)}$ in div (f) is positive. Then we have an exact sequence

$$0 \to M^{(i)} \to M^{(i-1)} \to \mathrm{Tr}_{C^{(i)}} M^{(i-1)} \to 0.$$

From this, we get $\dim M^{(i-1)} = \dim M^{(i)} + \dim \mathrm{Tr}_{C^{(i)}} M^{(i-1)}$. Using this and Lemma 16, we find $\dim M = \Sigma_0^{\nu} \dim \mathrm{Tr}_C M^{(i)}$, for a ν satisfying our inequality. $M^{(i)}$ is a submodule of $M = M^{(0)}$, and consequently $\mathrm{Tr}_C M^{(i)}$ is a submodule of $\mathrm{Tr}_C M$. Thus $\dim M \leq (\nu + 1) \dim \mathrm{Tr}_C M$, which proves our lemma.

LEMMA 18. *If $h^0(D_s, \mathcal{O}(K_V \cdot D_s - \Sigma_1^{s-1} D_i' \cdot D_s) > 0$,*

$$\dim \mathrm{Tr}_{C_s} H^0(D_s, \mathcal{O}(K_V \cdot D_s - \sum_1^{s-1} D_i' \cdot D_s)) \leq 3(2 + (\xi + 3d)r^{(e-1)/e})rI(K_V, D_s, X) + 1.$$

Proof. We shall omit s and let $Y = K_V \cdot D - \Sigma_1^{s-1} D_i' \cdot D$. Denote by c one of the constants c_1, c_2, c_3 of Lemma 15. Then for large $t \equiv 0$ (3), $\Lambda(tcrX \cdot D)$ contains $(t/3)C + F$, $F > 0$, by the lemma. Let Z be a generic divisor of $\Lambda(Y)_{\mathrm{red}}$ over k.

By definition, this means the following. Let (D^*, g) be the normalization of D over k. $\Lambda(Y)$ is the image of $\Lambda(g^{-1}(Y))$ and $\Lambda(Y)_{\mathrm{red}}$ is the image of $\Lambda(g^{-1}(Y))_{\text{red}}$ by g. Therefore, there is a generic divisor $Z^* + G^*$ of $\Lambda(g^{-1}(Y))$ over k such that G^* is k-rational, no component of Z^* is k-rational and that $Z = g(Z^*)$.

When that is so, we have $\dim \operatorname{Tr}_C H^0(D, \mathcal{O}(Y)) = \dim \operatorname{Tr}_C \Lambda(Z) + 1 \leq \deg C \cdot Z + 1$. As Z and F intersect properly on D by our choice of Z, $\deg C \cdot Z \leq \deg (C + (3/t)F) \cdot Z \leq I(3crX \cdot D, Z)_D$. It follows that $\dim \operatorname{Tr}_C H^0(D, \mathcal{O}(Y)) \leq I(Z, 3crX \cdot D)_D + 1 \leq I(Z + g(G^*), 3crX \cdot D)_D + 1 \leq I(Y, 3crX \cdot D)_D + 1 \leq I(K_V, D, 3crX) + 1$ by (II.1.4). As e is a positive integer satisfying $e \geq 2$, c_2 is the largest among the c_i (ξ is positive by our assumption on $\Lambda(mK_V)$). Our lemma is thereby proved.

COROLLARY 1. *With the same assumption as in our lemma,*

$$h^0(D_s, \mathcal{O}(K_V \cdot D_s - \sum_1^{s-1} D_i' \cdot D_s))$$
$$\leq 63(\xi + 3d)(2 + (\xi + 3d)r^{(e-1)/e})rI(K_V, D_s, X) + 21(\xi + 3d).$$

This is an immediate consequence of our lemma and Lemma 17.

COROLLARY 2.

$$\sum h^0(D_s, \mathcal{O}(K_V \cdot D_s - \sum_1^{s-1} D_i' \cdot D_s))$$
$$< 63(\xi + 3d)^2(2 + (\xi + 3d)r^{(e-1)/e})r^2 + 21(\xi + 3d)\, dr.$$

Proof. By Lemma 13, $I(K_V + 3X, D_i, X) \geq 0$ for all i. Moreover, $\xi > 0$ by our assumption on K_V. It follows that

$$\sum h^0(D_s, \mathcal{O}(K_V \cdot D_s - \sum_1^{s-1} D_i' \cdot D_s)) < 63(\xi + 3d)(2 + (\xi + 3d)r^{(e-1)/e})$$
$$\cdot \sum I(K_V + 3X, D_s, rX) + 21(\xi + 3d)\, dr.$$

as the number of components in $W + \sum_1^n D_i \sim rX$ is at most dr. From this, our corollary follows at once.

THEOREM 2. *If* $\Lambda(mK_V) \neq \emptyset$ *for large* m,

$$h^1(V, \mathcal{O}(K_V - rX)) < -h^0(V, \mathcal{O}(K_V)) + h^1(V, \mathcal{O}(K_V))$$

$$+ h^0(V, \mathcal{O}(K_V - rX))$$
$$+ 63(\xi + 3d)^2(2 + (\xi + 3d)r^{(e-1)/e})r^2$$
$$+ 21(\xi + 3d)\,dr.$$

where e is a prefixed positive integer, satisfying $e \geq 2$.

THEOREM 3. *If* $\Lambda(-mK_V)$ *has no fixed point for large m,*

$$h^1(V, \mathcal{O}(K_V - rX)) \leq h^1(V, \mathcal{O}(K_V)) + 1.$$

This follows at once from (II.1.5) and from (II.1.4).

BRANDEIS UNIVERSITY

REFERENCES

[1] S. Abhyankar, *Resolution of singularities of embedded algebraic surfaces,* Academic Press (1966).

[2] A. Grothendieck-J. Diedonné, "Eléments de géométrie algébrique," *Pub. Math.* IHES, Paris.

[3] K. Kodaira, "Pluricanonical systems on algebraic surfaces," *J. of the Math. Soc. Japan,* **20** (1968), pp. 1-23.

[4] T. Matsusaka, "Polarized varieties with a given Hilbert polynomial," *Amer. J. Math.* **94** (1972), pp. 1027-1077.

[5] ———, "On some cohomology groups of invertible sheaves," to appear in the Proceeding of the international conference, held in Kyoto, Japan, 1976.

[6] T. Matsusaka and D. Mumford, "Two fundamental theorems on deformations of polarized varieties," *Amer. J. Math.* **86** (1964), pp. 668-684.

[7] D. Mumford, "Enriques' classification of surfaces *I*," *Global Analysis,* Princeton University Press (1969), pp. 325-329.

[8] J.-P. Serre, *Groupes algébriques et corps de classes,* Hermann, Paris, 1959.

[9] G. Shimura, "Reduction of algebraic varieties with respect to a discrete valuation of the basic field," *Amer. J. Math.* **77** (1955), pp. 134-176.

[10] A. Weil, "Foundations of Algebraic Geometry," *Amer. Math. Soc. Coll. Publ.* **29** (1960).

[11] ———, *Sur les courbes algébriques et les variétés qui s'en deduisent,* Hermann, Paris (1948).

[12] O. Zariski, "Complete linear systems on normal varieties and a generalization of a lemma of Enriques-Severi," *Ann. of Math.* **55** (1952), pp. 552–592.

[13] ———, "Introduction to the problem of minimal models in the theory of algebraic surfaces," *Publ. of the Math. Soc. Japan*, **4** (1958).

[14] ———, "The theorem of Riemann-Roch for high multiples of an effective divisor on an algebraic surface," *Ann. of Math.* **76** (1962), pp. 560–615.

AN ALGEBRAIC SURFACE WITH K AMPLE, $(K^2) = 9$, $p_g = q = 0$

By D. MUMFORD

Severi raised the question of whether there existed an algebraic surface X homeomorphic to $\mathbf{P}^2$ but not isomorphic to it (as a variety), and conjectured that such a surface did not exist. The essential problem in proving this is to eliminate the possibility that the canonical class K, as a member of the infinite cyclic group $H^2(X, \mathbf{Z})$ might be a *positive* multiple (in fact, 3) of the ample generator of $H^2(X, \mathbf{Z})$ instead of a negative multiple (in fact, -3) as it ought to be. That this is the problem is clear from Castelnuovo's criterion for rationality, and was analyzed and generalized to higher dimensions in the paper [3] of Hirzebruch and Kodaira where it was shown that in *odd* dimensions, $\mathbf{P}^n$ is the only variety in its homeomorphism class. Severi's question was finally answered by S. Yau [8] as a Corollary of his result that all varieties X on which K is ample carry a unique Kähler-Einstein metric. In fact, this result shows that when X is a surface on which K is ample, then the Chern numbers satisfy $c_1^2 \leq 3c_2$, with equality if and only if X is isomorphic to D_2/Γ (D_2 = unit ball in $\mathbf{C}^2$, $\Gamma \subset SU(2, 1)/(\text{center})$ a discrete torsion-free co-compact subgroup; Hirzebruch in [2] had much earlier shown that the surfaces D_2/Γ did satisfy $c_1^2 = 3c_2$). However, the question arises: how close can we come to a surface with K ample which mimics the topology of $\mathbf{P}^2$? In particular, does there exist such a surface with the *same Betti numbers as* $\mathbf{P}^2$? By the standard results on algebraic surfaces, this means that we seek a surface X such that:

$$p_g = q = 0, \quad \text{hence } \chi(\mathcal{O}_X) = 1$$

$$(c_1^2) = (K^2) = 9$$

$$B_0 = B_2 = B_4 = 1, \qquad B_1 = B_3 = 0, \qquad \text{hence } c_2 = 3$$

I shall exhibit one such surface. My method is not by complex uniformization, as used by Shavel [6] and Jenkins (unpublished) in the construction of a surface with K positive and the same Betti numbers as $\mathbf{P}^1 \times \mathbf{P}^1$, but by the p-adic uniformization introduced recently by Kurihara [1] and Mustafin [5]. After looking for such an example at some length, I would hazard the guess that there are, in fact, very few such surfaces (combining Yau's results with Weil's theorem [7] that discrete co-compact groups $\Gamma \subset SU(2, 1)$ are rigid, it follows that there are in any case only *finitely* many such surfaces). But it seems a difficult matter to find some way of enumerating all such surfaces.

1. p-adic uniformizations in general. In this section we wish to summarize and extend somewhat the results of Kurihara and Mustafin cited above, restricting ourselves however to the 2-dimensional case. Let R be a complete discrete valuation ring with fraction field K and residue field $k = R/\pi R$. We assume k is finite. The basis of the construction is a beautiful scheme $\mathfrak{X}$, locally of finite type over R, which may be described by charts as follows:

$$\mathfrak{X} = \bigcup_{A \in GL(3,K)} \operatorname{Spec} R\left[\frac{l_0}{l_1}, \frac{l_1}{l_2}, \pi\frac{l_2}{l_0}\right] - (C_0 \cup C_1 \cup C_2)$$

where $l_i = \sum_{j=0}^{2} A_{ij}x_j$, $A = (A_{ij})$
C_0 = set of curves

$$\pi = \frac{l_0}{l_1} = 0, \qquad a\left(\frac{l_1}{l_2}\right)\cdot\left(\pi\frac{l_2}{l_0}\right) + b\left(\pi\frac{l_2}{l_0}\right) + c = 0$$

$a, b, c \in k$, $a \cdot c \neq 0$, plus the curves

$$\pi = \frac{l_0}{l_1} = 0, \qquad \left(\pi\frac{l_2}{l_0}\right) + c = 0$$

and $$\pi = \frac{l_0}{l_1} = 0, \qquad \frac{l_1}{l_2} + c = 0 \quad (c \in k^*).$$

C_1, C_2 = similar sets of curves where the role of l_0/l_1, l_1/l_2, $\pi(l_2/l_0)$ are permuted cyclically.

Here the glueing represented by the union sign is induced by the requirement that $\mathfrak{X}$ is irreducible and separated with function field

$$K\left(\frac{X_1}{X_0}, \frac{X_2}{X_0}\right),$$

which is the common fraction field of all affine rings.

The closed fibre $\mathfrak{X}_0$ of $\mathfrak{X}$ can be represented graphically by means of the Bruhat-Tits building Σ attached to PGL(3, K). In fact, the 3 sets:

Components E of $\mathfrak{X}_0$
Free rank 3 R-submodules $M \subset K\cdot X_0 \oplus K\cdot X_1 \oplus K\cdot X_2$, modulo $M \sim \pi^k\cdot M$
Vertices ν of Σ

are isomorphic. Moreover, the components of $\mathfrak{X}_0$ cross normally, and if E_i, M_i, ν_i, $i = 1, 2, 3$, correspond as above, then:

a) $E_1 \cap E_2$ is a curve $\Leftrightarrow M_1 \supsetneqq \pi^k M_2 \supsetneqq \pi M_1$, some $k \in \mathbf{Z}$

$\Leftrightarrow \nu_1, \nu_2$ are joined in Σ by a segment

b) $E_1 \cap E_2 \cap E_3$ is a triple point $\Leftrightarrow M_1 \supsetneqq \pi^k M_2 \supsetneqq \pi^l M_3 \supsetneqq \pi M_1$, some $k, l \in \mathbf{Z}$ (or same with 2, 3 interchanged)

$\Leftrightarrow \nu_1, \nu_2, \nu_3$ are the vertices of 2-simplex in Σ.

To describe $\mathfrak{X}$ in a Zariski-open neighborhood of some component E of $\mathfrak{X}_0$, we can proceed geometrically as follows: let E correspond to M and let Y_0, Y_1, Y_2 be an R-basis of M. Start with $\mathbf{P}_R^2$ based on homogeneous coordinates Y_0, Y_1, Y_2 (hence with function field $K(X_1/X_0, X_2/X_0)$ still). First, blow up all k-rational points of the closed fibre $\mathbf{P}_k^2$ of $\mathbf{P}_R^2$ (if k has q elements, there are $q^2 + q + 1$ of these). Second, blow up the proper transforms on this scheme of all k-rational lines on the original closed fibre $\mathbf{P}_k^2$ (again there are $q^2 + q + 1$ of these). Call this $\mathfrak{X}_M$ and let $E_M \subset \mathfrak{X}_M$ be the proper transform of $\mathbf{P}_k^2$. Then a suitable Zariski-neighborhood of E_M in $\mathfrak{X}_M$ is isomorphic to a neighborhood of E in $\mathfrak{X}$. In particular, all surfaces E are rational surfaces gotten by blowing up $\mathbf{P}_k^2 (q^2 + q + 1)$ times and they meet the $2(q^2 + q + 1)$ adjacent components in rational curves. These rational curves are either exceptional curves C of the first kind, in which case $(C^2) = -1$, or proper transforms of lines along which $q+1$ points have been blown up, in which case:

$$(C^2) = +1 - (q+1) = -q.$$

Thus geometrically, if $C = E_1 \cap E_2$, then $(C^2)_{E_1} = -1$ and $(C^2)_{E_2} = -q$ or vice versa; this asymmetry corresponds in (a) above to whether

$$\dim_k(M_1/\pi^k M_2) = 1 \quad \text{or} \quad \dim_k(\pi^k M_2/\pi M_1) = 1$$

and in Σ to the orientation on the segment from v_1 to v_2.

Now if $\Gamma \subset \mathrm{PGL}(3, K)$ is a discrete torsion-free co-compact group, we define first a formal scheme $\mathfrak{X}/\Gamma$ over R by dividing the formal completion of $\mathfrak{X}$ along $\pi = 0$ by Γ (this is possible because Γ acts freely and discontinuously in the Zariski-topology on $\mathfrak{X}_0$). Secondly, one verifies that the dualizing sheaf $\omega_{\mathfrak{X}}$ is ample on each component of $\mathfrak{X}_0$, hence it descends to an invertible sheaf $\omega_{(\mathfrak{X}/\Gamma)}$ on $\mathfrak{X}/\Gamma$ with the same property: this allows one to conclude that $\mathfrak{X}/\Gamma$ can be algebraized to true projective scheme over R, which, for simplicity, we denote $\mathfrak{X}/\Gamma$. Since the generic fibre $\mathfrak{X}_\eta$ of $\mathfrak{X}$ is smooth over K, the generic fibre $(\mathfrak{X}/\Gamma)_\eta$ is also smooth over K, hence

$$\omega_{(\mathfrak{X}/\Gamma)_\eta} = \Omega^2_{(\mathfrak{X}/\Gamma)_\eta}$$

hence $(\mathfrak{X}/\Gamma)_\eta$ is a surface of general type without smooth rational curves C with $(C^2) = -1$ or -2. It is not hard to compute the invariants of $(\mathfrak{X}/\Gamma)_\eta$: to do this, note that $(\mathfrak{X}/\Gamma)_0$ consists of finite set of rational surfaces, crossing each other (possibly crossing themselves) transversally in rational double curves and triple points. Let

E_α = normalizations of the components of $(\mathfrak{X}/\Gamma)_0$, $\quad 1 \le \alpha \le \nu_2$
C_β = normalizations of the double curves of $(\mathfrak{X}/\Gamma)_0$, $\quad 1 \le \beta \le \nu_1$
P_γ = triple points of $(\mathfrak{X}/\Gamma)_0$, $\quad 1 \le \gamma \le \nu_0$

We get an exact sequence:

$$0 \to \mathcal{O}_{(\mathfrak{X}/\Gamma)_0} \to \bigoplus_{\alpha=1}^{\nu_2} \mathcal{O}_{E_\alpha} \xrightarrow{\lambda} \bigoplus_{\beta=1}^{\nu_1} \mathcal{O}_{C_\beta} \xrightarrow{\mu} \bigoplus_{\gamma=1}^{\nu_0} \mathcal{O}_{P_\gamma} \to 0$$

hence

$$
\begin{aligned}
\chi(\mathcal{O}_{(\mathfrak{X}/\Gamma)_\eta}) &= \chi(\mathcal{O}_{(\mathfrak{X}/\Gamma)_0}) \\
&= \sum_\alpha \chi(\mathcal{O}_{E_\alpha}) - \sum_\beta \chi(\mathcal{O}_{C_\beta}) + \sum_\gamma \chi(\mathcal{O}_{P_\gamma}) \\
&= \nu_2 - \nu_1 + \nu_0.
\end{aligned}
$$

Let N be the number of orbits when Γ acts on the vertices of Σ. Clearly $N = \nu_2$. But each E_α contains $2(q^2 + q + 1)$ double curves, each on two E_α's, so

$$\nu_1 = N(q^2 + q + 1).$$

And each double curve passes through $(q + 1)$ triple points, each on three double curves, so

$$\nu_0 = N\frac{(q^2 + q + 1)(q + 1)}{3}.$$

Thus

$$
\begin{aligned}
\chi(\mathcal{O}_{(\mathfrak{X}/\Gamma)_\eta}) &= N\left[1 - (q^2 + q + 1) + \frac{(q^2 + q + 1)(q + 1)}{3}\right] \\
&= N\frac{(q - 1)^2(q + 1)}{3}.
\end{aligned}
$$

Next:

$$
\begin{aligned}
(c_{1,(\mathfrak{X}/\Gamma)_\eta}{}^2) &= (c_1(\omega_{(\mathfrak{X}/\Gamma)_0})^2) \\
&= \sum_\alpha (\mathrm{res}_{E_\alpha} c_1(\omega_{(\mathfrak{X}/\Gamma)_0})^2) \\
&= \sum_\alpha \Big((c_1(\omega_{E_\alpha}) + \sum_{\beta \neq \alpha} E_\alpha \cap E_\beta)^2\Big)
\end{aligned}
$$

All E_α's are just $B =$ (the blow-up of $\mathbf{P}_k{}^2$ at all $(q^2 + q + 1)$-rational points). Let $\pi: B \to \mathbf{P}_k{}^2$ be the blow-up map, let h be the divisor class of a line on $\mathbf{P}_k{}^2$, let $e_i \subset B$ be the exceptional divisors and let $l_j \subset B$ be the proper transforms of the lines. Then $c_1(\omega_{E_\alpha}) + \Sigma_{\beta \neq \alpha} (E_\alpha \cap E_\beta)$ corresponds on B to:

$$\begin{aligned} K_B + \sum_i e_i + \sum_j l_j &\equiv (\pi^{-1}(-3h) + \Sigma e_i) \\ &\quad + \Big(\sum_i e_i\Big) + \sum_j (\pi^{-1}(h) - \text{the } e_i \text{ meeting } l_j) \\ &\equiv \pi^{-1}((q^2 + q - 2)h) - (q - 1)\Big(\sum_i e_i\Big) \end{aligned}$$

with self-intersection $3(q - 1)^2(q + 1)$. Thus

$$(c_{1,(\mathfrak{X}/\Gamma)_\eta}{}^2) = 3N(q - 1)^2(q + 1).$$

By Riemann-Roch,

$$c_{2,(\mathfrak{X}/\Gamma)_\eta} = N(q - 1)^2(q + 1).$$

To determine the irregularity of $(\mathfrak{X}/\Gamma)_\eta$, we can use the relative Picard scheme $Pic_{\mathfrak{X}/\Gamma}{}^0$: its closed fibre is $Pic_{(\mathfrak{X}/\Gamma)_0}{}^0$, and since $(\mathfrak{X}/\Gamma)_0$ has normal crossings and rational components, this is an algebraic torus. In particular points of finite order l, $p \nmid l$, are dense: these correspond to l-cyclic coverings of $(\mathfrak{X}/\Gamma)_0$ and such coverings lift to $(\mathfrak{X}/\Gamma)_\eta$. Therefore the points of finite order l, $p \nmid l$, of $(Pic_{\mathfrak{X}/\Gamma}{}^0)_0$ lift to points of $(Pic_{\mathfrak{X}/\Gamma}{}^0)_\eta$ and hence $Pic_{\mathfrak{X}/\Gamma}{}^0$ is flat over R. On the other hand, a line bundle on $(\mathfrak{X}/\Gamma)_0$ is a line bundle on $\mathfrak{X}_0$ with Γ action: if it is in Pic^0, it is the trivial line bundle on $\mathfrak{X}_0$ and a Γ-action is just a homomorphism from Γ to $\mathbf{G}_m$. Thus finally, using Kajdan's theorem [4] that $\Gamma/[\Gamma, \Gamma]$ is finite, we deduce

$$\begin{aligned} \text{irregularity of } (\mathfrak{X}/\Gamma)_\eta &= \dim\, (Pic_{\mathfrak{X}/\Gamma}{}^0)_\eta \\ &= \dim\, (Pic_{\mathfrak{X}/\Gamma}{}^0)_0 \\ &= \dim \operatorname{Hom}(\Gamma, \mathbf{G}_m) \\ &= rk_{\mathbf{Z}} \Gamma/[\Gamma, \Gamma] \\ &= 0. \end{aligned}$$

Thus the numbers $h^{p,q}$ of $(\mathfrak{X}/\Gamma)_\eta$ fit into the pattern:

$$\begin{array}{c|ccc} q & & & \\ & M-1 & 0 & 1 \\ & 0 & M & 0 \\ & 1 & 0 & M-1 \\ \hline & & & \to p \end{array} \qquad M = N\frac{(q-1)^2(q+1)}{3}$$

In particular, if $N = 1$, $q = 2$, then $M = 1$ and $(\mathfrak{X}/\Gamma)_\eta$ is a surface of the desired type. In this case, in fact $(\mathfrak{X}/\Gamma)_0$ is one rational surface, $\mathbf{P}^2$ blown up 7 times, crossing itself in 7 rational double curves, themselves crossing in 7 triple points. The confusion arising from trying to draw the result brings vividly to mind Lewis Carroll's comment on the sandy shore—"If seven maids with seven brooms were to sweep it for half a year, do you suppose, the Walrus said, that they could get it clear?"

2. The Example. The example is based on the 7th roots of unity: fix the notation:

$$\zeta = e^{2\pi i/7}$$

$$\lambda = \zeta + \zeta^2 + \zeta^4 = \left(\frac{-1+\sqrt{-7}}{2}\right)$$

$$\bar{\lambda} = \zeta^3 + \zeta^5 + \zeta^6 = \left(\frac{-1-\sqrt{-7}}{2}\right)$$

We have the fields:

$$\begin{array}{ll} \mathbf{Q}(\zeta) & \\ \quad \big| \text{ deg } 3 & \text{Galois group } \mathbf{Z}/3\mathbf{Z}, \text{ generator } \sigma,\ \sigma(\zeta) = \zeta^2 \\ \mathbf{Q}(\lambda) & \\ \quad \big| \text{ deg } 2 & \text{Galois group } \mathbf{Z}/2\mathbf{Z}, \text{ generator } z \to \bar{z} \end{array}$$

Note $\mathbf{Q}(\lambda)$ is a UFD, $2 = \lambda \cdot \bar{\lambda}$ is the prime factorization of 2 and $7 =$

$-(\sqrt{-7})^2$ is the prime factorization of 7. We set $\mathbf{Q}(\zeta) = V$ and think of it only as a 3-dimensional vector space over $\mathbf{Q}(\lambda)$. We put the Hermitian form

$$h(x, y) = \mathrm{tr}_{\mathbf{Q}(\zeta)/\mathbf{Q}(\lambda)}(x\bar{y}) = [x\bar{y} + \sigma(x\bar{y}) + \sigma^2(x\bar{y})]$$

on V. Taking $1, \zeta, \zeta^2$ as a basis of V, we find that h has the matrix

$$H = \begin{pmatrix} 3 & \bar{\lambda} & \bar{\lambda} \\ \lambda & 3 & \bar{\lambda} \\ \lambda & \lambda & 3 \end{pmatrix}$$

so that h is positive definite with determinant 7. Note that V contains the lattice $L = \mathbf{Z}[\zeta]$, with basis $1, \zeta, \zeta^2$ over $\mathbf{Z}[\lambda]$. Define

$\Gamma_1 = \mathbf{Q}(\lambda)$-linear maps $\gamma: V \to V$ which preserve the form h and map $L[1/2]$ to $L[1/2]$

Since 2 splits in $\mathbf{Q}(\lambda)$, the λ-adic completion of $\mathbf{Q}(\lambda)$ is isomorphic to the 2-adic completion $\mathbf{Q}_2$ of $\mathbf{Q}$ (in fact, in $\mathbf{Q}_2$, we may take $\lambda = (\text{unit}) \cdot 2$, $\bar{\lambda} = \text{unit}$). So we have a canonical map $V \to$ (λ-adic completion of V) $\cong \mathbf{Q}_2 \cdot 1 \oplus \mathbf{Q}_2 \cdot \zeta \oplus \mathbf{Q}_2 \cdot \zeta^2$ and a canonical homomorphism

$$\Gamma_1 \to GL(3, \mathbf{Q}_2) \to PGL(3, \mathbf{Q}_2).$$

From standard results on the theory of arithmetic groups*, the image $\overline{\Gamma_1}$ of Γ_1 is discrete and co-compact. We introduce 3 elements of Γ_1: the first is σ itself; the second is

$$\tau(x) = \zeta \cdot x.$$

Note that $\sigma^3 = e$, $\tau^7 = e$ and $\sigma\tau\sigma^{-1} = \tau^2$, so together σ and τ generate

*Consider $U(V, h)$ as an algebraic group over $\mathbf{Q}$. Γ_1 is its $\mathbf{Z}[1/2]$-rational points. Since U is compact at the infinite place, Γ_1 is discrete and co-compact in $U(V, h)(\mathbf{Q}_2)$. But over $\mathbf{Q}_2$, $U(V, h) \cong GL(3)$, so mod scalars Γ_1 defines a discrete co-compact subgroup of $PGL(3, K)$.

a subgroup $\Gamma_2 \subset \Gamma_1$ of order 21. The third is a map ρ given by

$$\rho(1) = 1$$

$$\rho(\zeta) = \zeta$$

$$\rho(\zeta^2) = \lambda - \frac{\lambda^2}{\bar{\lambda}}\zeta + \frac{\lambda}{\bar{\lambda}}\zeta^2$$

It can be checked easily that $\rho \in \Gamma_1$. It can also be checked that

$$(\rho \cdot \tau)^3 = \text{multiplication by } \lambda/\bar{\lambda}.$$

Note that the scalar matrices in Γ_1 are exactly

$$\pm(\lambda/\bar{\lambda})^k \cdot I_3$$

PROPOSITION. *ρ, σ, τ and $-I$ generate Γ_1. All torsion elements in $\overline{\Gamma_1}$ are conjugate to either $\sigma^i \cdot \tau^j$ or to $(\rho \cdot \tau)^i$ (some $0 \leq i \leq 2$, $0 \leq j \leq 6$).*

Proof. Consider the action of Γ_1 on $\Sigma_0' =$ [the set of free rank $3\mathbf{Z}_2$-submodules of $\mathbf{Q}_2{}^3$]. Let M_0 be the submodule $\mathbf{Z}_2 \cdot 1 \oplus \mathbf{Z}_2 \cdot \zeta \oplus \mathbf{Z}_2 \cdot \zeta^2$ or $\mathbf{Z}_2{}^3$ for short. If an element $\alpha \in \Gamma_1$ maps M_0 to itself, then back in V, α is given by a 3×3 matrix with coefficients in $\mathbf{Z}[\lambda][1/\bar{\lambda}]$. Since α is H-unitary, its coefficients are also in $\mathbf{Z}[\lambda][1/\lambda]$, so α in fact has coefficients in $\mathbf{Z}[\lambda]$ and maps L to itself. But in L, it is easy to see that $\{\pm\zeta^i\}$ are the only elements $x \in L$ with $h(x, x) = 3$. So α permutes them. Then $\pm\tau^i \circ \alpha$ also carries the element $1 \in L$ to itself. Now the equations

$$h(x, x) = 3$$

$$h(1, x) = \bar{\lambda}$$

have only 3 solutions in L:$x = \zeta, \zeta^2$ or ζ^4. So $(\pm\tau^i \cdot \alpha)$ carries ζ to ζ, ζ^2 or ζ^4. Then $(\pm\sigma^j \circ \tau^i \circ \alpha)$ fixes 1 *and* ζ and it is easy to check that such a map must be the identity. Thus $\pm\Gamma_2$ is the stabilizer of M_0.

As in the Bruhat-Tits building, call $M, M' \in \Sigma_0'$ adjacent if $M \supset M'$ and $\dim_{\mathbf{Z}/2\mathbf{Z}} M/M' = 2$ or vice versa. Then $\rho(M_0) \subset M_0$ and is adjacent to M_0. Because $M/2M \cong (\mathbf{Z}/2\mathbf{Z})^3$, there are only 7 modules $M' \subset M_0$ adjacent to M_0. One checks easily that these are the modules $\tau^i\rho(M_0)$, $0 \leq i \leq 6$. Thus $(\tau^i\rho)^{\pm 1}(M_0)$ is the set of all $M \in \Sigma_0'$ adjacent to M_0. Since Σ_0' is connected under adjacency, this shows that *all* elements of Σ_0' can be expressed as:

$$(\tau^{i_1}\rho)^{\epsilon_1} \cdot \ldots \cdot (\tau^{i_k}\rho)^{\epsilon_k}(M_0), \qquad 0 \leq i_l \leq 5, \qquad \epsilon_l = \pm 1.$$

Thus the subgroup of Γ_1 generated by ρ, σ, τ and $-I$ acts as transitively as Γ_1 on Σ_0' and M_0 has the same stabilizer in both groups, so they are equal.

Now let $\alpha \in \Gamma_1$ be torsion in $\overline{\Gamma_1}$. If α is torsion in Γ_1, then α fixes e.g. the module

$$M_1 = \sum_{i=1}^{\text{order}(\alpha)} \alpha^i(M_0)$$

and, if $M_1 = \beta(M_0)$, then $\beta^{-1}\alpha\beta$ fixes M_0. Thus $\beta^{-1}\alpha\beta \in \pm\Gamma_2$ and $\bar{\alpha}$ is conjugate to $\sigma^j \circ \tau^i$, some i, j. In general, we consider $\det \alpha$. Then $|\det \alpha|^2 = 1$ and $\det \alpha \in \mathbf{Z}[\lambda][\frac{1}{2}]$, hence

$$\det \alpha = \pm(\lambda/\bar{\lambda})^i.$$

Replacing α by $\pm(\lambda/\bar{\lambda})^i\alpha^{\pm 1}$, we may assume $\det \alpha = \lambda/\bar{\lambda}$. Then

$$(\lambda/\bar{\lambda})^{-1}\alpha^3$$

has determinant 1 and is torsion in PGL(3), so it is torsion in GL(3). Now consider α and α^3/λ acting on $\mathbf{Q}_2{}^3$. Since $\mathbf{Z}_2[\alpha, \alpha^3/\lambda]$ is a finite ring over $\mathbf{Z}_2$, there is a free rank $3\mathbf{Z}_2$-module $M \subset \Phi_2{}^3$ such that $\alpha(M) \subset M$, $\alpha^3/\lambda(M) \subset M$. Since $\alpha^3/(\lambda/\bar{\lambda})$ is torsion and $\bar{\lambda}$ is a λ-adic unit, it follows that

$$M \supset \alpha(M) \supset \alpha^2(M) \supset \alpha^3(M) = 2M.$$

As before, replacing α by a conjugate, we can assume $M = M_0$. But now it is easily checked that the 21 maps $\sigma^i \circ \tau^j$ act simply transitively on the flags

$$M_0/2M_0 \supset (\text{2-dim}' \text{ subspace}) \supset (\text{1 dim}' \text{ subspace}).$$

So conjugating α by $\sigma^i \circ \tau^j$, we can assume

$$\alpha(M_0) = (\rho \circ \tau)(M_0)$$

$$\alpha^2(M_0) = (\rho\circ\tau)^2(M_0).$$

Then $(\rho\circ\tau)^{-1}\circ\alpha$ carries M_0 into itself and fixes a flag in $M_0/2M_0$. The former implies that $(\rho\circ\tau)^{-1}\circ\alpha = \pm\sigma^i\circ\tau^j$, some i, j, and the latter implies that $i = j = 0$. Thus in $\overline{\Gamma_1}$, $\alpha = \rho\circ\tau$. Q.E.D.

It remains to choose a suitable subgroup $\Gamma \subset \Gamma_1$ of finite index such that:

a) Γ/scalars is torsion-free
b) Γ acts transitively on Σ_0', (hence Γ/scalars acts transitively on Σ_0, the vertices of the Bruhat-Tits building).

It will then follow from the results of Section 1 that the corresponding surface $(\mathfrak{X}/\Gamma)_\eta$ is a surface of the desired type. To find Γ, it is convenient to use a congruence subgroup for the prime 7. In fact, consider the maps:

$$\mathbf{Z}[\lambda, \tfrac{1}{2}] \to \mathbf{Z}[\lambda, \tfrac{1}{2}]/(\sqrt{-7}) \cong \mathbf{Z}/7\mathbf{Z}$$
$$\lambda, \bar{\lambda} \mapsto 3$$
$$L[\tfrac{1}{2}] \to \underbrace{L[\tfrac{1}{2}]/(\sqrt{-7})L[\tfrac{1}{2}]}_{\text{call this } L_0} \cong (\mathbf{Z}/7\mathbf{Z})^3$$

The induced form h_0 on L_0 is of rank 1 and has null-space $L_1 \subset L_0$ spanned by $\zeta - 1$, $\zeta^2 - 1$. Taking 1, $\zeta - 1$, $(\zeta - 1)^2$ as a basis of L_0, it is easy to check that mod 7:

$$\sigma \mapsto \begin{pmatrix} 1 & 0 & 0 \\ 0 & 2 & 0 \\ 0 & 1 & 4 \end{pmatrix}, \qquad \tau \mapsto \begin{pmatrix} 1 & 0 & 0 \\ 1 & 1 & 0 \\ 0 & 1 & 1 \end{pmatrix}$$

$$\rho \mapsto \begin{pmatrix} 1 & 0 & 0 \\ 0 & 1 & 4 \\ 0 & 0 & 1 \end{pmatrix}, \qquad \rho\circ\tau \mapsto \begin{pmatrix} 1 & 0 & 0 \\ 1 & 5 & 4 \\ 0 & 1 & 1 \end{pmatrix}$$

In particular, considering the action of Γ_1 on L_1, we get a homomorphism

$$\pi:\Gamma_1 \to GL(2, Z/7\mathbf{Z}) \cap \{X \mid \det X = \pm 1\} \underset{\text{def}}{=} G.$$

The group G on the right has order $2^5 \cdot 3 \cdot 7$. Let H be a 2-Sylow subgroup and define $\Gamma = \pi^{-1}(H)$. Since all 21 elements $\sigma^i \tau^j$ and all 3 elements $(\rho \circ \tau)^i$ except e have non-zero images in G of orders 3 or 7, Γ is torsion-free. As the full group Γ_1 is set-theoretically $\Gamma \times \Gamma_2$, Γ acts transitively on $\Sigma_0{}'$. This completes the construction.

HARVARD UNIVERSITY

REFERENCES

[1] A. Kurihara, "On certain varieties uniformized by q-adic automorphic functions," to appear

[2] F. Hirzebruch, "Automorphe Formen und der Satz von Riemann-Roch," *Symp. Int. de Top. Alg.*, UNESCO, 1958

[3] F. Hirzebruch and K. Kodaira, "On the complex projective spaces," *J. Math. Pures et Appl.*, **36** (1957), p. 201.

[4] D. Kajdan, "Connection of the dual space of a group with the structure of its closed subgroups," *Funct. Anal. and its Applic.*, **1** (1967), p. 63 in English transl.

[5] G. A. Mustafin, "Non-Archimedean Uniformizations," *Mat. Sbornik*, **105** (1978), p. 207.

[6] Ira Shavel, "A class of algebraic surfaces of general type constructed from quaternion algebras," to appear, *Pac. J. Math.*

[7] A. Weil, "On discrete subgroups of Lie groups," *Ann. of Math.*, **72** (1960), p. 369.

[8] S. T. Yau, "Calabi's conjecture and some new results in algebraic geometry," *Proc. Nat. Acad. Sci.*, USA, **74** (1977), p. 1798

A GENERALIZATION OF THE NOTION OF A VALUATION

By Masayoshi Nagata

Under the classical definition, a valuation v of a field K is a mapping of K into the real number field $\mathbf{R}$ such that, for arbitrary $a, b \in K$, the following three conditions are satisfied:

(1) $va \geq 0$; $va = 0$ if and only if $a = 0$.
(2) $v(ab) = (va)(vb)$.
(3) $v(a + b) \leq va + vb$.

If v satisfies (3*) below, then v is *non-archimedean*; otherwise *archimedean.*

(3*) $v(a + b) \leq \max\{va, vb\}$.

The non-archimedean case was generalized by Krull [3] to the notion of additive valuations, which are very useful in the theory of commutative rings. The multiplicative expression of the generalization can be stated as follows:

> v is a mapping of a field K into an ordered group enjoying the properties (1), (2), (3*).

This last case can be realized if we take a suitable ordered field instead of the real number field (Section 2). Therefore we like to discuss mappings of a field K into an ordered field enjoying similar properties.

On the other hand, Artin [1] weakened (3) to the following:

(3°) There is a positive number C such that
$$v(a + b) \leq C \cdot \max\{va, vb\} \text{ for any } a, b \in K.$$

He showed that this modification produces only new valuations which are equivalent to the classical ones.

Therefore, in view of the fact that any ordered field contains the rational number field $\mathbf{Q}$ as an ordered subfield, we define as follows:

Definition 0.1. A *generalized valuation* v of a field K is a mapping of K into a suitable ordered field R enjoying the following three properties (1), (2), (3°).

(1) $va \geq 0$ for any $a \in K$; $va = 0$ if and only if $a = 0$.
(2) $v(ab) = (va)(vb)$ for any $a, b \in K$.
(3°) There is a positive rational number C such that $v(1 + b) \leq C$ whenever $vb \leq 1$.

Adapting the notions of archimedean and non-archimedean valuations, we classify generalized valuations into three classes:

Definition 0.2. A generalized valuation v of a field K is said to *be strongly non-archimedean* or *non-archimedean* if v satisfies (3*) or (3′) below, respectively.

In the other case, v is called *semi-archimedean.*

(3*) $v(1 + \mathrm{b}) \leq 1$ whenever $vb \leq 1$.
(3′) $v(1 + b) \leq 1 + \mathrm{r}$ whenever $vb \leq 1$ and r is a positive rational number.

Note that (3*) above is equivalent to (3*) before and that in the classical case (3′) is equivalent to (3*). Note also that, as we shall show in Section 2, strongly non-archimedean valuations are nothing but multiplicative expressions of additive valuations.

In the classical case or in the case of additive valuations, equivalence of valuations is defined by the coincidence of order. But, as will be shown in Section 4, immediate adaption of the definition is not nice (Remark 4.4), and we define first:

Definition 0.3. Two generalized valuations v, w of a field K are *semi-equivalent* to each other if it holds that $va > vb$ if and only if $wa > wb$.

Equivalence of generalized valuations will be defined later in Section 3, because we need some preliminaries for the definition. Under our definition, we shall show that every generalized valuation is equivalent to one enjoying the triangular inequality (3).

On the other hand, the notion of an absolute value defined below is an adaption of the absolute value in the complex number field $\mathbf{C}$, and gives an example of a semi-archimedean generalized valuation.

Let K be an arbitrary field of characteristic zero. Take the algebraic closure F of K. Then take a maximal formally real subfield R' of F. R' is then real closed and $F = R'(\sqrt{-1})$ (see Section 1). Every element of K is written as $b + c\sqrt{-1}$ with $b, c \in R'$, and we set $|b + c\sqrt{-1}| = \sqrt{b^2 + c^2}$. Now

Definition 0.4. The mapping $|\ |$ above (which depends on the choice of R') is called an *absolute value* on K.

We shall show that every semi-archimedean generalized valuation v of a field K with values in an ordered field R is a modification of an absolute value by a group homomorphism of $U(v) = \{a \in K \mid 1 - va \in P\}$ into $1 + P$, where $P = \{x \in R \mid -r < x < r$ for any positive rational number $r\}$.

1. Some Known Facts on Ordered Fields. The results we are stating in this section are well known (see for instance, Nagata [4]).

LEMMA 1.1 *A field R can be ordered so that R becomes an ordered field if and only if R is formally real (i.e.,* $X_1{}^2 + \cdots + X_n{}^2 = 0$ *has no non-trivial solution in K for any natural number n). In this case, R is of characteristic zero, hence contains the rational number field* $\mathbf{Q}$.

LEMMA 1.2. *A real closed field R has unique structure as an ordered field. In this case,* $a \geq 0$ *if and only if* $a = b^2$ *with some* $b \in R$.

LEMMA 1.3. *An ordered field R has its real closure (i.e., a real closed field containing R as an ordered subfield and algebraic over R), which is unique within isomorphisms.*

LEMMA 1.4. *A field R is real closed if and only if* (i) $\sqrt{-1} \notin R$ *and* (ii) $R(\sqrt{-1})$ *is algebraically closed.*

LEMMA 1.5. *If R is an ordered field, then* $V = \{x \in R \mid |x| < n$ *for some natural number* $n\}$ *is a valuation ring with maximal ideal* $P = \{x \in R \mid |x| < r$ *for any positive rational number* $r\}$. *(For an element a of R,* $|a|$ *denotes non-negative one among* $\pm a$.

Definition 1.6. We express the situation stated above by saying that (V, P) is the *canonical valuation ring* of the ordered field R.

LEMMA 1.7. *With* (V, P) *as above, let f be the natural homomorphism of V to V/P. Then V/P becomes an ordered field isomorphic to a subfield of the real number field* $\mathbf{R}$, *by defining*:

$$fx > fy \text{ if and only if } x - y \notin P \text{ and } x > y.$$

Note that this lemma can be generalized easily to the case of a valuation ring of R containing V. Conversely,

LEMMA 1.8. *Assume that W is a valuation ring of a field K. If f is a homomorphism of W into an ordered field R' and if the kernel of f is the maximal ideal of W, then K can be ordered so that K is an ordered field whose canonical valuation ring is the composite of W with the canonical valuation ring of fW.*

2. Ordered Groups. In this section, we are to prove:

THEOREM 2.1. *Assume that V is a valuation ring of a field K, G an ordered group and f an injection of a subgroup H of G into the value group of V. Then there is a valuation ring W of an extension field of K so that* (i) *W gives an extension of V (i.e., $W \cap K = V$) and* (ii) *f is extended to an injection of G into the value group of W.*

Note that, in view of Lemma 1.8, this assertion implies that the multiplicative expression of an arbitrary additive valuation is obtained as a strongly non-archimedean generalized valuation.

We begin with a kind of product of ordered additive groups. Let $\{G_\lambda \mid \lambda \in \Delta\}$ be a set of ordered additive groups G_λ indexed by a linearly ordered set Δ. For each well-ordered subset Ω of Δ, the direct product $\Pi_{\lambda\in\Omega} G_\lambda$ becomes an ordered group under lexicographical order. This is canonically embedded in the direct product $\Pi_{\lambda\in\Delta} G_\lambda$.

Definition 2.2. Under the circumstances, the union U of all such $\Pi_{\lambda\in\Omega} G_\lambda$ is again an ordered additive group. This U is called the *ordered product* of these G_λ. If all G_λ are isomorphic to an ordered group G, then we say that U is the *ordered product* of G along Δ.

Let G be an ordered group. We classify elements of G by an equivalence relation $\sim$ defined by $a \sim b$ if and only if there are natural numbers m and n so that $m|a| > |b|$, $n|b| > |a|$. Let us denote by C_a the equivalence class containing a. We introduce an order on the set Δ of these equivalence classes by: $C_a > C_b$ if and only if $-|a| > -|b|$ and $C_a \neq C_b$. Note that this definition does not depend on the particular choice of the representatives a, b. Note also that $C_0 = \{0\}$ is the largest. For each $0 \neq a \in G$, we set $G_a = \{b \in G \mid C_b \geq C_a\}$, $H_a = \{b \in G \mid C_b > C_a\}$. Then G_a and H_a are subgroups of G.

Definition 2.3. A subset B of the group G is said to be *strongly linearly independent* (over the rationals) if for any $a \in B$ and for any mutually distinct elements $a_1, \ldots, a_n$ of $B \cap C_a$, their residue classes modulo H_a are linearly independent over the rational number field $\mathbf{Q}$.

THEOREM 2.4. *Under the circumstances, G is isomorphic to a subgroup of the ordered product* $\mathbf{P}$ *of the additive group* $\mathbf{R}$ *of the real numbers along the linearly ordered set* Δ.

Proof. Take a maximal strongly linearly independent subset B of G. For each C_a ($a \neq 0$), we fix an injection f_a of G_a/H_a into $\mathbf{R}$. Let $\mathbf{R}_a$ be the direct factor of $\mathbf{P}$ corresponding to C_a. Then, for each $d \in C_a \cap B$, we define fb to be $f_a(b)$ in $\mathbf{R}_a$. This f is extended to an injection of the subgroup G_1 generated by B into $\mathbf{P}$. Then, consider the set F of pairs (G_λ, f_λ) of subgroups G_λ containing G_1 and injections f_λ of G_λ into $\mathbf{P}$ which extends f (as injections of ordered groups). Then F is not empty and has a maximal member, say (G^*, f^*). One sees easily that $G^* = G$. QED

Now our Theorem 2.1 follows from the following lemma in view of Theorem 2.4 above.

LEMMA 2.5. *If the value group of a valuation ring V is a subgroup of an ordered group G', then there is an extended valuation ring W so that its value group contains G'.*

Proof is easy and we omit it.

3. Structure of a Generalized Valuation. Assume that v is a generalized valuation of a field K with values in an ordered field R and that (V, P) is the canonical valuation ring of R. Then we see the following result immediately.

Proposition 3.1. $W = \{x \in K \mid vx \in V\}$ is a valuation ring of K with maximal ideal $M = \{y \in K \mid vy \in P\}$.

Definition 3.2. Under the circumstances, we say that (W, M) is the *structure ring* of v.

PROPOSITION 3.3. *The characteristic of W/M coincides with that of K. The mapping v* of W/M into V/P, naturally induced by v, is a valuation of the field W/M.*

Proof. Assume that $(v(1 + x) \bmod P) \neq 1$ with some $x \in M$. Then we may assume that $v(1 + x) > r > 1$ with a rational number r. Considering some power of $1 + x$, we have a situation so that $v(1 + x)$ is very large, contradicting our condition (3°). QED

Definition 3.4. The valuation v^* stated above is called the valuation *associated* to v.

As for the relationship between v and the valuation defined by W, we have:

PROPOSITION 3.5. *Let w be a multiplicatively expressed valuation defined by W. Then we have:*

(i) *$wx > wy$ implies $vx > vy$.*
(ii) *$wx = wy$ if and only if $vx \sim vy$ (in the sense of Section 2).*

Proof is easy and we omit it.
We now define the equivalence of generalized valuations.

Definition 3.6. Two generalized valuations v, w of a field are *equivalent* to each other if (i) they are semi-equivalent to each other, (ii) their structure rings coincide with each other and (iii) the valuations associated to them are equivalent to each other.

PROPOSITION 3.7. *A generalized valuation v is semi-archimedean if and only if the valuation v^* associated to v is archimedean.*

Proof. Each of the conditions (3*) and (3′) implies that $v^*(1 + x) \leq 1$ whenever $v^*x \leq 1$, namely that v^* is non-archimedean. QED

PROPOSITION 3.8. *If v is a non-archimedean generalized valuation of a field K, then $W^* = \{x \in K \mid vx < 1 + r$ for any positive rational number $r\}$ is a valuation ring of K with maximal ideal $M^* = \{x \in K \mid vx < 1 - r$ for some positive rational number $r\}$. Let w be a multiplicatively expressed valuation defined by W^*. Then, for any $x, y\ (\neq 0) \in K$, we have*

(i) *$vy \geq vx$ implies $wy \geq wx$.*
(ii) *$wy = wx$ implies $1 - r < v(x/y) < 1 + r$ for any positive rational number r.*

Proof. That W^* is a valuation ring follows from that v is non-archimedean. If $vy \geq vx$, then $v(x/y) \leq 1$ and $x/y \in W^*$ and $wy \geq wx$. If $wy = wx$, then x/y is a unit in W^* and we see (ii). QED

Remark 3.9. W^* above is the composite of W with the valuation ring of v^*.

Next, let us discuss what would determine a generalized valuation v within equivalence. By definition, the structure ring (W, M) and the associated valuation v^* are basic data. The following group $U(v)$ is also basic: $U(v) = \{x \in W \mid v^*(x \bmod M) = 1\}$.

Definition 3.10. $U(v)$ above is called the *unitary group* of v.

Remark 3.11. $U(v)$ contains $1 + M$ as a subgroup.

Furthermore, we consider a complete set of representatives of the residue class field W/M, which we denote by $r(W/M)$, and a complete set of representatives of non-zero principal ideals in W, which we denote by $p(W)$. (Namely, $p(W)$ is a subset of $W - \{0\}$ such that each non-zero principal ideal of W has unique generator in $p(W)$.)

THEOREM 3.12. *Let w be another generalized valuation of the field K. Then w is equivalent to v if and only if:*

(i) *The structure ring and the unitary group of w coincides with W and $U(v)$, respectively, and*

(ii) *it holds that $wx > wy$ if and only if $vx > vy$ for arbitrary two elements x, y of any one of $U(v)$, $r(W/M)$ and $p(W)$.*

Proof. The only if part is obvious and we assume that (i) and (ii) hold good. The equivalence of the associated valuations v^* and w^* follows from the coincidence of the unitary groups and the condition (ii) for $r(W/M)$. Take arbitrary $a, b \in K$. If $aW \neq bW$, say $aW \subset bW$, then $va < vb$, $wa < wb$ by Proposition 3.5. Assume now that $0 \neq aW = bW$. Set $c = a/b$. By the symmetry, it suffices to show that $vc > 1$ implies $wc > 1$, which follows immediately from (ii) if $c \in U(v)$; otherwise from the equivalence of v^* and w^*. QED

Now let us discuss how to construct a generalized valuation v of a field K with a given structure ring (W, M) and an associated valuation v^*. By these W and v^*, the unitary group $U(v)$ is well determined, because $U(v) = \{x \in W \mid v^*(x \bmod M) = 1\}$. Let G be any one of $U(v)$ and $1 + M$. We choose $r(W/M)$ and $p(W)$. Let R be an ordered field with canonical valuation ring (V, P). In order to define v with values in R, we have to decide values for elements of G, $r(W/M)$, $p(W)$ under obvious requirements:

(i) For $x \in r(W/M)$, $(vx \bmod P) = v^*(x \bmod M)$,

(ii) $v|_G$ is a homomorphism of the group G into $1 + P$,

(iii) $xW \mapsto (vx)V$ defines an injection of the semi-group of non-zero principal ideals of W into that of V,

(iv) if $x_1, \ldots, x_n \in r(W/M)$ and $\Pi x_i^{m_i} \in G$ $(m_i \in \mathbf{Z})$, then $\Pi(vx_i)^{m_i} = v(\Pi x_i^{m_i})$,

(v) if $y_1, \ldots, y_n \in p(W)$ and $\Pi y_i^{m_i} \in W - M$, then with $z \in r(W/M)$ such that $z - \Pi y_i^{m_i} \in M$, it holds that $\Pi(vy_i)^{m_i}/vz = v(\Pi y_i^{m_i}/z)$.

Now we prove:

THEOREM 3.13. *If* $v: G \cup r(W/M) \cup p(W) \to R$ *satisfies these five conditions, then* v *is uniquely extended to a generalized valuation of* K.

Proof. For $0 \neq a \in K$, first choose $y \in p(W)$ so that $ay^c \in W - M$ $(c = \pm 1)$, and $z \in r(W)$ so that $ay^c/z \in 1 + M$. Thus we define $va = v(ay^c/z)(vz)(vy)^{-c}$. QED

As the simplest case, we have:

THEOREM 3.14. *Assume that* (W, M), $r(W/M)$ *and* $p(W)$ *are given so that* (i) $r(W/M)$ *is a subfield (representing* W/M*) and* (ii) $P(W)$ *is a semi-group. We define* $v: r(W/M) \cup p(W) \cup (1 + M) \to R$ so *that* (i) $v|_{r(W/M)}$ *is a generalized valuation whose structure ring is* $r(W/M)$ *itself,* (ii) $v|_{p(W)}$ *is a semi-group injection of* $p(W)$ *into* V *such that* $xW \subset yW$ *if and only if* $(vx)V \subset (vy)V$, *and* (iii) $v|_{1+M}$ *is a homomorphism of the group* $1 + M$ *into* $1 + P$. *Then* v *is uniquely extended to a generalized valuation of* K.

Proof is easy and we omit it.

THEOREM 3.15. *Every generalized valuation* v *of a field* K *is equivalent to one which satisfies the triangular inequality* (3).

Proof. First, we may assume that the associated valuation v^* satisfies the inequality $v^*(a + b) < v^*a + v^*b$ if $ab \neq 0$ $(a, b \in W/M)$. In view of Theorem 3.13, we may modify v, preserving order on each of $U(v)$, $r(W/M)$, $p(W)$. We enlarge the ordered field R so that the canonical valuation ring contains a prime ideal Q such that (i) V/Q is isomorphic to the old V and (ii) V_Q is isomorphic to an extension of the old V. Then we may modify v so that $v(U(v)) \subseteq 1 + Q$ and such that $v|_{p(W)}$ is unchanged in the sense that $V/Q =$ (old V). Then we see that the modified v satisfies the triangular inequality. QED

4. Some Examples.

Let R be an ordered field with canonical valuation ring (V, P) such that $P \neq \{0\}$. For each prime number p, we take an element $m_p \in P$ and we define a generalized valuation v of the rational number field $\mathbf{Q}$ by $v(\pm p_1^{e_1} \cdots p_n^{e_n}) = \Pi\,(1 + m_{p_i})^{e_i}$ (p_i are prime numbers and e_i are integers).

Example 4.1. This v is a non-archimedean generalized valuation of $\mathbf{Q}$ such that the associated valuation v^* is trivial. v is strongly non-archimedean if and only if v is trivial.

Proof is easy and we omit it.

Note that:

PROPOSITION 4.2. *If v is a generalized valuation of $\mathbf{Q}$ and if the associated valuation v^* is trivial, then, v is obtained as in Example 4.1.*

THEOREM 4.3. *Assume that the multiplicative group $K^* = K - \{0\}$ of a field K is free and that f is a homomorphism of K^* into the additive group of the ordered field R. Then there is a generalized valuation v of K such that* (i) *the associated valuation is trivial, and* (ii) *$vx > vy$ if and only if $fx > fy$ (for $x, y \in K^*$).*

Proof. Multiplying a suitable element of R to fK^*, we may assume that $fK^* \subseteq P$ (the maximal ideal of the canonical valuation ring V of R). Extending R if necessary, we may assume that there is a non-zero prime ideal Q of V such that $fx - fy \in Q$ implies that $fx = fy$. Take a non-zero element z of Q and a free base $\{b_\lambda \mid \lambda \in \Delta\}$ of K^*, and we define $v(b_\lambda) = 1 + (fb_\lambda)z$. This v is extended to the required generalized valuation. QED

Remark 4.4. This result shows that semi-equivalence is quite different from equivalence; for instance there is a generalized valuation of $\mathbf{Q}$ as above which is semi-equivalent to the absolute value $|\ |$.

Next, we consider a non-archimedean generalized valuation v of $\mathbf{Q}$ such that the associated valuation v^* is not trivial. v^* is then a p-adic valuation with a prime number p. For other prime number q, $vq = 1 + m_q$ with $m_q \in P$. Conversely,

Example 4.5. $vp = p^{-1}$, $vq = 1 + m_q$ (p is a fixed prime number; q runs through all prime numbers other than p) defines a generalized valuation of $\mathbf{Q}$.

Thus one sees easily the following:

PROPOSITION 4.6. *Let v be a generalized valuation of $\mathbf{Q}$ with asso-*

ciated valuation v^*. If v is non-archimedean, then v is obtained as in Example 4.1 *or Example* 4.5. *If v is semi-archimedean, then v is equivalent to the absolute value.*

In order to discuss generalized valuations of a subfield of the complex number field **C**, we define a special case of the notion of a generalized absolute value (the general case will be defined in Section 6).

Definition 4.7. Assume that f is an injection of a field K into **C**. Let v be a generalized valuation of K and let v^* be the associated valuation. If it holds that $v^*(x \bmod M) = |fx|$ for all $x \in K$ (where M is the maximal ideal of the structure ring W of v), then we say that v is a *generalized absolute value* associated with the injection f.

Remark 4.8. By the definition, M must be 0 in this case, hence $W = K$.

THEOREM 4.9. *Let K and f be as above and let R be an ordered field with canonical valuation ring (V, P). Then a generalized absolute value v associated with f and with values in R is determined, within equivalence, by the group homomorphism $v|_{U(v)}: U(v) \to 1 + P$; here $U(v) = \{x \in K \mid |fx| = 1\}$*

Proof is immediate by Theorem 3.13.

Remark 4.10. Consider the case where $K = \mathbf{C}$ for simplicity. Then there are infinitely many group homomorphisms of $U(v)$ into $1 + P$. This means that even though $[\mathbf{C}:\mathbf{R}] = 2$, there are infinitely many extensions of the absolute value on **R**.

5. Extensions of a Generalized Valuation.

In this section, let v be a generalized valuation of a field K with values in an ordered field R, v^* the associated valuation, (W, M) the structure ring of v, and $p(W)$, $r(W/M)$ as in Section 3. Furthermore, (V, P) is the canonical valuation ring of R.

LEMMA 5.1. *Let L be a field containing K. If a valuation ring (W', M') of L dominates W, $W'/M' = W/M$, and if the value group of W' coincides with that of W, then there is a generalized valuation v' of L which extends v and with structure ring W'.*

Proof. $p(W)$ and $r(W/M)$ are good for W' (in order to apply Theorem 3.13), and we have to observe $1 + M'$. Since $1 + M'$ is a commutative group, we can use terminology for modules over $\mathbf{Z}$, and we take a maximal linearly independent subset B of $1 + M'$ over $1 + M$. Then, by defining $v'b = 1$ for any $b \in B$, $v|_{1+M}$ is naturally extended to a homomorphism $v'|_{1+M'}:1 + M' \to 1 + P$ (by enlarging R if necessary). Thus we have a mapping $v':(1 + M') \cup r(W/M) \cup p(W) \to R$, and v' satisfies the requirements stated in Theorem 3.13. QED

COROLLARY 5.2. *Let W^* be a maximal completion of W. Then v is extended to a generalized valuation of the field of fractions of W^* so that its structure ring coincides with W^*.*

The following three lemmata on maximally complete valuation rings are easily proved by virtue of the characterization of maximal completeness in terms of pseudo-convergent series (a result of Kaplansky [2]; cf. [5]):

LEMMA 5.3. *Let T be a maximally complete valuation ring with maximal ideal I and let k be a subfield of T/I. In the family of valuation rings dominated by T and with residue class fields contained in k, there is a maximal member. Every maximal member in the family is maximally complete.*

LEMMA 5.4. *With (T, I) as above, let $f(x)$ be a monic polynomial over T. If $f(X)$ modulo I is irreducible, then $T[X]/f(X)T[X]$ is a maximally complete valuation ring.*

LEMMA 5.5. *Assume that a valuation ring W' is dominated by T as above. Then there is a maximal completion W'' of W' which is dominated by T.*

Now we shall prove:

THEOREM 5.6. *Let L be a field containing K and let (W', M') be a valuation ring of L dominating W. If W'/M' is algebraic over W/M, then there is a generalized valuation v' of L which extends v and such that its structure ring coincides with W'.*

Proof. We first consider a maximal completion of W'. Then, applying Lemma 5.5 and Corollary 5.2, we may assume that W is maximally complete. On the other hand, if we know the result in the case where L is a simple extension of K, then we see the general case easily, and we assume that $L = K(b)$. Then we can reduce to the following three cases: (i) the value of b under the additive valuation defined by W' is linearly independent over the value group of W, (ii) b is algebraic over K and the monic irreducible polynomial $f(X)$ over W, which has b as a root, is irreducible modulo M and (iii) $W'/M' = W/M$, b is algebraic over K, and the value of b generates the value group of W' over that of W. Case (i) is proved easily. The other cases can be proved easily by virtue of Theorem 3.13. QED

6. Semi-archimedean Case.

We begin with the definition of a generalized absolute value on a field K of characteristic zero. As in Definition 0.4, the algebraic closure F of K is of the form $R(\sqrt{-1})$ with a real closed field R. Consider the canonical valuation ring (V, P) of R. Then V/P is represented by a subfield k of V and k is naturally identified with a subfield of the real number field $\mathbf{R}$. $V[\sqrt{-1}]$ is a valuation ring of F. Let v' be a generalized absolute value on $k(\sqrt{-1})$ associated with the natural embedding of $k[\sqrt{-1}]$ into $\mathbf{R}[\sqrt{-1}] = \mathbf{C}$. Then:

Definition 6.1. Let w be a generalized valuation of F such that its structure ring coincides with $V[\sqrt{-1}]$ and such that $w|_{k(\sqrt{-1})} = v'$. Then the restriction $w|_K$ is called a *generalized absolute value* on K.

THEOREM 6.2. *If v is a semi-archimedean generalized valuation of a field K, then v is equivalent to a generalized absolute value on K.*

Proof. Consider the structure ring (W, M) of v, and also v^*, $r(W/M)$, $p(W)$, etc. as in Section 3. By Theorem 5.6, we may assume that K is algebraically closed. By Corollary 5.2, we may assume that W is maximally complete. (Note that every maximal completion of valuation ring of an algebraically closed field is a valuation ring of an algebraically closed field, as is well known.) v^* is equivalent to an absolute value on W/M. This means that there is a subfield k of W/M such that k is real closed, $W/M = k(\sqrt{-1})$ and such that the referred absolute value is defined with respect to k. By Lemma 5.3, there is a subfield K'' of K such that $W'' = K'' \cap W$ is maximally complete, k coincides with W''/M'' ($M'' = M \cap W''$) and the value group of W'' coincides with that of W. Then by Lemma 5.4, $W''[\sqrt{-1}]$ is maximally complete. This means that $W = W''[\sqrt{-1}]$. Let k'' be a subfield of W'' which represents W''/M''. Then $k''(\sqrt{-1})$ represents W/M and $v|_{k''(\sqrt{-1})}$ is equivalent to a generalized absolute value on $k''(\sqrt{-1})$.

QED

Remark 6.3. By Definition 0.4, we have an absolute value on K (with respect to K''). The difference of v from the absolute value comes from the group homomorphism $v|_{U(v)}:U(v) \to 1 + P$. Thus we may say that every semi-archimedean generalized valuation is equivalent to a modification by a group homomorphism $U(v) \to 1 + P$ of a suitable absolute value on K.

KYOTO UNIVERSITY

REFERENCES

[1] E. Artin, "Algebraic numbers and algebraic functions, I," *Lecture notes.* Princeton Univ. & New York Univ., 1950-1951.
[2] I. Kaplansky, "Maximal fields with valuations," *Duke J. Math.* **9** (1942), pp. 303-321.
[3] W. Krull, "Allgemeine Bewertungstheorie," *Crelle J.* **167** (1932), pp. 160-196.
[4] M. Nagata, *Field theory,* Dekker, New York, 1977.
[5] O. F. G. Schilling, "The theory of valuations," *Math. Surveys AMS* **4**, 1950.

ON THE VALUE GROUP OF A DIFFERENTIAL VALUATION

By MAXWELL ROSENLICHT*

1. Introduction. If k is an ordinary differential field of characteristic zero whose subfield of constants is C, by a *differential valuation* of k is meant a valuation v of k that is trivial on C, for which C is a system of representatives of the residue class field, that satisfies the further condition that if a, b are nonzero elements of k and $v(a)$, $v(b) > 0$, then also $v(a'b/b') > 0$ [1]. The significance of this notion is that with it various versions of L'Hospital's rule hold and one can do a considerable amount of asymptotic analysis.

It is known [1, Theorem 4] that if k is a differential field and v a differential valuation of k with value group Γ, then there is a map ψ from $\Gamma^* = \Gamma - \{0\}$ into Γ such that for all $a \in k^*$ with $v(a) \neq 0$ we have $\psi(v(a)) = v(a'/a)$ and

(a) if $\alpha \in \Gamma^*$ and $n \in \mathbf{Z}$, $n \neq 0$, then $\psi(n\alpha) = \psi(\alpha)$
(b) for any $\gamma \in \Gamma$, the set $\{\alpha \in \Gamma : \alpha = 0 \text{ or } \psi(\alpha) \geq \gamma\}$ is a subgroup of Γ
(c) for any $\alpha, \beta \in \Gamma^*$, $\psi(\beta) < \psi(\alpha) + |\alpha|$.

Theorem 1 of the present paper shows that if Γ is an arbitrary ordered abelian group and $\psi : \Gamma^* \to \Gamma$ a function satisfying conditions (a), (b), (c), then the pair (Γ, ψ) comes from a differential valuation of a differential field, as above, at least if suitable extra hypotheses are imposed. Our other main result, Theorem 2, uses the pair (Γ, ψ) associated with a differential valuation v of a differential field k to handle the problem of asymptotic integration: if $a \in k$, does there exist some $b \in k$ that approximates $\int a$, or, which should mean roughly the same thing, does there exist $b \in k$ such that $a - b'$ is small compared to a, that is such that $v(a - b') > v(a)$? For example, consider the differential field $k = \mathbf{R}(x, e^{x^2})$ whose derivation is the usual differentiation with respect to x, which can be identified with a Hardy field of germs of dif-

* Research supported by National Science Foundation grant number MCS 77-03719 A02.

ferentiable real-valued functions on neighborhoods of $+\infty$ in $\mathbf{R}$, which has a natural differential valuation corresponding to the place that sends each function into its limit as $x \to +\infty$ [1, Example 1]. The element $e^{x^2} \in k$ has a nonelementary integral $\int_0^x e^{t^2}\,dt$, but this can be approximated by $e^{x^2}/2x \in k$ in the above sense and also in the sense that the ratio of the two functions approaches 1 as $x \to +\infty$. (In fact, these two senses are equivalent, since $k(\int_0^x e^{t^2}\,dt)$ is also a Hardy field.) Our Section 3 below handles some general aspects of this theory.

2. Differential Fields with Given Value Group. For any torsion free abelian group Γ, the map $\gamma \to \gamma \otimes 1$ is an embedding of Γ in the $\mathbf{Q}$-vector space $\mathbf{Q}\Gamma = \Gamma \otimes_{\mathbf{Z}} \mathbf{Q}$, and if Γ is an ordered abelian group then there is a unique ordering of $\mathbf{Q}\Gamma$ that preserves the ordering of Γ. Suppose now that Γ is an ordered abelian group and that $\psi:\Gamma^* \to \Gamma$ is a function satisfying properties (a), (b), (c). We claim that there is a unique function from $(\mathbf{Q}\Gamma)^*$ into $\mathbf{Q}\Gamma$, which we also denote ψ, that extends the given ψ on Γ^* and that also satisfies (a), (b), (c): Property (a) clearly gives a well-defined definition of ψ on $(\mathbf{Q}\Gamma)^*$, property (b) for $\mathbf{Q}\Gamma$ is an immediate consequence of (b) for Γ, while property (c) for $\mathbf{Q}\Gamma$ is an immediate consequence of [1, Theorem 5]. Note that $\psi(\Gamma^*) = \psi((\mathbf{Q}\Gamma)^*)$ and that the cardinality of the latter set is at most the vector space dimension $\dim_{\mathbf{Q}} \mathbf{Q}\Gamma$. If (Γ, ψ) comes from a differential valuation v of a differential field k with subfield of constants C, then $\dim_{\mathbf{Q}} \mathbf{Q}\Gamma$ is just the rational rank of v, which is known to be at most deg tr k/C; thus $\psi(\Gamma^*)$ is finite if k is a finite extension field of C, in particular if k is generated as a differential extension field of C by a finite number of elements that are differentially algebraic over C.

THEOREM 1. *Let Γ be an ordered abelian group and $\psi:\Gamma^* \to \Gamma$ a function with properties* (a), (b), (c). *Suppose that the ordered subset $\psi(\Gamma^*)$ of Γ, in the opposite ordering, is well-ordered. Let C be a field of characteristic zero such that the vector space dimension* $\dim_{\mathbf{Q}} C \geq \dim_{\mathbf{Q}} \mathbf{Q}\Gamma$. *Then there exists a differential field K whose subfield of constants is C and a differential valuation v of K whose value group is Γ such that for each $a \in K^*$ with $v(a) \neq 0$ we have $v(a'/a) = \psi(v(a))$.*

For each $\gamma \in \psi((\mathbf{Q}\Gamma)^*) = \psi(\Gamma^*)$ choose a set $\Sigma_\gamma \subset \Gamma$ whose canonical image in the $\mathbf{Q}$-vector space

$$\{\alpha \in \mathbf{Q}\Gamma : \alpha = 0 \quad \text{or} \quad \psi(\alpha) \geq \gamma\}/\{\alpha \in \mathbf{Q}\Gamma : \alpha = 0 \quad \text{or} \quad \psi(\alpha) > \gamma\}$$

is a **Q**-basis of the latter vector space. We shall prove the theorem for any field C of characteristic zero such that $\dim_{\mathbf{Q}} C \geq \operatorname{card} \Sigma_\gamma$ for each $\gamma \in \psi(\Gamma^*)$, a slight weakening of the stated hypothesis. We have $\psi(\Sigma_\gamma) = \{\gamma\}$. Let $\Sigma = \cup(\Sigma_\gamma : \gamma \in \psi(\Gamma^*))$. We claim that Σ is a **Q**-basis of $\mathbf{Q}\Gamma$. First, if Σ were linearly dependent we could find an integer $n \geq 1$ and elements $\alpha_1, \ldots, \alpha_n \in \Sigma$ and $c_1, \ldots, c_n \in \mathbf{Q}^*$ such that $c_1\alpha_1 + \cdots + c_n\alpha_n = 0$. Taking the canonical linear map of $\mathbf{Q}\Gamma$ with kernel

$$\{\alpha \in \mathbf{Q}\Gamma : \alpha = 0 \quad \text{or} \quad \psi(\alpha) > \min\{\psi(\alpha_1), \ldots, \psi(\alpha_n)\}\}$$

gives a relation of linear dependence among those α_i's for which $\psi(\alpha_i) = \min\{\psi(\alpha_1), \ldots, \psi(\alpha_n)\}$, a contradiction. Therefore Σ is linearly independent. Next, if $\alpha_0 \in (\mathbf{Q}\Gamma)^*$ then we can find an element β_0 in the **Q**-space spanned by $\Sigma_{\psi(\alpha_0)}$ such that either $\alpha_0 - \beta_0 = 0$ or $\psi(\alpha_0) < \psi(\alpha_0 - \beta_0)$. If $\alpha_0 - \beta_0 \neq 0$, we can find an element β_1 in the space spanned by $\Sigma_{\psi(\alpha_0 - \beta_0)}$ such that either $\alpha_0 - \beta_0 - \beta_1 = 0$ or $\psi(\alpha_0 - \beta_0) < \psi(\alpha_0 - \beta_0 - \beta_1)$. We can try to continue this process, thus getting $\beta_2, \beta_3, \ldots$, but by virtue of our well-ordering hypothesis this process must come to an end, giving $\alpha_0 = \beta_0 + \beta_1 + \cdots + \beta_n$ for some n, with each β_i in the **Q**-space spanned by Σ. Thus Σ is a **Q**-basis of $\mathbf{Q}\Gamma$.

Our desired field K will be an algebraic extension field of $C(X)$, where X is some fixed set of indeterminates having the same cardinality as Σ. Let $v: X \to \Sigma$ be a fixed bijection. Also fix a function $h: X \to C^*$ with the property that for each $\gamma \in \psi(\Gamma^*)$, h maps the set $\{x \in X : v(x) \in \Sigma_\gamma\}$ onto a set of elements of C that are linearly independent over **Q**; that such an h exists follows from our cardinality assumption. Now the multiplicative group of an algebraically closed field is divisible, hence an injective **Z**-module. Thus for each $x \in X$ we can find a homomorphism φ_x of the group **Q** into $\overline{C(X)}^*$, where $\overline{C(X)}$ is some fixed algebraic closure of $C(X)$, such that $\varphi_x(1) = x$. For $x \in X$ and $a \in \mathbf{Q}$, we define the symbol x^a to be $\varphi_x(a)$. Thus the set of all expressions $x_1^{a_1}x_2^{a_2} \cdots x_n^{a_n}$, where n is some positive integer, each $x_i \in X$ and each $a_i \in \mathbf{Q}$, forms a multiplicative subgroup U_1 of $\overline{C(X)}$ and $x_1^{a_1}x_2^{a_2} \cdots x_n^{a_n}$ takes on its usual meaning if each $a_i \in \mathbf{Z}$. For any $u = x_1^{a_1} \cdots x_n^{a_n}$ in U_1, we define $v(u) = \sum_{i=1}^n a_i v(x_i) \in \mathbf{Q}\Gamma$, which is consistent with the previous definition of v on X. Since the elements of X are algebraically independent over C and $v(X) = \Sigma$ is a **Q**-basis of $\mathbf{Q}\Gamma$, the map $v: U_1 \to \mathbf{Q}\Gamma$ is a group isomorphism. Now look at the ring $C[U_1] \subset \overline{C(X)}$. Any nonzero element of $C[U_1]$ can be written in the form $c_1u_1 + \cdots + c_ru_r$, where $u_1, \ldots, u_r$ are distinct elements of U_1 and each $c_i \in C^*$, and this representation is unique except for the order of the terms, again by the algebraic independence over C of the elements of X. We can therefore extend the map $v: U_1 \to \mathbf{Q}\Gamma$ to a well-defined map of the nonzero elements of $C[U_1]$ into $\mathbf{Q}\Gamma$ by setting $v(c_1u_1 + \cdots + c_ru_r) = \min \{v(u_1), \ldots, v(u_r)\}$, if $u_1, \ldots, u_r$ are distinct elements of U_1 and each $c_i \in C^*$, and we check immediately that for nonzero elements $y_1, y_2 \in C[U_1]$ we have $v(y_1y_2) = v(y_1) + v(y_2)$, $v(y_1 + y_2) \geq \min \{v(y_1), v(y_2)\}$. Thus the map $v: (C[U_1])^* \to \mathbf{Q}\Gamma$ extends to a valuation $v: (C(U_1))^* \to \mathbf{Q}\Gamma$. Thus there is a unique valuation v on $C(U_1)$ that is trivial on C and agrees with our original v on X, and the value group and residue class field of v are respectively $\mathbf{Q}\Gamma$ and C. Now let $T_1 = \{u \in U_1: v(u) \in \Gamma\}$. Then T_1 is a subgroup of U_1 and v induces an isomorphism between T_1 and Γ. Setting $K = C(T_1)$, v induces a valuation of K that is trivial on C and has value group Γ and residue class field C. We now go about defining a derivation on K. There is a well-defined function $\xi: X \to U_1$ such that for each $x \in X$ we have $v(\xi(x)) = \psi(v(x))$. Since ψ maps Γ^* into Γ, we have $\xi(X) \subset T_1$. Since X is a transcendence base for K over C, there is a unique derivation of K that is trivial on C and such that for each $x \in X$ we have $x' = h(x)\xi(x)x$. This derivation makes K into a differential field such that for each $x \in X$ we have $v(x'/x) = v(\xi(x)) = \psi(v(x))$. Now let $T = C^*T_1$, a subgroup of the multiplicative group of K. We have $C^* \subset T$, $K = C(T)$, and the kernel of the homomorphism $v: T \to \Gamma$ is C^*. We now wish to show that any constant in T is in C. It suffices for this to show that the only constant in T_1 is 1. So suppose that $x_1, \ldots, x_n$ are distinct elements of X, that $a_1, \ldots, a_n \in \mathbf{Q}^*$, that $x_1^{a_1} \cdots x_n^{a_n} \in T_1$ and that $(x_1^{a_1} \cdots x_n^{a_n})' = 0$. Then $(x_1^{a_1} \cdots x_n^{a_n})'/(x_1^{a_1} \cdots x_n^{a_n}) = a_1x_1'/x_1 + \cdots + a_nx_n'/x_n = a_1h(x_1)\xi(x_1) + \cdots + a_nh(x_n)\xi(x_n) = 0$. Now each $a_ih(x_i) \in C^*$ and each $\xi(x_i) \in U_1$, and there is a partition $\mathcal{P}$ of the set $\{1, \ldots, n\}$ such that for $i, j \in \{1, \ldots, n\}$ we have $\xi(x_i) = \xi(x_j)$ if and only if i and j are in the same set in $\mathcal{P}$. Therefore for any set $I \in \mathcal{P}$ we have $\Sigma_{i \in I}\, a_ih(x_i) = 0$. For $i \in I \in \mathcal{P}$ we have $\psi(v(x_i)) = v(\xi(x_i))$, so that $v(x_i) \in \Sigma_{v(\xi(x_i))}$, which depends only on I. But the function $h: X \to C^*$ was constructed so that the elements $\{h(x_i)\}_{i \in I}$ are linearly independent over $\mathbf{Q}$, which implies that $a_i = 0$ for all $i \in I$, a contradiction. Therefore each constant in T is in C. We now verify that for each $u \in T$ such that $v(u) \neq 0$, we have $v(u'/u) = \psi(v(u))$. Since $T = C^*T_1 \subset C^*U_1$, we need only prove this for $u \in U_1$. Write $u = x_1^{a_1} \cdots x_n^{a_n}$, with $x_1, \ldots, x_n$ distinct elements of X and $a_1, \ldots, a_n \in \mathbf{Q}^*$. Then $v(u'/u) = v(\Sigma_{i=1}^n a_i x_i'/x_i) = v(\Sigma_{i=1}^n a_i h(x_i)\xi(x_i))$. We know that for $i = 1, \ldots, n$, $\xi(x_i)$ depends only on $\psi(v(x_i))$, so by the way the function h was defined we get $\Sigma_{i \in I} a_i h(x_i) \neq 0$ if I is any maximal subset of $\{1, \ldots, n\}$ for which $\xi(x_i)$ assumes a constant value. We therefore obtain $v(u'/u) = \min_{i=1,\ldots,n} v(\xi(x_i)) = \min_{i=1,\ldots,n} \psi(v(x_i))$. By our construction of the set Σ and property (b), the latter value is just $\psi(\Sigma_{i=1}^n a_i v(x_i)) = \psi(v(u))$, as was to be shown. We can now verify that if $a, b \in T$ and $v(a), v(b) > 0$, then $v(a'b/b') > 0$, by noting that $v(a'b/b') = v(a'/a) + v(a) - v(b'/b) = \psi(v(a)) + v(a) - \psi(v(b))$, which is positive by property (c). Thus Theorem 2 of [1] can be applied to our current situation, using the present v, C, T, and K, with the k and $\mathcal{C}$ of the quoted theorem both taken to be C. We conclude that v is a differential valuation of K and C is its subfield of constants. We already know that the value group of v is Γ, and all that remains to be verified is the statement that if $a \in K^*$ and $v(a) \neq 0$, then $v(a'/a) = \psi(v(a))$. This statement is already known for $a \in T$ and it follows for any $a \in K^*$ such that $v(a) \neq 0$ by choosing, as we can, some $a_1 \in T$ such that $v(a) = v(a_1)$ and noting that since v is a differential valuation of K we have $v(a') = v(a_1')$. This completes the proof.

The procedure of the above proof is often easy to carry out in practice. Suppose for example that Γ is the subgroup of the lexicographically ordered group $\mathbf{R}^2$ given by

$$\Gamma = \{(m + n\pi, p)\colon m, n, p \in \mathbf{Z}\},$$

with ψ given by

$$\psi(m + n\pi, p) = \begin{cases} (-1, 0) & \text{if} \quad m \neq 0 \\ (-1, 1) & \text{if} \quad m = 0, (n, p) \neq (0, 0). \end{cases}$$

Properties (a), (b), (c) are easily checked. Following the procedure of the proof, we can choose, for example, $\Sigma_{(-1,0)} = \{(1, 0)\}$, $\Sigma_{(-1,1)} = \{(\pi, 0), (0, 1)\}$. Let $C = \mathbf{C}$ and let x_1, x_2, x_3 be indeterminates over $\mathbf{C}$ with $v(x_1) = (1, 0)$, $v(x_2) = (\pi, 0)$, $v(x_3) = (0, 1)$. The equation $v(\xi(x)) = \psi(v(x))$ gives $\xi(x_1) = x_1^{-1}$, $\xi(x_2) = \xi(x_3) = x_1^{-1}x_3$. Trying $h(x_1) = h(x_3) = 1$, $h(x_2) = a$, where a is an arbitrary element of

$\mathbf{C} - \mathbf{Q}$, and using the equation $x' = h(x)\xi(x)x$, we get $K = \mathbf{C}(x_1, x_2, x_3)$, with $x_1' = 1$, $x_2' = ax_1^{-1}x_2x_3$, $x_3' = x_1^{-1}x_3^2$. A specific complex variable solution of this system of equations is $x_1 = z$, $x_2 = (-1/\log z)^a$, $x_3 = -1/\log z$. Thus the given (Γ, ψ) is associated with a differential valuation of the field $\mathbf{C}(z, \log z, (-1/\log z)^a)$.

The well-ordering assumption on $\psi(\Gamma^*)$ made in the statement of Theorem 1 can be dropped if it turns out that the set Σ constructed in the proof spans $\mathbf{Q}\Gamma$. As an example where this happens, let $\Gamma = \bigoplus_{i\in \mathbf{Z}} \mathbf{Z}\,\epsilon_i$, where each $\epsilon_i > 0$ and each $\epsilon_i > n\epsilon_{i+1}$ for any $n \in \mathbf{Z}$ (so that Γ is isomorphic to the additive group of doubly infinite sequences of integers almost all of which are zero, lexicographically ordered) and let ψ be defined by $\psi(\epsilon_i) = \epsilon_0 + \epsilon_1 + \cdots + \epsilon_i$ if $i \geq 0$, $\psi(\epsilon_{-1}) = 0$, $\psi(\epsilon_i) = -\epsilon_{i+1} - \cdots - \epsilon_{-1}$ if $i < -1$. Choose $\Sigma = \{\epsilon_i\}_{i\in \mathbf{Z}}$, $C = \mathbf{R}$, indeterminates $\{x_i\}_{i\in\mathbf{Z}}$ over $\mathbf{R}$ such that each $v(x_i) = \epsilon_i$, and each $h(x_i) = -1$. We get $K = \mathbf{R}(\{x_i\}_{i\in\mathbf{Z}})$ with $x_i'/x_i = -x_0x_1 \cdots x_i$ if $i \geq 0$, $x_{-1}'/x_{-1} = -1$, $x_i'/x_i = -x_{i+1}^{-1} \cdots x_{-1}^{-1}$ if $i < -1$. A real variable solution of this system is $x_0 = 1/x$, $x_1 = 1/\log x$, $x_2 = 1/\log\log x, \ldots, x_{-1} = 1/e^x$, $x_{-2} = 1/\exp(e^x), \ldots$, so that (Γ, ψ) is associated with the differential valuation of the Hardy field $\mathbf{R}(\ldots, e^x, x, \log x, \log\log x, \ldots)$ as $x \to +\infty$.

3. Asymptotic Integration. Let k be a differential field with subfield of constants C and let v be a differential valuation of k. Given a nonzero element $a \in k$, we may want to find an integral of a in k, or, if this is not possible, an element $b \in k$ whose derivative is near a in an appropriate sense, which is here that $v(a - b') > v(a)$. If this can be done, we may be able to find a $c \in k$ such that $v((a - b') - c') = v(a - (b + c)') > v(a - b')$, and with luck we may be able to find a sequence $0 = b_0, b_1, b_2, \ldots$, in k such that the sequence $v(a - b_i')$ increases in $\Gamma = v(k^*)$, possibly without bound. Whether or not the sequence $v(a - b_i')$ is bounded, the various approximations b_i to $\int a$ may be useful for whatever task is at hand. Thus the principal question is whether, given $a \in k^*$, there is a $b \in k$ such that $v(a - b') > v(a)$. If we have such a b, then $v(a) = v(b')$. Conversely, if $b \in k$ is such that $v(a) = v(b')$, then $v(a/b') = 0$, so there exists some $c \in C$ such that $v((a/b') - c) > 0$, or $v(a - (cb)') > v(b') = v(a)$. Therefore the question is that of finding $b \in k$ such that $v(b') = v(a)$. If $v(b) = 0$, then for some $c_1 \in C$ we have $v(b - c_1) > 0$ while $(b - c_1)' = b'$, so the question is that of finding $b \in k$ such that $v(b) \neq 0$ and $v(b') = a$, or, since $v(b') = \psi(v(b)) + v(b)$, the question is of finding $\beta \in \Gamma^*$ such that $v(a) = \psi(\beta) + \beta$. Thus the nonzero elements of k which are "asymptotically integrable" are those elements whose values are of the form $\psi(\beta) + \beta$, for some $\beta \in \Gamma^*$.

THEOREM 2. *Let Γ be an ordered abelian group and $\psi: \Gamma^* \to \Gamma$ a function with properties* (a), (b), (c). *If $\psi(\Gamma^*)$ has a maximal element α, then there exists no $\beta \in \Gamma^*$ such that $\psi(\beta) + \beta = \alpha$. If the set $\psi(\Gamma^*)$ is well-ordered, then there is at most one element $\alpha \in \Gamma$ such that $\alpha \neq \psi(\beta) + \beta$ for any $\beta \in \Gamma^*$.*

If $\alpha_0 \in \Gamma^*$ is such that $\alpha = \psi(\alpha_0)$ is maximal in $\psi(\Gamma^*)$ and if $\beta \in \Gamma^*$ is such that $\psi(\beta) + \beta = \alpha = \psi(\alpha_0)$, then we must have $\psi(\beta) \leq \psi(\alpha_0)$, so that $\beta \geq 0$, implying $\beta > 0$ and contradicting $\psi(\alpha_0) < \psi(\beta) + |\beta|$. This proves the first contention. Now suppose that our well-ordering condition holds and that $\alpha \in \Gamma$ is not of the form $\psi(\beta) + \beta$ for any $\beta \in \Gamma^*$. If for some $\gamma \in \psi(\Gamma^*)$ we had $\alpha - \gamma \in \psi^{-1}(\gamma)$, then we would have $\psi(\alpha - \gamma) + (\alpha - \gamma) = \gamma + (\alpha - \gamma) = \alpha$, which is impossible. Hence $\alpha - \gamma \notin \psi^{-1}(\gamma)$ for each $\gamma \in \psi(\Gamma^*)$. We claim that if $\gamma_1, \gamma_2 \in \psi(\Gamma^*)$ and $\gamma_1 \leq \gamma_2$ then $\alpha - \gamma_2 \notin \psi^{-1}(\gamma_1)$. For otherwise we could find $\gamma_1, \gamma_2 \in \psi(\Gamma^*)$ with $\gamma_1 \leq \gamma_2$ and $\alpha - \gamma_2 \in \psi^{-1}(\gamma_1)$ such that γ_2 is minimal for all pairs (γ_1, γ_2) with these properties. Since $\alpha - \gamma_2 \notin \psi^{-1}(\gamma_2)$ we must have $\gamma_1 < \gamma_2$. By the minimality property of γ_2, for each $\gamma \in \psi(\Gamma^*)$ such that $\gamma \leq \gamma_1$ we have $\alpha - \gamma_1 \notin \psi^{-1}(\gamma)$. But either $\alpha - \gamma_1 = 0$ or $\alpha - \gamma_1 \in \psi^{-1}(\psi(\alpha - \gamma_1))$, so that

$$\alpha - \gamma_1 \in \cup\{\psi^{-1}(\gamma): \gamma \in \psi(\Gamma^*), \gamma > \gamma_1\} \cup \{0\}.$$

Recalling that $\alpha - \gamma_2 \in \psi^{-1}(\gamma_1)$ and using property (b) we obtain $\gamma_2 - \gamma_1 = (\alpha - \gamma_1) - (\alpha - \gamma_2) \in \psi^{-1}(\gamma_1)$. Therefore property (c) implies that $\gamma_2 < \psi(\gamma_2 - \gamma_1) + |\gamma_2 - \gamma_1| = \gamma_1 + (\gamma_2 - \gamma_1) = \gamma_2$, which is false. This proves that for each $\gamma_1 \leq \gamma_2$ in $\psi(\Gamma^*)$ we have $\alpha - \gamma_2 \notin \psi^{-1}(\gamma_1)$, so that

$$\alpha - \gamma_2 \in \cup\{\psi^{-1}(\gamma): \gamma \in \psi(\Gamma^*), \gamma > \gamma_2\} \cup \{0\}.$$

Thus if $\alpha_1, \alpha_2 \in \Gamma$ are neither of the form $\psi(\beta) + \beta$ for any $\beta \in \Gamma^*$, the statement $\alpha_1 - \alpha_2 = (\alpha_1 - \gamma_2) - (\alpha_2 - \gamma_2)$ and property (b) imply

$$\alpha_1 - \alpha_2 \in \cup\{\psi^{-1}(\gamma): \gamma \in \psi(\Gamma^*), \gamma > \gamma_2\} \cup \{0\}$$

for all $\gamma_2 \in \psi(\Gamma^*)$. If $\alpha_1 \neq \alpha_2$ we get a contradiction by taking $\gamma_2 = \psi(\alpha_1 - \alpha_2)$, which completes the proof.

COROLLARY. *Let k be a differential field, C its subfield of constants, and let v be a differential valuation of k. Suppose that $k \neq C$ and that each element of k is differentially algebraic over C. Then for any $a \in k^*$ there exists a $b \in k$ such that $v(a - b') > v(a)$ except in the case where* $\max\{v(u'/u): u \in k^*, v(u) \neq 0\}$ *exists and $v(a)$ is this maximum.*

For any $a, s \in k^*$ with $v(s) \neq 0$, the differential field $C\langle a, s\rangle$ generated by C, a and s is of finite transcendence degree over C, so that the value group $v((C\langle a, s\rangle)^*)$ is of finite rational rank. Thus the set $\{v(u'/u): u \in (C\langle a, s\rangle)^*, v(u) \neq 0\}$ is finite. The theorem tells us that for some nonzero $\beta \in v((C\langle a, s\rangle)^*)$ we have $v(a) = \psi(\beta) + \beta$, which implies the existence of some $b \in C\langle a, s\rangle$ such that $v(a - b') > v(a)$, except in the case where $v(a) = \max\{v(u'/u): u \in (C\langle a, s\rangle)^*, v(u) \neq 0\}$, which implies in particular that $v(a)$ is of the form $v(u'/u)$ for some $u \in k^*$ such that $v(u) \neq 0$ and that $v(a) \geq v(s'/s)$. Since s was an arbitrary element of k^* such that $v(s) \neq 0$, we are done.

For example, consider the sequence of Hardy fields

$$\mathbf{R}(x) \subset \mathbf{R}(x, \log x) \subset \mathbf{R}(x, \log x, \log\log x) \subset \cdots$$

of germs of differentiable real-valued functions on neighborhoods of $+\infty$ in $\mathbf{R}$, differentiation being the usual differentiation with respect to x. According to [1, (*)], in a Hardy field we get a maximal ψ from a minimal positive element of the value group. Elements of the above sequence of fields with minimal positive values are

$$1/x,\ 1/\log x,\ 1/\log\log x,\ \ldots$$

respectively, which have logarithmic derivatives

$$-1/x,\ -1/x\log x,\ -1/x\log x\log\log x,\ \ldots$$

respectively, whose values are precisely the values of those elements of the respective fields which cannot be asymptotically integrated in the same field, each element, however, having an integral in the next field.

In the union Hardy field $\mathbf{R}(x, \log x, \log\log x, \ldots)$, all elements can be asymptotically integrated.

UNIVERSITY OF CALIFORNIA, BERKELEY

REFERENCE

[1] M. Rosenlicht, "Differential valuations," *Pacific J. Math.* (to appear).

QUASIRATIONAL SINGULARITIES

By SHREERAM S. ABHYANKAR

In the fall of 1967, I heard a beautiful lecture by Hirzebruch, at M.I.T., in which he studied the singularity which the complex surface

$$X^a + Y^b + Z^c = 0$$

has at the origin, where a, b, c are pairwise coprime integers. By topological arguments he showed that, upon resolving the singularity, only rational curves come out of it. Actually, my recollection may be faulty and Hirzebruch might have only considered the case when $(a, b, c) = (2, 3, 5)$. At any rate, after the lecture, Wagreich asked me to explain the proof. In a couple of days I was able to make an algebraic argument.

Namely, first I noted that by projecting along the Z-axis, the above singularity can be regarded as a c-fold cyclic covering of the origin in the (X, Y)-plane and then the branch locus is the analytically irreducible plane curve

$$X^a + Y^b = 0$$

having only one characteristic pair (a, b). Next, from this hypothesis I deduced that only rational curves can come out of the singularity of the surface, no matter how we blow it up birationally. A surface singularity with the latter property may be called quasirational.

The prefix quasi is appended to avoid confusion with the rational singularities studied by DuVal and revisited by Artin, Brieskorn, Hirzebruch, Lipman, et al. Briefly, rational singularities are those singularities which upon resolving leave the arithmetic genus of the surface unchanged. Evidently, every rational singularity is quasirational but not conversely. To see that a quasirational singularity need not be rational, firstly recall that the quadratic transform of a normal rational singularity is again always normal. Secondly, consider the singularity at the origin of the surface

$$X^2 + Y^b + Z^c = 0 \qquad \text{with } 3 < b < c$$

where b and c are coprime odd integers. The surface is clearly devoid of multiple curves and so it is normal. Upon making the quadratic transformation

$$X = Y'X'$$
$$Y = Y'$$
$$Z = Y'Z'$$

and factoring out Y'^2, as part of the proper transform we get the surface

$$X'^2 + Y'^{b-2} + Z'^c Y'^{c-2} = 0.$$

The proper transform is met by the exceptional locus $Y' = 0$ in the line $X' = Y' = 0$. In other words, the origin of the original surface is blown up into this line. It is clearly a double line of the proper transform and hence the latter is not normal. So the said singularity is not rational. However, by the claim which was made earlier and which will be proved in a moment, it is quasirational.

I guess, in his lecture, Hirzebruch probably considered only those a, b, c for which one does get a rational singularity.

To discuss the above claim about the quasirationality of certain cyclic coverings, let us start with Jung's pioneering work on the local uniformization of algebraic surfaces which was first expounded in his 1908 memoir in volume 133 of the Crelle Journal and later exposed in his 1950 book on Algebraic Functions of Two Variables. Other than Zariski's own work, this 1908 memoir of Jung was the very first paper which Zariski had advised me to study in 1953 when, as a graduate student, I was working under his masterful guidance. Incidently, if I may also be allowed to reproduce the list of books which, at that time, Zariski had advised me to study: Neumann, Osgood, Severi, Stahl, and Weyl. I have still lovingly preserved the slip of paper on which Zariski wrote down the names of these books as well as the letter in which he described the said memoir of Jung.

In studying the above works of Jung, firstly I realized that a simple point of a surface is always quasirational; this was formalized in Proposition 3 of my Valuations paper in volume 78 of the American Journal of 1956. Secondly, Jung considers surface singularities which are local (tame) coverings of simple points of surfaces with branch loci having only normal crossings; let us designate such surface singularities as Jungian singularities. By an analytic argument, Jung shows that a Jungian singularity can be covered by a simple point; this aspect of Jungian singularities was algebracized in Theorem 3 of my Ramification paper in volume 77 of the American Journal of 1955. Now the classical theorem of Lüroth asserts the rationality of a curve which can be covered by a rational curve; from this we immediately deduce the quasirationality of any surface singularity which can be covered by a quasirational singularity. Thus we conclude that every Jungian singularity is quasirational.

To review the notion of characteristic pairs introduced by Halphen and Smith towards the end of the last century, consider an irreducible analytic plane curve given by

$$F(X, Y) = 0$$

where $F(X, Y)$ is an irreducible power series with complex coefficients, or, more generally, with coefficients in an algebraically closed field of characteristic zero. In other words,

$$F(X, Y) \text{ is an irreducible nonzero nonunit in } k[[X, Y]] \tag{1}$$

where

$$k \text{ is an algebraically closed field of characteristic zero} \tag{2}$$

and where, as usual, $k[[X, Y]]$ is the ring of all formal power series in indeterminates X and Y with coefficients in k. Let

$$n = \operatorname{ord} F(X, Y), \text{ i.e., the order of } F(X, Y) \tag{3}$$

and note that then

$$F(X, Y) = F_n(X, Y) + \sum_{i+j>n} F_{ij}X^iY^j \qquad \text{with } F_{ij} \text{ in } k$$

where $F_n(X, Y)$ is a nonzero homogeneous polynomial of degree n in X and Y with coefficients in k and it is called the initial form of $F(X, Y)$.

Clearly we can choose

$$\text{nonunits } A^*(X, Y) \text{ and } B^*(X, Y) \text{ in } k[[X, Y]]$$

with

$$k[[A^*(X, Y), B^*(X, Y)]] = k[[X, Y]] \tag{1*}$$

such that upon letting

$$F^*(X, Y) = F(A^*(X, Y), B^*(X, Y))$$

we have

$$\operatorname{ord} F^*(0, Y) = n. \tag{2*}$$

By the Weierstrass Preparation theorem we can uniquely write

$$F^*(X, Y) = D^*(X, Y)E^*(X, Y)$$

where

$$D^*(X, Y) \text{ in } k[[X, Y]] \text{ with } D^*(0, 0) \neq 0$$

and

$$E^*(X, Y) = Y^n + \sum_{i=1}^{n} E_i^*(X)Y^{n-i} \text{ with } E_i^*(X) \text{ in } k[[X]] \text{ and } E_i^*(0) = 0.$$

Now by Newton's Theorem we have

$$E^*(T^n, Y) = \prod_{i=1}^{n} (Y - y^*(w_iT))$$

where

$$y^*(T) = \Sigma\, y_j^*T^j \text{ in } k[[T]] \text{ with } y_j^* \text{ in } k$$

and

$$Y^n - 1 = \prod_{i=1}^{n} (Y - w_i) \text{ with } w_i \text{ in } k.$$

We get unique sequences

$$0 < m = m_1 < m_2 < \cdots < m_{h+1} = \infty \quad \text{and}$$

$$n = d_1 > d_2 > \cdots > d_{h+1} = 1$$

where $h, m_1, m_2, \ldots, m_h, d_1, \ldots, d_{h+1}$ are nonnegative integers such that for $1 \leq i \leq h + 1$ we have

$$d_i = GCD(n, m_1, \ldots, m_{i-1})$$

and

$$m_i = \text{the smallest integer for which } y^*_{m_i} \neq 0 \text{ and } m_i \not\equiv 0(d_i).$$

The nonnegative integer h is called the number of characteristic pairs of $F(X, Y)$, and the pairs

$$(m_i/d_{i+1}, d_i/d_{i+1})_{1 \leq i \leq h}$$

of coprime positive integers are called the characteristic pairs of $F(X, Y)$. Now the number as well as the values of the characteristic pairs depend only on the residue class ring

$$k[[X, Y]]/F(X, Y)k[[X, Y]]$$

and so in particular they are independent of the choice of the nonunits $A^*(X, Y)$ and $B^*(X, Y)$ in $k[[X, Y]]$ satisfying (1*) and (2*), and they are also independent of the choice of k. This is proved in my Inversion paper in volume 89 of the American Journal of 1967.

We can take

nonzero elements P^* and Q^* in k

such that

$$\text{if } h \neq 0 \text{ then } P^{*m} = y_m{}^* \text{ and } Q^{*mn} = D^*(0, 0)$$

whereas

$$\text{if } h = 0 \text{ then } P^* = 1 \text{ and } Q^* = D^*(0, 0).$$

Now let

$$P = P^{*-n}Q^{*-n}$$

$$Q = \begin{cases} Q^{*-m} & \text{if } h \neq 0. \\ Q^{*-1} & \text{if } h = 0 \end{cases}$$

$$C(X) = \sum_{j<m/n} y^*{}_{nj}X^j$$

and

$$y(T) = Q^{-1}[y^*(P^{*-1}Q^{*-1}T) - C(P^{*-n}Q^{*-n}T^n)].$$

Then clearly

$$y(T) = \Sigma\, y_j T^j \text{ in } k[[T]] \text{ with } y_j \text{ in } k \tag{4}$$

such that

$$y_j = 0 \text{ whenever } j < m, \text{ whereas if } h \neq 0 \text{ then } y_m = 1. \tag{5}$$

Upon letting

$$A(X, Y) = A^*(PX, QY + C(PX)) \quad \text{and}$$

$$B(X, Y) = B^*(PX, QY + C(PX))$$

we clearly have that

$$A(X, Y) \text{ and } B(X, Y) \text{ are nonunits in } k[[X, Y]] \tag{6}$$

such that

$$k[[A(X,\ Y), B(X,\ Y)]] = k[[X,\ Y]]. \tag{7}$$

Now upon letting

$$\overline{F}(X,\ Y) = F(A(X,\ Y), B(X,\ Y)) \tag{8}$$

we have

$$\operatorname{ord} \overline{F}(O,\ Y) = n \tag{9}$$

and upon letting

$$D(X,\ Y) = D^*(0, 0)^{-1}D^*(PX,\ QY + C(PX)) \quad \text{and}$$

$$E(X,\ Y) = D^*(0, 0)E^*(PX,\ QY + C(PX))$$

we have

$$\overline{F}(X,\ Y) = D(X,\ Y)E(X,\ Y) \tag{10}$$

$$D(X,\ Y) \text{ in } k[[X,\ Y]] \text{ with } D(0, 0) = 1 \tag{11}$$

$$E(X,\ Y) = Y^n + \sum_{i=1}^{n} E_i(X)Y^{n-i} \text{ with } E_i(X) \text{ in } k[[X]] \text{ and } E_i(0) = 0 \tag{12}$$

$$E(T^n,\ Y) = \prod_{i=1}^{n} (Y - y(w_i)) \tag{13}$$

and hence we also have

$$\operatorname{ord} \overline{F}(X,\ 0) = m. \tag{14}$$

We note that

$$h = 0 \text{ iff } n = 1 \text{ iff } m = \infty \text{ iff } E(X,\ Y) = Y \tag{15}$$

and that

$$\begin{cases} \text{if } h \neq 0 \text{ then } (m/GCD(m, n), n/GCD(m, n)) \text{ is the first} \\ \text{characteristic pair of } F(X, Y) \end{cases} \tag{16}$$

and finally that

$$h = 1 \text{ iff } n > 1 = GCD(m, n). \tag{17}$$

Now if $h \neq 0$ then we would have

$$E(T^n, Y) = \prod_{i=1}^{n} [(Y - w_i{}^m T^m) - \sum_{j>m} y_j w_i{}^j T^j] \qquad \text{by (5) and (13)}$$

$$= \overline{E}(T, Y) + \prod_{i=1}^{n} (Y - w_i{}^m T^m)$$

$$\text{where } \overline{E}(T, Y) = \sum_{i+mj>mn} \overline{E}_{ij} T^i Y^j \text{ with } \overline{E}_{ij} \text{ in } k$$

$$= \overline{E}(T, Y) + \prod_{i=0}^{n-1} (Y - w^{im} T^m)$$

where w is a primitive nth root of 1 in k

$$= \overline{E}(T, Y) + \prod_{i=0}^{n-1} (Y - \bar{w}^i T^m)$$

where $\bar{w} = w^m =$ a primitive (n/d)th root of 1 in k
where $d = GCD(m, n)$

$$= \overline{E}(T, Y) + \prod_{i=0}^{d-1} \prod_{j=0}^{(n/d)-1} (Y - \bar{w}^{i(n/d)+j} T^m)$$

$$= \overline{E}(T, Y) + (Y^{n/d} - T^{mn/d})^d$$

and hence we would have

$$E(X, Y) = (Y^{n/d} - X^{m/d})^d + \sum_{in+jm>mn} \overline{E}*_{ij} X^i Y^j \qquad \text{with } \overline{E}*_{ij} \text{ in } k$$

and therefore in view of (10) and (11) we would have

$$\overline{F}(X, Y) = (Y^{n/d} - X^{m/d})^d + \sum_{in+jm>mn} \overline{F}_{ij}X^iY^j \qquad \text{with } \overline{F}_{ij} \text{ in } k.$$

Thus we have shown that

$$\left\{\begin{array}{l}\text{if } h \neq 0 \text{ then } \overline{F}(X, Y) = (Y^{n/d} - X^{m/d})^d + \sum_{in+jm>mn} \overline{F}_{ij}X^iY^j \\ \\ \text{with } \overline{F}_{ij} \text{ in } k \text{ where } d = GCD(m, n)\end{array}\right. \tag{18}$$

and hence, in view of (15), (16) and (17), we see that

$$\left\{\begin{array}{l}\text{if } F(X, Y) \text{ has exactly one characteristic pair then that} \\ \text{characteristic pair is } (m, n) \text{ and we have} \\ \overline{F}(X, Y) = Y^n - X^m + \sum_{in+jm>mn} \overline{F}_{ij}X^iY^j \text{ with } \overline{F}_{ij} \text{ in } k.\end{array}\right. \tag{19}$$

Again, if $h \neq 0$ then upon letting

$$\hat{A}(T, U) = A(T^n, U^m) \text{ and } \hat{B}(T, U) = B(T^n, U^m)$$

by (6) and (7) we would see that $\hat{A}(T, U)$ and $\hat{B}(T, U)$ are nonunits in $k[[T, U]]$ which are devoid of nonunit common factors in $k[[T, U]]$, and upon letting

$$G(T, U) = F(\hat{A}(T, U), \hat{B}(T, U)) \quad \text{and} \quad d = GCD(m, n)$$

by (8) and (18) we would see that ord $G(T, U) = mn$ and the initial form of $G(T, U)$ is the dth power of a nonzero homogeneous polynomial of degree mn/d in $k[T, U]$ which factors into mn/d pairwise coprime homogeneous linear factors. Thus we have the following conclusion:

(20) If $h \neq 0$ then there exist nonunits $\hat{A}(T, U)$ and $\hat{B}(T, U)$ in $k[[T, U]]$ which are devoid of nonunit common factors and which are such that upon letting

$$G(T, U) = F(\hat{A}(T, U), \hat{B}(T, U)) \quad \text{and} \quad d = GCD(m, n)$$

we have that ord $G(T, U) = mn$ and that the initial form of $G(T, U)$ is the dth power of a nonzero homogeneous polynomial of degree mn/d in $k[T, U]$ which factors into mn/d pairwise coprime homogeneous linear factors.

Now given any nonzero $H(T, U)$ in $k[[T, U]]$ with ord $H(T, U) = e$, by Hensel's Lemma we see that if the initial form of $H(T, U)$ factors into e pairwise coprime homogeneous linear factors $H_1^*(T, U), \ldots, H_e^*(T, U)$, then $H(T, U)$ also factors into the product of e factors $H_1(T, U), \ldots, H_e(T, U)$ in $k[[T, U]]$ so that $H_i^*(T, U)$ is the initial form of $H_i(T, U)$, whence in particular ord $H_i(T, U) = 1$, for $1 \leq i \leq e$. Therefore, in view of (15), (16) and (17), by (20) we get the following

Proposition on Coverings of Curves. Consider an irreducible analytic curve given by $F(X, Y) = 0$ where $F(X, Y)$ is an irreducible nonzero nonunit element of $k[[X, Y]]$. Assume that $F(X, Y)$ has exactly one characteristic pair (m, n). Then the local (X, Y)-plane can be covered by the local (T, U)-plane so that the inverse image of the given curve has an ordinary multiple point of order mn, i.e., it consists of mn regular analytic curves whose tangent lines are pairwise distinct and so that each of the said mn regular analytic curves occurs in the said inverse image with multiplicity one. Equivalently, algebraically speaking, there exist nonunits $\hat{A}(T, U)$ and $\hat{B}(T, U)$ in $k[[T, U]]$ which are devoid of nonunit common factors in $k[[T, U]]$ and which are such that upon letting

$$G(T, U) = F(\hat{A}(T, U), \hat{B}(T, U))$$

we have

$$G(T, U) = v \prod_{p=1}^{mn} G_p(T, U) \qquad \text{with } v \neq 0 \text{ in } k$$

where

$$G_p(T, U) = U - v_p T + \sum_{i+j>1} G_{pij} T^i U^j \qquad \text{in } k[[T, U]]$$

with v_p and G_{pij} in k such that $v_1, \ldots, v_{mn}$ are pairwise distinct.

This suggests the following

Problem on Coverings of Curves. Consider an analytic curve given by $\tilde{F}(X, Y) = 0$ where $\tilde{F}(X, Y)$ is a nonzero nonunit in $k[[X, Y]]$. We can ask as to when it is possible to resolve the singularity of the curve into an ordinary multiple point, allowing multiple components, by means of a local covering of the plane by a plane, instead of by the usual birational method of quadratic transformations. Algebraically speaking, we are asking as to when it is possible to find nonunits $\tilde{A}(T, U)$ and $\tilde{B}(T, U)$ in $k[[T, U]]$ which are devoid of nonunit common factors and for which we have

$$\tilde{F}(\tilde{A}(T, U), \tilde{B}(T, U)) = \prod_{r=1}^{s} [U - \tilde{F}_r T + \sum_{i+j>1} \tilde{F}_{rij} T^i U^j]^{a(r)}$$

where the $a(r)$ are positive integers and where the $\tilde{F}_r$ and $\tilde{F}_{rij}$ are elements in k such that $\tilde{F}_1, \ldots, \tilde{F}_s$ are pairwise distinct. Now whenever this is possible, it would yield a nice method of computing the local fundamental group of the plane minus the curve. However, there is evidence to suggest that if this is possible and $\tilde{F}(X, Y)$ is irreducible then $\tilde{F}(X, Y)$ has at most one characteristic pair; if the said evidence is borne out then that would give an interesting characterization of having at most one characteristic pair.

With this discussion of characteristic pairs out of the way, let us return to the claim about the quasirationality of a normal surface singularity which can locally be realized as a cyclic covering of a simple point of a surface with analytically irreducible branch locus having exactly one characteristic pair whereby it is assumed that the covering degree is coprime with each of the two numbers occurring in the characteristic pair. By the usual tricks about suitable choices of primitive elements of cyclic extensions, such a surface singularity can be arranged to have an equation

$$Z^p - F(X, Y) = 0$$

where $F(X, Y)$ is an irreducible nonzero nonunit in $k[[X, Y]]$ having exactly one characteristic pair (m, n) so that p is coprime with mn. By the Proposition on Coverings of Curves we can cover the (X, Y)-plane locally by the (T, U)-plane so that the inverse image $G(T, U)$ of $F(X, Y)$

has an ordinary multiple point of order $q = mn$. Upon composing the two coverings, the given surface singularity gets covered by the normal surface singularity having the equation

$$Z^p - G(T, U) = 0$$

and now, in view of Lüroth's theorem, it suffices to show that the latter is quasirational.

So we are reduced to proving the auxilliary claim about the quasirationality of a normal surface singularity which can locally be realized as a special p-fold cyclic covering of a simple point of a surface with branch locus having a q-fold ordinary multiple point where it is assumed that p is coprime with q. Here special refers to the assumption that the surface singularity can be given by an equation of the form

$$Z^p - G(T, U) = 0$$

where $G(T, U)$ has a q-fold ordinary multiple point which is a priori assumed to be devoid of multiple components because now, unlike in the original claim in which the branch locus was irreducible, we cannot simultaneously rid ourselves of all the multiple components of the branch locus by the tricks about primitive elements.

To prove the auxilliary claim, we blow up the (T, U)-plane into the (T', U')-plane by the quadratic transformation centered at the origin, i.e., we make the substitution

$$T = T' \quad \text{and} \quad U = T'U'.$$

Also we take the corresponding blowing up of the given surface and normalize it. In other words, we take the normalization of the (T', U')-plane in the function field of the given surface in the sense which was first introduced by Zariski in his 1950 A.M.S. Memoir on Abstract Holomorphic Functions which was declared by Krull to be bahnbrechung. Now the original branch locus had an ordinary multiple point and so the new branch locus will have only normal crossings and hence, by the quasirationality of Jungian singularities, all the points of the new surface will be quasirational. Therefore any further birational blowing ups of the new surface will add only rational curves. So it only remains to consider the curves which were added when we passed from the given surface to the new surface. These however all lie above the exceptional line $T' = 0$ of the (T', U')-plane. Now the coprimality of p and q shows that this exceptional line is totally ramified. Therefore above it lies only one irreducible curve and that curve is in a birational correspondence with the exceptional line. Thus only one curve got added and it is rational. And so the auxilliary claim is established.

This paper could conveniently end right here if I were writing it only *to celebrate the eightieth birthday of Zariski* the geometer, i.e., Zariski up to his wertvolle 1935 Ergebnisse book on Algebraic Surfaces. But then Zariski has a dual personality; indeed, in my files I have always filed his reprints up to 1937 under C. Zariski and his reprints after 1938 under M. Zariski where obviously C is for Classical or the Geometer while M is for Modern or the Algebraist. So to complete *my affectionate tribute to the whole of Zariski,* I shall now reword the paper in algebraic terms full of local rings and such. Actually, in my own writings I have usually been overly algebraical which is probably due to the influence of Chevalley who was the second author recommended to me by Zariski when I was a graduate student; lately I have been trying to mollify these algebraic tendencies by the reading of Baker's Principles of Geometry.

So again consider a point of a surface whose quasirationality we want to discuss. The matter being local we may pass to the local ring of the point, i.e., the ring of all functions on the surface which are regular at the point. These functions may be rational or meromorphic or formal and accordingly we get an algebraic or analytic or complete local ring. It does not matter which one of these three local rings we take because, as the Remark at the end of the paper will show, a local ring and its completion are either both quasirational or neither is. Formally we introduce the

Definition of Quasirationality. We shall denote the maximal ideal of a (noetherian) local domain S by $M(S)$, and we shall denote the canonical epimorphism of S onto $S/M(S)$ by N_S, and we shall call $S/M(S)$ the residue field of S. We recall that a local domain V is said to dominate a local domain S if S is a subring of V and $M(S) = M(V) \cap S$. Given a local domain V and a subring H of V, we say that V is residually rational over H to mean that $N_V(V) = N_V(H_{M(V)\cap H})$ and we say that V is residually algebraic (resp. finite algebraic, finitely generated, transcendental, pure transcendental, simple transcendental, almost pure transcendental, almost simple transcendental) over H to mean that $N_V(V)$ is an algebraic (resp. finite algebraic, finitely generated, transcendental, pure transcendental, simple transcendental, pure transcendental extension of a finite algebraic, simple transcendental extension of a finite algebraic) extension of $N_V(H_{M(V)\cap H})$ and, finally, by the residual transcendence degree of V over H we mean the transcendence degree of $N_V(V)$ over $N_V(H_{M(V)\cap H})$. Given a noetherian domain H, by a prime divisor of H we mean a one-dimensional regular local overdomain V of H having the same quotient field as H such that the residual transcendence degree of V over H equals $(\dim H_{M(V)\cap H}) - 1$. Given a noetherian domain H, by a hidden prime divisor of H we mean a prime divisor of H which is residually transcendental over H, i.e., equivalently, a prime divisor V of H such that $\dim H_{M(V)\cap H} \geq 2$. Note that if S is a two-dimensional local domain then a hidden prime divisor of S is simply a prime divisor of S which dominates S. Also note that if H is a noetherian domain of dimension ≤ 2 then a hidden prime divisor of H is simply a one-dimensional regular local overdomain of H having the same quotient field as and residually transcendental over H; (for instance see Theorem 1 of my Valuations paper). A noetherian domain H of dimension ≤ 2 is said to be quasirational if every prime divisor of H is residually almost pure transcendental over H. Now if S is a one-dimensional local domain and V is a one-dimensional regular local domain dominating S and having the same quotient field as S then V is residually finite algebraic over S; (for instance see Theorem 1 of my Valuations paper). Therefore a noetherian domain of dimension ≤ 1 is always quasirational, whereas a noetherian domain H of dimension ≤ 2 is quasirational iff every hidden prime divisor of H is residually almost simple transcendental over H.

Definition of Order. For any local domain S and any $g \in S$ we put

$$\operatorname{ord}_S g = \max r \text{ such that } g \in M(S)^r.$$

Moreover, if S is regular then for any t in the quotient field of S, upon writing

$$t = g/g' \text{ with } g \in S \text{ and } 0 \neq g' \in S$$

we put

$$\text{ord}_S t = (\text{ord}_S g) - (\text{ord}_S g')$$

and we note that this independent of the way of writing t as quotient of two elements of S.

Lemma on the Quasirationality of a Simple Point. Every two-dimensional regular local domain is quasirational.

Proof. This is proved in Proposition 3 of my Valuations paper.

Lemma on the Quasirationality of a Domain and Its Integral Extensions. Let H be a noetherian domain of dimension ≤ 2 and let H' be an overdomain of H such that H' is a finite H-module. [Note that then H' is a noetherian domain of the same dimension as H.] Assume that H' is quasirational. Also assume that for every height-two prime ideal P in H, the residue field of H_P is algebraically closed. Then H is quasirational.

Proof. Let L and L' be the quotient fields of H and H' respectively. Given any hidden prime divisor V of H, we can find a one-dimensional regular local domain V' with quotient field L' such that $V' \cap L = V$. Now V' dominates and is residually finite algebraic over V. Also V' is a hidden prime divisor of H' and hence, by the first assumption, V' is residually almost simple transcendental over H'. Let

$$S = H_{M(V)\cap H} \quad \text{and} \quad S' = H'_{M(V')\cap H'}.$$

Then clearly S and S' are two-dimensional local domains such that S' dominates and is residually finite algebraic over S. However, by the second assumption we see that the residue field of S is algebraically closed and hence S' is residually rational over S while V' is residually simple transcendental over S'. Therefore by Lüroth's theorem V is residually simple transcendental over S. Thus H is quasirational.

Lemma on the Simultaneous Quasirationality of a Domain and Its Normalizations. Let H be a noetherian domain of dimension ≤ 2 with quotient field L and let H' be a subdomain of L such that H' contains H and H' is a finite H-module. Then H is quasirational iff H' is quasirational.

Proof. Now H' is a noetherian domain having the same dimension as H, and hence the proof follows by noting that for any one-dimensional regular local domain V with quotient field L we have that: $H \subset V$ iff $H' \subset V$, and that: if $H' \subset V$ then $H'_{M(V)\cap H'}$ dominates and is residually finite algebraic over $H_{M(V)\cap H}$.

Notation. Henceforth R will denote a two-dimensional regular local domain with quotient field K and with an algebraically closed residue field k of characteristic zero. If w is a homomorphism of a ring H into a ring H^*, then for any polynomial

$$G(X, Y, \ldots) = \Sigma\, G_{ij\ldots} X^i Y^j \cdots$$

in indeterminates $X, Y, \ldots$ with coefficients in H, by $w(G)(X, Y, \ldots)$ we shall denote the polynomial in $X, Y, \ldots$ with coefficients in H^* obtained by putting

$$w(G)(X, Y, \ldots) = \Sigma\, w(G_{ij\ldots}) X^i Y^j \cdots$$

Definition of an Ordinary Multiple Point. Given $f \in R$, we say that f has a q-fold ordinary multiple point in R, where q is a positive integer, if there exists a basis (x, y) of $M(R)$ and a homogeneous $G(X, Y) \in R[X, Y]$ of degree q such that $f = G(x, y)$ and such that $N_R(G)(X, Y)$ is nonzero and factors into q pairwise coprime homogeneous linear factors in $k[X, Y]$. We note that if $f \in R$ has a q-fold ordinary multiple point in R and if (x, y) is any basis of $M(R)$ then $f = G(x, y)$ for some homogeneous $G(X, Y) \in R[X, Y]$ of degree q, and moreover for every such $G(X, Y)$ we have that $N_R(G)(X, Y)$ is nonzero and factors into q pairwise coprime homogeneous linear factors in $k[X, Y]$.

First Theorem on the Quasirationality of Certain Special Cyclic Coverings with Branch Locus Having Ordinary Multiple Point. Let p and q be coprime positive integers and let

$$S = R[z] \text{ with } z^p = f \text{ where } f \in R$$

and

$$f \text{ has a } q\text{-fold ordinary multiple point in } R.$$

Then S is quasirational.

Proof. Clearly S is a two-dimensional local overdomain of R such that S is integral over R, S dominates R, and S is residually rational over R. Let L be the quotient field of S. Given any one-dimensional local domain V with quotient field L such that V dominates and is residually transcendental over R, we shall show that V is residually simple transcendental over R, and this will complete the proof. We can take a basis (x, y) of $M(R)$ and then relabelling it suitably we can suppose that $y/x \in V$. Now $N_V(R)$ is algebraically closed and hence if $N_V(y/x)$ is algebraic over $N_V(R)$ then $N_V((y - ex)/x) = 0$ for some $e \in R$ and so upon replacing (x, y) by $(x, y - ex)$ we could arrange that $N_V(y/x) = 0$. Thus, to begin with, we may assume that (x, y) is a basis of $M(R)$ such that $y/x \in V$ and such that either $N_V(y/x)$ is transcendental over $N_V(R)$ or $N_V(y/x) = 0$. Since f has a q-fold ordinary multiple point in R, we have $f = G(x, y)$ with homogeneous $G(X, Y) \in R[X, Y]$ of degree q such that $N_V(G)(X, Y)$ is nonzero and factors into q pairwise coprime homogeneous linear factors in $N_V(R)[X, Y]$. Upon letting $P(Y) = G(1, Y)$ we get

$$P(Y) = \Sigma P_i Y^i \in R[Y] \quad \text{with} \quad P_i \in R \tag{1.1}$$

such that

$$z^p = f = x^q P(y/x) \tag{1.2}$$

$$N_V(P)(Y) \neq 0 \tag{1.3}$$

and

$$\text{either } P_0 \notin M(R) \quad \text{or} \quad P_1 \notin M(R). \tag{1.4}$$

We also note that

$$\operatorname{ord}_V(x) \text{ and } \operatorname{ord}_V(y) \text{ are positive integers.} \tag{1.5}$$

We shall now divide the argument into two cases according as $N_V(y/x)$ is transcendental over $N_V(R)$ or $N_V(y/x) = 0$.

First consider the Case when $N_V(y/x)$ is transcendental over $N_V(R)$. Now in view of (1.3) we get $\operatorname{ord}_V P(y/x) = 0$ and hence in view of (1.2) we get

$$\operatorname{ord}_V z = (q/p)\operatorname{ord}_V x. \tag{1.1*}$$

Given any

$$r \in V \quad \text{with} \quad r \notin M(V)$$

we can write

$$r = s/t \text{ with nonzero elements } s \text{ and } t \text{ in } R[z] \tag{1.2*}$$

and then we have

$$\operatorname{ord}_V s = \operatorname{ord}_V t \tag{1.3*}$$

and now we can write

$$s = \sum_{i \in A} s_i z^i \quad \text{and} \quad t = \sum_{j \in B} t_j z^j \tag{1.4*}$$

where A and B are nonempty sets of integers between 0 and $p - 1$ and where

$$0 \neq s_i \in R \text{ for all } i \in A \text{ and } 0 \neq t_j \in R \text{ for all } j \in B.$$

For every $i \in A$ and $j \in B$, upon letting

$$u(i) = \operatorname{ord}_R s_i \quad \text{and} \quad v(j) = \operatorname{ord}_R t_j$$

we have

$$s_i = G_i(x, y) \text{ and } t_j = H_j(x, y)$$

where $G_i(X, Y)$ and $H_j(X, Y)$ are nonzero homogeneous polynomials of degree $u(i)$ and $v(j)$ in X, Y with coefficients in R such that

$$N_V(G_i)(X, Y) \neq 0 \quad \text{and} \quad N_V(H_j)(X, Y) \neq 0. \tag{1.5*}$$

Now for every $i \in A$ and $j \in B$ we have

$$s_i = x^{u(i)}G_i(1, y/x) \quad \text{and} \quad t_j = x^{v(j)}H_j(1, y/x) \tag{1.6*}$$

and, because $N_V(y/x)$ is transcendental over $N_V(R)$, by (1.5*) we also get

$$\operatorname{ord}_V G_i(1, y/x) = 0 \quad \text{and} \quad \operatorname{ord}_V H_j(1, y/x) = 0. \tag{1.7*}$$

By (1.1*), (1.6*) and (1.7*) we get

$$\operatorname{ord}_V(s_i z^i) = [u(i) + (q/p)i]\operatorname{ord}_V x \qquad \text{for all } i \in A \tag{1.8*}$$

and

$$\operatorname{ord}_V(t_j z^j) = [v(j) + (q/p)j]\operatorname{ord}_V x \qquad \text{for all } j \in B. \tag{1.9*}$$

And hence, because p and q are coprime and $\operatorname{ord}_V x$ is a nonzero integer, there exists a unique $a \in A$ and a unique $b \in B$ such that

$$\operatorname{ord}_V(s_a z^a) < \operatorname{ord}_V(s_i z^i) \quad \text{for all} \quad i \neq a \text{ in } A \tag{1.10*}$$

and

$$\operatorname{ord}_V(t_b z^b) < \operatorname{ord}_V(t_j z^j) \quad \text{for all} \quad j \neq b \text{ in } B. \tag{1.11*}$$

Now by (1.4*), (1.8*), (1.9*), (1.10*) and (1.11*) we see that

$$\operatorname{ord}_V(s) = \operatorname{ord}_V(s_a z^a) = [u(a) + (q/p)a]\operatorname{ord}_V x$$

and

$$\operatorname{ord}_V(t) = \operatorname{ord}_V(t_b z^b) = [v(b) + (q/p)b]\operatorname{ord}_V x$$

and hence, because $\operatorname{ord}_V x$ is a nonzero integer, by (1.3*) we get

$$u(a) + (q/p)a = v(b) + (q/p)b$$

and therefore, again because p and q are coprime, we must have

$$a = b \quad \text{and} \quad u(a) = v(b). \tag{1.12*}$$

By (1.2*), (1.3*), (1.4*), (1.10*) and (1.11*) we see that

$$N_V(r) = N_V(s_a z^a / t_b z^b)$$

and by (1.6*), (1.7*) and (1.12*) we see that

$$N_V(s_a z^a / t_b z^b) = N_V(G_a(1, y/x)/H_b(1, y/x)).$$

Therefore

$$N_V(r) = N_V(G_a(1, y/x)/H_b(1, y/x)).$$

Thus

$$N_V(V) = (N_V(R))(N_V(y/x))$$

and hence V is residually simple transcendental over R.

Next we consider the Case when $N_V(y/x) = 0$. Now $(x, y/x)R[y/x]$ is a maximal ideal in $R[y/x]$ and the quotient ring of $R[y/x]$ with respect to this maximal ideal is a two-dimensional regular local ring R^* such that $(x, y/x)$ is a basis of $M(R^*)$ and such that R^* dominates and is residually rational over R, (for instance see (1.4) of my Resolution book published by the Academic Press in 1966). Clearly V dominates R^*. So V is residually transcendental over R^* and what we want to show is that V is residually simple transcendental over R^*. Upon letting

$$(s, y^*, u) = \begin{cases} (P(y/x), y/x, 0) & \text{if } P_0 \notin M(V) \\ (1, P(y/x), 1) & \text{if } P_0 \in M(V) \end{cases}$$

by (1.1), (1.2) and (1.4) we see that (x, y^*) is a basis of $M(R^*)$ and u is a nonnegative integer and

$$z^p = f = sx^q y^{*u} \qquad \text{with } s \in R^* \text{ and } s \notin M(R^*). \tag{1.1'}$$

We can take elements x' and y' in an overfield of L such that

$$x'^p = x \quad \text{and} \quad y'^p = y^*. \tag{1.2'}$$

Let

$$K' = K(x', y') \quad \text{and} \quad R' = R[x', y'].$$

Then clearly R' is a two-dimensional regular local ring with quotient field K', (x', y') is a basis of $M(R')$, R' is the integral closure of R^* in K', and R' dominates and is residually rational over R^*; (for instance see Theorem 6 of my Jungian paper in volume 160 of Mathematische Annalen of 1965). We can take a one-dimensional regular local domain V' with quotient field L' such that $V' \cap L = V$. Now V' dominates and is residually finite algebraic over V. Also V' dominates and is residually transcendental over R'. We shall show that V' is residually simple transcendental over R'; by Luroth's theorem this will yield that V is residually simple transcendental over R^* which will complete the proof. Let

$$L' = L(x', y') \text{ and } W = \text{the integral closure of } R' \text{ in } L'.$$

Then $W \subset V'$ and upon letting

$$S' = W_{M(V') \cap W}$$

we have that S' is a two-dimensional local domain with quotient field L' and that S' dominates and is residually rational over R' and finally that V' dominates and is residually transcendental over S'. We shall show that V' is residually simple transcendental over S' and this will complete the proof. Upon letting

$$z' = z/x'^q y'^u$$

in view of (1.1′) and (1.2′) we have

$$L' = K'(z') \text{ and } z'^p = s \qquad \text{with } s \in M(R') \text{ and } s \notin M(R').$$

Since the residue field of R' is of zero characteristic, we see that the Z-discriminant of $Z^p - s$ is a unit in R'; therefore $M(R')S' = M(S')$; (for example see (10.17) of my Resolution book). Consequently S' is regular and hence by the above Lemma on the Quasirationality of Simple Points we conclude that V' is residually almost simple transcendental over S'. Since the residue field of S' is algebraically closed, we conclude that V' is residually simple transcendental over S'.

Second Theorem on the Quasirationality of Certain Special Cyclic Coverings With Branch Locus Having Ordinary Multiple Point. Let p and q be coprime positive integers, let $f \in R$ be such that f has a q-fold ordinary multiple point in R, and let z be an element in an overfield of K such that

$$z^p = sf^r g^p \qquad \text{where } 0 \neq g \in R \text{ and } s \in R \text{ with } s \notin M(R)$$

and where r is a nonnegative integer. Let $L = K(z)$ and let S be an overdomain of R with quotient field L such that S is integral over R; (for instance $S = R[z]$). Then S is quasirational.

Proof. Let there be given any one-dimensional regular local domain with quotient field L such that $S \subset V$ and V is residually transcendental over S. What we have to show is that then V is residually almost simple transcendental over S. Now clearly $S_{M(V)\cap S}$ dominates and is residually rational over R. So V dominates and is residually transcendental over R, and what we have to show is that V is residually simple transcendental over R. Let

$$d = GCD(p, r) \quad \text{and} \quad p' = p/d \quad \text{and} \quad r' = r/d.$$

Let

$$z^* = z^{p'} f^{-r'} g^{-p'}$$

and

$$K' = K(z^*) \text{ and } W = \text{the integral closure of } R \text{ in } K'.$$

Then $W \subset V$ and upon letting

$$R' = W_{M(V)\cap W}$$

we have that R' is a two-dimensional local domain with quotient field K' and that R' dominates and is residually rational over R and finally that V dominates and is residually transcendental over R'. So, what we want to show is that V is residually simple transcendental over R'. Now

$$z^{*d} = s \quad \text{with} \quad s \in R \quad \text{and} \quad s \notin M(R)$$

and so the Z-discriminant of $Z^d - s$ is a unit in R and hence $M(R)R' = M(R')$; (for instance see (10.17) of my Resolution book). Therefore R' is regular. The above equation also shows that

$$z^* \in R' \quad \text{and} \quad z^* \notin M(R').$$

Now $GCD(p', r') = 1$ and hence we can write

$$ar' + bp' = 1 \qquad \text{with integers } a \text{ and } b.$$

Let

$$z' = z^a f^b g^{-a} \quad \text{and} \quad s' = z^{*a}.$$

Then

$$z'^{r'} z^{*b} g = z$$

and hence

$$L = K'(z').$$

Also

$$z'^{p'} = s'f \qquad \text{where } f \in M(R') \text{ and } s' \in R' \text{ with } s' \notin M(R')$$

and so upon letting

$$S' = R'[z]$$

we see that S' is a two-dimensional local domain with quotient field L and that S' dominates R' and is residually rational over R' and finally that V is a hidden prime divisor of S'. Since f has a q-fold ordinary point in R, upon taking a basis (x, y) of $M(R)$ we can write $f = G(x, y)$ with homogeneous $G(X, Y) \in R[X, Y]$ of degree q such that $N_R(G)(X, Y)$ is nonzero and factors into q pairwise coprime homogeneous linear factors in $N_R(R)[X, Y]$; since R' dominates R and $M(R)R' = M(R')$, it follows that (x, y) is a basis of $M(R')$ and that $G(X, Y) \in R'[X, Y]$ is homogeneous at degree q and finally that $N_{R'}(G)(X, Y)$ is nonzero and factors into q pairwise coprime homogeneous linear factors in $N_{R'}(R')[X, Y]$. Thus f has a q-fold ordinary multiple point in R'; also clearly p' and q are coprime, and hence by the above First Theorem ... we conclude that S' is quasirational. Therefore V is residually almost simple transcendental over S'. Since S' is residually rational over R' and the residue field of R' is algebraically closed, we see that V is residually simple transcendental over R'.

Definition of Characteristic Pairs. Given $f \in R$, we say that f is analytically irreducible in R if $0 \neq f \in M(R)$ and $\hat{R}/f\hat{R}$ is a domain where $\hat{R}$ is the completion of R and where we regard $\hat{R}$ to be an overdomain of R. Recall that there exist isomorphisms u of $\hat{R}$ onto $k[[X, Y]]$, and moreover, for any such u and any $g \in R$, upon letting $u(g) = G(X, Y) \in k[[X, Y]]$ we have that $N_R(g) = G(0, 0)$. Given $f \in R$ such that f is analytically irreducible in R, by the characteristic pairs of f in R we mean the characteristic pairs of $u(f)$ in $k[[X, Y]]$ where u is an isomorphism of $\hat{R}$ onto $k[[X, Y]]$; from what we have said earlier, it follows that the characteristic pairs of f in R, and so also their number, are independent of the choice of u. Given any height-one prime ideal I in R, we say that I is analytically irreducible in R to mean that R/IR is a domain. Given any height-one prime ideal I in R which is analytically irreducible in R, by the characteristic pairs of I in R we mean the characteristic pairs of a generator of I in R; these are clearly independent of the particular generator.

Theorem on Covering a Curve with One Characteristic Pair by a

Curve With Ordinary Multiple Point. Let any $f \in R$ be given such that f is analytically irreducible in R and f has exactly one characteristic pair (m, n) in R. Then there exists a basis of (x, y) of $M(R)$ such that if x' and y' are any elements in any overfield of K with

$$x'^n = x \quad \text{and} \quad y'^m = y$$

then upon letting

$$K' = K(x', y') \quad \text{and} \quad R' = R[x', y']$$

we have that: R' is the integral closure of R in K', R' is a two-dimensional regular local domain, (x', y') is a basis of $M(R')$, R' dominates and is residually rational over R, and f has an (mn)-fold ordinary multiple point in R'.

Proof. Let $\hat{R}$ be the completion of R where we regard $\hat{R}$ to be an overdomain of R. Now there exists an isomorphism u of $\hat{R}$ onto $k[[X, Y]]$. Let

$$u(f) = F(X, Y) = \Sigma F_{ij} X^i Y^j \in k[[X, Y]] \qquad \text{with } F_{ij} \in k.$$

In view of (6), (7), (8) and (19) we can find

$$A(X, Y) \text{ and } B(X, Y) \text{ in } M(k[[X, Y]])$$

with

$$k[[A(X, Y), B(X, Y)]] = k[[X, Y]]$$

such that upon letting

$$\overline{F}(X, Y) = F(A(X, Y), B(X, Y))$$

we have

$$\overline{F}(X, Y) = Y^n - X^m + \sum_{in+jm>mn} \overline{F}_{ij} X^i Y^j \text{ with } \overline{F}_{ij} \in k.$$

We get an automorphism v of $k[[X, Y]]$ by putting

$$v(G(x, Y)) = G(A(X, Y), B(X, Y)) \quad \text{for all } G(X, Y) \in k[[X, Y]]$$

and then we get an isomorphism $\bar{u}$ of $\hat{R}$ onto $k[[X, Y]]$ by putting

$$\bar{u}(g) = v(u(g)) \quad \text{for all } g \in \hat{R}.$$

Now we have

$$\bar{u}(f) = \overline{F}(X, Y).$$

Since $\hat{R}$ is the completion of R, we can find

$$x \text{ and } y \text{ in } R$$

such that

$$\bar{u}(x) = X + P(X, Y) \quad \text{and} \quad \bar{u}(y) = Y + Q(X, Y)$$

with

$$P(X, Y) \text{ and } Q(X, Y) \quad \text{in } M(k[[X, Y]])^{mn}.$$

Now clearly

$$(x, y) \text{ is a basis of } M(R)$$

and there exist unique elements

$$P^*(X, Y) \text{ and } Q^*(X, Y) \quad \text{in } M(k[[X, Y]])^{mn}$$

such that

$$X + P^*(X, Y) + P(X + P^*(X, Y), Y + Q^*(X, Y)) = X$$

and

$$Y + Q^*(X, Y) + Q(X + P^*(X, Y), Y + Q^*(X, Y)) = Y.$$

We get an automorphism v^* of $k[[X, Y]]$ by putting

$$v^*(G(X, Y)) = G(X + P^*(X, Y), Y + Q^*(X, Y)) \qquad \text{for all } G(X, Y) \in k[[X, Y]]$$

and then we get an isomorphism u^* of $\hat{R}$ onto $k[[X, Y]]$ by putting

$$u^*(g) = v^*(\bar{u}(g)) \qquad \text{for all } g \in \hat{R}.$$

Now clearly

$$u^*(x) = X \quad \text{and} \quad u^*(y) = Y$$

and upon letting

$$u^*(f) = F^*(X, Y) \in k[[X, Y]]$$

we have

$$F^*(X, Y) = Y^n - X^m + \sum_{in+jm>mn} F_{ij}^* X^i Y^j \quad \text{with} \quad F_{ij}^* \in k.$$

We can rewrite the above equation in the form

$$F^*(X, Y) = \sum_{in+jm=mn} G_{ij}^*(X, Y) X^i Y^j \quad \text{with} \quad G_{ij}^*(X, Y) \in k[[X, Y]]$$

with

$$G_{ij}^*(0, 0) = \begin{cases} 1 & \text{if } (i, j) = (0, n) \\ -1 & \text{if } (i, j) = (m, 0) \\ 0 & \text{in all other cases} \end{cases}$$

which shows that $u^*(f)$ belongs to the ideal in $k[[X, Y]]$ generated by the elements $u^*(x^i y^j)$ as i and j range over all nonnegative integers with $in + jm = mn$; therefore, because $\hat{R}$ is the completion of R, we can conclude that f belongs to the ideal in R generated by the elements $x^i y^j$ as i and j range over all nonnegative integers with $in + jm = m$;

thus we can write

$$f = \sum_{in+jm=mn} g(i, j)x^i y^j \quad \text{with } g(i, j) \in R.$$

Upon letting

$$u^*(g(i, j)) = G_{ij}(X, Y) \in k[[X, Y]]$$

we get

$$\begin{aligned} Y^n - X^m + \sum_{in+jm>mn} F_{ij}^* X^i Y^j &= F^*(X, Y) \\ &= u^*(f) \\ &= \sum_{in+jm=mn} G_{ij}(X, Y) X^i Y^j \\ &= \left[\sum_{in+jm=mn} G_{ij}(0, 0) X^i Y^j\right] \\ &\quad + \sum_{in+jm>mn} F_{ij}^{**} X^i Y^j \quad \text{with } F_{ij}^{**} \in k \end{aligned}$$

and so by comparing coefficients we get

$$G_{ij}(0, 0) = \begin{cases} 1 & \text{if } (i, j) = (0, n) \\ -1 & \text{if } (i, j) = (m, 0) \\ 0 & \text{in all other cases} \end{cases}$$

and hence keeping in mind that

$$N_R(g(i, j)) = G_{ij}(0, 0)$$

we see that

$$N_R(g(i, j)) = \begin{cases} 1 & \text{if } (i, j) = (0, n) \\ -1 & \text{if } (i, j) = (m, 0) \\ 0 & \text{in all other cases.} \end{cases}$$

Thus we have found a basis (x, y) of $M(R)$ such that

$$f = \sum_{in+jm=mn} g(i, j)x^i y^j \quad \text{where } g(i, j) \in R$$

with

$$N_R(g(i, j)) = \begin{cases} 1 & \text{if } (i, j) = (0, n) \\ -1 & \text{if } (i, j) = (m, 0) \\ 0 & \text{in all other cases.} \end{cases}$$

Now let x' and y' be any elements in any overfield of K such that

$$x'^n = x \quad \text{and} \quad y'^m = y$$

and let

$$K' = K(x', y') \quad \text{and} \quad R' = R[x', y'].$$

Then clearly R' is a two-dimensional regular local domain, (x', y') is a basis of $M(R')$, R' dominates and is residually rational over R, and R' is the integral closure of R in K'; (for instance see Theorem 6 of my Jungian paper). Upon letting

$$G'(X, Y) = \sum_{i+j=mn} g'(i, j)X^i Y^j$$

where

$$g'(i, j) = \begin{cases} g(i/n, j/m) & \text{if } i \equiv 0(n) \text{ and } j \equiv 0(m) \\ 0 & \text{otherwise} \end{cases}$$

we get a homogeneous $G'(X, Y) \in R'[X, Y]$ of degree mn such that

$$f = G'(x', y')$$

and

$$N_{R'}(G')(X, Y) = Y^{mn} - X^{mn} = \prod_{i=1}^{mn} (Y - w_i X)$$

with pairwise distinct elements $w_1, \ldots, w_{mn}$ in k. Thus f has an (mn)-fold ordinary multiple point in R'.

First Theorem on the Quasirationality of Certain Cyclic Coverings With Branch Locus Having One Characteristic Pair. Let any $f \in R$ be given such that f is analytically irreducible in R and f has exactly one characteristic pair (m, n) in R. Let p be any positive integer which is coprime with mn. Let z be an element in an overfield of K such that

$$z^p = sf^r g^p \qquad \text{where } 0 \neq g \in R \text{ and } s \in R \text{ with } s \notin M(R)$$

and where r is some nonnegative integer. Let $L = K(z)$ and let S be any overdomain of R with quotient field L such that S is integral over R; (for instance $S = R[z]$). Then S is quasirational.

Proof. By the above Theorem on Covering a Curve . . . we can find a basis (x, y) of $M(R)$ such that upon taking elements x' and y' in an overfield of L with

$$x'^n = x \quad \text{and} \quad y'^m = y$$

and upon letting

$$K' = K(x', y') \quad \text{and} \quad R' = R[x, y]$$

we have that: R' is the integral closure of R in K', R' is a two-dimensional regular local domain, (x', y') is a basis of $M(R')$, R' dominates and is residually rational over R, and f has an (mn)-fold ordinary multiple point in R'. Let

$$L' = L(x', y') \quad \text{and} \quad S' = S[x', y'].$$

Then S' is an overdomain of R' with quotient field L' such that S' is integral over R'. Also clearly $L' = K'(z)$ and

$$z^p = sf^r g^p \qquad \text{where } 0 \neq g \in R' \text{ and } s \in R' \text{ with } s \notin M(R').$$

Therefore by the above Second Theorem ... we conclude that S' is quasirational. Also clearly S' is an overdomain of S such that S' is a finite S-module; consequently, by the Lemma on the Quasirationality of a Domain and Its Integral Extensions, we conclude that S is quasirational.

Definition of Branching. Given any height-one prime ideal I in R and given any finite algebraic extension L of K, we say that I is branched in L if there exists a prime ideal J in the integral closure W of R in L such that $J \cap R = I$ and $IW_J \neq JW_J$.

Second Theorem on the Quasirationality of Certain Cyclic Coverings With Branch Locus Having One Characteristic Pair. Let I be any height-one prime ideal in R such that I is analytically irreducible in R and I has exactly one characteristic pair (m, n) in R. Let p be a positive integer which is coprime with mn. Let L be a finite galois extension of K such that the galois group of L over K is a cyclic group of order p. Let S be an overdomain of R with quotient field L such that S is integral over R. Assume that no height-one prime ideal in R, other than I, is branched in L. Also assume that K contains a primitive pth root of 1. Then S is quasirational. [Although we shall not show it here, this theorem remains true without assuming that K contains a primitive pth root of 1.]

Proof. Now there exists $0 \neq z' \in L$ such that $L = K(z')$ and $z'^p \in K$. We can write $z'^p = a/b$ with nonzero elements a and b in R. Upon letting $z = bz'$ we then have

$$L = K(z) \text{ and } z^p = c \qquad \text{with } 0 \neq c \in R.$$

We can take a generator f of I in R and we can write

$$c = sf^r f_1^{r(1)} \cdots f_t^{r(t)} \qquad \text{where } s \in R \text{ with } s \notin M(R)$$

and where r, t, $r(1), \ldots, r(t)$ are nonnegative integers and $f_1, \ldots, f_t$ are elements R such that I, $f_1R, \ldots, f_tR$ are pairwise distinct height-one prime ideals in R. Now f_iR is unbranched in L and hence $r(i)$ must be divisible by p; this being so for $1 \leq i \leq t$, upon letting

$$g = f_1^{r(1)/p} \ldots f_t^{r(t)/p}$$

we get

$$z^p = sf^r g^p \quad \text{with} \quad 0 \neq g \in R.$$

Therefore by the immediately preceding First Theorem ... we see that S is quasirational.

Note. The principles discussed in the following two Remarks were essentially used in the proofs of the above theorems. However, we did not use these principles explicitly but preferred to give more direct, and ad-hoc, arguments. The same comment also applies to the other principles outlined in the geometric preamble.

Remark on Birational Blowing-Ups and the Finiteness of the Number of Hidden Prime Divisors Which are not Residually Almost Simple Transcendental. Let S be a two-dimensional local domain with quotient field L. Let W be a normal projective model of L over S, i.e., let W be a collection of normal local domains for which there exists a finite number of nonzero elements $x_1, \ldots, x_t$ in S, with $t \geq 1$, such that W is the union of $W_1, \ldots, W_t$ where W_i is the set of all localizations of $H_i = S[x_1/x_i, \ldots, x_t/x_i]$ with respect to the various prime ideals in H_i. Alternatively, we may call W the birational blowing-up of S by the nonzero ideal in S generated by $x_1, \ldots, x_t$. Clearly W is two-dimensional, i.e., the dimension of any member in W is ≤ 2 and W does have two-dimensional members. Also clearly every two-dimensional member of W dominates and is residually finite algebraic over S. Moreover, the one-dimensional members of W_i which dominate S are finite in number because they are the localizations of H_i with respect to the height-one prime ideals in H_i which contain the nonzero ideal $M(S)H_i$ and these height-one prime ideals are clearly finite in number. This being so for $1 \leq i \leq t$, we see that the one-dimensional members of W which dominate S are finite in number; amongst these some may be residually transcendental over S and they will precisely be the hidden prime divisions of S which are members of W; geometrically we may think of them as the curves which have come out of S by the said blowing-up. Finally, we note that if V is any one-dimensional local domain with quotient field L dominating S then V dominates a unique member of W and the said member dominates S; (for instance see (1.6) to (1.9) of my Resolution book). From all this discussion and in view of the Lemma on the Quasirationality of Simple Points we conclude with the following observations.

(2.1) The hidden prime divisions of S which belong to W are finite in number.

(2.2) Upon letting $B(S')$ to denote the set of all hidden prime divisors of any two-dimensional local domain S' which are not residually almost simple transcendental over S' and upon letting W^* to denote the set of all two-dimensional nonregular members of W, we have the following representation of $B(S)$ as a disjoint union:

$$B(S) = (B(S) \cap W) \cup \bigcup_{S^* \in W^*} B(S^*).$$

(2.3) The following three conditions are mutually equivalent.

(2.3.1) S is quasirational.

(2.3.2) Every member of W is quasirational and every member of W which is a hidden prime divisor of S is residually almost simple transcendental over S.

(2.3.3) Every two-dimensional nonregular member of W is quasirational and every member of W which is a hidden prime divisor of S is residually almost simple transcendental over S.

Recalling that a projective model is said to be nonsingular if every member of it is regular, by (2.1) and (2.2) we get the following:

(2.4) If S has a nonsingular birational blowing-up, i.e., if there exists a nonsingular projective model of L over S, then the hidden prime divisors of S which are not residually almost simple transcendental over S are finite in number.

Now it is known that if S is a localization of a finite ring extension of a field, and more generally if S is excellent, then S has nonsingular birational blowing-ups; see my Resolution book and also my Resolution paper in the Proceedings of the Bombay Colloquium On Algebraic Geometry of 1968 published by the Oxford University Press. So by (2.4) we get the following:

(2.5) If S is a localization of a finite ring extension of a field or, more generally, if S is excellent, then the hidden prime divisors of S which are not residually almost simple transcendental over S are finite in number.

Remark on the Simultaneous Quasirationality of a Local Domain and Its Completion. Let S be a two-dimensional local domain. Assume that S is analytically irreducible, i.e., its completion is a domain. Then S is quasirational iff its completion is quasirational.

Proof. Let L be the quotient field of S, let $\hat{S}$ be the completion of S, and let $\hat{L}$ be the quotient field of S where we regard $\hat{L}$ to be an overfield of L. Now $\hat{S}$ dominates and is residually rational over S, and hence the proof follows from the following two observations. Firstly, by Lemma 12 of my Uniformization paper in volume 63 of Annals of Mathematics of 1956, we see that if V^* is any hidden prime divisor of $\hat{S}$ then upon letting $V = V^* \cap L$ we have that V is a prime divisor of S, V^* dominates V, and V^* is residually rational over V. Secondly, by Lemma 13 of my Uniformization paper and Theorem 1 of my Valuations paper, we see that given any hidden prime divisor V of S, there exists a hidden prime divisor V^* of $\hat{S}$ such that $V = V^* \cap L$.

BHASKARACHARYA PRATISHTHANA, PUNE, INDIA

AND

UNIVERSITY OF POONA, PUNE, INDIA

A VERSION OF ZARISKI'S MAIN THEOREM FOR POLYNOMIAL IDENTITY RINGS

By M. ARTIN and W. SCHELTER

Introduction. Let k be an algebraically closed field. Unless the contrary is explicitly mentioned, the word *ring* in this paper will mean finitely generated k-algebra satisfying the identities of $n \times n$ matrices, and the notation Spec R will stand for the space of maximal ideals of R. Our purpose is to extend Zariski's Main Theorem to this class of rings, using the concept of integrality developed in [1].

When Zariski [8] formulated his theorem, he found it convenient to phrase it in terms of local algebra, as a theorem about normal local rings. For non-commutative rings, neither localization nor integral closure behave well, so we have gone back to a statement about homomorphisms of affine rings. In this context the theorem (cf. (2.1)) asserts the existence of integral elements with certain properties.

The homomorphisms which motivated us originally were the central extensions $R \to R[\overline{T}]$ of prime rings which are obtained by adjoining traces. However, we found that there were not enough central integral

The authors would like to thank the Université de Paris VII, Université de Paris Sud at Orsay, NRC and NSF for their generous support.

elements even for these homomorphisms (cf. (2.3)), and because of this we were led to consider a more general class of maps $R \xrightarrow{\varphi} S$. We consider maps which are compositions of central extensions and integral homomorphisms, and which satisfy a certain property, of being *geometric* (1.2). The assumption that S is integral over a central extension gives us a substitute for techniques of algebraic geometry related to Noether normalization, and is used strongly (cf. (1.7), (2.4)).

We prove the theorem by embedding Spec S into a scheme Y over projective space, thereby replacing the problem of finding integral elements by the one of finding global sections of $\mathcal{O}_Y$, to which cohomological methods can be applied. The proof is therefore more related to Zariski's proof of this theorem on holomorphic functions [9] than to his original proof [8] of the Main Theorem. But we have very poor control over the cohomology at present, and have not succeeded in finding a proof of the connectedness theorem in general.

There is one aspect of the argument which is perhaps worth mentioning. That is a substitute for Zariski's famous conductor lemma [8, p. 524]. We do use the conductor, but have been able to prove only that it is not the zero ideal. In particular, we do not know how to determine its support. So at the crucial point we resort to Noetherian induction and use the fact that the cohomology of $\mathbf{P}^1$ is trivial (cf. (7.7)).

1. Geometric homomorphisms. A ring homomorphism $R \xrightarrow{\varphi} S$ will be called *geometric* if for every maximal ideal $\mathfrak{m}$ of S, the map

$$R/\varphi^{-1}(\mathfrak{m}) \to S/\mathfrak{m} \tag{1.1}$$

is an isomorphism. (This definition is correct only when the ground field k is algebraically closed. In the general case, one should require that $\varphi^{-1}(\mathfrak{m})$ be maximal in R, and that $R/\varphi^{-1}(\mathfrak{m})$ and $S/\mathfrak{m}$ have the same pi degree.) We may also say that a homomorphism is geometric on a subset of Spec S if the above property holds at the corresponding subset of maximal ideals.

A geometric homomorphism induces a single valued map

$$Y = \operatorname{Spec} S \xrightarrow{f} \operatorname{Spec} R = X \tag{1.2}$$

which is continuous because the correspondence induced by any ring homomorphism is continuous. We will also refer to the map f as a geometric map if it is induced by a geometric homomorphism φ.

Examples of geometric homomorphisms are central extensions. Also, if R is a subring of a ring Q and $\alpha \in R$ is invertible in Q, then the subring $R\{\alpha^{-1}\}$ generated by α^{-1} and R is geometric over R, i.e., the inclusion $R \to R\{\alpha^{-1}\}$ is a geometric map.

We leave the proof of the following proposition to the reader:

PROPOSITION (1.3).

(1) *The composition of geometric homomorphisms is geometric.*
(2) *If* $R \xrightarrow{\varphi} S \xrightarrow{\psi} Q$ *are homomorphisms and* $\psi\varphi$ *is geometric, then so is* ψ.
(3) *If* φ *is geometric and* S *is semiprime (has Jacobson radical zero), then* φ *maps the center of* R *to the center of* S.
(4) *Suppose that* φ *is geometric, and let* $J = J(R)$ *be the Jacobson radical of* R. *Then* $\varphi(J) \subset J(S)$.
(5) φ *is geometric if and only if for every prime ideal* $\mathfrak{p} \subset S$ *of pi degree* n, $\varphi^{-1}(\mathfrak{p})$ *is a prime ideal of pi degree* n.
(6) *A ring homomorphism* $R \xrightarrow{\varphi} S$ *is geometric if and only if the correspondence it induces sends* $\mathrm{Spec}_i S$ *to* $\mathrm{Spec}_i R$ *for each* i.

PROPOSITION (1.4). *Let* $R \xrightarrow{\varphi} S$ *be a homomorphism of rings, where* R *is Azumaya of rank* n *and* S *satisfies the identities of* $n \times n$ *matrices. Then* φ *is a central extension, and* S *is Azumaya.*

Proof. We can assume $R \subseteq S$. Recall that R is Azumaya of rank n iff there exist evaluations of central polynomials in R which generate the unit ideal. Because our assumption says S and R both satisfy the identities of $M_n(\mathbf{Z})$, the same evaluations are evaluations of central polynomials for S and generate the unit ideal of S. It follows as in [6] that there are central polynomials f_i for R, and elements $r_i \in R$, so that for any x in R

$$x = \sum_i r_i f_i(x).$$

But because S also satisfies the identities of $n \times n$ matrices, we can take x to be any element of S, and then $f_i(s)$ will be central in S. This shows that S is a central extension, and is Azumaya.

We will often work with a geometric homomorphism φ which is a composition of a central map and an integral map, that is, has the form

$$R \to R[u] \xrightarrow{\psi} S, \tag{1.5}$$

where $R[u] = R[u_1, \ldots, u_n]$ denotes the ring of polynomials over R in central variables u_i, and where ψ is *integral* and *geometric.*

If $Y \xrightarrow{f} X$ is the geometric map corresponding to φ, then the *fibre* $f^{-1}(x)$ is a closed subset of Y which we denote by Y_x.

THEOREM (1.6). *Let $Y \xrightarrow{f} X$ be a geometric map which is a composition* (1.5) *of an integral map and a central map. Then $h(y) = \dim_y f^{-1}(f(y))$ is an upper semi-continuous function on Y.*

Remark. By $\dim_y F$ of a closed subset F of Y, we mean $\dim(F \cap U)$ for a small neighborhood U of y. Upper semi-continuity at $a \in Y$ means that there is a neighborhood U of a so that $h(y) \le h(a)$ for all $y \in U$. We may restrict X and Y to open sets containing $f(a)$ and a, without affecting upper semi-continuity at a.

The theorem fails for general geometric maps, and for arbitrary extensions, as the following examples show:

Example (1.7). A geometric homomorphism for which the fibre dimension is not upper semi-continuous: $A = k[u]$ and $B = k[u, v]$. Let $\mathfrak{b} = uB$ and let $\mathfrak{a} = \mathfrak{b} \cap A$. The example is

$$R = \begin{pmatrix} A & A \\ \mathfrak{a} & A \end{pmatrix} \xrightarrow{\varphi} \begin{pmatrix} B & B \\ \mathfrak{b} & k + \mathfrak{b} \end{pmatrix} = S.$$

The map f induced by φ on spectra is illustrated in Fig. 1. The rings R and S are finitely generated; for instance S is generated by the matrices

$$\begin{pmatrix} 1 & 0 \\ 0 & 0 \end{pmatrix}, \begin{pmatrix} v & 0 \\ 0 & 0 \end{pmatrix}, \begin{pmatrix} 0 & 0 \\ u & 0 \end{pmatrix}, \begin{pmatrix} 0 & 1 \\ 0 & 0 \end{pmatrix}.$$

Both rings are algebras over $k[u]$. Also $R[u^{-1}]$ and $S[u^{-1}]$ are full matrix rings over $A[u^{-1}]$ and $B[u^{-1}]$. So φ is geometric at points not containing u. On the other hand, rad uS contains the ideal $I = M_2(\mathfrak{b})$,

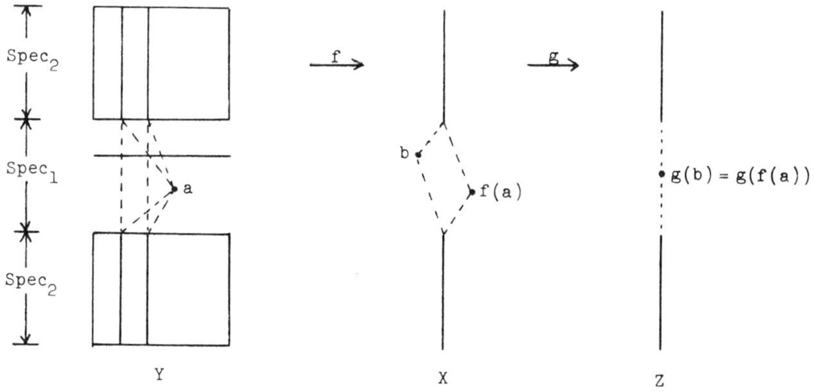

and we have

$$S/I \simeq \begin{pmatrix} k[v] & k[v] \\ 0 & k \end{pmatrix}.$$

Therefore the remaining points are in $\text{Spec}_1 S$, and have to be mapped to $\text{Spec}_1 R$. Let a be a point of Y corresponding to the maximal ideal

$$\begin{pmatrix} B & B \\ \mathfrak{b} & \mathfrak{b} \end{pmatrix}$$

of S. Then $f^{-1}(f(a)) = \{a\}$, so $h(a) = 0$, but $h(y) = 1$ for all $y \neq a$.

An example of an *extension* for which the fibre dimension is not semi-continuous is the map $A \to S$. This corresponds to the map $Y \xrightarrow{gf} Z$ in Figure 1. For this map, $h(a)$ is also zero.

LEMMA (1.8). *Let $Y \xrightarrow{f} X$ be a geometric map. Then the set $\{y\colon h(y) = n\}$ is constructible for each n, i.e., it is a finite union of locally closed subsets of Y.*

Proof. Let Y_i denote the constructible subset of Y of points of pi degree i. If $y \in Y_i$, then $f^{-1}(f(y)) \subseteq Y_i$ by Proposition (1.3)(6). But we know that the restriction of f to Y_i is a map of Azumaya schemes, so the results follow from the commutative case [3, p. 46].

Proof of Theorem (1.6). We have to show that for any integer d, the subset $V = \{y \mid h(y) \geq d\}$ is closed in Y. It is a constructible set, by (1.8), so if it is not closed there is an accessible point p in its boundary [1, (5.13)]. We may then choose a curve C in Y passing through p, with almost all points in V, that is, $h(y) \geq d$ for all but a finite number of points on C, but $h(p) < d$. Thus the fibre over all but a finite number of points of $f(C)$ is of dimension $\geq d$. We can restrict Y to an irreducible component of $f^{-1}(f(C))$ since the supremum of upper semicontinuous functions is upper semicontinuous. Also X can be restricted to the closure of $f(C)$. Then dim $X = 1$ and X is irreducible. The fibre over all but a finite number of points of X is of dimension greater than or equal to d. We know R is a finite module over its center E, but by hypothesis S is integral and geometric over a central extension $R[u]$ of R. Thus S is a finite module over its center, and in fact over $E[u]$. Let Z denote the integral closure of the center of S. Then $R \xrightarrow{g} S[Z]$ satisfies the hypothesis of the theorem. We replace S by $S[Z]$. In Spec $S[Z]$ there is a curve covering C [1, (1.10)], such that most points of it are on a fibre of dimension at least d. Let C denote this new curve and let p be a point on it over the old p. Now then on an open subset of Spec E the fibre dimension of Spec $E \xleftarrow{j}$ Spec Z is at least d. For $X \leftarrow Y$ is Azumaya on an open subset. By the commutative theorem [3, p. 46], the fibre dimension of j is at least d everywhere. Now use going down [7, Th. 3], applied to $Z \to S$, to show that there is a d-dimensional irreducible subset of Y containing p, which maps to a single point of Spec E and hence to an irreducible ((1.3)(5)) subset of X. Thus it maps to a single point of X, contradicting $h(p) < d$.

2. Statement of the Main Theorem. The affine version of the theorem is the following:

THEOREM (2.1). *Let $R \xrightarrow{\varphi} S$ be a geometric homomorphism which is made up of a central extension $R[u]$ followed by an integral map. There is a subring $\tilde{S}$ of S, integral and geometric (and finitely generated) over R:*

$$\begin{array}{ccc} R[u] & \xrightarrow{\psi} & S \\ \uparrow & & \uparrow \\ R & \xrightarrow{\bar{\psi}} & \tilde{S}, \end{array}$$

such that the fibres of the map Spec $S \to$ Spec $\tilde{S}$ either

(a) *consist of one point, or*
(b) *contain no zero-dimensional component.*

Note that the map $\tilde{S} \to S$ is automatically geometric (1.3)(2).

Even in the case that S is a central extension of R, there may be no central integral elements to separate points in a finite fibre; thus $\tilde{S}$ need not be a central extension:

Examples (2.2). (1) Let R be the ring of upper triangular 2×2 matrices with entries in k. Spec R consists of two points p, p', where say p corresponds to evaluation at the upper left entry of a matrix. Let $A = k[u]$, and $\overline{A} = A/(u^2 - u) \approx k \oplus k$. The ring

$$S = \begin{bmatrix} \overline{A} & \overline{A} \\ 0 & A \end{bmatrix},$$

is a quotient of $R[u]$, and is therefore a central extension of R. The fibre of Spec S over p consists of two points. The center of S consists of the diagonal matrices in S, and no such matrix is integral over R, unless it is constant. Thus there is no possibility of separating the points of the fibre by adjoining a central integral element. The required ring is generated over R by the non-central element

$$\alpha = \begin{pmatrix} \overline{u} & 0 \\ 0 & 0 \end{pmatrix}; \tilde{S} = R\{\alpha\} \approx \begin{bmatrix} k & 0 & k \\ 0 & k & k \\ 0 & 0 & k \end{bmatrix}. \tag{2.3}$$

(2) Another example: Let $A = k[u, u^{-1}, v]$,

$$R = \begin{pmatrix} k[u^2] + vA & A \\ A & k[u^{-1}] + vA \end{pmatrix},$$

$$M = \begin{pmatrix} (u^2 - a) + vA & A \\ A & k[u^{-1}] + vA \end{pmatrix},$$

and $S = R[u]$. Then the induced map is a homeomorphism on Spec_2, but for fixed $a \neq 0$, there are two points in Spec S over the one corresponding to the maximal ideal m. The center of S is $k[u] + vA$. If a central element $p(u) + vg$ is integral over R, with $g \in A$, then $p(u)$ is integral over

$$\begin{pmatrix} k[u^2] & 0 \\ A & k[u^{-1}] \end{pmatrix},$$

hence $p(u)$ is a constant.

There are various directions in which one may look for generalizations of this theorem, the most promising being to eliminate the hypothesis that ψ be geometric. We do not have a reasonable conjecture at this time. The particular form (1.5) for the map φ is however used in our proof, and in fact the following example, obtained from example (1.7), shows that it is essential.

Example (2.4). A geometric homomorphism $R \xrightarrow{\varphi'} S'$ for which the conclusion of (2.1) is false: Let $R \to S$ be the homomorphism introduced in example (1.7), and let S' be the central extension of S generated by a single element w, satisfying the relation

$$w^2 = uv + 1.$$

Then $R \xrightarrow{\varphi'} S'$ is the required example. The isolated point of $\mathrm{Spec}_1 S$ is covered by two points of $\mathrm{Spec}_1 S'$, corresponding to the two values $w = \pm 1$. This pair of points constitutes a fibre over Spec S. But there is no subring $\tilde{S}' \subset S'$ which is strictly larger than R and integral over R. To show this, it suffices to prove the same assertion for the map $R[u^{-1}] \to S'[u^{-1}]$, and clearly $S'[u^{-1}] \approx M_2(k[u, u^{-1}, w])$. An integral subring would be a central extension of $R[u^{-1}] = M_2(k[u, u^{-1}])$ (1.4), but $k[u, u^{-1}]$ is integrally closed in $k[u, u^{-1}, w]$.

Again, we can replace the ring R by A to get an example in which the map is an *extension* $A \to S'$.

To prove Theorem (2.1) we use induction on the number m of variables u_i, the bulk of the argument being for the case $m = 1$. As a preliminary reduction, we note that the variables u_i may be assumed central in S:

LEMMA (2.5). *It suffices to prove the theorem in the case that S is replaced by the semiprime ring S/J, where $J = J(S)$ is the Jacobson* *radical. Hence by* (1.3)(3) *we may assume that the elements* u_i *map to central elements of* S.

Proof. The rings S and S/J have homeomorphic spectra. Assume that an integral subring of S/J exists, as in the theorem. Call it $\overline{S}$, and choose generators $\overline{\alpha}_1, \ldots, \overline{\alpha}_n$ for $\overline{S}$ over R. Let $\alpha_i \in S$ be a representative for $\overline{\alpha}_i$, and let $R\{\alpha\}$ denote the subring of S generated by R and $\{\alpha_i\}$. Then the homomorphism $R\{\alpha\} \to R\{\overline{\alpha}\} = \overline{S}$ is surjective, with a nilpotent kernel. Therefore $R\{\alpha\}$ is integral over R, and Spec $R\{\overline{\alpha}\} \approx$ Spec $R\{\alpha\}$. Thus $\tilde{S} = R\{\alpha\}$ is the required subring of S.

3. The case $m = 1$: Reduction to the projective version. Let us consider the case that there is a single variable u in the presentation (1.5) of S. By Lemma (2.5), we may assume that u is central in S. We will embed Spec S as the affine part at finite distance of a scheme Y which is integral over the projective line $\mathbf{P}_R{}^1$, and will reduce the proof of Theorem (2.1) to a theorem about Y. The basic facts which we need about schemes over projective space are in the following sections.

To embed Spec S into Y, we use the following proposition.

PROPOSITION (3.1). *Let R be a ring containing a central element t. Let $R' = R[t^{-1}]$, and let S' be integral and geometric over R'. There exists a subring $S \subset S'$ integral and geometric over R, such that S' is R'-isomorphic to $R' \otimes_R S = S[t^{-1}]$.*

Example (3.2). An integral *extension* $R' \to S'$ with $R' = R[t^{-1}]$ and $t \in Z(R)$. The ring S' has no subring S which is integral over R, and such that $S' = S[t^{-1}]$: Let $B = k[u, v]$, and $C = k[u, uv^{10}, uv^{11}]$. Let $t = uv^{10}$. Then $B[t^{-1}] = C[t^{-1}] = k[u, v, u^{-1}, v^{-1}]$. Call this ring B'. The example is

$$R = \begin{bmatrix} B & 0 \\ 0 & C \end{bmatrix}, \quad R' = \begin{bmatrix} B' & 0 \\ 0 & B' \end{bmatrix}, \quad S = M_2(B').$$

To show that no subring S as above exists, it suffices to show that if

$$\alpha = \begin{pmatrix} 0 & 1 \\ v & 0 \end{pmatrix} \quad \text{and} \quad \beta = \begin{pmatrix} v & 0 \\ 0 & 0 \end{pmatrix},$$

then $R \to R[t^n\alpha]$ is not integral for any n. The over-ring generated by $t^n\alpha$ contains

$$(\beta^m t^n \alpha + t^n \alpha \beta^m)^2 = u^{2n} v^{20n+2m+1},$$

which is not integral over C, hence not over R, provided $m > n$.

Proof of Proposition (3.1). If $S' = R[\alpha_1, \ldots, \alpha_m]$, we will take $S = R[t^N \alpha_i]$, for a suitably large N, determined below. First note that S is geometric over R. For if $\mathfrak{m}$ is a maximal ideal of S then either $t \notin \mathfrak{m}$ and thus $t^{-1} \in S/\mathfrak{m}$, or else $t \in \mathfrak{m}$ in which case $S/\mathfrak{m} = R/\mathfrak{m}$. Next note that if R were a finite module over its center, we would only have to check integrality of a finite number of monomials, by Sirsov's theorem [4, p. 152], and hence we could choose N large enough to clear out any denominators in the monic polynomials (for each monomial) over $R[t^{-1}]$.

But to check that a geometric map $R \to S$ is integral, it suffices [1, (9.3)] to check that $R[T_m(R)] \to S[T_m[R]]$ is integral for each $m \leq$ pi deg S, where $R[T_m(R)]$ is the ring R modulo $m \times m$ identities, with the $m \times m$ coefficients of the characteristic polynomial adjoined. Thus we have only a finite number of conditions on N, and we take N big enough to satisfy them all.

We go back to our ring S. A scheme over $\mathbf{P}_R{}^1$ is given (see Section 4) by a pair of homomorphisms $R[u_i] \to S_i$ $(i = 0, 1)$ such that u_i is central in S_i, and an $R[u_0, u_1]$-isomorphism $S_0[u_1] \approx S_1[u_0]$ $(= S_{01})$, where u_i are related by the equation $u_0 u_1 = 1$. We take $u = u_0$ and $S = S_0$. Then we put $R = R[u_1]$, $t = u_1$, $S' = S_0[u_1]$ in Proposition (3.1) to find the required ring S_1, integral and geometric over $R[u_1]$. This leads to the basic diagram

$$\begin{array}{ccc} \mathbf{P}_R{}^1 & \xleftarrow{\quad g \quad} & Y \\ & {\scriptstyle \pi}\searrow \qquad \swarrow{\scriptstyle f} & \\ & X = \operatorname{Spec} R, & \end{array} \qquad (3.3)$$

in which g is *integral* and *geometric*.

THEOREM (3.4). *Given a diagram* (3.3), *let* $y \in Y$ *and* $x = f(y)$. *Let* Y_x *be the fibre of* Y *over* x. *If* y *is a component of* Y_x, *i.e.*, $Y_x =$

$\{y\} \amalg \Delta$ for some closed set Δ, there is an element $\alpha \in H^0(Y, \mathcal{O}_Y)$ such that $\alpha = 1$ at y and $\alpha = 0$ on Δ.

Here $H^0(Y, \mathcal{O}_Y)$ is the ring of globally defined elements of $\mathcal{O}_Y$, which is described by the exact sequence

$$0 \to H^0(Y, \mathcal{O}_Y) \to S_0 \oplus S_1 \xrightarrow{d_0 - d_1} S_{01} \to H^1(Y, \mathcal{O}_Y) \to 0 \tag{3.5}$$

(see (6.1)).

To apply this theorem we note that $H^0(Y, \mathcal{O}_Y)$ is a ring which, though not necessarily finitely generated, is integral over R (Proposition (6.5)). Therefore any finitely generated subring R' is integral over R. Moreover Y has a structure of scheme integral over $\mathbf{P}_{R'}{}^1$, given by the obvious homomorphisms $R'[u_i] \to S_i$. Therefore (5.2) the image of $Y \xrightarrow{f'} \operatorname{Spec} R' = X'$ is a closed set. Since $R' \subset S_0 \oplus S_1$, that closed set is all of Spec R', i.e., the map f' is surjective. From this and the fact that f is geometric, we obtain

LEMMA (3.6). *With the notation of (3.4), let R' be any finitely generated subring of $H^0(Y, \mathcal{O}_Y)$. The homomorphism $R \to R'$ is geometric, the map $Y \xrightarrow{g'} \mathbf{P}_{R'}{}^1$ is integral and geometric, and $Y \xrightarrow{f'}$ Spec R' is surjective.*

LEMMA (3.7). *Theorem (3.4) is equivalent with the following assertion: There is a finitely generated subring R' of $H^0(Y, \mathcal{O}_Y)$, so that $f'(y) = x'$ is not contained in $f'(\Delta)$, or equivalently, that $f'^{-1}(f'(y)) = \{y\}$.*

Proof. If Theorem (3.4) holds, it suffices to take for R' the ring $R\{\alpha\}$ generated by R and α in $H^0(Y, \mathcal{O}_Y)$. Conversely, suppose R' exists. Then since x' and $f'(\Delta)$ are disjoint closed sets of X', there is an $\alpha \in R'$ with $\alpha = 1$ at x' and $\alpha = 0$ on $f'(\Delta)$. This α is the element required by Theorem (3.4).

COROLLARY (3.8): (*of Theorem* (3.4)). *With the above notation, there is a finitely generated subring R' of $H^0(Y, \mathcal{O}_Y)$ so that the fibres of f' either*

(a) *consist of one point, or*
(b) *contain no zero-dimensional component.*

We will generalize this corollary to schemes over $\mathbf{P}^n$, $n > 1$, in section 9.

Proof. Assume the theorem. For any point $y \in Y$, which is isolated in its fibre Y_x, there is some ring R' which separates y from the rest Δ of Y_x (3.7). We need only show that a single finitely generated subring R' will work for every y. Now the fibres of f' are controlled by the fibred products $Y \times_{X'} Y = \{(y_1, y_2) | f'(y_1) = f'(y_2)\} \subset Y \times Y$, and an ascending chain of subrings R' leads to a descending chain of subsets $Y \times_{X'} Y$. Since f' is geometric, $Y \times_{X'} Y$ is the union of $Y_i \times_{X_i'} Y_i$ (1.3)(6), and these sets are easily seen to be closed subvarieties of the algebraic varieties $Y_i \times Y_i$. Since $Y_i \times Y_i$ is a noetherian topological space, any descending chain terminates. Thus $Y \times_{X'} Y$ achieves a minimum, and the corresponding R' is the required subring.

LEMMA (3.9). *Theorem* (2.1) *in the case* $m = 1$ *follows from Theorem* (3.4).

Proof. Given $R \to S$, we construct the scheme Y (3.3) and take for $\tilde{S}$ the ring R' occurring in Corollary (3.8). This ring is integral by (6.5) and geometric. Since Spec S is an open subset of Y, the corollary implies the Theorem (2.1) for Spec S.

4. Projective space over a ring. Let Z be a (commutative) scheme [2] with structure sheaf $\mathcal{O}_Z = \mathcal{O}$, and let $\mathcal{A}$ be a quasi-coherent sheaf of associative $\mathcal{O}$-algebras. One can introduce a topological space *Spec* $\mathcal{A}$ lying over Z, exactly as for commutative schemes [2, p. 128]: If $U \subset Z$ is an affine open set, then $\mathcal{A}(U)$ is an $\mathcal{O}(U)$-algebra, and *Spec* $\mathcal{A}$ is the union of the spectra Spec $\mathcal{A}(U)$. The spaces Spec $\mathcal{A}(U)$ glue together in a natural way, because for any pair U, V of affine open sets in Z, the map

$$\operatorname{Spec} \mathcal{A}(U \cap V) \to \operatorname{Spec} \mathcal{A}(U)$$

induced by the homomorphism $\mathcal{A}(U) \to \mathcal{A}(U \cap V)$ is a homeomorphism onto an open subset of Spec $\mathcal{A}(U)$:

PROPOSITION (4.1). *With the above notation, let* $U' \subset U$ *be affine open sets in* Z. *Then Spec* $\mathcal{A}(U') \to$ *Spec* $\mathcal{A}(U)$ *is an open immersion.*

The space *Spec* $\mathcal{A}$ comes equipped with a sheaf $\mathcal{A}$ of rings, defined on *certain* open subsets, namely on those which are inverse images of affine opens of Z. It is natural to ask for an extension of this sheaf to all open sets, but we will not enter into that question here; in fact we have not considered it in detail. For our purposes it is more convenient to consider the structure as consisting of the space Spec $\mathcal{A}$ lying over Z, and the sheaf of $\mathcal{O}$-algebras $\mathcal{A}$ on Z. We will refer to such a structure as a *scheme of algebras over Z,* or often just as a *scheme,* and will use notation such as $Y = \mathit{Spec}\ \mathcal{A}$, and $\mathcal{A} = \mathcal{O}_Y =$ sheaf of algebras defining Y. When we refer to an affine open of Y, it is to be understood to be an open set of the form Spec $\mathcal{A}(U)$, where U is an affine open of Z. Also, we work only with algebras $\mathcal{A}$ such that each $\mathcal{A}(U)$ is an affine pi ring.

The case in which we are mainly interested is that $Z = \mathbf{P}^n$ is projective space over k. To be specific, let us introduce homogeneous coordinates $y_0, \ldots, y_n$. Then $\mathbf{P}^n$ is the union of $n + 1$ affine spaces

$$U_j = \text{Spec}\, k[u_{0j}, \ldots, u_{\nu j}] = \text{Spec}\, P_j, \tag{4.2}$$

where $u_{\nu j}$ are central variables related by the formal substitutions

$$u_{\nu j} = y_\nu / y_j.$$

These affine spaces are open subsets of $\mathbf{P}^n$ whose intersections are $U_i \cap U_j = U_{ij} = \text{Spec}\, P_{ij}$, where $P_{ij} = P_i[u_{ji}^{-1}] = P_j[u_{ij}^{-1}] = P_{ji}$. For our purposes we may regard $\mathbf{P}^n$ as the union of the spectra obtained from the diagram of rings and localizations

$$\begin{array}{lll} P_i & & \\ & \searrow & \\ & & P_{ij} \\ & \nearrow & \\ P_j & & \end{array} \qquad (i, j = 0, \ldots, n). \tag{4.3}$$

Let $\mathcal{A}$ be a quasi-coherent sheaf of algebras on $\mathbf{P}^n$. Then $\mathcal{A}(U_i) = A_i$ is an algebra over P_i, and the property of being quasi-coherent implies that $\mathcal{A}(U_{ij}) = P_{ij} \otimes_{P_i} A_i = A_i[u_{ji}]$ [2, Ch. II, §5]. Thus $\mathcal{A}$ gives rise to a diagram

$$\begin{array}{lll} A_i & & \\ & \searrow & \\ & & A_{ij} \\ & \nearrow & \\ A_j & & \end{array} \qquad (i, j = 0, \ldots, n). \tag{4.4}$$

of algebras over the diagram (4.3), with the property that the arrows of (4.4) are the central localizations which are induced by tensor product from the arrows of (4.3). Conversely, any diagram (4.4) defines a quasi-coherent sheaf $\mathcal{A}$ [2, II, 5]. In terms of diagram (4.4), *Spec* $\mathcal{A}$ is the union of the schemes Spec A_i, which are identified along their open subsets Spec A_{ij}.

We define morphism of schemes of algebras in terms of algebra homomorphisms: Let $Z \xrightarrow{f} Z'$ be a map of commutative schemes, and let Y, Y' be schemes of algebras over Z and Z' respectively. Then a *morphism* $Y \to Y'$ *over* f is simply an $\mathcal{O}_{Z'}$-homomorphism $\mathcal{O}_{Y'} \to f_*\mathcal{O}_Y$, or equivalently, an $\mathcal{O}_Z$-homomorphism $f^*\mathcal{O}_{Y'} \to \mathcal{O}_Y$. Such a morphism induces a correspondence of topological spaces $Y \to Y'$, making a commutative square

$$\begin{array}{ccc} Y & \to & Y' \\ \downarrow & & \downarrow \\ Z & \to & Z'. \end{array}$$

We will consider mainly the case that $Z = Z'$ and $f =$ identity. But one should also take note of the case $Z' =$ Spec k. In this case Y' is simply the spectrum of a ring ($= k$-algebra). So, we have a notion of morphism from a scheme of algebras over Z to an *affine scheme* Spec R.

Note (4.5). Obviously the morphism $Y \to Y'$ could be described without reference to the map $Z \to Z'$, in terms of a map between suitable diagrams such as (4.4) of rings and central localizations. Moreover, the abstract notion of scheme Y can be defined by means of a diagram like (4.4), without reference to Z. But problems arise even in the case that the maps appearing in (4.4) are central localizations, because the center is not functorial. In particular, composition of such morphisms would not always be defined. It is conceivable that these difficulties could be avoided if the notion of morphism were restricted to be made up of *geometric* homomorphisms.

By *projective space* over a ring R we mean the scheme over $\mathbf{P}^n$:

$$\mathbf{P}_R{}^n = Spec(R \otimes_k \mathcal{O}_{\mathbf{P}^n}).$$

The rings (4.4) defining this scheme are

$$R \otimes_k P_i = R[u_{0i}, \ldots, u_{ni}]. \tag{4.6}$$

A *scheme Y over $\mathbf{P}_R^n$* is a morphism $Y \to \mathbf{P}_R^n$ of schemes over $\mathbf{P}^n$. It is defined by a map of $\mathcal{O}_{\mathbf{P}}$-algebras

$$R \otimes_k \mathcal{O}_{\mathbf{P}} \to \mathcal{O}_Y, \tag{4.7}$$

or equivalently, a k-homomorphism $R \to \mathcal{O}_Y$. Thus projective space is formally a product

$$\mathbf{P}_R^n = \operatorname{Spec} R \times \mathbf{P}^n.$$

If the maps (4.7) defining a scheme Y over $\mathbf{P}_R^n$ are *surjective,* Y identifies topologically with a closed subset of $\mathbf{P}_R^n$, and is called a *closed subscheme* of projective space over R. For example, $\mathbf{P}_R^1$ is defined by the diagram of rings

$$\begin{array}{lll} R[u_0] & & \\ & \searrow & \\ & & R[u_0, u_1] \qquad (u_0 u_1 = 1), \\ & \nearrow & \\ R[u_1] & & \end{array} \tag{4.8}$$

where, in the notation of (4.2), $u_0 = u_{10}$ and $u_1 = u_{01}$. A closed subscheme Y of $\mathbf{P}_R^1$ is therefore defined by ideals $I_\nu \subset R[u_\nu]$ ($\nu = 0, 1$) such that $I_0 R[u_0, u_1] = I_1 R[u_0, u_1]$. The corresponding algebras are $A_\nu = R[u_\nu]/I_\nu$.

We recall that a functorial property $\mathcal{P}$ of structures over commutative rings, C is said to be a *local property* if these conditions are satisfied:

(a) Let S be a structure over C, and $u \in C$. Let S_u denote the induced structure over $C_u = C[u^{-1}]$. If $S \in \mathcal{P}$, then $S_u \in \mathcal{P}$. (4.9)

(b) With the same notation, suppose $u_1, \ldots, u_r$ generate the unit ideal in C. If $S_{u_i} \in \mathcal{P}$ for all $i = 1, \ldots, r$, then $S \in \mathcal{P}$.

PROPOSITION (4.10). (i) *The properties semiprime, pi, pi degree n are local properties of algebras.* (ii) *The properties injective, surjective, extension, central extension, geometric, integral are local properties of homomorphisms of algebras.*

Proof. (i) To check (a) we must verify that if S is semiprime, pi or of pideg n then so is a central localization S_u. The property of being pi or pideg n can be phrased in terms of satisfying multilinear identities, and multilinear identities are clearly preserved under central localization. Now suppose S is semiprime. Then we have an inclusion $S \to \bigoplus^m_{i=1} S/p_i$, where the p_i are the minimal primes of S. So

$$S_u \hookrightarrow \bigoplus_{i=1}^{m} (S/p_i)_u$$

and $(S/p_i)_u$ is either 0 or prime. Now use the fact that a semiprime ideal of a central extension contracts to a semiprime ideal to conclude that S_u is semiprime.

Part (b) is also easy: just use the fact that $S \to \bigoplus S_{u_i}$ is a *central* inclusion.

(ii) Let us verify (b) for the property of being an extension. Thus we assume for u equal to any u_i, that $R_u \to S_u$ is an extension. Let $R = k\{x_1, \ldots, x_m\}$. Recall that $S^R = \{s \in S\colon sr = rs \text{ all } r \in R\}$. Then $s \in S_u{}^{R_u}$ iff $sx_j = x_js$ in S_u, all $j \leq m$ iff $(sx_j - x_js)u^n = 0$, all $j \leq m$, some n. Thus $su^n \in S^R$, and hence $S \otimes C_u = S^R R \otimes C_u$ for u equal to any u_i. Since the sum of the localizations of C is faithfully flat we conclude $S = S^R R$, so S is an extension of R. We omit the proofs of the remaining properties.

A local property extends to schemes of the type *Spec* $\mathfrak{A} = Y$ considered above. Thus we may speak of a *semiprime* scheme Y, or scheme Y *integral* over $\mathbf{P}_R{}^n$, etc.

5. Properness of projective space. Let Y be a scheme of algebras over a commutative scheme Z. In order to define the notion of curve in Y, we consider a curve Spec A [1, (5.1)], where A is an order over a dedekind ring D, as a scheme over Spec D. Thus a *curve in Y* is a central morphism Spec $A \to Y$ over a map Spec $D \to Z$.

Let Y be a scheme integral over $\mathbf{P}_R{}^n$. Then the morphism $Y \to$ Spec $R = X$ satisfies the following *curve criterion* for properness: Let A' be a D'order in $M_n(K)$, where $D' = D[t^{-1}]$ for some $t \neq 0$ in D, D is a dedekind ring and K is its field of fractions. Let $f'\colon$ Spec $A' \to Y$ be a curve in Y, and let $R[D]$ denote the subring of A' generated by the image of R and D.

PROPOSITION (5.1). *If* $t^{-1} \notin R[D]$, *there is a unique D-order A and central map f: Spec A → Y such that f′ is the localization of f with respect to t.*

In the case $y = \mathbf{P}_R{}^1$, this is related to a result of Procesi [4, p. 114].

Proof. First of all, projective space $\mathbf{P}^n$ over k is proper [2, p. 103], and this implies that the map g': Spec $D' \to \mathbf{P}^n$ underlying f' extends uniquely to a map g: Spec $D \to \mathbf{P}^n$. We may adjust coordinates so that the image under g of the finite set F where $t = 0$ lies in the affine open U_0. The problem of extending f' is local on Spec D at the set F, and so we may assume that the image of Spec D lies entirely in U_0. Then the extension problem is reduced to the affine situation

$$R[u_{10}, \ldots, u_{n0}] \xrightarrow{\varphi_0'} \mathcal{O}_Y(U_0) = S_0 \xrightarrow{\psi_0} A'.$$

Since $u_{10}, \ldots, u_{n0}$ map to D under ψ_0, the rings $R[D]$ and $R[u_{io}][D]$ are the same. Therefore the curve criterion [1, (6.3)], applied to the integral map φ_o', proves that the extension exists.

PROPOSITION (5.2). *Let Y be a scheme integral over* $\mathbf{P}_R{}^n$. *Then the correspondence* $Y \xrightarrow{f} X =$ *Spec R carries closed sets to closed sets.*

Proof. Since the correspondence Y $\to \mathbf{P}_R{}^n$ is integral, it is closed [1, (7.1)]. We may therefore replace Y by its image in $\mathbf{P}_R{}^n$, and this reduces us to showing that a closed set, say Y, in $\mathbf{P}_R{}^n$ has closed image in X. The image of Y is constructible [1, (4.11)], and so if it is not closed, there is an accessible point p in its boundary [1, (5.13)]. We may then find a curve φ: $C \to X$ passing through p, with all but a finite number of points lying in $f(Y)$. Reparametrizing C if necessary, we may assume that an open subset $C' \subset C$ lifts to Y, i.e., that there is a map ψ': $C' \to Y$ such that $f\psi' = \varphi'$. Apply proposition (5.1).

6. Cohomology. Let Y be a scheme of algebras over Z. By left (or right) $\mathcal{O}_Y$-module M, we mean a quasi-coherent sheaf of $\mathcal{O}_Z$-modules, such that on each affine open set U of Z, $M(U)$ is an $\mathcal{O}_Y(U)$ module. The *cohomology* of M is defined to be its cohomology as sheaf over Z:

$$H^q(Y, M) = H^q(Z, M). \tag{6.1}$$

It is a left (or right) module over the ring

$$H^0(Y, \mathcal{O}_Y) = \text{ring of global sections of } \mathcal{O}_Y. \tag{6.2}$$

We refer the reader to [2, III, 2] for a treatment of cohomology.

For example, if Y is a scheme over $\mathbf{P}_R{}^1$, given by a diagram of algebras

$$\begin{array}{ccccccc} R[u_0] & & & & A_0 & & \\ & \searrow & & & & \searrow & \\ & & R[u_0, u_1] & \rightarrow & & & A_{01} \\ & \nearrow & & & & \nearrow & \\ R[u_1] & & & & A_1 & & \end{array} \qquad (u_0 u_1 = 1),$$

where $A_i = \mathcal{O}_Y(U_i)$, then an $\mathcal{O}_Y$-module M is a diagram

$$\begin{array}{ccc} M_0 & & \\ & \searrow^{d_1} & \\ & & M_{01} \\ & \nearrow_{d_0} & \\ M_1 & & \end{array}$$

of A-modules, and the cohomology of M is given by the exact sequence

$$0 \rightarrow H^0(Y, M) \rightarrow M_0 \oplus M_1 \xrightarrow{d_0 - d_1} M_{01} \rightarrow H^1(Y, M) \rightarrow 0. \tag{6.3}$$

It is important to note that since our rings R are in general not noetherian, the cohomology of a scheme over $\mathbf{P}_R{}^n$ need not be finitely generated over R.

Example (6.4). Consider the ring

$$R = \begin{bmatrix} k[u] & k[u, v] \\ 0 & k[v] \end{bmatrix}.$$

A closed subscheme $Y \subset \mathbf{P}_R{}^2$ is defined by an ideal $\mathfrak{a} \subset \mathcal{O}_{\mathbf{P}_R{}^2}$, and any such ideal will have the form

$$\mathfrak{a} = \begin{bmatrix} I & J \\ 0 & I' \end{bmatrix},$$

where I, I', J are ideal of the rings $k[u]$, $k[v]$, $k[u, v]$ respectively, and

$J \supset I + I'$. Let I be the ideal generated by the homogeneous element y_0, and let I' be generated by $y_1 y_2$. The loci $\{y_0 = 0\}$, $\{y_1 y_2 = 0\}$ are connected, but their intersection consists of two points. Let J be the corresponding ideal. Then if $\mathcal{O}_Y = \mathcal{O}_{\mathbf{P}_R{}^2}/\mathfrak{a}$, we find

$$H^0(Y, \mathcal{O}_Y) \approx \begin{bmatrix} k[u] & k[u, v]^2 \\ 0 & k[v] \end{bmatrix},$$

which is not a finite left or right R-module.

However, we do have this fact:

PROPOSITION (6.5). *Let Y be integral over $\mathbf{P}_R{}^n$. Then every element $\alpha \in H^0(Y, \mathcal{O}_Y)$ is integral over R.*

Proof. We may assume Y semi-prime. Let $R' = R\{\alpha\}$ be the subring of $H^0(Y, \mathcal{O}_Y)$ generated by the image of R and by α. Then since R' maps to $\mathcal{O}_Y$, the scheme Y has a natural structure of scheme over $\mathbf{P}_{R'}{}^n$ (cf. (4.7)), and it is integral over that scheme. Let $Y' \subset \mathbf{P}_{R'}{}^n$, denote the image of Y. This is a closed subset which we give the canonical structure of semi-prime scheme. Since Y is semi-prime, it maps to Y', and we may replace Y by Y', hence suppose that Y is a closed subscheme of $\mathbf{P}_{R'}{}^n$.

We test the curve criterion [1, (6.3)] for the map $R \to R'$: Say that with the usual notation we are given a central homomorphism $R' \to A'$, an order over $D' = D[t^{-1}]$, and $t^{-1} \notin R[D]$. If the morphism Spec $A' \to$ Spec $R' = X'$ lifts to a curve Spec $A' \xrightarrow{f'} Y$, the properness (5.1) of Y over Spec R will imply that f' extends to a curve Spec $A \xrightarrow{f} Y$, and the composed map Spec $A \to Y \to X'$ will verify the criterion in that case. Now being a closed subscheme of $\mathbf{P}_{R'}{}^n$, Y is central over X'. Also, $Y \to$ Spec R' is a closed map (5.2), and hence is surjective. Therefore the curve can be lifted, after a change of parameter ring D [1, (5.16)]. This completes the proof.

Let us denote by L the hyperplane at infinity in $\mathbf{P}^n$, and by $V \subset Y$ its inverse image in a scheme Y which is integral over $\mathbf{P}_R{}^n$. Let f denote the morphism $Y \to X =$ Spec R. By Proposition (5.2), $f(V) = W$ is a closed subset of X. Let $\beta \in R$ be an element which vanishes on W. This implies that for every $y \in V$, $\beta = 0$ at each of the finite set [1, (4.1)] of points making up $f(y)$. Then if m is the pi degree of Y, $\beta^m = 0$ at y, and hence at every point of V.

PROPOSITION (6.6). *Let Y be a scheme integral over $\mathbf{P}_R{}^n$, and let $\beta \in R$ be an element which vanishes at the points at infinity, as above. Let M be a left $\mathcal{O}_Y$-module.*

(i) *If $\mu \in H^0(Y, M)$* and $q > 0$, *then $\beta^N\mu = 0$ if N is sufficiently large.*

(ii) *Let U_0 be the standard affine open set in $\mathbf{P}^n$, and let $M_0 = M(U_0)$. If $m \in M_0$, then $\beta^N m$ extends to a section of $H^0(Y, M)$ for large N.*

Proof. Consider an affine open $U_i = \operatorname{Spec} k[u_{oi}, \ldots, u_{ni}]$ of $\mathbf{P}^n$. Let $S_i = \mathcal{O}_Y(U_i)$. The hyperplane $L \cap U_i$ has equation $u_{oi} = 0$. Since some power of β vanishes at all points of $Y_i = \operatorname{Spec} S_i$ lying over L, the Nullstellensatz implies that u_{oi} divides a power of β in S_i. We may replace β by a high enough power so that u_{oi} divides β for each $i = 1, \ldots, n$: say $\beta = u_{oi}\gamma_i$ in S_i.

Let $I \subset \mathcal{O}_Y$ denote the sheaf of ideals of elements which have support on L. So, by definition

$$I(U_i) = \{\alpha \in S_i \mid u_{oi}{}^N\alpha = 0 \text{ for some } N\}.$$

Consider the exact sequence of $\mathcal{O}_Y$-modules

$$0 \to IM \to M \to M/IM \to 0.$$

Every section of IM is annihilated by a power of β, so the proposition is certainly true when M is replaced by IM. It follows from the exact sequence for cohomology that we need prove it only for $M/IM = \overline{M}$. This is a sheaf of $\mathcal{O}_Y/I$-modules, and we replace Y by *Spec* $\mathcal{O}_Y/I$. Thus we are reduced to the case that $\mathcal{O}_Y$ has no non-zero sections with support on L, i.e., $I = 0$. Then the division $\beta = u_{oi}\gamma_i$ in S_i is unique, and in S_{ij} we have the equation

$$\gamma_i = u_{ij}\gamma_j.$$

This allows us to factor multiplication by β^N in M as

$$\begin{array}{ccc} M & \xrightarrow{\beta^N} & M \\ & \searrow \quad \nearrow_{\gamma^N} & \\ & M(N) & \end{array},$$

where $M(N) = \mathcal{O}_{\mathbf{P}}(N) \otimes_{\mathcal{O}_{\mathbf{P}}} M$, and $\mathcal{O}_{\mathbf{P}}(N)$ is the sheaf of functions on $\mathbf{P}^1$ having a pole of order N on L. The map γ^N is multiplication by γ_i^N on U_i.

It is known that positive dimensional cohomology classes are annihilated by the map $M \to M(N)$, and that sections become extendable in $M(N)$, if N is sufficiently large. (Unfortunately we do not know a reference for this well-known fact. It can be deduced from [2, p. 228] by writing a quasi-coherent sheaf F on $\mathbf{P}_k^n$ as direct limit of coherent subsheaves. Cohomology commutes with this direct limit.) Thus the proposition follows.

7. Proof of Theorem (3.4).

LEMMA (7.1). *Let $N \subset \mathcal{O}_Y = \mathcal{O}$ be an ideal with square zero: $N^2 = 0$. Let $\overline{\mathcal{O}} = \mathcal{O}/N$, and $\overline{Y} =$ Spec $\overline{\mathcal{O}}$. If Theorem* (3.4) *is true for $\overline{Y}$, it is also true for Y.*

Proof. Note that the schemes Y, $\overline{Y}$ have homeomorphic underlying spaces. We denote by $\bar{y}$ the point which corresponds to $y \in Y$. Let $\bar{\alpha} \in H^0(\overline{Y}, \overline{\mathcal{O}})$ be the element required by the theorem for $\bar{y}$, and let $R' = R\{\bar{\alpha}\}$ be the subring of $H^0(\overline{Y}, \overline{\mathcal{O}})$ generated by R and $\bar{\alpha}$. We may view $\overline{Y}$ as a scheme over $X' =$ Spec R' (see section 3):

$$\begin{array}{ccccccc} \mathbf{P}_R{}^1 & \xleftarrow{\;g\;} & Y & & \mathbf{P}_{R'}{}^1 & \xleftarrow{\;g'\;} & \overline{Y} \\ & {\scriptstyle\pi}\searrow & \swarrow{\scriptstyle f} & \longleftarrow & {\scriptstyle\pi'}\searrow & & \swarrow{\scriptstyle f'} \\ & & X & & & X' & \end{array} \tag{7.2}$$

Let us write the fibre of $\overline{Y}$ over x as $\bar{y} \cup \bar{\Delta}$. By construction, $f'^{-1}(f'(\bar{y})) = \bar{y}$, hence $f'(\bar{y}) = x'$ is not contained in $f'(\bar{\Delta})$. We may choose projective coordinates so that $\bar{y}$ maps to a point at finite distance in $\mathbf{P}_{R'}{}^1$. Let $\overline{V}$ be the projection to X' of the locus at infinity of $\overline{Y} \to \mathbf{P}_{R'}{}^1$. Then $\overline{V}$ is closed and $x' \notin \overline{V}$. Therefore there is an element $\bar{\beta} \in R'$ which takes the values 1 at x' and 0 on $\overline{V} \cup f'(\bar{\Delta})$. We replace $\bar{\alpha}$ by $\bar{\beta}$, which we view as element of $H^0(\overline{Y}, \overline{\mathcal{O}})$, and we try to lift $\bar{\alpha}$ to $H^0(Y, \mathcal{O})$.

From the exact sequence

$$0 \to N \to \mathcal{O} \to \overline{\mathcal{O}} \to 0$$

of sheaves on $\mathbf{P}_k{}^1$, it follows that the obstruction to lifting $\bar{\beta}$ lies in $H^1(Y, N) = H^1(\mathbf{P}_k{}^1, N)$. Specifically, let a subscript $i = 0, 1$ denote the sections of a given sheaf on the standard affine U_i (4.2). Then we have the two exact sequences

$$0 \to N_i \to \mathcal{O}_i \to \overline{\mathcal{O}}_i \to 0.$$

We may lift $\overline{\alpha}_i$ to an element $\alpha_i \in \mathcal{O}_i$. The difference $\alpha_1 - \alpha_0 = h \in \mathcal{O}_{01}$ lies in N_{01}, and represents the cohomology class of the obstruction. Note that N is a 2-sided $\overline{\mathcal{O}}$-module. Since $\overline{\alpha}$ vanishes on $\overline{V}$, we may apply Proposition (6.6), to conclude that $\alpha^n h$ and $h\alpha^n$ represent zero in $H^1(Y, N)$, if $n \gg 0$.

Consider the lifting of $\overline{\alpha}^{2n}$:

$$\begin{aligned}\alpha_1{}^{2n} = (\alpha_0 + h)^{2n} &= \alpha_0{}^{2n} + \alpha^{2n-1}h + \alpha^{2n-2}h\alpha + \cdots + h\alpha^{2n-1} \\ &= \alpha_0{}^{2n} + p + q,\end{aligned}$$

where $p \in R'\alpha^n hR'$ and $q \in R'h\alpha^n R'$. Since $H^1(\overline{Y}, N)$ is a 2-sided R'-module, the obstruction vanishes. So $\overline{\alpha}^{2n}$ lifts to $H^0(Y, \mathcal{O})$, and is the required element.

LEMMA (7.3). (i) *It suffices to prove the theorem when Y is replaced by the semiprime scheme Spec $\mathcal{O}_Y/J(\mathcal{O}_Y)$.*

(ii) *To prove the theorem for $Y \xrightarrow{f} X$, it is permissible to replace R be any finitely generated subring of $H^0(Y, \mathcal{O}_Y)$ containing R.*

Proof. (i) Since the Jacobson radical of $\mathcal{O}_Y$ is nilpotent [5], it follows by induction from Lemma (7.1), that Y may be assumed semiprime.

(ii) Let R'' be such a subring: $R \subset R'' \subset H^0(Y, \mathcal{O}_Y)$. If a point $y \in Y$ is isolated in its fibre F over X, it is also isolated in its fibre F'' over $X'' = \text{Spec } R''$, because F'' is a closed subset of F. The assertion now follows from Lemma (3.7).

Assume now that X and Y are semiprime. An irreducible (or prime) component of Y is defined as *Spec* $\mathcal{O}_Y/\mathfrak{p}$, where $\mathfrak{p}$ is a minimal prime ideal of $\mathcal{O}_Y$. There is no difficulty in making this extension of the concept of component to schemes over $\mathbf{P}^n$ because a central localization of a prime ideal is prime. We can decompose Y as $Y = Y' \cup Z$, where Y' is the union of those components passing through the point y under consideration, and Z is the union of the other components.

LEMMA (7.4). *To prove the theorem for the point $y \in Y$, it suffices to prove it for Y', at the same point.*

Proof. Let W be the scheme-theoretic intersection $Y' \cap Z$, defined by the sheaf $\mathcal{O}_W = \mathcal{O}_Y/(\mathfrak{a} + \mathfrak{b})$, where $\mathcal{O}_{Y'} = \mathcal{O}_Y/\mathfrak{a}$ and $\mathcal{O}_Z = \mathcal{O}_Y/\mathfrak{b}$. This sheaf has the property that

$$\mathcal{O}_Y \approx \mathcal{O}_{Y'} \times_{\mathcal{O}_W} \mathcal{O}_Z. \tag{7.5}$$

The symbol on the right is the subsheaf of the direct product, such that both its projections agree on $\mathcal{O}_W$.

Assume that we can find an element $\alpha' \in H^0(Y', \mathcal{O}_{Y'})$ taking the value 1 at y and zero on the rest of the fibre of Y' over X. Let $R' = R\{\alpha\}$, and let f' denote the map $Y' \to \operatorname{Spec} R'$. Then $f'^{-1}(f'(y)) = \{y\}$. Therefore $f'(W)$ does not contain $f'(y)$, and so there is a $\beta' \in R'$ taking the value 0 on W and 1 at y. Replace α' by $\alpha'\beta'$. By the Nullstellensatz, the section $\alpha'^n \in H^0(Y, \mathcal{O}_Y)$ represents zero in $H^0(W, \mathcal{O}_W)$ if n is large. Replace α' by that power. Then α' extends by zero on Z to the required section $\alpha \in H^0(Y, \mathcal{O}_Y)$.

By Lemma (7.4), we may assume that every component of Y passes through y. Let $W \subseteq Y$ be the locus of points at which the fibre dimension is positive. It is closed by Theorem (1.6), and by assumption $y \notin W$. Therefore, no component of Y is contained in W. It follows that $\dim X = \dim Y$, and that it is finite-to-one on a non-empty open subset of X of the same dimension.

Let $X^* = g(Y)$ be the image of Y in $\mathbf{P}_R{}^1$, that is, the closed subscheme defined by the sheaf of algebras $\mathcal{O}_{X^*} = \mathcal{O}_{\mathbf{P}_R{}^1}/I$, where $I = \ker(\mathcal{O}_{\mathbf{P}_R{}^1} \to \mathcal{O}_Y)$. (Thus $\mathcal{O}_{X^*} \subset \mathcal{O}_Y$.)

LEMMA (7.5). *We may assume that there is a non-zero central element $c \in Z(R)$ which is in the conductor of $\mathcal{O}_{X^*}$ in $\mathcal{O}_Y$, i.e., such that $c\mathcal{O}_Y \subset \mathcal{O}_{X^*}$.*

Proof. Adjust coordinates so that y is not at 0 or infinity. Let $V \subset X$ be the projection of the locus at infinity of Y. Then $\dim V < \dim X$, since no component of Y is entirely at ∞. Choose $\beta \in R$ which vanishes on V but not on any component of maximal dimension of X. Then if $S_o = \mathcal{O}_Y(U_o)$ (where U_o is the standard affine in $\mathbf{P}^1$), and if α is any element of S_o, Proposition (6.6) implies that $\beta^n\alpha \in H^0(Y, \mathcal{O}_Y)$ for sufficiently large n. By Lemma (7.3) (ii), we may adjoin $\beta^n\alpha$ to R. We do this a finite number of times until the map $R \to S_o$ becomes birational on every component of X of maximal dimension. In other words, so that if $\mathfrak{p} \subset R$ is a prime ideal and $\dim R/\mathfrak{p} = \dim R$, then there is a unique prime $\mathfrak{p}' \subset S_o$ lying over $\mathfrak{p}$, and $R/\mathfrak{p} \to S_o/\mathfrak{p}'$ is birational. When this is done, the map $Y \to X^*$ is also birational on all components of maximal dimension, and it is integral.

Assume for the moment that Y is prime. Let $R^* = \mathcal{O}_{X^*}(U_o)$ and $k[T]$ be the ring generated by the coefficients of the characteristic polynomials of elements of R^*. Then $R^* \to S_o$ is integral and birational. We pass to $R^*[T] \subseteq S_o[T]$ which is also integral. We know that $S_o[T]$ is a finitely generated $k[T]$ module, and if K is the quotient field of the centre $Z(R)$, then we have $R^*K = S_oK$. Thus there is an element $d \in R \cap K$, such that $dS_o \subseteq R^*[T]$. If c is a large power of the evaluation of a central polynomial for R then by [7, Th. 1]

$$cdS_o \subseteq cR^*[T] \subseteq R^*.$$

If R is semiprime and has several minimal primes, we take the above c to lie in the intersection of all minimal primes other than a particular one we are considering. Then cd is still central in R. It kills all the other components of S_o, so replacing c by cd, we have $cS_o \subseteq R$. We can do the same for $S_1 = \mathcal{O}_Y(U_1)$ choosing an element which works for the same component. The product of the two elements will do the job.

We can now complete the proof of Theorem (3.4) by noetherian induction on the underlying topological space of Y. This allows us to assume the theorem known for any closed subscheme $Spec(\mathcal{O}_Y/I)$, and the reductions leading to Lemma (7.5) are valid. With the notation of that lemma, let $M = \mathcal{O}_Y/\mathcal{O}_{X^*}$, viewed as sheaf on $\mathbf{P}_k{}^1$, and let c be the element obtained there. Using the fact that c is in the conductor, we construct a diagram.

$$\begin{array}{ccccccc} 0 \to & \mathcal{O}_{X^*} & \to & \mathcal{O}_Y & \to M \to 0 \\ & \downarrow & & \downarrow & \| \\ 0 \to & \mathcal{O}_{X^*}/c\mathcal{O}_{X^*} & \to & \mathcal{O}_Y/c\mathcal{O}_Y & \to M \to 0. \end{array} \tag{7.6}$$

Case 1. The fibre Y_x is finite. In this case, choose coordinates so that the fibre is at finite distance over $\mathbf{P}_x{}^1$. Then if V is the projection to X of the locus at infinity of Y, we have $x \notin V$. Hence we can choose $\beta \in R$ to take the value 1 at x and 0 on V. Also, we can find an appropriate element $\alpha_o \in \mathcal{O}_Y(U_o)$ which is 1 at y and zero on Δ. By Proposition (6.6), $\alpha_o\beta^n = \alpha$ is the required section of $H^o(Y, \mathcal{O}_Y)$ if $n \gg 0$.

Case 2. The fibre $Y_x = y \sqcup \Delta$ has components of dimension 1. Then X_x^* is the whole projective line $\mathbf{P}_x{}^1$, and in fact Δ maps onto $\mathbf{P}_x{}^1$. Thus the map $Y \to X^*$ is not 1-1 at y. Clearly, this implies that c vanishes at y (i.e., at x). Let y' be a second point having the same image $x^* \in X^*$ as y.

We apply the induction hypothesis to conclude that the theorem is true for the closed subscheme $\overline{Y} = Spec(\mathcal{O}_Y / c\mathcal{O}_Y)$. Note that y and y' are points of $\overline{Y}$. Let $\overline{\alpha} \in H^0(Y, \mathcal{O}_Y) = H^0(\mathbf{P}^1, \mathcal{O}_Y / c\mathcal{O}_Y)$ be the required section at the point y, and denote by m its image in $H^0(\mathbf{P}^1, M)$. The obstruction to lifting m to a section of $H^0(Y, \mathcal{O}_Y)$ lies in $H^1(\mathbf{P}^1, \mathcal{O}_{X^*})$.

Recall that

(i) $H^1(\mathbf{P}^1, \mathcal{O}_{\mathbf{P}^1}) = 0$ [2, p. 225]. Also, $H^1(\mathbf{P}^1, \mathcal{O}_{\mathbf{P}_R{}^1})$ $= H^1(\mathbf{P}^1, \mathcal{O}_{\mathbf{P}^1}) \otimes_k R$ because R is k-flat [2, p. 255]. Therefore $H^1(\mathbf{P}^1, \mathcal{O}_{\mathbf{P}_R{}^1}) = 0$. (7.7)

(ii) H^1 is a right exact functor of quasi-coherent sheaves over $\mathbf{P}_k{}^1$, because $\mathbf{P}_k{}^1$ has dimension 1 [2, p. 208].

(iii) $\mathcal{O}_{X^*}$ is a quotient of $\mathcal{O}_{\mathbf{P}_R{}^1}$.

It follows that $H^1(\mathbf{P}^1, \mathcal{O}_{X^*}) = 0$. Therefore m lifts to a section $\gamma \in H^0(\mathbf{P}^1, \mathcal{O}_Y) = H^0(Y, \mathcal{O}_Y)$. Let $\overline{\gamma} \in H^0(\mathbf{P}^1, \mathcal{O}_{\overline{Y}})$ be its image. Then $\overline{\alpha} - \overline{\gamma} = \overline{\delta}$ is, by construction, zero in M, hence lies in $H^0(\mathbf{P}^1, \mathcal{O}_{X^*}/c\mathcal{O}_{X^*})$. Since $\overline{\alpha}$ takes different values at y and y', while $\overline{\delta}$ does not, γ is not constant on the fibre Y_x. Thus replacing R by $R\{\gamma\}$ results in a fibre which is a proper closed subset of Y_x. By noetherian induction on Y_x, we may make the fibre be y alone. Then Lemma (3.7) completes the proof.

8. End of the proof of Theorem (2.1). By Lemma (3.9) and Theorem (3.4), the proof is complete when $m = 1$ in the presentation (1.5). It remains to carry out the induction for $m > 1$: Let $R^* = R[u_1, \ldots, u_{m-1}]$. Replace R by R^*:

$$R^* \to R^*[u_m] \to S,$$

and construct a diagram (3.3)

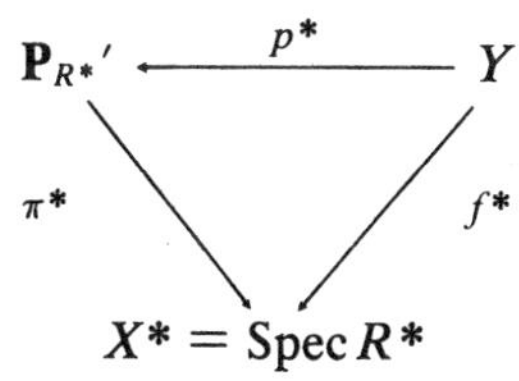

Let R' be the ring obtained in Corollary (3.8) for this diagram, so that if $X' = \operatorname{Spec} R'$, we have a chain of geometric maps

$$Y \xrightarrow{f'} X' \to X^* \to X.$$

Suppose y isolated in its fiber $Y_x = f^{-1}(f(y))$ over X. Then it is also an isolated point of the fibre $f^{*-1}(f^*(x))$ over X^*, and so by Corollary (3.8) it is the whole fibre $f'^{-1}(f'(x))$ over X'. Let F' be the fibre of X' over X at x. Then $F' = f'(Y)$ and $Y = f'^{-1}(F')$. If we write $Y_x = y \sqcup \Delta$ and use the facts that the map $Y \to X'$ is closed (5.2) and 1-1 at y, it follows that $x' = f'(y)$ is an isolated point of F'. That is what is needed. We apply Theorem (2.1) by induction, to the map

$$R \to R[u_1, \ldots, u_{m-1}] \to R',$$

to construct the required ring $R \subset \tilde{S} \subset R'$ ($\subset S$).

9. The projective version in dimension >1. We are now going to generalize Theorem (3.4) and its Corollary (3.8) to projective spaces of arbitrary dimension:

THEOREM (9.1). *Let Y be a scheme integral and geometric over $\mathbf{P}_R{}^m$. There is a finitely generated subring $R' \subset H^0(Y, \mathcal{O}_Y)$, so that the fibres of Y over Spec R' consist of a single point, or else contain no component of dimension zero.*

As in (3.4), it suffices to work at a single point $y \in Y$ which is isolated in its fibre Y_x, and to find an $\alpha \in H^0(Y, \mathcal{O}_Y)$ such that $\alpha = 1$ at y and 0 on the rest of Y_x.

If we start with any homomorphism $R \to S$ of the form (1.5), we can embed Spec S into a scheme Y integral and geometric over $\mathbf{P}_R{}^m$ exactly as in the case $m = 1$: Let u_i be the standard coordinates u_{oi}, and say that $S = S_o = R[u]\{\alpha_\nu\}$. Applying Proposition (3.1), we see that the ring $S_i = R[u_{io}, \ldots, u_{im}]\{u_{io}{}^N\alpha_\nu\}$ is integral and geometric over $R[u_{io}, \ldots, u_{im}]$ if N is sufficiently large. With this definition of S_i, we do have the property that $S_i[u_{ji}] \approx S_j[u_{ij}]$, and so these rings define the required sheaf $\mathcal{O}_Y$. Having Spec S embedded into Y, Theorem (2.1) follows from Theorem (9.1). However, the projective version seems to be the stronger one because there may be components of Y which are entirely at infinity, and it is not clear to us how to deduce (9.1) directly from (2.1). The following example shows the extra information contained in the projective version:

Example (9.2). This example shows that we may have a 1-1 correspondence between the points of R and the trace ring $R[t]$, but in the projective closure of Spec $R[t]$ there is a fibre with two points. Let $R = k\{x, y, z\}$ where $t\bar{t} = 1$ and

$$x = \begin{pmatrix} t(t-1) & 0 \\ 0 & \bar{t}(\bar{t}-1) \end{pmatrix},$$

$$y = \begin{pmatrix} t(t-1)(2t-1) & 0 \\ 0 & \bar{t}(\bar{t}-1)(2\bar{t}-1) \end{pmatrix},$$

$$z = \begin{pmatrix} 0 & 1 \\ v & 0 \end{pmatrix}.$$

Then $R[T] = R[u]$, where $u = 2 - (t + \bar{t}) = \det x$. Let p be the point of $\mathrm{Spec}_1 R$ corresponding to sending x, y, and z to 0. Then since $x^2 - x(u^2 - 3u) + u = 0$ in $R[u]$, the only point lying over p in $\mathrm{Spec}_1 R[u]$ is p_1 corresponding to u being sent to 0. Note that there is such a point, since it corresponds to setting $t = 1$ and $v = 0$. Now we introduce $\bar{u}$ with $u\bar{u} = 1$. Then in $\mathrm{Spec}_1 R[\bar{u}]$, we have a point p_2 over p also. For

$$\bar{u} = -t/(t-1)^2 = -\bar{t}/\bar{t} - 1)^2,$$

so that once we put $z = 0$ we have

$$\bar{u} = \begin{pmatrix} -t/(t-1)^2 & \\ & -\bar{t}/(\bar{t}-1)^2 \end{pmatrix}$$

and we can clearly send x, y, and $\bar{u}$ to 0. This is the only point in $\mathrm{Spec}_1 R[\bar{u}]$ over p, since

$$\bar{u}^2 x^2 - (1 - 3\bar{u})x - \bar{u} = 0$$

forces $\bar{u}$ to go to 0.

If Y denotes the scheme over $\mathbf{P}_R{}^1$ with local coordinate rings $R[u]$ and $R[\bar{u}]$, and $\operatorname{Spec} R \xleftarrow{f} Y$, then $f^{-1}(f(p)) = \{p_1, p_2\}$. Let

$$\alpha = \begin{pmatrix} t^3 & 0 \\ 0 & \bar{t}^3 \end{pmatrix}.$$

Then $\alpha = x(-u - 2^1/_2) - {}^1/_2 y + 1 = x^2\bar{u}$. So $\alpha \in H^0(Y, \mathcal{O}_Y)$. Also the above formulae for α show that α is 1 at p_1 and 0 at p_2.

Proof of Theorem (9.1). We reduce to the case $m = 1$ (3.4) by induction, by projection of $\mathbf{P}^m$ to $\mathbf{P}^1$ if $m > 1$. The standard diagram for doing this is

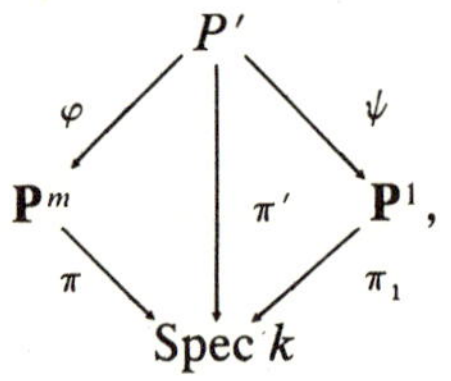

(9.3)

where φ is the blowing-up of the center L of projection and ψ is a fibre space with fibres $\mathbf{P}^{m-1}$. Here L is the linear space in $\mathbf{P}^m$ defined by $y_0 = y_1 = 0$. Coordinates should be chosen so that the point of interest y does not lie over L.

Given Y over $\mathbf{P}^m$, we form the pull-back $Y' = P' \times_{\mathbf{P}^m} Y$. Since φ is an isomorphism except above L, the fibre $f'^{-1}(x)$ of Y' over X has an isolated point y' lying over y. Therefore, y' is an isolated point of $\psi^{-1}(\psi(y'))$. The proof consists of two steps. First, the theorem is deduced for Y' by regarding the map π' as the composition $\pi' = \pi_1\psi$, and applying the induction hypothesis to the two maps π_1, ψ. Then the blowing-up φ is analyzed, to go from Y' to Y.

Consider the covering of $\mathbf{P}^1$ by the two affine opens $U_i = \operatorname{Spec} k[u_i]$ $(i = 0, 1)$. (These u_i are not the same as those defining Spec S above.) Note that $\psi^{-1}(U_i) \approx U_i \times \mathbf{P}^{m-1}$. Therefore the open subset Y_i' of Y' lying over U_i can be viewed as a scheme over $(\operatorname{Spec} R) \times U_i \times \mathbf{P}^{m-1} = \mathbf{P}^{m-1}_{R[u_i]}$. It is integral and geometric over this scheme, and hence we may apply induction to find a subring $\tilde{S}_i \subset H^0(Y_i', \mathcal{O}_{Y'})$ containing $R[u_i]$, and such that the point y' is the whole fibre over its image $\bar{y}_i$ in Spec $\tilde{S}_i$. The two rings $\tilde{S}_0$, $\tilde{S}_1$ can be adjusted so that $\tilde{S}_0[u_1]$

$= \tilde{S}_1[u_0]$, and then they define a scheme $\tilde{Y}$ integral and geometric over $\mathbf{P}_R{}^1$, and a morphism $Y' \to \tilde{Y}$. Corollary (3.8) applied to $\tilde{Y}$ then guarantees the existence of a subring $R' \subset H^0(\tilde{Y}, \mathcal{O}_{\tilde{Y}})$ with the property that $\tilde{y}$ is the fibre of $\tilde{y}$ over its image x' in $X' = \operatorname{Spec} R'$. Then also y' is the fibre of Y' over x', and this shows that the conclusion of Theorem (9.1) is valid when the scheme Y is replaced by Y'. So, there is an element $\alpha' \in H^0(Y', \mathcal{O}_{Y'})$ with the property that the fibre of Y' over Spec $R\{\alpha'\}$ is the single point y'. We take $R\{\alpha\}$ for R', and denote by f' the map $Y' \to X' = \operatorname{Spec} R'$.

Let $W' \subset Y'$ be the inverse image of the center of projection L. Then $x' = f(y')$ is not contained in $f'(w)$. So we may adjust α' to take the value 1 at y' and 0 on W' and on the rest of the fibre $Y_x{}'$. It suffices to show that then some power α' is the image of an element $\alpha \in H^0(Y, \mathcal{O}_Y)$. That element α will obviously be the required element, and so the proof will be complete. In fact, it suffices to show this in case Y is semiprime:

LEMMA (9.4). *To prove Theorem* (9.1) *for* Y, *it suffices to prove it for the semiprime subscheme* Spec$(\mathcal{O}_Y/J(\mathcal{O}_Y))$.

The proof of this lemma is the same as that of (7.3).

Assume now that Y is semiprime. The map φ is surjective, and therefore $Y' \to Y$ is also surjective, and it follows from the Nullstellensatz that the map of sheaves $\mathcal{O}_Y \to \varphi_* \mathcal{O}_{Y'}$ is injective. This means that a section α mapping to α' is unique, if it exists, and therefore the problem of finding α is local on Y, i.e., can be carried out separately on affine open subsets Spec S_i of Y. By (6.5), we know that α' generates an integral over-ring S_i' of S_i. Moreover, the map $Y' \to Y$ is an isomorphism except on W', and α' vanishes on W'. The following lemma therefore completes the proof:

LEMMA (9.5). *Let* $R \xrightarrow{\varphi} S$ *be an injective integral homomorphism and let* $I \subset R$ *be a centrally generated, finitely generated, ideal. Suppose that the map* Spec $S \to$ Spec R *induced by* φ *is an isomorphism except on* $V(I)$. *This means that for every* $u \in Z(R) \cap I$, *the map* $R[u^{-1}] \to S[u^{-1}]$ *is an isomorphism. Then some power of* IS *is contained in* R.

Proof. Let $\{u_1, \ldots, u_r\}$ be central generators for I, and let $\{\alpha_1, \ldots, \alpha_r\}$ be generators for S. By assumption, given any $\beta \in S$ and u_i, there is an integer so that $u_i{}^m \beta \in R$. By [1, (8.1)], every β can be expressed as a sum of generalized monomials in the $\{\alpha_i\}$, with coefficients in R and of bounded degree. Therefore the integer n can be bounded independently of β and i. Then $I^nS = (IS)^n \subset R$, as required. This completes the proof of Theorem (9.1).

MASSACHUSETTS INSTITUTE OF TECHNOLOGY

REFERENCES

[1] M. Artin and W. Schelter, "Integral ring homomorphisms." (In preparation.)
[2] R. Hartshorne, *Algebraic geometry,* Springer, New York, 1977.
[3] D. Mumford, *Algebraic geometry I: Complex projective varieties,* Springer, Berlin 1976.
[4] C. Procesi, *Rings with polynomial identities,* Marcel Dekker, New York, 1973.
[5] Y. Razmyslov, "Jacobson radical in pi algebras," Algebra i Logika **13** (1975) 192–204 (translation).
[6] W. Schelter, "Azumaya algebras and Artin's theorem," *J. of Alg.* **46** (1977) 303–304.
[7] W. Schelter, "Noncommutative affine pi rings are catenary," *J. of Alg.* **51** (1978) 12–17.
[8] O. Zariski, "Foundations of a general theory of birational correspondences," *Transactions A.M.S.*, **50** (1941) 48–70.
[9] O. Zariski, "Theory and applications of holomorphic functions on algebraic varieties over arbitrary ground fields," *Memoirs A.M.S.* vol. 69 (1951).

REMARKS ON ADJOINTS AND ARITHMETIC GENERA OF ALGEBRAIC VARIETIES

By PIOTR BLASS and JOSEPH LIPMAN[1]

Let $\mathbf{P}^{n+1}$ be projective $(n + 1)$-space over a fixed algebraically closed field k, let $F \subset \mathbf{P}^{n+1}$ be a (reduced) hypersurface of degree d, and let $f:\tilde{F} \to F$ be a desingularization. A classical characterization of the arithmetic genus of $\tilde{F}$:

$$p_a(\tilde{F}) = (-1)^n[\chi(\mathcal{O}_{\tilde{F}}) - 1] = h^n(\mathcal{O}_{\tilde{F}}) - h^{n-1}(\mathcal{O}_{\tilde{F}}) + \cdots + (-1)^{n-1}h^1(\mathcal{O}_{\tilde{F}})$$

is that

(#) $p_a(\tilde{F}) - 1$ *is the virtual* (*or "postulated"*) *dimension of the linear system of hypersurfaces in* $\mathbf{P}^{n+1}$ *having degree* $d - n - 2$ *and adjoint to* F.

(For the case $n = 2$, cf. [**10**, p. 73]; for $n = 3$ see [**11**, p. 590, footnote 13].)

Our main observation is that this characterization is an immediate corollary of duality theory and the Grauert-Riemenschneider vanishing theorem, the latter being valid when the characteristic of k is zero [**2**], and, if $n \leq 2$, also for characteristic > 0 (combine [**8**, Proposition 2.6] with [**7**, Theorem 2.3]).

To see this, we restate the characterization in a form which depends only on F (*not its embedding in* $\mathbf{P}^{n+1}$), and is in fact meaningful for *any* complete algebraic variety (reduced and pure n-dimensional).

Let $\omega_{\tilde{F}}$ be a dualizing sheaf on $\tilde{F}$, and let ω_F be a dualizing sheaf on F (ω_F is determined up to isomorphism by its property of representing the functor $\mathrm{Hom}_k(H^n(\mathcal{L}), k)$ of coherent $\mathcal{O}_F$-modules $\mathcal{L}$, cf. [**4**, Chapter III, §7]. Also, $\omega_F = H^{-n}(\mathcal{R}_F)$, where $\mathcal{R}_F$ is a *residual complex* on F [**3**, Chapter VI].) Since $\mathcal{O}_{\mathbf{P}^{n+1}}(-n - 2)$ is a dualizing sheaf on $\mathbf{P}^{n+1}$,

[1] Supported by NSF grant MCS76-8134.

the "adjunction formula" gives

$$\omega_F \cong \mathcal{E}xt^1_{\mathbf{P}^{n+1}}(\mathcal{O}_F, \mathcal{O}_{\mathbf{P}^{n+1}}(-n-2)) \cong \mathcal{O}_F(d-n-2).$$

There is a natural *injective* trace map

$$\tau: f_*(\omega_{\tilde{F}}) \to \omega_F,^2$$

and the adjoint ideal $\mathcal{A} \subseteq \mathcal{O}_F$ may be defined to be the annihilator of the cokernel $\mathcal{C}$ of τ, i.e.

$$\begin{aligned}\mathcal{A} = \mathcal{H}om_{\mathcal{O}_F}(\omega_F, f_*(\omega_{\tilde{F}})) &= \mathcal{H}om_{\mathcal{O}_F}(\mathcal{O}_F(d-n-2), f_*(\omega_{\tilde{F}})) \\ &= (f_*(\omega_{\tilde{F}}))(-d+n+2).\end{aligned}$$

Let $\pi: \mathcal{O}_{\mathbf{P}^{n+1}} \to \mathcal{O}_F$ be the natural map, and let $\mathcal{A}' = \pi^{-1}(\mathcal{A})$, so that we have an exact sequence

(*) $$0 \to (\text{kernel of } \pi) = \mathcal{O}_{\mathbf{P}^{n+1}}(-d) \to \mathcal{A}' \to \mathcal{A} \to 0.^3$$

---

[2] τ can be obtained from [**3**, p. 369, Theorem 2.1]. In more down-to-earth terms, τ can be described (sketchily) as follows. Factor f as $\tilde{F} \xrightarrow{g} F_1 \xrightarrow{h} F$, where F_1 is the normalization of F. Over any affine open set $U \subseteq F$, the sections $\Gamma(U, \omega_F)$ can be identified with certain meromorphic n-differentials on F, cf. [**6**] from which one also sees that

$$\Gamma(U, h_*\omega_{F_1}) \subseteq \Gamma(U, \omega_F).$$

Furthermore, $\Gamma(U, h_*\omega_{F_1})$ consists of meromorphic n-forms having no polar divisors in $h^{-1}(U)$, while $\Gamma(U, f_*\omega_{\tilde{F}})$ consists of meromorphic n-forms without poles in $g^{-1}h^{-1}(U)$; thus

$$\Gamma(U, f_*\omega_{\tilde{F}}) \subseteq \Gamma(U, h_*\omega_{F_1}) \subseteq \Gamma(U, \omega_F),$$

and τ is just the resulting inclusion map.

[3] So the geometric genus $p_g(\tilde{F}) = H^0(f_*(\omega_{\tilde{F}}))$ satisfies

$$p_g(\tilde{F}) - 1 = H^0(\mathcal{A}(d-n-2)) - 1 = H^0(\mathcal{A}'(d-n-2)) - 1$$

(cf. (*)), which is the *actual* dimension of the linear system of hypersurfaces of degree $d-n-2$ adjoint to F.

What (#) says then is that

$$\begin{aligned}(-1)^n[\chi(\mathcal{O}_{\tilde{F}}) - 1] &= \chi(\mathfrak{A}'(d-n-2)) \\ &= \chi(\mathfrak{A}(d-n-2)) + \chi(\mathcal{O}_{\mathbf{P}^{n+1}}(-n-2)) \\ &= \chi(\mathfrak{A}(d-n-2)) + (-1)^{n+1} \\ &= \chi(f_*(\omega_{\tilde{F}})) + (-1)^{n+1};\end{aligned}$$

and finally Serre duality gives

$$(-1)^n\chi(\mathcal{O}_{\tilde{F}}) = \chi(\omega_{\tilde{F}}).$$

Thus, (#) simply says that

$$\chi(\omega_{\tilde{F}}) = \chi(f_*(\omega_{\tilde{F}})). \tag{\#\#}$$

But (##) holds for *any* reduced pure n-dimensional variety F proper over k, with $n \leq 2$ when k has characteristic > 0 (for such F a desingularization $f:\tilde{F} \to F$ always exists [**5**], [**1**]). For, the Leray spectral sequence for f gives

$$\chi(\omega_{\tilde{F}}) = \sum_{i=0}^{n-1} (-1)^i\chi(R^if_*(\omega_{\tilde{F}}));$$

and the vanishing theorem says that $R^if_*(\omega_{\tilde{F}}) = 0$ for $i > 0$. (In fact the spectral sequence degenerates, so that $H^i(\tilde{F}, \omega_{\tilde{F}}) = H^i(F, f_*\omega_{\tilde{F}})$ for all i). Q.E.D.

Remarks. (1) The forms of degree j adjoint to F form a finite dimensional k-vector space

$$A_j = H^0(\mathbf{P}^{n+1}, \mathfrak{A}'(j)) \subseteq H^0(\mathbf{P}^{n+1}, \mathcal{O}_{\mathbf{P}^{n+1}}(j)) = S_j \qquad (0 \leq j < \infty)$$

and $A = \bigoplus_{j=0}^{\infty} A_j$ is an ideal in the polynomial ring

$$k[X_0, X_1, \ldots, X_{n+1}] = \bigoplus_{j=0}^{\infty} S_j.$$

The dimension of A_j is given, for sufficiently large j, by the polynomial

$$H_A(j) = \chi(\mathcal{A}'(j));$$

and so (#) says that

$$p_a(\tilde{F}) = H_A(d - n - 2).$$

All this begs the question: how to calculate A, or at least H_A?

In general, there is no easy answer. In case $\tilde{F}$ is the normalization of F (for example in the classical situation where F is a generic projection of $\tilde{F}$ into $\mathbf{P}^{n+1}$) we have

$$\begin{aligned} f_*(\omega_{\tilde{F}}) &= \mathcal{H}om_{\mathcal{O}_F}(f_*(\mathcal{O}_{\tilde{F}}), \omega_F) \\ &= [\mathcal{H}om_{\mathcal{O}_F}(f_*(\mathcal{O}_{\tilde{F}}), \mathcal{O}_F)](d - n - 2) \end{aligned}$$

and τ is "evaluation at 1" (cf. [**3**, p. 319, (c)]; or use [**6**] as in footnote 2 above). Hence

$$\mathcal{A} = \mathcal{H}om_{\mathcal{O}_F}(f_*(\mathcal{O}_{\tilde{F}}), \mathcal{O}_F)$$

which is the *conductor* of $\mathcal{O}_{\tilde{F}}$ in $\mathcal{O}_F$. Thus, in this case, adjoints coincide with *subadjoints*, which are more or less calculable [cf. [**10**, pp. 71–72]).

(2) For any reduced pure m-dimensional Cohen-Macaulay variety F, proper over k, by Serre-Grothendieck duality $\chi(\mathcal{O}_F) = (-1)^m\chi(\omega_F)$. So, given a desingularization $f: \tilde{F} \to F$, we have, using Grauert-Riemenschneider as above,

$$\begin{aligned} p_a(F) - p_a(\tilde{F}) = (-1)^m(\chi(\mathcal{O}_F) - \chi(\mathcal{O}_{\tilde{F}})) &= \chi(\omega_F) - \chi(\omega_{\tilde{F}}) \\ &= \chi(\omega_F) - \chi(f_*\omega_{\tilde{F}}) \\ &= \chi(\mathcal{C}). \end{aligned}$$

($\mathcal{C}$ is the cokernel of the above injective map τ). This generalizes formula (**) on p. 153 of [**7**].

If, in particular, $F \subseteq \mathbf{P}^{n+1}$, and F is such that the dualizing sheaf ω_F is $\mathcal{O}_F(D - n - 2)$ for some D (for example if the vertex of the projecting cone over F is Gorenstein; in particular if F is a complete intersection of hypersurfaces of degrees $d_1, \ldots, d_{n+1-m}$, we can take $D =$

$d_1 + \cdots + d_{n+1-m}$) then, proceeding as above, we have

$$\mathcal{C} = \mathcal{O}_{\mathbf{P}^{n+1}}(D - n - 2)/\mathcal{A}'(D - n - 2)$$

so that

$$(\#)' \ p_a(F) - p_a(\tilde{F}) = \chi(\mathcal{C})$$
$$= \chi(\mathcal{O}_{\mathbf{P}^{n+1}}(D - n - 2)) - \chi(\mathcal{A}'(D - n - 2))$$
$$= \binom{D-1}{n+1} - \chi(\mathcal{A}'(D - n - 2)).$$

(If F is a hypersurface of degree D then

$$p_a(F) = \binom{D-1}{n+1},$$

and $(\#)'$ reduces to $(\#)$.)

(3) Let F be as in (2), with $m \leq 2$ if k has characteristic > 0. If F is normal and has only isolated singularities, then $\mathcal{C}$ has zero-dimensional support (viz. on the singular points), and

$$\chi(\mathcal{C}) = \sum_{x \text{ singular}} \dim_k(\mathcal{C}_x).$$

Here is another "dual" description of $\dim_k(\mathcal{C}_x)$. Let U be an affine neighborhood of a singular point x, containing no other singular point, let $\tilde{U} = f^{-1}(U)$, $E = f^{-1}(x)$, and assume (without real loss of generality) that f induces an isomorphism $\tilde{U} - E \xrightarrow{\sim} U - x$. We have an exact sequence

$$0 \to H^0_E(\omega_{\tilde{F}}) \to H^0(\tilde{U}, \omega_{\tilde{F}}) \to H^0(U - x, \omega_{\tilde{F}})$$
$$\to H^1_E(\omega_{\tilde{F}}) \to H^1(\tilde{U}, \omega_{\tilde{F}}) = 0$$

(the last equality by Grauert-Riemenschneider). Now

$$H^0(U - x, \omega_{\tilde{F}}) = H^0(U - x, \omega_F) = H^0(U, \omega_F)$$

because, F being normal, ω_F is the sheaf of meromorphic m-forms with no polar divisors on F.

We conclude that

$$\mathcal{C}_x \cong H^1_E(\omega_{\tilde{F}}).$$

But by [7, p. 188], $H^1_E(\omega_{\tilde{F}})$ is dual to the stalk $R^{m-1}f_*(\mathcal{O}_{\tilde{F}})_x$. Thus

$$\mathcal{C}_x \cong \mathrm{Hom}_k(R^{m-1}f_*(\mathcal{O}_{\tilde{F}})_x, k)$$

and so

$$\dim_k(\mathcal{C}_x) = \dim_k(R^{m-1}f_*(\mathcal{O}_{\tilde{F}})_x).$$

(This is essentially Theorem A in [**9**].)

PURDUE UNIVERSITY

REFERENCES

[1] S. S. Abhyankar, "Local uniformization on algebraic surfaces over ground fields of characteristic $p \neq 0$," *Annals of Math.*, **63** (1956), pp. 491–526.

[2] H. Grauert and O. Riemenschneider, "Verschwindungssätze für analytische Kohomologiegruppen auf komplexen Räumen," *Inven. Math.*, **11** (1970), pp. 263–292.

[3] R. Hartshorne, *Residues and duality*, Lecture Notes in Math. **20**, Springer-Verlag, 1966.

[4] ______, *Algebraic Geometry*, Springer-Verlag, 1977.

[5] H. Hironaka, "Resolution of singularities of an algebraic variety over a field of characteristic zero," *Annals of Math.*, **79** (1964), pp. 109–326.

[6] E. Kunz, "Holomorphe Differentialformen auf algebraischen Varietäten mit Singularitäten I," *Manuscripta Math.*, **15** (1975), pp. 91–108.

[7] J. Lipman, "Desingularization of two-dimensional schemes," *Annals of Math.*, **107** (1978), pp. 151–207.

[8] J. Wahl, "Vanishing theorems for resolutions of surface singularities," *Inven. Math.*, **31** (1975), pp. 17–41.

[9] S. S. T. Yau, "Two theorems on higher dimensional singularities," *Math. Ann.*, **231** (1977), pp. 55–59.

[10] O. Zariski, *Algebraic Surfaces* (second supplemented edition), Springer-Verlag, 1971.

[11] ______, "Complete linear systems on normal varieties and a generalization of a lemma of Enriques-Severi," *Annals of Math.*, **55** (1952), pp. 552–592.

APPLICATIONS OF ALGEBRAIC GEOMETRY IN SYSTEM THEORY

By Christopher I. Byrnes* and Peter L. Falb†

1. Introduction. System theory is concerned with the modelling and analysis of phenomena both natural and man-made. It is a discipline whose formal beginnings go back at least to Watt and Maxwell and much of its motivation stems from engineering problems. Before World War II, system design and analysis were primarily an art. During and after the war, techniques based on complex variable theory were developed and applied primarily to single input, single output systems represented by a rational function, called the transfer function. The theory of servomechanisms developed rapidly from the end of the war to the early fifties and time-domain methods were applied. The representation of transfer functions via linear, constant-coefficient, differential equations led to a renewed interest in so-called state space methods. The rapid development of the theory followed and continues today. However, the increasing complexity of the engineering and economic problems considered required greater mathematical sophistication. Little did one of the authors dream while he was a student of Professor Zariski, that the techniques of modern algebraic geometry could (and would) be applied in a critical and essential way in system theory. Our purpose, in this paper, is to illustrate several of the many applications of algebraic geometry to linear system theory. These applications are both mathematically and practically non-trivial.

We begin, in Section 2, with an analysis of two (classical) questions; namely, (i) when is a meromorphic function of the form $f(s) = \Sigma^{\infty}_{l=1} h_l s^{-l}$ rational?, and, (ii) when is such a rational function stable (i.e.

*This work was supported by NASA under grant NSG-2265 and by the Office of Naval Research under JSEP-N00014-75-C-0648.

†This work was supported by the Air Force Office of Scientific Research under AF-AFOSR 77-3182.

has poles in the left-half plane)? The first question was answered by Hankel ([20]) and the second was answered by Cauchy, Hermite, Routh and Hurwitz ([15], [29], [30]). Our treatment serves to motivate the subsequent development. The concept of a time-invariant, finite-dimensional linear system is introduced in Section 3. Such systems admit an external description as either a matrix, $T(s)$, of rational functions or as an input-output map $f(s) = \sum_{l=1}^{\infty} L_l s^{-l}$ with the L_l constant matrices and rank $H_f < \infty$ where $H_f = (L_{i+j-1})$ is the Hankel matrix of f. Corresponding to each external description is an internal description or representation. Thus, a pair of polynomial matrices $(R(s), P(s))$ realizes $T(s)$ if $T(s) = R(s)P^{-1}(s)$ and $R(s)$, $P(s)$ are relatively right prime. The group $\mathcal{U}$ of polynomial matrices with determinant a unit acts on such realizations. For an input-output map f, a triple (A, B, C) of constant matrices represents f if $H_f = (CA^{i+j-1}B)$. The general linear group G (of appropriate dimension) acts on such triples via $g \cdot (A, B, C) = (gAg^{-1}, gB, Cg^{-1})$. Three main topics are treated in Section 4. The first is the construction of a moduli space for linear systems under the action of G and a proof of a theorem on "realization with parameters," based on Zariski's Main Theorem and Mumford's geometric invariant theory. It turns out that the "exceptional points" are precisely the representations that are undesirable from a practical viewpoint. Thus, the mathematical and physical considerations are exactly congruent. The second topic is a brief study of the biregular and topological invariants of the moduli space with an emphasis on the impact these invariants have on system theoretic questions analogous to the questions raised in Section 2. The third topic is the consideration of moduli spaces for systems with richer symmetries such as Hamiltonian or symmetric systems. In fact, we sketch the equivalence between an old problem in network theory ([53]) and what might be called the "Serre conjecture for quadratic modules." Recent results on quadratic modules ([1]) are precisely what is needed for the solution of the network theory problem. The vital concept of feedback is considered in Section 5 together with its relation to the problem of pole placement (stabilization) or coefficient assignment. Equivalence under feedback involves the action of a group which is not reductive. Nonetheless, a moduli space can be constructed and its invariants calculated. For controllable systems, the result can be found in [7], [28], [32] and [42]. The general case is completely analyzed in [19]. It should also be possible to extend Mumford's geometric invariant theory to groups of feedback type. A critical portion of the feedback invariants is the Kronecker set of indices $(\kappa_1, \ldots, \kappa_n)$ where the κ_i are non-negative integers. This set was used to define an ordering for systems by Rosenbrock ([42]). It turns out that this ordering is precisely the Harder-Narasimhan ordering ([45]) and we conclude with an indication of the use of this complex of ideas in system theoretic problems. We hope to indicate throughout the richness of the relationship between algebraic geometry and system theory both in terms of the application of algebraic geometry to system theory and in terms of the generation of problems in algebraic geometry from the practical considerations of system theory.

2. Routh-Hurwitz Theory. We begin by considering two questions, which were posed and solved in the 19th century by Cauchy, Hankel, Hermite, Hurwitz, Maxwell, Routh and others, and which now form part of the folklore of linear system theory.

Question 1. When is a proper (i.e., vanishing at ∞) meromorphic function f on **C** rational?

Question 2. If such an f is real, when do its poles lie in the left-hand plane?

In fact, we can ask the same questions for proper, matrix-valued meromorphic functions and, as a motivating example, consider the resolvent or transfer function, $f(s) = (sI - F)^{-1}$, of a differential equation, $(d/dt - F)x = 0$. This is still more interesting in the control system case (see Section 3), but serves to explain Maxwell's interest ([38]) in such questions. Question 1 can be thought of in terms of an algebro-geometric classification of differential equations, whereas Question 2 characterizes asymptotic stability.

One approach to Question 1, due to Hankel ([20]), is to form the matrix, $H_f = (h_{i+j-1})$, where h_l is given by

$$f(s) = \sum_{l=1}^{\infty} h_l s^{-l}. \tag{2.1}$$

Hankel's Theorem then asserts: f is rational if, and only if, rank $H_f = n < \infty$, as one can plainly see. In this case, f induces a holomorphic map

$$f\colon \mathbf{CP}^1 \to \mathbf{CP}^1 \tag{2.2}$$

whose degree is given by

$$\deg_{\mathbf{C}} f = \operatorname{rank}(H_f). \tag{2.3}$$

Now, in case f is real, we can restrict

$$f: \mathbf{RP}^1 \to \mathbf{RP}^1 \tag{2.4}$$

and, since $f \mapsto H_f$ is bijective, we can ask for $\deg_{\mathbf{R}} f$ in terms of H_f. In this way, we obtain the beautiful result which underlies the Routh-Hurwitz conditions:

THEOREM 2.5 (*Hermite-Hurwitz*) *$\deg_{\mathbf{R}} f$ = signature (H_f).*

This was discovered by Hermite in 1856 ([29]) in case f has distinct poles, and later extended by Hurwitz. Of course, $\deg_{\mathbf{R}} f$ was expressed in a different way:

Definition 2.6 (Cauchy) The local index of a real, rational f at a real pole x_0 is $+1$ if f changes from $+\infty$ to $-\infty$, -1 if the alternate change occurs, and 0 if f does not change sign while passing through x_0. The index of f, $C(f)$, is the sum of the local indices.

$C(f)$, which is clearly the winding number of the map in (2.4), was defined by Cauchy in [15]. In part I of [15], he uses the Cauchy index to compute the number of real roots of a real polynomial (generalizing, among other things, Descartes' rule of signs), the number of negative real roots, and related questions. In part II, he uses the Cauchy index to define and evaluate the local index of a non-degenerate plane vector field at an equilibrium point. Of course, this latter application was generalized by Kronecker, for $n \geq 2$, and others. On the other hand, the computation of the number of negative real zeroes of a polynomial is a bit harder than Question 2 and this computation was, in fact, the starting point for Hurwitz in [30]. Following [30], we assume for the moment that f has no poles on the imaginary axis. Thus, if g is the monic denominator of f, having no roots in common with the associated numerator h, computing the change in arg $g(-is)$, s between $+\infty$ and $-\infty$, is equivalent to computing $p - q$, the number of zeroes in the left-half plane minus the number of zeroes in the right-half plane. This is, by trigonometry, the Cauchy index of the rational function v/u

where

$$cg(-is) = (u + iv)(s),$$

and c is a complex constant rendering v/u proper. By the Hermite-Hurwitz Theorem,

$$p - q = \text{sign}\, H_{u/v}. \tag{2.7}$$

Thus, by the Jacobi-Frobenius algorithm, one obtains numeric criteria (in terms of the coefficients of g!) for Question 2; i.e., for g to be a Hurwitz polynomial. Explicitly, this is positive-definiteness of a quadratic form, and one should stress that it is a known fact in the folklore that $H_{u/v}$ is intimately related to a Liapunov function for $g(d/dt)x = 0$. In fact, this is very much the spirit of Hermite's treatment.

We close the section by sketching a moduli-theoretic proof of Theorem 2.5, due to R. W. Brockett ([2]). With the notation as above, consider

$$\text{Rat}(n) = \{(g, h)\colon g/h \text{ is proper and } (g, h) = 1 \text{ in } \mathbf{R}[s]\}.$$

As the colocus of the resultant, $\text{Rat}(n)$ is Zariski open in $\mathbf{R}^{2n}$, the Euclidean topology being the same as the compact-open topology on the maps (2.4). In particular, the Cauchy index $p - q$ is constant on Euclidean components of $\text{Rat}(n)$ and there are at least $n + 1$ of these. By continuity, $\text{sign}(H_f)$ is constant on components of $\text{Rat}(n)$, in light of (2.3). Thus, already, a general position argument reduces Theorem 2.5 to the result obtained by Hermite. Indeed, if $\text{Rat}(p, q)$ is the submanifold on which the Cauchy index is $p - q$, then, by a somewhat tedious but intuitive canonical form argument based on divisors, Brockett proved

THEOREM 2.8. *$Rat(p, q)$ is connected.*

It is entirely trivial to evaluate the induced bijection between values of the Cauchy index and the signature of H_f. This yields Theorem 2.5 as well as some insight into the global structure of $\text{Rat}(n)$ (i.e., of scalar input-output linear systems). It is very natural then to study the topology of the submanifolds, $\text{Rat}(p, q) \subset \Omega^1{}_{p-q}(S^1)$. G. Segal has recently shown the inclusion is an approximation in homotopy in certain degrees. For example, based on published work of R. Brockett and G. Segal, one can show that each Rat(p, q) can be given the structure of a Stein manifold and, in particular, deduce the vanishing of its higher cohomology groups. Their work also shows that Rat(p, q) is simply connected only in case either $p = n$, or $q = n$, in which case Rat$(n, 0) \simeq$ Rat$(0, n) \simeq \mathbf{R}^{2n}$. It is also known that Rat$(n - 1, 1) \simeq$ Rat$(1, n - 1) \simeq S^1 \times \mathbf{R}^{2n-1}$ (see [2]). In another direction ([13]), from all of the above, we can construct a vector bundle V_+ on Rat(p, q), obtained by assigning the positive eigenspace of H_f to $f \in$ Rat(p, q). This is, in general, non-trivial. For example, on Rat(1, 1), V_+ induces a classifying map

$$T_+ \colon \text{Rat}(1, 1) \to \mathbf{RP}^1$$

which is not homotopic to a constant map. Indeed ([21]), under the isomorphism Rat$(1, 1) \simeq S^1 \times \mathbf{R}^3$, T_+ is just projection on the first factor. In section four, we will interpret these obstructions in a system-theoretic context.

3. Linear Systems. The class of systems which we consider is the class of finite-dimensional, linear time-invariant multivariable systems. Such systems can be characterized via an external intrinsic description using either the transfer matrix or an input-output mapping. Associated with each description is an internal representation and an appropriate notion of equivalence which is defined by the action of an algebraic group. The critical link between these descriptions is provided by (a generalization of) Hankel's Theorem on the rationality of proper matrix valued "meromorphic" functions.

Now let R be a Noetherian integral domain which is integrally closed in its quotient field K and let $R[s]$ be a polynomial ring in s over R. We call an element $n(s)/d(s)$ of $K(s)$ *proper* if degree $n(s) <$ degree $d(s)$. Let $M_{p,m}(\cdot)$ denote the set of $p \times m$ matrices with entries in $\cdot$. Then we have:

Definition 3.1. Let $\Sigma_{p,m} = \{T(s) \in M_{p,m}(K(s)) \mid$ the entries $t_{ij}(s)$ of $T(s)$ are proper$\}$. Elements of $\Sigma_{p,m}$ are called *proper transfer matrices.*

Definition 3.2. Let $T(s)$ be an element of $\Sigma_{p,m}$. A pair $(R(s), P(s))$ with $R(s) \in M_{p,m}(K[s])$ and $P(s)$ an invertible element of $M_{m,m}(K[s])$ is called a *representation* of $T(s)$ if $T(s) = R(s)P^{-1}(s)$. A representation is called a *realization* (or minimal representation) of $T(s)$ if $R(s)$ and $P(s)$ are relatively right prime.

If $T(s) \in \Sigma_{p,m}$, then, letting $\Delta(s)$ to be the (monic) least common multiple of the denominators of the entries $t_{ij}(s)$ of $T(s)$, we have $T(s) = R(s)P^{-1}(s)$ where $R(s) = T_0 + T_1 s + \cdots + T_{n-1}s^{n-1}$, $P(s) = \Delta(s)I_m$ and so, representations exist. Moreover, let $\mathfrak{U}_m = \{U(s) \in M_{m,m}(K[s])$: det $U(s)$ is a unit i.e., a non-zero element of $K\}$. Note that $\mathfrak{U}_m$ is an algebraic group which acts on pairs $(R(s), P(s))$ via right multiplication and preserves the property of being relatively right prime. We now have:

THEOREM 3.3. *Given $T(s) \in \Sigma_{p,m}$, there exist realizations of $T(s)$ and any two realizations are equivalent under the action of $\mathfrak{U}_m$. (For a proof, see* [18], [49].)

Let $T(s)$ be an element of $\Sigma_{p,m}$ and let $\sigma_T(s)$ be an element of $M_{p+m,m}(K[s])$ which corresponds to $T(s)$. In other words, $\sigma_T(s)$ is an element of $M_{p+m,m}(K[s])$ such that

$$\sigma_T(s) = \begin{bmatrix} R_T(s) \\ P_T(s) \end{bmatrix}$$

$R_T(s)$, $P_T(s)$ relatively right prime and $T(s) = R_T(s)P_T^{-1}(s)$. Any such $\sigma_T(s)$ is called a *linear system with transfer matrix* $T(s)$. If $\Sigma_{p,m}$ is viewed in this way, then $\Sigma_{p,m}$ is stable under the action of $\mathfrak{U}_m$ and the orbits correspond to the transfer matrices. More precisely, if $S_{p,m} \subset M_{p+m,m} \cdot (K[s])$ is the set of all linear systems, then $S_{p,m}$ is stable under the action of $\mathfrak{U}_m$ and the transfer matrix is a complete invariant for this action. These results can be interpreted in still another way which is frequently useful in applications. Namely, let $\sigma(s)$ be an element of $S_{p,m}$ and let $\sigma_j(s)$ be the jth column of $\sigma(s)$. Then $\sigma_1(s), \ldots, \sigma_m(s)$ are free over $K[s]$ and we let M_σ be the free module with generators $\sigma_1(s), \ldots, \sigma_m(s)$. M_σ is called the *system module of* $\sigma(s)$ and is also a complete invariant for the action of $\mathfrak{U}_m$. If

$$\sigma(s) = \begin{bmatrix} R_\sigma(s) \\ P_\sigma(s) \end{bmatrix},$$

then n_σ = degree det $P_\sigma(s)$ is called the *McMillan degree of* $\sigma(s)$ and n_σ is also an invariant for the action of $\mathfrak{U}_m$. It is rather tempting to heuristically view $(R_\sigma(s), P_\sigma(s))$ as "homogeneous coordinates" relative to multiplication by $U(s)$ in $\mathfrak{U}_m$. Indeed, in the scalar case ($p = m = 1$), this is expressed in (2.2). More precisely, $\sigma(s)$ induces a map $\mathbf{T}_\sigma$ of $\mathbf{P}^1$ into $\text{Grass}_K(m, m + p)$ given by

$$T_\sigma(s) = \{(R_\sigma(s)w, P_\sigma(s)w) \mid w \in K^m\} \tag{3.4}$$

for $s \neq \infty$ and by K^m for $s = \infty$. $\mathbf{T}_\sigma$ is essentially the graph of $T_\sigma(s)$ by virtue of the fact that $R_\sigma(s)$, $P_\sigma(s)$ are relatively right prime ([27]). In addition, if K is algebraically closed, then (2.3) generalizes also to the following

$$\deg_K \mathbf{T}_\sigma = n_\sigma \tag{3.5}$$

which was first proved over $\mathbf{C}$ by Hermann and Martin ([27]). We shall soon see that n_σ is also the rank of a Hankel matrix.

We now turn our attention to the alternate external description of linear systems. Consider the (formal) Laurent series

$$f(s) = \sum_{l=1}^{\infty} L_l s^{-l} \tag{3.6}$$

where $L_l \in M_{p,m}(K)$. We associate with f the (generalized) Hankel matrix

$$H_f = (L_{i+j-1}) \tag{3.7}$$

and we have:

Definition 3.8. f is *admissible* if rank $H_f = n_f < \infty$. Let $\Sigma^f{}_{p,m} = \{f(s) \mid f \text{ is admissible}\}$. Elements of $\Sigma^f{}_{p,m}$ are called *proper input-output maps* and n_f is called the *dimension of* f.

Definition 3.9. Let $f(s)$ be an element of $\Sigma^f{}_{p,m}$. A triple (A, B, C) with $A \in M_{n,n}(K)$, $B \in M_{n,m}(K)$, $C \in M_{p,n}(K)$ is called a *representation* of $f(s)$ if $CA^{l-1}B = L_l$ for $l = 1, \ldots$. A representation is called a *realization* (or minimal representation) of $f(s)$ if $n = n_f$.

Definition 3.10. A representation (A, B, C) of $f(s)$ is called *con-*

trollable if rank $\mathcal{C}(A, B, C)$ = rank $[B, AB, \ldots, A^{n-1}B] = n$ and a representation (A, B, C) of $f(s)$ is called *observable* if rank $\mathcal{O}(A, B, C)$ $= \text{rank}[C', A'C', \ldots, (A')^{n-1}C'] = n$.

If $f(s)$ is a proper input-output map, then it is well-known that representations and realizations exist ([33]). Moreover, if (A, B, C) is a representation of $f(s)$, then $C(sI - A)^{-1}B = T_f(s)$ is an element of $\Sigma_{p,m}$ and (A, B, C) can be called a *state-space representation* of $T_f(s)$. Similarly, if (A, B, C) is a realization of $f(s)$, then $C(sI - A)^{-1}B = T_f(s)$ is an element of $\Sigma_{p,m}$ and (A, B, C) is called a state-space realization of $T_f(s)$. Thus, there is a natural mapping of $\Sigma^f{}_{p,m}$ into $\Sigma_{p,m}$ and it can be shown that this mapping is bijective by virtue of Hankel's Theorem ([20]).

Now, if (A, B, C) is a triple with $A \in M_{n,n}(K)$, $B \in M_{n,m}(K)$ $C \in M_{p,m}(K)$, then $GL(n, K)$ acts on (A, B, C) in the following way: $(A, B, C) \to (gAg^{-1}, gB, Cg^{-1})$, $g \in GL(n, K)$. It is clear that controllability and observability are preserved under this action. We now have:

THEOREM 3.11. *Given $f(s) \in \Sigma^f{}_{p,m}$ (or, equivalently, $T_f(s) \in \Sigma_{p,m}$), there exist realizations of $f(s)$ (state-space realizations of $T_f(s)$) and any two realizations are equivalent under the action of $GL(n_f, K)$. Moreover, any realization of $f(s)$ is both controllable and observable. Finally, if (A, B, C) and (A_1, B_1, C_1) are realizations of $f(s)$, then the $g \in GL(n_f, K)$ such that $(gAg^{-1}, gB, Cg^{-1}) = (A_1, B_1, C_1)$ is unique.*

(For a proof, see [3], [17]).

The final part of Theorem 3.11 is often referred to as the state-space isomorphism theorem.

So, let $f(s) \in \Sigma^f{}_{p,m}$ and let (A, B, C) be an element of $\mathbf{A}^{n_f^2+n_f(m+p)}(K)$ $= \mathbf{A}^{n^2+n(m+p)}$ which corresponds to $f(s)$. In other words, (A, B, C) is an element of $\mathbf{A}^{n^2+n(m+p)}$ such that (A, B, C) is both controllable and observable and $T_f(s) = C(sI - A)^{-1}B$ and $f(s) = \Sigma^\infty{}_{l=1} (CA^{l-1}B)s^{-l}$. Any such triple (A, B, C) is called a *linear system with input-output map* $f(s)$. Let $S^n{}_{p,m}$ be the set of all linear systems of dimension n. Then $S^n{}_{p,m}$ is stable under the action of $GL(n, K)$ and the orbits correspond to the proper input-output maps of dimension n. In the next section, we shall characterize the moduli space of $S^n{}_{p,m}$ under this action. Finally, we remark that the dimension of $f(s)$ is precisely the McMillan degree of $T_f(s)$.

So far, we have considered notions of equivalence based on classical groups such as $\mathfrak{U}_m$ and $GL(n, K)$. However, in Section 5, we will examine the critical concept of feedback equivalence. This notion requires, amongst other things, an additional set of invariants known as the Kronecker indices ([14], [32]) which we now introduce in two ways.

If $P(s) \in M_{m,m}(K[s])$, then let $\kappa_j(P) = \max \text{degree}\{p_{ij}(s)\ i = 1, \ldots, m\}$ be the jth *column degree* of $P(s)$ so that $P(s)$ can be written in the form

$$P(s) = \Delta_c(P) \operatorname{diag}[s^{\kappa_1}, \ldots, s^{\kappa_n}] + P_1(s) \tag{3.12}$$

where $\kappa_j(P_1) < \kappa_j(P)$ and $\Delta_c(P) \in M_{m,m}(K)$. $P(s)$ is called *column proper* if $\Delta_c(P) \in GL(m, K)$. If

$$\sigma = \sigma(s) = \begin{bmatrix} R_\sigma(s) \\ P_\sigma(s) \end{bmatrix}$$

is an element of $S_{p,m}$, then it is well-known that there is a $U(s) \in \mathfrak{U}_m$ such that $P_\sigma(s)U(s)$ is column proper ([49]). We now have:

Definition 3.13. Let σ be an element of $S_{p,m}$ and let $P(s)$ be a column proper element of $M_{m,m}(K[s])$ such that $P(s) = P_\sigma(s)U(s)$ for some $U(s) \in \mathfrak{U}_m$. Then the set of integers $\kappa_\sigma = \{\kappa_1(P), \ldots, \kappa_m(P)\}$ is called the *Kronecker set of* σ.

THEOREM 3.14. *Let σ be an element of $S_{p,m}$. Then* (i) *κ_σ is well-defined, and* (ii) *if $\tau = \sigma U$ for some $U \in \mathfrak{U}_m$, then $\kappa_\sigma = \kappa_\tau$ (as sets). (i.e., κ_σ is an invariant for the action of $\mathfrak{U}_m$ on $S_{p,m}$).*

Alternatively, in state-space form, let (A, B, C) be controllable so that $\operatorname{rank}[B, AB, \ldots, A^{n-1}B] = n$. Then it is possible to lexicographically order the first n linearly independent columns of $[B, AB, \ldots, A^{n-1}B]$ as follows: $b_1, Ab_1, \ldots, A^{\kappa_1-1}b_1, b_2, \ldots, A^{\kappa_2-1}b_2, \ldots, A^{\kappa_m-1}b_m$ where $b_1, \ldots, b_m$ are the columns of B. The set of integers $\kappa = (\kappa_1, \ldots, \kappa_m)$ is called the *Kronecker set of* (A, B, C). It can be shown that κ is an invariant under the action of $GL(n, K)$ ([7]). For example, if we consider the pencil $[A - sI, B]$, then equivalence under $GL(n, K)$ induces a strict equivalence (in the classical sense of Kronecker [32]) of pencils and the integers $(\kappa_1, \ldots, \kappa_m)$ are a complete invariant for strict equivalence of pencils. Explicitly, following Kronecker, the

κ_i may be computed by the process: choose $\phi_1(s) \in \operatorname{Ker}[A - sI, B] \otimes K$ of minimal degree κ_1 in s then choose $\phi_2(s) \in \operatorname{Ker}[A - sI, B] \otimes K$, independent of $\phi_1(s)$, and of minimal degree κ_2 among all such $\phi(s)$, etc. (cf. [10]). We also have:

THEOREM 3.15. *Let (A, B, C) be a state-space realization of $T(s)$ and let $\sigma_T(s)$ be a realization of $T(s)$. Then the Kronecker sets are the same.*

(For a proof, see [18] and [19].)

We note that, for ease of exposition, we have worked over the quotient field K of R. However, if the data were defined over R, then the realization problem is essentially a question of rationality which accounts for the assumption of integral closure. Thus, the results presented here hold with data in R (see [43], [44], [47]).

4. Classical Groups of Symmetry and Moduli for Linear Systems. There are three main topics covered here. The first is the construction, in external and in internal terms, of the moduli space for linear systems and a proof of the theorem on "realization with parameters," which is based on Zariski's Main Theorem. The second is a brief survey of biregular and topological invariants of the moduli space, with emphasis on the impact these invariants have on system theory, as in Section 2. We close by considering moduli for systems having richer external symmetries, such as Hamiltonian systems or systems describing symmetric electrical networks. Here we see a more arithmetic side come to the foreground, and in fact we sketch the equivalence between an old problem in network synthesis and what might be called the "Serre conjecture for quadratic modules." In particular, recent results on quadratic modules ([1]) turn out to be sufficient for the classical formulation of this problem.

Now, over $\mathbf{R}$ or C, it is natural to view the set of transfer functions with m inputs, p outputs, and fixed McMillan degree n as a (complete) metric space. Indeed, this topology ([13]) is simply the compact-open topology on transfer functions, regarded as maps

$$\mathbf{T}_\sigma : \mathbf{P}^1 \to \operatorname{Grass}(m, m + p),$$

as in (3.4). Although such a description is in fact useful in certain contexts, it is not entirely clear that this space is an algebraic variety, nor is it even clear what its dimension should be. For this reason, the first problem we consider is to construct, as explicitly as possible, a moduli space for the action of $GL(n, k)$ on $S^n{}_{p,m}$. In fact, since it is rather important as a classifying space for families of linear systems, it is rather crucial that we construct a fine moduli space. This is done in several stages, it turns out (by universality) to be sufficient to construct a geometric quotient for the action of $GL(n, k)$ on controllable pairs (A, B) in $\mathbf{A}^{n^2+nm}$. This is done within the context of geometric invariant theory, the rather remarkable result being that the properly stable points in $\mathbf{A}^{n^2+nm}$, relative to the standard character linearization, is precisely the space of controllable pairs! We start birationally.*

First of all, the corresponding space $\mathbf{H}$ of Hankels is clearly a quasi-affine variety, contained in $\mathbf{A}^N$, for $N = mp(n + 1)$. This was shown to be non-singular by J. M. C. Clark ([16]) and, in addition, his proof shows that $\mathbf{H}$ is a special variety, in the sense of Chevalley. In particular, $\mathbf{H}$ is rational and one can easily write down generators for the function field $K_{\mathbf{H}}$. On the other hand, if $R = K[A, B, C]$, K as the field of fractions of R, then Falb has shown in [18]:

PROPOSITION 4.1. $K_{\mathbf{H}} = K^{GL(n,k)}$.

This follows from the statement that the ring of invariant regular functions defined on the Zariski open subspace of semi-simple A is generated by the entries of the Hankel. This is proved, for B and C vectors, by Gurevitch—as an illustration of the symbolic method! It is proved in general in [18].

The biregular situation involves a bit more care—this is due to several people (e.g. [14], [26]) and our treatment follows ([14]), relying on [40]. It can be shown that, if $\tilde{\Sigma}_0$ is the Zariski open subspace of $\mathbf{A}^{n^2+nm}$ of controllable pairs and if R_0 is the ring of regular functions on $\tilde{\Sigma}_0$, then in contrast to Proposition 4.1, $R_0{}^{GL(n,k)}$ does not separate orbits in $\tilde{\Sigma}_0$, unless $m = 1$. In fact, if k is algebraically closed,

$$R_0{}^{GL(n,k)} = k[c_i], \tag{4.2}$$

where (c_i) denote the characteristic coefficients of A. In particular,

*The birational situation using these ideas was sketched in [18] and the entire theory fully developed in [10].

although Theorem 3.9 implies that the orbits in $\tilde{\Sigma}_0$ are closed, an affine quotient does not exist, unless $m = 1$. However, we do have:

THEOREM 4.3. *$\tilde{\Sigma}_0 \to \Sigma_0$ is a universal geometric quotient, with Σ_0 a quasiprojective, non-singular, irreducible variety of dimension nm.*

Sketch of Proof. Following [40], since (4.2) implies that the set of properly stable points for the trivial character ($\mathrm{Pic}(\tilde{\Sigma}_0) = (0)!$) is empty, we compute the properly stable points for the relative invariants of weight 1. Here we find a surprisingly pleasant result:

PROPOSITION 4.4. *The set of properly stable points, relative to $det(\cdot)$, in $\mathbf{A}^{n^2+nm}$ is exactly $\tilde{\Sigma}_0$.*

This is very much related to the fact that $\tilde{\Sigma}_0$ is the principal orbit type for the action. In the scalar input case, this can be proved by linear algebra, one direction being already implied by the proof of the state-space isomorphism theorem. For the other half, to say (A, B) is in the principal orbit type is to say, by differentiating the action,

$$n^2 = \dim \mathcal{O}_{(A,B)} = \dim \mathcal{O}_A + \dim Z(A)B, \tag{4.5}$$

where $Z(A)$ is the centralizer of A and $\mathcal{O}_.$ denotes the orbit of $\cdot$. Now, (4.5) can be refined since

$$\dim \mathcal{O}_A \leq n^2 - n \quad \text{and} \quad \dim Z(A)B \leq n \tag{4.6}$$

holds for any A. Thus, $\dim \mathcal{O}_A = n^2 - n$ is maximal and A therefore possesses a cyclic vector. By Frobenius's Theorem, $Z(A) = k[A]$ and thus

$$\dim Z(A)B = \dim k[A]B = \mathrm{rank}(B, AB, \ldots, A^{n-1}B) = n,$$

which was to be shown. The multivariable case also follows from linear algebra (see [14]).

Assuming Proposition 4.4, Theorem 4.3 follows from Mumford's Theorem ([40], Theorem 1.19).

By universality, a geometric quotient for the twisted action of $GL(n, k)$ on $\tilde{\Sigma}_p = \{(A, B, C): (A, B) \text{ controllable}\}$ exists and shares many of the same properties. In fact, $\Sigma_p \to \Sigma_0$ is an algebraic vector bundle with base Σ_0 and fiber k^{np}. Since the quotient, $\tilde{\Sigma}_p \to \Sigma_p$ is geometric, we also obtain a quotient $\Sigma^n{}_{p,m}$ for the Zariski open subspace $S^n{}_{p,m} \subset \tilde{\Sigma}_p$.

THEOREM 4.7. *A fine moduli space for state-space equivalence in $S^n{}_{p,m}$ exists. Moreover, $\Sigma^n{}_{p,m}$ is a rational, smooth, irreducible quasi-projective variety of dimension $n(m+p)$.*

The construction of a universal family of systems on Σ_0 was carried out in [26] and extended to $\Sigma^n{}_{p,m}$ in [10]. Thus, any map $f\colon X \to \Sigma^n{}_{p,m}$ induces a linear system defined over the coordinate ring R of X. This, of course, extends to the study of differentiable or continuous families as well.

THEOREM 4.8. *If $H(x)$ is a regular Hankel matrix, of locally constant rank on X, then $H(x)$ is relizable over R, with finitely generated, projective state module.*

Proof. After all of the above, including the existence of a universal family, i.e. a vector bundle Q—the state bundle, an endomorphism $\mathcal{A}$ of Q, and sections and co-sections $\mathcal{B}_i$, $\mathcal{C}_j$, resp., this amounts to proving that realization is, in fact, algebraic in its parameters (see [25], [14]). That is, if η is the natural map

$$\eta\colon \Sigma^n{}_{p,m} \to \mathbf{H} \tag{4.9}$$

$$[A, B, C] \to (CA^{i+j-1}B)$$

we claim η is biregular. By Theorem 3.9, η is bijective and by Proposition 4.1 η is birational. By Clark's Theorem, $\mathbf{H}$ is normal and thus η is biregular by Zariski's Main Theorem. Finally, the assumption on $H(x)$ is simply that H define a map to $\mathbf{H}$.

Example 4.10. Given $(h_{ij}) \in l^1(\mathbf{Z})^{(n+1)}$ of rank n, we seek a realization $(A(z), B(z), C(z))$ summable in $z \in \mathbf{Z}$. If we ask that the dual system be minimal too, the following sufficient condition (see [10], [46]) becomes necessary.

(T) the closed linear span of $\det(h_{ij})$ and its translates is $l^1(\mathbf{Z})$.

Indeed, Fourier transforming and using the Tauberian Theorem, we obtain a continuous map

$$(\hat{h}_{ij}): S^1 \to \mathbf{H} \tag{4.11}$$

and, thus, a linear system $(\hat{A}, \hat{B}, \hat{C})$ defined over $\widehat{l^1(\mathbf{Z})}$, since η^{-1} is analytic. Now, the state module is trivial over $C(S^1)$, for topological reasons, and the Docquier-Grauert Theorem asserts that the corresponding bundle can even be trivialized over $\widehat{l^1(Z)}$! In particular, we obtain a realization, in matrix form, by inverting the Fourier transform. Recent work by M. Hazewinkel on constructing partial compactifications for $\Sigma^n{}_{p,m}$, suggests that we may be able to drop the hypothesis on the Hankel.

We stress the fact that the computation made above actually depends on the topology of S^1; for example, it can be shown that there exists $\mathbf{P}^N \subset \Sigma^n{}_{p,m}$, of codimension $n(p+1)$, over which the universal family is non-trivial. This is, of course, very much related to the existence of algebraic (or continuous) canonical forms—a problem which has received a great deal of attention.

Now, Mumford's Theorem yields an actual imbedding of Σ_0 in $\mathbf{P}^M$; i.e. a very ample line bundle L on Σ_0. L also arises in another way, viz. $L = \Lambda^n \Sigma_1$, as one can see by checking cocycles (c.f. [14], Section 4), and thus L is associated to the principal $GL(n, k)$-bundle, $\hat{\Sigma}_0 \to \Sigma_0$. Furthermore, L cannot be trivial if $m > 1$, by (4.2). If $m = 1$, $\Sigma_0 \simeq \mathbf{A}^n$ by the existence of the rational canonical form. The more precise results for $m > 1$ are:

THEOREM 4.12. *If k is algebraically closed, $Pic(\Sigma_0) = (L) \simeq \mathbf{Z}$ and there exist no algebraic canonical forms. If $k = \mathbf{C}$, then the Betti numbers β_{2i} do not vanish, for $i = 0, \ldots, nm - n$ and continuous canonical forms do not exist.*

Sketch of Proof. Since $\hat{\Sigma}_0 \to \Sigma_0$ is an algebraic principal $GL(n, k)$-bundle, one can compute Pic(Σ_0) by descent ([40], p. 32):

$$\mathrm{Pic}^{GL(n,k)}(\hat{\Sigma}_0) \simeq \mathrm{Pic}(\Sigma_0).$$

Since Pic$(\hat{\Sigma}_0) = (0)$, we have a surjection (by the rigidity lemma),

$$\widehat{GL(n, k)} \to \mathrm{Pic}^{GL(n,k)}(\tilde{\Sigma}_0)$$

which is injective by (4.2). In terms of group cohomology, this is just the map,

$\widehat{GL(n, k)} \to H^1(GL(n, k), k^*) \to H^1(GL(n, k), \mathcal{O}_{\Sigma_0}{}^*$. If $m = 1$, the unit $\det(B, AB, \ldots, A^{n-1}B)$ is a coboundary of the character $\det(\cdot)^l$, regarded as a cocycle with value in $\mathcal{O}_{\Sigma_0}{}^*$.

As for the topological obstructions, notice that (4.2) asserts that $\mathbf{A}^n$ is the categorical quotient for the action, in the category of affine varieties. In particular, consider the natural foliation

$$\Sigma_0 \to \mathbf{A}^n.$$

In the codimension 1 case, $n = 1$ and thus $\Sigma_0 = \mathbf{A}^1 \times \mathbf{P}^{m-1}$ while $\tilde{\Sigma}_0 \to \Sigma_0$ is the pullback of the universal fibration. This is quite general, on a Zariski open subspace of $\mathbf{A}^n$, the leaves are codimension n projective spaces, $\mathbf{P}^N$! Over $\mathbf{C}$, these can be imbedded non-trivially, giving a proof of the second statement. Q.E.D.

Over $\mathbf{R}$, M. Hazewinkel ([24]) has shown that there exist no continuous canonical forms, by a clever imbedding of S^1 into Σ_0. It is possible to prove non-vanishing of higher homotopy groups in the proper ranges over $\mathbf{R}$ and $\mathbf{C}$, by making some crude computations along the lines in [11], and applying Bott Periodicity. These computations also show that the natural foliation on Σ_0 is not, in general, a product of $\mathbf{A}^n$ with a typical leaf. Of course, the situation remains the same for Σ_p, but $\Sigma^n{}_{p,m}$ is a bit more elusive. However, the statements about canonical forms still hold, answering in the negative—by topological and geometric means—a question which many engineers had assumed to be evident. This global form of the question was originally raised by R. E. Kalman [31].

For the remainder of the section, we consider the real field $\mathbf{R}$, and discuss an application, following [13], of a theorem of Harder to a problem in network synthesis, posed by Youla in [53]. It should be stressed that, even a decade ago, this problem was known to be algebro-geometric in nature and, in fact, the partial results obtained by Koga ([36]) were based on his use of the Riemann-Roch theorem for curves to analyze coprime factorizations of transfer functions in two variables.

Now, a result of Glover ([21]) asserts that if $m > 1$ or $p > 1$, the space $\mathbf{H}$ is connected in contrast to Theorem 2.7. This is due largely to the fact that, even in case $m = p$, the Hankel is not necessarily symmetric and one has more room to move in. However, symmetric Hankels or, equivalently, symmetric transfer functions are important in network theory (see [6], [37], [54]). This symmetry is displayed in state space-form as:

$$I_{p,q}A = A'I_{p,q} \qquad (4.13)$$
$$I_{p,q}B = C',$$

where $I_{p,q}$ is the standard form of signature $p - q$. Here $p - q$ is the signature of the Hankel and has the same physical significance ([4]) as the signature of the indefinite metric in the Brayton-Moser equations. In fact, the state-space isomorphism theorem applies here to give the group of symmetrices $\mathcal{O}(p, q) \subset GL(n, \mathbf{R})$ (see [54]). Moduli for these linear dynamical systems have been studied in [13] where, in particular, Theorem 2.7 is shown to hold. The invariant playing the role of the Cauchy index is the Maslov index where the real Grassmann in (3.4) is replaced by a Lagrangian Grassmann.

Now, the problem alluded to above is just a question of realization with parameters, subject to the symmetry constraints. That is, given a symmetric transfer function involving delays—which is simply a transfer function over $R[x_1, \ldots, x_N]$ ([34], [35])—find a symmetric realization. By the validity of the Serre conjecture, the state module can be assumed free and the corresponding question is equivalent ([13]) to:

> given a symmetric Hankel with constant rank, defined over $R = \mathbf{R}[x]$, find a symmetric realization (4.13) over R.

Since, over $\mathbf{R}$, this depends only on Sylvester's Theorem, we can invoke the following to finish off the problem.

THEOREM (Harder [1]). *Any non-degenerate symmetric bilinear form defined over $k[x]$ is equivalent to a form defined over k.*

We close by remarking that the general theory of quadratic modules arises in several instances, especially in connection with families of Hamiltonian systems. Here, again, over $\mathbf{R}$ we have a Hamiltonian realization, with symmetry group $Sp(n, \mathbf{R})$ and thus the moduli problem is already "in canonical form."

5. Stabilization and Feedback Groups. To be sure, one of the most important results in deterministic control theory is the pole-placement theorem, Theorem 5.2, which is concerned with stabilization of linear systems by means of "state feedback"—a topic very much related to the material developed in Section 2. It is fair to add, however, that a much deeper understanding of the whole feedback question comes only with the interpretation of feedback substitutions as part of a group action on $\mathbf{A}^N$, $N = n^2 + nm + np$. Explicitly, for (n, m) fixed we define $\mathfrak{F} = \mathfrak{F}(n, m)$ as the semi-direct product of 3 groups, generated by the actions on triples (A, B, C):

(i) $GL(n, k)$ acting via change of basis in the state space,
(ii) $GL(m, k)$ acting via change of basis in the input space,
(iii) $\mathrm{Hom}(k^n, k^m)$ acting via $(A, B) \mapsto (A + BF, B)$.

This action is more complicated than the action in (i), for both mathematical and control-theoretic reasons. That is, $\mathfrak{F}$ is a non-reductive extension of $GL(n, k)$ and, although $\mathfrak{F}$ preserves controllability, and hence acts on $\tilde{\Sigma}_p$, $\mathfrak{F}$ does not leave the space of linear systems, $S^n{}_{p,m}$, invariant. Thus, we begin this section by studying the action

$$\mathfrak{F} \times \tilde{\Sigma}_0 \to \tilde{\Sigma}_0 \tag{5.1}$$

and extending to the case $p > 0$. This extension is possible, in the sense of constructing moduli, since each of the isotropy subgroups arising in (5.1) is the semidirect product of a reductive group with the unipotent radical. We reproduce the theorem of Hermann and Martin ([28]) which states that the classical invariants for (5.1)—which are arithmetic, (being the Kronecker set introduced in Section 3)—arise as the Birkhoff-Grothendieck invariants of a bundle on $\mathbf{P}^1$. We extend these considerations to families of controllable pairs and, following Harder, Narasimhan and Shatz ([45]), introduce what might be called the "ubiquitous ordering" on the set of partitions of an integer n. Indeed, this ordering is well-known in system theory, having been introduced by Rosenbrock ([42]) in his study of the generalized pole-placement theorem, and we briefly indicate (based on joint work with C. Martin) how this circle of ideas yields some powerful new tools in the analysis of state feedback.

First of all, recall that the poles of a transfer function $T(s)$ are precisely the roots of χ_A—the characteristic polynomial of A, where

(A, B, C) is a linear system realizing $T(s)$. Now, if $F \in \mathrm{Hom}(R^n, R^m)$, ($R$ a commutative ring with identity) we may regard the action (iii) of F as feeding back the state of the system as a component of the input. In this setting, we ask (over $\mathbf{R}$) whether there exists an F such that χ_{A+BF} is a Hurwitz polynomial. More generally, over an arbitrary field k we consider ([51]):

THEOREM 5.2. *Any monic $p(s)$ can be expressed as χ_{A+BF} if, and only if, (A, B) is controllable.*

In the scalar-input case, this result was classically derived from the existence of the rational canonical form, and thus holds for any ring R. However, for families of systems with $m > 1$, technicalities arising from the non-existence of the rational canonical form, and thus holds for any ring R. However, for families of systems with $m > 1$, technicalities arising from the non-existence of canonical forms and Theorem 4.12, are non-trivial. In fact, a counterexample for the general situation has recently been announced by Bumby and Sontag ([9]). In general, our knowledge for arbitrary coefficient domains is very scant and is complete only under strong assumptions on the ring and the type of monic polynomial considered ([39]) or on the type of system ([12]).

Nevertheless, we can still glean some insight from the scalar input case. In terms of the action (5.1), which is transitive in this case, we fix (A, B) and consider

$$\tilde{\alpha}_{(A,B)}: \mathfrak{F} \to R^n, \tag{5.3}$$

$$\tilde{\alpha}_{(A,B)}(\bar{g}, s, F) = (c_i(\bar{g}(A + BF)\bar{g}^{-1})),$$

(where $c_i(\cdot)$ is the ith characteristic coefficient). Following E. Kamen, we note that the image of $\tilde{\alpha}_{(A,B)}$ is the translation by the vector χ_A, of a linear subspace, so that if we reduce modulo (χ_A), and suppress $\bar{g}$ and s, the corresponding map

$$\alpha_{(A,B)}: \mathrm{Hom}(R^n, R) \to R^n \tag{5.4}$$

is linear! If $\mathcal{C}^{(A,B)} = [B, AB, \ldots, A^{n-1}B]$ is the controllability operator, then the relationship between $\mathcal{C}$ and α, i.e. Theorem 5.2, is brought out by the beautiful identity due to B. F. Wyman ([52]):

$$\operatorname{coker} \alpha \simeq \operatorname{coker}(\mathcal{C}^*) \tag{5.5}$$

If the pair (A, B) is controllable over the fraction field of R, the right hand side can be computed in terms of $\mathcal{C}$, i.e.:

$$\mathrm{Ext}_R{}^1(\mathrm{coker}\,\mathcal{C}, R) \simeq \mathrm{coker}\,\alpha,\ \mathrm{Ext}_R{}^1(\mathrm{coker}\,\alpha, R) \simeq \mathrm{coker}\,\mathcal{C}. \quad (5.6)$$

In particular, if R is a field and if A is semi-simple, (5.5) contains the folklore result that one cannot change the uncontrollable modes by state feedback. That is, if A is diagonal, then any eigenvalue λ_i for which b_i is zero must persist in χ_{A+BF}.

In the case, $m > 1$, if k is algebraically closed and of characteristic 0, we can replace α by its Jacobian and thus obtain the corresponding generic result from the dominant morphism theorem. The computation of the Jacobian, over **C**, is due to Hermann and Martin ([27]) and relies on a differential analogue of Proposition 4.4.

Thus far, the only invariants which we have attached to a linear system have been topological, viz. the degree of the associated transfer function,

$$\mathbf{T}: \mathbf{P}_k{}^1 \to \mathrm{Grass}_k(m, m+p). \quad (5.7)$$

For $k = \mathbf{R}$, this defines the Cauchy index if $m = p = 1$ or else the McMillan degree mod (2). For $k = C$, this defines the McMillan degree. However, **T** is an algebraic map, and as such determines finer invariants, namely: the isomorphism class of T^*U, where U is the (algebro-geometric) universal vector bundle on $\mathrm{Grass}_k(m, m+p)$. By the well-known result of Birkhoff—Grothendieck (and probably others!), T^*U is determined by a partition of $\deg_k T$ into non-negative integers (n_i), $n_1 \geq n_2 \geq \cdots \geq n_m \geq 0$, if k is algebraically closed. (If $k = \mathbf{R}$, a theorem of Serre's still gives a decomposition of T^*U into a sum of line bundles, but these in turn are only determined by integers mod(2).) We owe to Hermann and Martin the identification of these geometric invariants ([28]):

THEOREM 5.8. *The Birkhoff-Grothendieck invariants of T^*U are the Kronecker set of any minimal realization.*

Sketch of Proof. Since one set of invariants is defined in the frequency domain and the other via state-space techniques, the key step will involve the Laplace transform. Note that, for constant coefficient differential operators, the Laplace transform has its range in a field of transcendence degree 1 and genus 0. This theorem very clearly indicates the relevance of the genus in such questions.

The key step is due to Rosenbrock: over **C**, consider the Laplace transforms of the solutions to the initial value problem,

$$\dot{x}(t) = Ax(t) + Bu(t), \qquad x(0) = 0.$$

These are elements of the $\mathbf{C}[s]$-module, defined as the kernel of the following pencil of matrices,

$$[A - sI, B]: (\mathbf{C}^n \oplus \mathbf{C}^m) \otimes \mathbf{C}[s] \to \mathbf{C}^n \otimes \mathbf{C}[s]. \tag{5.9}$$

The theory of such pencils, particularly the problem of equivalence, is due to Weierstrass and Kronecker (see [20]) and applies remarkably well to the situation at hand. Explicitly, controllability of (A, B) is equivalent to right surjectivity of (5.9) and the notion of strict equivalence of pencils is precisely that of feedback equivalence of (A, B) with (A', B'). Indeed, strict equivalence is an equivalence of $k[s]$-module maps, with the isomorphisms independent of s. By writing the definition of strict equivalence and comparing degrees, we find that the group of strict equivalence of pencils of the form (5.9) is a matrix representation of $\mathfrak{F}$! Since independence of s connotes projectivity, we follow another clue from Kronecker: we projectivize the pencil (5.9). In this setting, right surjectivity guarantees that we obtain a homogeneous bundle on $\mathbf{A}^2 - \{0\}$, and hence an m-bundle on $\mathbf{P}_k{}^1$. According to the Kronecker recipe, given at the end of Section 3, this bundle decomposes into a sum $\bigoplus^m{}_{i=1} \mathcal{O}(\kappa_i)$, $\Sigma^m{}_{i=1} \kappa_i = n$. That this bundle, $\ker[tA - sI, tB]$, on $\mathbf{P}_k{}^1$ is isomorphic to T^*U, follows at once from observability: we simply map $[x, u]$ to $[u, Cx]$, sending the kernel above to the graph of T. Observability is then just the statement that this map is injective.

Thus, Kronecker's classification also shows more, as noted in [32], for controllable pairs:

THEOREM 5.10. *$(A, B) \equiv (A', B')$ mod $\mathfrak{F}$ if, and only if, the Kronecker sets are the same.*

In fact, Kronecker's proof (see [20]) is to construct a canonical form for the pencil, given the data (κ_i). This canonical form coincides with the form discovered by P. Brunovsky ([7]) by other methods. We remark, however, that as an immediate application of the construction of this form, pole-placement becomes trivial. Indeed, the invariants $n_1 \geq n_2 \geq \cdots \geq n_m \geq 0$ are much more refined, they give a "generalized pole-placement theorem."

THEOREM 5.11 (*Rosenbrock* [42]). *The invariant factors of $A + BF$ can be made arbitrary, subject only to the constraints.*

$$\deg \phi_1 \geq n_1, \deg \phi_2 \geq n_1 + n_2, \ldots, \deg \phi_l \geq \sum_{i=1}^{l} n_i.$$

Another way of interpreting T^*U as a complete feedback invariant for the action (5.1) is to count sections. Thus, if $h^0(V)$ is the dimension of $H^0(\mathbf{P}^1; V)$, we have

$$h^0(T^*U) = h^0\left(\bigoplus^{m} \mathcal{O}(n_i)\right) = \sum_{i=1}^{m} (n_i + 1) = n + m, \tag{5.12}$$

On the other hand, evaluating T^*U at ∞ gives, canonically, the input space. Thus, if ϕ: $T_1{}^*U \to T_2{}^*U$ is an isomorphism, ϕ induces an isomorphism ϕ_*: $H^0(\mathbf{P}^1;\ T_1{}^*U) \to H^0(\mathbf{P}^1;\ T_2{}^*U)$, commuting with evaluations at ∞! Thus, ϕ_* is a triangular element of $GL(n + m, k)$, corresponding to the matrix representation of ϕ_* obtained above.

This point of view has several applications. For instance, if $(A, B) \in \tilde{\Sigma}_0$, then $\mathfrak{F}_{(A,B)}$—the isotropy subgroup for the action (5.1)—is just $\mathrm{Aut}(V)$, where V is the bundle defined by the kernel of $[tA - sI, tB]$. In particular,

$$\begin{aligned} \dim \mathfrak{F}/\mathfrak{F}_{(A,B)} &= \dim \mathfrak{F} - h^0(V \otimes V^*) \\ &= \dim \mathfrak{F} - \left(\sum_{n_i \geq n_j} (n_i - n_j + 1)\right), \end{aligned} \tag{5.13}$$

a formula originally derived by R. Brockett in [5], by a more explicit representation of $\mathfrak{F}_{(A,B)}$ together with a heuristic argument using Young diagrams corresponding to the partitions. Based on this description, it can also be shown ([5]) that connectivity (over $\mathbf{R}$) of the orbits is determined by the Kronecker indices mod(2).

Now, $\mathfrak{F}_{(A,B)}$ has also been written down explicitly by Falb and Wolovich in [19], in order to study the action of $\mathfrak{F}$ (and an extension of $\mathfrak{F}$)

on $\tilde{\Sigma}_p$. Then, one studies, in addition to the Kronecker set of (A, B), the action of $\mathfrak{F}_{(A,B)}$ on $\mathbf{A}^{np}$. In [19], it is shown that a moduli space for this action exists, using the fact that $\mathfrak{F}_{(A,B)}$ is always the semi-direct product of a reductive group with its unipotent radical.

For the remainder of the section we consider only $\mathfrak{F}$ acting on controllable pairs (A, B) and, as in Section 4, we begin to introduce parameters. Thus, we suppose (A, B) is an algebraic family, defined on an irreducible affine variety X. If (A, B) arises from a delay-differential system, it can be shown that we can take $X = \mathbf{A}^N$ ([10]) and that equivalence modulo $\mathfrak{F}(R)$ is not just a formal equivalence ([35], [10]). In this case, one may assign geometric data to (A, B), viz.

$$V_{(A,B)} \to X \times \mathbf{P}_k{}^1,$$

as before, by homogenizing the Rosenbrock pencil. If $X = \mathbf{A}^N$, it may be shown ([10]), based on the alternative view presented above, that $V_{(A,B)}$ is actually a complete invariant for (A, B) under $\mathfrak{F}(R)$. This situation is an improvement, as quite a bit more is known about deformations of bundles on $\mathbf{P}^1$ than about pencils over R. For example, if the deformation is point-wise trivial, i.e. if the Birkhoff-Grothendieck invariants are constant in X, then a theorem of C. Hanna ([23]) asserts:

$$V \simeq \Sigma\, p_1{}^*(V_i) \otimes p_2{}^*(W_i) \tag{5.14}$$

where the V_i are bundles on X, and the W_i are bundles on $\mathbf{P}^1$. Thus, if $X = \mathbf{A}^N$, such a deformation is globally trivial, since V_i is necessarily trivial. Applying this to $V_{(A,B)}$, we find:

PROPOSITION 5.15. *If the pointwise Kronecker invariants are constant, (A, B) is feedback equivalent to a system defined over k.*

This has the immediate corollary that such a system is coefficient-assignable, i.e. Theorem 5.2 holds. This latter fact generalizes to arbitrary R, although the Proposition fails to hold in general ([12]). It is not hard to show that, in this case, the pointwise Kronecker invariants are equal to the "global" Kronecker invariants, i.e. the Kronecker invariants computed over the fraction field of R. Indeed, in general, there is a Zariski open set on which the pointwise invariants are constant and equal to the global Kronecker invariants. What is more, the exceptional values must bear some relation to the global invariants. Explicitly, if

$V_i = \bigoplus^{m}{}_{i=1}\, \mathcal{O}(n_i)$, $V' = \bigoplus^{m'}{}_{i=1}\, \mathcal{O}(n_i')$ are bundles on $\mathbf{P}^1$, then $V \geq V'$ in the Harder-Narasimhan ordering just in case

$$\sum_{i=1}^{l} n_i \geq \sum_{i=1}^{l} n_i', \quad \text{for} \quad l = 1, \ldots, \max(m, m'). \tag{5.15}$$

THEOREM 5.16 (*Shatz,* [45]). *The Harder-Narasimhan ordering is upper semi-continuous in algebraic families of vector bundles on* $\mathbf{P}^1$.

This theorem has several applications in problems of computing feedback invariants. Now, any algebraic family of controllable systems induces a map,

$$X \to \tilde{\Sigma}_0,$$

and Theorem 5.16 thus gives a relation between the quotient topology on the finite set $\tilde{\Sigma}_0/\mathfrak{F}$ and the Harder-Narasimhan ordering or, equivalently, the Rosenbrock ordering of Theorem 5.11. We call the ordering induced by specialization, the geometric ordering, while we refer to the "ubiquitous ordering" as the natural ordering. The natural ordering is also present in combinatorics and in number theory, where it is studied as an ordering on $\mathfrak{L}_n$—the set of all partitions of the integer n. Now, it is an unpublished piece of folklore, not too hard to prove, that a stronger version of Theorem 5.16 holds for algebraic families constructed via system theory, viz. the geometric ordering is dual to the natural ordering. In, this context, the following theorem ([8]) is rather remarkable:

PROPOSITION 5.17. *$\mathfrak{L}_n$ is a lattice under the natural ordering.*

We close by indicating extensions of this situation which are motivated by control theory but seem likely to contribute some new examples to geometry. First, it is of much more practical significance to study the action of the output feedback group; i.e. $\mathrm{Hom}(k^n, k^m)$ is replaced by $\mathrm{Hom}(k^p, k^m)$ in (iii). In this setting, the proper generalization of Theorem 5.2 is not even conjectured—although a counter-example to the dominant morphism approach has been constructed by J. C. Willems ([48]). Second, pencils may also be defined by control systems governed by partial differential equations. In this setting, the generalization of the Rosenbrock pencil leads to coherent sheaves on $\mathbf{P}^N$ but whether these all split, as in Grothendieck's theorem, is unknown at present. Finally, there are several computations which suggest that, for non-constant coefficient differential equations, the transfer function ought to be, in the scalar input-output case, a meromorphic function on a Riemann surface of higher genus. Each of these questions deserves further study.

HARVARD UNIVERSITY
BROWN UNIVERSITY

REFERENCES

[1] H. Bass, "Quadratic modules over polynomial rings," *Contributions to Algebra* (H. Bass, P. Cassidy, J. Kovacic, eds.), Academic Press, New York, 1977.
[2] R. Brockett, "Some geometric questions in the theory of linear systems," *IEEE Trans. Aut. Contr.*, AC-21, 1976.
[3] R. Brockett, "Finite-Dimensional Linear Systems," Wiley-Interscience, New York, 1970.
[4] R. Brockett, "Lie algebras and rational functions, some control theoretic connections," *Lie Theories and Their Applications* (W. Rossmann, ed.), Queens University, Ontario.
[5] R. Brockett, "The geometry of the set of controllable linear systems," *Res. Repts. of Aut. Contr. Lab.*, Nagoya Univ., **24,** 1977.
[6] R. Brockett and R. Skoog, "A new perturbation theory for the synthesis of nonlinear networks," *Math. Aspects of Elec. Network Analysis, SIAM AMS* Proc., **III,** 1970.
[7] P. Brunovsky, "A classification of linear controllable systems," *Kybernetika*, **3,** 1970.
[8] T. Brylawski, "The lattice of integer partitions," *Discrete Mathematics*, **6,** 1973.
[9] R. Bumby and E. Sontag, "Reachability does not imply coefficient assignability," *Notices AMS*, 1978.
[10] C. Byrnes, "On the control of certain deterministic infinite dimensional systems by algebro-geometric techniques," to appear in *Amer. J. of Math.*
[11] C. Byrnes, "The moduli space for linear dynamical system," *Geometric Control Theory* (C. Martin, R. Hermann, eds.) Math. Sci. Press, Brookline, 1977.
[12] C. Byrnes, "Feedback invariants for linear system defined over rings," Proc. IEEE Conf. on Decision and Contr., San Diego, 1979.
[13] C. Byrnes and T. Duncan, "Topological and geometric invariants arising in control theory," to appear.
[14] C. Byrnes and N. Hurt, "On the moduli of linear dynamical systems," *Adv. in Math.: Studies in Analysis*, **4,** 1978.
[15] A-L. Cauchy, "Calcul de indices des fonctions," J. Ecole Polytechnique, 1835.
[16] J. M. C. Clark, "The consistent selection of local coordinates in linear system identification," *Proc. JACC*, Purdue, 1976.

[17] S. Eilenberg, *Automata, Languages, and Machines,* Vol. A, Academic Press, New York, 1978.

[18] P. Falb, "Linear Systems and Invariants," Lecture Notes, Control Group, Lund University, Sweden, 1974.

[19] P. Falb and W. Wolovich, "Invariants and canonical forms under feedback," to appear [also, Center for Dynamical Systems, Rep't. 1978, Brown University].

[20] F. R. Gantmacher, *Theory of Matrices,* Vols. I, II, Chelsea, New York, 1959.

[21] K. Glover, "Some geometrical properties of linear systems with implications in identification," *Proc. IFAC,* Boston, 1975.

[22] G. Gurevitch, *Foundations of the Theory of Algebraic Invariants,* Noordhoff Ltd., Netherlands, 1964.

[23] C. Hanna, "Decomposing algebraic vector bundles on the projective line," *Proc. Amer. Math. Soc.,* **61,** 1976.

[24] M. Hazewinkel, "Moduli and canonical forms for linear dynamical systems II: the topological case, *Math. Sys. Theory,*

[25] M. Hazewinkel, "Moduli and canonical forms for linear dynamical systems III: the algebraic-geometric case," in *Geometric Control Theory* (C. Martin, R. Hermann, eds.), Math. Sci. Press, Brookline, 1977.

[26] M. Hazewinkel and R. E. Kalman, "On invariants, canonical forms and moduli for linear constant finite-dimensional, dynamical systems," *Lecture Notes Econ.-Math. System Theory,* **131,** Springer-Verlag, Berlin, 1976.

[27] R. Hermann and C. Martin, "Applications of algebraic-geometry to system theory—Part I," *IEEE Trans. Aut. Cont.,* AC-22, 1977.

[28] R. Hermann and C. Martin, "Applications of algebraic geometry to systems theory: the McMillan degree and Kronecker indices of transfer functions as topological and holomorphic invariants," to appear in *SIAM J. Cont. and Optimization.*

[29] C. Hermite, "Sur les nombres des racines d'une equation algebrique comprise entre des limites donnes," *J. Reine Angewandte Math.,* **52,** 1856.

[30] A. Hurwitz, "Uber die bedingungen unter welchen eine gleichung nur wurzeln mit negativen reelen teilen besitz," *Mathem. Annalen,* **46,** 1895.

[31] R. E. Kalman, "Global structure of classes of linear dynamical systems," *Lectures,* NATO Adv. Study Inst., London, 1971.

[32] R. E. Kalman, "Kronecker invariants and feedback," *Ordinary Differential Equations* (L. Weiss, ed.), Academic Press, New York, 1972.

[33] R. E. Kalman, P. Falb and M. Arbib, *Topics in Mathematical System Theory,* McGraw-Hill, New York, 1969.

[34] E. Kamen, "On an algebraic theory of systems defined by convolution, *Math. System Theory,* **9**(1975).

[35] E. Kamen, "An operator theory of linear functional differential equations," *J. Diff. Eqns.* **27**(1978).

[36] T. Koga, "Synthesis of finite passive n-parts with prescribed two-variable reactance matrices, *IEEE Trans. Cir. Theory,* CT-13, 1966.

[37] S. Lefschetz, *Applications of algebraic topology,* Springer-Verlag, New York, 1976.

[38] J. C. Maxwell, "On governors," *Proc. Royal Soc. of London,* **16,** 1868.

[39] A. S. Morse, "Ring models for delay-differential systems," *Proc. IFAC Symp. on Multivariable Systems,* Manchester, 1974.

[40] D. Mumford, *Geometric Invariant Theory,* Springer-Verlag, Berlin, 1965.

[41] D. Mumford, *Algebraic Geometry I Complex Projective Varieties,* Springer-Verlag, Berlin, 1976.

[42] H. Rosenbrock, *State-space and Multivariable Theory,* Wiley Interscience, N. Y., 1970.

[43] Y. Rouchaleau, "Linear, discrete-time, finite dimensional, dynamical systems over some classes of commutative rings,

[44] Y. Rouchaleau, B. Wyman, and R. Kalman, "Algebraic structure of linear dynamical systems III. Realization theory over a c commutative ring." *Proc. Nat. Acad. Sci.,* **69**(1972).

[45] S. Shatz, "The decomposition and specialization of algebraic families of vector bundles," *Comp. Math.* **35**(1977).

[46] E. Sontag, "On split realizations of response maps over rings," *Inf. and Control,* **37**(1978).

[47] E. Sontag, "Linear Systems over Commutative Rings: A Survey," *Richerche di Automatica,* **7**(1976).

[48] J. C. Willems and W. H. Hesselink, "Generic properties of the pole-placement problem, *Proc. IFAC,* Helsinki, 1978.

[49] W. Wolovich, *Linear Multivariable Systems,* Springer-Verlag, New York, 1974.

[50] W. Wolovich and P. Falb, "Invariants and canonical forms under dynamic compensation," *SIAM J. on Cont. and Optimization,* **14,** 1976.

[51] W. Wonham, "On pole assignment in multi-input controllable linear systems, *IEEE Trans. Aut. Contr.* AC-12(1967).

[52] B. Wyman, "Pole placement over integral domains," to appear in *Comm. in Alg.*

[53] D. Youla, "The synthesis of networks containing lumped and distributed elements," *Networks and Switching Theory,* **11,** 1968.

[54] D. Youla and P. Tissi, "*n*-port synthesis via reactance extraction," *I. IEEE Int. Conv.* (1966).

[55] O. Zariski, "Theory and applications of holomorphic functions on algebraic varieties over arbitrary ground fields," *Mem. Amer. Math. Soc.,* **5,** 1951.

[56] O. Zariski and P. Samuel, *Commutative Algebra,* Vols. I, II. Van Nostrand, Princeton, 1960.

ON THE ALGEBRAICITY OF CERTAIN RINGED SPACES*

By Wei-Liang Chow

Many years ago we proved a theorem which says that a closed analytic subspace in the projective space is algebraic. Since then the theorem has received a number of other proofs. Of particular interest is the proof given by Serre in his well-known paper entitled "Géométrie algébrique et géométrie analytique" (often quoted as GAGA), in which this theorem appears as a corollary to a systematic treatment of the relations between analytic and algebraic sheaves. Under the circumstances, it would seem slightly absurd to offer here yet another proof of this old theorem. A few words of explanation are therefore in order.

We shall offer here a proof which depends essentially on a simple estimate of the "postulation" type. While this method is in spirit closely related to the geometric proof used in our original paper, it is more elementary in that very little use is made of the topological properties of the space. However, what makes this new proof worthy of attention is the fact that, with a few appropriate modifications, it applies also to the case of a non-archimedean valued field. In fact, we shall derive the result for both the complex-analytic case and the non-archimedean case from a general theorem, which states that if a ringed space of a certain type is immersed (see below) in a projective space, then it is essentially algebraic. In view of the radically different nature of the topology of the ground field in the two cases, it will be a basic feature of our treatment that direct use of this topology be avoided as much as possible. For this reason, we have replaced the usual concept of embedding by a definition of "immersion" which makes no use of the topology of the ground field. Such an immersion is a weaker concept than the usual embedding, but it is sufficient for our purpose.

Our approach can be used to prove other results of a similar nature. These will be presented in a subsequent paper.

We should like to mention that the possibility of extending the GAGA-type results to the non-archimedean case was indicated by Kiehl

* Research partially supported by a grant from National Science Foundation.

(Invent. Math., vol. 2, p. 191). This has been carried out in all generality by Ursula Koepf in her 1973 Muenster thesis.

I.

The object of our study is a G-ringed space $(X, \mathfrak{T}, \mathcal{O})$. We recall that such a space consists of the following objects: (1) A topological space X, which serves as the underlying space for the G-ringed space. (2) A Grothendieck topology $\mathfrak{T}$ on the category of certain open sets in X, called the admissible sets, and the "coverings" in $\mathfrak{T}$ are certain open coverings of admissible sets in X, called admissible coverings. (3) A sheaf of rings $\mathcal{O}(\mathfrak{T})$ on $\mathfrak{T}$, called the structural sheaf. More specifically, the structural sheaf $\mathcal{O}(\mathfrak{T})$ of X is defined by a contravariant functor $U \to \mathcal{O}(U)$ from admissible sets in $\mathfrak{T}$ to rings, which satisfies the usual sheaf condition F1 and F2 (in the notation of R. Godement: Théorie des Faisceaux) with respect to the admissible coverings in $\mathfrak{T}$. We note that the conditions F1 and F2 are not required to be satisfied with respect to coverings which are not admissible.

From the structural sheaf $\mathcal{O}(\mathfrak{T})$ we construct a sheaf space $\mathcal{O}_X$ over X as follows, whereby we observe that the construction is independent of the particular choice of the admissible coverings. By the definition of a sheaf, there is a canonical homomorphism $\rho_U{}^V$ of $\mathcal{O}(V)$ into $\mathcal{O}(U)$, for any pair of admissible sets U, V such that $U \subset V$. For every point x in X, we define the "stalk" $\mathcal{O}_x$ over x to be the inductive limit $\mathcal{O}_x = \lim_{x \in U} \mathcal{O}(U)$ of the family of rings $\mathcal{O}(U)$ with respect to the ordered system of all admissible sets containing the point x; then, for every admissible set U containing x, there is a canonical homomorphism $\rho_x{}^U$ of $\mathcal{O}(U)$ into $\mathcal{O}_x$. The sheaf space $\mathcal{O}_X$ is defined to be the union $\mathcal{O}_X = \bigcup_{x \in X} \mathcal{O}_x$ of the stalks over all points in X, endowed with the topology which has as a base for open sets all sets of the form $\{\rho_x{}^U(f) \mid x \in U\}$, for every admissible set U and every element f in $\mathcal{O}(U)$. To be sure, in order that the projection map of $\mathcal{O}_X$ onto X be continuous, we have to replace the original topology of X by the possibly coarser topology which has the admissible sets in $\mathfrak{T}$ as a base for open sets. However, to avoid unnecessary complications which are irrelevant for our present purpose, we shall assume that the admissible sets in $\mathfrak{T}$ already form a base for the original topology of X. Following the usual notations, we shall denote by $\Gamma(U, \mathcal{O}_X)$ the ring of all sections of $\mathcal{O}_X$ over the admissible set U. If f is any element in $\mathcal{O}(U)$, the correspondence $x \to f_x =$

$\rho_x{}^U(f)$ defines a map of U into $\mathcal{O}_X$, which is easily seen to be a section of $\mathcal{O}_X$ over U. There is therefore a canonical homomorphism of $\mathcal{O}(U)$ into $\Gamma(U, \mathcal{O}_X)$ for every admissible set U in T. We shall assume that this canonical homomorphism is injective, so that $\mathcal{O}(U)$ can be considered as a subring in $\Gamma(U, \mathcal{O}_X)$; and we shall express this by saying that the structural sheaf $\mathcal{O}(U)$ is Γ-*injective.*

From now on we shall make the following assumptions (without any further mention): (1) There is a fixed ground field k such that every ring $\mathcal{O}(U)$ is an algebra over k and every homomorphism between two such rings is a k-homomorphism. It follows then that the rings $\mathcal{O}_x$ are also k-algebras, and the homomorphisms $\rho_x{}^U$ are also k-homomorphisms. In fact, unless mentioned explicitly otherwise, all homomorphisms between k-algebras are understood to be k-homomorphisms. Furthermore, for simplicity, we shall also assume that the ground field k is algebraically closed, although much of our results are valid without this assumption or with appropriate modifications. On the other hand, we shall avoid any direct use of any topology that might exist in the field k, even though such a topology may have played a role in the original definition of the G-ringed space X. (2) Every stalk $\mathcal{O}_x$ is a local ring. More specifically, the situation is as follows. For any point x in an admissible set U, let $\mathcal{O}(U)_x$ be the quotient ring of $\mathcal{O}(U)$ with respect to the multiplicative set of all elements f such that $f_x = \rho_x{}^U(f)$ is a unit in $\mathcal{O}_x$. Then, for any admissible set V containing U, the canonical homomorphism $\rho_U{}^V$ of $\mathcal{O}(V)$ to $\mathcal{O}(U)$ can be extended to a homomorphism of $\mathcal{O}(V)_x$ to $\mathcal{O}(U)_x$, which we shall denote by the same symbol. If $\mathcal{O}_x$ is a local ring, then $\mathcal{O}(U)_x$ is also a local ring for every admissible set U containing x, and the canonical homomorphism of $\mathcal{O}(V)_x$ to $\mathcal{O}(U)_x$ is also local. The local ring $\mathcal{O}_x$ itself can be identified with the inductive limit of the family of rings $\mathcal{O}(U)_x$ with respect to the ordered system of all admissible sets containing x.

Let $\mathfrak{J}$ be a sheaf of ideals on a G-ringed space X. We shall associate with $\mathfrak{J}$ a G-ringed space $X(\mathfrak{J})$, which is called the *subspace* in X defined by the sheaf $\mathfrak{J}$. The underlying topological space of $X(\mathfrak{J})$ is the subspace of X (as an ordinary topological space) consisting of all points in X such that $\mathfrak{J}_x \neq \mathcal{O}_x$. Next, an open set in $X(\mathfrak{J})$ is said to be admissible if and only if it is the restriction $U_{X(\mathfrak{J})}$ to $X(\mathfrak{J})$ of an admissible set U in X. Similarly, an open covering of $X(\mathfrak{J})$ is said to be admissible if and only if it is the restriction to $X(\mathfrak{J})$ of an admissible covering in X. This defines then a Grothendieck topology $\mathfrak{T}_{\mathfrak{J}}$ on $X(\mathfrak{J})$. Finally, in order to define the structural sheaf $\mathcal{O}(\mathfrak{T}_{\mathfrak{J}})$ of the G-ringed space $X(\mathfrak{J})$, we first consider the presheaf $\mathcal{O}'(\mathfrak{T}_{\mathfrak{J}})$ defined by the functor $U_{X(\mathfrak{J})} \to \mathcal{O}(U)/\mathfrak{J}(U)$ on admissible sets in $\mathfrak{T}_{\mathfrak{J}}$, the canonical homomorphisms being the restrictions of the maps $\rho_U{}^V$ to the admissible sets $V_{X(\mathfrak{J})}$ and $U_{X(\mathfrak{J})}$ in $\mathfrak{T}_{\mathfrak{J}}$. The structural sheaf $\mathcal{O}(\mathfrak{T}_{\mathfrak{J}})$ of $X(\mathfrak{J})$ is then the sheaf on $\mathfrak{T}_{\mathfrak{J}}$ generated by the presheaf $\mathcal{O}'(\mathfrak{T}_{\mathfrak{J}})$. It is easily seen that for any point x in $X(\mathfrak{J})$, the stalk at x of the sheaf space $\mathcal{O}_{X(\mathfrak{J})}$ is isomorphic to the residue ring $\mathcal{O}_{X,x}/\mathfrak{J}_x$. The canonical homomorphism of $\mathcal{O}'(\mathfrak{T}_{\mathfrak{J}})$ into $\mathcal{O}(\mathfrak{T}_{\mathfrak{J}})$ is in general not injective. In order to give a criterion for the injectivity of this homomorphism, we introduce the following definition. Let $\mathfrak{A}$ be an ideal in $\mathcal{O}(U)$, and consider the ideal $\mathfrak{A}_x = \mathfrak{A} \otimes \mathcal{O}_x$ in $\mathcal{O}_x$ for any point x in U. The ideal $\mathfrak{A}$ is said to be *complete* if an element f in $\mathcal{O}(U)$ is in $\mathfrak{A}$ if (and only if) f_x is in $\mathfrak{A}_x$ for every point x in U. A sheaf $\mathfrak{J}$ of ideals is said to be *complete* if the ideal $\mathfrak{J}(U)$ in $\mathcal{O}(U)$ is complete for every admissible set U in $\mathfrak{T}$. One verifies readily that if the sheaf of ideal $\mathfrak{J}$ is complete, then the homomorphism of $\mathcal{O}'(\mathfrak{T}_{\mathfrak{J}})$ into $\mathcal{O}(\mathfrak{T}_{\mathfrak{J}})$ is injective.

In accordance with our aim of minimizing the use of the topology of our space, we introduce here a slightly different definition of the dimension of a G-ringed space, called the r-dimension (short for ring-dimension). The *r-dimension* of a (local) G-ringed space X is defined to be $\max_{x \in X} (\dim \mathcal{O}_{X,x})$, where $\dim \mathcal{O}_{X,x}$ is the Krull dimension (i.e. altitude) of the local ring $\mathcal{O}_{X,x}$. We shall assume that our space X has a finite r-dimension. It is clear that the r-dimension of a subspace $X(\mathfrak{J})$ is at most equal to the r-dimension of X; furthermore, if the sheaf $\mathfrak{J}$ has the property that $\mathfrak{J}_x$ is a non-zero-divisor (i.e. $\mathfrak{J}_x$ does not consist entirely of zero-divisors) in $\mathcal{O}_{X,x}$ for every x, then the r-dimension of $X(\mathfrak{J})$ is less than the r-dimension of X.

A complex-analytic space in the usual sense is of course an example of our G-ringed space. Another example is a holomorphic space over a complete non-archimedean valued field, and it is for the purpose of including this case in our treatment that we have introduced a Grothendieck topology on our space. The theory of holomorphic spaces over a non-archimedean valued field has been studied with great success by Grauert and Remmert (and their disciples) in recent years, following the initial fundamental work of Tate. It will be seen that the properties we shall impose on our space all hold for a holomorphic space, both in the complex-analytic case and in the non-archimedean case.

Following the usual practice, we shall from now on use the notation

$\mathcal{O}_X$ to denote both the structural sheaf $\mathcal{O}(\mathfrak{T})$ and the sheaf space $\mathcal{O}_X$ as defined above, and we shall refer to the whole thing as the structural sheaf of the G-ringed space X. The subscript will sometimes be dropped in case there is no ambiguity about the space under consideration.

A *morphism* φ of a G-ringed space $(X, \mathfrak{T}, \mathcal{O}_X)$ into a G-ringed space $(X', \mathfrak{T}', \mathcal{O}_{X'})$ consists of a pair of maps (φ_0, φ_1), where (1) φ_0 is a continuous map of X into X' such that the Grothendieck topology ${\varphi_0}^{-1}(\mathfrak{T}')$ is contained in $\mathfrak{T}$, and (2) φ_1 is a homomorphism (= local k-homomorphism) of the sheaf space $\mathcal{O}_{X'} \circ \varphi_0$ into the sheaf space $\mathcal{O}_X$ which carries $\mathcal{O}_{X'}(U')$ into $\mathcal{O}_X({\varphi_0}^{-1}(U'))$ for every admissible set U' in $\mathfrak{T}'$. This is of course the usual definition of a morphism of a G-ringed space into another. On the other hand, for our present purpose, we shall introduce the following definition of "immersion," which is a slight modification of the usual definition of "injective immersion," in that the condition that φ_1 be a surjection is replaced by a weaker one.

Definition 1. A morphism $\varphi = (\varphi_0, \varphi_1)$ of $(X, \mathfrak{T}, \mathcal{O}_X)$ into $(X', \mathfrak{T}', \mathcal{O}_{X'})$ is said to be an *immersion* of X in X' if (1) φ_0 is injective and (2) for every point x in X there exists a system of parameters in $\mathcal{O}_{X,x}$ consisting of elements in $\varphi_1(\mathcal{O}_{X'} \circ \varphi_0)_x$.

As we shall see later, this definition enables us to speak of an immersion of a holomorphic space in a projective space without having to make the latter itself into a holomorphic space. An immersion φ of X into X' is said to be an *embedding* if ${\varphi_0}^{-1}(\mathfrak{T}') = \mathfrak{T}$ and the homomorphism of $\mathcal{O}_{X'}(U')$ into $\mathcal{O}_X({\varphi_0}^{-1}(U'))$ is surjective for every U' in $\mathfrak{T}'$. It will turn out later that although we are primarily interested in embedded spaces (in a projective space), it is nevertheless sufficient to consider the embedding as a mere immersion.

Let φ be a morphism of X into X'. If $\mathfrak{F}$ is a sheaf of $\mathcal{O}_{X'}$-modules on X', we can define a sheaf $\mathfrak{F}_X$ of $\mathcal{O}_X$-modules on X which is the inverse image of $\mathfrak{F}$ under φ (or the lift of $\mathfrak{F}$ to X). We define first a presheaf $\hat{\mathfrak{F}}$ as follows. For any admissible set V in the Grothendieck topology ${\varphi_0}^{-1}(\mathfrak{T}')$, we set $\hat{\mathfrak{F}}(V) = \mathcal{O}_X(V) \otimes \mathfrak{F}(\varphi_0(V))$, where the tensor product is taken over $\mathcal{O}_{X'}(\varphi_0(V))$. We have then a presheaf $V \to \hat{\mathfrak{F}}(V)$ on the Grothendieck topology ${\varphi_0}^{-1}(\mathfrak{I}')$, and the canonical map $\mathfrak{F}(\varphi_0(V)) \to \hat{\mathfrak{F}}(V)$ is a sheaf homomorphism, which we shall denote also by φ_1. Next, we extend this presheaf $\hat{\mathfrak{F}}$ to a presheaf over the Grothendieck topology $\mathfrak{T}$, to be denoted by the same symbol $\hat{\mathfrak{F}}$. For any admissible set U in $\mathfrak{T}$, we define $\hat{\mathfrak{F}}(U)$ to be the inductive limit $\hat{\mathfrak{F}}(U) = \lim_{U \subset V}$

$\hat{\mathfrak{F}}(V)$ with respect to the ordered system of all admissible sets V in ${\varphi_0}^{-1}(\mathfrak{T}')$ containing U. The presheaf $U \to \hat{\mathfrak{F}}(U)$ on $\mathfrak{T}$ is then the desired extension. Finally, the sheaf $\mathfrak{F}_X$ is defined to be the sheaf associated with the presheaf $\hat{\mathfrak{F}}$. The homomorphism φ_1 of $\mathfrak{F}$ into $\hat{\mathfrak{F}}$ can be considered in a natural way as a homomorphism of $\mathfrak{F}$ into $\mathfrak{F}_X$. In case $\mathfrak{F}$ is an invertible sheaf, it is readily seen that $\mathfrak{F}_X$ is then also an invertible sheaf. In this case, the homomorphism φ_1 can be described locally as follows. Let V be an admissible set in ${\varphi_0}^{-1}(\mathfrak{T}')$ such that $\mathfrak{F}(\varphi_0(V))$ is a free $\mathcal{O}(\varphi_0(V))$-module of rank 1 and hence isomorphic to the ring $\mathcal{O}_{X'}(\varphi_0(V))$ as a module. The map φ_1: $\mathfrak{F}(\varphi_0(V)) \to \hat{\mathfrak{F}}(V)$ is then essentially the same as the map φ_1: $\mathcal{O}_{X'}(\varphi_0(V)) \to \mathcal{O}_X(V)$. We shall find this remark useful later.

II.

Let $S = S^N$ be the projective space of dimension N over the ground field k. We shall endow S with its Zariski topology, which also serves as its Grothendieck topology, and the structural sheaf of S is the usual one associated with the Zariski topology. Thus, the stalk $\mathcal{O}_{S,p}$ at a point p in S is the ring consisting of all quotients of two homogeneous elements of the same degree in the polynomial ring $k[u] = k[u_0, u_1, \ldots, u_N]$, with denominators not vanishing at p. In other words, the G-ringed structure of S is its usual Zariski structure. More generally, we can consider any algebraic scheme over k as a G-ringed space in the same way; this applies in particular to the subschemes in S.

Let φ be a morphism of a G-ringed space X into S. For every point x in X, the restriction $\varphi_{1,x}$ to x of φ_1 is a homomorphism of $(\mathcal{O}_S \circ \varphi_0)_x$ into $\mathcal{O}_{X,x}$, which can also be considered as a homomorphism of $\mathcal{O}_{S,\varphi_0(x)}$ into $\mathcal{O}_{X,x}$. Consider a homogeneous polynomial $f(u)$ in $k[u]$ with the following property: For every x in X there exists a homogeneous polynomial $g(u)$ of the same degree as $f(u)$ such that the image of $f(u)/g(u)$ under $\varphi_{1,x}$ is zero. The set of all polynomials having this property generates a homogeneous ideal $A(\varphi)$ in $k[u]$, and the subscheme in S defined by this ideal $A(\varphi)$ is called the *algebraic hull* of X in S (relative to the morphism φ) and will be denoted by $A(X)$, or $A_\varphi(X)$ in case it is advisable to specify the morphism involved. If we denote by $\iota = (\iota_0, \iota_1)$ the injection morphism of $A(X)$ into S, then we can factorize $\varphi = \iota \circ \varphi'$, where φ' is a morphism of X into $A(X)$. If φ is an immersion (embedding), then φ' is also an immersion (embedding), and we shall then say that X is an immersed (embedded) space in S. The following definition of algebraicity seems to be appropriate.

Definition 2. An immersed G-ringed space X in S is said to be *algebraic* if the dimension of (the scheme) $A(X)$ is equal to the r-dimension of X.

It is easily seen that the dimension of $A(X)$ is at least equal to the r-dimension of X. Hence, for an immersed G-ringed space X to be algebraic, it is sufficient to show that the dimension of $A(X)$ is at most equal to the r-dimension of X. To see this, we observe first that the homomorphism $\varphi_{1,x}$ of $\mathcal{O}_{S,\varphi_0(x)}$ into $\mathcal{O}_{X,x}$ induces a homomorphism of $\mathcal{O}_{A(X),\varphi_0(x)}$ into $\mathcal{O}_{X,x}$, which for simplicity we shall also denote by $\varphi_{1,x}$. Let n be the altitude of $\mathcal{O}_{A(X),\varphi_0(x)}$, and let $h_1, \ldots, h_n$ be a system of parameters of $\mathcal{O}_{A(X),\varphi_0(x)}$; then the ideal generated in $\mathcal{O}_{X,x}$ by the elements $\varphi_{1,x}(h_1), \ldots, \varphi_{1,x}(h_n)$ will have a radical which contains a system of parameters in $\mathcal{O}_{X,x}$ and hence must coincide with the maximal prime ideal in $\mathcal{O}_{X,x}$. It follows (from the Altitude Theorem of Krull) that the altitude of $\mathcal{O}_{X,x}$ is at most equal to n, which proves our assertion.

When we refer to a holomorphic subspace X in a projective space S, whether in the complex case or in the non-archimedean case, we must of course first endow S with a holomorphic structure. Then X as a subspace will be defined by a holomorphic sheaf of ideals $\mathcal{J}$, which then defines the holomorphic structure on X, and the injection map of X into S is then an embedding. The subspace X is usually said to be algebraic if the sheaf $\mathcal{J}$ defining X is the extension of an algebraic sheaf of ideals, and it is easily seen that this is the case (assuming that X is irreducible or equi-dimensional) if and only if the algebraic hull of X has the same dimension as X itself. On the other hand, the algebraic hull of X is defined even if we consider the embedding of X as a mere immersion into S; and we observe that any immersion of X into the holomorphic space S is also an immersion of X into the algebraic space S (i.e. endowed with its Zariski structure), and the algebraic hull of X is the same in either case.

Let $\mathcal{L}$ be the canonical invertible sheaf in S, and denote by $\mathcal{L}_X{}^m$ the lift to X of the sheaf $\mathcal{L}^m$. The map φ_1 of $\mathcal{L}^m$ into $\mathcal{L}_X{}^m$ induces on S a homomorphism of $\mathcal{L}^m(S)$ into $\mathcal{L}_X{}^m(X)$ as k-modules. We shall determine the kernel of this homomorphism. We recall first that $\mathcal{L}^m(S)$ can be identified with the module $k[u]_m$ of all homogeneous polynomials of degree m in $k[u]$. More generally, for any (Zariski) open set W in S, the $\mathcal{O}_S(W)$-module $\mathcal{L}^m(W)$ can be represented by the set of all quotients g/h of two homogeneous polynomials in $k[u]$, with $\deg g - \deg h = m$ and h not vanishing anywhere in W. Let x be a point in X, and let $h^{(x)}$ be an element in $k[u]_m$ which does not vanish at $\varphi_0(x)$; such a polynomial $h^{(x)}$ certainly exists. If we choose W so that $\varphi_0(x)$ is contained in W and $h^{(x)}$ does not vanish anywhere in W, then $h^{(x)}$ can serve as a generator of $\mathcal{L}^m(W)$, and the map $g/h \to g/hh^{(x)}$ defines an isomorphism between $\mathcal{L}^m(W)$ and $\mathcal{O}_S(W)$ as $\mathcal{O}_S(W)$-modules. As we have shown before, the map φ_1: $\mathcal{L}^m(W) \to \mathcal{L}_X{}^m({\varphi_0}^{-1}(W))$ can then be represented through this isomorphism by the map φ_1: $\mathcal{O}_S(W) \to \mathcal{O}_X({\varphi_0}^{-1}(W))$. Let f be an element in $\mathcal{L}^m(S)$; then its restriction to W can be represented by the element $f/h^{(x)}$ in $\mathcal{O}_S(W)$. If $\varphi_1(f) = 0$ in $\mathcal{L}_X{}^m(X)$, then $\varphi_1(f/h^{(x)}) = 0$ in $\mathcal{O}_X({\varphi_0}^{-1}(W))$; this implies that in the map $\varphi_{1,x}$: $\mathcal{O}_{S,\varphi_0(x)} \to \mathcal{O}_{X,x}$ the element $(f/h^{(x)})_x$ is mapped into zero. Since this is true for any point x in X, this implies that f is in the kernel $A(\varphi)$ of φ. On the other hand, since the structural sheaf of X is assumed to be Γ-injective, every element in $A(\varphi)$ is mapped by φ_1 into zero. Thus we have shown that the kernel of the homomorphism φ_1: $\mathcal{L}^m(S) \to \mathcal{L}_X{}^m(X)$ is the submodule $k[u]_m \cap A(\varphi)$, and hence the image $\varphi_1(\mathcal{L}^m(S))$ of $\mathcal{L}^m(S)$ in $\mathcal{L}_X{}^m(X)$ is isomorphic to the quotient module $k[u]_m/(k[u]_m \cap A(\varphi))$.

Now, the residue ring $k[u]/A(\varphi)$ is simply the graded ring associated with the projective scheme $A(X)$, and $k[u]_m/(k[u]_m \cap A(\varphi))$ is isomorphic to the module $(k[u]/A(\varphi))_m$ of elements of degree m in $k[u]/A(\varphi)$. It is well-known in algebraic geometry that if t is the dimension of $A(X)$, then $\dim_k(k[u]/A(\varphi))_m$ is given, for sufficiently high m, by a polynomial $\chi(m)$ of degree t in m, called the Hilbert polynomial. Therefore, if we can show that $\dim_k L_X{}^m(X) \leq Cm^s$ for an infinite number of positive integers m, for some constant C and an integer $s \leq r$-dim X, then dim $A(X) \leq r$-dim X and hence X must be algebraic. Thus we have proved the following lemma.

LEMMA 1. *Let X be an immersed G-ringed space. If there is a constant C and an integer $s \leq r$-dim X such that $\dim_k \mathcal{L}_X{}^m(X) \leq Cm^s$ for an infinite sequence of integers m, then X is algebraic.*

The next question is naturally how one can get an estimate of the dimension of $\mathcal{L}_X{}^m(X)$ as described in Lemma 1. An answer will be given by what we shall call a distinct system of sections. For the purpose of later generalization, we shall treat the case of an arbitrary invertible sheaf on a G-ringed space X, not necessarily the lift of an invertible from an immersion into S^N. We begin with an observation on a possible extension of the structural sheaf, which is often desirable in our context.

We shall say that an element f in a ring R is *rationally dependent* on a subring R_1 in R, if there are elements g and h in R_1 such that $gf = h$. A subring R_1 in a ring R is said to be *rationally closed* in R. if every element in R which is rationally dependent on R_1 is already contained in R_1. In general, for any subring R_1 in R, there is a smallest subring R_2 in R which contains R_1 and is rationally closed in R. We shall call this ring R_2 the *rational closure* of R_1 in R. Applying these concepts to the subring $\mathcal{O}(U)$ in $\Gamma(U, \mathcal{O}_X)$, we obtain the rational closure of $\mathcal{O}(U)$ in $\Gamma(U, \mathcal{O}_X)$, which we shall call the Γ-*closure* of $\mathcal{O}(U)$ and shall denote by $\tilde{\mathcal{O}}(U)$. It is easily verified that the functor $U \to \tilde{\mathcal{O}}(U)$ defines a sheaf on $\mathfrak{T}$, which has the same sheaf space $\mathcal{O}_X$ as the structural sheaf $\mathcal{O}(\mathfrak{T})$; we shall call this sheaf the Γ-closure of $\mathcal{O}(\mathfrak{T})$ and shall denote it by $\tilde{\mathcal{O}}(\mathfrak{T})$. The sheaf $\mathcal{O}(\mathfrak{T})$ is said to be Γ-*closed* if it coincides with its Γ-closure. Since $\mathcal{O}(\mathfrak{T})$ have the same sheaf space $\mathcal{O}_X$, the r-dimension of a space X is not changed if we replace $\mathcal{O}(\mathfrak{T})$ by $\tilde{\mathcal{O}}(\mathfrak{T})$ as the structural sheaf of X. The G-ringed space X is said to be Γ-*closed* if its structural sheaf is Γ-closed.

Let $\mathcal{L}$ be an invertible sheaf on X, and let f be an element in $\mathcal{L}(X)$. We shall associate with f a sheaf of ideals $\mathfrak{J}_f$ as follows. Let U be an admissible set in $\mathfrak{T}$. If $\mathcal{L}(U)$ is a free $\mathcal{O}_X(U)$-module of rank 1 and if we fix a basis element h for $\mathcal{L}(U)$, then the restriction f_U of f to U can be represented by an element $[f_U]$ in $\mathcal{O}_X(U)$, defined by the equation $f_U = [f_U]h$. We denote by $(f)_U$ the ideal in $\mathcal{O}_X(U)$ generated by this element $[f_U]$, and we note that this ideal $(f)_U$ is independent of the choice of the basis element h. If U is such that $\mathcal{L}(U)$ is not a free $\mathcal{O}_X(U)$-module of rank 1, then we let $(f)_U$ be the zero ideal in $\mathcal{O}_X(U)$. The map $U \to (f)_U$ then defines a presheaf of ideals on $\mathfrak{T}$, and the sheaf $\mathfrak{J}_f$ is defined to be the sheaf generated by this presheaf. We observe that $\mathfrak{J}_f(U) = (f)_U$ whenever $\mathcal{L}(U)$ is a free $\mathcal{O}_X(U)$-module of rank 1. If x is any point in such an admissible set U and if we take h_x to be the basis for $\mathcal{L}_x$, then $f_x = [f_U]_x h_x$, so that f_x can be represented by the element $[f_U]_x$ in $\mathcal{O}_{X,x}$. It is easily seen that the ideal $(f)_x$ generated by the element $[f_U]_x$ in $\mathcal{O}_{X,x}$ is independent of the choice of the element h as well as the choice of the admissible set U; in fact, the map $x \to (f)_x$ represents the sheaf space of $\mathfrak{J}_f$.

Consider the subspace $X_1 = X(\mathfrak{J}_f)$ in X defined by the sheaf $\mathfrak{J}_f$, and let $\psi = (\psi_0, \psi_1)$ be the injection morphism of X_1 into X. We recall that the homomorphism ψ_1 first carries $\mathcal{O}_X(U)$ into $\mathcal{O}_X(U)/\mathcal{J}_f(U)$ and then follows with the canonical homomorphism of $\mathcal{O}_X(U)/\mathcal{J}_f(U)$ into $\mathcal{O}_{X_1}(U)$, the structural sheaf $\mathcal{O}_{X_1}$ of X_1 being the sheaf generated by the presheaf $U \to \mathcal{O}_X(U)/\mathcal{J}_f(U)$. Let $\mathcal{L}_{X_1}$ be the lift to X_1 of $\mathcal{L}$ through the morphism ψ; the homomorphism ψ_1: $\mathcal{L} \to \mathcal{L}_{X_1}$ then induces a homomorphism of $\mathcal{L}^m(X)$ into $\mathcal{L}_{X_1}{}^m(X_1)$, which we shall also denote by ψ_1 for simplicity. On the other hand, the operation of multiplication by the element f induces a homomorphism of $\mathcal{L}^{m-1}(X)$ into $\mathcal{L}^m(X)$. We shall need the following lemma.

LEMMA 2. *Let $\mathcal{L}$ be an invertible sheaf on a G-ringed space X, and let f be an element in $\mathcal{L}(X)$. If $\mathcal{O}_X$ is Γ-closed and if f is nowhere a zero-divisor (i.e. f_x is not a zero-divisor for any point x in X), then the sheaf $\mathcal{J}_f$ is complete and the sequence*

$$\mathcal{L}^{m-1}(X) \xrightarrow{f} \mathcal{L}^m(X) \xrightarrow{\psi_1} \mathcal{L}_{X_1}{}^m(X_1)$$

is exact, where $X_1 = X(\mathcal{J}_f)$.

Proof. We first show that $\mathcal{J}_f$ is complete. Let g be an element in $\mathcal{O}_X(U)$, U being an admissible set, and assume that g_x is in $(f)_x$ for every point x in U. Then, for any point x in U, there is an element $h^{(x)}$ in $\mathcal{O}_{X,x}$ such that $g_x = h^{(x)}[f_U]_x$; this element $h^{(x)}$ is uniquely determined, since $[f_U]_x$ is not a zero-divisor. Further, there exist an admissible set V with $x \in V \subset U$ and an element h in $\Gamma(V, \mathcal{O}_X)$ with $h_x = h_x = h^{(x)}$ such that $g_y = h_y[f_U]_y$ for every point y in V. Since $[f_U]_y$ is not a zero-divisor in $\mathcal{O}_{X,y}$, it follows that $h_y = h^{(y)}$ for every point y in V. This shows that the map $x \to h^{(x)}$ is an element h in $\Gamma(U, \mathcal{O}_X)$, and we have the equation $g = h[f_U]$.

To prove the exactness of the sequence, let $U = \{U_i\}$ be an admissible covering in $\mathfrak{T}$ such that $\mathcal{L}(U_i)$ (and hence also $\mathcal{L}^m(U_i)$) is a free $\mathcal{O}_X(U_i)$-module of rank 1. For every i, we take a basis element b_i for $\mathcal{L}(U_i)$, and take $b_i{}^m$ to be the basis element for $\mathcal{L}^m(U_i)$. If g is an element in $\mathcal{L}^m(X)$, then g_{U_i} is represented by the element $[g_{U_i}]$ in $\mathcal{O}_X(U_i)$ defined by the equation $g_{U_i} = [g_{U_i}]b_i{}^m$. The homomorphism ψ_1: $\mathcal{L}^m(U_i) \to \mathcal{L}_{X_1}{}^m(U_{i,X_1})$ can then be represented by the homomorphism ψ_1: $\mathcal{O}_X(U_i) \to \mathcal{O}_X(U_i)/\mathcal{J}_f(U_i) = \mathcal{O}_X(U_i)/(f)_{U_i} \to \mathcal{O}_{X_1}(U_{i,X_1})$. Since $\mathcal{J}_f$ is complete, the second homomorphism here is injective; hence $\psi_1(g) = 0$ implies that $[g_{U_i}]$ is contained in the ideal $(f)_{U_i}$ for every i. There exists then an element h_i in $\mathcal{O}_X(U_i)$ such that $[g_{U_i}] = h_i[f_{U_i}]$. If U_j is another set in the covering U, then there are elements c_{ij} and c_{ji} in $\mathcal{O}_X(U_i \cap U_j)$ such that $b_i = c_{ij}b_j$, $b_j = c_{ji}b_i$, $c_{ij}c_{ji} = 1$ in $U_i \cap U_j$. Then we have the equations $[g_{U_i}] = c_{ij}{}^m[g_{U_j}]$ and $[f_{U_i}] = c_{ij}[f_{U_j}]$, and hence $h_i[f_{U_j}] = h_i c_{ji}[f_{U_i}] = c_{ji}[g_{U_i}] = c_{ij}{}^{m-1}[g_{U_j}] = c_{ij}{}^{m-1}h_j[f_{U_j}]$; since $[f_{U_j}]$ is not a zero-divisor, we have the equation $h_i = c_{ij}{}^{m-1}h_j$ in $U_i \cap U_j$. If we now assign to every U_i the element $h_i b_i{}^{m-1}$ in $\mathcal{L}^{m-1}(U_i)$, we obtain an element h in $\mathcal{L}^{m-1}(X)$ such that $g = hf$. This concludes the proof of our lemma.

For any invertible sheaf $\mathcal{L}$ on X, we denote by $\mathcal{L}^+(X) = \bigoplus_{m>0} \mathcal{L}^m(X)$ the graded ring of the sections of $\mathcal{L}^m$ over X, for all positive integers m. If $f_1, \ldots, f_s$ are homogeneous elements in $\mathcal{L}^+(X)$, we denote by $(f_1, \ldots, f_s)_U$ the ideal in $\mathcal{O}_X(U)$ generated by the ideals $(f_1)_U, \ldots, (f_s)_U$, U being an admissible set. As before, we denote by $\mathcal{J}(f_1, \ldots, f_s)$ the sheaf of ideals generated by the presheaf defined by the map $U \to (f_1, \ldots, f_s)_U$, and we denote by $(f_1, \ldots, f_s)_x$ the stalk of $\mathcal{J}(f_1, \ldots, f_s)$ at the point x. The subspace of X defined by this sheaf of ideals $\mathcal{J}(f_1, \ldots, f_s)$ will be denoted by $X(f_1, \ldots, f_s)$, or in short X_s (in case there is no ambiguity about the system $f_1, \ldots, f_s$ involved), and we note that at each point x in X_s its structural ring $\mathcal{O}_{X_s,x}$ is the residual ring $\mathcal{O}_{X,x}/(f_1, \ldots, f_s)_x$.

Definition 3. A system of homogeneous elements $f_1, \ldots, f_t$ in $\mathcal{L}^+(X)$ is said to be a *distinct system* (of length t) if for every i ($1 \leq i \leq t$) and for every point x in $X_{i-1} = X(f_1, \ldots, f_{i-1})$, the ideal $(f_i)_x$ is not a zero-divisor modulo the ideal $(f_1, \ldots, f_{i-1})_x$. A distinct system $f_1, \ldots, f_t$ is said to be *maximal* if X_t is empty.

It is easily seen that if $f_1, \ldots, f_t$ is a distinct system, then the r-dimension of $X(f_1, \ldots, f_t)$ is at most equal to r-dim $X - t$. It follows that if a distinct system has a length greater than the r-dimension of X, then it must be maximal. We note that if $f_1, \ldots, f_t$ is a maximal distinct system, then $(f_t)_x$ is the unit ideal for every point x in X_{t-1}. If the structural sheaf of X_{t-1} is Γ-closed, this implies that the restriction $f_t(X_{t-1})$ of f_t to X_{t-1} has an "inverse" element in $\mathcal{L}_{X_{t-1}}{}^{-m_0}(X_{t-1})$, if m_0 is the degree of f_t. Then the multiplication by the element $f_t(X_{t-1})^m$ induces an isomorphism of $\mathcal{O}_{X_{t-1}}(X_{t-1})$ onto $\mathcal{L}_{X_{t-1}}{}^{mm_0}(X_{t-1})$ for every positive integer m.

The following lemma gives an answer to the question posed after Lemma 1. In the proof we shall find it convenient to replace the structural sheaf $\mathcal{O}_{X_i}$ of each X_i by its Γ-closure $\tilde{\mathcal{O}}_{X_i}$. Although this extension of the structural sheaf will result in a corresponding extension of the sheaf $\mathcal{L}_{X_i}$, we shall for the sake of simplicity retain the same symbol for the so extended sheaf. We note that since the sheaf space of every X_i remains unchanged, a distinct system in the original $\mathcal{L}^+(X)$ remains such after the extension of $\mathcal{L}$ and its properties will not be essentially changed by the subsequent extensions of the other $\mathcal{L}_{X_i}$.

LEMMA 3. *Let X be a G-ringed space and let $\mathcal{L}$ be an invertible sheaf on X. If there exists a maximal distinct system $f_1, \ldots, f_{t+1}$ of length $t + 1$ in $\mathcal{L}^+(X)$, and if $\dim_k \tilde{\mathcal{O}}_{X_i}(X_i)$, $i = 1, \ldots, t$, are all finite, then there exist positive integers m_0 and C such that $\dim_k \mathcal{L}^{mm_0}(X) \leq Cm^t$ for all positive integers m.*

Proof. Let m_i be the degree of f_i, and set $m_0 = m_1 \cdots m_{t+1}$. We shall prove our lemma by showing that for every $i = 0, 1, \ldots, t$, we have a relation $\dim_k \mathcal{L}_{X_i}{}^{mm_0}(X_i) \leq C_i m^{t-i}$ for some positive integer C_i. For $i = 0$ this is the desired result with $C = C_0$, while for $i = t$ this follows from the above mentioned isomorphism between $\tilde{\mathcal{O}}_{X_t}(X_t)$ and $\mathcal{L}_{X_t}{}^{mm_0}(X_t)$, the isomorphism being induced by multiplication by $f_{t+1}(X_t)^{n_{t+1}}$, with $n_{t+1} = m_0/m_{t+1}$. We shall prove our relation by induction on the integer i in the reversed order from t to 0; that is, we assume its validity for $i + 1$ and shall show its validity for i. We set $n_i = m_0/m_i$, and denote by f_{i+1} the homomorphism of $\mathcal{L}_{X_i}{}^{(m-1)m_0}(X_i)$ into $\mathcal{L}_{X_i}{}^{mm_0}(X_i)$ obtained by multiplication with the element $f_{i+1}(X_i)^{n_i}$. We denote by $\psi^{(i+1)}$ the injection morphism of X_{i+1} into X_i; the homomorphism $\psi_1{}^{(i+1)}$: $\mathcal{L}_{X_i}{}^{mm_0} \to \mathcal{L}_{X_{i+1}}{}^{mm_0}$ then induces a homomorphism of $\mathcal{L}_{X_i}{}^{mm_0}(X_i)$ into $\mathcal{L}_{X_{i+1}}{}^{mm_0}(X_{i+1})$, which we shall denote also by $\psi_1{}^{(i+1)}$. By lemma 2, we have the exact sequence of k-modules

$$\mathcal{L}_{X_i}{}^{(m-1)m_0}(X_i) \xrightarrow{f_{i+1}} \mathcal{L}_{X_i}{}^{mm_0}(X_i) \xrightarrow{\psi_1{}^{(i+1)}} \mathcal{L}_{X_{i+1}}{}^{mm_0}(X_{i+1}),$$

and hence the inequality

$$\dim_k \mathcal{L}_{X_i}{}^{mm_0}(X_i) \leq \dim_k \mathcal{L}^{(m-1)m_0}(X_i) + \dim_k \mathcal{L}_{X_{i+1}}{}^{mm_0}(X_{i+1}).$$

Summing over all m down to 0, we obtain the inequality

$$\dim_k \mathcal{L}_{X_i}{}^{mm_0}(X_i) \leq \dim_k \tilde{\mathcal{O}}_{X_i}(X_i) + \sum_{j=1}^{m} \dim_k \mathcal{L}_{X_{i+1}}{}^{jm_0}(X_{i+1}),$$

and hence by induction hypothesis

$$\dim_k \mathfrak{L}_{X_i}{}^{mm_0}(X_i) \leq \dim_k \tilde{\mathfrak{O}}_{X_i}(X_i) + C_{i+1}m^{t-1} \leq C_i m^{t-i},$$

where we set $C_i = \dim_k \tilde{\mathfrak{O}}_{X_i}(X_i) + C_i$. This proves our assertion and hence the lemma.

III.

The next question is of course the existence of a distinct system. For this purpose we shall introduce a few finiteness properties about our space X. In fact, we have up to now assumed very little about our space X, and it is clear that we cannot expect to proceed much further without any additional assumptions. We therefore introduce the following definitions.

(1) An admissible set U in a G-ringed space is said to be *flat* if $\mathfrak{O}_x$ is a flat $\mathfrak{O}(U)_x$-module for every point x in U. An admissible covering of an admissible set in X is said to be *flat* if every set in the covering is flat. A G-ringed space X is said to be *flat* if every admissible covering of X admits a flat admissible refinement.

(2) If U is an admissible set in a G-ringed space X, then a subring R in $\mathfrak{O}(U)$ is said to be *quasi-Noetherian* if every complete ideal in R is the intersection of a finite number of primary ideals. If V is an admissible set in X which contains U, we shall denote by $\mathfrak{O}(V, U)$ the image ring of $\mathfrak{O}(V)$ in $\mathfrak{O}(U)$ under the homomorphism $\rho_U{}^V$. Then V is said to be *quasi-Noetherian relative to U* if the ring $\mathfrak{O}(V, U)$ is quasi-Noetherian. Let $\mathfrak{V} = \{V_j\}$ ($j \in J$, an indexing set) be an admissible covering of X, and let $\mathfrak{U} = \{\mathfrak{U}_i\}$ ($i \in I$) be an admissible refinement of $\mathfrak{V}$; we have then a map $\tau: I \to J$ such that $U_i \subset V_{\tau(i)}$ for every $i \in I$. We shall say that the admissible covering $\mathfrak{U}$ is a *quasi-Noetherian* refinement of $\mathfrak{V}$ if every $V_{\tau(i)}$ is quasi-Noetherian relative to U_i. The space X itself is said to be locally *quasi-Noetherian* if every admissible covering of X admits a quasi-Noetherian refinement.

We observe that if an admissible covering $\mathfrak{U}$ is a quasi-Noetherian refinement of an admissible covering $\mathfrak{V}$, then every admissible refinement of $\mathfrak{U}$ is also a quasi-Noetherian refinement of $\mathfrak{V}$. To prove this statement, we introduce another way to express the quasi-Noetherian property, whereby U and V are as before two admissible sets such that $U \subset V$. An ideal $\mathfrak{A}$ in $\mathfrak{O}(V)$ is said to be *U-complete* if it contains every element f in $\mathcal{O}(V)$ with the property that $f_x \in \mathfrak{A} \otimes \mathcal{O}_x$ for every point x in U. Then $\mathcal{O}(V, U)$ is quasi-Noetherian if and only if every U-complete ideal in $\mathcal{O}(V)$ is the intersection of a finite number of primary ideals. If W is an admissible set contained in U, then a W-complete ideal in $\mathcal{O}(V)$ is clearly also U-complete. It follows then that if V is quasi-Noetherian relative to U, then it is also such relative to W; this proves our assertion.

(3) A G-ringed space X is said to be *c-finite* (short for covering-finite) if every admissible covering of X admits a finite admissible refinement. This is of course the usual compactness definition for a topological space. However, in our present context where we have to deal with non-archimedean holomorphic spaces, it is advisable to give this property a different name. Similarly, the space X is said to be *c-countable,* if every admissible covering of X admits a countable admissible refinement.

LEMMA 4. *Let X be a G-ringed space which is flat, quasi-Noetherian, and c-finite (c-countable). Let $\mathcal{L}$ be an invertible sheaf on X, and let $\mathcal{J}$ be a quasi-coherent sheaf of ideals on X. Then there is a finite (countable) set of points in X such that an element in $\mathcal{L}^+(X)$ is a non-zero-divisor modulo $\mathcal{J}$ everywhere in X if and only if it is such at every point in this set.*

Proof. Let $\mathfrak{V} = \{V_j\}$ $(j \in J)$ be an admissible covering of X such that each $\mathcal{L}(V_j)$ is a free $\mathcal{O}(V_j)$-module of rank 1. Then, for every integer $m > 0$, each $\mathcal{L}^m(V_j)$ is also a free $\mathcal{O}(V_j)$-module of rank 1; hence, if f is an element in $\mathcal{L}^m(X)$, it can be represented in each V_j by an element $[f_{V_j}]$ in $\mathcal{O}(V_j)$, with respect to a module basis for $\mathcal{L}^m(V_j)$. If x is any point in V_j, then $[f_{V_j}]_x$ is a zero-divisor modulo $\mathcal{J}_x$ as an element in $\mathcal{O}_x$ if and only if f_x is a zero-divisor modulo $\mathcal{J}_x$ as an element in the $\mathcal{O}_x$-module $\mathcal{L}_x{}^m$. Since the space X is flat, we can assume that the covering $\mathfrak{V}$ is flat, replacing it by a refinement if necessary. Since X is quasi-Noetherian, there is a quasi-Noetherian refinement $\mathfrak{U} = \{U_i\}$ $(i \in I)$ of $\mathfrak{V}$, and since X is c-finite (c-countable), we can assume that $\mathfrak{U}$ is finite (countable). In view of the quasi-coherence of the sheaf $\mathcal{J}$, our lemma therefore will follow from the following assertion: If U and V are admissible sets in X such that $U \subset V$ and if V is flat and quasi-Noetherian relative to U, and if $\mathfrak{A}$ is an ideal in $\mathcal{O}(V)$, then there is a finite set of points x_i such that an element f in $\mathcal{O}(V)$ is a non-zero-divisor modulo $\mathfrak{A}$ everywhere in U if and only if f_{x_i} is such in $\mathcal{O}_{x_i}$ for every i.

To prove this assertion, consider the U-complete ideal $\tilde{\mathfrak{A}}$ generated by $\mathfrak{A}$ in $\mathcal{O}(V)$. Since $\mathcal{O}(V)$ is quasi-Noetherian relative to U, we have the representation $\tilde{\mathfrak{A}} = \bigcap^{s}{}_{i=1} \mathfrak{q}_i$, where the $\mathfrak{q}_i$ are primary ideals, and we can assume that the intersection on the right hand side is irredundant. Let $\mathfrak{p}_1, \ldots, \mathfrak{p}_s$ be the primary ideals belonging to $\mathfrak{q}_1, \ldots, \mathfrak{q}_s$ respectively. We maintain that, for every $i = 1, \ldots, s$, there is a point x_i in U such that every element in $\mathfrak{p}_i$ vanishes at x_i. In fact, assume that one of the $\mathfrak{p}_i$, say $\mathfrak{p}_1$, does not have this property, so that there is no point in U at which every element in $\mathfrak{p}_1$ vanishes. Then, for every point x in U, there is an element $h^{(x)}$ in $\mathfrak{p}_1$ which is a unit at x. If g is any element in the intersection $\bigcap_{i \neq 1} \mathfrak{q}_i$, then $g_x h_x{}^{(x)}$ is contained in $\tilde{\mathfrak{A}} \otimes \mathcal{O}_x$, and hence g_x must be contained in $\tilde{\mathfrak{A}} \otimes \mathcal{O}_x$, for every point x in U. Since $\tilde{\mathfrak{A}}$ is U-complete, this implies that g is contained in $\tilde{\mathfrak{A}}$; this would imply that $\mathfrak{q}_1$ is actually redundant in the intersection $\tilde{\mathfrak{A}} = \bigcap^{s}{}_{i=1} \mathfrak{q}_i$, in contradiction to our assumption. Therefore, there exists a set of points $x_1, \ldots, x_s$ such that every element in $\mathfrak{p}_i$ vanishes at x_i, and we shall show that this set has the desired property.

Let f be an element in $\mathcal{O}(V)$, and let $\mathfrak{n}$ be the ideal in $\mathcal{O}(V)$ consisting of all elements g such that gf is contained in $\tilde{\mathfrak{A}}$. Then we have the exact sequence of $\mathcal{O}(V)$-modules: $\mathfrak{n} \to \mathcal{O}(V) \xrightarrow{f} \mathcal{O}(V)/\tilde{\mathfrak{A}}$, where the first map is injection and the second map is induced by the multiplication by f followed by going over to the residue ring. Assume that f_{x_i} is not a zero-divisor modulo $\tilde{\mathfrak{A}} \otimes \mathcal{O}_{x_i}$ for any i, so that f is not contained in any $\mathfrak{p}_i$. Then every element in $\mathfrak{n}$ must be contained in every $\mathfrak{q}_i$, and hence $\mathfrak{n}$ is contained in $\tilde{\mathfrak{A}}$. Now, for every point x in U, $\mathcal{O}(V)_x$ is a flat $\mathcal{O}(V)$-module, by a well-known property of the quotient ring; and since V is flat, $\mathcal{O}_x$ is a flat $\mathcal{O}(V)_x$-module. It follows that $\mathcal{O}_x$ is a flat $\mathcal{O}(V)$-module, and hence the above exact sequence remains exact upon tensorisation by $\mathcal{O}_x$. Noting that $\mathcal{O}_x/\tilde{\mathfrak{A}} \otimes \mathcal{O}_x$ is canonically isomorphic to the tensorisation of $\mathcal{O}(V)/\tilde{\mathfrak{A}}$ by $\mathcal{O}_x$, we have then the exact sequence $\mathfrak{n} \otimes \mathcal{O}_x \to \mathcal{O}_x \xrightarrow{f_x} \mathcal{O}_x/\tilde{\mathfrak{A}} \otimes \mathcal{O}_x$, where the second homomorphism is induced by multiplication by f_x followed by going over to the residue ring. Since $\mathfrak{n} \otimes \mathcal{O}_x \in \tilde{\mathfrak{A}} \otimes \mathcal{O}_x$, it follows that f_x is not a zero-divisor modulo $\tilde{\mathfrak{A}} \otimes \mathcal{O}_x$. This concludes the proof of our lemma.

We are in a position to prove our main result.

THEOREM. *Let X be a flat, quasi-Noetherian, c-finite (c-countable if k is non-countable) G-ringed space which is immersed in a projective space, and assume that $\dim_k \tilde{\mathcal{O}}_{X(\mathfrak{J})}(X(\mathfrak{J}))$ is finite for every subspace defined by a coherent sheaf of ideals $\mathfrak{J}$ on X. Then X is algebraic.*

Proof. Let $\mathfrak{L}$ be the canonical invertible sheaf of the projective space S^N in which X is immersed by an immersion φ, and let $\mathfrak{L}_X$ be the lift of $\mathfrak{L}$ to X. In view of Lemma 1 and Lemma 3, we have only to show the existence of a maximal distinct system in $\mathfrak{L}_X{}^+(X)$ of length at most equal to r-dim $X + 1$. We shall construct such a maximal distinct system by induction, assuming that a distinct system $f_1, \ldots, f_i$ has already been constructed. If this system is already maximal, then we have nothing more to prove; we therefore assume that it is not maximal and proceed to extend it by another element f_{i+1}. Let x_j be the finite (countable) set of points given by Lemma 4, for the sheaf $\mathfrak{L}_X$ with reference to the ideal sheaf $\mathfrak{J}(f_1, \ldots, f_i)$, and consider the points $\psi_0(x_j)$ in S^N. We recall that the k-module $\mathfrak{L}(S)$ can be represented through a coordinate system in S^N by the system of all linear forms in N indeterminates. Since the ground field k is infinite (non-countable), there is at least one linear form g which does not vanish at any one of the finite (countable) set of points $\varphi_0(x_j)$. Let f_{i+1} be the element in $\mathfrak{L}_X(X)$ which is the image of g under φ_1; then f_{i+1} is a non-zero-divisor modulo $\mathfrak{J}(f_1, \ldots, f_i)$ at every point x_j and hence by Lemma 4 is a non-zero-divisor modulo $\mathfrak{J}(f_1, \ldots, f_i)$ everywhere in $X(f_1, \ldots, f_i)$. Then the system $f_1, \ldots, f_{i+1}$ is a distinct system in $\mathfrak{L}_X{}^+(X)$. Thus we can proceed until we reach a maximal system, but this must happen at the latest when the length becomes r-dim $X + 1$, for every distinct system of this length must be maximal.

COROLLARY. *If X is a closed holomorphic subspace in a projective space, either over the complex field or over a complete non-archimedean valued field, then it is algebraic.*

THE JOHNS HOPKINS UNIVERSITY

COMPLETE INTERSECTIONS IN CHARACTERISTIC $p > 0$

By Robin Hartshorne

A variety V of dimension r in a projective space $\mathbf{P}^n$ is a *set-theoretic complete intersection* if there are $n - r$ hypersurfaces $H_1, \ldots, H_{n-r}$ whose intersection (as subsets of $\mathbf{P}^n$) is V. For varieties of dimension ≥ 2 there are various necessary conditions to be a set-theoretic complete intersection. For example, V must be locally analytically connected in codimension 1 [3]. Thus a surface in $\mathbf{P}^4$ with an isolated double point with distinct tangent planes cannot be a set-theoretic complete intersection. If V is nonsingular and the groundfield is $\mathbf{C}$, then the Picard group of V must be finitely generated [4, p. 449]. So the quintic elliptic scroll in $\mathbf{P}_\mathbf{C}^4$ is not a set-theoretic complete intersection. Beyond this, there are necessary local and global conditions on the cohomology of V (using algebraic De Rham cohomology in characteristic zero, and étale $\mathbf{Z}/p\mathbf{Z}$ cohomology in characteristic $p > 0$) for V to be a set-theoretic complete intersection. These have been worked out in a series of papers [5, Ch. III], [6], [7]. However, these results say nothing in case V is a curve (except for the elementary result that V must be connected). The problem whether every irreducible (or even nonsingular) curve in $\mathbf{P}^3$ is a set-theoretic complete intersection remains open.

In this note we show that certain rational curves of degree $d \geq 4$ in $\mathbf{P}^3$ are set-theoretic complete intersections in all prime characteristics $p > 0$. However, even for $d = 4$ it is not known whether these curves are set-theoretic complete intersections in characteristic zero.

While the result given here is quite special, it gains interest in relation to some other recent results. Cowsik and Nori [1] have shown that every curve in *affine* n-space over a field of characteristic $p > 0$ is a set-theoretic complete intersection. Stagnaro [8] has given some examples of other rational quintic curves, not contained in a quadric surface like the ones here, which are set-theoretic complete intersections in characteristic 2. Finally Ferrand [2] has found a class of curves, including those here, which are set-theoretic complete intersections in every prime characteristic. Also he has given a nice conceptual proof of this result, and a new proof of the theorem of Cowsik and Nori.

For any $d \geq 4$, let C_d be the curve in $\mathbf{P}_k{}^3$ given parametrically by

$$x = tu^{d-1}$$
$$y = t^{d-1}u$$
$$z = t^d$$
$$w = u^d.$$

This is a nonsingular rational curve of degree d.

THEOREM.* *If k is an algebraically closed field of characteristic* $p > 0$, *then for any* $d \geq 4$, C_d *is a set-theoretic complete intersection in* $\mathbf{P}_k{}^3$.

Note: Of course for $d = 3$ this curve is the twisted cubic curve, which is a set-theoretic complete intersection in any characteristic, using e.g. the equations $y^2 = xz$ and $x^3 - 2xyw + zw^2 = 0$.

Proof of theorem. Given d and p, choose integers $n > 0$ and r such that, if $q = p^n$, then

$$\left(1 + \frac{1}{(d-1)}\right) q < r < \left(1 + \frac{1}{(d-2)}\right) q$$

and

$$r \leq \left(1 + \frac{1}{d-1}\right) q + 1.$$

This is possible, by choosing n large enough so that

$$\left(1 + \frac{1}{d-1}\right) q + 1 < \left(1 + \frac{1}{d-2}\right) q$$

for example, and then taking r to be the largest integer $\leq (1 + 1/(d-1))q + 1$.

---

*These calculations were first made on 6/30/64, and the result (for $d = 4$) was mentioned as Exercise 5.16, p. 125 in [5].

From these inequalities it follows that

$$r(d-1) - dq > 0 \tag{1}$$

$$q(d-1) - r(d-2) > 0 \tag{2}$$

$$r(d-1) - qd \leq d - 1. \tag{3}$$

Now we consider the two surfaces with equations

$$y^r = x^{r(d-1)-dq} z^q w^{q(d-1)-r(d-2)}$$

and

$$z^q w^{q(d-1)} - r y^{r-1} x^{dq-(r-1)(d-1)} w^{(r-1)(d-2)} + \cdots + (-1)^k \binom{r}{k} y^{r-k} x^{dq-(r-k)(d-1)} w^{(r-k)(d-2)} \pm \cdots + (-1)^r x^{dq} = 0.$$

First note these equations make sense, because all the exponents are nonnegative by choice of q and r. We will show that the intersection of these two surfaces is C_d. If $w = 0$, then $x = 0$ and $y = 0$, so there is only one point with $w = 0$, and that is on the curve C_d. Similarly, if $z = 0$, then $y = 0$ and $x = 0$, so there is only that one point, which is on C_d. Thus to show that C_d is the intersection of these two surfaces, we can set $w = 1$, $x = t$, and, assuming $t \neq 0$, it is sufficient to show that the only common solution of those two equations is $y = t^{d-1}$ and $z = t^d$.

After substituting $w = 1$ and $x = t$, multiply the second equation by $t^{r(d-1)-dq}$ and then replace the first term by y^r. It then becomes

$$y^r - r y^{r-1} t^{d-1} + \cdots + (-1)^k \binom{r}{k} y^{r-k} t^{k(d-1)} + \cdots + (-1)^r t^{r(d-1)} = 0.$$

This is simply

$$(y - t^{d-1})^r = 0,$$

so we conclude that $y = t^{d-1}$.

Then going back to the first equation, it says

$$t^{r(d-1)} = t^{r(d-1)-dq} z^q.$$

Since we are assuming $t \neq 0$, this implies

$$z^q - t^{dq} = 0.$$

But since $q = p^n$ is a power of the characteristic, this is just

$$(z - t^d)^q = 0.$$

So we conclude that $z = t^d$. Q.E.D.

Note. The curve C_d is obtained as an intersection of multiplicity rq. Thus, for fixed d, this multiplicity will grow with the characteristic p.

UNIVERSITY OF CALIFORNIA, BERKELEY

REFERENCES

[1] Cowsik, R. C., Nori, M. V., Affine curves in characteristic p are set-theoretic complete intersections. Invent. math. **45** (1978) 111-114.

[2] Ferrand, D., Set theoretical complete intersections in characteristic $p > 0$, in Algebraic Geometry, Copenhagen, Aug. 1978, ed. K. Lønsted, Lecture Notes in Math. Springer, to appear.

[3] Hartshorne, R. Complete intersections and connectedness, Amer. J. Math. **84** (1962) 497-508.

[4] ———, Cohomological dimension of algebraic varieties, Annals of Math., **88** (1968) 403-450.

[5] ———, Ample subvarieties of algebraic varieties, Lecture Notes in Math. **156,** Springer Verlag, Heidelberg (1970) xiii and 256 pp.

[6] ———, Speiser, R., Local cohomological dimension in characteristic p, Annals of Math. **105** (1977) 45-79.

[7] Ogus, A. Local cohomological dimension of algebraic varieties, Annals of Math. **98** (1973) 327-365.

[8] Stagnaro, E., On Vahlen's quintics in char. 2, 10-min. talk at ICM Helsinki (1978).

ON ZARISKI DIMENSIONALITY TYPE

By Heisuke Hironaka

Introduction. This article contains my own modest contribution to the beautiful work [Z] of Oscar Zariski on equisingular stratifications of algebraic hypersurfaces. When the theory of equisingular stratifications was first initiated by Zariski more than ten years ago, it appeared to be a rather experimental attempt on the topic, complete and excellent in small dimensions but somewhat uncomfortably artificial in higher dimensions and codimensions. Quite recently, however, a solid foundation was established all by Zariski himself for a general theory of equisingularity for hypersurfaces of any dimensions. The main role is played by the notion of *dimensionality type* of singularity, introduced and developed by Zariski, in his theory of equisingular stratification in which a stratum consists of the points of equal dimensionality type. The dimensionality type of an algebraic hypersurface at a point is defined by considering successive non-special projections of the ambient space to lower dimensional spaces and then by looking at the multiplicities at the corresponding points of the successive discriminant loci in those lower dimensional spaces. In order to make the notion intrinsic, or independent of the choice of coordinate system in the ambiant space, Zariski takes the "generic projection" $x = (x_1, \ldots, x_n) \to x^* = (x_1^*, \ldots, x_{n-1}^*)$ by means of independent generic formal power series

$$x_i^* = \sum_{A \in \mathbf{Z}_0^n \text{ with } |A| \geq 1} u_{iA} (x - \xi)^A, \qquad 1 \leq i \leq n-1,$$

with independent variables u_{iA} with respect to the given coordinate system $x = (x_1, \ldots, x_n)$ in the ambiant space and with respect to the given point ξ of the hypersurface. To apply the notion to the ultimate global stratification of the given hypersurface, we need the semi-continuity (in Zariski topology) of dimensionality types when the point ξ in question varies along any given subvariety of the hypersurface. The main (and sole) contribution of this article is a proof of this semi-continuity property of dimensionality types. The only difficulty in any proof is due to the fact that those variables x_i^* used in the definition are transcendental, or non-algebraic, with respect to the given variables x. Zariski himself proved that for a given point ξ there can be found an integer $H(\xi) > 0$ such that those u_{iA} with $|A| > H(\xi)$ may be all replaced by zero in x^* which are used to define projection and discriminant locus in regards to the definition of the dimensionality type at ξ. This is certainly useful in comparing the dimensionality types at any pair of points, and provides us a convincing evidence for the said semi-continuity. But there we must face the difficulty of proving that the number $H(\xi)$ (chosen to be the smallest for each ξ, for instance) is uniformly bounded when ξ varies along any given subvariety. In this article, we prove the semi-continuity rather directly from the definition of dimensionality type by means of "generic projections" as above. Incidentally, the uniform boundedness of $H(\xi)$ can be rather easily deduced from the semi-continuity as well as it is read out from the proof of the semi-continuity. The question of semi-continuity naturally carries a certain similarity to Oka's coherency theorem in the theory of several complex variables. As a matter of fact, just like in the proof of Oka's coherency theorem, Weierstrass preparation theorem plays the key role. From the technical point of view, the essence of this work is in fact to formulate, to prove and to make good use of Weierstrass theorem for elements of the ring $K[x][[u]]$ viewed as a subring of the formal power series ring $K((u))[[x]]$, where K is a suitable base field.

§1. Completion of topological rings. Let $\mathbf{A}$ be a commutative ring with unity. Let F be a set of ideals in $\mathbf{A}$. F is said to be *directed* if for every pair $I, I' \in F$ there exists $I'' \in F$ such that $I'' \subset I \cap I'$. If F is directed, we obtain a *projective system* of rings $\mathbf{A}/I$, $I \in F$, with respect to the natural homomorphisms $\mathbf{A}/I \to \mathbf{A}/I'$ for pairs $I, I' \in F$ such that $I' \supset I$. Thus we obtain the *projective limit* ring:

$$\mathbf{A}_F = \varprojlim_{I \in F} \mathbf{A}/I.$$

A directed set F of ideals in $\mathbf{A}$ defines a topology in $\mathbf{A}$ such that for every $a \in \mathbf{A}$, $\{a + I, I \in F\}$ forms a basis of the open neighborhoods of a in the topology. With such a topology, $\mathbf{A}$ becomes a topological ring. Let $\mathbf{A}, \mathbf{A}'$ be two rings with directed sets of ideals F, F' respectively. Then a homomorphism $\varphi: \mathbf{A} \to \mathbf{A}'$ is continuous with respect to the topologies in $\mathbf{A}, \mathbf{A}'$ defined by F, F' respectively, if and only if for every $I' \in F'$ there exists $I \in F$ such that $I \subset \varphi^{-1}(I')$. If this is the case, then φ induces a homomorphism of the completions $\varphi: \mathbf{A}_F \to \mathbf{A}'_{F'}$.

Let $u = \{u_s\}_{s \in S}$ be a set of elements $u_s \in \mathbf{A}$. We then let Δ be the set of monomials in u (each monomial, by definition, involves only a finite number of u_s, $s \in S$). Then for every finite subset δ of Δ, we obtain an ideal I_δ in $\mathbf{A}$ which is generated by the monomials belonging to $\Delta \setminus \delta$. The set F of the ideals I_δ for all the finite subsets δ of Δ, is then directed and defines a topology in $\mathbf{A}$. In this manner, any set u of elements in $\mathbf{A}$ gives rise to a definite notion of completion $\mathbf{A}_F$ of $\mathbf{A}$. We shall call $\mathbf{A}_F$ the *completion of* $\mathbf{A}$ *with respect to the subset* u of $\mathbf{A}$. For instance, let $u = \{u_s\}_{s \in S}$ be a set of indeterminates (not necessarily finite) over a ring $\mathbf{B}$ (commutative with unity, as always throughout this paper), and let $\mathbf{A} = \mathbf{B}[u]$, the ring of polynomials in u with coefficients in $\mathbf{B}$ (each involving only finitely many u_s, $s \in S$). Then the completion of $\mathbf{A}$ with respect to u will be denoted by $\mathbf{B}[[u]]$.

Remark (1.1). We can have two different sets of indeterminates, say u and v, over $\mathbf{B}$ which are related by $\mathbf{B}[u] = \mathbf{B}[v]$ and $(u)\mathbf{B}[u] = (v)\mathbf{B}[v]$. If u and v are finite sets (so that they have the same cardinality), then u and v define the same topology in $\mathbf{A} = \mathbf{B}[u] = \mathbf{B}[v]$ and the same completion $\mathbf{B}[[u]] = \mathbf{B}[[v]]$. But if u and v are infinite, then the topologies and the completions may differ from each other. For example, let $u = \{u_j\}_{j \in \mathbf{Z}_+}$ ($\mathbf{Z}$ denotes the set of positive integers) be a given set of indeterminates over B. Let us set $v_{r,i} = u_i - u_{i-r}$ for i, $r \in \mathbf{Z}_+$ where we use the convention $u_j = 0$ for all $j \leq 0$. Let $v_r = \{v_{r,i}\}_{i \in \mathbf{Z}_+}$. For each $r \in \mathbf{Z}_+$, we have $\mathbf{B}[u] = \mathbf{B}[v_r]$ and $(u)\mathbf{B}[u] = (v_r)\mathbf{B}[v_r]$. Let T_r be the topology in $\mathbf{A} = \mathbf{B}[u] = \mathbf{B}[v_r]$ defined by the set v_r. Then for r, $r' \in \mathbf{Z}_+$, we can prove that $T_{r'}$ is weaker (strictly) than T_r if and only if r' divides r (and $\neq r$). Moreover the topology in $\mathbf{A}$ defined by u is strictly weaker than any of those T_r, $r \in \mathbf{Z}_+$. For instance, therefore, we have a natural (continuous) homomorphism φ_r from $\mathbf{B}[[v_r]]$ to $\mathbf{B}[[u]]$ which is surjective but not injective. Ker(φ_r) is generated by

$$v_{r,j} + v_{r,j+r} + v_{r,j+2r} + \cdots$$

for $1 \leq j \leq r$.

Remark (1.2). Let $\mathbf{B}$ and u be the same as in (1.1). Let us assign a positive integer $\omega(u_i)$ to each indeterminate u_i of the set u, $i \in \mathbf{Z}_+$, in such a way that $\lim_{i\to\infty} \omega(u_i) = \infty$. Then we can extend ω to a map $\mathbf{B}[u] \to \mathbf{Z}_+ \cup \{\infty\}$ (also denoted by ω) as follows. For each monomial in $\mathbf{B}[u]$, say

$$P = \prod_i u_i^{a_i}, \qquad a_i \in \mathbf{Z}_0$$

we define $\omega(P) = \sum_i a_i\omega(u_i)$. For each polynomial

$$f = \sum_\alpha b_\alpha P_\alpha, \qquad b_\alpha \in \mathbf{B}$$

with distinct monomials P_α, we define

$$\omega(f) = \min\{\omega(P_\alpha) \mid b_\alpha \neq 0\} \qquad (= \infty \text{ if } f = 0).$$

This ω is a *quasi-valuation* in $\mathbf{A} = \mathbf{B}[u]$ in the sense that we have

(i) $\omega(fg) \geq \omega(f) + \omega(g)$

(ii) $\omega(f + g) \geq \min\{\omega(f), \omega(g)\}$

for all f and $g \in \mathbf{A}$. Now, for each integer $i \in \mathbf{Z}_0$, we have an ideal $I(i) = \{f \in \mathbf{A} \mid \omega(f) \geq i\}$ in $\mathbf{A}$. The set of ideals $F = \{I(i) \mid i \in \mathbf{Z}_0\}$ is directed and hence defines a topology T_ω in $\mathbf{A}$. This T_ω is then equal to the topology in $\mathbf{A}$ defined by the set u, which is in fact independent of the choice of ω as above. Thus $\mathbf{B}[[u]]$ is the completion of $\mathbf{A}$ with respect to T_ω. Moreover ω extends uniquely to a quasi-valuation in the completion $\mathbf{B}[[u]]$.

Remark (1.3). Let $\mathbf{B}[[u]]$ and ω be the same as in (1.2). Let $x = (x_1, \ldots, x_n)$ be a finite set of indeterminates over $\mathbf{B}[[u]]$. Then we can easily prove that

(i) $\mathbf{B}[[x]][[u]]$ is canonically isomorphic to $\mathbf{B}[[u]][[x]]$ and hence $\mathbf{B}[x][[u]]$ can be identified as a subring of $\mathbf{R}[[x]]$ where $\mathbf{R} = \mathbf{B}[[u]]$

(ii) we have

$$\mathbf{B}[x][[u]] = \left\{ \sum_{A\in\mathbf{Z}_0{}^n} r_A x^A \,\middle|\, r_A \in \mathbf{R}, \qquad \lim_{|A|\to\infty} \omega(r_A) = \infty \right\}$$

as a subring of $\mathbf{R}[[x]]$.

§2. Weierstrass Division Theorem. Let $\mathbf{R}$ be a commutative ring with unity. Let $v: \mathbf{R} \to \mathbf{Z}_0 \cup \{\infty\}$ be a *quasi-valuation* in $\mathbf{R}$ in the sense of (1.2).

Let us define $I(\nu) = \{b \in \mathbf{R} \mid v(b) \geq \nu\}$ for every $\nu \in \mathbf{Z}_0$. In this section, we assume

(2.1) $\mathbf{R}$ is v-complete for the quasi-valuation v in $\mathbf{R}$, i.e., the canonical homomorphism

$$\mathbf{R} \to \varprojlim_{\nu} \mathbf{R}/I(\nu)$$

is an isomorphism.

Let us take a system of n indeterminates $x = (x_1, \ldots, x_n)$ over $\mathbf{R}$, and pick a *weight function* $e: \mathbf{Z}_0{}^n \to \mathbb{R}_0$ $(= \{r \in \mathbb{R} \mid r \geq 0\})$ which simply means a linear function

$$e(u_1, \ldots, u_n) = \sum_{i=1}^{n} e_i u_i$$

with non-negative real numbers $e_i \in \mathbb{R}_0$, $1 \leq i < n$. We do not exclude the case in which $e_i = 0$ for all i. Given v and e as above, we obtain a *quasi-valuation* ω in the formal power series ring $\mathbf{R}[[x]]$ as follows. For

$$f = \sum_{A \in \mathbf{Z}_0{}^n} c_A x^A, \qquad c_A \in \mathbf{R}$$

we define

$$\omega(f) = \min_{A \in \mathbf{Z}_0{}^n} \left\{ v(c_A) + e(A) \right\}.$$

Here it should be noted that if $c_A = 0$ then $v(c_A) + e(A) = \infty$, and that by (2.1) we have $\omega(f) = \infty$ if and only if $f = 0$.

With reference to ω (i.e., to v and e), we define

$$\mathbf{R}\langle x \rangle^{\omega} = \left\{ f \in \mathbf{R}[[x]] \,\middle|\, \lim_{|A| \to \infty} \omega(c_A x^A) = \infty \right\}$$

which is clearly an $\mathbf{R}$-subalgebra of $\mathbf{R}[[x]]$. We have two special cases as follows.

Example (a). Let $\mathbf{R}$ be any commutative ring with unity and let $v(a) = 0$ for $a \in \mathbf{R} \setminus \{0\}$ while $v(0) = \infty$. Assume that the weight function is *positive*, i.e., $e_i > 0$ for every i. In this case, we have $\mathbf{R}\langle x \rangle^{\omega} = \mathbf{R}[[x]]$.

Example (b). Let $e = 0$. Then

$$\mathbf{R}\langle x \rangle^{\omega} = \left\{ f \in \mathbf{R}[[x]] \,\middle|\, \lim_{|A| \to \infty} v(c_A) = \infty \right\}$$

which is strictly smaller than $\mathbf{R}[[x]]$ (unless $\mathbf{R} = (0)$). In this case, we write $\mathbf{R}\langle x \rangle^{v}$ for $\mathbf{R}\langle x \rangle^{\omega}$.

Let us consider a finite system of non-zero elements $f_1, \ldots, f_m$ in $\mathbf{R}\langle x \rangle^{\omega}$ and, for each i, let us pick $p_i \in \mathbf{R}[x]$ (a polynomial) such that

(2.2) $\omega(f_i) = \omega(p_i) < \omega(h_i)$ where $h_i = f_i - p_i$, $1 \leq i \leq m$. Let E be a subset of $\mathbf{Z}_0{}^n$ such that

(2.3) for every $A \in E$, there exist $q_{Ai} \in \mathbf{R}[x]$ and $r_A \in \mathbf{R}[x]$ such that

$$x^A = \sum_{i=1}^{m} q_{Ai} p_i + r_A$$

where 1) $\omega(q_{Ai}) \geq \omega(x^A) - \omega(p_i)$ for all i, 2) r_A has no term with exponents in E, i.e.,

$$r_A \in \sum_{B \in \mathbf{Z}_0{}^n \setminus E} \mathbf{R} x^B$$

Remark (a). Consider the situation in Example (a). Let us assume that for each i there exists $A_i \in \mathbf{Z}_0{}^n$ such that

$$f_i = c_i x^{A_i} + h_i$$

where c_i is a unit in $\mathbf{R}$ and $\omega(h_i) > e(A_i)$. Then let $p_i = c_i x^{A_i}$, $1 \leq i \leq m$, and let

$$E = \bigcup_{i=1}^{m} A_i + \mathbf{Z}_0{}^n$$

then the conditions (2.2) and (2.3) are satisfied.

Remark (b). Consider the situation in Example (b). Pick any $p_i \in \mathbf{R}[x]$ satisfying (2.2) for each i. For instance, let

$$f_i = \sum_{A \in \mathbf{Z}_0{}^n} c_{iA} x^A \quad \text{with} \quad c_{iA} \in \mathbf{R}$$

and let

$$p_i = \sum_{\nu(c_{iA})=0} c_{iA} x^A.$$

Then let G be any subset of $\mathbf{Z}_0{}^n$ such that

$$\mathbf{R}[x]/(p_1, \ldots, p_m)\mathbf{R}[x]$$

is generated as $\mathbf{R}$-module by the images of those monomials x^A, $A \in G$. Then $E = \mathbf{Z}_0{}^n \setminus G$ satisfies the condition (2.3).

Remark (c). In Remark (b), let us consider the special case in which $m = 1$ and p_1 is a monic polynomial of degree $d > 0$ in x_n (with coefficients in $\mathbf{R}[x_1, \ldots, x_{n-1}]$). Then

$$E = \{(a_1, \ldots, a_n) \in \mathbf{Z}_0{}^n \mid a_n \geq d\}$$

satisfies the condition (2.3).

PROPOSITION 1. *Let* $\mathbf{R}$, ν, e, ω, $(f_1, \ldots, f_m)$, $f_i \in \mathbf{R}\langle x \rangle^\omega$, *and* E *satisfy the conditions* (2.1), (2.2) *and* (2.3). *Then for every* $g \in \mathbf{R}\langle x \rangle^\omega$ *there exist* $g_i \in \mathbf{R}\langle x \rangle^\omega$, $1 \leq i \leq m$, *and* $r \in \mathbf{R}\langle x \rangle^\omega$ such that

$$g = \sum_{i=1}^{m} g_i f_i + r$$

where if $r = \sum b_A x^A$ *with* $b_A \in \mathbf{R}$ *then* $b_A = 0$ *for all* $A \in E$.

Proof. Let us write

$$g = \sum_{A \in \mathbf{Z}_0{}^n} a_A x^A.$$

Let us pick and fix q_{Ai} and r_A for $A \in E$ as in (2.3). Let us define

$$\sigma_i(g) = \sum_{A \in E} a_A q_{Ai}$$

$$\rho(g) = \sum_{A \in \mathbf{Z}_0{}^n \setminus E} a_A x^A + \sum_{A \in E} a_A r_A.$$

Since $\lim_{|A| \to \infty} (\nu(a_A) + \omega(x^A)) = \infty$ and $\omega(q_{Ai}) \geq \omega(x^A) - \omega(p_i)$, we have

$$\lim_{|A| \to \infty} (\omega(a_A q_{Ai})) = \infty.$$

It follows that $\sigma_i(g) \in \mathbf{R}\langle x \rangle^\omega$ for every i. Moreover

$$g = \sum_{i=1}^{m} \sigma_i(g) p_i + \rho(g)$$

by the equalities of (2.3). It follows that $\rho(g) \in \mathbf{R}\langle x \rangle^\omega$. Let us also define

$$s(g) = -\sum_{i=1}^{m} \sigma_i(g) h_i$$

with those h_i defined in (2.2). Then we have

$$g = \sum_{i=1}^{m} \sigma_i(g) f_i + \rho(g) + s(g).$$

Let us remark that

(2.4) $\sigma_i(g)$ is *additive* in g, i.e., $\sigma_i(g + g') = \sigma_i(g) + \sigma_i(g')$, and

$$\omega(\sigma_i(g)) \geq \omega(g) - \omega(p_i)$$

(2.5) $\rho(g)$ is additive in g, and

(2.6) $s(g)$ is additive in g and

$$\omega(s(g)) \geq \omega(g) + \alpha$$

where $\alpha = \min_{l \leq i \leq m} \{\omega(h_i) - \omega(p_i)\}$ which is positive by (2.2).

In fact, the additivity is clear from the definition in all cases.

$$\begin{aligned} \omega(g) &= \min\{\omega(a_A x^A) \mid A \in \mathbf{Z}_0{}^n\} \\ &\leq \min\{\omega(a_A x^A) \mid A \in E\}. \end{aligned}$$

But $\omega(a_A x^A) = v(a_A) + \omega(x^A) \leq v(a_A) + \omega(q_{Ai}) - \omega(p_i) \leq \omega(a_A q_{Ai}) - \omega(p_i)$. Hence

$$\begin{aligned} \omega(g) &\leq \min\{\omega(a_A q_{Ai}) \mid A \in E\} - \omega(p_i) \\ &\leq \omega(\sigma_i(g)) - \omega(p_i) \end{aligned}$$

which proves (2.4). The inequality of (2.5) follows (2.4) by the equality

$$\rho(g) = g - \sum_{i=1}^{m} \sigma_i(g) p_i.$$

Finally

$$\begin{aligned} \omega(s(g)) &\geq \min\{\omega(\sigma_i(g)) + \omega(h_i) \mid 1 \leq i \leq m\} \\ &\geq \min\{\omega(g) - \omega(p_i) + \omega(h_i) \mid 1 \leq i \leq m\} \\ &\geq \omega(g) + \alpha, \text{ which proves (2.6).} \end{aligned}$$

Let us denote by $s^d(g)$ the result obtained by applying s d-times to g. Then (2.6) implies

$$\omega(s^d(g)) \geq \omega(g) + d\alpha \qquad \text{for all } d \in \mathbf{Z}_0.$$

Since $\alpha > 0$, the series $g + s(g) + s^2(g) + \cdots + s^d(g) + \cdots$ is convergent in $\mathbf{R}\langle x \rangle^\omega$. ($\mathbf{R}\langle x \rangle^\omega$ is ω-complete.) Call this infinite sum $\tilde{g}$.

Then, thanks to (2.4) and (2.5),

$$\sigma_i(\tilde{g}) = \sigma_i(g) + \sigma_i(s(g)) + \cdots$$

$$\rho(\tilde{g}) = \rho(g) + \rho(s(g)) + \cdots$$

which converge in $\mathbf{R}\langle x\rangle$. It is then easy to check that

$$g = \sum_{i=1}^{m} \sigma_i(\tilde{g}) f_i + \rho(\tilde{g}).$$

By the definition of $\rho(\tilde{g})$ and by 2) of (2.3), $\rho(\tilde{g})$ has the property of r as is claimed in Proposition 1. So with $g_i = \sigma_i(\tilde{g})$, $1 \le i \le m$, and $r = \rho(\tilde{g})$, the assertion of Proposition 1 is verified.

§3. Applications of Weierstrass Theorem. Throughout this section, we will have the following data. K will be any field and $x = (x_1, \ldots, x_n)$ will be a finite system of indeterminates over K. We then consider a system of countably many indeterminates u over $K[x]$. We let $\mathbf{R} = K[[u]]$. The topology of $\mathbf{R}$ is defined by a valuation v in $\mathbf{R}$ (with values in $\mathbf{Z}_0 \cup \{\infty\}$), which is obtained by assigning value 0 to every non-zero element in K and a suitable positive value to each indeterminate in the system u as we did in Remark (1.2) of §1. We fix such a valuation v throughout this section. We shall write $\mathbf{R}\langle x\rangle$ for $K[x][[u]]$ which is viewed as an $\mathbf{R}$-algebra in a natural way. Recall that we have

$$\mathbf{R}\langle x\rangle = \left\{ \sum a_A x^A \,|\, a_A \in \mathbf{R}, \lim_{|A|\to\infty} v(a_A) = \infty \right\}$$

(cf. Ex. (b), §2, in which $\mathbf{R}\langle x\rangle = \mathbf{R}\langle x\rangle^v$.) $K[x]$ is canonically identified as the homomorphic image of $\mathbf{R}\langle x\rangle$ whose kernel is the closure of $\max(\mathbf{R})\mathbf{R}\langle x\rangle$ by means of the canonical inclusion $K[x] \to \mathbf{R}\langle x\rangle$. We shall be interested in some applications of Weierstrass Theorem of §2 to an element $f \in \mathbf{R}\langle x\rangle$ and to a system of elements $y = (y_1, \ldots, y_n)$, $y_i \in \mathbf{R}\langle x\rangle$.

Example (3.1). In the later application to questions of dimensionality type, we shall have the following situation. Let $z = (z_1, \ldots, z_n)$ be a system of indeterminates over a field k. We then take a system of indeterminates over $k[z]$, which are named as follows: $t = (t_1, \ldots, t_n)$, $s = (s_{ij})$, $1 \leq i \leq n - 1$, $1 \leq j \leq n$, $r = (r_{iA})$, $1 \leq i \leq n - 1$, $A \in \mathbf{Z}_0{}^n$, $1 < |A| \leq N$, $u = (u_{iA})$, $1 \leq i \leq n - 1$, $A \in \mathbf{Z}_0{}^n$, $|A| > N$, where N is a fixed positive integer. We then pick

$$K = \text{any field containing } k(t, s, r)$$

$$x_n = y_n = z_n, \text{ and for } 1 \leq i \leq n - 1,$$

$$x_i = t_i{}^q + \sum_{j=1}^{n} s_{ij}{}^q z_j + \sum_A r_{iA}{}^q z^A, \qquad 1 < |A| \leq N,$$

$$y_i = x_i + \sum_A u_{iA}{}^q z^A, \qquad |A| > N$$

where q is a fixed power of the characteristic p of k if $p > 0$ but $q = 1$ if $p = 0$. The valuation ν in $\mathbf{R} = K[[u]]$ is defined as follows: For a monomial M in u, say

$$M = (u_{i_1 A_1})^{e_1} \cdots (u_{i_d A_d})^{e_d}$$

we let

$$\nu(M) = \sum_{j=1}^{d} e_j |A_j|,$$

and for $\varphi = \Sigma_M a_M M$ with $a_M \in K$ we let

$$\nu(\varphi) = \{\min \nu(M) \mid a_M \neq 0\}.$$

Note that for any integer $b > 0$, there are only a finite number of M such that $\nu(M) < b$, that $\mathbf{R}$ is the completion of $K[u]$ with respect to the restriction ν_0 of ν in $K[u]$, and that ν is the canonical extension of ν_0 by completion.

Now, back to the general set-up of this section, if y is a finite system of elements in $\mathbf{R}\langle x \rangle$, then we denote by $\mathbf{R}\langle y \rangle$ the subring of $\mathbf{R}\langle x \rangle$ consisting of those elements in $\mathbf{R}[[x]]$ which can be expressed as where $y = (y_1, \ldots, y_m)$, $c_A \in \mathbf{R}$ and $\lim_{|A| \to \infty} v(c_A) = \infty$. If we take a system of indeterminates $Y = (Y_1, \ldots, Y_m)$ over $\mathbf{R}$, then there exists a unique $\mathbf{R}$-homomorphism (continuous)

$$\lambda: \mathbf{R}\langle Y \rangle \to \mathbf{R}\langle x \rangle$$

which maps Y_i to y_i for all i. $\mathbf{R}\langle y \rangle$ is then the image of λ.

$\mathbf{R}$ is a local ring whose residue field is canonically isomorphic to K. The maximal ideal is not generated by u but is the closure of the ideal $(u)\mathbf{R}$ in the topology of $\mathbf{R}$. We shall denote bt $((*))\mathbf{R}$ the closure of the ideal generated by $(*)$, so that for instance $\max(\mathbf{R}) = ((u))\mathbf{R}$.

In what follows, we shall often identify

$$\mathbf{R}\langle x \rangle / ((u))\mathbf{R}\langle x \rangle = K[x]$$

by the natural inclusion $K[x] \subset \mathbf{R}\langle x \rangle$. In this sense, for $f \in \mathbf{R}\langle x \rangle$, we shall denote by $\overline{f} \in K[x]$ the image of $[f \bmod ((u))\mathbf{R}\langle x \rangle]$.

Proposition 1. *Let* $y = (y_1, \ldots, y_m)$ *with* $y_i \in \mathbf{R}\langle x \rangle$, *we have* $\mathbf{R}\langle y \rangle = \mathbf{R}\langle x \rangle$ *if and only if* $K[\overline{y}] = K[x]$.

Proof. The "only if" is clear. Conversely, assume $k[\overline{y}] = k[x]$. Let us introduce a system of indeterminates $Y = (Y_1, \ldots, Y_m)$ over $\mathbf{R}[[x]]$. Let

$$f_i = Y_i - y_i \in R\langle Y, x \rangle$$
$$p_i = Y_i - \overline{y}_i \in K[Y, x], \qquad 1 \leq i \leq m.$$

We extend v to a valuation ω in $\mathbf{R}[[Y, x]]$ by zero weight function. (Cf. Example (b) of §2.) Let

$$E = \mathbf{Z}_0{}^n \times \mathbf{Z}_0{}^n \setminus \mathbf{Z}_0{}^n \times (0).$$

We want to show that the conditions (2.2) and (2.3) of §2 are satisfied and hence that Proposition 1 of §2 (Weierstrass Theorem) is applicable to our situation. We have $h_i = f_i - p_i \in ((u))\mathbf{R}\langle x \rangle$ and hence

$$\omega(h_i) > 0 = \omega(p_i) \text{ for all } i$$

which proves (2.2). Take any $A \times B \in E$ with $A, B \in \mathbf{Z}_0{}^n$. Since $K[\overline{y}]$

$= K[x]$, we have a polynomial φ with coefficients in K such that $x^B = \varphi(\bar{y})$. Then

$$Y^A x^B - Y^A \varphi(Y) = Y^A(\varphi(\bar{y}) - \varphi(Y))$$

which belongs to the ideal $(p_1, \ldots, p_m)L[Y, x]$. In other words, there exist $q_{A\times B,i} \in K[Y, x]$ such that

$$Y^A x^B = \sum_{i=1}^{m} q_{A\times B,i}\, p_i + Y^A \varphi(Y)$$

which proves the condition (2.3). Now, applying the Weierstrass Theorem (Prop. 1, §2), we obtain that for every $g \in \mathbf{R}\langle Y, x\rangle$, there exist $g_i \in \mathbf{R}\langle Y, x\rangle$, $1 \leq i \leq m$, such that if

$$g - \sum_{i=1}^{m} g_i f_i = \sum_{A, B \in \mathbf{Z}_0{}^n} b_{A,B} Y^A x^B$$

with $b_{A,B} \in \mathbf{R}$, then $b_{A,B} = 0$ for all $A \times B \in E$. This means that

$$g - \sum_{i=1}^{m} g_i f_i \in \mathbf{R}\langle Y\rangle.$$

In particular, for every $g \in \mathbf{R}\langle x\rangle$, there exists $g^* \in \mathbf{R}\langle Y\rangle$ such that

$$g - g^* \in (f_1, \ldots, f_m)\mathbf{R}\langle Y, x\rangle.$$

In other words, the homomorphism λ: $\mathbf{R}\langle Y\rangle \rightarrow \mathbf{R}\langle x\rangle$ by $\lambda(Y_i) = y_i$, $1 \leq i \leq m$, is surjective. Namely, $\mathbf{R}\langle y\rangle = \mathbf{R}\langle x\rangle$. Q.E.D.

COROLLARY 1. *For every* $\xi = (\xi_1, \ldots, \xi_n) \in (K[[u]])^n$, *we have*

$$\mathbf{R}\langle x\rangle = \mathbf{R}\langle x - \xi\rangle.$$

COROLLARY 2. *Let us consider the case of* $N = 1$ *in Example* (3.1) *in which* $\mathbf{R}$, x *and* y *are defined. Then we have* $\mathbf{R}\langle x\rangle = \mathbf{R}\langle z\rangle = \mathbf{R}\langle y\rangle$.

In the next proposition we consider a slightly more general situation than the one described in the beginning of this section. Namely we let $\tilde{\mathbf{R}}$

be any local ring (not necessarily noetherian) with a quasi-valuation ν as in §2 such that $I(1) = \max(\tilde{\mathbf{R}})$ and $\tilde{\mathbf{R}}$ is ν-complete in the sense of (2.1), §2. Moreover we assume that $\tilde{\mathbf{R}}$ contain a field K which is canonically isomorphic to the residue field $\tilde{\mathbf{R}}/I(1)$. (Refer to the beginning paragraph of §2 for the definition of the ideals $I(\nu)$.) We are interested in the cases of $\tilde{\mathbf{R}} = \mathbf{R}$ (of this section) or $= \mathbf{R}/I$ with some closed ideal I in $\mathbf{R}$.

PROPOSITION 2. *Let $f \in \tilde{\mathbf{R}}\langle x\rangle$ and let $\bar{f}$ be the image of f in $K[x]$, where $\tilde{\mathbf{R}}\langle x\rangle/((max(\tilde{\mathbf{R}}))\tilde{\mathbf{R}}\langle x\rangle$ is canonically identified with $K[x]$. Suppose that $\bar{f}$ is a monic polynomial of degree $d > 0$ in x_n with coefficients in $K[x_1, \ldots, x_{n-1}]$. Then there exists a monic polynomial P of x_n of degree d with coefficients in $\tilde{\mathbf{R}}\langle x_1, \ldots, x_{n-1}\rangle$ such that $f = UP$ with an invertible element U in $\tilde{\mathbf{R}}\langle x\rangle$. This P is uniquely determined by f.*

Proof. The Weierstrass Theorem, Prop. 1 of §2, applies to the case in which $m = 1, f = f_1, p_1 = \bar{f}$ and

$$E = \{(a_1, \ldots, a_n) \in \mathbf{Z}_0{}^n \,|\, a_n \geq d\}$$

(cf. Remark (C), §2). Therefore for every $g \in \tilde{\mathbf{R}}\langle x\rangle$ there exists $g_1 \in \tilde{\mathbf{R}}\langle x\rangle$ such that

$$g - g_1 f = \sum_{A \in \mathbf{Z}_0{}^n} c_A x^A \in \tilde{\mathbf{R}}\langle x\rangle$$

where $c_A = 0$ for all $A \in E$. In other words,

$$g - g_1 f \in \sum_{0 \leq i \leq d-1} x_n{}^i\, \tilde{\mathbf{R}}\langle x_1, \ldots, x_{n-1}\rangle.$$

Take $g = x_n{}^d$. Then

$$g_1 f = x_n{}^d + \sum_{i=0}^{d-1} \varphi_i x_n{}^i$$

where $\varphi_i \in \tilde{\mathbf{R}}\langle x_1, \ldots, x_{n-1}\rangle$ for all i. Let P be the right hand side of this equality. Similarly, applying the Weierstrass Theorem to $f_1 = P$, we get $g_2 \in \tilde{\mathbf{R}}\langle x\rangle$ and $\psi_i \in \tilde{\mathbf{R}}\langle x_1, \ldots, x_{n-1}\rangle$ such that

$$f - g_2 P = \sum_{i=1}^{d-1} \psi_i x_n{}^i.$$

Since $P = g_1 f$, we must have

$$\sum_{i=1}^{d-1} \psi_i x_n{}^i = hf$$

with $h = 1 - g_1 g_2 \in \tilde{\mathbf{R}}\langle x\rangle$. We claim that $h = 0$. Suppose $h \neq 0$. Then we should have $\nu > 0$ such that $h \notin ((I(\nu)))\tilde{\mathbf{R}}\langle x\rangle$. To show that this leads to a contradiction, we may replace $\tilde{\mathbf{R}}$ by $\tilde{\mathbf{R}}/I(\nu)$ in the reasoning. Thus we assume that $\tilde{\mathbf{R}}$ is discrete in its topology. We then have a free base $\{M\}$, including $1 \in \tilde{\mathbf{R}}$, of $\tilde{\mathbf{R}}$ as K-module such that for every $\mu \geq 0$ a suitable subset of $\{M\}$ induces a free base of $I(\mu)/I(\mu + 1)$. Write $h = \Sigma\, h_M M$ with $h_M \in K[x]$ for all M. Let us pick one M, say M^*, such that $h_{M^*} \neq 0$ and $\nu(M^*) = \min\{\nu(M) \mid h_M \neq 0\}$. Then, when hf is similarly expressed, the coefficient of M^* is equal to $h_{M^*}\bar{f}$. If ψ_{iM^*} is the coefficient of M^* in the expression of ψ_i, then

$$\sum_{i=1}^{d-1} \psi_{iM^*}(x_1, \ldots, x_{n-1})x_n{}^i = h_{M^*}(x_1, \ldots, x_n)\bar{f}(x_1, \ldots, x_n).$$

This is impossible because of the difference of the degrees in x_n. Thus $h = 0$ or $g_1 g_2 = 1$, and $f - g_2 P = 0$. Thus $U = g_2$ is a unit in $\tilde{\mathbf{R}}\langle x\rangle$ and $f = UP$. Finally, if $f = U'P'$ is another expression subject to the same conditions as in Prop. 2, then with $\tilde{h} = 1 - U^{-1}U'$ we obtain $\tilde{h}P'$ is a polynomial of degree $< d$ in x_n. The same reasoning as above, once again, proves $\tilde{h} = 0$. This proves the uniqueness of P. Q.E.D.

COROLLARY. *Assume that $f = P$ in the proposition. Then for every $g \in \tilde{\mathbf{R}}\langle x\rangle$, we have a unique expression: $g = hP + r$ where $h \in \tilde{\mathbf{R}}\langle x\rangle$ and r is a polynomial in x_n of degree $< d$* with coefficients in $\tilde{\mathbf{R}}\langle x_1, \ldots, x_{n-1}\rangle$. (*The uniqueness is the point of assertion.*)

Proof. If there were two different expressions, then by taking their difference we obtain a division by P of a non-zero polynomial D of x_n with coefficients in $\tilde{\mathbf{R}}\langle x_1, \ldots, x_{n-1}\rangle$ whose degree is less than d. This means $P + D = UP$ with $U \in \tilde{\mathbf{R}}\langle x\rangle$ which is necessarily invertible. This contradicts the uniqueness of the proposition. Q.E.D.

§4. Dimensionality type (formal case). Let k be an algebraically closed field and let $\mathbf{A} = k[[x]]$ with a system of n indeterminates $x = (x_1, \ldots, x_n)$. Let f be a non-zero element with no multiple factors in $(x)\mathbf{A}$. Let us recall Zariski's definition of *dimensionality type of* $\hat{\mathbf{X}} = \mathrm{Spec}(\mathbf{A}/f\mathbf{A})$ at $P \in \hat{\mathbf{X}}$ (and in particular at $\max(\mathbf{A}) = (x)\mathbf{A}$). For this purpose, we take a system of *independent generic* power series

$$x_i^* = \sum_{j=1}^{n} s_{ij}^* x_j + \sum_A u_{iA}^* x^A$$

where $1 \leq i \leq n - 1$, A ranges over $\{A \in \mathbf{Z}_0{}^n \,|\, |A| > 1\}$ and s_{ij}^*, u_{iA}^* are independent variables over $\mathbf{A}$. Let k^* be the algebraic closure $k(s^*, u^*)$ where $s^* = \{s_{ij}^* \,|\, 1 \leq i \leq n - 1, 1 \leq j \leq n\}$ and $u^* = \{u_{iA}^* \,|\, 1 \leq i \leq n - 1, A \in \mathbf{Z}_0{}^n, |A| > 1\}$. Let $\mathbf{A}^* = k^*[[x^*]]$ with $x^* = (x_1^*, \ldots, x^*_{n-1})$ and let f^* be a reduced discriminant of f with respect to $\mathbf{A}^*$. Namely, f^* is a generator (unique up to a unit multiple) of the radical of the ideal

$$(f, f')\mathbf{A}^*[[x_n]] \cap \mathbf{A}^*$$

where f' is $\partial f/\partial x_n$ with reference to the system of parameters (x^*, x_n) of $\mathbf{A}^*[[x_n]]$. Such f^* is also obtained as follows. By Weierstrass preparation theorem, we have a unique factorization $f = UF$ in $k^*[[x^*, x_n]]$ such that U is a unit and

$$F = x_n{}^m + \varphi_1(x^*)x_n{}^{m-1} + \cdots + \varphi_m(x^*)$$

with $m = \mathrm{ord}(f)$ (i.e., the *multiplicity* of $\mathbf{X}$ at $\max(\mathbf{A})$) and $\varphi_i \in \mathbf{A}^*$ for all i. It then follows that $\mathrm{ord}(\varphi_i) \geq i$ for all i. Let $D(x^*)$ be the resultant of F and

$$\frac{\varphi F}{\partial x_n} = mx_n{}^{m-1} + (m - 1)\varphi_1 x_n{}^{m-1} + \cdots + \varphi_{m-1}$$

or

$$D = \det \begin{vmatrix} 1 & \varphi_1 & \cdots & \varphi_m & 0 & \cdots & 0 \\ 0 & 1 & \varphi_1 & \cdots & \varphi_m & \cdots & 0 \\ \cdots & \cdots & \cdots & \cdots & \cdots & \cdots & \cdots \\ m & (m-1)\varphi_1 & \cdots & \varphi_{m-1} & 0 & \cdots & 0 \\ 0 & m & \cdots & \cdots & \varphi_{m-1} & \cdots & 0 \\ \cdots & \cdots & \cdots & \cdots & \cdots & \cdots & \cdots \end{vmatrix} \begin{matrix} \left.\vphantom{\begin{matrix}1\\0\\ \cdots\end{matrix}}\right\} m-1 \\ \left.\vphantom{\begin{matrix}m\\0\\ \cdots\end{matrix}}\right\} m \end{matrix}$$

Then f^* is the product of distinct irreducible factors that appear in the factorization of D in $\mathbf{A}^*$. Since k is algebraically closed, $Pk^*[[x]]$ is also a prime ideal. Here P is viewed as a prime ideal in $\mathbf{A}$ by the natural inclusion $\operatorname{Spec}(\mathbf{A}/f\mathbf{A})$ into $\operatorname{Spec}(\mathbf{A})$. As $k^*[[x]] = \mathbf{A}^*[[x_n]]$, we obtain a prime ideal $P^* = P\mathbf{A}^*[[x_n]] \cap \mathbf{A}^*$. Zariski's dimensionality type, $\text{d.t.}(\hat{\mathbf{X}})_P$ for short, is inductively defined by the formula

$$\text{d.t.}(\hat{\mathbf{X}})_P = \text{d.t.}(\hat{\mathbf{X}}^*)_{P^*} + 1$$

where $\hat{\mathbf{X}}^* = \operatorname{Spec}(\mathbf{A}^*/f^*\mathbf{A}^*)$. Here if $P^* \notin \hat{\mathbf{X}}^*$, then $\text{d.t.}(\hat{\mathbf{X}}^*)_{P^*} = -1$. In other words, if P is a smooth point of $\hat{\mathbf{X}}$ (or equivalently $\mathbf{A}_P/f\mathbf{A}_P$ is regular), then $\text{d.t.}(\hat{\mathbf{X}})_P = 0$.

Remark (4.1). In the above inductive definition of dimensionality type, we may replace k^* by any algebraically closed field containing $k(s^*, u^*)$. This is easily proven by induction, by the fact that if K is an algebraically closed field and if $y = (y_1, \ldots, y_r)$ is a system of r indeterminates over K, then an irreducible $g \in K[[y]]$ stays to be irreducible in $L[[y]]$ for every field extension L of K.

Remark (4.2). Let K be any algebraically closed field containing $k(s^*, u^*)$. Let $\mathbf{A}_i^* = K[[x_1^*, \ldots, x^*_{n-i}]]$ for $i = 1, \ldots, n-1$. Let us choose $f_i^* \in \mathbf{A}_i^*$ as follows:

(1) f_i^* is a generator of the radical of the ideal $(f^*_{i-1}, \partial f^*_{i-1}/\partial x^*_{n-i+1})\mathbf{A}^*_{i-1} \cap \mathbf{A}_i^*$ for every i, where $f = f_0^*$, $x_n = x_n^*$ and $\mathbf{A}_0^* = \mathbf{A}_1^*[[x_n]]$. Now let $P \in \operatorname{Spec}(\mathbf{A})$ be such that $f \in P$ and let us define $P_i^* \in \operatorname{Spec}(\mathbf{A}_i^*)$, $0 \leq i \leq n-1$, as follows:

(2) $P_0{}^* = P\mathbf{A}_0{}^*$ and $P_i{}^* = P^*{}_{i-1} \cap \mathbf{A}_i{}^*$ for all $i \geq 1$. Then, under the conditions (1) and (2), we have

(3) $\text{d.t.}(\hat{\mathbf{X}})_P$ is the index d, $0 \leq d \leq n-1$, such that ($f_d{}^* \in P_d{}^*$ and) $P_d{}^*$ is a smooth point of $\text{Spec}(\mathbf{A}_d{}^*/f_d{}^*\mathbf{A}_d{}^*)$.

Proof. We shall prove (3) by induction on d. It is clearly true for $d \leq 1$. Let us assume that $d > 1$ and the assertion (3) is true for smaller d. Let us take independent variables $\tilde{s}_{\alpha i}$ and $\tilde{u}_{\alpha B}$ over $\mathbf{A}_0{}^*$, where $1 \leq \alpha \leq n-2$, $1 \leq i \leq n-1$ and $B \in \mathbf{Z}_0{}^{n-1}$ with $|B| > 1$. Let us define

$$y_\alpha = \sum_{i=1}^{n-1} \tilde{s}_{\alpha i} x_i{}^* + \sum_B \tilde{u}_{\alpha B} x^{*B} \tag{4}$$

for $1 \leq \alpha \leq n-2$. Let L be any algebraically closed field containing $K(\tilde{s}, \tilde{u})$. Let $\mathbf{B}_0 = L[[y, x^*{}_{n-1}]]$ with $y = (y_1, \ldots, y_{n-2})$, and let $\mathbf{B}_j = L[[y_1, \ldots, y_{n-1-j}]]$ for $j = 1, 2, \ldots, n-2$. Let $g_0 = f_1{}^*$ and $y_{n-1} = x^*{}_{n-1}$, and choose a generator g_j of the radical of the ideal $(g_{j-1}, \partial g_{j-1}/\partial y_{n-j})\mathbf{B}_{j-1} \cap \mathbf{B}_j$ for $j \leq 1$. Let $Q_0 = P_1{}^*\mathbf{B}_0$ and $Q_j = Q_{j-1} \cap \mathbf{B}_j$ for all $j \geq 1$. Then by induction assumption, $\text{d.t.}(\mathbf{X}^*)_{P^*}$ is the smallest e, $0 \leq e \leq n-2$, such that ($g_e \in Q_e$ and) Q_e is a smooth point of $\text{Spec}(\mathbf{B}_e/g_e\mathbf{B}_e)$. By (4), we get

$$y_\alpha = \sum_{j=1}^{n} S_{\alpha j} x_j + \sum_A U_{\alpha A} x^A, \qquad 1 \leq \alpha \leq n-2,$$

where A ranges over $\{A \in \mathbf{Z}_0{}^n \,\|\, |A| > 1\}$. Here $S_{\alpha j}$ and $U_{\alpha A}$ are certain polynomials in $k[s^*, u^*, \tilde{s}, \tilde{u}]$. Clearly $s^*{}_{n-1j}$, $u^*{}_{n-1A}$, $S_{\alpha j}$, $U_{\alpha A}$ are independent variables over $\mathbf{A}$, where $1 \leq j \leq n$, $1 \leq \alpha \leq n-2$ and $A \in \mathbf{Z}_0{}^h$, $|A| > 1$. Therefore we can find an automorphism σ of L *over* k such that $\sigma(s^*{}_{n-1j}) = s^*{}_{n-1j}$, $\sigma(u^*{}_{n-1A}) = u^*{}_{n-1A}$, $\sigma(s_{\alpha j}{}^*) = S_{\alpha j}$ and $\sigma(u_{\alpha A}{}^*) = U_{\alpha A}$ for all j, A, α. This σ extends to an automorphism of $L[[x^*, x_n]] = L[[y, x^*{}_{n-1}, x_n]]$ in such a way that $\sigma(x_\alpha{}^*) = y_\alpha$, $1 \leq \alpha \leq n-2$, $\sigma(x^*{}_{n-1}) = x^*{}_{n-1}$ and $\sigma(x_n) = x_n$. This σ induces the identity in $k[[x]] = \mathbf{A}$. For instance $\sigma(f) = f$. Since $L[[x^*]] = L[[y, x^*{}_{n-1}]]$, $\partial f/\partial x_n$ is the same whether it is with respect to (x^*, x_n) or it is to $(y, x^*{}_{n-1}, x_n)$. Therefore $\sigma(\partial f/\partial x_n) = \partial f/\partial x_n$. (Write f as a power series in x_n with coefficients in $L[[x^*]] = L[[y, x^*{}_{n-1}]]$. Then these coefficients are σ-invariant.) Hence the ideal $f_1{}^*\mathbf{B}_0 = g_0\mathbf{B}_0$ is σ-invariant. From this, it readily follows that, if $\mathbf{A}_i' = L[[x_1{}^*, \ldots, x^*{}_{n-i}]]$ for $i \geq 1$, then σ induces an isomorphism from $\mathbf{A}'{}_{i+1}$ to $\mathbf{B}_i$ which maps $f^*{}_{i+1}\mathbf{A}'{}_{i+1}$ to $g_i\mathbf{B}_i$. On the other hand, σ maps $PL[[x^*, x_n]] = PL[[y, x^*{}_{n-1}, x_n]]$ into itself because σ is identity in A. Hence σ maps $P_1{}^*\mathbf{A}_1' = Q_0$. Therefore σ maps $P_{i+1}\mathbf{A}'{}_{i+1}$ to Q_i for all $i \geq 1$. Now it is clear, that $g_i \in Q_i$ and $\mathrm{Spec}(\mathbf{B}_i/g_i\mathbf{B}_i)$ is smooth at Q_i if and only if $f^*{}_{i+1} \in P^*{}_{i+1}\mathbf{A}'{}_{i+1}$ and $\mathrm{Spec}(\mathbf{A}'{}_{i+1}/f^*{}_{i+1}\mathbf{A}'{}_{i+1})$ is smooth at $P^*{}_{i+1}\mathbf{A}'{}_{i+1}$ for all $i \geq 1$. The last condition is not affected by replacing $\mathbf{A}'{}_{i+1}$ by $\mathbf{A}^*{}_{i+1}$ because k^* is algebraically closed. We can readily conclude that the number in (3) is equal to $\mathrm{d.t.}(\hat{\mathbf{X}}^*)_{P^*} + 1$, which is $\mathrm{d.t.}(\hat{\mathbf{X}})_P$.

Remark (4.3). In Remark (4.2), we have

$$K[[x_1, \ldots, x_n]] = K[[x_1{}^*, \ldots, x^*{}_{n-1}, x_n]] = \mathbf{A}_0{}^*.$$

This means that we can write

$$x_i = \sum_{j=1}^{n} s_{ij}x_j{}^* + \sum_A u_{iA}x^{*A}, \qquad 1 \leq i \leq n-1,$$

where $x_n{}^* = x_n$ and $A \in \mathbf{Z}_0{}^n$ with $|A| > 1$. By symmetry between x and x^*, it is clear that $s = (s_{ij})$ and $u = (u_{iA})$ are independent variables over $\mathbf{A} = k[x]$ (as well as over $k[x^*]$). K is an algebraically closed field containing $k(u, s)$. In other words, given $f \in \mathbf{A} = k[x]$, we could have introduced indeterminates $s = (s_{ij})$, $1 \leq i \leq n-1$, $1 \leq j \leq n$, and $u = (u_{iA})$, $1 \leq i \leq n-1$, $A \in \mathbf{Z}_0{}^{\mathrm{n}}$, $|A| > 1$, over $\mathbf{A}$ and then define the $x_j{}^*$, $1 \leq i \leq n-1$, by the above equalities. K being any algebraically closed field containing $k(s, u)$, the criterion of $\mathrm{d.t.}(\hat{\mathbf{X}})_P$ in (4.2) holds exactly as it is stated. When we deal with dimensionality types in the algebraic case (rather than the formal case), we will find it more convenient to introduce indeterminates s, u as above than to introduce s^*, u^* as in the beginning of this section.

§5. Dimensionality type (algebraic case). Let $x = (x_1, \ldots, x_n)$ be a system of n indeterminants over an algebraically closed field k. Let f be a non-constant polynomial without multiple factors in $k[x]$. Then $f = 0$ defines a hypersurface $\mathbf{X}$ (an algebraic variety of pure dimension

$n - 1$, possibly reducible) in the affine n-space. Let $\mathbf{Y}$ be an irreducible subvariety of $\mathbf{X}$ and let $\mathbf{Z}$ be the singular locus of $\mathbf{Y}$ which is nowhere dense in $\mathbf{Y}$. We are interested in the dimensionality types of $\mathbf{X}$ at various points ξ of $\mathbf{Y} \setminus \mathbf{Z}$. We are particularly interested in the set of those ξ at which the dimensionality type is the same as that at the generic point of $\mathbf{Y}$.

For this purpose, we first introduce a set of indeterminates which are later used in defining dimensionality types at various points ξ. Let p denote the characteristic of the field k, if it is positive, and let $p = 1$ if the characteristic is zero. Let $t = (t_1, \ldots, t_n)$, $s = (s_{ij})$, $1 \leq i \leq n - 1$ and $1 \leq j \leq n$, and $u = (u_{iA})$, $1 \leq i \leq n - 1$ and $A \in \mathbf{Z}_0{}^n$ with $|A| > 1$, be indeterminates (independent) over $k[x]$. For every power $q = p^e$ with $e > 0$, let $u_q = (u_{qiA})$ with $u_{qiA} = (u_{iA})^{1/q}$. For a positive integer N, we let $r_q(N) = (u_{qiA}, 1 \leq i \leq n - 1, A \in \mathbf{Z}_0{}^n$ with $|A| \leq N)$ and $u_q(N) =$ the difference of u_q and $r_q(N)$. We let $r_q(N, N') =$ the difference of $r_q(N')$ and $r_q(N)$. Let K be the algebraic closure of $k(t, s)$ and let K_{N+1} be the algebraic closure of the field of fractions of $K_N[[r_q(N, N + 1)]]$ for every $N \geq 1$ where $K_1 = K$. These K_N are independent of q. We define a new system of n variables $z = (z_1, \ldots, z_n)$ over K_N (for every N) by the following equalities:

$$(1) \quad x_i = t_i + \Sigma^n{}_{j=1} s_{ij} z_j + \Sigma_A u_{iA} z^A, 1 \leq i \leq n - 1, \text{ and } x_n = z_n.$$

Remark (5.1). Let σ_0 be any non-singular k-linear transformation on $(z_1, \ldots, z_{n-1})$, and let σ be the transformation of z by $\sigma(z_i) = \sigma_0(z_i)$ for $i \leq n - 1$ and $\sigma(z_n) = z_n$. We then have a new expression of the same x_i in terms of $z' = \sigma(z)$:

$$x_i = t_i' + \sum_{j=1}^{n} s_{ij}' z_j' + \sum_A u_{iA}' z'^A, \qquad 1 \leq i \leq n - 1,$$

where t', s' and u' are again independent indeterminates over $k[x]$. The new situation with z', t', s' and u' is completely similar to the old one with z, t, s and u. To be precise, σ extends to a continuous k-automorphism $\rho_q(N)$ of $K_N[z][[u_q(N)]]$ in such a way that $\rho_q(N)$ maps z, t, s, u to z', t', s', u' respectively and it induces continuous k-automorphisms of $K_M[[r_q(M, M + 1)]]$ for $1 \leq M < N$. Such an extension $\rho_q(N)$ of σ is unique up to automorphisms of K_{M+1} over $K_M[[r_q(M, M + 1)]]$, $1 \leq M < N$. It induces the identity in $k[x]$.

Step 1. The defining polynomial f of $\mathbf{X}$ will be viewed as an element of

$$A_q(N) = K_N[z][[u_q(N)]]$$

by the obvious inclusion $k[x] \to A_q(N)$ under (1). If we let

$$\bar{f}_N = [f \bmod ((u_q(N)))A_q(N)]$$

where ((**)) denotes the closure of (*) and [ ] denotes the residue class, then

(2) $\bar{f}_N$ is a non-zero constant multiple of a monic polynomial in $\bar{z}_n$ of degree $= N \deg_x f$ with coefficients in $K_N[\bar{z}_1, \ldots, \bar{z}_{n-1}]$, where $\bar{z}_i$ denotes the residue class of z_i modulo $((u_q(N)))A_q(N)$ for all i.

This is easily deduced from the fact that s and $r_q(N)$ are independent variables over $k[x]$.

Now we can apply Weierstrass theorem, Prop. 2 of §3, to f and find

(3) a monic polynomial $f_N{}^{(1)}$ in z_n of degree $= N \deg_x f$ with coefficients in

$$A_q{}^{(1)}(N) = K_N[z_1, \ldots, z_{n-1}][[u_q(N)]]$$

such that $(f)A_q(N) = (f_N{}^{(1)})A_q(N)$.

It is clear that $\bar{f}_N$ has no multiple factors. Hence the same is true for $f_N{}^{(1)}$. Therefore

(4) the discriminant $D_N{}^{(1)}$ of $f_N{}^{(1)}$ with respect to $A_q{}^{(1)}(N)$ is a non-zero element of $A_q{}^{(1)}(N)$, and hence there exists $N' > N$ (N' depending upon N) such that

$$D_N{}^{(1)} \notin ((u_q(N')))A_q{}^{(1)}(N).$$

We shall use the following notation and the terminology: For $g \in A_q{}^{(1)}(N)$ and $N' > N$, $g(N')$ denotes the residue class of g modulo $(u_q(N')))A_q{}^{(1)}(N)$. We shall say that g is *reduced* if $g(N')$ does not have any multiple factors for all $N' \gg N$. Here it should be noted that we have a canonical isomorphism

$$(5)\quad A_q^{(1)}(N)/((u_q(N')))A_q^{(1)}(N) \rightleftharpoons K_N[z_1(N'), \ldots, z_{n-1}(N')][[r_q(N, N')]]$$

where $r_q(N, N')$ is the difference of the two systems $r_q(N')$ and $r_q(N)$. This ring is clearly a unique factorization domain for every $N' > N$, as $r_q(N, N')$ is a finite system of indeterminates. By taking $N' \to \infty$, we can even prove that $A_q^{(1)}(N)$ itself is a unique factorization domain (though it is non-Noetherian).

(6) For every $\eta = (\eta_1, \ldots, \eta_n)$, $\eta_i \in K_N[[u_q(N)]]$, we have $K_N[z_1 - \eta_1, \ldots, z_n - \eta_n][[u_q(N)]] = K_N[z_1, \ldots, z_n][[u_q(N)]] = A_q(N)$. (cf. Cor. 1 of Prop. 1, §3). Similarly, we have $K_N[z_1 - \eta_1, \ldots, z_{n-1} - \eta_{n-1}][[u_q(N)]] = A_q^{(1)}(N)$. If we let L be the algebraic closure of the field of fractions of

$$\bigcup_q \bigcup_{N' \geq 1} K_{N'}[[u_q(N')]]$$

then we have the canonical inclusions:

$$\begin{array}{ccc} A_q(N) & \subset & L[[z - \eta]] \\ \cup & & \cup \\ A_q^{(1)}(N) & \subset & L[[z_1 - \eta_1, \ldots, z_{n-1} - \eta_{n-1}]] \end{array}$$

(7) Given N, if q is a sufficiently high power of p (depending upon N), then the reduction (i.e., the result of replacing each multiple factor by the corresponding simple factor) of $D_N^{(1)}$ in $A_q^{(1)}(N)$ (which depends upon q) is reduced also in $L[[z_1 - \eta_1, \ldots, z_{n-1} - \eta_{n-1}]]$ for all η of (6).

This follows from the fact that the field of fractions of

$$\bigcup_q K_N[[r_q(N, N')]]$$

(for any fixed pair of N and N') is a perfect field. Here we should refer to (5) in which $z_i(N')$ may be replaced by $z_i(N') - \eta_i$, $1 \leq i \leq n - 1$.

Next we want to compare $D_N^{(1)}$ and $D_{N'}^{(1)}$, for $N' > N$. As in (3), $f_{N'}^{(1)}$ is defined as an element of $K_{N'}[z][[u_q(N')]]$. But actually it is an element of

$$K_N((r_q(N, N')))[z][[u_q(N')]] = \Gamma$$

where $K_N((*))$ denotes the field of fractions of $K_N[[*]]$, because the Weierstrass theorem is applicable if we replace $K_{N'}$ by any field containing $k(t, s, r_q(N, N'))$. On the other hand, $f_N^{(1)}$ is contained in

$$\begin{aligned} A_q(N) &= K_N[z][[u_q(N)]] \\ &= K_N[z][[r_q(N, N')]][[u_q(N')]] \\ &\subset (K_N[z][[r_q(N, N')]])_\Delta[[u_q(N')]] = \Gamma^* \end{aligned}$$

where $\Delta = K_N[[r_q(N, N')]] \setminus \{0\}$ and $(\ )_\Delta$ denotes the localization by the multiplicatively closed subset Δ. We have a natural inclusion $\Gamma \subset \Gamma^*$ and we have

$$(f)\Gamma^* = (f_N^{(1)})\Gamma^* = (f_{N'}^{(1)})\Gamma^*$$

so that $f_N^{(1)}$ divides $f_{N'}^{(1)}$ in Γ^*. As a matter of fact, we have the stronger division as follows.

(8) $f_N^{(1)}$ divides $f_{N'}^{(1)}$ as a polynomial in z_n, i.e., the division is true in $\Gamma_1^*[z_n]$ where

$$\Gamma_1^* = (K_N[z_1, \ldots, z_{n-1}][[r_q(N, N')]])_\Delta[[u_q(N')]]$$

so that $f_{N'}^{(1)}/f_N^{(1)}$ is a monic polynomial in z_n of degree $= (N' - N)$ $\deg_x f$.

To prove (8), let us take any open ideal J' in the topological ring $K_N[[u_q(N')]]$. Then we have $\Gamma^*/((J'))\Gamma^* = (A_q(N)/((J'))A_q(N))_\Delta$ so that, with bars denoting modulo $((J'))\Gamma^*$, we have $\delta \in \Delta$ such that $\delta \bar{f}_{N'}^{(1)}$ is in $A_q(N)/((J'))A_q(N)$. Moreover $\delta \bar{f}_{N'}^{(1)}$ is divisible by $\bar{f}_N^{(1)}$ in $A_q(N)/((J'))A_q(N)$. By applying the Euclidian method of division we can write

$$\bar{f}_{N'}^{(1)} = Q \cdot \bar{f}_N^{(1)} + R$$

where Q is a monic polynomial in $\bar{z}_n$ of degree $= (N' - N) \deg_x f$ with coefficients in $\Gamma_1^*/((J'))\Gamma_1^*$ and R is a polynomial in $\bar{z}_n$ of degree less than $N \deg_x f$ (= the degree of $f_N^{(1)}$ in z_n). We then invoke the uniqueness of Weierstrass division (cf. §2 and §3) applied to

$$\bar{f}_N^{(1)} \text{ in } A_q(N)/((J'))A_q(N) = \mathbf{B}\langle z \rangle$$

where $\mathbf{B} = K_N[[u_q(N)]]/((J'))K_N[[u_q(N)]]$. The uniqueness holds because

$$\deg_{\bar{z}_n}[\bar{f}_N^{(1)} \bmod ((\max(\mathbf{B})))\mathbf{B}\langle z \rangle] = \deg_{\bar{z}_n}(\bar{f}_N^{(1)}) = N \deg_x f.$$

It proves that R would not be divisible by $\bar{f}_N^{(1)}$ in $\mathbf{B}\langle z \rangle$ if it were non-zero. Therefore we conclude $R = 0$. This being true for every choice of J', (8) is now verified.

Now (8) implies that, as polynomial equations for z_n, the roots of $f_N^{(1)} = 0$ are also roots of $f_{N'}^{(1)} = 0$. It follows that

(9) $D_N^{(1)}$ divides $D_{N'}^{(1)}$ in $\Gamma_1{}^*$.

Furthermore, $f_{N'}^{(1)}$ is monic in z_n with coefficients in $A_q^{(1)}(N')$ and hence the roots of $f_{N'}^{(1)} = 0$ are algebraic and integral over $A_q^{(1)}(N')$. It follows that

(10) $D_N^{(1)}$ is algebraic and integral over $A_q^{(1)}(N')$ and its reduced norm in $A_q^{(1)}(N')$ divides $D_{N'}^{(1)}$ in $A_q^{(1)}(N')$.

(11) Given any fixed N (for instance $N = 1$), let us choose $N' \gg N$ according to (4) and then choose a sufficiently high power q of p according to (7) with respect to N' (instead of N as in (7)). Then we define $g^{(1)}$ to be the reduction of the norm of $D_N^{(1)}$ in $A_q^{(1)}(N')$.

PROPOSITION 1. *The element $g^{(1)} \in A_q^{(1)}(N')$ multiplied by a suitable non-zero constant $\in K_{N'}$, has the following properties:*

a) *if $\bar{g}^{(1)}$ denotes the residue class of $g^{(1)}$ mod$((u_q(N')))A_q^{(1)}(N')$, then $\bar{g}^{(1)}$ is a non-zero monic polynomial in $\bar{z}_{n-1}$ with coefficients in $K_{N'}[\bar{z}_1, \ldots, \bar{z}_{n-2}]$,*

b) *for every $\eta = (\eta_1, \ldots, \eta_n) \in (K_N[[u_q(N)]])^n$, $g^{(1)}$ is reduced in $L[[z_1 - \eta_1, \ldots, z_{n-1} - \eta_{n-1}]]$ with reference to the inclusion of $A_q^{(1)}(N')$ into this ring by (6), and*

c) *$g^{(1)}$ is divisible by the reduced discriminant of the Weierstrass* *polynomial in* $z_n - \eta_n$ *of* f *with respect to* $L[[z_1 - \eta_1, \ldots, z_{n-1} - \eta_{n-1}]]$.

Proof. By (4), $D_N^{(1)}$ is not in $((u_q(N')))A_q^{(1)}(N)$ and hence not in $((u_q(N')))\Gamma_1^*$ which is a prime ideal (cf. (8)). Therefore $g^{(1)}$ is not in $((u_q(N')))A_q^{(1)}(N')$, i.e., $\bar{g}^{(1)} \neq 0$. Remark (5.1) shows that a suitable non-zero constant ($\in K_N$) multiple of $\bar{g}^{(1)}$ is a monic polynomial in $\bar{z}_{n-1}$ as is claimed in a). The claim b) follows (7). As for c), it is clear that the Weierstrass polynomial in $z_n - \eta_n$ of f divides the monic polynomial $f_N^{(1)}$ in $L[[z_1 - \eta_1, \ldots, z_{n-1} - \eta_{n-1}]][z_n - \eta_l]$. Hence its discriminant in question divides $D_N^{(1)}$ in this ring. By the definition (11) of $g^{(1)}$, 3) follows immediately. Q.E.D.

We now proceed to *Step II*, starting with the chosen $g^{(1)} \in A_q^{(1)}(N')$ as above and ending with a selection of $g^{(2)} \in A_q^{(2)}(N'')$ with some $N'' > N'$ (and perhaps with a new higher power q of p). It should be noted that q may be replaced by a higher power, if necessary, without affecting the definition of $g^{(1)}$ or its properties a), b) and c) of Prop. 1. We then proceed to *Step III* and so on. Let us now describe the general step in this inductive procedure.

General Step. We begin with $g^{(i)} \in A_q^{(i)}(N^{(i)})$, subject to certain conditions, and end with $g^{(i+1)} \in A_q^{(i+1)}(N^{(i+1)})$ after replacing q by a suitable power, where

$$A_q^{(j)}(N) = K_N[z_1, \ldots, z_{n-j}][[u_q(N)]].$$

We shall use the following symbols:

$$y(j) = (z_1, \ldots, z_{n-j})$$

$$K_{N,q}(N') = (K_N[[r_q(N, N')]])_\Delta \text{ with the multiplicatively closed}$$

$$\Delta = K_N[[r_q(N, N']] \setminus \{0\}$$

$$\Gamma_{N,q}^{(j)}(N') = K_{N,q}(N')[y(j)][[u_q(N')]]$$

$$\Gamma_{N,q}^{(j)}(N')^* = (K_N[y(j)][[r_q(N, N')]])_\Delta[[u_q(N')]]$$

Note that we have inclusions

$$\Gamma_{N'',q}^{(j)}(N) \subset A_q^{(j)}(N) \subset \Gamma_{N,q}^{(j)}(N')^* \supset \Gamma_{N,q}^{(j)}(N')$$
$$\supset \Gamma_{N'',q}^{(j)}(N') \quad \text{for} \quad N'' \leq N \leq N'.$$

Remark (5.2). For every non-singular k-linear transformation τ of $y(j) = (z_1, \ldots, z_{n-j})$, we first extend τ to a linear transformation σ of z by letting $\sigma(z_a) = z_a$ for all $a > n - j$. We then extend σ, according to (5.1), to an automorphism $\rho_q(N)$ of $A_q(N)$. This $\rho_q(N)$ induces an automorphism of $\Gamma_{M,q}{}^{(j)}(N)$ for every M, $1 \leq M < N$ and maps the ideals $((u_q(N')))\Gamma_{M,q}{}^{(j)}(N)$ onto themselves for all $N' \geq N$.

Definition (5.3). We shall say that $h \in \Gamma_{M,q}{}^{(j)}(N)$ is *τ-invariant* (resp. *τ-quasi-invariant*) for a given non-singular k-linear transformation τ of $y(j)$, if we have $\rho_q(N)(h) = h$ (resp., $\rho_q(N)(h)$ is a unit multiple of h in $\Gamma_{M,q}{}^{(j)}(N)$) where $\rho_q(N)$ denotes every extension of τ according to (5.1) and (5.2).

Remark (5.4). Let $\mathbf{R} = \Gamma_{M,q}{}^{(j)}(N)$ and $\mathbf{E} = K_{M,q}(N)$ for some j, M, q and N, so that we have $\mathbf{R}/((u_q(N)))\mathbf{R} = \mathbf{E}[\bar{z}_1, \ldots, \bar{z}_{n-j}]$ where bars denote the residue classes $\mathrm{mod}((u_q(N)))\mathbf{R}$. Now if $h \in \mathbf{R}$ is τ-quasi-invariant for every non-singular k-linear transformation τ of $y(j)$, then it has the following properties: For every index a, $1 \leq a \leq n - j$,

(i) the residue class $\bar{h}$ of h is a constant ($\in \mathbf{E}$) multiple of a monic polynomial in $\bar{z}_a$ of degree $= \deg \bar{h}$,

(ii) every common factor of h and $\partial h/\partial z_a$ in $\mathbf{R}$ is also a factor of $\partial h/\partial z_b$ for all b, $1 \leq b \leq n - j$.

Proof. Pick any extension $\rho_q(N)$ of any τ according to (5.1) and (5.2). It induces an automorphism α of E and an automorphism ρ of the polynomial ring $\mathbf{E}[\bar{y}]$ where $\bar{y} = (\bar{z}_1, \ldots, \bar{z}_{n-j})$. We have $\rho(\bar{y}) = \tau(\bar{y}), = \bar{y}'$ say. Write $\bar{h}$ as a polynomial $p(\bar{y})$ and let α act on the coefficients of p. The τ-quasi-invariance of h implies that $\bar{h}$ is a non-zero constant multiple of $\rho(\bar{h})$ because the non-zero constants are the only units in a polynomial ring. This means that $p(\bar{y})$ is a non-zero constant multiple of $\alpha(p)(\tau(\bar{y}))$. For a suitable choice of τ, it has the property of (i). As for (ii), each $\rho_q(N)$ induces a permutation of the irreducible factors (finitely many) of h and hence we obtain a homomorphism from the group of all those $\rho_q(N)$ into the finite group of such permutations. We then see that for almost all τ we can choose extensions $\rho_q(N)$ which induce trivial permutation on the irreducible factors of h. Then (ii) follows because when we replace h by a unit multiple of h the common factors of h and $\partial h/\partial z_b$ stay unchanged.

We are now ready to describe the general step of our inductive procedure of selecting $g^{(j)} \in A_q{}^{(j)}(N^{(j)})$ for all j.

Let us fix an index $i \geq 0$ and assume that we are given the following data:

S_i = an infinite set of positive integers
$M = \min\{N \in S_i\}$, which we shall later call $N^{(i)}$
$q(i, N)$ = a positive power of p depending upon (i, N) with $N \in S_i$
$g_N{}^{(i)} \in \Gamma_{M,q}{}^{(i)}(N)$, one for each $N \in S_i$, where q denotes any positive power of $q(i, N)$
($g_M{}^{(i)}$ will later be denoted by $g^{(i)}$, viewed as an element of $A_q{}^{(i)}(N^{(i)})$.)

We then assume that these data satisfy the following conditions for every $N \in S_i$, for every positive power q of $q(i, N)$, for every $N < N' \in S_i$ and for every common positive power q' of $q(i, N)$ and $q(i, N')$:

a)$_i$ $g_N{}^{(i)} \notin ((u_q(N)))\Gamma_{M,q}{}^{(i)}(N)$

b)$_i$ For every $\eta(i) = (\eta_1, \ldots, \eta_{n-i})$ with $\eta_j \in K_N[[u_q(N)]]$, the image of $g_N{}^{(i)}$ is *reduced* with respect to the injection

$$\Gamma_{M,q}{}^{(i)}(N) \to L[[y(i) - \eta(i)]]$$

which is induced by the inclusion of (6).

d)$_i$ $g_N{}^{(i)}$ divides $g_{N'}{}^{(i)}$ when they are viewed as elements of $\Gamma_{N,q'}{}^{(i)}(N')^*$ which contains $\Gamma_{M,q'}{}^{(i)}(N)$ and $\Gamma_{M,q'}{}^{(i)}(N')$.

e)$_i$ For every non-singular k-linear transformation τ of $y(i)$, $g_N{}^{(i)}$ is τ-quasi-invariant in the sense of (5.3).

Under these assumptions, we assert and shall prove

PROPOSITION 2. *There exists an infinite subset S_{i+1} of S_i and there exists a pair $(q(i+1, N), g_N{}^{(i+1)})$ for each $N \in S_{i+1}$, where $q(i+1, N)$ is a positive power of $q(i, N)$ and $g_N{}^{(i+1)}$ is an element of $\Gamma_{M',q}{}^{(i+1)}(N)$ with $M' = \min\{N \in S_{i+1}\}$ for every positive power q of $q(i+1, N)$, such that these new data satisfy the conditions* a)$_{i+1}$, b)$_{i+1}$, d)$_{i+1}$, e)$_{i+1}$ *and the following*

c)$_{i+1}$ *for every $\eta(i) = (\eta_1, \ldots, \eta_{n-i})$, $\eta_j \in K_M[[u_q(M)]]$, $g_{M'}{}^{(i+1)}$ is divisible by the reduced discriminant of $g_M{}^{(i)}$ (viewed as an element of $L[[y(i) - \eta(i)]]$) with respect to $L[[y(i+1) - \eta(i+1)]]$, where the divisibility is meant in this last power series ring.*

Proof. The proof is quite similar to that of Proposition 1 and we shall follow more or less the same procedure described in the sequence of those reasonings from (2) through (10). To simplify the notation, we shall write g_N for $g_N{}^{(i)}$. By (5.4), $\bar{g}_N$ is a non-zero constant multiple of a monic polynomial in $\bar{z}_{n-i}$ of degree $= \deg_{\bar{y}(i)}(\bar{g}_N)$ and moreover g_N has no non-trivial common factor with $\partial g_N/\partial z_{n-i}$ in $\Gamma_{M,q}{}^{(i)}(N)$, for if otherwise then g_N should have a non-trivial common factor with all the $\partial g_N/\partial z_a$ with $1 \leq a \leq n - i$ while this is impossible by b)$_i$. Now by Weierstrass theorem, Prop. 2 of §3, we have a unique monic polynomial h_N in z_{n-i} with coefficients in $\Gamma_{M,q}{}^{(i+1)}(N)$, which is a unit multiple of g_N in $\Gamma_{M,q}{}^{(i)}(N)$. Clearly h_N and $\partial h_N/\partial z_{n-i}$ have no non-trivial common factor. Therefore if D_N denotes the discriminant of h_N as a polynomial in z_{n-i} with respect to $\Gamma_{M,q}{}^{(i+1)}(N)$, then $D_N \neq 0$. Hence for all $N' \gg N$,

$$(*) \qquad D_N \notin ((u_q(N')))\Gamma_{M,q}{}^{(i+1)}(N)$$

Let us pick and fix one $N' \in S_i$ having this property. Let q' be any common positive power of $q(i, N)$ and $q(i, N')$. With these N' and q', we define $h_{N'}$ and $D_{N'}$ in the same way as we did h_N and D_N. By d)$_i$, h_N divides $h_{N'}$ in $\Gamma_{N,q'}{}^{(i)}(N')^*$. Actually we claim a stronger division that h_N divides $h_{N'}$ as polynomials in z_{n-i} with coefficients in $\Gamma_{N,q'}{}^{(i+1)}(N')^*$. By Euclidian division of polynomials, we can write $h_{N'} = Qh_N + R$ with polynomials Q, R in z_{n-i}, where Q has the degree equal to the difference of those of $h_{N'}$ and h_N while R has a degree less than that of h_N. We claim $R = 0$. It is enough to show that for every open ideal J' of $K_N[[u_{q'}(N')]]$ we have $R \in ((J'))\Gamma_{N,q'}{}^{(i)}(N')^*$ which denotes the closed ideal generated by J'. If Δ denotes the multiplicative set $K_N[[r_{q'}(N, N')]] \setminus \{0\}$, then we have

$$\Gamma_{N,q'}{}^{(i)}(N')^*/((J'))\Gamma_{N,q'}{}^{(i)}(N')^* = (\mathbf{B}\langle w\rangle)_\Delta, \qquad \text{localization by } \Delta,$$

where $\mathbf{B} = K_N[[u_{q'}(N)]]/((J'))K_N[[u_{q'}(N)]]$ and w denotes the residue class of $y(i)$ in the left hand side of the equality. (cf. the notation of §3) Note that w is a system of indeterminates over $\mathbf{B}$. Let us denote by $(\ )^*$ the residue class modulo $((J'))\Gamma_{N,q'}{}^{(i)}(N')^*$. We have $(h_{N'})^* = Q^*(h_N)^* + R^*$ and we have $\delta \in \Delta$ such that the multiplication by δ maps $(h_{N'})^*$, Q^* and R^* into $\mathbf{B}\langle w\rangle$. We have $(h_N)^* \in \mathbf{B}\langle w\rangle$ because $\Gamma_{M,q'}{}^{(i)}(N) \subset A_{q'}{}^{(i)}(N)$. With $w = (w_1, \ldots, w_{n-i})$, $(h_N)^*$ is a monic polynomial in w_{n-i} with coefficients in $\mathbf{B}\langle w_1, \ldots, w_{n-i-1}\rangle$, and its degree is equal to the degree of $\bar{h}_N$ in $\bar{z}_{n-i}$, where bars denote residue classes modulo

$((\max(\mathbf{B})))\mathbf{B}\langle w\rangle$ (or equivalently, modulo $((u_{q'}(N)))\Gamma_{M,q'}{}^{(i)}(N))$. Since h_N divides $h_{N'}$ in $\Gamma_{N,q'}{}^{(i)}(N')^*$, $\delta(h_{N'})^*$ should be divisible by $(h_N)^*$ in $\mathbf{B}\langle w\rangle$. Hence δR^* should be divisible by the same in $\mathbf{B}\langle w\rangle$. But δR^* has degree less than that of $(h_N)^*$ in w_{n-i} and hence, by the uniqueness of Weierstrass division, we must have $\delta R^* = 0$ and so $R^* = 0$. (Note here that the elements of Δ are non-zero divisors in $\mathbf{B}\langle w\rangle$.) We conclude that $R = 0$, i.e.,

(**) $\quad h_{N'} = Qh_N$, divisibility as polynomials in z_{n-i}

Let us consider the roots of the polynomials $h_{N'}$ and h_N in a fixed algebraic closure of the field of fractions of $\Gamma_{N,q'}{}^{(i+1)}(N')^*$ which contains $\Gamma_{M,q'}{}^{(i+1)}(N)$ and $\Gamma_{M,q'}{}^{(i+1)}(N')$. Then, by (**), the roots of $h_N = 0$ are also roots of $h_{N'} = 0$ and therefore we have

(***) all the roots of $h_N = 0$ are algebraic and integral over $\Gamma_{M,q'}{}^{(i+1)}(N')$, D_N is algebraic and integral over the same, and moreover the norm $D_{N,N'}$ of D_N into $\Gamma_{M,q'}{}^{(i+1)}(N')$ divides $D_{N'}$ in this ring. Furthermore

$$D_{N,N'} \notin ((u_{q'}(N')))\Gamma_{M,q'}{}^{(i+1)}(N')$$

This last assertion follows (*). Now let us choose a sufficiently high power $q(i + 1, N')$ of the above q' (which is a positive power of $q(i, N')$) so that if q'' is any positive power of $q(i + 1, N')$, then the reduction $g_{N'}{}^{(i+1)}$ of $D_{N,N'}$ in $\Gamma_{M,q''}{}^{(i+1)}(N')$ satisfies the condition $\mathrm{b})_{i+1}$ (which is the analogy of $\mathrm{b})_i$ in which i, N, q are replaced by $i + 1$, N', q''). This condition $\mathrm{b})_{i+1}$ is clearly held true even if we change M by any M', $M < M' \leq N'$ (but keep the same $g_{N'}{}^{(i+1)}$). Let us next prove that $g_{N'}{}^{(i+1)}$ is τ'-quasi-invariant for every non-singular k-linear transformation τ' of $y(i + 1)$, i.e., the condition $\mathrm{e})_{i+1}$ in which i, N, τ are replaced by $i + 1$, N', τ'. In fact, let us extend τ' to a k-linear transformation τ of $y(i)$ by letting $\tau(z_{n-i}) = z_{n-i}$. With this τ, g_N is τ-quasi-invariant by the assumption $\mathrm{e})_i$ and hence so is h_N. Now h_N is a monic polynomial in z_{n-i} and this z_{n-i} is invariant by τ. Therefore any extension $\rho_{q''}(N')$ of τ in the sense of (5.1)-(5.2) must transform h_N into a monic polynomial in z_{n-i} of the same degree. The transform is a unit multiple of h_N by the τ-quasi-invariance of h_N, and hence it is equal to h_N by the uniqueness in the Weierstrass theorem, Prop. 2

of §3. In other words, the coefficients of the polynomial h_N are all τ'-invariant and so is D_N. Namely $\rho_{q''}(N')(D_N) = D_N$ which shows that the minimal monic polynomial of D_N over the ring $\Gamma_{M,q''}{}^{(i+1)}(N')$ is invariant by such $\rho_{q''}(N')$ and so is $D_{N,N'}$ in particular. As $D_{N,N'}$ is τ'-invariant, its reduction $g_{N'}{}^{(i+1)}$ is at least τ'-quasi-invariant as was claimed. Now our proof of Prop. 2 is just about complete. As a matter of fact, all we need is to choose the set S_{i+1} systematically by induction. Firstly we choose N' as above for $N = M$. Call it N_1. Suppose we have chosen N_s for an integer $s \geq 1$. Then we apply the same argument as above to $N = N_s$, pick an integer N' accordingly and call it N_{s+1}. We let S_{i+1} be the subset of S_i consisting of those chosen integers N_s for all positive integers s, so that in particular $M' = N_1$. a)$_{i+1}$ follows from (***) by the definition of $g_{N'}{}^{(i+1)}$ in the above reasoning. b)$_{i+1}$ and e)$_{i+1}$ are actually proven there. d)$_{i+1}$ follows from the relationship between $N = N_s$ and $N' = N_{s+1}$ in (***). Finally c)$_{i+1}$ is derived from (***) in which $N = M$ and $N' = N_1 = M'$. In fact the Weierstrass polynomial in $z_{n-i} - \eta_{n-i}$ for $g_N = g_M{}^{(i)}$ divides $h_N(z_{n-i} - \eta_{n-i} + \eta_{n-i})$ (with the monic polynomial h_N of (***)) as polynomials in $z_{n-i} - \eta_{n-i}$ with coefficients in the last power series ring of c)$_{i+1}$. It follows that the reduced discriminant of c)$_{i+1}$ divides D_N and hence $D_{N,N'}$. It then divides any reduction of $D_{N,N'}$ and in particular $g_{M'}{}^{(i+1)}$. Q.E.D.

Note that the assumptions a)$_0$, b)$_0$, d)$_0$ are obviously satisfied by the following initial data: $S_0 = \mathbf{Z}_+$, $g_N{}^{(0)} = f$ and $q(0, N) = p$ for all $N \in S_0$. By repeatedly applying Prop. 2 for $0 \leq i \leq n - 1$, starting with this initial data, we obtain S_i, $q(i, N)$, $g_N{}^{(i)}$ for all $N \in S_i$ and for all i, $0 \leq i \leq n - 1$, which satisfy all the conditions a)$_i$, b)$_i$, c)$_{i+1}$, d)$_i$ and e)$_i$ for all $i < n - 1$. Among these data, what we will need later are: $N^{(i)} = \min\{N \in S_i\}$ and $g^{(i)} = g_N{}^{(i)}$ for $N = N^{(i)}$ for each i, $0 \leq i \leq n - i$. Let us pick and fix one common positive power q of all the $q(i, N^{(i)})$. We then have:

Proposition 3. *To simplify the notation, let us write $K(i)$ and $A(i)$ for K_N and $A_q{}^{(i)}(N)$ with $N = N^{(i)}$, respectively. Then the following are true:*

1) $g^{(0)} = f$ *and* $N^{(0)} = 1 \leq N^{(1)} \leq \cdots \leq N^{(n-1)}$.

2) *For every $i \geq 0$, $g^{(i)}$ is an element of $A(i)$ but is not in $((u_q(N^{(i)})))A(i)$. Moreover $\bar{g}^{(i)}$ is a constant multiple of a monic poly-*

nomial of $\bar{z}_{n-i}$ *in the polynomial ring*

$$K(i)[\bar{z}_1, \ldots, \bar{z}_{n-i}] = A(i)/((u_q(N^{(i)})))A(i)$$

where bars denote the residue classes.

3) *For every* $i > 0$ *and for every* $\eta(i) = (\eta_1, \ldots, \eta_{n-i})$ *with* $\eta_j \in K(i-1)[[u_q(N^{(i-1)})]]$, $g^{(i)}$ *is divisible by the reduced discriminant of* $g^{(i-1)}$ *with respect to* $L[[z_1 - \eta_1, \ldots, z_{n-i} - \eta_{n-i}]]$ *where* $g^{(j)}$ *is viewed as an element of* $L[[z_1 - \eta_1, \ldots, z_{n-j} - \eta_{n-j}]]$ *for* $j = i, i-1$, *by the canonical injection from* $A(j)$ *into this power series ring.*

The proof is immediate from a)$_i$, b)$_i$, d)$_i$, e)$_i$ and c)$_{i+1}$ of Prop. 2.

Now let us recall that K denotes the algebraic closure of $k(t, s)$ and is contained in all those $K(i)$.

PROPOSITION 4. *Let us pick any* $\eta = (\eta_1, \ldots, \eta_n)$ *with* $\eta_j \in K[[u]]$ *and let* $z_j^* = z_j - \eta_j$, $1 \leq j \leq n$. *Let* $z^*(i) = (z_1^*, \ldots, z^*_{n-i})$. *Viewing* f *as an element of* $L[[z^*(0)]]$, *let us define* $f^{(i)} \in L[[z^*(i)]]$ *by the following procedure*: $f^{(0)} = f$, $f^{(i+1)} =$ *reduced discriminant of* $f^{(i)}$ *with respect to* $L[[z^*(i+1)]]$ *for all* $i \geq 0$. *Then* $f^{(i)}$ *divides* $g^{(i)}$ in $L[[z^*(i)]]$ *for all* $i \geq 0$.

Proof. The proof is a simple induction on i. Assume that $f^{(i)}$ divides $g^{(i)}$ for some $i \geq 0$. Then the reduced discriminant of $f^{(i)}$ divides the same of $g^{(i)}$ with respect to $L[[z^*(i+1)]]$. But this divides $g^{(i+1)}$ by 3) of Prop. 3. Q.E.D.

Let us now go back to the geometric situation in which **X** is the hypersurface defined by $f = 0$, **Y** is an irreducible subvariety of **X** and **Z** is the singular locus of **Y**. Our objective is to find an open dense subset **U** of **Y**$\setminus$**Z** such that the dimensionality type of **X** at $\xi \in$ **U** is independent of ξ. Such **U** will be explicitly constructed by means of those selected elements: $g^{(i)} \in A(i) = A_q^{(i)}(N^{(i)}) = B(i)\langle y(i)\rangle$, where $B(i)$ denotes $K(i)[[u_q(N^{(i)})]]$. (cf. the notation of §2-3) Let us define

$$E_q(N) = K_N[x][[u_q(N)]]$$

which is canonically identified as a subring of $A_q(N)$.

LEMMA (5.5). $A_q(N)$ *is a finite integral extension of* $E_q(N)$. *For every prime ideal* Q *in* $k[x]$, *the ideal* $QE_q(N)$ *is prime and there exist* *only a finite number of prime ideals Q' in $A_q(N)$ such that $Q' \cap E_q(N) = QE_q(N)$ and $Q' \supset QA_q(N)$.*

Proof. We can identify $A_q(N) = B_N\langle z\rangle$ with the residue class ring of $B_N\langle x, Z\rangle$ modulo the ideal generated by $F(x, Z) = (F_1(x, Z), \ldots, F_n(x, Z))$, where $B_N = K_N[[u_q(N)]]$, $Z = (Z_1, \ldots, Z_n)$ is a system of indeterminates and $F_j(x, Z) = -x_j + \Sigma^n{}_{a=1} s_{ja}Z_a + \Sigma_A u_{iA}Z^A$, $j < n$, $F_n(x, Z) = -x_n + Z_n$. (Note that Z_j corresponds to z_j in view of (1) in the beginning of this section.) Bars denoting the residue classes $\text{mod}((u_q(N)))A_q(N)$,

$$K_N[\bar{x}, \overline{Z}]/(\overline{F(\bar{x}, \overline{Z})})K_N[\bar{x}, \overline{Z}]$$

is a finite $K_N[\bar{x}]$-module. In fact it is generated by those monomials in $\overline{Z}$ whose degrees in the $\overline{Z_j}$ are less than N. By Remark (b) and Prop. 1 (Weierstrass theorem) of §2, $A_q(N) = B_N\langle x, Z\rangle/(F(x, Z))B_N\langle x, Z\rangle$ is finitely generated as a module over $B_N\langle x\rangle = E_q(N)$. Since k is algebraically closed, $QK_N[x]$ is prime and so is $QE_q(N)$. The last assertion then follows the first. Q.E.D.

Notation (5.6). Let P be the prime ideal of $\mathbf{Y}$ in $k[x]$. Then, for each i, $1 \leq i \leq n - i$, we denote by $P(i)_b{}^*$ with $1 \leq b \leq e(i)$ those prime ideals in $A_q(N^{(i)})$ which contain and contract to $PE(i)$ where $E(i)$ denotes $E_q(N^{(i)})$.

Let $\text{Diff}^m(A(i))$ denote the $A(i)$-module of *differential operators in $A(i)$ of orders $\leq m$ with reference to the variables $y(i) = (z_1, \ldots, z_{n-i})$*. To be explicit, it is the free $A(i)$-module freely generated by those *elementary differential operators ∂_B, $B \in \mathbf{Z}_0{}^{n-i}$* with $|B| \leq m$, which are defined as follows: Express each element h of $A(i)$ in the form

$$h = \sum_{C \in \mathbf{Z}_0{}^{n-i}} h_C y(i)^C \qquad \text{where } h_C \in B(i),$$

then we have

$$\partial_B(h) = \sum_{C-B \in \mathbf{Z}_0{}^{n-i}} \binom{C}{B} h_C y(i)^{C-B}$$

It is clear that ∂_B maps $A(i)$ into itself. If the characteristic of k is zero,

$$\partial_B = \left(\prod_j (b_j!)\right)^{-1} \frac{\partial^{|B|}}{(\partial z_1)^{b_1} \cdots (\partial z_{n-i})^{b_{n-i}}}$$

where $B = (b_1, \ldots, b_{n-i})$. In general, none of these elementary differential operators are affected at all, as endomorphisms of $A(i)$, if the variables $y(i)$ for reference are replaced by $y(i) - \eta(i)$ with any $\eta(i) = (\eta_1, \ldots, \eta_{n-i})$, $\eta_j \in B(i)$. (Recall that $B(i)\langle y(i)\rangle = B(i)\langle y(i) - \eta(i)\rangle$ by Cor. 1 of Prop. 1, §3.) Moreover with respect to the canonical inclusion $A(i) \to L[[y(i) - \eta(i)]]$, differential operators in $A(i)$ extend in a natural way to such in $L[[y(i) - \eta(i)]]$, as is clear from the above expression for ∂_B. Thus we have the canonical injection $\text{Diff}^m(A(i)) \to \text{Diff}^m(L[[y(i) - \eta(i)]])$, which is merely the extension of the coefficient ring from $A(i)$ to $L[[y(i) - \eta(i)]]$.

Now pick any i and b as in Remark (5.6). Let $m(i, b)$ be the smallest integer $m \geq 0$ such that

$$\text{(5.7)} \qquad \text{Diff}^m(A(i))(g^{(i)}) \not\subset P(i)_b = P(i)_b{}^* \cap A(i)$$

where $g^{(i)}$ is the element selected by Prop. 3 and $P(i)_b{}^*$ is that of (5.6). We have $0 \leq m(i, b) \leq \deg(\bar{g}^{(i)})$ where the bar denotes the residue class modulo $((u_q(N^{(i)})))A(i)$. We then let

$$\text{(5.8)} \qquad J^*(i, b) = \text{Diff}^m(A(i))(g^{(i)}) \text{ with } m = m(i, b)$$

which is an ideal in $A(i)$. And we let

$$\text{(5.9)} \qquad J'(i, b) = J^*(i, b)A_q(N^{(i)}) \cap E(i) \text{ with } E(i) = E_q(N^{(i)})$$

Each element h of $E(i) = K(i)[x][[u_q(N^{(i)})]]$ can be uniquely expressed as follows:

$$h = \sum_M{}^* \left(\sum_\gamma \gamma h_{M,\gamma}\right) M \text{ with } h_{M,\gamma} \in k[x]$$

where Σ^* denotes an infinite sum for all monomials M in $u_q(N^{(i)})$, $\{\gamma\}$ is any fixed free base of $K(i)$ as a k-module and the sum for the γ's is finite.

We finally define an ideal in $k[x]$ as follows:

$$J = \prod_{i,b} J(i, b) \tag{5.10}$$

where $0 \leq i \leq n - 1$, $1 \leq b \leq e(i)$ and $J(i, b)$ denotes the ideal in $k[x]$ which is generated by all those $h_{M,\gamma}$ for all $h \in J'(i, b)$.

Notation (5.11). Let $\mathbf{U}$ be the open (in Zariski topology) subset $\mathbf{Y} \setminus (\mathbf{Z} \cup \mathbf{Z}')$ of $\mathbf{Y}$ where $\mathbf{Z}$ is the singular locus of $\mathbf{Y}$ and $\mathbf{Z}'$ is the closed algebraic set in $\mathbf{X}$ defined by the ideal J in $k[x]$ of (5.10).

PROPOSITION 5. *The open subset* $\mathbf{U}$ *is dense in* $\mathbf{Y}$.

Proof. In other words, J is not contained in P. For this, we shall prove that $J(i, b) \not\subset P$ for every pair (i, b). In fact, if this were false, we should have $J'(i, b) \subset PE_q(N^{(i)})$. By (5.6) and (5.9), we must have $J^*(i, b)A_q(N^{(i)}) \subset P(i)_b{}^*$ and hence $J^*(i, b) \subset P(i)_b$ which contradicts (5.7)-(5.8). Q.E.D.

THEOREM. *The open dense subset* $\mathbf{U}$ *of* $\mathbf{Y}$, *defined by* (5.11), *has the following property*: *The dimensionality type of* $\mathbf{X}$ *at* $\xi \in \mathbf{U}$ *is independent of* ξ *and is equal to the dimensionality type of* $\mathbf{X}$ *at the generic point of* $\mathbf{Y}$.

Proof. Let us pick any point $\xi \in \mathbf{U}$, $= (\xi_1, \ldots, \xi_n)$ with $\xi_j \in k$. Let $\hat{\mathbf{R}} = k[[x - \xi]]/(f)k[[x - \xi]]$ and $\hat{\mathbf{X}} = \operatorname{Spec}(\hat{\mathbf{R}})$ which is the algebroid variety associated with $\hat{\mathbf{X}}$ at ξ. Let $\hat{M} = \max(\hat{\mathbf{R}})$ and $\hat{P} = P\hat{\mathbf{R}}$ which is a prime ideal in $\hat{\mathbf{R}}$. Then, by the definition and theorems of Zariski [Z, §6],

$$\text{d.t.}(\hat{\mathbf{X}})_{\hat{M}} = \text{the dimensionality type of } \mathbf{X} \text{ at } \xi$$

$$\text{d.t.}(\hat{\mathbf{X}})_{\hat{P}} = \text{the dimensionality type of } \mathbf{X} \text{ at the generic point of } \mathbf{Y}$$

Let us call these numbers d and d' respectively. We want to prove $d = d'$. Let Q be the maximal ideal corresponding to ξ in $k[x]$, i.e., $Q = (x - \xi)k[x]$. By Cor. 1 of Prop. 1, §3, we have an isomorphism $\mathbf{E} = K[x][[u]] \xrightarrow{\sim} K[z][[u]] = \mathbf{A}$, which induces isomorphisms $K[[u]] \xrightarrow{\sim} (K[x]/Q)[[u]] \xrightarrow{\sim} \mathbf{E}/Q\mathbf{E} \xrightarrow{\sim} \mathbf{A}/Q\mathbf{A}$. Let η_j be the element of $K[[u]]$ which corresponds to the residue class of z_j. Let $z_j{}^* = z_j - \eta_j$, $z^* = (z_1{}^*, \ldots, z_n{}^*)$ and $z(i)^* = (z_1{}^*, \ldots, z_{n-i}{}^*)$. Now with respect to the canonical injections $A(i) \to L[[z(i)^*]]$, if we define the successive reduced discriminants $f^{(i)} \in L[[z(i)^*]]$ for $f^{(0)} = f$ as we did in Prop. 4, then $f^{(i)}$ divides $g^{(i)}$ for all i by the same proposition. By Remark (4.3), §4, or by the isomorphism $\mathbf{E} \xrightarrow{\sim} \mathbf{A}$ as above, we can write

$$z_i^* = \sum_j S_{ij}(x_j - \eta_j) + \sum_A U_{iA}(x - \eta)^A, \qquad 1 \leq i \leq n - 1,$$

then those coefficients S_{ij} and U_{iA} are algebraically independent over $k[x]$. Moreover, if $\hat{P}(i)$ denotes the contraction of $\hat{P}$ to $L[[z(i)^*]]$, then by (4.2)–(4.3), §4,

$$d' = \text{the index } i \text{ such that } \hat{P}(i) \text{ is a smooth point of } \hat{\mathbf{X}}(i)$$

where $\hat{\mathbf{X}}(i)$ denotes $\operatorname{Spec}(L[[z(i)^*]]/f^{(i)}L[[z(i)^*]])$. Similarly if $\hat{M}(i) = \max(L[[z(i)^*]])$, then

$$d = \text{the index } i \text{ such that } \hat{M}(i) \text{ is a smooth point of } \hat{\mathbf{X}}(i)$$

Now for each i, $0 \leq i \leq n - i$, $\hat{Q}$ is a smooth point of $\hat{\mathbf{X}}(i)$ if and only if ($f^{(i)} \in \hat{Q}$ and) the order of $f^{(i)}$ in the local ring $L[[z(i)^*]]_{\hat{Q}}$ is equal to one, where $\hat{Q}$ is either $\hat{M}(i)$ or $\hat{P}(i)$. Therefore to prove $d = d'$, it is enough that $f^{(i)}$ has the same order either in $L[[z(i)^*]]$ or $L[[z(i)^*]]_{\hat{P}(i)}$ for all i. Since $g^{(i)}$ is non-zero and is divisible by $f^{(i)}$, it is then enough that $g^{(i)}$ has the same order either in $L[[z(i)^*]]$ or in $L[[z(i)^*]]_{\hat{P}(i)}$ for all i. Note that $L[[z(i)^*]]/\hat{P}(i)$ is regular because ξ is a smooth point of $\mathbf{Y}$. Therefore we can choose a regular system of parameters, say w, of $L[[z(i)^*]]$ so that $L[[z(i)^*]] = L[[w]]$ and $\hat{P}(i)$ is generated by a suitable subsystem of w. $\operatorname{Diff}^m(L[[z(i)^*]])$ stays to be the same if we replace the reference variables from $z(i)^*$ to w. With the help of this w, it is then easy to verify that the order of $g^{(i)}$ in $L[[z(i)^*]]_{\hat{P}(i)}$ is equal to the minimum of those integers $m \geq 0$ such that $\operatorname{Diff}^m(L[[z(i)^*]](g^{(i)})$ is not contained in $\hat{P}(i)$. Call it $m^{(i)}$. With respect to the canonical injection $E(i) = K(i)[x - \xi][[u_q(N^{(i)})]] \to L[[z^*]]$, $PL[[z^*]]$ contracts to $PE(i)$. Hence with respect to the injection $A_q(N^{(i)}) \to L[[z^*]]$, $PL[[z^*]]$ contracts to one of those ideals defined in (5.6), say $P(i)_b^*$. This implies that $\hat{P}(i)$ contracts to $P(i)_b$ of (5.7) by the injection $A(i) \to L[[z(i)^*]]$. Since $\operatorname{Diff}^m(L]]z(i)^*]])$ is merely the coefficient ring extension of $\operatorname{Diff}^m(A(i))$, the number $m^{(i)}$ defined above must be equal to the number $m(i, b)$ defined in (5.7). Now the point ξ is in $\mathbf{U}$ and hence is not in $\mathbf{Z}'$ in the notation of (5.11). Hence, in view of (5.10), there exists at least one $h_{M,\gamma}$ with $h \in J'(i, b)$ such that $h_{M,\gamma}(\xi) \neq 0$. This shows that $(x - \xi)E(i)$ does not contain $J'(i, b)$ and therefore $(z - \eta)A_q(N^{(i)})$ does not contain $J^*(i, b)A_q(N^{(i)})$ in the notation of (5.8)-(5.9). Now $(z^*) \cdot A_q(N^{(i)})$ with $z^* = z - \eta$ is a prime ideal as is easily seen by $A_q(N^{(i)}) = K(i)[z^*][[u_q(N^{(i)})]]$, and it contracts to the prime ideal $(z(i)^*)A(i)$ where $A(i) = K(i)[z(i)^*][[u_q(N^{(i)}]]$. Hence the above result implies that $(z(i)^*)A(i)$ does not contain $J^*(i, b) = \mathrm{Diff}^m(A(i))(g^{(i)})$ with $m = m(i, b) = m^{(i)}$. It follows that $(z(i)^*)L[[z(i)^*]]$ $(= \hat{M}(i))$ does not contain $\mathrm{Diff}^m(A(i))(g^{(i)})L[[z(i)^*]] = \mathrm{Diff}^m(L[[z(i)^*]])(g^{(i)})$ with the same m, because the prime ideal $(z(i)^*)A(i)$ is clearly the contraction of $(z(i)^*)L[[z(i)^*]]$ to $A(i)$. It follows that the order of $g^{(i)}$ in $L[[z(i)^*]]$ is at most $m^{(i)}$ and hence is equal to $m^{(i)}$, which was the order of $g^{(i)}$ in $L[[z(i)^*]]_{\hat{P}(i)}$. Q.E.D.

HARVARD UNIVERSITY

REFERENCE

[Z] Oscar Zariski: "*Foundations of a general theory of equisingularity on r-dimensional algebraoid and algebraic varieties of embedding dimension r + 1*," pages 380–419.

SUR LA GEOMETRIE DES SURFACES COMPLEXES I. TANGENTES EXCEPTIONNELLES

By Lê Dũng Tráng et Bernard Teissier

Introduction. Dans ce travail, nous commençons l'étude de l'équisingularité du point de vue des limites d'espaces tangents; il s'agit de comprendre à quel point la structure de l'espace des limites d'espaces tangents détermine la géométrie d'une singularité de surface. Il se trouve que du point de vue de l'équisingularité la partie essentielle de l'espace des limites d'espaces tangents d'une surface dans $\mathbf{C}^3$ est composée de pinceaux de plans passant par certaines génératrices du cône tangent à la surface au point singulier, dites *génératrices exceptionnelles*.

Nous donnons ici plusieurs caractérisations des génératrices exceptionnelles et montrons en particulier que pour toute projection assez générale de notre surface sur $\mathbf{C}^2$ il existe une composante du lieu critique de la projection qui est tangente à une génératrice exceptionnelle donnée: elles apparaissent ainsi comme tangentes fixes du lieu critique, indépendantes de la projection. Nous analysons ici le cas des surfaces qui n'ont pas de tangentes exceptionnelles et montrons que dans ce cas si le cône tangent est réduit la surface se déforme équisingulièrement sur son cône tangent. La motivation de ce travail est de définir un concept d'équisingularité faisant comprendre le lien entre le critère discriminant de Zariski pour une projection générale et la structure de l'espace des limites d'espaces tangents, théorie que nous développerons dans un travail ultérieur. Les résultats présentés ici, tout en étant de nature préliminaire, sont d'une part techniquement essentiels pour la théorie que nous nous proposons d'étudier et d'autre part encourageants en ce qui concerne l'étude de l'équisingularité de ce point de vue. Comme le lecteur le verra nous utilisons constamment des idées de la théorie de l'équisingularité introduite par O. Zariski ainsi que les résultats qu'il a démontrés. C'est pour nous un grand plaisir de lui dédier ce travail et ceux qui suivront.

1. Tangentes Exceptionnelles. Dans [5], J. P. G. Henry et Lê D. T. ont introduit la notion de génératrices exceptionnelles du cône tangent d'un germe de surface analytique complexe de $(\mathbf{C}^3, 0)$. Dans ce paragraphe nous allons donner plusieurs caractérisations de ces génératrices exceptionnelles. Ceci nous sera utile dans la suite pour l'étude de la géométrie des surfaces.

(1.1) *Notations*: Soit $(X, 0)$ une surface réduite dans $(\mathbf{C}^3, 0)$. Soit $C_{X,0}$ son cône tangent. On appellera *projection locale* de $(\mathbf{C}^3, 0)$ dans $(\mathbf{C}^2, 0)$ un germe de morphisme analytique $p: (\mathbf{C}^3, 0) \to (\mathbf{C}^2, 0)$ de rang 2 à l'origine. On appellera *projection locale de* $(X, 0)$ *sur* $(\mathbf{C}^2, 0)$ la restriction $\pi = p|_X$ à X d'une projection locale p de $(\mathbf{C}^3, 0)$ sur $(\mathbf{C}^2, 0)$. L'espace critique de π, défini par le 0-ième idéal de Fitting du faisceau $\Omega^1{}_\pi$ des formes différentielles relatives de π (cf. [18]), sera noté C_π. On remarque que l'on peut choisir les coordonnées locales de $\mathbf{C}^3$ en 0 de telle façon que $p(x, y, z) = (x, y)$. Dans ce cas si $f(x, y, z) = 0$ est une équation de $(X, 0)$ dans $(\mathbf{C}^3, 0)$, le $\mathcal{O}_X$-module $\Omega^1{}_\pi$ est isomorphe à $\mathcal{O}_X/f_z'\mathcal{O}_X$, où f_z' est la dérivée partielle de f par rapport à z, par conséquent le 0-ième idéal de Fitting du $\mathcal{O}_X$-module $\Omega^1{}_\pi$ n'est autre que l'idéal de $\mathcal{O}_X$ engendré par f_z'.

Si π induit sur $(C_\pi, 0)$ un germe de morphisme fini, on peut définir l'espace discriminant de π à l'aide du 0-ième idéal de Fitting du $\mathcal{O}_{\mathbf{C}^2,0}$ module $(\pi_*\mathcal{O}_{C_\pi})_o$ (cf. [18] (2.6)).

(1.2) *Rappels sur l'équisingularité*: Nous allons rappeler ici celles des nombreuses caractérisations équivalentes des familles équisingulières de courbes localement planes réduites que nous utiliserons dans la suite.

Soit $\rho: Z \to Y$ un morphisme plat d'espaces analytiques réduits à fibres réduites de dimension 1. On suppose que pour tout $y \in Y$ la dimension de plongement de la fibre Z_y en chacun de ses points est au plus égale à 2 (i.e. les fibres sont des courbes localement planes). Etant donnés $y \in Y$ et $z \in \rho^{-1}(y)$ les conditions suivantes sont équivalentes:

1) Il existe un voisinage ouvert U de z dans Z tel que, notant B l'ouvert $\rho(U)$ et notant, pour tout $y' \in B$, par z_i' $(1 \leq i \leq k)$ les points singuliers de la courbe $Z_{y'} \cap U$, on ait $\mu(Z_{y'}, z) = \sum^k{}_{i=1} \mu(Z_{y'}, z_i')$ où $\mu(Z_y, z)$ désigne le nombre de Milnor de Z_y en z.

2) Il existe un voisinage U de z dans Z, et une section $\sigma: B \to U$ de ρ (où $B = \rho(U)$) tels que

α) ρ induise une submersion d'espaces non singuliers

$$U - \sigma(B) \to B$$

β) On a $\mu(Z_{y'}, \sigma(y')) = \mu(Z_y, z)$ pour tout $y' \in B$.

2′) Comme 2) mais avec en plus:

γ) On a $m(Z_{y'}, \sigma(y')) = m(Z_y, z)$, où $m(Z_y, z)$ désigne la multiplicité dans l'anneau local $\mathcal{O}_{Z_y,z}$ de son idéal maximal.

3) Il existe un voisinage ouvert U de z dans Z, une section σ: $B \to U$ de ρ, avec $B = \rho(U)$ et une projection η: $B \times \mathbf{C}^2 \to B \times \mathbf{C}$ tels que:

α) on ait un plongement local i: $(U, z) \to (Y \times \mathbf{C}^2, y \times 0)$ faisant commuter le diagramme:

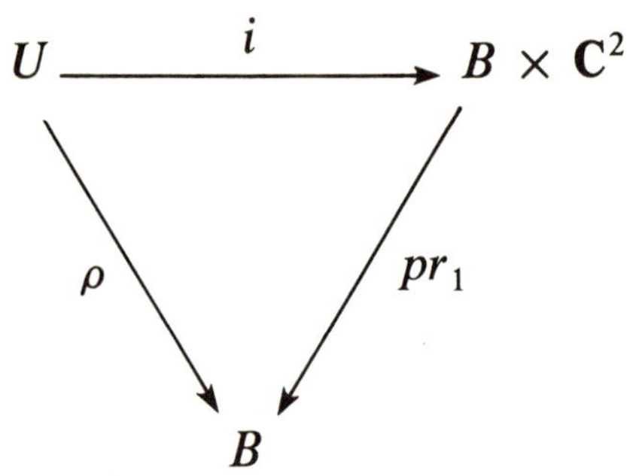

pour lequel on a $i(\sigma(B)) = B \times \{0\}$.

β) le lieu de ramification de la restriction de η à U (i.e. l'ensemble réduit sous-jacent au discriminant de cette restriction) coïncide avec $B \times \{0\}$.

4) Il existe un voisinage U de z dans Z et une section σ: $B \to U$ de ρ, avec $B = \rho(U)$, tels que, pour chaque $y' \in B$, le type topologique du germe de courbe plane $(Z_{y'}, \sigma(y'))$ soit le même.

5) Il existe un voisinage U de z dans Z et une section σ: $B \to U$ de ρ, avec $B = \rho(U)$, telles que, pour tout morphisme analytique

$h: \tilde{B} \to B$, avec $\tilde{B}$ non singulier, dans le diagramme cartésien:

$$\begin{array}{ccc} \tilde{U} & \longrightarrow & U \\ {\scriptstyle \tilde{\rho}}\downarrow & & \downarrow{\scriptstyle \rho} \\ \tilde{B} & \xrightarrow{h} & B \end{array}$$

on ait une section $\tilde{\sigma}$ de $\tilde{\rho}$ qui provient de σ et le couple de strates $(\tilde{U} - \tilde{\sigma}(\tilde{B}), \tilde{\sigma}(\tilde{B}))$ satisfait les conditions de Whitney en tout point de $\tilde{\sigma}(\tilde{B})$.

Remarques. 1. Dans la pratique nous utiliserons ce résultat dans le cas où la base Y est non singulière.

2. On rappelle que tout morphisme plat étant ouvert (cf. [2]), U étant ouvert, $B = \rho(U)$ est ouvert.

3. L'ensemble des points de y de Y où les conditions équivalentes ci-dessus sont réalisées est un ouvert analytique partout dense de Y quand la restriction de ρ à son espace critique C_ρ est finie.

4. Si les conditions équivalentes précédentes sont satisfaites pour $y \in Y$ et $z \in \rho^{-1}(y)$, on dit que Z est *équisingulier le long de $\sigma(Y)$ en z,* si σ est une section de ρ comme en 2), telle que $\sigma(y) = z$.

5. Soit $(Z, 0)$ un germe de surface réduite dans $(\mathbf{C}^3, 0)$. On peut montrer que, si Z est équisingulière le long de $\sigma(Y)$ en 0, cela implique que $\sigma(Y)$ est non singulière en 0.

L'équivalence des conditions ci-dessus est conséquence de résultats antérieurs. De façon précise:

1) $\Rightarrow$ 2) résulte de [19] et [11];

2) $\Rightarrow$ 1) est évident;

2) $\Rightarrow$ 2′) résulte de [12] ou [20] (pour une démonstration algébrique);

2′) $\Rightarrow$ 3) résulte des travaux de O. Zariski [27] (cf. aussi [21]);

3) $\Rightarrow$ 4) résulte aussi de [27] (cf. aussi [21]);

4) $\Rightarrow$ 5) est aussi un résultat de O. Zariski [29] (c'est aussi un cas particulier des résultats de [22] Chap. 2 §3).

5) $\Rightarrow$ 2) résulte de ce que les conditions de Whitney impliquent la constance du type topologique d'après un théorème de Mather-Thom (cf. [15]) et de ce que le nombre de Milnor est un invariant du type topologique (cf. [22] Chap. 1, théorème 1.4 ou [10] théorème (3.3)). On aurait pu utiliser les résultats de J. Briançon, J. P. G. Henry, J. P. Spéder dans [1] qui donne une réciproque aux résultats cités ci-dessus ([22] Chap. 2 §3).

(1.3) *Rappels sur les limites d'espaces tangents.* Soient X un espace analytique réduit équidimensionel de dimension d, $\Omega^1{}_X$ le faisceau des différentielles de X. Dans [18], on considère l'espace analytique en grassmanienne $\mathrm{Grass}_d(\Omega^1{}_X)$ au-dessus de X:

$$g\colon \mathrm{Grass}_d(\Omega^1{}_X) \to X$$

Sur l'ouvert X_o des points non singuliers de X on a une section $\tau\colon X_o \to \mathrm{Grass}_d(\Omega^1{}_X)$ de g (car g est un isomorphisme au-dessus de X_o). Soit $\tilde{X}$ la fermeture de $\tau(X_o)$ dans $\mathrm{Grass}_d(\Omega^1{}_X)$. On peut montrer que $\tilde{X}$ est un espace analytique réduit (cf. [18]). On appelle *modification de Nash* la restriction γ de g à $\tilde{X}$:

$$\gamma\colon \tilde{X} \to X$$

Remarque (1.3.1). Dans le cas où X est un sous-espace analytique d'un ouvert U de $\mathbf{C}^N$, on a une application analytique ψ de X_o dans $\mathrm{Grass}_d(\mathbf{C}^N)$ qui à $x \in X_o$ fait correspondre $T_x(X_o)$, l'espace tangent de X_o en x. On peut montrer que $\tilde{X}$ est isomorphe à l'adhérence du graphe $\mathrm{Gr}\,\psi$ de ψ dans $X \times \mathrm{Grass}_d(\mathbf{C}^N)$ et que, pour cet isomorphisme, la restriction à cette adhérence $\overline{\mathrm{Gr}\,\psi}$ de la première projection donne le diagramme commutatif:

$$\begin{array}{ccc} \tilde{X} & \xrightarrow{\ \sim\ } & \overline{\mathrm{Gr}\,\psi} \\ & {\scriptstyle\gamma}\searrow \quad \swarrow & \\ & X & \end{array}$$

Dans ce qui suit on s'intéresse au cas d'un germe de surface dans $(\mathbf{C}^3, 0)$. Le morphisme γ est alors isomorphe à l'éclatement de l'idéal engendré dans $\mathcal{O}_{X,0}$ par les dérivées partielles d'une fonction $f(x, y, z)$ qui définit $(X, 0)$ dans $(\mathbf{C}^3, 0)$.

On appellera *espace des limites d'espaces tangents à X en x* l'espace analytique $\gamma^{-1}(x)$. L'espace analytique réduit sous-jacent sera appelé *ensemble des limites d'espaces tangents à X en x* et ses éléments sont *les limites d'espaces tangents à X en x.*

Remarque (1.3.2). Quand X est une surface analytique dans un ouvert U de $\mathbf{C}^3$, la dimension de $\gamma^{-1}(x)$ est 1 ou 0 quand $x \in X$.

Dans ce cas $\gamma^{-1}(x)$ est isomorphe à un sous-schéma algébrique de $\check{\mathbf{P}}^2$, espace projectif de dimension 2 des directions de plans de $\mathbf{C}^3$.

Dans [9] nous montrons que l'ensemble $|\gamma^{-1}(x)|$ sous-jacent à $\gamma^{-1}(x)$, est une variété algébrique union de la courbe duale $(\mathrm{Proj}\,|C_{X,x}|)^v$ a la courbe projective réduite Proj $|C_{X,x}|$ associée au cône tangent $C_{X,x}$, et de droites $\mathcal{L}_1, \ldots, \mathcal{L}_k$ de $\check{\mathbf{P}}^2$ représentant des pinceaux de directions de plans de $\mathbf{C}^3$ dont les axes sont des génératrices de $C_{X,x}$. Ces droites sont donc tangentes à $|\mathrm{Proj}\, C_{X,x}|^v$ ou passent par une de ses composantes de dimension 0. Nous appelons ces génératrices les *génératrices exceptionnelles* de X en x; on dira aussi que ce sont les *tangentes exceptionnelles de X en x.* La structure de $\gamma^{-1}(x)$ et donc les tangentes exceptionnelles ne dépend que du type analytique du germe (X, x).

Exemples (1.3.3). Il peut n'y avoir aucune tangente exceptionnelle de X en x: par exemple quand X est non singulier en x. Si X est singulier en x, quand (X, x) est isomorphe au cône de $\mathbf{C}^3$ sur une courbe non singulière de $\mathbf{P}^2$, on remarquera que grâce à l'identité d'Euler X n'a pas de tangentes exceptionnelles en x. Nous verrons plus bas une réciproque "à équisingularité près" de ce résultat.

Remarquons que si $|C_{X,x}|$ est un plan, l'ensemble des limites d'espaces tangents $|\gamma^{-1}(x)|$ de X en x est constitué uniquement des pinceaux $\mathcal{L}_1, \ldots, \mathcal{L}_k$.

Nous allons maintenant décrire d'après [9] le cas où $\gamma^{-1}(x)$ est de dimension 0.

Remarquons que toute submersion analytique $l: (\mathbf{C}^3, 0) \to (\mathbf{C}, 0)$ telle que $l^{-1}(0) \cap X$ soit de dimension 1 nous permet de considérer X comme la famille de courbes $\lambda: (X, 0) \to (\mathbf{C}, 0)$, restriction de l à X.

On dira que l *est assez générale pour* X *en* 0 si $l^{-1}(0) \cap X$ est réduite en 0 et son nombre de Milnor en 0 est minimum parmi toutes les sections de X par des surfaces non singulières passant par 0. Sous ces hypothèses la restriction de l au lieu singulier Σ de X est un morphisme fini et de rang 1 en dehors de 0. Par conséquent pour $t \neq 0$ assez petit $l^{-1}(t)$ coupe transversalement le lieu singulier Σ de X en des points $x_1(t), \ldots, x_l(t)$ qui sont les points singuliers de $X \cap l^{-1}(t)$ quand le représentant X du germe (X, x) est convenablement choisi. En vertu de la remarque 3 de (1.2) ci-dessus, X est équisingulier en $x_i(t)$, pour $1 \leq i \leq l$, le long de Σ et, d'après O. Zariski [29], l est assez générale pour X en $x_i(t)$ $(1 \leq i \leq l)$, car en un point d'équisingularité une surface lisse transverse au lieu singulier donne nécessairement une intersection avec le nombre de Milnor minimum (en fait dans [29], O. Zariski démontre que la somme du nombre de Milnor et de la multiplicité moins 1, est minimum, mais ceci entraîne évidemment que le nombre de Milnor et la multiplicité sont minima). (1.3.3.1) Rappelons que pour un germe de courbe analytique $(C, 0)$ on appelle *discriminant* de $(C, 0)$ le nombre:

$$\Delta(C, 0) = \mu(C, 0) + m(C, 0) - 1$$

(la terminologie provient de ce que $\Delta(C, 0)$ est la multiplicité du discriminant d'une projection assez générale (i.e. de degré $m(C, 0)$) sur une droite complexe, (cf. [21] et comparer à [29]).

Dans [9], on a établi:

THÉORÈME (1.3.4). *Soit* l: $(C^3, 0) \to (\mathbf{C}, 0)$ *une submersion analytique assez générale pour* X *en* 0. Soient $x_1(t), \ldots, x_l(t)$ *les points d'intersection de* $l^{-1}(t)$ *avec le lieu singulier* Σ *de* X, *où* X *est un représentant convenable assez petit de* $(X, 0)$. *L'ensemble des limites d'espaces tangents à* X *en* 0 *est fini si et seulement si*

$$\Delta(l^{-1}(0) \cap X, 0) = \sum_{i=1}^{l} \Delta(l^{-1}(t) \cap X, x_i(t))$$

pour $t \neq 0$ *assez petit.*

Dans le cas où *le lieu singulier* Σ *de* X *est non singulier en* 0, on a $l = 1$, et le critère discriminant de O. Zariski (cf. [27]) donne:

COROLLAIRE (1.3.5). *La surface X est équisingulière en* 0 *le long de* Σ *si et seulement si les limites d'espaces tangents de X en* 0 *sont finies. Dans ce cas ces limites sont précisément les plans du cône tangent.*

Remarque (1.3.6). Il est équivalent de dire que les limites d'espaces tangents de X en 0 sont finies et que la modification de Nash $\gamma: \tilde{X} \to X$ est dominée par la normalisation $n: \overline{X} \to X$ de X. Dans l'hypothèse où le lieu singulier Σ est non singulier en 0, c'est le critère jacobien d'equisingularité de O. Zariski (cf. [27]).

(1.4) Caractérisation des tangentes exceptionnelles

(1.4.1) Soit $(X, 0)$ un germe de surface analytique définie dans $(\mathbf{C}^3, 0)$ par $f = 0$. Soit $f = f_m + f_{m+1} + \cdots$ le développement de Taylor de f en 0, avec $m = m(X, 0)$ multiplicité de X en 0. Soit U un voisinage ouvert de 0 où la série de Taylor de f ci-dessus converge. Si U est convenablement choisi la série $F = f_m + tf_{m+1} + t^2 f_{m+2} + \cdots = t^{-m}f(tx, ty, tz)$ converge pour tout $(x, y, z) \in U$ et $t \in \mathbf{D}$, où $\mathbf{D} = \{t / |t| < 2\}$. Soit $\mathfrak{X}$ l'hypersurface de $U \times \mathbf{D}$ définie par $F = 0$. La projection sur $\mathbf{D}$ restreinte à $\mathfrak{X}$ définit un morphisme analytique plat $\varphi: \mathfrak{X} \to \mathbf{D}$, muni d'une section $\sigma: \mathbf{D} \to \mathfrak{X}$ d'image $\{0\} \times \mathbf{D}$. Cette construction est un cas particulier d'une construction générale (cf. [3] et [23]).

(1.4.2) Soit $g: Z \to \mathbf{D}$ un morphisme analytique à valeurs dans un disque ouvert centré en 0 de $\mathbf{C}$. Soit $Z = \bigcup_{\alpha \in A} Z_\alpha$ une stratification de Z, i.e. une partition localement finie de Z en sous-espaces analytiques non singuliers localement fermés qu'on appellera strates de Z. Soient Z_α et Z_β deux strates de Z telles que $Z_\beta \subset Z_\alpha$. On suppose que les restrictions de g à Z_α et Z_β sont de rang constant. Soit $z \in Z_\beta$, on dit que le couple de strates (Z_α, Z_β) satisfait la *condition de Thom en* z relativement à g, si pour toute suite $(z_n)_{n \in \mathbf{N}}$ de points de Z_α qui converge vers z et pour laquelle la suite des espaces tangents $T_{z_n}(g^{-1}(g(z_n)) \cap Z_\alpha)$ a une limite T, on a:

$$T \supset T_z(g^{-1}(g(z)) \cap Z_\beta)$$

la limite et l'inclusion ci-dessus étant comprises dans une grassmannienne convenable grâce à un plongement local de Z dans $\mathbf{D} \times \mathbf{C}^N$. Si (Z_α, Z_β) satisfait la condition de Thom, si (Z_α, Z_β) satisfait la condition de Thom en tout point z de Z_β.

Dans [6], H. Hironaka a montré que, sous la condition que, pour tout $t \in \mathbf{D}$, $g^{-1}(t)$ soit contenu dans l'adhérence dans Z de $Z - g^{-1}(t)$, il existe une stratification comme ci-dessus telle que, pour tout couple $(\alpha, \beta) \in A \times A$ avec $Z_\beta \subset \overline{Z_\alpha}$, (Z_α, Z_β) satisfait la condition de *R*. Thom relativement à g et la condition de H. Whitney. Remarquons que la condition précédente est satisfaite, quand g est plat.

Dans le cas de la déformation ci-dessus dans (1.4.1), $\varphi\colon \mathfrak{X} \to \mathbf{D}$ est plat. On peut stratifier $\mathfrak{X}$ de telle sorte que $\varphi^{-1}(0) = \mathfrak{X}$ soit une union de strates, que la partie non singulière de $\mathfrak{X}$, la partie singulière de $\mathfrak{X}$ moins $\sigma(\mathbf{D}) = \{0\} \times \mathbf{D}$ et $\sigma(\mathbf{D}) - \{0\}$ soient des strates et que cette stratification satisfasse la condition de Thom relativement à φ et celle de Whitney.

Au cours de la démonstration du théorème, nous donnerons une relation intéressante entre la condition de Thom et l'équisingularité.

(1.4.3) Nous allons réaliser $\mathfrak{X}$ comme un ouvert dans l'éclatement de l'origine dans la famille triviale de $U \times \mathbf{D}$ définie par $f(x, y, z) = 0$.

Décrivons d'abord l'éclatement $\epsilon\colon E \to U \times \mathbf{D}$ de l'origine dans $U \times \mathbf{D}$. Soient A, B, C, D les coordonnées homogènes de $\mathbf{P}^3$ associées aux coordonnées (x, y, z, t) de $U \times \mathbf{D}$. Dans $U \times \mathbf{D} \times \mathbf{P}^3$ l'éclatement de l'origine dans $U \times \mathbf{D}$ est donné par:

$$\begin{cases} Dx - At = 0 \\ Dy - Bt = 0 \\ Dz - Ct = 0 \end{cases}$$

ou par

$$\begin{cases} Cx - Az = 0 \\ Cy - Bz = 0 \\ Ct - Dz = 0 \end{cases}$$

etc....

Notons U_D la carte de $\mathbf{P}^3$ où $D \neq 0$. L'intersection $E \cap V_D$, avec $V_D = U \times \mathbf{D} \times U_D$, est isomorphe à un ouvert de $\mathbf{C}^4$ et on peut choisir comme coordonnées locales $a = A/D$, $b = B/D$, $c = C/D$ et $t' =$

$t \circ \epsilon$. Dans cette carte l'image inverse de la famille triviale $X \times \mathbf{D}$ a pour équation:

$$f(at', bt', ct') = 0$$

et par conséquent la transformée stricte de $X \times \mathbf{D}$ par l'éclatement ϵ intersecte cette carte en l'hypersurface d'équation:

$$t'^{-m} f(at', bt', ct') = 0$$

où $m = m(X, 0)$. Dans un ouvert approprié de cette carte on retrouve bien un sous-espace analytique isomorphe à $\mathfrak{X}$. La projection sur $\mathbf{D}$ composée avec l'éclatement ϵ donne le morphisme analytique qui correspond à φ.

Dans les cartes V_A, V_B, V_C, un calcul analogue montre que la trace de la transformée stricte par ϵ de $X \times \mathbf{D}$ est isomorphe au produit par $\mathbf{D}$ de la trace de la transformée stricte par ϵ de $X \times \{0\}$. Par exemple $E \cap V_C$ est isomorphe à un ouvert de $\mathbf{C}^4$ et on peut choisir comme coordonnées $a = A/C$, $b = B/C$, $C = z \circ \epsilon$ et $d = D/C$. La transformée stricte de $X \times \mathbf{D}$ dans cette carte est donnée par:

$$c^{-m} f(ac, bc, c) = 0$$

dont l'intersection avec $d = 0$ donne bien la trace de la transformée stricte de $X \times \{0\}$ dans V_C. L'équation précédente ne dépendant pas de d, cette transformée stricte de $X \times \mathbf{D}$ par ϵ admet une structure de produit analytique.

Cette structure de produit dans V_C n'est autre que celle qui provient par éclatement de la structure de produit naturelle que possède l'espace total de la famille des sections de $X \times \mathbf{D}$ par les hyperplans d'équations $t - \lambda z = 0$ avec $\lambda \in \mathbf{C}$: on remarquera que ces hyperplans sont ceux du pinceau $\delta t + \gamma z = 0$ qui sont transverses à $\{0\} \times \mathbf{D}$.

En conclusion au voisinage de chacun des points x_o de $\mathfrak{X}_o - \{0\}$, où $\mathfrak{X}_o = \varphi^{-1}(0)$, $(\mathfrak{X}, x_o)$ est isomorphe à $(X' \times \mathbf{C}, (l_o, 0))$ où X' est l'espace total de l'éclatement $e: X' \to X$ de centre $\{0\}$ et l_o est le point de $e^{-1}(0)$ qui représente la génératrice de $C_{X,0}$ qui passe par x_o.

Le dessin ci-dessous aidera peut être le lecteur:

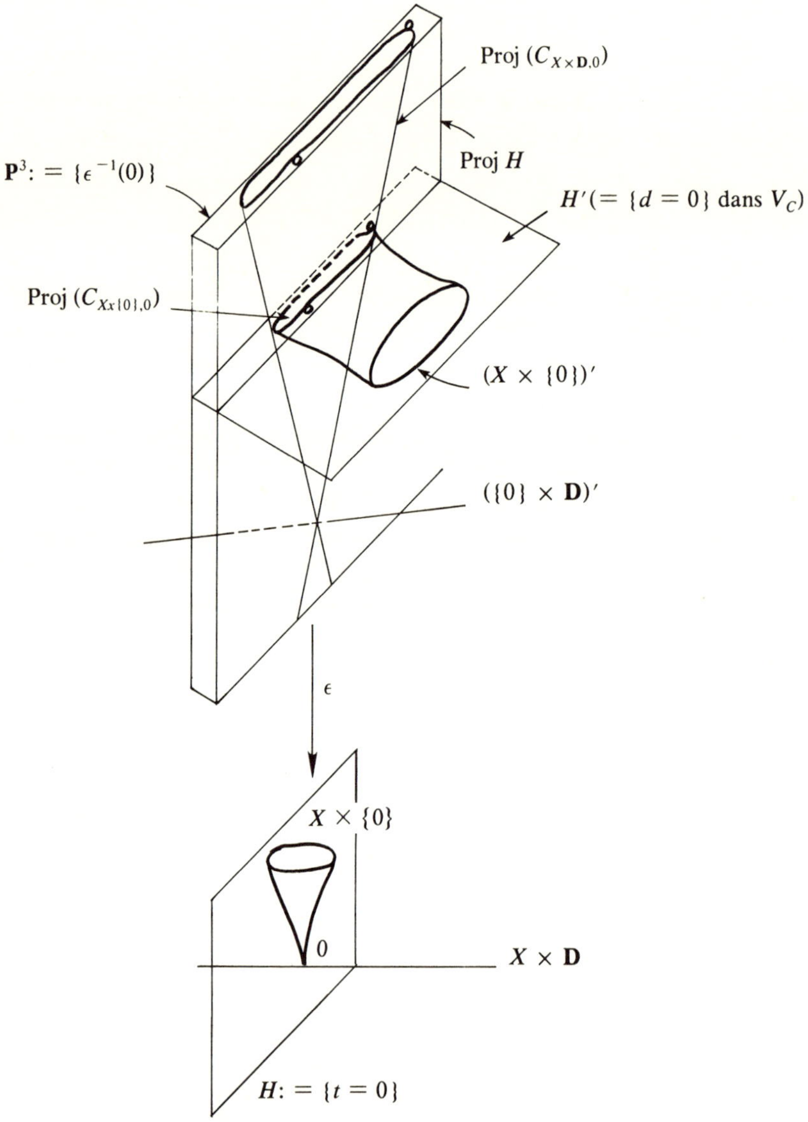

Légende. Le $'$ désigne la transformée stricte par ϵ.

(1.4.4) Nous allons donner diverses caractérisations des tangentes exceptionnelles. On reprend les notations précédentes du paragraphe (1.4).

THÉORÈME (1.4.4.1). *Soit* $(X, 0)$ *un germe de surface analytique de* $(\mathbf{C}^3, 0)$ *défini par* $f = 0$. *Soit* l *une génératrice du cône tangent* $C_{X,0}$ *de* X *en* 0, *qui n'est pas tangente en* 0 *au lieu singulier* Σ *de* X. *Les assertions suivantes sont équivalentes*:

1) *La génératrice* l *est une tangente exceptionnelle de* X *en* 0.

2) *Soit* $e\colon X' \to X$ *l'éclatement de* X *de centre* 0 *et soit* $J(f)\mathcal{O}_{X'}$ *l'idéal de* $\mathcal{O}_{X'}$ *engendré par* $(\partial f/\partial x)\circ e$, $(\partial f/\partial y)\circ e$, $(\partial f/\partial z)\circ e$. *L'éclatement de l'idéal* $J(f)\mathcal{O}_{X'}$, *n'est pas fini au-dessus d'un voisinage du point de* $e^{-1}(0) = \mathrm{Proj}\, C_{X,0}$ *qui représente* l.

3) *Pour toute projection locale* $\pi\colon (X, 0) \to (\mathbf{C}^2, 0)$ (*cf.* (1.1)) *de degré* $m(X, 0)$, *la génératrice* l *est tangente à une composante de la partie du lieu critique de* π *qui n'est pas le lieu singulier.*

4) *La surface éclatée* X' *de* X *en* 0 *n'est pas équisingulière le long de* $|e^{-1}(0)| = \mathrm{Proj}\, |C_{X,0}|$ *au point de* $e^{-1}(0)$ *qui représente* l.

5) *La déformation* $\varphi\colon \mathfrak{X} \to \mathbf{D}$ *définie dans* (1.4.1), *le couple de strates formé par la partie non singulière* $\mathfrak{X}_{\mathrm{reg}}$ *de* $\mathfrak{X}$ *moins* $\varphi^{-1}(0)$ *et de la partie non singulière de* $|\varphi^{-1}(0)| = |C_{X,0}| \times \{0\}$ *ne satisfait la condition de Thom relativement à* φ *en aucun point de* $l \times \{0\} - 0$.

Démonstration. Montrons que 1) entraîne 2). Pour cela on considère le diagramme d'éclatement du produit $\mathfrak{M}J(f)$ de l'idéal maximal définissant 0 dans X et de l'idéal $J(f) = (\partial f/\partial x, \partial f/\partial y, \partial f/\partial z)\mathcal{O}_X$.

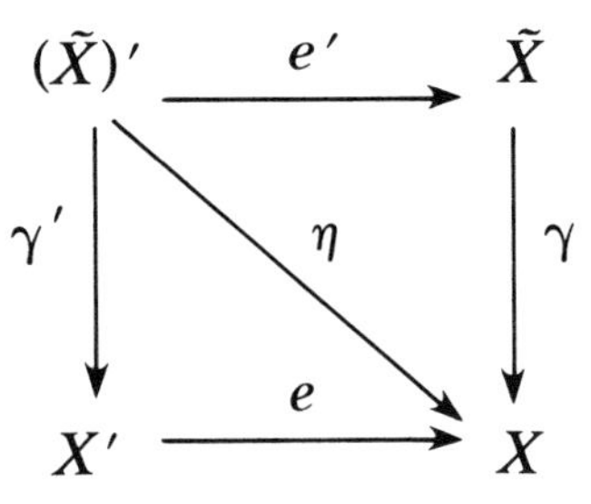

où e' est l'éclatement de $\gamma^{-1}(0)$ dans $\tilde{X}$ et γ' est l'éclatement de $J(f)\mathcal{O}_{X'}$. D'après la propriété universelle de l'éclatement, ce diagramme est commutatif et $\gamma \circ e' = e \circ \gamma'$ est l'éclatement de $\mathfrak{M}J(f)$. On notera $\eta = \gamma \circ e' = e \circ \gamma'$.

Rappelons que le morphisme d'éclatement de l'idéal maximal en 0 (resp. la modification de Nash) peut être décrit en considérant l'adhérence dans $X \times \mathbf{P}^2$ (resp. dans $X \times \check{\mathbf{P}}^2$) du graphe du morphisme canonique de $X - \{0\}$ dans $\mathbf{P}^2$ qui à $x \in X - \{0\}$ fait correspondre la direction de la droite $\overline{0x}$ (resp. du graphe du morphisme de $X - \Sigma$ dans $\check{\mathbf{P}}^2$ qui à $x \in X - \Sigma$ fait correspondre la direction du plan tangent a X en x). De la même façon considérons l'application de $X - \Sigma$ dans $\mathbf{P}^2 \times \check{\mathbf{P}}^2$ obtenue en considérant simultanément les deux précédentes. L'adhérence dans $X \times \mathbf{P}^2 \times \check{\mathbf{P}}^2$ du graphe de cette application n'est autre que $(\tilde{X})'$, et la projection sur X induit le morphisme η.

On peut exprimer le lemme de H. Whitney (cf. [25] ou [6]) en disant que $\eta^{-1}(0)$ est contenu dans l'hypersurface H de $\mathbf{P}^2 \times \check{\mathbf{P}}^2$ des couples (l, P) tels que $l \subset P$.

Soit l une tangente exceptionnelle. Montrons que $\gamma'^{-1}(l)$ est formé des couples (l, P) où P appartient au pinceau des plans qui contiennent l, i.e. $\dim \gamma'^{-1}(l) = 1$. Pour cela nous démontrons que $\gamma'^{-1}(l)$ contient les (l, P) pour P dans un ouvert dense du pinceau des plans qui contiennent l. Soit P un plan qui contient l, ne passe par aucune tangente exceptionnelle, et de plus n'est pas tangent à $C_{X,0}$, alors $e'^{-1}(P) = \{(l, P)\}$. En effet $e'^{-1}(P) \subset \{(l, P), (l_1, P), \ldots, (l_k, P)\}$ où $l, l_1, \ldots, l_k$ sont les génératrices de $C_{X,0}$ contenues dans P. Comme $l_1, \ldots, l_k$ sont des génératrices qui ne sont pas exceptionnelles, $\gamma'^{-1}(l_i) = \{(l_i, P_{ij})\}$ $(i = 1, \ldots, k)$ où P_{ij} est une limite le long de l_i de plans tangents à $C_{X,0}$ (cf. [9] théorème (2.3.7)).

Comme P n'est pas tangent à $C_{X,0}$ en l_i $(i = 1, \ldots, k)$, on a bien $e'^{-1}(P) = \{(l, P)\}$. Donc pour tous les plans P du pinceau qui contient l qui ne passent par aucune génératrice exceptionnelle autre que l et qui ne sont pas tangents à $C_{X,0}$, i.e. pour un ouvert dense du pinceau des plans qui contiennent l, on a $(l, P) \in \gamma'^{-1}(l)$. Donc on a bien $\dim \gamma'^{-1}(l) = 1$.

En fait nous pouvons montrer de la même façon que l'image inverse par e' d'un point P non singulier de $|\gamma^{-1}(0)|$ ne contient qu'un seul point. Les points P de l'ouvert dense ci-dessus sont les points de la droite duale à l qui sont non singuliers dans $|\gamma^{-1}(0)|$.

Remarquons que, si l n'est pas exceptionnel, nous avons vu ci-dessus (cf. [9] (2.3.7)) que $\gamma'^{-1}(l) = \{(l, P_j)\}$, où P_j est une limite d'espaces tangents à $C_{X,0}$ le long de l. Or, comme $C_{X,0}$ est un cône, cet ensemble est bien fini parce qu'il correspond aux droites tangentes en l à la courbe projective $|\mathrm{Proj}\, C_{X,0}|$. Ceci démontre que 2) entraîne 1).

Pour démontrer l'équivalence de 1) et de 3) nous remarquons que l'espace critique d'une projection locale peut se décrire de la façon suivante: Quitte à changer les coordonnées locales, on peut supposer que π: $(X, 0) \to (\mathbf{C}^2, 0)$ est donnée par une projection linéaire p de $\mathbf{C}^3$ sur $\mathbf{C}^2$. Le modifié de Nash $\tilde{X}$ de X en 0 est plongé dans $X \times \check{\mathbf{P}}^2$. Considérons $\gamma^{-1}(0)$ dans $\{0\} \times \check{\mathbf{P}}^2$. On a vu dans (1.3) que $|\gamma^{-1}(0)|$ est l'union de $(\text{Proj}\,|C_{X,0}|)^{\vee}$ variété duale de Proj $|C_{X,0}|$, et d'un nombre fini de droites, duales des tangentes exceptionnelles. Soit $L \subset \check{\mathbf{P}}^2$ la droite duale de la droite $p^{-1}(0)$ de $\mathbf{C}^3$, qui représente le pinceau des plans de $\mathbf{C}^3$ qui contiennent $p^{-1}(0)$. On a un morphisme analytique $\tilde{g}$: $\tilde{X} \to \check{\mathbf{P}}^2$ (morphisme de Gauss) restriction à $\tilde{X}$ de la projection de $X \times \check{\mathbf{P}}^2$ sur $\check{\mathbf{P}}^2$. On remarque que l'image par γ de $\tilde{g}^{-1}(L)$ est contenue dans le lieu critique de π et contient les composantes du lieu critique qui ne sont pas dans le lieu singulier Σ de X. On appellera (cf. [9] (2.2.3)) *partie verticale du lieu critique* la partie du lieu critique égale à $\gamma(\tilde{g}^{-1}(L))$.

Le dessin ci-dessous aidera peut être le lecteur:

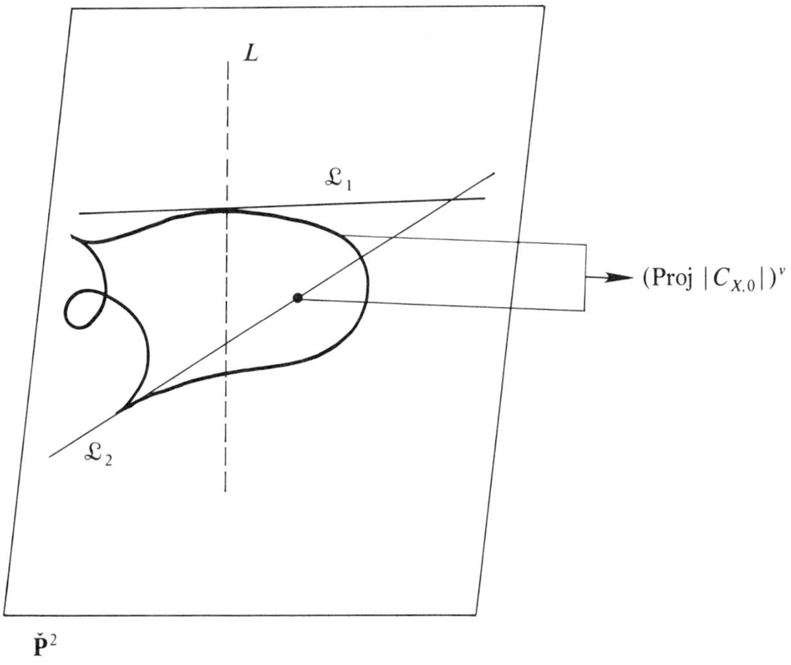

On a:

LEMME (1.4.4.2). *En un point h de $L \cap \gamma^{-1}(0)$ la dimension de $\tilde{g}^{-1}(L) = \tilde{X} \cap (X \times L)$ est au moins égale à 1. De plus si la projection locale de $(X, 0)$ sur $(\mathbf{C}^2, 0)$ définie par L est finie, la dimension en h de $\tilde{g}^{-1}(L)$ égale 1 et si de plus L n'est pas dual d'une tangente exceptionnelle $\tilde{g}^{-1}(L)$ n'est pas contenue dans $\gamma^{-1}(0)$.*

En effet $X \times \check{\mathbf{P}}^2$ est une intersection complète de dimension 4, $\tilde{X}$ est de dimension 2 et $X \times L$ est de dimension 3. Par conséquent $\tilde{X} \cap (X \times L) = \tilde{g}^{-1}(L)$ est au moins de dimension 1 en h. Si la projection locale π définie par L est finie, le lieu critique de π est de dimension 1. Comme $\gamma(\tilde{g}^{-1}(L))$ est une union de composantes irréductibles de ce lieu critique, et que γ est un morphisme fini en dehors de $\gamma^{-1}(0)$, en vertu du corollaire (1.3.5) et de la remarque 3 de (1.2) sur l'ouverture de l'équisingularité, la dimension de $g^{-1}(L)$ en $h \in L \cap \gamma^{-1}(0)$ est égale à 1 (ne pas oublier que $L \cap \gamma^{-1}(0)$ peut être vide; cf. 1.3.4).

On remarque:

(1.4.4.3) Les seules droites projectives dans $\gamma^{-1}(0)$ sont les duales des tangentes exceptionnelles, car une droite projective de $\check{\mathbf{P}}^2$ n'est pas duale d'une composante de Proj $|C_{X,0}|$ qui est équidimensionel de dimension 1 (bidualité pour les courbes projectives planes).

Ceci montre que, si L n'est pas dual à une tangente exceptionnelle $\{0\} \times L$ n'est pas contenue dans $\gamma^{-1}(0)$. Or $\tilde{g}^{-1}(L) \cap \gamma^{-1}(0) = (\{0\} \times L) \cap \gamma^{-1}(0)$. Ceci démontre notre lemme.

(1.4.4.4) Comme dans la démonstration de l'équivalence de 1) et de 2) ci-dessus, on remarque que, si $h \in \gamma^{-1}(0)$ n'est pas une composante de $(\text{Proj } |C_{X,0}|)^{\vee}$, i.e. si h n'est pas le dual d'un plan qui est dans $C_{X,0}$, au-dessus d'un voisinage de h dans $\tilde{X}$ le morphisme $e': (\tilde{X})' \to \tilde{X}$ est fini. En fait on a vu que l'image réciproque par e' d'un point non singulier h de l'ensemble $|\gamma^{-1}(0)|$ est réduite à un point. Donc si L coupe transversalement $|\gamma^{-1}(0)|$, alors le morphisme e' induit une bijection de $e'^{-1}(\tilde{g}^{-1}(L)) \cap \eta^{-1}(0)$ avec $\tilde{g}^{-1}(L) \cap \gamma^{-1}(0)$. Comme dans la démonstration ci-dessus de l'équivalence de 1) et de 2), ceci démontre que les tangentes à la partie verticale $\gamma(\tilde{g}^{-1}(L))$ du lieu critique de π sont les tangentes exceptionnelles et les droites du lieu critique de l'application tangente à π en 0 restreinte à $|C_{X,0}|$.

Si la projection locale $\pi: (X, 0) \to (\mathbf{C}^2, 0)$ associée à L est de degré $m(X, 0)$, le point de $\mathbf{P}^2$ dual de L n'est pas dans Proj $C_{X,0}$. Ceci signifie que L n'est pas tangente à $(\mathrm{Proj}\ |C_{X,0}|)^\vee$ (dans le cas d'un point singulier de $(\mathrm{Proj}\ |C_{X,0}|)^\vee$, ceci signifie que L n'est pas dans le cône tangent de $(\mathrm{Proj}\ |C_{X,0}|)^\vee$ en ce point et dans le cas d'une composante de dimension 0 de $(\mathrm{Proj}\ |C_{X,0}|)^\vee$, ceci signifie que L n'y passe pas). La seule difficulté nouvelle par rapport au cas considéré précédemment est que L peut passer par un point singulier de $|\gamma^{-1}(0)|$ dont l'image réciproque par e' est finie mais n'est pas réduite à un point.

Pour traiter ce cas nous faisons la construction suivante:

Il existe un pinceau $(L_t)_{t\in\mathbf{P}^1}$ de droites dans $\check{\mathbf{P}}^2$ tel que $L_o = L$ et pour tout $t \neq 0$ suffisamment proche de 0, L_t coupe transversalement $|\gamma^{-1}(0)|$. Considérons l'hypersurface Z de $\check{\mathbf{P}}^2 \times \mathbf{P}^1$ formée des points (x, t) tels que $x \in L_t$. Notons $\lambda\colon Z \to \mathbf{P}^1$ la restriction à Z de la projection de $\check{\mathbf{P}}^2 \times \mathbf{P}^1$ sur $\mathbf{P}^1$. On définit $\mathfrak{X}$ dans $(\tilde{X})' \times \mathbf{P}^1$ par le diagramme cartésien:

$$\begin{array}{cccl} \mathfrak{X} \subset & (\tilde{X})' \times \mathbf{P}^1 & & \\ \big\downarrow \quad \square & \big\downarrow & & (g\circ e')\times Id_{\mathbf{P}^1} \\ Z \subset & \check{\mathbf{P}}^2 \times \mathbf{P}^1 & & \end{array}$$

On notera Λ le morphisme composé de $\mathfrak{X} \to Z$ et de $\lambda\colon Z \to \mathbf{P}^1$. On remarque que $\Lambda^{-1}(t) = e'^{-1}(\tilde{g}^{-1}(L_t)) \times \{t\}$.

Soit h un point de $L \cap |\gamma^{-1}(0)|$ qui est singulier sur $|\gamma^{-1}(0)|$. Soient $\mathcal{L}_1, \ldots, \mathcal{L}_r$ les droites projectives de $|\gamma^{-1}(0)|$, duales de tangentes exceptionnelles, qui passent par h et soient $C_1, \ldots, C_s$ les composantes de $(\mathrm{Proj}\ |C_{X,0}|^\vee$ qui passent par h. Comme dans la démonstration de l'équivalence de 1) et 2) on fait la remarque suivante:

(1.4.4.5) On a $e'^{-1}(h) = \{(l_1, h), \ldots, (l_k, h)\}$ où les l_i sont les génératrices de $C_{X,0}$ duales des tangentes à $|\gamma^{-1}(0)|$ en h.

Si l_i, pour un i entre 1 et k, est une génératrice exceptionnelle duale de $\mathcal{L}_i$, le point (l_i, h) est limite quand t tend vers 0 des points (l_i, h_t) où $h_t = L_t \cap \mathcal{L}_i$. Or le point h_t, quand $t \neq 0$ est assez petit, est un point non singulier de $|\gamma^{-1}(0)|$. D'après le cas transversal vu plus haut, $\Lambda^{-1}(t) = e'^{-1}(\tilde{g}^{-1}(L_t)) \times \{t\}$ est de dimension 1 en (l_i, h_t, t), quand $t \neq 0$ est assez petit. Or $(l_i, h, 0) = \lim_{t\to 0} (l_i, h_t, t)$ appartient à $\Lambda^{-1}(0)$. La semi-continuité de la dimension des fibres de Λ donne que $\Lambda^{-1}(0)$ est de dimension 1 en $(l_i, h, 0)$. Ceci montre que $e'^{-1}(\tilde{g}^{-1}(L_o))$ est de dimension 1 en (l_i, h) et que l'image par η du germe

$(e'^{-1}(\tilde{g}^{-1}(L_o)), (l_i, h))$ est une composante de la partie verticale du lieu critique qui est tangente à l_i. Dans le cas où l_j, pour un j entre 1 et k, n'est pas une génératrice exceptionnelle, l_j est une droite du contour apparent de l'application tangente à π en 0 restreinte à $|C_{X,0}|$. Comme ci-dessus on montre que $e'^{-1}(\tilde{g}^{-1}(L_o))$ est de dimension 1 en (l_j, h) et l'image par η du germe $((e'^{-1}(\tilde{g}^{-1})(L_o)), (l_j, h))$ est une composante de la partie verticale du lieu critique qui est tangente à l_j. Ceci montre que, si l est une génératrice exceptionnelle de X en 0, pour toute projection locale π: $(X, 0) \to (\mathbf{C}^2, 0)$ de degré $m(X, 0)$, l est tangente au lieu critique de π. Les autres tangentes du lieu critique qui ne sont pas exceptionnelles sont les tangentes au lieu singulier et les droites du contour apparent de la restriction à $|C_{X,0}|$ de l'application tangente à π en 0. Ces dernières droites ne sont plus tangentes au lieu critique d'une projection voisine. Ceci démontre l'équivalence de 1) et de 3).

L'équivalence de 1) et 4) est démontrée dans [9] (théorème (2.3.7) et remarque (2.3.8)).

La preuve de l'équivalence de 4) et 5) repose sur le résultat suivant:

(1.4.4.5) Lemme. *Soit* $f(t_1, \ldots, t_k, x, y) = 0$ *l'équation d'une hypersurface réduite* X *de* $\mathbf{C}^k \times \mathbf{C}^2$ (*considérée comme une famille de courbes planes* (*que l'on suppose toutes réduites*) *paramétrée par* $\mathbf{C}^k$, *au voisinage de* $0 \in \mathbf{C}^k \times \mathbf{C}^2$) *et soit* $\alpha: X \to \mathbf{C}$ *la restriction à* X *d'une submersion* A: $(\mathbf{C}^k \times \mathbf{C}^2, 0) \to (\mathbf{C}, 0)$, *telle que* $(\alpha^{-1}(0)$ *contienne le lieu singulier* Σ *de* X, *que l'on suppose réduit à* $\mathbf{C}^k \times \{0\}$. *Alors si le couple de strates* $(X - \Sigma, \Sigma)$ *satisfait la condition* A_α *de Thom au voisinage de* 0 *on a*: *l'hypersurface* X *est équisinguliere le long de* $\mathbf{C}^k \times \{0\}$ *au voisinage de* 0.

Démonstration. Considérons la projection π: $X \to \mathbf{C}^k \times \mathbf{C}$ définie par $t_i \circ \pi = t_i\,(1 \leq i \leq k)$ et $u \circ \pi = \alpha$. Nous allons montrer que le lieu de bifurcation Δ_π de π est réduit à $\mathbf{C}^k \times \{0\}$. S'il n'en était pas ainsi, soit Δ_i une composante de Δ_π qui n'est pas $\mathbf{C}^k \times \{0\}$ et soit $p \in \Delta_i$. Puisque le lieu singulier est $\mathbf{C}^k \times \{0\}$, $d_\pi^{-1}(T_{\Delta_i,p})$ est un hyperplan de $\mathbf{C}^k \times \mathbf{C}^2$ tangent à X en au moins un point c du lieu critique C_π de π. L'intersection de ce plan avec l'espace tangent $T_{\alpha^{-1}(\alpha(c)),c}$ est nécessairement transversale pour $c \in X - \Sigma$ assez proche de 0 (car sinon $T_{\Delta_i,p}$ coinciderait avec l'hyperplan de $\mathbf{C}^k \times \mathbf{C}$ parallèle à $\mathbf{C}^k \times \{0\}$ et passant par p, pour tout p assez proche de 0 ce qui est absurde).

Ainsi, $d\pi^{-1}(T_{\Delta_i,p}) \cap T_{\alpha^{-1}(\alpha(c)),c)}$ coincide avec l'image inverse par $d\pi$ de l'intersection de $T_{\Delta_i,p}$ avec l'hyperplan parallèle à $\mathbf{C}^k \times \{0\}$

passant par p. Cette image inverse est de dimension k. On remarque qu'elle est verticale dans la projection de $\mathbf{C}^k \times \mathbf{C}^2$ sur $\mathbf{C}^k \times \mathbf{C}$ définie par la projection de $\mathbf{C}^k \times \mathbf{C}^2$ sur $\mathbf{C}^k$ et A. En effet c'est aussi l'image réciproque par l'application tangente à cette projection de l'intersection de $T_{\Delta_i,p}$ et de l'hyperplan parallèle à $\mathbf{C}^k \times \{0\}$ passant par p. Ceci montre que quand p tend vers zéro $[d\pi^{-1}(T_{\Delta_i,p})] \cap T_{\alpha^{-1}(\alpha(c)),c}$ tend vers une limite qui ne peut pas contenir $\mathbf{C}^k \times \{0\}$. Ceci contredit la condition A_α de Thom pour $(X - \Sigma, \Sigma)$ en 0. Donc il faut que $\Delta_\pi = \mathbf{C}^k \times \{0\}$, ce qui d'après le critère discriminant de O. Zariski (cf. (1.2)3)) implique l'équisingularité.

Remarque (1.4.4.6). Le lemme précédent admet la réciproque partielle suivante: si l'hyperplan $dA^{-1}(0)$ n'est pas dans le cône tangent à X en 0, si X est équisingulière le long de Σ en 0, la condition A_α de Thom est satisfaite pour $(X - \Sigma, \Sigma)$ sur un voisinage de 0 dans Σ. En effet choisissons les coordonnées x et y pour que $A(t_1, \ldots, t_k, x, y) = x$. La condition A_α de Thom pour $(X - \Sigma, \Sigma)$ en 0 s'exprime alors en disant que dans l'algèbre locale $\mathcal{O}_{X,0}$ on a:

$$\bar{\nu}_{\partial f/\partial y \cdot \mathcal{O}_{X,0}}(\partial f/\partial t_i) > 1 \qquad \text{pour } 1 \leq i \leq k.$$

(ceci résulte de la définition de la fonction d'ordre $\bar{\nu}_{\partial f/\partial y \cdot \mathcal{O}_{X,0}}$ cf. [22] et [14]; comparer ce résultat avec la remarque 3.10 de [22] et [13]).

Or l'équisingularité de X le long de Σ en 0 équivaut à 2) de (1.2), i.e. le nombre de Milnor des courbes de la famille X est constant. Dans 3.10 de [22] ceci se traduit par:

$$\bar{\nu}_{(\partial f/\partial x, \partial f/\partial y) \cdot \mathcal{O}_{\mathbf{C}^k \times \mathbf{C}^2, 0}}(\partial f/\partial t_i) > 1 \text{ pour } 1 \leq i \leq k$$

Par restriction à X le $\bar{\nu}$ ne peut qu'augmenter d'où:

$$\bar{\nu}_{(\partial f/\partial x, \partial f/\partial y) \cdot \mathcal{O}_{X,0}}(\partial f/\partial t_i) > 1 \text{ pour } 1 \leq i \leq k$$

La transversalité de x implique que $|\partial f/\partial x| \leq C|\partial f/\partial y|$ avec $C > 0$ dans un voisinage de 0 dans X. Or ceci se traduit (cf. [22] et [14]) par $\partial f/\partial x$ dans $\mathcal{O}_{X,0}$ est entier sur $\partial f/\partial y \cdot \mathcal{O}_{X,0}$. Les propriétés de la fonction d'ordre $\bar{\nu}$ (cf. [22] et [14]) impliquent alors:

$$\bar{\nu}_{(\partial f/\partial x, \partial f/\partial y) \cdot \mathcal{O}_{X,0}} = \bar{\nu}_{\partial f/\partial y \cdot \mathcal{O}_{X,0}}$$

d'où la condition A_α de Thom pour $(X - \Sigma, \Sigma)$ en 0.

Montrons maintenant que le lemme (1.4.4.5) donne: 4) implique 5) dans le théorème de (1.4.4). Dans (1.4.3) nous avons vu qu'en tout point x_o de $\mathfrak{X}_o - \{0\}$, où $\mathfrak{X}_o = \varphi^{-1}(0)$, $(\mathfrak{X}, x_o)$ est isomorphe à $(X' \times \mathbf{C}, (l_o, 0))$, l_o étant le point de $e^{-1}(0)$ qui représente la droite $0x_o$.

Pour démontrer 4) implique 5), d'après la formulation de 5) il suffit de considérer un point x_o non singulier dans $\mathfrak{X}_o$. Si X' n'est pas équisingulière en l_o le long de $e^{-1}(0)$, $\mathfrak{X}$ n'est pas équisingulière en x_o le long de $\mathfrak{X}_o$, à cause de la structure de produit. Le lemme (1.4.4.5) montre alors que $(\mathfrak{X} - \mathfrak{X}_o, \mathfrak{X}_o)$ ne satisfait pas la condition A_φ de Thom en x_o.

Montrons que 5) implique 1). Soit x un point de $l \times \{0\} - 0$, non singulier sur $|\mathfrak{X}_o - \{0\}|$. Si $\mathfrak{X}_{\text{reg}} - \mathfrak{X}_o$ ne satisfait pas la condition A_φ de Thom en x, il existe une suite de points $(x_n)_{n\in\mathbf{N}}$ de $\mathfrak{X}_{\text{reg}} - \mathfrak{X}_o$ qui tend vers x pour laquelle la suite de plans $T_{x_n}(\varphi^{-1}(\varphi(x_n)))$ converge vers un plan T qui ne contient pas $T_x(|\mathfrak{X}_o|)$. Cette suite de points donne une suite de points $(x_n')_{n\in\mathbf{N}}$ de X qui tend vers 0 telle que la suite des plans tangents $T_{x_n'}(X)$ tende vers T et la suite de droites $0x_n'$ tende vers $l = 0x$. Le plan T contient l et n'est pas tangent à $|C_{X,0}|$. Ceci n'est possible que si l est une tangente exceptionnelle. Ceci achève la démonstration du théorème de (1.4.4).

Remarque (1.4.4.7). Dans le cas où $(X, 0)$ a une singularité isolée en 0, en vertu du 4) du théorème (1.4.4.1) et de la remarque 5 de (1.2), les génératrices singulières de $|C_{X,0}|$ sont nécessairement exceptionnelles (comparer à [5], theoreme (3.1)).

2. Surfaces Sans Tangentes Exceptionnelles. Dans ce paragraphe nous allons étudier les surfaces sans tangentes exceptionnelles. Dans le cas des surfaces avec singularité isolée ou celui des surfaces dont le cône tangent est réduit, les surfaces sans tangentes exceptionnelles se déforment équisingulièrement sur leur cône tangent. Nous donnerons des exemples de surfaces sans tangentes exceptionnelles dont le cône tangent n'est pas réduit.

(2.1) *Cas d'un cône tangent réduit.*

(2.1.1) Théorème. *Supposons que le cône tangent $C_{X,0}$ de X en 0 soit réduit. Alors les deux conditions suivantes sont équivalentes:*

1) *la surface analytique X n'a pas de tangentes exceptionnelles en* 0.

2) *la déformation de $(X, 0)$ sur son cône tangent $\varphi\colon \mathfrak{X} \to \mathbf{D}$ (cf.* 1.4.1 *ci-dessus) est équisingulière en* 0 *le long de $\{0\} \times \mathbf{D}$ dans l'un des sens suivants:*

a) *la partie non singulière de l'espace $\mathfrak{X}'$ obtenu par éclatement de $\{0\} \times \mathbf{D}$ dans $\mathfrak{X}$ satisfait les conditions de Whitney le long de sa partie singulière.*

b) *La partition suivante de $\mathfrak{X}$: $(\mathfrak{X} - Sing\ \mathfrak{X}, Sing\ \mathfrak{X} - \{0\} \times \mathbf{D}, 0 \times \mathbf{D})$ est une stratification de Whitney de $\mathfrak{X}$.*

c) (*définition de Zariski*) *Pour une projection linéaire assez générale $\pi: \mathfrak{X} \to \mathbf{C}^2 \times \mathbf{D}$ compatible avec φ et la projection $\mathbf{C}^2 \times \mathbf{D} \to \mathbf{D}$, notant Δ_π le discriminant de π, on a: Δ_π est une famille de courbes planes équisingulières le long de $\{0\} \times \mathbf{D}$.*

Démonstration. Montrons que 1) entraîne 2) a). Examinons d'abord le cas où Proj $C_{X,0}$ est une courbe non singulière (i.e. $C_{X,0}$ est en fait un cône avec une singularité isolée).

Soit $x_o \in \mathfrak{X}_o - \{0\}$. La propriété A_φ de $(\mathfrak{X} - \mathfrak{X}_o, \mathfrak{X}_o)$ en x_o montre que $(\mathfrak{X}, x_o)$ est non singulier. D'après la structure de produit de $\mathfrak{X}$ au voisinage de x_o (cf. (1.4.3)), X' est non singulier en l_o, point de $\mathbf{P}^2$ qui représente la droite $0x_o$, et transverse à ce $\mathbf{P}^2$. Ceci est vrai en tout point du diviseur exceptionnel de X'. On obtient donc en éclatant $(0) \times \mathbf{D}$ dans $\mathfrak{X}$ un espace analytique non singulier $\mathfrak{X}'$ tel que le morphisme composé $\mathfrak{X}' \to \mathfrak{X} \xrightarrow{\varphi} \mathbf{D}$ soit une submersion d'espaces analytiques non singuliers, que pour chaque $t \in \mathbf{D}$ le morphisme $\mathfrak{X}_t' \to \mathfrak{X}_t$, induit sur les fibres de $\mathfrak{X}'$ et $\mathfrak{X}$ au-dessus de t, soit une résolution des singularités de $\mathfrak{X}_t$ et que l'image inverse de $\{0\} \times \mathbf{D} \subset \mathfrak{X}$ dans $\mathfrak{X}'$ soit égale à Proj $C_{X,0} \times \mathbf{D}$ et donc analytiquement triviale au-dessus de $\mathbf{D}$ (résolution simultanée forte cf. [20]). Dans ce cas où Proj $C_{X,0}$ est non singulier, on a donc bien 2) a). (Comparer ces résultats à ceux de [22] Chap. II, 2.7, et de [12] Cor. 2.4).

(2.1.1.1) Supposons donc que Proj $(C_{X,0})$ soit singulière en l_o. Comme la génératrice l_o de $C_{X,0}$ n'est pas une tangente exceptionnelle, on a vu dans [9] (remarque (2.3.8)) que l_o est nécessairement tangente à une union Σ_{l_o} de composantes irréductibles en 0 du lieu singulier Σ de X. Si cette union Σ_{l_o} des composantes irréductibles de Σ tangentes à l_o en 0 est non singulière, en tout point $x_o \neq 0$ de $l_o \times \{0\}$, le lieu singulier $\mathfrak{S}$ de $\mathfrak{X}$ est non singulier en x_o.

Soit H un plan de dimension 2 dans $\mathbf{C}^3$ tel que $H \times \mathbf{D}$ coupe $l_o \times \{0\}$ transversalement en x_o. L'intersection de $H \times \mathbf{D}$ avec $\mathfrak{X}$ est une famille de courbes planes paramétrée par $\mathbf{D}$ dont le lieu singulier est $\mathfrak{S} \cap (H \times \mathbf{D})$. Remarquons que $H \times \{0\}$, étant transverse à $l \times \{0\}$ dans $\mathbf{C}^3 \times \{0\}$, n'est pas limite de plans tangents à $\mathfrak{X}_t$ quand t tend vers 0 Il en résulte que les limites des droites tangentes aux courbes

$\mathfrak{X}_t \cap (H \times \mathbf{D})$ quand t tend vers 0 sont les tangentes à $\mathfrak{X}_o \cap (H \times \mathbf{D})$. Si nous projetons $\mathfrak{X} \cap (H \times \mathbf{D})$ sur $\mathbf{C} \times \mathbf{D}$ par une projection dont l'axe est transverse à $\mathfrak{X}_o \cap (H \times \mathbf{D})$ d'après ce que l'on vient de voir le discriminant de cette projection coïncide avec l'image du lieu singulier $\mathfrak{S} \cap (H \times \mathbf{D})$. Ce discriminant est donc non singulier en 0 et d'après le critère discriminant de O. Zariski (cf. (1.2) 3)), cette famille de courbe $\mathfrak{X} \cap (H \times \mathbf{D})$ est équisingulière en x_o le long de son lieu singulier $\mathfrak{S} \cap (H \times \mathbf{D})$. D'après la structure de produit de $(\mathfrak{X}, x_o)$ (cf. (1.4.3)) l'éclaté X' de X est équisingulier en l_o le long de la transformée stricte de Σ_{l_o} par l'éclatement $e: X' \to X$. Ainsi dans le cas où Σ_{l_o} est non singulier, nous avons bien 2)a) à cause de (1.2) 5).

Il nous reste à montrer que sous nos hypothèses, Σ_{l_o} est bien non singulier en 0.

Soit $E: Z \to U \times \mathbf{D}$ l'éclatement de $U \times \mathbf{D}$ le long de $\{0\} \times \mathbf{D}$. Soit $\mathfrak{X}'$ la transformée stricte de $\mathfrak{X}$. On note $E_{\mathfrak{X}}: \mathfrak{X}' \to \mathfrak{X}$ l'éclatement de $\mathfrak{X}$ le long de $\{0\} \times \mathbf{D}$, induit par E. On rappelle que $Z = Z_o \times \mathbf{D}$ où Z_o est l'éclaté de U en 0. On a également $E^{-1}(\{0\} \times \mathbf{D}) = \operatorname{Proj}(C_{X,0}) \times \mathbf{D}$. On appellera $l_1, \ldots, l_k$ les points singuliers de Proj $C_{X,0}$ (qui est réduit par hypothèse). On note $\Delta_1, \ldots, \Delta_k$ les discriminants de Proj $C_{X,0}$ en $l_1, \ldots, l_k$ (cf. (1.3.3.1)).

Nous allons montrer que sous nos hypothèses, $\mathfrak{X}$ en 0 et $\mathfrak{X}'$ en les points $(l_i, 0)$ $(i = 1, \ldots, k)$ ont leur courbe polaire par rapport à $t = 0$ vide (cf. [8]). Nous verrons par un argument topologique que ceci implique la non singularité de Σ_{l_i} $(i = 1, \ldots, k)$.

Montrons que l'hyperplan $t = 0$ n'est pas limite d'espaces tangents à $\mathfrak{X}$ en 0. D'après un résultat de [22] (chap. 1 (2.9)) ceci est équivalent à montrer que $\partial F/\partial t$ est entier dans $\mathcal{O}_{\mathfrak{X},0}$ sur l'idéal $(\partial F/\partial x, \partial F/\partial y, \partial F/\partial z)\mathcal{O}_{\mathfrak{X},0}$. Remarquons que l'hypothèse qu'il n'y a pas de tangentes exceptionnelles implique que chaque composante du diviseur exceptionnel de l'éclatement dans $\mathfrak{X}$ de l'idéal $(\partial F/\partial x, \partial F/\partial y, \partial F/\partial z)\mathcal{O}_{\mathfrak{X}}$ s'envoie surjectivement sur $\mathbf{D}$ par le morphisme composé de l'éclatement et de $\varphi: \mathfrak{X} \to \mathbf{D}$. En effet la propriété A_φ de Thom pour $\mathfrak{X}$ implique que le diviseur exceptionnel de la transformée stricte de $\mathfrak{X}_o$ coïncide avec la fibre au-dessus de 0 du diviseur exceptionnel de l'éclatement précédent (on utilise bien sûr (1.3.2)). En utilisant l'argument de [22] (chap. 2 (3.2)) (cf. Aussi [20]), App. II) on montre que pour tout $g \in \mathcal{O}_{\mathfrak{X},0}$ qui donne un germe $g \in \mathcal{O}_{\mathfrak{X},(0,t)}$ pour t assez petit, on a:

avec

$$J\mathcal{O}_{\mathfrak{X}} = (\partial F/\partial x,\ \partial F/\partial y,\ \partial F/\partial z)\mathcal{O}_{\mathfrak{X}}$$

Comme on a pour tout $t \neq 0$, $t \in \mathbf{D}$:

$$\bar{\nu}_{J\mathcal{O}_{\mathfrak{X},(0,t)}}(\partial F/\partial t) > 1$$

à l'aide de [22] (chap. 2 (3.6)) car $\mathfrak{X}_{\text{reg}} - \{0\} \times \mathbf{D}$ satisfait la condition A de Whitney en tout point de $\{0\} \times \mathbf{D} - \{0\}$ puisque la famille $\mathfrak{X}$ est analytiquement triviale au voisinage de $(0, t)$ avec $t \neq 0$, on obtient:

(2.1.1.2) $$\bar{\nu}_{J\mathcal{O}_{\mathfrak{X},0}}(\partial F/\partial t) > 1$$

ce qui exprime que $t = 0$ n'est pas limite d'espaces tangents à $\mathfrak{X}$ en 0.

(2.1.1.3) Ceci entraîne qu'en un point de $l_i \times \{0\} - \{0\}$ $(i = 1, \ldots, k)$ le lieu singulier de $\mathfrak{X}$ est transverse à $t = 0$. D'après la structure de produit de $\mathfrak{X}$ en un point de $\mathfrak{X}_o - \{0\}$ (cf. (1.4.3)) on obtient que la transformée stricte de la partie Σ_{l_i} du lieu singulier de X qui est tangente à l_i a la même multiplicité que Σ_{l_i} et que Proj $C_{X,0}$ en l_i.

(2.1.1.4) Un autre conséquence de ce que $t = 0$ ne soit pas limite d'espaces tangents à $\mathfrak{X}$ en 0 est que l'éclatement de $(\partial F/\partial x, \partial F/\partial y, \partial F/\partial z, \partial F/\partial t)\mathcal{O}_{\mathfrak{X}}$ a une fibre au-dessus de 0 qui est de dimension 1. L'éclatement de l'idéal $(\partial F/\partial x, \partial F/\partial y, \partial F/\partial z, \partial F/\partial t)$ dans $U \times \mathbf{D}$ a donc une fibre au-dessus de 0 de dimension au plus égale à 2. Ceci entraîne d'après un argument de [4] (2.2. p. 329 et 330) que la courbe polaire de $\mathfrak{X}$ relativement à $t = 0$ est vide. Ce qu'on peut aussi vérifier très facilement par un calcul direct.

Considérons maintenant une projection p de $U \times \mathbf{D}$ sur $V \times \mathbf{D}$, où V est un ouvert de $\mathbf{C}^2$, induite par une projection p_o: $U \to V$. Soit π la restriction de p à $\mathfrak{X}$. On note $D(\pi)$ le discriminant de π, qui est une surface de $V \times \mathbf{D}$. Quand p est assez général, remarquons qu'en tout point de $\{0\} \times \mathbf{D}$ la multiplicité de $D(\pi)$ égale $m(m - 1)$ où m est la multiplicité de x en 0. En effet cette multiplicité en $(0, t)$ est le discriminant de la courbe image inverse par π d'une droite générale de $V \times \mathbf{D}$ passant par $(0, t)$. Si $t \neq 0$ il suffit de considérer une droite de la forme $L \times \{t\}$. Comme le cône tangent est réduit, le germe de courbe $\pi^{-1}(L \times \{t\}, (0, t))$ a le type d'équisingularité de m droites qui passent par 0 et distinctes. Le discriminant d'une telle singularité (ordinaire) vaut $m(m-1)$. D'autre part on a vu dans (1.4.4.4) que les composantes du lieu critique d'une projection locale assez générale d'une surface sur $\mathbf{C}^2$ autres que celles du lieu singulier sont tangentes aux génératrices exceptionnelles, aux droites du contour apparent de la restriction de l'application tangente de Zariski de cette projection au cône tangent. Par conséquent, si p est assez générale, l'espace critique de π est composé du lieu singulier de $\mathfrak{X}$ et de $m^\vee$ surfaces non singulières qui contiennent $\{0\} \times \mathbf{D}$ et $m^\vee$ génératrices de $\mathfrak{X}_o$. En fait $m^\vee$ égale le degré de la courbe duale de Proj $C_{X,0}$. La formule de Plücker (cf. [7] ou [20] Appendice 2 remarque (2.5)) nous donne:

$$m^\vee = m(m-1) - \sum_{i=1}^{k} \Delta_i$$

où Δ_i est le discriminant de Proj $C_{X,0}$ en l_i. Soient $\Sigma_i^{(1)}, \ldots, \Sigma_i^{(r_i)}$ les composantes analytiques de Σ_{l_i} en 0 et soient $\nu_{1i}, \ldots, \nu_{r_i i}$ leurs multiplicités respectives. On considère une section de X en un point non singulier de $\Sigma_i^{(j)}$ proche de 0 par un plan transverse à $\Sigma_i^{(j)}$. Soit $\Delta_i^{(j)}$ le discriminant de cette section en ce point. La contribution de l'image de $\Sigma_i^{(j)}$ dans le discriminant d'une projection sur $\mathbf{C}^2$ transverse à $\Sigma_i^{(j)}$ est $\nu_{ji}\Delta_i^{(j)}$. La contribution totale du lieu singulier dans le discriminant d'une projection de $(X, 0)$ sur $(\mathbf{C}^2, 0)$ transverse à ce lieu singulier est donc: $\sum_{i,j} \nu_{ji}\Delta_i^{(j)}$.

On obtient donc que la multiplicité du discriminant d'une projection assez générale de $(X, 0)$ sur $(\mathbf{C}^2, 0)$ égale $\sum_{ij}\nu_{ji}\Delta_i^{(j)} + m^\vee$.

D'où:

$$\text{(2.1.1.5)} \qquad \sum_{i=1}^{k} \sum_{j=1}^{r_i} \nu_{ji}\Delta_i^{(j)} = \sum_{i=1}^{k} \Delta_i$$

Par un argument de semi-continuité on obtient en fait:

$$\text{(2.1.1.6)} \qquad \sum_{j=1}^{r_i} \nu_{ji}\Delta_i^{(j)} = \Delta_i$$

(2.1.1.7) Ceci entraîne avec le résultat de (2.1.1.3) que le germe de surface (X', l_i) n'a qu'un nombre fini de limites d'espaces tangents en l_i d'après le théorème (1.3.4).

Appelons $F_1(x', y', z', t) = 0$ l'équation de la transformée stricte $\mathfrak{X}'$ au voisinage de $\{l_i\} \times \mathbf{D}$. Nous remarquons, de façon analogue au cas de $\mathfrak{X}$, que $t = 0$ n'est pas limite d'espaces tangents à $\mathfrak{X}'$ en $(l_i, 0)$. En effet on peut stratifier $\mathfrak{X}'$ par sa partie non singulière, son lieu singulier moins sa partie dans $t = 0$ et $\{l_i\} \times \mathbf{D}$, la partie du lieu singulier de $\mathfrak{X}'$ dans $t = 0$ moins $\{0\}$, $\{l_i\} \times \mathbf{D} - \{0\}$ et $\{0\}$. Ceci donne une stratification de $\mathfrak{X}'$ qui satisfait à la condition $A_{\varphi \circ E}$. Comme dans le cas de $\mathfrak{X}$, chaque composante du diviseur exceptionnel de l'éclatement de l'idéal $(\partial F_1/\partial x', \partial F_1/\partial y', \partial F_1/\partial z')\mathcal{O}_{\mathfrak{X}'}$ s'envoie surjectivement sur $\mathbf{D}$ par le morphisme composé de l'éclatement $E_{\mathfrak{X}}$ et de φ. On termine comme on a fait dans le cas de $\mathfrak{X}$ et on trouve que $\partial F_1/\partial t$ est entier dans $\mathcal{O}_{\mathfrak{X}',(l_i,0)}$ sur $(\partial F_1/\partial x', \partial F_1/\partial y', \partial F_1/\partial z')\mathcal{O}_{\mathfrak{X}',(l_i,0)}$. Par un raisonnement analogue à (2.1.1.4), comme la fibre de l'éclatement de $(\partial F_1/\partial x', \partial F_1/\partial y', \partial F_1/\partial z', \partial F_1/\partial t)\mathcal{O}_{\mathfrak{X}'}$ est de dimension 0, la fibre au-dessus de 0 de l'éclatement de $(\partial F_1/\partial x', \partial F_1/\partial y', \partial F_1/\partial z', \partial F_1/\partial t)$ dans $(Z, (l_i, 0))$ a au plus la dimension 1. Dans ce cas on obtient donc par un argument de [4] (2.2, p. 330) que la courbe polaire de $\mathfrak{X}'$ relativement à $(t = 0)$ est vide en $(l_i, 0)$.

Soit $V_i \times \{0\}$ une boule fermée de $Z_o \times \{0\}$ de centre $(l_i, 0)$ et de rayon $\epsilon > 0$ assez petit définie à l'aide des coordonnées locales x', y', z', t. Pour λ assez petit et pour tout $t \in \mathbf{D}$, $(V_i \times \{t\}) \cap \{F_1 = \lambda\}$ est la fibre de Milnor de $\mathfrak{X}_t'$ en (l_i, t) (cf. [16] ou [10] théorème (2.3)): ceci résulte de ce que tous les $\mathfrak{X}_t'$ pour $t \in \mathbf{D}$, $t \neq 0$, sont isomorphes avec $\mathfrak{X}_1'$ par l'isomorphisme: $(x', y', z', t) \to (x'/t, y', z', 1)$. Donc si pour $\epsilon > 0$ assez petit les sphères $S_{\epsilon'} \times \{1\}$ de $Z_o \times \{1\}$ centrées en $(0, 1)$, avec $0 < \epsilon' \leq \epsilon$, coupent transversalement $\mathfrak{X}_1 = X \times \{1\}$ dans $Z_o \times \{1\}$, $S_{\epsilon'} \times \{t\}$ coupe transversalement $\mathfrak{X}_t$ dans $Z_o \times \{t\}$ pour tout $t \neq 0$, $t \in \mathbf{C}$, $|t| \leq 1$.

La fibre de Milnor de $\mathfrak{X}_o'$ en $(l_i, 0)$ est difféomorphe au produit de la fibre de Milnor de Proj $C_{X,0}$ en l_i et d'un disque de $\mathbf{C}$, car $\mathfrak{X}_o'$ est un cylindre (cf. (1.4.3)). La caractéristique d'Euler de cette fibre de Milnor est donc $1 - \mu_i$ où μ_i est le nombre de Milnor de Proj $C_{X,0}$ en l_i.

La fibre de Milnor de $\mathfrak{X}_t'$ en (l_i, t) pour $t \neq 0$ a une section générique difféomorphe à la fibre de Milnor de Proj $C_{X,0}$ en l_i. Comme la courbe polaire de $\mathfrak{X}'$ relativement à $t = 0$ est vide, le résultat de [8] montre que la fibre de Milnor de $\mathfrak{X}_t'$ a la caractéristique d'Euler de celle de $\mathfrak{X}_o'$. Celle-ci égale donc $1 - \mu_i$. Or d'après [8], ceci n'est possible que, si la courbe polaire de $\mathfrak{X}_t'$ en (l_i, t) par rapport à $x' = 0$, équation locale du diviseur exceptionnel de $Z_0 \times \{t\}$, est également vide. Ceci implique immédiatement que

$$\sum_{j=1}^{r_i} \nu_{ji}\mu_i^{(j)} = \mu(\mathrm{Proj}\,(C_{X,0}), l_i),$$

où $\mu_i^{(j)}$ est le nombre de Milnor de la section plane de X par un plan transverse è $\Sigma_i^{(j)}$ en un point non singulier de $\Sigma_i^{(j)}$ voisin de O. D'après [11], ceci n'est possible que si $r_i = 1$ et $\nu_{1i} = 1$. Donc Σ_{l_i} est non singulière en 0.

Nous avons ainsi démontré l'implication 1) entraîne 2) a) du théorème (2.1.1).

Montrons que 2) a) entraîne 2) b). D'après (2.1.1.2) on a la condition A de Whitney pour la partie non singulière de $\mathfrak{X}$ le long de $\{0\} \times \mathbf{D}$ en 0. La condition B de Whitney provient du lemme de Whitney (cf. [25] ou [5]), du fait que φ donne une déformation triviale des cônes tangents de $\mathfrak{X}_t$ en $(0, t)$ et que l'on a la propriété A_φ de Thom. Le lieu singulier étant equisingulier le long de $\{0\} \times \mathbf{D}$ d'après la démonstration ci-dessus, on a la propriété de Whitney le long de $\{0\} \times \mathbf{D}$, et, pour la partie non singulière de $\mathfrak{X}$ le long du lieu singulier en dehors de $\{0\} \times \mathbf{D}$, on a en fait équisingularité en codimension 1 et on conclut grâce à (1.2) 5).

Remarquons que 2) b) entraîne 2) a) parce que 2) b) entraîne 1). En effet si on a une stratification de Whitney comme dans 2) b) les limites d'espaces tangents de $\mathfrak{X}$ en 0 sont les plans qui contiennent $T_0(\{0\} \times \mathbf{D})$ et un plan tangent au cône tangent $C_{X,0}$. Ceci montre que les limites d'espaces tangents à X en 0 ne sont que les plans tangents au cône tangent.

Enfin 2) c) entraîne 1) parce qu'un cône n'a pas de tangentes exceptionnelles et l'équisingularité du discriminant le long de $\{0\} \times \mathbf{D}$ d'une projection linéaire assez générale de $\mathfrak{X}$ sur $\mathbf{C}^2 \times \mathbf{D}$ implique que pour $t \neq 0$, $\mathfrak{X}_t$ n'a pas de tangentes exceptionnelles (cf. (1.4.4.4)). En fait 1) entraîne 2) c) puisque l'on a vu dans la démonstration de 1) entraîne 2) a) que pour toute génératrice l singulière sur $C_{X,0}$ la partie Σ_l du lieu singulier de X en 0 tangente à l est elle-même non singulière. Le discriminant d'une projection générale de $\mathfrak{X}$ sur $\mathbf{C}^2 \times \mathbf{D}$ est la réunion de $m^\vee + k$ surfaces non singulières contenant $\{0\} \times \mathbf{D}$ non tangentes entre elles, où $m^\vee$ est le degré de la courbe duale de Proj $C_{X,0}$ et k le nombre de génératrices singulières de $C_{X,0}$. Ce discriminant est évidemment équisingulier le long de $\{0\} \times \mathbf{D}$. Ceci achève la démonstration du théorème (2.1.1).

(2.2) Cas d'une singularité isolée

(2.2.1) THÉORÈME. *Soit $(X, 0) \subset (\mathbf{C}^3, 0)$ un germe de surface à singularité isolée en* 0. *Les propriétés suivantes sont équivalentes:*

i) *La surface éclatée $X' \to X$ est équisingulière le long de la courbe* $|e^{-1}(0)|$.

ii) *Le cône tangent $C_{X,0}$ de X en* 0 *a une singularité isolée en* 0. (*et donc $e^{-1}(0)$ est une courbe réduite*).

iii) *Le nombre de Milnor $\mu^{(3)}$ de $(X, 0)$ et celui $\mu^{(1)} = m(X, 0) - 1$ de sa section par une droite générique satisfont l'égalité*

$$\mu^{(3)} = (\mu^{(1)})^3$$

iv) *La déformation $\varphi: \mathfrak{X} \to \mathbf{D}$ de X sur son cône tangent* (*cf.* (1.4.1)) *est équisingulière au sens de la résolution simultanée forte de* ([20]) (*ou dans l'un des sens équivalents* 2) a), 2) b), 2) c) *du théorème* (2.1.1)).

v) *L'éclatement e de l'idéal maximal de $(X, 0)$ est une résolution des singularités et* $|e^{-1}(0)|$ est non singulier.

vi) *La surface $(X, 0)$ n'a pas de génératrice exceptionnelle, i.e. la fibre $\gamma^{-1}(0)$ de la modification de Nash $\gamma: \tilde{X} \to X$ de X est exactement la courbe duale de* $|Proj\ C_{X,0}|$.

Démonstration. Montrons i) implique ii). On remarque que dans l'éclatement $E_o: Z_o \to U$ de U de centre $\{0\}$, on peut calculer le nombre de Milnor de X en 0 de la façon suivante: soit T un voisinage tubulaire du diviseur exceptionnel $E_o^{-1}(0)$ dans Z_o; soit T_o un voisinage tubulaire du diviseur exceptionnel de X' dans Z_o. On considère $\{f \circ E_o = \lambda\} \cap (T - T_o) = Y_1$ et $\{f \circ E_o = \lambda\} \cap T_o = Y_2$ quand $\lambda \neq 0$ est assez petit. La fibre de Milnor de X en 0 est isomorphe à $Y_1 \cup Y_2$. Par conséquent la caractéristique d'Euler de la fibre de Milnor F de X en 0 vaut:

$$\chi(F) = 1 + \mu(X, 0) = \chi(Y_1) + \chi(Y_2) - \chi(Y_1 \cap Y_2)$$

On peut montrer que l'on a: $\chi(Y_1 \cap Y_2) = 0$. De plus Y_1 est homéomorphe à un revêtement de $E_o^{-1}(0) - E_o^{-1}(0) \cap X'$ de degré $m(X, 0)$

$= m$, multiplicité de X en 0, le cône tangent réduit de X en 0 est un cône sur une courbe plane non singulière de degré d on a:

$$\chi(E_o^{-1}(0) - E_o^{-1}(0) \cap X') = 1 + (d-1)(d-2)$$

Donc: $\chi(Y_1) = m(1 + (d-1)(d-2))$.

On remarque que Y_2 est une fibration sur $E_o^{-1}(0) \cap X' = \operatorname{Proj} C_{X,0}$ localement triviale de fibre homéomorphe à la fibre de Milnor d'une courbe non réduite composée d'un germe de courbe $(C', 0)$ et d'une courbe non réduite d'ensemble sous-jacent non singulier $(D, 0)$ compté avec la multiplicité m et ayant un nombre d'intersection avec C' égal à k en 0. En effet soit k la multiplicité de $|C_{X,0}|$. On a:

$$kd = m$$

Soit $\mathcal{C}$ la courbe section de $(X, 0)$ par un plan assez général de $\mathbf{C}^3$ qui passe par 0. Cette courbe a d tangentes distinctes et l'union C des composantes de $\mathcal{C}$ ayant la même tangente a la multiplicité l. La courbe $(C', 0)$ a le type d'équisingularité en 0 de la transformée stricte de C par E_o. On a donc la relation (cf. [17] appendice, ou [29]).

$$\mu_1 = \mu(C', 0) = \mu(C, 0) - k(k-1)$$

D'autre part:

$$\chi(Y_2) = \chi(\operatorname{Proj} |C_{X,0}|) \cdot \chi(F')$$

où F' est la fibre de Milnor de $(C' \cup D)$ en 0.

Calculons $\chi(F')$. Soit $x^m g = 0$ une équation locale de $C' \cup D$ en 0. En utilisant les résultats de [8], on trouve:

$$\chi(F') - m - m(C', 0) = -(m+1)k - \mu(C', 0) - m(C', 0) + 1$$

D'où:

$$\chi(F') = -\mu_1 - mk + 1 - k + m.$$

Comme

$$\chi(\mathrm{Proj}\,|C_{X,0}|) = 2 - (d-1)(d-2)$$

On obtient

$$\chi(Y_2) = (2 - (d-1)(d-2))(1 - \mu_1 - mk - k + m)$$

Donc:

$$1 + \mu(X, 0) = m + m(d-1)(d-2)$$
$$+ (mk + \mu_1 + k - m - 1)((d-1)(d-2) - 2)$$

La formule de [17] (Appendice) (ou [29]) donne pour un plan D de $\mathbf{C}^3$ assez général:

$$\mu^{(2)} = \mu(X \cap D, 0) = \mu(C, 0) = m(m-1) + d\mu_1 - (d-1)$$

Donc:

$$\mu_1 = \frac{\mu^{(2)} - m(m-1) + (d-1)}{d}$$

On a alors:

$$\frac{\mu(X, 0)}{\mu(X \cap P, 0)} = \frac{\mu^{(3)}}{\mu^{(2)}}$$

$$= (\mu^{(2)})^{-1}\Bigg(m + m(d-1)(d-2)$$

$$+ \left(mk + \frac{\mu^{(2)} - m(m-1) + (d-1)}{d}\right.$$

$$\left. + k - m - 1\right)((d-1)(d-2) - 2) - 1\Bigg)$$

Or $m = kd$. On peut donc écrire:

$$\frac{\mu^{(3)}}{\mu^{(2)}} = \frac{kd + kd^3 - 3kd^2 + 2kd + (k^2d^2 + \mu^{(2)} - k^2d^2 + kd - 1 + kd - kd^2)(d-3) - 1}{\mu^{(2)}}$$

$$= \frac{3kd + kd^3 - 3kd^2 + \mu^{(2)}(d-3) + 2kd^2 - 6kd - d + 3 - kd^3 + 3kd^2 - 1}{\mu^{(2)}}$$

$$= \frac{-3kd + \mu^{(2)}(d-3) + 2kd^2 - d + 2}{\mu^{(2)}}$$

$$= \frac{-3kd + 2kd^2 - d + 2}{\mu^{(2)}} + d - 3$$

Or on a toujours (cf. [22]): $\mu^{(2)} \geq (m-1)^2 = (kd-1)^2$. Donc comme $m > 1$:

$$\frac{\mu^{(3)}}{\mu^{(2)}} \leq \frac{2kd^2 - 2d - (3kd - 3)}{(kd-1)^2} + d - 3 + \frac{d-1}{(kd-1)^2}$$

$$= \frac{2d-3}{(kd-1)} + \frac{d-1}{(kd-1)^2} + d - 3$$

$$= \frac{2(d-1)}{kd-1} + \frac{d-1}{(kd-1)^2} - \frac{1}{kd-1} + d - 3$$

Or

$$\frac{2(d-1)}{(kd-1)} \leq 2 \text{ et } \frac{d-1}{(kd-1)^2} - \frac{1}{kd-1} = \frac{d(1-k)}{(kd-1)^2} \leq 0.$$

D'où:

$$\frac{\mu^{(3)}}{\mu^{(2)}} \leq d - 1$$

Or on a toujours

$$\frac{\mu^{(3)}}{\mu^{(2)}} \geq kd - 1 \qquad \text{(cf. [22] Chap. (2.1))}$$

D'où:

$$0 = (k - 1)d$$

i.e. $k = 1$.

L'équivalence de ii) et iii) est faite dans [22] Chap. 2 (2.7).

L'implication iii) implique iv) résulte immédiatement de la description de φ dans (1.4.1).

L'implication iv) donne v) est immédiate, car l'éclatement de 0 dans un cône à singularité isolée est une résolution des singularités, l'hypothèse de résolution simultanée implique qu'il ne peut pas y avoir de tangentes exceptionnelles, et en vertu de la remarque (1.4.4.7), $|e^{-1}(0)|$ est non singulière.

v) implique vi) résulte de l'équivalence de 1) et 4) dans le théorème (1.4.4.1).

Enfin vi) implique i) provient aussi du théorème (1.4.4.1).

Nous remercions J. Giraud de nous avoir signalé une imprécision dans l'énoncé du théorème (2.2.1).

(2.3) Quelques exemples

(2.3.1) La queue d'aronde, discriminant de la déformation semi-universelle de $X^4 = 0$ d'équation:

$$256x^3 - 27y^4 - 128x^2z^2 + 144xy^2z + 16xz^4 - 4y^2z^3 = 0$$

n'a pas de tangentes exceptionnelle et son cône tangent n'est pas qu'un nombre fini de limites d'espaces tangents en 0: car dans [18] (5.5.1)) on démontre que sa modification de Nash est sa normalisation qui est lisse. On remarque aussi qu'une section plane générique passant par 0 a le type d'équisingularité de $x^3 + y^4 = 0$ et qu'une section plane générique proche de 0 mais ne passant pas par 0 est:

On remarque que dans ce cas le cône tangent de la queue d'aronde est donné par $x^3 = 0$ et n'est pas réduit.

(2.3.2) La surface d'équation

$$(x^2 + y^2 + z^2)^2 + z^5 = 0$$

n'a pas de tangentes exceptionnelle et son cône tangent n'est pas réduit. Les tangentes au lieu critique d'une projection générique contiennent les génératrices du contour apparent du cône tangent dont la contribution de chacune dans la multiplicité de l'espace critique est 2 parce que la section plane générique passant par l'origine de cette surface a le type d'équisingularité d'une courbe plane composée de deux cusps transverses:

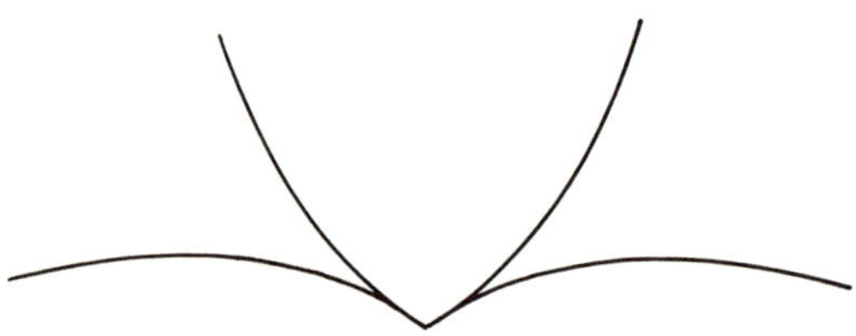

On montrera dans un prochain article que l'union des composantes du lieu critique d'une projection générique dont la tangente est une génératrice non exceptionnelle a le type d'équisingularité de l'union des composantes d'une section plane générique ayant la même tangente.

La contribution du lieu singulier au lieu critique peut être calculée égale à 10 dans ce cas et on trouve ainsi qu'il ne peut pas y avoir de tangentes exceptionnelles, car la contribution totale de l'espace critique est $14 = 10 + 2.2$.

Centre de Mathématiques de l'Ecole Polytechnique Laboratoire de Recherche Associé Au C.N.R.S.

BIBLIOGRAPHIE

---

[1] J. Briançon, J. P. G. Henry, J. P. Speder, Note aux C. R. Acad. Sc. Paris, t. 282, série A, p. 279 (1976).

[2] Adrien Douady, "Flatness and privilege," *L'enseignement mathématique,* n° **14** (1968).

[3] Murray Gerstenhaber, "On the deformation of rings and algebras II," *Annals of Math.,* **84,** n° 1 (1966).

[4] Helmut Hamm et Lê D. T., "Un théorème de Zariski du type de Lefschetz," *Annales de l'E. N. S.* 1973, p. 317-366.

[5] J. P. G. Henry et Lê Dũng Tráng, "Limites d'espaces tangents in Fonctions de plusieurs variables complexes II," *Séminaire F. Norguet,* Springer Lecture Notes, n° 482.

[6] Heisuke Hironaka, "Stratification and flatness. Real and complex singularities," Nordic Summer School Oslo 1976, Noordhoff 1977.

[7] Steve Kleiman, "The enumerative theory of singularities," *Real and complex singularities,* Oslo 1976, Noordhoff 1977.

[8] Lê Dũng Tráng, "Calcul du nombre de cycles évanouissants d'une hypersurface complexe," *Annales de l'Institut Fourier,* t. XXIII, Fasc. 4 (1973).

[9] Lê D. T., "Limites d'espaces tangents, à paraître Proc. Symp. on Algebraic singularities," Rheinhardtsbrunn 1978, Academia Leopoldina, Preprint Centre de Mathématiques, Ecole Polytechnique, 1978.

[10] Lê D. T., "Topologie des singularités des hypersurfaces complexes," Singularités à Cargèse 1972. Astérisque, n° 7-8, S. M. F., Paris.

[11] Lê D. T., "Une application d'un théorème d'A'Campo à l'équisingularité," *Indag. Math.* **35,** n° 5 (1973).

[12] Lê D. T. et C. P. Ramanujam, "The invariance of Milnor's number implies the invariance of the topological type." *Amer. Journ. of Math.* **98** (1976), p. 67-68.

[13] Lê D. T. et Kyoji Saito, "La constance du nombre de Milnor donne de bonnes stratifications," Note aux C. R. Acad. Sc., t. 277 n° 16 (22 Octobre 1973) p. 793.

[14] Monique Lejeune-Jalabert et Bernard Teissier, "Clôture intégrale des idéaux et équisingularité," Séminaire Ecole Polytechnique 1973-74, Chap. I, publié par l'Institut Fourier, St. Martin d'Hères, 38 (France).

[15] John Mather, *Notes on topological stability.* Harvard Notes 1970.

[16] John Milnor, "Singular points of complex hypersurfaces," *Annals of Math.* Studies n° 61 (1968 (Princeton University Press.

[17] Frédéric Pham, "Singularités planes d'ordre 3. Singularités à Cargèse 1972," Astérisque n° 7-8, S. M. F. Paris.

[18] Bernard Teissier, "The hunting of invariants in the geometry of discriminants,"

Real and complex singularities, Nordic Summer School Oslo 1976, Noordhoff, 1977.

[19] B. Teissier, "Déformations à type topologique constant II," Séminaire Douady-Verdier E. N. S. 1971-72, Astérisque n° 16, S. M. F. Paris.

[20] B. Teissier, "Cycles évanescents et résolution simultanée," Séminaire Demazure-Pinkham-Teissier sur les Singularités des surfaces, Ecole Polytechnique, Centre de Mathématiques 1976-77.

[21] B. Teissier, "Introduction to equisingularity problems," Summer school in Algebraic geometry, Arcata 1974, Proc. A.M.S., Symp. in Pure Math. n° 29, A.M.S. 1975.

[22] B. Teissier, "Cycles évanescents, sections planes, et conditions de Whitney," Singularités à Cargèse 1972. Astérisque n° 7-8, S.M.F. Paris.

[23] B. Teissier, "Appendice au cours de O. Zariski: Modules de branches planes," Ecole Polytechnique Centre de Mathématiques 1973.

[24] A. N. Varchenko, "Theorems of topological equisingularity," Isv. Ak. Nauk. SSSR Ser. Mat. 36 (1972).

[25] Hassler Whitney, *Local Properties of Analytic Varieties in Differential and Combinatorial Topology.* Princeton University Press, 1965.

[26] H. Whitney, "Tangents to an anlytic variety," *Annals of Maths.,* t. 81 (1965) n° 3, p. 496-549.

[27] Oscar Zariski, "Studies in equisingularity I, II." *Amer. Journ. of Math.* **87** (1965).

[28] O. Zariski, "Studies in equisingularity III." *Amer. Journ. of Math.* **90** (1968).

[29] O. Zariski, "Contributions to the problem of equisingularity," *Problems in Algebraic geometry,* C. I. M. E. Varenna 1969. Edizioni Cremonese, Roma.

FOUNDATIONS OF A GENERAL THEORY OF EQUISINGULARITY ON r-DIMENSIONAL ALGEBROID AND ALGEBRAIC VARIETIES, OF EMBEDDING DIMENSION r + 1.*

By Oscar Zariski

Introduction. We develop in this paper a general (local) theory of equisingularity on algebroid hypersurfaces V, and a general (global) theory of equisingularity on r-dimensional algebraic varieties V, defined over an algebraically closed ground field k (of characteristic zero) and having local embedding dimension $r + 1$ at each of their singular k-rational points. In preceding papers, especially in [1], we dealt only with equisingularity of an algebraic variety V, of dimension r, along singular irreducible subvarieties W of V, of codimension 1. In the present paper we deal with singular subvarieties of any codimension s $(1 \leq s \leq r)$.

In the absence of any general theory of equivalence of singularities of varieties of dimension >1, any definition of equisingularity of V along singular subvarieties W of codimension >1 is bound to be to some extent a matter of individual choice and perception. We cannot, nor do we intend to, claim that our definition is a definitive one. We can say, nevertheless, that our definition is not trivially too strong, as for instance one based on analytical equivalence would be, nor is it trivially too weak as a definition based on equimultiplicity would be. A definitely positive aspect of our theory is that it leads to a substantial collection of results which are both non-trivial and geometrically sound, including a complete theory of stratification of our varieties V into equisingularity strata.

Since one of the principal conditions which we wanted our concept of equisingularity to satisfy is that it be analytically intrinsic (and not only biregularily invariant), we decided to develop first a *local* theory of equisingularity of *algebroid* hypersurfaces. The greatest part of our paper (five out of a total of six sections) is dedicated to this local theory.

* This work was supported by a National Science Foundation Grant.

After this local theory is developed, it is applied in section 6 to the "global" theory of equisingularity of algebraic varieties, including the stratification of these varieties into equisingularity strata. Such a stratification is also developed for algebroid varieties (in section 5), but in that case the strata are easier to introduce and have a more formal character. Nevertheless, some of the properties of the formal strata in the algebroid case are essential for the theory of stratification of algebraic varieties [see, for instance, the use we make, in section 6, of part (d) of Theorem 5.1].

Our theory of equisingularity is based on the concept of the *dimensionality type* s of a singular point of V; here $1 \leq s \leq r$ (but *simple* points has dimensionality type zero). Our singularity strata are maximal open irreducible subvarieties S of V all points of which have the same dimensionality type s, and the closure of S (i.e., the smallest algebraic subvariety $\overline{S}$ of V containing S) is an irreducible variety of dimension $r - s$, while S is a Zariski dense open subset of $\overline{S}$ (i.e., $\overline{S} - S$ is a proper subvariety of $\overline{S}$). An informal preparatory discussion of the concept of dimensionality type and of the nature of the strata S is to be found in section 3. We advise the reader to read section 3 before reading the more technical sections 1 and 2.

We call attention of the reader to the fact that one crucial result in section 6, namely the proof that each equisingularity stratum S is a Zariski dense open subset of its algebraic closure $\overline{S}$ has been proved by H. Hironaka in his paper [5] published in the present issue of the American Journal of Mathematics.

1. Algebroid hypersurfaces, their projections and critical loci. Let V be a formal (reduced) algebroid variety, of pure dimension r, defined over a field k of characteristic zero. Let P be the center (i.e., the closed point) of V and let $\mathfrak{o}$ be the (complete) local ring of V at P. Thus $\mathfrak{o}$ is a noetherian local ring, of dimension r, free from nilpotent elements different from zero, and having k as coefficient field. We assume that *the embedding dimension* of V is $r + 1$, i.e., that the maximal ideal $\mathfrak{m}$ of $\mathfrak{o}$ has a minimal basis of $r + 1$ elements (equivalently: the k-vector space $\mathfrak{m}/\mathfrak{m}^2$ has dimension $r + 1$). Thus V can be represented (up to an arbitrary analytical isomorphism over k) as a hypersurface in an $(r + 1)$-dimensional affine $(r + 1)$-space $\mathbf{A}_{r+1}/k$, and P is necessarily a *singular* point of V. We have thus excluded the case in which $\mathfrak{o}$ is a regular local ring, in which case every minimal basis of $\mathfrak{m}$ would consist of r elements and P would be a simple point of V.

The variety V is to be thought of as being the set *Spec* $(\mathfrak{o})$, i.e., the set of all prime ideals $\mathfrak{p}$ of $\mathfrak{o}$. We shall refer to any such ideal $\mathfrak{p}$ as a *point of* V. In particular, the closed P of V is the maximal ideal $\mathfrak{m}$ of $\mathfrak{o}$.

Given two points R_1 and R_2 of V, corresponding to two prime ideals $\mathfrak{p}_1$ and $\mathfrak{p}_2$ of $\mathfrak{o}$, we say that R_2 is an *analytic specialization of* R_1 if $\mathfrak{p}_2 \supset \mathfrak{p}_1$. We shall usually omit the word "analytic" and will say simply that R_2 is a specialization R_1 if $\mathfrak{p}_2 \supset \mathfrak{p}_1$. In particular, the closed point P of V is a specialization of any point R of V.

The set of all points which are specializations of a given point R of V shall be referred to as the *irreducible* (algebroid) *subvariety* W *of* V *having* R *as general point.* This algebroid variety W can be canonically identified with the algebroid (irreducible) variety Spec $(\mathfrak{o}/\mathfrak{p})$, where $\mathfrak{p}$ is the prime ideal associated with the point R, the local ring $\mathfrak{o}/\mathfrak{p}$ being a complete local *domain,* having k as coefficient field. The customary notation for this variety W is $V(\mathfrak{p})$ (*the variety of* $\mathfrak{p}$). More generally, if $\mathfrak{A}$ is any ideal in $\mathfrak{o}$ then the set of prime ideals $\mathfrak{p}$ in $\mathfrak{o}$ which contain $\mathfrak{A}$ is the variety $V(\mathfrak{A})$ of the ideal $\mathfrak{A}$. We have $V(\mathfrak{A}) = V(Rad\ \mathfrak{A})$, and $V(Rad\ \mathfrak{A})$ is canonically identifiable with the algebroid variety Spec $(\mathfrak{o}/Rad\ \mathfrak{A})$, which is a reduced variety since the local ring $\mathfrak{o}/Rad\ \mathfrak{A}$ is free from non-zero nilpotent elements. The ideal $Rad\ \mathfrak{A}$ is an irredundant (finite) intersection of prime ideals in $\mathfrak{o}$, and the associated varieties of these prime ideals are the irreducible components of $V(Rad\ \mathfrak{A})$. In particular, the zero ideal (0) in $\mathfrak{o}$ coincides with its own radical (since V is reduced), and if

$$(0) = \mathfrak{p}_1 \cap \mathfrak{p}_2 \cap \cdots \cap \mathfrak{p}_q \quad (\mathfrak{p}_i \not\supset \mathfrak{p}_j \quad \text{if} \quad i \neq j) \tag{1}$$

is the irredundant representation of (0) as an intersection of prime ideals, then the q irreducible varieties $V_i = V(\mathfrak{p}_i)$ are the irreducible components of V, and we have

$$V = V_1 \cup V_2 \cup \cdots \cup V_q. \tag{2}$$

Our assumption that V is of pure dimension r signifies that each V_i has dimension r, or—equivalently—that each of the q prime ideals $\mathfrak{p}_i$ in (1) has dimension r (or—what is the same—depth r). These prime ideals are, of course, all of height zero, i.e., they are the minimal prime ideals of $\mathfrak{o}$.

The (well known) terms and concepts introduced above are independent of the choice of the coefficient field k of $\mathfrak{o}$. From now on, however, the chosen coefficient field k will begin to play an essential role. In fact, the entire theory of equisingularity which we develop in this paper depends *a priori* on the choice of the coefficient field k of $\mathfrak{o}$. It is a basic open question whether our equisingularity concept is or is not actually independent of the choice of the coefficient field k of our complete local ring $\mathfrak{o}$.

Given a minimal basis $\{x_1, x_2, \ldots, x_{r+1}\}$ of the maximal ideal $\mathfrak{m}$ of $\mathfrak{o}$, it is well known that $\mathfrak{o}$ is a homomorphic image of the power series ring $\mathfrak{O} = k[[X_1, X_2, \ldots, X_{r+1}]]$ in $r + 1$ independent variables over k and that the kernel of this homomorphism is a principal ideal generated by a power series $f(X)(=f(X_1, X_2, \ldots, X_{r+1}))$ which is free from multiple factors (since V is reduced) and which is uniquely determined by the basis $\{x_1, x_2, \ldots, x_{r+1}\}$ of $\mathfrak{m}$ and by the coefficient field k up to an arbitrary unit factor in $\mathfrak{O}$. The elements $x_1, x_2, \ldots, x_{r+1}$ are the $f(X)$-residues of $X_1, X_2, \ldots, X_{r+1}$. If $f^{(1)}(X), f^{(2)}(X), \ldots, f^{(q)}(X)$ are the irreducible factors of $f(X)$ then the q prime ideals $\mathfrak{p}_i$ in (1) are the principal ideals $\mathfrak{o}\cdot f^{(i)}(x)$. We refer to the equation $f(X_1, X_2, \ldots, X_{r+1}) = 0$ as the equation of V (relative to the given minimal basis $\{x_1, x_2, \ldots, x_{r+1}\}$ of $\mathfrak{m}$ and the given coefficient field k):

$$V: f(X_1, X_2, \ldots, X_{r+1}) = 0. \tag{3}$$

The equation of the irreducible component V_i of V is then $f^{(i)}(X_1, X_2, \ldots, X_{r+1}) = 0$, relative to the minimal basis $[x_1{}^{(i)}, x_2{}^{(i)}, \ldots, x_{r+1}{}^{(i)}\}$ of the maximal ideal $\mathfrak{m}/\mathfrak{p}_i$ of the local domain $\mathfrak{o}/\mathfrak{p}_i$; here $x_j{}^{(i)}$ is the $\mathfrak{p}_i$-residue of x_j, or—what is the same thing—the $f_i(X)$-residue of X_j.

A set of r elements $x_1, x_2, \ldots, x_r$ of $\mathfrak{m}$ constitute a set of *parameters* of $\mathfrak{o}$ if these elements generate in $\mathfrak{o}$ an ideal which is primary for $\mathfrak{m}$. *We shall deal from now on only with such sets of elements* $x_1, x_2, \ldots, x_r$ *of* $\mathfrak{m}$ *that are parameters and that can be extended to a minimal basis* $\{x_1, x_2, \ldots, x_r, x_{r+1}\}$ *of* $\mathfrak{m}$. If $\{x_1, x_2, \ldots, x_r, x_{r+1}\}$ is a minimal basis of $\mathfrak{m}$ and if (3) is the equation of V relative to that basis, then $x_1, x_2, \ldots, x_r$ are parameters of $\mathfrak{o}$ if and only if $f(0, 0, \ldots, 0, X_{r+1})$ is not identically zero (equivalently: if the line $X_1 = X_2 = \cdots = X_r = 0$ does not belong to V).

If $f_n(X_1, X_2, \ldots, X_{r+1})$ is the leading form of the power series $f(X)$ in (3), then $n > 1$ (since P is a singular point of V), and this integer n is independent of the choice of the coefficient field k and of the basis $\{x_1, x_2, \ldots, x_{r+1}\}$ of $\mathfrak{m}$. The integer n is the *multiplicity* of the local ring $\mathfrak{o}$ and is also the multiplicity of the singular point P of V. The r elements $x_1, x_2, \ldots, x_r$ of the minimal basis $\{x_1, x_2, \ldots, x_r, x_{r+1}\}$ are said to be *transversal* parameters if the leading term of the non-zero power series $f(0, 0, \ldots, 0, X_{r+1})$ is of degree n, or—equivalently—if the monomial $X_{r+1}{}^n$ actually occurs in the leading form $f_n(X_1, X_2, \ldots, X_r, X_{r+1})$ of $f(X)$, i.e., if $f_n(0, 0, \ldots, 0, X_{r+1}) \neq 0$. It is clear that if $r + 1$ elements $x_1', x_2', \ldots, x_r', x_{r+1}'$ of $\mathfrak{o}$ are defined implicitly by equations of the form

$$x_i = \sum_{j=1}^{r+1} c_{ij}x_j' + \text{terms of degree} > 1 \text{ in the } x\text{'s},$$

$$(i = 1, 2, \ldots, r + 1;\ c_{ij} \in k)$$

where we assume that the determinant $|c_{ij}|$ is different from zero, then $\{x_1', x_2', \ldots, x_r', x_{r+1}'\}$ is a basis of $\mathfrak{m}$ (necessarily minimal), and $x_1', x_2', \ldots, x_r'$ are transversal parameters if and only if $f_n(c_{1,r+1}, c_{2,r+1}, \ldots, c_{r+1,r+1}) \neq 0$.

We assume now that $x_1, x_2, \ldots, x_r$ are parameters of $\mathfrak{o}$ (not necessarily transversal) and that $\{x_1, x_2, \ldots, x_r, x_{r+1}\}$ is a basis of $\mathfrak{m}$. Let $X_{r+1}{}^m$ be the least power of X_{r+1} which actually occurs in $f(0, 0, \ldots, 0, X_{r+1})$. We have them $m \geq n$, with equality if and only if $x_1, x_2, \ldots, x_r$ are *transversal* parameters. Since the power series $f(X)$ is only defined to within an arbitrary unit factor in the power series ring $k[[X_1, X_2, \ldots, X_r, X_{r+1}]]$, we may assume, by the Weierstrass preparation theorem, that $f(X)$ is a monic polynomial in X_{r+1}, of degree m, with coefficients in $k[[X_1, X_2, \ldots, X_r]]$:

$$f(X_1, X_2, \ldots, X_r, X_{r+1}) = X_{r+1}{}^m + a_1(X_1, X_2, \ldots, X_r)X_{r+1}{}^{m-1} + \cdots + \alpha_m(X_1, X_2, \ldots, X_r), \quad (4)$$

where the leading form of each power series $a_i(X_1, X_2, \ldots, X_r)$ is of positive degree, and, furthermore, for $i = m - n + 1, m - n + 2,$

..., m, the leading form of $a_i(X)$ is of degree $\geq n - m + i$, with equality for at least one of these n values of i.

The parameters $x_1, x_2, \ldots, x_r$ are analytically independent over k, and thus the subring

$$A = k[[x_1, x_2, \ldots, x_r]] \tag{5}$$

of $\mathfrak{o}$ is canonically isomorphic to the power series ring $k[[X_1, X_2, \ldots, X_r]]$. We denote by $\hat{\mathbf{A}}_r/k$ the algebroid (irreducible) r-dimensional variety Spec (A); this variety is merely the r-dimensional affine space $\hat{\mathbf{A}}_r$ over k, considered locally at the origin P_0, as a formal algebroid variety (having at P_0 a simple point). We have a natural projection

$$\pi_A : V \to \hat{\mathbf{A}}_r/k \tag{6}$$

of V onto $\hat{\mathbf{A}}_r/k$, defined by

$$\pi_A(\mathfrak{p}) = \mathfrak{p} \cap A = \mathfrak{p}_0; \; (\mathfrak{p} \in \text{Spec}\,(\mathfrak{o}),\ \mathfrak{p}_0 \in \text{spec}\,(A)). \tag{7}$$

Using the geometric language of points, we shall write (7) in the form

$$\pi_A(R) = R_0, \tag{7'}$$

where R and R_0 are the points of V and of $\hat{\mathbf{A}}_r/k$ represented respectively by the prime ideals $\mathfrak{p}$ and $\mathfrak{p}_0$ of $\mathfrak{o}$ and A. That the mapping π_A is surjective follows from the fact that, in view of (4), $\mathfrak{o}$ is integral over A, and that consequently over every prime ideal $\mathfrak{p}_0$ of A there lies at least one prime ideal $\mathfrak{p}$ of $\mathfrak{o}$. Furthermore, the fact that $\mathfrak{o}$ is a finite A-module implies that the number of such prime ideals $\mathfrak{p}$ is *finite* (in fact, not greater than m). For this reason we shall refer to the projection π_A as a *finite* projection.

We observe that the projection π_A depends only on the (regular) local ring A and not on the choice of the parameters $x_1, x_2, \ldots, x_r$ of $\mathfrak{o}$ which form a basis of the maximal ideal of A. On the other hand, if the parameters $x_1, x_2, \ldots, x_r$ are given then the projection π_A depends on the choice of the coefficient field k, since $A = k[[x_1, x_2, \ldots, x_r]]$. We say that the projection π_A is *transversal at P* if the integer m in (4) is equal to n, i.e., if the parameters $x_1, x_2, \ldots, x_r$ are transversal. The coördinates of P being all zero, to say that the projection π_A is transversal at P is the same as saying that $X_{r+1} = 0$ is an n-field root of $f(0, 0, \ldots, 0, X_{r+1})$; or — equivalently — that the line $X_1 = X_2 = \cdots = X_r = 0$ has exactly an n-fold intersection with V at P (and therefore does not belong to the tangent hypercone of V at P). More generally, if R is any point of V and if $\xi_1, \xi_2, \ldots, \xi_r, \xi_{r+1}$ are the coördinates of R (i.e., the $\mathfrak{p}$-residues of R in the local domain $\mathfrak{o}/\mathfrak{p}$, where $\mathfrak{p}$ is the prime ideal of $\mathfrak{o}$ associated with the point R), then we say that π_A is transversal at R if the multiplicity of the root $X_{r+1} = \xi_{r+1}$ of the polynomial $f(\xi_1, \xi_2, \ldots, \xi_r, X_{r+1})$ is equal to the multiplicity of the point R of V [The multiplicity of R at V is the multiplicity of the local ring of quotients $\mathfrak{o}_\mathfrak{p}$; it can be proved (and is well known) that ν is the multiplicity of V at R if and only if all the partial derivatives of $f(X_1, X_2, \ldots, X_{r+1})$, of order $\leq \nu - 1$, vanish at R, but not all partial derivatives of f, of order ν, vanish at R]. The following observation is obvious: we have always $\nu \leq n$ for any point R of V, and if $\nu = n$ then every projection π_A which is transversal at P is also transversal at R (the converse is not necessarily true).

Let R and R_0 be as in (7′) and let $\xi_1, \xi_2, \ldots, \xi_r$ be the coördinates of R, i.e., the $\mathfrak{p}_0$-residues of $x_1, x_2, \ldots, x_r$ in the local (regular) *domain* $A/\mathfrak{p}_0$. It follows from (7) that $\xi_1, \xi_2, \ldots, \xi_r$ can be identified also with the $\mathfrak{p}$-residues of $x_1, x_2, \ldots, x_r$ in the local domain $\mathfrak{o}/\mathfrak{p}$. If then ξ_{r+1} is the $\mathfrak{p}$-residue of x_{r+1} then $\xi_1, \xi_2, \ldots, \xi_r, \xi_{r+1}$ are the coördinates of R, and we must have

$$f(\xi_1, \xi_2, \ldots, \xi_r, \xi_{r+1}) = 0.$$

Conversely, if the point $R_0 = (\xi_1, \xi_2, \ldots, \xi_r)$ is given and if ξ_{r+1} is any root of the polynomial $f(\xi_1, \xi_2, \ldots, \xi_r, X_{r+1})$ in the algebraic closure of the field $k((\xi_1, \xi_2, \ldots, \xi_r))$ of meromorphic functions of $\xi_1, \xi_2, \ldots, \xi_r$ (over k), then the point R whose coördinates are $\xi_1, \xi_2, \ldots, \xi_r, \xi_{r+1}$ belongs to V, and R_0 is the π_A-projections of R. Two distinct roots ξ_{r+1}, ξ_{r+1}' of $f(\xi_1, \xi_2, \ldots, \xi_r, X_{r+1})$ yield the same point R of V if and only if these roots are conjugate over the field $k((\xi_1, \xi_2, \ldots, \xi_r))$, for it is in this and only in this case that the ideal $\mathfrak{p}$ of those elements $g(x_1, x_2, \ldots, x_r, x_{r+1})$ of $\mathfrak{o}$ which yield true relations $g(\xi_1, \xi_2, \ldots, \xi_r, \xi_{r+1}) = 0$ coincides with the ideal $\mathfrak{p}'$ of those elements $g(x_1, x_2, \ldots, x_{r+1})$ of $\mathfrak{o}$ which yield true relations $g(\xi_1, \xi_2, \ldots, \xi_r, \xi_{r+1}') = 0$. If μ is the number of distinct roots of $f(\xi_1, \xi_2, \ldots, \xi_r, X_{r+1})$ which are conjugate to ξ_{r+1}, we call μ *the relative degree of the point R* *over its π_A-projection R_0*. This integer μ is also the relative degree $[k((R)): k((R_0))]$, where $k((R))$ and $k((R_0))$ are respectively the fields of fractions $k((\xi_1, \xi_2, \ldots, \xi_r, \xi_{r+1}))$, $k((\xi_1, \xi_2, \ldots, \xi_r))$ of the complete local domains $\mathfrak{o}/\mathfrak{p}$ and $A/\mathfrak{p}_0$. Let then $R_1, R_2, \ldots, R_g$ be the *distinct* points of V which have the same projection R_0, let μ_j be the relative degree of R_j over R_0, let $\xi_{r+1}{}^{(j)}$ be a root of $f(\xi_1, \xi_2, \ldots, \xi_r, X_{r+1})$ which yields the point R_j, and let e_j be the multiplicity of that root. Then we have

$$e_1\mu_1 + e_2\mu_2 + \cdots + e_g\mu_g = m, \tag{8}$$

where—we recall—m is the degree of $f(X_1, X_2, \ldots, X_r, X_{r+1})$ in X_{r+1} [see (4)]. It is clear, from the very definition of the integer g and of the g integers μ_j, that these integers are independent of the choice of the element x_{r+1} which together with the elements $x_1, x_2, \ldots, x_r$ gives a basis of $\mathfrak{m}$. It is also easy to see that the integers e_j are also independent of the choice of the element x_{r+1}; the proof is left to the reader.

We call the integer e_j the *ramification index* of R_j relative to the projection π_A, and we write $e_j = e(R_j; \pi_A)$. We say that a point R of V is a *ramified* point of the projection π_A if $e(R; \pi_A) > 1$. We call a point R_0 of $\hat{\mathbf{A}}_r/k$ a *critical* point of π_A if at least one of the points of V which project into R_0 under π_A is a ramified point of π_A. The set of critical points of π_A is an algebroid hypersurface in $\hat{\mathbf{A}}_r/k$, for this set is defined by the equation

$$D(X_1, X_2, \ldots, X_r) = 0,$$

where D is the X_{r+1}-discriminant of $f(X_1, X_2, \ldots, X_r, X_{r+1})$. We call this hypersurface the *critical variety* of the projection π_A of V and we denote it by $\Delta(\pi_A)$. Since we wish to regard $\Delta(\pi_A)$ as a *reduced* variety, we replace the power series $D(X_1, X_2, \ldots, X_r)$ by the product of its distinct (i.e., non-associated) factors in $k[[X_1, X_2, \ldots, X_r]]$. We denote this product by $\Delta_A(X_1, X_2, \ldots, X_r)$. Thus the reduced critical variety $\Delta(\pi_A)$ is defined by the equation

$$\Delta_A(X_1, X_2, \ldots, X_r) = 0. \tag{9}$$

The critical variety $\Delta(\pi_A)$ is uniquely determined by the two local rings

$\mathfrak{o}$ and A (and hence depends also on the choice of the coefficient field k of $\mathfrak{o}$).

The set of ramified points R of the projection π_A is an algebroid subvariety of V, of pure dimension $r - 1$, since this set is the intersection of V with the hypersurface $F_{X_{r+1}}'(X_1, X_2, \ldots, X_r, X_{r+1}) = 0$, this latter hypersurface having no irreducible components in common with V, since $F(X)$ is free from multiple factors. In particular, all *singular* points of V are ramified points of *any* projection π_A.

A *simple* point R of V, having coördinates $\xi_1, \xi_2, \ldots, \xi_{r+1}$, is a ramified point of π_A if and only if the "line" $X_i = \xi_i$ $(i = 1, 2, \ldots, r)$ is tangent to V at R. Actually, the term "line" is somewhat misleading in the present context, unless R is the closed point P of V (in which case all the ξ_i are zero). If $R \neq P$ and if ρ is the dimension of the irreducible subvariety W_ρ of V having R a general point, then the irreducible subvariety $W_\rho^{(0)}$ of $\Delta(\pi_A)$, having as general point the projection R_0 of R, is also of dimension ρ, and in the ambient space $\hat{\mathbf{A}}_{r+1}/k$ of V the equations $X_i = \xi_i$ $(i = 1, 2, \ldots, r)$ define an algebroid *cylinder* $L_{\rho+1}$ of dimension $\rho + 1$, having $(\xi_1, \xi_2, \ldots, \xi_r, t)$ as general point over k, where t is an analytical transcendental over the field $k((\xi_1, \xi_2, \ldots, \xi_r))$ of meromorphic functions of $\xi_1, \xi_2, \ldots, \xi_r$. The variety $W_\rho^{(0)}$ is the base of this cylinder $L_{\rho+1}$, while the X_{r+1}-axis is its axis. If $R_1 (= R)$, $R_2, \ldots, R_g$ are the points of V which project into the point R_0, then the intersection of V with $L_{\rho+1}$ is the union of g irreducible subvarieties $W_{\rho,1} (= W_\rho)$, $W_{\rho,2}, \ldots, W_{\rho,g}$, of dimension ρ, R_j being the general point of $W_{\rho,j}$. Since we have assumed that R is a simple point of V and a ramified point of π_A, the variety W_ρ is a *simple* subvariety of V, and the cylinder $L_{\rho+1}$ is tangent to V at the point R and, consequently, also at each point of W_ρ which is simple for V (note that since R is a simple point of V, *almost all points* of W_ρ are simple points of V; namely, if V_{sing} denote the set of singular points of V then the intersection $W_\rho \cap V_{sing}$ is a proper subvariety of the irreducible variety W_ρ).

For any prime ideal $\mathfrak{p}_0$ of A, or for the corresponding point of R_0 of Spec (A), the primary irredundant decomposition of the extended ideal $\mathfrak{o}\mathfrak{p}_0$ is essentially well known and is completely determined by the factorization of the polynomial $f(\xi_1, \xi_2, \ldots, \xi_r, X_{r+1})$ into irreducible factors over the field $k((\xi_1, \xi_2, \ldots, \xi_r))$ of meromorphic functions of $\xi_1, \xi_2, \ldots, \xi_r$; here $\xi_1, \xi_2, \ldots, \xi_r$ are the $\mathfrak{p}_0$-residues of $x_1, x_2, \ldots, x_r$. Namely, if $R_1, R_2, \ldots, R_g$ are the points of V whose π_A-projection is the point $R_0(=\mathfrak{p}_0)$, and if μ_j and e_j are respectively the relative degree and the ramification index of R_j over R_0 ($j = 1, 2, \ldots, g$), then the monic polynomial $f(\xi_1, \xi_2, \ldots, \xi_r, X_{r+1})$ in X_{r+1} factors into irreducible factors as follows:

$$f(\xi_1, \xi_2, \ldots, \xi_r, X_{r+1}) = \prod_{j=1}^{g} [F_j(\xi_1, \xi_2, \ldots, \xi_r, X_{r+1})]^{e_j}.$$

Here $F_j(\xi_1, \xi_2, \ldots, \xi_r, X_{r+1})$ is a monic polynomial in X_{r+1}, of degree μ_j, with coefficient in $k[[\xi_1, \xi_2, \ldots, \xi_r]]$. The coördinates of the point R_j are $\xi_1, \xi_2, \ldots, \xi_r, \xi_{r+1}^{(j)}$, where $\xi_{r+1}^{(j)}$ can be any of the μ_j roots of the polynomial $F_j(\xi_1, \xi_2, \ldots, \xi_r, X_{r+1})$. The prime ideal $\mathfrak{p}_j$ in $\mathfrak{o}$ which is associated with the point R_j is the ideal

$$\mathfrak{p}_j = \mathfrak{o}\cdot\mathfrak{p}_0 + \mathfrak{o}\cdot f_j(x_1, x_2, \ldots, x_r, x_{r+1}). \tag{10}$$

The primary irredundant decomposition of the ideal $\mathfrak{o}\cdot\mathfrak{p}_0$ is then the following:

$$\mathfrak{o}\cdot\mathfrak{p}_0 = \bigcap_{j=1}^{g} \mathfrak{q}_j, \tag{11}$$

where

$$\mathfrak{q}_j = \mathfrak{o}\cdot\mathfrak{p}_0 + \mathfrak{o}\cdot(f_j(x_1, x_2, \ldots, x_r, x_{r+1})^{e_j}). \tag{12}$$

2. The generic projection of the variety V. If $\{x_1, x_2, \ldots, x_{r+1}\}$ is a (necessarily) minimal basis of the maximal ideal $\mathfrak{m}$ of our local ring $\mathfrak{o}$, every element z of $\mathfrak{m}$ is a power series in $x_1, x_2, \ldots, x_{r+1}$ with coefficients in k and constant term zero:

$$z = \sum_{j=1}^{r+1} c_j x_j + \sum_{1\le j_1\le j_2\le r+1} c_{j_1 j_2} x_{j_1} x_{j_2} + \cdots + \sum_{1\le j_1\le j_2\le\cdots\le j_N\le r+1} c_{j_1 j_2\cdots j_N} x_{j_1} x_{j_2}\cdots x_{j_N} + \cdots,$$

where the $c_j, c_{j_1 j_2}, \ldots, c_{j_1 j_2\cdots j_N}, \cdots$ are elements of k. The basic idea of this section is to consider power series x^* in $x_1, x_2, \ldots, x_{r+1}$ with *indeterminate* coefficients u_j, $u_{j_1 j_2}$, ... Actually we shall introduce r such power series x_1^*, x_2^*, ..., x_r^*, with indeterminate coefficients $u_j^{(i)}$, $u_{j_1 j_2}^{(i)}$, ..., $u_{j_1 j_2 \cdots j_N}^{(i)}$, ... ($i = 1, 2, \ldots, r$; $1 \leq j_1 \leq j_2 \leq \cdots \leq j_N \leq r + 1$):

$$\begin{aligned} x_i^* &= \varphi_i(x_1, x_2, \ldots, x_{r+1}; \{u\}) \\ &= \sum_{j=1}^{r+1} u_j^{(i)} x_j + \sum_{1 \leq j_1 \leq j_2 \leq r+1} u_{j_1 j_2}^{(i)} x_{j_1} x_{j_2} + \cdots \\ &\quad + \sum_{1 \leq j_1 \leq j_2 \leq \cdots \leq j_N \leq r+1} u_{j_1 j_2 \cdots j_N}^{(i)} x_{j_1} x_{j_2} \cdots x_{j_N} + \cdots . \end{aligned} \tag{1}$$

We denote by k^* the field generated over k by the infinitely many indeterminates $u_{\{j\}}^{(i)}$ and introduce the local ring $\mathfrak{o}^* = k^*[[x_1, x_2, \ldots, x_{r+1}]]$ obtained from $\mathfrak{o}$ by the coefficient field extension $k \to k^*$. When we say that the u's are indeterminates we use this term not merely in an algebraic sense (i.e., that the u's are algebraically independent over $\mathfrak{o}$) but in a stronger analytic sense. Namely, we mean to stipulate that if R is any point of V, the associated prime ideal of $\mathfrak{o}$ being $\mathfrak{p}$, and if $\xi_1, \xi_2, \ldots, \xi_{r+1}$ are the coördinates of R (i.e., the $\mathfrak{p}$-residues of x_1, $x_2, \ldots, x_{r+1}$), then any analytic relation $G^*(\xi_1, \xi_2, \ldots, \xi_{r+1}) = 0$, with coefficients in k^*, is a consequence of analytic relations $g(\xi_1, \xi_2, \ldots, \xi_{r+1}) = 0$ between the ξ's *over* k (where therefore $g(x_1, x_2, \ldots, x_{r+1}) \in \mathfrak{p}$). That means that we must have

$$G^*(x_1, x_2, \ldots, x_{r+1}) \in \mathfrak{o}^*\mathfrak{p} \quad \text{if} \quad G^*(\xi_1, \xi_2, \ldots, \xi_{r+1}) = 0, \tag{2}$$

where $\mathfrak{o}^* = k^*[[x_1, x_2, \ldots, x_{r+1}]]$.

A formal and intrinsic definition of the ring $\mathfrak{o}^*$ is the following: We consider the tensor product

$$\mathfrak{o}_0^* = \mathfrak{o} \bigotimes_k k^*.$$

The ring $\mathfrak{o}_0^*$ is generated by its two subrings $\mathfrak{o}$ and k^*, and these two subrings are linearly disjoint over k (see [3], CA1, Ch. III, Section 14). In the ring $\mathfrak{o}_0^*$ the extended ideal $\mathfrak{o}_0^*\mathfrak{m}$ of the maximal ideal $\mathfrak{m}$ of $\mathfrak{o}$ is a maximal ideal, and $\mathfrak{o}^*$ is defined as the $\mathfrak{o}_0^*\mathfrak{m}$-adic completion of $\mathfrak{o}_0^*$:

$$\mathfrak{o}^* = \mathfrak{o}_0^*\mathfrak{m}\text{-}\textit{adic completion of } \mathfrak{o}_0^*. \tag{3}$$

It is then clear that if $\{x_1, x_2, \ldots, x_{r+1}\}$ is *any* basis of $\mathfrak{m}$ in $\mathfrak{o}$, then $\mathfrak{o}^*$ can be canonically identified with $k^*[[x_1, x_2, \ldots, x_{r+1}]]$:

$$\mathfrak{o}^* = k^*[[x_1, x_2, \ldots, x_{r+1}]]$$

$$\cong k^*[[X_1, X_2, \ldots, X_{r+1}]]/(f(X_1, X_2, \ldots, X_{r+1})). \quad (3')$$

The ring $\mathfrak{o}^*$ is a complete local ring, of dimension r, free from non-zero nilpotent elements, and $\mathfrak{o}^*\mathfrak{m}$ is the maximal ideal $\mathfrak{m}^*$ of $\mathfrak{o}^*$. Furthermore, every minimal basis of $\mathfrak{m}$ is also a minimal basis of $\mathfrak{o}^*$. All these statements follow directly from well known properties of tensor products and of completions.

The algebroid variety $V^* = \text{Spec}\,(\mathfrak{o}^*)$ is defined by the same equation (3) (Section 1) by which V is defined, and since k^* is a pure transcendental extension of k, the irreducible factors of $f(X_1, X_2, \ldots, X_{r+1})$ remain irreducible over k^*. Hence V^* is still of pure dimension r and has the same number q of irreducible components as V. While V^* can be regarded as being obtained from V by the ground field extension $k \to k^*$, it is important to point out that *V is canonically a subset of V^**. Namely, if $\mathfrak{p}$ is any prime ideal of $\mathfrak{o}$, then it follows from results of Serre and Grothendieck on "excellent schemes" (see [4], Ch. IV, second part, Section 7.9 and especially Corollary 7.9.9, p. 222) that the extended ideal $\mathfrak{o}^*\mathfrak{p}$ is a prime ideal in $\mathfrak{o}^*$ and that $\mathfrak{o}^*\mathfrak{p} \cap \mathfrak{o} = \mathfrak{p}$. It is clear that the multiplicity n of the closed point P of V is also the multiplicity of the closed point P of V^*. From now on we shall identify any point R of V, represented by a prime ideal $\mathfrak{p}$ of $\mathfrak{o}$, with the point of V^* represented by the prime ideal $\mathfrak{o}^*\mathfrak{p}$ of $\mathfrak{o}^*$.

We now consider the elements $x_1^*, x_2^*, \ldots, x_r^*$ of $\mathfrak{o}^*$ introduced in (1). Since all the r-rowed determinants of the r by $r+1$ matrix $|u_j^{(i)}|$ of the coefficients of the linear terms in (1) are different from zero and since each coefficient $u_j^{(i)}$ is different from zero, it follows, for any $j = 1, 2, \ldots, r+1$, that the elements of $x_1^*, x_2^*, \ldots, x_r^*, x_j$ form a basis of $\mathfrak{m}^*$. Furthermore, it is easily seen that $x_1^*, x_2^*, \ldots, x_r^*$ are *transversal parameters* of $\mathfrak{o}^*$. To see this we consider, say, the basis $x_1^*, x_2^*, \ldots, x_r^*, x_{r+1}$ of $\mathfrak{m}^*$. Upon solving the r equations (1) with respect to $x_1, x_2, \ldots, x_r$ we get

$$x_j = \sum_{i=1}^{r+1} U_j^{(i)} x_i^* + \text{terms of degree} > 1 \text{ in the } x_i^*,$$

$$(j = 1, 2, \ldots, r;\ x_{r+1}^* = x_{r+1}) \quad (4)$$

where the $U_j^{(i)}$ are the cofactors of the $(r+1)$-rowed square matrix obtained from the matrix $\|u_j^{(i)}\|$ by adding to this r-rowed matrix an extra $(r+1)$-th row consisting of the elements $0, 0, \ldots, 0, 1$ (note that we have then $U_{r+1}^{(i)} = 0$ for $i = 1, 2, \ldots, r$, while $U_{r+1}^{(r+1)} = 1$).

It is obvious that the $r(r+1)$ elements $U_j^{(i)}$ in (4) are also generators of the field $k(\{u_j^{(i)}\})$ and therefore are algebraically independent over k. Now if $f^*(X_1^*, X_2^*, \ldots, X_r^*, X_{r+1}) = 0$ is the equation of V^* relative to the basis $x_1^*, x_2^*, \ldots, x_r^*, x_{r+1}$ of $\mathfrak{m}^*$, then it follows from (4) that the coefficient of X_{r+1}^n in the leading form $f_n^*(X_1^*, X_2^*, \ldots, X_r^*, X_{r+1})$ of f^* is $f_n(U_1^{(r+1)}, U_2^{(r+1)}, \ldots, U_r^{(r+1)}, 1)$ and is therefore different from zero. This proves our assertion that $x_1^*, x_2^*, \ldots, x_r^*$ are transversal parameters of $\mathfrak{o}^*$. Furthermore, we recall that these parameters belong to a minimal basis of $\mathfrak{m}^*$, namely that for any $j = 1, 2, \ldots, r+1$ the set $\{x_1^*, x_2^*, \ldots, x_r^*, x_j\}$ is a basis of $\mathfrak{m}^*$. Thus the parameters $x_1^*, x_2^*, \ldots, x_r^*$ belong to the category of parameters of $\mathfrak{o}^*$ similar to the one we agreed to deal with exclusively in the preceding section (in the case of the local ring $\mathfrak{o}$).

The r transversal parameters $x_1^*, x_2^*, \ldots, x_r^*$ of $\mathfrak{o}^*$ will play an important role in the sequel. For this reason we shall show at once that these parameters are *essentially* independent of the choice of the basis $\{x_1, x_2, \ldots, x_{r+1}\}$ of $\mathfrak{m}$ which served to define $x_1^*, x_2^*, \ldots, x_r^*$ in (1). To see this, let $\{y_1, y_2, \ldots, y_{r+1}\}$ be any other basis of $\mathfrak{m}$. We will have then

$$x_j = \psi_j(y_1, y_2, \ldots, y_{r+1}) = \sum_{\nu=1}^{r+1} c_\nu^{(j)} y_\nu$$

$$+ \text{ terms of degree } >1 \text{ in the } y_\nu\text{'s} \quad (j = 1, 2, \ldots, r+1),$$

where the ψ_j are power series with coefficients in k and where the $(r+1)$-rowed determinant $|c_\nu^j|$ is different from zero. Substituting these expressions of the x_j into (1) we find that the expressions of the r elements x_i^* as power series in $y_1, y_2, \ldots, y_{r+1}$ can be obtained by replacing in (1) the indeterminates $\{u_j^{(i)}\}$, $\{u_{j_1j_2}^{(i)}\}$ $\cdots$ (i-fixed) by quantities $\{v_j^{(i)}\}$, $\{v_{j_1j_2}^{(i)}\}$, $\cdots$ which are polynomials in the $\{u_j^{(i)}\}$, $\{u_{j_1j_2}^{(i)}\}$, $\cdots$ with coefficients in k, and that for each N *the polynomial ring* $k[\{v_j^{(i)}\}, \{v_{j_1j_2}^{(i)}\}, \ldots, \{v_{j_1j_2\cdots j_N}^{(i)}\}]$ *coincides with the polynomial ring* $k[\{u_j^{(i)}\}, \{u_{j_1j_2}^{(i)}\}, \ldots, \{u_{j_1j_2\cdots j_N}^{(i)}\}]$. Hence also the v's are indeterminates, and generate over k the same field k^* which is generated by the u's.

We now set

$$A^* = k^*[[x_1^*, x_2^*, \ldots, x_r^*]] \tag{5}$$

and we apply to the algebroid hypersurface V^*/k^* the considerations developed in Section 1, replacing the rings $\mathfrak{o}$ and A of Section 1 by the rings $\mathfrak{o}^*$ and A^*, with k^* playing the role of the coefficient field k. We therefore have now a projection π_{A^*}, which we shall denote by $\pi_u{}^*$:

$$\pi_u{}^*: V^* \rightarrow \hat{\mathbf{A}}_r/k^*, \tag{6}$$

defined by

$$\pi_u{}^*(\mathfrak{p}^*) = \mathfrak{p}^* \cap A^* = \mathfrak{p}_0{}^*, \qquad (p^* \in Spec\,(\mathfrak{o}^*), \mathfrak{p}_0{}^* \in Spec\,(A^*)). \tag{7}$$

Since V is canonically a subset of V^*, the restriction of $\pi_u{}^*$ to V maps V *into* $\hat{\mathbf{A}}_r/k^*$. We call this restriction of $\pi_u{}^*$ *the generic projection of* V and we denote it by π_u. This generic projection π_u of V is defined explicitly as follows:

$$\pi_u(\mathfrak{p}) = \mathfrak{o}^*\mathfrak{p} \cap A^* = \mathfrak{p}_0{}^*, \qquad (\mathfrak{p} \in \mathrm{Spec}\,(\mathfrak{o}), \mathfrak{p}_0{}^* \in \mathrm{Spec}\,(A^*)), \tag{8}$$

or—using the geometric language of points:

$$\pi_u(R) = R_0{}^*, \qquad (R \in V, R_0{}^* \in \hat{\mathbf{A}}_r/k^*), \tag{8$'$}$$

where R and $R_0{}^*$ are the points represented by the prime ideals $\mathfrak{p}$ and $\mathfrak{p}_0{}^*$ respectively.

Since the generic projection π_u depends only on the two local rings $\mathfrak{o}^*$ and A^*, we can replace the regular parameters $x_1{}^*, x_2{}^*, \ldots, x_r{}^*$ of A^* [see (5)] by any other set of regular parameters of A^*, without affecting the generic projection π_u. We proceed to replace the $x_i{}^*$ by other regular parameters of A^* which have a somewhat simple expression as power series in $x_1, x_2, \ldots, x_{r+1}$ (with coefficients in k^*). This will somewhat simplify the algebra of some proofs given in the sequel.

From the definition of the cofactors $U_j{}^{(i)}$ which appear in (4) it follows that the cofactors $U_j{}^{(i)}$ in which $1 \leq i \leq r$ are also the cofactors of the *square* matrix $\|u_j{}^{(i)}\|$ in which both j *and* i take the values 1, 2, $\ldots, r$. It follows that if we set

$$\tilde{x}_i{}^* = U_i{}^{(1)}x_1{}^* + U_i{}^{(2)}x_2{}^* + \cdots + U_i{}^{(r)}x_r{}^*, \qquad (i = 1, 2, \ldots, r) \tag{9}$$

then also the r elements $\tilde{x}_1{}^*, \tilde{x}_2{}^*, \ldots, \tilde{x}_r{}^*$ are regular parameters of A^* (since the determinant of the coefficients of the elements $x_1{}^*, x_2{}^*,$

$\ldots, x_r^*$ in (9) is different from zero), and the expression of each element $\tilde{x}_i^*$ as a power series $\tilde{\varphi}_i$ in $x_1, x_2, \ldots, x_r, x_{r+1}$ has $x_i + v^{(i)}x_{r+1}$ as leading form of degree 1, where $v^{(i)} = -u_r^{(r+1)}$. It is easy to see that the r quantities $v^{(i)}$ and all the coefficients v of the terms of degree ≥ 2 in $\bar{\varphi}_i$, for $i = 1, 2, \ldots, r$, are algebraically independent over $\mathfrak{o}$. We now replace the elements x_i^* by the elements $\tilde{x}_i^*$, and we use the old letters x_i^* to denote the new generators $\tilde{x}_i^*$ of A^*. We shall use the letters u to denote the coefficients v in the terms of degree ≥ 2 in $\bar{\varphi}_i$, and we shall use the letter $u^{(i)}$ to denote the coefficient $v^{(i)}$ of x_{r+1}. Thus, the definition (1) of the x_i^* is now as follows:

$$x_i^* = x_i + u^{(i)}x_{r+1} + \sum_{1 \leq j_1 \leq j_2 \leq r+1} u_{j_1 j_2}{}^{(i)} x_{j_1} x_{j_2} + \cdots + \sum_{1 \leq j_1 \leq j_2 \leq \cdots \leq j_N} u_{j_1 j_2 \cdots j_N}{}^{(i)} x_{j_1} x_{j_2} \cdots x_{j_N} + \cdots \qquad (i = 1, 2, \ldots, r). \quad (10)$$

From now on we shall denote by k^* the field generated over k by the indeterminates u which appear in (10).

From (10) it follows that $x_1, x_2, \ldots, x_r$ are power series in $x_1^*, x_2^*, \ldots, x_r^*, x_{r+1}$, with coefficients which are elements of the *polynomial ring* $k[\{u\}]$. Therefore the equation of V^* is now of the form

$$V^*: f^*(X_1^*, X_2^*, \ldots, X_r^*, X_{r+1}) = 0, \qquad (11)$$

where f^* is a power series with coefficients in the polynomial ring $k[\{u\}]$. The leading form f_n^* of f^* is still of degree n, and the coefficient of $X_{r+1}{}^n$ in f_n^* is equal to $f_n(-u^{(1)}, -u^{(2)}, \ldots, -u^{(r)}, 1)$ and is therefore different from zero. This coefficient is a polynomial in the $u^{(i)}$.

From the fact that the coefficient $a^*(u)$ of $X_{r+1}{}^n$ in f^* is different from zero and that all the coefficients of the power series f^* belong to the polynomial ring $k[\{u\}]$ follows, by the Weierstrass preparation theorem, that f^*, after multiplication by a suitable unit factor in $k^*[[X_1^*, X_2^*, \ldots, X_r^*, X_{r+1}]]$, *having coefficients in* $k[\{u\}]_{(a^*)}$, where $k[\{u\}]_{(a^*)}$ denotes the ring of quotients of $k[\{u\}]$ with respect to the multiplicative system consisting of all the positive powers of $a^*(u)$, may be assumed to be a monic polynomial in X_{r+1} of degree n, with coefficient in $k[\{u\}]_{(a^*)}[[X_1^*, X_2^*, \ldots, X_r^*]]$:

$$f^*(X_1^*, X_2^*, \ldots, X_r^*, X_{r+1}) = X_{r+1}{}^n + a_1^*(X_1^*, X_2^*, \ldots, X_r^*)X_{r+1}{}^{n-1} + \cdots + a_n^*(X_1^*, X_2^*, \ldots, X_r^*), \quad (12)$$

where $a_\nu^*(X_1^*, \ldots, X_r^*)$ is a power series in $k[\{u\}]_{(a^*)}[[X_1^*, X_2^*, \ldots, X_r^*]$ having a leading form of degree $\geq \nu$.

Let $D^*(X_1^*, X_2^*, \ldots, X_r^*)$ be the X_{r+1}-discriminant of f^*. The coefficients of the power series D^* belong in $k[\{u\}]_{(a^*)}$, and we have $D^*(0, 0, \ldots, 0) = 0$ since the origin P is a singular point of V^*. Apart from the fact that D^* may have multiple factors, the equation $D^*(X_1^*, X_2^*, \ldots, X_r^*) = 0$ yields the critical hypersurface of the projection π_u^* of V^* onto $\hat{\mathbf{A}}_r/k^*$ [see (6)]. We shall denote this critical hypersurface by Δ_u^*.

We consider this hypersurface as defined over k^*. Let m be the degree of the leading form of D^*. Upon replacing, if necessary, the variables $X_1^*, X_2^*, \ldots, X_r^*$ by suitable linear form in these variables, with coefficients in k, we may assume that the monomial X_r^{*m} actually occurs in D^*. If b^* is the coefficient of X_r^{*m} in D^* (where $b^* \in k[\{u\}]_{(a^*)}$), then by the Weierstrass preparation theorem it follows that after multiplying D^* by a suitable unit factor in $k[\{u\}]_{(\nu^*)}[[X_1^*, \ldots, X_r^*]]$, where $\nu^* = a^*b^*$, we may assume that D^* is a monic polynomial in X_r^*, of degree m, with coefficients in $k[\{u\}]_{(\nu^*)}[[X_1^*, X_2^*, \ldots, X_{r-1}]]$.

The roots of the polynomial D^* in X_r^* are therefore integral over the integral domain $k[\{u\}]_{(\nu^*)}[[X_1^*, X_2^*, \ldots, X_{r-1}^*]]$. Now, the polynomial ring $k[\{u\}]$ is an integrally closed domain, and therefore also the ring of quotients $k[\{u\}]_{(\nu^*)}$ is integrally closed. *It is known that this implies that also the ring* $k[\{u\}]_{(\nu^*)}[[X_1^*, X_2^*, \ldots, X_{r-1}^*]]$ *is integrally closed in its quotient field.* For this ring is obviously a Krull ring, and it is known that any power series ring $A[[X_1, X_2, \ldots, X_n]]$ over a Krull ring A is itself a Krull ring and is therefore integrally closed (See Bourbaki, *Algèbre Commutative,* Chapter 7, p. 83, exercise 9(b)).

Now, let $D_1^*(X_1^*, X_2^*, \ldots, X_r^*)$ be any *monic* polynomial in X_r^* which is an *irreducible* factor of D^* (over k^*) and let g_1 be the degree of D_1^* on X_r^*. The roots $x_r^{(1)}, x_r^{(2)}, \ldots, x_r^{(g_1)}$ of D_1^* are then integral over $k[\{u\}]_{(\nu^*)}[[X_1^*, \ldots, X_{r-1}^*]]$, and therefore the elementary symmetric functions of these roots must be elements of the ring $k[\{u\}]_{(\nu^*)}$ $[[X_1^*, X_2^*, \ldots, X_{r-1}^*]]$ (since they belong to the field of fractions of that *integrally closed* ring). We have therefore shown that D_1^* *is a monic polynomial in* X_r^* *with coefficients in* $k[\{u\}]_{(\nu^*)}$. Since this is true for any irreducible factor of D^*, it follows that the product $\Delta_u^*(X_1^*, X_2^*, \ldots, X_r^*)$ of the *distinct* irreducible factors of D^* is a monic polynomial in X_r^* with coefficients in the ring $k[\{u\}]_{(\nu^*)}[[X_1^*, X_2^*, \ldots, X_{r-1}^*]]$. Thus, the (reduced) equation of the (reduced) critical hypersurface $\Delta_u{}^*$ of the generic projection $\pi_u{}^*$ is given by

$$\Delta_u{}^*\colon \Delta_u{}^*(X_1{}^*, X_2{}^*, \ldots, X_r{}^*) = 0, \tag{13}$$

where

$$\Delta_u{}^*(X_1{}^*, X_2{}^*, \ldots, X_{r-1}{}^*, X_r{}^*) = X_r{}^{*g} + \sum_{\nu=1}^{g} b_\nu(X_1{}^*, X_2{}^*, \ldots, X_{r-1}{}^*)X_r{}^{*g-\nu}, \tag{14}$$

and where

$$b_\nu(X_1{}^*, X_2{}^*, \ldots, X_{r-1}{}^*) \in k[\{u\}]_{(\nu^*)}[[X_1{}^*, X_2{}^*, \ldots, X_{r-1}{}^*]]. \tag{15}$$

We note also that since D^* is a monic polynomial in $X_r{}^*$ of degree m, where m is the degree of the leading form of D^*, it follows that also the degree g of $\Delta_u{}^*(X_1{}^*, X_2{}^*, \ldots, X_{r-1}{}^*, X_r{}^*)$ in $X_r{}^*$ is equal to the degree of the leading form of $\Delta_u{}^*(X_1{}^*, X_2{}^*, \ldots, X_r{}^*)$. In other words, the elements $x_1{}^*, x_2{}^*, \ldots, x_{r-1}{}^*$ are transversal parameters of $\Delta_u{}^*$, and the origin P_0 is a g-fold point of the hypersurface $\Delta_u{}^*$.

3. Introductory informal remarks on equisingularity and on the dimensionality type of the points of V. We shall introduce informally in this section a basic concept of our theory of equisingularity: *the dimensionality type* of any point Q of our algebroid r-dimensional variety V. This will be a numerical character of the point Q which we shall denote by $d.t.(V, Q)$. It will be an *integer* σ satisfying the inequalities $0 \leq \sigma \leq r$. Since this integer will depend *a priori* not only on V and Q but also on the choice of the coefficient field k of the complete local ring $\mathfrak{o}$ of V at its origin P, we should actually use the notation $d.t.(V, Q)/k$, but since the coefficient field k for V is assumed in this work to have been fixed once and for always, we shall omit the reference to k and use the notation $d.t.(V, Q)$. However, in the course of the exposition other algebroid varieties (related to V) will come to the fore, with coefficient fields which are field extensions of k, and in those cases the notation will have to include the reference to the associated coefficient fields, as indicated in the above notation $d.t.(V, Q)/k$.

We give in the next section the formal definition of the integer $d.t.(V, Q)$. In this section we shall develop the intuitive idea which lies behind this concept. We begin with simple points of V.

If Q is a simple point of V, then Q is represented by a prime ideal $\mathfrak{p}$ of $\mathfrak{o}$ such that the localization $\mathfrak{o}_\mathfrak{p}$ of $\mathfrak{o}$ at $\mathfrak{p}$ is a *regular* local ring. It is well known that this is equivalent to assuming that not all the $r + 1$ partial derivatives of $f(X_1, X_2, \ldots, X_{r+1})$ vanish at the point Q. For that it is necessary and sufficient that if $f_1, f_2, \ldots, f_q$ are the irreducible factors of f then only one of the q polynomials f_i vanish at Q and that if, say, f_i vanishes at Q then not all the $r + 1$ partial derivatives of f_i vanish at Q. Geometrically speaking, this signifies that *Q belongs to only one of the q irreducible components $V_1, V_2, \ldots, V_q$ of V* [see (2), Section 1] *and is a simple point of the component.** In particular the general point of any component V_i of V is a simple point of V. We have thus, in the special case of a *simple* point Q of V, that the following is true: there exists one and only one *maximal irreducible* algebroid subvariety $W(Q)$ of V containing Q and *such that the type of "singularity" which V has at Q is the same as the type of "singularity" which V has at the general point of $W(Q)$.* The variety $W(Q)$ is that irreducible component of V which contains the point Q. (We have put the term "singularity" in quotation marks, because in the present case both Q and the general point of $W(Q)$ are not singular points at all).

Since the codimension of each V_i is zero we prefer to attach to any *simple* point Q of V the *dimensionality type zero* rather than r (which is the dimension of V_i), the idea being that the smaller the dimensionality type of a point the "simpler" is the singularity type of that point (simple points being non-singular all).

Our geometric intuition, or—at least—our geometrically motivated expectation, suggests that the following be true *for any point Q of V* (this time we assume that Q is a singular point): there exists a *uniquely determined maximal irreducible* algebroid subvariety $W(Q)$ of V containing Q such that the singularity which V has at Q is of the same type

*From an ideal-theoretic point of view this conclusion follows from well-known properties of quotient rings $\mathfrak{o}_\mathfrak{p}$. Namely, assuming, if possible, that $\mathfrak{p}$ contains more than one of the q prime ideals $\mathfrak{p}_i$ of the zero ideal of $\mathfrak{o}$, say $\mathfrak{p} \supset \mathfrak{p}_i$, $i = 1, 2, \ldots, \rho$, $\rho > 1$, $\mathfrak{p} \not\supset \mathfrak{p}_i$ if $i > \rho$, then it is well known [see C.A. vol. I, Ch. IV, Theorems 18 and 19, pp. 225–228) that the ρ extended ideals $\mathfrak{o}_\mathfrak{p} \cdot h(\mathfrak{p}_i)$ (where h is the canonical homomorphism of $\mathfrak{o}$ into $\mathfrak{o}_\mathfrak{p}$), $i = 1, 2, \ldots, \rho$, are distinct and are the prime ideals of the zero ideal in $\mathfrak{o}_\mathfrak{p}$. Since $\rho > 1$, this would imply that $\mathfrak{o}_\mathfrak{p}$ is not an integral domain, which is in contradiction with the regularity of $\mathfrak{o}_\mathfrak{p}$.

Furthermore, if, say, $\mathfrak{p} \supset \mathfrak{p}_1$ (whence $\mathfrak{p} \not\supset \mathfrak{p}_i$, $i > 1$), then the two local rings $\mathfrak{o}_\mathfrak{p}$ and $(\mathfrak{o}/\mathfrak{p}_1)_{\mathfrak{p}/\mathfrak{p}_1}$ are canonically isomorphic. Since $\mathfrak{o}/\mathfrak{p}_1$ is the local ring of V_1 and the prime ideal $\mathfrak{p}/\mathfrak{p}_1$ in $\mathfrak{o}/\mathfrak{p}_1$ corresponds to the point Q, considered as a point of the algebroid variety V_1, it follows that Q is also a simple point of V_1.

as, or equivalent (in some sense) to, the singularity which V has at the general point of $W(Q)$ (see Proposition 5.4). The difficulty of giving this intuitive statement a strictly logical and formal foundation lies in the absence, in the theory of singularities, of any general criterion of equivalence of two singular points of a given variety of dimension greater than 1 (or of two varieties of that dimension). We would also expect Q to be a *simple* point of the variety $W(Q)$, because at any singular point Q' of $W(Q)$ one would *expect* V to have (*because of the maximality property of* $W(Q)$) a more complicated singularity than it has at the general point of $W(Q)$. Again here we are dealing with a statement which is not obvious but which will be proved in this paper (see Section 5, Theorem 5.4). For the moment we were able to illustrate this statement only in the trivial case of a simple point Q of V. In that case $W(Q)$ is the irreducible component V_i of V which contains Q, and Q is then certainly a simple point of V_i. If, however, $r > 1$, then there is nothing to prevent a simple point Q from being a singular point of some *proper* subvariety of V_i, say, of a curve contained in V_i and passing through Q. This shows the essential role which the maximality property of $W(Q)$ is bound to play in any proof of the claim that Q must be a simple point of $W(Q)$.

Once a strictly logical and formal foundation is provided for this intuitive outline of our theory, it will be possible to define the dimensionality type of any singular point Q of V and to define also the concept of equisingularity on V. Informally speaking, this could be done as follows:

If the dimension of the irreducible variety $W(Q)$ is $r - \sigma$, then we can say that the dimensionality type of the point Q is σ. Here, if Q is a singular point, then necessarily σ is *positive,* because the irreducible variety $W(Q)$ must be contained in the singular locus V_{sing} of V (since the general point of $W(Q)$ must also be singular for V), whence $r - \sigma < r$. (This shows incidentally that a point Q is simple for V if and only if its dimensionality type is zero). On the other hand, we must have also $r - \sigma \geq 0$, i.e., $\sigma \leq r$, and the maximum $\sigma = r$ is reached if and only if $W(Q)$ consists of the point Q only. In this case we are dealing with an *isolated* singularity Q of V. Here the term "isolated" is not used only in the set-theoretic sense, i.e., we do not mean to say that Q is necessarily a *component* of singular locus of V. We only mean to say that Q is isolated as far as the complexity of the singularity which V possesses at Q. On the other hand, it is clear that for any point Q of V which *is* a component of the singular locus of V we will have $W(Q) = Q$ and hence $d.t.(V, Q) = r$.

In the actual treatment given in this paper the procedure will be just *the opposite* of the one just outlined above. We will first *define* the integer $\sigma = d.t.(V, Q)$ and then we shall prove the existence and uniqueness of the irreducible variety $W(Q)$. We shall then also prove that $W(Q)$ has dimension $r - \sigma$ and that Q is a simple point of $W(Q)$.

Once the integer $\sigma = d.t.(W, Q)$ is defined and the existence of the variety $W(Q)$ is proved, one can consider the subset of $W(Q)$ consisting of points of dimensionality type σ. We denote this subset by $S(Q)$. We shall prove in the case of algebroid varieties that the set $S(Q)$ is a Zariski dense open subset of $W(Q)$, i.e., that the complement of $S(Q)$ in $W(Q)$ is a *proper subvariety of* $W(Q)$. In the case of algebraic varieties this has been proved by Hironaka; see Introduction and Section 6. Hence $W(Q)$ is the closure of the set $S(Q)$. If Q' is any point of $S(Q)$ then from the unicity of the variety $W(Q')$ it follows at once that the two (open) varieties $S(Q)$ and $S(Q')$ coincide (and we have $W(Q) = W(Q')$), and that every point on the boundary $W(Q) - S(Q)$ of the open variety $S(Q)$ has dimensionality type *greater* than σ.

We shall call the open variety $S(Q)$ an *equisingularity stratum* S of V, and we shall say that V *is equisingular along* $W(Q)$ *at each point* Q' *of* $S(Q)$. We shall also prove (both in the algebroid and in the algebraic case) that the boundary of a stratum is a union of strata. This will give a *covering* of V by a *finite* number of equisingularity strata.

More generally, given any irreducible subvariety Z of V, we can consider the unique equisingularity stratum S of V which contains the general point of Z. Then $S \cap Z$ is a Zariski dense open subset of Z, and we shall say that V *is equisingular along* Z *at a point* Q *of* Z *if and only if* $Q \in S \cap Z$.

As we have mentioned earlier, the difficulty of giving out intuitive program of work a strictly logical and formal foundation lies in the absence of a general definition of equivalent singularities. This difficulty is not present only in the case of plane curves. For this reason, the notion of equisingularity can be introduced quite rigorously in an intuitive manner in the *case* $s = 1$. In this special case, using the above informal introduction of the variety $W(Q)$, we require that there exist a unique $(r - 1)$*-dimensional* irreducible subvariety $W(Q)$ of V such that the singularities which V has at Q and at the general point R of $W(Q)$ be equivalent in some sense. In this special case we can consider the section of V with a generic plane L_2 through the point Q, or—to be even more specific—a section with a plane L_2 which is *transversal* to the tangent linear $(r - 1)$-space T_{r-1} of $W(Q)$ at Q (we assume here that Q is a simple point of $W(Q)$). The section of L_2 with V is a plane algebroid curve $C_1(L_2)$ in L_2, having at Q a singular point. In a similar manner we obtain a plane algebroid curve $C_1(L_2^*)$, having a singular point at the general point R of $W(Q)$. Since we know what it means to say that the two plane algebroid curves $C_1(L_2)$ and $C_1(L_2^*)$ have equivalent singularities at Q and R respectively, the irreducible variety $W(Q)$ is well defined *if the dimensionality type of Q is 1*. Namely, $W(Q)$ must be an $(r - 1)$-dimensional irreducible component of V_{sing} which has at Q a simple point and is such that the two generic plane sections $C_1(L_2)$ and $C_1(L_2^*)$ have equivalent singularities. (Actually, it must be required that Q be a simple point of V_{sing}, so that Q must lie on *only one* $(r - 1)$-dimensional irreducible components of V_{sing}). We have developed a complete theory of the singularities of V which have dimensionality type 1 in our three papers SES I, II, III, especially in [1]. In these papers, more precisely in SES II, we prove that $C_1(L_2)$ and $C_1(L_2^*)$ have equivalent singularities (and that consequently $d.t.(V, Q) = 1$) if and only if there exists a projection π of V onto affine $\mathbf{A}_r$ which is, locally at Q, transversal to V (i.e., the direction of the projection is not tangent to the tangent hypercone of V at Q) and which is such that the point $\pi(Q)$ in $\mathbf{A}_r$ is a *simple* point of the critical hypersurface Δ_π of π in $\mathbf{A}_r$. In other words, the point $\pi(Q)$ of Δ_π must have dimensionality type 0, hence *one* less than the dimensionality type of Q on V. We have also proved in SES II that if $\pi(Q)$ is a simple point of Δ_π for *one* locally transversal projection π, at Q, then the same will also be true for *any* projection π', locally transversal to V at Q.

Our idea is to generalize this method of projections, involving the singularities of the critical hypersurface Δ of the projection π. Since Δ has dimension $r - 1$, an induction procedure becomes available in the general case. But, as can be shown by examples, it is not sufficient, in the case of points of dimensionality of type ≥ 2, to use arbitrary transversal projections. We find that in the *first* step, meaning in the *definition* of dimensionality type, we must use the *generic* projections introduced in Section 2. This we proceed to do in the next section.

4. Formal definition and some preliminary properties of the concept of dimensionality type of points of V. Let π_u be the generic projection of our r-dimensional algebroid variety V onto the affine space $\mathbf{A}_r/k^*$ [see (8) and (8′), Section 2] and let Q be a singular point of V. We may assume, by induction on r, that the dimensionality type of points of algebroid varieties of dimension ρ less than r (and of embedding dimension $\rho + 1$) has already been defined, since for $r = 1$ everything is straightforward. Namely, if the origin P of the *plane* algebroid curve V is a simple point of V, then V is irreducible and consists of two points only: the point P and the general point of V. Both these points are *simple* points of V and therefore have dimensionality type zero (as was stipulated in Section 3). If, however, P is a singular point of V, it is an isolated singular point of V and therefore has dimensionality type 1 ($=r$), while the remaining q points of V (where q is the number of the irreducible branches V_i of V) are simple points of V (they are the general points of $V_1, V_2, \ldots, V_q$) and have dimensionality type zero.

Let Δ_u^* be the (reduced) critical hypersurface of the generic projection π_u of V onto $\mathbf{A}_r/k^*$ and let $Q_0^* = \pi_u(Q)$, $P_0^* = \pi_u(P)$. Since Q was assumed to be a singular point of V, the point Q_0^* belongs to Δ_u^*, while P_0^* is the origin of Δ_u^*. We take the field $k^*(=k(\{u\}))$ as coefficient field of the local ring of Δ_u^* at P_0^* (see Section 2). By our induction hypothesis, the integer $d.t.(\Delta_u^*, Q_0^*)/k^*$ is well defined.

Definition 4.1. The dimensionality type σ of the point Q of V is equal to $1 + d.t.(\Delta_u^*, Q_0^*)/k^*$:

$$\sigma = d.t.(V, Q)/k = 1 + d.t.(\Delta_u^*, Q_0^*)/k^*. \tag{1}$$

Since, by this definition, we have that $d.t.(\Delta_u^*, Q_0^*)k^* = \sigma - 1$, and since, by induction on r, we have $0 \leq d.t.(\Delta_u^*, Q_0^*)/k^* \leq r - 1$ for any point Q_0^* of Δ_u^* (these inequalities being trivially true if Δ_u^* is a curve, i.e., if $r = 2$), it follows from (1) that

$$1 \leq d.t.(V, Q) \leq r \tag{2}$$

for any *singular* point of V. We recall, from Section 3, that simple points of V have dimensionality type 0. It is easy to see that our convention of assigning to simple points of V of dimensionality type 0 is in agreement with (1) of Definition 4.1, if we agree to assign to any point of $\mathbf{A}_r$, which *does not belong to the critical hypersurface Δ_u^*, the dimensionality type* -1. This follows from the (intuitively obvious) fact that if Q is a simple point of V then the π_u projection Q_0^* of Q does not belong to the critical hypersurface $\Delta_u{}^*$. The proof is straightforward, and is as follows:

Let $\xi_1, \xi_2, \ldots, \xi_{r+1}$ be the coördinates of the point Q. The indeterminates u are to be regarded as indeterminates over the field $k((\xi_1, \xi_2, \ldots, \xi_{r+1}))$ of meromorphic functions of the ξ's (except for the single relation $u^{(1)}u^{(2)} \cdots u^{(r)} = 1$ [see (11), Section 2]. The X^*-coördinates of $Q_0{}^*$ are $\xi_1{}^*, \xi_2{}^*, \ldots, \xi_r{}^*$, where the ξ^*'s are obtained by replacing in (10) of Section 2 the x's by the ξ's. We have to show two things: (1) the projecting "line" $QQ_0{}^*$ is not tangent to V at any simple point of V and (2) that this line does not pass through any singular point of V. The prove (1) it will be sufficient to prove that the "line" $QQ_0{}^*$ is not tangent to V at Q. Now, the equations of the "line" $QQ_0{}^*$ in the coördinate system of $X_1, X_2, \ldots, X_{r+1}$ are the following [see (10), Section 2]:

$$X_{r+1} - \xi_{r+1} + u^{(i)}(X_i - \xi_i) + (\text{terms of degree} \geq 2$$
$$\text{in } X_1 - \xi_1, X_2 - \xi_2, \ldots, X_{r+1} - \xi_{r+1}) = 0 \qquad (i = 1, 2, \ldots, r). \quad (3)$$

The curve defined by (3) has at Q a simple point and its tangent line at Q is the line $X_{r+1} - \xi_{r+1} + u^{(i)}(X_i - \xi_i) = 0$, $i = 1, 2, \ldots, r$. Thus, along this tangent line the quantities $X_1 - \xi_1, X_2 - \xi_2, \ldots, X_{r+1} - \xi_{r+1}$ are proportional to $1/u^{(1)}, 1/u^{(2)}, \ldots, 1/u^{(r)}, -1$. The tangent hyperplane of V at Q is given by the equation

$$\sum_{i=1}^{r+1} \frac{\partial f}{\partial \xi_i}(X_i - \xi_i) = 0.$$

Since $u^{(1)}, u^{(2)}, \ldots, u^{(r)}$ satisfy only the relation $u^{(1)}u^{(2)} \cdots u^{(r)} = 1$, it follows that

$$\sum_{i=1}^{r} \frac{1}{u^{(i)}} \frac{\partial f}{\partial \xi_i} - \frac{\partial f}{\partial \xi_{r+1}}$$

is different from zero (since the $r + 1$ quantities $\partial f/\partial \xi_i$ are not all zero). This proves that the line $QQ_0{}^*$ is not tangent to V at Q.

To prove part (2) of our assertion it is sufficient to observe that the dimension of V_{sing} is $\leq r - 1$, and that V_{sing} (and also $V_{sing}{}^*$) is defined over k, while the curve defined by the equations (3) is essentially a generic "line" through R in the affine space $\mathbf{A}_{r+1}$ and therefore does not meet V_{sing}. As an immediate corollary of Definition 4.1, we have the following

PROPOSITION 4.2. *If Q, Q' are two points of V and Q' is a specialization of Q over k, then $d.t.(V, Q') \geq d.t.(V, Q)$.*

Proof. Let $\sigma = d.t.(V, Q)$. The assertion is trivial if $\sigma = 0$ (i.e., if Q is a simple point of V). It is also trivial if $\sigma = 1$, for in that case Q is a singular point of V, whence also Q' is a singular point of V, and therefore $d.t.(V, Q') \geq 1$. We shall therefore use induction on σ, assuming that the proposition is true *for any dimension r* whenever $d.t.(V, Q)$ is less than σ.

Let $\mathfrak{p}$ and $\mathfrak{p}'$ be the prime ideals of $\mathfrak{o}$ which are associated with the points Q and Q' respectively. Let $\pi_u(Q) = Q_0{}^*$, $\pi_u(Q') = Q_0{}'^*$, and let $\mathfrak{p}_0{}^*$ and $\mathfrak{p}_0{}'^*$ be the prime ideals of A^* associated with $Q_0{}^*$ and $Q_0{}'^*$ respectively. We have then [see (8) and (8′), Section 2]: $\mathfrak{p}_0{}^* = \mathfrak{o}^*\mathfrak{p} \cap A^*$, $\mathfrak{p}_0{}'^* = \mathfrak{o}^*\mathfrak{p}' \cap A^*$. Since Q' is a specialization of Q over k, we have $\mathfrak{p} \subset \mathfrak{p}'$. Therefore $\mathfrak{p}_0{}^* \subset \mathfrak{p}_0{}'^*$, i.e., $Q_0{}'^*$ is a specialization of $Q_0{}^*$ over k^*. Since $d.t.(\Delta_u{}^*, Q_0{}^*)/k^* = \sigma - 1$, it follows by our induction hypothesis, that $d.t.(\Delta_u{}^*, Q_0{}'^*)/k^* \geq d.t.(\Delta_u{}^*, Q_0{}^*)/k^*$. Since $d.t.(V, Q') = 1 + d.t.(\Delta_u{}^*, Q_0{}'^*)$ and $d.t.(V, Q) = 1 + d.t.(\Delta_u{}^*, Q_0{}^*)$, the proposition follows.

Another consequence of our Definition 4.1 is the following

PROPOSITION 4.3. *The dimensionality type of any point of V is invariant under any extension of the coefficient field k.*

Proof. Let k' be any field containing k. By the complete local ring $\mathfrak{o}'$ which is obtained from $\mathfrak{o}$ by the coefficient field extension $k \to k'$ we mean the analog of the local ring $\mathfrak{o}^*$ defined in Section 2 by means of the coefficient field extension $k \to k^*$. Namely, $\mathfrak{o}'$ is the completion of the ring $\mathfrak{o} \otimes_k k'$ with respect to the (maximal) ideal in $\mathfrak{o} \otimes_k k'$ which is the extension of the maximal ideal $\mathfrak{m}$ of $\mathfrak{o}$. We now introduce the set of indeterminates $\{u\}$ and we set (as in Section 2) $k^* = k(\{u\})$. We consider the u's also as indeterminates over k' and we set $k'^* = k'(\{u\})$. It is clear that k'^* is also an extension $k^{*\prime}$ of the field k^*, obtained by extending the subfield k of k^* to k'. We can write therefore:

$$k'^* = k'(\{u\}) = k^{*\prime}. \tag{4}$$

We set

$$\mathfrak{o}^{*\prime} = \widehat{\mathfrak{o}^* \bigotimes_{k^*} k^{*\prime}}, \tag{5}$$

$$\mathfrak{o}'^{*} = \widehat{\mathfrak{o}' \bigotimes_{k'} k'^{*}}, \tag{5'}$$

where the symbol $\widehat{\ }$ indicates completion with respect to the maximal ideals in $\mathfrak{o}^* \otimes_{k^*} k^{*\prime}$ and $\mathfrak{o}' \otimes_{k'} k'^*$ which are the extensions of the maximal ideal $\mathfrak{m}^*$ of $\mathfrak{o}^*$ and of the maximal ideal $\mathfrak{m}'$ of $\mathfrak{o}'$ respectively. Now, since both $\mathfrak{m}^*$ and $\mathfrak{m}'$ are the extensions (in $\mathfrak{o}^*$ and $\mathfrak{o}'$ respectively) of the maximal ideal $\mathfrak{m}$ in $\mathfrak{o}$, it follows that two completions in (5) and (5′) are both with respect to the extensions of $\mathfrak{m}$ in the two rings $\mathfrak{o}^* \otimes_{k^*} k^{*\prime}$ and $\mathfrak{o}' \otimes_{k'} k'^*$. Now it is easy to see that these two completions coincide and that consequently

$$\mathfrak{o}^{*\prime} = \mathfrak{o}'^{*} \tag{6}$$

To see this, we observe that we have

$$\begin{aligned} \mathfrak{o}^* \bigotimes_{k^*} k^{*\prime} &= (\widehat{\mathfrak{o} \bigotimes_{k} k^*}) \bigotimes_{k^*} k^{*\prime} \\ &= k^*[[x_1, x_2, \ldots, x_{r+1}]] \bigotimes_{k^*} k^{*\prime}, \end{aligned}$$

whence

$$\mathfrak{o}^{*\prime} = k^{*\prime}[[x_1, x_2, \ldots, x_{r+1}]], \tag{7}$$

where the defining relation between the $+ 1$ basis elements x_1, x_2, $\ldots$, x_{r+1} of $\mathfrak{m}$ is the original relation $f(x_1, x_2, \ldots, x_{r+1}) = 0$ which defines our variety V. On the other hand, we have

$$\mathfrak{o}' \bigotimes_{k'} k'^* = (\widehat{\mathfrak{o} \bigotimes_{k} k'}) \bigotimes_{k'} k'^* = k'[[x_1, x_2, \ldots, x_{r+1}]] \bigotimes_{k'} k'^*,$$

and hence

$$\mathfrak{o}'^* = k'^*[[x_1, x_2, \ldots, x_{r+1}], \tag{7'}$$

where the defining relation between the elements $x_1, x_2, \ldots, x_{r+1}$ is still $f(x_1, x_2, \ldots, x_{r+1}) = 0$. Now (6) follows from (7), (7′) and (4). Let now V^* be the algebroid variety Spec ($\mathfrak{o}^*$) and let $V^{*\prime}$ = Spec ($\mathfrak{o}^{*\prime}$). Similarly, let V' = Spec ($\mathfrak{o}'$), V'^* = Spec ($\mathfrak{o}'^*$). We have then by (6): $V^{*\prime} = V'^*$. Since Proposition 4.3 is trivially true in the case $r = 1$, we can use induction on r. We assume therefore that the proposition is true for algebroid varieties of dimension $r - 1$. We observe now that the critical hypersurface $\Delta_u'^*$ of the generic projection π_u' of V' is obviously obtainable from Δ_u^* by the coefficient field extension $k^* \to k^{*\prime}$. Let now Q be any point of V. The proposition is trivially true if Q is a simple point of V, for in that case Q is also a simple point of V', whence $d.t.(V, Q)/k = d.t.(V', Q)/k' = 0$. Assume now that Q is a singular point of V (and therefore also a singular point of V'). Then $\pi_u(Q) = Q_0^*$ is a point of $\Delta_u'^*$; therefore, by our induction hypothesis, we have $d.t.(\Delta_u'^*, Q_0^*)/k^{*\prime} = d.t.(\Delta_u^*, Q_0^*)/k^*$. Since $d.t.(V, Q)/k = 1 + d.t.(\Delta_u^*, Q_0^*)/k^*$ and $d.t.(V', Q)/k' = d.t.(\Delta_u'^*, Q_0^*)/k'^*$, and since $k^{*\prime} = k'^*$[by (4)], the equality $d.t.(V, Q)/k = d.t.(V', Q)/k'$ follows, and the proposition is proved.

5. The basic theorem on the equisingularity stratification of *V*. We shall denote by s the dimensionality type of the closed point P of V. Since P is a specialization of every point of V, it follows from Proposition 4.2 that the dimensionality type σ of any point Q of V satisfies the inequality $0 \leq \sigma \leq s$. We shall denote by $V(\sigma)$ the set of points of V which have dimensionality type σ (for a given σ this set may be empty if σ is positive). We have therefore

$$\dot{\bigcup_{\sigma=0}^{s}} V(\sigma) = V; \qquad \dot{\bigcup_{\sigma=1}^{s}} V(\sigma) = V_{sing}. \tag{1}$$

We shall denote by $V\{\sigma\}$ the set of points of V which have dimensionality type $\geq \sigma$. We have therefore

$$V\{\sigma\} = V(\sigma) \cup V\{\sigma + 1\}; \tag{2}$$

$$V\{0\} = V, \qquad V\{1\} = V_{sing}. \tag{2'}$$

The set $V(0)$ is the set of simple points of V. If $V_1, V_2, \ldots, V_q$ are the irreducible components of V, then $V(0)$ is the disjoint union of q sets

$S_{i,0}$, where $S_{i,0} = V_i - (V_i \cap V_{sing})$ $(i = 1, 2, \ldots, q)$. The algebroid closure $\overline{S_{i,0}}$ of $S_{i,0}$ (i.e., the least algebroid subvariety of V which contains $S_{i,0}$) is the irreducible variety V_i, of dimension r, and $S_{i,0}$ is a Zariski open (dense) subset of V_i, i.e., the boundary $\overline{S_{i,0}} - S_{i,0}$ of $S_{i,0}$ is a *proper* algebroid subvariety of V_i, namely $V_i \cap V_{sing}$. Moreover, each point of $S_{i,0}$ is a simple point of $\overline{S_{i,0}}$.

Our study of the case $\sigma = 1$ (see [1], SEQ, II) shows that the set-up in this case is quite similar to the above set-up in the case $\sigma = 0$, except that $V(1)$ *may be empty*. If $V(1)$ is non-empty, the closure $\overline{V(1)}$ of $V(1)$ is a subvariety of V, of pure dimension $r - 1$. The set $V(1)$ is non-empty if and only if V_{sing} has dimension $r - 1$. When that is so, then $V(1)$ is the disjoint union of q_1 subsets $S_{i_1,1}$, where q_1 is the number of irreducible $(r - 1)$-dimensional components of V_{sing}, and these components are precisely the algebroid closures $\overline{S_{i_1,1}}$ of the q_1 sets $S_{i_1,1}$. Each set $S_{i_1,1}$ is a Zariski dense open subset of $\overline{S_{i_1,1}}$, the boundary $\overline{S_{i,1}} - S_{i,1}$ being a proper algebroid subvariety of $\overline{S_{i_1,1}}$ (and therefore of dimension $< r - 1$). Moreover, *each point of* $S_{i_1,1}$ *is a simple point of* $\overline{S_{i_1,1}}$, and in SES II we have shown that each such point is even a *simple point of* V_{sing} (hence, in particular, $\overline{S_{i_1,1}}$ is the only $(r - 1)$-dimensional component of V_{sing} which contains that point).

We call the $q + q_1$ open (irreducible) smooth varieties $S_{i,0}$, $S_{i_1,1}$ the *equisingularity strata* of V, of dimensionality type 0 and 1 respectively. We repeat that the strata $S_{i_1,1}$ are present if and only if V_{sing} has dimension $r - 1$.

Since the boundary of $S_{i,0}$ is the variety $V_i \cap V_{sing}$ and since the union of these q varieties is V_{sing}, it follows that each of the q_1 $(r - 1)$-dimensional irreducible components $\overline{S_{i_1,1}}$ of V_{sing}, and therefore also each stratum $S_{i_1,1}$, belongs to the boundary $V_i \cap V_{sing}$ of at least one r-dimensional stratum $S_{i,0}$. Moreover, since each point $S_{i_1,1}$ is a simple point of V_{sing}, it follows easily that *if a point Q of a stratum $S_{i_1,1}$ belongs to the boundary of a stratum $S_{i,0}$ then the entire stratum $S_{i_1,1}$ belongs to the boundary of $S_{i,0}$*. For, let say Q be a common point of a stratum $S_{1,1}$ and of the boundary $V_1 - S_{1,0}$ of a stratum $S_{1,0}$. Assuming, if possible, that $S_{1,1} \not\subset V_1 - S_{1,0}$, we must have $S_{1,1} \subset V_i \cap V_{sing}$ for some $i \neq 1$, say $S_{1,1} \subset V_2 \cap V_{sing}$. The intersection $V_1 \cap V_2$ must have at least one irreducible component $\overline{S_{i,1}}$ of dimension $r - 1$, passing through Q, and this component is different from $\overline{S_{1,1}}$, since $\overline{S_{1,1}} \not\subset V_1$, while $S_{i,1} \subset V_1$. Hence Q lies on at least *two* distinct irreducible $(r - 1)$-dimensional components of V_{sing} and is therefore *not* a simple point of V_{sing}, in contradiction with what was said above (namely, that the points of each stratum $S_{i,1}$ are simple points of V_{sing}).

The main object of this work is to prove the following basic theorem, which generalizes to any value of σ the results stated above for the cases $\sigma = 0, 1$:

THEOREM 5.1.

(*a*) *For any integer* σ, such that the set $V(\sigma)$ *is non-empty* ($0 \leq \sigma \leq s$), $V(\sigma)$ *is a disjoint union of a certain number* q_σ *of sets* $S_{i_\sigma,\sigma}$ ($i_\sigma = 1, 2, \ldots, q_\sigma \geq 1$), *called strata in the sequel, where each* $S_{i_\sigma,\sigma}$ *is an irreducible* $(r - \sigma)$*-dimensional Zariski dense open subset of its algebroid closure* $\overline{S}_{i_\sigma,\sigma}$ (*equivalently,* $\overline{S}_{i_\sigma,\sigma}$ *is an irreducible* $(r - \sigma)$*-dimensional algebroid subvariety of V, and* $\overline{S}_{i_\sigma,\sigma} - S_{i_\sigma,\sigma}$ *is a proper subvariety of* $\overline{S}_{i_\sigma,\sigma}$).

(*b*) *Every point of each* $S_{i_\sigma,\sigma}$ *is a simple point of* $\overline{S}_{i_\sigma,\sigma}$, *and—furthermore—V is equimultiple along* $S_{i_\sigma,\sigma}$.

(*c*) *The stratum W(P) which contains the closed point P of V is a closed subvariety of V (i.e., its boundary is empty) and coincides with the set V(s) of all points of V which have the maximum dimensionality type s.*

(*d*) *The boundary of each stratum is a union of strata.*

We call each stratum $S_{i_\sigma,\sigma}$ (σ being such that $V(\sigma)$ is non-empty) an *equisingularity stratum* of V, of *dimensionality type* σ. Theorem 5.1 yields a stratification of V by equisingularity strata, satisfying the usual properties of a stratification, in view of the disjointness of any two distinct strata (part (*a*) of the theorem) and in view of parts (*b*) and (*d*).

This section is devoted largely to the proof of Theorem 5.1. The proof will require a preliminary basic lemma which we shall state and prove below and which concerns an algebroid hypersurface $V_t{}^*$ whose equation has coefficients in a field $k^* = k(\{t\})$ generated over k by a (finite or infinite) set $\{t\}$ of parameters. This lemma will then be applied to the critical hypersurface $\Delta_u{}^*$ of the generic projection π_u of our original r-dimensional algebroid variety V/k, the parameters t being in that case the indeterminates u. The reason why we use the letter t for the parameters which appear in the equation of the hypersurface $V_t{}^*$ of the lemma (instead of using the letter u) is that, in the course of the proof of the lemma, we shall have to consider the generic projection of $V_t{}^*$, and, following the notations which we have used in the preceding sections, we will want to continue using the letter u for the parameters of the generic projection $\pi_u{}^*$ of $V_t{}^*$, the u's being then indeterminates over the coefficient field $k(\{t\})$ of the local ring of $V_t{}^*$. We recall (see Section 2) that the coefficients of the equation of $\Delta_u{}^*$ were not arbitrary elements of the field $k(\{u\})$ but were elements of the ring of quotients $k[\{u\}]_{(v^*)}$ of the *polynomial ring* $k[\{u\}]$ with respect to the multiplicative system (v^*) consisting of all the positive integral powers of some non-zero element v^* of the polynomial ring $k[\{u\}]$ [see (15), Section 2]. We also recall that the equation of $\Delta_u{}^*$ was given by (14), Section 2, the left-hand side of that equation being a monic *polynomial* in one of variables $X_1{}^*, X_2{}^*, \ldots, X_r{}^*$ (namely in $X_r{}^*$; see (14) and (15), Section 2). For this reason we impose a similar restriction on the class of r-dimensional hypersurfaces $V_t{}^*$ which our lemma will deal with. We assume namely that $V_t{}^*$ is defined by an equation

$$V_t{}^*: f^*(X_1, X_2, \ldots, X_{r+1}; \{t\}) = 0, \tag{3}$$

where f^* is a monic polynomial in X_{r+1} and a power series in $X_1, X_2, \ldots, X_r$ (zero at the origin $X_1 = X_2 = \cdots = X_{r+1} = 0$), and that the coefficients of f^* belong to the ring of quotients $k[\{t\}]_M$ of the polynomial ring $k[\{t\}]$ with respect to the multiplicative system M consisting of the positive powers of some non-zero element $a(\{t\})$ of $k[\{t\}]$:

$$M = \text{set of positive powers of a non-zero polynomial } a(\{t\}). \tag{4}$$

In the applications of our lemma to the case in which $V_t{}^*$ is the critical hypersurface $\Delta_u{}^*$ (and therefore r is to be then replaced by $r - 1$) we shall be dealing only with those points of $\Delta_u{}^*$ which are π_u-projections of (necessarily singular) points of V. Now, the coördinates $\xi_1{}^*, \xi_2{}^*, \ldots, \xi_r{}^*$ of the $\pi_u{}^*$-projection $Q_0{}^*$ of a point $Q(\xi_1, \xi_2, \ldots, \xi_{r+1})$ of V are given by (1), Section 2, where $x_1, x_2, \ldots, x_{r+1}$ are to be replaced by $\xi_1, \xi_2, \ldots, \xi_{r+1}$. The coördinates $\xi_1{}^*, \xi_2{}^*, \ldots, \xi_r{}^*$ of $Q_0{}^*$ are therefore elements of the tensor product $k[\{u\}] \otimes_k k[[\xi_1, \xi_2, \ldots, \xi_{r+1}]]$, namely power series in the ξ's, with coefficients in $k[\{u\}]$ and constant term zero, the u's being indeterminates over the complete local domain $k[[\xi_1, \xi_2, \ldots, \xi_{r+1}]]$. We shall therefore be interested only in a special class of points of $V_t{}^*$, namely in the points Q^* whose coördinates $\xi_1{}^*, \xi_2{}^*, \ldots, \xi_{r+1}{}^*$ belong to the tensor product (over k) of $k[\{t\}]$ with a complete local domain $k[[\xi_1, \xi_2, \ldots, \xi_N]]$; here, $\xi_1, \xi_2, \ldots, \xi_N$ are the coördinates of a variable point (ξ) of a *fixed* algebroid variety Z embedded in some affine N-space and defined over k. We shall denote the associated point of V_t^* by Q_ξ^*. We furthermore assume that the coördinates ξ_i^* of Q_ξ^* are given by the following expressions:

$$\xi_i^* = \varphi_i(\xi_1, \xi_2, \ldots, \xi_N; \{t\}), \qquad (i = 1, 2, \ldots, r+1) \tag{5}$$

where the φ_i are power series in the ξ's, with *fixed* coefficients in $k[[t]]$ (i.e., independent of the variable point (ξ) of Z) and constant term zero. Here the t's are considered as indeterminates over the local domain $k[[\xi_1, \xi_2, \ldots, \xi_N]]$. More precisely (and in complete agreement with the definition of tensor products), let $\mathfrak{p}$ be the prime ideal in $k[[Y_1, Y_2, \ldots, Y_N]]$ which is associated with the point $(\xi_1, \xi_2, \ldots, \xi_N)$, and let $\mathfrak{p}^*$ be the prime ideal in $k^*[[X_1, X_2, \ldots, X_{r+1}]]$ associated with the point Q_ξ^*. Let h be the homomorphism $k^*[[X_1, X_2, \ldots, X_{r+1}]] \to k^*[[Y_1, Y_2, \ldots, Y_N]]$ defined by $h(X_i) = \varphi_i(Y_1, Y_2, \ldots, Y_N; \{t\})$. Then $h(\mathfrak{p}^*)$ is the extension to $k^*[[Y_1, Y_2, \ldots, Y_N]]$ of the ideal $k[\{t\}] \otimes_k \mathfrak{p}$ in the subring $k[\{t\}] \otimes_k k[[Y_1, Y_2, \ldots, Y_N]$ of $k^*[[Y_1, Y_2, \ldots, Y_N]]$.

We shall denote by $S_k(V_t^*)$ the set of points Q_ξ^* described above.

We shall consider specializations $\{t\} \to \{\bar{t}\}$ of the parameters t, where the $\bar{t}$'s are arbitrary elements of $k[\{t\}]$. For the proof of Theorem 5.1 the important case will be the one in which the $\bar{t}$'s belong to k. We shall also consider specializations $\{\xi_1, \xi_2, \ldots, \xi_N\} \xrightarrow{k} \{\zeta_1, \zeta_2, \ldots, \zeta_N\}$, where $(\zeta_1, \zeta_2, \ldots, \zeta_N)$ is therefore an arbitrary point of the (irreducible) algebroid subvariety of Z having $(\xi_1, \xi_2, \ldots, \xi_N)$ as general point. It is clear that the point Q_ζ^* is a specialization of the point Q_ξ^*, over k^*, and belongs to $S_k(V_t^*)$.

We assume, of course, that V_t^* is a *reduced* hypersurface, i.e., that the power series f^* in (3) has no multiple factors. Furthermore, we assume that if n is the degree of the leading form of f^* $(n \geq 1)$ then f^* is a monic polynomial in X_{r+1}, of degree n (this agrees with the condition which is satisfied by Δ_u^*; see the end of Section 2).

We shall say that a given property, which may or may not hold true for some specializations $\{t\} \to \{\bar{t}\}$, *is true for almost all specializations* $\{t\} \to \{\bar{t}\}$, if there exists a non-zero polynomial $g(\{t\})$ in $k[\{t\}]$ such that that property holds true for all specializations $\{t\} \to \{\bar{t}\}$ such that $g(\{\bar{t}\}) \neq 0$.

If Q_ξ^* is any point in $S_k(V_t^*)$ and $\{t\} \to \{\bar{t}\}$ is any specialization of the parameters t, we shall denote by $Q_{\xi,\bar{t}}{}^*$ the point whose coördinates $\bar{\xi}_1{}^*, \bar{\xi}_2{}^*, \ldots, \bar{\xi}_{r+1}$ are obtained from (5) by replacing the parameters t by the $\bar{t}$'s. Thus, we have:

$$\bar{\xi}_i{}^* = \varphi_i(\xi_1, \xi_2, \ldots, \xi_N; \{\bar{t}\}). \qquad (i = 1, 2, \ldots, r + 1) \tag{6}$$

Before we state our basic lemma we introduce some notations. Given an algebroid variety $V = \text{Spec}\ (\mathfrak{o})$ and given a point Q of V, *we denote by* V_Q *the set of points* R *of* V *such that* Q *is a specialization of* R. As was explained earlier in Section 1, if $\mathfrak{p}$ is the prime ideal in $\mathfrak{o}$ which is associated with Q then V_Q is the set of prime ideals of $\mathfrak{o}$ which are contained in $\mathfrak{p}$. Since there is a canonical (1, 1)-correspondence between the prime ideals in $\mathfrak{o}$ which are contained in $\mathfrak{p}$ and the prime ideals of the quotient ring $\mathfrak{o}_{\mathfrak{p}}$, it follows that V_Q *can be canonically identified with the variety* Spec $(\mathfrak{o}_{\mathfrak{p}})$. However, this latter variety is not, strictly speaking, an algebroid variety (in the sense in which we use this term), since, in general, the local ring $\mathfrak{o}_{\mathfrak{p}}$ is not complete (if the dimension r of $\mathfrak{o}$ is ≥ 3 and the dimension ρ of the prime ideal $\mathfrak{p}$ satisfies the inequalities $0 < \rho \leq r - 2$, then $\mathfrak{o}_{\mathfrak{p}}$ is not complete). *We next denote by the symbol* $\text{EQS}_k(V, Q)$ (where k is the fixed coefficient field of $\mathfrak{o}$) the set of points R of V_Q such that $d.t.(V, R) = d.t.(V, Q)$.

We can now state our basic lemma.

LEMMA 5.2. *Let* $V_t{}^*$ *be the* (*reduced*) *algebroid variety defined by* (3) *and let* $Q_\zeta{}^*$ *be a point of* $S_k(V_t{}^*)$. *Then for almost all specializations* $\{t\} \to \{\bar{t}\}$, *where the* $\bar{t}$'s *are in* $k[\{t\}]$ (*in particular, for almost all specializations* $\{t\} \to \{\bar{t}\}$, *the* $\bar{t}$'s *in* k) *the equation*

$$V_{\bar{t}}{}^*: f^*(X_1, X_2, \ldots, X_{r+1}; \{\bar{t}\}) = 0 \tag{7}$$

is meaningful [*i.e., we have* $a(\{\bar{t}\}) \neq 0$, *see* (4)], *defines a reduced algebroid variety* $V_{\bar{t}}{}^*$ (*over the field* $k(\{\bar{t}\})$ *which contains the specialization* $R_{\xi,\bar{t}}{}^*$ *of any point* $R_\xi{}^*$ *of* $S_k(V_t{}^*)$, *and furthermore the following two conditions are satisfied*:

(*a*) $d.t.(V_{\bar{t}}{}^*, Q_{\zeta,\bar{t}}{}^*)/k(\{\bar{t}\}) = d.t.(V_t{}^*, Q_\zeta{}^*)/k^*$;

(*b*) *if* $R_\xi{}^*$ *is any point of* $S_k(V_t{}^*)$ *such that the point* $(\zeta_1, \zeta_2, \ldots, \zeta_N)$ *is a specialization, over* k, *of the point* $(\xi_1, \xi_2, \ldots, \xi_N)$, *then*

$$R_{\xi,\bar{t}}^* \in \mathrm{EQS}_{k(\{\bar{t}\})}(V_{\bar{t}}^*, Q_{\xi,\bar{t}}^*) \leftrightarrow$$
$$R_\xi^* \in \mathrm{EQS}_{k^*}(V_t^*, Q_\xi^*).$$

Proof. One condition that we must impose on our specializations $\{t\} \to \{\bar{t}\}$ is the one mentioned in the statement of the lemma, namely the inequality

$$a(\{\bar{t}\}) \neq 0. \qquad (8)$$

This insures that the power series $f^*(\{X\}; \{\bar{t}\})$ in $X_1, X_2, \ldots, X_{r+1}$ is well defined and has coefficients in the ring $k[\{\bar{t}\}]_{\bar{M}}$, where $\bar{M}$ is the multiplicative system of the non-negative powers of $a(\{\bar{t}\})$. In particular, if the $\bar{t}$'s belong to k, then the coefficients of $f^*(\{X\}; \{\bar{t}\})$ belong to k. Furthermore, the power series $f^*(\{X\}; \{\bar{t}\})$ is a monic polynomial in X_{r+1}, of degree n, and its leading form is also of degree n.

Let $D^*(X_1, X_2, \ldots, X_r; \{t\})$ be the X_{r+1}-discriminant of $f^*(X_1, X_2, \ldots, X_{r+1}; \{t\})$. This discriminant is not identically zero (since $f^*(X_1, X_2, \ldots, X_{r+1}; \{t\})$ has no multiple factors) and is a power series in $X_1, X_2, \ldots, X_r$ with coefficients in the ring of quotients $k[\{t\})_M$ (see (4)). If (8) is satisfied then $D^*(X_1, X_2, \ldots, X_r; \{\bar{t}\})$ is well defined and is the X_{r+1}-discriminant of $f^*(X_1, X_2, \ldots, X_{r+1}; \{\bar{t}\})$ (since $f^*(X_1, X_2, \ldots, X_{r+1}; \{\bar{t}\})$ is a monic polynomial in X_{r+1}, of the same degree n as $f^*(X_1, X_2, \ldots, X_{r+1}; \{t\})$). Since $f^*(X_1, X_2, \ldots, X_{r+1}; \{t\})$ has no multiple factors, $D^*(X_1, X_2, \ldots, X_r; \{t\})$ is not identically zero. Furthermore, $f^*(X_1, X_2, \ldots, X_{r+1}; \{\bar{t}\}$ has no factors independent of X_{r+1} (since the degree of the leading form of $f^*(X_1, X_2, \ldots, X_{r+1}; \{\bar{t}\}$ is equal to the degree n of $f^*(X_1, X_2, \ldots, X_{r+1}; \{\bar{t}\})$ in X_{r+1}). Therefore, the power series $f^*(X_1, X_2, \ldots, X_{r+1}; \{\bar{t}\})$ will have no multiple factors if (and only if) $D^*(X_1, X_2, \ldots, X_r; \{\bar{t}\})$ is not identically zero. To insure that this is the case it is sufficient to consider any non-zero coefficient $b(\{t\})/a(\{t\})^\mu$ of the power series $D^*(X_1, X_2, \ldots, X_r; \{t\})$ and impose on the specialization $\{t\} \to \{\bar{t}\}$ the inequality

$$b(\{\bar{t}\}) \neq 0. \qquad (9)$$

If then both (8) and (9) are satisfied, then the equation (7) will define a *reduced* algebroid hypersurface $V_{\bar{t}}^*$ having at the origin P an n-fold point.

It is easily seen that for any point Q_ξ^* of $S_k(V_t^*)$ and for any specialization $\{t\} \rightarrow \{\bar{t}\}$ (satisfying (8)), the point $Q_{\xi,\bar{t}}^*$ belongs to the hypersurface $V_{\bar{t}}^*$. This follows from the fact that the t's are indeterminates over the local domain $k[[\xi_1, \xi_2, \ldots, \xi_N]]$. Namely, from what was pointed out earlier about the relationship between $h(\mathfrak{p}^*)$ and the ideal $k[\{t\}] \otimes_k \mathfrak{p}$, where $\mathfrak{p}^*$ is the prime ideal of Q_ξ^* in $k^*[[X_1, X_2, \ldots, X_{r+1}]]$ and $\mathfrak{p}$ is the prime ideal of the point $(\xi_1, \xi_2, \ldots, \xi_N)$ in $k[[Y_1, Y_2, \ldots, Y_N]]$, it follows that since $f^*(\xi_1^*, \xi_2^*, \ldots, \xi_{r+1}^*; \{t\}) = 0$, we must have an identity of the form

$$f^*(\varphi_1(\{Y\}; \{t\}), \varphi_2(\{Y\}; \{t\}), \ldots, \varphi_{r+1}(\{Y\}; \{t\})) = \sum_{\mu=1}^{m} A_\mu(\{Y\}; \{t\}) g_\mu(\{Y\}), \quad (10)$$

where the A_μ are power series in $Y_1, Y_2, \ldots, Y_N$, with coefficients in $k[\{t\}]_M$, and the $g_\mu(\{Y\}) \in \mathfrak{p}$, i.e., $g_\mu(\xi_1, \xi_2, \ldots, \xi_N) = 0$. From (10) it follows therefore that

$$f^*(\varphi_1(\{\xi\}; \{\bar{t}\}), \varphi_2(\{\xi\}; \{\bar{t}\}), \ldots, \varphi_{r+1}(\{\xi\}; \{\bar{t}\}); \{\bar{t}\}) = 0,$$

i.e., $f^*(\bar{\xi}_1^*, \bar{\xi}_2^*, \ldots, \bar{\xi}_{r+1}^*; \{\bar{t}\}) = 0$ [see (6)], whence $Q_{\xi,\bar{t}}^* \in V_{\bar{t}}^*$, as was asserted.

We first consider the special case in which Q_ζ^* is a *simple* point of V_t^*. To prove part (*a*) of the Lemma in this special case we have to prove that *for almost all specializations* $\{t\} \rightarrow \{\bar{t}\}$, *the point* $Q_{\zeta,\bar{t}}$ *is a simple point of* $V_{\bar{t}}^*$. Since Q_ζ^* is a simple point of V_t^*, one of the $r + 1$ partial derivatives $\partial f^*/\partial X_i$ is different from zero at Q_ζ^*. Let, say, $\partial f^*/\partial X_1$ be different from zero at Q_ζ^*. We shall show that for almost all specializations $\{t\} \rightarrow \{\bar{t}\}$ the partial derivative $\partial f^*(\{X\}; \{\bar{t}\})/\partial X_1$ is different from zero at $Q_{\xi,\bar{t}}^*$, and this will complete the proof of part (*a*) of the lemma in the special case in which Q_ζ^* is a simple point of V_t^*.

By a suitable linear homogeneous transformation of the coördinates $\zeta_1, \zeta_2, \ldots, \zeta_N$, with coefficients in k, we may assume that if ρ is the dimension of the local domain $\mathfrak{o} = k[[\zeta_1, \zeta_2, \ldots, \zeta_N]]$ then $\zeta_1, \zeta_2, \ldots, \zeta_\rho$ are parameters of that local domain. Since we are dealing with a *complete* (noetherian) local domain it follows that the local domain $k[[\{\zeta\}]]$ is a finite module of the (regular) ρ-dimensional local domain $R = k[[\zeta_1, \zeta_2, \ldots, \zeta_\rho]]$. Let $\omega_1, \omega_2, \ldots, \omega_h$ be an R-module basis of $\mathfrak{o}$, and let d be the relative degree of the quotient field K of $\mathfrak{o}$ over to the quotient field K_0 of R (the former being obviously a finite algebraic extension of the latter). We have then necessarily $h \geq d$, since obviously $\omega_1, \omega_2, \ldots, \omega_h$ is also a K_0-basis of K. We may assume that $\omega_1, \omega_2, \ldots, \omega_d$ are linearly independent over K_0 and therefore form a linearly independent basis of K over K_0. There exists therefore an element $c(\zeta_1, \zeta_2, \ldots, \zeta_\rho)$ in R, different from zero, such that

$$c(\zeta_1, \zeta_2, \ldots, \zeta_\rho)\cdot\omega_{d+j} \in \sum_{\nu=1}^{d} R\cdot\omega_\nu, \qquad j = 1, 2, \ldots, h-d.$$

Therefore

$$c(\zeta_1, \zeta_2, \ldots, \zeta_\rho)\cdot\mathfrak{o} \subset \sum_{\nu=1}^{d} R\cdot\omega_\nu. \tag{11}$$

We denote the partial derivative $\partial f^*(X_1, X_2, \ldots, X_{r+1}; \{t\})/\partial X_1$ by $f_1^*(X_1, X_2, \ldots, X_{r+1}; \{t\})$. If $\zeta_1^*, \zeta_2^*, \ldots, \zeta_{r+1}^*$ are the X-coördinates of Q_ζ^*, then, by assumption, we have

$$f_1^*(\zeta_1^*, \zeta_2^*, \ldots, \zeta_{r+1}^*; \{t\}) \neq 0. \tag{12}$$

Upon replacing the ζ^*'s by their expressions in terms of $\zeta_1, \zeta_2, \ldots, \zeta_N$ and the t's we have that

$$f_1^*(\zeta_1^*, \zeta_2^*, \ldots, \zeta_{r+1}^*; \{t\}) = F(\zeta_1, \zeta_2, \ldots, \zeta_N; \{t\}),$$

where F is a power series in $\zeta_1, \zeta_2, \ldots, \zeta_N$ with coefficients in $k[\{t\}]_M$. From (11) it follows therefore at once that

$$c(\zeta_1, \zeta_2, \ldots, \zeta_\rho)f_1^*(\zeta_1^*, \zeta_2^*, \ldots, \zeta_{r+1}^*; \{t\}) = \sum_{\nu=1}^{d} F_\nu(\zeta_1, \zeta_2, \ldots, \zeta_\rho; \{t\})\omega_\nu, \tag{13}$$

where the F_ν are power series in $\zeta_1, \zeta_2, \ldots, \zeta_\rho$, with coefficients in $k[\{t\}]_M$. From (12) it follows that we must have $F_\nu(\zeta_1, \zeta_2, \ldots, \zeta_\rho; \{t\}) \neq 0$, for some value of ν. Let, say, $F_1(\zeta_1, \zeta_2, \ldots, \zeta_\rho; \{t\}) \neq 0$, and let $\alpha_{i_1 i_2 \cdots i_n}(\{t\})/a(\{t\})^q$ be a coefficient of the power series F_1 which is different from zero. Here $\alpha_{i_1 i_2 \cdots i_n}(\{t\})$ is a polynomial in the t's, and the above quotient $\alpha_{(i)}/a^q$ is the coefficient of the term $\zeta_{i_1}\zeta_{i_2} \cdots \zeta_{i_n}$, where $1 \le i_1 \le i_2 \le \cdots \le i_n \le \rho$, and n is some positive integer. We impose on our specializations $\{t\} \to \{\bar{t}\}$ the following inequality (in addition to the inequalities (8) and (9)):

$$\alpha_{i_1 i_2 \cdots i_n}(\{\bar{t}\}) \neq 0. \tag{14}$$

Since $\zeta_1, \zeta_2, \ldots, \zeta_\rho$ are analytically independent over k they are also analytically independent over $k(\{t\})$ (always by the usual argument based on our assumption that the t's are indeterminates over $k[[\zeta_1, \zeta_2, \ldots, \zeta_N]]$). Since $k(\{\bar{t}\})$ is a subfield of $k(\{t\})$, it follows that $\zeta_1, \zeta_2, \ldots, \zeta_\rho$ are also analytically independent over the field $k(\{\bar{t}\})$. It follows therefore from (14) that $F_1(\zeta_1, \zeta_2, \ldots, \zeta_\rho; \{\bar{t}\}) \neq 0$. Since $\omega_1, \omega_2, \ldots, \omega_d$ are linearly independent over the quotient field K_0 of $k[[\zeta_1, \zeta_2, \ldots, \zeta_\rho]]$, they are also linearly independent over $k(\{t\})[[\zeta_1, \zeta_2, \ldots, \zeta_\rho]]$ and therefore also over $k(\{\bar{t}\})[[\zeta_1, \zeta_2, \ldots, \zeta_\rho]]$. It follows therefore from (13) and from the inequality $F_1(\zeta_1, \zeta_2, \ldots, \zeta_\rho; \{\bar{t}\}) \neq 0$ that $f_1^*(\zeta_1^*, \zeta_2^*, \ldots, \zeta_{r+1}^*; \{\bar{t}\}) \neq 0$. This completes the proof of part (a) of the lemma in the case of a simple point Q_ζ^* of V_t^*. As to part (*b*) of the lemma, it is trivial in the case of a simple point Q_ζ^*. Namely, from the assumption that the point $(\zeta_1, \zeta_2, \ldots, \zeta_N)$ is a specialization, over k, of the point $(\xi_1, \xi_2, \ldots, \xi_N)$, it follows that the point Q_ζ^* is a specialization of R_ξ^*, over k^*. Hence, since Q_ζ^* is a simple point of V_t^* also R_ξ^* is necessarily a simple point of V_t^*. Therefore in this case we always have $R_\xi^* \in \mathrm{EQS}_{k^*}(V_t^*, Q_\zeta^*)$, i.e., the implication $\Rightarrow$ in part (*b*) of the lemma is trivially satisfied. On the other hand, from the fact that Q_ζ^* is a specialization of R_ξ^* over k^* follows trivially that $Q_{\zeta,\bar{t}}^*$ is a specialization of $R_{\xi,\bar{t}}^*$ over $k(\{\bar{t}\})$. Since $Q_{\zeta,\bar{t}}^*$ is a simple point of $V_{\bar{t}}^*$ if (14) holds true, also $R_{\xi,\bar{t}}^*$ is then necessarily a simple point of $V_{\bar{t}}^*$, i.e., $R_{\xi,\bar{t}}^* \in \mathrm{EQS}_{k(\{\bar{t}\})}(V_{\bar{t}}^*, Q_{\zeta,\bar{t}}^*)$. We have therefore proved that, if Q_ζ^* is a simple point of V_t^* then all the conditions stated in the lemma are satisfied for all the specializations $\{t\} \to \{\bar{t}\}$ such that

$$a(\{\bar{t}\}) \cdot b(\{\bar{t}\}) \cdot \alpha_{i_1 i_2 \cdots i_n}(\{\bar{t}\}) \neq 0.$$

From now on we assume that Q_ζ^* is a singular point of V_t^*, whence $\sigma = d.t.(V_t^*, Q_\zeta^*)/k^* \ge 1$. In this case we shall use induction with respect to r. If $r = 1$ V_t^* is a plane (reduced) algebroid curve, defined over k^*. The only point of the curve V_t^* which can be singular is its closed point P (the origin $X_1 = X_2 = 0$). Therefore Q_ζ^* is necessarily the point P. The remaining points of V_t^* are the q general points of the q irreducible branches of V_t^*, and these are simple points of V_t^*. Therefore, in the present case the set $(V_t^*)_{Q_\zeta^*}$ coincides with $(V_t^*)_P$, i.e., with V_t^*, and the set $\mathrm{EQS}_{k^*}(V_t^*, P)$ consists of the point P only. If then the inequalities (8) and (9) are satisfied then, as was pointed out earlier (for arbitrary values of r), the reduced curve $V_{\bar{t}}^*$ has at P a singular point (in fact, a point of the *same* multiplicity n as the multiplicity of P for V_t^*). Thus $d.t.(V_{\bar{t}}^*, P)/k(\{\bar{t}\}) = d.t.(V_t^*, P)/k^*(= 1)$, and condition ($a$) of the lemma is satisfied for all the specializations $\{t\} \rightarrow \{\bar{t}\}$ such that $a(\{\bar{t}\}) \cdot b(\{\bar{t}\}) \neq 0$. As to condition ($b$) of the lemma, it is vacuous, since the set $\mathrm{EQS}_{k(\{\bar{t}\})}(V_{\bar{t}}^*, P)$ consist of the point P only.

From now on we shall assume that the lemma is true in dimension $r - 1$. We introduce the generic projection π_u^* of V_t^*/k^* onto the affine variety $\mathbf{A}_r/k^{**}$, where $k^{**} = k^*(\{u\})$ [see Section 2, where k is to be replaced by $k^*(= k(\{t\})]$. Let $\Delta_{t,u}^{**}$ be the critical hypersurface of π_u^*, and let $\pi_u^*(Q_\zeta^*) = Q_{\zeta,u}^{**}$. Since Q_ζ^* is a singular point of V_t^*, the point $Q_{\zeta,u}^{**}$ belongs necessarily to $\Delta_{t,u}^{**}$, and we have, by the definition of dimensionality type:

$$d.t.(\Delta_{t,u}^{**}, Q_{\zeta,u}^{**})/k^{**} = d.t.(V_t^*, Q_\zeta^*)/k^* - 1 = \sigma - 1. \tag{15}$$

We apply our inductive hypothesis to the $(r - 1)$-dimensional variety $\Delta_{t,u}^{**}$. In applying our lemma in this case, we replace k by the field $k(\{u\})$, the t's being indeterminates over this field. It follows that for almost all specializations $\{t\} \rightarrow \{\bar{t}\} \subset k(\{u\})[\{t\})]$—*and here we shall restrict ourselves to specializations in which the* $\bar{t}$'s *belong to* $k[\{t\}]$—the hypersurface $\Delta_{\bar{t},u}^{**}$ is well defined and is reduced, and that, by part (a) of the lemma, and by (15) we have

$$d.t.(\Delta_{\bar{t},u}^{**}, Q_{\zeta,\bar{t},u}^{**}) = \sigma - 1. \tag{16}$$

Here $Q_{\zeta,\bar{t},u}^{**}$ is the specialization of the point $Q_{\zeta,u}^{**}$ under $\{t\} \rightarrow \{\bar{t}\}$. Now, under this specialization the projection π_u^* is specialized to the generic projection $\pi_{\bar{t},u}^*$ of $V_{\bar{t}}^*$, as follows at once from the definition of the generic projection given in Section 2. It is also clear that $\Delta_{\bar{t},u}^{**}$ is the (generic) critical hypersurface of the projection $\pi_{\bar{t},u}^{*}$ of $V_{\bar{t}}^{*}$ and that $\pi_{\bar{t},u}^{*}(Q_{\zeta,\bar{t}}^{*}) = Q_{\zeta,\bar{t},u}^{**}$. It follows therefore from (16) and from the definition of dimensionality type that

$$d.t.(V_{\bar{t}}^{*}, Q_{\zeta,\bar{t}}^{*})/k(\{\bar{t}\}) = \sigma,$$

and this proves part (*a*) of the lemma.

To prove part (*b*) of the lemma, we set $R_{\xi,u}^{**} = \pi_u^{*}(R_\xi^{*})$. Since $(\zeta_1, \zeta_2, \ldots, \zeta_N)$ is a specialization of $(\xi_1, \xi_2, \ldots, \xi_N)$ over k, it follows that *$Q_{\zeta,u}^{**}$ is a specialization of $R_{\xi,u}^{**}$ over $k^{*}(\{u\})(= k^{**})$*. From the definition of dimensionality type we have

$$d.t.(\Delta_{t,u}^{**}, R_{\xi,u}^{**})/k^{**} = d.t.(V_t^{*}, R_\xi^{*})/k^{*} - 1,$$

and hence, by (15), it follows that

$$R_{\xi,u}^{**} \in \mathrm{EQS}_{k^{**}}(\Delta_{t,u}^{**}, Q_{\zeta,u}^{**}) \Leftrightarrow R_\xi^{*} \in \mathrm{EQS}_{k^{*}}(V_t, Q_\zeta^{*}). \qquad (16')$$

In applying our induction hypothesis to $\Delta_{t,u}^{**}$, we replace V_t^{*} by $\Delta_{t,u}^{**}$, the fields k and k^{*} of the lemma by the fields $k(\{u\})$ and $k^{**}[= k(\{u\}, \{t\})]$ respectively, and the point Q_ζ^{*} by the point $Q_{\zeta,u}^{**}$. Since $(\zeta_1, \zeta_2, \ldots, \zeta_N)$ is a specialization of $(\xi_1, \xi_2, \ldots, \xi_N)$ over k, hence also over $k(\{u\})$, it follows that for almost all specializations $\{t\} \to \{\bar{t}\} \subset k^{**}$, and hence also for almost all specializations $\{t\} \to \{\bar{t}\} \subset k^{*}(= k(\{t\})$, we have, for all points $R_{\xi,u}^{**}$ such that $(\xi_1, \xi_2, \ldots, \xi_N) \xrightarrow{k} (\zeta_1, \zeta_2, \ldots, \ldots, \zeta_N)$:

$$R_{\zeta,\bar{t},u}^{**} \in \mathrm{EQS}_{k(\{u\},\{\bar{t}\})}(\Delta_{\bar{t},u}^{**}, Q_{\zeta,\bar{t},u}^{**}) \Leftrightarrow R_{\xi,u}^{**} \in \mathrm{EQS}_{k^{**}}(\Delta_{t,u}^{**}, Q_{\zeta,u}^{**}). \qquad (17)$$

Hence, by (16′), we have for almost all specializations $\{t\} \to \{\bar{t}\} \subset k(\{t\})$:

$$R_{\xi,\bar{t},u}^{**} \in \mathrm{EQS}_{k(\{u\},\{\bar{t}\})}(\Delta_{\bar{t},u}^{**}, Q_{\zeta,\bar{t},u}^{**}) \Leftrightarrow R_\xi^{*} \in \mathrm{EQS}_{k^{*}}(V_t^{*}, Q_\zeta^{*}). \qquad (18)$$

Now each of our specializations $\{t\} \to \{\bar{t}\}$ defines a generic projection $\pi_{\bar{t},u}^{*}$ of $V_{\bar{t}}^{**}$, and the critical hypersurface of $\pi_{\bar{t},u}^{*}$ is the hypersurface

$\Delta_{\bar{t},u}$**. We have also

$$\pi_{\bar{t},u}{}^*(Q_{\zeta,\bar{t}}{}^*) = Q_{\zeta,\bar{t},u}{}^{**},$$

$$\pi_{\bar{t},u}{}^*(R_{\xi\bar{t}}{}^*) = R_{\xi,\bar{t},u}.$$

Hence, by (18), and using the very definition of the dimensionality type (based on the use of generic projections), we find that part (b) of the lemma is valid.

This concludes the proof of our basic lemma.

Before we proceed to the proof of the basic Theorem 5.1 we draw some preliminary consequences of Lemma 5.2. We go back to the r-dimensional algebroid (reduced) hypersurface V/k dealt with in the previous sections.

PROPOSITION 5.3. *Given a singular point Q of V, of dimensionality type σ, the following two conditions are satisfied for almost all k-rational projections $\pi_{\bar{u}}$ of V onto $\mathbf{A}_r/k$ and for all points R of V such that Q is a specialization of R over k:*

(a) $d.t.(\Delta_{\bar{u}}, \overline{Q}_0)/k = \sigma - 1$,

(b) $\pi_{\bar{u}}(R) \in \mathrm{EQS}_k(\Delta_{\bar{u}}, \overline{Q}_0) \Leftrightarrow R \in \mathrm{EQS}_k(V, Q)$,

where $\overline{Q}_0 = \pi_{\bar{u}}(Q)$ and $\Delta_{\bar{u}}$ is the critical hypersurface of $\pi_{\bar{u}}$.

Proof. We apply Lemma 5.2 to the case in which the r-dimensional variety V_t is replaced by the $(r-1)$-dimensional critical hypersurface Δ_u of the generic projection π_u of V, the indeterminates u now playing the role of the parameters t. It follows from Lemma 5.2 that for almost all specializations $\{u\} \to \{\bar{u}\} \subset k$ the specialized hypersurface $\Delta_{\bar{u}}$ is well defined and is reduced. Furthermore, it is clear that for almost all specializations $\{u\} \subset \{\bar{u}\} \subset k$, the specialized quantities $\bar{x}_i{}^*$ (where the r elements $x_i{}^*$ are defined in (1) of Section 2) form a set of parameters of the local ring $\mathfrak{o}$ of V (equivalently: the determinant $|\bar{u}_j{}^{(i)}|$ is different from zero) and hence define a finite projection $\pi_{\bar{u}}$ of V onto $\mathbf{A}_r(k)$.

The hypersurface $\Delta_{\bar{u}}$ is then the critical hypersurface of $\pi_{\bar{u}}$. We take as set $S_k(\Delta_u)$ the set of points of Δ_u which are π_u-projections R_0 of (necessarily singular) points R of V. The $r+1$ power series $\varphi_i(\xi_1, \xi_2, \ldots, \xi_N; \{t\})$ of Lemma 5.2 [see (5)] and the algebroid variety Z/k are now respectively the r power series given in (1), Section 2 (where $x_1, x_2, \ldots, x_{r+1}$ are to be replaced by $\xi_1, \xi_2, \ldots, \xi_{r+1}$, the integer N now being equal to $r+1$) and Z being now our original variety V. Our proposition is then a direct consequence of Lemma 5.2, the conditions (*a*) and (*b*) of the proposition being the analogues of the conditions (*a*) and (*b*) of the lemma.

PROPOSITION 5.4 *Let Q be a point of V, of dimensionality type σ, and let $W(Q)$ denote the algebroid closure in V of the set* $\mathrm{EQS}_k(V, Q)$ *(i.e., $W(Q)$ is the smallest algebroid subvariety of V containing the set* $\mathrm{EQS}_k(V, Q)$*). Then $W(Q)$ is an irreducible subvariety of V, of dimension $r - \sigma$, whose general point R belong to the set* $\mathrm{EQS}_k(V, Q)$, *and it is the greatest irreducible subvariety of V which contains Q and which, in addition, has the property that its general point has dimensionality type σ. Furthermore, $W(Q)$ is smooth at Q, and V is equimultiple at Q along $W(Q)$. In particular, we have $W(P) =$* $\mathrm{EQS}_k(V, P)$, *where P is the closed point of V.*

Proof. The proportion is trivial if Q is a simple point of V. In that case there is only one irreducible component V_i of V which contains Q. If, say, V_1 is that component then it is clear that $\mathrm{EQS}_k(V, Q)$ is a subset of $V_1 - (V_1 \cap V_{sing})$ containing the general point R of V_1/k and that therefore $W(Q) = V_1$. Hence the dimension of $W(Q)$ is in this case equal to r, while σ is equal to zero. The general point R of V_1 is of course a simple point of V, and Q is a specialization of R over k. The rest of the proposition is obvious in this case (note that if P is a simple point, then $\mathrm{EQS}_k(V, P) = V$). We assume from now on that Q is a singular point of V. We shall use induction with respect to r, the proposition being trivial in the case $r = 1$, for in that case the point Q, being a singular point of the plane curve V, must be the closed point P of V and hence $W(P) = \mathrm{EQS}_k(V, P) = P$. We assume now that the proposition is true in the case of $(r - 1)$-dimensional algebroid varieties.

We shall say that a *k-rational projection* $\pi_{\bar{u}}$ of V (i.e., a $\pi_{\bar{u}}$ such that the $\bar{u}$'s belong to k) is *Q-permissible* if it satisfies the conditions (*a*) and (*b*) of Proposition 5.3. We fix a Q-permissible k-rational projection $\pi_{\bar{u}}$ of V and use the notations of Proposition 5.3, except that we write π and Δ instead of $\pi_{\bar{u}}$ and $\Delta_{\bar{u}}$. Since Q is a singular point of V, of dimensionality type σ, and since π is a Q-permissible projection of V, it follows that the π-projection $\overline{Q}_0$ of Q belongs to Δ and has dimensionality type $\sigma - 1$ $(\sigma \geq 1)$. Since Δ has dimension $r - 1$ it follows from our induction hypothesis that if we set $Z_0 = \mathrm{EQS}_k(\Delta, \overline{Q}_0)$ then the algebroid closure $\overline{Z}_0$ of Z_0 in Δ is an irreducible subvariety of Δ, of dimension $r - \sigma$ $(= (r - 1) - (\sigma - 1))$, having at $\overline{Q}_0$ a simple point, and that Δ is equimultiple at $\overline{Q}_0$ along $\overline{Z}_0$. It follows therefore from a generalized version of our PNAS Note [2] (see Appendix) that the full inverse image $\pi^{-1}\{\overline{Z}_0\}$ of $\overline{Z}_0$ on V has a simple point at Q and that V is equimultiple at Q along $\pi^{-1}\{\overline{Z}_0\}$. Since π is a finite projection, the set $\pi^{-1}\{\overline{Z}_0\}$ is an algebroid subvariety of V, of pure dimension $r - \sigma$. Since Q is a simple point of that subvariety, there is only one irreducible component of $\pi^{-1}\{\overline{Z}_0\}$ which contain the point Q. We denote by Z that component. Let R be the general point of Z. Then R is a point of V, of dimension $r - \sigma$, and the equimultiplicity of V at Q along $\pi^{-1}\{\overline{Z}_0\}$ signifies that R and Q are points of V *of the same* multiplicity. From the implication $\Leftarrow$ in condition (*b*) of Proposition 5.3 it follows that the set $\mathrm{EQS}_k(V, Q)$ is contained in $\pi^{-1}\{Z_0\}$. If Q' is any point of that set then Q' must belong to at least one irreducible component of $\pi^{-1}\{Z_0\}$, and any such component must also contain Q, since Q is a specialization of Q', over k. It follows that *the set* $\mathrm{EQS}_k(V, Q)$ *is contained in* Z. Since R is the general point of Z, the point $\overline{R}_0(= \pi(R))$ must be the general point of $\overline{Z}_0$ (since R and R_0 have the same dimension over k, whence the dimension of $\overline{R}_0$ is $r - \sigma$, equal to the dimension of $\overline{Z}_0$). By our induction hypothesis, the general point $\overline{R}_0$ of $\overline{Z}_0$ belongs to the set $\mathrm{EQS}_k(\Delta, \overline{Q}_0)$, i.e., we have $d.t.(\Delta, \overline{R}_0)/k = d.t.$ $(\Delta, \overline{Q}_0)/k = \sigma - 1$. Since Q is a specialization of R over k, it follows from the implication $\Rightarrow$ in Proposition 5.3 that $R \in \mathrm{EQS}_k(V, Q)$, whence $d.t._k(V, R) = \sigma$. From the two facts just established above, namely that 1) $\mathrm{EQS}_k(V, Q) \subset Z$ and that 2) the general point R of Z is contained in $\mathrm{EQS}_k(V, Q)$, follows that Z is the algebroid closure of the set $\mathrm{EQS}_k(V, Q)$, i.e., *we have* $Z = W(Q)$. Thus the variety $W(Q)$ of the proposition has all the properties stated in that proposition, except that we still have to show that $W(Q)$ is the greatest irreducible subvariety of V which contains Q and whose general point has dimensionality type σ. Let then Z' be any irreducible subvariety of V which contains Q and whose general point R' has dimensionality type σ. Since Q is a specialization of R', it follows that $R' \in \mathrm{EQS}_k(V, Q)$. Hence $R' \in W(Q)$, showing that $Z' \subset W(Q)$. This proves the maximality property of the variety $W(Q)$.

Finally, if Q is the point P, then the set $\mathrm{EQS}_k(V, P)$ is the set of *all* points of V which have dimensionality type s, since Q is a specialization of any point of V. Hence $\mathrm{EQS}_k(V, P) = V(s)$. Now, let Q' be any point of $W(P)$. We have then that Q' is a specialization of R and P is a specialization of Q'. Hence, by Proposition 4.2, we have $d.t._k(V, R)$

$\leq d.t._k(V, Q') \leq d.t._k(V, P)$. Since $d.t._k(V, R) = d.t._k(V, P) = s$, it follows that $d.t._k(V, Q') = s$, whence $Q' \in \text{EQS}_k(V, P)$. This shows that $W(P) = \text{EQS}_k(V, P)$.

This completes the proof of the proposition.

We shall explicitly state the following inequality which holds for any point Q of V and which is a direct consequence of Theorem 5.4:

$$\dim Q/k + d.t.(V, Q) \leq r. \tag{19}$$

For, if $d.t.Q = \sigma$ then we have $Q \in W(Q)$ and $\dim (W(Q)) = r - \sigma$, whence $\dim Q/k \leq r - \sigma$, and this is precisely (19).

The inequality (19) will be used very frequently in the next section.

We now proceed to the proof of the basic Theorem 5.1. We begin by observing that *for any integer* σ, $0 \leq \sigma \leq s$, such that $V(\sigma) \neq \emptyset$, *there is only a finite number of points R on V such that $d.t._k(V, R) = \sigma$ and* $\dim R/k = r - \sigma$. This assertion is trivial in the case $r = 1$, for in that case V itself is a finite set of points. It is also trivial for any value of r if $\sigma = 0$. Assuming that this assertion is true for algebroid varieties of dimension $r - 1$, we consider the generic projection π_u of V and the critical hypersurface Δ_u of π_u. By induction, we have that Δ_u has only a finite number of points of a given dimensionality type $\sigma - 1$ ($\sigma \geq 1$) (over the ground field $k^* = k(\{u\})$) and of dimension $r - \sigma$ $(= (r-1) - (\sigma - 1))$ over k^*. Now, since π_u is a finite projection, any point of Δ_u is the projection of only a finite number of points of V^*, where V^* is the algebroid variety obtained from V by the ground field extension $k \to k^*$. If R is any point of V, of dimensionality type σ over k and of dimension $r - \sigma$ over k, then R is also a point of V^*, of dimensionality type σ over k^* and of dimension $r - \sigma$ over k^* (see Proposition 4.3), and $\pi_u(R)$ is a point of Δ_u of dimensionality type $\sigma - 1$ and of dimension $r - \sigma$ over k^*. From this our assertion follows.

As a consequence it follows that as the point Q ranges over the entire set $V(\sigma)$ of points of dimensionality type σ (assuming of course that $V(\sigma)$ is non-empty) the variety $W(Q)$ ranges over a *finite set of varieties*. This finite set of varieties consists of all irreducible $(r - \sigma)$-dimensional algebroid subvarieties of V having as general point a point of dimensionality type σ. (Note that if W is any such variety and if R is the general point of W/k then $\text{EQS}(V, R)$ consists of the single point R and $W(R)$ is W.) Let $\{W_1, W_2, \ldots, W_{q_\sigma}\}$, $q_\sigma \geq 1$, be this finite set of varieties and let R_i be the general point of W_i. Since every point

Q of $V(\sigma)$ belongs to one of these q_σ varieties it follows that if we set

$$S_i = W_i \cap V(\sigma), \qquad (i = 1, 2, \ldots, q_\sigma) \tag{20}$$

then

$$V(\sigma) = \bigcup_{i=1}^{q_\sigma} S_i. \tag{21}$$

It is easy to see that $S_i \cap S_j = \emptyset$ if $i \neq j$, and that consequently (21) *is a decomposition of* $V(\sigma)$ *into disjoint sets* S_i. This follows directly from the fact that, by Proposition 5.4, every point Q of $V(\sigma)$ belongs to one and only one irreducible $(r - \sigma)$-dimensional subvariety of V whose general point has dimensionality type σ.

We recall that we agreed to denote by $V\{\sigma\}$ the set of all points of V which have dimensionality type $\geq \sigma$. *We prove now that* $V\{\sigma\}$ *is a subvariety of* V. We use induction from $\sigma + 1$ to σ, since $V\{s\}$ *coincides with* $V(s)$, and $V(s) = \mathrm{EQS}_k(V, P) = W(P)$, by Proposition 5.4. We have $V\{\sigma\} = V(\sigma) \cup V\{\sigma + 1\}$. Since by our induction hypothesis $V\{\sigma + 1\}$ is a variety, we have $\overline{V\{\sigma\}} \subset \overline{V(\sigma)} \cup V\{\sigma + 1\}$. Now, by (21), we have $\overline{V(\sigma)} = \bigcup_{i=1}^{q_\sigma} W_i$ since obviously $\overline{S_i} = W_i$ (the *subset* S_i of the *variety* W_i contains the general point R_i of W_i). Since every point of W_i is a specialization of R_i and since $d.t.(V, R_i) = \sigma$, it follows from Proposition 4.2 that $W_i \subset V\{\sigma\}$. Hence $\overline{V(\sigma)} \subset V\{\sigma\}$, and therefore from the above inclusion $\overline{V\{\sigma\}} \subset \overline{V(\sigma)} \cup V\{\sigma + 1\}$ follows the inclusion $\overline{V\{\sigma\}} \subset V\{\sigma\} \cup V\{\sigma + 1\} = V\{\sigma\}$, and this proves that $\overline{V\{\sigma\}} = V\{\sigma\}$ and that consequently $V\{\sigma\}$ is a variety.

It was pointed out in the above proof that $\overline{S_i} = W_i$. We prove next that $W_i - S_i$ *is a proper subvariety of the* (*irreducible*) *variety* W_i (*and that consequently each set* S_i *is a Zariski dense open subset of its algebroid closure* $\overline{S_i}$). We have $\overline{S_i} \not\subset V\{\sigma + 1\}$ (since $R_i \in S_i \subset \overline{S_i}\}$. Since we have just proved that $V\{\sigma + 1\}$ is a variety it follows that $\overline{S_i} \cap V\{\sigma + 1\}$ is a proper subvariety of $\overline{S_i}$. Since $S_i \subset V(\sigma)$ we have $S_i \cap V\{\sigma + 1\} = \emptyset$, and therefore $\overline{S_i} \cap V\{\sigma + 1\} \subset \overline{S_i} - S_i$. On the other hand, if Q' is any point of $W_i - S_i (= \overline{S_i} - S_i)$ then $Q' \in W_i$ and $Q' \neq S_i (= W_i \cap V(\sigma))$, whence $d.t._k(V, Q') > \sigma$ (by Proposition 4.2 we have $d.t._k(V, Q') \geq \sigma$ for any point Q' of W_i, since the general point R_i of W_i has dimensionality type σ). This shows that $Q' \in V\{\sigma + 1\}$, and consequently $\overline{S_i} - S_i \subset V\{\sigma + 1\}$. This, in conjunction with the above inclusion $\overline{S_i} \cap V\{\sigma + 1\} \subset \overline{S_i} - S_i$, implies that $\overline{S_i} - S_i = \overline{S_i} \cap V\{\sigma + 1\}$ = a proper subvariety of $\overline{S_i}$, as was asserted.

This completes the proof of part (*a*) of Theorem 5.1.

Parts (*b*) and (*c*) of Theorem 5.1 are included in the statement of Theorem 5.4 which we have already proved. We now proceed to the proof of part (*d*) of Theorem 5.1.

Let S be a stratum of V, let R be the general point of S (i.e., the general point of the irreducible subvariety $\overline{S}$ of V which is the algebroid closure of S) and let $\sigma = d.t.(V, R)$. Let Q' be any point of the boundary $\overline{S} - S$ of S and let $\sigma' = d.t.(V, Q')$. We have to prove that if S' denotes the stratum containing the point Q' then $S' \subset \overline{S} - S$.

We first observe that, by Proposition 4.2, we have $\sigma' \geq \sigma$ (since Q' is a specialization of R over k). We know that $S = \overline{S} \cap V(\sigma)$ [see (20)]. Since $Q' \in \overline{S} - S$ it follows that $Q' \notin V(\sigma)$. Hence $d.t.(V, Q') \neq \sigma$, and this implies that $\sigma' > \sigma$. Let S' denote the stratum containing the point Q' and let R' be the general point of S'. *We have to prove that* $R' \in \overline{S}$, for once this is proved it will follow that $R' \in \overline{S} - S$ (since $d.t.(V, R') = d.t.(V, Q') = \sigma' > \sigma$) and that therefore the entire stratum S' belongs to the boundary $\overline{S} - S$ of the stratum S.

To prove part (*d*) of Theorem 5.1 we shall use induction on the dimension r of V, since for $r = 1$ part (*d*) of that theorem is trivial (the only point Q' of a plane algebroid curve V which can (and does) belong to the *boundary* of a stratum is *the closed point* P of V, provided P is a singular point of V; in this case the stratum S' containing Q' is the point $Q'(= P)$ itself, and therefore $R' = Q'(= P)$). We therefore assume that part (*d*) of Theorem 5.1 is true in dimension $r - 1$.

We first consider the case $\sigma = 0$, in which case S is a stratum of non-singular points. Therefore, as has been pointed out in the beginning of this section, we have $S = V_1 - (V_1 \cap V_{sing})$, where V_1 is some irreducible component of V. The boundary of S is the variety $V_1 \cap V_{sing}$. We have to show that $R' \in V_1 \cap V_{sing}$. By Theorem 5.4 we know that R' and Q' are points of V of the same multiplicity. Since the multiplicity of R' on V is the sum of the multiplicities $e_{R'}(V_i)$ of R' as a point of V_i ($i = 1, 2, \ldots, q$), and since we have $e_{R'}(V_i) \leq e_{Q'}(V_i)$ for each i (since Q' is a specialization of R'), the equality

$$\sum_{i=1}^{q} e_{R'}(V_i) = \sum_{i=1}^{q} e_P(V_i)$$

implies that $e_{R'}(V_i) = e_{Q'}(V_i)$, for $i = 1, 2, \ldots, q$. Since the point Q' belongs to V_1 it follows that also R' belongs to V_1, and therefore $R' \in V_1 \cap V_{sing}$ (since $\sigma' > 0$), showing that $R' \in \overline{S}(= V_1 - V_{sing})$.

We now deal with the case $\sigma > 0$. In this case the points R, Q' and R' are singular points of V. We consider the generic projection π_u of V onto $\mathbf{A}_r$. Let Δ_u^* be the critical hypersurface of π_u and let

$$\pi_u(R) = R_0^*, \qquad \pi_u(Q') = Q_0'^*, \qquad \pi_u(R') = R_0'^*.$$

The points R_0^*, $Q_0'^*$ and $R_0'^*$ belong to Δ_u^*, since R, Q' and R' are singular points of V. We have

$$d.t.(\Delta_u^*, R_0^*)/k^* = \sigma - 1,$$
$$d.t.(\Delta_u^*, Q_0'^*) = d.t.(\Delta_u^*, R_0'^*) = \sigma' - 1, \tag{22}$$

and since Q' is a specialization of R over k, the point $Q_0'^*$ is a specialization of R_0^* over k^*; here k^* is the field $k(\{u\})$. We have $\dim R_0^*/k^* = \dim R/k = r - \sigma, =(r-1) - (\sigma - 1)$, and since $d.t.(\Delta_u^*, R_0^*)/k^* = \sigma - 1$, it follows that $R_0'^*$ is the general point of the stratum S_0^* containing R_0^*. Since $Q_0'^*$ is a specialization of R_0^* over k^* and since the dimensionality type of $Q_0'^*$ on Δ_u^* is $\sigma' - 1$ [see (22)], hence different from $\sigma - 1$, it follows that $Q_0'^*$ belongs to the boundary $\overline{S_0}^* - S_0^*$ of the stratum S_0^*. By our induction hypothesis it follows then that the entire stratum $S_0'^*$ containing the point $Q_0'^*$ is contained in $\overline{S_0}^* - S_0^*$. Since $R_0'^*$ belongs to $S_0'^*$, and is in fact the general point of $S_0'^*$, we have then that $R_0'^* \in \overline{S_0}^* - S_0^*$. We proceed to show that this last inclusion implies the inclusion $R' \in \overline{S} - S$, and this will complete the proof of part (*d*) of Theorem 5.1.

We consider the full inverse image $\pi_u^{-1}\{R_0'^*\}$ of the point $R_0'^*$. This is a finite set of points, and the singular point R' is one point of that set. We shall show now that *all the remaining points of the set* $\pi_u^{-1}\{R_0'^*\}$ *are simple points of the variety* V/k^*. Once this is proved, the inclusion $R' \in \overline{S} - S$ can be proved as follows:

Since $R_0'^* \in \overline{S_0}^* - S_0^*$, $R_0'^*$ is a specialization of R_0^* over k^*. Therefore some point R_i' of the set $\pi_u^{-1}\{R_0'^*\}$ must be a specialization of the point R. Since R is a singular point of V, also R_i' must be a singular point of V. Therefore R_i' must be the point R'. We have therefore that R' is a specialization of R over k^*, and therefore also over k. This proves the desired inclusion $R' \in \overline{S} - S$.

Let $\xi_1, \xi_2, \ldots, \xi_{r+1}$ be the coördinates of R. The r coördinates $\xi_i{}^*$ of the point $R_0{}^*$ are the quantities $\varphi_i(\xi_1, \xi_2, \ldots, \xi_{r+1}; \{u\})$ [see (1), Section 2]. We recall that each function $\varphi_i(X_1, X_2, \ldots, X_{r+1}; \{u\})$ depends only on the indeterminates $\{u^{(i)}\}$, each indeterminate $u_{j_1 j_2 \cdots j_N}{}^{(i)}$ being the coefficient of the monomial $X_{j_1} X_{j_2} \cdots X_{j_N}$, N being an arbitrary positive integer, and $1 \leq j_1 \leq j_2 \leq \cdots \leq j_N \leq r + 1$. The points of the set $\pi_u{}^{-1}\{R_0{}^*\}$ are the intersections of V/k^* with the *algebraic* curve $\Gamma_u{}^*$ in $\mathbf{A}_{r+1}/k^*$ defined by the r equations

$$\varphi_i(X_1, X_2, \ldots, X_{r+1}; \{u^{(i)}\}) - \xi_i{}^* = 0, \qquad i = 1, 2, \ldots, r. \qquad (23_i)$$

Since $\xi_i{}^* = \varphi_i(\xi_1, \xi_2, \ldots, \xi_{r+1}; \{u^{(i)}\})$, it follows that the lefthand side of (23_i) is a power series in the $X_j - \xi_j$, $j = 1, 2, \ldots, r + 1$, with coefficients in $k[[\{\xi\}]][\{u^{(i)}\}]$ (which are linear in the $u^{(i)}$). Therefore each of the equations (23_i) defined an algebroid hypersurface F_i having R as its closed point, with $k((\{\xi\}))(\{u^{(i)}\})$ as coefficient field. It is clear that each F_i is the general algebroid hypersurface in $\mathbf{A}_{r+1}/k$ having R as its closed point and that, moreover, each F_i is also the general hypersurface in $\mathbf{A}_{r+1}/K_{i-1}{}^*$ having R as its closed point; here $K_{i-1}{}^*$ denotes the field $k((\{\xi\}))(\{u^{(1)}\}; \{u^{(2)}\}, \ldots, \{u^{(i-1)}\})$. The points of the set $\pi_u{}^{-1}\{R_0{}^*\}$ are the intersections of $V/K_r{}^*$ with the algebroid curve $\Gamma_u{}^*$ defined by the r equations (23_i). Since V is defined over k, also V_{sing} is an algebroid variety defined over k. Let ρ be the dimension of V_{sing}. The intersection $V_{sing} \cap F_1$ is clearly an algebroid variety Z_1 of dimension $\rho - 1$, defined over K_1. Since the $u^{(2)}$ are indeterminates over the field K_1, it follows that the intersection $Z_1 \cap F_2$ is an algebroid variety Z_2 of dimension $\rho - 2$, defined over K_2. Proceeding in this fashion we find that the intersection $V_{sing} \cap F_1 \cap F_2 \cap \cdots \cap F_\rho$ is a finite set of points, one of which is the point R. Since $\rho \leq r - 1$, it follows that the intersection $V_{sing} \cap F_1 \cap F_2 \cap \cdots \cap F_\rho \cap F_{\rho+1}$ is the single point R. Therefore $V \cap F_1 \cap F_2 \cap \cdots \cap F_r$ consists of a *finite* set of points all of which, except R, are simple points of V. This completes the proof of part (d) of Theorem 5.1.

6. Foundations of a general (global) theory of equisingularity on r-dimensional algebraic varieties, of embedding dimension $= r + 1$ at each of their singular k-rational points.

Let V be an algebraic variety (in projective or affine space), of pure dimension r, defined over an algebraically closed ground field k of characteristic zero, and of embedding dimension $\leq r + 1$, locally, at each of its k-rational points. If P is any such point of V we consider the completion $\hat{\mathfrak{o}}$ of the local ring $\mathfrak{o}$ of V at P and we denote by $\hat{V}_P$ the algebroid variety Spec $(\hat{\mathfrak{o}})$ having origin at P. The ground field k is a coefficient field of $\hat{\mathfrak{o}}$, and it will be with reference to this ground field k that we shall speak of the dimensionality type of points of $\hat{V}_P$. The point P is singular for V if and only if it is singular for $\hat{V}_P$. If P is singular for V, any minimal basis $\{x_1, x_2, \ldots, x_{r+1}\}$ of the maximal ideal $\mathfrak{m}$ of $\mathfrak{o}$ will be a minimal basis of the maximal ideal $\hat{m}$ of $\hat{\mathfrak{o}}$, and thus also $\hat{V}_P$ is of embedding dimension $r + 1$. We define *the dimensionality type of P on V*, in symbols: $d.t.(V, P)/k$ (or simply $d.t.(V, P)$) as being the dimensionality type of P on $\hat{V}_P$. We have then:

$$0 \leq d.t.(V, P) \leq r,$$

with $d.t.(V, P) = 0$ if and only if P is a simple point of V. This definition is, by its very nature, analytically intrinsic.

If we wish (as we do) to define the dimensionality type of points R of V which are not k-rational (and hence of positive dimension over k), the following procedure suggests itself quite naturally:

We consider the irreducible subvariety W of V having R as general point and we pick a k-rational point P on W at which P is analytically irreducible (in particular, we can take for P any simple point of W). The analytical irreducibility of W at P implies that W determines a unique irreducible *algebroid* subvariety $\hat{W}$ of $\hat{V}_P$, and hence R determines a unique point of $\hat{V}_P$, namely the general point of $\hat{W}$. We continue to denote by R the general point of $\hat{W}$. *Then we can define the dimensionality type $d.t.(V, R)$* (relative to the given ground field) *as being the dimensionality type of R on $\hat{V}_P$*. This immediately raises a question, which will be one of the various questions that we shall study and answer in this section, namely: *is the above definition of $d.t.(V, R)$ independent of the choice of the k-rational point P on W? We shall answer this question later on in the affirmative.* The proof is not straightforward and will require a considerable amount of preparatory analysis (unless R is a simple point of V, in which case we obviously have $d.t.(V, R) = 0$).

Another, less important question, is the following: if the irreducible variety W is not analytically irreducible at P and has, say, g irreducible analytical branches at P, do the g distinct points $R_1, R_2, \ldots, R_g$ which correspond on $\hat{V}_P$ to R have the same dimensionality type?

Another paramount question which will be studied and answered (affirmatively) in this section is the following: *Is the irreducible algebroid variety* EQS$(\hat{V}_P, P)$ *algebraic* (this variety has been defined in Section 5, just before Lemma 5.2 was stated; see also the last statement of Theorem 5.4).

We assume from now on that P is a singular point of V (in the case of a simple point P the questions formulated above are trivial). We fix a minimal basis $x_1, x_2, \ldots, x_r, x_{r+1}$ of the maximal ideal $\mathfrak{m}$ of the local ring $\mathfrak{o}$ of V at P. The $r + 1$ elements are algebraically dependent over k. We have then

$$f(x_1, x_2, \ldots, x_r, x_{r+1}) = 0, \tag{1}$$

where f is a polynomial with coefficients in k and is uniquely determined (up to an arbitrary non-zero factor in k) by the condition of being of least possible degree. The polynomial f is free from multiple factors, its leading form f_n has degree n, where n is the multiplicity of the singular point P of V. The local ring $\mathfrak{o}$ of V at P is the ring of quotients of the ring $k[X_1, X_2, \ldots, X_{r+1}]/(f(X_1, X_2, \ldots, X_{r+1}))$ with respect to its maximal ideal generated by the f-residues $x_1, x_2, \ldots, x_{r+1}$ of the X_i. As to the completion $\hat{\mathfrak{o}}$ of $\mathfrak{o}$ we have

$$\hat{\mathfrak{o}} = k[[X_1, X_2, \ldots, X_{r+1}]]/(f(X_1, X_2, \ldots, X_{r+1})). \tag{2}$$

Let V_1 be the algebraic hypersurface in affine $\mathbf{A}_{r+1}$ defined by the equation

$$V_1 \colon f(X_1, X_2, X_{r+1}) = 0. \tag{3}$$

The hypersurface V_1 is birationally equivalent to V, and the birational transformation of V into V_1 is biregular at P, the corresponding point of V_1 being the origin P_1: $X_1 = X_2 = \cdots = X_{r+1} = 0$. We therefore have $\hat{V}_P = \hat{V}_{1,P_1}$.

In the present case of *algebraic* varieties we shall deal primarily with projections $\pi_{\bar{u}}$ of V_1 *which are of finite degree N*. By this we mean a projection $\pi_{\bar{u}}$ of V_1 onto an r-dimensional affine space $\mathbf{A}_r{}^*$:

$$\pi_{\bar{u}}(X_1, X_2, \ldots, X_{r+1}) = (X_1{}^*, X_2{}^*, \ldots, X_r{}^*),$$

where the $X_i{}^*$ are polynomials of degree N in $X_1, X_2, \ldots, X_{r+1}$ which are zero at P and have coefficients $(\bar{u})$ in some extension field $k^* = k(\bar{u})$:

$$X_i^* = \varphi_{i,(\bar{u})}(X_1, X_2, \ldots, X_{r+1}) = \sum_{j=1}^{r+1} \bar{u}_j^{(i)}(X_j + \sum_{1 \le j_1 \le j_2 \le r+1} \bar{u}_{j_1 j_2}^{(i)} X_{j_1} X_{j_2}$$
$$+ \cdots + \sum_{1 \le j_1 \le \cdots \le j_N \le r+1} \bar{u}_{j_1 j_2 \cdots j_N}^{(i)} X_{j_1} X_{j_2} \cdots X_{j_N}. \quad (4)$$

The equations (4) define a finite projection $\pi_{\bar{u}}$ of V_1 onto the space $\mathbf{A}_r^*$, in the neighborhood of the origin P_0^* of $\mathbf{A}_r^*$, if and only if the r elements

$$x_{i,(\bar{u})}^* = \varphi_{i,(\bar{u})}(x_1, x_2, \ldots, x_{r+1}), \qquad (i = 1, 2, \ldots, r) \qquad (4')$$

form a set of parameters of the local ring obtained from $\mathfrak{o}$ by the coefficient field extension $k \to k^*$.

We shall say that the projection $\pi_{\bar{u}}$ is a generic projection of V_1, of degree N, if all the coefficients $\bar{u}$ of the r polynomials $\varphi_{i,\bar{u}}$ are indeterminates. In the case of a generic projection we shall use, as in the preceding sections, the letter u instead of $\bar{u}$, and the projection will be therefore denoted by π_u. The r polynomials in (4) will be denoted in that case by $\varphi_{i,(u)}$.

We shall call a projection $\pi_{\bar{u}}$ *k-rational* if all the coefficients $\bar{u}$ belong to the ground field k.

If $\Delta_{\bar{u}}$ is the critical hypersurface of a projection $\pi_{\bar{u}}$ of V_1 and if Q is a singular point of $\hat{V}_P$ we shall say that $\pi_{\bar{u}}$ is *quasi-permissible at Q* if the following equality holds true:

$$d.t.(\hat{V}_P, Q) = d.t.(\Delta_{\bar{u}}, \pi_{\bar{u}}(Q))/k(\{\bar{u}\}) + 1, \qquad (5)$$

or—equivalently, if condition (a) of Proposition 5.3 (where V is to be replaced by $\hat{V}_P$) is satisfied by the projection $\pi_{\bar{u}}$. As in section 5, we shall say that $\pi_{\bar{u}}$ is *permissible at Q* if $\pi_{\bar{u}}$ is k-rational and satisfies both condition (a) and (b) of Proposition 5.3. We know, by Proposition 5.3, that for every sufficiently high integer N almost all projections $\pi_{\bar{u}}$ of degree N are quasi-permissible at Q, and that almost all k-rational projections of degree N are permissible at Q.

For almost all projections $\pi_{\bar{u}}$ of degree N it is true that the inverse image $\pi_{\bar{u}}^{-1}\{P_0^*\}$ of the origin P_0^*: $X_1^* = X_2^* = \cdots = X_r^* = 0$, is a finite set of points of V_1, or—more precisely—a zero-dimensional cycle $Z_{\bar{u}}$ on V_1. The cycle $Z_{\bar{u}}$ is the intersection of V_1 with the curve $\Gamma_{\bar{u}}$ defined by the r equations

$$\Gamma_{\bar{u}}\colon \varphi_{i,(\bar{u})}(X_1, X_2, \ldots, X_r, X_{r+1}) = 0, \qquad i = 1, 2, \ldots, r. \tag{6}$$

As the $\bar{u}$'s vary, the curve $\Gamma_{\bar{u}}$ varies in an irreducible algebraic system Σ of curves, the general member of this system being the curve Γ_u associated with the generic projection π_u of degree N, and the cycle $Z_{\bar{u}}$ varies in an irreducible algebraic system $\{Z_{\bar{u}}\}$ of zero-dimensional cycles. The system $\{Z_{\bar{u}}\}$ is defined over k, and the cycles $Z_{\bar{u}}$ are of degree mN^r, where m is the degree of the hypersurface V_1 (since N^r is the order of the curve $\Gamma_{\bar{u}}$). The general member of this system is the intersection cycle Z_u of V_1 with the general member Γ_u of Σ.

PROPOSITION 6.1. *The general cycle Z_u of the system $\{Z_{\bar{u}}\}$ consists of the point P_1 counted n times and of $mN^r - n$ distinct* (*simple*) *points of V_1* (*all counted to multiplicity* 1).

Proof. All the curves $\Gamma_{\bar{u}}$ pass through the origin P_1: $X_1 = X_2 = \cdots = X_r = 0$, and the tangent of the general curve Γ_u at P_1 is the line

$$\sum_{j=1}^{r+1} u_j^{(i)} X_j = 0, \qquad i = 1, 2, \ldots, r.$$

This line obviously does not lie on the tangent hypercone $f_n(X_1, X_2, \ldots, X_{r+1}) = 0$ of V_1 at P_1 (here f_n is the leading form of the polynomial f). Thus P_1 occurs in the general cycle Z_u to multiplicity exactly equal to n.

To prove the proposition it is sufficient to exhibit a particular cycle in the system $\{Z_{\bar{u}}\}$ which consists of the point P_1 counted n times and of $mN^r - n$ distinct (simple) points of V_1, all counted to multiplicity 1. We take for each of the polynomials $\varphi_{i,(\bar{u})}$ the product of N *linear generic* polynomials $\varphi_i^{(\nu)}$ in $X_1, X_2, \ldots, X_{r+1}$ ($\nu = 1, 2, \ldots, N$) subject to the only condition that the hyperplane $H_i^{(1)}$: $\varphi_i^{(1)} = 0$ pass through P_1. With such a choice of the parameter $\bar{u}$ we find that the corresponding curve $\Gamma_{\bar{u}}$ consists of one *generic* line of $\mathbf{A}_{r+1}$ through P_1 and of $N^r - 1$ *generic* lines of $\mathbf{A}_{r+1}$. This proves our proposition.

COROLLARY 6.2. *For almost all k-rational specializations $\{u\} \to \{\bar{u}\} \subset k$ the cycle $Z_{\bar{u}}$ consists of the point P_1 counted n times and of $mN^r - n$ distinct* (*simple*) *points of V_1* (*all counted to multiplicity* 1).

Obvious.

We now consider the r elements $x_{i,(u)}^*$ introduced in (4′), which define the *generic* projection π_u of degree N. Each of the $r+1$ elements $x_1, x_2, \ldots, x_r, x_{r+1}$ is algebraic over the field $K^* = k(u)(x_{1,(u)}^*, \ldots, x_{r,(u)}^*)$, and by Proposition 6.1 the field $K^*(x_1, x_2, \ldots, x_r, x_{r+1})$ is of relative degree mN^r over K^*. It can be easily verified, using the proof of Proposition 6.1, that the $mN^r - n$ points of the general cycle Z_u, different from P_1, have, for each $i = 1, 2, \ldots, r, r+1$, distinct X_i-coördinates. Let us consider the minimal polynomial $f_{(u)}(x_{1,(u)}^*, x_{2,(u)}^*, \ldots, x_{r,(u)}^*; X_{r+1})$ of X_{r+1} over K^*. Here $f_{(u)}(X_1^*, X_2^*, \ldots, X_r^*; X_{r+1})$ is a polynomial in $X_1^*, X_2^*, \ldots, X_r^*, X_{r+1}$ of degree mN^r in X_{r+1} and with coefficients in the polynomial ring $k[\{u\}]$. (Note that x_{r+1} is integrally dependent on $k^*[x_{1,(u)}^*, x_{2,(u)}^*, \ldots, x_{r,(u)}^*]$). We denote by $a(u)$ the coefficient of $X_{r+1}{}^{mN^r}$ in $f_{(u)}$. Since 0 is an n-fold root of the polynomial $f_{(u)}(0, 0, \ldots, 0; X_{r+1})$, it follows from the Weierstrass preparation theorem that we have an identity of the form

$$\begin{aligned} &f_{(u)}(X_1^*, \ldots, X_r^*; X_{r+1}) \\ &= F_{(u)}(X_1^*, \ldots, X_r^*; X_{r+1})G_{(u)}(X_1^*, \ldots, X_r^*; X_{r+1}), \quad (7) \end{aligned}$$

where $F_{(u)}$ is a monic polynomial in X_{r+1}, of degree n, with coefficients in the power series ring $k[\{u\}]_{(a(u))}[[X_1^*, \ldots, X_r^*]]$, while $G_{(u)}$ is a power series in $X_1^*, \ldots, X_r^*, X_{r+1}$ with coefficients in $k[\{u\}]_{(a(u))}$, and we have $G_{(u)}(0, \ldots, 0; 0) \neq 0$. Here $k[\{u\}]_{(a(u))}$ denotes the ring of quotients of the polynomial ring $k[\{u\}]$ with respect to the multiplicative system consisting of the non-negative powers of $a(u)$.

The algebroid hypersurface $F_{(u)}(X_1^*, X_2^*, \ldots, X_r^*; X_{r+1}) = 0$ is the original algebroid variety $\hat{V}_P(= V_{1,P_1})$ with $k(\{u\})$ as new (extended) ground field, while the equation $f_{(u)}(X_1^*, \ldots, X_r^*; X_{r+1}) = 0$ defines an algebraic hypersurface $V_{(u)}^*$ in $\mathbf{A}_{r+1}/k(\{u\})$ which is a rational transform of $V_1/k(\{u\})$ (and hence also of $V/k(\{u\})$) and which has the property that if P^* is the origin $X_1^* = X_2^* = \cdots = X_r^* = X_{r+1} = 0$ then $V/k(\{u\})$ and $V_{(u)}^*/k(\{u\})$ are analytically equivalent at P and P^* respectively.

For any specialization $\{u\} \to \{\bar{u}\} \subset k$ such that $a(\bar{u}) \neq 0$ the specialized polynomial $f_{(\bar{u})}$ and the specialized power series $F_{(\bar{u})}$ and $G_{(\bar{u})}$ are well defined, and for almost all such specializations the equation $F_{(\bar{u})} = 0$ defines our original algebroid variety $\hat{V}_P(= \hat{V}_{1,P_1})$, while the equation $f_{(\bar{u})} = 0$ defines a rational transform $V_{(\bar{u})}^*$ of V/k, and the two algebraic varieties $V_{(u)}^*$ and V are analytically equivalent at P.

Let $\Delta_{(\bar{u})}$ and $\hat{\Delta}_{(\bar{u})}$ denote respectively the critical hypersurfaces of the $\pi_{\bar{u}}$-projection of the algebraic hypersurface $V_{(\bar{u})}{}^*$ and of the algebroid hypersurface $\hat{V}_P(= \hat{V}_{1,P_1})$. The two hypersurfaces $\Delta_{(\bar{u})}$ and $\hat{\Delta}_{(\bar{u})}$ are therefore the *reduced* hypersurfaces which are defined by the equations $D_{X_{r+1}}(f_{(\bar{u})}) = 0$ and $D_{X_{r+1}}(F_{(\bar{u})}) = 0$ respectively, where $D_{X_{r+1}}(f_{(\bar{u})})$ and $D_{X_{r+1}}(F_{(\bar{u})})$ are the X_{r+1}-discriminants of $f_{(\bar{u})}$ and $F_{(\bar{u})}$ respectively.

PROPOSITION 6.3. *For almost all specializations $\{u\} \to \{u\} \subset k$ the algebraic critical hypersurface $\Delta_{(\bar{u})}$ coincides, locally at the origin $P_0{}^*$ of the image space $\mathbf{A}_r{}^*$ of $\pi_{\bar{u}}$, with the algebroid hypersurface $\hat{\Delta}_{(\bar{u})}$.*

Proof. Let $\bar{x}_{r+1}{}^{(1)}, \bar{x}_{r+1}{}^{(2)}, \ldots, \bar{x}_{r+1}{}^{(mN^r)}$ be the branches of the mN^r-valued algebraic function $\bar{x}_{r+1}$ of $X_1{}^*, X_2{}^*, \ldots, X_r{}^*$ defined by the equation $f_{(\bar{u})}(X_1{}^*, X_2{}^*, \ldots, X_r{}^*; X_{r+1}) = 0$. We consider the values of these branches at the origin $P_0{}^*$. By Corollary 6.2 we know that for almost all specializations $(u) \to (\bar{u}) \subset k$, n and only n of these branches, say $\bar{x}_{r+1}{}^{(1)}, \bar{x}_{r+1}{}^{(2)}, \ldots, \bar{x}_{r+1}{}^{(n)}$, have value zero, while the remaining $mN^r - n$ branches have values which are distinct from each other and different from zero. Since we have

$$D_{X_{r+1}}(f_{(\bar{u})}) = \prod_{1 \le i < j \le n} (\bar{x}_{r+1}{}^{(i)} - \bar{x}_{r+1}{}^{(j)})^2$$

$$\cdot \left(\prod_{\nu=1}^{mN^r} \cdot \prod_{\mu(>\nu)=n+1}^{mN^r} (\bar{x}_{r+1}{}^{(\nu)} - \bar{x}_{r+1}{}^{(\mu)}) \right)^2,$$

where the second factor (the double product) is different from zero at the origin P^*, while the first factor is equal to $D_{X_{r+1}}(F_{(\bar{u})})$, the proposition is proved.

We are now in a position to answer affirmatively the questions raised in the beginning of this section.

The first question concerns the definition of the dimensionality type of a *singular* point R of V which is not k-rational. We consider the irreducible algebroid subvariety W/k of V having R as general point. We consider on W any two k-rational points P, P', such that W is analytically irreducible at P and also at P' (in particular, we can take for P and P' any two simple points of W). *We have to prove that*

$$d.t.(\hat{V}_P, R)/k = d.t.(\hat{V}_{P'}, R)/k. \tag{8}$$

To prove this we make use of part (a) of Proposition 5.3 and of Proposition 6.3 just proved above. By Proposition 5.3 almost all k-rational projections $\pi_{\bar{u}}$ of $\hat{V}_P$ are quasi-permissible at P *and* at R. Similarly, almost all k-rational projections $\pi_{\bar{u}}'$ of $\bar{V}_{P'}$ are quasi-permissible at P' *and* at R. It follows that for all sufficiently high integers N almost all k-rational projections $\pi_{\bar{u}}$ of $\hat{V}_P$, of degree N, are quasi-permissible at P and R, while also almost all k-rational projections $\pi_{\bar{u}}'$ of $\hat{V}_{P'}$, of the same degree N, are quasi-permissible at P' and R. We fix an affine model of our algebraic variety V such that both P and P' belong to that model. We continue to denote by V that affine model.

Let $k[z_1, z_2, \ldots, z_M]$ be the coördinate ring of V; here M is the dimension of ambient affine space of V. Let $a_1, a_2, \ldots, a_M$ and $a_1', a_2', \ldots, a_M'$ be respectively the coördinates of P and P'. The M elements $z_1 - a_1, z_2 - a_2, \ldots, z_M - a_M$ generate the maximal ideal $\mathfrak{m}$ of the local ring $\mathfrak{o}$ of V at P; and, similarly, $z_1 - a_1', z_2 - a_2', \ldots, z_M - a_M'$ generate the maximal ideal $\mathfrak{m}'$ of the local ring $\mathfrak{o}'$ of V at P'. For almost all sets of $M(r+1)$ elements c_{ij} in k ($i = 1, 2, \ldots, r+1$; $j = 1, 2, \ldots, M$) the $r+1$ elements

$$x_i = \sum_{j=1}^{M} c_{ij}(z_j - a_j), \qquad (i = 1, 2, \ldots, r+1) \tag{9}$$

form a minimal basis of $\mathfrak{m}$, and the $r+1$ elements

$$x_i' = \sum_{j=1}^{M} c_{ij}(z_j - a_j'), \qquad (i = 1, 2, \ldots, r+1) \tag{9$'$}$$

form a minimal basis of $\mathfrak{m}'$. We have

$$x_i - x_i' = b_i, \qquad (i = 1, 2, \ldots, r+1) \tag{10}$$

where

$$b_i = \sum_{j=1}^{M} c_{ij}(a_j' - a_j). \qquad (i = 1, 2, \ldots, r+1)$$

As in (3) we introduce the (reduced) hypersurface V_1 in $\mathbf{A}_{r+1}$ such that the element $x_1, x_2, \ldots, x_{r+1}$ in (9) are the f-residues of $X_1, X_2, \ldots, X_{r+1}$. The origin P_1 belongs to V_1. Since $f(x_1, x_2, \ldots, x_{r+1}) = 0$, it follows from (10) that $f(x_1' + b_1, x_2' + b_2, \ldots, x_{r+1}' + b_{r+1}) = 0$.

Passing to the $\mathfrak{m}'$-residues we conclude that $f(b_1, b_2, \ldots, b_{r+1}) = 0$, i.e., that the point $P_1'(b_1, b_2, \ldots, b_{r+1})$ also belongs to V_1. It is clear that the birational transformation between V and V_1 is biregular at the corresponding points P, P_1 and is also biregular at the corresponding points P', P_1'. Since P (as well as P') is a specialization of R over k, it follows that the birational transformation of V onto V_1 is also biregular at R. To prove (8) it is therefore permissible to replace in (8) the variety V by the variety V_1, and the points P, P', R be the corresponding points P_1, P_1' and R_1 of V_1. We shall therefore identify from now on our original variety V with the hypersurface V_1, and we shall therefore denote by V the hypersurface defined by (3) and by P, P', R the points P_1, P_1', R_1.

We now fix an integer N sufficiently high so that the following be true: *almost all k-rational projections $\pi_{\bar{u}}$ of V, of degree N, are quasi-permissible at P, P' and R.* In terms of the coördinates $X_1, X_2, \ldots, X_{r+1}$ which are zero at P the equations of $\pi_{\bar{u}}$ are: $\pi_{\bar{u}}(\{X\}) = \{X^*\}$, where $\{X\} = (X_1, X_2, \ldots, X_{r+1})$, $\{X^*\} = (X_1^*, X_2^*, \ldots, X_r^*)$ and (see (4)):

$$\pi_{\bar{u}}: X_i^* = \varphi_{i,(\bar{u})}(X_1, X_2, \ldots, X_{r+1}). \qquad (i = 1, 2, \ldots, r) \tag{11}$$

If we use the coördinates $X_1', X_2', \ldots, X_{r+1}'$ which are zero at P' (where $X_i' = X_i - b_i$; see (10)) we must also make a change of coördinates $\{X^*\} \to \{X'^*\}$ of the image space $\mathbf{A}_r$ of $\pi_{\bar{u}}$, namely $X_i'^* = X_i^* - \varphi_{i,(\bar{u})}(b_1, b_2, \ldots, b_{r+1})$, $i = 1, 2, \ldots, r$. Hence in terms of the new coördinate systems in $\mathbf{A}_{r+1}$ and $\mathbf{A}_r^*$, the equations of $\pi_{\bar{u}}$ will be as follows:

$$\pi_{\bar{u}} = \pi_{\bar{v}}': X_i'^* = \varphi_{i,(\bar{v})}(X_1', X_2', \ldots, X_{r+1}'), \qquad (i = 1, 2, \ldots, r) \tag{11'}$$

where

$$\varphi_{i,(\bar{v})}(X_1', X_2', \ldots, X_{r+1}') = \varphi_{i,(u)}(X_1' + b_1, X_2' + b_2, \ldots, X_{r+1}' + b_{r+1}) - \varphi_{i,(u)}(b_1, b_2, \ldots, b_{r+1}).$$

It is clear that for almost all k-rational specializations $(u) \to (\bar{u})$ also the specialization $(v) \to (\bar{v})$ will yield a projection $\pi_{\bar{v}}'$ (locally at P') which is quasi-permissible at P'. It is in this sense that we have used the expression "$\pi_{\bar{u}}$ is quasi-permissible at P *and* P'". When we say that $\pi_{\bar{u}}$ is also quasi-permissible at the point R, we actually mean to imply that $\pi_{\bar{u}}$ is quasi-permissible at R when R is considered as a point of $\hat{V}_P$ and that $\pi_{\bar{v}}'$ is quasi-permissible at R when R is considered as a point of $\hat{V}_{P'}$.

We now consider the *algebraic* critical hypersurface $\Delta_{(\bar{u})}(= \Delta_{(\bar{v})})$ of the (global) projection $\pi_{\bar{u}}(= \pi_{\bar{v}})$ of the algebraic hypersurface V. By Proposition 6.3 we can assume that the specialization $(u) \to (\bar{u}) \subset k$ has also the following two properties: $\Delta_{(\bar{u})}$ coincides, locally at the point $P_0 = \pi_u(P)$, with the critical algebroid hypersurface $\hat{\Delta}_{(\bar{u})}$ of the $\pi_{(\bar{u})}$-projection of $\hat{V}_P$, and coincides, locally at the point $P_0' = \pi_{\bar{v}}(P')$, with the critical algebroid hypersurface $\hat{\Delta}_{(\bar{v})}$ of the $\pi_{(\bar{v})}$-projection of $\hat{V}_{P'}$:

$$\widehat{\Delta_{(\bar{u}),P_0}} = \hat{\Delta}_{(\bar{u})}, \qquad (P_0 = \pi_{(\bar{u})}(P)); \tag{12}$$

$$\widehat{\Delta_{(\bar{v}),P_0'}}(= \widehat{\Delta_{(\bar{u}),P_0'}}) = \hat{\Delta}_{(\bar{v})}, \qquad (P_0' = \pi_{\bar{v}}(P')). \tag{12'}$$

We set

$$R_0 = \pi_{(\bar{u})}(R)(= \pi_{\bar{v}}(R)).$$

By definition of the integers $d.t.(\hat{V}_P, R)$, $d.t.(\hat{V}_{P'}, R)$ we have

$$d.t.(\hat{V}_P, R) = d.t.\widehat{(\Delta_{(\bar{u})}, R_0)} + 1, \tag{13}$$

$$d.t.(\hat{V}_{P'}, R) = d.t.\widehat{(\Delta_{(\bar{v})}, R_0)} + 1. \tag{13'}$$

Hence, by (12) and (12′), we have

$$d.t.(\hat{V}_P, R) = d.t.(\widehat{\Delta_{(\bar{u}),P_0}}, R_0) + 1, \tag{14}$$

$$d.t.(\hat{V}_{P'}, R) = d.t.(\widehat{\Delta_{(\bar{u}),P_0'}}, R_0) + 1. \tag{14'}$$

Now, $\Delta_{(\bar{u})}$ is an *algebraic* hypersurface, and the points P_0, P_0' and R_0 belong to $\Delta_{(\bar{u})}$ (R_0 belongs to $\Delta_{(\bar{u})}$ since R was assumed to be a singular point of V), and both P_0 and P_0' are specializations of R_0, over k. We can assume, by induction on r, that (8) is true if V is replaced by the $(r-1)$-dimensional hypersurface $\Delta_{(\bar{u})}$ and P, P', R are replaced by P_0, P_0', R_0. For, if $r = 1$, then (8) is trivial (any point R of an algebraic curve, which is not k-rational, is obviously a simple point of the curve). We have therefore the equality

$$d.t.(\widehat{\Delta_{(\bar{u}),P_0}}, R_0) = d.t.(\widehat{\Delta_{(\bar{u}),P_0'}}, R_0).$$

From this the equality (8) follows in view of (14) and (14′).

For the sake of completeness we shall deal also with the case in which W is analytically reducible at one of the points P, P', say at P, while being analytically irreducible at P'. Let $\hat{W}_1, \hat{W}_2, \ldots, \hat{W}_\rho$ be the irreducible analytical components of W at P, and let R_ν be the general point of $\hat{W}_\nu$ ($\nu = 1, 2, \ldots, \rho$). The ρ points R_ν are distinct points of $\hat{V}_P$. We shall continue to denote by R the unique point of $\hat{V}_{P'}$ which is associated with the general point R of W. We fix our k-rational projection $\pi_{\bar{u}}$ of sufficiently high degree N in such a manner that it be quasi-permissible at P and at each point R_ν of $\hat{V}_P$ and that the resulting k-rational projection $\pi_{(\bar{v})}$ of $\hat{V}_{P'}$ be quasi-permissible at P' and R. Furthermore, we assume that the condition of Proposition 6.3 is satisfied both at P and P', or—equivalently—that $\pi_{(\bar{u})}$ and $\pi_{(\bar{v})}$ satisfy (12) and (12′) respectively.

The $\pi_{\bar{u}}$-transform W_0 of W will be a reducible algebroid subvariety of $\widehat{\Delta_{(\bar{u}),P_0}}$ of the same dimension as W, and its ρ irreducible components $W_{0,1}, W_{0,2}, \ldots, W_{0,\rho}$ have $R_{0,1}, R_{0,2}, \ldots, R_{0,\rho}$ as general points, where $R_{0,\nu} = \pi_{(\bar{u})}(R_\nu)$. Then (13′) is satisfied, and also (13) is satisfied if R is replaced by any of the ρ points R_ν and R_0 is replaced by the corresponding point $R_{0,\nu}$. Note that since R is a singular point, the points $R_{0,\nu}$ belong to $\Delta_{(\bar{u}),P_0}$. By induction on r, we have for each $\nu = 1, 2, \ldots, \rho$:

$$d.t.(\widehat{\Delta_{(\bar{u}),P_0}}, R_{0,\nu}) = d.t.(\widehat{\Delta_{(\bar{u}),P_0'}}, R_0),$$

and from these equalities follow the desired equalities which generalize (8), namely

$$d.t.(\hat{V}_P, R_\nu) = d.t.(\hat{V}_{P'}, R), \qquad \nu = 1, 2, \ldots, \rho.$$

We shall now deal with the last question raised in the beginning of this section, namely we shall prove the following

THEOREM 6.4. *If V is our original algebraic variety and P is a k-rational point of V, the algebroid variety $W(P) = \mathrm{EQS}_k(\hat{V}_P, P)$ is algebraic, irreducible, of dimension $r - s$, and P is a simple point of $W(P)$. Furthermore, V is equimultiple at P along $W(P)$. Here s is the dimensionality type of P on V.*

Proof. We recall that we have proved in section 5 (see last statement of Theorem 5.4) that the subset $\mathrm{EQS}_k(\hat{V}_P, P)$ of $\hat{V}_P$ is a (closed) algebroid subvariety of $\hat{V}_P$. For the sake of clarity we shall make precise the meaning we attach to a statement that *an algebroid variety W, embedded in an affine space $\mathbf{A}_M$ and defined over k, is algebraic.* Let $X_1, X_2, \ldots, X_M$ be affine coördinates in $\mathbf{A}_M$ and let P: $X_1 = X_2 = \cdots = X_M = 0$ be the origin of W. We say that W is *algebraic* if W is the variety of some ideal $\mathfrak{A}$ in the power series ring $k[[X_1, X_2, \ldots, X_M]]$ such that $\mathfrak{A}$ has a basis consisting of *polynomials* in $X_1, X_2, \ldots, X_M$ (with coefficients in k).

We now consider the hypersurface V_1 in $\mathbf{A}_{r+1}$, defined by the equation (3). We know that the local ring of V_1 at the origin P_1: $X_1 = X_2 = \cdots = X_{r+1} = 0$ coincides with the local ring of V at P. We now consider a k-rational projection $\pi_{(\bar{u})}$ of V_1 onto $\mathbf{A}_r$, of a sufficiently high degree N, such that $\pi_{(\bar{u})}$ is P-permissible (and not only quasi-permissible). Then, by Proposition 5.3, we have that $W(P) = \pi_{(\bar{u})}^{-1}(W_0(P_0))$, where P_0 is the origin $X_1^* = X_2^* = \cdots = X_r^* = 0$ of the image space $\mathbf{A}_r^*$ of $\pi_{(\bar{u})}$ and where $W_0(P_0) = \mathrm{EQS}_k(\hat{\Delta}_{(\bar{u})}, P_0)$. We can now use induction on r, since the theorem is trivial in the case $r = 1$ (in the case $r = 1$, $W(P)$ is either the point P, if P is a singular point of the curve V, and $W(P) = V$ is P is a simple point of V). We can therefore assume that $W_0(P_0)$ is an algebraic variety, since by Proposition 6.3 $\hat{\Delta}_{(\bar{u})}$ is an algebraic hypersurface $\Delta_{\bar{u}}$. Therefore $W_0(P_0)$ can be defined by a set of *polynomial* equations $\psi_\nu(X^*) = 0$ in $X_1^*, X_2^*, \ldots, X_r^*$. Since $W(P) = \pi_{\bar{u}}^{-1}(W_0(P_0))$, it follows that $W(P)$ is the variety defined by the equation $\psi_\nu(X^*) = 0$ and $f(X_1, X_2, \ldots, X_{r+1}) = 0$. Since the X^*'s are polynomials in $X_1, X_2, \ldots, X_{r+1}$, and since the birational transformation of V onto V_1 is given by the equations $X_i = \sum_{j=1}^{M} c_{ij}Z_j$ (where we assume that P is the origin in affine space $\mathbf{A}_M$ in which V is embedded), it follows that $W(P)$ (which we now regard as being embedded in $\mathbf{A}_M$) can be defined by polynomial equations in $Z_1, Z_2, \ldots, Z_M$, and hence is algebraic.

The rest of the theorem is merely a repetition of Theorem 5.4, with the closed point P replacing the point Q.

Our main purpose in the remainder of this section is to introduce the *equisingularity strata* on V and prove that the resulting covering of V by these strata enjoys the usual properties of a stratification. Given any point P of V, of dimensionality type s, we consider the irreducible subvariety $W(P)$ (see Theorem 6.4). The general point R of $W(P)$ also has dimensionality type s (since $W(P) = \text{EQS}(V_P, P)$ and has dimension $r - s$). The first thing that has to be proved is that the set $W(P) \cap V(s)$, i.e., the set of points of $W(P)$ which have the same dimensionality type s as P and R is a Zariski dense open subset of $W(P)$, or—equivalently—that the set of points of $W(P)$ which have dimensionality type $>s$ is a *proper* algebraic subvariety of $W(P)$. This essential and important result is proved by H. Hironaka in the paper [5] cited by us in the Introduction.

We use the result of Hironaka and we denote by $S(P)$ the set of points of $W(P)$ which have dimensionality type s. By Theorem 5.1, the point P is a simple point of $W(P)$. If P' is any other k-rational point contained in $S(P)$, then P' is also a specialization of R, and therefore $W(P)$ coincided with $W(P')$, whence $S(P) = S(P')$. We see therefore that all the points of $S(P)$ are simple points of $W(P)$. We call $S(P)$ the *equisingularity stratum* containing the point P. Thus each equisingularity stratum S, of dimensionality type s, is a smooth Zariski dense open subset of the $(r - s)$-dimensional irreducible subvariety $\overline{S}$ of V ($\overline{S} = W(P)$, for any point P in S). Any two equisingularity strata are obviously disjoint in view of the fact just proved, namely that if $P' \in S(P)$ then $S(P') = S(P)$.

We now prove that *the number of equisingularity strata of given dimensionality type s is finite.* We know that this is trivially true for the set of strata of dimensionality type zero; and also of dimensionality 1 (see our paper [1], SES II). We shall use induction from s to $s + 1$. We use the notations $V(s)$ and $V\{s\}$ introduced in section 5, where now V is our *algebraic* variety (and not an algebroid variety). Let Σ_s denote the (finite) set of equisingularity strata of dimensionality type $\leq s$. We have then

$$V\{s + 1\} = V_{sing} - \bigcup_{S_i \in \Sigma_s} S_i.$$

It is easily seen that $V\{s + 1\}$ is an algebraic subvariety of V. For, if R is any point of $V\{s + 1\}$ then by Proposition 4.2 every specialization of R over k belongs to $V\{s + 1\}$. On the other hand, if $R \notin V\{s + 1\}$, then $d.t.(V, R) = \sigma \leq s$, and hence, by Hironaka's result, almost all specializations of R over k have dimensionality type σ and therefore do not belong to $V\{s + 1\}$. This proves the algebraic character of the set $V\{s + 1\}$. No point R of $V\{s + 1\}$ can have dimension (over k) $\geq r - s$, for in the contrary case we would have necessarily $d.t.(V, R) \leq s$. Therefore the variety $V\{s + 1\}$ has dimension $\leq r - s - 1$. We may assume that $V(s + 1) \neq \emptyset$ (in the contrary case, we would have $V\{s + 1\} = V\{s + 2\}$, and we could replace in our induction hypothesis the integer s by the integer $s + 1$). From the assumption that $V(s + 1) \neq \emptyset$ follows that $V\{s + 1\}$ must have dimension exactly equal to $r - s - 1$ (and not *less* than $r - s - 1$), for $V(s + 1) \neq \emptyset$ implies the existence of equisingularity strata S of dimensionality type $s + 1$, and any such stratum S has dimension $r - s - 1$, and the closure $\overline{S}$ of S must be an irreducible component of $V\{s + 1\}$. Conversely, every irreducible component W of $V\{s + 1\}$, of dimension $r - s - 1$, gives rise to a stratum S of dimensionality type $s + 1$; it is the stratum containing the general point R of W, for the inequality $d.t.(V, R) \geq s + 1$ together with the equality $dim\ R/k = r - s - 1$ imply that $d.t.(V, R) = s + 1$. We conclude therefore that the number of strata of dimensionality type $s + 1$ is equal to the number of $(r - s - 1)$-dimensional components of the variety $V\{s + 1\}$, and is therefore finite. This completes our inductive proof of the statement that the set of equisingularity strata of V is finite.

We shall conclude this section (and the paper) by proving that *the boundary of any equisingularity stratum is a union of strata.*

Let S be an equisingularity stratum, of dimensionality type s, let R be the general point of the closure $\overline{S}/k$ of S, and let P' be a k-rational point of the boundary $\overline{S} - S$ of S. We know then that if $d.t.(V, P') = s'$ then $s' > s$. Let $S' = S(P')$ be the equisingularity stratum containing P' and let R' be the general point of $\overline{S'}/k$. What we have to show is that $R' \in \overline{S} - S$, because if that is shown then it will follow that $S' \subset \overline{S} - S$. Let $\hat{V}_{P'}$ be the *algebroid* r-dimensional variety which is defined by V, locally at the k-rational point P', i.e., let $\hat{V}_{P'} = Spec\ (\mathfrak{o}')$, where $\mathfrak{o}'$ is the completion of the local ring of V at P'. We set $W = \overline{S}$ and $W' = \overline{S'}$. The irreducible algebraic variety W', of dimension $r - s'$, is analytically irreducible at P' since P' is a simple point of W'. Therefore the general point R' of W'/k defines a unique point of $\hat{V}_{P'}$. We shall continue to denote that point by R'. Since P' is also a specialization of point R, over k, the point R is associated with a finite number of points $\hat{R}_1, \hat{R}_2, \ldots, \hat{R}_g$ of $\hat{V}_{P'}$, where g is the number of irreducible analytical branches of W at P'. Consider any of these g points $\hat{R}_i$ and denote it by $\hat{R}$. The stratum S having R as general point yields a well-defined stratum $\hat{S}$ of $\hat{V}_{P'}$ having $\hat{R}$ as general point, for we know that $d.t.(\hat{V}_{P'}, \hat{R}) = d.t.(V, R) = s$ (by the definition of the dimensionality type of R given earlier in this section) and we know also that the (analytic) dimension of the point $\hat{R}$ is equal to the (algebraic) dimension of the point R, i.e., is equal to $r - s$. Since $P' \in \overline{S}$, it follows that $P' \in \overline{\hat{S}}$. On the other hand we have $P' \notin \hat{S}$ since $d.t.(V, P') = s' > s = d.t.(V, R) = d.t.(\hat{V}_{P'}, \hat{R})$. Hence P' belongs to the *boundary* of $\hat{S}$. Since the *closed* stratum $\hat{S}'$ of $\hat{V}_{P'}$ (i.e., the stratum of $\hat{V}_{P'}$ containing the closed point P' of $\hat{V}_{P'}$; see Theorem 5.1, part (*c*)) contains the point R', it follows from Theorem 5.1, part (*d*), that $R' \in \overline{\hat{S}}$. We have found therefore that R' is an (analytic) specialization of the general point $\hat{R}$ of $\hat{S}$, and therefore R' is also an (algebraic) specialization of R. This completes the proof of the stratification property stated above.

Appendix

A generalization of a theorem on equimultiple subvarieties of an algebroid hypersurface (proved in [2]).

Our PNAS note [2] deals only with the special case in which the point Q is the closed point of V. This special case is the only case which needs to be considered in our theory of equisingularity of *algebraic* varieties (see section 6). Although our primary purpose in dealing with algebroid varieties is to prepare the ground for an *analytically intrinsic* theory of equisingularity of algebraic varieties, we nevertheless shall give here the generalization of the proof given in the PNAS note to the general case in which Q is an arbitrary point of V. This generalization consists essentially in replacing V by its localization $\tilde{V}_Q$ of V at Q; here $\tilde{V}_Q = \text{Spec}\,(\hat{\mathfrak{o}}_{\mathfrak{p}})$, where $\hat{\mathfrak{o}}_{\mathfrak{p}}$ is the completion of the (local) ring of quotients $\mathfrak{o}_{\mathfrak{p}}$ of $\mathfrak{o}$ with respect to the prime ideal $\mathfrak{p}$ associated with the point Q. The point Q is now the *closed* point of $\tilde{V}_Q$ and we shall show that the statement used in the text in the case of the pair $\{V, Q\}$ can be derived from our PNAS note, as applied to the closed point Q of $\tilde{V}_Q$.

Our k-rational projection π is defined by a set of parameters $x_1, x_2, \ldots, x_r$ belonging to a minimal basis $\{x_1, x_2, \ldots, x_r, x_{r+1}\}$ of the maximal ideal $\mathfrak{m}$ of $\mathfrak{o}$. Since x_{r+1} is integral over the (regular) local domain $k[[x_1, x_2, \ldots, x_r]]$, it follows that if $\xi_1, \xi_2, \ldots, \xi_r, \xi_{r+1}$ are the coördinates of Q (i.e., the $\mathfrak{p}$-residues of $x_1, x_2, \ldots, x_r, x_{r+1}$), then ξ_{r+1} is integral over the integral domain $k[[\xi_1, \xi_2, \ldots, \xi_r]]$. If ρ is the (analytic) dimension of Q over k, then $\tilde{V}_Q$ is an algebroid variety of dimension ρ defined over the field of fraction $K = k((\xi_1, \xi_2, \ldots, \xi_{r+1}))$ of the integral domain $\mathfrak{o}/\mathfrak{p}$ (K is a coefficient field of $\hat{\mathfrak{o}}_\mathfrak{p}$). Necessarily ρ of the r elements $\xi_1, \xi_2, \ldots, \xi_r$, say $\xi_1, \xi_2, \ldots, \xi_\rho$, are analytically independent over k, while K is an algebraic extension of the field $k((\xi_1, \xi_2, \ldots, \xi_\rho))$ of meromorphic functions of $\xi_1, \xi_2, \ldots, \xi_\rho$ (we could, if necessary, identify $\xi_1, \xi_2, \ldots, \xi_\rho$ with $x_1, x_2, \ldots, x_\rho$).

The equation of $\tilde{V}_Q$ is given by

$$\tilde{V}_Q : \tilde{f}(Y_1, Y_2, \ldots, Y_{r-\rho+1}) = 0, \tag{1}$$

where

$$\tilde{f}(Y_1, Y_2, \ldots, Y_{r-\rho+1}) = f(\xi_1, \xi_2, \ldots, \xi_\rho, Y_1 + \xi_{\rho+1}, \ldots, Y_{r-\rho+1} + \xi_{r+1}),$$

and the maximal ideal of $\hat{\mathfrak{o}}_\mathfrak{p}$ has the minimal basis $y_1, y_2, \ldots, y_{r-\rho+1}$, where $y_i = h(x_{\rho+i}) - \xi_{\rho+i}$, h being the canonical homomorphism of $\mathfrak{o}$ into $\mathfrak{o}_\mathfrak{p}$. Thus, $\tilde{f}(Y)$ is a power series in $Y_1, Y_2, \ldots, Y_{r-\rho+1}$, with coefficients in K, and the Y-coördinates of Q are all zero. The set of points of $\tilde{V}_Q$ consists of those points of V which have Q as a specialization over k. If R is any such point then it is easy to see that the multiplicity $e(V, R)$ of V at R is equal to the multiplicity $e(\tilde{V}_Q, R)$ of $\tilde{V}_Q$ at R:

$$e(\tilde{V}_Q, R) = e(V, R). \tag{2}$$

To see this we first observe that from the very definition of the multiplicity of Q on $\tilde{V}_Q$ (i.e., the multiplicity of the *complete* local ring $\hat{\mathfrak{o}}_\mathfrak{p}$; see C.A.II, p. 214) it follows that the multiplicity $e(\tilde{V}_Q, Q)$ is equal to the multiplicity of Q on V_Q (i.e., to the multiplicity of the local ring $\mathfrak{o}_\mathfrak{p}$). By a similar argument it follows that the multiplicity of R on $\tilde{V}_Q$ is equal to the multiplicity of R on V_Q. Hence in order to prove (2) we have only to show that

$$e(V_Q, R) = e(V, R). \tag{2'}$$

We set $\tilde{\mathfrak{o}} = \mathfrak{o}_{\mathfrak{p}}$. Let $\mathfrak{P}$ be the prime ideal in $\mathfrak{o}$ which is associated with the point R of V. Then $e(V, R)$ is the multiplicity of the local ring $\mathfrak{o}_{\mathfrak{p}}$, while $e(V_Q, R)$ is the multiplicity of the local ring $\tilde{\mathfrak{o}}_{\mathfrak{p}}$, where $\tilde{\mathfrak{P}} = \tilde{\mathfrak{o}}.h(\mathfrak{P})$, h being the canonical homomorphism of $\mathfrak{o}$ into $\tilde{\mathfrak{o}}$. From $\tilde{\mathfrak{P}} = \tilde{\mathfrak{o}}.h(\mathfrak{P})$ follows at once that the local rings $\tilde{\mathfrak{o}}_{\mathfrak{p}}$ and $\mathfrak{o}_{\mathfrak{P}}$ have the same completion and therefore have the same multiplicity. This proves (2′) and hence also (2).

If $D(X_1, X_2, \ldots, X_r)$ denotes the X_{r+1}-discriminant of $f(X_1, X_2, \ldots, X_r, X_{r+1})$ and $\tilde{D}(Y_1, Y_2, \ldots, Y_{r-\rho})$ denotes the $Y_{r-\rho+1}$-discriminant of $\tilde{f}(Y_1, Y_2, \ldots, Y_{r-\rho+1})$, then

$$\tilde{D}(Y_1, Y_2, \ldots, Y_{r-\rho}) = D(\xi_1, \xi_2, \ldots, \xi_\rho, Y_1 + \xi_{\rho+1}, \ldots, Y_{r-\rho} + \xi_{r-\rho}). \quad (3)$$

Our k-rational projection π of V onto $\hat{A}_r/k$ determines a unique projection $\tilde{\pi}$ of $\tilde{V}_Q$ onto $\hat{A}_{r-\rho}/K$ defined by $\tilde{\pi}(\{y_1, y_2, \ldots, y_{r+\rho+1}\}) = \{y_1, y_2, \ldots, y_{r-\rho}\}$. More specifically, if $\tilde{\mathfrak{P}}$ is any prime ideal of $\hat{\mathfrak{o}}_{\mathfrak{p}}(= K[[y_1, y_2, \ldots, y_{r-\rho+1}]]$ and if $\tilde{\mathfrak{P}}_0 = \tilde{\mathfrak{P}} \cap K[[y_1, y_2, \ldots, y_{r-\rho}]]$, then $\tilde{\pi}(R) = R_0$, where R and R_0 are the points of $\tilde{V}_Q$ and of $\hat{A}_{r-\rho}/K$ associated respectively with the prime ideals $\tilde{\mathfrak{P}}$ and $\tilde{\mathfrak{P}}_0$. From this and from (3) it follows therefore that the critical variety $\tilde{\Delta}$ of $\tilde{\pi} \mid \tilde{V}_Q$ is obtained from the critical variety Δ of $\pi \mid V$ by first passing from Δ to the localization $\tilde{\Delta}_{Q_0}$ of Δ at the point Q_0, where $Q_0 = \pi(Q)$, and then by extending the coefficient field $K_0(= k((\xi_1, \xi_2, \ldots, \xi_r))$ of $\tilde{\Delta}_{Q_0}$ to the field K. Since a coefficient field extension does not affect the multiplicity of any point of $\tilde{\Delta}_{Q_0}$, we therefore have, by applying the general result (2) to Δ/k and $\tilde{\Delta}_{Q_0}/K$, the following equality for any point R_0 of $\tilde{\Delta}_{Q_0}/K$:

$$e(\tilde{\Delta}_{Q_0}/K, R_0) = e(\Delta/k, R_0), \qquad (R_0 \in \tilde{\Delta}_{Q_0}). \quad (4)$$

Now, let W_0 be any irreducible subvariety Δ, passing through the point Q_0 (in the text, W_0 is the variety $\overline{Z_0}$). The dimension of W_0 is ρ', where $\rho' \geq \rho$. The case $\rho' = \rho$ being trivial, we may assume that $\rho' > \rho$. The variety W_0 determines an irreducible subvariety $\tilde{W}_0$ of $\tilde{\Delta}_{Q_0}$, of dimension $\rho' - \rho$, and if Q_0 is a simple point of W_0, Q_0 is also a simple point of $\tilde{W}_0$. By assumption (in the text) we have that if R_0 is the general point of W_0/k then $e(\Delta, R_0) = e(\Delta, Q_0)$. Applying (4) to both sides of this equality, where we replace R_0 by Q_0 in applying (4) to the right-hand of the equality, we find that

$$e(\tilde{\Delta}_{Q_0}/K, R_0) = e(\tilde{\Delta}_{Q_0}/K, Q_0). \tag{5}$$

Since Q_0 is the closed point of $\tilde{\Delta}_{Q_0}$ and is a simple point of $\tilde{W}_0$, our theorem of the PNAS note yields the following result:

$\tilde{\pi}^{-1}\{\tilde{W}_0\}$ *has a unique irreducible component* $\tilde{W}$ *at the point* Q, *and* $\tilde{V}_Q$ *is equimultiple at* Q *along* $\tilde{W}$, i.e., *we have*

$$e(\tilde{V}_Q, Q) = e(\tilde{V}_Q, R), \tag{6}$$

where R *is the general point of* $\tilde{W}$, *and clearly* $\tilde{\pi}(R) = R_0$. *Furthermore,* Q *is a simple point of* $\tilde{W}$.

Now $\tilde{W}$ is the Q-localization of a unique irreducible subvariety W of V, and R is also the general point of W. Furthermore, since $\tilde{\pi} = \pi \mid \tilde{V}_Q$, it follows that $\pi(R) = R_0$. Thus W is the irreducible component of $\pi^{-1}\{W_0\}$ which contains the point Q. Since Q is a simple point of $\tilde{W}$, it is also a simple point of W. Applying (2) to (6) we find that $e(V, Q) = e(V, R)$, i.e., V is equimultiple at Q along W. This completes the proof of the generalization of the result established in our PNAS note.

HARVARD UNIVERSITY

REFERENCES

[1] O. Zariski, "Studies in equisingularity II. Equisingularity in codimension 1 (and characteristic zero)," *Amer. J. Math.*, **87** (1965) pp. 972–1006. This paper will be referred to in the text as SEQ II.

[2] ______, "On equimultiple subvarieties of algebroid hypersurfaces," *Proc. Nat. Acad. Sci. U.S.A.*, **72** (1975) pp. 1425–1426.

[3] O. Zariski and P. Samuel, *Commutative Algebra*, v. I(1958) and v. II(1960). These two books will be referred to as C.A.I, and C.A.II respectively.

[4] A. Grothendieck, *Elements de géometrie algébrique*, Chapter IV.

[5] H. Hironaka, *On Zariski dimensionality type.* Amer. J. Math. **101**, No. 2(1979) pp. •••-•••.

Respiration of Amphibious Vertebrates

Respiration of Amphibious Vertebrates

Edited by

G. M. HUGHES

Research Unit for Comparative Animal Respiration
University of Bristol, Woodland Road, Bristol, England

1976

ACADEMIC PRESS
LONDON NEW YORK SAN FRANCISCO
A Subsidiary of Harcourt Brace Jovanovich, Publishers

ACADEMIC PRESS INC. (LONDON) LTD.
24/28 Oval Road,
London NW1

United States Edition published by
ACADEMIC PRESS INC.
111 Fifth Avenue,
New York, New York 10003

Library of Congress Catalog Card Number: 75-19651
ISBN: 0 12 360750-7

Printed in Great Britain by
Willmer Brothers Limited, Birkenhead

Contributors

CRAWFORD, E. C., JR., *Abteilung Physiologie, Max-Planck-Institut für experimentelle Medizin, Göttingen, West Germany*

DEHADRAI, P. V., *Central Inland Fisheries Research Institute, Barrackpore, West Bengal, India*

DEJOURS, P., *Laboratoire de Physiologie Respiratoire, Centre National de la Recherche Scientifique, Strasbourg, France*

FÄNGE, R., *Department of Zoophysiology, University of Göteborg, Göteborg, Sweden.*

GANS, C., *Department of Zoology, The University of Michigan, Ann Arbor, Michigan, U.S.A.*

GATZ, R. N., *Abteilung Physiologie, Max-Planck-Institut für experimentelle Medizin, Göttingen, West Germany*

GRAHAM, J. B., *Smithsonian Tropical Research Institute, Box 2072, Balboa, Canal Zone*

GUIMOND, R. W., *Department of Biology, Boston State College, Boston, Massachusetts, U.S.A.*

HEATWOLE, H., *Department of Zoology, University of New England, Armidale, New South Wales, Australia*

HOWELL, B. J., *Department of Physiology, State University of New York at Buffalo, Buffalo, New York, U.S.A.*

HUGHES, G. M., *Research Unit for Comparative Animal Respiration, Bristol University, Bristol, England.*

HUTCHISON, V. H., *Department of Zoology, University of Oklahoma, Norman, Oklahoma, U.S.A.*

LENFANT, C. J. M., *Division of Lung Diseases, National Heart and Lung Institute, Bethesda, Maryland, U.S.A.*

MUNSHI, J. S. DATTA, *Post-Graduate Department of Zoology, Bhagalpur University, Bihar, India.*

PATTLE, R. E., *Chemical Defence Establishment, Porton Down, Salisbury, Wiltshire, England*

PIIPER, J., *Abteilung Physiologie, Max-Planck-Institut für experimentelle Medizin, Göttingen, West Germany*

RAHN, H., *Department of Physiology, State University of New York at Buffalo, Buffalo, New York, U.S.A.*

SATCHELL, G. H., *Department of Zoology, University of Otago, Dunedin, New Zealand*

SCHEID, P., *Abteilung Physiologie, Max-Planck-Institut für experimentelle Medizin, Göttingen, West Germany*

SEYMOUR, R. S., *Department of Zoology, Monash University, Clayton, Victoria, Australia*
SINGH, B. N., *Department of Zoology, Banaras Hindu University, Varanasi, India*
TRIPATHI, S. D., *Central Inland Fisheries Research Institute, Barrackpore, West Bengal, India*
ULTSCH, G. R., *Department of Biology, University of Alabama, Tuscaloosa, Alabama, U.S.A.*
WEIBEL, E. R., *Anatomical Institute, University of Berne, Berne, Switzerland*
WOOD, S. C., *University of New Mexico, School of Medicine, Albuquerque, New Mexico, U.S.A.*

List of Participants (Registered)

Dr. C. M. Ballintijn (Gröningen, Netherlands)
Sri B. P. Bhagat (Bhagalpur)
Dr. J. N. Cameron (Alaska, U.S.A.)
Sri T. K. Chakaravarty (Monghyr)
Dr. B. J. Choubey (Bhagalpur)
Sri Devendra Choudhury (Begusarai)
Dr. P. V. Dehadrai (Barrackpore)
Professor P. Dejours (Strasbourg, France)
Professor R. Fänge (Göteborg, Sweden)
Md. Fasiuddin (Bhagalpur)
Professor C. Gans (Ann Arbor, U.S.A.)
Dr. J. B. Graham (Balboa, Panama)
Dr. R. W. Guimond (Boston, U.S.A.)
Sri Abdul Hakim (Arrah)
Sri Niz amul Haque (Darbhnaga)
Professor H. Heatwole (Armidale, Australia)
Professor G. M. Hughes (Bristol, England)
Professor V. H. Hutchison (Oklahoma, U.S.A.)
Dr. Kazuko Ishii (Fukushima, Japan)
Dr. Kosei Ishii (Fukushima, Japan)
Dr. A. G. Jha (Bhagalpur)
Dr. A. P. Jha (Bhagalpur)
Sri U. G. Jha (Bhagalpur)
Sri S. C. Joshi (Ujjain)
Dr. G. L. Kooyman (La Jolla, U.S.A.)
Dr. B. K. Mandal (Bhagalpur)
Dr. Ira Mazumdar (Bhagalpur)
Sri Narendra Mishra (Godda)
Dr. A. K. Mittal (Varanasi)
Sri L. K. Moitra (Bhagalpur)
Professor J. S. Datta Munshi (Bhagalpur)
Sri S. A. K. Nasar (Bhagalpur)
Dr. J. Ojha (Bhagalpur)
Dr. B. N. Pandey (Gaya)
Dr. P. K. Pandey (Bhagalpur)
Dr. R. E. Pattle (Salisbury, England)
Professor H. M. Peters (Tübingen, W. Germany)
Professor J. Piiper (Göttingen, W. Germany)

Dr. S. M. Z. Rahman (Bhagalpur)
Professor H. Rahn (Buffalo, U.S.A.)
Sri S. P. Roy (Bhagalpur)
Sri M. P. Saha (Bhagalpur)
Professor G. H. Satchell (Otago, New Zealand)
Dr. P. Scheid (Göttingen, W. Germany)
Dr. R. S. Seymour (Monash, Australia)
Sri G. S. Sharma (Bhagalpur)
Dr. B. N. Singh (Varanasi)
Dr. B. R. Singh (Muzaffarpur)
Sri N. K. Singh (Bhagalpur)
Sri A. K. Sinha (Bhagalpur)
Mrs. Asha Lata Sinha (Bhagalpur)
Sri N. D. P. Sinha (Bhagalpur)
Dr. S. P. Sinha (Bhagalpur)
Dr. V. Sinha (Bhagalpur)
Sri B. K. Tiwary (Patna)
Sri N. P. Tiwary (Bhagalpur)
Dr. G. R. Ultsch (Alabama, U.S.A.)

Preface

This volume contains some of the proceedings of a satellite symposium held at the Post-graduate Department of Zoology, Bhagalpur University, Bihar, India, October 28th—November 1st, 1974 following the XXVI International Congress of Physiological Sciences at New Delhi. It was the second satellite symposium concerned with comparative physiology of respiration, the first having been held at Göttingen in 1971. On this occasion a topic was chosen which was particularly relevant to the local conditions in India and especially with the work done at Professor Munshi's department. Indian air-breathing fishes have long attracted the study of biologists but it is only in recent years that modern physiological and morphometric methods have been applied to them. By choosing a topic broader than air-breathing fish, an attempt was made to place these particular animals in perspective and in relation to current general problems of respiratory physiology. Research has developed quite rapidly in this field during the past ten years and the papers presented summarized results obtained during this period.

Invitations to speakers were given so as to bring together scientists not only with specialist interest in respiratory physiology, but also biologists more concerned with the ecological conditions, morphology, and ecophysiology of vertebrates which can breathe both in water and air. The meeting provided many lively discussions but perhaps the success of the symposium will be best judged by a future assessment of the stimulus that it provided not only to biologists in India but to those in other parts of the world where animals await investigation using comparable techniques.

With such a variety of backgrounds it was inevitable that some problems should arise in communication and at this particular time especially in relation to the symbols and units that were adopted. In editing the volume, I have taken the view that the time is not quite ripe for complete uniformity, although some approach to this has been made. It is hoped, however, that each individual paper is self-consistent. In order to assist readers who are used to different units, equivalents have been provided, sometimes within individual papers, but also in appendices at the beginning of this volume.

I should like to thank all those people who helped in the conception and planning of the symposium and all who gave their time and energies most willingly at the time of the meeting in Bhagalpur. In particular, however, I thank Johannes Piiper and Hermann Rahn for their constant support during the planning stages. At Bhagalpur, Professor Munshi performed many miracles by completing arrangements for the first ever international symposium to be held in that city and I know that he has felt most strongly a great indebtedness to his Vice-Chancellor and the Deputy Minister of Education for their unfailing support. To all our Indian hosts, the visitors from overseas will long feel indebted for the great efforts they made and which contributed so greatly to the success of a meeting that will always be remembered.

GEORGE M. HUGHES
Bristol, 4th November 1975

Contents

Introduction

INAUGURAL SPEECH OF SHRI D. P. YADAV, UNION DEPUTY MINISTER FOR EDUCATION AND SOCIAL WELFARE AND CULTURE AT THE INTERNATIONAL SATELLITE SYMPOSIUM AT BHAGALPUR ON 29th OCTOBER, 1974

FRIENDS,

I consider myself privileged to be present in the midst of this distinguished gathering of scholars and scientists assembled from the different countries and to inaugurate this satellite symposium on "Respiration of Amphibious Vertebrates" under the sponsorship of the XXVI International Congress of Physiological Sciences. The subject matter is highly technical and I am sure that many specialists who are participating in the discussions would be going into the various aspects of this interesting problem. Apart from the academic interests of this particular subject, I consider that this problem has its applied aspects which, I hope, will be gone into by the participants. In this connection I understand that the Zoology Department of the Bhagalpur University is specially interested in the physiological studies on air-breathing fishes and that a scheme on the study of certain aspects of this problem is under consideration for a financial grant from the Indian council of Agricultural Research. As a layman, I am more interested in the applied aspects of these studies, especially in relation to the development of fish culture in this country.

Perhaps many of you here are aware of the existence of such varieties of fish as the "Singhi", "Kzoi" and the "Magur" which are held in some esteem as food fishes and which inhabit the swamps in Bihar and elsewhere. I am given to understand that these fishes are endowed with remarkable powers of respiration so as to lead a kind of amphibious life which help them to survive the adverse conditions in swampy areas. The extensive swamps, subjected to the ravages of the seasons, permit the survival of only a limited number of species. Reclamation of swamps for carp culture would entail considerable expenditure but at the same time it would be a waste if such vast areas were not better utilized for the culture of air-breathing fishes. Recognizing the importance of the problem, I am happy to say that the Indian council of Agricultural Research has initiated during the Fourth Plan an all-India coordinated Research Project on the culture and propagation of air-breathing fishes in swamps, which is being continued, so as to evolve a package of practices in utilizing the swamps for fish culture purposes. This has considerable significance in the augmentation of fish production in rural areas.

In the Government of India's approach to the Fifth Five Year Plan programme in Fisheries, our broad objectives include increasing the availability of fish, so as to help in bridging the protein gap in the Indian diet, and at the same time improve the socio-economic conditions of the fishermen, who are economically regarded as belonging to the weaker section of the society. Thus, fish culture would be one of the obvious measures that could be adopted to provide more effective and remunerative methods of production and also for finding means of generating employment. Fish culture operations are proposed to be developed in an intensive manner by bringing over 3·0 lakh hectares of tanks and ponds under this programme which is expected to bring the level of annual inland fish production to the order of 11·25 lakh tonnes by the end of the Fifth Plan. In attaining this target it is the intention of the Government to render necessary assistance to the fish culturists by organizing supplies of necessary inputs and the flow of technology.

At present, control of the majority of tanks and ponds in this country is with the local bodies or State Governments who adopt a three year leasing system of fisheries rights. Realizing the many drawbacks of the existing leasing practices and to encourage private farmers to undertake fish culture, the Government is considering improvements in leasing policies which would extend leases for a sufficiently reasonable period so that farmers can reap fuller benefits from their investments and at the same time enable them to repay any loans over this period. It is also essential to create progressive fish farmers all over the country by introducing suitable training programmes. The farmers are to be provided with technical guidance in regard to the improvement of water areas and cultural practices. Assistance also will have to be rendered in regard to securing adequate loans from commercial banks, supply of other necessary inputs like fish feed, manure, fish seed, etc. For this purpose the Government of India is proposing to set up Fish Farmers' Development Agencies in the various States/Union Territories. The main functions of the agencies will be to assist intensive fish culture in derelict watersheds by coordinating and integrating the work relating to the various components of the developmental activity. While conversion of the marginal areas of silted up water systems into ponds in States like West Bengal and Assam will constitute an important component of the scheme, development of the beels as well as the mane and jheels of the States of Bihar and Utter Pradesh will be taken up. In this process, fish farmers would be encouraged to set up nursery units to meet their own seed requirements and to sell seed to other needy fish farmers.

While the Fish Farmers' Development Agency is envisaged as a Central Scheme, other State level schemes are also contemplated in the V plan to aid in fish culture activities. These would mainly cover such aspects as development of district level fish seed farms and block level nurseries, development of fisheries in silted-up water systems, river fishery management and spawn production, introduction of the technique of cage fisheries and extending necessary assistance to fish culturists. For developing the district level fish seed farms and block level nursery units the area will be determined with reference to the needs and availability of land. Necessary fish farm construction cells also will be set up or the existing facilities for the purpose will be suitably strengthened. The Government also has in mind at present the setting up of suitable beel, mane jheel development agencies in those States where there are a sizeable number of such bodies of water and to coordinate these activities. Spawn collection from the rivers is proposed to be intensified besides formulating and enforcing measures to lend protection to brood fish in the rivers. In addition to riverine spawn collections, this will be augmented through the adoption of induced breeding and/or bund breeding at the existing District level Fish Seed Farms and nursery units at block level. About 7200 million numbers of spawn are expected to be produced under this scheme which should meet a great portion of the fish farmer's requirements.

In the matter of large-scale developments in fish culture, water can become a serious constraint. But unlike agricultural crops, fishes are essentially non-consumptive of water, requiring water only as their living medium. It is in this context that cage-culture, pen-culture and other similar techniques assume importance. We have large river systems and irrigation canals which could be used for the cage-culture of a variety of fish. Employing the available technology for the purpose, it is proposed to make a beginning in this direction so as to attain an annual production of about 500 tonnes in the coming years.

A suitable scheme is proposed to be evolved to provide financial assistance to fish culturists so as to enable them to organize supply of necessary inputs in their venture. In order to ensure timely transfer of technology the extension machinery would be suitably strengthened with setting up of National Demonstrations Centres and Operational Research Projects.

Although perhaps not quite relevant to Bihar, I may mention in this connection the proposals to develop fish culture in brackish waters for which the available technology is relatively limited. Nevertheless, it is proposed to make a beginning by initially taking up about 800 hectares of brackish water for this purpose. Fish culture possibilities in purely marine environments also exist and experiments on mariculture operations also have proved rewarding.

I have hitherto only tried to indicate the broad outlines in our approach for increasing fish production through culture operations which constitute an important aspect of our fisheries developmental programmes. In all these

I am happy to find that adequate research support is being given by the various organizations, notably the fisheries research institutes under the Indian Council of Agricultural Research. I am also aware of the significant research contributions being made by other agencies like the Fisheries Departments of the States, the research Departments of various universities and the Fishery Faculty of some of the Agricultural Universities. During the recent past, greatly as a result of the researches and experimental work conducted by the Central Inland Fisheries Research Institute in Barrackpore, we may consider ourselves on the threshold of a major breakthrough in inland fish culture. A fairly well tested system of composite culture practices has been evolved by which the production of fish in small bodies of water could be raised substantially compared to the yield obtained through traditional methods. These techniques have been successfully demonstrated in the farms located in the various States centres of the all India Coordinated Project on Composite culture of Indian and exotic fishes.

In stressing fish culture at this Symposium, it has been my intention to highlight one of the aspects which has some relevance to the larger topic to be discussed during the coming days and at the same time to bring in to focus the national values involved. During the Fifth Five Year Plan the Planning Commission has approved a total provision of nearly Rs.167 crores including Central and State Sectors for the purpose of development and research in the various branches of fisheries.

On going through the papers coming up for discussion, I am happy to find that a number of them are devoted to the respiratory studies on fish. It is my earnest hope that the forthcoming deliberations will throw added light on our existing knowledge particularly on the physiology and anatomy of respiratory mechanisms of the amphibious type of fishes, frogs and other vertebrates which will have an impact on our programmes for development in aquaculture. I have, therefore, great pleasure in inaugurating this Symposium.

JAI HIND

Appendix I—Symbols

Symbols used in respiratory physiology and mainly adopted in this volume are based on those given by Dejours in "Principles of Comparative Respiratory Physiology". The general principles for writing these symbols are to be found in Pappenheimer (1950), *Federation Proc.*, **9**, 602–603. A glossary of terms on respiration and gas exchange is given by Bartels *et al.* (1973), *J. Appl. Physiol.*, **34**, 549–558.

w	water	I	inspired
g	gas	E	expired
b	blood	A	alveolar
a	arterial	L	pulmonary
v	venous	D	dead space
$\bar{v}$	mixed venous	T	tidal
c	capillary	B	barometric
t	tissue	G	gill

STPD	standard temperature, pressure, dry (0°C, 760 mmHg)
BTPS	body temperature, pressure, saturated with water
ATPS	ambient temperature, pressure, saturated with water
D	coefficient of diffusion
K	Krogh permeation coefficient
G	conductance
$G_{I,A}$	conductance between inspired and alveolar gas
$D_{L_{O_2}}$	lung diffusing capacity (conductance) for O_2
W or B	body mass
f_H	heart frequency
f or f_R	respiratory frequency
C_{O_2} diss	concentration of dissolved oxygen
C_{Hb}	concentration of haemoglobin
C_{HbO_2}	concentration of oxyhaemoglobin (HbO_2)
$C^{max}_{HbO_2}$	maximal concentration of oxygen as HbO_2
C_{O_2}tot	total oxygen concentration
P_{50}	partial pressure of oxygen at which 50% of the Hb is oxygenated
M	mass of substance
$\dot{M}$	quantity of substance per unit time
$\dot{M}_{O_2}$ or $\dot{V}_{O_2}$	oxygen consumption (quantity of O_2 per unit time, e.g. $mmol.min^{-1}$, ml $O_2STPD.min^{-1}$)
P	gas pressure (may be partial pressure)
C	concentration
S	saturation
F	fractional concentration in dry gas phase
R	the gas constant
R	respiratory exchange ratio
$R_{a,\bar{v}}$	respiratory exchange ratio between arterial and mixed venous blood
V	volume
V_T	tidal volume
V_S	stroke volume

V_L	lung volume
$\dot{V}$	volume per unit of time
$\dot{V}\beta$	conductance or capacity rate
$\dot{V}_{O_2}$	oxygen consumption (volume of O_2 per unit time)
$\dot{V}w$	water flow rate
$\dot{V}air$	air flow rate
$\dot{V}b$	blood flow rate
$\dot{V}I$	rate of inspired air flow
U	coefficient of utilisation or extraction
Ew_{O_2}	effectiveness of O_2 extraction from water
Eb_{O_2}	effectiveness of O_2 uptake by blood
β	capacitance coefficient ($\equiv \alpha$ or solubility coefficient)
βb_{O_2}	capacitance coefficient of oxygen in blood

Other symbols are defined by individual contributors

Appendix II—Equivalent units

Some useful conversion factors are as follows:

Amount of gas
1 mmol O_2 = 22·4 ml O_2 (STPD)
1 ppm O_2 in water = 1 mg O_2/litre H_2O
150 mmHg P_{O_2} in water = 150 × solubility coefficient (for that particular water) at particular temperature—ml O_2/litre
Oxygen consumption ($\dot{V}_{O_2}$ or $\dot{M}_{O_2}$)
1 mmol O_2/kg/hr. = 22·4 ml O_2/kg/hr.
1 ml O_2/kg/hr. = 1·43 mg O_2/kg/hr.
1 mg O_2/kg/hr. = 0·70 ml O_2/kg/hr.
Pressure
1 mm Hg = 1 torr.
1 cm H_2O = 0·7355 mmHg
1 millibar = 1·02 cm H_2O
Surface tension
1 dyne/cm = 1 millinewton/metre
Length
1 inch = 2·54 cm
1 micron (μm) = 10^{-4} cm
1 Angstrom (A) = 10^{-8} cm
Number
1 lakh = 100,000
1 million = 1,000,000
1 crore = 100 lakhs (usually of Rupees)
1 billion = 1 million million
1 billion (U.S.) = 1,000 million

The table of energy and power equivalents given in the table below is based upon that in Dejours "Principles of Comparative Respiratory Physiology"

Conversion factors between various physical units of power and the two physiological units of rate of energy expenditure (volume STPD of O_2)(time)$^{-1}$ and (mmol O_2)(time)$^{-1}$. The energy equivalents of 1 ml STPD and of 1 mmol of oxygen are respectively 4·8 and 108 calories (1 mmol — 22·4 ml STPD).

		W	J/min	cal/min	ml O_2STPD/min	mmol O_2/min
1 W	=	—	60	14·3	2·99	0·133
1 J.min^{-1}	=	0·0167	—	0.239	0·0498	0·00222
1 cal.min^{-1}	=	0·0698	4·186	—	0·208	0·0093
1 ml O_2STPD.min^{-1}	=	0·335	20·1	4·8	—	0·0446
1 mmol O_2.min^{-1}	=	7·50	450	108	22·4	—

Water Versus Air as the Respiratory Media

PIERRE DEJOURS

Laboratoire de Physiologie Respiratoire,
Centre National de la Recherche Scientifique, Strasbourg, France

According to the present concepts of palaeontology, primitive animals lived in water. The first aerial animals, land scorpions, are recorded as living 450 millions years ago, in the late Ordovician. A few millions years later, in the Silurian, one finds fossils of the Dipnoi, the first air-breathing vertebrates. In our times, certain animals which are primarily aquatic during their development, such as the tadpole, metamorphose to become almost perfect air-breathers as in some terrestrial amphibians. Dual-breathing fish, amphibians or reptiles pass more or less frequently from one medium to the other. Animals moving from water- to air-breathing meet fundamentally different conditions for respiration. The differences concern not only oxygen and carbon dioxide but also other ambient factors related to the mechanics of breathing and to heat loss balance (Table I).

To clarify the differences between water and air as to respiratory gases, the conductance equations (Piiper, Dejours, Haab and Rahn, 1971) may be used as a guide line:

$$\dot{M}_{O_2} = G \cdot \Delta P_{O_2} \qquad (1)$$

where $\dot{M}_{O_2}$ is the rate of oxygen transfer between locations 1 and 2, G the conductance of the medium between 1 and 2, and ΔP_{O_2} the difference of partial pressure of O_2 between the two locations.

For *diffusive transfer*, eq. (1) becomes

$$\dot{M}_{O_2} = D_{O_2} \cdot \beta_{O_2} \cdot \frac{A}{E} \cdot \Delta P_{O_2} \qquad (2)$$

where D_{O_2} is the diffusion coefficient of the medium interposed between 1

TABLE I

Summary of the main physical differences between water and air which affect external respiration (from Dejours, 1975)

			Distilled water	Air	$\frac{\text{Water}}{\text{Air}}$
Gas transfer factors					
Diffusion	D_{O_2}	$\frac{cm^2}{sec}$	0·000025	0·178	$\simeq \frac{1}{7000}$
coefficients of O_2 and CO_2	D_{CO_2}	$\frac{cm^2}{sec}$	0·000018	0·139	$\simeq \frac{1}{8000}$
Capacitance	β_{O_2}	$\frac{nmol}{ml \cdot torr}$	1·82	54·73	$\simeq \frac{1}{30}$
coefficients of O_2 and CO_2 at 20°C	β_{CO_2}	$\frac{nmol}{ml \cdot torr}$	51·89	54·73	$\simeq 1$
Krogh's constants of O_2 and CO_2	K_{O_2}	$\frac{nmol}{cm \cdot sec \cdot torr}$	0·000046	$\simeq$ 9·7	$\simeq \frac{1}{2 \cdot 10^5}$
	K_{CO_2}	$\frac{nmol}{cm \cdot sec \cdot torr}$	0·00093	$\simeq$ 7·6	$\simeq \frac{1}{8000}$
Fluid transport factors					
Viscosity at 15°C	η	poise	0·0114	0·000180	$\simeq$ 60
Density at 15°C	ρ	$\frac{kg}{L}$	0·999	0·001226 at 1 atm	$\simeq$ 800
Kinematic viscosity at 15°C	$\frac{\eta}{\rho}$	$\frac{poise \cdot L}{kg}$	0·0114	0·147	$\simeq \frac{1}{13}$
Heat transfer factors					
Heat capacity		$\frac{kcal}{L \cdot °C}$	1 00	$\simeq$ 0·175	$\simeq$ 6
Heat conductivity at 15°C		$\frac{\mu cal}{sec \cdot °C \cdot cm}$	1440	60·3	$\simeq$ 24

and 2, β_{O_2}, the capacitance ("solubility") coefficient of this medium, A, the surface area of the cross section between 1 and 2, E, the distance between 1 and 2.

The product $(D_{O_2} \cdot \beta_{O_2})$ is Krogh's diffusion constant, K_{O_2}; thus eq. (2) becomes

$$\dot{M}_{O_2} = K_{O_2} \cdot \frac{A}{E} \cdot \Delta P_{O_2} \qquad (2b)$$

In eqs. (2) and (2b) the ratio A/E takes care of the geometry of the volume through which diffusion takes place. For spherical or cylindrical organisms a particular geometrical factor is substituted for A/E.

For *convective transfer*, eq. (1) takes the following form

$$\dot{M}_{O_2} = \beta_{O_2} \cdot \dot{V} \cdot \Delta P_{O_2} \qquad (3)$$

where $\dot{V}$ is the flow rate of water, $\dot{V}w$, or of gas, $\dot{V}g$, between 1 and 2.

Similar equations may be written for CO_2 transfer.

I. Water

Table I gives some values for O_2 and CO_2 diffusion coefficients, capacitance coefficients and Krogh's constants in distilled water. These values are valid for freshwater of low salinity, of low carbonate alkalinity and for a given temperature.

The physiological consequences of the enormous difference in the capacitance coefficients (Fig. 1, left) will be studied further.

Many factors may modify the respiratory conditions of natural waters.

(1) *Salinity.* An increase in salinity decreases solubility as can be seen in Fig. 2 which compares distilled water and sea water (salinity 35 $g \cdot kg^{-1}$).

(2) *Temperature.* The solubilities of O_2 and CO_2 decrease with the increase in temperature (Fig. 2). Since temperature-dependency is more marked for

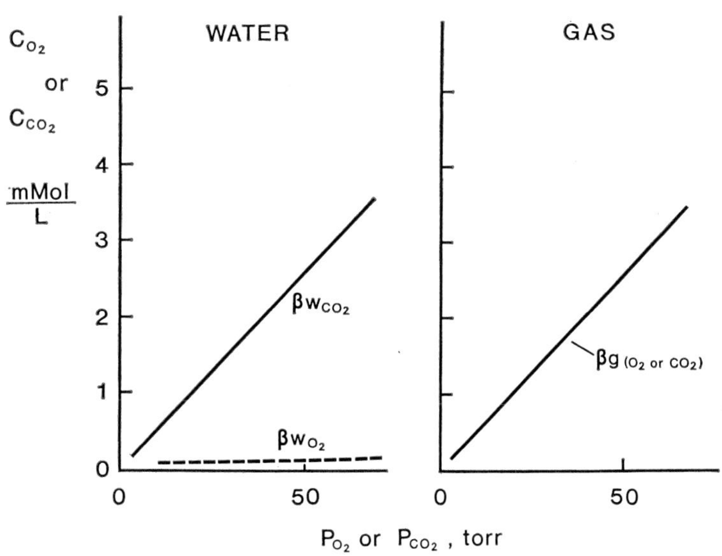

FIG. 1. Concentration of O_2 and CO_2 in distilled water at 15°C and in gas at 37°C, plotted against P_{O_2} or P_{CO_2}. The slopes in water and in air correspond to the O_2 and CO_2 capacitance in the two media. For distilled water, capacitance coefficients are synonymous with solubility coefficients. Note that O_2 and CO_2 capacitance coefficients are identical.

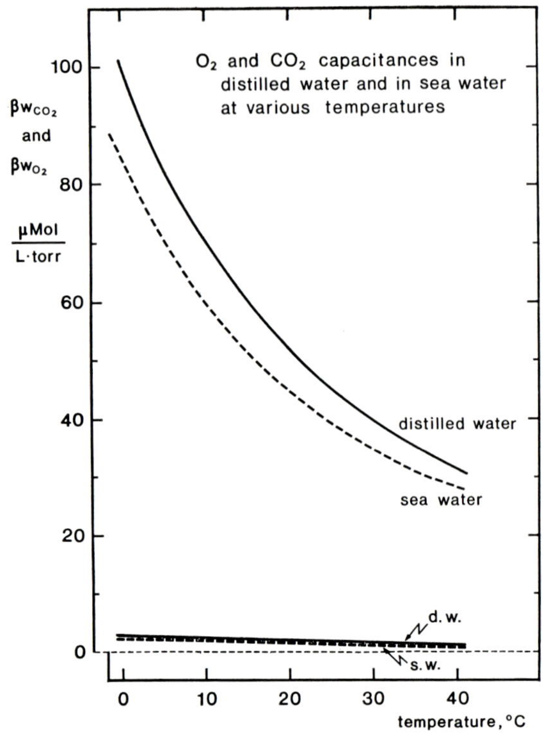

FIGURE 2.

CO_2 than for O_2, the ratio CO_2 capacitance/O_2 capacitance is not constant (Fig. 3). Because the increase in temperature increases the diffusion coefficient, the product of the diffusion and capacitance coefficient, i.e. Krogh's constant, is not very temperature-dependent (Fig. 4).

(3) *Carbonate alkalinity*. In highly carbonated water, as in well-aerated sea water, $CO_3^=$ and HCO_3^- ions exist concomitantly, their relative quantities depending on the pH value. In such water, added CO_2 (hence H_2CO_3) is buffered by the transformation of some $CO_3^=$ into HCO_3^-. Figure 5 shows the relation between the CO_2 concentration and P_{CO_2} in distilled water and three carbonated waters, including sea water.

A. EFFECT OF LIVING PROCESSES

Figure 6 shows with an enlarged ordinate the relation between $C_{CO_2 tot}$ and P_{CO_2} of the upper curve of Fig. 5. Point 1 corresponds to inspired water; point E″ to a "virtual" expired water, assuming that all CO_2 produced by the animal goes into solution (the hypothetical line 1 – E″ has the slope of the

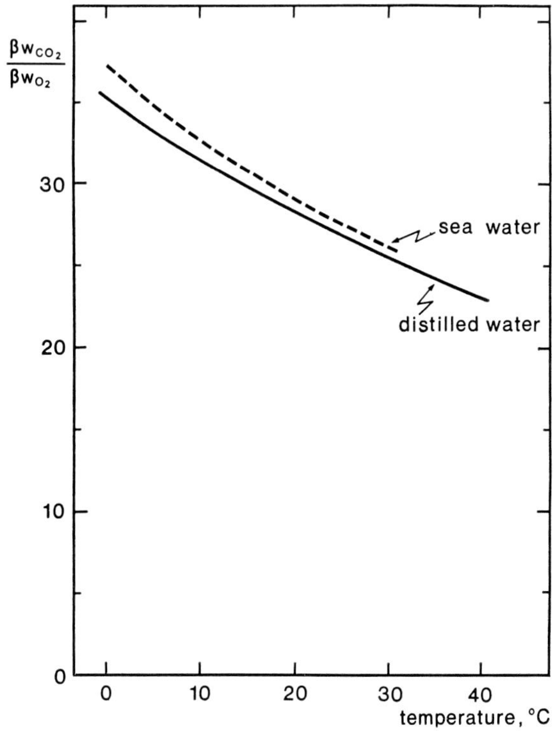

FIG. 3. Ratio of the capacitance coefficient of CO_2, βw_{CO_2}, to the capacitance coefficient of O_2, βw_{O_2}, as a function of temperature.

CO_2 solubility curve in distilled water as on Fig. 4). Due to the curvilinearity of the relation in carbonated water at high pH, i.e. at low P_{CO_2}, expired water should be located at E′ instead of E″. Actually for two teleost fish, the freshwater goldfish and the seawater sculpin, the expired water point is still more to the left because their expired water has a higher titration alkalinity than inspired water (Dejours, Armand and Verriest, 1968). The study of the ionic balance shows that there are more cations, mainly NH_4^+, in the expired water of the ammonotelic teleost than in the inspired water. The hypothesis that the increase in titration alkalinity is mainly due to NH_4^+ ion in the ammonotelic teleost is supported by the observation that in the ureotelic elasmobranch, the dogfish *Scyliorhinus canicula*, the CO_2 dissociation curves of the inspired and expired water are not different (Fig. 7).

The examples of carbonated water of Fig. 5 are by no means extreme; there exist more highly carbonated waters, e.g. those of the lakes of the Basin Range, California, and of British Columbia (Livingstone, 1963). For example, in the water of Borax Lake, California, the total CO_2 concentration varies

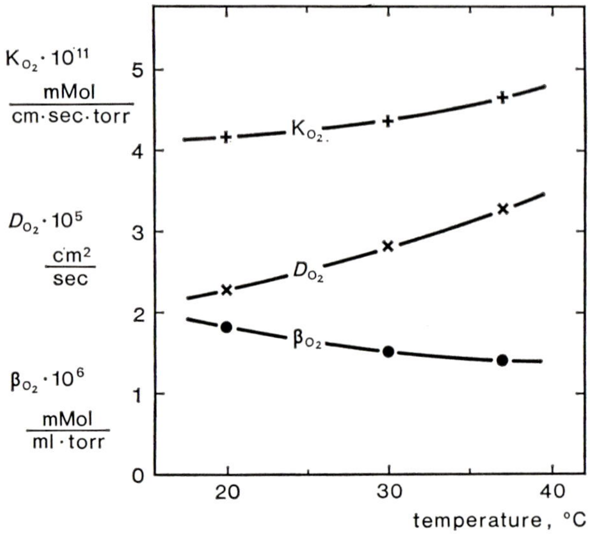

FIG. 4. The O_2 diffusion coefficient, D_{O_2}, the O_2 capacitance of water (solubility), β_{O_2}, and Krogh's constant of diffusion of O_2 in water, K_{O_2}, as functions of temperature changes (Grote, 1967).

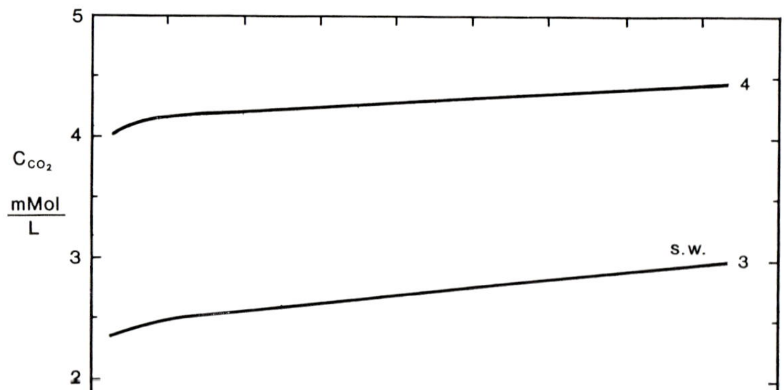

FIG. 5. Carbon dioxide absorption curves of different types of water. Curve 1, distilled water at 25°C. Curve 2, low-carbonate tap water at 25°C. Curve 3, sea water at 20°C. Curve 4, tap water at 25°C. Note the curvatures of the waters with carbonate alkalinities (from Dejours *et al.*, 1968).

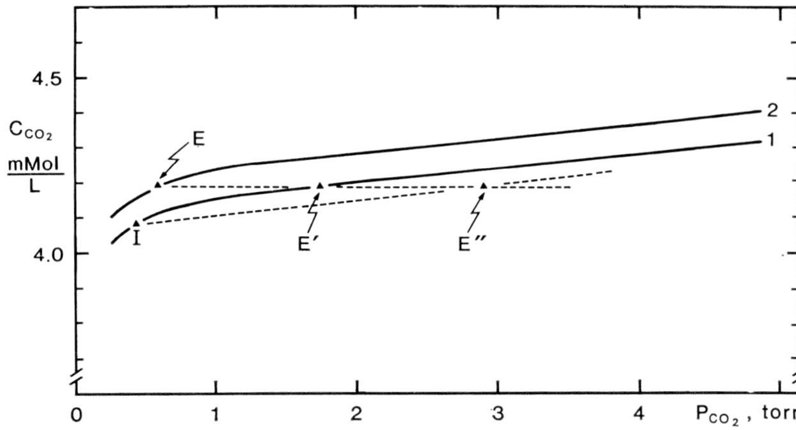

FIG. 6. Carbon dioxide absorption curves of tap water inspired (1) and expired (2) by goldfish. Triangles I and E indicate inspired and expired waters. For significance of triangles E′ and E″, see text (from Dejours *et al.*, 1968).

between 100 and 200 mmol·L^{-1} and pH between 9·5 and 9·8, yet two species of Crustacea live in these conditions (Wetzel, 1964). Another extreme example is the water of the springs and of the lagoon of Lake Magadi, Kenya; here the temperature is near 40°C, the pH is greater than 10, and the main salts are bicarbonate and carbonate in concentrations approaching 600 mosmol·kg^{-1}; the fish *Tilapia grahami* live successfully in these waters (Reite, Maloiy and Aasehaug, 1974).

Two biological processes, respiration and photosynthesis, act in opposite direction to modify O_2 and CO_2 concentrations and pressures in water. At night, respiration is the only process: O_2 decreases and CO_2 increases. During the day, photosynthesis of green plants may supersede the animal and vegetal respiration and lead to a net uptake of CO_2 and a net production of O_2. Consequently daily cycles of O_2 and CO_2 concentration and pressure occur. During the hot days of summer, temperature and O_2 concentration were measured each hour in the surface water of a ditch in June (Fig. 8, from Jones, 1961); at 1.30 p.m., O_2 concentration reached 750 μmol·L^{-1} and P_{O_2} about 480 torr, whereas at 3.30 a.m. the corresponding values were 18 μmol·L^{-1} and about 10 torr. Similar observations have been reported by others, in particular by Garey and Rahn (1970).

Diurnal variations of CO_2 and pH have rarely been examined; but Carter (1935) reported value of P_{CO_2} above 30 torr (≡ mm Hg), of pH below 5·0 and of P_{O_2} of a few torr in the water of a British Guiana swamp sampled at 9.00 a.m. More data were reported by Carter (1955) on the variation of O_2, CO_2 and acidity in the water of the swamps of Uganda.

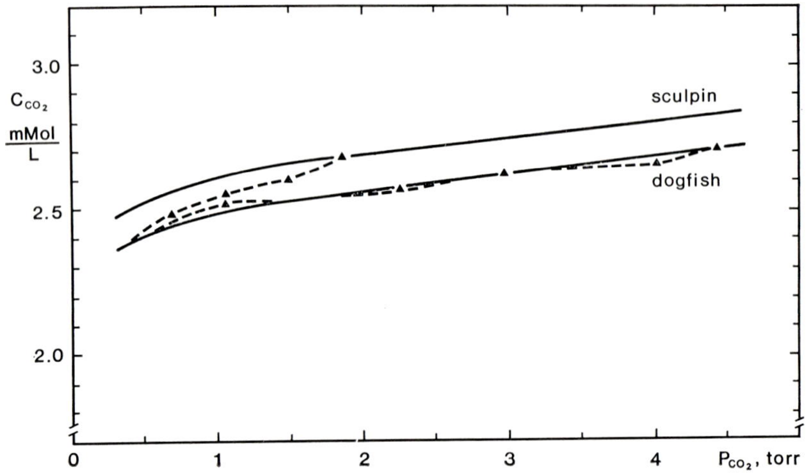

FIG. 7. The two solid lines are the CO_2 absorption curves of CO_2 of sea water (concentration of CO_2, C_{CO_2}, vs the partial pressure of CO_2, P_{CO_2}). The lower curve corresponds to normal sea water; the ureotelic dogfish (*Scylorhinus canicula*) does not modify the CO_2 absorption curve of the water which it breathes in a confinement. The upper curve corresponds to the CO_2 absorption curve of the water at the end of the confinement of a teleost, the sculpin *Cottus bubalis*. This ammonotelic fish modifies the titration alkalinity of the water it breathes (Dejours and Armand, unpublished observations). The triangles represent the C_{CO_2} vs P_{CO_2} values at different moments of confinement period.

II. AIR

Table I gives values of diffusion coefficients, capacitance coefficients and Krogh's constants for O_2 and CO_2.

The diffusion coefficients are inversely proportional to the barometric pressure. Thus if D^1, D^2, P^1, P^2 are the diffusion coefficients and the barometric pressures at altitudes 1 and 2, we have, for oxygen as an example,

$$\frac{D^1}{D^2} = \frac{P^2}{P^1} \tag{4}$$

This pressure-dependency of the diffusion coefficient has enormous consequences for the incubation of eggs since high altitude favours not only the diffusion of O_2, but also the loss of water vapour (Wangensteen, Rahn, Burton and Smith, 1974).

The capacitance coefficients are the same for all gases. It may be demonstrated (Piiper *et al.*, 1971) that

$$\beta g = \frac{1}{RT} \tag{5}$$

in which R is the gas constant, and T the absolute temperature.

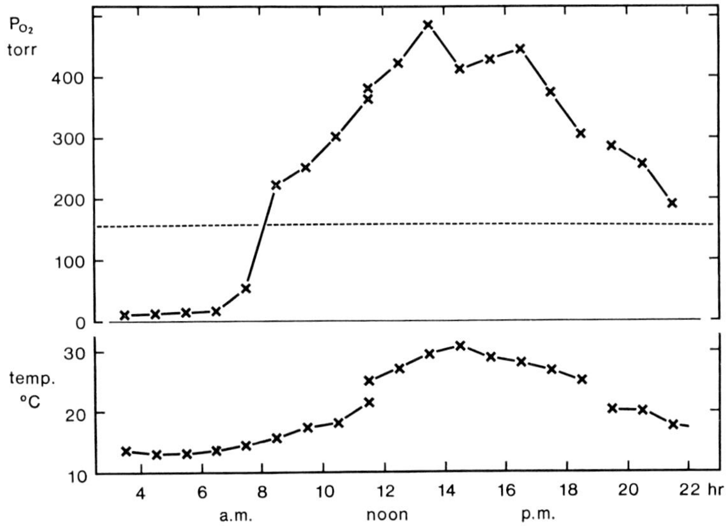

FIG. 8. Changes of the partial pressure of oxygen and of temperature in the surface water of a ditch filled with clumps of aquatic plants on a hot sunny June day (data from Jones, 1961). The lowest value of P_{O_2} is observed at 3.30 a.m. before sunrise. The horizontal dashed line indicates the presence of O_2 in the air; for any point above or below that line the water is said to be O_2-oversaturated or O_2-undersaturated.

This equation shows also that the capacitance coefficients are (1) independent of the barometric pressure (Figs. 1 and 9); (2) inversely proportional to the absolute temperature, but nearly constant in the biological range of temperatures (Fig. 9).

It follows from eqs. (3) and (5), and dropping the time dimension appearing on both sides of eq. (3), that the concentration of O_2, C_{O_2}, in air is defined by

$$C_{O_2} = \frac{M_{O_2}}{Vg} = \frac{1}{RT}(P_B - P^T_{H_2O}) \cdot F_{I_{O_2}} \quad (6)$$

Vg is the volume of gas in BTPS conditions, P_B the barometric pressure, $P^T_{H_2O}$ the saturating water vapour at temperature T, and $F_{I_{O_2}}$ the fraction of O_2 in dry air which is normally 0·2095. This equation is a form of the gas law (Dejours, 1974).

It is not customary to speak of concentration of a substance in a gas phase, but this concept permits the use of a Fick type equation for O_2 uptake from gas, as may be written for O_2 uptake from water or blood.

As an example, in a man at sea level, at 37°C, the O_2 concentration in BTPS inspired gas (tracheal gas) is 173 ml STPD/L BTPS of air, or 7·72 mmol O_2/L BTPS.

The concentration of O_2 depends (1) on the barometric pressure P_B and (2) on the temperature which acts twice in eq. (6), i.e. on the capacitance

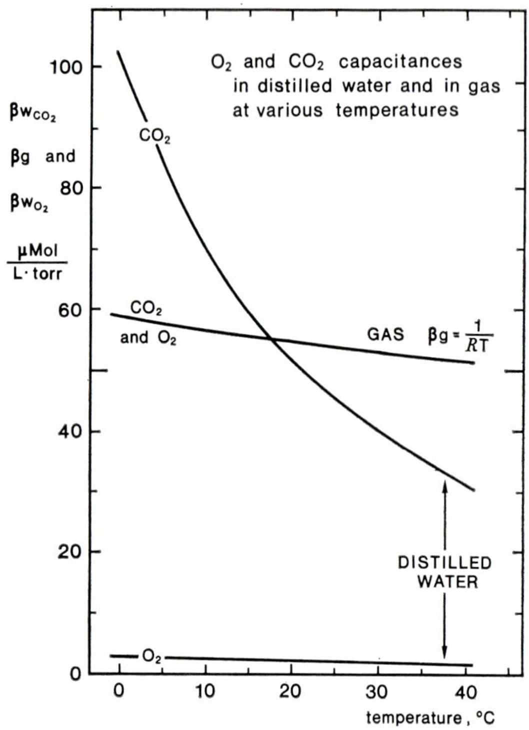

FIG. 9.

$1/RT$ and on the water vapour tensions. Since $P^T_{H_2O}$ is constant for a given temperature, it follows that the lower the barometric pressure, the higher the fractional volume occupied by water vapour. At the theoretical altitude of about 19·2 km, the barometric pressure is 47 torr; then all the alveolar gas of a mammal at 37°C would be occupied by water vapour (Fig. 10).

Since O_2 and N_2 diffuse over large distances, and since air is an enormous mass submitted to powerful convective currents, its composition in O_2 and in CO_2 is not much affected by respiratory processes or even by urban life. Only in special circumstances does O_2 decrease below and CO_2 increase above their normal fractions, 0·2095 and 0·00033 respectively; for example, a CO_2 fraction of 0·135 and an O_2 fraction of 0·04 have been measured in the dens of hibernating mammals (Williams and Rausch, 1973).

III. Main Differences between Water and Air as Respiratory Media

1. *Comparison of the diffusion and capacitance coefficients and the Krogh's constants for* O_2 *and* CO_2

Table I, Figs. 1, 2 and 9 show that the diffusion and capacitance coefficients and the Krogh's constants of O_2 and CO_2 are much higher in air than in water.

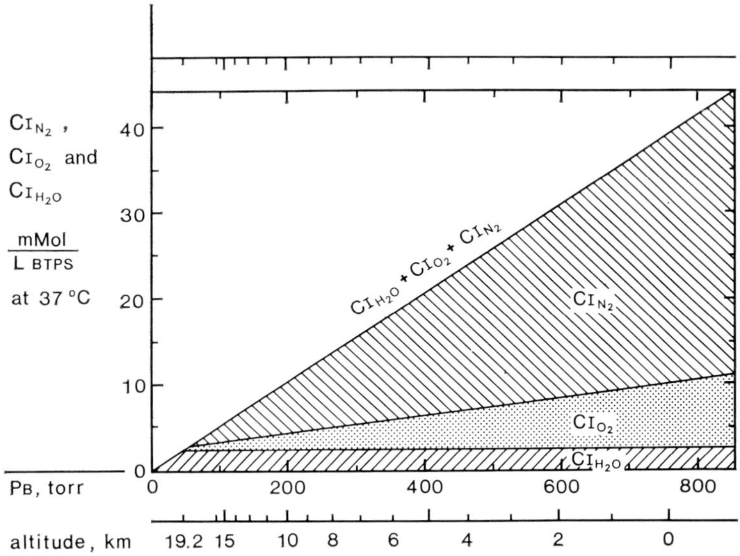

FIG. 10. Concentrations of water vapour (CI_{H_2O}), of O_2 (CI_{O_2}) and of nitrogen (CI_{N_2}) in the BTPS conditions (tracheal gas) of a man at 37°C as functions of barometric pressure. Note that the proportion occupied by water vapour increases with altitude and that at the theoretical altitude of 19·2 km, at which the water boils, the alveolar spaces would be completely occupied by water vapour.

If we do not find the size of aerial animals which exchange by diffusion much larger than that of aquatic animals, it is because the aerial animals, by exposing their integuments to open air, would quickly dry out. However, those aerial animals which confine themselves to a wet environment or which bathe from time to time exchange O_2 and CO_2 by diffusion; this is the case for some terrestrial plethodontids (Urodeles) like *Hydromantes genei*. Also the aerial eggs of reptiles and birds are protected against dessication by a hard shell; only over a limited surface, i.e. the pores, can diffusion of O_2, CO_2 and water vapour take place.

An extremely important difference in the properties of water and air is the fact that the O_2 capacitance is much lower in water than in air whereas CO_2 has about the same capacitance in both media. Actually at 18°C, the CO_2 capacitance in pure water and in air are the same, whereas the O_2 capacitance in water is 29 times less in water than in air. Consequently, to take up a given quantity of oxygen an air-breather can breathe much less than a water breather. This behaviour is best expressed by the ventilatory requirement per unit of O_2 consumed, that is the ratio Vg/M_{O_2} for the air breather and Vw/M_{O_2} for the water breather (Dejours, Garey and Rahn, 1970). It has been found for oxygen that the ratio is ten times lower in man than in a goldfish. Analogous differences are found for all water-breathers and air-breathers which have been studied. Consequently, since the air-breathers breathe much less than the water-breathers, and since the CO_2 capacitance of both media are nearly the same, the CO_2 partial pressure is much higher in the air-breathers than in the water-breathers (Rahn, 1966). Since the pH's of aerial and aquatic poikilotherms are similar, it follows that blood $[HCO_3^-]$ is much higher in the former than in the latter (Rahn, this volume).

2. *Influence of respiration on the respiratory properties of water and of air*

Except in very special environments, the composition of air is not noticeably affected by living processes, because diffusion in air is very high and because enormous convective currents renew the medium. It is simple, knowing barometric pressure and temperature, to predict the changes air will undergo when it is breathed. Furthermore, the absolute changes of O_2 and CO_2 partial pressure are the same (for a respiratory quotient of 1).

But there exists a variety of waters, and respiration and photosynthesis may modify considerably the partial pressures and concentrations of O_2 and CO_2 (Fig. 8). For O_2, as long as temperature and salinity are known, it is easy to calculate the possible changes of P_{O_2} and C_{O_2} caused by respiration. But these data are not sufficient for predicting changes of the carbon dioxide, because the actual pH, the titration alkalinity and a possible change of the ionic balance between inspired and expired water must also be taken into account.

3. *Comparison of viscosity and density of water and air*

The work of breathing depends on the viscosity and density of the respired medium. Since viscosity and density are respectively 60 times and 800 times higher in water than in air, it follows that the work of breathing, which depends on these rheological factors, must be much higher in aquatic animals than in aerial animals. In addition, as seen above, an aquatic animal must breathe much more than an aerial animal in order to extract a given quantity of oxygen. Figure 11 shows the change of the rheological factors of water and air with temperature. One will note that an increase in temperature decreases the density and kinematic viscosity of water whereas in air the effect is the opposite. This differing effect of temperature must have consequences, little studied as yet, on the work of breathing of aquatic animals (see Hughes and Roberts, 1970).

4. *Comparison of heat capacity and heat conductivity of water and air*

The fact that the heat capacity and the heat conductivity are much lower in air than in water (Table I) seems to be a prerequisite for the existence of homeotherms. It may be argued that some homeothermic mammals like the whales and the seals live in water; but, in reality, only their core temperature

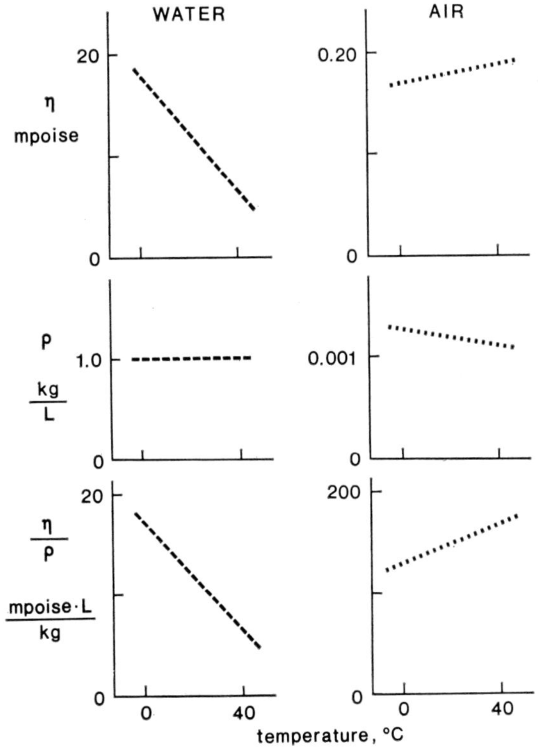

FIG. 11. Viscosity, η, density, ρ, and kinematic viscosity η/ρ of water and air as functions of temperature.

is actually at high temperature; certainly the relative body mass actually maintained at a temperature above 35°C is much smaller in aquatic mammals than in terrestrial mammals.

IV. Conclusion

For air it is sufficient to know the barometric pressure and the body temperature in order to calculate the partial pressures and concentrations of all gases in inspired air or in any inspired artificial gas mixture whose dry fractional concentrations are known. It is easy to predict which changes O_2 and CO_2 may undergo in the process of respiration of aerial animals. Furthermore, O_2 uptake and CO_2 production of living beings do not alter the composition of ambient air, except in some rare instances.

In water, the two gases O_2 and CO_2 behave quite differently; their solubilities are extremely different and are not affected in the same proportion by changes of temperature and salinity. Except for water analysed *ad hoc*, it is impossible to calculate P_{CO_2} from its known concentration and to predict the change it may undergo in the process of respiration. Furthermore, life processes (respiration, photosynthesis) markedly change concentrations and pressures of the ambient O_2 and CO_2; seasonal and nycthemeral cycles affect these variables. There are no two identical waters as far as gas exchanges are concerned.

Also the rheological factors (viscosity and density) and the heat transfer factors (caloric capacity and conductivity) of air and water are markedly different.

References

Carter, G. S. (1935). The fresh waters of the rain-forest areas of British Guiana. *J. Linn. Soc. London, Zoology* **39**, 147–193.

Carter, G. S. (1955). "The Papyrus Swamps of Uganda". Cambridge, W. Heffer, 25p.

Dejours, P. (1974). Recent concepts in the physiology of gas exchange. Proc. of IUPS congress (New Delhi). Volume X, pp. 25–26.

Dejours, P. (1975). "Principles of Comparative Respiratory Physiology". Amsterdam, North-Holland Publishing Company, 253p.

Dejours, P., Armand, J. and Verriest, G. (1968). Carbon dioxide dissociation curves of water and gas exchange of water-breathers. *Resp. Physiol.* **5**, 23–33.

Dejours, P., Garey, W. F. and Rahn, H. (1970). Comparison of ventilatory and circulatory flow rates between animals in various physiological conditions. *Resp. Physiol.* **9**, 108–117.

Garey, W. F. and Rahn, H. (1970). Gas tensions in tissues of trout and carp exposed to diurnal changes in oxygen tension of the water. *J. exp. Biol.* **52**, 575–582.

Grote, J. (1967). Die Sauerstoffdiffusionskonstanten im Lungengewebe und Wasser und ihre Temperaturabhängigkeit. *Pflügers Arch. ges. Physiol.* **295**, 245–254.

Hughes, G. M. and Roberts, J. L. (1970). A study of the effect of temperature changes on the respiratory pumps of the rainbow trout. *J. exp. Biol.* **52**, 177–192.

Jones, J. D. (1961). Aspects of respiration in *Planorbis corneus* L. and *Lymnaea stagnalis* L. (Gastropoda: Pulmonata). *Comp. Biochem. Physiol.* **4**, 1–29.

Livingstone, D. A. (1963). Chemical composition of rivers and lakes. (pp. 1–64). Chapter G of Data of Geochemistry – Geological Survey Professional paper 440 G – US Government Printing Office, Washington, D.C.

Piiper, J., Dejours, P., Haab, P. and Rahn, H. (1971). Concepts and basic quantities in gas exchange physiology. *Resp. Physiol.* **13**, 292–304.

Rahn, H. (1966). Aquatic gas exchange: Theory. *Resp. Physiol.* **1**, 1–12.

Reite, O. B., Maloiy, G. M. O and Aasehaug, B. (1974). pH, salinity and temperature tolerance of Lake Magadi Tilapia. *Nature, Lond.* **247**, 315.

Wangensteen, O. D., Rahn, H., Burton, R. R. and Smith, A. H. (1974). Respiratory gas exchange of high altitude adapted chick embryos. *Resp. Physiol.* **21**, 61–70.

Wetzel, R. G. (1964). A comparative study of the primary productivity of higher aquatic plants, periphyton, and phytoplankton in a large, shallow lake. *Int. Rev. ges. Hydrobiol.* **49**, 1–61.

Williams, D. D. and Rausch, R. L. (1973). Seasonal carbon dioxide and oxygen concentrations in the dens of hibernating mammals (Sciuridae). *Comp. Biochem. Physiol.* **44A**, 1227–1235.

Quantitative Functional Analysis of Branchial Gas Transfer: Theory and Application to *Scyliorhinus stellaris* (Elasmobranchii)

PETER SCHEID and JOHANNES PIIPER

Max-Planck-Institut für experimentelle Medizin, Abteilung Physiologie, Göttingen (Germany)

I. INTRODUCTION

Many studies have been devoted to analysis of gas exchange data obtained in fish (cf. Randall, 1970; Johansen, 1971). In most studies attention has been focussed on one of a few factors governing gas exchange. Recently Hill and Hughes (1970) and Scheid and Piiper (1971) have assessed the role of diffusion resistance in interlamellar water in limiting gas exchange in gills. However, at that time a full analysis was impossible due to the lack of exact morphometrical data on gills of those fish that were used in physiological studies on gas exchange. Recently gills of *Scyliorhinus stellaris* have been investigated morphometrically (G. M. Hughes, S. Perry and J. Piiper, unpublished observations). Using the physiological measurements on the same species reported by Baumgarten–Schumann and Piiper (1968), an attempt is made in this paper to analyse quantitatively the factors limiting gas exchange in the gills. Although provisionally the application is limited to one elasmobranch species, the models used and the quantitative relationships derived may be considered to be valid and useful for analysis of gas exchange in other fishes, including teleosts.

II. Structure of Fish Gills

Structure of gills of various fish species has been reviewed recently (Hughes and Morgan, 1973). A schema of gills of an elasmobranch fish is shown in Fig. 1. Each gill filament carries a great number of *secondary* lamellae that form a sieve through which respiratory water passes from the cavity to the outside. The secondary lamellae are perfused by blood from the afferent (venous blood) to the efferent artery (arterial blood). This blood flow in the secondary lamella may be considered as a "sheet flow" (Fung and Sobin, 1969) in the blood lacunae that are partitioned by pillar cells.

It is the structure of the gill apparatus in elasmobranchs that is specifically depicted in Fig. 1. In teleosts, the septa supporting the filaments are absent; however, the general arrangement is the same. The following analysis is, therefore, applicable to teleost as well as to elasmobranch gills.

Figure 2 shows a schema of the smallest functional unit of gas exchange. It consists of the *interlamellar space* bounded by adjacent secondary lamellae. A thin tissue layer, *the secondary lamellar membrane*, separates the water in the interlamellar space from blood in the blood lacunae. Both, this membrane and part of the interlamellar space have to be traversed, by diffusion, by the respiratory gases.

Most important for quantitation of gas exchange is the relative flow direction of interlamellar water and lacunar blood. Both anatomical (cf. Hughes and Morgan, 1973) and physiological measurements (Piiper and Baumgarten-Schumann, 1968a,b; Hughes and Hills, 1971; Grigg and Read, 1972) are in favour of opposite directions of water and blood flow in the gas exchange unit, meaning presence of a *counter-current gas exchange system* in elasmobranch fish as is also generally assumed for the teleosts.

For a quantitative analysis of gas exchange, which is attempted in this paper, detailed knowledge of the quantitative dimensions of the gills is needed. The parameters required include height, width and length of the gas

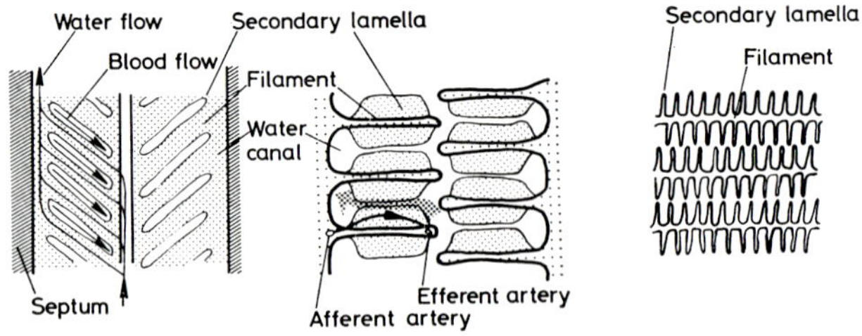

Fig. 1. Schematic representation of the structure of gills in elasmobranchs. Left: view, parallel to the secondary lamellae, onto the filament. Middle: cross section of filaments, parallel to secondary lamellae (the slant of which is neglected). Right: cross section in a plane perpendicular to both filaments and secondary lamellae, visualizing the "gill sieve".

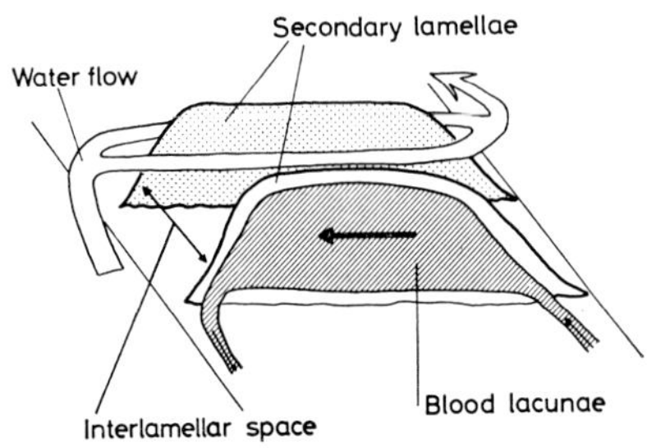

FIG. 2. Schema of the gas exchange unit in fish gills. Either an interlamellar space bounded by two secondary lamellae or a secondary lamella with adjacent two interlamellar hemi-spaces may be taken as a gas exchange unit. Interlamellar water flow and interlacunar blood flow oppose each other thus forming the basis for counter-current gas exchange.

exchange unit and the number of such units in a fish. The shape of the secondary lamella would determine the distribution of water flow along the height of the gas exchange unit. Furthermore, area and thickness of the secondary lamellar membrane are important parameters for its diffusing capacity. These parameters are considered in this study.

III. THEORY

A. BASIC FEATURES OF THE COUNTER-CURRENT SYSTEM

1. *Relative partial pressure differences*

As has been shown by Piiper and Scheid (1975), gas exchange in a counter-current system quantitatively can be assessed using the *relative partial pressure differences*, $\Delta p'$, associated with ventilation (Δp_{vent}), perfusion (Δp_{perf}) and transfer of gas between blood and medium (= water) (Δp_{tr}), as defined below:

Let,

$$\Delta p_{vent} = \frac{P_I - P_E}{P_I - P_v}, \quad (1)$$

$$\Delta p_{tr} = \frac{P_E - P_a}{P_I - P_v}, \quad (2)$$

$$\text{and } \Delta p_{perf} = \frac{P_a - P_v}{P_I - P_v}. \quad (3)$$

[P_I, P_E, partial pressures (of O_2 or CO_2) in inspired and expired water; P_a, P_v, partial pressures in arterial and venous blood].

In general, the partial pressure differences can be expressed as functions of 3 conductances: a *ventilation conductance*, $G_{vent} = \dot{V}_m \cdot \beta_m$; a *perfusion conductance*, $G_{perf} = \dot{V}_b \cdot \beta_b$; and a *diffusion conductance*, $G_{diff} = D$. $\dot{V}_m$ and $\dot{V}_b$ represent gill water ventilation and blood perfusion, respectively; β_m and β_b, *capacitance coefficients* in medium (= water) and blood (Piiper, Dejours, Haab and Rahn, 1971; see also Dejours, this volume); D, *diffusing capacity* or transfer factor. The capacitance coefficient β equals the slope of the respective dissociation curve; for a substance that is not chemically bound, β equals the physical solubility, α. As has been shown by Piiper and Scheid (1975), eqs. (1) to (3) can be expressed as simple functions of the conductances provided the following idealizing assumptions are met:

(a) the system is functionally homogeneous;

(b) $\dot{V}_m$ and $\dot{V}_b$ are constant;

(c) β_m and β_b are constant over the range of P_I to P_E and of P_a to P_v, respectively;

(d) diffusion and chemical reaction of the gases under study in blood and medium are infinitely rapid processes.

Under these conditions eqs. (1) to (3) become

$$\Delta P_{vent} = \frac{1 - e^{-Y(1-1/X)}}{X - e^{-Y(1-1/X)}} \tag{4}$$

$$\Delta P_{tr} = \frac{Xe^{-Y(1-1/X)} - 1}{X - e^{-Y(1-1/X)}} \tag{5}$$

$$\Delta P_{perf} = \frac{X(1 - e^{-Y(1-1/X)})}{X - e^{-Y(1-1/X)}} \tag{6}$$

$$X = G_{vent}/G_{perf} \qquad Y = G_{diff}/G_{perf} \tag{7}$$

These equations are useful for determination, for instance, of $D = G_{diff}$ when partial pressures in blood and water as well as both ventilation and perfusion are known, as will be shown later. For any given set of values for the conductances and for given P_I and P_v, P_a and P_E can be calculated using any two of the eqs. (4) to (6). Figure 3 shows for some values of G_{diff}/G_{perf} and G_{vent}/G_{perf} the levels of P_E and P_a relative to P_I and P_v.

It is evident from Fig. 3 that the *maximum possible* equilibration of the counter-current system, *viz.* $P_E = P_v$ and $P_a = P_I$, can only be obtained when $G_{diff}/G_{perf} \to \infty$ (i.e. in the absence of a diffusion barrier) and when at the same time $G_{perf} = G_{vent}$ (i.e. perfect matching between ventilation and perfusion conductances). On the other hand, both mismatching at infinitely high G_{diff} (upper row in Fig. 3) as well as diffusion resistance in the perfectly matched system (middle column in Fig. 3) will result in equilibration deficiencies in P_E or P_a or both.

Similarly, the *overlap* of partial pressures, $(P_E - P_a)/(P_I - P_v) < 0$, alleged to be characteristic for the counter-current, is linked to both high G_{diff}/G_{perf} ratio and matching of G_{vent} to G_{perf}. For instance, for ideally matched G_{vent}

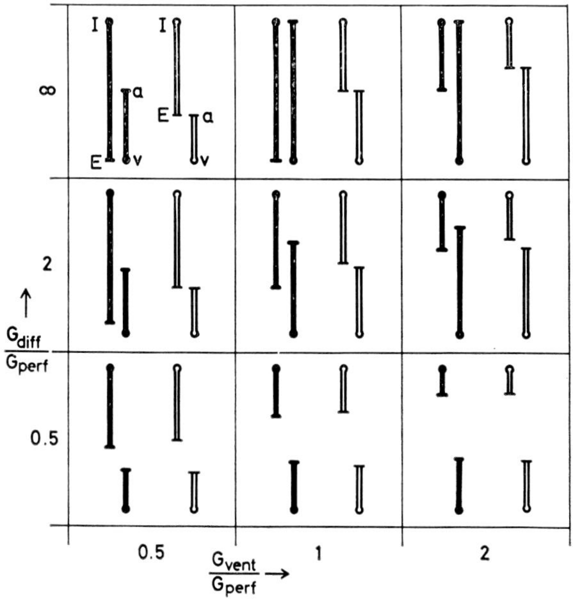

FIG. 3. Gas exchange efficacy of counter-current and co-current systems. Gas partial pressures in outflowing medium and blood (E, a) relative to their values in inflowing medium and blood (I,v) in counter-current system (closed bars) and in co-current system (open bars). Dependence on G_{vent}/G_{perf} shown in the rows, dependence on G_{diff}/G_{perf} depicted in columns.

and G_{perf}, $G_{vent}/G_{perf} = 1$, the overlap will occur for values of G_{diff}/G_{perf} exceeding unity. However, if $G_{vent}/G_{perf} = 2$, G_{diff}/G_{perf} will have to increase to beyond 1·39 for the overlap to occur.

Figure 3 shows the differences between counter-current and co-current systems in respect of partial pressures P_E and P_a. It can be seen that the differences are pronounced at high values of G_{diff}/G_{perf}; in particular, no partial pressure overlap is possible in the co-current system, the maximum equilibration being that of P_E to P_a at $G_{diff}/G_{perf} \rightarrow \infty$. However, at low values of G_{diff}/G_{perf}, the differences between the two systems virtually disappear.

In conclusion, the counter-current shows its alleged superior gas exchange performance only when ventilation and perfusion conductances are well matched and, further, when diffusion conductance is large compared with both ventilation and perfusion conductances.

2. *Limitations to gas exchange*

The limitation imposed on gas exchange by either ventilation, perfusion or transfer between medium and blood can be calculated as the relative deficit in transfer rate caused by the respective limiting process. It is appropriate to use the index L based on the following relationship between the amount of gas actually transferred, $\dot{M}_{act}$, and the amount that could be transferred if one of the three conductances were infinite, $\dot{M}_{max}$:

$$L = \frac{\dot{M}_{max} - \dot{M}_{act}}{\dot{M}_{max}} \tag{8}$$

Thus a ventilation limitation, L_{vent}, can be defined by introducing into eq. (8) $\dot{M}_{max} = \dot{M}(G_{vent \to \infty})$. Similarly, L_{perf}, and L_{diff} may be defined. As has been shown by Piiper and Scheid (1975), the L values can be expressed as functions of the two conductance ratios of eq. (7),

$$1 - L_{vent} = \frac{X(1 - e^{-Y(1-1/X)})}{(X - e^{-Y(1-1/X)})(1 - e^{-Y})} \tag{9}$$

$$1 - L_{perf} = \frac{1 - e^{-Y(1-1/X)}}{(X - e^{-Y(1-1/\ )})(1 - e^{-Y/X})} \tag{10}$$

$$1 - L_{diff} = \frac{A \cdot (1 - e^{-Y(1-1/X)})}{X - e^{-Y(1-1/X)}} \tag{11}$$

$A = 1$ for $X \leq 1$, $A = X$ for $X > 1$.

3. *Partial pressure profiles along the gill exchange surface*

The values of partial pressures in medium and blood, $P_m(L)$ and $P_b(L)$, along the length, L, of the exchange surface in the secondary lamella can be expressed as simple functions of the conductances. With the convention of eq. (7)

$$\frac{P_m(L) - P_v}{P_I - P_v} = \frac{X - e^{-Y(1-1/X)\cdot L/L_0}}{X - e^{-Y(1-1/X)}} \tag{12}$$

$$\frac{P_b(L) - P_v}{P_I - P_v} = \frac{X(1 - e^{-Y(1-1/X)\cdot L/L_0})}{X - e^{-Y(1-1/X)}} \tag{13}$$

L, distance from the blood inflowing end of the secondary lamella; total length, L_o.

Figure 4 shows $P_m(L)$ and $P_b(L)$ for three sets of values of conductances. It can be seen that only for perfect matching ($G_{vent} = G_{perf}$) are the partial pressure profiles straight lines. In this case, the mean water-to-blood partial pressure difference $P_E - P_v = P_I - P_a$ = const, its actual values being determined only by G_{diff}. For any mismatching of G_{vent} and G_{perf} the partial pressure profiles in medium and blood are curved lines, and neither $P_E - P_v$ nor $P_I - P_a$ nor their arithmetic mean are, in general, representative of the mean water-to-blood difference which is needed for calculating D (see below).

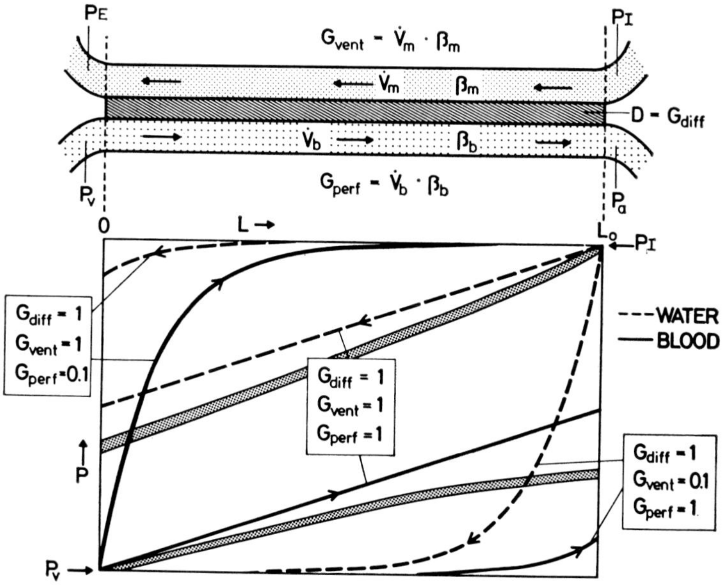

FIG. 4. Partial pressure profiles in medium and blood of a counter-current system. Gas transport in medium and blood is effected solely by convection (no diffusion resistance in both medium and blood), gas transfer from medium to blood, by diffusion. Theoretical partial pressure profiles calculated for 3 sets of conductances (dashed lines, in water; continuous lines, in blood). Profiles calculated for experimental values of Table I are shown as a band.

B. RESISTANCE TO O_2 DIFFUSION IN FISH GILLS

1. *Overall O_2 diffusing capacity*

The overall diffusing capacity, D, is defined as the amount of gas transferred, $\dot{M}$, per mean medium-to-blood partial pressure difference,

$$D = \frac{\dot{M}}{(\overline{P_m - P_b})} \tag{14}$$

Whereas $\dot{M}$ can be measured by physiological techniques, $(\overline{P_m - P_b})$ is not accessible to direct measurement. When solving eq. (4) for $G_{diff} = D$, use being made of the following relationships,

$$G_{vent} = \dot{M}/(P_I - P_E) \tag{15}$$

$$G_{perf} = \dot{M}/(P_a - P_v) \tag{16}$$

D can be expressed as

$$D = \dot{M} \cdot \frac{\ln[(P_I - P_a)/(P_E - P_v)]}{(P_I - P_a) - (P_E - P_v)} \tag{17}$$

Combination of eqs. (14) and (16) yields the relationship between the spatial mean, $\overline{(P_m - P_b)}$, $P_I - P_a$ and $P_E - P_v$,

$$\overline{(P_m - P_b)} = \frac{(P_I - P_a) - (P_E - P_v)}{\ln[(P_I - P_a)/(P_E - P_v)]} \tag{18}$$

Thus D can be determined from the partial pressures in inflowing and outflowing medium and blood.

The assumptions underlying eq. (17) are the same as those listed above for the calculation of the counter-current system. In particular, the blood and water dissociation curves are assumed to be linear. Piiper and Baumgarten-Schumann (1968a) have used a modified Bohr integration method for determining D_{O_2} in *Scyliorhinus stellaris*. In practice, however, the D_{O_2} value thus obtained did not differ much from that calculated according to eq. (17).

The overall diffusion resistance, $R_{O_2} = 1/D_{O_2}$ (see below), thus calculated is expected to result mainly from 2 different diffusion barriers—that offered by the interlamellar membrane and that offered by the interlamellar water. Both factors cannot easily be separated applying physiological measurements. Thus in the following text membrane diffusion resistance will be calculated from morphometrical measurements and the resistance offered by the interlamellar water will be assessed from calculations based on simple models.

2. *Diffusing capacity of secondary lamellar membrane, D_m*

The *membrane diffusing capacity*, D_m, can be estimated from morphometrical data and Krogh's diffusion constant, K, of the gas under study in the membrane,

$$D_m = K \cdot \frac{F}{\overline{x_h}} \tag{19}$$

$\overline{x_h}$ denotes the harmonic mean thickness of the blood-water barrier. The total area of secondary lamellae that contribute to gas exchange, F, can be calculated from the area of a single secondary lamella, f, and the total number of secondary lamellae, N

$$F = N \cdot f \tag{20}$$

3. *Diffusion in interlamellar water*

(*a*) *Inefficiency of equilibration in interlamellar water.* Unlike air breathers, water breathing animals are expected to face pronounced diffusion resistances in their breathing medium due to the low diffusivities of gases in water (10^5 to 10^6 times smaller than in air). The resulting functional diffusion barrier has to be considered together with the diffusion resistance offered by the secondary lamellar membrane in order to assess the whole diffusion resistance.

Scheid and Piiper (1971) have investigated the *diffusion in interlamellar water* and its dependence on the shape of the secondary lamellae and on the gill water velocity. To isolate the diffusion resistance from other factors affecting gas exchange, e.g. membrane diffusion and blood perfusion, these authors assumed rather artificial boundary conditions, viz. a constant partial pressure at the contact of water with gill membrane.

Figure 5 summarizes the essentials of the analysis for their Model IIIB with tapering secondary lamellae, ventilated with gill water of laminar flow profile, the mean water velocity at any given height, i.e. vertical distance from

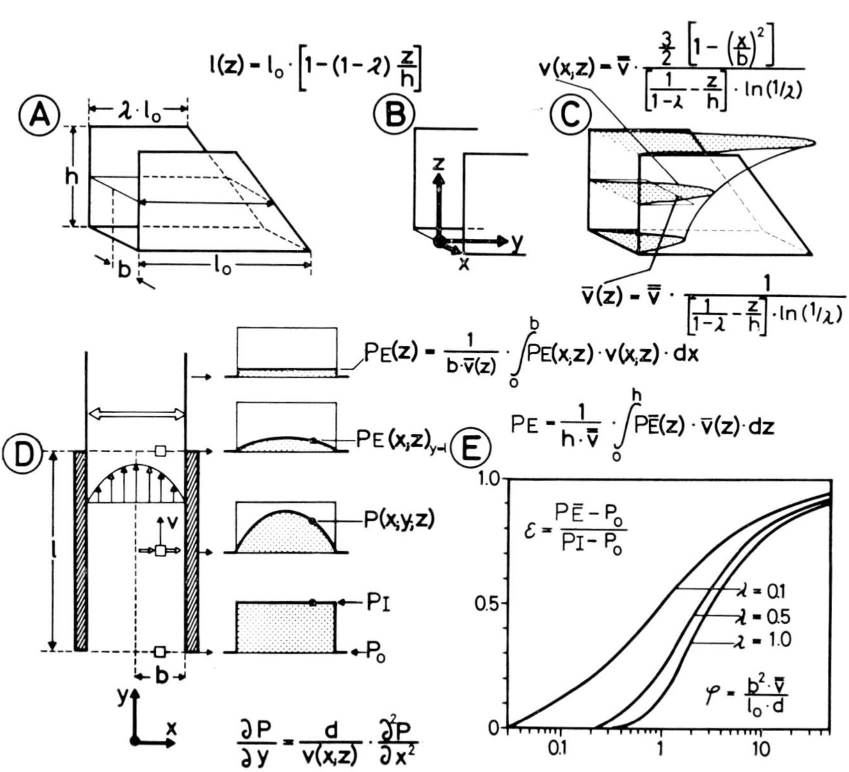

FIG. 5. Models and basic equations for calculating partial pressure equilibration in water flowing through an interlamellar space. A. Model for interlamellar space with terminology of pertinent parameters. B. Definition of space coordinates. C. Distribution of interlamellar water velocity, assuming laminar flow (parabolic flow profiles in x direction) and mean velocity at each value of z inversely related to secondary lamellar length at z. $\bar{v}$ (z), mean water velocity at a given height, z; $\bar{\bar{v}}$, overall mean water velocity. D. View from above (z direction) into interlamellar space. Convective transport solely in y direction (black arrow); diffusive transport solely in x direction (white arrow); no gas transport in z direction. To the right, partial pressure profiles at various points along interlamellar space; $P_{\bar{E}}$ (z), mean partial pressure at height z. Overall mean partial pressure, $P_{\bar{\bar{E}}}$, flow weighted mean over slit height. E. Dependence of inefficiency, ε, on parameter, φ, for different shape parameters, λ.

the base of the interlamellar space being inversely proportional to the length at this height. The Fick diffusion equation can be solved to give partial pressure of gas under study for any given point in the interlamellar space. The flow-weighted mean value of the partial pressure, averaged over both the space width and height, results in a mean expired partial pressure, $P\bar{E}$. The relative location of $P\bar{E}$ between the inspired partial pressure, PI, and the partial pressure at the boundary, P_o, can be taken as a measure for the inefficiency, ε, of partial pressure equilibration ascribed to diffusion resistance in water

$$\varepsilon = \frac{P\bar{E} - P_o}{PI - P_o} \tag{21}$$

For the case of complete equilibration of $P\bar{E}$ with P_o, ε and thus the equilibration inefficiency become o; on the contrary, if no equilibration occurs at all, $P\bar{E}$ stays at PI, and ε becomes unity.

ε can be shown to depend on a parameter, φ,

$$\varphi = \frac{b^2 \cdot \bar{v}}{l_o \cdot D} \tag{22}$$

b, half-width of the space; l_o, its length; $\bar{v}$, average linear water velocity in interlamellar space; D, diffusion coefficient.

The dependence of ε on φ for three different shapes of secondary lamellae, characterized by the factor λ, is shown in Fig. 5. The factor λ is the ratio of seondary lamella length at its top to the length at its basis.

$\bar{v}$ can be obtained by dividing the gill ventilation, $\dot{V}_m$, by the overall area offered by the gill sieve, i.e. the area perpendicular to the water velocity

$$\bar{v} = \dot{V}_m/(N \cdot 2b \cdot h)$$

h, height of the secondary lamella; 2b, width of the interlamellar space; N, total number of secondary lamellae. N, b, h as well as l_o and the shape factor λ can be obtained from morphometrical determinations.

(*b*) *Equivalent diffusing capacity*, D_{eq}. In order to assess quantitatively the limitation to partial pressure equilibration and to compare the diffusion resistance with the resistance offered by the interlamellar membrane, an *equivalent diffusing capacity*, D_{eq}, can be calculated from ε using a simple model, shown in Fig. 6. In this model the water ventilating the interlamellar space is subdivided into a central, well-mixed component with homogeneous partial pressure, and a stagnant surface layer that offers a diffusion resistance, much like the membrane resistance. In this model the partial pressure in expired water, PE, is a simple exponential function of the ventilation conductance, $G_{vent} = \dot{V}_m \cdot \beta_m$, and the equivalent diffusing capacity, D_{eq},

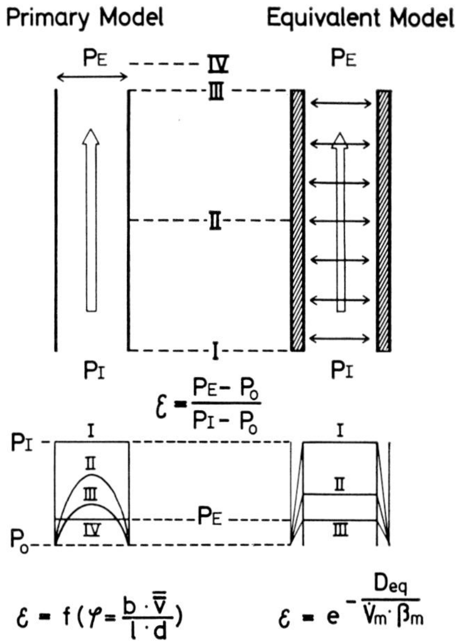

FIG. 6. Equivalent diffusing capacity for equilibration of interlamellar water. Left, situation of model of Fig. 5. Right, diffusion resistance concentrated into a stagnant surface layer (shaded) of diffusing capacity D_{eq}. Inefficiency parameter, ε, for this equivalent model is a simple function of D_{eq} and $G_{vent} = \beta_m \cdot \dot{V}_m$.

$$\frac{P_E - P_o}{P_I - P_o} = \varepsilon = e^{-D_{eq}/\dot{V}_m \cdot \beta_m} \tag{23}$$

Eq. (23) permits calculation of D_{eq} from known values of ε and $\dot{V}_m$.

4. *Balance of resistance to O_2 transfer in fish gills*

Both diffusion barriers—that of the secondary lamellar membrane and that of the surface layer—are in series with respect to O_2 transfer. Thus the diffusion resistance offered by each should be additive to give the total resistance to O_2 transfer.

The resistance to O_2 diffusion, R_{O_2}, can be defined as the reciprocal of the diffusional conductance (= diffusing capacity),

$$R_{O_2} = 1/D_{O_2} \tag{24}$$

Hence the total diffusion resistance imposed on O_2 transfer by interlamellar water and lamellar membrane, R_{O_2},

$$R_{O_2} = 1/D_{eq,O_2} + 1/D_{m,O_2} \tag{25}$$

can be compared to the overall resistance that results from introducing overall diffusing capacity into eq. (24).

IV. Application to Experimental Data

A. Physiological Measurements

Table I shows data obtained using physiological techniques in the larger spotted dogfish (Baumgarten–Schumann and Piiper, 1968; Piiper and Baumgarten–Schumann, 1968a, b).

Figure 7 summarizes the dissociation curves for O_2 and CO_2 in both water and blood. As a result of the much steeper dissociation curves of CO_2 in

Table I

Gas exchange in unanaesthetized Scyliorhinus stellaris *at rest at 17°C. Mean values (After Baumgarten-Schumann and Piiper, 1968; Piiper and Baumgarten-Schumann 1968, a, b)*

	Value	Unit
Body mass	2·2	[kg]
O_2 uptake, $\dot{M}_{O_2}$	60	[μmol·min^{-1}]
Ventilation, $\dot{V}_m$	440	[ml·min^{-1}]
Cardiac output, $\dot{V}_b$	48	[ml·min^{-1}]
Partial pressure for O_2:		
inspired water, P_I	149	[torr]
expired water, P_E	57	[torr]
arterial blood, P_a	49	[torr]
venous blood, P_v	10	[torr]
Partial pressure for CO_2:		
inspired water, P_I	0·7	[torr]
expired water, P_E	1·3	[torr]
arterial blood, P_a	2·0	[torr]
venous blood, P_v	3·0	[torr]
Capacitance coefficient:		
O_2 in water, β_{mO_2}	0·0014	[mmol·l^{-1}·torr^{-1}]
O_2 in blood, β_{bO_2}	0·034	[mmol·l^{-1}·torr^{-1}]
CO_2 in water, β_{mCO_2}	0·24	[mmol·l^{-1}·torr^{-1}]
CO_2 in blood, β_{bCO_2}	2·2	[mmol·l^{-1}·torr^{-1}]
Conductance for O_2:		
ventilation, G_{vent}	0·65	[μmol·min^{-1}·torr^{-1}]
perfusion, G_{perf}	1·53	[μmol·min^{-1}·torr^{-1}]
diffusion, $G_{diff} = D_{O_2}$	0·85	[μmol·min^{-1}·torr^{-1}]
Ratio of conductances for O_2:		
G_{vent}/G_{perf}	0·42	[−]
G_{diff}/G_{perf}	0·56	[−]
Limitation for O_2:		
ventilation, L_{vent}	0·33	[−]
perfusion, L_{perf}	0·09	[−]
diffusion, L_{diff}	0·36	[−]

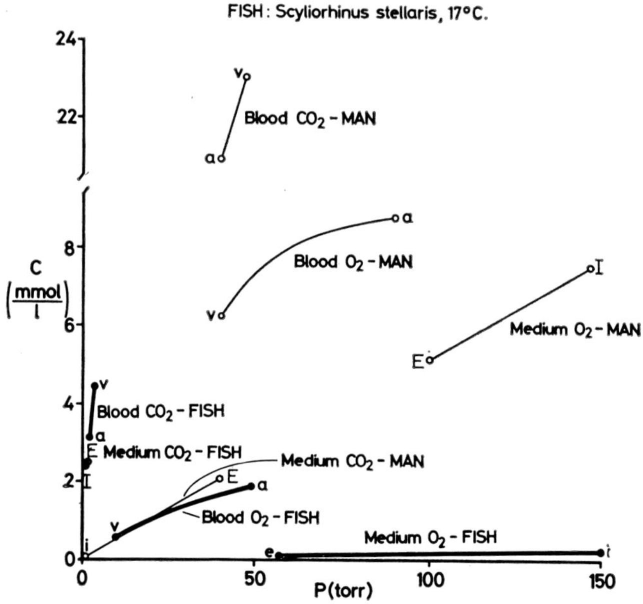

FIG. 7. Dissociation curves for O_2 and CO_2 in medium and blood. Comparison between *Scyliorhinus stellaris* (at 17°C) and man (at 37°C). Dissociation curves for O_2 and CO_2 in blood for both fish and man are curved. Pronounced differences between fish and man result from dissociation curves in medium. For man straight lines of equal slope for both O_2 and CO_2. For fish, the O_2 dissociation curve for medium is also straight (= physical solubility in water), but is much less than the slope for the blood O_2 dissociation curve, this being the main reason for high $\dot{V}_m$ compared with $\dot{V}_b$. The much greater slope of the CO_2 dissociation curve of medium in fish results in small $(P_E - P_I)_{CO_2}$ and thus in very low $P_{a_{CO_2}}$ and $P_{v_{CO_2}}$.

water and blood as compared with those for O_2, the partial pressures for CO_2 cover an extremely narrow range of little over 2 torr. The accuracy of the P_{CO_2} values in relation to their narrow range does not permit a meaningful quantitative analysis on the basis outlined in the section "Theory" which will thus be restricted to the analysis of the P_{O_2} values.

B. CONDUCTANCES, LIMITATIONS, AND PARTIAL PRESSURE PROFILES

Table I shows that $\dot{V}_m$ exceeds $\dot{V}_b$ by almost a factor of 10. However, due to the much steeper slope of the blood O_2 dissociation curve compared with the O_2 solubility in water, G_{vent} and G_{perf} are much less different, the ratio G_{vent}/G_{perf} being 0·42. Since this ratio, rather than the ratio of $\dot{V}_m/\dot{V}_b$, has to be considered for analysis of gas exchange, ventilation and perfusion are reasonably matched in the animal of Table I.

It can be seen from Table I that the ventilation limitation, 33%, is considerably higher than that for perfusion, 9%. Thus an increase in ventilation will increase O_2 exchange much more than an equal increase in perfusion. This is opposed to the situation in mammals where hyperventilation only has a slight effect on O_2 exchange. The diffusion limitation for O_2 is marked, 36%, a value that is much higher than that for mammals, even in hypoxic conditions.

The partial pressure profiles expected to prevail along the secondary lamella of *Scyliorhinus stellaris* can be calculated applying eqs. (12) and (13) to the conductance ratios of Table I. The result is shown in Fig. 4.

C. RESISTANCE TO O_2 DIFFUSION

1. *Overall diffusing capacity*

An overall D_{O_2} can be calculated from values of Table I using eq. (17). This value is listed in Table I. Piiper and Baumgarten–Schumann (1968a) have applied a Bohr integration technique to these values to obtain D_{O_2} taking into account the non-linear dissociation curve of O_2 in blood. Their value did not significantly differ from that calculated using eq. (17).

2. *Membrane diffusing capacity of secondary lamella*

G. M. Hughes, S. Perry and J. Piiper (unpublished observations) have determined morphometrical parameters in gills of *Scyliorhinus stellaris*. The values are listed in Table II, together with Krogh's diffusion constant for O_2. The membrane diffusing capacity for O_2, D_{m,O_2}, calculated using eq. (19) is about twice as large as the overall diffusing capacity (Table I). In calculating D_{m,O_2}, the total secondary lamellar surface area was used, without taking into account the fact that its fraction which is subtended by the pillar cells (possibly about 10%) should be rather ineffective in gas transfer.

TABLE II

Morphometrical data on Scyliorhinus stellaris (*Hughes, Perry and Piiper, unpublished observations*)

Body weight	2·2 kg
Area of a secondary lamella, f	1·1 mm^2
Number of secondary lamellae, N	224 600
Total area of secondary lamellae, F = N·f	0·248 m^2
Mean harmonic water-blood distance, $\bar{x}_h$	9·9 μm
Krogh's diffusion constant K_{O_2}	$6{\cdot}76 \times 10^{-10}$ $mmol \cdot min^{-1} \cdot cm^{-1} \cdot torr^{-1}$
D_{mO_2}	1·70 $mmol \cdot min^{-1} \cdot torr^{-1}$

3. *Equivalent diffusing capacity of interlamellar water*

Figure 8 shows the shape of a typical secondary lamella in *Scyliorhinus stellaris*. The top length is about 60% of the base length, l_o; thus λ (Fig. 5) is 0·6. Figure 8, moreover, shows morphometrical data on the gills of *Scyliorhinus stellaris* (G. M. Hughes, S. Perry and J. Piiper, unpublished observations) that are pertinent to the analysis of O_2 equilibration in the interlamellar water. Applying eq. (22) to these data, a value of $\varphi = 0{\cdot}44$ can be calculated for resting *Scyliorhinus stellaris*.

Figure 9 shows that under these conditions the inefficiency of O_2 equilibration is only 5%. The equivalent diffusing capacity for interlamellar water, D_{eq}, resulting from eq. (23) using the value of ε and $\dot{V}$ of Table I is 2·00 $\mu mol \cdot min^{-1} \cdot torr^{-1}$.

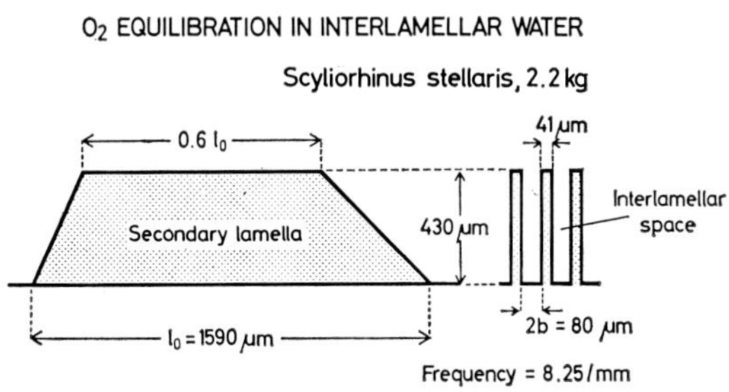

Number of secondary lamellae N = 224 600

Cross sectional area of an interlamellar space $a = 34400\ \mu m^2$

Total cross sectional area of interlamellar spaces $A = N \cdot a = 77.2\ cm^2$

Diffusion coefficient of O_2 in water at 17 °C: $d_{O_2} = 13 \cdot 10^{-4} \frac{cm^2}{min}$

Resting ventilation, 17 °C $V_m = 440$ ml/min

Mean water velocity $\bar{v} = \dot{V}_m / A = 5.7$ cm/min

$$\varphi = \frac{b^2 \cdot \bar{\bar{v}}}{l_0 \cdot d_{O_2}} = 0.44$$

FIG. 8. Dimensions of interlamellar space and physiological data used for calculating O_2 equilibration in interlamellar water for *Scyliorhinus stellaris* of 2·2 kg body mass.

4. *Balance of O_2 diffusing resistances*

Table III summarizes the diffusing capacities: the overall total diffusing capacity, obtained by analysis of O_2 partial pressures in water and blood on the basis of a perfect counter-current system; the equivalent interlamellar water diffusing capacity; and the membrane diffusing capacity derived from morphometrical data.

It can be seen that the diffusion resistances offered by interlamellar water and by blood-to-water membrane are of the same magnitude and add up to

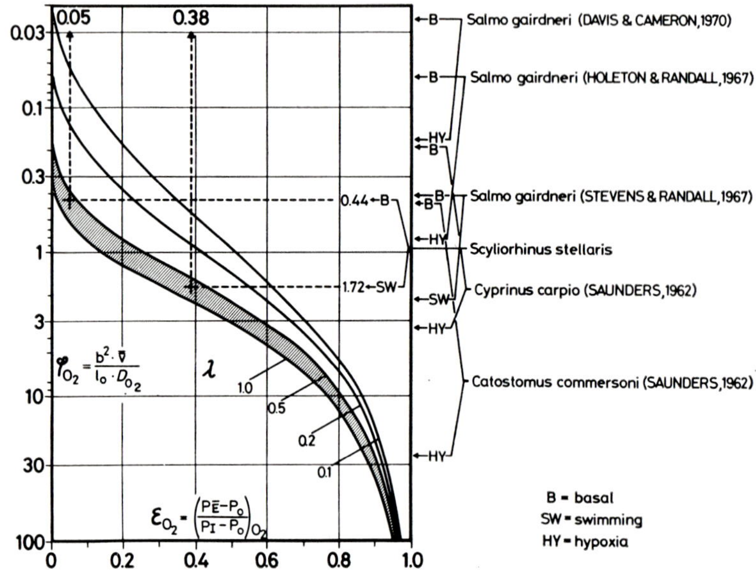

FIG. 9. Plot of parameter φ_{O_2} (ordinate) against ε_{O_2} (abscissa). The shaded area, λ between 0·5 and 1·0, gives range of shape parameters observed in fish. Aside from values calculated for *Scyliorhinus stellaris* in this study at rest and during swimming activity, there are values of φ calculated from literature data, and ε values resulting therefrom, for measurements both during physiological rest and during increased ventilation due to hypoxia or to swimming activity.

about 90% of the total resistance to O_2 transfer. This result is in agreement with the conclusions of Hills and Hughes (1970) who used a completely different method of analysis.

V. DISCUSSION

A. COUNTER-CURRENT VS. CO-CURRENT GAS EXCHANGE

It has been pointed out above that the differences in gas exchange performance of the counter-current and the co-current systems are pronounced only under conditions of high diffusion conductance. When applying the physiological measurements reported in Table I to Fig. 3 it can be seen that differences between both systems are virtually absent for this species under resting conditions.

B. PARTIAL PRESSURE EQUILIBRATION IN FISH GILLS

It has been claimed by some authors (cf. Randall, 1970, p. 261) that in the absence of diffusion resistance the partial pressure in expired water leaving

TABLE III

Balance of diffusion resistances to O_2 uptake in gills of unanaesthetized Scyliorhinus stellaris *at rest. Body mass 2·2 kg*

	D_{O_2} (μmol·min^{-1}·torr^{-1})	R_{O_2} (μmol^{-1}·min·torr)	R_{rel} (%)
Overall	0·85	1·18	100
Interlamellar water	2·00	0·50	42
Blood water barrier	1·70	0·59	49
Other		0·09	9

an ideal counter-current system should be equilibrated to the partial pressure in venous blood. That this generally is not so can easily be seen from the top row of Fig. 3 where for $G_{vent}/G_{perf} > 1$, P_E never equilibrates to P_v, not even in an ideal counter-current of no diffusion resistance. Thus it is essential to consider the matching of G_{vent} and G_{perf} together with G_{diff} for a proper analysis of partial pressure equilibration.

For this reason, the formula given by Randall (1970) to calculate a water shunt for a counter-current in the absence of diffusion resistance from differences in $P_E - P_v$ has to be questioned when $G_{vent}/G_{perf} > 1$. We would suggest replacing the denominator of his formula for the water shunt on pp. 261 and 262 with $(P_I - P_v)_{O_2}$.

Hughes and Shelton (1962) have introduced the partial pressure ratios of eqs. (1) and (3), termed here Δp_{vent} and Δp_{perf}, to designate "effectiveness of gas exchange". It should be noted that both ratios consider only one side of the gas exchange organ, the ventilation or the blood perfusion. Piiper and Scheid (1971, 1975) have tried to define the gas exchange efficiency of vertebrate gas exchange organs in more general terms. Thus they have considered ΔP_{tr} of eq. (2) as a parameter useful to designate efficacy of partial pressure equilibration, considering both medium and blood.

C. PARTIAL PRESSURE PROFILE AND MEAN PARTIAL PRESSURE DIFFERENCE BETWEEN GILL WATER AND BLOOD

Randall, Holeton and Stevens (1967) have proposed the calculation of the overall diffusing capacity, which they called transfer factor of the gills, on the basis of a formula similar to our eq. (14) but using for mean medium-to-blood partial pressure difference the arithmetic mean between the partial pressure values on both ends of the secondary lamella:

$$(\overline{P_m - P_b})_{O_2} = \tfrac{1}{2}(P_I - P_a)_{O_2} + \tfrac{1}{2}(P_E - P_v)_{O_2} \qquad (26)$$

When this approximation is compared with the exact mean value of eq.

(18), applying to the counter-current system for linear dissociation curves, it can be seen that both agree only for the case of $P_I - P_a = P_E - P_v$. It is evident from eqs. (15) and (16) that this condition is fulfilled only for $G_{vent} = G_{perf}$, that is for perfect matching of the system. In this case eqs. (12) and (13) yield

$$\frac{P_m(L) - P_v}{P_I - P_v} = \frac{1 + Y \cdot L/L_o}{1 + Y} \tag{27}$$

$$\frac{P_b(L) - P_v}{P_I - P_v} = \frac{Y \cdot L/L_o}{1 + Y} \tag{28}$$

which are the mathematical expressions of straight, parallel lines. The relative water-to-blood partial pressure difference then becomes

$$\frac{\overline{(P_m - P_b)}}{P_I - P_v} = \frac{1}{1 + Y} \tag{29}$$

which is constant along the secondary lamella and depends only on $Y = G_{diff}/G_{perf} = G_{diff}/G_{vent}$ (since $G_{vent} = G_{perf}$).

The error introduced by calculating $\overline{(P_m - P_b)}$ according to eq. (26) rather than to eq. (18) increases with increasing mismatching of G_{vent} and G_{perf}. For the values of Table I the partial pressure profiles in water and blood are almost straight lines (Fig. 3), although not parallel. $\overline{(P_m - P_b)}$ calculated according to eq. (26) overestimates the value calculated from eq. (18) by 5%, and hence overall D_{O_2} would be underestimated by the same amount.

However, larger errors are introduced by more severe mismatching like those of Fig. 4. For the set of $G_{diff} = 1$, $G_{vent} = 0{\cdot}1$ and $G_{perf} = 1$, P_E virtually equilibrates to P_v. This example is instructive for two reasons:

(1) Practically complete partial pressure equilibration of P_E to P_v or of P_a to P_I but not of both is possible even for finite diffusion conductance, if only G_{vent} and G_{perf} are sufficiently different. On the other hand, as is demonstrated in Fig. 3, infinite diffusion conductance does not necessarily lead to partial pressure equilibration of P_E to P_v or P_a to P_I. Thus equilibration of P_E to P_v does not necessarily mean absence of diffusion resistance as has been proposed by Randall (1970, p. 262).

(2) $\overline{(P_m - P_b)}$ calculated according to eq. (26) yields a value that overestimated the true value [calculated by using eq. (18)] by a factor of 4·5 and thus would underestimate D_{O_2} by the same factor. However, in cases in which P_E or P_a equilibrate with P_v or P_I, respectively, the partial pressure profiles cannot be calculated using eq. (18) and thus $\overline{(P_m - P_b)}$ becomes a highly elusive quantity. Thus D, although finite, cannot be calculated with certainty. In practical cases, however, G_{vent} and G_{perf} are not far from being equal and overall D_{O_2} can be calculated from P_I, P_v, P_E and P_a. Whenever

possible, eq. (18) should be preferred to eq. (26), as suggested by Hughes (1972).

D. DIFFUSION RESISTANCES IN WATER AND MEMBRANE

The analysis of diffusion resistances in water is subject to several idealizing assumptions. Water flow was assumed to be laminar, which might be true for the low values of linear velocity of gill water. Nothing is known, however, about the distribution of water flow at different heights in the interlamellar space, which was assumed to be inversely related to the length of the secondary lamella at a particular height and thus to the resistance of water flow at that height. Moreover, for the analysis, diffusion in the z direction (Fig. 5) was neglected.

The most critical assumption, however, concerns the boundary condition, viz. the constancy of partial pressure at the secondary lamellar surface, which was chosen in order to study the effect of diffusion in interlamellar water in isolation. This condition assumes that both G_{diff} and G_{perf} are infinite, the partial pressure in blood being P_o. In a simple model, the inefficiency of water equilibration was ascribed to diffusion in a stagnant membrane and its diffusion resistance was added to that of the secondary lamellar membrane. It has to be shown in more sophisticated models if this addition of equivalent water resistance, $1/D_{eq}$, and membrane resistance, $1/D_m$, yields a reasonable approximation to the overall resistance to diffusion in a composite system, in which both water diffusion and membrane diffusion limit gas transfer.

Scheid and Piiper (1971) have used literature data on both physiological and morphometrical values in 3 species of teleost fish to calculate the parameter φ_{O_2} and therefrom the inefficiency parameter ε_{O_2} (Fig. 9). The authors have discussed the difficulties inherent in calculation of φ_{O_2} from literature values, the most pronounced one being that for most studies physiological and complete morphometrical data are not available for fishes of the same origin. Moreover, measurement of gill ventilation in teleosts is difficult for methodological reasons. This may contribute to the wide scatter in calculated φ_{O_2} and in ε_{O_2} resulting therefrom.

The values for basal conditions are low, the ε_{O_2} value of 5% calculated in this study being the upper limit. During swimming activity, gill ventilation increases and thus both φ_{O_2} and ε_{O_2} increase. The ε value for *Scyliorhinus stellaris* during swimming, calculated using values collected by Drees (unpublished observations), approaches 38%; a similar value can be calculated for the trout (Fig. 9). Similarly, ε_{O_2} increases with increased ventilation due to hypoxia.

Averaging the values listed in Fig. 9, ε_{O_2} is about 2% for basal conditions and approaches 40% for nonbasal conditions.

E. OTHER FACTORS THAT MAY LIMIT GAS EXCHANGE IN FISH GILLS

It can be seen from Table III that although $1/D_{eq}$ and $1/D_m$ constitute the main factors for limiting O_2 transfer in the gills of *Scyliorhinus stellaris* in basal conditions, they do not account quantitatively for the overall resistance. This may in part be due to the limited validity of the model analysis, particularly the validity of the additivity of $1/D_{eq}$ and $1/D_m$, and in part to the limited accuracy of the underlying measurements, both physiological and morphometrical. However, other factors that may limit O_2 exchange should be considered.

(a) *Water shunt* at both the upper edge of the secondary lamella and between adjacent filaments. It should be noted that an equivalent water shunt offered by diffusion resistance in interlamellar water is accounted for by the value of the equivalent water resistance, $1/D_{eq}$.

(b) *Blood shunt.* Various possible pathways have been described for blood shunt (Randall, 1970).

(c) *Unequal distribution* of water and blood flow. This condition leads to inefficiency of branchial gas exchange basically in the same manner as the well-studied unequal distribution of alveolar ventilation to pulmonary capillary blood flow produces O_2 and CO_2 equilibration deficit in mammalian lungs.

(d) *Diffusion in blood.* Since D_m was calculated from morphometrical data on the blood/water barrier only, diffusion in blood plasma, red cell membrane and interior was neglected.

(e) *Reaction with haemoglobin.* This factor is expected to play a considerable role at the low body (= water) temperatures usually encountered in fishes in contrast to mammals and birds.

(f) *Pulsatility* of water and blood flow. Unsteady water or blood flow or both is expected to cause inefficiency of gas exchange much like an unequal distribution in space of ventilation, blood flow and diffusing capacity. The disturbance would be least with synchronous pulsations, of equal magnitude and pattern, of both water and blood flow.

Each of these factors, and possibly many other factors, should be studied in some detail if a comprehensive analytical picture of branchial gas exchange is to be established.

VI. Summary

A theoretical approach to analysis of gas transfer in a model of fish gills was developed. Its applicability was exemplified by application to experimental data obtained in the elasmobranch *Scyliorhinus stellaris*.

The following aspects were specifically investigated because of their amenability to quantitative treatment:

(1) Water/blood counter-current flow, particularly the conditions for its superior gas exchange efficacy as compared to co-current flow systems;

(2) O_2 equilibration of water flowing through an interlamellar space with the bounding secondary lamellar surfaces;

(3) O_2 transfer through the tissue layer separating the intralemellar blood from surrounding water (water-blood barrier).

For the O_2 transfer in resting *Scyliorhinus stellaris* the effective resistance of the interlamellar water and of the water-blood barrier are estimated to be of about equal magnitude and to amount jointly to about 90% of the total resistance to O_2 transfer (= reciprocal diffusing capacity or transfer factor), thus leaving little scope to other factors expected to limit branchial O_2 transfer. However, these factors, comprising water and blood shunts; unequal distribution and pulsatility of water and blood flow; diffusion and chemical reaction in blood, may be more effective in limiting branchial gas exchange in other species or in other physiological conditions.

References

Baumgarten-Schumann, D. and Piiper, J. (1968). Gas exchange in the gills of resting unanesthetized dogfish (*Scyliorhinus stellaris*). *Resp. Physiol.* **5**, 317–325.

Davis, J. C. and Cameron, J. N. (1970). Water flow and gas exchange at the gills of rainbow trout, *Salmo gairdneri*. *J. exp. Biol.* **54**, 1–18.

Fung, Y. C. and Sobin, S. S. (1969). Theory of sheet flow in lung alveoli. *J. Appl. Physiol.* **26**, 472–488.

Holeton, G. F. and Randall, D. J. (1967). The effect of hypoxia upon the partial pressure of gases in the blood and water afferent and efferent to the gills of rainbow trout. *J. exp. Biol.* **46**, 317–327.

Grigg, G. C. and Read, J. (1972). Gill function in an elasmobranch. *Z. vergl. Physiol.* **73**, 569–574.

Hills, B. A. and Hughes, G. M. (1970). A dimensional analysis of oxygen transfer in the fish gill. *Res. Physiol.* **9**, 126–40.

Hughes, G. M. (1972). Morphometrics of fish gills. *Resp. Physiol.* **14**, 1–26.

Hughes, G. M. and Hills, B. A. (1971). Oxygen tension distribution in water and blood at the secondary lamella of the dogfish. *J. exp. Biol.* **55**, 399–408.

Hughes, G. M. and Morgan, M. (1973). The structure of fish gills in relation to their respiratory function. *Biol. Rev.* **48**, 419–475.

Hughes, G. M. and Shelton, G. (1962). Respiratory mechanisms and their nervous control in fish. *Adv. Comp. Physiol. Biochem.* **1**, 275–364.

Johansen, K. (1971). Comparative physiology: Gas exchange and circulation in fishes. *Ann. Rev. Physiol.* **33**, 569–612.

Piiper, J. and Baumgarten-Schumann, D. (1968a). Transport of O_2 and CO_2 by water and blood in gas exchange of the dogfish (*Scyliorhinus stellaris*). *Resp. Physiol.* **5**, 326–337.

Piiper, J. and Baumgarten-Schumann, D. (1968b). Effectiveness of O_2 and CO_2 exchange in the gills of the dogfish (*Scyliorhinus stellaris*). *Resp. Physiol.* **5**, 338–349.

Piiper, J. and Scheid, P. (1971). Maximum gas transfer efficacy of models for fish gills, avian lungs and mammalian lungs. *Resp. Physiol.* **14**, 115–124.

Piiper, J. and Scheid, P. (1975). Gas transport efficacy of gills, lungs and skin: theory and experimental data. *Resp. Physiol.* **23**, 209–221.

Piiper, J., Dejours, P., Haab, P. and Rahn, H. (1971). Concepts and basic quantities in gas exchange physiology. *Resp. Physiol.* **13**, 292–304.

Randall, D. J. (1970). Gas exchange in fishes. In: "Fish Physiology" (ed. W. S. Hoar and D. J. Randall), Vol. IV, pp. 253–292. Academic Press, New York and London.

Randall, D. J., Holeton, G. F. and Stevens, E. D. (1967). The exchange of oxygen and carbon dioxide across the gills of rainbow trout. *J. exp. Biol.* **46**, 339–348.

Saunders, R. L. (1962). The irrigation of the gills in fishes. II. Efficiency of oxygen uptake in relation to respiratory flow activity and concentration of oxygen and carbon dioxide. *Can. J. Zool.* **40**, 817–862.

Scheid, P. and Piiper, J. (1971). Theoretical analysis of respiratory gas equilibration in water passing through fish gills. *Resp. Physiol.* **13**, 305–318.

Stevens, E. D. and Randall, D. J. (1967). Changes of gas concentrations in blood and water during moderate swimming activity in rainbow trout. *J. exp. Biol.* **46**, 329–337.

Environment and Ecology of Freshwater Air-breathing Teleosts

P. V. DEHADRAI and S. D. TRIPATHI

Central Inland Fisheries Research Institute, Barrackpore, West Bengal, India

I. Introduction

In all tropical regions of the world, occurrence of derelict and swampy water masses with adverse ecological conditions is common. Survival of fishes in this environment is meagre. However, there exists a very interesting group of curious, amphibious teleosts, mostly confined to freshwater or partially saline environments in various parts of the tropical world, which are able to thrive in such ecologically adverse swamps. These fishes are different from others by virtue of adaptations such as development of accessory respiratory organs, modifications of their blood physiology, thyroid hormone specializations, occurrence of diving reflexes and adaptations in their food and feeding habits, reproduction and behaviour.

The air-breathing fishes have been studied considerably because of their peculiar double-respiratory mechanism. However, much remains to be understood about the other life processes of these fishes in relation to their environment.

II. Systematic Status

A wide variety of teleostean fishes which need to breathe air owing to the special conditions of their life have evolved specialized organs: in some, the

TABLE I

Systematic position, habitat and types of organs of air-breathing teleosts

Sl.	Species	Air-breathing organ	Taxonomic position: Order	Taxonomic position: Family	Habitat	Reference
1	*Electrophorus electricus*	Bucco-pharyngeal	Cypriniformes	Electrophoridae	Rivers and swamps —S. America	Carter (1935b) Johansen *et al.* (1968)
2	*Ophicephalus* (≡*Channa*) (*O. punctatus*, *O. marulius*, *O. striatus*)	Pharyngeal lung	Ophicephaliformes	Ophicephalidae	Tropical ponds and rivers —Asia and Africa	Rauther (1910) Das (1927) Qasim *et al.* (1960) Hughes and Munshi (1973a)
3	*Amphipnous cuchia*	Pharyngeal lung	Symbranchiformes	Amphipnoidae	Rivers, ponds and swamps —Southern Asia	Das (1940)
4	*Hypopomus brevirostris*	Opercular chamber	Cypriniformes	Sternarchidae	Swamps —S. America	Carter and Beadle (1931b)
5	*Symbranchus marmoratus*	Opercular chamber	Symbranchiformes	Symbranchidae	Swamps —S. America	Taylor (1913) Johansen (1966)
6	*Monopterus javanensis*	Opercular chamber	Symbranchiformes	Symbranchidae	Fresh waters —Southern Asia	Wu and Kung (1940)
7	*Pseudapocryptes lanceolatus*	Opercular chamber	Perciformes	Gobiidae	Pools and swamps —S. Asia	Das (1927)

8	*Heteropneustes* (≡*Saccobranchus*) *fossilis*	Opercular lung	Cypriniformes	Saccobranchidae	Ponds and swamps —Sri Lanka, India, Burma, Laos, Thailand, Vietnam	Das (1927)
9	*Clarias* (*C. lazera* and *C. magur*)	Opercular lung	Cypriniformes	Clariidae	Ponds and swamps —Africa and South and West Asia	Rauther (1910) Moussa (1956) Munshi (1961)
10	*Macropodus cupanus*	Opercular lung	Perciformes	Anabantidae	Tropical ponds —Asia	Rauther (1910) Das (1940)
11	*Colisa fasciata*	Opercular lung	Perciformes	Anabantidae	Freshwater —S. Asia	Munshi (1965)
12	*Betta*	Opercular lung	Perciformes	Anabantidae	Freshwater —S. Asia	Rauther (1910)
13	*Osphronemus* (*O. gorami*)	Opercular lung	Perciformes	Anabantidae	Freshwater —S. Asia	Rauther (1910) Das (1940)
14	*Anabas testudineus*	Opercular lung	Perciformes	Anabantidae	S. Asia, Indo-Malayan Archipelago; Tropical and S. Africa	Rauther (1910) Das (1927)
15	*Ancistrus anisitsi*	Stomach			Swamps —S. America	Carter and Beadle (1931b)
16	*Plecostomus plecostomus*	Stomach	Cypriniformes	Loricariidae	Swamps —S. America	Jobert (1878)
17	*Misgurnus fossilis*	Intestine	Cypriniformes	Cobitidae	Rivers and pools —Europe and Asia	Calugareanu (1907) Leiner (1938)

TABLE I (*continued*)

Sl.	Species	Air-breathing organ	Taxonomic position Order	Family	Habitat	Reference
18	*Lepidocephalichthys guntea*	Intestine	Cypriniformes	Cobitidae	Freshwaters —Asia	Das (1937)
19	*Doras*	Intestine	Cypriniformes	Doradidae	Rivers and swamps —S. America	Jobert (1878)
20	*Hoplosternum litorale*	Intestine			Swamps —S. America	Carter and Beadle (1931b)
21	*Arapaima gigas*	Swimbladder	Clupeiformes	Arapaimidae	Swamps and R. Amazon —S. America	Sawaya (1946)
22	*Gymnarchus*	Swimbladder	Mormyriformes	Gymnarchidae	Swamps and Rivers —S. Africa	Assheton (1907)
23	*Erythrinus unitaeniatus*	Swimbladder	Cypriniformes	Characinidae	Swamps —S. America	Carter and Beadle (1931b)
24	*Umbra*	Swimbladder	Clupeiformes	Umbridae	Stagnant waters —Europe and N. America	Geyer and Mann (1939)
25	*Notopterus* (*N. notopterus N. chitala*)	Swimbladder	Clupeiformes	Notopteridae	Freshwater —Asia	Dehadrai (1960 and 1962)
26	*Phractolaemus ansorgii*	Swimbladder	Clupeiformes	Phractolaemidae	Tropical —West Africa	Audenaerdae (1959)
27	*Anguilla* (*A. anguilla, A. bengalensis, A. japonicus*)	Skin	Anguilliformes	Anguillidae	Rivers —Europe, Asia, Africa and N. America	Krogh (1904) Berg and Steen (1965)

swimbladder; in others, altogether new organs; in a few, the gills; and in still others, the skin has taken up the respiratory function. These teleosts belong to different orders but, by virtue of their special adaptations to aerial respiration, they are generally grouped together (Table I).

The presence of aerial respiration in teleosts is a typical example of convergent adaptation towards a particular environment to which were exposed species of widely separated orders and families during the course of evolution. Respiratory adaptations alone could not have been adequate to enable these fishes to face the inimical conditions of life. Since the environment of shallow tropical waters is set with adversely changing patterns through their various phases of dereliction, it demands suitable overall changes in various life processes of air-breathing fishes.

III. Nature of the Environment

Air-breathing fishes are by no means restricted to derelict and swampy waters; most species are abundant in all types of waters. However, the incompatible air-breathing fishes have been systematically eliminated, as an ideal fishery practice, from clean water ponds and lakes which form easily manageable culture fishery resources. Thus, there remain shallow, derelict and swampy waters of the tropical world, which may be discussed as the habitual abode of air-breathing fishes.

The use of the term swamp in relation to an aquatic habitat may be deceiving. Water areas which become unusable for conventional human activities pertaining to commercial prospecting could be regarded as swampy or derelict. Ecologically, the swamps, through their different phases of dereliction, are considered to be the final stage of the evolution of land from lakes (Hora and Pillay, 1962). The occurrence of amphibious vertebrates in swamps is thus synchronous with the transitional phase of the environment. The gradual process of dereliction of water bodies is caused both by natural and man-made factors. Periodic exposure of small water areas to extreme drought conditions, failure of rains with high atmospheric temperature causing rapid evaporation of water and bringing it to a dangerously low level can easily turn the low water column derelict and hostile to the aquatic fauna, particularly fishes. Further, the alluvial soil system commonly occurring in the tropics is highly conducive to organic decay and siltation. Such biologically derelict water masses are automatically neglected by man and their reclamation postponed. Infestation of these waters with commonly occurring aquatic weeds—floating, emergent or submerged—in almost unmanageable proportions, further precludes their usefulness as a healthy aquatic habitat. On the other hand, for example, the flood control measures applied to the rivers in Bihar (India) have kept a large number of functional lakes and ponds devoid of their annual replenishment of water thus exposing them to the inevitable process of dereliction. There are records of many functional water bodies having become totally derelict swamps in north-east and southern India with heavy siltation, shallow water level, profuse infestation of aquatic weeds, grasses and other aquatic flora (Philipose, 1963; David, Rao and Ray, 1974).

However, many swamps retain a definite connection with the rivers, being themselves a discarded river course, and replenish their water (including fish) every year as in north-east India and Guiana whereas certain other swamps are completely enclosed and stagnant as in central and southern India, Paraguayan Chaco and Uganda.

Although the general ecology of tropical swamps has been characterized, there is a paucity of information on the detailed environmental features of these ecosystems. Certain reports worth recording are by Beadle (1932 and 1958), Carter (1935a, b, 1955), Carter and Beadle (1931a, b), Weber (1942), Migahid (1947), Talling (1957a, b), Ling (1960), Bishai (1962), Talling and Rzoska (1967), Kurdin (1968), Mizuno and Mori (1970). Earlier work on Indian swamps mainly pertained to ecology of aquatic macroflora (Philipose, 1963; Philipose, Ramachandran, Singh and Ramaprabhu, 1970; David *et al.*, 1974; Radhakrishnan and Bhuyan, MS). Recently, detailed studies on the environmental features of swamps have been made in Assam, Bihar and Karnataka States of India in connection with the understanding of the environment for culture of air-breathing fishes (Dehadrai, Thakur and Das, MS).

While surveying the swamps and low-lying derelict waters in north-east and south India, four different types of aquatic ecosystems were observed to be abundantly habited by air-breathing fishes during different periods of the year: (1) *Polluted swamps*, near or in townships are generally small water bodies receiving town sewage. The waters have a high BOD and exhibit permanent algal blooms. The bottom silt is very rich with organic matter and other nutrients. The balance of carbon dioxide and oxygen is usually far from normal. (2) *Semi-senescent* or *senescent* swamps on the basis of the degree and period of dereliction are other abodes of air-breathing fishes. Derelict shallow water bodies, in proximity with human habitation, are usually manipulated for commercial use by cultivating aquatic crops such as "Makhana" (*Euryale ferox*, Salisb.) and "Singhara" (*Trapa* spp.) (Fig. 1). Such activities prevent the water areas from becoming completely swampy and thus remain in a *semi-senescent* condition. (3) The *senescent swamps* are those which have reached a point of no return in regard to their dereliction. They apparently look like grasslands (Fig. 2) but with water 0·5–2 m deep below them. (4) Low-lying areas (*Chaurs*) and paddy fields get inundated during the rainy season either by the spilling over of neighbouring rivers and lakes or due to the rushing water from the catchment areas. Air-breathing teleosts migrating to breed in shallow waters during the rainy season, abound in such areas for a good part of the year till these regions dry up.

FIG. 1. Plantation of "makhana" (*Euryale ferox* Salisb.) in a semi-senescent swamp in North Bihar, India.

FIG. 2. A 9-hectare senescent swamp with a mat of floating and emergent macrophytes situated near Tamuria Railway Station, about 56 km from Darbhanga in North Bihar, India.

Decaying terrestrial vegetation under inundated areas provide the protection and food for air-breathing teleosts for considerable periods. In fact these shallow rainfilled areas form the most easily exploitable fishery grounds in rural areas. The practice of building coffer dams and dewatering the enclosed areas alternately yields considerable quantities of air-breathing fishes (Fig. 3).

Morphologically, the swamps are low-lying, waterlogged, shallow areas with a loose peaty bottom rich in decaying organic matter, retaining water either perennially, or shrinking or drying up completely during the summer

FIG. 3. Rain-filled, low-lying water areas divided by coffer dams to retrieve air-breathing fishes in North Bihar, India.

months. The water surface in direct contact with the air and sunlight is substantially reduced which influences the dissolved gases and productivity of water. The excessive accumulation of plant remnants shallows up the bottom. The hard parts of the plants do not mineralize quickly and reduce productivity. The accumulation of silt among the dense roots and stems leads to the formation of floating masses which grow rapidly in association with other plants to form floating islands. The water level rapidly decreases mainly due to transpiration by floating plants and the essential nutrients are utilized by the weeds at the expense of fish food organisms and fish.

Due to the accumulation of organic matter, the soil tends to be acidic which in turn retards mineralization and circulation of nutrients. Swamps situated in lateritic soils are all the more acidic. In such a situation, nitrogen fixation by bacteria is slow and the soil tends to bind nutrients (Hora and Pillay, 1962). Circulation of phosphates from such soils is also impaired as ferric and aluminium phosphate (Matida, 1956; Hepher, 1958). In most swamps which drain regularly, however, phosphates are not high but they accumulate in stagnant waters where they are not removed by currents or by animal and plant life. The normal metabolic cycle is essentially disrupted by the presence of a dense growth of vegetation (Ling, 1960). The shading effect of macro-vegetation mat causes the water body to be completely pocketed, adversely affecting the photosynthesis, nutrient circulation and the normal balance of metabolic gases (Fig. 4).

In general, the swamps have very poor oxygen conditions (Table II). The lower layers are often completely deoxygenated whereas this is only occasionally true of the surface waters. However, a great variety of oxygen conditions exist in Ugandan swamps. With standing waters and floating vegetation, supersaturation may also develop in the outer margins during sunny exposures (Rzoska, 1974). Carter (1935a) believes that swamps in other parts of the tropical world are less deoxygenated than those of Paraguayan Chaco and Guiana. In Nile swamps, the oxygen conditions deep inside the vegetation fringe are only at about 10% saturation (Rzoska, 1974).

Dissolved free carbon dioxide in tropical swamps is generally high (Table II). The weed-infested waters have 25–30 times more free CO_2 than weed-free

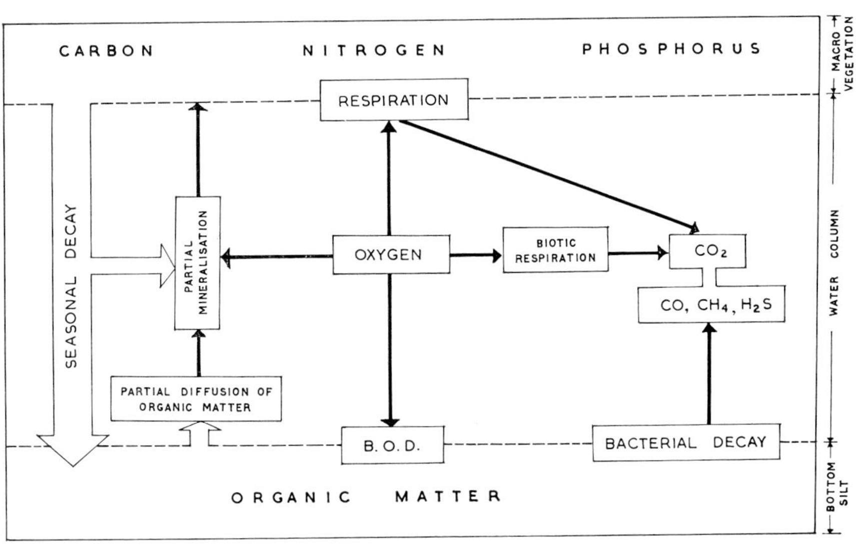

FIG. 4. Diagram of the ecological cycle of a pocketed water body of a typical senescent swamp with a thick macro-vegetation mat indicating absence of photosynthesis, low oxygen, high free carbon dioxide and soil-bound nutrients.

TABLE II

Environmental features of some representative swamps of the tropical world

Sl. No.	Name and locality of the swamp	Temp. °C	pH	Dissolved oxygen (mg/l)	Free CO_2 (mg/l)	Total alkalinity (10^{-4}N)	Conductivity (μmho/cm at 20°C)	Investigator
	NILE SWAMPS							
	Swamp rivers							
1	Bahr el Gabel	22–28	6·8–8·2	0·66–6·2	2·80–18	—	112–550	Modified after Rzoska (1974)
2	Bahr el Ghazal	22–28	6·4–8·4	1·3–8·0	—	—	40–370	
3	Bahr el Sobat	22–28	6·8–7·8	3·4–6·2	—	15·2–50·0	110–280	
4	Bahr el Zeraf	22–28	7·5–8·0	2·2–5·9	—	92·0–116·0	245–370	
	Standing waters							
5	Shambe Lagoon	22–28	6·95–8·0	2·4–6·0	—	31·7	—	Modified after Rzoska (1974)
6	Lagoon R.P. 12	22–28	7·25–8·4	6·1–10·2	—	21·4–31·4	200–250	
7	Lake Ambadi	22–28	6·4–7·2	8·0	—	25·4	40–55	
8	Khor Perboi	22–28	—	1·8–7·15	—	—	—	
9	Khor Atar	22–28	8·0–8·6	6·88–11·5	—	29	—	
	GUIANAN SWAMPS							
10	Grass swamp A	26–29·5	4·42–4·5	0·22–1·24	8·37–9·18	Absent	—	Carter (1935a)
11	Other swamps	26–29·5	4·53	0·65	13·48	Absent	—	
12	Pool B	26–29·5	4·07–4·15	0·60–1·45	5·02	Absent	—	
	EAST AFRICAN SWAMPS							
13	Lake Naivasha (Papyrus swamp)	25	7·8	2·51	—	—	—	Beadle (1932)
14	Kazinga Channel (Papyrus swamp)	26·7	5·9	nil-traces	—	7·5	—	
15	Kitoma (Papyrus swamp)	—	6·7	nil	—	71·6	—	
	INDIAN SWAMPS							
16	Polluted swamp	19–33	8·0–9·6	6·45–8·2	16·4–20	148 ppm	—	Dehadrai *et al.* (MS)
17	Semi-senescent swamp	19–33	8·0–8·6	0·6–3·2	20	143·4 ppm	—	
18	Senescent swamp	30·1–32·5	8·0	1·5–2·65	12	94–162 ppm	—	

waters (Table III). Up to 30 ppm free carbon dioxide is recorded in some Indonesian swamps (Vaas-van Oven and Vaas, 1960). Total alkalinity and pH are generally low in the swamp water and gases such as hydrogen sulphide, methane, carbon monoxide, free ammonia and chloride contents, all caused mainly by organic decay, are very high. The Guiana swamps have a very low pH (4·3–4·4) whereas the pH of the acid swamps in Assam (India) varies from 6·25 to 7·55 (Table III). In Bihar, the pH of alluvial swamps ranges from 8 to 9·6. The overall adverse nature of the swampy environment is highlighted when observed through its diurnal changes in relation to seasonal variations (Figs. 5, 6 and 7).

Primary productivity of swamps in India (Dehadrai, *et al.*, MS) appears to be considerably affected by the environment (Fig. 5). In polluted swamps, the primary productivity ranged from 31·25 to 1210·94 mgC/m^3/hr as against 11·46–281·25 mgC/m^3/hr in semi-senescent swamp and 11·46–23·44 mgC/m^3/hr in senescent swamp. The recorded high production from semi-senescent swamps during October–November is because of the raking of the bottom while harvesting the ripe aquatic crop of *Euryale ferox*. The release of nutrients into the water, availability of bright sun-light due to removal of floating "Makhana" leaves and general favourable ecological conditions of the spring are conducive to high primary productivity. The high primary productivity during March–April in polluted swamps without macro-vegetation conforms to the general ecological pattern of tropical ponds.

Although the trophic potential of swampy waters is low, the profusely growing macrophytes may be regarded as primary producers, and are thus the most important biotic forms of the swamps (Table IV). Surveys have revealed that the infestation of free-floating and emersed weeds is most extensive in the Indian swamps (Philipose *et al.*, 1970) and that the eastern zone comprising West Bengal, Bihar, Orissa, Assam and Manipur has the maximum incidence of 40–70% as compared to 20–25% for the rest of India (Philipose, 1966). The inference that "the clearer the water and the higher the weed density the poorer the plankton production", holds true in the case of swampy environments. In certain unshaded swamps, however, the phytoplankton occurs and varies seasonally but zooplankton is usually negligible. In Guiana swamps, rotifers are conspicuously rare (Carter, 1935a).

Rzoska (1974) reports that the vegetation fringe surrounding waters in the Nile swamps form a unique and biologically the richest habitat. The stems of emerging, submerged and floating vegetation are covered with epiphytes including the red alga and a varied animal life of tube-building Protozoa, Oligochaeta and Bryozoa. A short-term study on the occurrence of chironomid larvae in Indian swamps (Dehadrai, 1972a) indicated a high production of 20,000–40,000 units/sq m in 15 days to 40,000–80,000 units/sq m in 60 days.

Thick vegetation provides a good environment for aquatic insects (dip-

TABLE III

Physico-chemical conditions of water in weed infested ponds and ponds free of weeds in Sibsagar, Assam (India)
(*After Radhakrishnan and Bhuyan, MS*)

	Water temp. (°C)	pH	Free CO_2 (ppm)	Dissolved O_2 (ppm)	Total alkalinity (ppm)	Nitrate as NO_3 (ppm)	Phosphate as PO_4 (ppm)	Free ammonia (ppm)	Chloride (ppm)	Turbidity
Ponds infested with weeds	28·83 – 28·90	6·25 – 6·57	14·91 – 27·22	1·1 – 4·16	35·90 – 80·20	0·068 0·075	0·095 0·174	0·160 2·38	4·30 6·02	138·95 – 149·20
Ponds without weeds	27·23	7·55	0·99	8·74	30·96	0·065	0·051	0·02	0·65	272·24

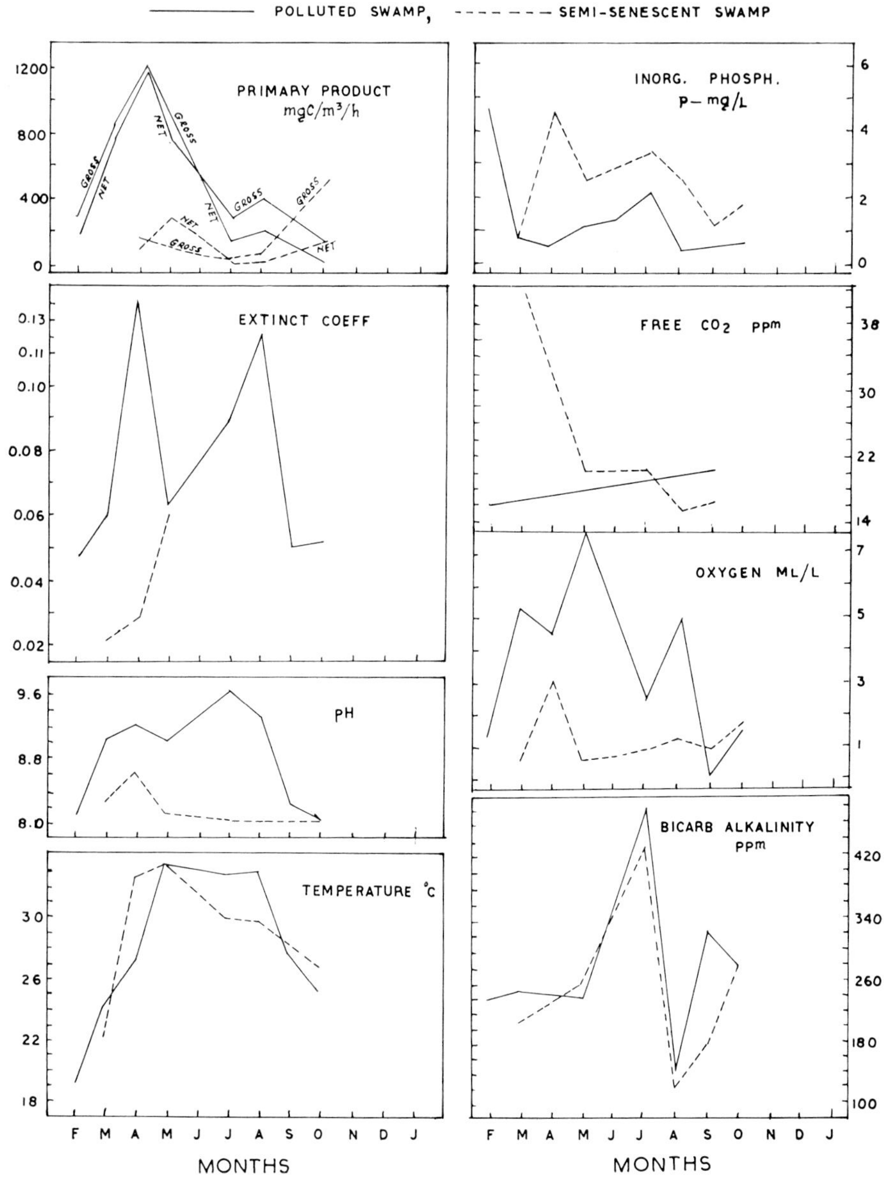

FIG. 5. Seasonal environmental features of polluted and semi-senescent swamps in North Bihar, India (Dehadrai *et al.*, MS).

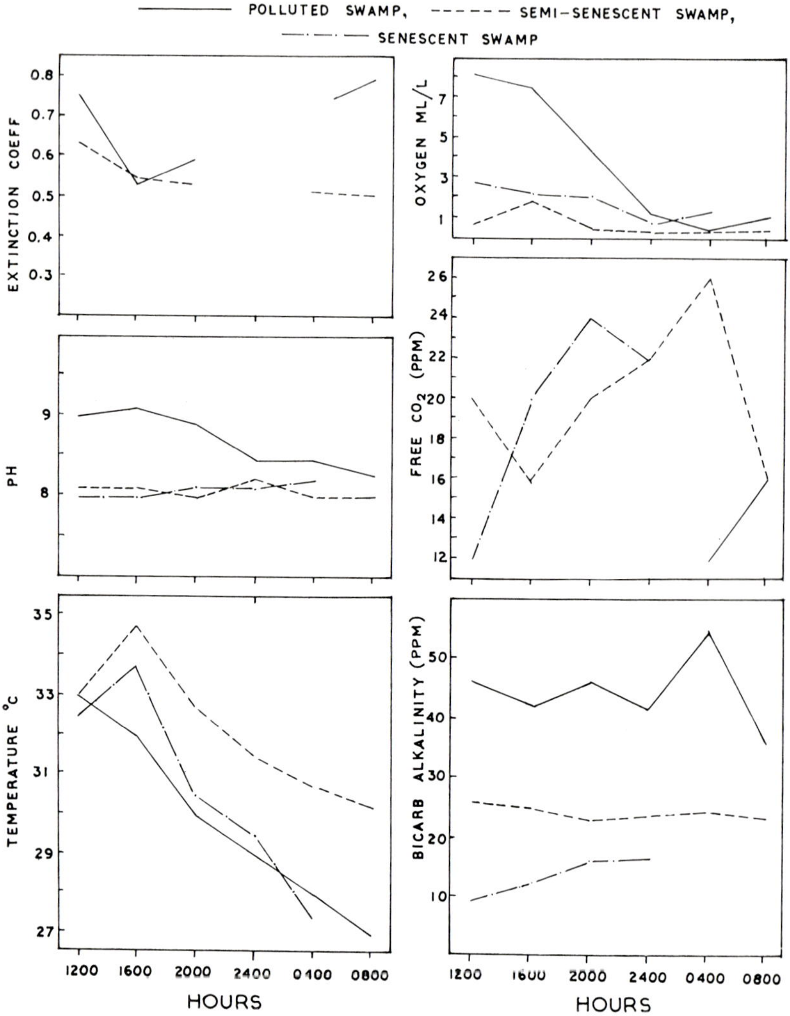

FIG. 6. Diurnal fluctuations of ecological factors in polluted, semi-senescent and senescent swamps during the summer of 1972 (Dehadrai *et al.*, MS).

terans, hemipterans, and coleopterans) which abound in swamps. Gastropods (*Viviparus*, *Planorbis*, *Pila*, larvae and adults of *Gyraulus* and *Limnaea*) are also common in swamps (David *et al.*, 1974).

Considerable quantities of fish food material including micro- and macrophytes, bottom fauna, insects and decaying organic matter may be available in different combinations in derelict and swampy waters but their fish fauna is limited. This is mainly attributed to the adverse physico-chemical conditions of these waters where fast growing, non-predaceous fishes with short

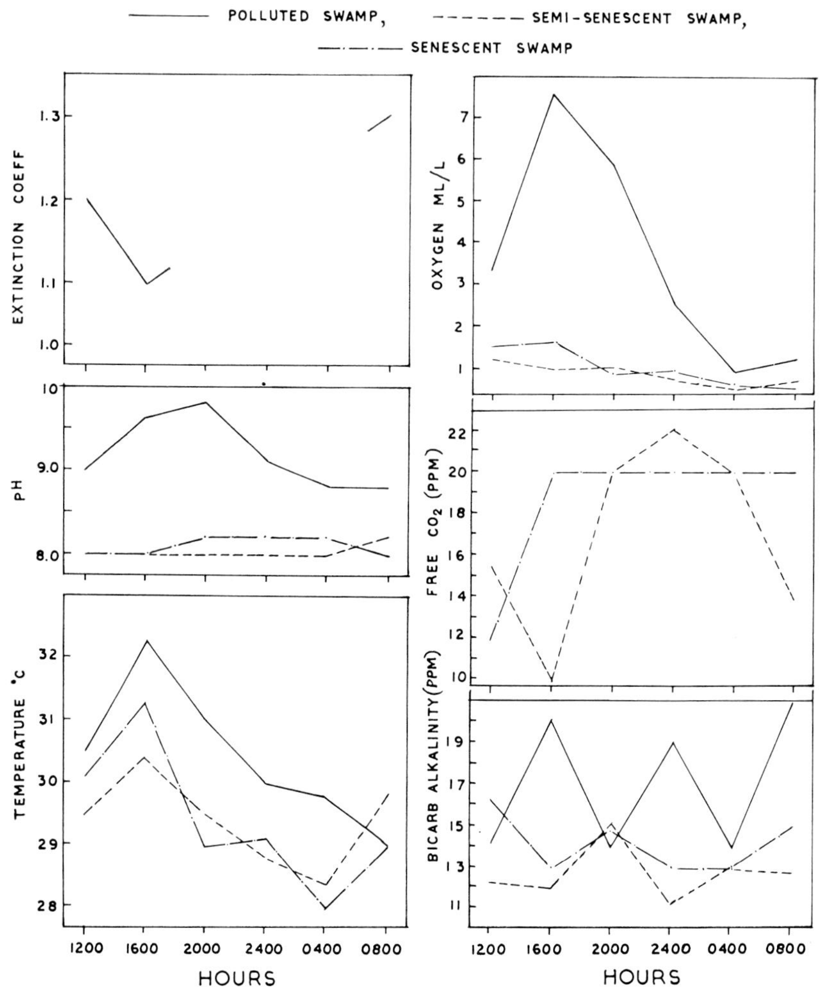

FIG. 7. Diurnal fluctuations of ecological factors in polluted, semi-senescent and senescent swamps during a monsoon (1972) (Dehadrai *et al.*, MS).

food cycles are not able to thrive and are susceptible to various diseases. However, certain slow-growing, hardy fishes with carnivorous food habits and low sensitivity to harmful gases survive in such inimical environments. Air-breathing teleosts appear to be adapted all the more with their predaceous food habits, accessory aerial respiration and other biological and physiological adaptations. It seems the air-breathing fishes and the swamps are made for each other.

TABLE IV

Common aquatic macrophytes in swamps of India

Littoral	Submersed	Emersed	Floating
Colocasia antiquorum	*Hydrilla verticillata*	*Trapa natans*	*Eichhornia crassipes*
Marsilea quadrifoliata	*Ceratophyllum demersum*	*Nymphoides cristatum*	*Salvinia cucullata*
Jussiaea repens	*Najas indica*	*Nymphaea* spp.	*Azolla pinnata*
Ipomoea spp.	*Utricularia stellaris*	*Monochoria* sp.	*Pistia stratiotes*
Polygonum sp.	*Chara* sp.	*Nelumbo nucifera*	*Lemna minor*
Typha sp.	*Nitella* sp.	*Nechamandra alternifolia*	*Spirodela polyrhiza*
Cyperus sp.	*Limnophila heterophylla*		
Scirpus sp.	*Vallisneria spiralis*	*Ottelia alismoides*	
Pseudoraphis sp.	*Lagarosiphon* sp.	*Euryale* sp.	
Phragmites sp.	*Potamogeton* sp.		
Acorus sp.			
Herpestes sp.			
Commelina sp.			

Compiled from Radhakrishnan and Bhuyan (MS) and David *et al.* (1974)

IV. Respiratory Adaptations to Environment

A. TYPES OF AERIAL RESPIRATORY ORGANS:

The accessory respiratory organs of air-breathing fishes show diverse structural adaptations. Details of the fine and gross structure of the aerial respiratory organs in different teleosts, their physiology and respiratory behaviour, are discussed by Munshi (p. 73) and Singh (p. 125) elsewhere in this volume.

While some species rely predominantly on aquatic breathing with gills and only supplement gas exchange with their accessory aerial respiratory organs under adverse respiratory conditions, others are obligatory air-breathers and "drown" if denied access to air for a short time.

The air-breathing organs may have evolved as a modification of an already existing structure or may be neomorphic. The *monoaccessory* type are the organs in the form of simple sacs, receptacles or diverticula without any additional structures present inside whereas the others are called *diaccessory* type and possess additional structures to increase the respiratory surface area.

An accessory respiratory organ is functional when it has direct connection with atmospheric air, its blood circulation is modified such that the aerated blood reaches the heart through the shortest path and the network of blood capillaries is separated from the lumen of the organ by a thin layer of epithelium interspersed with mucous or fatty cells (Carter, 1957).

In general, the occurrence of accessory aerial respiratory organs in fishes is accompanied by a reduction in gill surface area by almost one-half compared to that of other purely aquatic breathers (Dubale, 1955; Hughes, Dube and Munshi, 1973; Hughes, Singh, Guha, Dube and Munshi, 1974).

Seven different types of accessory aerial respiratory organs in teleosts can be recognized (Table I).

1. *Bucco-pharyngeal organ*

Probably as a constant reaction to the lack of oxygen in ambient water, the habit of surfacing to gulp air may have been instigated, acquired and later established as a mode of aerial respiration to be carried on in the mouth cavity and pharynx as in *Elec rophorus electricus* (Evans, 1929; Johansen, Lenfant, Schmidt-Nielsen and Petersen, 1968).

2. *Pharyngeal "lung"*

A definite pouch or air chamber or "lung" which could store the inhaled air for at least a short period evolved in the pharyngeal region probably as a response to more rigorous and pronounced aerial respiratory exercises in certain fishes. In Ophicephalidae, the air chamber or "lung" lies above the gills and is enclosed within the skull (Munshi, 1962; Hughes and Munshi,

1973a). In *Amphipnous cuchia*, the pharyngeal lung is free and extends beyond the region of the head (Das, 1940).

3. *Modified opercular chamber*

In certain fishes the inhaled air is stored in the opercular chamber which has become highly vascularized and modified to subserve aerial respiration. Air is taken in by the mouth, lodged for some time inside the opercular chambers which become bulged out like two little balloons behind the head and then by a sudden collapse of the walls of these chambers, the air is either passed out through the mouth or as a rule through the small opercular openings. Such modifications are observed in *Hypopomus brevirostris*, *Monopterus* sp. and several gobiids (*Pseudapocryptes lanceolatus*) (Wu and Kung, 1940).

4. *Opercular lung*

Yet another adaptation to air breathing is in the form of definite specialized receptacles developed as diverticula of opercular chambers between the hyoid and first branchial arches. These receptacles lie above the gills and are, in some cases, enclosed within the skull bones. They may extend backwards far beyond the region of the gills up to about the middle of the tail region as a long tubular sac-like structure embedded within the myotomes as in *Heteropneustes fossilis*. Its inner walls are vascular and the epithelium is provided with tongues and "alveoli" interspersed with fat cells (Hughes and Munshi, 1973a; Hughes *et al.*, 1974). In other forms, the diverticula of the air chambers are permanently located inside the skull and either form shelly, labyrinthine organs, as in Anabantidae, or coral-like, dendritic "air trees" as in *Clarias batrachus*. The fine and gross structural details of the labyrinthine organs in Anabantids and "air trees" in *Clarias* have been discussed in detail by Moussa (1956), Munshi (1961) and Hughes and Munshi (1973a,b).

5. *Gas exchange in the gastrointestinal tract*

In a few fishes, the inhaled air is swallowed and passed backwards inside the alimentary canal and it is lodged for some time either in the stomach (*Ancistrus anisitsi* and *Plecostomus plecostomus*) or intestine [(*Misgurnus fossilis*, *Lepidocephalichthys guntea* (Cobitidae); *Doras* sp. (Doradidae), *Hoplosternum* sp. (Callichthyidae)], which are modified to effect the gaseous exchange between the blood and swallowed air. The used-up air is passed out to the exterior either through the mouth (*Doras*, *Hoplosternum*, *Ancistrus* and *Plecostomus*) or through the anus (*Misgurnus* and *Lepidocephalichthys*).

6. *Swimbladder as a respiratory organ*

In fishes where the swimbladder functions as an accessory respiratory organ, its structure is more complex and it acts almost like a vertebrate lung.

Teleosts in which the swimbladder acts as a lung are *Arapaima* (Osteoglossidae), *Gymnarchus* (Mormyridae), *Erythrinus* (Characinidae), *Umbra* (Esocidae), *Notopterus* (Notopteridae) and *Phractolemus*. For detailed information on the structure and function see Johansen (1970), Dehadrai (1962), Audenaerdae (1959) and Fänge (this volume).

7. *Skin*

In many air-breathing teleosts (e.g. *Amphipnous*), the skin serves as a respiratory organ.

B. GAS EXCHANGE IN AIR-BREATHING FISHES

Air-breathing fishes largely avoid the problems of dessication and carbon dioxide accumulation in the blood by remaining essentially aquatic, exploiting the atmosphere mainly as a source of oxygen.

All air-breathing organs in fishes show a low gas exchange ratio indicating that their function is mainly oxygen absorption. Gills and skin play their most important role in the elimination of carbon dioxide. In *Misgurnus fossilis*, an intestinal air breather (Calugareanu, 1907), oxygen absorption is reported to be the dominant feature of air breathing. However, the capacity of the air-breathing organs to eliminate carbon dioxide in *Ophicephalus* (Qasim, Qayyum and Garg, 1960) and *Notopterus* (Dehadrai, 1962) under respiratory stress is reported to be considerable. Gas transport between the respiratory organs and metabolizing tissues depends on the metabolic rate, the pattern and rate of blood circulation and the respiratory properties of blood.

C. BLOOD OF AIR-BREATHING FISHES

Respiratory properties of the blood of air-breathing fishes have disclosed adaptive features which correlate with environmental factors and the relative importance of air and water breathing. A comparative study of the blood of tropical fishes by Willmer (1934) disclosed a definite tendency for higher oxygen carrying capacity in air-breathing forms and in some cases through certain circulatory modifications (Johansen, 1970). Dubale (1959) found the loading tension of oxygen to be higher in the blood of air breathers than that of water breathers. The amount of iron in the blood of water breathers ranged from 15·13 mg/100ml to 27·52 mg/100ml against 29·41 mg/100ml to 400·00 mg/100ml in the case of air breathers.

The internal air-breathing organs in fishes contain up to 3% carbon dioxide (Carter, 1957). Much higher carbon dioxide (6·5%) has been observed in the swimbladder of *Notopterus* (Dehadrai, 1962) when the fish was subjected to extreme respiratory stress with high (120 ppm) free carbon dioxide and low (0·5 ppm) dissolved oxygen in the ambient waters. At this level of carbon dioxide, oxygenation of fish haemoglobin would be greatly impaired even at atmospheric P_{O_2}'s. The air-breathing fishes have evolved a haemoglobin much less sensitive to carbon dioxide as in *Electrophorus* (Willmer, 1934) and *Clarias* (Fish, 1956; Hughes and Singh, MS.).

D. RESPIRATORY BEHAVIOUR

Air-breathing fishes exhibit different respiratory responses when subjected to hypoxic waters. The response patterns express an apparent relationship to the relative role of aquatic and air-breathing in overall gas exchange (Johansen, 1970). The role of aquatic and aerial respiration under different tensions of dissolved oxygen and free carbon dioxide in teleosts has been studied only in *Erythrinus* (Willmer, 1934), *Notopterus* (Dehadrai, 1962), *Symbranchus* (Johansen, 1966) and *Anabas* (Hughes and Singh, 1970a,b; Singh and Hughes, 1973). Deoxygenation of water elicits an increased rate of air-breathing while the rate and frequency of branchial breathing drops depending upon the carbon dioxide in the ambient water. The external and internal levels of carbon dioxide are thus important factors influencing aquatic and aerial respiration in the air-breathing fishes.

In derelict tropical swamps, the carbon dioxide levels could build up to an extent that would adversely affect the oxygenation of haemoglobin and disrupt the internal acid-base balance. Resorting to aerial respiration continuously also leads to the retention and increase of internal carbon dioxide tension. The air-breathing teleosts seem to possess a sensory mechanism that can probably differentiate between the external and internal carbon dioxide levels.

E. THYROID IN RELATION TO AERIAL RESPIRATION

Unlike other teleosts, the thyroid tends to become compact and organized in air-breathing fishes such as in *Ophicephalus* and *Heteropneustes* (Swarup and Belsare, 1959a,b); *Notopterus* (Dehadrai, 1972b); *Anabas* and *Clarias* (author's observations). Functionally, the thyroid is known to be more metamorphic than metabolic in its effect in poikilothermic vertebrates (Gorbman and Bern, 1962). But the structural modification of the thyroid, observed in air-breathing fishes, is noteworthy. Its role in the control of a double respiratory mechanism in air-breathing fishes needs to be understood in detail. Intraperitoneal injections of 25 μg of l-thyroxine into specimens of 70–80 gm every 24 hrs for three days brought about an accelerated air breathing associated with an activated thyroid in *Notopterus*. Treatment of the fish with 0·3% thiourea in water for 22 days caused gradual blocking of thyroid activity followed by a marked inhibition of aerial respiration. Fish subjected to forced aerial respiration for 15 days showed activated thyroid. The thyroid in *Notopterus notopterus* appeared to be functionally associated with the physiology of the secondarily developed aerial respiration in the fish (Dehadrai, 1972b). However, reservations in regard to the use of thiourea as an antithyroid drug may be kept in mind.

Harms (1929, 1935) succeeded to a considerable extent in his attempts to transform various amphibiotic fishes towards a more terrestrial life by feeding thyroid. Later Cagler (1950) successfully induced semi-terrestrial life by feeding thyroid in *Blennius pavo* and *B. montagui*, which are used to leaving water. But no change could be brought about by him in the completely aquatic species *Blennius sphinx* and *B. erythrocephalus*.

In certain cases a perfectly functional hormone exists which may have no target organ upon which to act. This could be the situation with thyroxine in some poikilothermic vertebrates. The target organs may develop independently in individual vertebrate species. Accordingly, in the air-breathing fishes, with the existence of thyroxine in the blood, the secondarily-evolved aerial respiratory organ became a responsive receptor site upon which the hormone could act. Cagler's (1950) failure to induce semi-terrestrial habits in purely aquatic species of *Blennius* by feeding thyroid was probably due to the absence of a target organ for the hormone to act upon in the fish.

V. Diving Reflexes

Certain air-breathing fishes such as *Ophicephalus* and *Periophthalmodon* are known to exhibit diving reflexes which are manifested in the form of bradycardia, reduced oxygen consumption and lactic acid changes in the blood and muscles during and after prolonged exposure to air similar to those observed in diving animals under and after submergence (Dehadrai, 1962; Garey, 1962; Scholander, Bradstreet and Garey, 1962). Increase of blood circulation to the accessory respiratory organ during acute respiratory stress is of common occurrence in *Notopterus* (Dehadrai, 1962).

The occurrence of diving reflexes and subsequent lowering of metabolism when exposed to unfavourable respiratory medium has been attributed to asphyxiation. In air-breathing teleosts such as *Ophicephalus* and *Periophthalmodon*, the occurrence of diving reflexes in spite of the access to oxygen through their air-breathing organs is noteworthy. Garey (1962) suggested that since the mudskipper is adapted to land life, "an aquatic existence may now constitute an asphyxic environment" for it. However, it may appear that in addition to the respiratory stress, which is known to lower the metabolic rate of fishes exposed to air, a change in the natural environment of fishes alone may probably also induce a shift towards a lowered metabolic rate. Drying of the skin of air-breathing fishes during air exposure and build up of carbon dioxide in the blood as a result of resorting to aerial respiration may initiate certain nervous reflexes which in turn may induce the physiological adjustments in the fish described above.

VI. Maturation and Spawning

Detailed studies on the reproductive biology of air-breathing fishes are few. The males are generally smaller than the females and are poorly represented in sex ratio (Table V). The Indian air-breathing teleosts attain maturity generally by the summer months of the first year but their sizes and weights at first maturity vary greatly (Table VI).

The fecundity of air-breathing teleosts (Table VII) appears low but when assessed in relation to their respective body sizes, the number of ova is considerable. However, the body size–fecundity relationship is also associated with the occurrence of parental care. Murrels (*Ophicephalus*) despite their large sizes, have a very low ratio of ova to their body size. On the contrary, *Anabas*, *Clarias* and *Heteropneustes* have a high ratio.

Most of the air-breathing fishes are known to breed in confined waters either before or during the rainy season. Breeding in *Anabas* starts with the onset of the monsoon and continues till the middle of the monsoon. During this period they migrate from their normal habitat to the nearby inundated pools and puddles for spawning. Under the spawning urge they are known to travel on wet land *en route* to the nearby puddles. The eggs are small, spherical, buoyant and yellowish or whitish in colour measuring 0·7 mm in diameter with a single large oil globule. The eggs are generally laid in marginal areas amongst aquatic vegetation so that they are not carried away by the wind or water currents. *Clarias* and *Heteropneustes* breed during the rainy season. *Clarias* lays its eggs in horizontal holes prepared in the embankment slightly below the water surface. The eggs of *Clarias* are round and

Table V

Sex ratio of certain Indian air-breathing teleosts

Sl. No.	Species	Sex ratio Male	Female	Authors
1.	*Heteropneustes fossilis*	1	: 2	Bhatt (1968)
2.	*Clarias batrachus*	1	: 2	Mookerjee, Mazumdar and Das Gupta (1941)
3.	*Anabas testudineus*	1	: 1·2	Mookerjee and Mazumdar (1946)
4.	*Ophicephalus marulius*	1	: 1·0535	Parameswaran (1974)
5.	*O. striatus*	1	: 1·0550	Parameswaran (1974)
6.	*O. punctatus*	1	: 1·0225	Parameswaran (1974)
7.	*Notopterus notopterus*	1	: 1·07	Parameswaran and Sinha (1966)

TABLE VI

Age, size and growth of certain air-breathing fishes of India

Species	Length-weight relationship	Size (in mm) at at different ages (in yrs) 1	2	3	4	5	First maturity at L. (mm)	Wt. (g)	Age (yrs)	Length-age/weight-age relationship	Authors
O. marulius	♂:Log W = −5·2191 + 2·9977 Log L ♀:Log W = −5·2696 + 3·0204 Log L	275	400	505	615	715	♂:300 ♀:320	265 282	2 2	$l_t = 1372{\cdot}8[1-e^{-0{\cdot}1260\,(t+0{\cdot}6870)}]$	Parameswaran (1974)
O. striatus	♂:Log W = −5·0076 + 2·9745 Log L ♀:Log W = −5·0734 + 3·0071 Log L	153	239	310	365	411	♂:173 ♀:179	36 39	1 1	$l_t = 867{\cdot}4[1-e^{-0{\cdot}1313\,(t+0{\cdot}8789)}]$	Parameswaran (1974)
		351	411	451	481	531	–	–	–	$l_t = 56{\cdot}5[1-e^{-0{\cdot}420\,(t+0{\cdot}981)}]$ $W_t = 3281{\cdot}71[1-e^{-0{\cdot}207\,(t+2{\cdot}4583)3}]$	Bhatt (1970)
O. punctatus	♂:Log W = −5·1610 + 3·0703 Log L ♀:Log W = −4·4901 + 2·7920 Log L	115	153	192	231	265	♂:103 ♀:105	18 21	1 1	$l_t = 345{\cdot}5[1-e^{-0{\cdot}1593\,(t+1{\cdot}0215)}]$	Parameswaran (1974)
N. notopterus	Log W = −4·9556 + 2·9677 Log L	–	–	–	–	–	–	–	–	–	Parameswaran and Sinha (1966)
C. batrachus	Log W = −5·4634 + 3·0882 Log L	–	–	–	–	–	–	–	–	–	Thakur (1974)
H. fossilis	♂:Log W = −4·2147 + 2·5271 Log L ♀:Log W = −5·0817 + 2·9418 Log L Juvenile:Log W = −5·6460 + 3·1646 Log L	–	–	–	–	–	♂: 85 ♀:120	–	2	–	Bhatt (1968) Thakur and Das (1974)
A. testudineus		149					73	8·4		–	Mookerjee and Mazumdar (1946), Banerji and Prasad (1974)

TABLE VII

Fecundity of the air-breathing fishes of India

Species	Length and weight of specimens studied	Total No. of ova (Fecundity)	Fecundity/Length-weight relationship	Authors
1. *Heteropneustes fossilis*	164–307 mm	2843–44723	$F = 10^{-6} \times 9647\ L^{*2 \cdot 65681}$ $F = 1040 \cdot 49149 + 207 \cdot 27833W\dagger$	Bhargava (1971)
2. *Clarias batrachus*	315 mm/252 gm	11612	$\text{Log } F = -7 \cdot 7219 + 3 \cdot 5613 \text{ Log } L^*$	Mookerjee and Mazumdar (1950);
	150 mm–350 mm/ 26·63 gm–359·90 gm	1000–20000	$\text{Log } F = -0 \cdot 5514 + 0 \cdot 0592 \text{ Log } W\dagger$	Thakur (1974)
3. *Anabas testudineus*	73 mm/8·4 gm– 171 mm/80·5 gm	4588–34993	—	Banerji and Prasad (1974)
4. *Ophicephalus marulius*	320 mm/252 gm– 994 mm/6260 gm	1799–38375	$F = 0 \cdot 0001972L^{*2 \cdot 76055}$ $F = 1 \cdot 03926W\dagger^{0 \cdot 9398}$	Parameswaran (1974)
5. *Ophicephalus striatus*	206 mm/80 gm– 405 mm/830 gm	2794–28046	$\text{Log } F = -4 \cdot 2819 + 3 \cdot 3203 \text{ Log } L^*$ $\text{Log } F = 1 \cdot 1328 + 1 \cdot 17625 \text{ Log } W\dagger$	Parameswaran (1974)
	248 mm/133 gm– 448 mm/794 gm	4422–20070	—	Alikunhi (1953)
6. *O. punctatus*	122 mm/23 gm– 287 mm/205 gm	2477–25483	$\text{Log } F = -3 \cdot 6903 + 3 \cdot 38183 \text{ Log } L^*$ $\text{Log } F = 1 \cdot 92999 + 1 \cdot 07501 \text{ Log } W\dagger$	Parameswaran (1974)
7. *Notopterus notopterus*	198 mm/57 gm– 294 mm/220 gm	175–1188	$\text{Log } Y\ddagger = -8 \cdot 6916 + 4 \cdot 7624 \text{ Log}\S$ $\text{Log } Y\ddagger = -0 \cdot 1118 + 1 \cdot 3635 \text{ Log}\S$	Parameswaran and Sinha (1966)

*L = Total length in mm, † W = Total weight in gm, Y‡ = Number of ova, § is the total length in mm/total weight in gm.

yellowish brown measuring about 1·8 mm in diameter whereas those of *Heteropneustes* are round and greenish or dark-brown measuring about 1·4 mm in diameter. Both the air-breathing cat fishes have adhesive and demersal eggs. The various species of *Ophicephalus* breed in ponds and tanks throughout the rainy season, the peak period being immediately preceding and during the rainy season. Murrels are known to make some sort of a nest by clearing a small water area amidst shallow marginal weeds. They also breed in ponds devoid of any vegetation. The eggs of murrels are pelagic, each having a single, large oil globule. In colouration they are straw, pale yellow or amber. The eggs as well as the brood of young fry are guarded both by the male and female in *Ophicephalus* spp. (Alikunhi, 1953). The parental care ceases when fingerlings attain a size of 150 mm in *O. marulius*, 57 mm in *O. striatus* and 35 mm in *O. punctatus* (Parameswaran, 1974). The spawn or broods of murrels occur in great abundance in shallow, rain-filled puddles, ditches as well as tanks and ponds infested with weeds. The early fry of murrels exhibit specific colouration such as iridescent lemon-yellow with black bands in *O. punctatus*, vermilion red in *O. striatus* and orange-yellow longitudinal bands on either side of the body in *O. marulius*. The eggs of *Notopterus* are large and adhesive and pale-golden yellow in colour.

Adaptations in regard to the spawning habits, nature of eggs and reproductive behaviour including the occurrence of parental care help to enhance their survival in the adverse environment.

VII. Survival in Inimical Swampy Environments

Special adaptations of air-breathing fishes in regard to reproductive behaviour include breeding migration to avoid predators in the habitual waters, nature of eggs ensures propagation, nesting habits and parental care to enhance survival of the progeny. Nevertheless, from their hatchling stage onwards these fishes have to face a continuous struggle for existence. Several biotic and abiotic factors determine their survival through various stages of their life history in the derelict swampy waters.

Most of the air-breathing fishes spawn in shallow waters, preferably newly-formed pools and puddles or the low-lying inundated areas. A low water column is essential for aerial respiration, a characteristic habit which develops at a very young stage, about the 5th to 8th day of their hatching in *Anabas*, *Heteropneustes*, *Clarias*, *Ophicephalus* spp. (Dehadrai, 1972a; Banerji, 1974; Parameswaran, 1974; Thakur, Pal and Khan, 1974). The energy loss by their vertical trips for aerial respiration could be considerable. A young *Ophicephalus punctatus*, kept in an aquarium containing water 40 cm, is reported to surface 1879 times/day travelling a distance of 1318m/day at an energy cost of 161 cal/day. By comparison those reared in 2·5 cm water depth surfaced 482 times/day travelling a distance of 45 m/day at an energy cost of 92 cal/day. Consequently, the former group converts 28 mg wet food/g/day though it consumes 246 mg wet food/g/day, whereas those exposed to 2·5 cm water depth consume only 160 mg wet food/g/day yet convert 34 mg wet food/g/day (Dr. T. J. Pandian, personal communication). Thus, the surfacing trips appear to tell upon the general metabolism in regard to maintenance and growth of the fishes.

Periodic exposure of air-breathing fishes to desiccation due to drying up of shallow puddles and pools in the tropics is not infrequent. Reported adaptations of certain air-breathing fishes, such as *Clarias* and *Heteropneustes* which take refuge in the soft mud, are helpful in this situation.

Paucity of food in the derelict environment is a major stress affecting survival of air-breathing fishes. The spawn and fry of air-breathing fishes being zooplankton feeders suffer on account of its poor availability in most swamps. The juveniles and adults have usually to cover a larger area in search of their prey, the latter being not so abundant and restricted in variety. The newly-hatched larvae less than 2 mm in size, sluggish and fragile, are very prone to destruction in nature, paradoxically, even through the attack of *Cyclops* which is otherwise a preferred item of food of the larvae. *Cyclops* at a density of 1500/litre of water could bring about 100% destruction of 5–7 mm spawn of *Anabas*. However, in *Anabas*, the fry beyond the 8 mm size can successfully ward off the Cyclops attack. (Banerji, 1974). Hatchlings of *Heteropneustes*, *Clarias* and *Ophicephalus* spp. have also been observed to suffer heavy damage due to *Cyclops*.

Occurrence of cannibalism or intra-specific belligerence is probably instigated by the general lack of food. Cannibalism in the younger stages of *Heteropneustes*, *Anabas* and *Ophicephalus* spp. is predominant. Specimens up to 60 mm in *Ophicephalus* spp. were found damaged at the posterior end. In *Anabas*, on the other hand, the incidence of cannibalism becomes less predominant after the fry attains a size of about 22 mm. The ultimate number of survivors in a progeny of any of these fishes is largely determined by the extent of a belligerent tendency marked by the fact that over 3–6% of the fry outgrow the others in the same brood and pose a danger to the stock. In an experiment, with 1068 10-day-old fry of *Anabas*, only 5·4% survival was recorded at the end of 302 days. However, in the above experiment the other hazards of nature were completely avoided (Banerji, 1974).

Survival of the air-breathing fishes at their younger stages is partially enhanced by parental care in *Ophicephalus* and *Osphronemus*. In *Heteropneustes*, *Clarias* and *Anabas*, the absence of parental care is compensated by the development of sharp pectoral and dorsal spines. The pectoral spine of *Heteropneustes* is poisonous (Bhimachar, 1944). In *Anabas*, the dorsal and opercular spines help in warding off the predators.

The fact that air-breathing fishes are carnivorous poses considerable problems of interspecific predation. The presence of other carnivorous and piscivorous, hardy fishes in the swampy environment are additional threats in the struggle for survival of air-breathing fishes. The survival of air-breathing fishes in their natural habitat is indeed poor.

VIII. Food and Feeding Habits

Air-breathing fishes are, in general, carnivorous, each species being well-adapted to seeking, seizing and digesting its food. Most species have a large mouth, numerous pointed teeth, few short gillrakers, a large stomach usually with a number of pyloric caecae, and short intestinal tract. No detailed work on the structure of the alimentary canal and digestive physiology of air-breathing fishes is available.

Heteropneustes fossilis and *Clarias batrachus* have more or less similar feeding habits. The larvae and young fry feed mainly on microcrustaceans while the juveniles and adults feed on insects and their larvae, shrimps, ostracods and other smaller crustaceans, worms, higher plants and organic debris. They subsist largely on a diet of mud from the swamp bottom, extracting sufficient nutrient in the form of decaying animal and plant matter, to which their small mouth and weak jaws armed with feeble teeth are suitably adapted. Contrary to general belief these fishes are thus at the very base of the food chain where they utilize the raw material in the form of organic matter. The barbels amply compensate for the lack of clear vision in the muddy and dark environment of the swamp and pond bottoms. The environment of the swamp suits them as the abundance of insects and rich organic debris coupled with hide-outs favours a fast growth. Chaudhuri (1971) has reported very high populations of *Heteropneustes* and *Clarias* in Pagoda ponds of Burma where these fishes are fed heavily by the visitors with rice and bundles of *Ipomoea aquatica* leaves and stems.

Anabas testudineus is a microplankton feeder in its larval and young fry stages but soon becomes insectivorous and a voracious feeder. *Anabas* (Fig. 8) can consume insects up to 16 times its body weight per day (Dehadrai and Banerji, 1973). The authors also noted the preference of *Anabas* for *Anisops* over *Corixa*. However, the fast rate in consumption of aquatic insects (*Anisops*) is perhaps not maintained throughout life and hence the poor growth of the species after two years of age. Though it is also piscivorous, it is not as highly predatory as the murrels.

Osphronemus gorami is primarily herbivorous in its natural habitat subsisting largely on submerged weeds, surface creepers, marginal grasses and algae. Occurrence of insects, shrimps, worms, etc. in its stomach has also been observed occasionally.

The murrels (*Ophicephalus* spp.) are highly predatory and markedly piscivorous. The fry in almost all cases feed heavily on zooplankton whereas the fingerlings feed on insects, insect larvae, shrimps and fish fry. However the young of *Ophicephalus punctatus* prefer phytoplankton (Banerji, 1974). The smaller species (*O. punctatus* and *O. gachua*) feed mainly on minnows and other small fishes, shrimps, insects and occasionally molluscs while the larger varieties (*O. marulius* and *O. striatus*) feed mainly on fish, frogs, insects and other live food. The mouth is well adapted to the highly predatory habits of murrels in being large, and provided with strong jaws and numerous small but strong teeth. The pyloric caecae are also present. The occurrence of trash fish in the swamps together with aquatic vegetation, which provides a

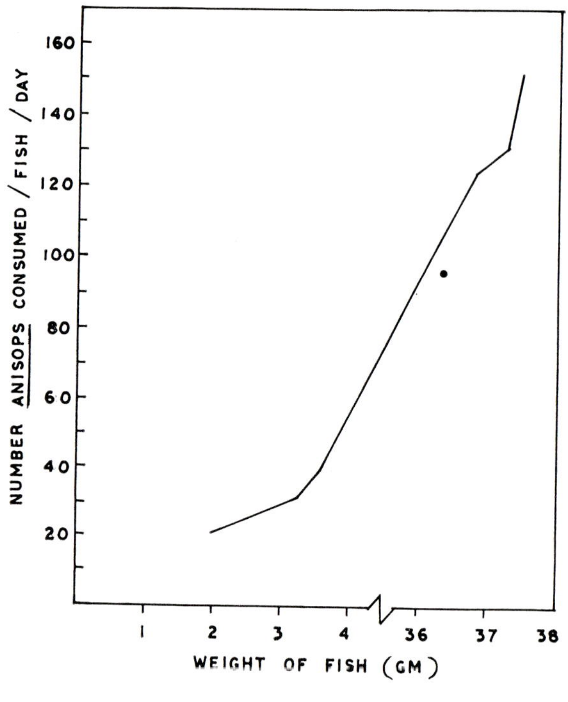

FIG. 8. Graph showing number of *Anisops* consumed/fish/day by different weight groups of *Anabas testudineus* in the aquarium.

cover for the murrels, thus seem an ideal environment for this highly predaceous group of freshwater fishes. The larvicidal habits of *Anabas*, *Osphronemus* and fingerlings of murrels are useful in environmental hygiene.

N. notopterus is also a planktophagous species up to 30 mm in length but thereafter it switches over to a predominantly shrimp and insect diet. Parameswaran and Sinha (1966) however observed that when its preferred food is in short supply, the species turns markedly predatory and takes fish fry and fish.

IX. Growth

The air-breathing fishes are generally small in size (the only exceptions in India being *O. marulius* and *O. striatus*). Paucity of proper food and the adverse environment affecting the metabolism of the fish also contribute to their genetically slow growth. While detailed studies on the growth of air-breathing fishes in relation to their environment have not been made, some studies on the biology of these fishes give an idea as to their growth at different ages (Table VI).

No systematic studies on the growth of *Heteropneustes*, *Clarias* and *Anabas* are available. [Thakur and Das (1974) have recently worked out the length-weight relationship of *Heteropneustes* and found that the value of the exponent (Table VI) in the male is less than 3 signifying that they become lighter for their length as they grow larger.]

Osphronemus is an extremely slow-growing fish and even in culture ponds its rate of growth could not be increased.

In general, poor growth in terms of size in most of the air-breathing fishes could be attributed to the stress they are subjected to in the adverse environments of derelict and swampy waters. Specific genetical limitations prevail in the growth processes of the fishes, though it has been possible at least to hasten the growth rate of *Clarias*, *Heteropneustes*, *Anabas* and certain murrels under controlled conditions with adequate food and healthy environment.

ACKNOWLEDGEMENT

The authors' grateful thanks to Dr. V. G. Jhingran, Director, Central Inland Fisheries Research Institute, Barrackpore for providing facilities and encouragement.

References

Alikunhi, K. H. (1953). Notes on the bionomics, breeding and growth of the murrel, *Ophicephalus striatus* (Bloch). *Proc. Indian Acad. Sci.* **38B**, 10–20.

Assheton, R. (1907). *In*: "Budgett Memorial Volume" (Graham Kerr, ed.), p. 293 Cambridge University Press, London.

Audenaerdae, F. E. T. v. D. (1959). Existence d'une vessie natatoire pulmonide chez *Phractolaemus ansorgei* Blgr. *Rev. Zool. Bot. Afr.* **59**, 364.

Banerji, S. R. (1974). Hypophysation and life history of *Channa punctatus* (Bloch). *J. Inland Fish Soc. India* **6**, 62–73.

Banerji, S. R. and Prasad, D. (1974). Observations on reproduction and survival of *Anabas testudineus* (Bloch) in Bihar region. *J. Inland. Fish. Soc. India* **6**, 6–17.

Beadle, L. C. (1932). Observations on the bionomics of some East African swamps. *J. Linn. Soc. London* (Zool.) **38**, 135–155.

Beadle, L. C. (1958). Hydrobiological investigations on tropical swamps. *Verhdlg. Int. Ver. Limnol.* **13**, 135–155.

Berg, T. and Steen, J. B. (1965). Physiological mechanisms for aerial respiration in the eel. *Comp. Biochem. Physiol.* **15**, 469–484.

Bhargava, R. M. S. (1971). The fecundity of *Heteropneustes fossilis* (Bloch). *J. Bombay nat. Hist. Soc.* **67**, 583–588.

Bhatt, V. S. (1968). Studies on the biology of some freshwater fishes. Part VII, *Heteropneustes fossilis* (Bloch.). *Indian J. Fish.* **15** (1 & 2), 99–115.

Bhatt, V. S. (1970). Studies on the growth of *Ophicephalus striatus* (Bloch). *Hydrobiologia* **36**, 165–177.

Bhimachar, B. S. (1944). Poison glands in pectoral spines of two catfishes *H. fossilis* and *Plotossus arab* with remarks on the nature of their venom. *Proc. Indian Acad. Sci.* **19** (B), 65–70.

Bishai, H. M. (1962). The water characteristics of the Nile in Sudan with a note on the effect of *Eichhornia crassipes* on the hydrobiology of the Nile. *Hydrobiologia* **19**, 357–382.

Cagler, M. (1950). Über die Wirkung des Thyreoidins auf einige Blenniusarten. *Istanbul Univ. Fen. Fak. Mecmuasi* (Rev. Fac. Sci. Univ. Istanbul), B **15**, 95–109.

Calugareanu, D. (1907). Die Darmatumung Von *Cobitis fossilis* II. Mitteilung: Über den Gaswechsel *Arch. Ges. Physiol.* **120**, 425–450.

Carter, G. S. (1935a). Results of the Cambridge expedition to British Guiana, 1933. The freshwaters of the rain forest areas of British Guiana. *J. Linn. Soc. London* **39**, 147–193.

Carter, G. S. (1935b). Reports of the Cambridge Expedition to British Guiana 1933. Respiratory adaptations of the fishes of the forest waters, with descriptions of the accessory respiratory organs of *Electrophorus electricus* L. and *Plecostomus plecostomus* L. *J. Linn. Soc. Lond.* (Zool.). **39**, 219–233.

Carter, G. S. (1957). Air-breathing. *In* "The Physiology of Fishes." (Ed. M. E. Brown) Vol. 1, p. 65, Academic Press Inc., New York, London and San Francisco.

Carter, G. S. and Beadle, L. C. (1931a). The fauna of the swamps of the Paraguayan Chaco in relation to its environment. I. Physico-chemical nature of the environment. *J. Linn. Soc. Lond.* (Zool.) **37**, 205–258.

Carter, G. S. and Beadle, L. C. (1931b). The fauna of the swamps of Paraguayan Chaco in relation to its environment. II. Respiratory adaptations in the fishes. *J. Linn. Soc. Lond.* (Zool.) **37**, 327–366.

Chaudhuri, H. (1971). Workshop Report I-Coordinated Project on Culture of Air-breathing Fishes in Swamps. Central Inland Fisheries Research Institute (ICAR). Cuttack.

Das, B. K. (1927). The bionomics of certain air-breathing fishes of India, together with an account of the development of their air-breathing organs. *Phil. Trans. Roy. Soc. Lond.* **216**, 183–219.

Das, B. K. (1937). Some further observations on the structure and physiology of an air-breathing Loach, *Lepidocephalus* (Syn. *Lepidocephalichthys*) *guntea* (Ham. Buch.) found in Hyderabad. *Proc. 24th Indian Sci. Cong.*, Hyderabad, 305.

Das, B. K. (1940). Nature and causes of evolution and adaptation of the air-breathing fishes. *Proc. 27th Indian Sci. Congr.* Presidential address, No. 2, 215–260.

David, A., Rao, N. G. S. and Ray, P. (1974). Tank fishery resources of Karnataka. *Bull. Cent. Inl. Fish. Res. Inst. Barrackpore*, (20), 87 p. (*Mimeo*).

Dehadrai, P. V. (1960). Respiratory function of the swim-bladder of *Notopterus* (Lacepede). *Nature* (*Lond.*) **185**, 929.

Dehadrai, P. V. (1962). Respiratory function of the swim-bladder of *Notopterus* (Lacepede) *Proc. Zool. Soc. Lond.* **139**, 341–357.

Dehadrai, P. V. (1972a). Workshop Report II, All India Coordinated Research Project on Air breathing Fish Culture in Swamps. Central Inland Fisheries Research Institute (ICAR), Patna, Bihar.

Dehadrai, P. V. (1972b). Thyroid in relation to aerial respiration in *Notopterus notopterus* Ham. (Teleostei). *J. Biol. Sci.* **14**.

Dehadrai, P. V. and Banerji, S. R. (1973). Capacity of *Anabas testudineus* (Bloch) to feed on insects. *J. Inland. Fish Soc. India* **5**, 137–140.

Dehadrai, P. V., Thakur, N. K. and Das, N. K. (MS). Environmental features of polluted, semi-senescent and senescent swamps in north Bihar, India.

Dubale, M. S. (1955). Paucity of oxygen in the medium as a factor in the evolution of air-breathing habit among teleosts. *J. Univ. Baroda* **4**, 63.

Dubale, M. S. (1959). A comparative study of the oxygen carrying capacity of the blood in water and air-breathing teleosts. *J. Anim. Morph. Physiol.* **6**, 48–54.

Evans, M. (1929). Some notes on the anatomy of the electric eel, *Gymnotus electrophorus*, with special reference to a mouth-breeding organ and the swimbladder. *Proc. Zool. Soc. Lond.* 17–23.

Fish, G. R. (1956). Some aspects of the respiration of six species of fish from Uganda. *J. exp. Biol.* **33**, 186–195.

Garey, W. F. (1962). Cardiac responses of fishes in asphyxic environments. *Biol. Bull.* **122**, 362–368.

Geyer, F. and Mann, H. (1939). Atmung des ungarischen Hundfisches (*Umbra lacustris*) Grossinger. *Zool. Anz.* **127**, 234–245, 305–312.

Gorbman, A. and Bern, H. A. (1962). "A Text Book of Comparative Endocrinology". John Willey and Sons, New York.

Harms, J. W. (1929). Die Realisation von Genen and die consecutive Adaptation. I. Phasen in der Differezierung der Anlagenkomplexe und die Frage der Landtierwerdung. *Z. wiss. Zool.* **133**, 211–397.

Harms, J. W. (1935). Die Realisation von Genen und die consecutive Adaptation. IV. Experimentell hervorgerufener Medianwechsel Wasser zu Feuchluft bzul zu Trockenluft bei Gobiiformes (*Gobius*, *Boleophthalmus* und *Periophthalmus*). *Z. wiss. Zool.* **146**, 417–462.

Hepher, B. (1958). On the dynamics of phosphorus added to fish ponds in Israel. *Limnol. Oceanogr.* **3**, 84–100.

Hora, H. L. and Pillay, T. V. R. (1962). Hand book of fish culture in the Indo-Pacific region. *FAO Fish. Biol. Tech. pap* **14**, 204p.

Hughes, G. M. and Munshi, J. S. Datta (1973a). Nature of the air-breathing organs of the Indian fishes *Channa*, *Amphipnous*, *Clarias* and *Saccobranchus* as shown by electron microscopy. *J. Zool. Lond.* **170**, 245–270.

Hughes, G. M. and Munshi, J. S. Datta (1973b). Fine structure of the respiratory organs of the Climbing perch, *Anabas testudineus* (Pisces Anabantidae). *J. Zool. Lond.* **170**, 201–225.

Hughes, G. M. and Singh, B. N. (1970a). Respiration in air-breathing fish, the climbing perch *Anabas testudineus* Bloch. I. Oxygen uptake and carbon dioxide release into air and water. *J. exp. Biol.* **53**, 265–280.

Hughes, G. M. and Singh, B. N. (1970b). Respiration in an air-breathing fish, the climbing perch, *Anabas testudineus* II. Respiratory patterns and the control of breathing. *J. exp. Biol.* **53**, 281–98.

Hughes, G. M., Dube, S. C. and Munshi, J. S. Datta (1973). Surface area of the respiratory organs of the climbing perch, *Anabas testudineus* (Pisces: Anabantidae) *J. Zool. Lond.* **170**, 227–243.

Hughes, G. M., Singh, B. R., Guha, G., Dube, S. C. and Munshi, J. S. Datta (1974). Respiratory surface areas of an air-breathing siluroid fish *Saccobranchus* (*Heteropneustes*) *fossilis* in relation to body size. *J. Zool. Lond.* **172**, 215–232.

Jobert, C. (1878). Recherches anatomiques et physiologiques pour servir à l'histoire de la respiration chez les poissons. *Ann. Sci. Nat.* (*Zool.*) 6 Ser., Vol. 7, 7p.

Johansen, K. (1966). Air-breathing in the teleost *Symbranchus marmoratus*. *Comp. Biochem. Physiol.* **18**, 383–395.

Johansen, K. (1970). Air-breathing in fishes. *In:* "Fish Physiology." (Eds W. S. Hoar and D. J. Randall) Vol. 4, pp. 361–411. Academic Press, New York, London and San Francisco.

Johansen, K., Lenfant, C., Schmidt-Nielsen, K. and Petersen, J. A. (1968). Gas exchange and control of breathing in the electric eel, *Electrophorus electricus*. *Z. vergleich. Physiol.* **61**, 137–163.

Krogh, A. (1904). Some experiments on the cutaneous respiration of vertebrate animals. *Skand. Arch. Physiol.* **16**, 348–357.

Kurdin, V. P. (1968) (in Russian). Data of hydrological and hydrochemical observations on the White Nile. *Biologia vnutrennich vod. Inform Biuletin. Acad. Sc. USSR.*

Leiner, M. (1938). *Die Physiologie der Fischatmung*. Akademische Verlagsges, Leipzig, 134 pp.

Ling, W. (1960). Control and alteration of biological environment. Control of aquatic vegetation. *In* Lectures presented at the Third International Training Centre conducted by the Government of Indonesia and the FAO of the UN with the cooperation of the Indo-Pacific Fisheries Council at Bogor, Indonesia, 31st October to 10th December, 1955, **1**, 12p.

Matida, Y. (1956). Study of farm pond fish culture 3. Fates of fertilized elements and the efficiency of fertiliser and biochemical environment in the pond. *Bull. Freshw. Res. Lab. Tokyo* **6**, 27–39.

Migahid, A. M. (1947). An ecological study of the "Sudd" swamps of the Upper Nile. *Proc. Egypt. Acad. Sci.* **3**, 57–86.

Mizuno, T. and Mori, S. (1970). Preliminary hydrobiological survey of some south east Asian inland waters. *Biol. J. Linn. Soc.* **2**, 77–117.

Mookerjee, H. K. and Mazumdar, S. R. (1946). On the life history, breeding and rearing of *Anabas testudineus* Bloch. *J. Dep. Sci. Calcutta Univ.* **2**, 101–140.

Mookerjee, H. K. and Mazumdar, S. R. (1950). Some aspects of the life history of *Clarias batrachus* (Linn.) *Proc. Zool. Soc. Bengal* **3**, 71–84.

Mookerjee, H. K., Mazumdar, S. R. and Das Gupta, B. (1941). On the nature of urinogenital papillae of *Clarias batrachus* (Linn.) and *Heteropneustes fossilis* (Bloch). *Indian J. vet. Sci. Anim. Hus.* **11**, 334–336.

Moussa, T. A. (1956). Morphology of the accessory air-breathing organs of the teleost, *Clarias lazera* (C & V). *J. Morph.* **98**, 125–160.

Munshi, J. S. Datta (1961). The accessory respiratory organs of *Clarias batrachus* (Linn). *J. Morph.* **109**, 115–140.

Munshi, J. S. Datta (1962). On the accessory respiratory organs of *Ophicephalus punctatus*, and *Ophicephalus striatus*. *J. Linn. Soc. Lond.* (*Zool.*) **44**, 616–626.

Munshi, J. S. Datta (1965). On the accessory respiratory organs of *Trichogaster fasciatus*. *Proceedings of seminar on recent trends of research in animal morphology. Vikram University, Ujjain.*

Parameswaran, S. (1974). Investigations on the biology of some fishes of the genus *Channa* Gronovius 1763. Ph.D. Thesis. Magadh University, Bodh-Gaya.

Parameswaran, S. and Sinha, M. (1966). Observations on the biology of the featherback, *Notopterus notopterus* (Pallas). *Indian J. Fish.* **13**, 232–250.

Philipose, M. T. (1963). Making fish ponds weed free. *Indian Livestock* **1**, 20–21.

Philipose, M. T. (1966). Present trends in the control of weeds in fish cultural waters of Asia and the Far East. *FAO Fish. Rep.*, (44) Vol. 5, 26–52.

Philipose, M. T., Ramachandran, V., Singh, S. B. and Ramaprabhu, T. (1970). Some observations on the weeds of cultivable freshwaters in Orissa. *J. Inland Fish. Soc. India* **2**, 61–83.

Qasim, S. Z., Qayyum, A. and Garg, K. K. (1960). The measurement of CO_2 produced by air-breathing fishes and evidence of the respiratory function of the accessory respiratory organs. *Proc. Indian Acad. Sci.* **52**, 19–26.

Radhakrishnan, S. and Bhuyan, B. R. (MS). Observations on the incidence of weeds and seasonal fluctuation in abundance of *Lemna minor* (L), *Azolla pinnata* (L) and *Salvinia cucullata* (L.) in ponds in Sibsagar, Assam.

Rauther, M. (1910). Die akzessorischen Atmungsorgane der Knochenfische. *Ergeb. Fortschr. Zool.* **2**, 517–585.

Rzoska, J. (1974). The Upper Nile swamps, a tropical wet land study. *Freshwater Biol.* **4**, 1–30.

Sawaya, P. (1946). Sobre abiologia de alguns peixes de respiracao aerea. *Univ. Sao Paulo, Fac. Filosof., Cienc. Letras, Zool.* **11**, 255–286.

Scholander, P. F., Bradstreet, E. and Garey, W. F. (1962). Lactic acid responses in the grunion. *Comp. Biochem. Physiol.* **6**, 201.

Singh, B. N. and Hughes, G. M. (1973). Cardiac and respiratory responses in the climbing perch *Anabas testudineus*. *J. comp. Physiol.* **84**, 205–226.

Swarup, H. and Belsare, D. K. (1959a). Thyroid gland in Ophicephalidae (Actinopterygii, Percomorphi). *Curr. Sci.* **28**, 369.

Swarup, H. and Belsare, D. K. (1959b). Thyroid gland in *Heteropneustes fossilis*. *Sci. & Cult.* **25**, 221.

Talling, J. F. (1957a). The longitudinal succession of water characteristics in the White Nile. *Hydrobiologia* **11**, 73–89.

Talling, J. F. (1957b). Diurnal changes of stratification and photosynthesis in some tropical African waters. *Proc. Roy. Soc. (B)* **147**, 57–83.

Talling, J. F. and Rzoska, J. (1967). The development of plankton in relation to hydrological regime in the Blue Nile. *J. Ecol.* **55**, 637–662.

Taylor, M. (1913). The development of *Symbranchus marmoratus*. *Quart. J. Micro. Sci.* **59**, 1–51.

Thakur, N. K. (1974). The fecundity of *Clarias batrachus* (Linn.). *Proc. Zool. Soc. Calcutta* (In press).

Thakur, N. K. and Das, N. K. (1974). Length-weight relationship of *Heteropneustes fossilis* (Bloch). *J. Inland Fish. Soc. India* **6**, 95–96.

Thakur, N. K., Pal, R. N. and Khan, H. A. (1974). Embryonic and larval development of *Heteropneustes fossilis* (Bloch). *J. Inland Fish. Soc. India* **6**, 33–44.

Thakur, R. (MS). Studies on the respiratory system of an eel fish *Amphipnous cuchia* (Ham.). Ph.D. Thesis, University of Bihar, Muzzafarpur.

Vaas-van Oven and Vaas, K. F. (1960). Appraisal of inland Fishery resources. Biological appraisal. *In* Lectures presented at the Third International Training Centre conducted by the Government of Indonesia and the FAO of the UN with the Cooperation of the Indo-Pacific Fisheries Council at Bogor, Indonesia, 31st October to 10th December, 1955, **1**, 19p.

Willmer, E. N. (1934). Some observations on the respiration of certain tropical freshwater fish. *J. exp. Biol.* **11**, 283–306.

Wu, H. W. and Kung, C. C. (1940). On the accessory respiratory organ of *Monopterus*. *Sinensia* **11**, 59–67.

Gross and Fine Structure of the Respiratory Organs of Air-breathing Fishes*

J. S. DATTA MUNSHI

Postgraduate Department of Zoology, Bhagalpur University, Bihar, India

I. INTRODUCTION

In India there are many fishes belonging to different families and genera which are amphibious in nature. These fishes mainly live in those areas where water pools and tanks dry up in the summer and become marshy. Due to the luxuriant development of macrophytes and the decomposition of organic matter there is depletion of oxygen and accumulation of carbon dioxide in these marshy waters. Consequently, they become uninhabitable for purely water-breathing fishes (Nasar and Munshi, 1971).

These marshy waters abound in the eastern zone of India which includes especially Gorakhpur division of Uttar Pradesh, Saharsa, Purnea, Darbhanga and Bhagalpur in Bihar and most parts of Assam, Bengal and Orissa. The Central Inland Fisheries Research Institute, Barrackpore has made a general survey of the areas where these air-breathing fishes could be cultured.

The following species of amphibious air-breathing fishes are common in these waters: *Anabas testudineus* (Bloch), local Name—*Kawai*; *Colisa* (= *Trichogaster*) *fasciatus* (Bl. & Schn.), local Name—*Khalisha*; *Amphipnous cuchia* (Ham.), local Name—*Kuchia*; *Clarias batrachus* (Linn.), local Name—*Maguri*; *Heteropneustes* (= *Saccobranchus*) *fossilis* (Bloch), local Name—*Singhi*; *Channa punctatus* (= *Channa punctata*) (Bloch), local Name—*Girai, Lata*; *Channa striatus* (Bloch), local Name—*Sauri*; *Channa marulius* (Ham.), local Name—*Saur*; *Channa gachua* (Ham.), local Name—*Chainga*.

There is another group of fishes belonging to the family Gobiidae which

* This work was supported by the University Grants Commission, New Delhi, Grant No. F.27–4/73 (Sc.I/SR), and the Nuffield Foundation.

are found in the ascending estuaries and tidal waters of India. The prominent species of these amphibious fishes are *Periophthalmus vulgaris* and *Boleophthalmus boddaerti*.

In this paper the gross and fine morphology of the gills and air-breathing organs of the common amphibious fishes of India will be discussed. Most of the findings summarized here are from previous studies which have been separately published.

II. Gills

A. GROSS STRUCTURE

The secondary lamellae of the gills of fishes are beautifully adapted for maintaining respiratory and osmoregulatory efficiency. This efficiency is directly related to the effective gill area and inversely proportional to the water-blood barrier which consists of three main elements—the epithelium, basement membrane and the pillar cell flange.

In *Amphipnous cuchia* the gill lamellae are present on the second gill arch. The third gill arch bears on its ventral side a continuous fleshy vascular flap-like structure which gives it a false filamentous appearance due to the folded nature of its epithelium and large number of capillary loops (Munshi and Singh, 1968b). Typical gill rays are absent. The gill filaments are blade like structures that divide into two or three sub-filaments. Each primary gill filament is supplied with an afferent and one or two efferent blood vessels running parallel to each other. The afferent and efferent vessels are connected with each other by means of coiled capillary loops which project laterally and take the place of secondary lamellae. Pillar cells are not present. The gill filamentar muscles are absent and as a result individual movement of the filaments has been lost.

In *Channa striatus* the fourth pair of gills are very much reduced in size and gill lamellae are mainly present on the ceratobranchials. The primary lamellae of the outer and inner hemibranchs of the first three holobranchs interdigitate among themselves to a great extent and as a result water pockets are found between them (Munshi, 1960). The primary lamellae of the outer hemibranch of the fourth gill arch are small finger-like structures and have no secondary lamellae. The free edges of the primary lamellae of each hemibranch fuse *inter se* towards their proximal end and form water-retaining pockets alternating with the primary lamellae. The epithelium lining these pockets is folded and many mucous glands are present. The following special features of the fourth holobranch may be noted:

(i) The main afferent and efferent blood vessels which run through the gill arch are directly connected with one another by means of short vascular loops. This means that deoxygenated blood of the afferent vessel passes directly into the efferent vessel without any oxygenation.

(ii) The primary afferent and efferent vessels which run along the margin of the primary gill lamellae are directly interconnected with each other, because of the partial development of the secondary lamellae.

(iii) The secondary lamellae have been suppressed to a great extent and if present are in the form of small vascular buds. Within these buds the secondary afferent vessels break up into an irregular plexus before opening into the efferent vessels. These degenerative features of the primary lamellae of the fourth gill suggest that it has lost its efficiency as a respiratory organ.

Each secondary lamella comprises a middle vascular layer surrounded on both sides by a thin layer of connective tissue and a thick external epithelium. Occasionally adjacent secondary lamellae fuse with one another at their tips. In some cases the fusion is so complete that even the vascular layers of adjacent lamellae become interconnected with one another.

In *Anabas testudineus* there is a gradual reduction in size of the four gills (Munshi, 1968). The third gill is relatively small in size and the fourth is very much reduced. No secondary lamellae are borne by the fourth gill. The capillaries joining the afferent and efferent blood vessels are connected directly by broad blood spaces within the primary lamellae.

In *Heteropneustes fossilis* and *Clarias batrachus* all four pairs of gills are provided with lamellae. However, the fourth pair are comparatively smaller in size and bear few primary lamellae. Like other fishes the primary lamellae are provided with secondary gill lamellae. In both species a few posterior primary lamellae of each gill take part in the formation of fan-like structures and they (Munshi, 1961, 1962a) migrate into the respiratory membrane of the supra-branchial chambers (Fig. 1).

The four pairs of gills of *Periophthalmus vulgaris* bear small and stumpy primary lamellae on epi- and ceratobranchials. The filaments of adjacent hemibranchs of each holobranch interdigitate among themselves and bear short and thick secondary lamellae (Singh and Munshi, 1969).

B. ULTRASTRUCTURE

1. *Epithelium*

In *Amphipnous cuchia* the surface of the gill filaments is covered by a multilayered epithelium on the outside of which many microvilli are present. The epithelial cells interdigitate with each other with finger-like processes. Many of these cells have a characteristic smooth endoplasmic reticulum with quite large spaces interconnected with each other and many mitochondria. In some of the EM preparations a ribosome-bearing endoplasmic reticulum is discernible near the Golgi zone and fine glycogen granules are found in the cytoplasm. The basal epithelial cells contain fine contractile fibrils which may be tonofibrils. The nuclei are either rounded or polymorph in shape. In *Channa striatus* the epithelial layer is formed of more than three layers and

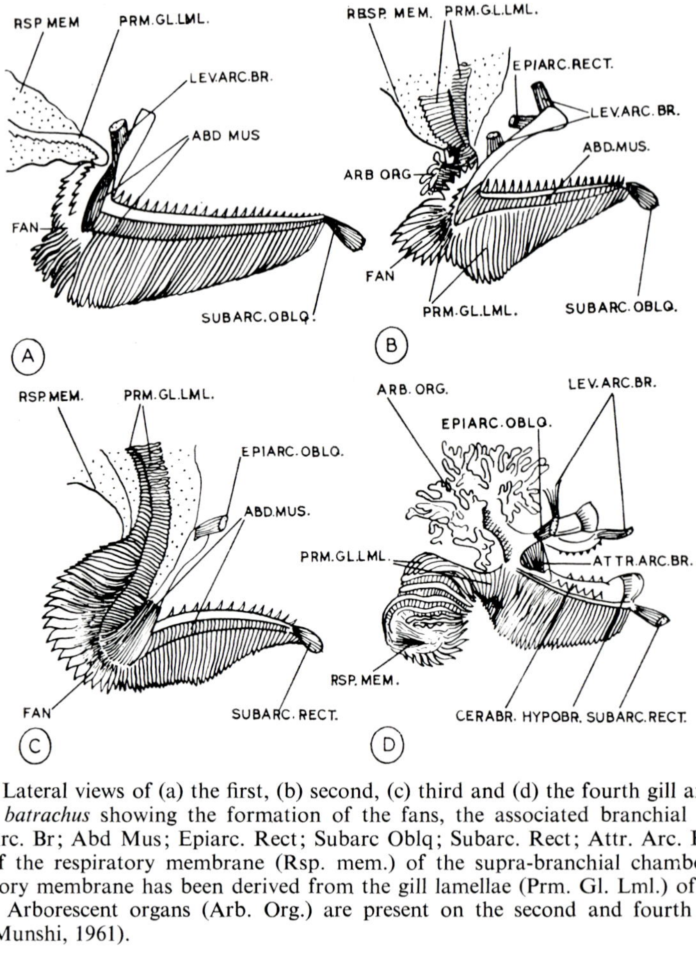

FIG. 1. Lateral views of (a) the first, (b) second, (c) third and (d) the fourth gill arches of *Clarias batrachus* showing the formation of the fans, the associated branchial muscles (Lev. Arc. Br; Abd Mus; Epiarc. Rect; Subarc Oblq; Subarc. Rect; Attr. Arc. Br.) and parts of the respiratory membrane (Rsp. mem.) of the supra-branchial chamber. This respiratory membrane has been derived from the gill lamellae (Prm. Gl. Lml.) of the gill arches. Arborescent organs (Arb. Org.) are present on the second and fourth arches. (After Munshi, 1961).

consists of cuboidal cells. Some of these epithelial cells seem to be the mitochondria-rich chloride cells. No microvilli could be detected on the epithelium.

The comparatively thinner epithelium (2–4 μm) in some specimens of *Anabas testudineus* which is much less than the values (15–20 μm) reported for *Anabas* based on earlier electron micrographs (Hughes and Munshi, 1968, 1973a) has led to the recognition of two varieties (Dube and Munshi, 1974a). In narrow-trunked variety of *Anabas* the thickness of the water–blood pathway increases with increasing body weight by a power of 0·227 which suggests that the efficiency of the gills decreases with increasing body weight which has been confirmed by the gill diffusing capacity of *Anabas testudineus* (Dube and Munshi, 1974b). The most effective surface for oxygen transfer in *Anabas* gills seems to be the free edge or margin of the secondary lamellae where the total water–blood distance is only 3·5–7 μm. Many mitochondria-rich cells having many membrane-bound vacuoles have been detected in the epithelium of this fish, which are thought to be the ion-regulating chloride cells. Amoebocytes containing lysosomes and vacuoles are very common.

In *Clarias batrachus* the cuboidal cells of the epithelium are covered by the processes spreading out from other epithelial cells (Fig. 2). These cells have microvilli on their outer surface. The microvilli are covered over by osmiophilic granules which may be mucus. The cuboidal cells characteristically have a system of smooth endoplasmic reticulum in between which mitochondria are found. In many such cells irregular-shaped empty vesicles are also found. The junction between the outer epithelial cells and the large cuboidal cell has a double membrane and one or other of these membranes seem to have pores. The cuboidal mitochondria-rich cells open to the surface through gaps present between thin outer epithelial cells. There is a suggestion that the endoplasmic reticulum is connected with the Golgi apparatus and is involved in the process of secretion. The apical region of several of these cells contains many small smooth-walled vesicles. At higher magnification the tubular nature of the endoplasmic reticulum is very clear. In *Saccobranchus* (= *Heteropneustes*) *fossilis* two layers of epithelial cells are distinguishable. Prominent microvilli are usually present. The thickness of the water–blood pathway varies from 3–6 μm but this becomes quite thin near the marginal channels of the secondary lamellae. Towards the base of the secondary lamellae there are again many cells which have extensive endoplasmic reticulum and mitochondria. In *Boleophthalmus boddaerti* the epithelium has become thick to the extent of 16–32 μm. The outer epithelial region of the gills is dense and quite fibrous. There are many spaces present in the epithelial layer which contain lymphocytes (Hughes and Munshi, 1975).

2. *Basement Membrane*

In *Amphipnous cuchia* the basement membrane is well defined and has an outer clear area beneath the epithelial cells and then a more dense layer. This is followed by a thick inner fibrous layer in which many fibrocytes and connective tissue cells and gill capillaries are found.

In *Channa striatus* the basement membrane has very characteristic peg-like processes on the epithelial side. The basement membrane is fairly thick and contains many fibres. Two layers are very clearly distinguished—the outer being more granular and the inner being thicker and having certain types of elongated osmiophilic pigment particles.

In *Anabas testudineus* the basement membrane is thin in the secondary lamellae of the first gill but thicker in the second, third and fourth pair of gills. It is differentiated into clear, fine and coarse fibrous layers.

In *Clarias batrachus* the basement membrane is very thin in comparison to the epithelium. Even at higher magnification it is not possible to distinguish an outer clear layer and inner dense layer. In *Heteropneustes fossilis* the basement membrane has the usual two layers of which the outer one is most dense and coarse, the inner being made up of fine fibrous material.

The basement membrane is rather thin in *Boleophthalmus boddaerti* and consists of a middle dense layer with two light areas.

3. *Pillar cells*

Due to the absence of the secondary lamellae from the gills of *Amphipnous cuchia*, the characteristic pillar cells are absent in this fish. The gill capillaries are separated from one another by basement membrane. Some of the capillaries are lined by single endothelial cells.

In *Channa striatus* the capillary that runs within the secondary lamellae are made up of a hyaline collagenous material containing the pillar cells and other endothelial cells. In several places connective tissue cells are found in the connective tissue wall of the capillaries. In the vascular buds of the fourth gill arch the normal disposition of the pillar cells is disturbed and they are quite difficult to identify.

In *Anabas testudineus* the pillar cell column sometimes remains at the

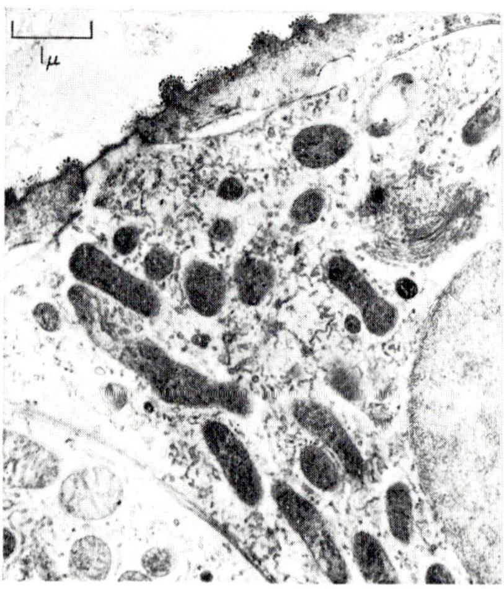

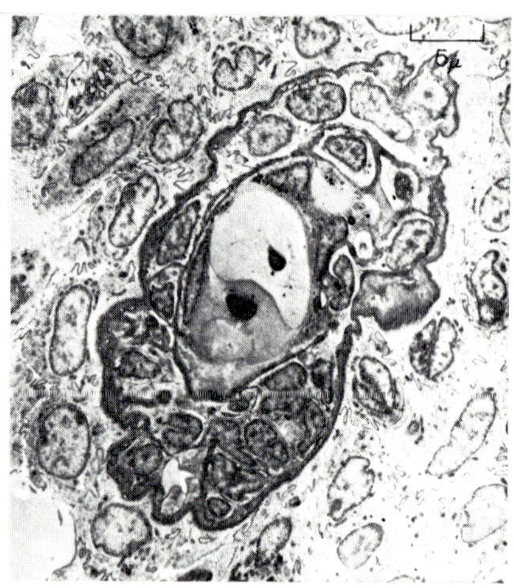

FIG. 2. Electron micrograph showing the structure of gill epithelium of *C. batrachus*. The covering epithelium bears stumpy microvilli with darkly stained osmiophillic granular material. Part of a mitochondria-rich cell is seen with its round nucleus, Golgi zone, mitochondria and smooth endoplasmic reticulum. (Hughes and Munshi, 1975).
Fig. 3. Section passing through primary gill lamella of *A. testudineus* showing the nest of embryonic mesenchymal cells covered by thick basement membrane. Columns of basement membrane penetrate in between the cells and also note development of small vascular spaces. (Hughes and Munshi, 1975).

surface where they occur in blocks. It is quite likely that the columns here occupy the original position and only where they are likely to come into contact with blood do they get incorporated within the infoldings of the plasma membrane. This might serve to reduce the danger of any thromboplastic effect that the collagen of the columns might have (Hughes and Weibel, 1972). In Fig. 3 we see a small blood capillary surrounded by groups of mesenchymal cells. These mesenchymal cells are enveloped by thick well-defined basement membrane. Strands of the connective tissue also penetrate in between these cells. On the right side of the figure there is a small vascular space which is clearer at higher magnification. The nuclei of the mesenchymal cells are very prominent with scanty cytoplasm. In contrast, on the left of the figure these cells become completely surrounded by connective tissue, having more cytoplasm, and develop all the characteristics of pericyte cells. It seems that with further development of the lamellae the vascular spaces increase in size and the pericyte-like cells take the position of pillar cells. Also seen in this remarkable micrograph are typical endothelial cells with osmiophilic granules and pericyte cells surrounding the capillaries and the so-called pillar cells.

In *Clarias batrachus* the basement membrane is clear and has the appearance of two pillar cells but it is quite possible that it is a single cell having two

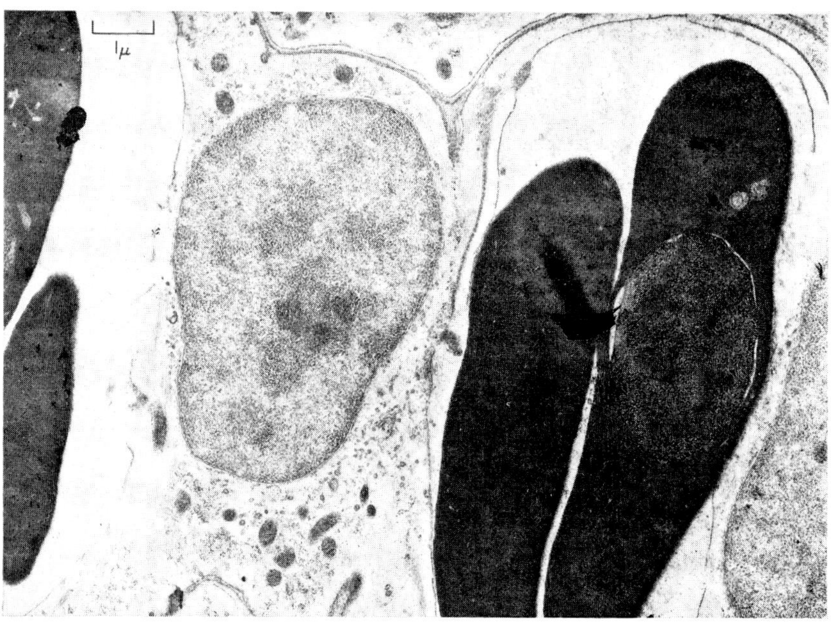

FIG. 4. Transverse section passing through secondary gill lamella of *C. batrachus* showing the structure of a pillar cell. Note overlapping flange of pillar cells. (Hughes and Munshi, 1975).

nuclei. The flange of the pillar cell is quite thick at certain points. There are many different types of vesicular bodies and the mitochondria are distributed in the different parts of the cytoplasm. Figure 4 shows the pillar cell at high magnification and there appears to be a double flange layer. This may be due to one cell overlying another for a greater distance. In the cytoplasm clear fibrillar material is discernible. The double arrangement of the flange layer is also distinguishable in this part of the pillar cell. The pillar cell body has a large nucleus and in its peripheral region there are many vesicles and microsomes. The epithelium adjacent to the pillar cells also possesses fine fibres which run at right angles to the pillar cell axis.

In *Heteropneustes fossilis* typical pillar cells are present and in this particular fish contractile fibrillae are vividly discernible in the cytoplasm (Fig. 5). When the gills were treated with collagenase and electron micrographs taken

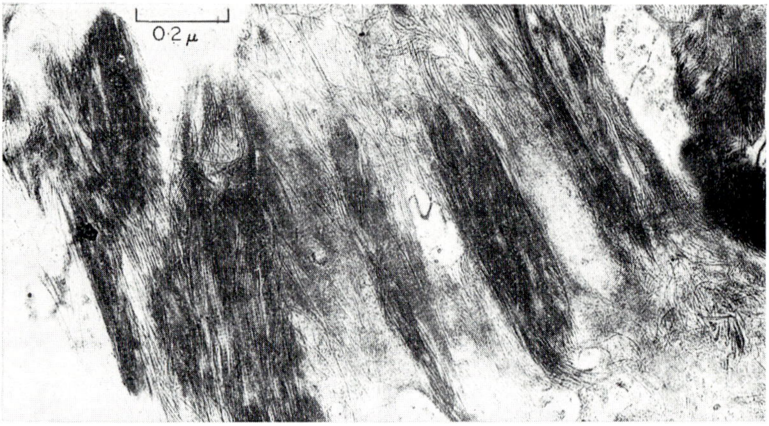

FIG. 5. Structure of basement membrane of *H. fossilis* showing bundles of collagen fibres in the columns of pillar cells. (Hughes and Munshi, 1975).

the collagen fibres of the column appear disrupted (Fig. 6) but the cytoplasmic fibres remain and become more distinct. This indicates that while the pillar cell columns contain collagen fibres, the pillar cell cytoplasm contain simple contractile fibres. Quite a good number of mitochondria are found in the cytoplasm.

In *Boleophthalmus boddaerti* the pillar cell flanges possess very characteristic darkly-staining material which suggests a supporting structure (Fig. 7). The elongate fibrillar structures are characteristically arranged in concentric layers. This specialization of pillar cells may be correlated with the amphibious habit of these fishes (Hughes and Munshi, 1975). Functionally the capillaries of the gills will not collapse when the fishes come out of water.

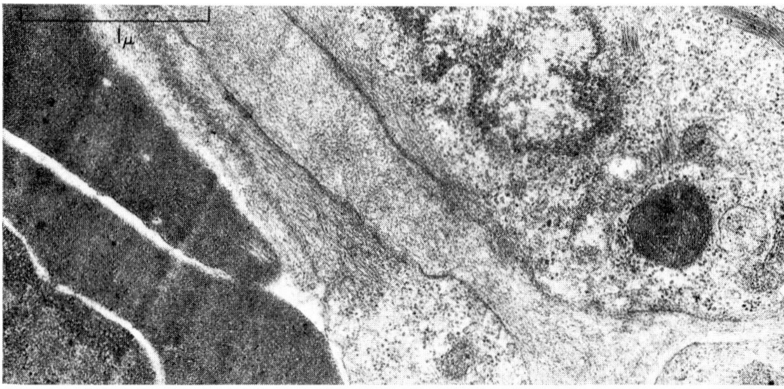

FIG. 6. Highly magnified view of part of a pillar cell of *H. fossilis* along with one of the columns after collagenase treatment. Note the digested collagen material in the columns and fine fibrillae in the cytoplasm of the pillar cells. (Hughes and Munshi, 1975).

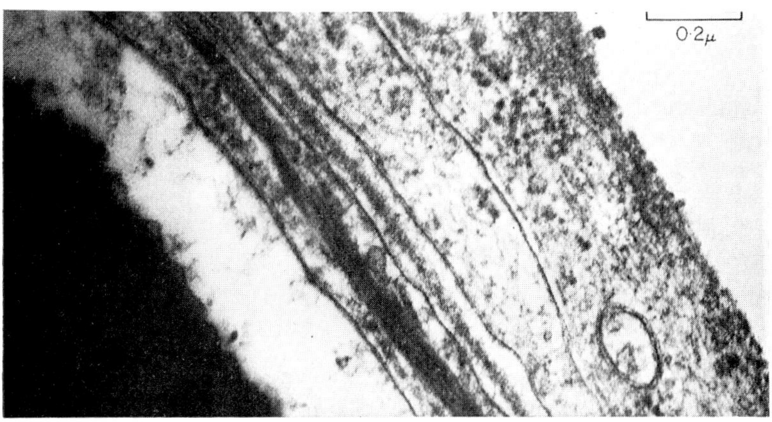

FIG. 7. Transverse section of a secondary gill lamella of *Boleophthalmus boddaerti* showing the characteristic darkly stained stiffened structural material in the cytoplasm of the pillar cell flange. Note also the structure of the basement lamella and epithelium. (Hughes and Munshi, 1975).

III. STRUCTURE OF AIR-BREATHING ORGANS

Air-breathing respiratory chambers have developed as extensions of the opercular (= extrabranchial), pharyngeal or branchial chambers.

A. OPERCULAR OR EXTRABRANCHIAL CHAMBERS MODIFIED FOR AIR BREATHING

Certain species of the family Gobiidae have their chambers modified for air-breathing purposes. Gobiidae include several air-breathing species all inhabiting estuaries. This led Hora (1935) and later Saxena (1963) to recognize estuarine fishes as the main associations of air-breathing species.

Based on the structure of the bucco-pharyngeal epithelium, Schöttle (1931) has described three categories of gobies namely, (i) those in which capillaries are sub-epithelial in position, e.g. aquatic *Gobius caninus* and *Periophthalmus koelreuteri*; (ii) those in which the blood capillaries are more numerous in the sub-epithelium and may even extend into the epithelium, e.g. *Boleophthalmus boddaerti*; and (iii) those in which the capillaries are mostly epithelial in position, e.g. *Periophthalmus schlosseri*. Todd and Ebeling (1966) have studied the structure of the bucco-pharyngeal epithelium of *Gillichthys mirabilis* and have placed it in the second category of the above scheme. These authors however do not say anything about the specialization of the opercular epithelium of these fishes. Das (1934) studied the air-breathing habit of *Pseudapocryptes lanceolatus* and found that the air is absorbed mainly by the lining of the operculum which is adapted to this function by a rich vascular supply and by the thinness and the elasticity of the opercular bones. Several other estuarine gobies of India, such as *Taenioides rubicundus*, *Apocryptes bato*, *Periophthalmus schlosseri* distend their gill-chambers with fresh air every few minutes (Hora, 1935). Amongst the Indian gobies *Periopthalmus vulgaris* and *Periophthalmus pearsei* and *Boleophthalmus viridis* and *Boleophthalmus boddaerti* may be taken as typical amphibious air-breathing fishes of estuaries.

Periophthalmus vulgaris, *Boleophthalmus viridis* and *B. boddaerti* spend a great part of their time skipping amongst the roots of mangrove trees of the Sunderban and Subarnarekha delta of Bengal and Orissa coast. At low tide they are found in burrows in the bank just above water level.

The accessory respiratory organs of *Periophthalmus vulgaris* provide an excellent example of adaptation by modification of the opercular chambers (Figs. 8 and 9) (Singh and Munshi, 1969). The following structural modifications have taken place:

(i) The opercular bones have become thin and elastic;
(ii) The opercular chambers have enlarged and extend below the basi-branchials and above the gill arches; air pockets have developed in the walls of the respiratory chambers;
(iii) The branchiostegal apparatus of both sides has developed a special type of safety valve which is workable by strips of muscles;
(iv) The epithelium lining the opercular chamber and the branchio-stegal membranes have become richly vascular and;
(v) Intricate mechanisms for opening and closing the inhalent and exhalent apertures have evolved.

The air–blood pathway along which the respiratory gases are exchanged is fundamentally the same as in the lungs of higher vertebrates. It is composed of an external squamous type of epithelium, a basement membrane and an internal lining of the blood capillary. The total thickness may vary from 1 μm–3·2 μm.

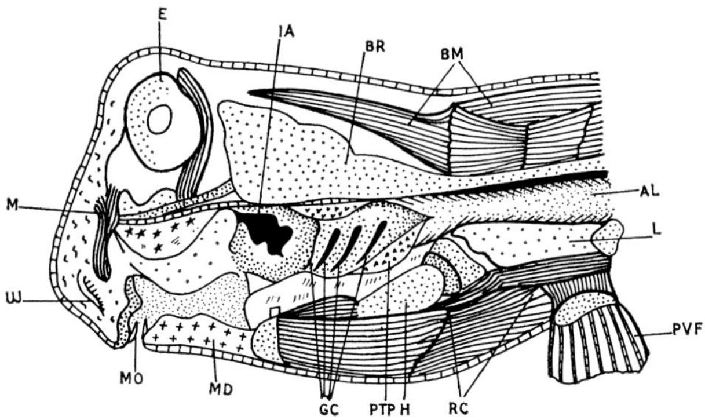

FIG. 8. A sagittal section of *P. vulgaris* showing the arrangement of the 4 gills and their gill clefts (= gill openings). The first gill cleft has become specialized to work as the inhalent (IA) aperture for air breathing. Other details of its anatomy, such as the position of the heart (H), brain (Br), alimentary canal (AL) and the various muscle systems (BM, M) are also shown. × 16 Nat. Size. (After Singh and Munshi, 1969).

B. PHARYNGEAL CHAMBERS MODIFIED FOR AIR BREATHING

In *Amphipnous cuchia*, and *Channa* spp. the respiratory air chambers are diverticular of the pharynx.

1. *Amphipnous cuchia*

The air sacs are in the form of a pair of lung-like structures situated along the lateral sides of the head partly covered over by the operculum (Fig. 10; Munshi and Singh, 1968a). Morphologically they are lateral extensions of the

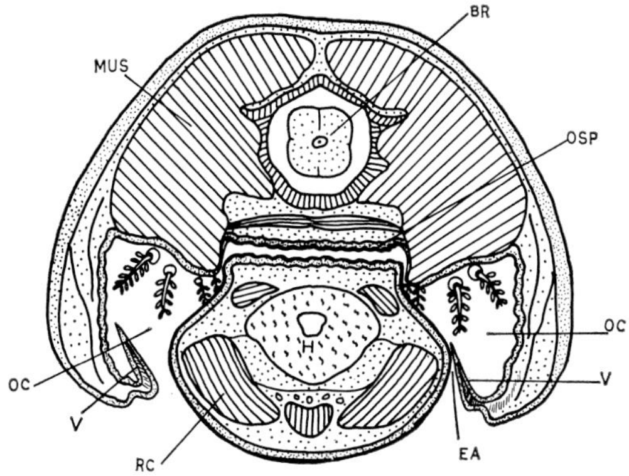

Fig. 9. Diagrammatic sketch of T. S. of *P. vulgaris* passing through the posterior part of the pharynx showing the disposition of four gill arches, the respiratory opercular chambers, Valves (V), the exhalent aperture (EA), branchial muscles and the heart (H). × 50 Nat. Size.

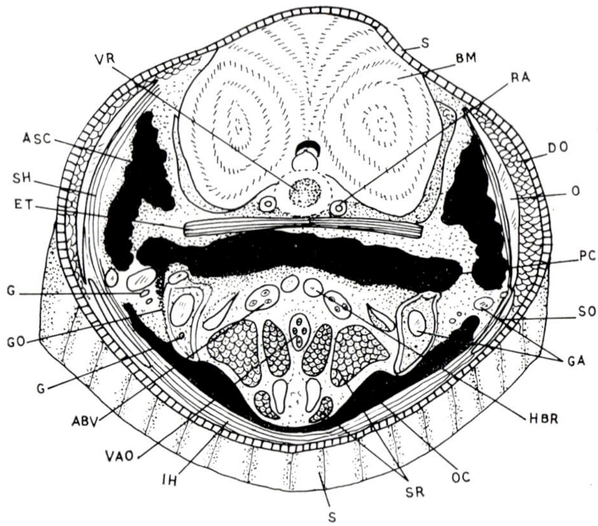

FIG. 10. A diagram of a T.S. of the head of *A. cuchia* passing through the middle part of the air sac showing the opercular cavity (OC), branchiostegal membrane, pharyngeal cavity (pc), dilator operculi (Do), the superior (SH) and the inferior hyohyoideus (IH), body myotomes (BM), the subarcualia recti (SR) and the epiarcualis transversus (ET). × 5 Nat. Size. (Modified after Munshi and Singh, 1968).

pharyngeal cavity situated one on each side dorsal to the gill arches. The pharynx opens through an inhalent aperture into the air sac on each side. There is no mechanism to close this opening. Each sac opens into the opercular cavity through a separate opening, the exhalent aperture situated above the ceratobranchial of the first arch.

(a) *Structure of the air sac.* The inner wall of the air sac is thrown into folds and ridges. Each fold is covered by the respiratory epithelium next to which are the elastic fibres of the connective tissue, a thin basement membrane and then a thick muscular coat derived from the hyohyoideus muscle.

(b) *Respiratory mucosa.* The mucosa lining the respiratory sac consists of vascular and non-vascular areas. The vascular area is comprised of small and large "islets" (Fig. 14e). These vascular islets become very distinct in injected preparation. Each islet is studded with hundreds of vascular rosettes, each of which comprises a group of vascular papillae. The vascular papillae composing each rosette of a respiratory islet varies from 20–30 in number arranged in a linear order. The non-vascular area of the respiratory mucosa is in the form of irregular spaces between the contiguous vascular respiratory islets. The epithelium of this region contains numerous mucous glands.

(c) *Blood circulation in respiratory islets.* The air sac of *Amphipnous cuchia* gets its blood supply from the first afferent branchial vessel. This branch of the afferent vessel enters the air sac of its own side and divides and redivides into a number of large and small blood vessels which ramify all over the sac to supply blood to the respiratory islets. The blood is collected from the rosette of the vascular papillae by an efferent vessel. All the efferent islets vessels open ultimately into large veins which fuse and open into the jugular vein.

(d) *Ultrastructure of vascular papillae.* Electron microscopic studies (Hughes and Munshi, 1973b) showed that individual vascular papillae are separated by micro air pockets. The arrangement is very characteristic (Fig. 11). Each micro air pocket is connected with the main cavity of the air chamber by a passage of 2 μm length and 0·1 μm in diameter. Each pocket has an irregular shape and is bordered by epithelial cells bearing microvilli.

The main part of each papilla which projects into the air sac has a dome-shaped appearance. The afferent and efferent capillaries within the papillae are separated from each other by means of valves. Each valve projects freely into the papilla lumen. The valve arises from the inner surface of each papilla and is formed of connective tissue material bounded by a

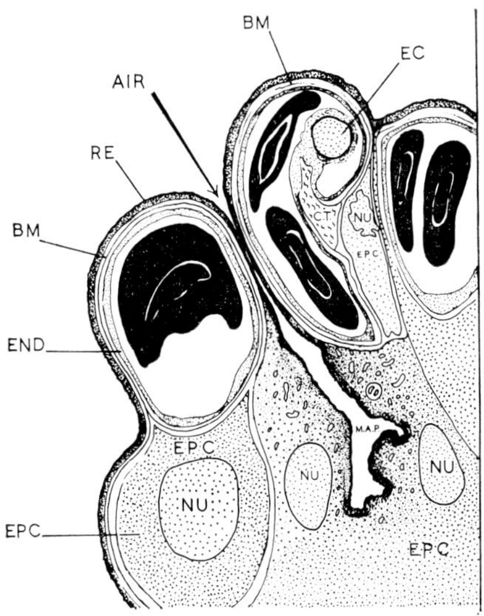

FIG. 11. Diagrammatic view of a vertical section through the respiratory membrane of an air sac of *A. cuchia* showing the relationship of micro-air pocket (M.A.P.) and the vascular *papillae*. Note also the position of the valve with its prominent nucleus situated inside the vascular *papilla*, also the position of the basement membrane (BM) and the epithelial cells. The respiratory epithelial cells (EPC) are also seen.

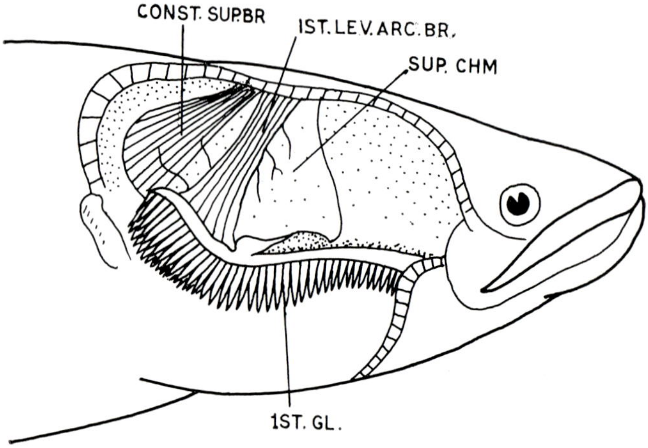

FIG. 12. Lateral view of the head of *Channa punctata* to show the relation of the supra-branchial chamber with the constrictor suprabranchialis (Const. supbr.), the first levator arcus branchialis (lev. arc. Br.) and the first gill-arch. × 2½ Nat. Size. (After Munshi, 1962).

basement membrane and covered by the endothelial lining. The terminal endothelial cells of the valve are characterized by a round nucleus. The stalk of the valve is mainly composed of connective tissue containing many contractile filaments. The valve seems to be concerned with regulating the flow of blood through individual vascular papilla. The air–blood pathway is extremely thin in *Amphipnous cuchia*. The total thickness being 0·435 μm. It is formed of single dense epithelial layer, a basal lamina and a thin endothelial layer.

2. *Channa punctatus and Channa striatus*

The main air-breathing organs are a pair of supra-branchial chambers which develop dorsal to the gill arches above the pharynx and lateral to the prootic bone of the auditory capsule (Fig. 12; Munshi, 1962b). Dorsally it is protected by parietal and pterotic bone. Posterolaterally the suprabranchial chamber extends into the opercular cavity. Each suprabranchial chamber is incompletely divided into compartments or recesses by a shelf-like transverse outgrowth from the hyomandibula which projects into it. The pharynx opens through the inhalent aperture into an anterior compartment (Fig. 13). There is no shutter to guard this opening. The anterior compartment communicates freely with the posterior compartment of the suprabranchial chamber. This posterior chamber also communicates with the pharynx through another inhalent aperture. This aperture however, is guarded by a shutter of dendritic plate borne by the first branchial arch. This shutter serves to close the inhalent aperture. The dorsal part of the first gill slit forms the exhalent aperture for the suprabranchial chamber.

The wall of the posterior compartment of the suprabranchial chamber is contractile due to the presence of the constrictor suprabranchialis muscle (Fig. 12). Both compartments of the suprabranchial chamber are lined by a respiratory membrane containing islets of vascularized tissue. Each respiratory islet is composed of small groups of vascular papillae where gaseous exchange takes place.

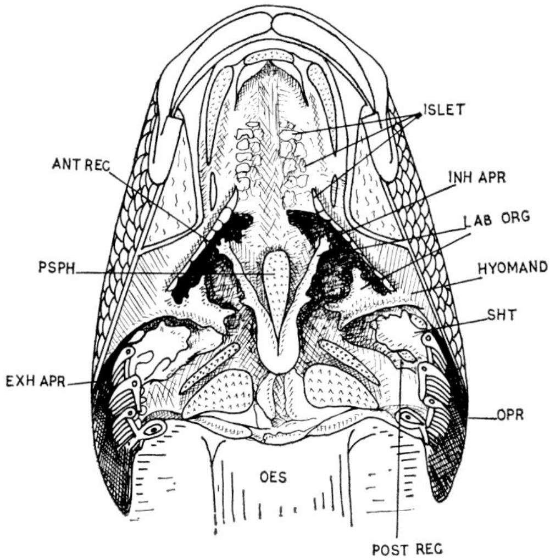

FIG. 13. A view of *Channa* (= *Ophicephalus*) *striatus* dissected from the ventral side to show the anterior and posterior recesses (Ant. Rec.) of the suprabranchial chamber, inhalent (Inh. Apr.) and exhalent aperture (Exh. Apr.), transverse process of the hyomandibula bearing the nodular labyrinthine organ (Lab. org.), shutter (SHT), gill arches and the vascular "islets" on the margin of the anterior recess and on the roof of the posterior section of the pharynx. $\times$ $1\frac{1}{2}$ Nat. Size. (After Munshi, 1962).

(a) *Arrangement of the vascular papillae.* Each vascular area (islet) is studded with hundreds of vascular rosettes each of which comprises of a group of papillae (Fig. 14f). The vascular papillae comprising each rosette varies from 25–30 in number. The afferent and efferent blood vessels which supply and collect blood from each rosette are situated at a somewhat deeper level. The rosettes are separated from each other by much folded micro air-pockets. In certain regions the air pockets are lined by a simple epithelium bearing microvilli.

(b) *Ultrastructure.* The sub-epithelial capillaries come to the surface of the suprabranchial chamber at several points. These minute blood channels terminate in the form of bulb-like structures called vascular papillae (Fig. 15). Epithelial cells separate the vascular papillae, each of which is formed by one or two endothelial cells whose nuclei generally lie at the base or side of the

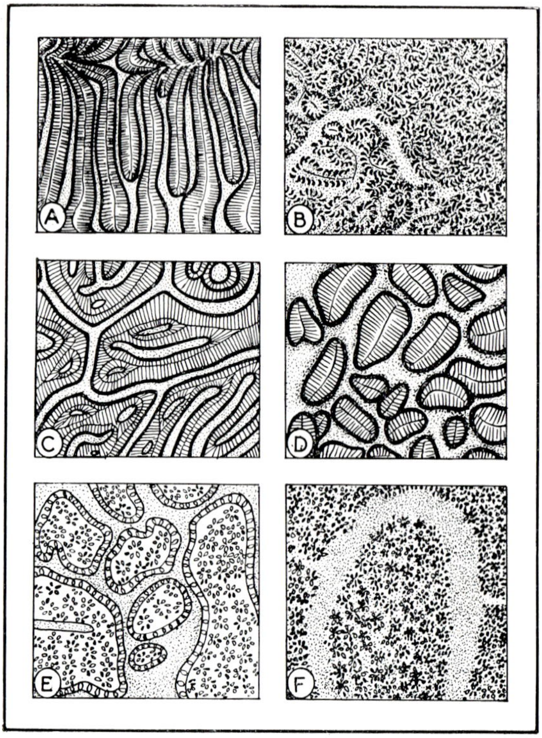

FIG. 14. Respiratory islets of (a) *Clarias batrachus* (b) *Heteropneustes fossilis* (c) *Anabas testudineus* (d) *Colisa fasciatus* (e) *Amphipnous cuchia* and (f) *Channa punctatus*.

bulb. The papilla wall is lined by extended flanges of the endothelial cells external to which the basement membrane forms the structural layer. The aerial surface of the papilla is covered by thin extensions of the epithelial layer which is single or double layered. At higher magnification the outer border of the epithelial flanges has a bushy appearance and its inner surface has many pores. The basal lamina is of uniform thickness and micro-pinocytotic vesicles from the endothelial cells open into it. In the cytoplasm of the endothelial cells are found a few mitochondria, small-walled vesicles, and aggregates of minute osmiophilic granules. The measurements of the air–blood distance give values varying from 0·56–0·99 μm (Hughes and Munshi, 1973b).

(c) *Blood circulation through the vascular papilla.* Blood enters and leaves each vascular papilla through two capillaries which are separated by a valve-like structure (Fig. 15a) which originates from the wall of the base of the vascular papilla. The body of the valve is formed of connective tissue bounded by a membrane which seems to give off column-like structures. The surface

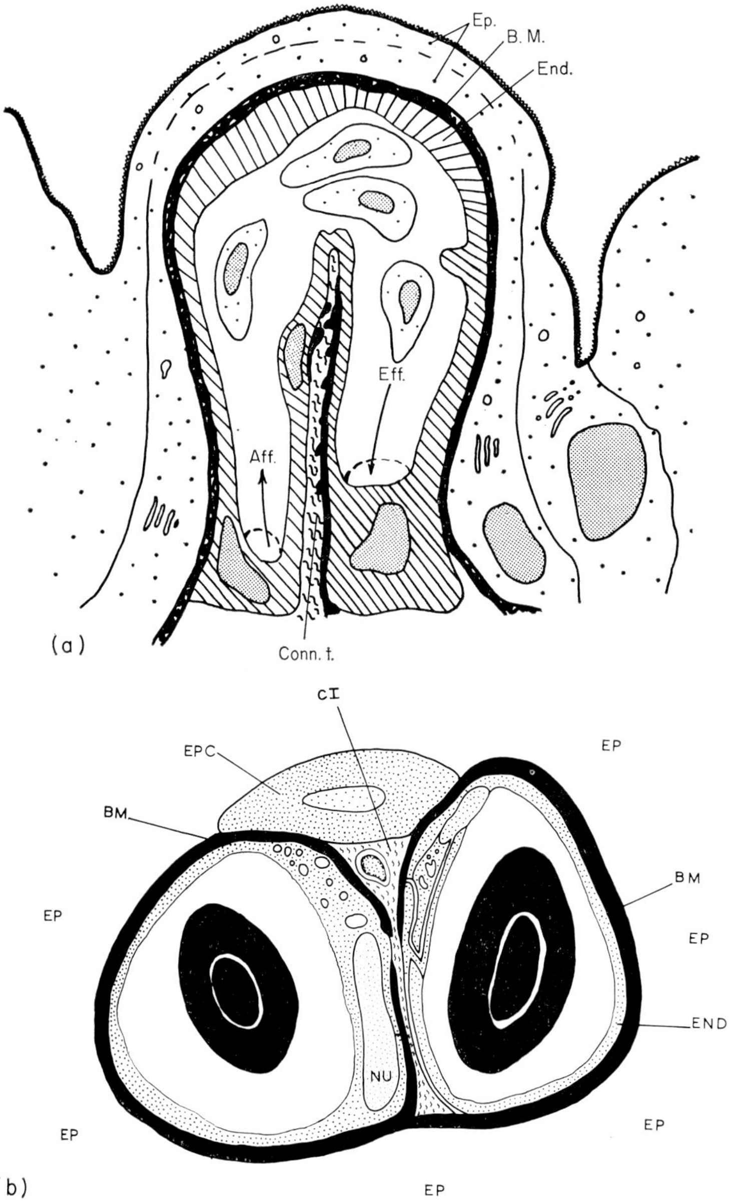

FIG. 15 (a) Diagrammatic view showing the ultrastructure of the vascular papilla of *Channa punctatus* showing the position of the valve, basement membrane (BM) and two layers of epithelial cells (EP). The direction of the movement of blood is shown by a thick arrow. (From Hughes and Munshi, 1973b). (b) Diagrammatic view of the ultrastructure at the base of one vascular papilla of *Channa* showing the afferent and efferent capillaries separated by a partition composed of connective tissue (CT) and endothelial lining (END). Under the light microscope this partition looks like a pillar cell with columns.

of the valve, however, is covered by the endothelial cells. Sections passing through the base of a papilla (Fig. 15b) show that there are two vascular channels separated from each other by a wall composed of connective tissue material bounded on both sides by endothelium. The basement membrane gives off several column-like structures into the connective tissue material of the valve. Under the light microscope it was this structure which appeared very much like a pillar cell (Munshi, 1962b).

In the case of *Channa striatus* the same sort of structure of the vascular papilla has been confirmed but the tip of each papilla has a characteristic epithelial cap cell. This increases the length of air–blood distance to the extent of 2·00–2·32 μm. At certain points however it is quite thin (0·525 μm).

(d) *Non-vascular area.* The non-vascular part of the lining of the suprabranchial chamber is formed of stratified columnar and polyhedral cells. The endoplasmic reticulum of some of these epithelial cells bear ribosomes. They are arranged in regular fashion about the centrally placed nuclei.

C. BRANCHIAL CHAMBERS MODIFIED FOR AIR BREATHING

In *Clarias batrachus*, *Heteropneustes* (= *Saccobranchus*) *fossilis*, *Anabas testudineus* and *Colisa fasciatus* the branchial chambers have been modified for air breathing.

1. *Clarias batrachus*

The suprabranchial chamber consists of two recesses—a dorsal and a ventroposterior chamber (Fig. 16a,b). The dorsal chamber is a cup-shaped recess formed by the approximation of the second and third fans. The first dendritic organ, developed on the second gill arch, lies within this recess. The ventroposterior chamber is situated behind the third fan and extends beyond the pectoral girdle into the trunk region. The second dendritic organ is situated in this chamber.

Well-defined inhalent and exhalent apertures are present (see Fig. 16b). The air from the pharynx enters the dorsal chamber through the inhalent aperture and then reaches the posterior chamber. Air escapes into the opercular cavities through the exhalent aperture of each side.

Four well-developed fans are present which are formed by the fusion of the primary gill lamellae.

The first dendritic organ is borne by the second gill arch and is smaller in size than the one which develops on the fourth gill arch (see Fig. 1). The dendritic organs are supported by an internal skeleton which remains cartilaginous throughout life. The terminal bulb of each dendritic organ consists of a core of cartilage, a cortex of connective tissue with a covering of respiratory mucosa having generally 8 folds. Each fold is dual in composition, each half representing one side of the primary gill lamella. The blood capillaries of the lamellae present in the bulb of the dendritic organs are supported by pillar cells.

The respiratory mucosa lining the suprabranchial chamber also comprises vascular and non-vascular areas. The former takes the form of numerous

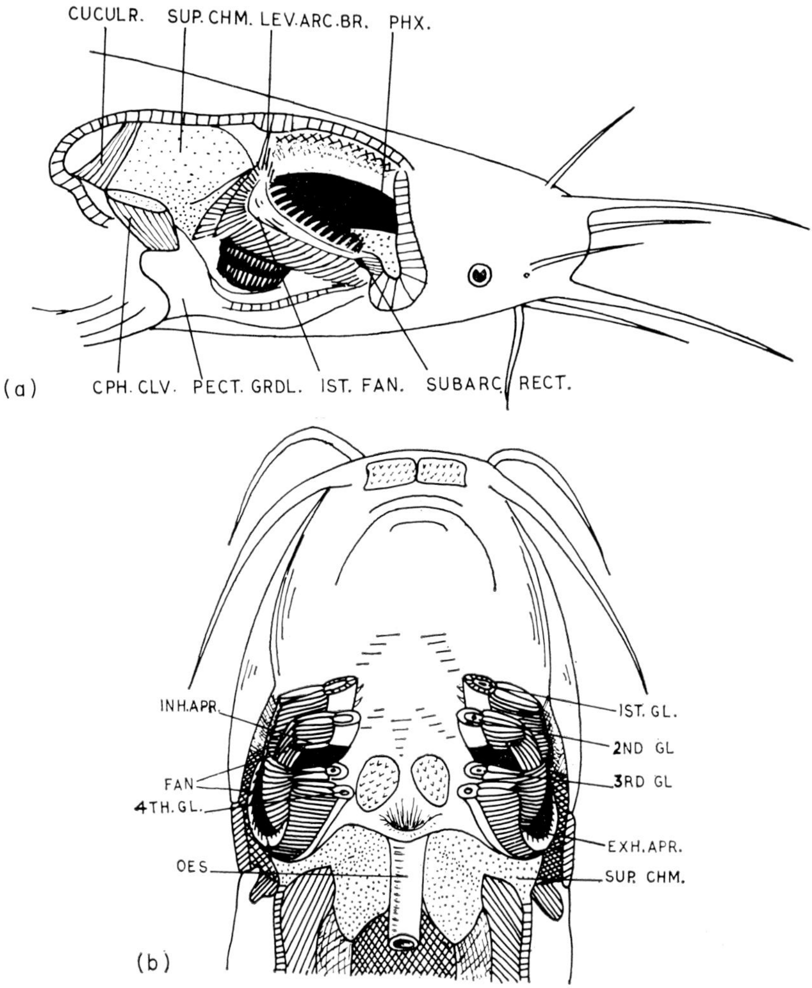

FIG. 16. (a) Lateral view of the respiratory membrane (dotted) covering the suprabranchial chamber (Sup. chm.) of *Clarias batrachus* as seen from outside and its relation with the Cucullaris (cuculr.) Cephalo-clavicularis (Cph. clv.), Levatores arcuum branchialium (Lev. arc. br.) and the gill arches. × $2\frac{1}{2}$ Nat. Size. (b) A view of *Clarias batrachus*, dissected from the ventral side to show the suprabranchial chamber (Sup. Chm.) (dotted membrane), the inhalent and exhalent aperture (Inh. Apr. and Exh. apr.) and their relationship with the branchial arches and the "fans". × 3 Nat. Size.

large and small respiratory islets in each of which two corresponding sets of secondary lamellae belonging to a primary lamella are incorporated (Fig. 14a). The vascular network existing within the lamellae and the general pattern and the configuration of the islets proclaim their derivation from the primary and secondary gill lamellae.

It has been found that after the formation of the fans the primary gill lamellae belonging to the first, second, third and fourth gill arches move away from the branchial arches and migrate on to the surface of the suprabranchial chambers where they undergo fusion and shortening of the primary lamellae, and abbreviation of the secondary lamellae to form a flattened respiratory membrane. The process is essentially similar in both *Clarias batrachus* and *H. fossilis*.

(a) *Ultrastructure*. Electron microscopic studies of the respiratory islets (Hughes and Munshi, 1973) confirmed the presence of typical pillar cells having columns surrounded by infoldings of the cell membrane. Sometimes several pillar cells are found adjacent to one another. The air–blood pathway comprises a single epithelial layer (0·38 μm), a very thin basement membrane (0·024 μm) and a thin lining of pillar cell flanges (0·146 μm).

2. *Heteropneustes fossilis*

The suprabranchial chambers extend backward into the trunk region in the form of sacs deeply embedded in the trunk myotomes. Each suprabranchial chamber is completely separated by the second and the third fans into two chambers. The anterior chamber thus formed, communicates freely with the posterior chamber which extends backwards in the form of a tubular sac (Fig. 17). Air first enters the anterior chamber through the inhalent slit and then goes into the posterior part. The respiratory mucosa of the air sac is thrown into folds or ridges. Each ridge is covered by a respiratory epithelium next to which, in order, are connective tissue, a thin membrane and externally a muscular coat derived from the cucullaris muscle. The respiratory membrane lining the respiratory sac consists of vascular and non-vascular areas. The vascular part comprises small and large islets distinguished irregularly over its surface. The islets make different types of configuration on account of the fusion of primary gill lamellae (Figs. 14b, 17). Each islet is formed by double rows of modified secondary lamellae. Associated with each islet are several transverse channels which interconnect them.

(a) *Ultrastructure*. Electron micrographs of the respiratory islets (Hughes and Munshi, 1973b) reveal a series of pillar cells having a basement membrane on either side. As the cells approach the surface, more and more vascular spaces appear between them. Invariably the vascular spaces bifurcate near the air surface and run parallel to them. This produces maximum contact

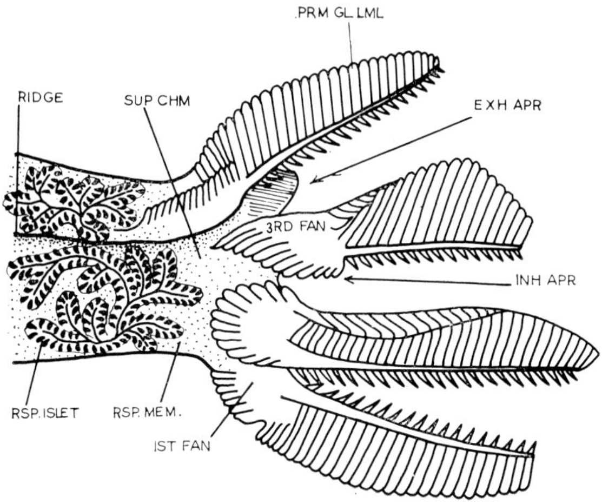

FIG. 17. The respiratory sac opened out by means of a mid-ventral incision to show the relation between the "fans", the inhalent (Inh. Apr.) and the exhalent apertures (Exh. Apr.), the gill arches and the suprabranchial chamber (Sup. chm.) and respiratory sac. ×4 Nat. size. (Modified after Munshi, 1962).

of the blood with the respiratory surface. The cells forming the vascular channels are certainly pillar cells as they contain columns of basement membrane material and closely resemble those of the gills. The nuclei of the pillar cells vary in shape and sometimes endothelial cells are also detectable. The outermost epithelial surface bears microvilli and is covered with a fuzzy material which seems to be mucus. The air–blood pathway is composed of a thin epithelial layer (1·36 μm), a very thin basement membrane (0·05 μm) and thin pillar cell lining (0·2 μm). The total thickness is about 1·6 μm. The non-vascular area of the respiratory membrane is covered by a multilayered epithelium. Lymphocytes are also found amongst the epithelial cells.

3. *Anabas testudineus and Colisa fasciatus*

While in *Anabas* the labyrinthine organs are complex structures (Fig. 18), in *Colisa fasciatus* they are of simpler nature (Fig. 19). The labyrinthine organ in *C. fasciatus* is a special structure possessing two leaf-like expansions situated on either side of the median septum (Fig. 19). The free and broad expansion of this organ projects into the two recesses of the suprabranchial chamber. The respiratory mucosa covering the labyrinthine organ and the suprabranchial chamber contains both vascular and non-vascular areas. The surface of the respiratory membrane shows irregularly arranged raised areas called respiratory islets.

The structure of the respiratory islets is more complex in *Anabas* than in *Colisa* (see Fig. 14c,d). However, in both cases each respiratory islet, besides

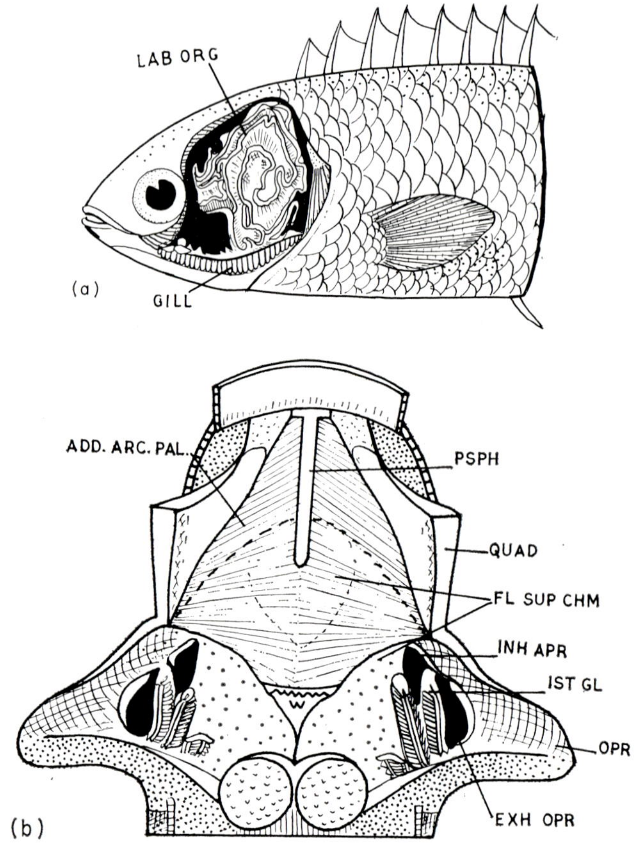

FIG. 18. (a) Lateral view of the head of *Anabas testudineus* showing the disposition of the labyrinthine organ and first gill arch. ×2 Nat. Size. (b) *Anabas testudineus* dissected from the ventral side to show the limits of the suprabranchial chamber (dotted line), adductor arcus palatini (Add. arc. pal.) and the inhalent (Inh. Apr.). and exhalent apertures (Exh. apr.). ×3 times. (Modified after Munshi, 1968).

other structural components, consists of double rows of parallel channels. Under low magnification these are parallel lines but at higher magnification their real nature is revealed. The contiguous blood channels of the islet are separated from each other by a series of epithelial cells.

(a) *Ultrastructure*. Electron microscopic studies were very informative (Hughes and Munshi, 1973a). The supporting epithelial cells give off long flanges that cover the adjacent capillaries externally and they also project into the intercapillary spaces (Fig. 20). The blood capillaries are essentially formed of a series of endothelial cells having large spherical nuclei and very characteristic tongue like processes which may act as small valves. These processes

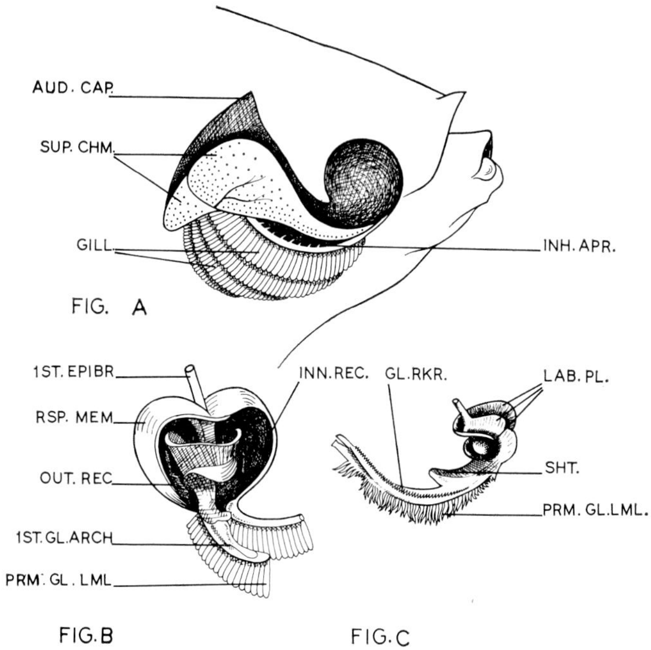

Fig. 19. (a) Lateral view of the head of *Colisa fasciatus* showing the suprabranchial chamber, gill arches and inhalent aperture. ×16 Nat. Size. (b) Fronto-lateral view of the outer and inner recesses of the suprabranchial chamber of *C. fasciatus* showing the disposition of the labyrinthine organ and the relation of the respiratory membrane with the first and second gill arches. ×10 Nat. Size. (c) Lateral view of the first gill arch of the *C. fasciatus* showing the labyrinthine organ, shutters and the primary gill lamellae. ×10 Nat. Size.

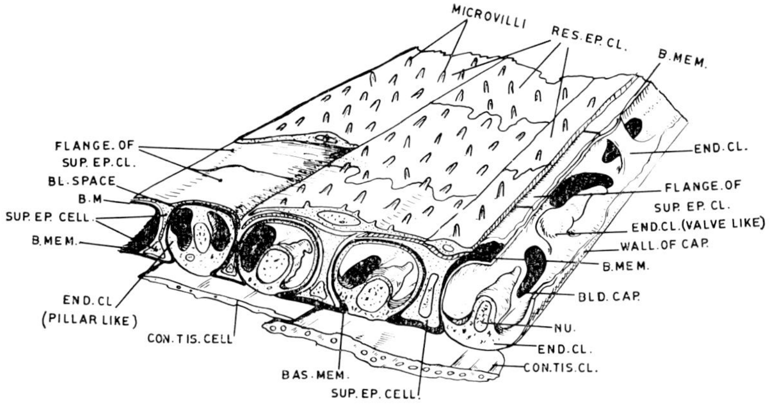

Fig. 20. A three-dimensional diagram to show the ultrastructure of a respiratory islet of *Anabas testudineus* as seen under the electron microscope. (Hughes and Munshi, 1973a).

project in the direction of blood flow and the valve may bend at different angles. A structural basis for contractility of the endothelial valves may be the fibriller material in their cytoplasm. These valves modify significantly the nature of the endothelial channels which looked like a system of ring-like spaces at regular intervals in indian ink preparations. The air–blood pathway is composed of a double layer of epithelial cells, a basement membrane and their extensions of endothelial cells. The total thickness measures about 0·2–0·3 μm.

Specialized chaemoreceptor cells have been discovered in the non-vascular part of the lining of the suprabranchial chamber (Hughes and Munshi, 1973a). They have all the characteristics of the integument chemoreceptor. Although the suprabranchial chambers are extensions of branchial chambers and the labyrinthine organs have developed on the epibranchials of Anabantidae, the respiratory islets in this group have evolved in different fashion.

The hypothesis that the labyrinthine organs and the respiratory islets of the supra-branchial chamber in *Anabas* represent modified gill capillaries was based on the presence of characteristic pillar cells in all three structures (Rauther, 1910; Munshi, 1968). But this is no longer tenable as EM studies (Hughes and Munshi 1968, 1973a) reveal that in Anabantidae the "pillar cells" of the air-breathing organs do not have the same relationship with that of the gills but are modified epithelial cells. Obviously we are dealing with analogous rather than homologous structures that have been used to serve the same function of supporting the contiguous vascular units which make up these respiratory organs. Such a supporting function is clearly of importance, and during evolution different structural arrangements have been selected to provide surfaces with a minimum diffusion barrier between the blood and the respiratory medium (Hughes and Munshi, 1968).

These recent studies reveal some remarkable convergence with the air-breathing organs of other vertebrates, but no evidence for lamellated bodies has been found. Hughes (1973) while describing the ultrastructure of the lung of *Neoceratodus* and *Lepidosiren*, has summarized the distribution of lamellated bodies in relation to surface tension reducing properties of washings from lung or homologous structures in all the groups of vertebrates. The gross and fine morphology of air-breathing organs of the fishes studied in this paper show that they have evolved independently in a different fashion.

In comparison to the water–blood pathway, the air–blood pathway is relatively thinner. Results on the air–blood pathway of Indian air-breathing organs are summarized in Table I.

IV. Morphometry of Respiratory Surface Areas

Fishes are primarily water breathers with gills taking care of the total metabolic gas exchange. However, bimodal gas exchange has also been

TABLE I

Fish sp.	Accessory respiratory organ	External epithelial layer (μm)	Basal lamina (μm)	Endothelial or pillar cell flange (μm)	Total thickness (μm)
Amphipnous cuchia	Air sac (Vascular papilla)	0·35	0·06	0·025	0·435
Anabas testudineus	Labyrinthine organ (respiratory islets)	0·15	0·05	0·01	0·21
Channa punctatus	Suprabranchial chamber (vascular papilla)	0·465	0·06	0·255	0·78
Channa striatus	Suprabranchial chamber (vascular papilla)	1·212	0·06	0·087	1·359
Clarias batrachus	Arborescent organ	0·38	0·024	0·146	0·55
Heteropneustes (=*Saccobranchus*) *fossilis*	Air sac (respiratory islet)	1·37	0·05	0·185	1·605

reported in many teleostean fishes, where gills are supplemented by skin and buccopharyngeal breathing. To evaluate the respiratory efficiency of the gills, the measurements of gill dimensions have been made for many water-breathing teleostean fishes (Putter, 1909; Price, 1931; Byczskowska–Smyk, 1957, 1958, 1959a,b; Gray, 1954; Hughes, 1966, 1970a,b,c, 1972a,b,c; Saunders, 1962; Muir and Hughes, 1969; Hughes and Morgan, 1973; Hughes and Gray, 1972). Of these, only Price (1931), Hughes and Morgan (1973), Muir and Hughes (1969) and Hughes and Gray (1972) made serious efforts to determine the relation between size and respiratory area in a single species.

Morphometric studies of the lungs of mammals and other tetrapods have been carried out by Tenney and Remmers (1966), Tenney and Tenney (1970), Weibel (1963, 1969), Hutchison, Whiteford and Kohl (1968) and others, but comparable studies on the air-breathing fishes are somewhat limited (see Hughes and Weibel, 1975). Recently Rahn, Rahn, Towell, Gans and Tenney

(1971) made morphometric measurements of the lungs of *Lepisosteus osseus* (Hughes and his colleagues, Hughes, Dube and Munshi, 1973; Hughes, Singh, Guha, Dube and Munshi, 1974; Hughes, Singh, Thakur and Munshi, 1974) have established a definite relationship between bimodal gas exchanger machinery (gills+accessory respiratory organs) and body weight in some air-breathing fishes of India. The relationship between the body weight and the respiratory surface areas of these air-breathing fishes have been summarized in Table II.

From the table it is evident that gill area in *Channa punctatus*, *Anabas testudineus* and *Heteropneustes* (= *Saccobranchus*) *fossilis* increases by a power of 0·592, 0·615 and 0·746 respectively. Because the power functions in all cases are less than 1, the weight specific gill area per unit body weight

TABLE II

Summary of the equations showing the relationship between the dimensions of the respiratory organs and body weight in air-breathing fishes of India

Fish species	Equation ($Y = aW^b$)	References
Anabas testudineus		
Gill area (mm^2)	$Y = 278{\cdot}3W^{0{\cdot}615}$	
Gill area per unit body wt. (mm^2/g)	$Y = 278{\cdot}0W^{-0{\cdot}383}$	
Labyrinthine organ area	$Y = 80{\cdot}69W^{0{\cdot}799}$	
Suprabranchial chamber	$Y = 55{\cdot}44W^{0{\cdot}574}$	Hughes *et al.* 1973
Total surface area of accessory resp. organs (mm^2)	$Y = 147{\cdot}4W^{0{\cdot}712}$	
Amphipnous cuchia		
Air sac area (mm^2)	$Y = 12{\cdot}46W^{0{\cdot}797}$	Hughes *et al.* 1974
Skin area (mm^2)	$Y = 877{\cdot}4W^{0{\cdot}707}$	
Heteropneustes (= *Saccobranchus*) *fossilis*		
Gill area (mm^2)	$Y = 186{\cdot}0W^{-0{\cdot}746}$	
Gill area per unit body wt.	$Y = 186{\cdot}1W^{0{\cdot}254}$	
Skin area (mm^2)	$Y = 851{\cdot}1W^{0{\cdot}684}$	Hughes *et al.* 1974
Air sac area (mm^2)	$Y = 145{\cdot}9W^{0{\cdot}662}$	
Gill+Skin+Air-sac area	$Y = 1181W^{0{\cdot}694}$	
Channa punctatus		
Gill area (mm^2)	$Y = 470{\cdot}4W^{0{\cdot}592}$	
Gill area (mm^2/g)	$Y = 470{\cdot}4W^{-0{\cdot}41}$	Hakim, Munshi and Hughes (1976)
Suprabranchial chamber (mm^2)	$Y = 159{\cdot}1W^{0{\cdot}696}$	
Total respiratory area (mm^2) (Gills+accessory organs)	$Y = 629{\cdot}8W^{0{\cdot}62}$	

will decrease with increasing body weight. From the table it is also clear that the weight specific gill area (mm^2/g) in *Channa punctatus* (b= −0·41) decreases more rapidly than those of *Anabus testudineus* (b−0·385) and *Heteropneustes fossilis* (b = −0·254).

Some existing data on the relationship of gill area and body weight in fishes as summarized in Table III show that the slope (b) of the regression line ranges from 0·5–1·0 in different teleostean fishes.

TABLE III

Summary of values for the regression coefficient (b) reported for the relationship between log gill area to log body weight for different fish species

Fish species	Value for (b)	References
Micropterus dolomieu	0·78	Price (1931)
Grays "Intermediates"	0·82	Ursin (1967)
Katsuwo.ius pelamis / *Thunnus albacares* / *T. thynnus*	0·85	Muir and Hughes (1969)
Coryphaena hippurus	0·71	Hughes (1970a)
Scomber scombrus	0·997	Hughes (1970b)
Tinca tinca	0·698	Hughes (1970b)
Blennius pholis	0·85	Milton (1971)
Opsanus tau	0·79	Hughes and Gray (1972)
Anabas testudineus	0·615	Hughes *et al.* (1973)
Saccobranchus fossilis	0·746	Hughes *et al.* (1974)
Macrognathus aculeatum	0·733	Ojha and Munshi (1974)
Channa punctatus	0·59	Hakim *et al.* (1976)

From the table it is evident that the power functions of the gills in relation to body weight of *Channa punctatus* (0·59), *Anabas testudineus* (0·615) and *Heteropneustes* (= *Saccobranchus*) *fossilis* (0·746) are less than those of aquatic water breathing active fishes like *Micropterus dolomieu* (0·78), *Scomber scombrus* (0·997) and *Blennius pholis* (0·85). The lower values in the regression line of gill area against body weight may be due to the amphibious nature of these fishes. The morphometric estimation of the gill diffusing capacity (Dg) reveals that Dg of the gills of *Anabas testudineus* (0·0071) is less than those of *Heteropneustes fossilis* (0·0242), *Salmo gairdneri* (0·060) and *Tinca tinca* (0·125) (Hughes, 1972a). The lower values of the diffusing capacity of the gills of air-breathing fishes (*Anabas* and *Heteropneustes*) are due to comparatively lower values of gill area and thicker water–blood pathway. Recently Dube and Munshi (1974) have reported in *Anabas testudineus* that the diffusing capacity of the gills decreases with increasing body weight which clarify the death of larger fishes when not allowed to breath atmospheric air.

The analysis of air-breathing organs in relation to body weight reveals that the power function of air sac (b = 0·797) of *Amphipnous cuchia* is greater than that obtained for the air sac of *Saccobranchus* (b = 0·662), the supra-branchial chambers of *Anabas* (b = 0·574) and *Channa punctatus* (b = 0·696) The greater slope of the regression line for *Amphipnous* is close to the value obtained for both respiratory surface area and the metabolic rate in relation to body weight for most vertebrates. It appears that the air sac of *Amphipnous* is quite capable of meeting the increased oxygen demands with increasing body weight. Other factors are also important, such as the thickness of the air-blood barrier and hence the diffusing capacity (Dt). In the case of *Heteropneustes*, Dt ($ml0_2$/min/mmHg/kg) was 0·0288 (Hughes *et al.*, 1974), for *Anabas* about 0·28 for the combined suprabranchial chamber and labyrinthine organs (Hughes *et al.*, 1973) and for *Amphipnous* 0·0165 (Hughes *et al.*, 1974). Thus, although the air sac area of a 100g. *Amphipnous* is about one-sixth that of *Heteropneustes* (= *Saccobranchus*), the diffusing capacity is more than half that of *Heteropneustes* because the air-blood barrier is thinner in *Amphipnous* (Table IV). The diffusing capacity in these air-breathing fishes reveals that the air-breathing organs of *Amphipnous* and *Saccobranchus* are clearly less well suited for O_2 uptake than those of *Anabas*.

TABLE IV

Diffusing capacity for the tissue barrier (Dt) of the respiratory organs of Amphipnous, Anabas *and* Saccobranchus *(from Hughes* et al., 1974)

Species	Respiratory surface for 1 g fish (mm^2)	Thickness of tissue barrier (μm)	Area/g 100 g fish (mm^2)	Diffusing cap. (100 g fish) $D_t = A.K/t$ (ml/min/ mmHg/kg)
Anabas testudineus:				
(a) gill	278·0	10·0	47·2	0·0071
(b) suprabranchial chamber	55·4	0·21	7·55	0·0539
(c) labyrinthine organ	80·7	0·21	32·0	0·2286
Amphipnous cuchia:				
air sac	12·9	0·44	4·84	0·0165
skin	877·0		227·5	
Saccobranchus fossilis:				
(a) gill	186·1	3·58	57·7	0·0242
(b) air sac	145·9	1·60	30·7	0·0288
(c) skin	851·1	98·0	200·0	0·0031

V. Conclusion

Morphological and electron microscopical studies have confirmed the earlier deduction (Greenwood, 1961; Munshi, 1961, 1962a) that the air-breathing organs of *Clarias batrachus* and *Saccobranchus* (= *Heteropneustes*) *fossilis* represent modified gill structures and that the suprabranchial chambers and air sacs represent extensions of the branchial chamber. The cells forming the vascular spaces in the dendritic organs are typical pillar cells of the gills whereas in *Anabas testudineus*, the fine capillaries of the respiratory membrane are in fact intra epithelial structures and pillar cells are not present (Hughes and Munshi, 1968). In *Amphipnous cuchia* and *Channa* sp. the suprabranchial chambers are extensions of the pharynx. The vascular mucosa of the air sacs and suprabranchial chambers have evolved independently and not from the gill lamellae as it looks in the histological preparation under light microscopy (Hughes and Munshi, 1973b).

Morphometric studies of the gills of air-breathing fishes reveal that the gills of *Saccobranchus fossilis* seem to be more efficient than those of *Channa punctatus* and *Anabas testudineus* so far as the higher weight groups are concerned. "Drowning" experiments on the other hand show that the lower weight groups of *Anabas* can survive for a long time whereas the larger fishes die when not allowed to breath atmospheric air. Specimens of *Channa punctatus*, *Saccobranchus* and *Clarias* (smaller as well as larger) can survive for a long time even when not allowed to breathe atmospheric air. From these drowning experiments it can be concluded that the gills of *Channa punctatus*, are quite capable of meeting the increase in O_2 demand with increasing body weight. But the gill area per unit body weight decreases at a comparatively fast rate ($b = -0{\cdot}4$). In this fish, therefore, it is suggested that the suprabranchial chamber may take part in gaseous exchange when the fish is inside the water and thus cope with the O_2 demand of the fish as it grows in size. Whereas in *Anabas testudineus* the gills alone fail to supply the oxygen demand of the fish as it grows in size and therefore dies when not allowed to breathe atmospheric air. In *Clarias batrachus* and *Heteropneustes fossilis* the gills and skin are quite capable of meeting the increasing oxygen demand of the fish as they grow in size and therefore they can sustain life inside the water alone.

The morphometry of the air-breathing organs in relation to body weight reveals that the power function of air sac ($b = 0{\cdot}799$) in *Amphipnous* is greater than that obtained for the air sac of *Saccobranchus* ($b = 0{\cdot}662$) and the suprabranchial chamber of *Anabas* ($b = 0{\cdot}574$) and *Channa punctatus* ($b = 0{\cdot}696$). From these findings it has been suggested that the air sac of *Amphipnous* is efficient in maintaining the increasing oxygen demand of the fish as it grows in size. However, the diffusing capacity of the air-breathing organs reveal that

Amphipnous and *Saccobranchus* are clearly less well suited for O_2 uptake than those of *Anabas*.

The morphometric studies of the dimensions of the respiratory areas give a very close estimation of their functional capacity.

References

Byczkowska-Smyk, W. (1957). The respiratory surface of gills in teleosts. 1. The respiratory surface of the gills in the flounder (*Pleuronectes platessa*) and perch (*Perca fluviatilis*). *Zoologica* **Pol.** 8, 91–111.

Byczkowska-Smyk, W. (1958). The respiratory surface of the gills in teleosts 2. The respiratory surface of the gills on the eel (*A. anguilla*), the loach (*Misgurnus fossilis* L.) *Acta Biol. Cracov.* **1**, 83–97.

Byczkowska-Smyk, W. (1959a). The respiratory surface of the gills in teleosts. 3. The respiratory surface of the gills in the tench (*Tinca tinca* L.), the silver bream (*Blicca bjoerkna* and the *Chondrostome nasus* L.). *Acta Biol. Cracov.* **2**, 73–88.

Byczkowska-Smyk, W. (1959b). The respiratory surface of the gills in teleosts. 4. The respiratory surface of the gills in the pike (*Esox lucius*) stone-perch (*Acerina cernus*) and the burbot (*Lota lota* L.). *Acta Biol. Cracov.* **2**, 113–127.

Das, B. K. (1934). The habits and structure of *Pseudopocryptes lanceolatus*, a fish in the first stages of structural adaptation to aerial respiration. *Proc. Roy. Soc. Lond.* **115**, 422–430.

Dube, S. C. and Munshi, J. S. D. (1974a). Studies on the blood water diffusion barrier of secondary gill lamellae of an air-breathing fish *Anabas testudineus. Zool. Anz.* **193**, 35–41.

Dube, S. C. and Munshi, J. S. D. (1974b). Diffusing capacity of gills of climbing perch *Anabas testudineus* (BL.) in relation to body weight. *Ind. J. exp. Biol.* **12(2)**, 207–208.

Gray, I. E. (1954). Comparative study of the gill area of marine fishes. *Biol. Bull. Mar. Biol. Lab. Woods Hole* **107**, 219–225.

Hakim, A., Munshi, J. S. D. and Hughes, G. M. (1976). Morphometrics of the respiratory organs of *Channa punctatus* (in preparation).

Hora, S. L. (1935). Physiology, bionomics and evolution of the air-breathing fishes of India. *Trans. Natn. Inst. Sci. India* **1**, 1–16.

Hughes, G. M. (1966). The dimensions of fish gills in relation to their function. *J. exp. Biol.* **45**, 177–195.

Hughes, G. M. (1970a). A comparative approach to fish respiration *Experientia* **26**, 113–122.

Hughes, G. M. (1970b). Morphological measurements on the gills of fishes in relation to their respiratory function. *Folia Morph. Praha* **18**, 78–95.

Hughes, G. M. (1970c). Gill dimensions in relation to other respiratory parameters. Proc. IXth. Int. Cong. Anat. Leningrad (unpublished manuscript).

Hughes, G. M. (1972a). Morphometrics of fish gills. *Resp. Physiol.* **14**, 1–26.

Hughes, G. M. (1972b). Distribution of oxygen tension in the blood and water along the secondary lamella of icefish gills. *J. exp. Biol.* **56**, 481–492.

Hughes, G. M. (1972c). Gills of a living coelacanth (*Latimeria chalumnae*). *Experientia* **28**, 1301–1302.

Hughes, G. M. (1973). Ultrastructure of the lung of *Neoceratodus* and *Lepidosiren* in relation to the lung of other vertebrates. *Folia Morphol. Prague* **21**, 155–161.

Hughes, G. M. and Gray, I. E. (1972). Dimensions and ultrastructure of toadfish gills. *Biol. Bull. Mar. Biol. Lab. Woods Hole* **143**, 150–161.

Hughes, G. M. and Morgan, M. (1973). The structure of the fish gills in relation to their respiratory function. *Biol. Rev.* **48**, 419–475.

Hughes, G. M. and Munshi, J. S. D. (1968). Fine structure of the respiratory surface of an air-breathing fish, the climbing perch. *Anabas testudineus* (Bloch). *Nature Lond.* **219**, 1382–1384.

Hughes, G. M. and Munshi, J. S. D. (1973a). Fine structure of the respiratory organs of the Climbing perch *Anabas testudineus* (Pisces: Anabantidae). *J. Zool. Lond.* **170**, 201–225.

Hughes, G. M. and Munshi, J. S. D. (1973b). Nature of the air-breathing organs of the Indian fishes *Channa*, *Amphipnous*, *Clarias*, and *Saccobranchus* as shown by electron microscopy. *J. Zool. Lond.* **170**, 245–270.

Hughes, G. M. and Munshi, J. S. D. (1975). Fine structure of the gills of some Indian air-breathing fishes (In preparation).

Hughes, G. M. and Weibel, E. R. (1972). Similarity of supporting tissue in fish gills and the mammalian reticulo-endothelium. *J. ultrastruct. Res.* **39**, 106–114.

Hughes, G. M. and Weibel, E. R. (1975). Morphometry of fish lungs. This volume p. 213.

Hughes, G. M., Dube, S. C. and Munshi, J. S. D. (1973). Surface area of the respiratory organs of the Climbing perch, *Anabas testudineus* (Pisces: Anabantidae) *J. Zool. Lond.* **170**, 227–243.

Hughes, G. M., Singh, B. R., Guha, G., Dube, S. C. and Munshi, J. S. D. (1974). Respiratory surface area of an air-breathing silurold fish *Saccobranchus* (= *Heteropneustes*) *fossilis* in relation to body size. *J. Zool. Lond.* **172**, 215–232.

Hughes, G. M., Singh, B. R., Thakur, R. N. and Munshi, J. S. D. (1974). Areas of the air-breathing surfaces of *Amphipnous cuchia* (Ham.). *Proc. Ind. Nat. Sci. Acad.* **40(4)**, 379–392.

Hutchison, V. H., Whiteford, W. S. and Kohl, M. (1968). Relation of body size and surface area to gas exchange in Anurans. *Physiol. Zool.* **41**, 63–85.

Milton, P. (1971). Oxygen consumption and osmoregulation in the shanny (*Blennius pholis*). *J. Mar. Biol. An. U.K.* **51**, 247–265.

Muir, B. S. and Hughes, G. M. (1969). Gill dimensions for three species of tunny. *J. exp. Biol.* **51**, 271–285.

Munshi, J. S. D. (1960). The structure of the gills of certain freshwater teleosts. *Ind. J. Zoot.* **1**, 135–174.

Munshi, J. S. D. (1961). The accessory respiratory organs of *Clarias batrachus*. *J. Morph.* **109**, 115–140.

Munshi, J. S. D. (1962a). On the accessory respiratory organs of *Heteropneustes fossilis* (BL.). *Proc. Roy. Soc. Edinb.* (*B*) **68**, 128–146.

Munshi, J. S. D. (1962b). On the accessory respiratory organs of *Ophiocephalus punctatus* (BL.) and *O. striatus* (BL.). *J. Linn. Soc. Lond.* (*Zoo*) **44**, 616–626.

Munshi, J. S. D. (1968). The accessory respiratory organs of *Anabas testudineus* (BL.) Anabantidae, Pisces. *Proc. Linn. Soc. Lond. (Zool).* **179**, 107–126.

Munshi, J. S. D. and Singh, B. N. (1968a). On the respiratory organs of *Amphipnous cuchia* (Ham. Buch). *J. Morph.* **124**, 423–444.

Munshi, J. S. D. and Singh, B. N. (1968b). On the micro-circulatory system of the gills of certain freshwater teleostean fishes. *J. Zool. Lond.* **154**, 365–376.

Nassar, S. A. K. and Munshi, Jayashree Datta (1971). Studies on the Macrophytic biomass production and fish population in an abandoned pond at Bhagalpur, Bihar. *J. Bh. U. (N.S.)* **IV(1)**, 8–16.

Ojha, J. and Munshi, J. S. D. (1974). Morphometric studies on the gill and skin Dimension in relation to body weight in a freshwater Mud-eel, *Macrognathus aculeatum* (BL.). *Zool. Anz. Jena* **193**, 364–381.

Price, J. W. (1931). Growth and gill development in the small-mouthed black bass. *Micropterus dolomieu* lacépède. *Ohio St. Univ., Stud.* **41**, 1–46.

Pütter, A. (1909). Die Ernahrung der. Wassertiere und der Staffhaushatt der Gewasser. *Z. allg. physiol.* **7**, 1–148.

Rahn, H., Rahn, K. B., Howell, J., Gans, C. and Tenney, S. M. (1971). Air breathing of the gar fish (*Lepisosteus osseus*) *Resp. Physiol.* **11**, 285–307.

Rauther, B. (1910). Die akzessorischen. Atmungorgane der Knochen fische *Ergeb. Fortschr Zool.* **(2)**, 517.

Saunders, R. L. (1962). The irrigation of the gills in fishes 2, efficiency of oxygen uptake in relation to respiratory flow activity and concentration of oxygen and carbon dioxide. *Can. J. Zool.* **40**, 817–862.

Saxena, D. B. (1963). A review on ecological studies and their importance in the physiology of air-breathing fishes. *Ichthyologica* **2**, 116–128.

Singh, B. N. and Munshi, J. S. D. (1969). On the respiratory organs and Mechanism of breathing in *Periophthalmus vulgaris.* (eggret) *Zool. Anz. Heft.* **183**, 92–110.

Todd, E. S. and Ebeling, A. W. (1966). Aerial respiration in the long jaw mud sucker (*Gillichthys mirabilis* (Teleostei-Gobiidae). *Biol. Bull. mar. biol. labs. Woods Hole* **130**, 265–288.

Tenney, S. M. and Remmers, J. E. (1966). Alveolar dimensions in the lungs of animal raised at high altitude. *J. Appl. Phys.* **21**, 1328–1330.

Tenney, S. M. and Tenney, J. B. (1970). Quantitative morphology of cold-blooded lungs: amphibia and reptilia. *Resp. Physiol.* **9**, 197–215.

Ursin, E. (1967). A mathematical model of some aspects of fish growth, respiration and mortality. *J. Fish Res. Bd. Can.* **24**, 2355–2453.

Weibel, E. R. (1963). "Morphometry of the Human Lung." Berlin, Gottingen-Heidelberg, Springer-Verlag.

Weibel, E. R. (1969). Stereological principles for morphometry in electron microscopic cytology. *Int. Rev. Cytol.* **26**, 235–302.

The Circulatory System of Air-breathing Fish

G. H. SATCHELL

Department of Zoology, University of Otago, Otago, New Zealand

It has been emphasized by both Munshi and Graham (this volume) that air breathing has evolved repeatedly in fish in many different ways. Even amongst the teleost fish air-breathing species are found in sixteen families (Bertin, 1958). Amongst them are genera ranging from fish which spend the greater part of their life on land to fish which resort to air breathing only in times of stress such as is caused by high temperature and low oxygen tension. Such fish are thus bimodal breathers possessing, in the one animal, mechanisms for ventilating and perfusing both types of respiratory organ. Their diverse strategies of air and water breathing are reflected in equally diverse arrangements to perfuse their two types of gas exchange organs. The modifications of their circulatory systems to this end will form the subject of this chapter.

I. Vascular Systems and Perfusion Pressures

The circulatory system of fish is an "in series" system in which the single ventricle of the heart pumps the blood successively through the gills and the peripheral circulation. Thus the resistance of the branchial circulation is imposed between the heart and the peripheral vessels and the blood pressure within the secondary gill lamellae exceeds that in the capillaries of the systemic circuit. The data on blood pressures in fish have recently been collated (Randall, 1970; Satchell, 1971). The fish heart develops a pressure from one-half to one-third that of a mammal. Approximately one-third to one-quarter of this is lost across the resistance of the gills. There are few records of both ventral and dorsal aortic pressures in teleost fish. If we assume that the average pressure within the small vessels of the secondary lamella is the mean of the dorsal aortic and ventral aortic pressures, then the salmon, *Oncorhynchus tshawytscha*, has a lamella pressure of 63 mmHg (Greene, 1904), and the rainbow trout, *Salmo gairdneri*, 31 mmHg (Stevens and Randall, 1967). In *Monopterus albus*, a synbranchoid air-breathing fish with vestigial gills and with the ventral and dorsal aortae joined by a wide shunt vessel, the pressure within this is 41·5 mmHg (Liem, 1961).

These relatively high pressures present within the small vessels of the fish gill may be contrasted with pulmonary capillary pressures in a mammal. Here the heart is completely divided and the lightly-muscled right ventricle develops systolic and diastolic pressures, at rest, of the order of 16 and 9 mmHg. Pulmonary capillary pressures have been variously estimated, but 8 mmHg is a probable figure. The low pressure pump of the right ventricle is matched to a vessel system which characteristically has low resistance, and high compliance, and these permit the two sides of the heart to pump equal volumes of blood at widely different pressures appropriate to the two circulations.

In mammals the plasma proteins of the blood, some 7g%, contribute a colloid osmotic pressure of 25 mmHg (Landis and Pappenheimer, 1963). This exceeds the pulmonary blood or hydrostatic pressure, and, in accordance with Starling's classic hypothesis, favours absorption rather than filtration. A minimum of fluid passes into the alveolar walls and alveoli, and gas exchange is facilitated. Thus the completely divided heart of a mammal permits not only the separation of the streams of oxygenated and venous blood: it also prevents the disruption of lung function by the outpouring of the liquid part of the blood into the alveoli.

The extent to which Starling's hypothesis is applicable to the gill micro-circulation and systemic capillaries of fish has been little discussed. The amount of protein in fish plasma has been measured in several species and Albritton (1951) collects together the data up to that year. Even in closely related species plasma protein levels differ widely. In the Perciformes they range from 1·68–6·19 g% (Sulya, Box and Gunter, 1961). Colloid osmotic pressures range from 5–9 mmHg (Keys and Hill, 1934), but are possibly higher in some freshwater fish such as eel, and in turbot. The published values, which are about one third to one half those of mammals, are thus of a magnitude to match blood pressures in fish and to suggest that Starling's concept operates here too. If this is so, the balance of hydrostatic over colloid osmotic pressure must favour an outward movement of water across the secondary lamellae of the gills. The problem is complicated by the osmotic forces between the blood and the environmental water and is different for marine and freshwater species. The gill is a multifunction organ engaged in the inward and outward transport of ions. Nevertheless it is likely that such an excess of hydrostatic over colloid osmotic pressure as occurs in gill lamellae would prove disadvantageous in a lung and that this has favoured, in evolution, changes which could bring about a lower pressure in the capillaries of the air-breathing organ.

II. The Efferent Outflow from the Air-breathing Organ

A. RETURN OF THE OXYGENATED BLOOD TO THE DORSAL AORTA

In several fish such as *Hypopomus brevirostris*, *Symbranchus marmoratus* and *Hoplosternum litorale* gills are used for air breathing and are little changed. In these genera the significant modification is an increase in the ratio of the breadth of the secondary lamellae to their thickness (Carter and Beadle, 1931; Johansen, 1966). Adaptations occur which prevent the lamellae collapsing and *Chirolophis* has them held apart by raised marginal ridges (Schmidt, 1942). Certain genera of gobiids have richly developed mucous cells which keep the gills moist. In these fish (Schöttle, 1931), as in normal water-breathing fish (Fig. 1a), oxygenated blood from the gill lamellae, passes to the lateral dorsal aortae by the four efferent branchial arteries.

The catfish *Clarias batrachus* has an air-filled suprabranchial chamber above the gills (Fig. 1c). In *Saccobranchus fossilis* this air-containing cavity extends far back to the level of the middle of the tail (Fig. 1b). These air-breathing organs appear at first sight to have little in common with gills but the recent fine-structural studies of Hughes and Munshi (1973) show that characteristic pillar cells can be recognized in their vascular areas. Pillar cells typically link together the two surfaces of secondary lamellae and presumably prevent them from ballooning apart in response to the high intralamellar blood pressure. These two air-breathing catfish reflect the clear derivation of their air-breathing organs from gills, also in their vascular connections. The blood system of *Clarias* can be derived from the descriptions of Munshi (1961) and Moussa (1956); that of *Saccobranchus* is described by Hyrtl (1854) and Burne (1896). In both (Fig. 1c and b) the oxygenated blood is returned from the air breathing organ to the dorsal aorta, by an efferent branchial artery or a derivative of one.

B. THE RETURN OF THE OXYGENATED BLOOD TO THE VEINS OF THE HEAD

The fine structure of the air-breathing organs of *Anabas*, *Channa* and *Amphipnous* contrasts with that of the previous two genera in that their small vessels are true capillaries and not blood spaces between pillar cells (Hughes and Munshi, 1968, 1973). Their air-breathing organs differ quite widely. *Anabas* has labyrinthine organs and a suprabranchial chamber (Munshi, 1968); *Channa* also has an extensive suprabranchial chamber

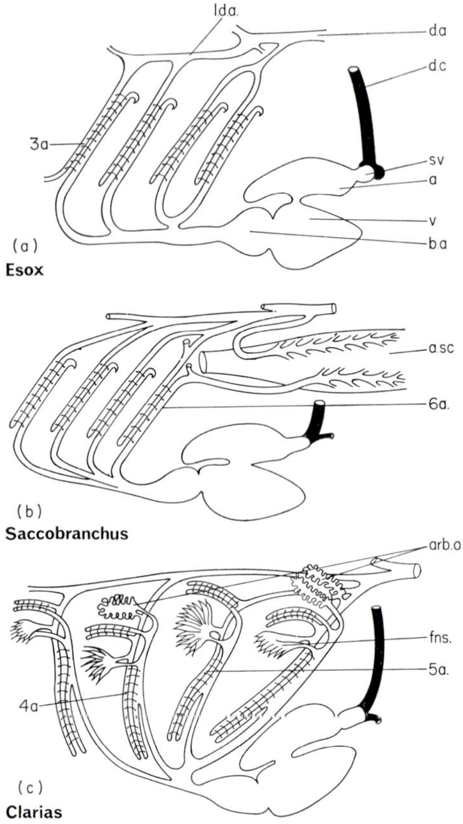

FIG. 1. Branchial vessels schematically drawn, of fish in which the oxygenated blood is returned to the dorsal aorta. (a) *Esox*, an unmodified water-breathing fish. (b) *Saccobranchus*, (c) *Clarias*, two air-breathing cat fish. a. atrium, arb. o. arborescent organ, a. sc. air sac, b.a. bulbus arteriosus, d.a. dorsal aorta, d.c. ductus cuvieri, fns. fans, l.d.a. lateral dorsal aorta, s.v. sinus venosus, v. ventricle, 3a, 4a, 5a, 6a, third fourth, fifth and sixth arterial arches. (a) redrawn after Allis (1912); (b) redrawn after Hyrtl (1854), and Burne (1896); (c) redrawn after Moussa (1956), Singh and Hughes (1973), and the account of Munshi (1961).

(Lele, 1932). *Amphipnous* has an elongate air sac extending down each side of the body (Munshi and Singh, 1968). In all three genera the oxygenated blood from the capillaries of the air-breathing organ is returned to the anterior cardinal vein or to branches of it (see Fig. 3a,b,c). In the gobiid genera *Pseudapocryptes*, *Boleophthalmus* and *Periophthalmus* air breathing by gills is supplemented by the vascularized linings of the pharynx, gill arches and operculum. Here too the oxygenated blood from the secondary lamellae of the gills passes to the dorsal aortae (Fig. 2a) by the efferent branchial arteries, whilst that from the capillaries of the supplementary air breathing surfaces passes to the anterior cardinal veins or inferior jugular veins (Das, 1934, Schöttle, 1931). These two veins (Fig. 2b) collect the oxygenated blood from the vascular papillae in the mouth of the electric eel, *Electrophorus electricus* (Carter, 1935). The lower pressure at the peripheral end of the respiratory capillaries caused by exposing them to the *vis a fronte* of the central veins, rather than the arterial pressure of the dorsal aorta must lower the average pressure prevailing along their length.

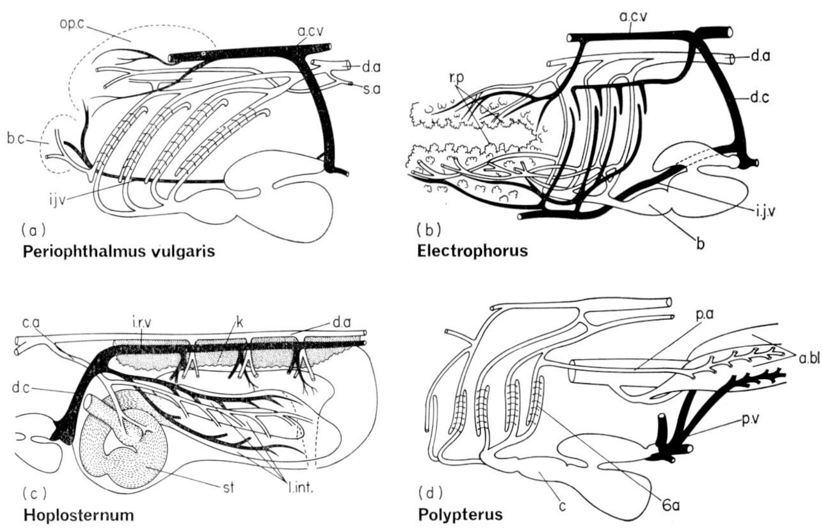

FIG. 2. Schematic figures illustrating the return of oxygenated blood from an air-breathing organ to a central vein. a.bl. air bladder, a.c.v. anterior cardinal vein, b.c. buccal cavity, c. conus arteriosus, c.a. celiac artery, i.j.v. internal jugular vein, i.r.v. inter renal vein, k. kidney, l. int. loops of intestine, op. c. opercular cavity, p.a. pulmonary artery, p.v. pulmonary vein, r.p. respiratory papillae, s.a. subclavian artery, st. stomach. Other letters as in Fig. 1. (a) redrawn after Schöttle (1931); (b) redrawn after Carter (1935) and Johansen *et al.* (1968b); (c) redrawn after Carter and Beadle (1931); (d) redrawn after Budgett (1901).

C. THE FORMATION OF AN INTERRENAL VEIN

Certain genera of air-breathing fish have parts of the alimentary canal modified as an air-breathing organ and air is swallowed into it from the pharynx and expelled from the cloaca. The wall of the region modified for gas exchange is exceedingly thin and transparent and richly supplied with capillaries (Carter and Beadle, 1931). In *Hoplosternum litorale*, (Fig. 2c) and *Callichthyes callichthyes* the greater part of the gut is modified in this way and loops back and forth within the abdomen. In *Ancistrus ancisti* the stomach forms a large U shaped air-containing sac. In these and similar forms the modified gut is supplied with arterial blood from the dorsal aorta by the coeliac artery as is the gut in other fish. The oxygenated blood is returned to the sinus venosus in a new vessel, the interrenal vein. It emerges into the abdominal cavity, from the trunk and tail, as the caudal vein, and runs forwards within the tissue of the kidneys; it is formed by the coalescence of the caudal vein, renal portal vein and right posterior cardinal vein (Carter and Beadle, 1931; Jobert, 1878). It follows the course of the subintestinal vein of the embryo (Hochstetter, 1888) and may be formed by its retention.

We can contrast this arrangement with the gut drainage of a normal fish. As in other vertebrates blood from the alimentary canal is returned to the liver by the hepatic portal vein to pass through the capillary bed of the hepatic portal system before it is delivered by the hepatic veins to the sinus venosus.

In *Erythrinus unitaeniatus* the swimbladder is provided with a broad pneumatic duct and its middle region is richly vascularized and used for gas exchange (Carter and Beadle, 1931). In this species too the outflow from the swimbladder is into an interrenal vein. In the majority of teleost fish the swimbladder veins also pass into the hepatic portal system.

The interrenal vein serves to free the capillaries of the air breathing parts of the viscera from the raised pressure imposed by the resistance of the hepatic portal system and exposes them instead to the *vis a fronte* suction of the heart. Stevens and Randall (1967) report that the subintestinal vein pressure of *Salmo gairdneri* was $+9{\cdot}6$ to $+13{\cdot}5$ cm H_2O whereas, in the eel, Mott (1950) reports that hepatic vein pressure was $-5{\cdot}4$ to $+6$ cm H_2O. Carter and Beadle (1931) point out that were the normal pattern of venous return to be retained the flow of blood through the air breathing section of the viscera would be dictated by changes in the vascular resistance of the liver capillary beds.

D. THE CLOSER APPROACH TO THE HEART OF THE VENOUS RETURN FROM THE AIR-BREATHING ORGAN

In the primitive genera *Polypterus*, *Amia* and *Lepisosteus*, and in the Dipnoi the air-breathing organ is a vascularized and primitively alveolated air bladder recalling in structure the amphibian lung. It receives its afferent blood supply either from the dorsal aorta, as in *Lepisosteus* or from the sixth aortic arch. These genera can be grouped in a series exhibiting a successively closer approach to the heart of the openings of the veins returning the oxygenated blood. In *Lepisosteus* (Potter, 1910) oxygenated blood is returned by numerous tributary veins arising along the length of the air bladder and opening separately into the right posterior cardinal vein. In *Amia* and in *Neoceratodus* a single pulmonary vein opens into the left ductus cuvieri. In *Polypterus* (Fig. 2d) and *Calamoichthyes* (Purser, 1926) the entrance is more medial, into the base of the hepatic vein. In *Protopterus* (Fig. 3) and *Lepiso-*

siren the pulmonary vein opens separately into the right side of the atrium. The mormyrid fish *Gymnarchus niloticus* curiously has a single pulmonary vein opening into the left posterior cardinal vein (Assheton, 1907).

It has been suggested (Bertin, 1958) that the perfusion of an air breathing organ can be judged to be functionally advanced according to the extent to which its blood supply arises from and is returned to the heart. This assemblage of largely unrelated genera shows an increasing approach to the amniote condition, culminating in *Protopterus* and *Lepidosiren*, which remarkably parallel the Amphibia in this respect.

III. The Formation of Shunt Vessels between the Ventral and Dorsal Aortae

Air-breathing fish which retain functional gills and are bimodal breathers face a problem if breathing from an air store when submerged in water of low oxygen tension. Oxygenated blood returned to the heart from the air breathing organ may lose its oxygen to the water as it passes through the gill lamellae. Vascular shunts which short circuit the lamellae can prevent this and enable the oxygenated blood to pass to the dorsal aorta for distribution to the tissues.

In *Anabas* (Fig. 3a) the last arch has the secondary lamellae absent and the

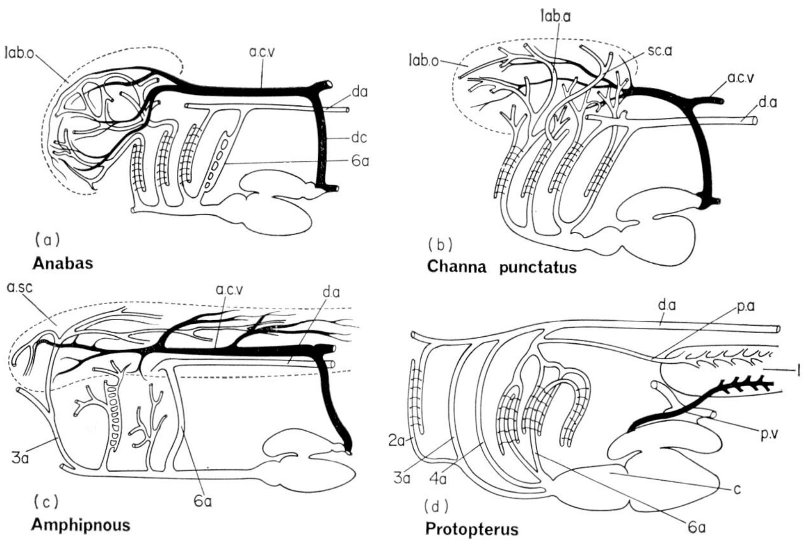

FIG. 3. Schematic figures illustrating the evolution of shunt vessels between ventral and dorsal aortae. l. lung, lab. o. labyrinthine organ, lab. a. labyrinthine artery, s.c.a. artery of suprabranchial chamber. Other letters as before. (a) drawn from account of Munshi (1968); (b) redrawn from Lele (1932); (c) drawn from account of Munshi and Singh (1968); (d) redrawn from Johansen *et al.* (1968b).

afferent and efferent arteries are joined by broad blood spaces (Munshi, 1968). In *Channa argus* the afferent and efferent vessels of this arch are connected by clustered loops of small blood vessels (Wu and Chang, 1947). In *Amphipnous* (Munshi and Singh, 1968) and in *Monopterus* (Liem, 1961) this arch forms a wide diameter shunt vessel which constitutes the only link between the two aortae (Fig. 3c). In *Channa striatus*, and *C. punctatus* (Hyrtl, 1853) the last two branchial vessels pass directly up to the dorsal aorta; it appears that this is not the case in the related chinese species *C. argus* (Wu and Chang, 1947). In *Protopterus* the arteries of the embryonic third and fourth arches lack filaments and form shunt vessels (Fig. 3d); these correspond to the carotid and systemic arches of the amniota. Lastly, in *Lepidosiren* (Foxon, 1950) and in *Electrophorus* (Carter, 1935) all of the afferent branchial arteries connect directly to the dorsal aortae.

It is likely that, even when the four pairs of gills are evident and well developed, shunt channels develop within them in air-breathing fish. Jobert (1878) when studying the intestinal breathers *Hoplosternum* and *Callichthyes*, and the swimbladder breather *Erythrinus* came to this conclusion. He noted how easily injection mass passed from the ventral aorta into the dorsal aorta in comparison with a normal fish: the subject needs to be re-examined.

IV. The Afferent Supply to the Air-breathing Organs

In such diverse air-breathing fish as *Polypterus*, *Amia*, *Lepisosteus*, *Anabas* and *Clarias* all of the blood perfusing the air-breathing organs must first pass through secondary lamellae. This provides a route whereby the fish can dispose of carbon dioxide if breathing air from an air store when submerged. In more advanced air breathers there is an increasing tendency for the air-breathing organ to receive its blood supply directly from an afferent branchial artery. In the gobiid fish *Boleophthalmus viridis* the ventral aorta continues forwards to bifurcate into the two hyomandibular arteries in the second arches; these vessels supply venous blood directly to the vascularized buccal and opercular surfaces which are thus a parallel circulation to the normal four pairs of gills (Schöttle, 1931). In the gobiid *Pseudapocryptes lanceolatus* a large bucco-opercular artery arises from the first afferent branchial artery in the third arch, and supplies the vascularized lining of the pharynx (Das, 1934). In *Channa punctatus* both the first and second afferent arteries (Fig. 3b) supply branches to the air chambers (Lele, 1932) and in *Amphipnous* (Munshi and Singh, 1968) and *Monopterus* (Liem, 1961) all of the first three afferent arteries are solely concerned in supplying blood to various air breathing surfaces. *Saccobranchus* and *Lepidosiren* use only the fourth artery, the vessel of the sixth embryonic arch, to supply the air sac.

It remains a curious and unexplained fact that the fourth afferent branchial artery, in the sixth embryonic arch, serves to carry the venous blood to the air breathing organ in such a diverse assemblage of fish. In the brachiopterygian genera *Polypterus* (Fig. 2d) (Budgett, 1901) and *Calamoichthys* (Purser, 1926), in the ganoid *Amia* (Allis, 1912), and in the Dipnoi the pulmonary artery arises from the dorsal end of the sixth arch vessel. It is by no means certain that these fish could all have derived the feature from some common ancestor of the tetrapods. The Dipnoi are a group of disputed affinity probably to be placed close to the Elasmobranchomorpha (Jarvik, 1968). The Brachiopterygii are an extraordinarily isolated group related perhaps both to the Crossopterygii and Actinopterygii (Schmalhausen, 1968). It is more curious that *Gymnarchus* (Assheton, 1907), a mormyrid fish classed amongst the Osteoglossids, and *Saccobranchus*, a clarid catfish commonly included amongst the Cypriniforms, should have their air-breathing organs so supplied. An intermediate stage is seen in *Clarias* (Fig. 1c) in which the fourth artery is of large diameter, almost as big as the aorta and supplies, in addition to its gill, the fan and arborescent organ of its arch (Moussa, 1956).

The origin of the sixth arch vessel from the ventral aorta is often very close to the outlet of the conus or bulbus and this may enable the muscle of these structures to regulate and direct the outflow so that it can be apportioned between the posterior and anterior afferent branchial arteries. This point will be discussed in the subsequent section on the heart.

V. The Regulation of Blood Flow through the Air-breathing Organs

In true water-breathing fish with their "in series" circulations, as in mammals with their completely divided hearts, flow through the respiratory and systemic circuits must in the long term be equal. In all air-breathing fish, other than those such as *Clarias* and *Symbranchus* where oxygenated blood passes to the dorsal aorta, the two circulations are either wholly or partly in parallel. There may be marked differences in the rate of flow in the two circuits. The various air-breathing teleosts must endure, it would seem, a degradation of the level of oxygenation of the blood leaving the air-breathing organ by its admixture with systemic venous blood in its passage through the heart.

Flow through the air-breathing organ varies widely during different phases of the breathing cycle, and increases markedly following an air breath. The long intervals between breaths in many air-breathing fish provide time for regional shifts of blood flow to occur. In the electric eel Johansen, Lenfant, Schmidt–Nielsen and Petersen (1968b) have shown that the systemic blood is normally only partly saturated and the mean arterial P_{O_2} averages 21 mmHg. Following an air breath the fraction of blood from the heart passing to the respiratory papillae in the mouth, $= QM/Q_T$, is 72% and only 28%

passes up the shunt vessels (Fig. 2b) to the aortae. After 75 seconds these values have changed to 52% and 48% respectively.

In the dipnoan *Protopterus* flow through the lung increases fourfold following an air breath (Johansen, Lenfant and Hanson, 1968a). In the gills of normal fish the vascular resistance diminishes when they are ventilated and at least part of this response is believed to be intrinsic to the vascular smooth muscle. In *Protopterus* the increased flow through the lung may also depend on an intrinsic relaxation of arteriolar smooth muscle but a reflex may also be involved as an increase sometimes occurs in anticipation of an air breath. The augmented lung flow is accompanied by an increase in cardiac output. A reflex control of lung blood flow is believed to occur in the amphibian *Xenopus* (Shelton, 1970).

The enforced partial unsaturation of the blood, resulting from the admixture of oxygenated and venous blood in its passage through the heart, may be the cause of the high level of haemoglobin occurring in the blood of many air-breathing fish. Banerjee (1966) reports the figure of 19·8 g% in *Anabas*, Johansen *et al.* (1968b) found values ranging between 6·2 and 11·2 g% in *Electrophorus*, and Munshi and Singh (1968) report a series of very high values for a range of air-breathing fish including *Amphipnous*, *Channa*, *Heteropneustes* and *Clarias*.

VI. The Role of the Heart in the Evolution of Air Breathing

The circulatory adaptations to air breathing described so far in this chapter have been essentially peripheral. The successively closer return of the oxygenated blood to the heart enabled the downstream end of the capillaries of the gas exchange organ to be increasingly exposed to the low pressure of the central veins. Concomitantly with these peripheral changes the hearts of certain genera have been modified in directions which have both favoured the achievement of low pulmonary capillary pressures, and also served to separate into two streams the returning oxygenated and venous blood.

In Amphibia, Jones and Shelton (1972) have emphasized the crucial role of the apical conal valves which guard the entrances to the caroticosystemic and pulmocutaneous circulations (Fig. 4c). Only when these have closed, and the two circulations are separated, can caroticosystemic and pulmocutaneous pressures diverge. The latter falls more steeply to a lower level than the systemic, because of the lower resistance and greater compliance of the pulmonary vessels. In addition the spiral valve (Fig. 4c) has the specific role in early diastole of occluding the entrance to the pulmocutaneous arch until pressure in the conus has risen to the level of that in the systemic circuit. The isolation of the pulmonary circulation during diastole must have occurred before it could evolve its specific haemodynamic properties of high compliance

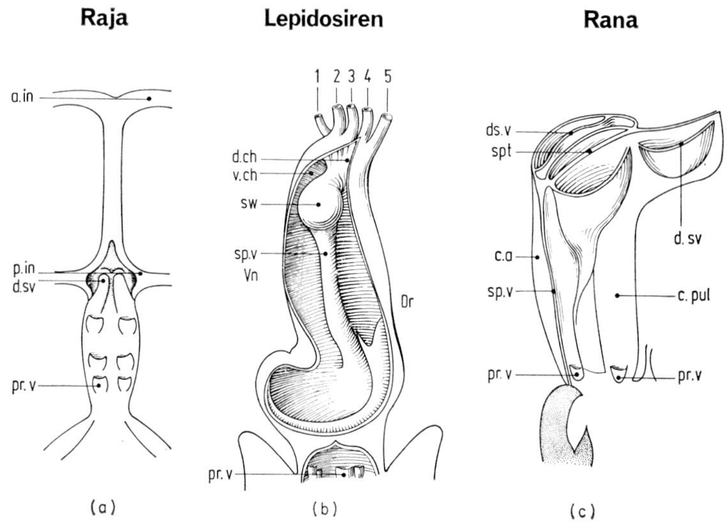

FIG. 4. The conus arteriosus in *Raja*, *Lepidosiren* and *Rana*, opened to show valves. a.in. anterior innominate artery, c.a. cavum aorticum, c.pul. cavum pulmonale, d.ch. dorsal chamber, Dr. dorsal, d.sv. distal semilunar valves, p.in. posterior innominate artery, pr.v. proximal semilunar valves, sp.v. spiral valve, s.w. swelling in spiral valve, v.ch. ventral chamber, Vn. ventral. (b) redrawn and simplified from Robertson (1914); (c) redrawn after Gaupp (1899).

and low resistance. Without the re-organization of the upper conal valves which made this possible a disproportionate amount of the stroke volume would have passed to the lung.

The hearts of lungfish show certain resemblances to those of Amphibia. A single pulmonary vein returns oxygenated blood to the left side of a partly divided atrium: a spiral valve subdivides an elongate muscular conus arteriosus. Recent studies of the physiology of the dipnoan heart make timely an assessment of the extent to which it can maintain separate the entering streams of oxygenated and venous blood, and contribute towards the attainment of a low pulmonary capillary pressure.

VII. The Dipnoan Heart

In the genus *Protopterus* (Lankester, 1879; Bugge, 1960) and *Lepidosiren* (Robertson, 1914) the atrium is partly divided by an atrioventricular plug, a type of valve unique to the Dipnoi, and by a partial septum called the pulmonalis fold. It is caused by the overlying pulmonary vein indenting the atrial wall; from it a well-developed trabecular band passes to the opposite wall and thus increases the separation of the right and left portions of the atrium during systole. Venous blood from the sinus venosus flows into the right of the atrium and oxygenated blood from the pulmonary vein passes to the left.

The almost symmetrical ventricle is also partly divided by a septum which extends from its apex towards the atrioventricular plug. An undivided part of the lumen extends from the septal edge to the origin of the conus arteriosus which is set to the right of the midline.

The conus has an investment of cardiac muscle limited to its proximal third. It bends through two right angles (Fig. 4b) and has a short transverse region between them. A median ventral ridge, formed from a fused row of valves, partly divides the proximal segment and a shorter second ridge arises on the opposite wall of the transverse segment. Both ridges spiral through 270° as Lankester (1879) noted; within the distal portion the two ridges fuse to form a transverse septum which divides the conus into dorsal and ventral chambers. Small valves at the base of the conus serve to separate it from the ventricle late in systole. There is virtually no ventral aorta; the five pairs of branchial arteries arise directly from the conus. Those to the through-channels in the third and fourth branchial arches of *Protopterus* (Fig. 3d) arise from the ventral chamber; those in the fifth and sixth (pulmonary) arches arise from the dorsal chamber.

A. THE SEPARATION OF OXYGENATED AND VENOUS BLOOD IN THE HEART OF *Protopterus*

In *Protopterus aethiopicus* Johansen and Hol (1968) and Johansen *et al.* (1968a) have presented evidence that the two streams of blood entering the left and right sides of the atrium remain largely separate in their passage through the heart to the branchial arteries. By simultaneously measuring blood flow and blood oxygen tension these authors were able to calculate the proportion of the flow through the anterior afferent branchial arteries, $= Q_{AB}$, derived from the systemic inflow from the vena cava, $= Q_S$. If complete separation had been achieved all vena caval blood would find its way into the pulmonary artery, $= Q_L$, and the ratio Q_S/Q_{AB} would be zero whilst Q_L/Q_{AB} would be 100%. In a specimen of *Protopterus* breathing under water following an air breath, Q_S/Q_{AB} was 9% and Q_L/Q_{AB} was 91%. In *Neoceratodus*, which has a less well developed spiral valve, values of 16% and 84% were achieved when breathing from its airstore in hypoxic water (Table I).

Johansen and Hol (1968) have presented X-ray radiographic evidence which helps to explain how this separation is achieved. Contrast media injected into the pulmonary vein is directed to the left posterior part of the atrium, due to the course and direction of the vein. Contrast media from the vena cava passes through the sinus venosus into the right side of the atrium. Separation of the two inflows is achieved in part by the atrial septum and in

TABLE I

*Composite table of blood oxygen tensions in mmHg and calculated shunt patterns in lungfish.**

Species	No. of Specimens	Condition	Systemic art. blood	Pulmonary art. blood	Pulmonary venous blood	Anterior branchial blood	Vena cava blood	Q_S/Q_{AB} (%)	Q_L/Q_{AB} (%)
Neoceratodus	8	In aerated water		38	36	20	14	67	33
	8	In hypoxic water		25	95	32	5	16	84
Protopterus Seattle series	5	In aerated water	27	20	40		2		
	5	Air exposed	30	22	35		2		
Protopterus Kampala series	3	In aerated water	30	25	46	38	2	9	91
Lepidosiren Amazon series	2	In aerated water	38	28	69				
	2	Air exposed	43	32	84				
Lepidosiren Seattle series	5	In aerated water	28	12					

* From Johansen *et al.* (1968a).

part by a residual volume of blood which inhibits mixing and tends to remain in the median undivided part of the atrium. Following atrial systole the two adjacent masses of blood pass into the ventricle and come to lie, largely unmixed, on the two sides of the ventricle in the posterior region where the septum is present.

Ventricular systole ejects the two masses of blood into the conus. The oxygenated blood from the left side passes up the left of the conus and, by virtue of its spiralled course, enters the ventral channel and passes to the shunt vessels in the third and fourth arches. In similar manner the venous blood from the right of the ventricle is directed up the right of the conus and into the dorsal chamber from which arise the pulmonary arteries. (Fig. 4b).

Johansen and Hol's radiographs reveal the great importance of the trabeculated ventricular wall. The deep cylindrical crypts between the trabeculae are each emptied during myocardial contraction and their contents can be visualized as they proceed forwards as multiple laminal flows which remain separate in their passage across the undivided part of the ventricle.

B. THE CONAL VALVES OF DIPNOI

There has been no detailed experimental study of the dipnoan conus comparable with that of Shelton and Jones (1965) and Jones and Shelton (1972) in Amphibia. Nevertheless certain functional conclusions can be drawn from the anatomical descriptions of the conus; the account of Lankester (1879) forms a convenient starting point. His drawings of *Neoceratodus* show clearly the proximal row of small valves, reminiscent of those in the shark *Heterodontus portusjacksoni* (Satchell and Jones, 1967); in *Neoceratodus* it is evident that these valves could only be competent during conal systole when they are briefly opposed across the wide conal lumen. Johansen *et al.* (1968a) have published pressure records from the conus of *Neoceratodus*; these show clearly that conal pressure remains briefly elevated after ventricular pressure has fallen and during this time the lower conal valves must be apposed.

From the proximal row of valves a spiral valve runs forwards and its derivation from the ventral row of conal valves is evident in the figures of Boas (1880). At the conal apex, separate pocket valves guard the entrances to the upper and lower divisions and it appears possible that the upper division, from which the pulmonary arch takes origin, can be separated from the lower division during diastole.

The situation in *Protopterus* and *Lepidosiren* is obscure even in terms of its anatomy and needs further study. The accounts of Lankester (1879), Boas (1880), Robertson (1914) and Bugge (1960) all agree that, as in *Neoceratodus*, small pocket valves guard the entrance to the conus. They also agree that no normal pocket valves occur at the distal end of the conus. The implications of this have never been discussed; plainly the heart could not function unless some valvular mechanism was able to prevent blood from the afferent branchial arteries regurgitating into the ventricle, once conal systole was over and the proximal valves could no longer be apposed. In Robertson's figures of *Lepidosiren* a spherical thickening is shown in the distal end of the spiral valve, and this may serve as the upper conal valve system. It is reminiscent of the plug-like valve at the atrio-ventricular junction; during ventricular systole blood may flow past this thickening and into the afferent branchial arteries. As ventricular systole comes to an end it may move backwards and impact against the conal walls. As the conus is almost completely divided into upper and lower channels at this level it would seem that the pulmonary and systemic circulations might indeed be separated during diastole, as they are in Amphibia, where separate sets of pocket valves occur (Fig. 4c). The distal end of the conus of Dipnoi is a further example, it would seem, of the tendency for lungfish to achieve a level of functional complexity comparable with that of Amphibia, by different structural means. There is clearly need to test these speculations on conal function in Dipnoi, by experiment.

In *Protopterus*, Johansen *et al.* (1968a, 1970) conclude that pulmonary vascular resistance is lower than systemic and the double pulse in the flow pattern of the pulmonary vein suggests this. But these studies do not tell us whether the spiral valve has a specific role to play, as in Amphibia, in guarding the entrance to the pulmonary circuit until pressure has risen to the systemic level.

The ability of the conus arteriosus to regulate flow into the posterior branchial arches is already foreshadowed in bottom living elasmobranchs. In *Raja binoculata* the two afferent branchial arteries supplying blood to gill slits ventilated from the spiracle arise from the anterior innominate artery. The three arteries to the posterior gills, ventilated more predominantly from the mouth, arise from a single pair of posterior innominate arteries (Fig. 4a). Simultaneous records of flow through the two stem vessels show (Satchell, Hanson and Johansen, 1970) that a greater proportion of the initial outflow passes into the anterior innominate arteries. No explanation of this was forthcoming at the time the paper was published but subsequent study shows (Fig. 4a) that the rostral valves of the conus in *Raja nasuta* overlap the openings to the posterior innominate arteries and must partially occlude them in early systole. The valves guarding the separate entrances to the pulmonary and systemic circulations in *Neoceratodus* and Amphibia must have been derived from the distal ring of valves. Presumably, it is the ability of the conus to, in part, regulate the flow of blood into the posterior arches which has made it advantageous for the sixth arch to arise close to its apex. Herein, perhaps, lies the explanation of why the afferent supply to an air-breathing organ is so often derived from the sixth arch. Constriction of the smooth muscle of the bulbus arteriosus as in teleost fish, or charges in the activity of the cardiac muscle of the conus, may provide an auxillary means of regulating the perfusion of the air-breathing organ. The potential of this region of the heart as a regulator of outflow to the various arterial arches is seen repeatedly in the Reptilia and is indeed only lost when the heart is completely divided and the conus arteriosus is absorbed into the right ventricular outflow tract.

Summarizing we may say that the anatomy of the dipnoan conus suggests that, as in Amphibia, the carotico-systemic and pulmonary circulations can be separated in diastole and that this may augment the ability of the low resistance pulmonary circuit to achieve a low pulmonary capillary pressure. During systole however both circulations must have similar pressures as they are perfused by a common pump. Evolution, it seems, has pursued a three-stage strategy in the perfection of the pulmonary circulation. The air-breathing fish suggest that initially the outflow from the air-breathing organ was switched from the dorsal aorta to a central vein thus permitting the distal end of the capillaries to be subject to the *vis a fronte*. Secondly, the distal conal valves evolved in ways which permitted the separation of the systemic from the pulmonary circulation during diastole, and made advantageous its further evolution in the direction of low resistance and high compliance. Finally the complete division of the atrium and ventricle enabled quite different perfusion pressures to be generated on the two sides of the heart, and thus provided the pulmonary circuit with its own low pressure pump.

References

Albritton, E. C. (1951). "Standard Values in Blood." p. 199 W. B. Saunders Co., Philadelphia.

Allis, E. P. (1912). The pseudobranchial and carotid arteries in *Esox*, *Salmo*, and *Gadus*, together with a description of the arteries in the adult *Amia*. *Anat. Anz.* **41**, 113–142.

Assheton, R. (1907). The development of *Gymnarchus niloticus*. *In:* "Budgett Memorial Volume." pp. 293–432 (Ed. G. J. Kerr) Cambridge.

Banerjee, V. (1966). Blood of some Indian vertebrates. *Naturwissenschaften* **137**, 53.

Bertin, L. (1958). Organes de la respiration aérienne. *In* "Traite de Zoologie" (Ed. P. D. Grassé) Vol. 13 Pt. 2. pp. 1363–1398. Massonet Cie, Paris.

Boas, J. E. V. (1880). Über Herz und Arterienbogen bei *Ceratodus* und *Protopterus*. *Morph. Jb.* **6**, 321–354.

Budgett, J. S. (1901). On some points in the anatomy of *Polypterus*. *Trans. Zool. Soc. Lond.* **15**, 323–338.

Bugge, J. (1960). The heart of the African lung fish *Protopterus*. *Vidensk. Meddr dansk naturh. Foren.* **123**, 193–210.

Burne, R. H. (1896). On the aortic arch system of *Saccobranchus fossilis*. *J. Linn. Soc.* **25**, 48–55.

Carter, G. S. (1935). Respiratory adaptations of the fishes of the forest waters, with descriptions of the accessory respiratory organs of *Electrophorus electricus* (Linn) (= *Gymnotus electricus* auctt) and *Plecostomus plecostomus* (Linn)). *J. Linn. Soc.* **39**, 219–233.

Carter, G. S. and Beadle, L. C. (1931). The fauna of the swamps of the Paraguayan Chaco in relation to its environment. II Respiratory adaptations in the fishes. *J. Linn. Soc.* **37**, 327–368.

Das, B. K. (1934). The habits and structure of *Pseudapocryptes lanceolatus*, a fish in the first stages of structural adaptation to aerial respiration. *Proc. R. Soc. B.* **115**, 422–430.

Foxon, G. E. H. (1950). A description of the coronary arteries in Dipnoan fishes and some remarks on their importance from the evolutionary standpoint. *J. Anat.* **84**, 121–131.

Gaupp, E. (1899). Blutgefassystem. A. Herz. *In* "Anatomie des Frosches" (Ed. A. Ecker und R. Wiedersheim) Vol. 2 pp. 247–284. Friederich Viewegund Sohn. Braunschweig.

Greene, C. W. (1904). Physiological studies of Chinook salmon. I. Relation of blood pressure to functional activity. *Bull. U.S. Bur. Fisheries* **24**, 431–455.

Hochstetter, F. (1888). Beitrage zur vergleichenden Anatomie und Entwicklungsgeschite des Venensystems der Amphibien und Fische. *Morph. Jb.* **13**, 115–172.

Hughes, G. M. and Munshi, J. S. Datta (1973). Nature of the air-breathing organs of the Indian fishes *Channa*, *Amphipnous*, *Clarias*, and *Saccobranchus* as shown by electron microscopy. *J. Zool. Lond.* **170**, 245–270.

Hughes, G. M. and Munshi, J. S. Datta (1968). Fine structure of the respiratory surfaces of an air-breathing fish, the climbing perch *Anabas testudineus* (Bloch). *Nature, Lond.* **219**, 1382–1384.

Hyrtl, J. (1853). Uber das Labyrinth und die Aortenbogen der Gattung *Ophiocephalus*. *Sber. Akad. Wiss. Wien Math. Naturw. Kl.* **10**, 148–153.

Hyrtl, J., (1854) Zur Anatomie von *Saccobranchus singio*. *Sber. Akad. Wiss. Wien Math. Naturw. Kl.* **11**, 302–307.

Jarvik, E. (1968). The systematic position of the Dipnoi. *In* "Current problems of lower vertebrate phylogeny." (Ed. T. Ørvig) pp. 223–246. Almquist and Wiksell, Stockholm.

Jobert, M. (1878). Recherches anatomiques et physiologiques pour servir a l'histoire de la respiration chez les poissons. *Ann. Sci. nat. Zool.* **7**, 1–7.

Johansen, K. (1966). Air breathing in the teleost *Symbranchus marmoratus*. *Comp. Biochem. Physiol.* **18**, 383–395.

Johansen, K., and Hol, R. (1968). A radiological study of the central circulation in the lungfish *Protopterus aethiopicus*. *J. Morph.* **126**, 333–348.

Johansen, K., Lenfant, C. and Hanson, D. (1968a). Cardiovascular dynamics in lungfishes. *Z. vergl. Physiol.* **59**, 157–186.

Johansen, K., Lenfant, C., Schmidt Nielsen, K. and Petersen, J. A. (1968b). Gas exchange and the control of breathing in the electric eel *Electrophorus electricus*. *Z. vergl. Physiol.* **61**, 137–163.

Johansen, K., Lenfant, C., and Hanson, D. (1970). Phylogenetic development of pulmonary circulation. *Fedn Proc. Fedn Am. Socs exp. Biol.* **29**, 1135–1140.

Jones, D. R. and Shelton, G. (1972). Factors affecting diastolic blood pressures in the systemic and pulmocutaneous arches of anuran amphibia. *J. exp. Biol.* **57**, 789–803.

Keys, A. and Hill, R. M. (1934). The osmotic pressure of the colloids in fish sera. *J. exp. Biol.* **11**, 28–34.

Landis, E. M. and Pappenheimer, J. R. (1963). Exchange of substances through the capillary walls. *In* "Handbook of Physiology" II, 2. Circulation, 961–1034. (Ed. W. F. Hamilton). American Physiological Society, Washington.

Lankester, E. R. (1879). On the hearts of *Ceratodus*, *Protopterus* and *Chimaera* with an account of undescribed pocket valves in the conus arteriosus of *Ceratodus* and *Protopterus*. *Trans. zool. Soc. Lond.* **10**, 493–506.

Lele, S. H. (1932). The circulation of blood in the air breathing chambers of *Ophiocephalus punctatus* Bloch. *J. Linn. Soc.* **38**, 49–54.

Liem, K. F. (1961). Tetrapod parallelisms and other features in the functional morphology of the blood vascular system of *Fluta alba* Zuiew (Pisces: Teleostei). *J. Morph.* **108**, 131–143.

Mott, J. C. (1950). Radiological observations on the cardiovascular system in *Anguilla anguilla*. *J. exp. Biol.* **27**, 324–333.

Moussa, T. A. (1956). Morphology of the accessory air-breathing organs of the teleost, *Clarias lazera* (C. and V.). *J. Morph.* **98**, 125–160.

Munshi, J. S. Datta (1961). The accessory respiratory organs of *Clarias batrachus* (Linn). *J. Morph.* **109**, 115–140.

Munshi, J. S. Datta (1968). The accessory respiratory organs of *Anabas testudineus* (Bloch) (Anabantidae, Pisces). *Proc. Linn. Soc. Lond.* **179**, 107–126.

Munshi, J. S. Datta and Singh, B. N. (1968). On the respiratory organs of *Amphipnous cuchia* (Ham. Buch.). *J. Morph.* **124**, 423–444.

Potter, C. M. L. (1910). Respiratory function of the swim-bladder in *Lepidosteus*. *J. exp. Zool.* **49**, 45–67.

Purser, G. L. (1926). On *Calamoichthys calabaricus* I. The alimentary and respiratory systems. *Trans. R. Soc. Edinb.* **54**, 767–784.

Randall, D. J. (1970). The circulatory system. *In* "Fish Physiology". Vol. IV, 133–172. (Eds W. S. Hoar and D. J. Randall.) Academic Press, New York, London and San Francisco.

Robertson, J. I. (1914). The development of the heart and vascular system of *Lepidosiren paradoxa*. *Q. Jl microsc. Sci.* **59**, 53–132.

Satchell, G. H. (1971). "Circulation in Fishes." 131 pp. Cambridge University Press, Cambridge.

Satchell, G. H. and Jones, M. P. (1967). The function of the conus arteriosus in the Port Jackson shark, *Heterodontus portusjacksoni*. *J. exp. Biol.* **46**, 373–382.

Satchell, G. H., Hanson, D. and Johansen, K. (1970). Differential blood flow through the afferent branchial arteries of the skate *Raja rhina*. *J. exp. Biol.* **52**, 721–726.

Schmalhausen, I. I. (1968). "The Origin of Terrestrial Vertebrates" 314 pp. Academic Press, New York, London and San Francisco.

Schmidt, P. T. (1942). A fish with feather-like gills. *Copeia*. pp. 98–100.

Schöttle, E. (1931). Morphologie und Physiologie der Atmung bei Wasser-, Schlamm- und landlebenden Gobiiformes. *Z. wiss. Zool.* **140**, 1–114.

Shelton, G. (1970). The effect of lung ventilation on blood flow to the lungs and body of the Amphibian *Xenopus laevis*. *Resp. Physiol.* **9**, 183–196.

Shelton, G., and Jones, D. R. (1965). Central blood pressure and heart output in surface and submerged frogs. *J. exp. Biol.* **42**, 339–357.

Singh, B. N. and Hughes, G. M. (1973). Cardiac and respiratory responses in the climbing perch *Anabas testudineus*. *J. comp. Physiol.* **84**, 205–226.

Stevens, E. D. and Randall, D. J. (1967). Changes in blood pressure, heart rate and breathing rate during moderate swimming activity in rainbow trout. *J. exp. Biol.* **46**, 307–315.

Sulya, L. L., Box, B. E. and Gunter, G. (1961). Plasma proteins in the blood of fishes from the Gulf of Mexico. *Am. J. Physiol.* **200**, 152–154.

Wu, H. W. and Chang, H. W. (1947). On the arterial system of the gills and the suprabranchial cavities in *Ophiocephalus argus*, with special reference to the correlation with the bionomics of the fish. *Sinensia* **17**, 1–16.

Balance Between Aquatic and Aerial Respiration

B. N. SINGH*

Department of Zoology, Banaras Hindu University, Varanasi, India

I. INTRODUCTION

The freshwater air-breathing fishes of tropical countries inhabit waters of low O_2 and high CO_2 contents. Moreover, the pools, ponds, swamps, creeks and also the torrential streams inhabited by them often either dry out during summer or their water becomes muddy, highly hypoxic and hypercarbic which is unsuitable for gaseous exchange across the gills. Under such unfavourable environmental conditions, these fishes depend on aerial respiration alone until the commencement of the next rainy season. However, while some fishes resort to air breathing only under adverse conditions, others such as the climbing perch adopt bimodal respiration even in normoxic waters. The adaptive features of the air-breathing fishes inhabiting hypoxic and hypercarbic waters have been described by Willmer (1934), Carter (1957), Johansen (1970) and several others. The environment and ecology of Indian air-breathing fishes are dealt with in detail by Dehadrai and Tripathi (this volume). The gross and fine structure of their respiratory organs are described by Munshi (this volume).

Until late 1950, much attention was paid to the structural modifications and adaptations of air-breathing fishes (see Das, 1927, 1933; Schöttle, 1931; Munshi, 1961, 1962a,b). However, during the last decade or so, a large number of physiologists have become interested in the study of various

* Present address: Central Inland Fisheries Research Institute, Barrackpore, West Bengal, India.

adaptive features of the amphibious vertebrates (Hughes, 1966; Rahn, 1966a; Johansen, 1966; Johansen and Lenfant, 1968; Johansen, Lenfant and Grigg, 1967; Johansen, Lenfant, Schmidt–Nielsen and Petersen, 1968; Johansen, Hanson and Lenfant, 1970; Lenfant and Johansen, 1967, 1968, 1972; Lenfant, Johansen and Grigg, 1967; Farber and Rahn, 1970; Garey and Rahn, 1970; Rahn, Rahn, Howell, Gans and Tenney, 1971; Berg and Steen, 1965, 1966; Gordon, Boetius, Evans, McCarthy and Oglesby, 1965, 1969; Gordon, Fisher and Edurdo, 1970; Hughes and Singh, 1970a,b, 1971; Singh and Hughes, 1971, 1973; Howell, 1970; Howell, Baumgardner, Bondi and Rahn, 1970; Hutchison, Whitford and Kohl, 1968; Whitford and Hutchison, 1963, 1965; and several others). Although a sequence of extant forms (including air-breathing fishes) using both water and air might not represent the exact physiological changes and adaptations that might have occurred during the evolution of land vertebrate ancestors (closest to which are the dipnoans and crossopterygians), nevertheless they are of great value in understanding the physiological changes that would have occurred during evolution of life from water to land.

Gas exchange and respiratory patterns of some Indian air-breathing fishes such as *Anabas testudineus*, *Heteropneustes fossilis* and *Clarias batrachus* have been studied in detail by Hughes and Singh (1970a,b and 1971; Singh and Hughes, 1971, 1973). These studies clearly indicate that there are large interspecific variations in the metabolic rates of these fishes inhabiting more or less similar waters. Moreover, there are also significant differences in $\dot{V}_{O_2}$ of the same species when living in water and when exposed to air. Their ability to extract O_2 from water and air via aquatic and aerial routes varies a great deal. However, it has now become clear that the function of the accessory respiratory organs is mainly to absorb O_2 and most of the CO_2 is lost through gills and/or skin.

It has also been found that if O_2 and CO_2 concentrations in the ambient water and air are changed, then these fishes also augment their water and air respiration to maintain a balance between aquatic and aerial gas exchange. Such adaptations in the mode of respiration also require associated changes in the perfusion, and also in the respiratory properties of the blood. Therefore, in this presentation an attempt has been made to study the mechanisms for the balance between aquatic and aerial respiration and also to throw some light on various respiratory and circulatory adaptations of these fishes that help them not only in bimodal respiration but also in survival during adverse environmental conditions including aestivation.

II. Gas Exchange from Water and/or Air

The metabolic rates of the climbing perch *Anabas testudineus* and the catfishes *Clarias batrachus* and *Heteropneustes* (= Saccobranchus) *fossilis* have been studied in water and/or air. Adult fishes were used for such studies: body weight of *Anabas testudineus* was in the range 28·0–51·0 g, and that of *Clarias batrachus* between 76·0–157·0 g; and *Heteropneustes fossilis* between 45·5–69·0 g. All measurements were made at 25 ± 1°C.

A. OXYGEN CONSUMPTION ($\dot{V}_{O_2}$) AND CARBON DIOXIDE LIBERATION ($\dot{V}_{CO_2}$) FROM WATER AND AIR

During such experiments fishes were allowed access to air in closed respirometers. The water and air were in contact over a small surface area (2–3″ in diameter) since a small air chamber was fitted on one side of the respirometer allowing the fish to gulp air from this chamber only. The use of paraffin oil to separate the two respiratory media was avoided since Spurway and Haldane (1963) have found that paraffin oil interferes with bimodal breathing of *A. testudineus*. The intervals between air breaths vary in each species and the frequencies of air breathing are also related to the gas concentrations in water and air.

The rates of O_2 consumption of *Anabas* are shown in Fig. 1 and Table I. This fish consumes a little more O_2 from air (54%) than from water (46%) (see Table IV). However, it is interesting to find that *Anabas* consumes more O_2 from water when the O_2 tension in the water is above 100 mmHg whereas a nearly constant rate of O_2 consumption from water is found when WP_{O_2} (oxygen tension in water) is between 100–60 mmHg (Fig. 1). It is worth pointing out here that at high levels of WP_{O_2} the fishes consumed more O_2 from water even when they were acclimatized in the respirometer for a few days and observations made only on resting fish. Therefore, high $\dot{V}_{O_2}$ from water above 100 mmHg WP_{O_2} cannot be accounted for by their activity. In natural environmental conditions *Anabas* inhabits water with P_{O_2} usually between 40–100 mmHg. Therefore, this appears to be a homeostatic zone for *Anabas* and a constant level of $\dot{V}_{O_2}$ in this range would certainly be advantageous for this fish in nature. It is likely that above 100 mmHg WP_{O_2}, the ΔP_{O_2} between water and blood across the gills is increased hence more oxygen diffuses into the blood resulting in a high rate of O_2 consumption.

Total mean $\dot{V}_{O_2}$ of *Anabas* during bimodal respiration is 113·4 ml/Kg/h. In *C. batrachus* the total mean $\dot{V}_{O_2}$ is 93·4 ml/Kg/h. *Clarias* also consumes more oxygen from air (58%) than from normoxic waters (42%) Tables I and IV).

Figure 2 shows $\dot{V}_{O_2}$ from water and air in *H. fossilis*. This fish consumes more oxygen from water (59%) than from air (41%) when living in normoxic water (WP_{O_2} = 154·0 mmHg). However, this ratio is reversed when fishes are kept in water with less than 100 mmHg P_{O_2} (Fig. 2). Moreover, like *Anabas*, *Heteropneustes* also shows a nearly constant level of O_2 consumption

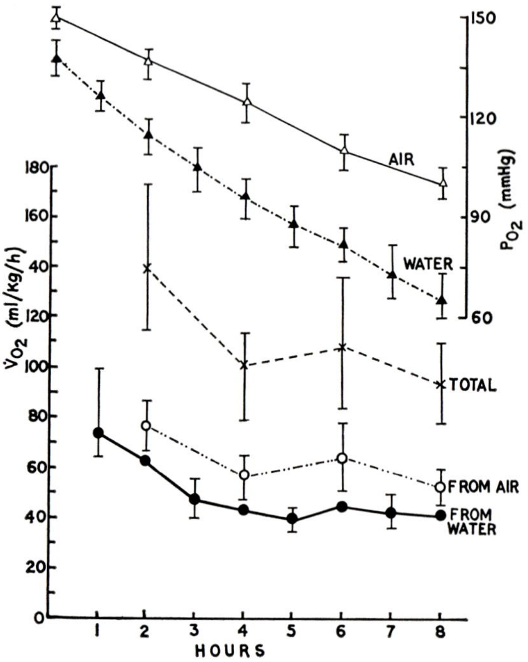

FIG. 1. Rate of O_2 consumption from water and air of *A. testudineus*. Changes in P_{O_2} of water and air clearly show that the two respiratory media did not come into full equilibrium during the experiments. Ranges of determinations are also given. (Data from Hughes and Singh, 1970a).

from water between WP_{O_2}, 100–60 mmHg; which is again the normal range of P_{O_2} in natural waters inhabited by them.

Thus air-breathing organs of *Anabas* and *Clarias* are of greater importance in oxygen uptake as compared to *Heteropneustes* (Tables I and IV). A trend similar to *Anabas* and *Clarias* is found in the adult forms of almost all of the air-breathing fishes studied so far at temperatures of 20–30°C (Table I). However, in the young forms of *Anabas*, *Clarias* and *Heteropneustes* the gills are better developed than the accessory respiratory organs (Munshi, this volume). Therefore, it is likely that a greater proportion of O_2 uptake will be via the aquatic routes in the young forms of all these fishes. However, in *Anabas* the skin is thick and scaly, and hence probably only the gills participate in O_2 absorption from water, whereas in *Clarias* and *Heteropneustes* both gills and skin participate in the absorption of oxygen (see Table IV).

In all three air-breathing teleosts much more carbon dioxide is eliminated via the gills and skin into the water than from the accessory respiratory organs into the air (see Tables III and IV). Total mean $\dot{V}_{CO_2}$ in *Anabas* is 132·7 ml/Kg/h. About 90% of this total CO_2 is eliminated via gills and the

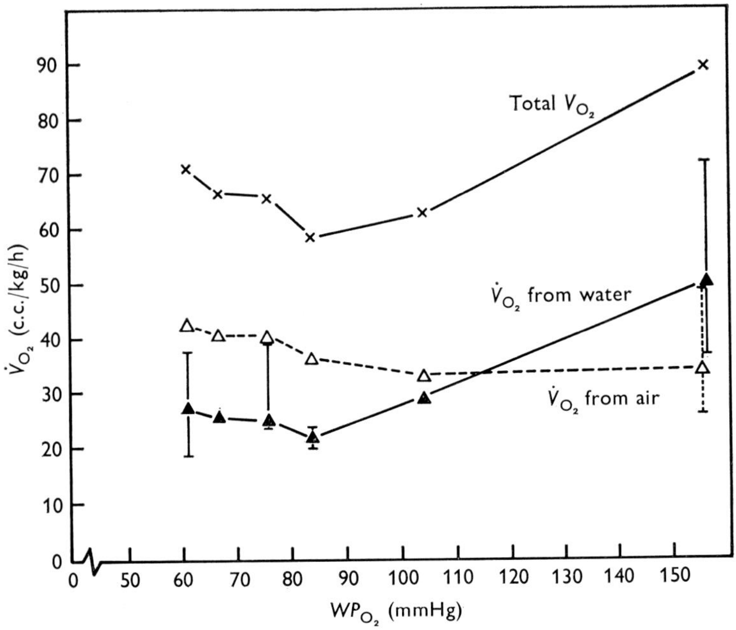

FIG. 2. Rate of O_2 consumption from water and air in *H. fossilis*. Effect of the lowering O_2 tension in water (WP_{O_2}) on the proportional changes in $\dot{V}_{O_2}$ from the two respiratory media are clearly shown. (From Hughes and Singh, 1971).

ratio for aquatic to aerial $\dot{V}_{CO_2}$ is 10:1 (see Tables IV and V). The overall gas exchange ratio (RQ) is about unity but aquatic RQ is 2·29 and aerial RQ, 0·2 only (see Table III). Since experiments on *Clarias* and *Heteropneustes* were conducted in continuously flowing water, it was not possible to measure directly the minute changes of CO_2 content in the water during such experiments. However, on examining the determined gas exchange ratio in a variety of bimodal fishes (see Table V), it becomes clear that total RQ from water and air is about 1·0. Therefore, assuming that the total RQ from water and air in *Clarias* and *Heteropneustes* is also 1·0; the measured aerial $\dot{V}_{CO_2}$ show clearly that in these species also about 90% of the metabolic CO_2 is eliminated into the water and that the accessory respiratory organs only account for about 10% of the total $\dot{V}_{CO_2}$ (see Table IV).

B. $\dot{V}_{O_2}$ FROM CONTINUOUSLY FLOWING NORMOXIC WATER

When these fishes are kept in continuously flowing normoxic waters (WP_{O_2} = 150 mmHg) and are prevented from surfacing; they show restlessness to begin with but settle down to rest after about ½ hour. During such experiments the frequency of the opercular movements is slightly elevated but the depth of breathing is normal in most cases.

TABLE I

Oxygen uptake in representative air-breathing fishes. Proportion of $\dot{V}_{O_2}$ from water and air and the quotient between aquatic and aerial $\dot{V}_{O_2}$ are also given

T °C	Species	O_2 uptake ($\dot{V}_{O_2}$, ml/*Kg*/h) Aquatic	Aerial	Total	Aquatic/Aerial	Reference
25	*Anabas testudineus*	52·62	60·80	113·42	0·865	Hughes and Singh (1970a)
25	*Clarias batrachus*	38·85	54·54	93·39	0·712	Singh and Hughes (1971)
25	*Heteropneustes* (*Saccobranchus*) *fossilis*	50·10	34·45	84·55	1·454	Hughes and Singh (1971)
26	*Electrophorus electricus*	6·92	23·08	30·00	0·30	Farber and Rahn (1970)
20	*Amia calva*	66·00	30·00	96·00	2·20	Johansen *et al.* (1970)
30	*A. calva*	30·00	84·00	114·00	0·36	Johansen *et al.* (1970)
22	*Lepisosteus osseus*	14·40	39·00	53·40	0·369	Rahn *et al.* (1971)
25	*Cobitis fossilis*	14·40	59·40	73·80	0·242	Calugareanu (1907)
20	*Protopterus aethiopicus*	1·26	10·14	11·40	0·123	Lenfant and Johansen (1968)
24	*P. aethiopicus*	5·80	56·70	62·50	0·102	McMahon (1970)
18	*Lepidosiren* (Juvenile, 150 g)	54·00	30·60	84·60	1·764	Johansen and Lenfant (1967)
20	*Lepidosiren* (Adult)	1·80	40·20	42·00	0·044	Sawaya (1946)
20	*Lepidosiren* (Adult)	1·00	21·20	22·20	0·04	Lenfant *et al.* (1970)

The rate of O_2 consumption is lowered compared to the total $\dot{V}_{O_2}$ from both water and air. $\dot{V}_{O_2}$ under such conditions is 75·5 ml/Kg/h in *Anabas*, 64·9 ml/Kg/h in *Clarias* and 66·4 ml/Kg/h in *Heteropneustes* (Table II). These values indicate about 30% reduction in overall rate of metabolism in *Anabas* and *Clarias* but *Heteropneustes* shows only about 20% reduction (Table II).

However, in *Clarias* and *Heteropneustes* the skin also contributes significantly to the absorption of O_2. In both species about 17% of the total $\dot{V}_{O_2}$ in normoxic water occurs via skin and the rest via gills (see Table VI). Similar rates of O_2 consumption via skin were also observed in *Anguilla vulgaris* (Berg and Steen, 1965) and *Monopterus albus* (Liem, 1967; see Table VI).

C. $\dot{V}_{O_2}$ FROM STILL WATER

The fishes were kept in closed respirometers and surfacing was prevented. Therefore, they rebreathed the same water, hence its P_{O_2} gradually declined, and slow hypoxia was imposed over a period of several hours.

$\dot{V}_{O_2}$ of *Clarias* and *Anabas* is reduced at lowering O_2 tension in water and they show dependent respiration throughout the range of 150·0–60·0 mmHg (Fig. 3). The lowering in the total $\dot{V}_{O_2}$ occurred even though, the fishes showed increased rate and depth of breathing (Fig. 3). However, $\dot{V}_{O_2}$ in *Anabas* was very high but this was due to the activity since they were allowed to take N_2-breaths from the gas chamber of the respirometer. These fishes often repeated N_2-breaths and they did so more frequently at lowered O_2 tensions. This indicates that in hypoxic waters they were unable to meet their metabolic demands via the gills alone and hence were searching for additional O_2 from air.

Unlike *Anabas* and *Clarias*, *H. fossilis* shows independent respiration between 150 and 40 mmHg O_2 tensions (Fig. 3). Below 40 mm $\dot{V}_{O_2}$ is sharply reduced. Both the rate and depth of breathing increased during independent respiration (Fig. 3). Munshi and Singh (1968a), and Hughes and Munshi (1968, 1973) have reported that the water/blood barrier of the secondary lamellae is thin in *H. fossilis* but thicker in *Clarias* and *Anabas* (5–29 mμ). It seems likely that this thin respiratory barrier together with a large gill surface area (Hughes *et al.*, 1974) are two very important factors that help *Heteropneustes* maintain a nearly constant level of O_2 transfer even at lower WP_{O_2} levels (40·0–100·0 mmHg).

D. EFFECT OF HYPOXIA IN CONTINUOUSLY FLOWING WATERS WHEN SURFACING IS PREVENTED

Figure 4 shows the effect of hypoxic waters in a continuously flowing system. *C. batrachus* again behaves like an O_2-conformer (dependent respira-

TABLE II

Rate of oxygen consumption of certain air-breathing fishes in different experimental respiratory media. A comparison is made between total $\dot{V}_{O_2}$ *from water and air with that from other respiratory media*

T °C	Species	Total $\dot{V}_{O_2}$ (ml/kg/h)				% of total $\dot{V}_{O_2}$ from air and water			Reference
		Air and water	Water (Cont. flow)	Air (Air-exposed)	Air (de-O_2 water)	Water (Cont. flow)	Air (Air-exposed)	Air (de-O_2 water)	
25	*Anabas testudineus*	113·42	75·50	104·78*	132·83*	66·6	92·4	117·1	Hughes and Singh (1970a)
25	*Clarias batrachus*	93·39	64·94	71·17	40·84	69·5	76·3	43·7	Singh and Hughes (1971)
25	*Heteropneustes fossilis*	84·50	66·35	54·50	35·09	78·6	64·5	41·5	Hughes and Singh (1971)
15	*Anguilla vulgaris*	—	26·6	11·54	—	—	43·4 (Air/water)	—	Berg and Steen (1965)
24	*Protopterus aethiopicus*	62·50	10·64	—	—	17·0	—	—	McMahon (1970)
20	*P. aethiopicus*	13·15	—	13·80	—	—	100·0	—	Lenfant and Johansen (1968)

* Active state

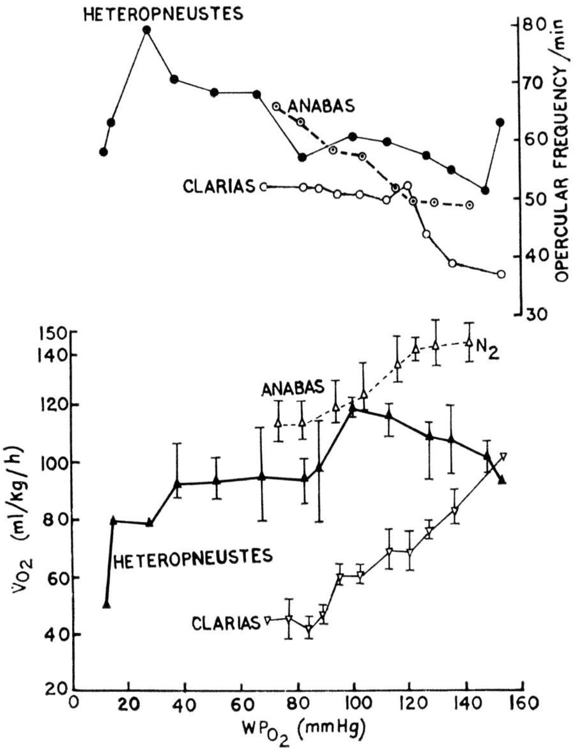

FIG. 3. Changes in the rate of O_2 consumption and opercular movements (gill ventilation frequencies) in *A. testudineus*, *C. batrachus* and *H. fossilis*, are plotted when they are subjected to gradual hypoxia in still water and prevented from surfacing. The range of variations in $\dot{V}_{O_2}$ at each WP_{O_2} level are also shown. (Data for *Anabas*—Hughes and Singh, 1970a; for *Clarias*—Singh and Hughes, 1971; and for *Heteropneustes*—Hughes and Singh, 1971).

tion) while *Heteropneustes* shows independent respiration at O_2 tensions between 100–40 mmHg. The $\dot{V}_{O_2}$ below 40 mmHg is not maintained constant in spite of a very steep rise in rate and depth of ventilation (Fig. 4).

E. EFFECT OF CARBON DIOXIDE ON $\dot{V}_{O_2}$

The rate of O_2 consumption in three individuals of *H. fossilis* was studied in normoxic (WP_{O_2} = 128·5–155·0 mmHg) but hypercarbic waters with free access to air. This fish normally does not surface for breathing air in normoxic waters. Neither did they surface for air in hypercarbic waters until 7·5–10% CO_2 in air was equilibrated with water. The rate and depth of breathing increased in hypercarbic waters but once the C_{CO_2} (total carbon dioxide content) in water was raised above 12 Vol.%, the depth of breathing decreased.

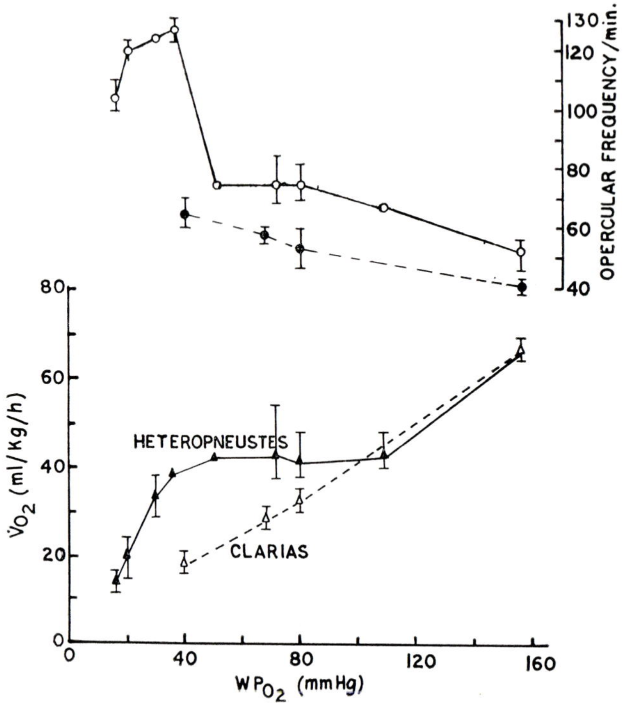

FIG. 4. Changes in $\dot{V}_{O_2}$ and opercular frequencies from water in continuously-flowing systems. (Data for *Clarias*—Singh and Hughes, 1971, and for *Heteropneustes*—Hughes and Singh, 1971).

$\dot{V}_{O_2}$ was reduced in hypercarbic waters (Fig. 5). The fishes switched over to bimodal respiration in hypercarbic waters having C_{CO_2} levels above 13·0–14·0 Vol.%. They consumed nearly equal amounts of O_2 from water and air (Fig. 5). The P_{O_2} in air was between 129·0–134·4 mmHg and P_{CO_2} between 11·0–42·0 mmHg.

F. $\dot{V}_{O_2}$ AND $\dot{V}_{CO_2}$ IN AIR-EXPOSED FISH

When these fishes are exposed to air they survive for several hours (4–8 or more) if protected from dehydration. They become a little restless and active on exposure to air but quickly settle down to resting conditions. The measurements were started at least $\frac{1}{2}$–1 hour after exposure to air when the fishes were apparently in resting conditions. But from Fig. 6 it appears that their O_2 consumption remained elevated for about 2 hours. Probably this resulted in the high rate of oxygen consumption during the first observations (Fig. 6). Later on, a constant level of $\dot{V}_{O_2}$ was established and this rate was maintained until P_{O_2} in the respirometer air was reduced to about 125·0 mmHg. As

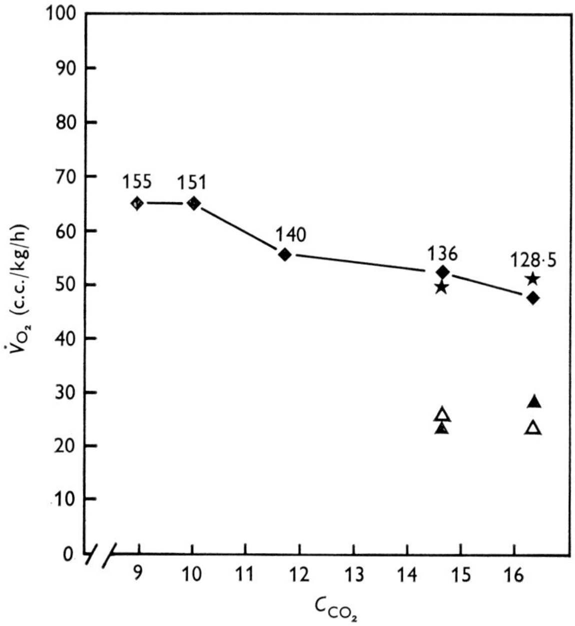

Fig. 5. Changes in $\dot{V}_{O_2}$ from hypercarbic waters in *H. fossilis* at different CO_2 contents and O_2 tensions in water when free access to air is not prevented. It exchanges gases with water alone until C_{CO_2} in water is raised above 14 vol%. Above this level of CO_2 this fish uses bimodal respiration and consumes nearly equal volumes of O_2 from water (▲) and air (△). Levels of WP_{O_2} in each case are given above the line for total $\dot{V}_{O_2}$ (From Hughes and Singh, 1971).

P_{O_2} in the ambient air was further lowered below this level and P_{CO_2} was raised above 20–30 mmHg, *Anabas* and *Heteropneustes* showed dependent respiration. However, *Clarias* showed independent respiration even when P_{O_2} in the air was lowered to about 70·0 mmHg and P_{CO_2} was raised above 40·0 mmHg (Fig. 5).

The total metabolic rate in air-exposed fishes is reduced (Table II). *C. batrachus* has a rate of 71·2 ml/Kg/h and *H. fossilis* 54·5 ml/Kg/h. These rates show about 25% reduction in total $\dot{V}_{O_2}$ of *Clarias* and about 35% reduction in *Heteropneustes* (Table II). *Anabas* also shows lowered $\dot{V}_{O_2}$ (104·9 ml/Kg/h). This is only about 8% lower than the overall $\dot{V}_{O_2}$ from water and air (Table II). But this reduction (8%) in $\dot{V}_{O_2}$ would have been still higher had these fishes not become active from time to time during the observations.

The loss of metabolic CO_2 was also at a lower level in air-exposed fishes. $\dot{V}_{CO_2}$ in air-exposed *A. testudineus* was 74·6 ml/Kg/h which gives a gas exchange ratio of 0·7. The rate of CO_2 elimination in *C. batrachus* and *H. fossilis* was 37·5 ml/Kg/h and 31·8 ml/Kg/h respectively. These values give

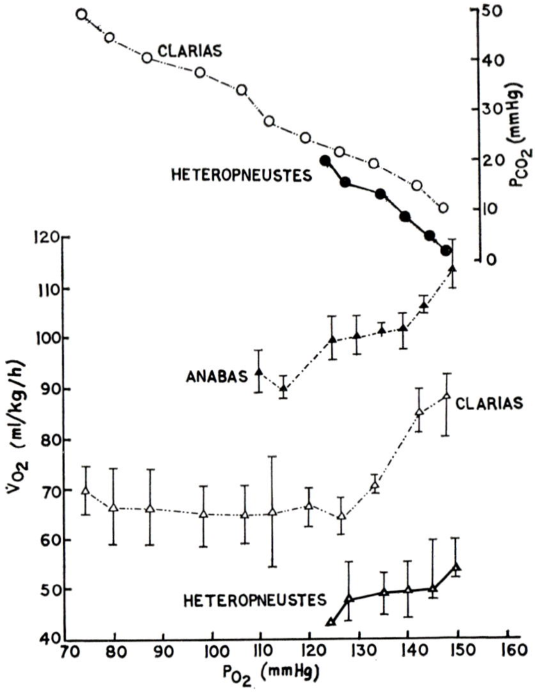

FIG. 6. Rates of O_2 consumption in *A. testudineus*, *C. batrachus* and *H. fossilis* when they are exposed to air in closed respirometers. Gradual increases in P_{CO_2} and decline in P_{O_2} of ambient air in the closed system are plotted for *Clarias* and *Heteropneustes*. Range of determinations for $\dot{V}_{O_2}$ are also given. (Data for *Anabas*—Hughes and Singh, 1970a; for *Clarias*—Singh and Hughes, 1971, and for *Heteropneustes*—Hughes and Singh, 1971).

gas exchange ratios of 0·52 in *Clarias* and 0·58 in *H. fossilis* (Table III). These ratios are very low and it seems that some of the metabolic CO_2 is retained during exposure to air.

It is worth noting here that the total amount of CO_2 released via the air-breathing organs has considerably increased. This will be advantageous since the gills are not suitable for CO_2 elimination in air. Some CO_2 might have been excreted via skin of *Heteropneustes* and *Clarias* but this needs further confirmation by direct measurements. However, it is clear that both the air-breathing organs and the skin are still insufficient to excrete all of the metabolic CO_2 (RQ about 0·6).

G. $\dot{V}_{O_2}$ AND $\dot{V}_{CO_2}$ FROM AIR IN FISHES LIVING IN DEOXYGENATED WATER

During the summer in tropical countries the oxygen content of the inhabited waters is sometimes depleted to almost zero and the air-breathing fishes are forced to live in almost deoxygenated waters Therefore, some

TABLE III

A comparison between the rate of CO_2 production and gas exchange ratio into water and/or air during different experimental conditions, in three air-breathing teleosts ($T = 25°C$)

Species		In water with access to air			In water (Access to N_2)	In air (Air exposed)	In de-O_2 water with access to air			Reference
		Water	Air	Total			Water	Air	Total	
Anabas testudineus	$\dot{V}_{CO_2}$ (ml/kg/h)	120·44	12·26	132·70	137·56	74·62	113·66*	19·17	132·83	Hughes and Singh (1970a)
	R	2·29	0·20	1·17	1·08	0·71	0·85*	0·15	1·00*	
Clarias batrachus	$\dot{V}_{CO_2}$ (ml/kg/h)	87·51*	5·88	93·39	—	37·47	23·60*	17·24	40·84	Singh and Hughes (1971)
	R	2·25*	0·11	1·00*	—	0·52	0·60*	0·40	1·00*	
Heteropneustes fossilis	$\dot{V}_{CO_2}$ (ml/kg/h)	79·10*	5·40	84·50	—	31·78	21·92*	13·17	35·09	Hughes and Singh (1971)
	R	1·58*	0·17	1·00*	—	0·58	0·62*	0·38	1·00*	

* Assuming overall R = 1·0

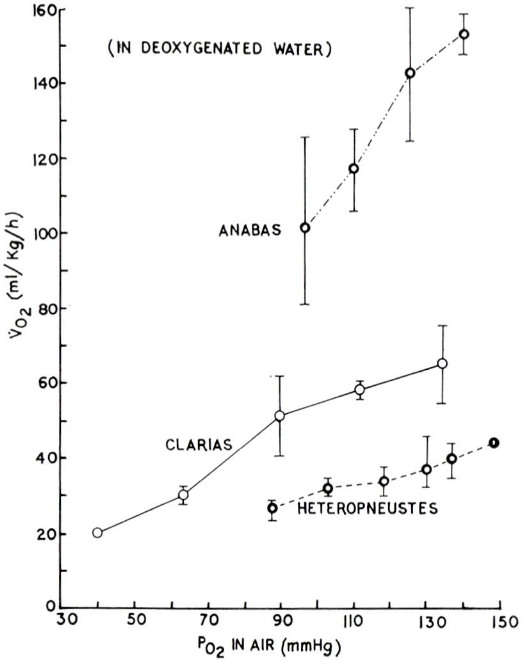

FIG. 7. Changes in $\dot{V}_{O_2}$ from still air in *Anabas*, *Clarias* and *Heteropneustes*, when fishes were kept in deoxygenated waters with free access to air.

experiments were designed to test the ability of these fishes to withstand similar conditions. Interestingly enough, all of them were found capable of surviving in deoxygenated waters for several days, breathing only air.

The rate of O_2 consumption of fishes living in deoxygenated waters with free access to air is plotted in Fig. 7. All of them showed dependent respiration under these conditions. This may be attributed largely to the fact that P_{O_2} in closed air chambers declined sharply below 100 mmHg, hence they were forced to breath hypoxic air while still spending extra energy to surface for air. *Clarias* and *Heteropneustes* show very low metabolic rates, when these are compared with the rate of fishes exchanging gases with both water and air or with fishes exposed to air (Table II). The lowering in overall metabolic rate is about 60%. Fishes in the still deoxygenated waters were very restive and frequently took air-breaths. The only movement made under such conditions was to surface for air. It seems as if they conserved energy during such adverse conditions.

However, under such conditions skin was not available for O_2 absorption, therefore a very high reduction in total $\dot{V}_{O_2}$ could also be due partly to the absence of O_2 exchange via the skin.

A. testudineus were placed in continuously flowing deoxygenated water with still air in the closed air chamber of the respirometer. They were very actively swimming during the observations and this is why they had a very high rate of O_2 consumption (Table II, Fig. 7).

The rate of CO_2 release via the aerial route was very low (R = 0·4). It is possible that some of the CO_2 eliminated into the air chamber might have diffused into the water. But the control experiments did not show this diffusion to reduce aerial RQ significantly in this particular experimental setup. Therefore, it appears that even in the deoxygenated water about 60% of the total metabolic CO_2 is lost via aquatic routes (Table III). *Anabas* were observed showing periodically a few opercular movements with very low amplitude, therefore most of the CO_2 might have been lost via gills during such short intervals. However, in *Clarias* and *Heteropneustes* skin would also contribute in the excretion of CO_2 probably at the expense of some O_2 loss into the deoxygenated water.

H. ROUTE OF CO_2 RELEASE IN AMPHIBIOUS FISHES

It is well-established that the two respiratory media (i.e. water and air) differ a great deal in their respiratory properties with particular reference to the solubility of O_2 and CO_2. (For details see Dejours, this volume.) O_2 becomes readily available in air but CO_2 is readily eliminated into the water. Therefore, the animals utilizing both water and air as respiratory media have to evolve better mechanisms to deal with the problems posed by CO_2 during exchange of gases with air (Hughes, 1966; Rahn, 1966a). CO_2 is readily lost during bimodal respiration but its level has to rise in the blood and tissues if the animal is living out of water. The retention of CO_2 in the blood leads to increased levels of HCO_3^-, P_{CO_2} and lowered pH in the blood and tissues (Howell, 1970; Howell *et al.*, 1970).

The problems raised by high CO_2 in the blood, tissues and lungs have been solved by terrestrial vertebrates. However, it is clear that the transitional forms using both water and air as respiratory media face an acute problem of acid-base balance (see Rahn, 1966a,b; Howell, 1970, Howell *et al.*, 1970; Rahn and Baumgardner, 1972; Lenfant and Johansen, 1972). However, Hughes (1966) suggested that in amphibious vertebrates the air-breathing organs chiefly function for absorption of O_2 while CO_2 will be easily lost into the water via gills and/or skin. He also suggested that in those amphibious forms which have lost gills in their adult forms, CO_2 will be excreted mainly through the skin until the lungs in the fully terrestrial vertebrates take over the function of total metabolic CO_2 elimination (Fig. 8). Recently this hypothesis has received wide support from a large number of physiological findings on respiration of amphibious vertebrates (see also Hutchison, this volume). This view holds true if we examine the situation among air-breathing fishes (Tables III, IV, V and VI).

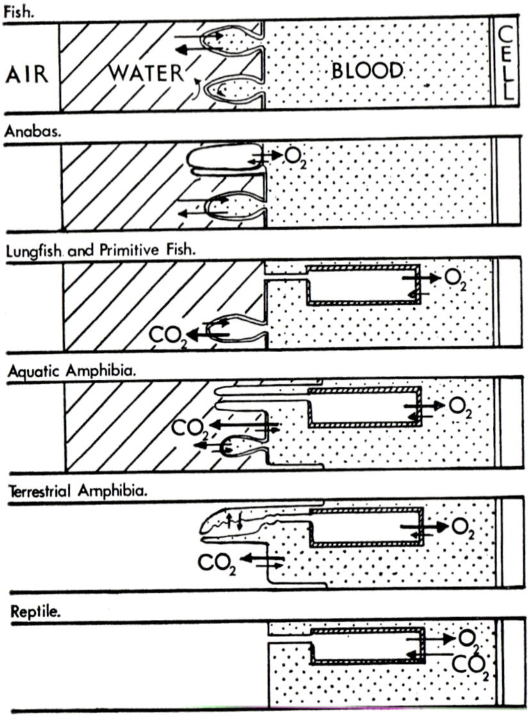

FIG. 8. Diagrammatic representation of pathway for gas exchange in a variety of vertebrates. The main sites for O_2 and CO_2 exchange are also indicated (From Hughes, 1966).

III. P_{O_2} AND P_{CO_2} IN THE SUPRABRANCHIAL CHAMBERS

Gas content of the suprabranchial chambers was measured in the climbing perch, *A. testudineus*, when living in water with access to air allowed and also on exposure to air. Patterns of changes in the P_{O_2} and P_{CO_2} of the suprabranchial chamber's gas following air-breaths are plotted in Fig. 9.

In bimodal respiration, P_{O_2} declines continuously over a period of 5–17 min during breath holding. However, the rate of decline in the suprabranchial chamber P_{O_2} (SBP_{O_2}) is slow if the fish continues gill ventilation for a few minutes following the air-breath, but it is faster if the fish ceases gill ventilation. P_{CO_2} rises quickly in the first few minutes following an air-breath (R = 0·33–0·1) but later on levels off at about 25·0 mmHg (R less than 0·1) before the next breath is repeated (Fig. 9).

P_{O_2} and P_{CO_2}, just prior to an air breath were not determined but the extrapolation of plots (Fig. 9 dotted lines) to the time when the spontaneous breaths actually occurred, indicate that *Anabas* usually repeats an air-breath when SBP_{O_2} is below 40 mmHg.

TABLE IV

Gas exchange partition between water-breathing organs (gill, G and Skin, S) and air-breathing organs of representative air-breathing fishes

T		% $\dot{V}_{O_2}$			% $\dot{V}_{CO_2}$			
°C	Species	From water	(Organ)	From air	Into water	(Organ)	Into air	Reference
20	*Amia calva*	63	(G)	37	75	(G)	25	Johansen *et al.* (1970)
25	*Heteropneustes fossilis*	59	(G, S)	41	94	(G, S)	6	Hughes and Singh (1971)
25	*Anabas testudineus*	46	(G)	54	91	(G)	9	Hughes and Singh (1970a)
25	*Clarias batrachus*	42	(G, S)	58	94	(G, S)	6	Singh and Hughes (1971)
22	*Lepisosteus osseus*	27	(G)	73	92	(G)	8	Rahn *et al.* (1971)
23	*Monopterus albus*	25	(G, S)	75	—	—	—	Liem (1967)
26	*Electrophorus electricus*	22	(S)	78	81	(S)	19	Farber and Rahn (1970)
25	*Cobitis fossilis*	20	—	80	34	—	66	Calugareanu (1907)
20	*Protopterus aethiopicus*	11	(G)	89	70	(G)	30	Lenfant and Johansen (1968)
24	*Protopterus aethiopicus*	10	(G)	90	68	(G)	32	McMahon (1970)
20	*Lepidosiren* (Juvenile)	64	(G, S)	36	76	(G, S)	24	Johansen and Lenfant (1967)
20	*Lepidosiren* (Adult)	4	(G, S)	96	41	(G, S)	59	Lenfant *et al.* (1970)

TABLE V

A comparison between the rate of CO_2 elimination into water and air in representative air-breathing fishes. Gas exchange ratio from water and air and total R are also given

T °C	Species	CO_2 production (ml/kg/h) Aquatic	Aerial	Total	Aquatic / Aerial	Gas Exchange Ratio (R) Aquatic	Aerial	Total	Reference
25	*Anabas testudineus*	120·44	12·26	132·70	9·82	2·29	0·20	1·17	Hughes and Singh (1970a)
25	*Clarias batrachus*	87·51*	5·88	93·39	14·88	2·25*	0·11	1·00*	Singh and Hughes (1971)
25	*Heteropneustes fossilis*	79·10*	5·40	84·50	14·66	1·58*	0·17	1·00*	Hughes and Singh (1971)
26	*Electrophorus electricus*	24·35	5·65	30·00	4·31	3·65	0·25	1·00	Farber and Rahn (1970)
20	*Amia calva*	80·50	23·50	104·00	3·44	1·22	0·78	1·08	Johansen *et al.* (1970)
30	*A. calva*	68·00	44·00	112·00	1·54	2·27	0·52	0·98	Johansen *et al.* (1970)
22	*Lepisosteus osseus*	39·00	3·60	42·60	10·86	2·70	0·09	0·80	Rahn *et al.* (1971)
20	*Cobitis fossilis*	23·40	48·00	71·40	0·49	1·60	0·81	0·98	Calugareanu (1907)
20	*Protopterus aethiopicus*	6·06	2·58	8·64	2·35	4·70	0·25	0·75	Lenfant and Johansen (1968)
24	*P. aethiopicus*	31·80	15·60	47·40	2·04	5·40	0·27	0·76	McMahon (1970)
18	*Lepidosiren paradoxa* (Juvenile, 150 g)	72·60	22·20	94·80	3·27	1·34	0·72	1·12	Johansen and Lenfant (1967)
20	*L. paradoxa* (Adult)	6·70	9·50	16·20	0·71	6·70	0·45	0·73	Lenfant *et al.* (1970)

* Assuming overall R = 1·0.

TABLE VI

A comparison between certain air-breathing teleosts showing partitioning of $\dot{V}_{O_2}$ from water via the skin and via the gills when the fish is submerged under water and not allowed to surface

T °C	Species	Experimental conditions	$\dot{V}_{O_2}$ in water (%) Via skin	Via gill	Reference
7–15	*Anguilla vulgaris*	Submerged under water	10–15	85–90	Berg and Steen (1965)
25	*Heteropneustes fossilis*		16·8	83·2	Hughes and Singh (1971)
25	*Clarias batrachus*		16·6	83·4	Singh and Hughes (1971)
23	*Monopterus albus*		14·0	86·0	Liem (1967)

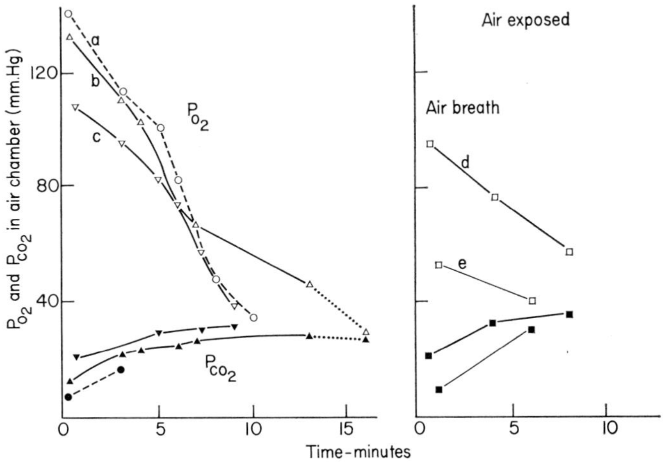

FIG. 9. Changes in P_{O_2} and P_{CO_2} of gas in the suprabranchial chambers of *A. testudineus* following air-breaths. Plots *a*, *b* and *c* are for fishes living in water with free access to air. Plots *d* and *e* are for air-exposed fish. Plot *e* is for a single air-breath when P_{O_2} and P_{CO_2} remained very low following an air-breath; this latter situation is observed only occasionally. (From Singh and Hughes, 1973).

P_{O_2} changes, in the suprabranchial gas following an air-breath in the air-exposed fishes, are similar to that of fish living in water. However, the mean SBP_{O_2}, following the air-breath, is generally low compared to the levels in a bimodal fish (Fig. 9). P_{CO_2} of the suprabranchial gas increases rapidly to about 34·0 mmHg in the first few minutes (R = 1·5–0·13) but later on starts to level off (R = 0·13–0·0). A very similar pattern of change in P_{O_2} and

P_{CO_2} of buccal cavity air has been observed in the electric eel, *Electrophorus electricus* (Farber and Rahn, 1970).

The results obtained for *Anabas* suggest that SBP_{CO_2} (P_{CO_2} of suprabranchial chamber gas) and the blood P_{CO_2} attain equilibrium at about 34·0 mmHg in air-exposed fish and at about 25·0 mmHg in fishes exchanging gases with both water and air. Such blood P_{CO_2} values are much higher than the actually measured blood P_{CO_2} values in purely water-breathing fishes (resting P_{CO_2} = 1·0–5·0 mmHg, Holeton and Randall, 1967, Piiper and Baumgarten–Schumann, 1968). Therefore, it is clear that to cope with such high blood P_{CO_2} values the air-breathing fishes should have blood less sensitive to CO_2 and with high buffering capacities.

IV. Respiratory Properties of the Blood

Oxygen dissociation curves of *C. batrachus* are drawn in Fig. 10 (Hughes and Singh, unpublished observations). The blood becomes fully saturated at P_{O_2}'s above 50 mmHg in the presence of 0·3 mmHg, P_{CO_2}. At this P_{CO_2}, P_{50} of *Clarias* blood is 16·2 mmHg. In the presence of higher P_{CO_2}'s, the O_2 dissociation curves shift to the right (Fig. 10). The magnitude of the Bohr effect is low (−0·30). However, it is interesting to find that the affinity for oxygen is not reduced much even in the presence of 23·0 mmHg P_{CO_2} (Fig. 10). If, we recall here that in *Anabas*, the suprabranchial gas and the blood attain

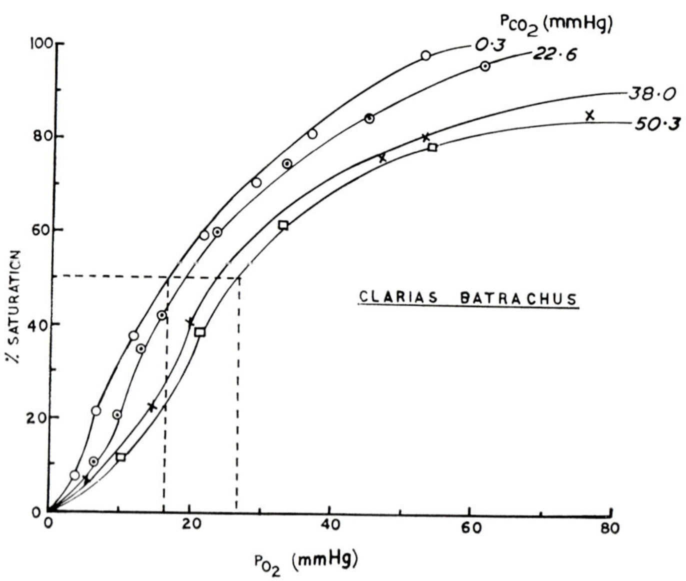

FIG. 10. Oxygen dissociation curves of *C. batrachus* at different P_{CO} levels. (Data from Hughes and Singh, unpublished observations).

equilibrium usually at P_{CO_2}'s above 25·0 mmHg (Fig. 9) and that these fishes inhabit CO_2-rich waters, the significance of the low sensitivity of their blood to CO_2 becomes very obvious.

However, the effect of CO_2 on the blood of *Anabas* has not been worked out as yet but the O_2 capacity and haemoglobin content of *Anabas* resemble values for *Clarias* and *Heteropneustes*. Haemoglobin content and O_2 capacity are higher in *C. batrachus* and *H. fossilis*, as is the case in most air-breathing teleosts (Table VII). Moreover, like other air-breathing fishes *C. batrachus* has a low Bohr effect, high buffering capacity, and the Root effect is very slight. All these respiratory properties of *Clarias* and also that of *H. fossilis* blood appear to favour their bimodal and aerial respiration (Table VII). It also becomes clear from this table that *Clarias* is better adapted for aerial life than *Heteropneustes*.

V. Patterns of Respiration and Control of Breathing

H. fossilis breathes from both water and air when it is kept in hypoxic waters. Air-breathing intervals are irregular and vary from a few minutes to half an hour. The intervals between air-breaths are related to the O_2 content in water and air. Air-breathing frequencies increase both in hypoxic water and air (Hughes and Singh, 1971). *Heteropneustes* ceases air breathing and exchanges gases with water alone for periods of 6–12 hours or more if it is placed in continuously-flowing air-saturated water. In normoxic but hypercarbic waters also, it continues aquatic respiration alone unless more than 7·5% CO_2 in air is bubbled into the water (C_{CO_2} more than 13 vol %).

Clarias shows bimodal respiration in tap waters (O_2 content below air saturation). The intervals between air-breaths usually vary between 2–20 min and are related to the O_2 and CO_2 content of water and air. The frequencies of air breathing increase in hypoxic water and also in hypoxic and hypercarbic air. Air-breathing intervals are very much reduced (between 20 seconds to 2 min) if the fish is allowed access to N_2 instead of air. It often repeats N_2-breaths as if "not satisfied" (Singh and Hughes, 1971). A very similar behaviour is shown by the climbing perch when forced to breathe nitrogen (Hughes and Singh, 1970a).

It is interesting to find that in normoxic waters, *C. batrachus* breathes quietly for 6–8 h exchanging gases with gills alone, without showing any signs of asphyxiation. However, in the African catfish, *Clarias lazera*, the gills alone are insufficient to meet the metabolic demand when exchanging gases with air-saturated water and they died of asphyxiation in a few hours (12–20 h) if surfacing was not allowed (Moussa, 1957). However, the Asian species, *C. batrachus* does not show any signs of restlessness for several hours (6–8 h) when exchanging gases with air-saturated water alone. Moreover, this fish does not even attempt air breathing for about 6–8 h, when living in normoxic water.

TABLE VII

A comparison of the respiratory properties of the blood of some representative air-breathing fishes. Respiratory properties of the blood of two water-breathing fishes are given for comparison (from Hughes and Singh unpublished observations)

T °C	Species	Hb. g%	RCV %	O_2 capacity Vol %	P_{50} (At pH P_{CO_2})	Bohr effect	Root effect	Haldane effect	HCOE (mM/L)	Buffering capacity	Reference
28·0	*Electrophorus electricus*	14·74*	—	19·75	13·0 (pH = 6·2)	Some	Some	—	—	—	Willmer (1934)
28·0	*E. electricus*	11·2	41·0	13·90	10·7 (pH = 7·57)	−0·68	None	4·8	33·5	4·7	Johansen *et al.* (1968)
28·0	*Hoplosternum littorale*	13·54*	—	18·14	15·0 (pH = 6·8)	Slight	Some	—	—	—	Willmer (1934)
25·0	*Symbranchus marmoratus*	13·0	47·0	17·30	7·2 (pH = 7·6)	−0·40	None	2·1	24·1	2·7	Lenfant *et al.* (1970)
27·0	*Amia calva*	5·82	22·8	7·8	24·5 (pH = 7·6)	—	Some	—	6·7	—	Johansen *et al.* (1970)
23·5	*C. mossambicus*	13·3	—	17·82*	5·5 (P_{CO_2} = 0)	Present	None	—	—	—	Fish (1956)
25·0	*C. batrachus*	12·59*	36·2	16·87	16·2 (pH = 7·58)	−0·30	Some	—	10·45 (At pH=6·98)	2·55	Hughes and Singh (Unpublished)
25·0	*H. fossilis*	11·58*	43·8	15·52	7·7 (pH = 7·105)	—	—	—	10·7 (At pH=7·10)	—	Singh and Hughes (Unpublished)
25·0	*P. aethiopicus*	7·26*	30·1	9·73	25·0 (pH = 7·64)	−0·28	None	2·3	34·2	2·6	Lahiri *et al.* (1968)
	P. aethiopicus	6·2	25·0	6·80	11·5 (P_{CO_2} = 6·2)	−0·47	None	1·0	20·0	1·52	Lenfant and Johansen (1968)
28·0	*Lepidosiren* (Adult)	6·64*	28·0	8·25	—	−0·295	None	1·5	27·2	2·43	Lenfant *et al.* (1970)
28·0†	*Myleus setiger*	8·04*	—	10·78	12·0 (P_{CO_2} = 0)	Large	Large	—	—	—	Willmer (1934)
28·0†	*Hoplias malabaricus*	4·87*	—	6·53	8·5 (P_{CO_2} = 0) (pH = 6·8)	Large	Large	—	—	—	Willmer (1934)

The climbing perch is an obligate air breather as it surfaces for air even in normoxic (P_{O_2} = 150·0 mmHg) and hyperoxic (P_{O_2} above 500·0 mmHg) waters (Hughes and Singh, 1970b). Gill ventilation continues for varying periods (a few seconds to a few minutes), after taking an air-breath. It then ceases for a few minutes but usually reappears before the next air-breath. The periods for which the fish shows gill ventilation, between successive air-breaths, appear to be related to the O_2 and CO_2 contents of water and air. The duration of gill ventilation is short in hypoxic and hypercarbic waters and the air-breaths are more frequent. But in normoxic waters, intervals between air-breaths are increased and gill ventilation occurs for comparatively longer periods (Hughes and Singh, 1970b). A similar pattern of respiration is shown by another obligatory air-breathing fish, Cuchia, *Amphipnous cuchia* (Munshi and Singh, 1968b and B. N. Singh, personal observation).

In normoxic waters (O_2 content about 5 ml/l and total CO_2 content 8–11 vol %), the mean interval between air breaths was about 10 min. However, the minimum and maximum intervals vary from 30 sec to 35 min or more.

Changes in opercular frequencies (gill ventilation) and air breathing of the climbing perch, living in experimental waters having different concentrations of O_2 and CO_2, are shown in Fig. 11. It is interesting to find that in highly hypercarbic waters (total CO_2 content above 20 vol %), climbing perch, tightly close their opercula and mouth, and only respire aerially. However, in hypercarbic waters this fish shows few opercular movements with very low amplitude of breathing for very short intervals (few seconds only). It appears that such behaviour enables the climbing perch to "test" the suitability of water for gill ventilation.

A. BREATHING PATTERNS IN RELATION TO pH OF WATER

The climbing perch changes its mode of respiration in alkaline and acid waters (Fig. 12). It shows both aquatic and aerial respiration in alkaline and neutral waters (pH 8·0 to 6·85). Aerial respiration predominates and aquatic respiration is depressed between pH 6·5–6·25. In highly acid waters (pH below 6·25), the climbing perch completely stops gill ventilation and only aerial respiration occurs. Thus it is clear that pH of water controls breathing in this fish. However, since a lowered water pH is correlated with increased CO_2 contents of water, it is difficult to decide whether this control is due to the changes in H^+ ions alone or is a combined effect of H^+ ions and the CO_2 content. Willmer (1934) also found that pH of water controls breathing in the yarrow, *Erythrinus erythrinus*, an air-breathing fish. He suggested that acidity of the water and the presence of carbonic anhydrase in the blood may assist the diffusion of CO_2 across the respiratory interface. It appears that in the climbing perch also acidity of the water enhances the transfer of CO_2 into the blood. Therefore it seems that at least low pH assists in the control of breathing by acting together with CO_2.

B. ROLE OF OXYGEN AND CARBON DIOXIDE IN THE CONTROL OF BREATHING

The patterns of respiration shown by *A. testudineus,* in waters having different concentrations of O_2 and CO_2 are plotted in Fig. 13. It is clear that this fish uses bimodal respiration in waters that have C_{CO_2} levels below 15

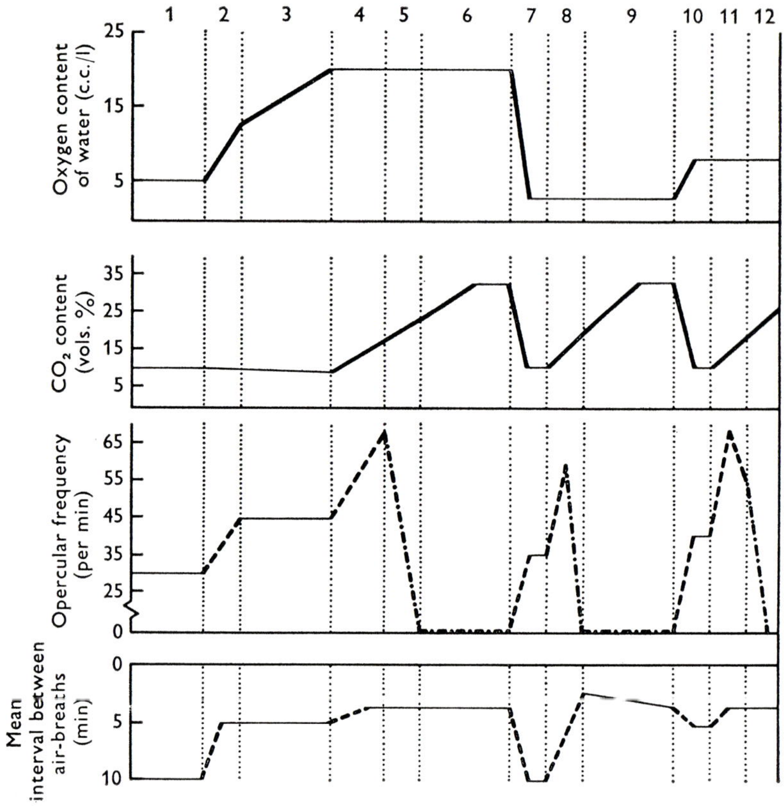

FIG. 11. Diagram to show the effect of changes in O_2 and CO_2 content of water (thick lines) on the frequencies of aquatic and aerial breathing in *A. testudineus*. The climbing perch shows a typical pattern of water and air breathing in water containing 0·7–5 ml O_2/l when the CO_2 content of water is normal (Sections 1, 7). Both water and air breathing frequencies increase when O_2 content of water is raised above 7 ml/l but the C_{CO_2} was lower than normal and free CO_2 was absent (Sections 2, 3, 10). A further rise in opercular and air-breathing frequencies occurs as total CO_2 content is raised above normal, hypercarbic waters (Sections 4, 8, 12). Fish shows "test water zone" if C_{CO_2} in water is raised between 18–22 Vol% (Sections 5, 8, 12). Mouth and opercula close tightly and only aerial respiration occurs if C_{CO_2} of the water is raised above 22 vol%, sections 6 and 9 (From Hughes and Singh, 1970b).

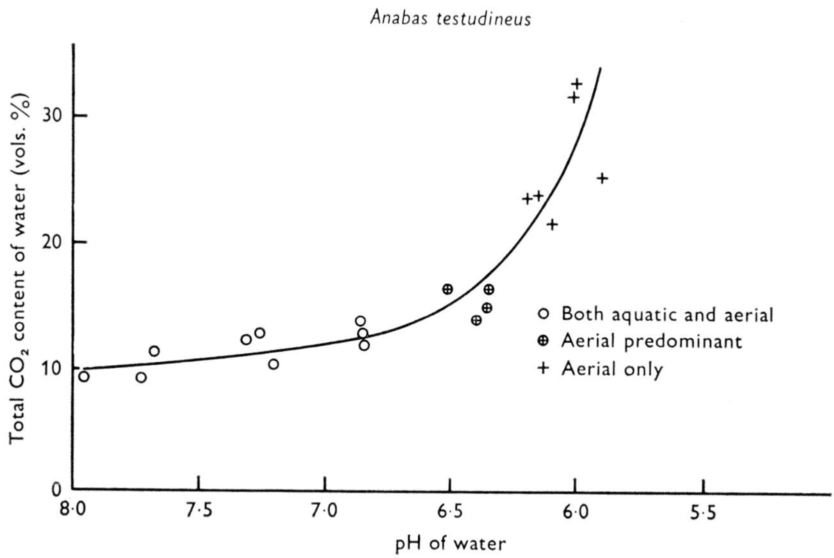

FIG. 12. Diagram to show the modes of respiration used by *Anabas* in relation to pH of water and C_{CO_2} (From Hughes and Singh, 1970b).

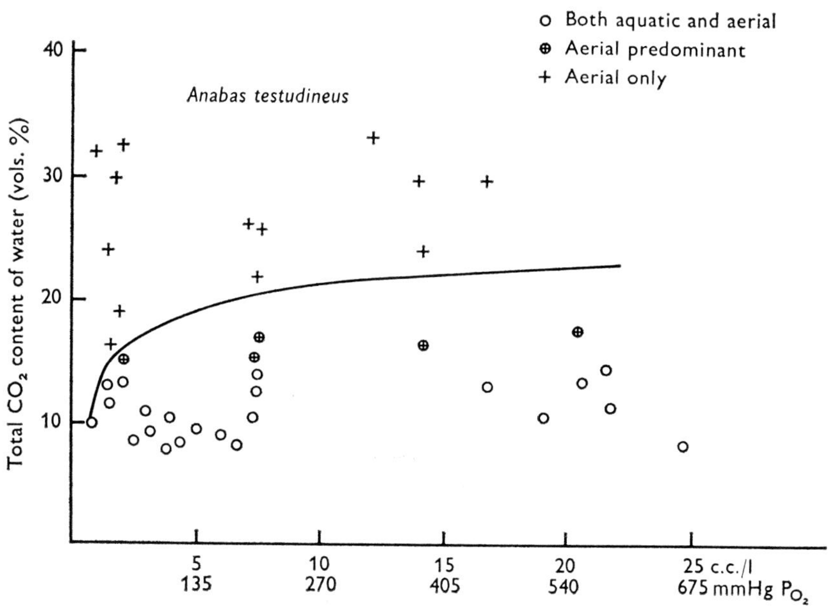

FIG. 13. Patterns of respiration used by *Anabas* in waters containing different O_2 and CO_2 concentrations. Aerial respiration predominates if CO_2 is raised above 15 vol% and only aerial respiration is seen if C_{CO_2} in water is raised above 22 vol%. For details see text (From Hughes and Singh, 1970b).

vol %. Above this level of CO_2 aerial respiration predominates (air-breathing frequencies increase) and gill ventilation is depressed. In waters containing C_{CO_2} above 20 vol %, only aerial respiration occurs. However, the level of total CO_2 content of water at which a complete switch over to aerial respiration occurs is also related to the O_2 contents of waters (Fig. 13). If P_{O_2} in the water is above 250 mmHg, the complete switch-over to aerial respiration occurs above 22 vol % C_{CO_2}, but in hypoxic water (P_{O_2} below 50 mmHg), the complete switch over to aerial respiration occurs at relatively lower CO_2 levels, 17 vol %.

Willmer (1934) found that yarrow (*E. erythrinus*) also stops gill ventilation altogether and continues only aerial respiration in hypercarbic waters (CO_2 content above 30 cc/1). Such a similarity in the respiratory behaviour of both yarrow and climbing perch, in highly hypercarbic waters, are to safeguard themselves from acidaemia because gill ventilation in such waters will certainly pose a very serious problem of the transfer of excess amounts of CO_2 from water into the blood across the respiratory epithelium. Therefore, it is advantageous for the fish to cease gill ventilation completely and exchange gases with air alone. In this way the climbing perch and yarrow can tolerate very high CO_2 in waters. However, yarrow is a facultative air-breather and it exchanges gases with water alone if P_{O_2} in the water is above 40 mmHg and CO_2 content is between 10–25 cc/l. Moreover, this fish used only aerial respiration if CO_2 content of water is below 5 cc/l. Like yarrow, another fresh water facultative air-breathing teleost, *Notopterus* (Indian fish inhabiting ponds, puddles and rivers) uses the air bladder for aerial respiration (Dehadrai, 1962). The respiratory patterns shown by this fish are very similar to the patterns used by yarrow in waters with different O_2 and CO_2 concentrations. *Notopterus* also uses only aerial respiration both in the highly hypercarbic (free CO_2 above 100 ppm) and also in hypocarbic (free CO_2 below 10 ppm) waters.

Willmer (1934) suggested that if the water CO_2 content is very low (below 5 cc/1) then yarrow ceases gill ventilation because a certain amount of CO_2 in the water is essential to stimulate movements of the gills. However, climbing perch did not cease gill ventilation completely in hypocarbic waters. But both the frequency and depth of gill ventilation were reduced and aerial respiration predominated if free CO_2 was absent. Moreover, these fishes often became restless in such waters and they even jumped out of the respirometer after some time. Therefore, it seems that for climbing perch also, the presence of a certain amount of free CO_2 in the water is essential to stimulate full ventilation of the gills. Also, in hypocarbic water which contain no free CO_2, there is a danger of excess CO_2 being lost from the blood into the water. This may clearly disturb the levels of CO_2 and pH in the blood and tissues.

From the foregoing account, it is clear that breathing in *A. testudineus* is controlled by O_2, CO_2 and pH of water. Like yarrow, and *Notopterus*, out of these three factors, CO_2 content of the water appears to be the most important in the control of breathing of the climbing perch living in water. C_{O_2}, C_{CO_2} and pH of the water also appear to control breathing in *Clarias* and *Heteropneustes*.

Therefore, a very interesting feature shown by the climbing perch, *A. testudineus* and other air-breathing fishes, is the balance between aquatic and aerial respiration for gas exchange. This is obviously advantageous for these fishes because they can satisfy their respiratory demands efficiently from the more suitable respiratory medium.

When *Anabas*, *Clarias* and *Heteropneustes* are exposed to air they breathe quietly for about 4–8h. On exposure to air, they are a little restless to begin with but quickly settle down to rest. The intervals between air breaths vary a great deal from species to species. However, they are usually shorter than when the fishes are living in waters with access to air. The frequency of air breathing increases when fishes become active. Air-breathing frequencies also increase if they are allowed an access to hypoxic air or N_2. Hypoxic gas and N_2 also act as a stimulus for increased air breathing in the lungfish, *Protopterus* (Johansen and Lenfant, 1968). Hypoxic gas also increases the rate of air breathing in the eel, *Electrophorus* (Johansen *et al.*, 1968). O_2 content of water and air also regulates breathing patterns in *Amphipnous cuchia* (Munshi and Singh, 1968b; and the mudskipper, *Periophthalmus vulgaris* (Singh and Munshi, 1968). In both the latter fishes, hypoxic gas and hypoxic water stimulate air breathing.

VI. Cardiac and Ventilatory Responses in Water and/or Air

It has been shown that *A. testudineus*, *H. fossilis* and *C. batrachus* regulate their breathing patterns and metabolic rates in response to environmental stresses. However, for regulating metabolism and the relative rates of gas exchange via gills and accessory respiratory organs, changes in the cardiac output may also be involved. Johansen (1971, 1972), and Johansen *et al.* (1970) have found that in the air-breathing fishes, blood is shunted past those parts of the respiratory system (i.e. gills, skin and accessory respiratory organs) which are not involved in exchange of gases during particular circumstances, because of the unavailability of the suitable respiratory medium. Singh and Hughes (1973) subjected the climbing perch to different environmental stresses by monitoring P_{O_2} and P_{CO_2} levels in water and air to study the effect on heart rate in relation to the changes in ventilation.

A. Cardiac Output and Ventilation Volume

Although the changes in heart rate and opercular frequencies accompanied by the changes in the amplitude of breathing are not certain guides to changes in cardiac output and ventilation volume, the balance of evidence (see Randall, 1970; Satchell, 1971; Johansen, 1971) suggests that changes in the above parameters certainly are correlated with changes in cardiac output and ventilation volume. Moreover, increases in the velocity of blood flow following an air breath has been observed in several air breathing fishes; for example in *Protopterus* (Johansen and Lenfant, 1968), *Electrophorus* (Johansen *et al.*, 1968) and *Amia* (Johansen *et al.*, 1970). Singh and Hughes (1973) observed marked tachycardia associated with air breaths in *Anabas* (Fig. 14). They suggested that in this fish tachycardia results in the increased cardiac output and marked bradycardia results in lowered cardiac output. Similar assumptions have also been made for the ventilation volume during increased and decreased rate and depth of breathing. Therefore, on the basis of these assumptions several derivations and possible conclusions have been drawn here for cardiac output and ventilation volume of the climbing perch. However, direct measurements of changes in cardiac output and ventilation volume during environmental stresses and in association with the air breaths would help elucidate the exact regulatory mechanisms of breathing and circulation in this fish.

1. *Cardiac and ventilatory responses of climbing perch living in water with free access to air*

The mean heart rate of the climbing perch during exchange of gases with normoxic water and air is 40 beats/min. The changes in the frequency of opercular movements and heart rate in relation to an air breath are shown in Fig. 15a. Tachycardia occurs within a few seconds following an air breath but heart rate returns to normal followed by bradycardia during the later phases of the inter-breath intervals. An air breath is usually repeated after the

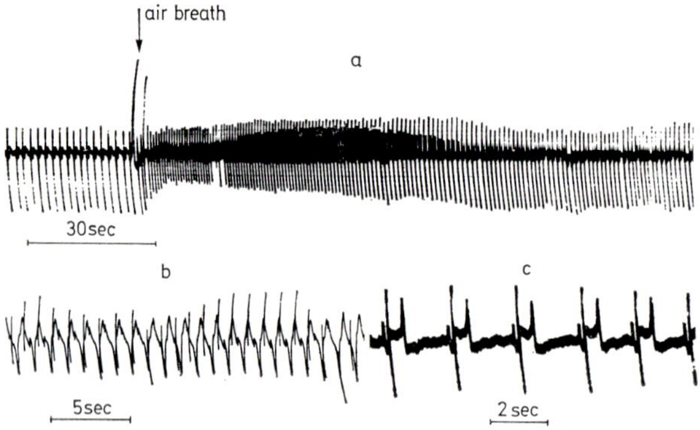

FIG. 14. Electrocardiogram (ECG) recording in *Anabas* showing occurence of tachycardia following an air-breath (a). ECG recording with gill ventilation movement artifacts showing changes in phase relationship between cardiac and ventilatory rhythms (b) and (c), faster recordings showing components of the ECG waveforms. (From Singh and Hughes, 1973).

heart rate has fallen to 15–30/min. The rate of decline of the heart beat from the initial tachycardia is related to the duration for which the fish continues gill ventilation following an air breath (Figs. 15a and 16).

The mean heart rate is reduced (lower than average 40 beats/min) but the mean opercular frequency is increased (more than 43/min) during exchange of gases with hypoxic water (P_{O_2} below 100 mmHg) and air (Figs. 15a and 16). The depth of breathing also increases in hypoxic waters specially when WP_{O_2} is reduced below 90 mmHg. However, it is interesting to find that in both normoxic and hypoxic waters air breath usually occurs once the heart rate has fallen between 15–30/min. Moreover, a 3–5 fold increase in the heart rate within a few seconds of taking an air breath is also similar in both normoxic and hypoxic waters (Fig. 16).

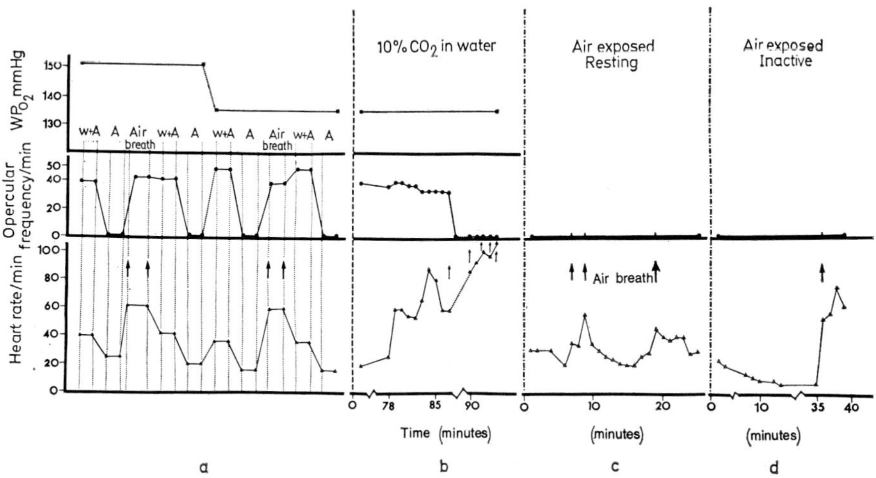

FIG. 15. Changes in cardiac and opercular frequencies of *Anabas* under different conditions. (a) Typical pattern of changes during different stages of gas exchange from water and air in normoxic water. Tachycardia is associated with the occurrence of an air-breath. Normal heart rate is found when exchanging gases with both water and air (w + A). Bradycardia occurs when aquatic breathing stops and gas exchange occurs only with hypoxic gases of the air chambers (A). (b) shows changes in heart rate and aquatic and aerial breathing in hypercarbic water. (c) and (d) show changes in heart rate in resting and inactive air-exposed climbing perch. Arrows indicate air-breaths. (From Singh and Hughes, 1973).

(a) *Effect of hypercarbic water and air*. The effect of hypercarbic waters are described elsewhere in this chapter. Figure 15b shows the changes in heart rate and breathing of one specimen when 10% CO_2 in air was equilibrated with water. It is interesting to find that after breathing hypercarbic water for some time, the fish stops gill ventilation and repeats an air breath almost every minute and tachycardia is maintained at a very high level (Fig. 15b).

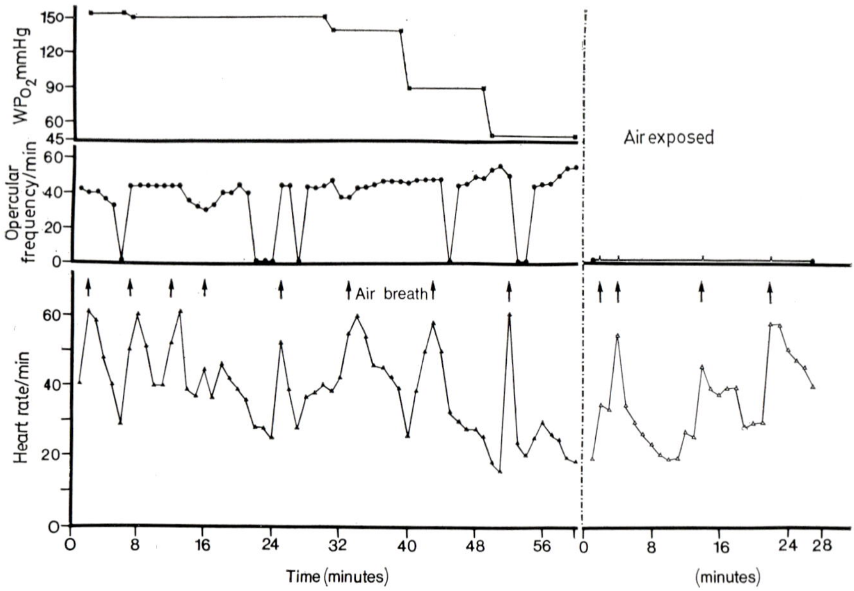

FIG. 16. Left section shows the time course and pattern of changes in heart rate of *Anabas* when living in normoxic and hypoxic waters with free access to air. The decline in cardiac rate following an air-breath is related with the duration of gill breathing. Pattern of changes in heart rate in resting air-exposed fish are shown for comparison. (From Singh and Hughes, 1973).

(b) *Heart rate and Ventilatory responses in water when surfacing is prevented.* In normoxic water (P_{O_2} = 154 mmHg) the ratio between breathing and perfusion is 1:1, since the mean cardiac and ventilatory frequencies are 42 and 43/min respectively. Bradycardia (19 beats/min) occurs in hypoxic water (P_{O_2} = 80 mmHg), but the ventilatory frequency (57/min) and also the depth of breathing increase considerably if surfacing is prevented. Hughes and Singh (1970a) found that $\dot{V}_{O_2}$ of *Anabas* is lower in hypoxic waters in spite of increases in both the frequency and amplitude of breathing. Therefore, the reduced $\dot{V}_{O_2}$ in hypoxic waters appears to be mainly due to the bradycardia (i.e. low cardiac output).

2. *Heart rate in air-exposed fish*

The average heart rate is lowered (30 beats/min) in resting, air-exposed climbing perch. However in air-exposed fish the heart rate varies a great deal in relation to the state of activity (Fig. 15c,d).

A comparison of the response of heart in air-breathing fishes, living in water and after exposure to air, is made in Table VIII. Like most of the fishes *Anabas* shows bradycardia on exposure to air.

TABLE VIII

A comparison of the heart rate, of some air-breathing fishes, when living in water and when exposed to air

T °C	Species	Heart rate in water (beats/min.)	Heart rate in air (beats/min.)	Reference
15	*Anguilla vulgaris*	45	25	Berg and Steen (1965)
21–24	*Anguilla japonica*	32	—	Itazawa (1970)
	Periophthalmus australis	46	108	Garey (1962)
	Leuresthes tenuis	84	16	Garey (1962)
	Cypseluneus californiens	84–92	26	Garey (1962)
24	*Periophthalmus sobrinus*	70–125	No indication of bradycardia	Gordon *et al.* (1969)
14	*Sicyases sanguineus*	60–80	40–50	Gordon *et al.* (1970)
25	*Anabas testudineus*	40 (20–70)	30 (15–70)	Singh and Hughes (1973)

The changes in cardiac rate following an air breath are very similar to the pattern of changes associated with an air breath in fish living in water (Figs. 15 and 16). Air-exposed fish normally repeat air breaths when the cardiac rate has dropped between 15–25/min. However, in air-exposed fish, the lowering in heart rate from the initial tachycardia following an air breath is usually slow and gradual compared to a fish living in water (Fig. 16). Hughes and Singh (1970a) have shown that the metabolic rate in resting air-exposed fish is lower compared to the fish exchanging gases with both water and air. Therefore, bradycardia observed in air-exposed fish during resting conditions suggests that lowering in $\dot{V}_{O_2}$ on exposure to air might be due to the low cardiac output (smaller amount of blood perfusing the air chambers).

B. RELATIONSHIP BETWEEN TACHYCARDIA, BRADYCARDIA AND O_2 AVAILABILITY IN AIR CHAMBERS

The three to five fold increase in heart rate immediately following an air-breath suggests that tachycardia is associated with the greater availability of oxygen in the air chambers. Tachycardia may increase the cardiac output so that a larger amount of the blood perfuses air-breathing organs for an efficient gaseous exchange. The lowering in the heart rate (bradycardia) during the interval between air breaths is associated with a decline in P_{O_2} and elevation of P_{CO_2} in the suprabranchial chambers (see Fig. 9). Bradycardia also occurs in hypoxic waters (P_{O_2} below 100 mmHg). However, tachycardia is observed in hypoxic water (P_{O_2} = 60 mmHg) during the outbursts of activity, when the fish is prevented from surfacing (Singh and Hughes, 1973). But after about 1h of active search for air, the heart rate fails to increase even after violent movements of fishes in search for air. This suggests that O_2 absorbed from hypoxic water alone is insufficient to meet the metabolic needs especially during activity and hence the lack of additional O_2 prevents the occurrence of tachycardia even after the violent movements. On the other hand tachycardia occurs in air-exposed active fishes for longer periods (several hours) and they never fail to show 2–5 fold increase in heart rate immediately following an air-breath (Singh and Hughes, 1973). This also suggests that tachycardia is associated with high O_2 availability for gaseous exchange.

Moreover, the dramatic 2–5 fold increase in heart rate following an air-breath further suggests that an initial tachycardia is due to the reflex action but after a few seconds it is mediated through nervous control following inputs from chemo/or baro-receptors. Although the nervous supply to the heart in the climbing perch has not been studied, it is well known that the teleost heart is supplied by branches of the vagus nerve (Randall, 1970). Moreover, Randall (1966) reported that the cardiac vagus contains cholinergic fibres. Therefore, the occurrence of tachycardia following an air-breath appears to be due to the reduced vagal inhibition whereas the bradycardia will result from increased vagal tone. Such a nervous control also appears to involve inputs from sources such as vascular baro- and chemoreceptors, located on the gills or in the buccal and suprabranchial cavities (Singh and Hughes, 1973).

C. POSSIBLE ROLE OF THE BLOOD P_{O_2}, P_{CO_2} AND pH IN CONTROL OF BREATHING AND PERFUSION

The vascular supply to the respiratory organs of *A. testudineus* is shown in Fig. 17. One of the very striking features of its circulatory arrangement is that the dorsal aorta is formed from the efferent branchial arteries of 3rd and 4th gill arches only. These gill arches have few secondary lamellae, hence they are not efficient in gaseous exchange (Munshi, 1968; Hughes and Munshi, 1973).

The labyrinthine organs and the respiratory membranes receive their afferent blood supply from the efferent branchial arteries of the 1st and 2nd gills only. Oxygenated blood from the accessory organs is returned to the heart via jugular veins. Therefore, the oxygenated blood gets mixed with the venous return before it enters the dorsal aorta for distribution to the body tissues. Thus, the tissues in *Anabas* receive mixed blood during both conditions, i.e. when living in water and exchanging gases with water and air, and when exposed to air. The circulation in climbing perch ensures that this mixed

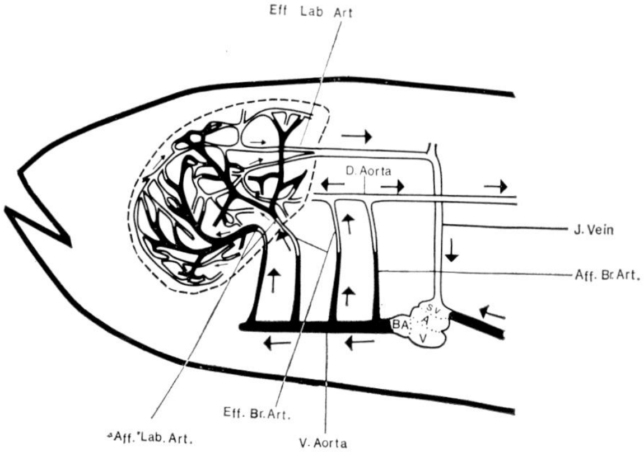

FIG. 17. Diagram of general circulation through gills and accessory respiratory organs (labyrinthine organs and respiratory membrane). Arrows indicate the direction of flow. (For details see the text). A, Atrium; BA, Bulbus aretriosus; SV, Sinus venosus; V, Ventricle, Aff. Br. Art. Afferent branchial artery, Aff. Lab. Art. Afferent labyrinthine artery. Eff. Br. Art. Efferent branchial artery; Eff. Lab. Art. Efferent labyrinthine artery; D.A. Dorsal aorta, V.A. Ventral aorta; J. Vein, Jugular vein. (From Singh and Hughes, 1973).

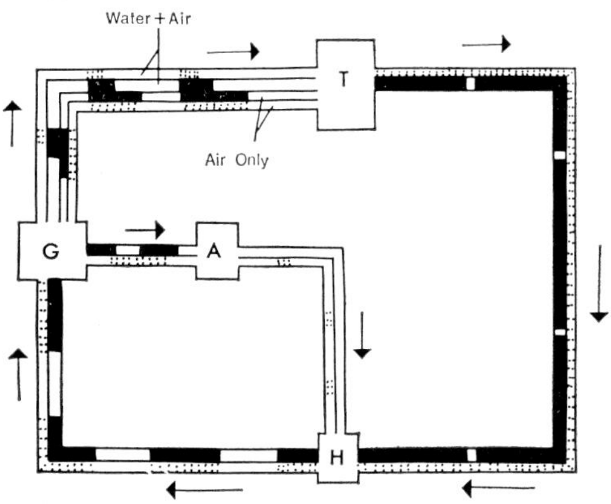

FIG. 18. Schematic representation of circulation in *Anabas* in relation to aquatic and aerial gas exchange. The approximate amount of oxygenation and deoxygenation of blood is represented with white and black respectively and the stippled area represent the relative amount of CO_2 content. Note that tissues are supplied with mixed blood which has different amounts of O_2 and CO_2 when exchanging gases with water and air (I) and when exposed to air only (II). Arrows indicate the direction of the blood flow. A. air chambers; G. gills; H. heart; T. tissues. (Fom Singh and Hughes, 1973).

blood can never reach full saturation *in vivo*. Moreover, it will have a high P_{CO_2} (Figs. 18 and 9).

Comparison of the gaseous composition in the suprabranchial chambers of *A. testudineus* and the alveoli of human lung shows striking similarities in P_{O_2} (Table IX). P_{CO_2} levels in the suprabranchial chambers also appear to show similar trends (Table IX and Fig. 9). We have already seen from Fig. 9

TABLE IX

Comparison between the partial pressure of the gases in the lung of man and the suprabranchial chambers of Anabas testudineus. *The values given for the climbing perch are very close to the average values obtained from analyses of several spontaneous air breaths of fishes living in water. However, when exposed to air* P_{CO_2} *in air chambers of the climbing perch usually rises to an average value of 34·0 mmHg; when air-chamber gas and the blood appear to be in equilibrium (for details see Fig. 9). Data for man are from Langley (1971). (From Singh and Hughes, 1973)*

Gas	Suprabranchial chamber air (climbing perch) at 25°C		Alveolar air (man) at 37°C		Atmospheric air dry	
	Vols %	mmHg	Vols %	mmHg	Vols %	mmHg
O_2	13·99	103·11	14·2	101·0	20·94	159·1
CO_2	2·64	19·50	5·5	39·0	0·04	0·3
N_2	83·37	614·25	80·3	573·00	79·02	600·6
	100·00	**736·86**	**100·00**	**713·00**	**100·00**	**760·00**

that P_{CO_2} in the blood and air chambers of *Anabas* reach equilibrium usually around 30·0 mmHg or more. Douglas, Haldane and others (Daly, 1968) have shown that in mammals hyperventilation results in an excessive washing-out of CO_2 from the lungs. This finally lowers blood P_{CO_2} with the consequent result that blood pH and the cerebrospinal fluid (CSF) pH are increased. It is well known that mammalian respiratory centres are highly sensitive to increased pH of the CSF which ultimately depresses breathing by either complete cessation of ventilation or by periodic breathing until normal levels of C_{CO_2}, P_{CO_2} and pH in the blood and CSF are restored. *Anabas* stops water breathing within a short time (seconds to a few minutes) following an air-breath. We have seen that in *Anabas* the gills are the main sites for CO_2 loss but during the early part of breath-holding appreciable amounts of CO_2 are also exchanged ($R = 0{\cdot}4-1{\cdot}5$) in the air chambers. Therefore it is suggested that the pattern of intermittent water and air breathing in *Anabas* is very similar to the forced periodic breathing of mammals since in both cases this results from an excessive loss of CO_2 during hyperventilation.

However, cessation of water breathing in *Anabas* partly appears to be due to the ready availability of O_2 in the air chambers. But from Figs. 17 and 18, it is clear that full saturation of the blood may not result in *Anabas* exchanging gases with air alone or with both water and air. Therefore, most probably greater availability of O_2 in the air chambers alone may not result in the complete cessation of water breathing. Moreover, climbing perch starts gill ventilation prior to an air-breath which appears mainly to remove accumulated CO_2 in the blood before the O_2 demand is satisfied by repeating the air-breath (see Fig. 19). This appears to suggest that *Anabas* has a double

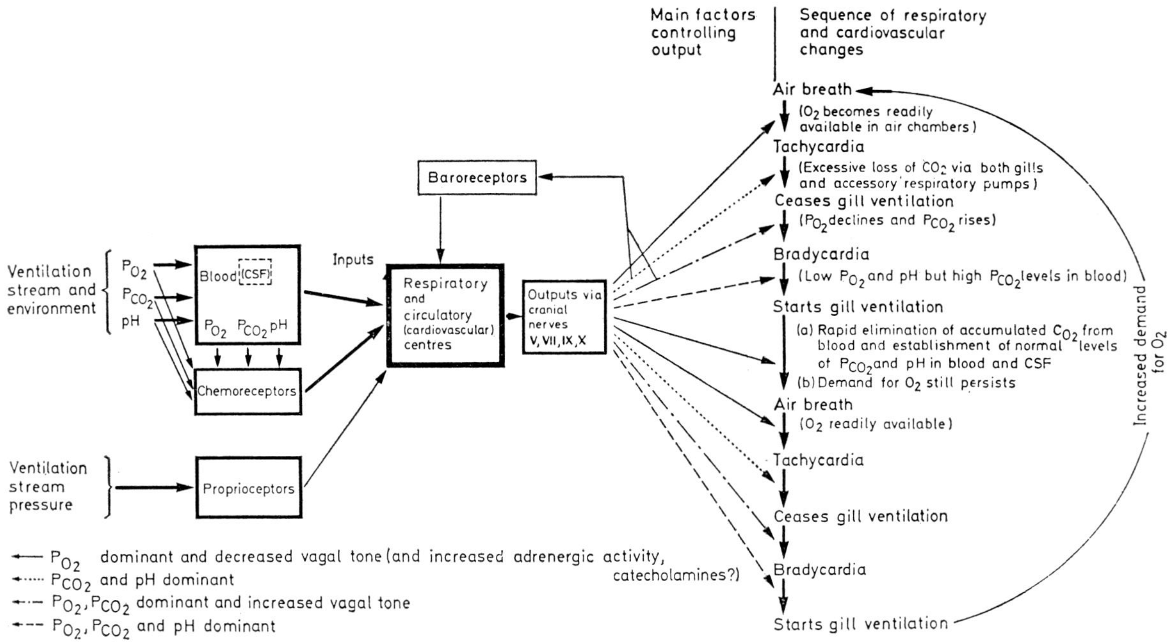

FIG. 19. Schematic representation of regulation of ventilation and perfusion in *Anabas*. Respiratory pattern is controlled by the respiratory centre and the heart rate by changes in vagal tone. Both these components in turn are stimulated by the levels of P_{O_2}, P_{CO_2} and pH in the blood, CSF and the chemo- and baro- receptors. The results produced by changes in breathing pattern and cardiac rate are also given for each step. For details see text. (Fom Singh and Hughes, 1973).

pathway to the brain, or elsewhere sensitive to C_{O_2}, P_{O_2}, C_{CO_2}, P_{CO_2} and pH to control both, the pattern of respiration and perfusion of blood in relation to variations in rates of gas exchange from water and air. Measurement of P_{O_2}, P_{CO_2} and pH of arterial venous blood of *Anabas*, in relation to air breath, would be most useful in solving such a complicated mechanism. The nervous control mechanisms of respiration in fish have been discussed in detail by Hughes and Shelton (1962) and Shelton (1970). Similar mechanisms may be involved in *Anabas* for the control of breathing. Tachycardia and bradycardia appear to be mediated via the vagus nerve as discussed earlier. Singh and Hughes (1973) proposed a diagram summarizing the possible mechanisms for the regulation of breathing and perfusion in *Anabas* (Fig. 19). This proposed scheme appears to be in very good agreement with the possible derivations on ventilation, perfusion and gas exchange in *Anabas* described in detail in this chapter.

Acknowledgements

I am very much indebted to Prof. G. M. Hughes, Professor of Zoology and Head of the Research Unit for Comparative Animal Respiration at University of Bristol, for his inspiration and constant encouragement which helped me in writing this chapter. The author also thanks Prof. J. P. Thapliyal, Professor of Zoology, at Banaras Hindu University, Varanasi, for his help and hospitality during my stay at B.H.U. while writing this work.

References

Berg, T. and Steen, J. B. (1965). Physiological mechanisms for aerial respiration in the eel. *Comp. Biochem. Physiol.* **15**, 469–484.

Berg, T. and Steen, J. B. (1966). Regulation of ventilation in eels exposed to air. *Comp. Biochem. Physiol.* **18**, 511–516.

Carter, G. S. (1957). Air breathing. *In* "The Physiology of Fishes". (Ed. M. E. Bown). Vol. I, pp. 65–79. Academic Press, New York, London and San Francisco.

Calugareanu, D. (1907). Die Darmatumung von *Cobitis fossiles*. II. Mitteilung: Uber den Gaswechsel. *Arch. Ges. Physiol.* **120**, 425–450.

Daly, M. De Burgh (1968). The regulation of breathing. *In* "Winton and Bayliss, Human Physiology." (Eds O. C. Lippold and F. R. Winton). pp. 188–215. J. and A. Churchill, London.

Das, B. K. (1927). The bionomics of certain air-breathing fishes of India, together with an account of the development of their air-breathing organs. *Phil. Trans. Roy. Soc. Lond.* **216**, 183–219.

Das, B. K. (1933). On the bionomics, structure and physiology of respiration in an estuarine air-breathing fish *Pseudapocryptes lanceolatus*. With special reference to a new mode of aerial respiration. *Curr. Sci.* **I**, 389–393.

Dehadrai, P. V. (1962). Respiratory function of the swimbladder of *Notopterus* (Lacépède). *Proc. Zool. Soc. Lond.* **139**, 341–357.

Farber, J. and Rahn, H. (1970). Gas exchange between air and water and the ventilation pattern in the electric eel. *Respir. Physiol.* **9**, 151–161.

Fish, G. R. (1956). Some aspects of respiration of six species of fish from Uganda. *J. exp. Biol.* **33**, 186–195.

Garey, W. F. (1962). Cardiac responses of fishes in asphyxic environments. *Biol. Bull.* **122**, 362–368.

Garey, W. F. and Rahn, H. (1970). Normal arterial gas tensions and pH and the breathing frequency of the electric eel. *Resp. Physiol.* **9**, 141–150.

Gordon, M. S., Boëtius, H., Evans, D. H., McCarthy, R. and Oglesby, L. C. (1965). Physiology of the amphibious habit in mudskipper fish (Periophthalmus sobrinus). Abstr. Paper, 23*rd Int. Congr. Physiol. Sci., Tokyo.* p. 169.

Gordon, M. S., Boëtius, J., Evans, D. H., McCarthy, R. and Oglesby, L. C. (1969). Aspects of the physiology of terrestrial life in amphibious fishes. I. The mudskipper *Periophthalmus Sobrinus. J. exp. Biol.* **50**, 141–149.

Gordon, M. S., Fisher, S. and Edurdo, T. S. (1970). Aspects of the physiology of terrestrial life in amphibious fishes. II. The Chilean clingfish *Sicyases sanguineus. J. exp. Biol.* **53**, 559 572.

Holeton, G. F. and Randall, D. J. (1967). The effect of hypoxia upon the partial pressure of gases in the blood and water afferent and efferent to the gills of rainbow trout. *J. exp. Biol.* **46**, 317–327.

Howell, B. J. (1970). Acid-base balance in transition from water to air breathing. *Federation Proc.* **29**, 1130–1134.

Howell, B. J., Baumgardner, W. F., Bondi, K. and Rahn, H. (1970). Acid-base balance in cold-blooded vertebrates as a function of body temperature. *Am. J. Physiol.* **218**, 600–606.

Hughes, G. M. (1966). Evolution between air and water. *In* "Ciba Foundation Symposium On Development of the Lung". (Eds A. V. S. de Reuck and R. Porter). pp. 64–80. Churchill Ltd., London.

Hughes, G. M. and Munshi, J. S. D. (1968). Fine structure of the respiratory surfaces of an air-breathing fish, the climbing perch *Anabas testudineus* (Bloch). *Nature, Lond.* **219**, 1382–1384.

Hughes, G. M. and Munshi, J. S. D. (1973). Nature of the air-breathing organs of the Indian fishes *Channa, Amphipnous, Clarias* and *Saccobranchus*, as shown by electron microscopy. *J. Zool. Lond.* **170**, 245–270.

Hughes, G. M. and Shelton, G. (1962). Respiratory mechanisms and their nervous control in fish. *In* "Advances in Comparative Physiology and Biochemistry". (Ed. O. Lowenstein) **1**, 275–364. Academic Press, London and New York.

Hughes, G. M. and Singh, B. N. (1970a). Respiration in an air-breathing fish, the climbing perch *Anabas testudineus* Bloch. I. Oxygen uptake and carbon dioxide release into air and water. *J. exp. Biol.* **53**, 265–280.

Hughes, G. M. and Singh, B. N. (1970b). Respiration in an air-breathing fish, the climbing perch *Anabas testudineus*. II. Respiratory patterns and the control of breathing. *J. exp. Biol.* **53**, 281–298.

Hughes, G. M. and Singh, B. N. (1971). Gas exchange with air and water in an air-breathing catfish *Saccobranchus fossilis. J. exp. Biol.* **55**, 667–682.

Hughes, G. M. and Singh, B. N. Respiratory properties of the blood of *Clarias batrachus* (Unpublished observations).

Hughes, G. M., Singh, B. R., Guha, G., Dube, S. C. and Munshi, J. S. D. (1974). Respiratory surface areas of an air-breathing siluroid fish *Saccobranchus fossilis* in relation to body size. *J. Zool. Lond.* **172**, 215–232.

Hutchison, V. H., Whitford, W. G. and Kohl, M. (1968). Relation of body size and surface area to gas exchange in anurans. *Physiol. Zool.* **41**, 65–85.

Itazawa, Y. (1970). Heart rate, cardiac output and circulation time of fish. *Bull. Jap. Soc. Sci. Fish.* **36**, 926–931.

Johansen, K. (1966). Air breathing in the teleost, *Symbranchus marmoratus*. *Comp. Biochem. Physiol.* **18**, 383–395.

Johansen, K. (1970). Air breathing in fishes. *In* "Fish Physiology" (Eds W. S. Hoar and D. J. Randall), Vol. IV, pp. 361–411. Academic Press, New York and London.

Johansen, K. (1971). Comparative physiology: Gas exchange and circulation in fishes. *Ann. Rev. Physiol.* **33**, 569–611.

Johansen, K. (1972). Heart and circulation in gill, skin and lung breathing. *Resp. Physiol.* **14**, 193–210.

Johansen, K. and Lenfant, C. (1967). Respiratory function in the south American lungfish, *Lepidosiren paradoxa* (Fitz). *J. exp. Biol.* **46**, 205–218.

Johansen, K. and Lenfant, C. (1968). Respiration in an African lungfish *Protopterus aethiopicus*. *J. exp. Biol.* **49**, 453–468.

Johansen, K., Lenfant, C. and Grigg, G. C. (1967). Respiratory control in the lungfish, *Neoceratodus forsteri* (Krefft). *Comp. Biochem. Physiol.* **20**, 835–854.

Johansen, K., Lenfant, C., Schmidt-Nielsen, K. and Petersen, J. A. (1968). Gas exchange and control of breathing in the electric eel, *Electrophorus electricus*. *Zeitschr. Vergl. Physiol.* **61**, 137–163.

Johansen, K., Hansen, D. and Lenfant, C. (1970). Respiration in a primitive air-breather, *Amia calva*. *Resp. Physiol.* **9**, 162–174.

Lahiri, S., Shub, C. and Fishman, A. P. (1968). Respiratory properties of blood of the African lungfish (Protopterus). *Proc. 24th Intern. Union Physiol. Sci.* Washington. **7**, 251.

Langley, L. L. (1971). Diffusion and transport in respiratory gases. *In* "Review of Physiology" pp. 405–421. McGraw Hill Book Co., New York.

Lenfant, C. and Johansen, K. (1967). Respiratory adaptations in selected amphibians. *Resp. Physiol.* **2**, 247–260.

Lenfant, C. and Johansen, K. (1968). Respiration in the African lungfish, *Protpterus aethiopicus*. I. Respiratory properties of blood and normal patterns of breathing and gas exchange. *J. exp. Biol.* **49**, 437–452.

Lenfant, C. and Johansen, K. (1972). Gas exchange in gill, skin and lung breathing. *Resp. Physiol.* **14**, 211–218.

Lenfant, C., Johansen, K. and Grigg, G. C. (1967). Respiratory properties of blood and pattern of gas exchange in the lungfish *Neoceratodus forsteri* (Krefft). *Resp. Physiol.* **2**, 1–12.

Lenfant, C., Johansen, K. and Hansen, D. (1970). Bimodal gas exchange and ventilation-perfusion relationship in lower vertebrates. *Federation Proc.* **29**, 1124–1129.

Liem, K. F. (1967). Functional morphology of the Integumentary, Respiratory and Digestive systems of the Synbranchoid Fish *Monopterus albus*. *Copeia*, **2**, 375–380.

McMahon, B. R. (1970). The relative efficiency of gaseous exchange across the lungs and gills of an African lungfish *Protopterus aethiopicus*. *J. exp. Biol.* **52**, 1–15.

Moussa, T. A. (1957). Physiology of the accessory respiratory organs of the teleost, *Clarias lazera* (C & V). *J. exp. Zool.* **136**, 419–454.

Munshi, J. S. D. (1961). The accessory respiratory organs of *Clarias batrachus* (Linn). *J. Morphol.* **109**, 115–140.

Munshi, J. S. D. (1962a). On the accessory respiratory organs of *Heteropneustes fossils*. *Proc. Roy. Soc. Edin.* **B68**, 128–146.

Munshi, J. S. D. (1962b). On the accessory respiratory organs of *Ophiocephalus punctatus* (bloch) and *Ophiocephalus striatus* (bloch). *Linn. Soc. Lond. Zool.* **44**, 616–626.

Munshi, J. S. D. (1968). The accessory respiratory organs of *Anabas testudineus* (BL), Anabantidae, Pisces, *Proc. Linn. Soc. Lond.* **179**, 107–127.

Munshi, J. S. D. and Singh, B. N. (1968a). A study of the gill-epithelium of certain freshwater teleostean fishes with special reference to the air-breathing fishes. *J. Zoot. Ind.* **9**, 91–107.

Munshi, J. S. D. and Singh, B. N. (1968b). On the respiratory organs of *Amphipnous cuchia* (Ham. Buch.). *J. Morph.* **124**, 423–444.

Piiper, J. and Baumgarten-Schumann, D. (1968). Carriage of O_2 and CO_2 by water and blood in gas exchange of the dogfish. (*Scyliorhinus stellaris*). *Resp. Physiol.* **5**, 326–337.

Rahn, H. (1966a). Gas transport from the external environment to the cell. *In* "Ciba Foundation Symposium on Development of the Lung" (Eds A. V. S. de Reuck and R. Porter) pp. 3–23. J. A. Churchill Ltd., London.

Rahn, H. (1966b). Aquatic gas exchange: Theory. *Resp. Physiol.* **1**, 1–12.

Rahn, H. and Baumgardner, F. W. (1972). Temperature and acid-base regulation in fish. *Resp. Physiol.* **14**, 171–182.

Rahn, H., Rahn, K. B., Howell, B. J., Gans, C. and Tenney, S. M. (1971). Air breathing of the garfish (*Lepisosteus osseus*). *Resp. Physiol.* **11**, 285–307.

Randall, D. J. (1966). The nervous control of cardiac activity in the tench *Tinca tinca* and the goldfish (*Carassius auratus*). *Physiol. Zool.* **34**, 185–192.

Randall, D. J. (1970). The Circulatory System. *In* "Fish Physiology" (Eds W. S. Hoar and D. J. Randall) Vol. IV, pp. 133–172. Academic Press, New York and London.

Satchell, G. H. (1971). "*Circulation in Fishes*" Cambridge Monographs in Experimental Biology, C.U.P.

Sawaya, P. (1946). Sobre abiologia de alguns peixes de respiracao aerea. *Univ. Sao Paulo, Fac. Filosof., Cienc., letras, Zool.* **11**, 255–286.

Schöttle, E. (1931). Morphologie und Physiologie der Atmung bei wasser schlam und landlebenden Gobüformes. *Z. Wiss. Zool.* **140**, 1–114.

Shelton, G. (1970). The regulation of breathing. *In* "Fish Physiology" (Eds W. S. Hoar and D. J. Randall) Vol. IV, pp. 293–359. Academic Press, New York and London.

Singh, B. N. and Hughes, G. M. (1971). Respiration of an air-breathing catfish *Clarias batrachus* (Linn.). *J. exp. Biol.* **55**, 421–434.

Singh, B. N. and Hughes, G. M. (1973). Cardiac and respiratory responses in the climbing perch, *Anabas testudineus*. *J. Comp. Physiol.* **84**, 205–226.

Singh, B. N. and Hughes, G. M. Respiratory properties of the blood of *Heteropneustes fossilis* (unpublished observations).

Singh, B. N. and Munshi, J. S. D. (1968). On the respiratory organs and mechanics of breathing in *Periophthalmus vulgaris*. *Zool. Anz.* **183**, 92–110.

Spurway, H. and Haldane, J. B. S. (1963). The regulation of breathing in a fish, *Anabas testudineus*. *In* "The Regulation of Human Respiration". (Eds D. J. C. Cunningham and B. B. Lloyd) pp. 431–434. Oxford: Blackwell.

Whitford, W. A. and Hutchison, V. H. (1963). Cutaneous and pulmonary gas exchange in the spotted salamander, *Ambystomia maculatum*. *Biol. Bull.* **124**, 344–354.

Whitford, W. A. and Hutchison, V. H. (1965). Gas exchange in salamanders. *Physiol. Zool.* **38**, 228–242.

Willmer, E. N. (1934). Some observations on respiration of certain tropical freshwater fishes. *J. exp. Biol.* **11**, 283–306.

Respiratory Adaptations of Marine Air-breathing Fishes

JEFFREY B. GRAHAM

Smithsonian Tropical Research Institute,
P.O. Box 2072, Balboa, Canal Zone

I. Introduction

Reviews of air-breathing in fishes generally place their major emphasis on descriptions of the diverse respiratory adaptations found among many freshwater species and, except for Hora's (1935) early paper, there have been no comparisons of the air-breathing adaptations of marine and freshwater species. The literature also lacks a substantive comparison of the requisite adaptations for gas exchange in amphibious and aquatic air-breathing fishes. (Amphibious air-breathing refers to the gas exchange of fish out of water. Aquatic air-breathers are those fish that remain in water but gulp air at the surface.)

The respiratory adaptations of marine air-breathing fishes have not been well studied and this is partly attributable to their small size, inconspicuous behaviour, and occurrence in isolated, sometimes formidable, habitats where observation, let alone capture, is difficult. Another factor that has discouraged research with marine air-breathers is that the little known about them suggests that their respiratory adaptations are simpler and less refined than those in even the least specialized freshwater air-breathers. Indeed, when air-breathing fishes are classified on the basis of their structural specializations for air breathing (Hora, 1935; Carter, 1957; Bertin, 1958; Saxena, 1963; Johansen, 1970) marine forms, in respiring aerially with no more than gills, modified buccal or pharyngeal epithelium, or possibly a vascularized skin, are invariably relegated to the ranks of the least specialized air-breathing fishes.

The objectives of this paper are to report on the variety of marine fishes that rely upon air-breathing, to review what is known about their respiratory specializations, and to indicate important ways that their morphological and physiological adaptations for air-breathing differ from those of freshwater species. Particular emphasis will be placed on amphibious marine air-breathers. Factors that have influenced the evolution of air breathing in marine and freshwater fishes are elucidated. Neglected areas of research and subjects that need to be developed in future studies are also discussed.

II. Occurrence of Air Breathing Among Marine Fishes

A survey of the literature on air-breathing and amphibious fishes has produced a list of 40 marine species, in six families, that are known to (or, as indicated by their behaviour, probably do) regularly breathe air (Table I).

This census cannot be considered complete as some species have doubtlessly been overlooked. Also, a few of the species' names may now be outdated, although an attempt was made to check the validity of each name, and some of the species names may actually refer to one and the same species. This seems particularly likely for the mudskippers (*Boleophthalmus*, *Periophthalmus*, and *Periophthalmadon*).

For the Blenniidae in Table I, there are many more species in each of the listed genera (e.g., in *Andamia* there are 4–5 species; *Alticus* 8–10; *Entomacrodus* 22–24; *Istiblennius* 25–30; and *Salarias* 8–12; cf. Smith–Vaniz and Springer, 1971) than are known to be amphibious. Further study of these fishes should indicate more air-breathing species and will most likely demonstrate graded degrees of amphibious adaptation and air breathing capability similar to that pointed out for *Andamia heteroptera*, *A. reyi*, and *Salarias* sp. by Rao and Hora (1938), and for several genera of blennies by Zander (1972).

The vast majority of air-breathing blennies and gobies in Table I are amphibious in mangroves or on rocky shores, although some occur on mudflats. Some species are more highly adapted for air breathing and amphibious life than others (MacNae, 1968b; Graham, 1973). The clingfish (Gobiesocidae) are secretive, inconspicuous inhabitants of the intertidal zone and commonly endure exposure to air during low tide. Only one species, the tarpon *Megalops atlantica* is pelagic and occurs in open water. Young tarpon are commonly observed to breathe air in estuaries, mangrove lagoons, and anoxic ponds (Böhlke and Chaplin, 1968).

Table I clearly indicates the unevenness of our knowledge about the physiology of marine air-breathing fishes. While much is known about some species, others, for example, the air-breathing Blenniidae, remain virtually unstudied. The tarpon, which has long been noted for its conspicuous surface-rolling behaviour during air-breathing, has not been studied since the account of social control of its air-breathing was given in 1940 by Shlaifer and Breder.

TABLE I

Marine fishes that have been determined to be air breathers through their natural behaviour (e.g., regular air gulping or amphibious), their ability to survive in hypoxic water, or through direct measurement. Only one reference, that dealing most directly with the species' respiratory adaptations for air breathing is given. (Many species were found to be listed without any discussion and thus only a general reference is available)

Family	Species	Habitat	Air-breathing Organs[a]	Remarks	Reference
Gobiidae	*Periophthalmus sobrinus*	mangrove swamps and rocky intertidal	G, BP, S	all species of *Periophthalmus* are amphibious	Gordon *et al.*, 1969
	P. vulgaris	mangroves, mud flats	G, BP, S		Singh and Munshi, 1968
	P. kalolo	mangroves, mud flats	G, BP, S	synonym with *koelreuteri*	Macnae, 1968a
	P. chrysospilos	mangroves, mud flats	G, BP, S	climbs trees at high tide to stay out of water	Macnae, 1968a
	P. papilio	mangroves, mud flats	G, BP, S	amphibious, also gulps at surface	Willem and Boelaert, 1937
	P. dipus	mangroves, mud flats	G, BP, S		Schöttle, 1932
	P. schlosseri	mangroves, mud flats	G, BP, S		Schöttle, 1932
	Boleophthalmus boddaerti	soft muds	G, BP, S	less amphibious than *Periophthalmus*	Schöttle, 1932
	B. viridis	mangrove channels	G, BP, S	less amphibious than *Periophthalmus*	Schöttle, 1932

TABLE I (*Continued*)

Family	Species	Habitat	Air-breathing Organs[a]	Remarks	Reference
	Periophthalmadon australis	mangroves	G, BP, S	highly amphibious, exhibits bradycardia in water and in hypoxic burrows	Garey, 1962
	Scartelaos viridis	mangroves	G, BP, S	amphibious	Macnae, 1968b
	Pseudapocryptes lanceolatus	estuaries	G, BP	gulps air at surface, occasionally out of water	Das, 1934
	Taenioides rubicundus	estuaries	G, BP	gulps at surface	Hora, 1935
	Apocryptes bato	estuaries	G, BP	gulps at surface	Hora, 1935
	Pisoodonophis boro	estuaries	G, BP	gulps at surface	Hora, 1935
	Gobionellus saggitula	mud flats	G, S	gulps at surface	Todd, in preparation
	Gillichthys mirabilis	tide lagoons, estuaries	BP, GB	amphibious	Todd and Ebeling, 1966
	G. seta	tide lagoons, estuaries	BP	less specialized for air-breathing than *G. mirabilis*	Todd and Ebeling, 1966
	Quietula guaymasiae	tidal flats	BP (weakly vascularized)	gulps at surface, holds air in mouth at surface	Todd and Ebeling, 1966
Blenniidae	*Andamia heteroptera*	rocky shores	G, S	amphibious	Rao and Hora, 1938
	A. reyi	rocky shores	G, S	less amphibious than heteroptera	Rao and Hora, 1938

	Salarias sp.	rocky shores	?	seldom amphibious	Rao and Hora, 1938
	Alticus kirkii	surf zone on coral reefs	G, S	amphibious, large cutaneous blood vessels	Zander, 1972
	A. saliens	surf zone on coral reefs	G, S		Magnus, 1963
	Istiblennius edentulus	supralittoral zone	G, S	less amphibious than *Alticus*	Zander, 1972
	I. rivulatus	supralittoral zone	G, S		Klausewitz, 1964
	Entomacrodus nigricans	high energy zone of coral reefs	G, S, BP	amphibious, especially at night	Graham *et al.*, in preparation
	E. chiostictus	high surge zones of rocky shores	G, S		Graham, personal observation
	Blennius galerita	high littoral zone	G, S	amphibious	Soljan, 1932
	B. pholis	high littoral zone	G, S	amphibious at low tide	Daniel, 1970
Gobiesocidae	*Sicyases sanguineus*	rocky intertidal	G, BP, S	amphibious but not too active	Gordon *et al.*, 1970
	Phallerodiscus funebris	rocky intertidal	G, S	exposed during low tide	Eger, 1971
	Gobiesox pinniger	rocky intertidal	G, S	exposed during low tide	Eger, 1971
	Tomicodon boehlkii	rocky intertidal	G, S	exposed during low tide	Eger, 1971
	T. humeralis	rocky intertidal	G, S	exposed during low tide	Eger, 1971
	T. petersi	rocky intertidal		exposed during low tide	C. L. Hubbs, personal communication
Clinidae	*Mnierpes macrocephalus*	intertidal	G, S	amphibious	Graham, 1973
	Dialommus fuscus	intertidal		amphibious	Curio, 1968
Megalopidae	*Megalops atlantica*	pelagic shallow water, and estuaries	GB	physostomous, breathes by rolling at surface	Shlaifer and Breder, 1940
Eleotridae	*Dormitator latifrons*	brackish water	S	gas bladder inflated to elevate vascular region into air	Todd, 1973

[a] G, gills; BP, buccal, pharyngeal or opercular epithelium; S, skin; GB, gas bladder.

The table also points out neglected areas of research. There have been few studies determining the relative importance of the different air-breathing surfaces to total aerial respiration. The marine air-breathing fishes' capacity to simultaneously regulate gas exchange in as many as three different tissues has not been investigated. Finally, a difficult yet crucial problem in the study of air-breathing in amphibious species is to determine when air-breathing commences. Some species can hold water in their branchial chambers or around their bodies and thus may still be respiring aquatically for part of the time they are out of the water (Stebbins and Kalk, 1961; Graham, 1970).

III. Air-breathing Organs of Marine Fishes

The following structures are known to function as air-breathing organs in marine fishes: gills; vascularized buccal, pharyngeal, or opercular epithelium; the skin; and, in a few cases, the gas bladder or swimbladder (Table I).

A. AIR BREATHING WITH GILLS

Among freshwater fishes only *Symbranchus* and *Hypopomus* use gills for aerial respiration (Carter, 1957). By contrast, Table I suggests that a number of air-breathing fishes utilize gill respiration in air. Experimental evidence has been obtained on this for *Sicyases* (Vargas and Concha, 1957), *Tomicodon* (Eger, 1971), and *Mnierpes* (Graham, 1973). However, in most other cases the function of gills in air has been presumed and no data are available. For species that possess vascularized buccopharyngeal epithelium, experiments which distinguish the respiratory activities of that tissue and gills, thus demonstrating both function in air, have not been performed.

Gills tend to collapse and dry in air and most studies have investigated their adaptations to prevent these occurrences. Thus, the view of gills has been more toward how they endure exposure rather than how or whether they function for gas exchange in air.

A reduction in the number of secondary lamellae per mm of primary lamella is an adaptation that prevents both gill collapse (greater spacing between secondary lamellae) and evaporation (lower total gill area) in air. In marine fishes this has been observed for *Sicyases* (Ebeling, Bernal and Zuleta, 1970), and *Periophthalmus vulgaris* (Singh and Munshi, 1968), and to a lesser extent in *Blennius* (Milton, 1971) and *Mnierpes* (Graham, 1973).

Total gill area is reduced in species that are more dependent on aerial respiration. Schöttle (1932) compared the ratio of total gill area to body surface area in gobies and her findings are summarized in Table II. In primarily aquatic species of *Gobius* she found that the ratio was slightly greater than 1. In progressively more terrestrial mudskippers the ratio decreased; about 0·7 for two species of *Boleophthalmus*, and about 0·35 for four species of *Periophthalmus* (Table II). Schöttle (1932) also observed that in *Periophthalmus schlosseri* the first gill arch was given over entirely to supporting a fold of vascularized epithelium and that the fish only had three functional pairs of gill arches. Todd and Ebeling (1966) noted that gills in *Gillichthys* were short and that total gill area was less than in species that did not breathe air. I compared the total gill area reported for *Blennius pholis* (Milton, 1971) to that of *Mnierpes* (Graham, 1973) and found that *Mnierpes*, which is more amphibious, has a slightly smaller total gill area. The gill area in *Mnierpes* (range 15·4–16·8 cm^2 in fish 7–8 cm, total length), however, is generally greater than that stated by Schöttle (1932) for mudskippers of a similar size range (Table II). This reflects the mudskippers' greater terrestrial capability (Graham, 1973).

TABLE II

Schöttles' (1932) data for the ratios of gill and body surface areas in gobies in relation to their degree of terrestriality. Gobius *is entirely aquatic,* Boleophthalmus *is terrestrial but less so than* Periophthalmus

Species	Length (cm)	Body Surface Area (cm^2)	Gill Surface Area (cm^2)	Gill Area/ Body Area
Gobius jozo	9·7	27·4	27·5	1·00
G. auratus	5·4	8·1	9·5	1·17
G. caninus	10·0	34·6	48·6	1·40
Boleophthalmus boddaerti	12·6	40·0	33·6	0·83
	12·1	34·2	23·4	0·68
B. viridis	6·8	9·2	6·6	0·72
Periophthalmus koelreuteri	10·7	46·6	21·6	0·46
P. dipus	9·8	35·1	12·3	0·35
P. chrysospilos	5·8	12·7	4·3	0·34
	8·8	24·0	8·7	0·36
P. vulgaris	6·7	14·7	4·7	0·32
	8·4	25·0	6·7	0·27

It has not been determined whether reduced gill area is related solely to the problems of collapse and evaporative water loss in air. In the species studied by Schöttle (1932) increased amounts of buccopharyngeal tissue were found to parallel the decreases in gill area (Section III, B). This suggests that air-breathers have evolved toward minimizing their total amount of gill tissue, but compensate this loss of gill area by adding amounts of tissue that function more efficiently in aerial respiration. Freshwater air-breathers, including some species that never leave water, also have low total gill areas (Carter, 1957; Hughes and Morgan, 1973). Lower gill area reduces the energetic cost of branchial ventilation in water and species that can breathe aerially may be opting towards this means of saving energy (Hughes and Morgan, 1973).

Another factor favouring reduced total gill area in aquatic air-breathers is the possible loss of oxygen from the gills as the blood is circulated from accessory air-breathing organs through the gills to body (see Satchell, this volume).

B. AIR-BREATHING WITH MODIFIED BUCCOPHARYNGEAL AND OPERCULAR EPITHELIUM

Increased vascularization in the walls of the buccal and pharyngeal cavities and the inside of the operculum is a common structural adaptation for aerial respiration in marine gobies and clingfish. In species that have been studied, vascularization in these tissues has occurred through the ramification of existing blood vessels (Schöttle, 1932; Carter, 1957; Todd and Ebeling, 1966; Singh and Munshi, 1968; also see Satchell, this volume), that are homologous to those in less specialized species.

Schöttle (1932) found that the trend for reduction in gill area with a greater dependence on aerial respiration that she observed in *Gobius*, *Boleophthalmus*, and *Periophthalmus* was paralleled by the development of larger areas of more heavily vascularized epithelium (particularly in the pharnyx), in the increasingly amphibious species. In *Gobius*, buccal capillaries were diffuse and buried under the epithelium while in *Boleophthalmus* they were more numerous and tended to bulge slightly up into the epithelium. In *Periophthalmus schlosseri* she observed that dense capillaries were actually inside the epithelium and made furrows, thus greatly increasing surface area (Schöttle, 1932). In *Periophthalmus vulgaris* the branchiostegal membrane is also vascularized (Singh and Munshi, 1968). Todd and Ebeling (1966) describe a respiratory epithelium in *Gillichthys*, which has a large capillary bed in the roof of its mouth. The vessels in the bed extend outward from the dermis into the epithelium giving it a ribbed appearance as in *Periophthalmus. Gillichthys* breathes air by gulping at the surface and holding an air bubble in its mouth directly under the capillary bed. *Pseudapocryptes* has vascular bundles on the floor and roof of its mouth and a richly vascularized operculum wall (Das, 1934). A small vascular patch on the inside of the operculum of *Entomacrodus* may function in aerial respiration (Graham, Flowers and Jones, in preparation).

C. CUTANEOUS RESPIRATION IN AIR

Aerial cutaneous respiration has been suggested for a number of marine air breathers (Table I), for most fishes this capability has been inferred on the basis of vascularized skin, rather than empirically determined.

The appearance and position of cutaneous respiratory blood vessels varies in different species. *Boleophthalmus* has vascularized papillae along its tail whereas *Periophthalmus* and *Periophthalmadon* have capillaries between their scales (Schöttle, 1932). Among the blennies, cutaneous vessels seem to be concentrated in the anterior part of the body near or on the operculum (Rao and Hora, 1938; Magnus, 1953; Zander, 1972).

A large number of capillaries occur on the ventral and lateral surfaces of *Entomacrodus* (Graham *et al.*, in preparation). In air-breathing *Dormitator* a thick capillary bed measuring 10% of the total body surface area is formed on the frontal part of the head (Todd, 1973).

Cutaneous blood vessels in *Mnierpes* are difficult to see although cutaneous aerial respiration at 40 to 48% of total oxygen uptake has been measured for this species (Graham, 1973). Teal and Carey (1967) found that aerial cutaneous respiration could account for 50 to 60% of total respiration in *Periophthalmus sobrinus*. They also found cutaneous respiration to occur in water at 40%, the rate of total aquatic respiration.

For *Tomicodon* which has dermal vascularization, Eger (1971) was able to detect low rates of cutaneous respiration in fish that had their mouths blocked to prevent branchial respiration. Ebeling *et al.* (1970) suggested that the skin, particularly in the frontal region, functioned for aerial respiration in *Sicyases*.

D. THE GAS BLADDER AND AERIAL RESPIRATION

Todd and Ebeling's (1966) data indicate that *Gillichthys* consumes oxygen from its physoclistous gas bladder to maintain metabolism during its short (average 17 min) transition period from aquatic to aerial respiration. The tarpon *Megalops* regularly ventilates its physostomous gas bladder, even in normoxic water, by a characteristic surface roll (Böhlke and Chaplin, 1968). Shlaifer and Breder (1940) describe the tarpon's gas bladder to be lung-like and to have alveolar tissue. Johansen (1970) has suggested that aerial respiration may augment the tarpon's aquatic respiration and increase its metabolic scope for activity. (See also Fänge, this volume.).

The physoclistous gas bladder in *Dormitator*, while not involved in gas exchange, is directly related to this species' ability to breathe air. In hypoxic waters *Dormitator* must achieve positive buoyancy in order to elevate its vascularized frontal skin into air (Todd, 1973).

IV. Physiological Adaptations for Amphibious Air breathing

The advantages of air over water as a respiratory medium have been frequently discussed; it is less dense, more O_2 is available, and the O_2 diffusion rate is higher in air than in water (Dejours, this volume; Johansen, 1970).

In spite of these advantages no marine species, unlike several freshwater forms, are known to be obligate air breathers. All species depend on water and retain gills for aquatic respiration although some, like mudskippers, certainly are less dependent upon aquatic existence than others.

The physiology of gas exchange in air-breathing fishes has been thoroughly reviewed by Johansen (1970). Most aquatic air breathers respire bimodally. Oxygen is removed from air gulped at the surface and held in the air-breathing organ while CO_2 is released into water from the gills or skin. Depending upon physical conditions, little or no O_2 is absorbed aquatically while very little CO_2 is excreted aerially. Carbon dioxide is highly soluble in sea water. Bimodal respiration has been demonstrated for *Gillichthys* (Todd and Ebeling, 1966) and probably occurs in all species that gulp air at the surface and remain in water.

Amphibious air-breathing and terrestrial activity, however, have resulted in special adaptations related to a species' metabolic and cardiac rates in air and its capacity for aerial CO_2 excretion that are somewhat different from the pattern typically found in aquatic air-breathing fishes.

A. METABOLIC RATES IN AIR

Among most aquatic air breathers, O_2 uptake is divided between air and water—the fraction given to each mode varies depending on physical conditions such as aquatic O_2 tension and temperature (Johansen, 1970; Hughes and Singh; 1970a; 1971). The major limitation imposed by amphibious air breathing is that bimodal respiration cannot be utilized. Thus, amphibious aerial respiration must replace rather than supplement aquatic respiration and amphibious fishes must be able to switch rapidly and exclusively to aerial respiration when they leave water.

An indication of a species' capability for air breathing can be gained from a comparison of its metabolic rate in air with its normal rate in water. Table III summarizes all known comparative measurements of the metabolic rates of amphibious teleosts in and out of water. Most of these data were actually obtained for the purpose of comparing aerial and aquatic respiration, and investigators generally adopted uniform procedures for handling the fish in air and water so that comparisons would be valid. Stress and activity in response to handling usually required that measurements in air be made over longer periods of time than the species are normally emersed. Ideally, respiration during the first few minutes of emersion or submersion should be measured, but this is technically difficult and fish usually become more active when they are disturbed.

Table III indicates that a majority of the marine species maintain the same metabolic rates when taken from water and placed in air while most of the freshwater species reduce metabolism in air. Most of the freshwater species

TABLE III

Comparison of the aerial and aquatic metabolic rates of air-breathing teleost fishes

Species	Habitat[a]	Reference
I. Aerial Metabolic Rate $\gtrsim$ Aquatic		
Periophthalmus sobrinus	m	Gordon *et al.*, 1969
Sicyases sanguineus (small fish only)	m	Gordon *et al.*, 1970
Tomicodon humeralis	m	Eger, 1971
Blennius pholis	m	Daniel, 1970
Mnierpes macrocephalus	m	Graham, 1973
Gillichthys mirabilis	m	Todd and Ebeling, 1966
Entomacrodus nigricans	m	Graham, *et al.*, in preparation
II. Aerial Metabolic Rate < Aquatic		
Ophiocephalus (= *Channa*) *striatus*	fw	Dehadrai, 1962
Anguilla vulgaris	fw	Berg and Steen, 1965
Anabas testudineus	fw	Hughes and Singh, 1970a
Sicyases sanguineus (larger fishes)	m	Gordon *et al.*, 1970
Clarias batrachus	fw	Singh and Hughes, 1971
Saccobranchus fossilis	fw	Hughes and Singh, 1971

[a] m, marine; fw, fresh water.

are inactive in air and their lower metabolism is expected. Interestingly, marine species such as *Blennius* and *Tomicodon* that are not particularly active during emersion do not lower their metabolism whilst still in air.

Other data suggests that the response to emersion seems to be fundamentally different in marine and freshwater air-breathers. Gordon, Boetius, Evans, McCarthey and Oglesby (1969) showed that air-breathing *Periophthalmus* does not concentrate lactic acid in its blood, whereas air-breathing *Anguilla* develops an O_2 debt (Berg and Steen, 1965).

Large *Sicyases* are the only exception to the trend in Table III and this seems related to these fishes' very low activity during exposure (Ebeling *et al.*, 1970). In air large *Sicyases* breathe intermittently, accumulate lactic acid in their blood, and have 35 to 45% bradycardia (Gordon, Fischer and Tarifeño, 1970). Gordon *et al.* (1970) suggested that elevated aerial repiration in small *Sicyases* may be related to their thinner skin which permits cutaneous respiration. Certainly the surface area:volume ratio in small fish favours higher cutaneous gas exchange. Ebeling *et al.* (1970) observed that the gills of large fish seem structurally better suited for air-breathing but also that total gill area is reduced (Section III, A). Gordon *et al.* (1970) interpreted the intermittent respiration of the larger fish to be evidence of no aerial cutaneous respiration. Smaller gill area with no cutaneous respiration might therefore limit metabolism in large fish. It is of interest to note that Vargas and Concha (1957) did not record intermittent respiration in *Sicyases*.

B. CARDIAC RESPONSES TO EMERSION

Heart-rate measurements during the transition from aquatic to aerial respiration are useful because, unlike metabolism, they can monitor a fishes' instantaneous response to emersion. Also, a correlation between metabolic and heart rates in air could be expected because cardiac output in fishes is related to O_2 availability (Singh and Hughes, 1973 and references). Examples of this are the bradycardia associated with partial anoxia that has been recorded for aquatic species of fish in air (Garey, 1962) and tachycardia just prior, during, and for short periods after an air breath in some species (Johansen, 1966; 1968; Singh and Hughes, 1973).

Heart-rate measurements for air-breathing fishes in air and water are summarized in Table IV, and a good correlation exists between aerial heart rate (cardiac output) and metabolism. Predictably in *Sicyases* and *Anguilla*, which both have lower metabolic rates in air, bradycardia occurs with emersion. In *Symbranchus*, bradycardia occurs with emersion but is followed by tachycardia once a breath of air is taken, suggesting that aerial metabolism might be accelerated, but this has not been studied. *Anabas* is bradycardic in air but exhibits tachycardia just before and right after it breathes. Unlike

TABLE IV

Observed cardiac response following emersion in amphibious air-breathing teleost fishes

Species	Cardiac Response	Reference
Marine		
Periophthalmadon australis	tachycardia (bradycardia with submersion)	Garey, 1962
Periophthalmus sobrinus	no change	Gordon, *et al.*, 1969
Sicyases sanguineus	bradycardia	Gordon, *et al.*, 1970
Blennius pholis	short bradycardia, then resume, normal heart rate in air	Daniel, 1970
Gillichthys mirabilis	tachycardia	Todd, 1971
Freshwater		
Anguilla vulgaris	bradycardia	Berg and Steen, 1965
Symbranchus marmoratus	bradycardia until air breath, then tachycardia	Johansen, 1966
Anabas testudineus	bradycardia, tachycardia with air breath, but heart rate still less than in water	Singh and Hughes, 1973

Symbranchus however the tachycardic heart rate is still less than its normal heart rate in water.

Blennius pholis and *Periophthalmus sobrinus* do maintain the same rate of oxygen consumption as they are transferred to air from water (Table III) and, as expected, also have the same heart rate in the two media. *Blennius*, however, like *Symbranchus*, is bradycardic for about five minutes following emersion. This indicates that the transition to air breathing in *Blennius* is probably not as rapid as in *Periophthalmus*, and that *Blennius* may be partially asphyxiated when first emersed.

Some of the mudskippers appear to be so highly terrestrial that aquatic life is alien to them. Garey (1962) found that when he submerged *Periophthalmadon australis* its heart rate dropped from 108 to 46 beats per minute. This is the only known occurrence of diving bradycardia in fishes. It has subsequently been confirmed and further studied by Bandurski, Bradstreet and Scholander (1968), who also investigated other aspects of *Periophthalmadon's* diving response. Diving bradycardia seems to correlate well with the behaviour of some mudskippers which pass high tides in their submerged and partially hypoxic burrows (Garey, 1962; MacNae, 1968a; 1968b). Gas exchange in *Periophthalmadon* has not been studied.

There are several important conclusions that can be drawn from this discussion of the cardiac and metabolic rates of air-breathing fish in air. First, based on existing data, the response to air exposure seems basically different in marine and freshwater air breathers. Most amphibious marine fish maintain normal metabolic and heart rates in air and this seems adaptive in view of their need to remain active while out of water. Most freshwater species show an asphyxic response (lowered cardiac output and metabolism) but this also seems adaptive in enabling them to conserve energy and endure prolonged aerial exposure such as during over-land migrations or seasonal droughts. Finally, it is emphasized that the physiological response to emersion is consistent for all species. Metabolic rate correlates well with cardiac output. In cases where air-breathing is periodic, cardiac output increases when a new supply of O_2 is available.

C. RELEASE OF CARBON DIOXIDE INTO AIR

Another problem that arises with unimodal gas exchange in air is the excretion of CO_2. This has been discussed by Hughes (1966) and Johansen (1970).

Aquatic air breathers have low aerial gas exchange ratios because they release virtually all CO_2 into water through their gills and skin (Johansen, 1970, Table III; Hughes and Singh, 1970a; 1971; Singh and Hughes, 1971). With exposure to air normal avenues of CO_2 release are interrupted and total release of this gas decreases and does not balance total O_2 consumption.

Emersion has been shown to lower total CO_2 loss in *Anabas*, *Clarias*, *Electrophorus* and *Saccobranchus* (Farber and Rahn, 1970; Hughes and Singh, 1970a; 1971; Singh and Hughes, 1971). It also lowers total CO_2 loss and causes an increase in blood CO_2 tension in *Protopterus* and *Neoceratodus* (Lenfant and Johansen, 1968) and *Lepidosiren* (Johansen and Lenfant, 1967). In *Anguilla* the blood CO_2 level is normally high and actually remains fairly constant, but does increase relative to arterial O_2 tension and O_2 consumption which both decrease upon exposure to air (Berg and Steen, 1965).

In *Symbranchus* aerial respiration through the gills blocks CO_2 release by that tissue and blood CO_2 tension increases during a breathing phase (Johansen, 1966). Apparently the effect is the same whether the fish is in water or air, although ventilation is faster in air. (The interesting possibility that this fish may be capable of cutaneous gas exchange has not been investigated.)

From this discussion it seems that because of their inability to excrete CO_2 into air, freshwater air breathers, even forms that may be amphibious, endure increases in blood CO_2 tension and blood pH changes while emersed. This may be tolerated, or alleviated, through blood-buffering systems, (Johansen, 1970) or in some cases oxidative metabolism might subsequently be reduced (Table III) in order to minimize CO_2 production and restore a more normal gas exchange ratio. However, full balance of CO_2 may not occur in some fishes until they return to water (Berg and Steen, 1965).

There are no data comparing blood-gas parameters of amphibious marine fishes in air and water. Such data would be extremely useful, but the small size and blood volume of these fishes precludes catheterization or repeated sampling under the necessary controlled conditions.

There are data, however, for amphibious marine air-breathers in which both aerial O_2 uptake and CO_2 release were measured. These indicate, in contrast to the situation in freshwater air breathers, that marine amphibious fishes have evolved the capacity for highly efficient CO_2 excretion into air. Schöttle (1932; Table IV) measured aerial respiratory quotients (RQ) of from 0·66 to 0·94 for three species of *Periophthalmus*. For *Mnierpes* the aerial RQ is above 0·70 and remains high for periods much in excess of the time this species normally spends out of water (Graham, 1973; and unpublished observations).

Admittedly the comparative evidence available on this extremely important point is sparse and more data are needed, but none of the existing data are equivocal. As additional evidence for efficient CO_2 excretion in amphibious marine fishes, it can be noted that in two cases where high rates of aerial metabolism were reported (*Blennius*, Daniel, 1970, and *Tomicodon*, Eger, 1971) CO_2 release and not O_2 uptake was actually measured.

On the basis of the evidence available, it can be concluded that amphibious marine air breathers generally appear to be more proficient in the aerial release of CO_2 than their freshwater counterparts. This capability seems most likely related to the active marine fish's need for high oxidative metabolism while amphibious. The physiological and biochemical bases for aerial CO_2 excretion remain uninvestigated. Amphibious marine air breathers may be able to better ventilate in air because their skin may normally function in gas exchange and their air-breathing organs are not as sequestered as those of most freshwater air breathers (Section VI, Table V). There may also be large concentrations of carbonic anhydrase in the skin or gills of these fish.

Finally it is interesting to point out the paradoxical separation that seems to exist among modern air-breathing fishes for two specialized steps that were linked and probably crucial to the first terrestrial vertebrates: a specialized air-breathing organ and the ability to release CO_2 into air (Hughes, 1966; Thomson, 1969). Amphibious marine air-breathers do not have highly specialized respiratory structures but have developed the capacity to release CO_2 into air. On the other hand, modern air breathers with lungs or special lung-like structures are primarily aquatic and remain bound to water for the release of CO_2.

D. ADAPTATIONS TO RESIST DESICCATION IN AIR

Fish in air lose water mainly from their gills and skin, which impairs the function of these tissues in aerial gas exchange and limits terrestrial activity. To minimize water loss in air, amphibious marine fish regulate the amount of time they stay out of water and avoid prolonged exposure to direct sunlight (Graham, 1970; 1973). Additionally some species such as *Periophthalmus* and *Sicyases* can change skin-pigment concentrations to regulate heat absorption (Stebbins and Kalk, 1961; Ebeling *et al.*, 1970).

Species of *Andamia*, *Salarias*, *Periophthalmus*, and *Alticus* (see references in Table I) have all been seen to remoisten their skins by rolling on their sides on moist substrates. Another variation of rolling, in which only the head and anterior part of the body alternately are tipped to either side, is often seen in *Periophthalmus* (Stebbins and Kalk, 1961). This must function to remoisten branchial respiratory surfaces with water inside that chamber (Section V, B).

When out of water, *Mnierpes* does not roll, but its skin is protected from desiccation by a complete covering of scales that are distally loosened from the body and have a slightly superior orientation, thus forming cups that hold water during emergence (Graham, 1970).

To protect their gills, most species reduce the frequency of opercular pumping in air (Section V, B). Also, some species (*Tomicodon*, *Pseudapocryptes*, and *Gillichthys*) have the ability to produce mucus in or around their air-breathing surfaces and most probably this seems to be related to slowing desiccation. In *Pseudapocryptes*, mucus also enhances the sealing of the opercula so that tidal ventilation can be employed (Das, 1934). *Tomicodon* produces mucus in its branchial chambers and has a special mucus-secreting pad at the base of its pectoral fin. During aerial respiration this substance partially blocks and reduces the opercular apertures (Eger, 1971).

V. Behavioural Patterns Related to Air Breathing

A. Factors Inducing Air Breathing

Johansen (1970) reviewed the various factors that elicit and control air breathing in fishes. Three naturally occurring conditions: hypoxia, stagnation, and emersion have been shown to cause air breathing in marine fishes. The role of CO_2 in air breathing has not been investigated in these forms.

Todd's (1971) experiments with *Gillichthys* demonstrated that a decrease in this fish's oxidative metabolism while in hypoxic water was necessary to cause surface gulping. Hypoxia also causes air breathing in *Pseudapocryptes* (Das, 1934).

For *Tomicodon*, Eger (1971) observed that water stagnation elicited surface orientation and air breathing, either by emersion or gulping. Eger suggested that the absence of water motion is a signal to the organism that the tide has receded. The fishes' response could allow it to endure fluctuations such as increased temperature or lowered aquatic oxygen while it is exposed in tide pools or under rocks during low tide.

B. Behaviour

Tarpon breathe air while making a slow surface roll (Böhlke and Chaplin, 1968). Shlaifer and Breder (1940) found that air breathing in young tarpon is socially controlled and that groups of fish often breathe at or close to the same time.

Both *Gillichthys* and *Pseudapocryptes* may be amphibious but they usually remain in water and gulp surface air at intervals of from 1 to 10 min, then expel respired air through their mouths (Das, 1934; Todd and Ebeling, 1966; Todd, 1968). *Pseudapocryptes* keeps its opercular margins tightly sealed to its body and uses a flexible opercular wall to "inhale" air in to its mouth and branchial chamber. *Gillichthys* uses gular pumping and holds air in its mouth. Excess air is purged from the opercula which are not tightly sealed. *Gillichthys* becomes positively buoyant when air breathing and clings to the substrate with its sucker-like pelvic fin (Todd and Ebeling, 1966). This fish has a slightly smaller gas bladder (3·6–4% of body volume) than that generally found in other fishes and this apparently compensates for added buoyancy during air breathing.

Air-breathing *Pseudapocryptes* were observed by Das (1934) to float at the surface with their opercular chambers fully inflated. This was also seen for *Symbranchus* (Johansen, 1966; 1968). The adaptive significance of floating in open water is unclear as it would seem to make these animals extremely vulnerable to predation.

Air-breathing *Sicyases* usually position themselves head downward, which concentrates water in the branchial chamber near the mouth to moisten incoming air (Ebeling *et al.*, 1970).

Stebbins and Kalk (1961) reported that amphibious *P. sobrinus* in mangroves carried measureable amounts of water in its opercular chambers and did not move its opercula in air. They also describe an array of behaviour patterns, such as rolling (see Section IV, D), which aid the fish in mixing air and water in the chamber. When the fish feeds, water is lost from the chamber and Stebbins and Kalk (1961) observed that after ingestion the fish restored water to its chamber either by drinking or submerging. Some of these observations were not confirmed by Gordon, Boetius, Evans and Oglesby (1968), who studied *P. sobrinus* on rocky shores. *Periophthalmus vulgaris* seals its branchial chambers in air but, at intervals of 30–120 seconds it makes brief ventilatory motions to recharge branchial air (Singh and Munshi, 1968). A similar pattern with different time intervals has been observed for *Mnierpes* and *Entomacrodus* (Graham, 1970; Graham *et al.*, in preparation).

C. CHAMBER BULGING AND GULPING IN AIR

Chamber-bulging during aerial respiration has been observed in a number of air breathers: *Pseudapocryptes* (Das, 1934); *Andamia* (Rao and Hora, 1938); *Periophthalmus* (Stebbins and Kalk, 1961; Singh and Munshi, 1968) and *Entomacrodus* (Graham *et al.*, in preparation).

In *Entomacrodus* this action is characterized by a marked jerking and an audible gasp. The lower jaw is thrust up inside the upper jaw to close the mouth while the opercula and branchiostegal rays seal the branchial chamber. After it has gulped, the fish's opercula, throat, and cheeks are all bulged with air. This posture is held for 30–60 sec, relaxed for a brief period of ventilation, then re-established. If small amounts of water are introduced around the fish's head, it pumps water. This suggests a possible air-breathing reflex—with emersion the fish does not initiate air breathing until water around its body can no longer be sucked up and pumped through the gills.

Rao and Hora (1938) attempted unsuccessfully to determine the significance of bulging in *Andamia*. It may reduce evaporative water loss. Our dissections of *Entomacrodus* preserved in the bulged position show that the gill arches of these fish are erect and separate. This indicates that bulging is a way of making room for expanded gills (favouring gas exchange and reducing the probability of collapse) during air breathing. However, the occurrence of bulging in *Periophthalmus*, which has buccopharyngeal epithelium for aerial respiration, suggests that gill expansion is not the sole reason for bulging. One intriguing possibility that deserves investigation is that the increased

TABLE V

General ways that the morphological and physiological adaptations for air-breathing in freshwater and marine fishes may be contrasted

	Marine	Freshwater
1. Habitat and the Nature of the Air-Breathing Response	The majority are amphibious.[a] This behaviour, or tidal exposure makes air breathing a continuous requirement from early in a fish's ontogony. Most fish make rapid and complete transitions between aerial and aquatic respiration.	Some are amphibious.[b] Stagnation and anoxia associated with tropical dry seasons make facultative air-breathing and amphibious behaviour a seasonal occurrence. Some species are obligate air-breathers and others breathe aerially to supplement aquatic respiration when they are active.[c]
2. Structure and Placement of Air-Breathing Organs	Organs have simple structure and involve only gills, buccopharyngeal epithelium, and skin.[d] The primary reason for this is that marine air-breathers usually inhabit air-saturated waters and in air switch completely to air-breathing.	Because fresh-water air-breathers usually inhabit hypoxic waters, and must depend upon their gills for aquatic carbon dioxide release, air-breathing organs are enclosed and usually sequestered away from the gills (e.g., alimentary canal, branchial diverticula, or the air bladder) to prevent oxygen loss into water.[d]
3. Physiological Requirements Associated with Air-Breathing	Aerial oxygen uptake occurs at rates sufficient to sustain normal activities and carbon dioxide is efficiently excreted into air. Data for most species that are active on land shows that heart rates do not decrease in air.[e]	Most are sluggish and respire at lower rates when emersed. Carbon dioxide excretion, while normally efficiently accomplished through the gills and skin in water, is slowed in air. Most forms that have been studied show bradycardia in air.[e]
4. Ecological Basis for Air-Breathing	Primarily evolved as species become adapted for terrestrial life. Amphibious fish seem to avoid competition and successfully exploit an open habitat.	Air breathing in majority of species either supplements aquatic respiration or enables species to survive and carry on a normal existence in a hypoxic habitat. In the case of amphibious species it permits migration between bodies of water or in some cases, a search for food.
5. Parallel Adaptations	In addition to effective aerial gas exchange, amphibious life has resulted in new suites of adaptations as resistance to desiccation and water conservation through modifications in the pathways of nitrogen metabolism, terrestrial locomotion, and the emphasis of some sensory modalities (e.g., vision) over others (vibration detection).	No amphibious species are as highly adapted for terrestrial life as some marine species. Except for aerial oxygen uptake, non-amphibious air breathers remain largely adapted for life in water.

Table notes on facing page.

atmospheric pressure that results inside the chamber when it is bulged increases O_2 diffusion into the blood.

VI. Summary—Comparisons among Air-breathing Fishes

Breathing adaptations of freshwater and marine fishes have been compared throughout this paper and some general statements that contrast air breathing in these two groups are presented in Table V.

As would be expected, there are exceptions to the generalizations in the table and these are discussed. It remains, however, that the type of air-breathing adaptations found in the two groups can be contrasted and differences observed between them related to both the physical conditions which require air breathing and the importance of this adaptation to a particular species.

The ecological circumstances surrounding air breathing in the two groups are different. Air breathing enables a number of marine species to exploit habitats that are not saturated with competitors and predators for food and shelter. With aerial respiration species like clingfish can endure tidal exposure and hold their position in the intertidal zone, thus avoiding continuous and energetically costly vertical migrations. Tarpon, like some freshwater species, may breathe air to increase their metabolic scope.

Air breathing in all marine fishes seems to be an integral part of their natural history from early ontogeny. By contrast sub-adults of some freshwater species cannot breathe air as efficiently as adults (Johansen, 1966).

Most freshwater air breathers are not amphibious and breathe air in response to seasonally recurring conditions of stagnation and hypoxia. Amphibious freshwater forms may occasionally feed out of water but none seem to exploit terrestrial resources and amphibious behaviour seems primarily to be a means of locating new aquatic habitats. Thus, the presence of amphibians and a host of other terrestrial organisms in and around freshwaters seems to have been important in preventing the evolution of a highly amphibious freshwater fish fauna like that of marine habitats.

The complex array of aerial respiratory organs commonly seen in fresh-

---

Notes for Table V opposite

a. Exceptions are *Gillichthys*, *Pseudapocryptes*, and *Dormitator* which are estuarine and breathe aerially in response to hypoxia.

b. *Symbranchus*, *Monopterus*, *Anguilla*, *Clarias*, *Anabas*, see discussion; existing data on these species does not contradict comparison 2, also see comparisons 4 and 5.

c. *Clarias*, Magid, 1970; *Anabas*, Hughes and Singh, 1970b.

d. Two marine exceptions are *Gillichthys* which removes oxygen from its gas bladder during its transition from aquatic to aerial respiration, and the tarpon which regularly ventilates its gas bladder. Two freshwater exceptions are *Symbranchus* and *Hypopomus* which breathe aerially with gills (Johansen, 1970).

e. See discussion in text.

water air breathers is not found in marine species. The reason for this again seems related to the presence of some freshwater forms in habitats that are hypoxic for perhaps long periods of time. These fishes must insulate their supply of aerial oxygen from ambient water. By contrast, the habitats of most of the marine air breathers usually are normoxic, and probably never hypoxic for extended periods.

In general, air-breathing adaptations in the two groups do not seem to directly relate to the different physical properties of salt and fresh water. It could be argued that amphibious marine species have greater osmotic stress and thus must be better adapted both to conserve water while in air and to replenish water lost by evaporation; however, this has not been studied. Attempts to generalize completely about the differences in the physiological adaptations for air breathing in the two groups are frustrated because a number of marine air breathers are not amphibious, and some freshwater species are. Also, both groups have species that breathe air in response to aquatic hypoxia. Anadromous species such as *Anguilla* also cloud the marine–freshwater distinction.

Clearly *Symbranchus* and *Periophthalmus*, and all amphibious species, have a number of common physiological problems related to their gas exchange in air. Future comparative investigations of air breathing that place their emphasis on a contrast of amphibious as opposed to aquatic air-breathing will no doubt increase our understanding of the physiology of air breathing in fishes.

Acknowledgements

This work was supported by the Smithsonian Tropical Research Institute and by the Smithsonian Institution's Office of Academic Studies through the W. R. Bacon Scholarship. I especially thank Dr. S. Dillon Ripley, Secretary of the Smithsonian Institution, for granting me special funds to conduct research for this paper. Travel and support to attend the XXVI International Congress of Physiologists in New Delhi and the Satellite Symposium on the Respiration of Amphibious Vertebrates was granted by the United States National Academy of Sciences—National Research Council and the Smithsonian Institution's Foreign Currency Program. I thank Drs. D. Kramer, J. McCosker, R. Rosenblatt, and R. Warner for discussions and advice in the preparation of this paper.

References

Bandurski, R. S., Bradstreet, E. and Scholander, P. F. (1968). Metabolic changes in the mud-skipper during asphyxia or exercise. *Comp. Biochem. Physiol.* **24**, 271–274.

Berg, T. and Steen, J. B. (1965). Physiological mechanisms for aerial respiration in the eel. *Comp. Biochem. Physiol.* **15**, 469–484.

Bertin, L. (1958). Organes de la respiration aérienne. *In* "Traite de Zoologie: Agnathes et Poissons, Anatomie, Ethologie Systematique." **13**, 1363–1398. (Ed. P. P. Grassé) L'Académie de Médecine, Paris.

Böhlke, J. E. and Chaplin, C. C. G. (1968). "Fishes of the Bahamas and Adjacent Tropical Waters." Academy of Natural Sciences, Philadelphia.

Carter, G. S. (1957). Air-breathing. *In* "The Physiology of Fishes". Vol. I, 65–79. (Ed. M. E. Brown) Academic Press, New York and London.

Curio, E. (1968). Some observations on the "four-eyed" blenny of the Galápagos Islands, *Dialommus fuscus* (Pisces: Clinidae). *Notic. Galapagos.* **12**, 13–17.

Daniel, M. J. (1970). Ph.D. Dissertation. University of London.

Das, B. K. (1934). The habits and structure of *Pseudapocryptes lanceolatus*, a fish in the first stages of structural adaptation to aerial respiration. *Proc. Roy. Soc. Lond.* **115**, 422–430.

Dehadrai, P. V. (1962). Observations on certain physiological reactions in *Ophiocephalus striatus* exposed to air. *Life Sci.* **11**, 653–657.

Ebeling, A. W., Bernal, P. and Zuleta, A. (1970). Emersion of the amphibious Chilean clingfish *Sicyases sanguineus*. *Biol. Bull.* **139**, 115–137.

Eger, W. H. (1971). Ecological and Physiological adaptations of intertidal clingfishes (Teleostei; Gobisocidae) in the northern Gulf of California. Ph. D. Dissertation, University of Arizona, Department of Biological Sciences.

Farber, J. and Rahn, H. (1970). Gas exchange between air and water and the ventilation pattern in the electric eel. *Resp. Physiol.* **9**, 151–161.

Garey, W. F. (1962). Cardiac responses of fishes in asphyxic environments. *Biol. Bull.* **122**, 362–368.

Gordon, M. S., Böetius, J., Evans, D. H. and Oglesby, L. C. (1968). Additional observations on the natural history of the mud-skipper, *Periophthalmus sobrinus*. *Copeia* (1968) 853–857.

Gordon, M. S., Böetius, I., Evans, D. H., McCarthey, R. and Oglesby, L. C. (1969). Aspects of the physiology of terrestrial life in amphibious fishes. I. The mud-skipper *Periophthalmus sobrinus*. *J. exp, Biol.* **50**, 141–149.

Gordon, M. S., Fischer, S. and Tarifeño, S., E. (1970). Aspects of the physiology of terrestrial life in amphibious fishes. II. The Chilean clingfish, *Sicyases sanguineus*. *J. exp. Biol.* **53**, 559–572.

Graham, J. B. (1970). Preliminary studies on the biology of the amphibious clinid *Mnierpes macrocephalus*. *Mar. Biol.* **5**, 136–140.

Graham, J. B. (1973). Terrestrial life of the amphibious fish *Mnierpes macrocephalus*. *Mar. Biol.* **23**, 83–91.

Hora, S. L. (1935). Physiology, bionomics and evolution of the air-breathing fishes of India. *Trans. natn. Inst. Sci. India* **1**, 1–16.

Hughes, G. M. (1966). Evolution between air and water. *In* "Ciba Foundation Symposium on Development of the Lung". 64–80. (Eds A. V. S. de Reuck and R. Porter) Churchill, London.

Hughes, G. M. and Morgan, M. (1973). The structure of fish gills in relation to their respiratory function. *Biol. Rev.* **48**, 419–475.

Hughes, G. M. and Singh, B. N. (1970a). Respiration in an air-breathing fish, the climbing perch *Anabas testudineus*. Bloch. I. Oxygen uptake and carbon dioxide release into air and water. *J. exp. Biol.* **53**, 265–280.

Hughes, G. M. and Singh, B. N. (1970b). Respiration in an air-breathing fish, the climbing perch *Anabas testudineus*. II. Respiratory patterns and the control of breathing. *J. exp. Biol.* **53**, 281–298.

Hughes, G. M. and Singh, B. N. (1971). Gas exchange with air and water in an air-breathing catfish, *Saccobranchus* (*Heteropneustes*) *fossilis*. *J. exp. Biol.* **55**, 667–682.

Johansen, K. (1966). Air-breathing in the teleost *Symbranchus marmoratus*. *Comp. Biochem. Physiol.* **18**, 383–395.

Johansen, K. (1968). Air-breathing fishes. *Sci. Am.* **219**, 102–111.

Johansen, K. (1970). Air-breathing in fishes. *In* "Fish Physiology". Vol. 4, 361–411. (Eds W. S. Hoar and D. J. Randall) Academic Press, New York and London.

Johansen, K. and Lenfant, C. (1967). Respiratory function in the South American lungfish, *Lepidosiren paradoxa* (Fitz.) *J. exp. Biol.* **46**, 205–218.

Johansen, K. and Lenfant, C. (1968). Respiration in the African lungfish. II. Control breathing. *J. exp. Biol.* **49**, 453–468.

Klausewitz, W. (1964). Taxonomische und okologische Untersuchungen an einigen Fischarten der Kustenzone. *Senckenbergiana Biol.* **45**, 123–144.

Lenfant, C. and Johansen, K. (1968). Respiration in the African lungfish, *Protopterus aethiopicus*, I. Respiratory properties of blood and normal patterns of breathing and gas exchange. *J. exp. Biol.* **49**, 437–452.

Macnae, W. (1968a). A general account of the fauna and flora of mangrove swamps and forests in the Indo-west Pacific region. *Adv. Mar. Biol.* **6**, 73–270.

Macnae, W. (1968b). Mudskippers. *African Wildlife* **22**, 241–248.

Magid, A. M. A. (1971). The ability of *Clarias lazera* (Pisces) to survive without air-breathing. *J. Zool. Lond.* **163**, 63–72.

Magnus, D. B. E. (1963). *Alticus saliens*, ein amphibisch lebender Fisch. *Natur U. Museum* **93**, 128–132.

Milton, P. (1971). Oxygen consumption and osmoregulation in the shanny, *Blennius pholis*. *J. Mar. Biol. Ass. U.K.* **51**, 247–265.

Rao, H. S. and Hora, L. S. (1938). On the ecology, bionomics and systematics of the blenniid fishes of the genus *Andamia* Blyth. *Rec. Indian Mus.* **40**, 377–401.

Saxena, D. B. (1963). A review on ecological studies and their importance in the physiology of air-breathing fishes. *Ichthyologica* **2**, 116–128.

Schöttle, E. (1932). Morphologie und Physiologie der Atmung bei wasser-, schlamm- und landlebenden Gobiiformes. *Z. Wiss. Zool.* **140**, 1–114.

Shlaifer, A. and Breder, C. M., Jr. (1940). Social and respiratory behaviour of small tarpon. *Zoologica* **25**, 493–512.

Singh, B. N. and Hughes, G. M. (1971). Respiration of an air-breathing catfish *Clarias batrachus* (Linn.) *J. exp. Biol.* **55**, 421–434.

Singh, B. N. and Hughes, G. M. (1973). Cardiac and respiratory responses in the climbing perch *Anabas testudineus*. *J. Comp. Physiol.* **84**, 205–226.

Singh, B. N. and Munshi, J. S. D. (1968). On the respiratory organs and mechanics of breathing in *Periophthalmus vulgaris*. *Zool. Anz.* **183**, 92–110.

Smith-Vaniz, W. F. and Springer, V. G. (1971). Synopsis of the tribe Salariini, with description of five new genera and three new species (Pisces Blenniidae). *Smithsonian Contrib. Zool.* **73**, 1–72.

Soljan, T. (1932). *Blennius galerita* L., poisson amphibien des zones supralittorale et littorale exposées de l'Adriatique. *Acta Adriatica* **2**, 1–14.

Stebbins, R. C. and Kalk, M. (1961). Observations on the natural history of the mud-skipper, *Periophthalmus sobrinus*. *Copeia* (1961), 18–27.

Teal, J. M. and Carey, F. G. (1967). Skin respiration and oxygen debt in the mud-skipper, *Periophthalmus sobrinus*. *Copeia* (1967), 677–679.

Thomson, K. S. (1969). The biology of lobe-finned fishes. *Biol. Rev.* **44**, 91–154.

Todd, E. S. (1968). Terrestrial sojourns of the longjaw mudsucker, *Gillichthys mirabilis*. *Copeia* (1968), 192–194.

Todd, E. S. (1971). Respiratory control in the longjaw mudsucker, *Gillichthys mirabilis*. *Comp. Biochem. Physiol.* **39A**, 147–164.

Todd, E. S. (1973). Positive buoyancy and air-breathing: A new piscine gas bladder function. *Copeia* (1973) 461–464.

Todd, E. S. and Ebeling, A. W. (1966). Aerial respiration in the longjaw mud-sucker *Gillichthys mirabilis* (Teleostei Gobiidae). *Biol. Bull.* **130**, 265–288.

Vargas, F., F. and Concha, B., J. (1957). Metabolismo respiratario del teleosteo *Sicyases sanguineus*. *Invest. Zool. Chilena.* **3**, 146–154.

Willem, V. and Boelaert, R. (1937). Les manoeuvres respiratorie de *Periophthalmus*. *Bull. Acad. Roy. Sci. Belg.* (Ser. 5). **23**, 942–959.

Zander, C. D. (1972). Beziehungen zwischen Körperbau und Lebensweise bei Blenniidae (Pisces) aus dem Roten Meer. I. ÄuBere Morphologie. *Mar. Biol.* **13**, 238–246.

Gas Exchange in the Swimbladder

RAGNAR FÄNGE

Department of Zoophysiology, University of Göteborg, Göteborg, Sweden

I. Phylogenetic and Structural Background

Two main lines of early vertebrates may have existed in paleozoic times, one giving rise to modern fishes or Actinopterygii, the other to lungfishes, crossopterygians and tetrapods (Romer, 1972). Comparative anatomical and paleontological evidence indicate that the early vertebrates were provided with lung-like oxygen resorbing structures. Whether air-breathing vertebrates first appeared in a fresh water or in a marine or brackish water environment has been debated (Packard, 1974). The dominating hypothesis is that lungs evolved in freshwater vertebrates during the Devonian time (Barrel, 1916).

In the forerunners of modern fishes ventral paired lungs are supposed to have transformed into the unpaired dorsal swimbladder characteristic of many fishes. In the tetrapods the development of lungs made possible the development of an intense oxygen-requiring metabolism. Carbon dioxide during early terrestrial life was probably eliminated mainly through the skin (Hughes, 1967).

Both swimbladders and lungs emerge embryologically from a single or paired pouch of the anterior part of the digestive tube. As a consequence the swimbladder, and to some extent the lungs, contain the same principal tissue elements as the gut wall: internal epithelium, connective tissue, smooth and sometimes striated muscles, blood vessels and nerves. The swimbladder may lie relatively free or be fixed dorsally to parapophyses, blood vessels etc. (Hagman, 1921).

Swimbladders may possess processes or chambers, such as lateral diverticula in Sciaenidae (Dutt and Thankam, 1968), anterior and posterior chambers in Cyprinidae, Gymnotidae, Characinidae etc. (Evans, 1929, 1930; Evans and Damant, 1928; Rowntree, 1907), connections with the inner ear (as in Clupeidae, Notopteridae etc.: Dehadrai, 1957, 1959). The enormous structural variability is correlated with a large functional variability: the swimbladder may serve a hydrostatic function, in production and reception of sound, in pressure reception and as an accessory respiratory organ (Fänge, 1966). Probably other physiological roles could be added to this list.

Swimbladders are never found in cyclostomes, elasmobranchs or holocephalans but occur in about half of the true fishes or Teleostomi. Fishes in which the swimbladder communicates with the digestive tract via a duct are termed *physostomes*, while *physoclists* possess a closed swimbladder.

II. Respiratory Swimbladders and Lungs

Among about 20,000 or more existing species of fishes only a few have the capacity to breathe air with lungs or swimbladders. Lungs are found in lungfish (Dipnoi) and bichirs (Polypteroidei), and swimbladders adapted for respiration in a few teleosts and holosteans (Table I). Fish lungs and swimbladders are supplied by blood, which has already passed through the gill-apparatus. From this region the blood reaches fish lungs through pulmonary arteries, while swimbladders are supplied either from the coeliac (coeliaco-mesenteric) artery or by intercostal arteries branching off from the dorsal aorta. Pulmonary veins drain the blood from fish lungs to the heart. A similar venous drainage into the heart or into large veins (cardinal or intercostal) also characterize respiratory swimbladders. This is in contrast to the mode of venous return from the gas secretory part of the swimbladder of most fishes, and from the intestine and other parts of the digestive canal. From these structures the venous blood has to pass through the capillaries of the hepatic portal system before reaching the heart.

Considering lungs and respiratory swimbladders it should be remarked that these structures by no means represent the only adaptations for uptake of atmospheric oxygen found in fish. The diversity of gas exchange mechanisms with this function is indeed very great (Johansen, 1970; Johansen, Hansen and Lenfant, 1970; Lomholt and Johansen, 1974; Munshi, this volume; other authors).

Fishes with lungs or respiratory swimbladders are faced with the problem of how to avoid loss of oxygen through the gills into surrounding oxygen-poor water. It is not clear how such loss is minimized or avoided in different species. Vascular reflexes and shunts, or stop of gill ventilation, may be involved. In lungfish, the specific vascular anatomy and the reduction of the gas exchange surface of the gills are important in this connection (Wood and Lenfant, this volume).

TABLE I

Lungs and respiratory swimbladders in fish

Fishes	Reference
DIPNOI (lungfish)	Johansen (1970)
Neoceratodus	McMahon (1969)
Lepidosiren	
Protopterus	
HOLOSTEI (bowfin, garfish)	Ballantyne (1927)
Amia	Johansen *et al.* (1970)
Lepisosteus	Rahn *et al.* (1971)
POLYPTEROIDEI (bichirs)	Kerr (1909), Gérard (1931)
Polypterus	Johansen (1970)
Calamoichthys	
TELEOSTEI (bony fish)	
Osteoglossidae:	Rauther (1910)
Arapaima (*Sudis*)	Nysten (1962)
Pantodon	Thys van den Audenaerde (1959)
Heterotis	
Mormyridae:	Assheton (1907), Ballantyne
Gymnarchus	(1927)
Esocidae:	Rauther (1914)
Umbra	Geyer and Mann (1939)
	Safford Black (1944)
Characidae:	Jobert (1878)
(*Hoplo*) *erythrinus*	Rowntree (1907)
Lebiasina	Carter and Beadle (1931)
	Böker (1933), Lüling (1964)
Notopteridae:	Müller (1950)
Notopterus	Dehadrai (1960)
Xenomystus	Greenwood (1963)
Phractolaemidae	
Phractolaemus	Thys van den Audenaerde (1959)

A. SWIMBLADDERS USED FOR RESPIRATION

In *Hoploerythrinus unitaeniatus*, a small tropical fish, the swimbladder in the intervals between the respiratory movements contained 7% oxygen and 0·8% carbon dioxide (Carter and Beadle, 1931). This indicates that oxygen is efficiently resorbed, but only small amounts of carbon dioxide are given off. Carbon dioxide probably is eliminated from the gills or the skin. Safford Black (1944) observed that the mudminnow (*Umbra limi*) is able to resorb oxygen from the swimbladder at very high tensions of carbon dioxide in the water. In this case high carbon dioxide values were found in the swimbladder.

Johansen *et al.* (1970) investigated the function of the respiratory swimbladder in *Amia calva*. They observed that inflation of the swimbladder with air, oxygen or nitrogen causes the fish to interrupt its ascent to the water surface for air intake. They compared this response with the Hering–Breuer reflex in mammals. The latter feedback mechanism is based on vagal stretch receptors of the tracheobronchial tree. Afferent discharge following inflation of the lungs causes the brain to terminate inspiration.

Air breathing of the garfish, *Lepisosteus*, has been investigated by Potter (1927) and Rahn, Rahn, Howell, Gans and Tenney (1971).

The neuromuscular mechanisms of intake and expulsion of air into respiratory swimbladders have not been much investigated. Inspiration undoubtedly is due to buccal force pumps, but activity of smooth muscles of the pneumatic duct and the swimbladder may also be of some importance. In *Hoploerythrinus* (Table I) the pneumatic duct is reported to be remarkably wide and muscular (Böker, 1933; Lüling, 1964). Expiration may be produced by similar release mechanisms as the gasspuckreflex of ordinary physostomes (cf. p. 194).

In the eel (*Anguilla anguilla*) the pneumatic duct (Rauther, 1937; Fänge, 1953) is adapted for gas exchange by diffusion. Structurally it resembles very much a primitive lung, for example that of an amphibian. The venous drainage from the eel pneumatic duct passes directly to the sinus venosus of the heart or a vein close to the sinus venosus (Mott, 1951). Jacobs (1898) suggested that the eel may be able to use the pneumatic duct of the swimbladder for respiration. However, there seems to exist no observations showing that when the eel is outside water it is able to use the swimbladder for uptake of atmospheric oxygen. Further, in the eel, as in amphibians, at low temperature (below 15°C) cutaneous respiration may suffice for survival (Krogh, 1904).

B. FISH LUNGS

The lungs of lungfish (Dipnoi) are provided with internal septa containing elastic connective tissue and smooth muscles. Histologically and ultrastructurally the richly vascularized wall of the lung of the African lungfish (*Protopterus*) largely resembles that of an amphibian lung (DeGroodt, Lagasse and Sebruyns, 1958; Klika and Lelek, 1967). A muscular sphincter controls the entrance to the lungs from the pharynx (the glottis region). The lungs are innervated by branches from the vagus nerve. Scattered ganglion cells are found in the lung tissue (Kerr, 1909; Gérard, 1931; Poll, 1962). The smooth muscles of the lung and the glottal sphincter are contracted by acetylcholine and relaxed by adrenaline and noradrenaline. The response to acetylcholine is blocked by atropine (Johansen and Reite, 1967).

In lungfish inspiration is brought about by a buccal force mechanism related to and perhaps phylogenetically derived from that used for ventilation of the gills (McMahon, 1969). The function of the pulmonary smooth muscles is not clear. Expiration is brought about mainly by elastic forces, although contractions of the pulmonary smooth muscles may also be of some importance. Further, tonus changes of the smooth muscles and blood vessels of the respiratory mucosa may influence the rate of gas exchange.

The lungs of bichirs (*Polypterus*, *Calamoichthys*) (Table I) contain two layers of striated muscles crossing each other diagonally. The muscles are innervated by the vagus nerve and supposedly perform respiratory movements (Kerr, 1909). The right lung is connected with the oesophagus via the slit-like glottis surrounded by a muscular sphincter.

C. AMPHIBIAN AND REPTILIAN LUNGS

In amphibians ventilation of the lungs occurs under positive pressure, air being pressed into the lungs by a buccal force mechanism and retained by closure of the glottis sphincter. The smooth muscles of amphibian lungs are extremely stretch-sensitive (Price and Pierce, 1963). Reflex movements of the smooth muscles of anuran lungs were studied by Carlson and Luckhardt (1920). The muscles receive predominantly inhibitory fibres but also excitatory fibres from the vagosympathetic trunks. The inhibitory response to nerve stimulation is blocked by nicotine indicating the presence of ganglionic synapses along the nervous pathways. Section of the pulmonary nerves causes hypertonus of the muscles. Wood and Burnstock (1967) found that in the toad (*Bufo*) sbout 80% of the inhibitory response to nerve stimulation is blocked by the β-blocking drug dichloroisoprenaline. The lungs of the frog (*Rana*) are relaxed by adrenaline, noradrenaline and 5-hydroxytryptamine (Schnizer, Hoang and Brecht, 1968). The transmitter of the inhibitory fibres could be adrenaline, but Campbell (1971a,b) holds the view that the fibres are non-adrenergic.

Pulmonary smooth muscles have been studied in a few reptiles. In the lizard, *Uromastix*, vagotomy caused the lungs to relax and overfill with air, indicating that the vagus nerve carries excitatory fibres. The smooth muscles were contracted by pilocarpine but did not react to adrenaline (Saalfeldt, 1933). In another lizard, *Tiliqua* (*Trachysaurus*), the pulmonary smooth muscles receive excitatory and inhibitory fibres from the vagosympathetic trunks. Adrenaline, noradrenaline and dopamine cause relaxation. Acetylcholine produces contraction, which is potentiated by eserine and inhibited by atropine. Noradrenaline may be the inhibitory transmitter (McLean and Burnstock, 1967; Burnstock and Wood, 1967).

The fact that vagotomy strongly changes the tonus of the lungs, although in opposite ways in amphibians and reptiles, indicates that the vagus nerve is active in continuous reflex control of the pulmonary smooth muscles.

However, the importance of the smooth muscles is poorly understood. They may play a role in expiratory release of air, in reflexes hindering entrance of foreign objects into the airways, or in regulation of the permeability of the respiratory membranes etc. Reflexes of the pulmonary vascular bed may also be important.

III. Non-respiratory Physostome Swimbladders

A. Function of the Pneumatic Duct

Physostomes have been observed to be able to inflate the swimbladder with air gulped in at the water surface. Probably the air is pressed into the swimbladder by a buccal force mechanism. Some physoclists are known to use a pneumatic duct during their larval stage for the first filling of the swimbladder (Ledebur and Wunder, 1937; Jacobs, 1938).

1. *The pneumatic sphincter*

The entrance of the pneumatic duct into the oesophagus is guarded by smooth and striated muscle sphincters, as in ganoids (Kudrjaschoff, 1926; smooth muscles), notopterids (Müller, 1950), *Electrophorus* (the electric eel) (Evans, 1929), cyprinids (carpfishes) (Guyénot, 1909; Evans and Damant, 1928; Plattner, 1941) and salmonids (Stärk, 1924; Fahlén, 1971). The sphincters may function in a similar way to corresponding sphincters in respiratory physostome swimbladders, and the glottal sphincters of primitive lungs (lungfish etc.).

2. *The gasspuckreflex*

Release of gases through the pneumatic duct leading to deflation of the swimbladder takes place by a reflex action, the gasspuckreflex (or gas spitting reflex). This is initiated by reduction of the external pressure, making the fishes too light (e.g. by intraperitoneal injection of air), or by nervous excitation (Table II). Most studies have been made on cyprinids, which perhaps is a pity because they possess a somewhat special "high pressure" swimbladder. In carpfishes the swimbladder, like a pneumatic tyre (bicycle etc.), has an inner gas pressure considerably exceeding the external hydrostatic pressure. The excess pressure may be as high as 80–90 mmHg or more (Evans and Damant, 1928). If the tonus of the pneumatic sphincter decreases, release of gases through the duct may occur, probably due to the combined effects of elasticity of the swimbladder wall, contraction of smooth swimbladder muscles, and contractions of body wall muscles.

3. *Nervous control*

Sensory nerve endings have been demonstrated in the wall of the cyprinid swimbladder histologically and electrophysiologically (Quotob, 1962).

TABLE II

Factors influencing the reflex opening of the pneumatic sphincter (= release of gas bubbles via the pneumatic duct). Mainly from Franz (1937), *Plattner* (1941), *Harvey* (1963) *and Srivastava* (1964)

Fishes	Facilitation	Inhibition
Cyprinidae, Salmonidae, Clupeidae	Slow reduction of external pressure 10–100 mmHg	Increased rate of pressure reduction
	Nervous excitation (frightening)	Removal of brain
		Death
	Intraperitoneal injection of air	Spinal section
	Adrenaline, ephedrine	Narcotics
	(Atropine)	Vagotomy
		Curare
		α-Adrenergic blocking agents

Probably the gasspuckreflex is initiated by increased discharge from vagal nerve endings in the swimbladder functioning as stretch receptors, but other receptors (eyes, inner ear) may also be of importance. According to Franz (1937) removal of the weberian ossicles connecting the swimbladder with the inner ear does not influence the gasspuckreflex. If the inner ear is involved, the labyrinth rather than vibration sensitive receptors may be of importance.

That the central nervous system has an important role for the gasspuck-reflex is obvious, because decerebration, spinal section or administration of narcotics inhibit opening of the pneumatic sphincter (Table II). A centre for gas spitting is probably localized in the brain, perhaps in the medulla oblongata.

In physostomes the smooth muscles of the swimbladder wall and the pneumatic sphincter are innervated by noradrenergic fibres. This has been found in the trout (*Salmo gairdneri*) and the herring (*Clupea harengus*) (Fahlén, Falck and Rosengren, 1965; Fahlén, 1967). In cyprinids (Franz, 1937), as probably in other physostomes, the pneumatic sphincter receives fibres from the vagus nerve. Unpublished work (Nilsson and Fänge, 1975) shows that the bulb-like pneumatic sphincter of the goldfish (*Carassius auratus*) contains a rich plexus of varicose adrenergic nerve endings. Similar nerve endings are found in the swimbladder itself. The histochemical fluorescence observations indicate that the transmitter may be noradrenaline. Pharmacologic experiments show that the smooth muscles of cyprinids (Fänge, 1953), salmonids (Harvey, 1963), and the secretory mucosa of the swimbladder of the eel (*Anguilla anguilla*) (Fänge, 1953; Nilsson and Fänge 1967) contract in the presence of adrenaline. The response is of the α-adrenergic type (Nilsson and Fänge, 1967). On the other hand the pneumatic sphincter of salmonids (Harvey, 1963), clupeids (Srivastava, 1964), and the pneumatic duct of the eel (*Anguilla anguilla*) react by relaxation, showing a response of the β-adrenergic type (Nilsson and Fänge, 1967; Fänge, 1973).

The gasspuckreflex requires a certain minimum of time in order to be put in action. This may be the reason why "Trommelsucht" may occur in physostomes caught from deep water. Siebold (1858) described the netting of the salmonid *Coregonus acronius* from a depth of 65 metres in the Swiss lake Bodensee. On arrival at the surface the fishes were highly swollen by expanded swimbladder gases. A similar phenomenon is commonly observed in physoclists. In experiments with *Phoxinus* (a cyprinid) Franz (1937) found that increasing the rate with which decompression is made tends to inhibit or delay opening of the pneumatic sphincter. Moehres (1940) made similar observations.

B. GAS EXCHANGE BY DIFFUSION

Because the pneumatic duct may function as a safety valve, exchange of gases by diffusion are probably not of great importance in physostomes. Diffusion barriers may exist as collagen layers in the swimbladder of sterlets (*Acipenser*) (for further facts see p. 198). In salmonids, in which the swimbladder had been filled with air through the pneumatic duct, the percentage of oxygen within 3–4 days decreased to 3·3–12·7 (Ledebur, 1929). Safford Black (1944) observed a decrease of oxygen in the swimbladder of physostomes during asphyxiation. With the exception of eels (*Anguilla*, *Conger*) and a few other forms, physostomes generally do not possess any richly vascularized wall structures specially adapted for gas resorption. However, so far only a few forms (cyprinids, salmonids, clupeids etc.) have been investigated in detail.

C. GAS SECRETION

In general the rate of gas secretion is slower among physostomes than in physoclists (Table III). But the eel (*Anguilla anguilla*) has a large rete mirabile and well developed gas gland epithelium (Krogh, 1936; Dorn, 1962) and in accordance with this a high rate of gas secretion (Steen, 1963). In other physostomes counter-current vascular systems and gas gland cells often are

TABLE III

Species differences in efficiency of gas secretion

Fishes	Type	Time	Reference
		Fast secretion:	
Pomatomus	Physoclist	2–6 h	Fänge (1966; cited from several authors)
Stenotomus	Physoclist	6–12 h	
Gadus	Physoclist	3–6 h	
Ctenolabrus	Physoclist	6–8 h	
Monacanthus	Physoclist	6 h	
Opsanus	Physoclist	18–24 h	
Perca	Physoclist	24 h	
Anguilla	Physostome	—	Steen (1963)
		Slow secretion:	
Salmo	Physostome	11–13 days	Wittenberg (1958)
Coregonus	Physostome	probably several weeks	Sundnes (1963)
Tinca, *Carassius*	Physostome	a few days to several weeks	Evans and Damant (1928), Jacobs (1934), Wittenberg (1958)
Esox	Physostome	many days	Dreser (1892), Jacobs (1934)
Clupea	Physostome	not measured; probably many days	Sundnes and Bratland (1972)
Acipenser	Physostome	No secretion (?):	Kudrajaschoff (1926)
Fishes with lungs or respiratory swimbladders* (Table I)	Physostome	No secretion	

* Exceptions: *Umbra* (Rauther, 1914) and *Gymnarchus* swimbladder-derived auditory vesicles (Stipetič, 1939).

poorly developed or have not been found. Fahlén (1971) discovered a well-developed rete mirabile in *Coregonus*, but not in other salmonids, nor in clupeids (herring etc.). However, counter-current systems consisting of a few vessels only and of low functional efficiency could very well exist without having been noticed by the investigators. This may be the case in some forms with a slow rate of gas secretion.

Many structural and functional problems remain concerning gas secretion in physostomes. Histological work has shown that in cyprinids (Rauther, 1937; Fänge and Mattisson, 1956) cells supposed to function as gas gland cells are separated from the swimbladder by a thick smooth muscle layer.

Similar situations may exist in the esocid *Umbra* (Rauther, 1914) and in the anterior chamber of the salmonoid physoclist *Argentina silus* (Fänge, 1958). Furthermore, in cyprinids the high gas pressure probably exceeds the blood pressure in the dorsal aorta, which may result in an extremely restricted capillary blood circulation in the inner layers of the swimbladder (Evans and Damant, 1928). In many physotomes there remains uncertainty concerning the structural identity or existence of gas gland cells.

Theoretical solutions have been presented to explain the remarkably high concentration of nitrogen in the swimbladder of the deep water salmonid *Coregonus*, as well as the secretion of inert gases into fish swimbladders generally (Abernethy, 1972). However, the high nitrogen content in the coregonid swimbladder is not due to a primary deposition of a gas mixture containing a high percentage of nitrogen, but may be mainly due to secondary loss of oxygen by diffusion (Sundnes, Bratland and Strand, 1969; Sundnes and Bratland, 1972).

Almost nothing is known about the biochemistry of gas secretion in the swimbladder of physostomes. Probably in principle it resembles that of physoclists. Cellular gas secretion without involvement of counter-current circulatory systems is not very plausible, but cannot be completely eliminated from the discussion (Fänge, 1973).

IV. Physoclist Swimbladders

A. GAS EXCHANGE BY DIFFUSION

1. *Diffusion barriers*

In the swimbladder counter-current vessels function as a lock against loss of gases (Scholander, 1958). Further, the swimbladder wall may contain diffusion-hindering layers. In the *Conger* eel removal of a silvery guanine-layer increases the permeability to gases about 100 times. This layer is especially well developed in deep water eels (Denton, Liddicoat and Taylor, 1972) but is also found in the European eel (*Anguilla anguilla*) (Kutchai and Steen, 1971). Tightly packed fibrils or crystals of collagen (Fauré–Fremiet and Cougny, 1937) probably have a similar function as guanine crystals. The swimbladder of sterlets (*Acipenser*) and other fishes due to richness of collagen is a source of isinglass (ichthyocolla). These fishes are physostomes. In physoclist swimbladders guanine (*Argentina silus*; Fänge, 1958) or collagen fibrils (Fig. 1a) are found primarily in the poorly vascularized tunica externa. The mucosa, in most physoclists, consists of a secretory and a resorbent part. Because of its relative thickness and counter-current vascularization the secretory mucosa is poorly permeable to gases. The very thin resorbent mucosa is supplied by a vascular network, structurally resembling that of a primitive lung, e.g. a frog lung. Both parts of the mucosa consist of

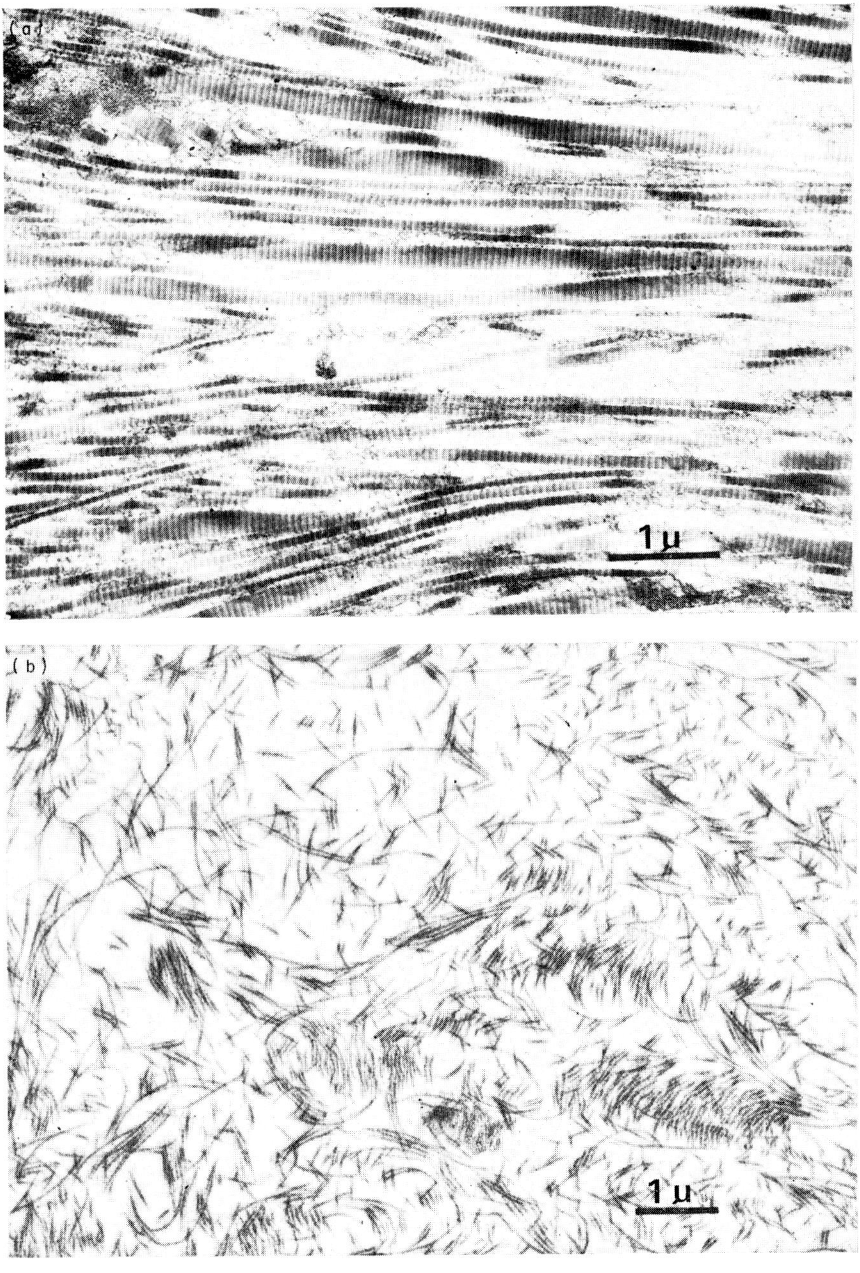

FIG. 1. (a) *Ctenolabrus rupestris*. Part of the swimbladder wall containing densely arranged collagen fibrils. (b) *Gadus morhua*, swimbladder. Submucosal loose connective tissue. (Prep. and electr. micr. photos, Dr. A. Mattisson).

epithelium and a thin smooth muscle layer, the muscularis mucosae. Because the middle layer of the swimbladder wall usually consists of loose and jelly-like connective tissue (Fig. 1b), contractile activity of the muscularis mucosae may result in movements of the mucosa independent of the external shape of the swimbladder (Fig. 2).

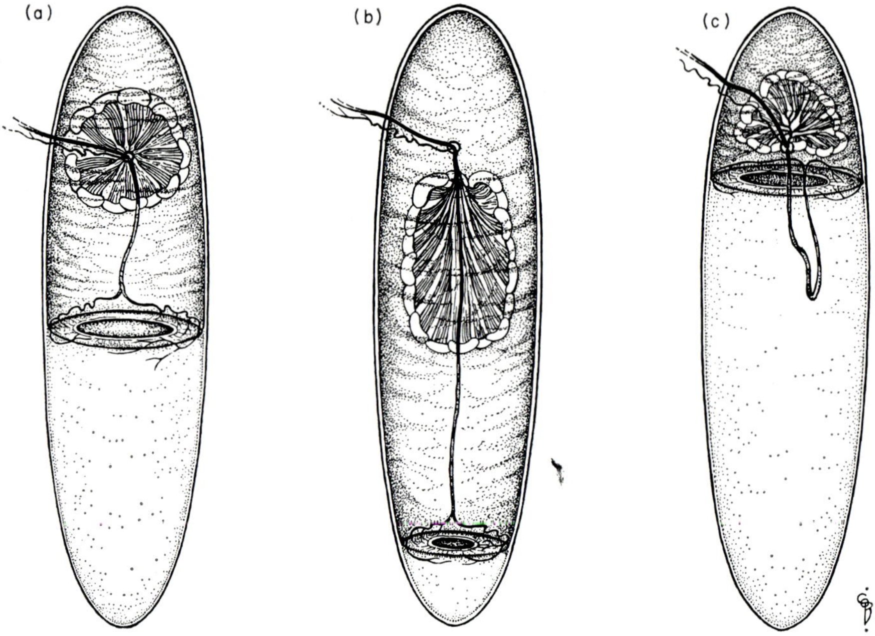

FIG. 2. *Ctenolabrus rupestris*. Musocal movements of the swimbladder. (a) Normal steady state. (b) Gas exchange by diffusion is minimal (inflatory reflex). (c) Gas exchange by diffusion is maximal (deflatory reflex). The swimbladder is shown as if it were fully transparent. The stippled part respresents the anteriorly situated secretory mucosa. Ventral view. (Drawing by Cynthia O'Brian).

2. *Autonomic nervous control of gas diffusion*

Work by Fänge (1953, 1966, 1973) and others has shown that diffusion of gases is controlled by autonomic nerve reflexes determining the tonus of the muscularis mucosae and the blood vessels of the two parts of the mucosa. The system is supplied by noradrenergic vagal nerve fibres (Nilsson, 1971, 1972) (Fig. 3). But whereas the secretory mucosal structures show mainly α-adrenergic responses resulting in contractions of smooth muscles, the resorbent mucosal structures are characterized by β-adrenergic relaxing responses (Fänge, 1973).

Elimination of gases from the physoclist swimbladder resembles release of gases by the gasspuckreflex of physostomes. In fact the resorbent mucosa

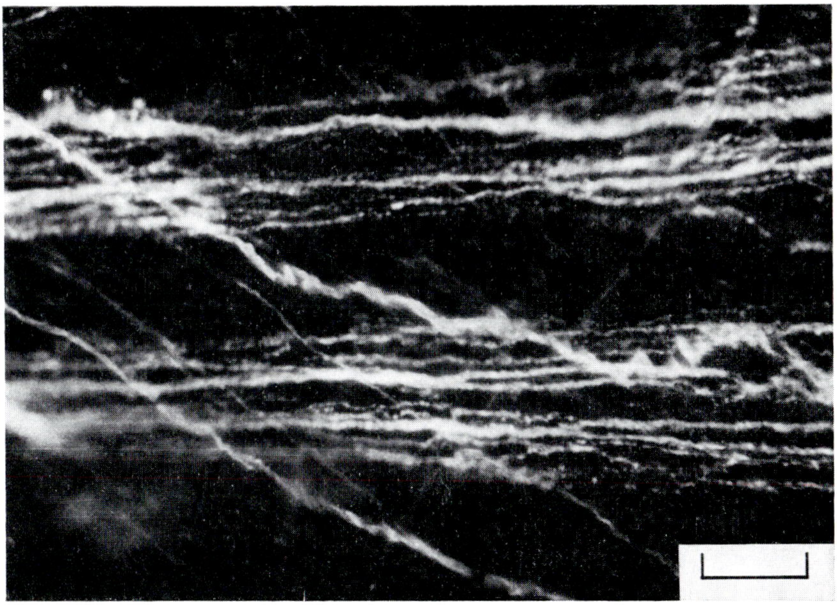

Fig. 3. *Gadus morhua*. Fluorescence histochemistry of catecholamines in the swimbladder secretory mucosa. Bundles of nerve fibres follow radial smooth muscles near the oval. Falck-Hillarp method. Calibration bar — 100 μm. (Prep. and microphoto Dr. S. Nilsson).

is thought to be anatomically homologous to the pneumatic duct and sphincter. The deflatory reflex of physoclists is initiated by decrease of external pressure, making the fishes too light, or by nervous excitation etc. (Cf. Table II: factors causing opening of the pneumatic sphincter). The afferent nerve impulses starting the deflatory reflex probably do not originate from receptors of the swimbladder but from other sensory organs (eyes, inner ear). However, pressure stretch receptors may exist in the swimbladder wall of some physoclists (*Gadus*: Tytler and Blaxter, 1973). During the deflatory reflex an increased discharge of vagal noradrenergic fibres in the swimbladder wall results in mucosal movements and vasomotor reactions, that serve to increase the permeability of the resorbent mucosa (Fänge, 1966; Fänge, 1973) (Fig. 2c).

Large variations are found in the permeability controlling mucosal mechanisms in the swimbladder of different species of euphysoclists. The resorbent part of the mucosa may form a dorsal "oval" (*Perca*, *Lucioperca*, *Gadus*, *Raniceps*, *Sciaena*, *Diodon*, *Caranx* etc.) or a "posterior chamber" or "posterior vascular organ" (*Labrus*, *Ctenolabrus*, *Tautogolabrus*, *Ophiocephalus*, *Syngnathus*, *Gasterosteus*, *Zeus*, *Opsanus*, *Trigla* etc.) (Rauther, 1937; Fänge, 1953; George, 1956; Peter, 1963). Remarkably large mucosal movements are observed in the swimbladder of labrids such as *Ctenolabrus rupestris* (Fig. 2a,b,c). In other forms the permeability of the resorbent mucosa probably is more influenced by vasomotor responses than by mucosal movements. For example in the toadfish (*Opsanus tau*) the diaphragm at the boundary between the secretory and resorbent mucosa is fixed to the gas gland and the tunica externa, and the mucosal movements are not very prominent (Fänge and Wittenberg, 1958).

3. *Diffusion rates of swimbladder gases*

Krogh (1919) showed that carbon dioxide diffuses through living membranes many times faster than oxygen. Nitrogen shows the slowest rate of diffusion. In the marine physoclist *Ctenolabrus rupestris* injected carbon dioxide disappeared from the swimbladder within a few minutes, while oxygen even after administration of adrenaline required several hours to decrease from about 60% to 15% (Fänge, 1953). Akita (1936) showed in another species (*Monacanthus*) that nitrogen disappears by diffusion from the swimbladder at a much slower rate than oxygen. Moore (1970) observed that in *Mugil*, in water with low content of oxygen, the swimbladder oxygen decreased within 1 hour from 18·7% to about 10%.

These and other results obtained on gas resorption in the swimbladder are comparable with results obtained in studies on sealed-off gas pockets during atelectasis or pneumothorax in mammals (Rist and Strohl, 1922; Dale and Rahn, 1952), or other systems (Campbell, 1931; Piiper, 1965; Hughes, 1967). However, the situation in the physoclist swimbladder is slightly more complicated, because one has to consider also the effect of gas secretion. Synchronously with gas resorption various swimbladder gases are transported from the blood into the swimbladder at individually different rates. The two main processes, resorption (= diffusion) and secretion, are continuously controlled by autonomic nervous reflexes. The resulting steady-state keeps the swimbladder volume constant. The control mechanisms function even in deep water because fishes with functional swimbladders live at depths of several thousand metres (Nielsen and Munk, 1963; Marshall, 1960; 1972).

The presence of a gas-filled closed swimbladder poses problems for mesopelagic species like lantern-fishes (myctophids), which perform diurnal vertical migrations of hundreds of metres. According to Bone (1973) it seems improbable that these fish would be able to secrete and reabsorb gas quickly enough to preserve neutral buoyancy at all depths. At greater depth they probably are more dense than sea water and obliged to swim actively to ascend or avoid sinking.

The amount of oxygen in the physoclist swimbladder probably is too small to be of any significance as an oxygen store during asphyxiation. Furthermore, any oxygen resorbed from the swimbladder by fish, occasionally swimming in water poor in oxygen, would be lost by diffusion from the gills.

B. GAS SECRETION

1. *Biochemistry*

Gas secretion in fish is supposed to be due to a counter-current multiplication of physico-chemical effects (Bohr and/or Root effects, salting out effects) produced by substances released into the blood by special cells (Enns, Douglas and Scholander, 1967). The importance of the buffer systems and the gas transporting capacity of the blood has been discussed by Steen (1963) and Berg and Steen (1968). Ball, Strittmatter and Cooper (1955) studied the metabolism of isolated gas gland tissue from *Stenotomus*, a fast secreter (Table III). This tissue was found to convert glucose to lactic acid at a very rapid rate. The conversion was followed by release of carbon dioxide from the medium, which contained sodium bicarbonate. Later biochemical works have confirmed that the gas gland cells have an enzymic equipment making them suited for the production of lactic acid at a rapid rate (D'Aoust, 1970).

Recent studies of metabolic enzyme ratios have contributed to give a fuller picture of the metabolism of the gas gland cells. Activities of hexokinase, glucose-6-phosphate dehydrogenase, 6-phosphogluconate dehydrogenase, aldolase and lactate dehydrogenase are remarkably high, while activities of phosphofructokinase, fumarase and cytochromoxidase are relatively low. Comparisons were made between gas gland tissue and skeletal muscle (Boström, Fänge and Johansson, 1971; Fänge, 1973). The lactate dehydrogenase of the gas gland is of the same isozymic type as found in anaerobically metabolizing tissue (Gesser and Fänge, 1971). The high activities of the shunt enzymes (glucose-6-phosphate dehydrogenase and 6-phophogluconate dehydrogenase) indicate that the glycolysis may largely bypass the reaction catalysed by the allosteric enzyme phosphofructokinase. This may explain why the gas gland cells, as shown by several investigators, show little or no Pasteur effect. They produce lactic acid even at high partial pressure of oxygen. Synthesis of large amounts of cholesterol and other lipids is observed in the gas gland of the deep water fish *Coryphaenoides* (Phleger, 1971). The specific metabolism of gas gland cells may in part be a biochemical response to the high pressure of oxygen prevailing in the swimbladder (D'Aoust, 1970).

The newly secreted gas consists mostly of oxygen and carbon dioxide. As much as 40% or more of carbon dioxide has been measured in rapidly secreting physoclists (*Perca*: Jacobs, 1930; *Pomatomus*: Wittenberg, Schwend and Wittenberg, 1964). The carbon dioxide emanates from two sources: bicarbonates of the blood and the metabolism of the gas gland, especially the hexosemonophosphate shunt activity. In many species the production of carbon dioxide is easy to observe during the early phase of the gas secretion, when loss of carbon dioxide due to diffusion is at minimum because of the mucosal movements (Fig. 2b).

The enzyme carbonic anhydrase occurs in the gas gland cells, the erythrocytes of the blood and perhaps the vascular walls of the counter-current vessels. There are different theories about its function. Several details of the gas secretory mechanism are still unexplained (Berg and Steen, 1968; Fänge, 1973). After administration of ^{3}H-histidine, significant amounts of labelled histamine appear in the eel swimbladder. However, preliminary studies *in vitro* of the histidine decarboxylase activity in the gas gland of *Gadus* showed only traces of histamine-forming enzyme. At present no definite statement on the role of histamine in the swimbladder can be made (B. Holstein, personal communication).

2. *Nervous control*

Bohr (1894) showed that in the codfish (*Gadus morhua*) vagotomy abolishes gas secretion. Several later authors have confirmed that the vagus nerve carries secretory fibres to the gas gland cells. There is some evidence, although mostly indirect, that the secretory fibres may be cholinergic (Fänge, 1973). Dehadrai (1962) observed that a vagal reflex producing bradycardia in the Indian fish *Ophiocephalus* increases the concentration of oxygen in the swimbladder.

The transmitter of the secretory nerve fibres, whether acetylcholine or some other substance, probably increases the glycolytic activity of the gas gland cells. The vasodilatation observed in the gas gland during gas secretion may be secondary, produced by substances released from the gas gland cells into the blood, for example lactic acide. Vasodilatation alone produced by yohimbine, an α-adrenergic blocking agent, does not lead to gas secretion (Fänge and Nilsson, 1972). Deck (1970) showed that the gas gland cells are stimulated by glucagon but not by adrenaline. Perfusion experiments of the gas gland of the codfish (*Gadus morhua*) combined with nerve stimulation showed that the blood vessels receive constrictor fibres from both the vagus nerve and the splanchnic nerve. The vessels showed α-adrenergic types of response (Nilsson, 1972).

V. Discussion and Conclusions

Considerable resemblances exist between swimbladders and primitive lungs in structure and function. Both types of organs contain smooth muscle layers, which are controlled by the autonomic nervous system. The controlling nerve, i.e. the vagus nerve or the vagosympathetic trunk, is a mixed nerve containing different kinds of fibres. The efferent innervation of the smooth muscles of swimbladders and maybe even lungs tends to be mainly adrenergic. In physostome fishes the gas content of the swimbladder is regulated either mechanically through the pneumatic duct, or by secretion and diffusion through the wall. The first mode of regulation may correspond to expiration and inspiration of air in primitive lungs. Relaxation of the pneumatic sphincter by adrenergic impulses may play a role in release of gases through the duct.

The volume homeostasis of the physoclist swimbladder is maintained solely by gas exchange through the wall. The permeability of the membranes, through which diffusion takes place, is controlled by efferent vagal fibres acting on an adrenergic receptor mechanism. It is tempting to assume that β-adrenergic effects generally increase the permeability for substances of smooth muscle containing membranes (e.g. in lung, swimbladder, blood-vessels) by producing relaxation. In the physoclist swimbladder this effect may be strengthened by the simultaneous contraction of the adjacent secretory mucosa, which shows an α-adrenergic response. The functionally important specific distribution of two different types of adrenergic responses probably is phylogenetically very old in vertebrate organisms.

Gas secretion implies transport of gases against pressure gradients. The gases are mainly oxygen and carbon dioxide, but inert gases (nitrogen, argon) are involved too. Gas secretion is best developed in closed swimbladders (physoclist type). It may exist in respiratory swimbladders but has not been found in lungs, although it seems to be based on common physiological principles realized in many systems (counter-current circulation, glycolytic activity etc.). One may speculate about resemblances between gastric acid secretion in the stomach and the production of lactic acid by gas gland cells.

The volume homeostasis of the swimbladder permits the fish to keep buoyancy, saving energy by not having to swim continuously to avoid sinking. This function probably is of enough biological importance to explain why, during phylogenesis, primitive lung structures of the ancestors to modern fishes evolved into highly specialized swimbladders.

References

Abernethy, J. D. (1972). The mechanism of secretion of inert gases into the fish swimbladder. *Aust. J. exp. Biol. med. Sci.* **50**, 365–374.

Akita, Y. K. (1936). Studies on the physiology of the swimbladder. *J. Fac. Sci. Imp. Univ. Tokyo*, Sect. IV. **4**, 111–135.

Assheton, R. (1907). The development of *Gymnarchus niloticus*. "John Samuel Budgett Memor. Vol." 293–422. Cambridge.

Ball, E. G., Strittmatter, C. F. and Cooper, O. (1955). Metabolic studies on the gas gland of the swimbladder. *Biol. Bull.* **108**, 1–17.

Ballantyne, F. M. (1927). Air-bladder and lungs. *Trans. Roy. Soc. Edinb.* **55**, 371–394.

Barrel, J. (1916). Influence of Silurian-Devonian climates on the rise of air-breathing vertebrates. *Bull. Geol. Soc. Am.* **27**, 387–436.

Berg, T. and Steen, J. B. (1968). The mechanism of oxygen concentration in the swimbladder of the eel. *J. Physiol.* **195**, 631–638.

Bohr, C. (1894). The influence of section of the vagus nerve on the disengagement of gases in the air bladder of fishes. *J. Physiol.* **15**, 494–500.

Böker, H. (1933). Über einige neue Organe bei luftatmenden Fischen. *Anat. Anz.* **76**, 148–155.

Bone, Q. (1973). A note on the buoyancy of some lantern-fishes (Myctophoidei). *J. mar. Biol. Ass. U.K.* **53**, 619–633.

Boström, S. L., Fänge, R. and Johansson, R. G. (1972). Enzyme activity patterns in gas gland tissue of the swimbladder of the cod (*Gadus morrhua*). *Comp. Biochem. Physiol.* **43B**, 473–478.

Burnstock, G. and Wood, M. J. (1967). Innervation of the lungs of the sleepy lizard, *Trachysaurus rugosus:* II. Physiology and pharmacology. *Comp. Biochem. Physiol.* **22**, 815–831.

Campbell, G. (1971a). Autonomic innervation of the lung musculature of a toad, *Bufo marinus. Comp. Gen. Pharmacol.* **2**, 281–294.

Campbell, G. (1971b). Autonomic innervation of the pulmonary vascular bed in a toad, *Bufo marinus. Comp. Gen. Pharmacol.* **2**, 287–294.

Campbell, J. A. (1931). Gas tension in the tissues. *Physiol. Rev.* **11**, 1–40.

Carlson, A. J. and Luckhardt, A. B. (1920). Studies on the visceral sensory nervous system. I. Lung automatism and lung reflexes in the frog (*Rana pipiens* and *Rana catesbiana*) *Am. J. Physiol.* **54**, 55–95.

Carter, G. S. and Beadle, L. C. (1931). The fauna of the swamps of the Paraguyan Chaco in relation to its environment. I. Physicochemical nature of the environment. II. Respiratory adaptations in fishes. *J. Linn. Soc. Lond.* **37**, 205–258; 327–366.

Dale, W. A. and Rahn, H. (1952). Rate of gas absorption during atelectasis. *Am. J. Physiol.* **170**, 606–615.

D'Aoust, B. G. (1970). The role of lactate ion in gas secretion in the teleost swimbladder. *Comp. Biochem. Physiol.* **32**, 637–668.

Deck, J. E. (1970). Lactic acid production by the swimbladder gas gland *in vitro* as influenced by glucagon and epinephrine. *Comp. Biochem. Physiol.* **34**, 317–324.

DeGroodt, M., Lagasse, A. and Sebruyns, M. (1958). Elektronenmikroskopische Morphologie der Lungenalveolen des *Protopterus* und *Amblystoma* Proc. 4th Int. Conf. Electon Microscopy. Berlin **1**, 418–421. Springer, Berlin. (Cit. fr. Johansen, 1970, p. 372).

Dehadrai, P. V. (1957). On the swim-bladder and its relation with the internal ear in genus *Notopterus* (Lacapede). *J. Zool. Soc. India* **9** (1), 50–61. (Calcutta 1958).

Dehadrai, P. V. (1959). The swimbladder and its connection with the internal ear in family Cichlidae. *Proc. Natl. Inst. Sci. India*, **25B**, 254–261.

Dehadrai, P. V. (1960). Respiratory function of the swimbladder of *Notopterus* (Lacapede). *Nature, Lond.* **185**, 929.

Dehadrai, P. V. (1962). Observations on certain physiological reactions in *Ophicephalus Striatus* exposed to air. *Life Sci.* **11**, 653–657.

Denton, E. J., Liddicoat, J. D. and Taylor, D. W. (1972). The permeability to gases of the swimbladder of the conger eel (*Conger conger*). *J. mar. biol. Ass. U.K.* **52**, 727–746.

Dorn, E. (1962). Über den Feinbau der Schwimmblase von *Anguilla vulgaris*. *Z. Zellforsch mitrosk Anat.* **55**, 894–912.

Dreser, H. (1892). Notiz über eine Wirkung des Pilokarpins. *Arch. exp. Path. Pharmak.* (Leipzig) **30**, 159–160.

Dutt, S. and Thankam, V. (1968). The Sciaenidae of the coastal waters of Visakhapatnan. *J. Bombay nat. Hist. Soc.* **65**, 335–347.

Enns, T., Douglas, E. and Scholander, P. F. (1967). Role of the swimbladder rete of fish in secretion of inert gas and oxygen. *Adv. Biol. Med. Physics.* **11**, 231–244.

Evans, H. M. (1929). Some notes on the anatomy of the Electric eel, *Gymnotus electrophorus*, with special reference to a mouth-breathing organ and the swimbladder. *Proc. Zool. Soc. Lond.* **57**, 17–23.

Evans, H. M. (1930). The swimbladder and Weberian ossicles and their relation to hearing in fishes. *Proc. Roy. Soc. of Med.* **23**, 11.

Evans, H. M. and Damant, G. C. C. (1928). Observations on the physiology of the swimbladder in cyprinoid fishes. *Br. J. exp. Biol.* **6**, 42–55.

Fahlén, G. (1967). Morphological aspects on the hydrostatic function of the gas bladder of *Clupea Harengus* L. *Acta Univ. Lund.* II. **1**, 1–49; **2**, 1–37.

Fahlén, G. (1971). The functional morphology of the gas bladder of the genus *Salmo*. *Acta Anct.* **78**, 161–184.

Fahlén, G., Falck, B. and Rosengren, E. (1965). Monoamines in the swimbladder of *Gadus callarias* and *Salmo irideus*. *Acta Physiol. Scand.* **64**, 119–126.

Fänge, R. (1953). The mechanism of gas transport in the euphysoclist swimbladder. *Acta Physiol. Scand.* **30**, suppl. 110, 1–133.

Fänge, R. (1958). The structure and function of the gas bladder in *Argentina silus*. *Q. Jl. Microsc. Sci.* **99**, 95–102.

Fänge, R. (1966). Physiology of the swimbladder. *Physiol. Rev.* **46**, 299–322.

Fänge, R. (1973). The physiology of the swimbladder. *In* "Comparative Physiology" (Eds L. Bolis, K. Schmidt-Nielsen and S. H. P. Maddrell) 135–159. North-Holland Publ. Co.

Fänge, R. and Mattisson, A. (1956). *In* "Bertil Hanström Zool. Papers in Honour of His Sixty-Fifth Birthday" (Ed. K. G. Wingstrand) pp. 153–162. Lund, Sweden: Zool. Inst.

Fänge, R. and Nilsson, S. (1972). (Preliminary results); Cit. fr. Fänge (1973).

Fänge, R. and Wittenberg, J. B. (1958). The swimbladder of the toadfish (*Opsanus tau* L.) *Biol. Bull.* **115**, 172–179.

Fauré-Fremiet, E. and Cougny, A. (1937). Propriétés de l'ichtyocolle provenant de diverses sortes de vessies natatoires. *Bull. Mus. Hist. Nat. Paris.* (2) **9**, 188–192.

Franz, G. (1937). Über den Reflex des Gasspuckens bei Fischen und die Funktion des Weberschen Apparates. *Z. vergl. Physiol.* **25**, 193–238.

George, M. G. (1956). The anatomy and histology of the swimbladder in *Ophiocephalus* (Block) or (*Channa* Scopoli). *J. Zool. Soc. India*, **8**, 181–194.

Gérard, P. (1931). Les sacs aériens des Crossoptérygiens et les poumons des Dipneustes. Etudes anatomique et histologique. *Arch. Biol. Liège* **42**, 251–278.

Gesser, H. and Fänge, R. (1971). Lactate dehydrogenase and cytochrome oxidase in the swimbladder of fish. *Int. J. Biochem.* **2**, 163–166.

Geyer, F. and Mann, H. (1939). Atmung des ungarischen Hundfisches (*Umbra lacustris*) Grossinger. *Zool. Anz.* **127**, 234–245; 305–312.

Greenwood, P. H. (1963). The swimbladder in African Notopteridae (Pisces) and its bearing on the taxonomy of the family. *Bull. Brit. Mus.* (*Nat. Hist.*) *Zoology* **2**, No. 5, 377–412.

Guyénot, E. (1909). Les fonctions de la vessie natatoire de poissons téléosteens. *Bull. Sci. France Belg.* **7**, 203–297.

Hagman, N. (1921). Studien über die Schwimmblase einiger Gadiden und Macruriden. Dr. thesis, Lund, Sweden, pp. 1–124.

Harvey, H. H. (1963). Pressure in the early life history of Sockeye Salmon. Ph.D. thesis, Zoology, Univ. Brit. Columbia, pp. 1–267.

Hughes, G. M. (1967). Evolution between air and water. *In* "Development of the Lung" (Eds A. V. S. de Reuck and R. Porter) Ciba Found. Symp. Churchill, London, pp. 64–84.

Jacobs, C. (1898). Über die Schwimmblase der Fische. *Tübinger Zool. Arb. Lpz.* **3**, 385–411.

Jacobs, W. (1930). Untersuchungen zur Physiologie der Schwimmblase der Fische. I. Über die "Gassekretion" in der Schwimmblase von Physoclisten. *Z. vergl. Physiol.* **11**, 565–629.

Jacobs, W. (1934). Untersuchungen zur Physiologie der Schwimmblase der Fische. III. Luftschlucken und Gassekretion bei Physostomen. *Z. vergl. Physiol.* **20**, 674–678.

Jacobs, W. (1938). Untersuchungen zue Physiologie der Schwimmblase der Fische. IV. Die erste Gasfüllung der Schwimmblase bei jungen Seepferdchen. *Z. vergl. Physiol.* **25**, 379–388.

Jobert, C. (1878). Recherches anatomiques et physiologiques pour servir à l'histoire de la respiration chez les Poissons. *Ann. Sci. nat.* (6). Zool. **7**, 1–7.

Johansen, K. (1970). Air-breathing in fish. *In* "Fish Physiology" Vol. 4, (Eds W. S. Hoar and D. J. Randall). Academic Press, New York. pp. 361–411.

Johansen, K. and Reite, O. (1967). Effects of acetylcholine and biogenic amines on pulmonary smooth muscle in the African lungfish, *Protopterus aethiopicus*. *Acta Physiol. Scand.* **71**, 248–252.

Johansen, K., Hanson, D. and Lenfant, C. (1970). Respiration in a primitive air-breather, *Amia calva*. *Resp. Physiol.* **9**, 162–174.

Kerr, J. G. (1909). Note on swimbladder and lungs. *Proc. Roy. Phys. Soc. Edinb.* **17**, 170–174.

Klika, E. and Lelek, A. (1967). A contribution to the study of the lungs of the *Protopterus annectens* and *Polypterus* senegalensis. *Folia Morphologica* (Chech.), **15**, 168–175.

Krogh, A. (1904). Some experiments on the cutaneous respiration of vertebrate animals. *Skand. Arch. Physiol.* **16**, 348–357.

Krogh, A. (1919). Rate of diffusion of gases through animal tissues. *J. Physiol.* **52**, 391–408.

Krogh, A. (1936). "The Anatomy and Physiology of Capillaries" Sec. ed. Yale Univ. Press, New Haven. 422 pp.

Kudrjaschoff, B. A. (1926). Zur Frage des Baues der Schwimmblase von *Pseudoscaphirhymchus kaufmanii* (Bogdanov). *Anat. Anz.* **62**, 163–180.

Kutchai, H. and Steen, J. B. (1971). The permeability of the swimbladder. *Comp. Biochem. Physiol.* **39A**, 119–123.

Ledebur, J. von (1929). Beiträge zur Physiologie der Schwimmblase der Fische II. Versuch einer experimentellen Sonderung des Gassekretions- und Gasresorptionsorganes in der Schwimmblase von Physoclisten. *Z. vergl. Physiol.* **10**, 431–439.

Ledebur, J. von and Wunder, W. (1937). Beiträge zur Physiologie der Schwimmblase. IV. Beobachtungen an Stichlingen, die ihre Schwimmblase nicht mit Gas füllen könnten. *Z. vergl. Physiol.* **25**, 149–155.

Lomholt, J. P. and Johansen, K. (1974). Control of breathing in *Amphipnous cuchia*, an amphibious fish. *Resp. Physiol.* **21**, 325–340.

Lüling, K. H. (1964). Uber die Atmung des *Hoploerythrinus unitaeniatus* (Pisces, Erythrinidae). *Bonner Zool. Beiträge* **15**, 90–102.

McLean, J. R. and Burnstock, G. (1967). Innervation of the lungs of the lizard, *Trachysaurus rugosus*: I, Fluorescent histochemistry of catecholamines. *Comp. Biochem. Physiol.* **22**, 809–813.

McMahon, B. R. (1969). A functional analysis of the aquatic and aerial respiratory movements of an African lungfish, *Protopterus aethiopicus*, with reference to the evolution of the lung-ventilation mechanism in vertebrates. *J. exp. Biol.* **51**, 407–430.

Marshall, N. B. (1960). Swimbladder structure of deep-sea fishes in relation to their systematics and biology. *Discovery Rep.* **31**, 1–122.

Marshall, N. B. (1972). Swimbladder organization and depth ranges of deep-sea teleosts. *In* "The Effects of Pressure on Organisms" S.E.B. Symp. 26. Cambridge Univ. Press. pp. 261–272.

Moehres, F. P. (1940). Untersuchungen über die Frage der Wahrnemung von Druckunterschieden des Mediums (Versuche an Bodenfischen). *Z. vergl. Physiol.* **28**, 1–42.

Moore, R. H. (1970). Changes in the composition of the swimbladder gas of the striped mullet, *Mugil Cephalus*, during hypoxia. *Comp. Biochem. Physiol.* **34**, 895–899.

Mott, J. C. (1950). Radiological observations on the cardiovascular system in *Anguilla anguilla*. *J. exp. Biol.* **27**, 324–333.

Müller, W. (1950). Die Mündung des Luftganges beim Messerfisch und ihre konstruktive Gestaltung. *Zool. Anz. Ergänz.* **145**, 635–642.

Nielsen, J. C. and Munk, O. (1964). A hadal fish (*Bassogigas profundissimus*) with a functional swimbladder. *Nature, Lond.* **204**, 302–315.

Nilsson, S. (1971). Adrenergic innervation and drug responses of the oval sphincter in the swimbladder of the cod (*Gadus morhua*). *Acta Physiol. Scand.* **83**, 446–453.

Nilsson, S. (1972). Autonomic vasomotorinnervation in the gas gland of the swimbladder of the teleost (*Gadus morhua*). *Comp. Gen. Pharmacol.* **3**, 371–375.

Nilsson, S. and Fänge, R. (1967). Adrenergic receptors in the swimbladder and gut of a teleost (*Anguilla anguilla*). *Comp. Biochem. Physiol.* **23**, 661–664.

Nilsson, S. and Fänge, R. (1975). In preparation.

Nysten, M. (1962). *Ann. Mus. Roy. de l'Afrique Centrale*, Tervuren, Belg., Ser. 8, Sci. Zool. (No. 108), 186–220.

Packard, G. C. (1974). The evolution of air-breathing in paleozoic gnathostome fishes. *Evolution* **28**, 320–325.

Peter, K. J. (1963). Studies on the swimbladder of fishes. I. Morphology, Histology and gas content of the swimbladder of *Diodon hystrix L. Bull. Deptm. Marine Biol. Oceanogr.* Univ. of Kerala, Ernakulum, **1**, 133–146.

Phleger, C. F. (1971). Pressure effects on cholesterol and lipid synthesis by the swimbladder of an abyssal *Coryphaenoides* species. *Am. Zool.* **11**, 559–570.

Piiper, J. (1965). *In* "Handbook of Physiology" Sect. Resp. (3), Vol. 2, pp. 1205–1218. (Fenn, W. O. and Rahn, H., eds.) Washington, D.C.: *Amer. Physiol. Soc.*

Plattner, W. (1941). Etudes sur la fonction hydrostatique de la vessie natatoire des poissons. *Rev. Suisse Zool.* **48**, 201–338.

Poll, M. (1962). Etude sur la structure aduate et la formation des sacs pulmonaires des Protoptères. *Ann. Mus. Roy. de l'Afrique Centrale.* Tervuren, Belg., Sér. 8, Sci. Zool. (No. 108), XI, 128–172.

Potter, G. E. (1927). Respiratory function of the swimbladder in *Leopidosteus. J. exp. Zool.* **49**, 45–67.

Price, J. A. and Pierce, J. A. (1963). The physical properties of the frog lung. *Proc Soc. Expl. Biol. Med.* **112**, 538–541.

Quotob, Z. (1962). The swimbladder of fishes as a pressure receptor. *Archs. neérl. Zool.* **15**, 1–67.

Rahn, H., Rahn, B., Howell, B. J., Gans, C. and Tenney, S. M. (1971). Air-breathing of the garfish (*Lepisosteus osseus*). *Resp. Physiol.* **11**, 285–307.

Rauther, M. (1910). Die akzessorischen Atmungsorgane der Knochenfische. *Ergebn. u. Fortschr. d. Zool. Jena* **2**, 517–585.

Rauther, M. (1914). Über die respiratorische Schwimmblase von *Umbra. Zool. Jahrb. Allg. Zool. Physiol.* **34**, 339–364.

Rauther, M. (1937). Die pneumatischen Darmanhänge. *In* "Bronns Klassen und Ordnungen des Tierreiches" 6, Abt. I. Buch 2, pp. 759–826. Leipzig: Akad. Verl. M.B.H. 1940.

Rist, E. and Strohl, A. (1922). Sur le role de la diffusion dans la résorption gazeuse et le maintien de la pression sous-atmosphérique dans la plèvre. *La presse medicale*, 69–71.

Romer, A. S. (1972). Skin breathing – primary or secondary? *Resp. Physiol.* **14**, 183–192.

Rowntree, W. S. (1907). On some points in the visceral anatomy of the Characinidae. *Trans. Linn. Soc. Lond.*, (2) Zool. **9**, 47–81.

Saalfeldt, E. von (1933). Die nervöse Regulierung der Atembewegungen bei *Uromastix. Pflügers Arch.* **233**, 449–468.

Safford Black, V. (1944). Gas exchange in the swimbladder of the mudminnow. *Umbra limi* (Kirtland). *Proc. Nova Scotia Inst. Sci.* **21**, (1,2), 1–22.

Schnizer, W., Hoang, D. N. and Brecht, K. (1968). Transmitter in der Froschlunge. *Pflügers Arch. ges. Physiol.* **304**, 271–283.

Scholander, P. F. (1958). Counter current exchange. A principle in biology. *Hvalrådets Skr. Norske Videnskaps Akad.* Oslo, **44**, 1–24.

Siebold, C. T. von (1858). Ueber den Kilch des Bodensees (*Coregonus acronius*). *Z. wiss. Zool.* **9**, 295–299.

Srivastava, V. M. B. (1964). Some functions of the swimbladder and its ducts in Atlantic and Pacific herring. Ph.D. Thesis, Univ. Brit. Columbia, (Cit. fr. Diss. Abstr. 1967–8. **28**, 1736-B).

Stärk, F. (1924). Histologische Studien an der Schwimmblase der Renken. *Z. Anat. Entw. Gesch.* **70**, 308–312.

Steen, J. B. (1963). The physiology of the swimbladder of the eel, *Anguilla vulgaris*. *Acta Physiol. scand.* **59**, 221–241.

Stipetic, E. (1939). Über das Gehörorgan der Mormyriden. *Z. vergl. Physiol.* **26**, 740–752.

Sundnes, G. (1963). Studies on the High Nitrogen content in the Physostome swimbladder. *Rep. Norweg. Fish. Invest.* **13**, (No. 5), 1–8.

Sundnes, G. and Bratland, P. (1972). Notes on the gas content and neutral buoyancy in physostome fish. *Rep. Norweg. Fish. Invest.* **16**, 89–97.

Sundnes, G., Bratland, P. and Strand, E. (1969). The gas content in the coregonid swimbladder. *Rep. Norweg. Fish. Invest.* **15**, 274–278.

Thys van den Audenaerde, D. F. E. (1959). Existence d'une vessie natatoire pulmonoide chez *Phractolaemus ansorgei* Blgr. *Rev. Zool. Bot. Afr.* **59**, 364–366.

Tytler, P. and Blaxter, J. H. S. (1973). Adaptation by cod and saithe to pressure changes. *Netherlands J. Sea Res.* **7**, 31–45.

Wittenberg, J. B. (1958). The secretion of inert gas into the swimbladder of fish. *J. Gen. Physiol.* **41**, 783–804.

Wittenberg, J. B., Schwend, M. J. and Wittenberg, B. A. (1964). The secretion of Oxygen into the swimbladder of Fish. III. The role of carbon dioxide. *J. Gen. Physiol.* **48**, 337–355.

Wood, J. G. and Burnstock, G. (1967). Innervation of the lungs of the toad, *Bufo marinus*: I. Physiology and pharmacology. *Comp. Biochem. Physiol.* **22**, 755–766.

Morphometry of Fish Lungs

G. M. HUGHES and E. R. WEIBEL

Research Unit for Comparative Animal Respiration, Bristol University, England
and
Anatomical Institute, University of Berne, Switzerland

I. Introduction

As discussed elsewhere in this volume, a wide variety of organs concerned with gas exchange between air and blood are found among fishes. During recent years, a number of morphometric studies of these organs have been made, in particular those of *Anabas* and other Indian air-breathing teleosts (see pp. 96–102). There is, however, a lack of corresponding investigations on structures homologous with the mammalian lung on which a considerable amount of morphometric information has been accumulated over the past years (Weibel, 1963, 1973a).

Morphometry of fish lungs is of interest from three main points of view:

(i) comparison with the air-breathing organs of other fishes
(ii) as a model for the earliest stages in evolution of the vertebrate lung, and
(iii) in relation to the physiology of these fish.

Dipnoan lungs were first described by Gunther (1871)—*Neoceratodus*, Hyrtl (1845)—*Lepidosiren*, and Owen (1840) and Parker (1892), *Protopterus*. In the present century some more detailed anatomical studies have been made, e.g. Poll (1962), Grigg (1965). These have included good light microscopical studies, and in recent years the electron microscope has also revealed interesting structural features of these lungs; see De Groodt, Lagasse and

Sebruyns (1958); Klika and Lelek (1967); Hughes (1970, 1973); Creasey, Pattle and Schock (1974).

However, apart from gross topographical measurements, there have been very few measurements of parameters relating to their respiratory physiology.

This paper tries to provide an example of the way in which current morphometric methods used with mammalian lungs and fish respiratory systems can be applied to these lungs. Some of the methods used have been modified in relation to the particular anatomy, and perhaps will be of value in studies on the lungs of other lower vertebrates.

II. Materials and Methods

Two specimens of *Lepidosiren paradoxa* weighing about 500 g were available for this work. They were anaesthetized by injection of urethane, the heart exposed with minimal dissection and prepared for perfusion of the pulmonary circulation using the method of Gil and Weibel (1971). A cannula was inserted into the conus arteriosus and the auricles cut so as to allow the escape of blood. Perfusion was started immediately following insertion of the cannula. The lung remained in an inflated condition throughout the perfusion. It was decided not to attempt periodic artificial ventilation because the lung normally remains in a fully expanded condition for long periods.

III. Morphometric Model and Methods

Following their dissection from the rest of the body, the lungs were photographed and one of them was cut transversely at six different levels. At each of these levels, one thin slice was cut into smaller portions for electron microscopy, samples being taken separately from the alveolar and the duct regions. The other major pieces were then available for light microscopic studies. Photographs of the cross sections show the basic structure of the lung (Fig. 1) with a large air duct which is surrounded asymmetrically by the alveolar regions. The main duct becomes smaller distally but less so than the alveolar regions which are relatively greater in the more anterior parts. This subdivision into duct and alveolar regions is illustrated diagrammatically in Fig. 2a, where the intersections and point counts used for morphometry are also indicated. Sections such as this were used from each of the main regions and systematically scanned using a Wild automatic sampling microscope (Weibel, 1970). The grid used had a set of coarse lines between which the space was subdivided by a finer grid (Fig. 2a); points falling on the air space of these two regions were recorded using the coarse grid (P_a, P_d). Coarse grid points were also recorded for the main blood vessels (P_v) and septal regions of the lung (P_s). Fine grid points were recorded where they fell on capillary (P_c) and tissue (P_t) parts of the lung. Thus the total area of a given section

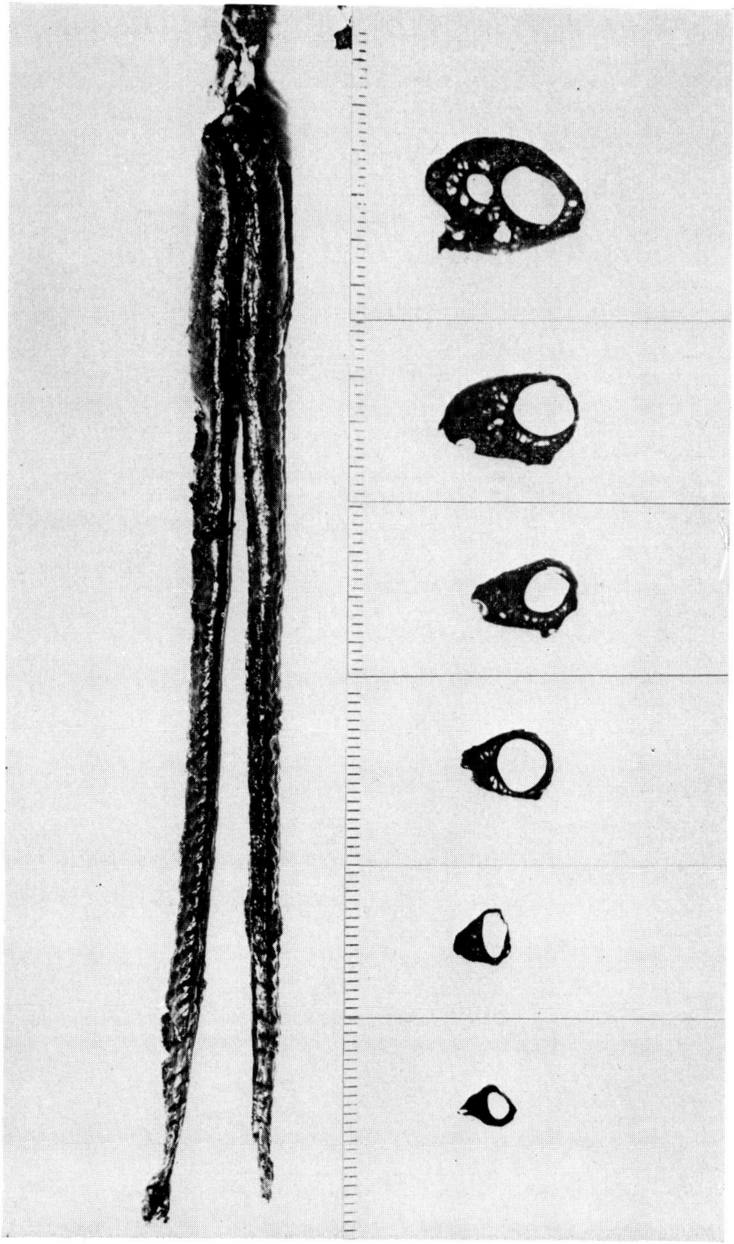

FIG. 1. Photographs of the excised perfused lung from a specimen of *Lepidosiren paradoxa* (body weight = 500 g). Left: The two lungs in ventral view (total length 30 cm). Right: Transverse sections (mm scale) at six different positions along the length of the lung.

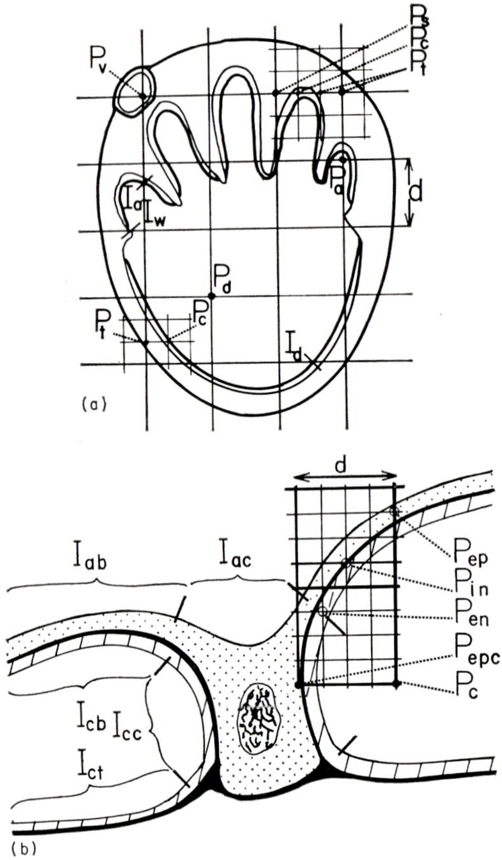

FIG. 2. Morphometric models. (a) For light microscopy, the section shows diagrammatically the duct and alveolar regions. For purposes of this diagram the grid is shown covering the whole section—in fact it was only used over a small part at any one time. Points of the coarse grid falling on these regions (P_d, P_a) are indicated. Coarse grid point counts were also carried out for P_v and P_S. The fine grid was used for counting P_c and P_t. Intersections with the alveolar surface and duct surface overlying capillaries were distinguished from surfaces where there were no capillaries (I_w). (b) Model used for electron micrographs showing the three barrier layers and a single epithelial cell body. The regions distinguished for intersections are shown, with the barrier alveolar surface (I_{ab}) and capillary surface (I_{cb}). Intersections with the alveolar and capillary regions adjacent to the cell body are indicated by I_{ac} and I_{cc} and the remaining capillary regions by I_{ct}. A coarser grid was used for point counts of the epithelial cell body and capillary, whereas the find grid is used for the three layers of the air/blood barrier.

was covered by the following points: P_d, duct air; P_a, alveolar air; P_c, capillary space; P_t, tissue: P_v, major vessels. Thus air-containing areas were $P_d + P_a$ and the total non-air area is made up of $P_v + P_t + P_c$. Points falling on the septa alone were also counted as P_p, but the same points were also either P_t or P_c.

Intersections of the grid with different surfaces were recorded for the duct respiratory surface (I_d) and the alveolar respiratory surface (I_a); intersections with non-capillary wall regions were recorded as I_w.

Measurements were made at two levels of magnification using this double lattice grid.

A. MODEL FOR ANALYSIS OF E.M. SECTIONS

This model (Fig. 2b) was primarily designed to obtain information about

(i) area and components of the respiratory surface;
(ii) the proportion of surface comprising epithelial cell bodies, and
(iii) the relative size of the capillaries.

The relative sizes of components of the barrier were estimated from points falling on the epithelial layer (P_{ep}), the interstitium (P_{in}) and the endothelium (P_{en}). Intersections with the outer epithelial surface were subdivided into those which fell over the capillaries (I_{ab}) and those which fell over the epithelial cell bodies (I_{ac}). A similar distinction was made for the capillary surfaces, i.e. those intersections of the air-blood barrier in the strict sense and those with the epithelial cell bodies. The line separating these two parts is not absolute but was usually taken as the region where the barrier thickness becomes more even. Intersections of the capillary surface on the septal side (I_{ct}) were also counted; this surface is not involved in gas exchange.

Analysis of the electron microscope sections made by systematic sampling

TABLE I

Lepidosiren *Lung. Summary of data from EM morphometry. For meaning of symbols, see text. Values given are unweighted means with their standard error for the six regions*

Scf/Sa	1·4259	0·2574
Sepc/Sa	0·5061	0·0358
Sc/Sa	2·527	0·584
Vc/Sa	3·302	0·5164
Vepc/Vc	0·3536	0·0874
(a)	0·4215	0·1049
(b)	0·2857	0·0749
Tb arith	3·7595	0·3354
Tb harm	0·8651	0·0476
Vep/Vb	0·0810	0·0106
Vin/Vb	0·2103	0·0105
Ven/Vb	0·1539	0·0246
Vep*/Vb	0·6307	0·0261

(Vep* includes Vep and Vepc)

of large sections was carried out by projecting 35 mm films onto a test system (Weibel, 1973b); a modified version of a computer programme used with mammalian lung was employed for the calculations. Data obtained from this analysis are summarized in Table I.

IV. Results

A. Gross Morphology

Each lung is wider anteriorly where the two lungs are fused and have a common cavity which opens into the posterior region of the pharynx at the glottis. This main airway continues throughout the whole length of the lung. The internal lung surface is much folded, especially in the dorsal regions. Folding takes the form of septa and trabeculae which are clearly visible when the lung is split open longitudinally (Fig. 3).

The alveolar sacs are about 1 mm in diameter; there are some variations in the different lung regions. The whole of the surface, and the main duct, is lined by respiratory epithelium except for certain areas, especially at the junction between the main duct and the dorsal alveolated region. Under the

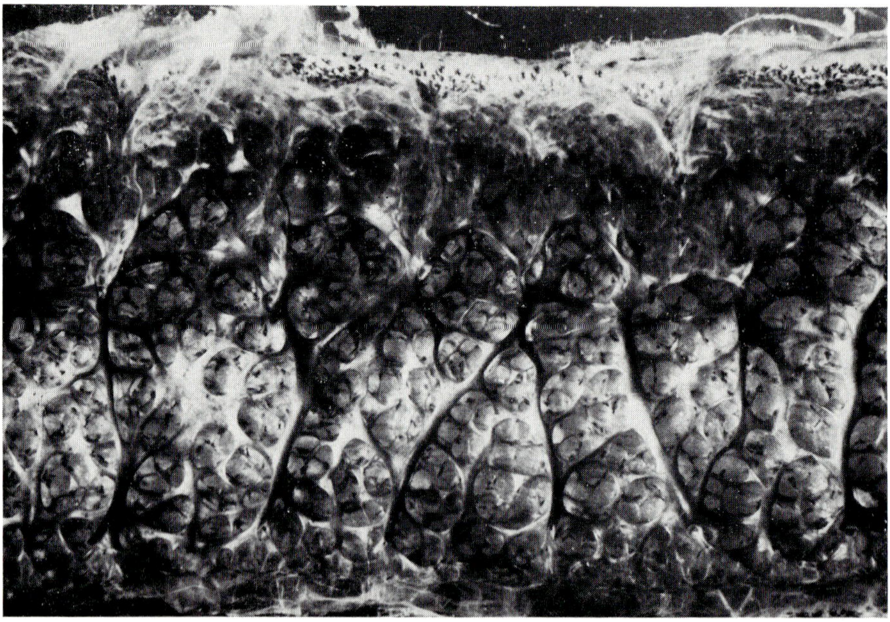

Fig. 3. View of the internal surface of *Lepidosiren* lung when opened up by a longitudinal section. The septa and trabeculae are clearly visible. (×7·5).

E.M. it is seen that the cells forming this epithelium have many lamellated bodies in their cell bodies. Over the surface of the lung there is a lung lining film which was very extensive in these perfused preparations (Fig. 4a). This film tends to become aggregated over the cell bodies themselves with the result that the whole alveolar surface is fairly smooth and continuous. The capillaries do not continue at the edges of some of the septa, and consequently

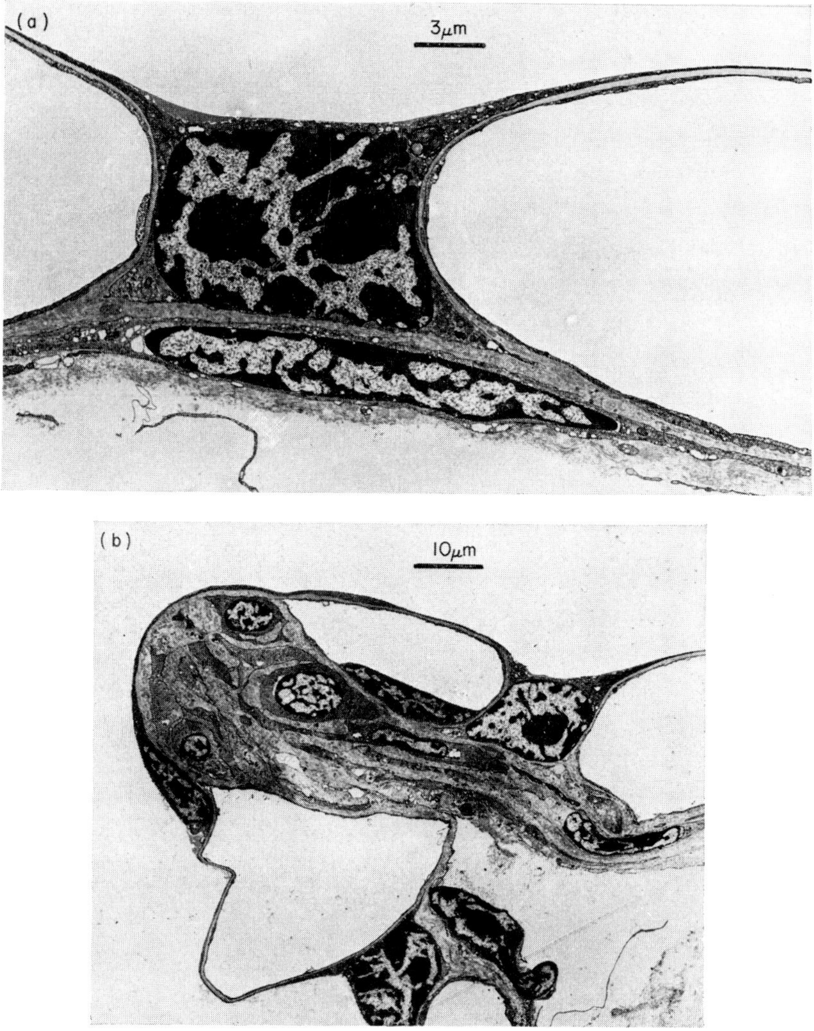

FIG. 4. Electron micrographs showing (a) single epithelial cell. The three layers of the barrier are visible and the extensive interstitial layer on the septal surface (×4380). (b) Section through the terminal edge of an interalveolar septum around which the epithelial layer is continuous but without any underlying capillary. (×1340).

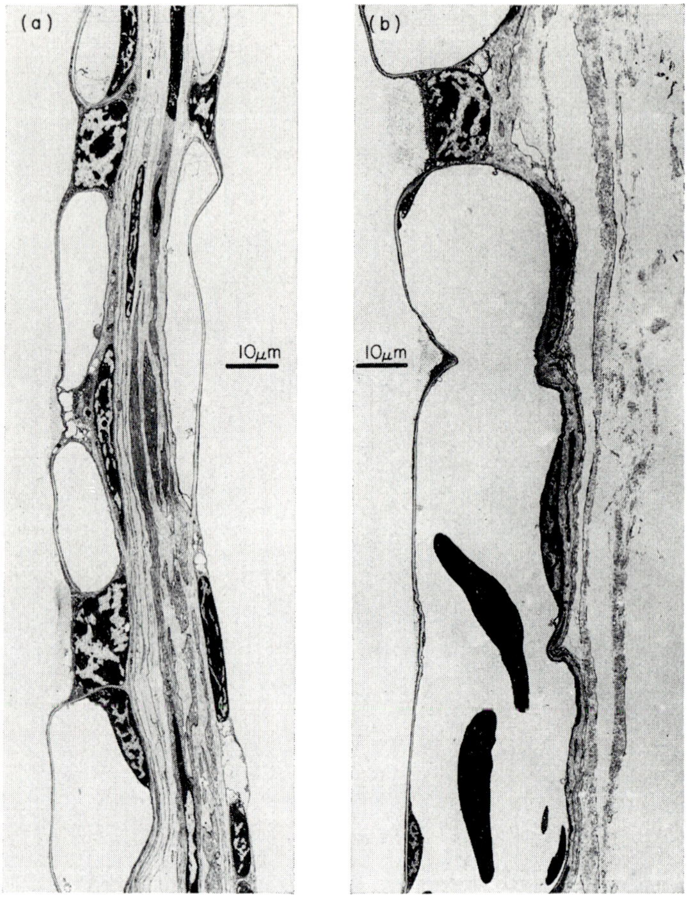

FIG. 5. Electron micrographs to show structure of the respiratory epithelium (×1094). In (a) capillaries are visible on both sides of the septum. Notice the aggregation of surfactant-like material in the spaces above the cell bodies. In (b) some red blood cells remain after the perfusion and show their large size.

non-respiratory surfaces are present (Fig. 4b). The septa are relatively thick, and, unlike the mammalian lung, there are capillaries only on one side. The septa contain much connective tissue and also smooth muscle (Fig. 5).

B. MORPHOMETRY

Broadly speaking, point counts give information about relative volumes of different structures, and intersections enable estimates to be made of surface areas. To account for the elongated shape of the *Lepidosiren* lung, we have subdivided it into six regions and have determined various parameters for each region. Thus from the relative areas covered by different tissues in a cross-section and multiplication by the length of the region, the relative surface or volume for the whole region is obtained (Fig. 6). Thus if in a cross-section, the area of a particular structure is A_i then the volume of this in the whole region $= A_i L_r$, where L_r is the length of the region. A_i is obtained by point counting so that $A_i = P_i \cdot d^2$, where d is the spacing of the points in the grid used. For the finer grid then $A_i = P_i \cdot d^2/q^2$ where q^2 is the number of points in the fine grid for each coarse grid point.

Thus using point counting we can readily determine the area of the alveolar air (A_a), of duct air (A_d), of capillaries (A_c), and of septa (A_s) etc. in a given cross section, so that the volumes for each of the six regions can be calculated.

By stereological methods (Weibel, 1973b) we can also estimate the volume density of the components from the relative number of test points which fall on profiles of that component. Thus the volume density of capillaries

$$V_{Vc} = \frac{P_c}{9(P_d+P_a)+P_t+P_c} = \frac{P_c}{P_{T(L)}}$$

where the denominator is the total number of points falling onto the lung; the factor 9 ($= q^2$) applied to P_a and P_d is necessary because P_d and P_a were counted with the coarse grid, whereas all other points were counted with the fine grid which had 9 times as many points.

The capillary volume in a given region is then obtained by multiplying V_{Vc} by the regional volume which is equal to the product of cross sectional area (A_r) and the length of the region:

$$V_c = V_{Vc} \cdot A_r \cdot L_r \cdot$$

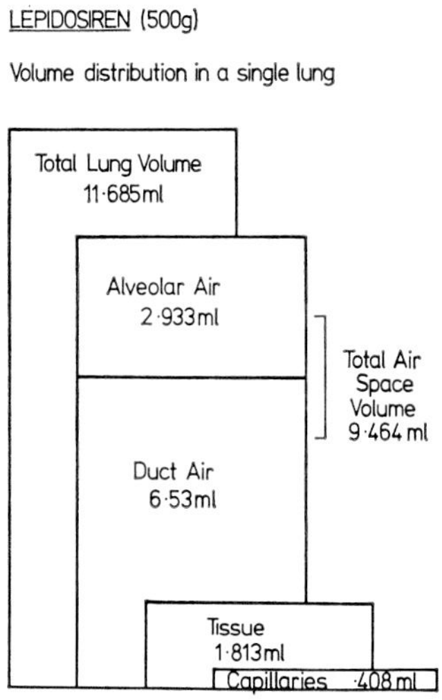

FIG. 6. Histogram indicating the distribution of volumes within a single lung.

A_r is obtained directly from the total number of points falling on the lung cross section:

$$A_r = P_{T(L)} \cdot \frac{d^2}{9}$$

Corresponding calculations can be applied to the measurement of surfaces using the method of intersection counting. The surface density of the alveolar air space is obtained from:

$$S_{Va} = 2I_{La}/L_{T(L)}$$

where

$$L_{T(L)} = P_{T(L)} \cdot 2d$$

and is the test line length enclosed within the lung; (if $P_{T(L)}$ is taken, as defined above, as the total number of fine points, then d must be replaced by d/3). By combining these two equations we obtain

$$S_{Va} = I_a/P_{T(L)} \cdot (d/3)$$

The total alveolar surface in one region of length L_r and cross-section A_r is then obtained by

$$S_a = S_{Va} \cdot A_r \cdot L_r$$

By introducing the above definition of A_r, we finally obtain

$$S_a = I_a \cdot L_r \cdot d/_3$$

This relationship has been used in this study for the calculation of Sa and Sd, i.e. the area of the alveolar and duct surfaces for a given lung region. Calculations were done separately for each region and summed to give values for the whole lung (Fig. 7). The data summarized in Table II and plotted in Fig. 8 indicates some of the differences between different lung regions. Calculations made from the primary results were carried out for a number of parameters which are of particular interest in relation to the function of these structures.

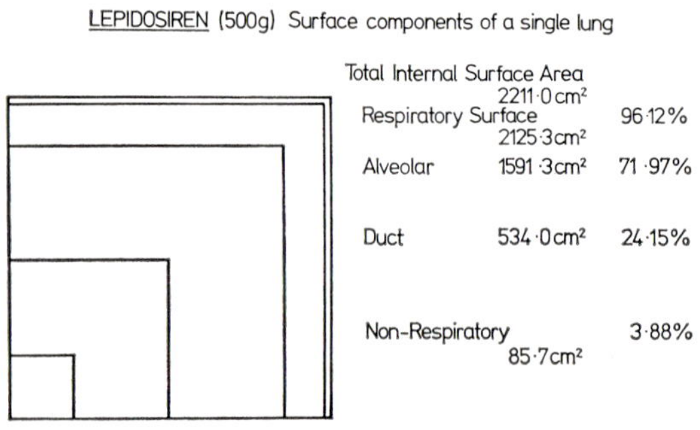

FIG. 7. Diagram summarizing the distribution of surface areas within a single lung. It is clear that a large proportion of the surface covers the alveolar regions of the lung.

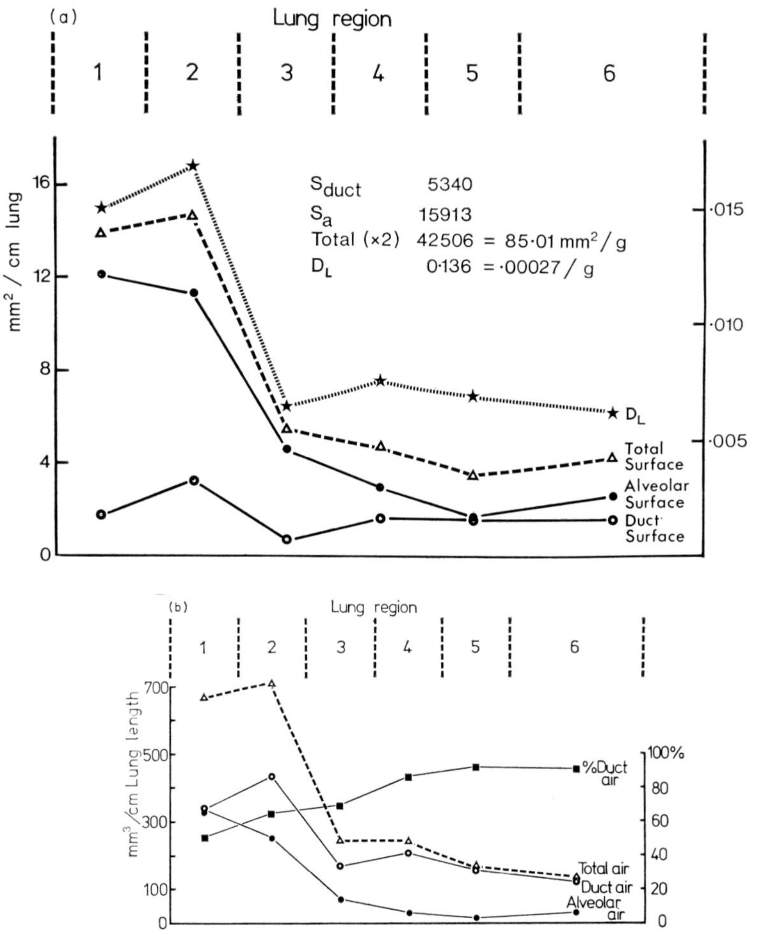

FIG. 8. Graphs indicating regional differences in area (a) and volume (b) along the length of a single lung. Measurements are given separately for each of the six sections. Notice that the final section was twice the length of the others and that the data is plotted per cm of lung length.

1. *Capillary loading*

Capillary loading (V_c/S_a) gives some indication of the surface area available for gas exchange with a given volume of blood. Morphometrically it is equivalent to the ratio of the respective volume and surface densities. Hence:

$$V_c/S_{Va} = V_{Vc}/S_{Va} = \frac{P_c}{P_{T(L)}} \div \frac{I_{ab}+I_{ac}}{P_{T(L)}\cdot d}$$

$$= P_c \cdot d/I_{ab}+I_{ac}$$

TABLE II

Dimensions of Lepidosiren *lung showing differences between the six regions*

Region	1	2	3	4	5	6	Totals for one lung
Length (L_r)	4 cm	4 cm	4 cm	4 cm	4 cm	10 cm	
Surface area (mm^2)							
alveolar	4852·8	4542·9	1863·4	1208·8	707·0	2738·4	15913·3
duct	720·1	1296·1	296·8	672·1	707·0	1647·4	5339·5
Total respiratory	5572·9	5839·0	2160·2	1880·9	1414·0	4385·8	21252·8 mm^2
Barrier thickness (μm) Harmonic mean	0·874	0·882	0·691	0·792	1·037	0·916	0·865 μm
Volumes (mm^3)							
alveolar	1320	1008·6	293·2	132·3	52·8	126·1	2933
duct	1335·7	1822·1	674·5	840·1	619·3	1238·6	6530·5
capillary (from light microscopy)	71·7	97·1	44·6	41·5	42·1	111·4	408·4

Capillary loading was calculated for each region separately. It appears to be less in the anterior parts of the lung, e.g. average for regions 1–3 is 8·446, but for regions 4–5 16·286. The overall average being 12·366 ml/m^2. From this value and determinations of total respiratory area we obtain an estimate of the volume of blood in the lung of 0·26–0·4 ml.

2. *Functional surface of capillary*

Functional surface of capillary relative to total capillary surface (S_{cf}/S_c) indicates the proportion of the pulmonary capillary surface available for gas exchange. It is obtained from the ratio $I_{cb}+I_{cc}$: $I_{cb}+I_{cc}+I_{ct}$ or as the ratio between S_{cf}/S_a and S_c/S_a. These latter ratios also enable us to obtain absolute values for S_{cf} and S_c, knowing S_a from the light microscopic investigation.

The mean of the values obtained for S_{cf}/S_c for each of the regions was 58·4%.

3. *Volume of barrier per unit surface area*

Volume of barrier for unit surface area (V_b/S_a) is obtained from the fine grid point and the intersection counts.

$$\frac{V_b}{S_a} = \frac{P_{ep}+P_{in}+P_{en}+P_{epc}}{I_{ab}+I_{ac}} \cdot \frac{d}{16}$$

A mean of values for each of the six regions was 4·83 cm^3/m^2.

4. *Barrier thickness*

In the calculation of V_b given above the epithelial cell bodies were considered as part of the barrier and this was also true for the calculations of barrier thickness. Arithmetic mean barrier thickness was obtained as the ratio between the total volume of the barrier and the mean of the alveolar and functional capillary surfaces:

$$T_b \text{ arith} = \frac{P_{ep}+P_{in}+P_{en}+P_{epc}}{I_{ab}+I_{ac}+I_{cb}+I_{cc}} \cdot \frac{d}{16}$$

Harmonic mean barrier thickness was obtained by measuring the length of the lines of a standard test grid between their intersection with the alveolar surface and the functional capillary surface. As expected the harmonic mean thickness was much less (0·865 μm) than the arithmetic mean (3·76 μm), the former takes less account of the long distances crossing the epithelial cell bodies, which are relatively more numerous in *Lepidosiren* lungs than those of mammals.

5. *Diffusing capacity*

Calculations were carried out using basically the same method as that employed for the mammalian lung (Weibel, 1973a,b). The overall diffusing capacity (D_L) being divisible into components for the tissue barrier (D_t), for the plasma (D_p), and for the erythrocyte (D_e). The surface lining layer was considered as part of D_t.

$$\frac{I}{D_L} = \frac{1}{D_t} + \frac{1}{D_p} + \frac{1}{D_e}$$

Values for Dt, Dp, De were calculated for each of the regions separately as follows:

$$D_t = K_t \cdot \frac{S_a+S_{cf}}{2} \cdot \frac{1}{T_b\text{harm}}$$

The value of Krogh's permeation coefficient taken was $3{\cdot}3 \times 10^{-6}$ml O_2/mm Hg/min/mm^2/μm as used by Weibel (1973a), after Grote (1967)

$$D_p = K_p \cdot S_{cf}/T_b p$$

K_p was taken as $4{\cdot}3 \times 10^{-6}$ and the value for $T_b p$ was 0·25 μm.

$$D_e = \theta \cdot V_c$$

The value taken for θ, the rate of O_2 uptake by whole blood, was the maximum (= 2·7 ml/ml.min.mm Hg) given in the literature, again the value most frequently used in mammalian studies.

For a single lung, the sum of the values calculated for each of the regions were: $D_t = 0{\cdot}09048$; $D_p = 0{\cdot}4553$; $D_e = 0{\cdot}83871$. Consequently, $D_L = 0{\cdot}0692$. Hence overall $D_L/W = 0{\cdot}000277$ ml O_2/min/mmHg/g.

In these calculations a number of assumptions have been made, and corrections might be introduced in relation to differences between the *Lepidosiren* lung and the mammalian lung which has been taken as a model. As presented, the data are of value for comparisons between these lungs as morphometric entities, but for a closer approach to the diffusing capacity of the *Lepidosiren* lung under its normal physiological conditions, some modifications in detail are desirable.

Such modifications might include:

(1) The value for K_t taken here is based on determinations for mammalian lung tissue at 37°C (Grote 1967). In other calculations of diffusing capacity of air-breathing organs of fishes (Hughes and Munshi, 1970; Hughes, Singh, Guha, Dube and Munshi, 1973) values based on determinations by Krogh (1919) for frog connective tissue at 20°C have been used. The latter figure is about half that obtained for mammalian lung tissue. The effect of temperature on K is not so marked because its component parts (solubility coefficient, α, and diffusion coefficient, D) are affected in opposite directions. Over the physiological range for fish the change is probably less than 10% for a change of 10°C. (Hughes and Roberts, 1970, also see Dejours, this volume).

(2) Although harmonic mean barrier thickness is appropriate, the mode of determination takes no account of differences in curvature of the alveolar and capillary surfaces.

(3) Values of K for the plasma have only been determined for mammals. It seems unlikely that lungfish plasma has a markedly different value, though a reduction of about 10% should probably be made because of the temperature difference.

(4) The major difficulties arise from the differences in red blood cell size and their nucleated nature in lungfish. Whereas the erythrocytes of mammals are 5–7 μm in diameter, the ovoid cells of lungfish and amphibians are about 15–30 μm in their smallest axis. One effect of this is to increase distances within the plasma and secondly that the proportion of the capillaries occupied by red cells is less than in mammals where the capillary haematocrit (V_e/V_c) is about 0·5. In *Lepidosiren* the nuclei occupy part of V_e and hence the appropriate ratio here is V_{ecyt}/V_c. The volume of cytoplasm (V_{ecyt}) is about 84% of V_e.

Thus for a mammal: $D_e = \theta \cdot V_c \cdot H_c/\bar{H}$, where H_c and $\bar{H}$ are the haematocrits for capillary blood and mixed venous blood respectively.

And for animals with nucleated red cells,

$$D_e = \theta \cdot V_c \left(\frac{V_{ecyt}/V_c}{\bar{H}} \right) = \theta \cdot V_{ecyt}/\bar{H}$$

In the case of *Lepidosiren*, $D_e = \theta \cdot 0 \cdot 84 V_e/0{\cdot}28 = 3 \cdot V_e$. However, this assumes that θ has been determined for *Lepidosiren* blood. If we are to use the only available values, for mammalian blood, we must make some allowance for the lower haematocrit.

Thus a mammalian value for $\bar{H}$ must be taken and the appropriate H_c calculated for *Lepidosiren* blood:

$$D_e = \frac{\theta \cdot V_c \cdot 0{\cdot}84 \cdot 0{\cdot}28}{0{\cdot}45}$$

$$= \theta \cdot V_c \cdot 0{\cdot}52$$

V. Discussion

This study provides the first set of morphometric data for a fish lung but, because of their essentially similar structures, the methods discussed could be applied to the lungs of many urodeles, ichthyophids, ophidians and some other reptiles. In spite of their homogeneous appearance these tubular lungs of fish and other vertebrates show regional differences, on the nature and functional significance of which we have little information. In *Lepidosiren* there is a preponderance of gas exchange surface nearer to the connexion with the alimentary canal. The more posterior regions might, therefore, be considered relatively more important in producing airflow during the ventilatory cycle. This antero-posterior differentiation is also known from snake and lizard lungs but it has not yet been studied quantitatively.

During expiration, nearly all the air is removed from the lung of *Lepidosiren* (Bishop and Foxon, 1968) and for *Protopterus*, tidal volume was about 80% of the lung volume (Jesse, Shub and Fishman, 1967). Lung volume for these juvenile (57–71g) *Protopterus* was about 26 ml/kg as estimated from radiographs. The lung volume for *Lepidosiren* was about 20 ml/kg for each lung. This value is much lower than that expected for an amphibian of the same body weight and the respiratory surface area is about 1/10th (Tenney and Tenney, 1970). When compared with other air-breathing fishes, however, *Lepidosiren* has a greater respiratory area, e.g. about 4× that of the air-breathing organs of *Anabas*. Furthermore, the respiratory surface available for each ml of air contained in the lung is 44·53 cm^2 which is more than twice the corresponding value for *Anabas* and four times that for *Saccobranchus*.

In estimating diffusing capacity morphometrically the *Lepidosiren* lung was treated as though it were a mammal. Hence the figures given are directly comparable with those for mammalian lungs. However, to estimate the functional diffusing capacities, i.e. for *Lepidosiren* at 25°C, this needs to be modified, but by exactly how much is very uncertain, because all the necessary constants are not available. Making approximations as described on p. 226, the values for a single lung are as follows:

$$D_t = 0{\cdot}0814 \text{ or } 0{\cdot}0407 \text{ using Krogh's value for } K_t$$

$$D_p = 0{\cdot}4553$$

$$D_e = 0{\cdot}43613$$

Hence, $$D_L = 0{\cdot}05876$$

and for both lungs $D_L/W = 0{\cdot}000235\ O_2$/min/mmHg/g.

This value is the same order of magnitude as those of other cold-blooded vertebrates that have been studied but significantly less than the diffusing capacity of the lungs of mammals of the same body weight (Table III).

TABLE III

Diffusing capacities (ml· O_2/min/mmHg/kg)

Rat	(500 g)	4·3	D_L
Dog	(500 g)	1·75–4·0	D_L
Terrapin*	(1 Kg)	0·54	D_t
Tortoise*	(1 Kg)	0·26	D_t
Anabas (air)	(100 g)	0·28	D_t
Lepidosiren	(500 g)	0·27	D_L
Lepidosiren		0·235	D_L
(corrected)		0·163	D_t

* Unpublished data kindly given by Dr. S. V. Perry.

A. COMPARISON WITH THE MAMMALIAN LUNG

The fish lung is a simpler, sac-like structure with fewer internal foldings. Capillaries are found only on one side of the septa whereas this condition may be present only during development of the mammalian lung. The amount of air contained within the lung is about 80% of the total lung volume. Of the remainder, which can be called tissue volume, the proportion occupied by the capillaries is very much greater in mammals. In the rat, for example, it is 14·1% of the total lung volume, whereas in *Lepidosiren* the capillaries only constitute 3·5%. Thus the proportion of the non-air volume comprising the capillaries is 18·4% in *Lepidosiren* but much greater (61·8%) in the rat. Comparisons made here are generally with mammals about the size of a rat in order to reduce complications from scaling factors as the relationship between many of the components and body weight is not linear (Table IV).

Another marked difference in the structure of the capillary bed emerges from measurements of the ratio between the volume of the epithelial cell bodies to the total volume of the capillaries. In *Lepidosiren* this ratio is 0·354; the same computer programme gave 0·169 for measurements on dog lung.

TABLE IV

Comparison of lung parameters for Lepidosiren *and mammal of the same body weight*

	Lepidosiren 500 g	*Lepidosiren* / Mammal	Mammal (Rat) 500 g
Resp. S_a/W	85	1/25	2000 mm^2/g
Harmonic mean t	0·86	2	0·40 μm
D_L/W =	0·000271	1/16	0·0043 ml/min/mmHg/g
% lung vol = tissue	15·5	2	8·7%
V_b/S_a	4·1–4·88	4	1·171 cm^3/m^2
Capil volume	0·512–0·662	1/3	1·5 ml
S_c/S_a	1·99–2·5	2	0·8–1·05
V_c/S_a	12·37–18·87	10	1·2–1·3 ml/m^2
Capil functional surface	58·43% of total	1/2	nearly 100%
V_{epc}/V_c	0·354	2	0·178

Estimates from data on the development of the rat at 131 days (Burri, Dbaly and Weibel, 1974; Fig. 9) for the ratios between volume of the epithelial layer (including cell bodies) and capillaries was 0·178. This change is probably related to the further differentiation of type 1 cells in the mammal and associated with their more extensive prolongations over the capillary surface. Thus in *Lepidosiren* it is found that the proportion of the lung alveolar surface overlying epithelial cell bodies is about 50%. Another way of expressing this comparison is as the *capillary loading* of the surface, i.e. V_c/S_a. For

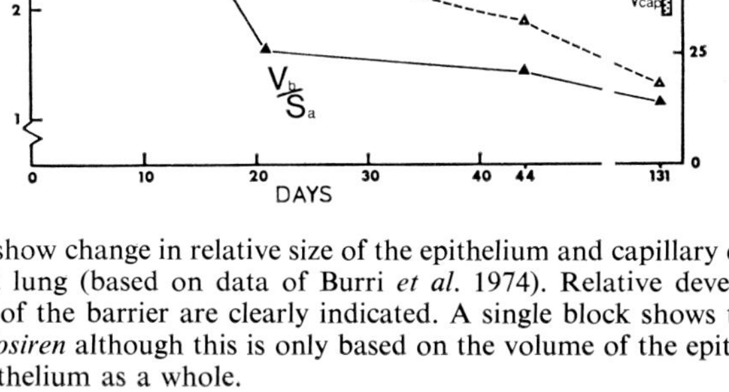

FIG. 9. Plot to show change in relative size of the epithelium and capillary during development of the rat lung (based on data of Burri *et al.* 1974). Relative development of the epithelium and of the barrier are clearly indicated. A single block shows the comparable figure for *Lepidosiren* although this is only based on the volume of the epithelial cell body and not the epithelium as a whole.

the dog, monkey, rabbit, rat and mouse this ranges from 1·20 to 1·30 ml/m^2. Corresponding figures for *Lepidosiren* are 12·37–18·87 ml/m^2. The ratio of capillary surface to the alveolar surface in dog and other mammals ranges between 0·8 and 1·05, whereas for *Lepidosiren* it is 1·99–2·5. One of the reasons for these differences is because the alveolar surface is only found on one side of the capillaries in *Lepidosiren*. Secondly, the high V_c/S_a in the lungfish is associated with very large capillary diameters between 22 and 28 μm (Fig. 4), as against diameters of 6–8 μm in the rat.

Calculation of the functional capillary surface including the surface adjacent to epithelial cell body, relative to total capillary surface gave values near to 60% whereas it is almost 100% in an adult mammalian lung.

Differences in the air/blood barrier are not so great but detailed comparisons are difficult because the figures obtained for *Lepidosiren* are for the barrier and epithelial cell bodies separately, whereas for the mammal both are included. In *Lepidosiren* 18·2% of the barrier itself, is made up of epithelium, 7·2% interstitium and 34·6% endothelium. For the mammal the percentages are 30, 40 and 30% respectively for the combined barrier and epithelial cell bodies.

VI. Summary

1. The gross morphology of the lung of *Lepidosiren* is described and a method of morphometric analysis is given.
2. Models for morphometry based on light and electron microscopy are described.
3. Measurements of the volume distribution indicate that anteriorly the proportion of air contained in the alveolar regions is greater than that in the more posterior parts of the lungs.
4. The total respiratory surface is about 85 mm^2/g body weight and, although greater than that of other air-breathing fish, it is significantly less than that of amphibians and higher vertebrates.
5. The ratio of alveolar surface to capillary volume is less than in lungs in which the septa contain a single layer of capillaries as in the mammal.
6. The harmonic mean thickness of the air/blood barrier is about 0·9 microns. A morphometric estimate is made of diffusing capacity based on the method used for mammals and gives values of 0·27 ml/min/mmHg/kg. Modifications to this calculation necessary to approach the physiological diffusing capacity are discussed.

Acknowledgement

We should like to thank Miss H. Claassen for her excellent technical assistance during all phases of this work. This work was supported by grants from the Swiss National Science Foundation (3.682.71), the Roche Foundation and the Medical Research Council.

References

Bishop, I. R. and Foxon, G. E. H. (1968). The mechanism of breathing in the South American lungfish, *Lepidosiren paradoxa*; a radiological study. *J. Zool. Lond.* **154**, 263–271.

Burri, P. H., Dbaly, J. and Weibel, E. R. (1974). The postnatal growth of the rat lung. I. Morphometry *Anat. Rec.* **178**, 711–730.

Creasey, J. M., Pattle, R. E. and Schock, C. (1974). Ultrastructure of inclusion bodies in Type II cells of lung, human and sub-simian. *J. Physiol.* **237**, 35–37P.

De Groodt, M., Lagasse, A. and Sebruyms, M. (1958). Elektronenmikroskopische Morphologie der Lungenalveolen des *Protopterus* und *Amblystoma*. 4th Int. Kongr. Elektronmicrosp. 418–421.

Gil, J. and Weibel, E. R. (1969/70). Improvements in demonstration of lining layer of lung alveoli by electron microscopy. *Resp. Physiol.* **8**, 13–36.

Grigg, G. C. (1965). Studies of the Queensland lungfish, *Neoceratodus forsteri* (Krefft) *Aust. J. Zool.* **13**, (2) 243–253.

Grote, J. (1967). Oxygen diffusion constants in lung tissue and water and its dependence on temperature. *Pflügers Arch. ges. Physiol.* **295**, 245–254.

Gunther, A. (1871). Description of *Ceratodus*. *Phil. Trans. Roy. Soc. Lond.* **CLXI**.

Hughes, G. M. (1970). Ultrastructure of the air-breathing organs of some lower vertebrates. 17th Congres Internat. de Microscopie Electronique, Grenoble, Vol. **3**, 599.

Hughes, G. M. (1973). Ultrastructure of the lung of *Neoceratodus* and *Lepidosiren* in relation to the lung of other vertebrates. *Folia Morph. Prague*, **21**, 155–161.

Hughes, G. M. and Roberts, J. L. (1970). A study of the effect of temperature changes on the respiratory pumps of the rainbow trout. *J. exp. Biol.* **52**, 177–192.

Hughes, G. M., Dube, S. C. and Munshi, J. S. Datta (1973). Surface area of the respiratory organs of the climbing perch, *Anabas testudineus* (Pisces: Anabantidae). *J. Zool. Lond.* **170**, 227–243.

Hughes, G. M., Singh, B. R., Guha, F., Dube, S. C. and Munshi, J. S. Datta (1974). Respiratory surface areas of an air-breathing Siluroid fish *Saccobranchus (Heteropneustes) fossilis* in relation to body size. *J. Zool. Lond.* **172**, 215–232.

Hyrtl, J. (1845). Monographie von *Lepidosiren paradoxa*. *Abhandlungen Böhmischen Gesellschaft der Wissenschaften. Prague*. **III**, 605–668.

Jesse, M. J., Shub, C. and Fishman, A. P. (1967). Lung and Gill ventilation of the African lungfish. *Resp. Physiol.* **3**, 267–287.

Klika, E. and Lelek, A. (1967). A contribution to the study of the lungs of *Protopterus annectens* and *Polypterus senegalensis*. *Folia. Morph. Prague*. **15**, 168–175.

Krogh, A. (1919). The rate of diffusion of gases through animal tissue. *J. Physiol. Lond.* **52**, 391–408.

Owen, R. (1840). Description of *Lepidosiren annectens*. *Trans. Linn. Soc.* **XVIII**.

Parker, W. N. (1892). On the anatomy and physiology of *Protopterus annectens*. *Trans. Roy. Irish. Acad.* **30**, 109–230.

Poll, M. (1962). Étude sur la structure adulte et la formation des sacs pulmonaires des Protoptères. *Ann. Reeks Zool. Wetenschap.* **108**, 131–172.

Tenney, S. M. and Tenney, J. B. (1970). Quantitative morphology of cold-blooded lungs: Amphibia and Reptilia. *Resp. Physiol.* **9**, 197–215.

Weibel, E. R. (1963). "Morphometry of the Human Lung". Heidelberg, Springer.

Weibel, E. R. (1970). Automatic sampling stage microscope for Stereology. *J. Microsc.* **91**, 1–18.

Weibel, E. R. (1973a). Morphological basis of alveolar-capillary gas exchange. *Physiol. Revs.* **53**, 419–495.

Weibel, E. R. (1973b). Stereological methods. *In* "Principles and Techniques of Electron Microscopy", vol. 3. (Ed. M. A. Hayat), Van Nostrand, Reinhold, New York.

The Lung Surfactant in the Evolutionary Tree

R. E. PATTLE

Chemical Defence Establishment, Porton Down, Salisbury, Wiltshire, England

I. Introduction

A. Scope of Review

The lungs of mammals are lined with a film containing phospholipids and other substances, whose main function is to reduce surface tension. The present review deals with this film, and its supposed homologues in various taxa of vertebrates.

The great interest taken in the surfactant arises mainly from its connection with the respiratory distress syndrome of the newborn human. The literature is mountainous, but in the non-mammalian field as yet fragmentary. The reviewer apologizes to those authors whom he does not mention.

The earlier work on the subject was reviewed by Pattle (1965). Other books, reviews, and symposia are those of Pattle (1966, 1975), Scarpelli (1968), Wolstenholme and O'Connor (1969), Sayers (1973), Weibel (1973), Villee, Villee and Zuckerman (1973), Comline, Cross, Dawes and Nathanielsz (1973), Tierney (1974) and Clements (1974). The reference "Pattle, unpublished" (P.U.) will be used for unpublished observations from the author's laboratory.

The reader should be warned at the outset that this is a difficult field, in which qualitative findings have in the past been fairly easy to make, but in which quantitative data are hard to get and as often as not untrustworthy (Pattle, 1965; Young and Tierney, 1972). There is also confusion as to the meaning of some of the terms used, and I shall deal with this first.

B. DEFINITIONS

1. *"Surpellic", a new term*

It is unfortunately the case that students of the lung use the words "surfactant" and "surface active" in a much narrower sense than do ordinary physical chemists. In the language of the latter, the words apply to any substance which tends to accumulate at a surface and lowers surface tension. Thus acetic acid would be described as surface active. In the ordinary terminology the statement that "unsaturated lecithins are not surface active" would be untrue, but lung specialists regard it as a statement of fact. In their language "surfactant" and "surface active" imply the capacity to form a metastable surface film capable of sustaining for a considerable time a surface pressure well in excess of its equilibrium value, and sufficient to produce a net surface tension well below that of ordinary liquids.

In this paper I propose to use the term "surpellic" to describe films with this property or substances which produce such films. I apologise to scholars for using together the French prefix "sur", the Latin root "pellere" to push, and the Greek suffix "ic". The resemblance of the latter part to the word "pellicle", from the Latin "pellis", skin, is accidental but useful.

The term "lung surfactant" will be retained as a noun; the substance is surfactant in the ordinary sense as well as being surpellic.

2. *Classification of surpellic properties from bubble and trough tests.*

Lord Rayleigh (1890) demonstrated that the freshly formed or continuously stretched surface of a colloidal solution had a surface tension ("dynamic surface tension") greater than that of the undisturbed solution. What is less well known is that when the surface area is continually reduced, the surface tension falls below its equilibrium value. The phenomenon is familiar to lung physiologists, but not, I think, to surface chemists in general.

When a bubble is placed in air-saturated water (Fig. 1), its excess internal pressure, due to surface tension, causes air to dissolve and the volume to shrink (Fig. 2). In a colloidal solution the surface tension (S.T.) falls off in a way, depending on the rate of contraction, which has not been extensively studied. The effective S.T. may be estimated from the rate of contraction. An approximation formula is:

$$\gamma = 0{\cdot}075(d_1^3 - d_2^3)/t$$

where γ is the S.T. in mN/m, d_1 and d_2 are the initial and final diameters in μm, and t is the time in seconds between observations (see Pattle, 1958). Bubbles initially about 60 μm in diameter may conveniently be used; the final measurement may be taken at about 20 μm or after 20 minutes' observation. The results should be regarded as qualitative only, as bubbles in the same batch differ from one another.

From the results for blood quoted by Pattle (1958) a S.T. of 15 mN/m may be deduced. I have found 20 mN/m for Tween 80 solution. While these are below the equilibrium values, they are not remarkable, and I propose to call these substances "poorly surpellic". I shall call films reducing the S.T. under these circumstances to less than 8 mN/m "partially surpellic", and those reducing it to 0·1 mN/m or less, "fully surpellic". Thus the lining film of the goldfish swimbladder will be classed as poorly surpellic, those of newt, *Amia*, and *Lepisosteus* lungs as partially surpellic, and those of the lungs of mammals and Dipnoi as fully surpellic (Fig. 2).

The S.T. may also be estimated from the contours of sessile or "clicking" bubbles (*vide infra*). Clicking usually implies potentially very low S.T. and fully surpellic properties (see Pattle, 1958).

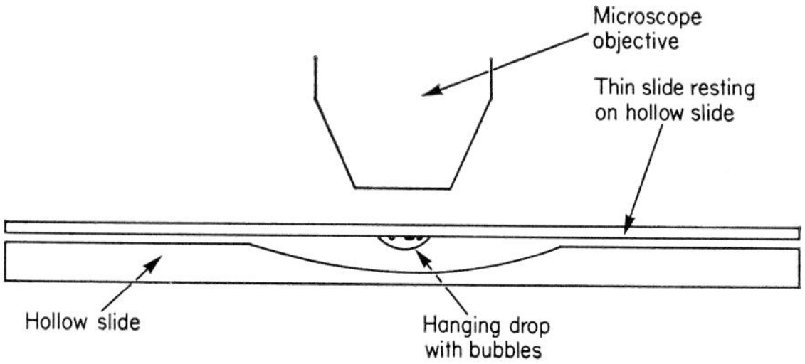

FIG. 1. Arrangement for viewing bubbles in hanging drop of air-saturated water.

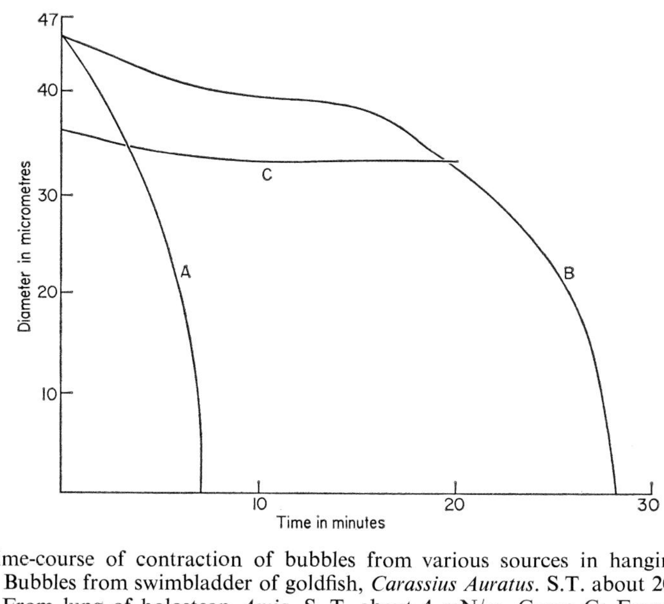

FIG. 2. Time-course of contraction of bubbles from various sources in hanging drop: Curve A: Bubbles from swimbladder of goldfish, *Carassius Auratus*. S.T. about 20 mN/m. Curve B: From lung of holostean *Amia*. S. T. about 4 mN/m. Curve C: From lung of dipnoan *Lepidosiren*. Final S.T. less than 0·1 mN/m.

Workers with the Wilhelmy trough commonly report "surfactant"—i.e. surpellic—properties if the minimum S.T. falls below 10 mN/m.

3. *Properties of di-palmitoyl lecithin: definition of "lung surfactant"*

The lung lining and lung washings contain di-palmitoyl lecithin (D.P.L.), which shows on the trough surpellic properties like those of lung washings. This has given rise to some controversy as to whether D.P.L. is "the" lung surfactant. Bubbles from slurries of D.P.L. contract slowly, often become non-spherical, and do not "click" (P.U.) (*vide infra*); in this way they differ from bubbles of pulmonary origin. In this paper the term "lung surfactant" will be applied to the substance, certainly a complex mixture, from which the lung surface film is derived.

4. *Lamellated bodies, lamellae and leaflets: definitions*

In lung it is usual to find osmiophilic bodies which in thin section have a lamellated appearance. The lamellae are often 20 to 100 nm thick. These bodies have been given various names in the literature. The present author has used "lamellated osmiophilic bodies" (LOPBs). There is now some indication that in non-simian mammals the lamellar appearance is artefactual, but, as this may not be the case in other groups, the reviewer proposes to keep this nomenclature.

The earlier authors referred to the coarser (20 to 100 nm) bands as "lamellae". Within the lamellae are systems of parallel sheets spaced at 4 to 5 nm, and lately some authors have referred to these as "lamellae", or failed to make it clear whether they are referring to the coarser or finer structures. It is proposed here to retain "lamellae" for the coarser structures, and "leaflets" for the structures spaced at 4 to 5 nm.

C. SOME DISPROVED HYPOTHESES

I do not propose to discuss the following hypotheses, which are to be found in the literature, but which I consider to have been already sufficiently confuted.

1. That the stability of bubbles from the lung is due to their possessing a surface film impermeable to gas.
2. That the lungs of birds and reptiles do not have a surface film like that of mammalian lung.
3. That the LOPBs are formed of material secreted by the Clara cells of the bronchi and phagocytosed by the Type II cells of the alveolus.
4. That the Type II cells of the alveolus are of mesodermal origin, and are not derived from the cuboidal cells present in the "glandular" stage of lung development.

5. That the lung surfactant is formed from the sloughed-off plasma membrane of the cells of the alveolar wall.
6. That the LOPBs of mammalian lung are normally derived from mitochondria.
7. That the LOPBs of human and simian lung are indistinguishable from those of non-simian mammals.
8. That surfactant passes through the blood-air barrier from the alveoli into the blood.
9. That surfactant passes through the blood-air barrier from the blood into the alveoli.

II. Lung Surfactant in Various Taxa

A. NON-SIMIAN MAMMALS

1. *Bubble observations*

The lung surfactant was first investigated in non-simian mammals. The lungs are lined with a surface film which is easily detached so as to form the lining of a bubble. This occurs when liquid enters the alveoli, either through a bronchus, or in acute pulmonary oedema. The bubbles, when placed in air-saturated water, show very low surface tension. This is manifested by the fact that the bubbles, say 50 μm or less in diameter, remain stable (Pattle, 1955, 1958). Bubbles of this size, in ordinary liquids, disappear in a few minutes (Fig. 2) under the influence of their own surface tension. If a fragment of lung is squeezed into a hanging drop of air-saturated water, stable bubbles small enough (2 μm or less) to show Brownian motion may be obtained. If bubbles derived from the lung are placed in de-aerated water, so that they dissolve, some of them execute a peculiar rhythmical change of shape; they slowly flatten, and suddenly resume spherical form. This is known as "clicking", and is a characteristic which has not been found in bubbles derived from non-pulmonary sources.

If an atelectatic lung is washed out with saline, and small bubbles are produced by injecting air into a hanging drop of the lavage fluid, the bubbles show stability, Brownian motion and sometimes clicking. This shows that there exists in lung a saline-dispersible substance, from which a surface film like that found in the inflated lung can be obtained.

2. *Trough observations*

The trough of Wilhelmy (1863) is an apparatus in which the surface tension of a liquid is measured by the force acting on a slide hanging in the liquid, and measurements can be made while the surface is being compressed or stretched. Extracts of lung or washings from the lung show very low surface tensions (down to 1 mN/m or less) when the surface is compressed (Clements, 1956, 1957).

3. *Surfactant chemistry*

The surface film obtained from bubbles has been analysed by Pattle and Thomas (1961) (pig) and by Klaus, Clements, and Havel (1961) (beef). The former found that its infra-red absorption spectrum resembled that of lecithin with about 1% of protein. The latter found 50%–70% of lipids, and 5% of nitrogen. The lipids contained 74% phospholipid, 8% cholesterol, 10% triglycerides, and 8% of fatty acids. It has been shown by Pattle (1958) that this film represents the original alveolar lining; it is not (as King and Clements, 1972, state) produced by foam fractionation of the lavage fluid. Much work has been done on the separation of surpellic fractions from lung lavage fluid; it is catalogued by King and Clements (1972). These authors carried out a careful fractionation of dog lung washings and isolated a fraction with 11% protein and 88% lipid; hexose, hexosamine, and nucleic acid were virtually absent. Of the lipid 78% was phosphatidyl choline, of whose fatty acids 79% were fully saturated (see Brown, 1964). Phosphatidyl ethanolamine, trigylcerides, and cholesterol made up most of the remainder of the lipid fraction.

More thorough analysis of the lining film is desirable to find out how far its composition corresponds to that of the fraction isolated by King and Clements (1972).

4. *Morphology of the lining film*

Although the presence of the lining film is easily deduced from the behaviour of lung bubbles (Pattle, 1958), its morphological demonstration is very difficult. This seems to be because it is thin (?10 or 20 nm), easily dissolved (because of its lipid composition), and highly surpellic. The latter property causes it to adhere to the surface and be washed off as a bubble lining. This makes it impossible to be sure that it is not washed off when the alveolus fills with fixative or embedding fluid. One must avoid confusion between "fixation" in the histological sense and "fixing" meaning permanent mechanical attachment. Bubbles which have a surpellic lining film can be squeezed from lungs histologically "fixed" with glutaraldehyde, as easily as from fresh lungs. Lung collapsed *in vivo* or *in vitro* fails to show, in the electron micrograph, the former alveolar lining film. The presence of a line, coloured or electron-dense, at the alveolar surface is no proof that it represents the surpellic lining film. Among workers who have visualised some, but probably not the whole, of the film are Dermer (1969), Gil and Weibel (1969/70), Pattle, Schock and Creasey (1972), and Hughes and Weibel (this volume).

Gil and Weibel (1969/70) also showed a layer of liquid, probably proteinaceous, but containing osmiophilic "myelin figures", under the lining film. The Type II cells have microvilli which would be squashed by surface tension forces, if they projected into air. This suggests that a little liquid is permanently present. The fact that bubbles squeezed out of a lung fragment bring the lining film with them also suggests a mobile layer between alveolar wall and lining film. The origin and composition of this layer await full elucidation.

Roth, Winkelmann and Meyer (1973) found in freeze-etched specimens a layer of 4 to 5 leaflets 4 nm thick lining the alveoli in places, and this is perhaps the best visualization of the surface film to date.

5. *Morphology of intracellular surfactant*

Macklin (1954) drew attention to the secretory properties of the Type II cell of the lung and suggested that its products might affect surface tension. Klaus, *et al.* (1961) proposed that the lamellated osmiophilic bodies (LOPBs) of the Type II cells were the source of the surfactant.

The evidence remained mainly circumstantial until Gil and Reiss (1973) obtained a fraction from homogenized lung which was shown by electron micrography to consist mainly of isolated LOPBs. This fraction was found to contain large amounts of fully saturated lecithin together with other lipids and protein. Its surface properties, as demonstrated by the Wilhelmy balance, resembled those of other preparations of lung surfactant. The evidence that the LOPBs consist mainly of lung surfactant is thus very strong. In the past pieces of lung have usually been prepared for electron microscopy by fixation in glutaraldehyde and osmium tetroxide, and embedded in Araldite with alcohol as dehydrator and epoxypropane as thinner. Sections from blocks prepared by this method often show the LOPBs badly eroded, and sometimes show vacuoles instead of LOPBs. Heavy metal fixation and staining give better results. Schock, Pattle and Creasey (1973) devised a solution containing 1 vol. M/40 lead nitrate and 1·05 vol. M/40 potassium ferricyanide, which helps to preserve the LOPBs. If chemical equivalents of the two salts are used blotchy precipitation takes place in the tissue. Uranyl acetate solution preserves LOPBs if it is administered by perfusion (Gil and Thurnheer, 1971) but not (Gil and Weibel, 1969/70) if the tissue is simply immersed in it. As seen in thin section under the electron microscope, the LOPBs (Fig. 3) are 0·5 to 1·5 μm across and display straight, arcuate, or concentric lamellae typically 20 nm to 100 nm thick. Within these lamellae, and running parallel to them, is a system of lines spaced at about 4·4 nm. As stated earlier, there is some confusion in the nomenclature. Serial sectioning shows that the lamellae in non-simians are sometimes domed or bell-shaped and sometimes flat throughout the body. The LOPBs appear to be formed by deposition of phospholipid in "multivesicular bodies". This process has been observed in the rat and the opossum (Sorokin, 1967) and in the hamster (Meban, 1972). Examination of freeze-etched specimens, however, makes it doubtful whether the interlamellar spaces exist *in vivo*. Illustrations of freeze-etched specimens of LOPBs are somewhat deceptive, as the most prominent features are the faces of 4 nm leaflets, where fracture has taken place between the leaflets

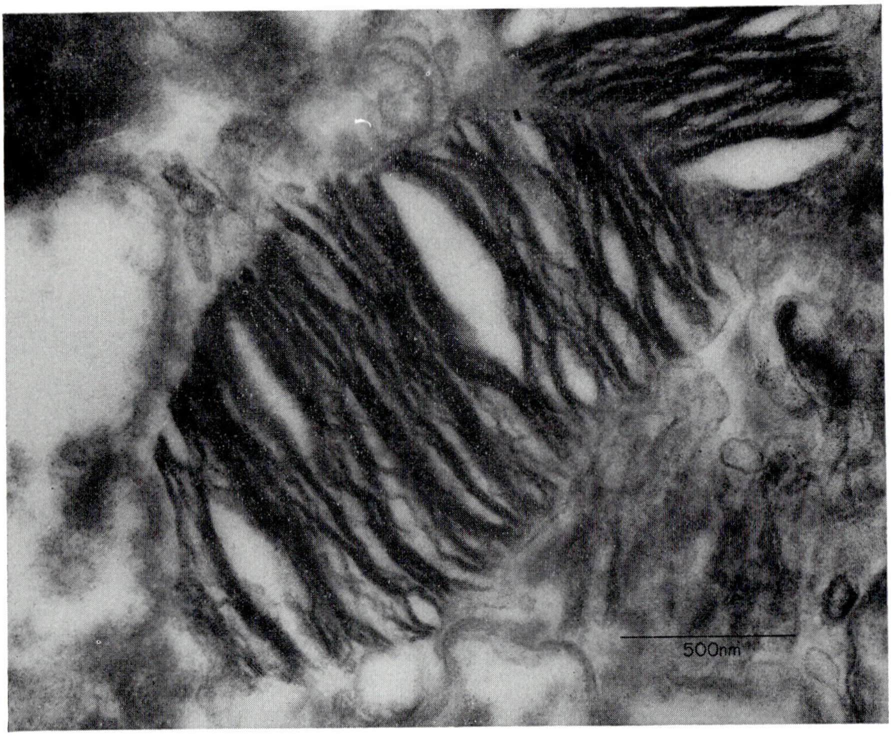

FIG. 3. Cross-barred lamellated bodies from Type II cell of guinea-pig lung. Scale bar 500 nm.

or by parting of their lipid layers; in thin section the leaflets are most prominent when they are at right angles to the section. In freeze-etched specimens where fracture has taken place approximately at right angles to the leaflets their spacing appears to be about 5 nm, and division of the block of leaflets into lamellae is (in non-simians) not obvious (P.U.). The appearances would be compatible with the leaflets having shrunk because of extraction of some components (perhaps triglyceride and cholesterol) by the embedding fluid, and come apart in places so as to give a lamellar appearance. It may be that the planes of splitting are determined by some real inter-leaflet structure, such as that described by Smith, Smith and Ryan (1973). Hughes, Ryan and Ryan (1974, and unpublished) however found that these structures were absent in *Protopterus* and *Lepidosiren*, whose LOPBs in thin section show a coarsely lamellated structure.

So far the LOPBs of marsupials, and of members of the eutherian orders Insectivora, Lagomorpha, Rodentia, Carnivora, Artiodactyla, and Perissodactyla have been examined and found to conform to the same pattern. The nonsimian pattern has also been found in the tree-shrew *Tupaia belangeri*, which some consider to have prosimian status, and in the mouse lemur, *Microcebus murinus rufus*, which ranks as a prosimian primate (Creasey, Pattle and Schock, 1974; Pattle, Gandy, Schock and Creasey, 1974). The LOPBs of monotremes have not been examined. Kikkawa and Spitzer (1969) claim that there are species differences between the LOPBs of different groups of non-simians. The LOPBs of the foetal rabbit and sheep, the dog, rat and mouse were, according to their findings, heavily osmiophilic and coarsely lamellar, those of the hamster and guinea-pig (but see Fig. 3) lightly osmiophilic and not lamellar, and those of the post-natal rabbit and sheep vacuolated with only a peripheral osmiophilic rim. We have, using tricomplex fixation (Schock *et al.*, 1973), examined the LOPBs of all these types, except the foetal sheep, and we find no systematic difference between them or the other non-simians examined. We therefore attribute the results of Kikkawa and Spitzer (1969) to erosion by embedding fluids.

6. *Quantitative aspects*

Quantitative work on surfactant is, as has already been pointed out, difficult. Gil and Thurnheer (1971) found, by morphometric measurements, that 1·22% by volume of the tissue of normal rat lung consists of LOPBs. Clements, Nellenbogen and Trahoun (1970), using the surface balance, studied extracts of lungs of various mammalian species. They found that these contained sufficient surpellic material to cover the internal surface area of the lung, at a surface pressure of 12 mN/m, between three and seven times over. The amounts of fully saturated lecithin per gram of wet tissue ranged from 2·30 mg (cow) to 7·95 mg. (guinea-pig).

7. *Embryology of surfactant*

The full functional development of surfactant in the foetus appears to occur in mice and rats just after the changeover from the "glandular" to the "canalicular" stage in lung anatomy. (Pattle, 1961). In the sheep the surfactant is only fully developed some 10 to 15 days after the change is complete (Howatt, Avery, Humphreys, Normand, Reid and Strang, 1965). In the thoroughbred foal the morphological change takes place around 210 days, while the surfactant does not mature till about 300 days, and is sometimes inadequate at full term (340 days) (Pattle, Rossdale, Schock and Creasey, 1975). Traces of surfactant can however be found, by the bubble "clicking" method, in the foal at 150 days, long before the morphological change has started, and while the lung tubules are fully lined with cuboidal epithelium. In the newborn foal, even at full term, signs referrable to surfactant deficiency sometimes appear, and constitute the "barker" or "neonatal maladjustment" syndrome (Rossdale, Pattle and Mahaffey, 1967; Rossdale, 1969, 1970, 1972). The corresponding "respiratory distress syndrome" in the human newborn is discussed in the next section. It is highly probable that careful search will reveal similar syndromes in other animals.

Surfactant is present in the lung of the pouch possum, a marsupial (Ellison, McPherson and Ellison, 1969).

8. *Surfactant reaction to silica*

The surfactant's best understood role is lowering the surface tension of the alveolar wall. It also appears to have protective properties. Heppleston and Young (1972) have found that under certain circumstances the lungs of rats and of man show a peculiar reaction to inhaled silica dust. Instead of the lung forming collagenous silicotic nodules, the Type II cells multiply and fill the lung with lipoprotein.

B. DIFFERENCES BETWEEN SIMIAN AND NON-SIMIAN MAMMALS

1. *Morphological*

The pattern of the LOPBs of man, apes, and monkeys (both old and new world) is quite different from that in non-simians (Creasey *et al.*, 1974; Pattle *et al.*, 1974; P.U.). The LOPBs in simians have thinner lamellae disposed in a concentric pattern (Fig. 4). It is not at all clear whether there is a bounding membrane. Freeze-etched specimens suggest that the inter-lamellar spaces are genuine and contain amorphous material (P.U.). Multi-vesicular bodies are rare in simians, and the LOPBs do not appear to be associated with them.

We have found this pattern in man, *Macaca mulatta* and *Erythrocebus Patas* (old world), and *Saimiri sciurea* (new world). The appearances would be compatible with direct formation of LOPBs from the endoplasmic reticulum (Balis and Conen, 1964). Van Golde, Fleischer and Fleischer (1971) found that enzymes responsible for de novo synthesis of lecithin were localized in the rough and smooth endoplasmic reticulum of rat and bovine liver. Auto-radiographic studies of lung cells (Chevalier and Collet, 1972; Massaro and Massaro, 1972; Dickie and Massaro, 1974) have suggested a transit of phospholipid and protein from rough endoplasmic reticulum to lamellar bodies and so to the extra-cellular space. It may be that in non-simians the phospholipid is stored in separate lamellar bodies, while in simians the endo-plasmic reticulum itself undergoes a change and stores phospholipid.

The morphology of the LOPBs in the tarsier, which in the evolutionary tree is more closely associated with the simian line than are the lemurs (Osman Hill, 1972), is unknown.

2. *Biochemical*

A critical review of this field has been given by Masoro (1973). Gluck, Kulovich, Eidelman, Cordero and Khazin (1972) claim that in the human,

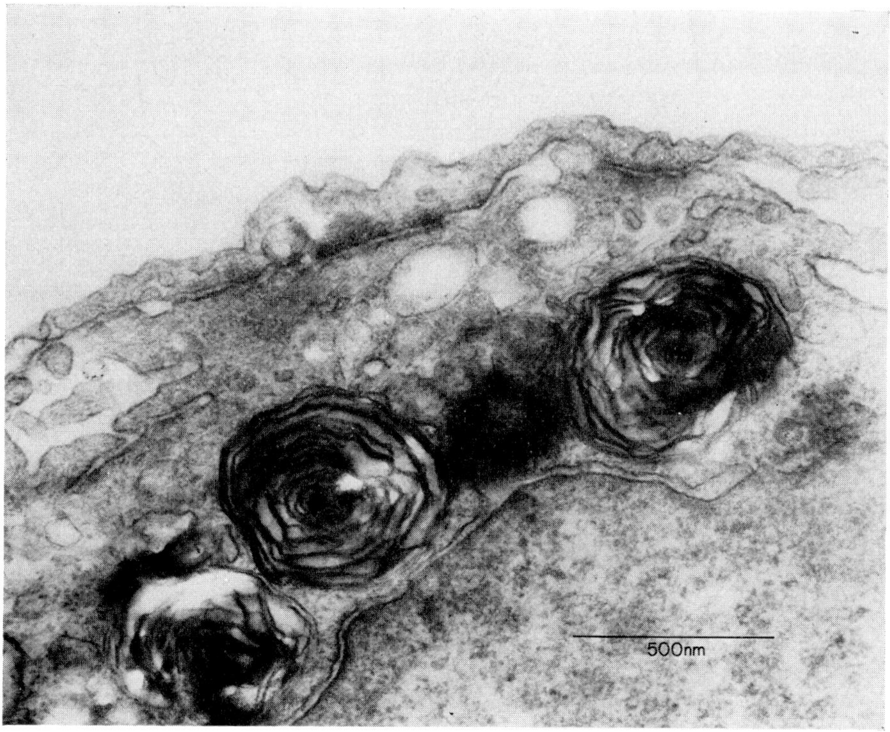

FIG. 4. Concentric lamellated bodies from Type II cell of lung of Patas monkey, *Erythrocebus Patas*. Scale bar 500 nm.

there is a biochemical mechanism (methylation of phosphatidyl ethanolamine) for the synthesis of lecithin, which does not exist in the sheep or rabbit; and that this mechanism is prominent in premature humans, whose survival it facilitates. They claim also that the lecithin synthesized by this mechanism is mainly α-palmitic/β-myristic. The other main mechanism, prominent only after 35–36 weeks, is a reaction between cytidine diphosphate-choline and diglyceride, and synthesizes di-palmitic lecithin. In the human adult, according to these workers, about 40% of the β acid of the lecithins of alveolar wash is myristic; in the adult rabbit the amount is less than 10% (Gluck, Landowne and Kulovich, 1970). In this respect monkeys resemble humans (Gluck, Chez, Kulovich, Hutchinson and Niemann, 1974).

3. *Reproductive and neonatal*

The tarsier and the simians have haemochorial placentas, in which the chorionic villi are bathed in maternal blood, while lemurs and other mammals have other forms of placenta. The connections, if any, between the various differences between simians and non-simians have not yet been traced.

In the premature newborn human, a serious disease, the "hyaline membrane disease", or "respiratory distress syndrome", is liable to occur. The surfactant is deficient in infants dying of this disease (Avery and Mead, 1959; Pattle, Claireaux, Davies and Cameron, 1962). It seems likely (Gluck, *et al.*, 1972) that the disease is mainly a consequence of prematurity—i.e. of the infant being born before its lungs have developed a sufficient supply of surfactant. The infant has usually died or recovered within three or four days. The statistics suggest that premature birth produces very rapid maturation of the lung.

Considerable success has attended a method of therapy in which oxygen or air at a constant excess pressure of about 10 cm of water is supplied to the infant's airway, while the infant himself is left to make the rhythmical movements of respiration (Gregory, Kitterman, Phibbs, Tooley and Hamilton, 1971; Dunn, Speidel and Storrs, 1973). The disease seems analogous to similar trouble in the foal (Dunn, 1972). It has not been proved to exist in laboratory animals. It must be stressed that respiratory distress and post-mortem atelectasis usually arise from causes other than lack of surfactant.

C. BIRDS

Bubbles squeezed from bird lung show the same stability and clicking as do those from mammalian lung (Pattle, 1958; Pattle and Hopkinson, 1963). Washings from bird lung form stable bubbles (Pattle and Hopkinson, 1963); Fujiwara, Adams, Nozaki and Dermer (1970) found that washings from turkey lung resembled those from rabbit lung both in their surface properties and in their containing di-palmitoyl lecithin.

The atria of bird lung have in their walls cells containing LOPBs like those of mammals. The lamellae of these bodies consist of leaflets 4 to 5 nm thick. In origin they appear to be connected, as in non-simian mammals, with multivesicular bodies (Jones and Radnor, 1972). Both these workers and Petrik (1967) showed electron micrographs of their extrusion from cells. LOPBs may be found in the atrial cells after 17 days of incubation. I have found (P.U.) that the capacity to form stable bubbles is fully established by 18 days. The "trilaminar substance" (*vide infra*) appears only at the time of hatching (21–22 d.) (Jones and Radnor, 1972). It therefore seems that the surpellic lining film of bird lung, like that of mammalian lung, is derived from the LOPBs.

The surpellic surface layer, if it resembles that of mammals, is also difficult to demonstrate in electron micrographs. The lungs of birds, however, contain relatively enormous quantities of a peculiar "trilaminar substance" (Jones and Radnor, 1972) which is not attacked by embedding fluids. It is secreted in the lining cells of the atrium, in which it appears as skeins or compact bodies in the cytoplasm. It appears in the micrographs of Lambson and Cohn (1968; goose lung), and probably in those of Tyler and Pangborn (1964; chicken) and Fujiwara *et al.* (1970; turkey).

The basic unit of trilaminar substance, when stained by the method of Schock *et al.* (1973) is a streak showing a pair of osmiophilic lines about 4·5 nm apart, measured from centre to centre (P.U.). When the streaks are close-packed, the 4·5 nm spaces alternate with spaces of about 7·4 nm; thus the 4·5 nm structures must be flanked by non-osmiophilic layers 3·7 nm thick. It may be surmised that the former consists of a bilayer of phospholipid molecules, and the latter consists of protein or polysaccharide.

For a function for the "trilaminar substance" (which appears to be neither mucus nor ordinary surfactant) we may perhaps look to the peculiar architecture of bird lung. In birds the passages in which gas exchange takes place are tubular, as opposed to cup-shaped, and have diameters down to 30 μm or less. The alveoli of shrew and bat lung are about 30 μm in diameter (Tenney and Remmers, 1963). Small amounts of liquid in mammalian alveoli probably accumulate in niches and among the microvilli (see Fig. 4 of Gil and Weibel, 1969/70), where they can have little adverse effect. In a tubular, as opposed to cup-shaped, space, any free liquid tends to accumulate at one place and block the tube. We may therefore speculate that the "trilaminar substance" is in some way a substitute for the free liquid of mammalian alveoli, or perhaps acts so as to immobilize any free liquid which appears. There is room for plenty of work on this very interesting substance.

D. REPTILIA

Pattle and Hopkinson (1963) showed that stable bubbles could be obtained from the lung of the legless lizard *Anguis fragilis* (slow-worm). Clicking, though present, was less marked than in birds or mammals. They were however unable to find any reserve of saline-dispersible surfactant. Tortoise lung also produces stable bubbles (P.U.). Brooks (1970b) states that in the lung of the rat snake (*Elaphe obsoleta obsoleta*) there are two types of cell, resembling the Type I and Type II of mammalian lung. Cells of the second type contain certain bodies resembling LOPBs, but their internal configuration is not clear. Okada, Ishiko, Daido, Kim and Ikeda (1962b) found LOPBs in lung epithelial cells of snakes of the genus *Elaphe*, and in those of the lizard *Gekko japonicus*. Putting this rather fragmentary information together, one may conclude that most reptiles have a surpellic lining film in their lungs, that they have little extra-cellular saline-dispersible surfactant, and that the surfactant in the Type II cells forms lamellar bodies rather like those of mammals. The surface properties of the surfactant, as evidenced by "clicking", are, however, somewhat different.

E. AMPHIBIA

1. *Anura*

The lung of the clawed toad (*Xenopus laevis*) produces stable bubbles which at room temperature do not "click" (Pattle and Hopkinson, 1963). Bubbles from the lungs of *Rana temporaria* are rather less stable, but "click" at 3°C; LOPBs are present (P.U.).

Bargmann and Knoop (1956) found LOPBs in the lung cells of *X. laevis* and *R. esculenta*; Brooks (1970b) in *R. pipiens*; and Okada *et al.* (1962a) in *R. nigromaculata* and *Bufo vulgaris japonicus*.

As far as can be seen from these data, the lining film of anuran lungs is rather less surpellic than that of reptile lungs; there may be species differences.

2. *Urodeles*

The lungs of the newt *Triturus vulgaris* produce bubbles which are unstable in air-saturated water; the bubbles are often non-spherical and their lining is only partially surpellic. We have not found LOPBs in the epithelial cells (P.U.). Okada *et al.* (1962a) found no LOPBs in *T. pyrrhogaster*. The permanently larval salamander *Proteus* lacks both surpellic lining film and LOPBs (Hughes, 1970, 1973) but according to Brooks (1970b) the lung alveolar cells of the closely related *Necturus* contain LOPBs. They also contain masses of fibrils, spaced at about 12 nm, reminiscent of the "trilaminar substance" of birds. De Groodt, Lagasse and Sebruyns (1960) and Sorokin (1963) found LOPBs in the lungs of *Amblystoma* salamanders.

The information available is thus not self-consistent and more work is needed to establish the facts about urodele lungs.

3. *Apoda*

The lung lining film of *Ichthyophis* resembles that of *R. temporaria* (P.U.). LOPBs are present (P.U. and Hughes, Personal communication).

F. FISHES

1. *General*

It is generally considered that the lungs of Dipnoi and of primitive actinopterygian fish which possess lungs are homologous with the swim-bladders of other species of fish. Arguments as to whether the original diverticula from the oesophagus were single or double, etc., based on classical comparative anatomy, are given by Goodrich (1930). Ultrastructural studies of the embryonic stages of development might illuminate these arguments, just as they have helped to settle the question of whether the alveolar cells of mammals are of endodermal or mesodermal origin. The lung or swim-bladder of the living coelacanth *Latimeria* is represented by a rudiment whose ultrastructure has not been studied (Hughes, 1973). Its lipids are mainly saturated wax esters (Nevenzel, Rodegker, Mead and Gordon, 1966).

2. *Lungfishes* (*Dipnoi*)

Bubbles from the lungs of the African lungfish *Protopterus*, and the South American *Lepidosiren*, are stable and show "clicking" (Hughes, 1967; Pattle, 1973). De Groodt *et al.* (1960) and Hughes *et al.* (1974) found LOPBs in *Protopterus*, though Klika and Lelek (1967) did not. *Lepidosiren* and the Australian *Neoceratodus* have LOPBs; the lung lining film of the latter has perhaps inferior surpellic properties (Hughes, 1973).

3. *Actinopterygian fish with lungs*

The primitive bony fishes *Polypterus*, *Lepisosteus* (garpike), and *Amia* (bowfin) have lungs, that of *Lepisosteus* being somewhat more convoluted than that of *Amia*; the interior of the lung of *Polypterus* is smooth (Klika and Lelek, 1967).

I am not aware of any surface studies done on *Polypterus*. Klika and Lelek (1967) report electron-dense granules in rather rare cells in the secretory crypts of the lung, but state that there is no cell resembling the granular pneumocyte of mammals. *Amia* and *Lepisosteus* show slightly surpellic linings (Hughes, 1967); surface tensions of 2 to 6 mN/m are found (P.U.). Hughes (1970) found well-marked osmiophilic bodies, of cross-barred and concentric forms, in the lungs of *Amia*. Hughes and Ryan (unpublished observations) found LOPB's which have a similar appearance to those from dipnoan lungs following freeze-fracture.

4. *Ordinary fish with swimbladders*

The swimbladder of the goldfish (*Carassius auratus*) does not show a surpellic film (S.T. 10 to 20 mN/m), and similar results have been obtained from a roach–dace hybrid (*Leuciscus leuciscus* × *Rutilis rutilis*) (Hughes, 1967).

There are dense lamellated concentric osmiophilic bodies in the non-ciliated cells of the physostomatous swimbladder of the rainbow trout (*Salmo gairdneri*) (Brooks, 1970a). Brooks (1970b) enumerates similar findings by other workers.

5. *Air-breathing fish without lungs*

I know of no surface studies on the respiratory organs of these fish. No mention is made of osmiophilic bodies in the suprabranchial chamber of the climbing perch (*Anabas testudineus*) by Hughes and Munshi (1973a). Nor are they mentioned by Hughes and Munshi (1973b) as being found in the respiratory organs of *Channa, Amphipnous, Clarias, or Saccobranchus*. These authors describe "micro air pockets" down to 1 μm in size in *Amphipnous* and in *Channa punctatus*, and it seems that if these are really filled with air *in vivo*, some means of counteracting surface tension must be present. It is the view of Hughes and Munshi (1973b) that the air-breathing organs of *Amphipnous* and *Channa* do not represent modified gills. Surface studies on these would be of great interest.

III. Discussion

A. General

The record of surfactant in the evolutionary tree is so far somewhat fragmentary. Many investigations have covered either the ultrastructure of the lungs of a particular species, or its pulmonary surface properties, but not both aspects of the same species. It is possible however to trace certain general trends, from the swimbladders of ordinary fish (which have poorly surpellic lining films), through *Amia* and *Lepisosteus* (whose lung lining is partially surpellic) the anurans (whose lining is surpellic at 3°C but does not show "clicking" at room temperature), the Dipnoi, and the reptiles, to birds and mammals. The lungs of the last two classes have a fully surpellic lining and reserves of surpellic matter; the simians show morphological differences from other mammals, and the significance of these is at present unknown.

It is known that the lung linings of mammals and birds contain much fully saturated lecithin which, in its pure form, on the Wilhelmy trough, is strongly surpellic. Some evidence that the linings of other lungs contain less saturated substances may be gained from illustrations in the literature. Osmium tetroxide does not react with fully saturated lecithin and does not fix it (Chapman and Fluck, 1966) but fixes unsaturated phospholipids well (Cope and Williams, 1969). In the illustrations of Brooks (1970a,b) the LOPBs of mouse, chicken, and snake lung show signs of serious erosion, while those of LOPBs from fish swimbladders show highly developed concentric bodies which look remarkably well preserved. Hughes' (1970) figure from *Amia* lung also suggests good preservation of the LOPBs. Brooks' (1970b) figures of frog lung are perhaps intermediate.

Thus it appears that in the gradation from fish swimbladder to mammalian lung the phospholipids become more fully saturated, in line with the development of the surpellic properties of the lining film. One gets the impression (P.U.) that the LOPBs of simians are even more easily eroded by embedding fluids than those of non-simians; this would be in accord with the findings of Gluck *et al.* (1970, 1972) that the fatty acids in the β position in lung lecithin were more saturated in the adult human (85%) than in the adult rabbit (75%). If this difference were found to be general, it would suggest that the simian lung had taken a further step in evolution.

B. Urodele Amphibia

Such facts as are known about the surface film and lamellated bodies of urodele amphibians do not fit in with one another or with the corresponding features of other lungs. Further examination of the surface films of other urodele species, and EM examination using the improved methods for fixing LOPBs (Schock *et al.*, 1973) are required. If the LOPBs of urodeles should be found absent, or quite different from those of other taxa, this might support the view of Jarvik (1966, 1968) that the urodeles arose from a line of fish (*Porolepiformes*) different from that (*Osteolepiformes*) from which the anurans and other tetrapods developed.

The view of Parsons and Williams (1963), that all living amphibia are descended from a "protolissamphibian" ancestor, is more widely held. It is not unlikely that the lung, in the course of evolution, readily adapts itself to circumstances, and does not retain features—e.g. a surpellic lining film—which it does not require. Partisans of either view might draw conclusions from the fact that in plethodontid salamanders the lung is absent (Tenney and Tenney, 1970). The existence of LOPBs and a "clicking" surpellic lining film in lungfish is a point against the suggestion of Jarvik (1968) that the lungs of Dipnoi evolved independently.

C. FUNCTION OF OSMIOPHILIC BODIES IN FISH SWIMBLADDERS

The evidence for the role of surfactant in maintaining alveolar patency and stability in convoluted lungs, and for the role of the LOPBs in the secretion of surfactant, seems conclusive (Pattle, 1965; Gil and Reiss, 1973). There is however no obvious role for a surfactant in fish swimbladders. The respiratory organs of lungless air-breathing fish do not contain LOPBs (Hughes, 1973). If the swimbladder has evolved from an earlier convoluted lung, the LOPBs may represent a vestigial remnant of a surfactant-secreting mechanism. Brooks (1970b) however suggests that they have an anti-oxidant role.

The swimbladders of deep-sea fishes often contain oxygen at pressures of many atmospheres (see Fänge, 1966). D'Aoust (1969) however found that if the juvenile pacific rockfish, *Sebastodes miniatus*, was kept in sea water saturated with oxygen at partial pressures (up to 5 atm) which in nature exist in its swimbladder, it suffered convulsions and died. In nature, the ambient oxygen tension in the sea does not differ much from 0·2 atm.

Oxygen at 0·35 atm pressure is harmless to mammals, but at 0·8 atm causes damage to the pulmonary capillary endothelium; the damage is much worse at 1·0 atm (see Back and Thomas, 1970). The Type I and endothelial cells of the lung are attacked, but the Type II cells proliferate (Kapanci, Weibel, Kaplan and Robinson, 1969).

Thus the phenomena of oxygen poisoning appear to be very odd indeed, and they have not yet been explained (see review by Menzel, 1970). They suggest the existence of some agency giving positive protection against oxygen up to a certain concentration. While there is no positive evidence for an anti-oxidant role for the LOPBs, it cannot be ruled out.

D. GAPS IN OUR KNOWLEDGE

Important gaps in our knowledge are:

(i) The exact chemical composition of the lining films of the lung and of homologous organs.

(ii) The surface properties of the air respiratory organs of air-breathing fish.

(iii) The surface properties of the lung of *Polypterus*.

(iv) The ultrastructure of the lungs of monotreme mammals.

(v) The ultrastructure of the lung of the tarsier.

(vi) The function of the osmiophilic bodies in fish swimbladders.

E. SUMMARY

(i) Some points of confusion in nomenclature are discussed. The word "surpellic" is coined to describe metastable films of high surface pressure and substances capable of producing them.

(ii) Evidence is assembled for a spectrum of properties, from the swimbladders of ordinary fish, with large radii of curvature, no surpellic lining film, and bodies probably containing unsaturated phospholipids, to the lungs of mammals, with small radii of curvature, a highly surpellic lining film, and a secretion containing mainly saturated phospholipids.

(iii) There is a morphological difference between the LOPBs of simian mammals and those of non-simians.

(iv) The evidence relating to urodele amphibians is not self-consistent. Further work is required.

Acknowledgement

We are grateful to the Zoological Society of London for specimens of mammalian lung tissue.

References

Avery, M. E. and Mead, J. (1959). Surface properties in relation to atelectasis and hyaline membrane disease. *A.M.A. J. Dis. Child.* **97**, 517–523.

Back, K. C. and Thomas, A. A. (1970). Aerospace problems in pharmacology and toxicology. *Ann. Rev. Pharmacol.* **10**, 395–412.

Balis, J. U. and Conen, P. E. (1964). The role of alveolar inclusion bodies in the developing lung. *Lab. Invest.* **13**, 1215–1229.

Bargmann, W. and Knoop, A. (1956). Vergleichende elektronenmikroskopische Untersuchungen der Lungenkapilläre. *Z. Zellforsch.* **44**, 263–281.

Brooks, R. E. (1970a). Ultrastructure of the physostomatous swimbladder of rainbow trout (*Salmo Gairdneri*). *Z. Zellforsch.* **106**, 473–483.

Brooks, R. E. (1970b). Lung alveolar cytosomes: a consideration of their significance. *Z. Zellforsch.* **106**, 484–497.

Brown, E. S. (1964). Isolation and assay of dipalmityl lecithin in lung extracts. *Am. J. Physiol.* **207**, 402–406.

Chapman, D. and Fluck, D. J. (1969). Physical studies of phospholipids. III. E.M. studies of some fully saturated 2–3 diacyl DL phosphatidyl ethanolamines and phosphatidyl cholines. *J. Cell Biol.* **30**, 1–11.

Chevalier, G. and Collet, A. J. (1972). *In vivo* incorporation of choline ^{3}H, leucine ^{3}H, and galactose ^{3}H in alveolar Type II pneumocytes in relation to surfactant synthesis. A quantitative radiautographic study in mouse by electron microscopy. *Anat. Rec.* **174**, 289–310.

Clements, J. A. (1956). Dependence of pressure volume characteristics of lungs on intrinsic surface-active material. *Am. J. Physiol.* **187**, 592.

Clements, J. A. (1957). Surface tension of lung extracts. *Proc. Soc. exp. Biol. Med.* **95**, 170–172.

Clements, J. A. (1974) (chairman). Symposium on biochemical aspects of pulmonary function. *Fed. Proc.* **33**, 2231–2262.

Clements, J. A., Nellenbogen, J. and Trahan, H. J. (1970). Pulmonary surfactant and evolution of the lungs. *Science, N.Y.* **169**, 603–604.

Comline, K. S., Cross, K. W., Dawes, G. S. and Nathanielsz, P. W. (1973). 'Foetal and Neonatal Physiology: Proceedings of Sir Joseph Barcroft centenary symposium", 25–27 July 1972. Cambridge U.P.

Cope, G. H. and Williams, M. A. (1969). Quantitative studies on the preservation of choline and ethanolamine phosphatides during tissue preparation for electron microscopy. *J. Microsc.* **90**, 31–46.

Creasey, J. M., Pattle, R. E. and Schock, C. (1974). Ultrastructure of inclusion bodies in Type II cells of lung, human and sub-simian. *J. Physiol.* **237**, 35P–37P.

D'Aoust, B. G. (1969). Hyperbaric oxygen toxicity to fish at pressures present in their swimbladders. *Science, N.Y.* **163**, 576–578.

De Groodt, M., Lagasse, A. and Sebruyns, M. (1960). Elektronenmikroscopische Morphologie der Lungenalveolen des *Protopterus* und *Amblystoma*. *Proc. Internat. Cong. Electron Microscopy*, 1958. Springer-Verlag. pp. 418–421.

Dermer, G. B. (1969). The fixation of pulmonary surfactant for electron microscopy. 1. The alveolar surface layer. *J. Ultrastruct. Res.* **27**, 88–104.

Dickie, K. and Massaro, D. (1974). Protein transport by lung. *Proc. Soc. exp. Biol. Med.* **145**, 154–156.

Dunn, P. M. (1972). Human "barkers". *Equine Vet. J.* **4**, 128–131.

Dunn, P. M., Speidel, B. D. and Storrs, C. N. (1973). In defence of the Gregory box (CPAP). *Lancet* (ii), 853–854.

Ellison, L. T., McPherson, J. C., Jr. and Ellison, R. G. (1969). Alveolar surfactant in small laboratory animals, fowl, snakes, and adult and pouch possums. *Fed, Proc.* **28**, 400.

Fänge, R. (1966). Physiology of the swimbladder. *Physiol. Rev.* **46**, 299–322.

Fujiwara, T., Adams, F. H., Nozaki, M. and Dermer, G. B. (1970). Pulmonary surfactant phospholipids from turkey lung: comparison with rabbit lung. *Am. J. Physiol.* **218**, 218–225.

Gil, J. and Reiss, O. K. (1973). Isolation and characterisation of lamellar bodies and tubular myelin from rat lung homogenates. *J. Cell Biol.* **58**, 152–171.

Gil, J. and Thurnheer, U. (1971). Morphometric evaluation of ultrastructural changes in Type II alveolar cells of rat lung produced by bromhexine. *Resp.* **28**, 438–456.

Gil, J. and Weibel, E. R. (1969–70). Improvements in demonstration of lining layer of lung alveoli by electron microscopy. *Resp. Physiol.* **8**, 13–36.

Gluck, L., Kulovich, M. V., Eidelman, A. I., Cordero, L. and Khazin, A. F. (1972). Biochemical development of surface activity in mammalian lung. IV. Pulmonary lecithin synthesis in the human fetus and newborn and etiology of the respiratory distress syndrome. *Pediat. Res.* **6**, 81–99.

Gluck, L., Landowne, R. A. and Kulovich, M. V. (1970). Biochemical development of surface activity in mammalian lung. III. Structural changes in lung lecithin during development of the rabbit fetus and newborn. *Pediat. Res.* **4**, 352–364.

Gluck, L., Chez, R. A., Kulovich, M. V., Hutchinson, D. L. and Niemann, W. H. (1974). Comparison of phospholipid indicators of fetal lung maturity in the amniotic fluid of the monkey (*Macaca mulatta*) and the baboon (*Papio Papio*) *Am. J. Obst. Gynecol.* **120**, 527–530.

Goodrich, E. S. (1930). "Studies on the Structure and Development of Vertebrates". pp. 578–612. Macmillan, London.

Gregory, G. A., Kitterman, J. A., Phibbs, R. H., Tooley, W. H. and Hamilton, W. K. (1971). Treatment of the idiopathic respiratory distress syndrome with continuous positive airway pressure. *New England J. Med.* **284**, 1333–1340.

Haugaard, N. (1968). Cellular mechanisms of oxygen toxicity. *Physiol. Rev.* **48**. 311–373.

Heppleston, A. G. and Young, A. E. (1972). Alveolar lipo-proteinosis: an ultrastructural comparison of the experimental and human forms. *J. Path.* **107**, 107–117.

Howatt, W. F., Avery, M. E., Humphreys, P. W., Normand, I. C. S., Reid, L. and Strang, L. B. (1965). Factors affecting pulmonary surface properties in the foetal lamb. *Clin. Sci.* **29**, 239–248.

Hughes, G. M. (1967). Evolution between air and water. *In* "Development of the Lung" (Ciba foundation symposium) (Eds A. S. de Reuck and R. Porter), pp. 64–80. Churchill, London.

Hughes, G. M. (1970). Ultrastructure of the air-breathing organs of some lower vertebrates. *7th Internat. Cong. Electron Microsc.*, Grenoble, pp. 599–600. Soc. Franc. de Microsc. Electronique, Paris.

Hughes, G. M. (1973). Ultrastructure of the lung of *Neoceratodus* and *Lepidosiren* in relation to the lung of other vertebrates. *Folia Morphol. Prague* **21**, 155–161.

Hughes, G. M. and Munshi, J. S. Datta (1973a). Fine structure of the respiratory organs of the climbing perch, *Anabas testudineus. J. Zool. Lond.* **170**, 201–225.

Hughes, G. M. and Munshi, J. S. Datta (1973b). Nature of the air-breathing organs of the Indian fishes *Channa*, *Amphipnous*, *Clarias*, and *Saccobranchus* as shown by electron microscopy. *J. Zool. Lond.* **170**, 245–270.

Hughes, G. M., Ryan, J. W. and Ryan, U. (1974). Freeze-fractured lamellate bodies of *Protopterus* lung: a comparative study. *J. Physiol.* **236**, 15P–17P.

Jarvik, E. (1966). Remarks on the structure of the snout in *Megalichthys* and certain other rhipidistid crossopterygians. *Ark. Zool.* **19**, 41–98.

Jarvik, E. (1968). Aspects of vertebrate phylogeny. *In* "Nobel Symposium 4.

Current problems of lower vertebrate phylogeny". (Ed. T. Ørvig) pp. 497–527. Interscience, London.

Jones, A. W. and Radnor, C. (1972). The development of the chick tertiary bronchus. II. The origin of the surface lining system. *J. Anat.* **113**, 325–340.

Kapanci, Y., Weibel, E. R., Kaplan, H. P. and Robinson, F. R. (1969). Pathogenesis and reversibility of the pulmonary lesions of oxygen toxicity in monkeys. II. Ultrastructural and morphometric studies. *Lab. Invest.* **20**, 101–118.

Kikkawa, Y. and Spitzer, R. (1969). Inclusion bodies of Type II alveolar cells: species differences and morphogenesis. *Anat. Rec.* **163**, 525–542.

King, R. J. and Clements, J. A. (1972). Surface active materials from dog lung. II. Composition and physiological correlations. *Am. J. Physiol.* **223**, 715–726.

Klaus, M. H., Clements, J. A. and Havel, R. J. (1961). Composition of surface-active material isolated from the lung. *Proc. Nat. Acad. Sci. U.S.A.* **47**, 1858–1859.

Klaus, M. H., Reiss, O. K., Tooley, W. H., Piel, C. and Clements, J. A. (1962). Alveolar epithelial cell mitochondria as source of the surface-active lung lining. *Science, N.Y.* **137**, 750–751.

Klika, E. and Lelek, A. (1967). A contribution to the study of the lungs of the *Protopterus annectens* and *Polypterus senegalensis*. *Folia Morphol., Prague* **15**, 168–175.

Lambson, R. O. and Cohn, J. E. (1968). Ultrastructure of the lung of the goose and its lining of surface material. *Am. J. Anat.* **122**, 631–649.

Macklin, C. C. (1954). The pulmonary alveolar mucoid film and the pneumocytes. *Lancet* (i), 1099–1104.

Massaro, G. D. and Massaro, D. (1972). Granular pneumocytes: electron microscopic radioautographic evidence of intracellular protein transport. *Am. Rev. Resp. Dis.* **105**, 927–931.

Masoro, E. J. (1973). Biosynthesis in the fetus. *In* "Respiratory Distress Syndrome" (Eds C. A. Villee, D. B. Villee and J. Zuckerman) pp. 7–27, Academic Press, New York, London and San Francisco.

Meban, C. (1972). The inclusion bodies in granular pneumocytes of hamster lung: a combined histochemical and ultrastructural study. *J. Anat.* **112**, 195–206.

Menzel, D. B. (1970). Toxicity of ozone, oxygen, and radiation. *Ann. Rev. Pharm.* **10**, 379–394.

Nevenzel, J. C., Rodegker, W., Mead, J. F. and Gordon, M. S. (1966). Lipids of the living coelacanth, *Latimeria Chalumnae*. *Science. N.Y.* **152**, 1753–1755.

Okada, Y., Ishiko, S., Daido, S., Kim, J. and Ikeda, S. (1962a). Comparative morphology of the lung with special reference to the alveolar epithelial cells. 1. Lung of the amphibia. *Acta Tuberc. Jap.* **11**, 63–72.

Okada, Y., Ishiko, S., Daido, S., Kim, J. and Ikeda, S. (1962b). Comparative morphology of the lung with special reference to the alveolar epithelial cells. 2. Lung of the reptilia. *Acta Tuberc. Jap.* **12**, 1–10.

Osman Hill, W. C. (1972). "Evolutionary Biology of the Primates". Academic Press, London, New York and San Francisco.

Parsons, T. S. and Williams, E. E. (1963). The relationships of modern Amphibia: a re-examination. *Q. Rev. Biol.* **38**, 26–53.

Pattle, R. E. (1955). Properties, function, and origin of the alveolar lining layer. *Nature, Lond.* **175**, 1125.

Pattle, R. E. (1958). Properties, function and origin of the alveolar lining layer. *Proc. Roy. Soc. Lond.* **B148**, 217–240.

Pattle, R. E. (1961). The formation of a lining film by foetal lungs. *J. Path. Bact.* **82**, 333–343.

Pattle, R. E. (1965). Surface lining of lung alveoli. *Physiol. Rev.* **45**, 48–79.

Pattle, R. E. (1966). Surface tension and the lining of the lung alveoli. *In* "Advances in respiratory Physiology" (Ed. C. G. Caro), pp. 83–105. Arnold, London.

Pattle, R. E. (1973). Inter-species differences in surface properties of the lung. *Proc. Roy. Soc. Med.* **66**, 385–386.

Pattle, R. E. (1975). The lung surfactant: an introduction. *In* Proc. First internat. Symposium upon equine Reproduction (Eds W. R. Allen and P. D. Rossdale) *J. Reprod. Fert. Supp.* 23.

Pattle, R. E. and Hopkinson, D. A. W. (1963). Lung lining in bird, reptile, and amphibian. *Nature, Lond.* **200**, 894.

Pattle, R. E. and Thomas, L. C. (1961). Lipoprotein composition of the film lining the lung. *Nature, Lond.* **189**, 844.

Pattle, R. E., Claireaux, A. E., Davies, P. A. and Cameron, A. H. (1962). Inability to form a lung-lining film as a cause of the respiratory-distress syndrome of the newborn. *Lancet* (i), 469–473.

Pattle, R. E., Schock, C. and Creasey, J. M. (1972). Electron microscopy of the lung surfactant. *Experientia* **28**, 286–287.

Pattle, R. E., Gandy, G., Schock, C. and Creasey, J. M. (1974). Lung inclusion bodies: different ultrastructure in simian and non-simian mammals. *Experientia,* **30**, 797–798.

Pattle, R. E., Rossdale, P. D., Schock, C. and Creasey, J. M. (1975b). The development of the lung and its surfactant in the foal and in other species. *In* Proc. First internat. Symposium upon equine Reproduction (Eds W. R. Allen and P. D. Rossdale). *J. Reprod. Fert. Supp.* 23.

Petrik, P. (1967). The ultrastructure of the chicken lung in the final stages of embryonic development. *Folia Morphol., Prague* **15**, 176–186.

Rayleigh, Lord (J. W. Strutt) (1890). On the tension of recently formed liquid surfaces. *Proc. Roy. Soc. Lond.* **47**, 281–287. Also *in* "Scientific Papers", vol. III, pp. 341–346. Cambridge U.P. (1902).

Rossdale, P. D. (1969). Clinical studies on 4 newborn thoroughbred foals suffering from convulsions, with special reference to blood gas chemistry and pulmonary ventilation. *Res. Vet. Sci.* **10**, 279–290.

Rossdale, P. D. (1970). Some parameters of respiratory function in normal and abnormal newborn foals with special reference to levels of p_a O_2 during air and oxygen inhalation. *Res. Vet. Sci.* **11**, 270–276.

Rossdale, P. D. (1972). Modern concepts of neonatal disease in foals. *Equine Vet. J.* **4**, 117–128.

Rossdale, P. D., Pattle, R. E. and Mahaffey, L. W. (1967). Respiratory distress in a newborn foal with failure to form lung lining film. *Nature, Lond,* **215**, 1498–1499.

Roth, J., Winkelmann, H. and Meyer, H. W. (1973). Electron microscopic studies in mammalian lungs by freeze-etching. IV. Formation of the superficial layer of the surfactant system by lamellar bodies. *Exp. Path.* **8**, 354–362.

Sayers, B. A. McA. (president) (1973). Lung surfactant: its contribution to lung distensibility. *Proc. Roy. Soc. Med.* **66**, 381–389.

Scarpelli, E. M. (1968). "The Surfactant System of the Lung". Lea and Febiger, Philadelphia.

Schock, C., Pattle, R. E. and Creasey, J. M. (1973). Methods for electron microscopy of the lamellated osmiophilic bodies of the lung. *J. Microsc.* **97**, 321–330.

Smith, U., Smith, D. S. and Ryan, J. W. (1973). Tubular myelin assembly in Type II alveolar cells. Freeze-fracture studies. *Anat. Rec.* **176**, 125–127.

Sorokin, S. P. (1963). Histochemistry in embryonic lungs. *In* "Neonatal Respiratory Adaptation" (Ed. T. K. Oliver, Jr.), pp. 79–93. U.S. Public Health Publication no. 1432.

Sorokin, S. P. (1967). A morphologic and cytochemical study of the great alveolar cell. *J. Histochem. Cytochem.* **14**, 884–897.

Tenney, S. M. and Remmers, J. E. (1963). Comparative quantitative morphology of the mammalian lung: diffusing area. *Nature, Lond.* **197**, 54–56.

Tenney, S. M. and Tenney, J. B. (1970). Quantitative morphology of cold-blooded lungs: amphibia and reptilia. *Resp. Physiol.* **9**, 197–215.

Tierney, D. F. (1974). "Lung metabolism and biochemistry". *Ann. Rev. Physiol.* **36**, 209–231.

Tyler, W. S. and Pangborn, J. (1964). Laminated membrane surface and osmiophilic inclusion bodies in avian lung epithelium. *J. Cell Biol.* **20**, 157–164.

Van Golde, L. M. G., Fleischer, B. and Fleischer, S. (1971). Some studies on the metabolism of phospholipids in Golgi complex from bovine and rat liver in comparison to other subcellular fractions. *Biochem. Biophys. Acta* **249**, 318–330.

Villee, C. A., Villee, D. B. and Zuckerman, J. eds. (1973). "Respiratory Distress Syndrome". Academic Press, New York, London and San Francisco.

Weibel, E. R. (1973). Morphological basis of alveolar-capillary gas exchange. *Physiol. Rev.* **53**, 419–495.

Wilhelmy, L. (1863). Ueber die Abhängigkeit der Capillaritäts-Constanten des Alkohols von Substanz und Gestalt des benetzten festen Körpers. *Poggendorfs Ann. der Physik* **119**, 177.

Wolstenholme, G. E. W. and O'Connor, M. eds. (1969). "Foetal Autonomy: a Ciba Foundation Symposium." J. and A. Churchill, London.

Young, S. L. and Tierney, D. F. (1972). Dipalmitoyl lecithin secretion and metabolism by the rat lung. *Am. J. Physiol.* **222**, 1539–1544.

Physiology of Fish Lungs

STEPHEN C. WOOD and CLAUDE J. M. LENFANT

The University of New Mexico, School of Medicine, Albuquerque, New Mexico, U.S.A. and *Division of Lung Diseases, National Heart and Lung Institute, Bethesda, Maryland, U.S.A.*

I. Introduction

Any structural adaptation for air breathing may be called a lung if it is ventilated with air and perfused with blood. In fishes these adaptations are diverse and the terms "air sac," "air bladder" and "lung" have often been used interchangeably in descriptions of air-breathing species. The present discussion, however, will be restricted to fish lungs, defined as air-breathing organs which are perfused from vessels homologous to the pulmonary arteries of higher vertebrates. Although such lungs were present in most or all late Paleozoic fishes (Romer, 1972), they are today found only in the three genera of Dipnoi and in the chondrosteans *Polypterus* and *Calamoichthys*. Also, although it is not clear whether lungs evolved from the air bladder (Johansen, 1970) or vice versa (Romer, 1972), the holosteans *Amia* and *Lepisosteus* have lungs as defined above.

The present article reviews the physiology of these lungs with emphasis on the mechanics and control of ventilation and the patterns of gas exchange.

II. Mechanics of Lung Ventilation

Although quantitative information on fish lung ventilation is quite fragmentary, the basic features have been described for most species. In all cases, inspiration requires positive pressure (buccal) pumping (McMahon, 1969; Gans, 1970 and this volume). The method of expiration, however, is quite variable. In the garfish (*Lepisosteus*), a facultative air breather, expiration is passive, brought about by hydrostatic pressure which forces the air sac gas into the buccal cavity. Retraction of the buccal cavity then forces gas through the opercular opening. When the snout is above the water surface the buccal floor is actively depressed and filled with air. Resubmergence then forces this air through the glottis into the air sac while the opercular openings are kept closed (Rahn, Rahn, Howell, Gans and Tenney, 1971). In the bowfin (*Amia*), expiration is at least partially active. Johansen, Hanson and Lenfant (1970) found that at higher temperatures the tidal volume was equal to total lung volume, i.e. at 30°C the air sac was completely emptied by contraction of surrounding striated muscles.

In the case of Dipnoi, more detailed information is available. McMahon (1969) used pressure, electromyographic and radiological measurements to describe the mechanics of ventilation in *Protopterus*. Figure 1 summarizes the pressure and EMG data obtained during a typical air-breathing cycle. This breathing pattern is essentially the same as that reported earlier for *Neoceratodus* (Grigg, 1965; Johansen, Lenfant and Grigg, 1967). As the lungfish approaches the surface (Fig. 1), the jaw muscles and buccal floor muscles are active, closing the mouth and increasing pressure in the buccal cavity causing water to be expelled through the operculum. Upon surfacing, the mouth opens, buccal and opercular pressure declines, and the operculum closes. Relaxation of glottal sphincter muscles produces expiration of com-

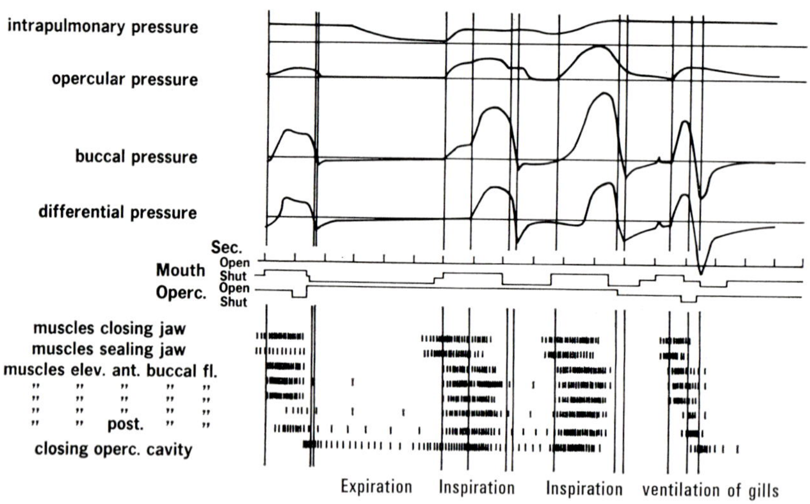

FIG. 1. Respiratory muscle activity and corresponding pressure changes during an air-breathing cycle of the African lungfish, *Protopterus aethiopicus*. Note that intrapulmonary pressure exceeds atmospheric pressure throughout the breathing cycle. (after McMahon, 1969.)

pressed lung gas. This is accomplished by contraction of smooth muscle and normal lung elasticity (Poll, 1962; Grigg, 1965; Johansen *et al.*, 1967); ribs play no part in expiration or inspiration.

Inspiration occurs as the jaw and opercular closing muscles become active resulting in buccal pressures up to 20 mmHg. Figure 1 shows a second inspiratory effort which was not uncommon in *Protopterus*. Bishop and Foxon (1968) report essentially the same breathing mechanism for *Lepidosiren* except that two, three or four inspiratory efforts were always required to fill the lung. Relaxation of the buccal cavity occurs after submergence, forcing water over the gills and expelling any residual air.

As reported for *Neoceratodus* by Johansen *et al.* (1967), the intrapulmonary pressure in *Protopterus* is positive throughout the air-breathing cycle. This compression of lung gases would favorably affect gas exchange by increasing the alveolar P_{O_2} (thus promoting HbO_2 "loading") and by increasing the utilization of lung O_2 during breath holding. This compression increase of pulmonary P_{O_2} would be particularly adaptive in lungfishes since the conditions which elicit air breathing and accompany breath holding usually involve a decreased blood pH, thereby a lowered HbO_2 affinity. The high intrapulmonary pressure in dipnoan lungs is also important in opposing the high pulmonary perfusion pressure resulting from incomplete separation of pulmonary and systemic circuits (Johansen and Reite, 1967), thereby preventing oedema.

The tidal volume of all these species is large, ranging from 40 to 100% of total lung volume (Table I).

III. Pulmonary Gas Exchange

Table I summarizes some features of gas exchange in fish lungs. Since all of these species also possess gills, the total gas exchange is bimodal. Figure 2 shows the relative contributions of aquatic and aerial breathing in dipnoans. The phylogenetic transition from water to air breathing is paralleled in modern Dipnoi where there is close correlation between gill development, habitat, and dependence on pulmonary gas exchange. As shown in Fig. 3, the Australian lungfish (*Neoceratodus*), an obligate water breather, succumbs quite rapidly if exposed to air, while the African (*Protopterus*) and South American (*Lepidosiren*) lungfish (both obligate air breathers) maintain arterial saturation and oxygen uptake throughout air exposure.

The relative importance of pulmonary gas exchange is further shown by comparing gas tensions and saturation in pulmonary and systemic vessels (Table I). Thus, in *Neoceratodus* the pulmonary arterial blood is in equilibrium with pulmonary venous blood at P_{O_2} of approximately 39 mmHg. Furthermore, because of the high HbO_2 affinity, the arterial blood is 95%

TABLE I

Gas exchange in fish lungs

Species (temp.)		Total $\dot{V}_{O_2}$ ml/min/kg	Lung $\dot{V}_{O_2}$ % of total	Lung $\dot{V}_{CO_2}$ % of total	Tidal Vol. % of lung Vol.	Blood Gas Values						
						Pulm. P_{O_2}	Art. P_{CO_2}	Pulm. P_{O_2}	Vein P_{CO_2}	Syst. P_{O_2}	Art. % S	References
Neoceratodus	(18–20°C)	·25	0	0	40–60	39	4	36	4	39	95	(1), (2)
Protopterus	(20°C)	·33	89	40	60–80	20	25	40	22	27	78	(3), (4), (5)
Lepidosiren	(18–20°C)	·37	96	60	≤ 100	10		49		31	80	(6)
	(25–28°C)					28		69		38	86	(7)
Lepisosteus	(22°C)	·65	73	8	40							
Amia	(10°C)	·3	0	0						≥ 100	100	(8)
	(20°C)	1·5	35	25	≤ 100					30	80	(9)
	(30°C)	2·0	75	40								

(1) Johanesen, Lenfant and Grigg (1967); (2) Lenfant, Johansen and Grigg (1966); (3) Johansen and Lenfant (1968); (4) Lenfant and Johansen (1968); (5) McMahon (1969); (6) Johansen and Lenfant (1967); (7) Lenfant, Johansen and Hanson (1970); (8) Rahn, Rahn, Howell, Gans and Tenney (1971); (9) Johansen, Lenfant and Hanson (1970).

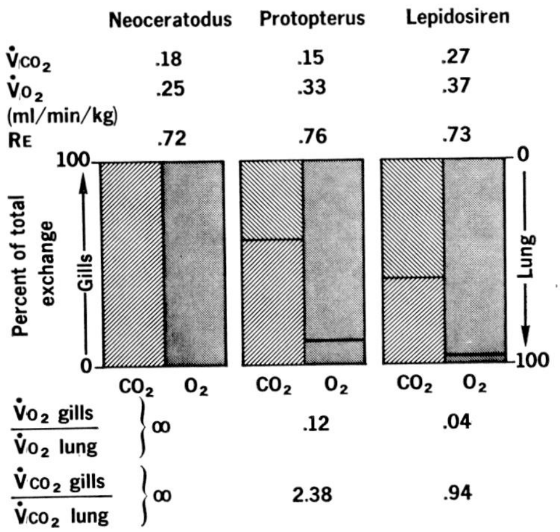

FIG. 2. Relative contributions of aquatic and aerial breathing to total gas exchange in the three genera of Dipnoi (Temperature = 20°C). (From Lenfant *et al.*, 1970.)

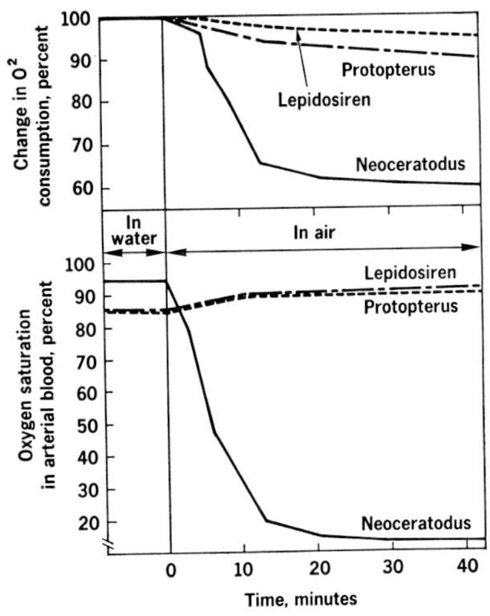

FIG. 3. Oxygen uptake and arterial oxygen saturation in Dipnoi during air exposure. Note the failure of adequate pulmonary O_2 uptake in *Neoceratodus*. (From Lenfant *et al.*, 1970.)

saturated at this P_{O_2} and the lung could contribute little additional oxygenation. Likewise in *Amia* (at lower temperatures), arterial blood is fully saturated by the gills and the air bladder plays no role in gas exchange (Johansen *et al.*, 1970). In fact, perfusion of the air bladder results in a decrease of arterial P_{O_2}, but this is minimized by a 60–70% shunt past the bladder exchange surface. At higher temperatures (discussed below) and in hypoxic water, air breathing becomes frequent, shunt flow past the gills occurs, and the air bladder becomes functional in gas exchange.

The African and South American lungfishes, which depend on pulmonary gas exchange, have pulmonary venous P_{O_2} values reflecting this dependence. Figure 4 shows blood gas values in unanaesthetized free-swimming *Protopterus*, where the importance of the lung (increasing O_2 saturation to 90%) and the high degree of functional separation of pulmonary and systemic blood flow is readily apparent. Similarly, in *Lepidosiren*, pulmonary venous blood is almost 100% saturated compared with 40–50% saturated pulmonary arterial blood (Johansen and Lenfant, 1967). The relative efficiency of pulmonary gas exchange in *Protopterus* and *Neoceratodus* is compared in Fig. 5 using a P_{O_2}–P_{CO_2} diagram. During air exposure, *Protopterus* does well but *Neoceratodus* is unable to maintain systemic arterial saturation. Although *Neoceratodus* can oxygenate the blood flowing through the lung (pulmonary venous P_{O_2} increases from 40 to 90 mmHg), systemic hypoxia ensues because of inadequate perfusion. CO_2 retention, inevitable during air exposure, is pronounced in *Neoceratodus* but in *Protopterus* it is minimized by the relatively high tidal volumes and greater efficiency of pulmonary CO_2 elimination. The latter is evidenced by the P_{CO_2} difference between the pulmonary artery and dorsal aorta.

If, instead of being totally exposed, hypoxic water induces the fish to surface intermittently for air, a potentially serious problem exists, i.e., oxygen gained by the lungs could be lost at the gills. In *Neoceratodus* this problem is reduced by the high HbO_2 affinity and in other dipnoans the degenerate gills provide a diffusion limitation. In *Amia* and *Protopterus* physiological and anatomical shunts prevent O_2 loss at the gills (Johansen, 1972).

A low pulmonary gas exchange ratio (R), characteristic of all transitional species, results mainly from the high solubility of CO_2 in water. The resulting cutaneous CO_2 elimination is essential for species with intermittent ventilation since breath-holding permits O_2 extraction but CO_2 elimination requires rapid ventilation. If cutaneous or gill CO_2 elimination does not minimize

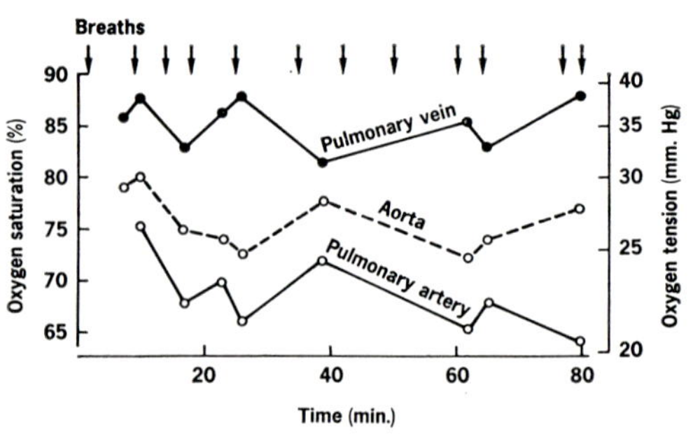

FIG. 4. Blood gas values during undisturbed breathing in the African lungfish, *Protopterus*. Arrows indicate air breaths. (From Lenfant and Johansen, 1968.)

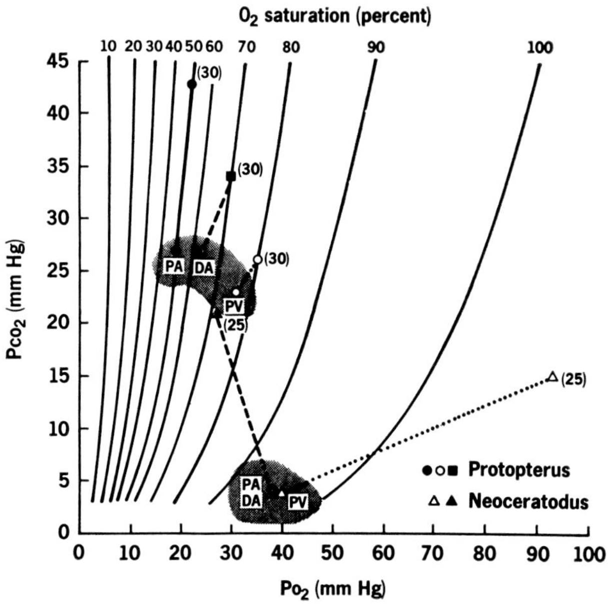

FIG. 5. $P_{O_2} - P_{CO_2}$ diagram showing blood gas values in *Neoceratodus* and *Protopterus* during undisturbed breathing in well-aerated water (values on stippled background) and following air exposure of (25) or (30) minutes. PA, pulmonary artery; DA, dorsal aorta; and PV, pulmonary vein. (From Lenfant and Johansen, 1968.)

pulmonary values of R, respiratory acidosis ensues. This did not occur during the evolutionary transition from water to air breathing because CO_2 retention was balanced by increasing plasma bicarbonate.

In facultative air breathers, the relative importance of lungs and gills to total gas exchange must be greatly affected by the temperature as well as the gaseous composition of the water. This was clearly demonstrated for *Amia* by Johansen *et al.* (1970) but remains to be shown for other species. As shown in Fig. 6, there is a complete reversal of the relative contribution of gills and air bladder to total $\dot{V}_{O_2}$ as temperature increases from 20° to 30°C. In concluding this section it is important to note that although the lungs of obligate air breathers such as *Protopterus* function well, they suffice only because of the much lower (10 to 50 fold) aerobic requirements compared with higher vertebrates. The inefficient gas exchange of *Protopterus* lungs relative to those of higher vertebrates is clearly shown in Fig. 7 by the very large gradient of P_{O_2} between pulmonary gas and pulmonary venous blood. A difference in oxygen pressure across the pulmonary surface may result from three factors: (1) diffusion limitation; (2) shunting of blood flow; and (3) unequal ventilation to perfusion ratios ($\dot{V}/\dot{Q}$). The diffusing capacity of the lung depends on the area and thickness of the alveolar-capillary mem-

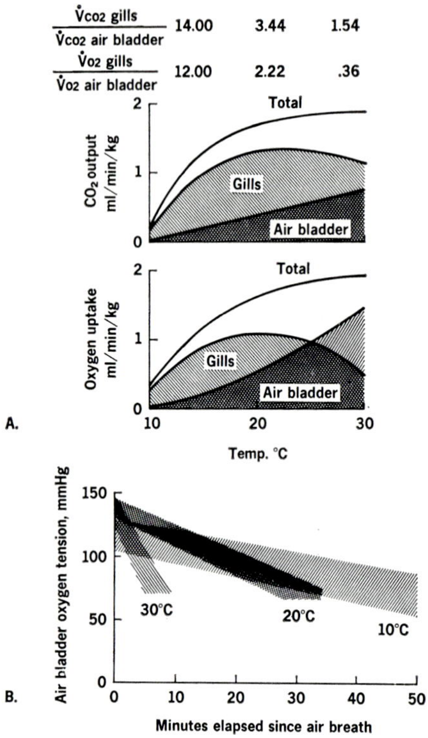

FIG. 6. (A) Relative roles of gills and lung to total $\dot{V}_{CO_2}$ and $\dot{V}_{O_2}$ in *Amia calva* as a function of temperature. (B) The rate of utilization of lung oxygen as a function of temperature. (From Johansen *et al.*, 1970.)

brane. Since in *Protopterus* the alveolar-capillary membrane is, at 0·5 microns, about the same thickness as in mammals (Klika and Lelek, 1967), a more likely source of diffusion limitation would be a small pulmonary surface area per unit oxygen uptake. This was found by Rahn *et al.* (1971) in *Lepisosteus*, which has an estimated pulmonary surface area of 800–1000 cm^2 per unit $\dot{V}_{O_2}$, several orders of magnitude lower than terrestrial vertebrates where pulmonary surface area per unit $\dot{V}_{O_2}$ ranges from 2000 to 4000 cm^2 (Tenney and Tenney, 1970). Venous admixture from anatomical shunts would also result in the large alveolar-arterial P_{O_2} difference seen in *Protopterus* when pulmonary P_{O_2} is high. As pulmonary P_{O_2} decreases, the magnitude of the P_{O_2} grandient decreases because the steep portion of the Hb–O_2 dissociation curve becomes operative and a given degree of shunt causes a smaller change in pulmonary venous P_{O_2}. Finally, areas of the lung with $\dot{V}/\dot{Q}$ values lower than 0·8–1·0 produce physiological shunts, thus contributing to the alveolar–arterial P_{O_2} difference.

IV. Control of Breathing

Bimodal breathers are faced with two distinct problems in respiratory control: (1) detection of respiratory stimuli in both the external and internal environments, and (2) integrated control of both aquatic and aerial breathing. The severity of these problems depends on the relative importance of each breathing mode to total gas exchange. In obligate air breathers such as *Protopterus* and *Lepidosiren* it would, *a priori*, seem likely that internally elicited control mechanisms would be more important. *Protopterus*, however, has a tremendously labile breathing pattern which seems only loosely related to arterial gas tensions or pH (Lenfant and Johansen, 1968; Johansen and Lenfant, 1968). The apparent low sensitivity of arterial chemoreceptors is consistent with the relationship between pulmonary and arterial P_{O_2} shown in Fig. 7, i.e. because of the large P_{O_2} gradient, cyclic variations of arterial P_{O_2} are much less than those of pulmonary P_{O_2}. Because of this, and the diffusion limitation in the degenerate gills of this species, other stimuli have assumed dominance in respiratory control, i.e. mechanical agitation of the water stimulates branchial respiration while air exposure is the most potent stimulus for air breathing.

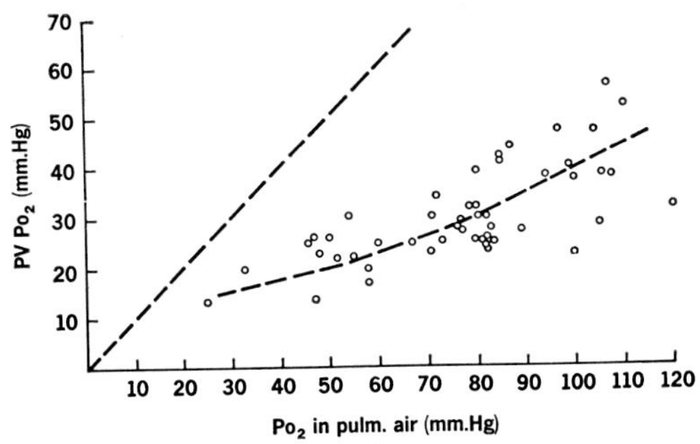

FIG. 7. Relationship of pulmonary venous P_{O_2} to lung P_{O_2} (alveolar-arterial P_{O_2} difference) in *Protopterus aethiopicus*. See text for discussion. (From Johansen and Lenfant, 1968.)

A. Oxygen as a Respiratory Control Factor

The concentration and partial pressure of oxygen in the water often varies considerably in the habitats of air-breathing fish. In the obligate gill breather, *Neoceratodus*, water ventilation increases sharply as inspired P_{O_2} falls below 75 mmHg. As shown in Fig. 8a, the increased ventilation results from an increase in the depth of ventilation with no change in the respiratory rate. The increased branchial ventilation induced by hypoxia is not capable of maintaining aerobic homeostasis, as indicated by decreasing O_2 extraction and O_2 uptake by the gills (Fig. 8b). In fact, aerial respiration was commonly

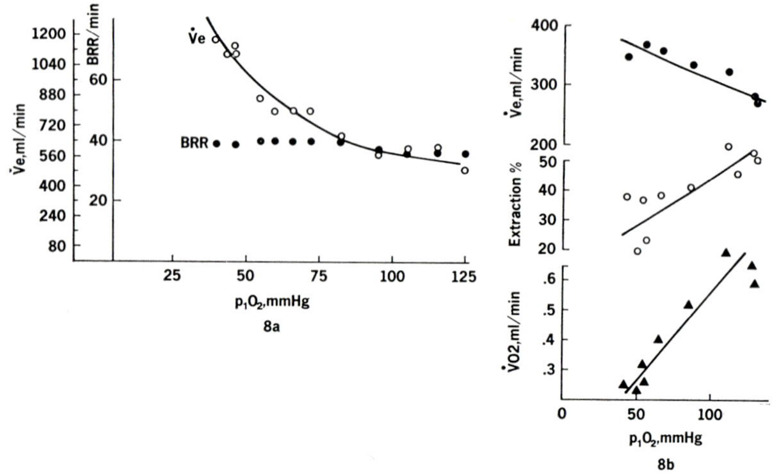

FIG. 8. (a) Water minute volume ($\dot{V}_E$) and branchial breathing rate of *Neoceratodus forsteri* as a function of the P_{O_2} of inspired water. (b) Water minute volume ($\dot{V}_E$), percent of O_2 extracted from water, oxygen uptake ($\dot{V}_{O_2}$) of *Neoceratodus* as a function of the P_{O_2} of inspired water. (From Johansen *et al.*, 1967.)

induced at water P_{O_2} levels of 85 mmHg or more, well above the levels at which branchial ventilation increased. The chemoreceptors responsible for these respiratory responses are apparently external (branchial), since injection of nitrogen into the lung of *Neoceratodus* did not elicit any compensatory response.

The response of *Amia* to hypoxic water is particularly interesting since this is the only species for which data are available showing the interaction of temperature and P_{O_2}. Thus, as seen in Fig. 9A, air breathing is unimportant even in very hypoxic water as long as the temperature is low (10°–18°C). However, at higher temperatures (20°), pulmonary O_2 uptake assumes an increasingly greater role when water P_{O_2} drops below 90 mmHg, and at 30°C there is a high frequency of air breathing even in well aerated water (Fig. 9B). Coordination of aquatic and aerial breathing is indicated in Fig. 9B,C which shows branchial breathing rate being inversely related to air-breathing rate.

In adult *Protopterus*, water P_{O_2} does not influence branchial or aerial breathing (Johansen and Lenfant, 1968), but decreased lung P_{O_2} stimulates aerial breathing. Lahiri, Szidon and Fishman (1970) also showed that decreased lung P_{O_2} (and systemic arterial P_{O_2}) increased the frequency of air breathing in *Protopterus*. They showed that the chemoreceptors responsible for the ventilatory response to hypoxia are located mainly in the first three afferent gill arteries. Juvenile *Protopterus*, in contrast, depend more on aquatic respiration and respond to low water P_{O_2} (8% O_2) by a 50% increase

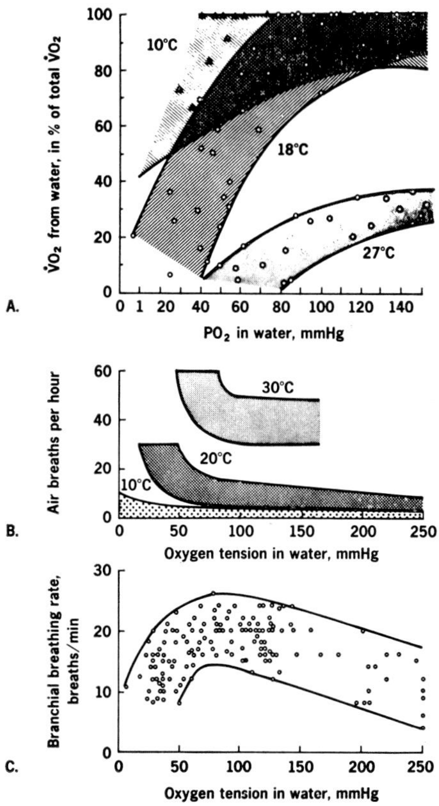

FIG. 9. *Amia calva*. (a) Contribution of aquatic to total $\dot{V}_{O_2}$, (b) frequency of air breathing and (c) water breathing as a function of temperature and P_{O_2} of inspired water. (From Johansen *et al.*, 1970).

in branchial and a 300% increase in aerial breathing (Jesse, Shub and Fishman, 1967). A similar distinction exists between juvenile and adult *Lepidosiren* in that juvenile fish display regular branchial movements (Johansen and Lenfant, 1967), whereas the adults do not (Bishop and Foxon, 1968).

B. CARBON DIOXIDE AS A RESPIRATORY CONTROL FACTOR

Air-breathing fishes often inhabit water which is CO_2-enriched as well as O_2-deficient. Consequently, influx of CO_2 and respiratory acidosis is a real problem if gill ventilation occurs at rates sufficient to obtain oxygen. Thus it is not surprising that bimodal breathers have external CO_2 receptors which, as shown in Fig. 10, reflexively inhibit branchial respiration and stimulate air breathing. In *Neoceratodus* this reflex (abolished by atropine) occurs

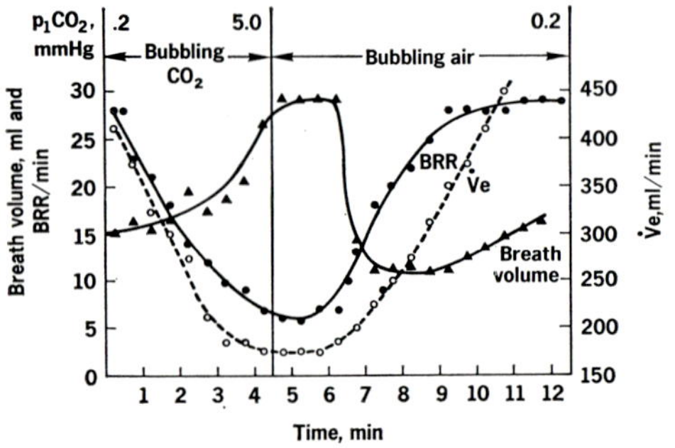

FIG. 10. Air breath volume, water minute volume ($\dot{V}_E$) and branchial breathing rate (BRR) in *Neoceratodus forsteri* during and after exposure to elevated CO_2 in the inspired water. (From Johansen *et al.*, 1967.)

with increased gill CO_2 but not from CO_2 introduced directly into the lung (Johansen, 1966). Further evidence that the CO_2 receptors are located in the branchial region (or arterial side of the branchial circulation) is provided by data for *Protopterus* (Johansen and Lenfant, 1968) showing that increased gill P_{CO_2} inhibits branchial breathing before arterial P_{CO_2} increases. Jesse *et al.* (1967) found that juvenile *Protopterus* (mean weight 65g compared with 3 kg for *Protopterus* studied by Johansen and Lenfant, 1968) showed a biphasic response to CO_2 similar to that of terrestrial mammals. At physiological levels of water CO_2 (0·5% or less), both gill and lung ventilation were stimulated; at higher levels of CO_2 (1–5%) gill ventilation declined steadily. In contrast, lung ventilation decreased initially but remained constant at 1% to 5% CO_2.

V. CONCLUSION

In the late 1960's Johansen, Lenfant and others applied modern physiological instrumentation and techniques to studies of air-breathing fishes. The resulting data, in many instances the first available for unanaesthetized, free-swimming fishes, revealed a fascinating parallel to the events which must have accompanied the evolution of tetrapod vertebrates. It is clear, however, that much remains to be learned about the respiratory functions of fish lungs. Some particularly obvious gaps in our understanding include: (1) the nature and mechanism of the ontogenetic changes in respiration; (2) the influence of temperature on bimodal gas exchange in Dipnoi; and (3) the localization and characterization of both sensory and motor neurones involved in respiratory control. These problems, and many others, provide challenging projects for future studies of this key group of vertebrates.

References

Bishop, I. R. and Foxon, G. E. H. (1968). The mechanism of breathing in the South American lungfish, *Lepidosiren paradoxa*; a radiological study. *J. Zool. Lond.* **154**, 263–271.

Gans, C. (1970). Strategy and sequence in the evolution of the external gas exchangers of ectothermal vertebrates. *Formio et Functio* **3**, 66–104.

Grigg, G. C. (1965). Studies of the Queensland lungfish *Neoceratodus forsteri*. I. Anatomy, histology and functioning of the lung. *Aust. J. Zool.* **13**, 243–253.

Jesse, M., Shub, C. and Fishman, A. (1967). Lung and gill ventilation of the African lungfish. *Resp. Physiol.* **3**, 267–287.

Johansen, K. (1966). Chemoreception in respiratory control of lungfish, *Neoceratodus*. *Fed. Proc.* **25**, 389.

Johansen, K. (1970). Air-breathing in fishes. *In* "Physiology of Fishes". Vol. 4 (Eds W. S. Hoar and J. D. Randall). Academic Press, New York, London and San Francisco.

Johansen, K. (1972). Heart and circulation in gill, skin and lung breathing. *Resp. Physiol.* **14**, 193–210.

Johansen, K. and Lenfant, C. (1967). Respiratory function in the South American lungfish, *Lepidosiren paradoxa* (Fitz). *J. exp. Biol.* **46**, 205–218.

Johansen, K. and Lenfant, C. (1968). Respiration in the African lungfish, *Protopterus aethiopicus*. II. Control of Breathing. *J. exp. Biol.* **49**, 453–468.

Johansen, K. and Reite, O. B. (1967). Effects of acetylcholine and biogenic amines on pulmonary smooth muscle inthe African lungfish, *Lepidosiren paradoxa*. *Acta Physiol. Scand.* **71**, 248–252.

Johansen, K., Lenfant, C. and Grigg, G. C. (1967). Respiratory control in the lungfish, *Neoceratodus forsteri* (Krefft). *Comp. Biochem. Physiol.* **20**, 835–854.

Johansen, K., Lenfant, C. and Hanson, D. (1968). Cardiovascular dynamics in the lungfishes. *Z. vergl. Physiol.* **59**, 157–186.

Johansen, K., Hanson, D. and Lenfant, C. (1970). Respiration in a primitive air breather, *Amia calva*. *Resp. Physiol.* **9**, 162–174.

Klika, E. and Lelek, A. (1967). A contribution to the study of the lungs of *Protopterus annectans* and *Polypterus senegalensis*. *Folia Morphol.* **15**, 168–175.

Lahiri, S., Szidon, J. P. and Fishman, A. P. (1970). Potential respiratory and circulatory adjustments to hypoxia in the African lungfish. *Fed. Proc.* **29**, 1141–1148.

Lenfant, C. and Johansen, K. (1967). Gas exchange and response to environment in the lungfish, *Protopterus aethiopicus*. *Fed. Proc.* **26**, 441.

Lenfant, C. and Johansen, K. (1968). Respiration in the African lungfish, *Protopterus aethiopicus*. I. Respiratory properties of blood and normal patterns of gas exchange. *J. exp. Biol.* **49**, 437–452.

Lenfant, C. and Johansen, K. (1972). Gas exchange in gill, skin and lung breathing. *Resp. Physiol.* **14**, 211–218.

Lenfant, C., Johansen, K. and Grigg, G. C. (1966). Respiratory properties of blood and pattern of gas exchange in the lungfish *Neoceratodus forsteri* (Krefft). *Resp. Physiol.* **2**, 1–21.

Lenfant, C., Johansen, K. and Hanson, D. (1970). Bimodal gas exchange and ventilation-perfusion relationship in lower vertebrates. *Fed. Proc.* **29**, 1124–1129.

McMahon, B. R. (1969). A functional analysis of the aquatic and aerial respiratory movements of an African lungfish, *Protopterus aethiopicus*, with reference to the evolution of the lung-ventilation mechanism in vertebrates. *J. exp. Biol.* **51**, 407–430.

Poll, M. (1962). Etude sur la structure adulte et la formation des sacs pulmonaires des *Protopters. Ann. Reeks Zool. Wetenschap.* **108**, 131–172.

Rahn, H., Rahn, K. B., Howell, B. J., Gans, C. and Tenney, S. M. (1971). Air breathing of the garfish, *Lepisosteus osseus. Resp. Physiol.* **11**, 285–307.

Romer, A. S. (1972). Skin breathing – Primary or secondary? *Resp. Physiol.* **14**. 183–192.

Tenney, S. M. and Tenney, J. B. (1970). Quantitative morphology of cold-blooded lungs: *Amphibia* and *Reptilia. Resp. Physiol.* **9**, 197–215.

Bimodal Gas Exchange

HERMANN RAHN and BARBARA J. HOWELL*

Department of Physiology, State University of New York at Buffalo, Buffalo, New York, U.S.A.

The transition from water to air breathing required many modifications in the structural and physiological adaptation of the external gas exchange system, and these modifications took place by trial and error over an extended period in vertebrate evolution. One may speculate that those animals which succeeded in surmounting all the hurdles became the stem forms of our reptilian ancestors and are no longer found today. Those that did not succeed adapted themselves to a particular ecological niche and were caught in the air–water interface where many of their modern offspring can still be found today. These latter forms can be easily recognized by their bimodal gas exchange mechanism and their large variety of approaches of exploiting air as a respiratory exchange medium. Furthermore, it becomes obvious that not all air-breathing fish today are ancient forms that have survived, and that attempts to obtain oxygen directly from the air is a continuing evolutionary process.

In this review we have tried to bring together data for the quantitative partition of the gas exchange between the gill-skin system on the one hand and the primitive lung or the air organ on the other, not only in aquatic but also in terrestrial species. This evidence suggests that in present day forms the lung or air organ can easily cope with serving the oxygen demand but is unable to function efficiently in the elimination of carbon dioxide. It is the handling of carbon dioxide by the lung system that appears to have been the greatest obstacle in the transition from water breathing to the modern lung, which is first encountered in our present day reptiles.

* Supported in part by NIH Grant 5 PO1 HL 14414.

Bimodal gas exchange

It has long been recognized that many aquatic air-breathing fish and aquatic as well as terrestrial Amphibia have a dual system of gas exchange where the air organ or primitive lung exchanges with the air, while the gill and/or skin exchanges gases with water. The term bimodal gas exchange for this dual system was probably first used by Lenfant, Johansen, and Hanson (1970) and serves as a useful description not only for aquatic forms but also for the terrestrial Amphibia where both the skin and the lung exchange with air.

However, only in recent years have reliable quantitative assessments of the partition of O_2 and CO_2 between the dual gas exchangers become available, and Lenfant *et al.* (1970) have provided an extensive review on this subject. In our approach we have expressed the bimodal gas exchange not in terms of the ratio of the total gas exchange between the two systems, but have rather looked at the O_2 and CO_2 of each exchange system as a percent of the total gas exchange. This focuses our attention upon the exchange ratio, R, of each system and allows one to evaluate the relative contribution of each system to the elimination of CO_2 or the acquisition of O_2.

Table 1 shows us the values of the O_2 and CO_2 exchange, expressed as a percentage of the total metabolic activity, for the "lung" (air organ or primitive lung) and for the gill–skin (which in the aquatic species should read gill and/or skin, and in the terrestrial forms pertains to skin only). In addition, the exchange ratio, R, is given for each system as well as the overall exchange ratio, R_{total}. When the CO_2 and O_2 exchange values are expressed as a percentage of the total, the exchange ratio for a particular system can be calculated; for example, R of the "lung" system $= R_{total}\,(\%\,\dot{V}_{CO_2}/\%\,\dot{V}_{O_2})$ lung. All the values given in Tables I and II have been rounded off.

In Table I aquatic Amphibia are included with the air-breathing fish. It also includes *Cryptobranchus* and *Necturus* which have been added because of their special interest although they do not fit into the scheme because their lung systems appear to serve no longer as a gas exchange organ but most likely as a hydrostatic organ (Guimond and Hutchison, 1972 and 1973). For this reason their values are not included in the overall average at the bottom of Table I. It will be noted that in terms of bimodal gas exchange the two aquatic amphibian species, *Siren* and *Amphiuma*, are very similar to the air-breathing fishes. The aquatic forms in general use their lung system for about 65% of the O_2 demand but their gill and/or skin for eliminating nearly 75% of their CO_2.

The terrestrial forms are shown in Table II. Here again we see that two-thirds of the O_2 is supplied by the lung, while three-fourths of the CO_2 is eliminated by the skin. It is of interest to note that the relative distribution of capillaries between the lung and the skin in a large number of amphibians was established by Czopek (1965). We have averaged his values for eighteen

TABLE I

Partition of gas exchange between lung and gill-skin system in aquatic forms
*(*R value assumed)*

Species		T	"Lung"—%			Gill-Skin—%			Total	Reference
		°C	$\dot{V}_{CO_2}$	$\dot{V}_{O_2}$	R	$\dot{V}_{CO_2}$	$\dot{V}_{O_2}$	R	R	
FISH										
Dipnoi:	*Lepidosiren paradoxa*	20	52	96	0·40	48	4	8·8	0·73	Lenfant *et al.* (1970)
	Protopterus aethiopicus	20	30	89	0·26	70	11	5·0	0·76	Lenfant and Johansen (1968)
Holostei:	*Amia calva*	25	30	50	0·60	70	50	1·4	1·00	Johansen, Hanson and Lenfant (1970)
	Lepisosteus osseus	22	8	73	0·09	92	27	2·7	0·80	Rahn *et al.* (1971)
Teleostei:	*Electrophorus electricus*	25	19	78	0·25	81	22	3·6	1·01	Farber and Rahn (1970)
	Anabas testudineus	25	9	54	0·20	91	46	2·3	1·17	Hughes and Singh (1970)
	Saccobranchus fossilis	25	6	40	0·17	94	60	1·6	1·00*	Hughes and Singh (1971)
	Clarias batrachus	25	6	58	0·11	94	42	2·3	1·00*	Singh and Hughes (1971)
AMPHIBIA										
Urodela:	*Amphiuma means*	25	23	56	0·36	77	44	1·6	0·88	Guimond and Hutchison (1974)
	Siren lacertina	25	22	50	0·40	78	50	1·5	0·91	Guimond and Hutchison (1973a)
	Cryptobranchus alleganiensis	25	2	8	0·23	98	92	0·9	0·86	Guimond and Hutchison (1973b)
	Necturus maculosus	25	1	4	0·20	99	96	0·9	0·82	Guimond and Hutchison (1972)
Mean (excluding last 2 species) n = 10			**20**	**65**	**0·28**	**80**	**35**	**3·0**	**0·91**	
S.D.			15	19	0·15	15	19	2·3	0·15	

TABLE II

Partition of gas exchange between lung and skin system in terrestrial amphibia

Species		T °C	Lung—% $\dot{V}_{CO_2}$	Lung—% $\dot{V}_{O_2}$	Lung—% R	Skin—% $\dot{V}_{CO_2}$	Skin—% $\dot{V}_{O_2}$	Skin—% R	Total R	Reference
Urodela:	*Ambystoma maculatum*	25	29	69	0·31	71	31	1·7	0·72	Whitford and Hutchison (1965)
	Taricha granulosa	25	32	68	0·40	68	32	2·1	0·86	Whitford and Hutchison (1965)
Anura:	*Rana clamitans*	25	24	74	0·28	76	26	2·6	0·89	Hutchison, Whitford and Kohl (1968)
	Rana pipiens	25	43	68	0·46	57	32	1·3	0·73	Hutchison *et al.* (1968)
	Hyla gratiosa	25	19	68	0·24	81	32	2·1	0·85	Hutchison *et al.* (1968)
	Hyla versicolor	25	16	67	0·19	84	33	2·1	0·81	Hutchison *et al.* (1968)
	Bufo americanus	25	17	66	0·24	83	34	2·3	0·95	Hutchison *et al.* (1968)
	Bufo boreas	25	31	69	0·42	69	31	2·1	0·93	Hutchison *et al.* (1968)
	Bufo cognatus	25	30	44	0·57	70	56	1·1	0·84	Hutchison *et al.* (1968)
	Bufo marinus	25	24	50	0·44	76	50	1·4	0·91	Hutchison *et al.* (1968)
	Bufo terrestris	25	25	78	0·30	75	22	3·1	0·92	Hutchison *et al.* (1968)
Mean (n = 11)			**26**	**65**	**0·35**	**74**	**35**	**2·0**	**0·86**	
S.D.			8	10	0·12	8	10	0·6	0·08	

European species of the genera *Rana*, *Hyla*, and *Bufo*, which indicates that their relative distribution is similar to the distribution of O_2 uptake, namely, the lungs have 62% and the skin 38% of the total respiratory capillaries.

I. CO_2 Conductance in Water and Air

Since the skin continued to be the major organ for CO_2 transport after the amphibians emerged upon land, it is of interest to compare the relative efficiency of the convective CO_2 transport in air and water. The conductance of CO_2 has the dimensions of $\dot{V}\beta_{CO_2}$ (Piiper, Dejours, Haab and Rahn, 1971) and is equal to the quantity of CO_2 transported per unit pressure difference of CO_2. $\dot{V}$ in our example is the volume of convective flow of air or water over the skin per unit time and βCO_2 is the capacitance coefficient (or solubility) of CO_2 in air or water and has the dimension of volume of CO_2 (·STPD) per litre of convection medium per torr.

In Fig. 1 the ratio of β water/β air are compared over the biological temperature range. We note that at about 17°C the capacity coefficient of water or air is the same, which indicates that water and air are equally efficient in removing CO_2 from the skin. Below this temperature water becomes more effective than air as a CO_2 transport system but since the solubility of CO_2 in water decreases more than that of air above 17°C, air now becomes a more effective medium. If we assume a similar convective flow over the skin in air and water and that diffusion of CO_2 in the medium plays a minor role, then little if any morphological skin changes were needed when aquatic animals emerged upon land. In this sense the new environment did not require any major adjustments.

Thus in terms of the gas partition between the two gas exchange systems,

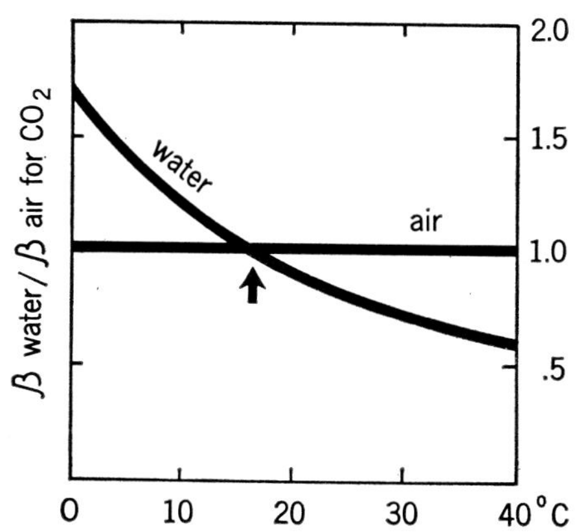

FIG. 1. The ratio of the capacity coefficients of CO_2, β, for water and air as a function of temperature. (See text.)

there is no great difference between the aquatic and terrestrial species: the lung is the primary O_2 acquisitor, the gill–skin or skin is the primary CO_2 eliminator. Whether the natural environment of the animal is water or air, the basic mechanism for the exchange of respiratory gases is the same.

II. CO_2 Elimination

An overall view of CO_2 elimination by the gill–skin system and the lung system is shown in Fig. 2 where the data of Tables I and II are combined and arranged according to decreasing O_2 uptake of the gill–skin system and increasing O_2 uptake of the lung systems. Columns with a dot indicate aquatic forms. *Cryptobranchus* and *Necturus* are omitted for reasons stated above.

What is impressive is the fact that percentage contribution of each system to the CO_2 elimination is rather constant in all the forms and averages 76% in the gill–skin system and 24% for the lung system. On the other hand, the percentage contribution to the oxygen demand varies with the species within each system. It is therefore of interest to plot the exchange ratio of the gill–skin system as a function of the relative amount of O_2 that this system provides. This is shown in Fig. 3. As the gill–skin system becomes less and less useful as an O_2 exchanging organ, the exchange ratio, R, increases to rather enormous values, indicating that the CO_2 elimination by this system remains unaltered. In this figure the values for *Cryptobranchus* and *Necturus* are plotted. Here the gill-skin system accounts for nearly all the O_2 and CO_2 exchange and therefore yields an essentially normal R value of 0·9. These two animals are from this point of view indistinguishable from a water breathing fish.

III. O_2 and CO_2 Tension

As has been pointed out previously (Rahn, 1966; Howell, 1970; Dejours, this volume), the transition from water to air exchange necessitates an

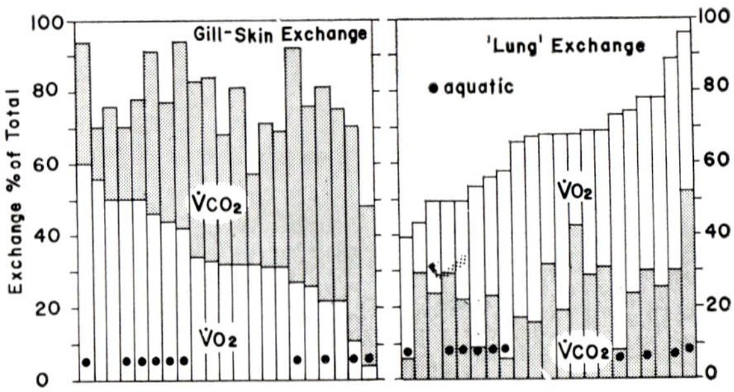

FIG. 2. The O_2 and CO_2 exchange (expressed as % of total exchange) for the gill–skin system and the "Lung" system for 21 species listed in Table I and II. The columns of the aquatic animals are identified by a dot. They are arranged in order of decreasing O_2 consumption (gill–skin system) and increasing O_2 consumption ("Lung" system).

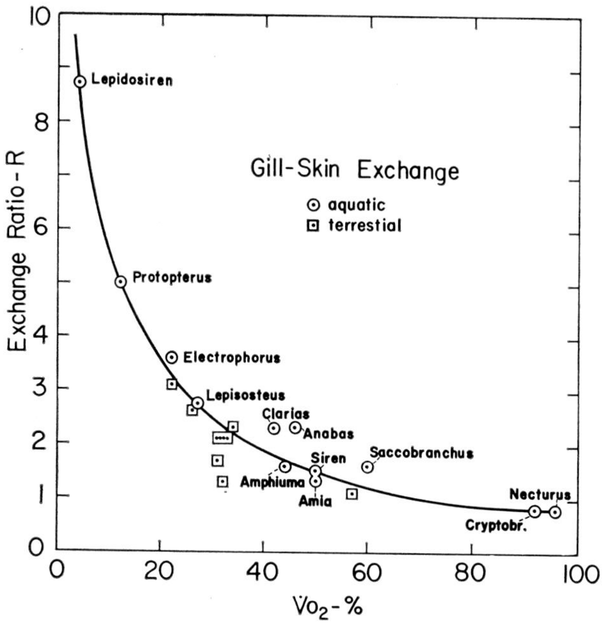

FIG. 3. The exchange ratio, R, of the gill–skin system as a function of the % O_2 uptake of the gill–skin system in all the species of Table I and II including *Necturus* and *Cryptobranchus*.

increase in blood CO_2 tension and a commensurate increase in plasma HCO_3^- to maintain a normal acid-base equilibrium. This is illustrated in Fig. 4, where A shows the simultaneous O_2 and CO_2 tensions which must occur at the surface of a water-exchanging organ and an air-exchanging organ if the exchange ratio is to be 0·85 (if the inspired medium has a P_{O_2} of 150 mmHg and a P_{CO_2} of 0 mm). The exact point along the R.Q. line, that is the exact O_2 and CO_2 value that will be found at the exchange surface, depends upon the ratio of the ventilation to the metabolic rate.

The curved arrow in Fig. 4 suggests the evolutionary change that must have occurred in the transition from a typical gill to a modern lung. It is of interest now to describe the intermediary steps that one finds in the bimodal gas exchange of a typical aquatic and terrestrial form. These are shown in Fig. 4b,c. The aquatic air breather now has a gill–skin exchange ratio of about 2·0 since most of the CO_2 but little O_2 is exchanged in the water. This fact alone would increase the CO_2 level above that of the water breather (Rahn, 1966). The air-breathing organ or primitive lung, however, has a typical exchange ratio of 0·35 (which is not significantly different from 0·28, the mean value of Table I), and for a given P_{O_2} must have a P_{CO_2} five times that of the gill–skin system. The curved arrow indicates that the primitive lung is still

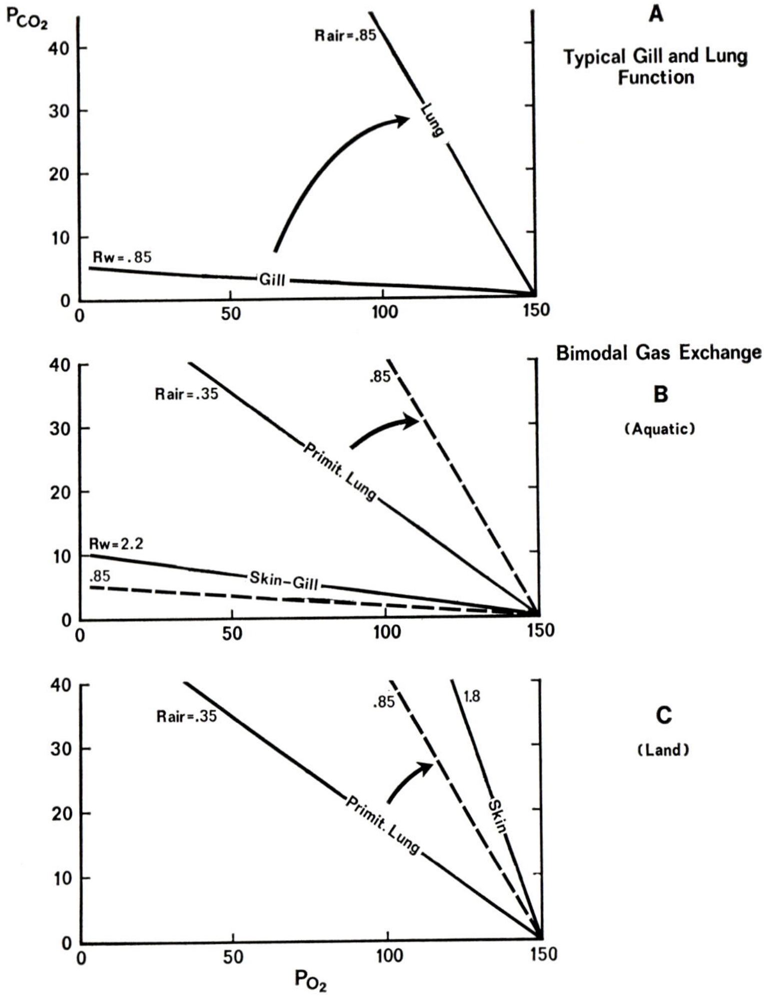

FIG. 4. (a) The R lines for an exchange ratio of 0·85 for water breathers or air breathers. The arrow suggests the general course of direction during the evolution from aquatic to air breathing. (b) Typical R lines for aquatic animals which have bimodal gas exchange. (See text.) (c) Typical R lines for terrestrial animals which have bimodal gas exchange. (See text.)

far removed from that of the modern lung which exchanges at an R.Q. = 0·85. In the terrestrial amphibians both skin and lung exchange with air. The high exchange ratio of the skin places the air R line to the right of the 0·85 line, while the primitive lung with the R = 0·35 is essentially on the same R line as that of the aquatic forms and equally far removed from the R = 0·85 line of the modern lung.

It should be pointed out that the actual gas composition of the two exchange systems are not indicated and known only for some species and then only for the lung air organ system. The P_{CO_2} values need not be the same in the two exchangers for any particular species. Much needs to be explored to fully describe the gas exchange, but nevertheless the R lines give us some clues since the particular O_2 and CO_2 value must fall somewhere on a particular R line if the exchange ratio is established. The description in Fig. 4 is only another approach in pointing out the inadequate performance of the primitive lung in its ability to eliminate CO_2.

IV. The P_{CO_2}–HCO_3^- Levels

The P_{CO_2} values of the air exchanging organ have been measured in a few aquatic forms. In *Lepisosteus* it is about 12 torr (Rahn, Rahn, Howell, Gans and Tenney, 1971) and in *Electrophorus* it is 35 (Farber and Rahn, 1970). Additional values have been obtained from the arterial blood in several species as well as the HCO_3^- concentration. But these blood values do not tell us whether they represent the "lung" or the gill–skin system or a mixture from both. Table III lists the values which are known and indicates the large differences which exist.

If P_{CO_2} values between 0·5 and 3 are characteristic for aquatic organisms depending entirely upon gill–skin function, then values higher than this do

Table III

Blood P_{CO_2} *and plasma* HCO_3^- *concentrations in various aquatic and terrestrial vertebrates with bimodal gas exchange at a body temperature near 28°C and average pH of 7·61. From the data of Rahn and Garey (1973)*

Common Name	Scientific Name	HCO_3^- mmol./L	P_{CO_2} torr	Reference
Walking catfish	*Clarias batrachus*	4	3·5	Howell (1970)
Garfish	*Lepisosteus osseus*	8	9	Rahn *et al.* (1971)
Marine toad	*Bufo marinus*	16	18	Rahn and Garey (1973)
Bullfrog	*Rana catesbiana*	28	17	Howell, Baumgardner, Bondi and Rahn (1970)
Electric eel	*Electrophorus electricus*	30	29	Farber and Rahn (1970)
African lungfish	*Protopterus aethiopicus*	34	22	Lenfant and Johansen (1968)

not necessarily indicate the degree to which the primitive lung system has developed. The P_{CO_2} and HCO_3^- level may actually reflect a particular niche and whether or not long periods of anaerobiosis are encountered which would benefit from a high HCO_3^- level and therefore high P_{CO_2} (Rahn and Garey, 1973). The arterial P_{CO_2} and HCO_3^- values for some aquatic and terrestrial species are given in Table III and illustrate the large differences that may exist. While in general the P_{CO_2} level in air breathers is substantially higher than in water breathing forms, it is of interest to trace this change during transition from water to air during metamorphosis of an amphibian or the seasonal transitions in the garfish, a facultative air breather.

The change in blood P_{CO_2} that occurs as an animal moves from an aquatic to a terrestrial environment can be illustrated in the present day amphibian, *Rana catesbiana*. During the early tadpole stage the animal depends entirely on its gills (and probably skin) for gas exchange. As shown in Fig. 5, the early tadpole is a typical aquatic organism, having a P_{CO_2} of about 2 torr (Erasmus, Howell and Rahn, 1970/71). During the later stages, before metamorphosis, a small air bubble can be seen in the developing lung tissues. To what extent this gas bubble contributes to the gas exchange is unknown, but its presence is associated with a P_{CO_2} of 5 mm; and when such an animal is denied access to air, the P_{CO_2} value will fall again to 2 torr (Just, Gatz and Crawford, 1973). During metamorphosis the gills disappear and the lung assumes more and more importance as an air exchange organ; its inadequacy as a CO_2 eliminator becomes apparent, as we observe the rise in blood P_{CO_2}. The adult bullfrog has an arterial P_{CO_2} some five times that of the tadpole. In order to avoid an appreciable respiratory acidosis, the blood bicarbonate must also increase during the 25 days of metamorphosis, as shown in the inset in Fig. 5. The adult bullfrog has now established the steady-state condition around which its blood pH will be maintained for the rest of its life.

Another present day animal, the garfish, *Lepisosteus osseus*, does not have even such a quasi-steady state as the adult bullfrog. Each year as the waters warm in the spring and the amount of dissolved oxygen decreases, but the metabolism increases, the gar is forced to utilize its lung for the acquisition of sufficient oxygen. In Fig. 6 we see that the blood P_{CO_2} must rise to attain the summer CO_2 tensions typical of an air breather. With the approach of winter, and the reduction in metabolism and increase in the O_2 content of the water, water is now the most hospitable environment for the gas exchange. The gar fish again becomes a typical fish, no longer breaks the surface to renew its O_2 supply and exchanges both O_2 and CO_2 via its gills under a heavy layer of ice in the northern latitudes (Rahn *et al.*, 1971).

In retrospect, it is easy to see how water-breathing animals became easily addicted to breathing air since its O_2 was easily accessible, always constant, and at a very high concentration compared with that dissolved in water. As long as they could eliminate the CO_2 through the gill–skin system, it would

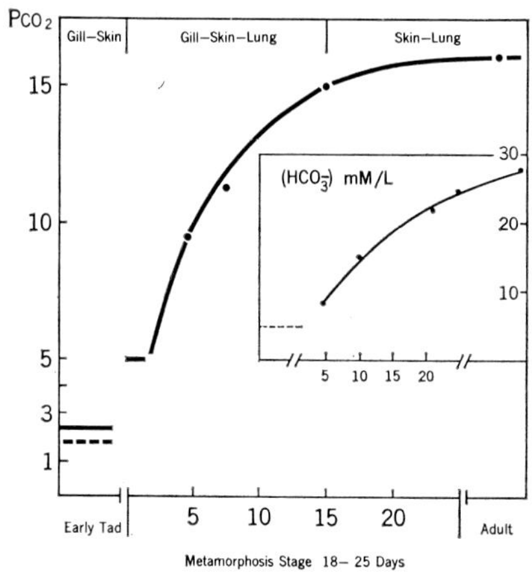

FIG. 5. Changes in blood P_{CO_2} and plasma bicarbonate during metamorphosis of *Rana catesbeiana*. Adapted from Just *et al.* (1973) and Erasmus *et al.* (1970/71).

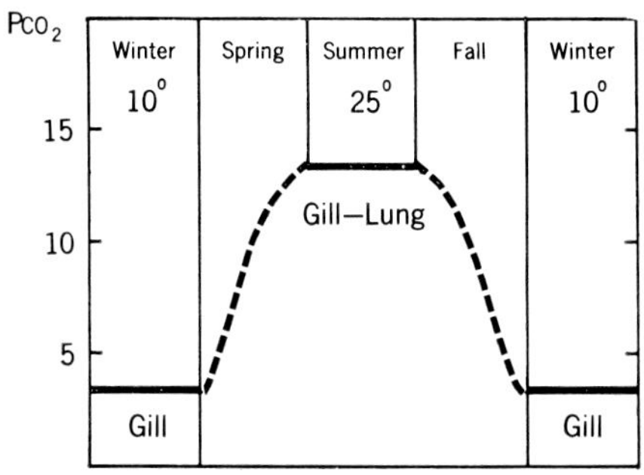

FIG. 6. Seasonal changes in blood P_{CO_2} of the garfish, *Lepisosteus*. Adapted from Rahn *et al.* (1971).

not build up in the air organ or primitive lung, and thus they could extract most of the inhaled O_2 by holding their breath for long periods and only occasionally replenishing their air supply. After a typical inspiration the O_2 in the air organ is reduced from 21% to 5% and 3%, respectively, in *Lepisosteus* and *Electrophorus* before the next inspiration. However, during this interval the CO_2 increases only to 2·5 and 5%, respectively. Thus it would appear that the gill–skin system served a very important function in that it allows the lung to ventilate only at rather protracted intervals.

The advantage of intermittent air breathing without producing a respiratory acidosis is obvious for an aquatic organism since it need surface only occasionally. This advantage can be illustrated by comparing the breath-holds of man and electric eel. Figure 7 shows the air organ or alveolar O_2 values for both species during four minutes of breath-hold.

There is a striking parallelism in the composition of man's lung gases and the eel's buccal cavity gases as they hold their breaths. The eel has virtually no FRC when it inhales and, therefore, its gas composition starts essentially with inspired air (Farber and Rahn, 1970). The single maximal breath of man provides an initial alveolar value of 135 mmHg and 24 mmHg for O_2 and CO_2 respectively (Hong, Lin, Lally, Yim, Kominami, Hong and Moore, 1971). Thereafter, the O_2 changes with time remain relatively parallel. One difference is the fact that after two minutes the P_{CO_2} of the eel reaches a plateau, while that of man continues to rise. The other striking difference is revealed in their behaviour when they must return periodically to renew their air since both are obligatory air breathers. The behavioural difference is

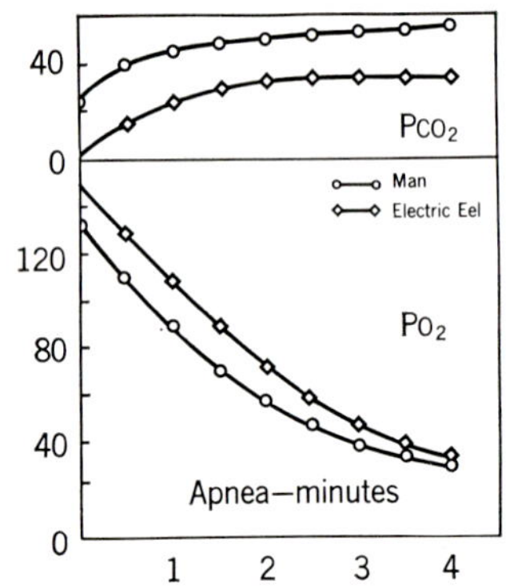

FIG. 7. Comparison of the P_{O_2} and P_{CO_2} changes during a four minute apnoea in the electric eel, which has a bimodal gas exchange, and man. Based upon the buccal gas concentrations in the electric eel (Farber and Rahn, 1970) and the alveolar air concentration in man (Hong *et al.*, 1971). For explanation see text.

seen in the time spent at the surface. The eel surfaces for only a fraction of a second. The single breath which is exchanged renews his oxygen while the CO_2 is being continuously excreted through the skin into the water, and that is why a CO_2 plateau is reached (Fig. 7). Probably the best *repeat* divers in the world are the Ama of Korea and Japan. Their recovery time at the surface is equal to their time under water. When they break the surface, the O_2 is quickly replenished in the lung, but the CO_2 stored in the blood and tissues is only slowly released. It must first be transported to the lung before being exhaled and this takes time.

V. SUMMARY

The mean values and standard error of $\dot{V}_{CO_2}$, $\dot{V}_{O_2}$ and R for all 21 species are shown in Table IV which indicates the narrow limits in which bimodal

TABLE IV

n = 21	"Lung"—%			Gill–Skin—%			R_{tot}
	$\dot{V}_{CO_2}$	$\dot{V}_{O_2}$	R	$\dot{V}_{CO_2}$	$\dot{V}_{O_2}$	R	
Mean	24	65	0·32	76	35	2·5	0·88
S.D.	12	14	0·14	12	14	1·7	0·11
S.E.	3	3	0·07	3	3	0·4	0·02

gas exchange vertebrates function today. Rather than looking at the divergence of adaptation of the many air-breathing fish and aquatic and terrestrial amphibians, it is of interest to look at them as a functional group that has adapted itself to an ecological niche which one might describe as the water–air interface.

The typical bimodal gas exchanging vertebrate uses its air organ or primitive lung to obtain two-thirds of its O_2 but eliminates only one-fourth of the metabolic CO_2 with an exchange ratio of 0·32, while the overall R.Q. is 0·88. What is of particular interest is the fact that the non-pulmonary exchange eliminates three-fourths of the metabolic CO_2 whether the animal lives in air or water. This would indicate that none of the present forms can be considered typical stem forms or "missing links" between the gill breathers

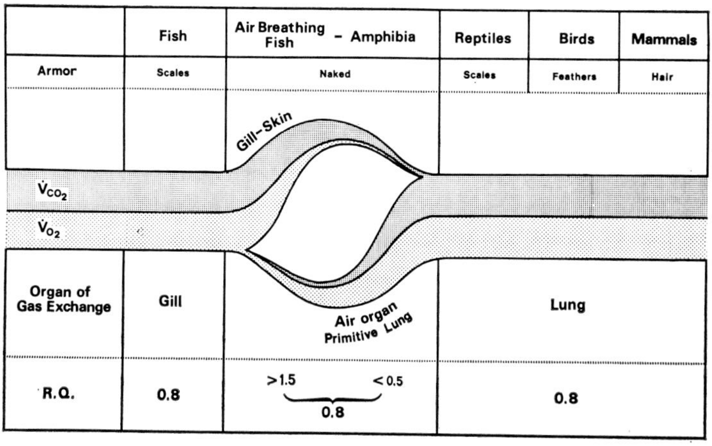

FIG. 8. Overall view of the evolution of the gas exchange from the gill system in fish to the modern lung in the higher vertebrates. Transition forms went through stages of bimodal gas exchange with a large exchange ratio initially in the gill–skin system and a low ratio initially in the air organ or primitive lung. During this stage the skin was essentially naked and subject to dehydration on land. Scales, feathers and hair had to await the development of the modern lung.

and animals with modern lungs, the present reptiles. While some species show that they can obtain up to 96% of their O_2 from the lung, their dependence upon the skin for most of their CO_2 elimination confines them to a water–air interface where the exposed vascularity of their skin can function in this capacity.

The ideal stem forms that led to the modern reptilian lung we might assume followed the path as indicated in Fig. 8. Here the shaded bands show the whole O_2 and CO_2 exchange occurring either in the gill of a typical fish or the modern lung of the higher vertebrates. In between we have pictured the dual path via the gill–skin and the air organ or primitive lung, the former showing a large and the latter a small exchange ratio. While early stem forms in this transition might be difficult to recognize, the later stem forms should reveal a primitive lung that would in addition to supplying the major O_2 demand also eliminate the major metabolic CO_2. As we have seen, such forms do not exist today. Only by switching the major CO_2 elimination to the primitive lung could an animal become independent of exposing its capillary in a naked skin and thus redevelop sufficient armour to avoid dehydration (Czopek, 1965). Only by such a device could our stem forms abandon the moist atmosphere of the water–air interface.

Obviously, some of the extinct stem forms made this transition that allowed them to rearmour their skin, but only after the lung had taken over all the functions of O_2 and CO_2 exchange. This we have called the modern lung, which is today represented by the reptiles and higher vertebrates. From the reptilian scales, feathers and hair evolved and allowed for the development of homeothermy. In this sense one can argue that temperature regulation had to await the development of the modern lung.

References

Czopek, J. (1965). Quantitative studies of the morphology of respiratory surfaces in amphibians. *Acta Anat.* **62**, 296–323.

Erasmus, B. deW., Howell, B. J. and Rahn, H. (1970/71). Ontogeny of acid-base balance in the bullfrog and chicken. *Resp. Physiol.* **11**, 46–53.

Farber, J. and Rahn, H. (1970). Gas exchange between air and water and the ventilation pattern in the electric eel. *Resp. Physiol.* **9**, 151–161.

Guimond, R. W. and Hutchison, V. H. (1972). Pulmonary, branchial and cutaneous gas exchange in the mud puppy, *Necturus maculosus maculosus* (Rafinesque). *Comp. Biochem. Physiol.* **42A**, 367–392.

Guimond, R. W. and Hutchison, V. H. (1973a). Trimodal gas exchange in the large aquatic salamander, *Siren lacertina* (Linnaeus). *Comp. Biochem. Physiol.* **46A**, 249–268.

Guimond, R. W. and Hutchison, V. H. (1973b). Aquatic Respiration: An unusual strategy in the hellbender *Cryptobranchus alleganiensis alleganiensis* (Daudin). *Science N.Y.* **182**, 1263–1264.

Guimond, R. W. and Hutchison, V. H. (1974). Aerial and aquatic respiration in the congo eel, *Amphiuma m. means*. *Resp. Physiol.* **20**, 147–159.

Hong, S. K., Lin, Y. C., Lally, D. A., Yim, B. J. B., Kominami, N., Hong, P. W. and Moore, T. O. (1971). Alveolar gas exchanges and cardiovascular function during breathholding with air. *J. Appl. Physiol.* **30**, 540–547.

Howell, B. J., Baumgardner, F. W., Bondi, K. and Rahn, H. (1970). Acid-base balance in cold-blooded vertebrates as a function of body temperature. *Am. J. Physiol.* **218**, 600–606.

Howell, B. J. (1970). Acid-base balance in transition from water-breathing to air-breathing. *Fed. Proc.* **29**, 1130–1134.

Hughes, G. M. and Singh, B. N. (1970). Respiration in an air-breathing fish, the climbing perch, *Anabas testudineus* (Bloch). *J. exp. Biol.* **53**, 265–280.

Hughes, G. M. and Singh, B. N. (1971). Gas exchange with air and water in an air-breathing catfish, *Saccobranchus fossilis*. *J. exp. Biol.* **55**, 667–682.

Hutchison, V. H., Whitford, W. G. and Kohl, M. (1968). Relation of body size and surface area to gas exchange in anurans. *Physiol. Zool.* **41**, 65–85.

Johansen, K., Hanson, D. and Lenfant, C. (1970). Respiration in a primitive air-breather, *Amia calva*. *Resp. Physiol.* **9**, 162–174.

Just, J. J., Gatz, R. N. and Crawford, E. C., Jr. (1973). Changes in respiratory function during metamorphosis of the bullfrog, *Rana catesbeiana*. *Resp. Physiol.* **17**, 276–282.

Lenfant, C. and Johansen, K. (1968). Respiration in the African lungfish, *Protopterus aethiopicus*. *J. exp. Biol.* **49**, 437–452.

Lenfant, C., Johansen, K. and Hanson, D. (1970). Bimodal gas exchange and ventilation-perfusion relationships in lower vertebrates. *Fed. Proc.* **29**, 1124–1129.

Piiper, J., Dejours, P., Haab, P. and Rahn, H. (1971). Concepts and basic quantities in gas exchange physiology. *Resp. Physiol.* **13**, 292–304.

Rahn, H. (1966). Aquatic gas exchange: theory, *Resp. Physiol.* **1**, 1–12.

Rahn, H., Rahn, K. B., Howell, B. J., Gans, C. and Tenney, S. M. (1971). Air-breathing of the garfish, *Lepisosteus osseus*. *Resp. Physiol.* **11**, 443–466.

Rahn, H. and Garey, W. F. (1973). Arterial CO_2, O_2, pH and HCO_3^- values of ectotherms living in the Amazon. *Am. J. Physiol.* **225**, 735–738.

Singh, B. N. and Hughes, G. M. (1971). Respiration of an air-breathing catfish, *Clarias batrachus* (Linn.) *J. exp. Biol.* **55**, 421–434.

Vinegar, A. and Hutchison, V. H. (1965). Pulmonary and cutaneous gas exchange in the green frog, *Rana clamitans*. *Zoologica* (N.Y. Zool. Soc.) Vol. **50**, 47–53.

Whitford, W. G. and Hutchison, V. H. (1965). Gas exchange in salamanders. *Physiol. Zool.* **38**, 228–242.

Eco-physiological Studies of some Metabolic and Respiratory Adaptations of Sirenid Salamanders

GORDON R. ULTSCH

University of Alabama, Department of Biology, University, Alabama, U.S.A.

I. INTRODUCTION

One of the most important physicochemical conditions to which aquatic organisms must adapt is the concentration of dissolved respiratory gases. In waters that contain little aquatic vegetation of any form dissolved oxygen is the gas of importance, since the high rate of diffusion of carbon dioxide in water results in very low concentrations of this gas as free CO_2. Consequently, much of the literature dealing with the adaptations of organisms to dissolved gas levels deals only with oxygen. However, in situations where the dissolved oxygen is low the dissolved carbon dioxide is often high, and this poses some serious physiological problems for water breathers. Such problems may include a decrease in the affinity of the blood for oxygen at the site of gas exchange with the external medium (Bohr effect), a decrease in the oxygen capacity of the blood (Root effect), disturbance of the acid-base balance by the decrease in pH of the water associated with an increase in P_{CO_2}, and a decrease in the ability to eliminate CO_2 to the water because of a reduction in the partial pressure gradient between the blood within the external gas exchanger and the surrounding water.

The presence of any type of vegetation in the water will usually induce a hypoxic and hypercarbic aquatic environment to some degree, except during periods of peak photosynthesis. Figure 1 illustrates this effect. The data points are those of Lynch, King, Chamberlain and Smith (1947) from a series of measurements they conducted during the summer in Louisiana and Florida at a depth of about 30 cm. Oxygen saturation for the temperatures encountered by these authors would be approximately 8* ppm for open, well-mixed water; carbon dioxide levels under such conditions would be neglible from a physiological viewpoint (less than 5 ppm). The obvious effect of the presence of vegetation is to lower the level of dissolved O_2 and increase the level of dissolved CO_2. This is the case for emergent, submerged and floating types of vegetation when dissolved oxygen is considered; but only the floating forms were found to both reduce dissolved oxygen to low levels and raise the dissolved carbon dioxide to levels higher than 30 ppm. *Eichhornia crassipes*, the water hyacinth, exhibited this combined effect to the greatest degree, frequently inducing levels of dissolved O_2 of zero, and causing by far the highest levels of dissolved carbon dioxide; as high as 80 ppm. Such conditions would obviously impose a serious respiratory stress upon most aerobic water-breathing organisms.

Since the water hyacinth community is one of the most restrictive environments for water-breathing organisms from the viewpoint of gas exchange, a study of respiration and related physiological functions of animals inhabiting that community could reasonably be expected to elucidate various strategies of respiratory adaptation. In order to identify the various microhabitats available to larger aquatic organisms, measurements of pH, temperature, dissolved O_2 and dissolved CO_2 were conducted throughout a representative hyacinth community and compared with open water in the same pond (Ultsch, 1973b). Once the parameters of physiological interest were defined, it was possible to conduct meaningful laboratory experiments, the results of which could be related to animals in the field.

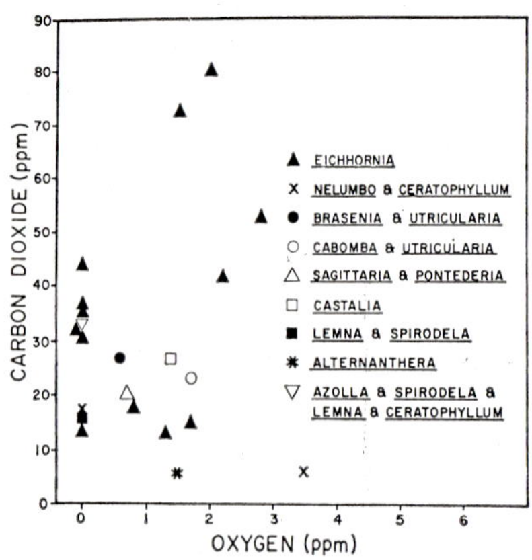

FIG. 1. Oxygen and carbon dioxide concentrations in aquatic environments containing various types of vegetation. Data points are selected from Lynch *et al.* (1947). Original measurements were taken in Louisiana and Florida from July–October (1946) at a depth of approximately 30 cm.

* For most freshwaters 1 ppm $\cong$ 15–20 mmHg P_{O_2} at 15–30°C respectively.

Within the water hyacinths, the most abundant vertebrates were three species of salamanders of the family Sirenidae. These three represent all of the known species of this family, but they are well separated by body size in this particular study area. The maximim weights of adults collected during the five years encompassed by various aspects of the study were: *Siren lacertina*, 1698 g; *S. intermedia intermedia*, 45 g; and *Pseudobranchus striatus*, 5·0 g. As is the case with many of the aquatic inhabitants of this community, the Sirenidae are bimodal breathers capable of respiring both air and water. All three forms possess external gills and have lungs. In captivity, they are observed to breathe air frequently, even in air-saturated water. For *Siren lacertina*, numerous studies (Czopek, 1962; Guimond and Hutchison, 1973; Shield and Bentley, 1973) have shown that the gills are relatively unimportant in gas exchange and that most of the aquatic gas exchange is carried on via the integument. In fact, *Siren lacertina* up to about 700 g and all sizes of the other two species can survive indefinitely in air-equilibrated water at 25°C without access to air (Ultsch, 1973a, 1974). So the capacity of the skin for gas exchange is appreciable in this group.

The intent of this paper is to examine the adaptations of all three species of sirenids for meeting the stresses imposed by (1) low levels of dissolved oxygen (2) high levels of dissolved carbon dioxide and (3) restrictions on the attainment of a large body size as related to gas exchange capacities. All considerations will then be related to variations in the microenvironment of the animals in terms of seasonal and daily fluctuations in dissolved gases.

II. Materials and Methods

The field location used for the study of the microenvironments of water hyacinth communities and for the collection of many of the sirenids was a pond in Gainesville, Alachua County, Florida (Fig. 2). The pond was almost entirely covered with water hyacinths, with only about $\frac{1}{2}$ acre (of approximately 10 total acres) of open water. This "open" water contained other vegetation, mostly *Ceratophyllum* sp., a submerged aquatic plant. Details of techniques for measuring O_2, CO_2, pH and temperature throughout the pond are given elsewhere (Ultsch, 1973b).

Sirenids and fish collected in the study area were captured with a dredge (often referred to as a Coleman dredge) composed of a 3 foot by 3 foot section of $\frac{1}{8}$ inch hardware cloth fastened to the bottom of a three-sided frame. This open front dredge is similar in appearance to a fireplace screen (Fig. 2). It is pushed rapidly under the hyacinths and lifted upward, entrapping the section of hyacinths above it. As the hyacinths are removed from the dredge, the various animals in the hyacinth mat fall onto the screen. By pushing the dredge through the mud at the bottom before lifting it, one can collect both bottom-dwelling and hyacinth mat-dwelling organisms. The open water

FIG. 2. Photograph of the study area taken during April, 1971. Shown are hyacinth-covered water, the open water (background) and the dredge used to collect animals under and within the hyacinth areas. The aerial portions of the plants are just beginning their annual regrowth and appear relatively short. Photograph (and student help) courtesy of Dr. Archie Carr.

area was sampled with both the dredge and with $\frac{1}{8}$ inch mesh seines. Because of the relative difficulty of collecting large *Siren lacertina* in the hyacinths by the arduous dredging method, many additional *S. lacertina* were collected elsewhere in Alachua County at night with dip nets.

Methods for the determination of surface area of the skin, metabolic rates of submerged animals, critical oxygen tensions, and permeability of the skin to oxygen are given elsewhere (Ultsch, 1973a, 1974).

Gas exchange partitioning experiments were conducted in two-section plexiglas chambers. The lower portion of a chamber contained the animal and was filled with water; the upper portion was filled with air. A small opening connected the two sections to allow the animal to breathe air. The surface of the water at the opening was covered with a 1 cm layer of heavy duty paraffin oil to minimize gas exchange between the air and water. Observation showed that the animals located the breathing hole much more readily when it was the only source of light; therefore the chamber was painted a dull grey except for the area above the breathing hole. In practice, the animals usually stayed near the opening, and air breathing merely necessitated raising the upper third of the body to be able to protrude the mouth into the aerial chamber. Gas concentrations were set at the desired levels by bubbling CO_2, O_2, or N_2 through the water. Diffusion between the air and water in controls was nil. The animals were acclimatized to the experimental conditions for at least 24 hours before the start of a series of experiments by placing them in the chamber at 25°C with the appropriate set of concentrations of dissolved gases and leaving the aerial portion open to the atmosphere. Immediately prior to an experiment, the water was changed and the desired gas concentrations re-established in the water. One to two hours were allowed before the aerial portion was sealed from the atmosphere and connected by a manometer to a thermobarometer chamber, after which the tank was completely submerged in the water bath at 25°C. Generally, two experiments were run per day, after which the gas concentrations were reset to the desired levels, and the animals were left in the chamber overnight. The same procedure was followed for the next day and thereafter until a series of experiments for a particular set of conditions was complete. Any acclimatization was apparently complete at the end of 24 hours, since there was no significant change in metabolic rates or partitioning between the first and final days of experimentation for a particular set of conditions. Volume control in the air phase was maintained by the addition of N_2 to re-level the manometer connecting with the thermobarometer chamber. Changes in the vol % of CO_2 and O_2 in the aerial phase were measured with a Scholander $\frac{1}{2}$ cc gas analyser. Dissolved O_2 was measured with the Radiometer PHA 27 Gas Monitor. CO_2 elimination to the water was determined by calculation from an RQ of 0·91 (SE = 0·03) reported by Guimond and Hutchison (1973) for *S. lacertina* at 25°C. The conditions of experimentation dealing with the effects of varying body size and dissolved gases are given in Tables I and II respectively.

The effect of high levels of carbon dioxide was studied by submerging animals in water with an oxygen tension near or above air saturation. After a period of submergence (17–23 hr) at nil levels of carbon dioxide, the P_{CO_2} was raised intermittently over a period of days. High oxygen levels were re-established whenever the P_{CO_2} was increased. Gas levels were usually adjusted twice daily by bubbling O_2 and CO_2 through the upper portion of the chamber while the animals were located in the lower portion. The experiments were terminated with the death of the animal or when the P_{CO_2} reached 200 mmHg, whichever came first. The latter figure corresponds to the upper end of the scale of the Radiometer Gas Monitor and is equivalent to about 390 ppm CO_2. This was far beyond the maximum field concentration.

All experimental animals were acclimatized at 25 ± 2°C for at least two weeks prior to use. All gas volumes are reported at STPD.

TABLE I

Conditions of gas exchange partitioning experiments for two large and one small S. lacertina. *The duration of the runs was 2·75–5·50 hrs for large animals, and 2.00–6.33 hrs for the small animal. All experiments were at 25°C. Values are given as* $\bar{x} \pm 2$ *S.E.*

Dates of Experiment	Weight (g)	Final Air O_2 (vol%)	Mean Water P_{O_2} (mmHg)	Mean Water P_{CO_2} (mmHg)
26th Feb. – 4th Mar. 1972	1489 ±26	15·6 ±1·0	145 ±8	less than 7
22nd Feb. – 2nd Mar. 1972	1182 ±9	15·2 ±1·4	144 ±3	less than 7
28th Mar. – 3 Apr. 1972	89·6 ±0·7	18·3 ±0·8	143 ±4	less than 7

TABLE II

Conditions of gas exchange partitioning experiments in a single, large S. lacertina. *The duration of the runs was 2.75–5.50 hrs. All experiments were at 25°C. Values are given as* $\bar{x} \pm 2$ *S.E.*

Dates of Experiment	Weight (g)	Final Air O_2 (vol%)	Mean Water P_{O_2} (mmHg)	Mean Water P_{CO_2} (mmHg)
26th Feb. – 4th Mar. 1972	1489 ±26	15·6 ±1·0	145 ±8	less than 7
17th Mar. – 20th Mar. 1972	1454 ±10	15·7 ±0·5	145 ±2	40 ±2
8th Mar. – 11th Mar. 1972	1464 ±4	15·3 ±1·2	24 ±5	39 ±2
13th Mar. – 15th Mar. 1972	1453 ±2	15·5 ±0·9	22 ±4	less than 7

III. RESULTS AND DISCUSSION OF FIELD STUDIES

Determinations of the levels of dissolved oxygen and carbon dioxide were made in both open water and hyacinth-covered water as a function of depth, time of day, extent of coverage by water hyacinths, and time of year (Ultsch, 1973b). There are two effects that are relevant to this study. The first is that there existed a distinct difference in the levels of dissolved respiratory gases in the open water and in the hyacinth areas. Regardless of the time of day or the time of year, there is less dissolved oxygen and more dissolved carbon dioxide in the water covered by hyacinths than in the open water. Figure 3 illustrates this effect of the hyacinths. This makes the hyacinth-covered portion of the pond distinctly less suitable than the open water portion for a water breathing animal. The levels of dissolved respiratory gases are given as a function of the month of the year in Fig. 4, from which the high CO_2–low O_2 levels of the sub-hyacinth water are evident, especially during the warmer months. The second effect is that there is a vertical difference in the dissolved gases within the hyacinth microhabitats that is especially pronounced during the growing season of the plants (March–October in Gainesville, Florida). To more clearly illustrate this effect, gas envelopes were constructed for surface and bottom conditions within the hyacinths. Such envelopes are represented graphically by plotting all measure-

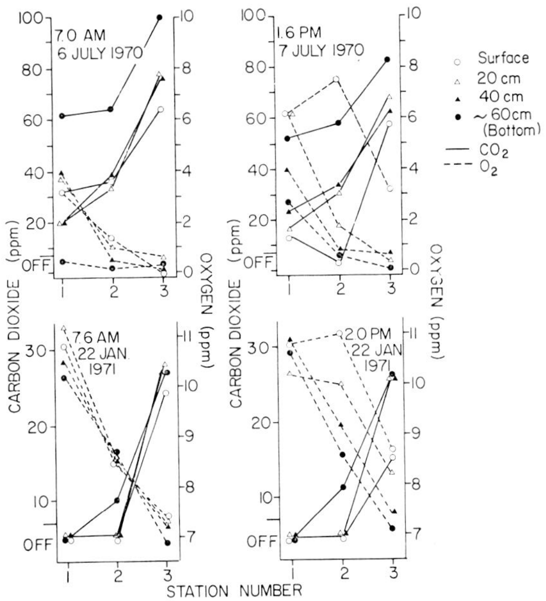

FIG. 3. Dissolved oxygen and carbon dioxide levels in the study area as a function of the time of day, time of year, and extent of coverage by *Eichhornia*. On the x-axis, (1) represents water not covered by water hyacinths, but often with a considerable crop of *Ceratophyllum* during the summer; (2) is water covered by a "thin" hyacinth mat (plants are small without an accumulation of detritus in the root mass) characteristic of the fringes of the main mat; (3) is the mature section of the mat with larger plants that have been present for at least several years and whose root mass has collected considerable organic debris that results in a much denser and thicker coverage of the water. The instrumentation used could detect CO_2 down to a P_{CO_2} of 7 mmHg; values below that were assigned a level of 3·5 mmHg, or approximately 7 ppm. Ecologically, this is a nil level for a water breather. Such levels are indicated as "off", meaning off the scale of the instrument.

ments of O_2 and CO_2 within a particular area of the pond for a particular time span and then enclosing these data with a polygon. An example is given in Fig. 5 for a month during the peak of the growing season (July) and for one during the non-growing season (January). These envelopes constitute the entire set of combinations of dissolved O_2 and CO_2 available for microhabitat selection by animals inhabiting the hyacinth portions of the pond. It should be pointed out that the rapid decrease in oxygen with depth in the hyacinths restricts the size of organisms that can effectively occupy the surface niche shown for July. Most of the water in the hyacinths at that time of year actually falls in or near the polygon enclosing the bottom measurements (see Ultsch, 1973b for more details).

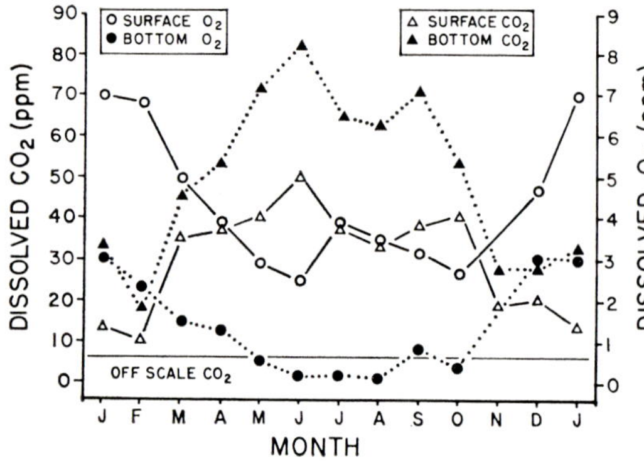

FIG. 4. Concentrations of dissolved carbon dioxide and oxygen in a water hyacinth community in Gainesville, Florida (U.S.A.). Each point is a 24-hr average of a sunny day within the given month. Water depth was approximately 60–80 cm; surface samples were from the upper 5 cm and bottom samples from the lower 5 cm.

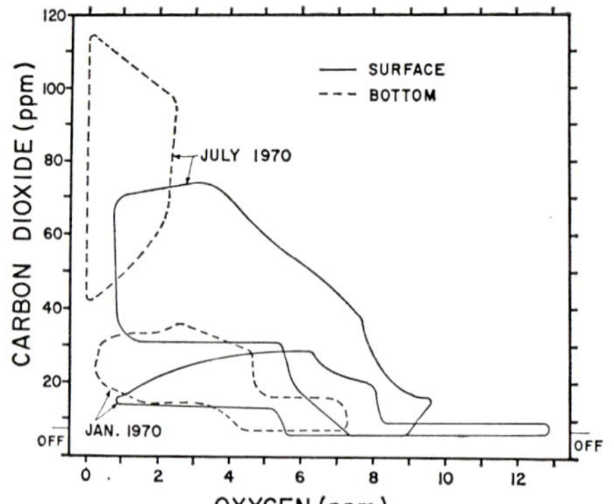

FIG. 5. Gas envelopes for January and July at the surface and bottom of water covered by water hyacinths. The depth was about 60 cm. Each envelope contains all combinations of O_2 and CO_2 measured at intervals throughout a 24 hr day and therefore represents the sum of available niches in terms of dissolved respiratory gases.

Some of the consequences of these two general effects are reflected in the distribution of fishes within the study area. There were four relatively distinct vegetational zones in the pond which represented an ecological transect from the shore to the open water. These zones can be characterized as follows:

I. Algae and emergent grasses; no hyacinths. Depths approximately 0–50 cm.
II. "Thin" hyacinths (see Fig. 3). Depths 30–100 cm.
III. "Thick" hyacinths (see Fig. 3). Depths 50 cm and greater.
IV. Open water. Depths 75 cm and greater. Some *Ceratophyllum*.

Fishes were collected in these four zones during a one day trial in April of three consecutive years (1970–72). The species distribution is given in Table III. This table is undoubtedly not exhaustive; collections were attempted for several hours until no new species were obtained. However, it should reflect the relative abundance of species in each area.

The paucity of fish in the dense hyacinths characteristic of zone three is related to the low levels of oxygen and high levels of carbon dioxide. The levels of these two gases expressed as the average daily values for April are

TABLE III

The occurrence of various fishes within the four vegetational zones of the study area (see text for details). Collections were made in April of 1970–72. Any occurrence in any year is indicated by an X; an O means no occurrence

Species	Zone I	Zone II	Zone III	Zone IV
Esox niger	X	O	O	O
Leptolucania ommata	X	X	O	O*
Enneacanthus gloriosus	X	X	O	X
Enneacanthus obesus	X	X	O	X
Gambusia affinis	X	X	X	X
Fundulus chrysotus	X	X	X	X
Lucania goodei	X	O	O	X
Elassoma zonatum	X	X	X	O
Aphredoderus sayanus	X	X	O	O
Notemigonus crysoluecas	X	O	O	O
Amia calva	X	O	O	O*
Jordanella floridae	X	O	O	O
Etheostoma fusiforme	X	O	O	O
Lepomis macrohirus	X	O	O	X
Ictalurus sp.	X	X	O	X
	15/15	8/15	3/15	7/15

* Probably occurs, but collections were unsuccessful.

given in Table IV. Only data for the main portion of the hyacinth mat (which is at least 95% of the total mat) and for open water (zone four) are presented, since they represent most of the area of the pond and are the only areas for which a 24 hour average is available. The data of Table IV are helpful in explaining the distributions shown in Table III. Some individual daytime measurements indicated that O_2 is higher and CO_2 lower in zone one than in any other zone, which helps to explain the occurrence of all species in that area. The reduction of species number as one encounters the hyacinth area is expected. What fish occurred were in the relatively oxygenated upper water layer. Furthermore, two of these (*Gambusia* and *Fundulus*) are cyprinodontoids, which have been found to be morphologically adapted for breathing the thin, oxygenated, topmost layer of water (Lewis, 1970). Some of the species reappear in the open water area, but the interposition of the hyacinths between zone one and zone four apparently prevents the reappearance of all of the species that might be expected, although this difference may be partially due to unsuccessful collections and is certainly related to other microhabitat differences in addition to CO_2 and O_2 levels. The fish distributions do, however, suggest that the hyacinths effect the distributions of aquatic organisms. The data reflects a definite choice of microhabitats available to water-breathing organisms between open water and hyacinth covered water and, to a degree that is dependent upon body size and the ability to use the upper few centimeters of water for respiration, within the hyacinth covered water itself. With regard to the sirenid salamanders that were so common in the pond, their relatively large body size (*Siren* is much larger than most of the fish listed in Table III) effectively limited their choice of microenvironments to the open water or to the subsurface hyacinth water. In terms of gas exchange, this would be particularly important in the summer, when dissolved oxygen under the hyacinths was near zero. When this decreased oxygen availability is considered in com-

TABLE IV

Dissolved respiratory gases in the open water and in water covered with a mature hyacinth mat. Data are the mean daily values for a representative sunny day in April, 1970

	LOCATION			
	Hyacinths (upper 5 cm)	Hyacinths (lower 5 cm)	Open water (upper 5 cm)	Open water (lower 5 cm)
Dissolved O_2 (ppm)	3·9	1·3	7·1	4·0
Dissolved CO_2 (ppm)	38	53	10	22

bination with the increased oxygen demand associated with the higher body temperatures and metabolic rates of these poikilotherms, such a micro-habitat difference would be significant for an animal dependent on aquatic gas exchange.

Numerous attempts to collect sirenids were made at various times of the day and year from 1969–72. Each successful attempt was accompanied by a determination of dissolved O_2 and CO_2. An unsuccessful attempt can not be interpreted to mean that no animals were present at a particular location, so dissolved gas determinations were not made in those instances. However, it soon became apparent that there were no sirenids in the open water, or that they were so scarce that none were collected. At the other extreme, there were no unsuccessful collections of *Pseudobranchus striatus* in the hyacinths. This animal was by far the most common vertebrate in the hyacinth community. As for the two species of *Siren*, the possibility remained that they were restricted to certain portions of the hyacinth mat that did not have gas concentrations as unfavourable for aquatic respiration as other portions. It was also possible that the distribution of the *Siren* might be size dependent, particularly for *S. lacertina.* Finally, the season might influence the distributions.

To test all of these possibilities, composite gas envelopes were constructed for the hyacinth areas for the growing and non-growing seasons for *Eichhornia.* These envelopes would then represent all possible combinations of dissolved O_2 and CO_2 available for selection by the various sirenids. Each successful collection can then be scored within the appropriate polygon by considering concentrations of dissolved gases where the animal was collected. Any avoidance, considering a large number of successful collections, should appear as a "hole" on the graph; that is, if a particular sirenid is avoiding an area that has especially low O_2 and/or high CO_2, there would be no scores in that portion of the envelope regardless of the number of collections made. The results of this portion of the field work are shown in Figs 6–8. The interpretation is conclusive: regardless of species, size, surface or bottom concentrations of dissolved O_2 and CO_2, or time of the year, none of the sirenids were excluded from microhabitats on the basis of dissolved respiratory gases.

Since the sirenids avoid the open water and are apparently indifferent to the levels of dissolved respiratory gases in the hyacinth covered water, I must conclude that all of their respiratory and metabolic adaptations are physiological or morphological, rather than behavioural.

IV. Results and Discussion of Experimental Studies

The strategies of adaptation for the exchange of gases by the Sirenidae inhabiting the water hyacinth community have to be complex. The animals can breathe either air or water, and both the metabolic demand for oxygen and the oxygen and carbon dioxide tensions in the water fluctuate on a seasonal basis. A reasonable approach is to first examine adaptations relating to aquatic gas exchange, and then investigate adaptations relating to aerial gas exchange. Body size must be considered when both of these

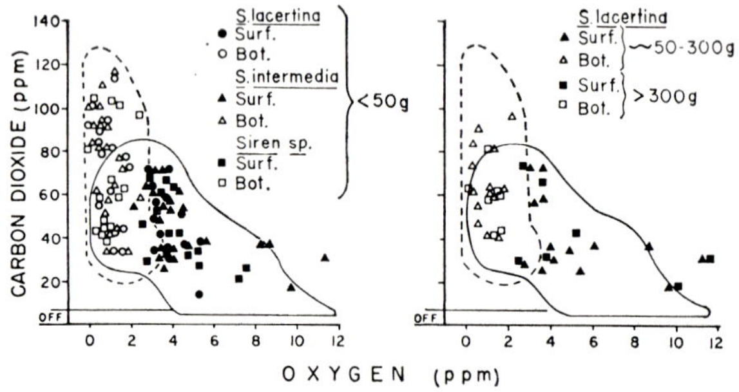

FIG. 6. Location of various size classes of *S. lacertina* and *S. intermedia* within the hyacinth area of the pond during the hyacinth growing season (March–October), as a function of concentrations of dissolved respiratory gases. Areas enclosed by solid and dashed lines represent concentrations of dissolved O_2 and CO_2 at the surface (Surf.) and bottom (Bot.), respectively, for all March–October (1970) field determinations (exclusive of determinations made concurrently with collections). Some scores are outside the 1970 polygon because collections were made from 1969–1972.

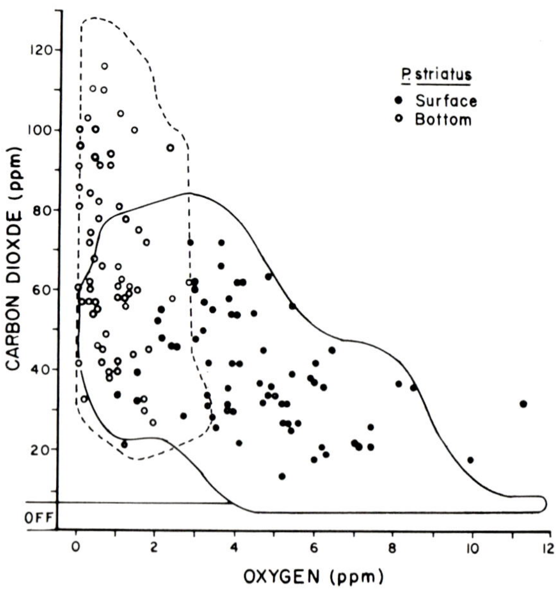

FIG. 7. Location of *P. striatus* within the hyacinth area of the pond during the growing season of *Eichhornia* (March–October), as a function of concentrations of dissolved respiratory gases. Enclosed areas are as in Fig. 6.

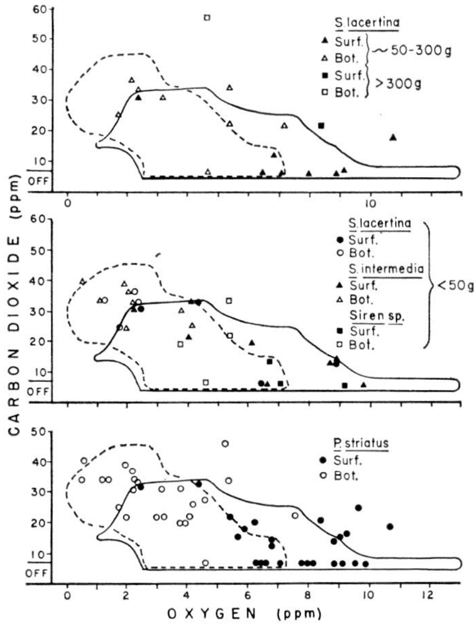

FIG. 8. Location of *Siren* and *Pseudobranchus* within the hyacinth area of the pond during the non-growing season of *Eichhornia* (November–February), as a function of concentrations of dissolved respiratory gases and body size. Enclosed areas are as in Fig. 6, except for data being for November–February rather than March–October.

modes are being investigated, for an adaptation suitable for a small animal may not be suitable for a large one.

A. AQUATIC GAS EXCHANGE

The ability of a sirenid to obtain oxygen from the water will vary with the oxygen tension of the water and the capacity of the external gas exchanger (the skin, in the case of the Sirenidae) to transport oxygen into the animal. The latter, herein referred to as the oxygen exchange capacity of the external gas exchanger (E_c), will depend upon the surface area of the gas exchanger and its permeability to oxygen. Furthermore, metabolic rate, the permeability to oxygen of the skin, and the surface area of the skin are all functions of the body size.

The detailed relationships between metabolic rate, oxygen exchange capacity, and body size are presented elsewhere (Ultsch, 1973a). In summary, there are two allometric equations that are relevant to this discussion:

$$\text{SMR} = k\ \text{W}^a, \text{(or } \dot{V}_{O_2} = a\text{W}^b\text{, of some authors) and} \qquad (1)$$

$$E_c = k_1\ \text{W}^b; \qquad (2)$$

where SMR = standard metabolic rate and E_c = oxygen exchange capacity of the skin. The exponent b in Eq. (2) would have to be larger than the exponent a in Eq. (1) to prevent a water breather from literally running out of gas as he increases in size. Equation (1) relates the minimal demand for oxygen to body weight, and Eq. (2) relates the maximal ability to supply that oxygen to body weight. If a is larger than b in this set of equations, the supply and demand curves for oxygen will intersect and impose an upper limit on the size of an animal. Therefore any adaptation that would lower the demand for oxygen or increase the ability to supply it would be useful in terms of aquatic gas exchange.

The demand for oxygen is usually considered in terms of the SMR at a particular temperature. The SMR is used comparatively in two ways: (1) the SMR at unit body weight and (2) the rate of change of SMR with increasing body weight. Whitford and Hutchison (1967) have studied the SMR in both lunged and lungless salamanders, and their results are compared to those for the sirenids in Table V. The sirenids were submerged

TABLE V

A comparison of important metabolic parameters in sirenids and other salamanders. The data for the sirenids is for submerged animals at $P_{O_2} = 155$ *mmHg. (Ultsch, 1974)*

Species or group	SMR at unit body weight (1 g) (μl O_2/hr)	Exponent of W in Eq. (1)
Lunged salamanders	110	0·86
Lungless salamanders	120	0·72
Sirenidae	33*	0·66
P. striatus	29*	0·74
S. intermedia	39*	0·63
S. lacertina	33*	0·65

* SMR is corrected to the 15° experimental temperature used by Whitford and Hutchison (1967) for the lunged and lungless forms. The correction is made by using a $Q_{10} = 2{\cdot}19$ as reported by Guimond and Hutchison (1973) for *S. lacertina* from 15–25°C.

and therefore unable to use their lungs, but within the size range used by Whitford and Hutchison (3–30 g) comparisons are valid, since at this size all three species of sirenids can live indefinitely at 25°C when submerged in air-equilibrated water and utilizing only aquatic respiration. However, there is some indication that sirenids show an increase in the SMR if allowed to breathe both air and water (Ultsch, 1973a, 1974 and unpublished data), so it is with some reservation that it is concluded from Table V that sirenids have an intrinsically lower SMR than other salamanders. In the context of looking only at the adaptations for aquatic gas exchange, however, it is true that these animals do have both a low SMR at unit body weight and a slower increase in oxygen consumption with size than other salamanders. These are certainly adaptations that are useful in diminishing the relative demand for oxygen while submerged.

The counterpart of the demand for oxygen is the ability to supply it as expressed in Eq. (2). Here several factors influence both the level of the E_c and its relation to body weight.

Equation (2) can be approximated for sirenids submerged at 25°C in air-saturated water as

$$E_c = P_{max}(SA)(\Delta P_{O_2}) \tag{3}$$

where P_{max} is the maximum permeability of the skin to oxygen ($\mu l\ O_2/hr \cdot cm^2 \cdot mmHg$), SA is the surface area of the skin (cm^2), and ΔP_{O_2} (mmHg) is the difference in the partial pressure of oxygen in the water and the mean capillary oxygen tension in the blood at the site of cutaneous gas exchange. By assuming that the loading tension of the blood is near 25 mmHg (Ultsch, 1973a), ΔP_{O_2} becomes a constant at $(P_e - 25)$ mmHg where P_e = the O_2 tension of the water. For a fixed oxygen tension in the water the ability of an animal to supply the oxygen necessary to meet its metabolic demand will be a function of the surface area and permeability to oxygen of the skin. Permeability with the units given is similar to transport capacity and can be modified by changes in heart rate, dilation of the peripheral vessels, and the shunting of blood to the skin; but there must be a maximum permeability after all physiological adjustments are made. This maximum permeability can be indirectly measured by the following considerations of the effects of declining oxygen tensions on the metabolic rate of a submerged sirenid.

All three of the sirenids have been found to be metabolic regulators (Ultsch, 1974); that is, they can maintain the level of their standard metabolic rate over a range of oxygen tensions. The lower end of this range is denoted as the critical oxygen tension (P_c), with the animals becoming oxygen conformers at tensions below this point; that is, their rate of oxygen consumption now becomes dependent upon the oxygen tension in the water, with lower tensions resulting in lower rates of oxygen consumption. It is reasonable to assume that all physiological methods of increasing the permeability of the skin to oxygen are maximized at P_c and that the P_{max} is realized at this point. Since O_2 consumed = O_2 taken up through the skin at all tensions for these submerged animals, P_{max} can be expressed as

$$P_{max} = SMR/SA(P_c - 25), \tag{4}$$
$$(\mu l\ O_2/hr \cdot cm^2 \cdot mmHg)$$

where P_c is the critical oxygen tension for the submerged sirenid. The SMR is fixed by the physiology of the animal, but appropriate changes in SA or P_c with increasing body size could be strategies that would effect E_c. It must be noticed that P_{max} in Eq. (4) is expressed on a per cm^2 basis; obviously if the animal had a greater surface area than actually observed for a given weight, this would enable more oxygen to enter the body. Generally, the equation that relates surface area to volume is:

$$SA = k_2 W^{0 \cdot 67} \tag{5}$$

for objects of similar proportions but of different volumes (weight is here considered equivalent to volume). If a water breathing sirenid disproportionately increased the surface area of the skin as he grew, perhaps by the use of skin folds or a tendency to become long and thin, this would help to alleviate the morphological problem associated with an increase in the surface area of the skin lagging behind the increase in volume. However, this turns out to be another unused strategy for adaptation to water breathing (Fig. 9); notably so in *S. lacertina* which has the largest range of body size. The only other way that the animals can increase the permeability to oxygen with increasing size is to reduce P_c with increasing body weight. The relationship of P_c and weight is shown in Fig. 10. Over the small size ranges of *P. striatus* and *S. intermedia* no trend is obvious, but over the large size range attained by *S. lacertina*, there is definitely an increase in P_c with body weight, which is maladaptive with regard to aquatic gas exchange.

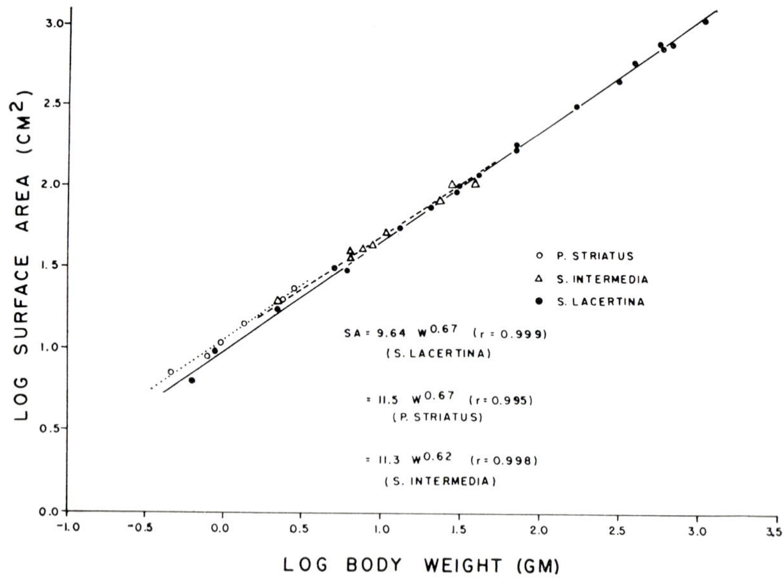

FIG. 9. The relationship between the surface area of the skin and body size in the three species of Sirenidae.

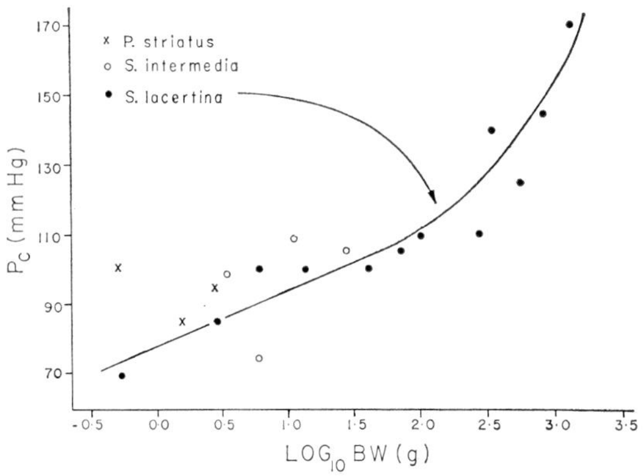

FIG. 10. Critical oxygen tension (P_c) as a function of body size in the Sirenidae. The curve for *S. lacertina* is fitted by eye.

The measurements of surface area, P_c and SMR enable one to make a comparison between the demand for oxygen (SMR) and the ability to supply it (E_c) for all three species. The conditions of comparison are chosen arbitrarily as $P_e = 155$ mmHg of O_2 at 25°C (air saturation), which makes $\Delta P_{O_2} = 130$ mmHg. The SMR and E_c can then be expressed as functions of body weight according to Eqs (1) and (3), respectively. The results are given in Fig. 11 (see Ultsch 1973a for more details on the mathematics and a sample calculation of E_c). Only in *S. lacertina* does the maximum size of the animal approach the theoretical upper size limit for submerged animals. For the sizes actually attained by *P. striatus* and *S. intermedia*, the results suggest that these animals should be able to survive indefinitely utilizing only aquatic respiration under the given set of conditions, and this is borne out experimentally. The maximum theoretical size of 2908 grams for *S. lacertina* is derived by considering the minimal metabolic rate and the maximal exchange capacity, a situation likely to be intolerable over any period of time since this would leave the animal no scope for activity. Death of submerged animals would be expected to occur at a smaller size and in fact first occurs at about 700–800 g. If the utilization of E_c is expressed as per cent, this per cent approaches 100 more rapidly for larger animals than for smaller ones, and it can be shown (Ultsch, 1973a) that 955 g is a better estimate of the maximum attainable size for breathing only water. This means that *S. lacertina* at large body sizes must be an obligate air breather during the summer in the hyacinth communities studied, regardless of the oxygen tension in the water, since summer water temperatures closely approximate 25°C.

The extrapolation of the curves of SMR and E_c for *S. intermedia* predict essentially the same upper size as for *S. lacertina*. Yet, in Gainesville, the

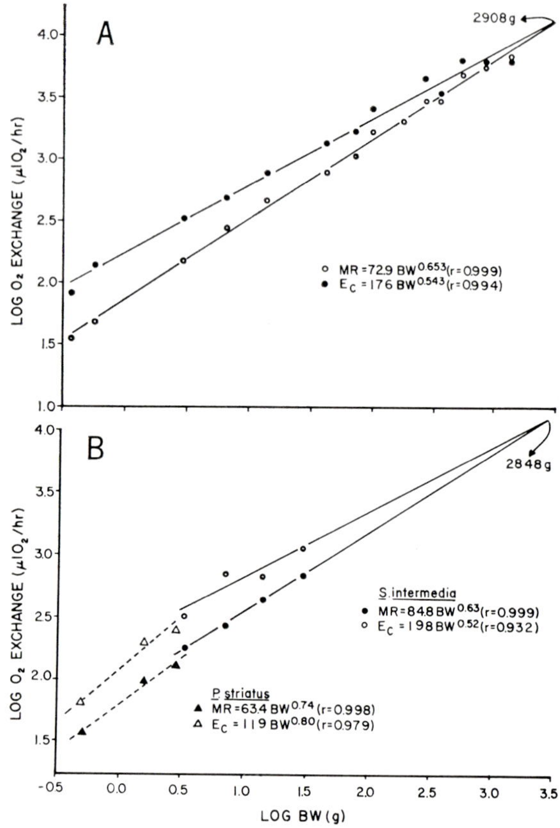

FIG. 11. The relationships of the standard metabolic rate (MR) and the maximum ability to take up oxygen (E_c) as functions of body size in three species of sirenid salamanders. A = *S. lacertina*; B = *S. intermedia* and *P. striatus*. Data are based on submerged animals at 25°C with P_{O_2} = 155 mmHg.

former reaches only 45 g while the latter grows to over 1700 g. Outside the range of *S. lacertina*, *S. intermedia* reaches a much larger size. Bishop (1943) gives the maximum length of *S. intermedia* as 660 mm in Texas, while I have never collected one over 250 mm in Gainesville. Since there seems to be an indication that gas exchange is not limiting the attainable size of *S. intermedia*, it is quite possible that this is yet another example of character displacement based upon food particle size.

B. AERIAL GAS EXCHANGE

The oxygen consumption of submerged sirenids indicated that the metabolic demand for oxygen was low. At 155 mmHg of O_2 (25°C), animals up to 700–800 g survived. But the field data indicated that at the time of the year when the natural habitat is near 25°C, the oxygen content of the water is very low. Saturation at this temperature is 8·2 ppm for oxygen, but this level is never realized anywhere under the mat during the growing season of the hyacinths. The mean value for the bottom 5 cm of water under the hyacinth mat during the growing season is 0·6 ppm O_2, and for the upper 5 cm it is 3·5 ppm O_2. Since the lowest P_c observed for any of the sirenids is about 4·0 ppm at 25°C, all three species are obligate air breathers in the field during the warmer months. This conclusion would only be strengthened by considerations of carbon dioxide, since it would be reasonable to expect that levels of carbon dioxide above the nil levels used in determining P_c would, if there is any effect at all, result in a higher P_c. Since the animals would be excluded from the hyacinth community on the basis of aquatic respiration, a major part of their adaptation must be based on aerial respiration, at least during the warmer months. During the winter, when the oxygen level under the hyacinths reaches 2·5 ppm and the CO_2 drops to 26 ppm, the lower oxygen demand associated with lower temperatures may be met largely by aquatic gas exchange.

The mean temperature for March–October at the surface and bottom of the hyacinths was 25·5°C and 23·0°C, respectively. Since the summer would be the most critical time in terms of respiratory stress on a sirenid, both in terms of increased metabolism and low levels of O_2 with high levels of CO_2 in the water, experimentation dealing with animals breathing both air and water was done at 25°C. Two experiments are pertinent. The first deals with the effects of body size on the partitioning of the exchange of O_2 and CO_2 between the air and water, and the second deals with the effects of variations in the dissolved O_2 and CO_2 partitioning for a given body size.

To study the effects of body size upon gas exchange partitioning, the animals were placed in an air–water respirometer with O_2 tensions near air saturation and nil levels of CO_2 in both air and water. This would be the most favourable microenvironment that a sirenid could naturally occur in and should reflect a maximum degree of aquatic respiration. Since *S. lacertina* attains the largest range of body size, this species was chosen for the investigation. The partitioning is given in Table VI. Under forced submergence, a 90 g animal is capable of completely aquatic respiration without any ill effects if kept forcibly submerged; the two larger animals would die within a day or two. Obviously access to air breathing is utilized by both large and small sizes, since even the 90 g individual obtained 42% of its oxygen from the air; but it is especially important to the larger animals (75%), which are obligate air breathers at 25°C even in air-equilibrated water. The elimination of CO_2 was mainly via the skin for both sizes, as is true for most aquatic amphibians. The increase from 6% to 26% in aerial CO_2 elimination by the large animals may be mainly a byproduct of the increase in aerial $\dot{V}_{O_2}$. The results are in agreement with Guimond and Hutchison

(1973) who found that *S. lacertina* of a body weight between the sizes used here (they used individuals averaging 628 g) had % aerial partitions also intermediate between those reported here; 50% for aerial $\dot{V}_{O_2}$ and 22% for aerial $\dot{V}_{CO_2}$. Generalizing, smaller animals obtain the majority of their oxygen from the water while larger ones obtain the majority of their oxygen from the air; all sizes eliminate most of their carbon dioxide to the water.

TABLE VI

Gas exchange partitioning between air and water for large and small S. lacertina *under conditions of high* O_2 *and low* CO_2 *in the water (see Table I). Absolute values are* μl *gas/g·hr* ($\bar{x} \pm 2$ *S.E.); values below these are percentages for each partition expressed as the mean and 95% confidence interval. N = 8 for 1489 g, N = 9 for 90 g, N = 6 for aquatic and total results and 7 for aerial results for 1182 g*

		LARGE 1489 g	LARGE 1182 g	SMALL 90 g
Aerial	$\dot{V}_{O_2}$	19·10 ± 3·00 75(61–89)	20·15 ± 4·20 75(56–94)	19·68 ± 7·94 42(22–60)*
	$\dot{V}_{CO_2}$	5·93 ± 1·88 26(16–35)	4·91 ± 1·72 20(12–29)	2·61 ± 1·98 6(1–11)*
Aquatic	$\dot{V}_{O_2}$	6·22 ± 0·62 25(22–27)	6·68 + 0·88 25(21–29)	27·25 ± 4·94 58(46–70)*
	$\dot{V}_{CO_2}$	17·16 ± 1·28 74(68–81)	19·49 ± 3·22 80(63–97)	40·05 ± 6·12 94(77–100)
Total	$\dot{V}_{O_2}$	25·32 ± 3·02	27·22 ± 5·34	46·93 ± 7·78
	$\dot{V}_{CO_2}$	23·09 ± 2·74	24·77 ± 4·86	42·57 ± 6·98

* Significantly different from large animals at the 95% level of confidence.

The second aspect of importance in gas exchange partitioning is the tensions of the dissolved respiratory gases in the water. This should have profound effects upon the extent to which each medium is used for gas exchange. Since a large body size would reflect the greatest disadvantage for aquatic respiration, the largest animal available was used in this experiment. The tensions used as low for O_2 (about 25 mmHg) and high for CO_2 (about 40 mmHg) were purposely chosen to closely reflect the actual gas tensions in the hyacinth portion of the study area during the summer. Therefore the gas exchange partitioning under the low O_2–high CO_2 conditions should be indicative of the actual respiratory behaviour in the hyacinth community and should provide an indication of the aerial gas exchange adaptations used under natural conditions. Absolute amounts of gas exchange in each partition are shown for the various conditions of dissolved gases in the water in Fig. 12. There was no difference (at the 95% level) in the aerial $\dot{V}_{O_2}$ for any set of aquatic conditions, while low O_2 tensions in the water did result in a significant lowering of the aquatic uptake of oxygen. The aquatic $\dot{V}_{O_2}$ was lower for high levels of CO_2 in the water than for low levels of CO_2. While the latter differences were not significant, they suggest the presence of a Bohr shift in *Siren* blood. Such a shift would be maladaptive for an animal that breathed water only and inhabited a hypoxic and hypercarbic environment, and has been found to be absent or even reversed in some animals found in such habitats. Apparently the ability to breathe air has made such an adaptation unnecessary for the Sirenidae.

The results dealing with the elimination of CO_2 can best be appreciated with reference to Table VII. High tensions of CO_2 in the water did not have nearly as large an effect upon the partitioning of CO_2 elimination as was

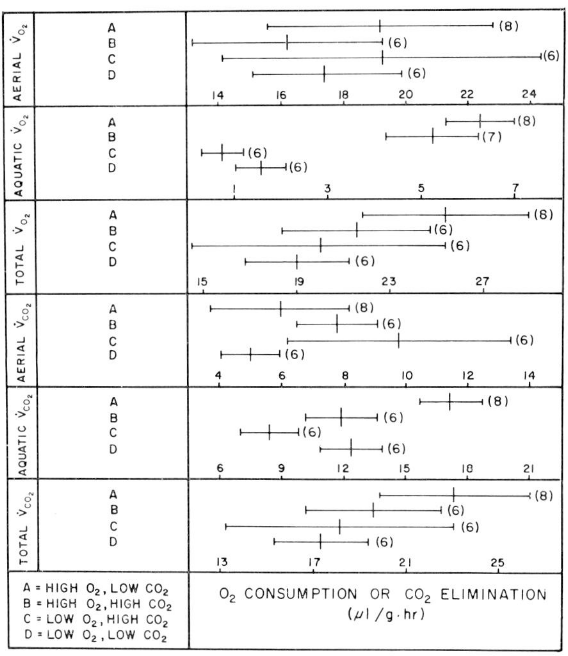

FIG. 12. Aerial-aquatic gas exchange partitioning in an individual large *S. lacertina* (1453–1489 g) under various conditions of dissolved O_2 and CO_2 (see Table II). Results are shown as $\bar{x}$ and the 95% confidence interval. Non-overlap of confidence intervals yields statistical difference in the mean values at the 95% level. Numbers in parentheses are sample sizes.

TABLE VII

Gas exchange partitioning between air and water for an individual, large S. lacertina *(1453–1489 g) as a function of concentrations of dissolved O_2 and CO_2 in the water phase. Values are expressed as in Table VI. N = 8 at all high O_2-low CO_2 values, N = 7 for aquatic $\dot{V}_{O_2}$ at high O_2-high CO_2, N = 6 for all other conditions. All experiments were at 25°C*

		HIGH O_2		LOW O_2	
		Low CO_2	High CO_2	Low CO_2	High CO_2
Aerial	$\dot{V}_{O_2}$	19·10±3·00 75(73–78)	16·16±2·34 76(61–89)	17·44±1·86 92(79–100)	19·21±3·94 96(71–100)
	$\dot{V}_{CO_2}$	5·93±1·88 26(16–35)	7·78±1·02 40(33–46)	5·00±0·78 26(23–29)	9·76±2·80 54(34–74)
Aquatic	$\dot{V}_{O_2}$	6·22±0·62 25(22–27)	5·21±0·80 24(20–29)	1·56±0·42 8(5–11)	0·75±0·34 4(2–6)
	$\dot{V}_{CO_2}$	17·16±1·28 74(65–84)	11·82±1·40 60(51–59)	12·32±1·16 74(63–80)	8·41±1·12 46(38–54)
Total	$\dot{V}_{O_2}$	25·32±3·02	21·54±2·48	19·00±1·74	19·96±4·20
	$\dot{V}_{CO_2}$	23·09±2·74	19·60±2·26	17·32±1·58	18·17±3·82

expected. Of particular interest are the results for CO_2 elimination to the water under the conditions of a high O_2 tension–low CO_2 tension and a low O_2 tension–high CO_2 tension. In spite of the small amount of aquatic O_2 uptake in the latter situation, the aquatic CO_2 elimination had only dropped from 74% to 46%. To eliminate CO_2 to water that has a P_{CO_2} = 40 mmHg, *Siren* must be able to tolerate a P_{CO_2} in the blood greater than 40 mmHg. Whether *Siren* has a particularly high level of HCO_3^- in the blood to buffer this abnormally high (for a water breather) level of carbon dioxide in the blood or whether it simply is tolerant to acidosis remains to be seen. Whatever the mechanism, the ability to tolerate a high blood P_{CO_2} is certainly an important adaptation enabling *Siren* (and presumably *Pseudobranchus*) to inhabit the water hyacinth community.

The ability to tolerate a high level of dissolved carbon dioxide suggests that *S. lacertina* is relatively insensitive to CO_2 tension. Tensions as high as 60 mmHg were measured in the field, which should result in a blood CO_2 tension that is high even by mammalian standards. A preliminary investigation was conducted to determine if there was a lethal carbon dioxide tension for this animal. Recall that a critical oxygen tension is defined as that level (usually measured with nil CO_2) at which the animal can no longer

TABLE VIII

The tolerance of submerged S. lacertina *to high tensions of* CO_2. *The* P_{O_2} *range refers to the highest and lowest oxygen levels in the water during the experiment (not to beginning and end pressures)*

Weight (g)	Final P_{CO_2} (mmHg)	Elapsed time (hr)*	P_{O_2} range (mmHg)	Final condition of animal
7·6	197†	139	129–205	good
9·0	202	130	144–240	good
11·8	54	71	145–173	dead
27·5	205	130	153–215	good
38	205	150	140–193	good
43	200	151	132–220	good
128	195	70	147–230	good
230	210	70	134–280	good

* Total time since the first introduction of CO_2. Does not include the initial submergence period in aerated water during which the animals were becoming adapted to the chamber.

† Final pH = 6·1. Apparently the well water used was naturally buffered.

maintain its SMR. Theoretically, this reduction of the SMR would be fatal eventually, since the animal should build up an oxygen debt that it cannot pay. Analogously, there should be a tension of CO_2, given a normally adequate O_2 tension, that would be lethal. This lethality might be the result of respiratory acidosis, a pronounced Root effect or Bohr effect, or some combination of these and other factors. Furthermore, submerged animals of small body size should be most susceptible to the effects of a high CO_2 tension, since their high weight-specific metabolic rate and large surface-to-volume ratio should combine to saturate any buffering capacity within a short period of time. It would follow that whatever the P_{CO_2} of the water, the P_{CO_2} of the blood would be at least equal to it.

Table VIII indicates that *S. lacertina* of relatively small size is apparently insensitive to high carbon dioxide tensions. These animals were kept forcibly submerged and were therefore utilizing aquatic respiration only; their lungs were not available for the elimination of the CO_2 that initially must have been entering the body through the highly permeable skin. How high a level they would have been able to tolerate is an interesting academic question, but of no consequence ecologically, as a P_{CO_2} of this magnitude is not encountered naturally by *Siren*. The one individual that died apparently was not killed by the high levels of CO_2.

In summary, the Sirenidae are very efficient bimodal gas exchangers for oxygen uptake, having the ability to respire only air or only water depending upon the O_2 tensions of the water (except for the larger sizes of *S. lacertina*). With regard to CO_2, they are essentially skin breathers at all sizes when in water of low CO_2 tensions, and are roughly equally divided between air and water at high CO_2 tensions. When their relatively low metabolic rate when submerged is also taken into consideration, these animals are well adapted for inhabitation of the hyacinth community. This community is a relatively new one in the southeastern United States (post Civil War), and this family of salamanders has rapidly assumed the position of dominant vertebrates within it, quite possibly as the result of their pre-adaptation for gas exchange in the new microenvironment induced with the advent of *Eichhornia* introduction.

V. Summary

1. The aquatic environment in the water hyacinth community is characterized by low levels of dissolved oxygen and high levels of dissolved carbon dioxide.

2. The distribution of water breathers (fish) reflects the differences in dissolved O_2 and CO_2 which occur in sub-hyacinth and open water.

3. The distribution of sirenid salamanders in the hyacinth community does not appear to be effected by the low oxygen and high carbon dioxide levels of hyacinth-covered water. These animals select the sub-hyacinth water in spite of the unfavourable conditions for aquatic respiration.

4. Submerged sirenids have a relatively low metabolic rate which is sufficient for survival at 25°C, $P_{O_2} = 155$ mmHg, P_{CO_2} nil, for sizes up to 700–800 g.

5. Sirenid salamanders have little or no change in body proportions with increasing body weight.

6. The critical oxygen tension (P_c) of sirenids increases with increasing body size, particularly in *S. lacertina*. In this animal, the ability to exchange oxygen through the skin (E_c) increases with weight at a slower rate than the metabolic demand for oxygen. For this reason, larger animals are obligate air breathers.

7. Under field conditions, all sirenid salamanders in hyacinth communities require air-breathing for survival. The dependence upon aerial respiration is a function of body size and of the dissolved O_2 and CO_2 levels of the water.

8. Submerged sirenids are extremely tolerant of high levels of dissolved carbon dioxide.

9. Bimodal gas exchange and insensitivity to carbon dioxide have enabled the sirenid salamanders to assume the role of dominant vertebrates in many of the water hyacinth communities of the southeastern United States.

Acknowledgements

The work reported here was part of that conducted during five years as a graduate student and one year as an assistant professor in the Department of Zoology at the University of Florida, Gainesville. Practically everyone on the staff gave some form of assistance at one time or another, either mental, physical, or financial. I thank all of them, particularly Drs Brian McNab, Frank Nordlie, Stephen Zam, John Anderson, Henry Prange, Carmine Lanciani, and Archie Carr. Most of the figures were drawn by Mr Paul Laessle of that department; some were done by Mr Gary Hellermann of the Center for Aquatic Sciences. Mr John Wakeman of the Department of Biology at the University of Alabama (Tuscaloosa) read and commented upon the manuscript.

This study was financed by the Department of Zoology at the University of Florida and by a Sigma Xi Grant-in-Aid of Research. The presentation was financed by the College of Arts and Sciences of the University of Alabama.

References

Bishop, S. (1943). "Handbook of Salamanders", Cornell University Press, Ithaca, New York.

Czopek, J. (1962). Vascularization of respiratory surfaces in some Caudata, *Copeia*, 576–587.

Guimond, R. W. and Hutchison, V. H. (1973). Trimodal gas exchange in the large aquatic salamander, *Siren lacertina* (Linnaeus). *Comp. Biochem. Physiol.* **46A**. 249–268.

Lewis, W. M., Jr. (1970). Morphological adaptations of cyprinodontoids for inhabiting oxygen deficient waters, *Copeia*, 319–326.

Lynch, J. J., King, J. E., Chamberlain, T. K. and Smith, A. L. (1947). Effects of aquatic weed infestations on the fish and wildlife of the Gulf States, *U.S. Dept. Int. Spec. Sci. Rept.* **39**, 1–71.

Penfound, W. T. and Earle, T. T. (1948). The biology of the water hyacinth. *Ecol. Monogr.* **18**, 447–472.

Shield, J. W. and Bentley, P. J. (1973). Respiration of some urodele and anuran Amphibia – II. In air, role of the skin and lungs. *Comp. Biochem. Physiol.* **46A**, 29–38.

Ultsch, G. R. (1971). The relationship of dissolved carbon dioxide and oxygen to microhabitat selection in *Pseudobranchus striatus*. *Copeia*, 1971: 247–252.

Ultsch, G. R. (1973a). A theoretical and experimental investigation of the relationships between metabolic rate, body size, and oxygen exchange capacity. *Resp. Physiol.* **18**, 143–160.

Ultsch, G. R. (1973b). The effects of water hyacinths (*Eichhornia crassipes*) on the microenvironment of aquatic communities. *Arch. Hydrobiol.* **72(4)**, 460–473.

Ultsch, G. R. (1974). Gas exchange and metabolism in the Sirenidae (Amphibia:

Caudata) I. Oxygen consumption of submerged sirenids as a function of body size and respiratory surface area. *Comp. Biochem. Physiol.* **47A**, 485–498.

Whitford, R. W. and Hutchison, V. H. (1967). Body size and metabolic rate in salamanders. *Physiol. Zool.* **40**, 127–133.

Gas Exchange of the Giant Salamanders of North America

ROBERT W. GUIMOND and VICTOR H. HUTCHISON

Department of Biology, Boston State College, Boston, Massachusetts, U.S.A. and *Department of Zoology, University of Oklahoma, Norman, Oklahoma, U.S.A.*

I. INTRODUCTION

Amphibians utilize the integument, buccal cavity, lungs and gills to various extents in their external respiration and were the first tetrapods to make the successful transformation from life in an aquatic medium to one in a terrestrial environment. Such a transition was necessarily associated with a number of physiological modifications, in particular an efficient method of extracting oxygen from the air. Although little can be gleaned from the fossil evidence in this respect, the study of existing amphibian species can provide information on these adaptive changes since amphibians span the spectrum of habitat preferences from those forms which are exclusively aquatic to those that are largely terrestrial.

The literature on the relative participation of the varied respiratory surfaces in the gas exchange of amphibians is extensive (Krogh, 1904; Dolk and Postma, 1927; Whitford and Hutchison, 1967; Guimond and Hutchison, 1968; Hutchison, Whitford and Kohl, 1968). Until recently, however, there was a group of amphibians whose gas exchange patterns were hardly examined. These animals are collectively referred to as the "giant salamanders" (Conant, 1958) and comprise four diverse salamander families. Some members of this assemblage attain lengths greater than three feet and weights up to two kilograms. These animals are aquatic, generally nocturnal and secretive, exhibit various degrees of paedomorphosis, and with few exceptions are largely confined to the drainage systems of eastern and southern United States.

We recently completed a series of investigations on the gas exchange patterns of selected members from this group (Guimond and Hutchison, 1972, 1973a, b, 1974, 1975). In addition, several recent publications have also considered the various aspects of the respiration in some of these animals (Toews, 1971a, b, 1973, Toews, Shelton and Randall, 1971; Ultsch, 1972, 1973a b, 1974; Shield and Bentley, 1973; Bentley and Shield, 1973; Reno, Gehlbach and Turner, 1972; and Gehlbach, Gordon and Jordan, 1973).

II. Experimental Animals and Design

A representative species from each of the four families was selected for our investigation (Fig. 1). The hellbender, *Cryptobranchus alleganiensis alleganiensis* (Daudin), Family Cryptobranchidae, is dorso-ventrally flattened, with conspicuous folds of skin hanging from the margins of the body and legs. *Siren lacertina* Linnaeus, Family Sirenidae, is known as the greater siren or mud eel and has a long eel-like body, small external gills and no hind limbs. The mud puppy, *Necturus maculosus maculosus* (Rafinesque), Family Proteidae, has a generally stout body, four well-developed limbs and elaborate gills. The southern congo eel, *Amphiuma means means* Garden, Family Amphiumidae, has an eel-like body form, four vestigial limbs and lacks external gills, although a spiracle is present.

These four species constitute a transition ranging from an obligate aquatic

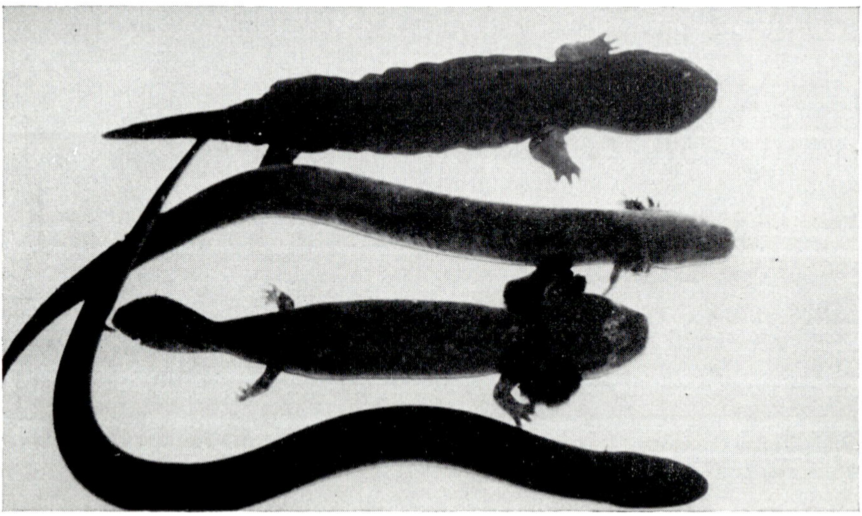

Fig. 1. Four species of Giant Salamanders: from top to bottom, *Cryptobranchus alleganiensis alleganiensis*, *Siren lacertina*, *Necturus maculosus maculosus* and *Amphiuma means means*.

animal, *Necturus*, that has large functional gills and rather simple sac-like lungs, to *Amphiuma* which has no gills and possesses long and elaborate lungs that enable this animal to survive out of water for extended periods. These four species represent various stages of metamorphosis. For example, Noble (1931) considered *Siren* as a form that has differentiated very little beyond a very early stage of metamorphosis; *Necturus* and *Cryptobranchus* have reached later stages; *Amphiuma* is practically metamorphosed.

We examined the relative contribution of the aerial and aquatic respiration of these four species, and evaluated the relative participation of the pulmo-buccopharyngeal, gill and cutaneous surfaces in the gas exchange of *Necturus* and *Siren*. Animals were acclimated and tested at 5°, 15°, and 25°C, to establish the roles of the different respiratory surfaces under varied environmental conditions and metabolic rates. Investigations on the morphology of the lungs and cineradiographic observations of the ventilatory mechanism of these forms were conducted for clarification of the inconsistencies in the literature concerning the mechanism of pulmonary ventilation.

Individuals of each species were submerged in air-saturated water or exposed to an aerial environment for two weeks at different temperatures to observe their viability under those conditions. The respiration of these aquatic animals kept in air was measured and comparisons were made with the other metabolic data. Pulmonary and branchial ventilatory frequencies were also noted and tidal volumes were obtained for *Siren*. Comparisons were also made to determine if differences in respiratory patterns were correlated with the distribution and ecology of these species.

III. Respirometry

A detailed description of the respirometric techniques employed in our investigations were described and figured earlier (Guimond and Hutchison, 1972). Three different closed system respirometers made of plexiglas were used, and only brief descriptions are given below.

A. Aerial-aquatic respirometry (AA)

The AA respirometer consisted of an aquatic compartment and a superadjacent aerial chamber; the aerial and aquatic phases were separated by a layer of paraffin oil which prevented diffusion between the two respiratory media. The animals were unrestrained and were free to surface for air at will. A combination of manometric, titrametric and polarographic techniques were utilized to monitor the oxygen consumption ($\dot{V}_{O_2}$) and carbon dioxide release ($\dot{V}_{CO_2}$).

B. Pulmonary-branchial-cutaneous respirometry (PBC)

The separate contributions of the lungs, gills and skin in the gas exchange of *Necturus* and *Siren* were determined in respirometers of three separate compartments arranged in tandem and containing the head, gill and body regions, respectively. These regions were separated by two sets of tailored rubber-dam partitions. The methods for determining $\dot{V}_{O_2}$ and $\dot{V}_{CO_2}$ were the same as in the AA experiments .

C. TOTAL AERIAL RESPIROMETRY (TA)

The TA respirometer was used for measurement of the metabolism of these aquatic animals when kept in an aerial environment. This respirometer was essentially that used and described by Whitford and Hutchison (1963).

IV. Results and Discussion

Gas exchange values are given in microlitres per gram per hour ($\mu l\ g^{-1}hr^{-1}$) STPD and are presented in Tables I–III and summarized in Figs 2–3. Animals in which metabolism was measured in the AA respirometer are referred to as AA animals; PBC animals and TA animals are those in which the gas exchange was determined in the PBC and TA respirometers, respectively.

The respiratory mechanism in *Amphiuma*, *Cryptobranchus*, *Necturus* and *Siren* is considerably different from that of most anurans and urodeles. These salamanders do not exhibit regular buccal oscillations interrupted by intermittent lung ventilations: only pulmonary ventilations are in evidence. The buccal cavities in *Amphiuma* and *Siren* contain less than 1% of their total respiratory capillaries (Czopek, 1965). Consequently, it is unlikely that this area contributes much to the total gas exchange because of the infrequent buccal flushings combined with the small proportion of respiratory capillaries located in the buccal region.

We made detailed observations to establish whether these animals used aquatic-buccopharyngeal ventilation. There were isolated instances when the mouth of *Siren* and *Necturus* opened and closed in a rhythmic fashion with water leaving via the gill slits. However, these movements were infrequent and were never persistent. On the basis of the findings, aquatic respiration in this paper represents cutaneous or combined cutaneous and branchial gas exchange only.

A. AERIAL AND AQUATIC GAS EXCHANGE (AA)

1. *Cryptobranchus and Necturus*

Guimond and Hutchison (1972, 1973b) reported that the aquatic respiration amounted to over 90% of the $\dot{V}_{O_2}$ and 97% of $\dot{V}_{CO_2}$ in the AA *Cryptobranchus* at all temperatures; the AA *Necturus* had an aquatic $\dot{V}_{O_2}$

and $\dot{V}_{CO_2}$ that represented over 96% and 99% of the total gas exchange (Table I, Fig. 2). These findings concur with observations of others that these species are principally water-breathers. (*Cryptobranchus*: Gage and Gage, 1886; Smith, 1907; Willey, 1920; Noble, 1925; Bishop, 1941; Baker, 1949; Robin, in discussion in Hughes, 1966. Nickerson and Mays, 1973. *Necturus*: Mitchill, 1822, 1824; Holbrook, 1842; Gibbes, 1853; Kneeland, 1859; Miller, 1900, 1905; Whipple, 1906; Bruner, 1914; Willey, 1919, 1920; Luckhardt and Carlson, 1921; Noble, 1925; Bishop, 1926; Baker, 1949; Harris, 1959a, b; Lenfant and Johansen, 1967; Shield and Bentley, 1973; Bentley and Shield, 1973).

2. *Amphiuma and Siren*

Guimond and Hutchison (1973a, 1974) noted the importance of the lungs in permitting *Amphiuma* and *Siren* to meet their oxygen requirements at higher temperatures; over 87% of the $\dot{V}_{O_2}$ and 96% of the $\dot{V}_{CO_2}$ was aquatic for both species at 5°C; however, at 25°C the aerial $\dot{V}_{O_2}$ accounted for over 50% of the total, while aerial $\dot{V}_{CO_2}$ represented approximately 23% of the total (Table I, Fig. 2). These findings are in opposition to the earlier beliefs of Noble (1925, 1931), Baker (1945, 1949) and Darnell (1949) who dis-

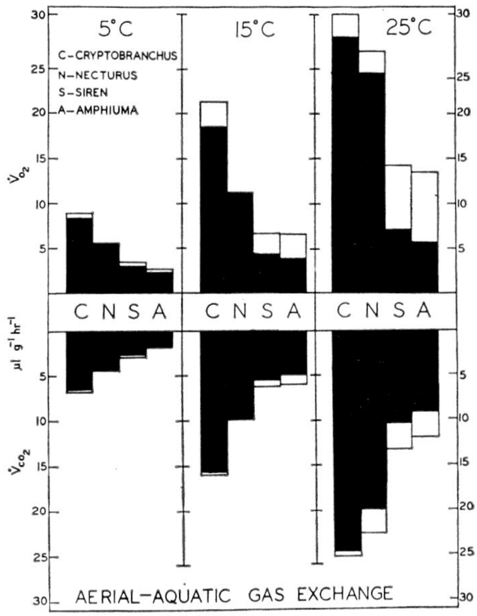

FIG. 2. Summary histogram of the aerial and aquatic gas exchange for *Amphiuma*, *Cryptobranchus*, *Necturus* and *Siren* at different temperatures. The dark portion of the bar represents the aquatic gas exchange; the white portion, the aerial gas exchange; the black plus the white portions equal the total respiratory exchange.

TABLE I

Aerial and aquatic gas exchange in four species of giant North American Salamanders, oxygen consumption ($\dot{V}_{O_2}$) and carbon dioxide production ($\dot{V}_{CO_2}$) are given in $\mu l\ g^{-1}\ hr^{-1}$

					$\dot{V}_{O_2}$			$\dot{V}_{CO_2}$		
	Temp. (°C)	$\bar{x}$ Body Wt. (g)	Hrs.	N	Aerial $\bar{x}$ ± SE	Aquatic $\bar{x}$ ± SE	Aerial $\%O_2$	Aerial $\bar{x}$ ± SE	Aquatic $\bar{x}$ ± SE	Aquatic $\%CO_2$
Amphiuma	5	352	80	16	0·21 ± 0·06	2·30 ± 0·21	8·37	0·00 ± 0·00	1·96 ± 0·17	100·00
	15	375	130	26	2·78 ± 0·15	3·97 ± 0·22	41·19	1·08 ± 0·17	4·11 ± 0·22	79·19
	25	376	140	28	7·52 ± 0·27	5·88 ± 0·39	56·12	2·70 ± 0·26	9·13 ± 0·46	77·18
Cryptobranchus	5	514	120	24	0·53 ± 0·10	8·15 ± 0·65	6·11	0·19 ± 0·07	6·40 ± 0·58	97·12
	15	511	115	23	2·29 ± 0·27	8·57 ± 0·93	10·98	0·26 ± 0·08	15·84 ± 0·99	98·39
	25	423	115	23	2·40 ± 0·17	28·52 ± 0·97	7·76	0·54 ± 0·10	24·54 ± 1·15	97·85
Necturus	5	214	40	8	0·00 ± 0·00	5·46 ± 0·92	0·00	0·00 ± 0·00	4·21 ± 0·58	100·00
	15	206	40	8	0·00 ± 0·00	11·18 ± 0·59	0·00	0·00 ± 0·00	9·99 ± 0·45	100·00
	25	212	40	8	1·13 ± 0·45	25·39 ± 0·81	4·26	0·22 ± 0·10	21·72 ± 1·45	99·00
Siren	5	601	55	11	0·48 ± 0·08	2·99 ± 0·27	13·83	0·14 ± 0·07	2·74 ± 0·23	95·14
	15	581	60	12	2·21 ± 0·54	4·35 ± 0·20	33·69	0·74 ± 0·12	5·50 ± 0·29	88·14
	25	628	55	11	7·19 ± 0·52	7·16 ± 0·53	50·10	2·91 ± 0·30	10·20 ± 1·01	77·80

counted the significance of cutaneous gas exchange in both *Amphiuma* and *Siren*, but are in agreement with the suppositions of Vaillant (1863) and Czopek (1962b, 1965), who noted the probable importance of cutaneous respiration. Numerous other investigators (Krogh, 1904; Whitford and Hutchison, 1967; Hutchison, *et al.*, 1968; Guimond and Hutchison, 1968) have also demonstrated the importance of the cutaneous surfaces in amphibian gas exchange.

The relatively thin epidermal layer of the skin with its numerous glands that keep it moist, along with the extensive cutaneous circulation, materially reduce the diffusion barrier between the respiratory medium and the blood, and confers upon most amphibians a considerable capacity for integumentary gas exchange, particularly in CO_2 elimination. The latter aspect is largely the result of the substantial CO_2 gradient between the cutaneous capillaries and the respiratory medium. The lungs of amphibians, on the other hand, have higher CO_2 concentrations than air or normoxic waters. This fact in conjunction with the reduced CO_2 in the blood due to continued cutaneous release establishes a negligible CO_2 gradient between the blood and pulmonary air, and results in the lower aerial $\dot{V}_{CO_2}$ ordinarily observed in amphibians.

Other anatomical and physiological investigations lend further support to our findings that the cutaneous surfaces of *Amphiuma* and *Siren* can account for considerable gas exchange. Czopek (1965) stated that approximately one-third of the respiratory capillaries of these two species are located in the skin; Lenfant and Johansen (1967) found that $P_{a_{CO_2}}$ in submerged *Amphiuma* did not rise as precipitously as it did in *Necturus* maintained in air; they suggested that the skin of *Amphiuma* must take part in gas exchange, especially in CO_2 elimination. The small arterio-venous O_2 differences on *Amphiuma* emphasized the importance of cutaneous respiration (Johansen, 1963). Similarly, Toews *et al.* (1971) found small fluctuations in CO_2 tensions in both lung and blood samples of *Amphiuma* even during a prolonged dive. Toews (1971a) observed that CO_2 injected into the lungs in doses 3–5 times the normal concentrations was rapidly removed from the lungs; he also found that an artificial increase of the P_{CO_2} in the surrounding water rapidly increased $P_{a_{CO_2}}$ in the dorsal aorta and Toews concluded that at 15°C, the skin is capable of considerable CO_2 gas exchange. In other experiments Toews (1973) assumed that little CO_2 is eliminated from the lungs at 15°C and thereby discounted any pulmonary $\dot{V}_{CO_2}$. We note, however, that about 21% and 23% of all respiratory CO_2 is removed via the lungs of unrestrained *Amphiuma* at 15° and 25°C, respectively (Table I); this contribution should not be ignored.

Shield and Bentley (1973) and Bentley and Shield (1973) observed that the cutaneous respiration in submerged *Siren* (at 20°C) represented approximately 50% of the total $\dot{V}_{O_2}$ of animals allowed to use their lungs. Other experiments with anaesthetized *Siren* also indicated that the cutaneous surfaces can account for as much as 50% of the total gas exchange of the animal. These findings are similar to our observations (Table I) that unrestrained *Siren* also utilize their cutaneous surfaces for about 50% of the total $\dot{V}_{O_2}$ at 25°C.

At lower temperatures, therefore, the integumentary surfaces (approximately $\frac{1}{3}$ of the respiratory capillaries) of *Amphiuma* and *Siren* account for the requisite gas exchange requirements, with occasional assistance from the lungs. At elevated temperatures, however, diffusion across the skin is insufficient to meet the accelerated metabolic demands, and the lungs assume a more prominent role in providing oxygen to the tissues. The results of the aerial and aquatic experiments (AA) agree with the conclusions of Foxon (1964) and Lenfant, Johansen and Hanson (1970) that in amphibians and other air-breathing lower vertebrates, the more oxygen absorbed through the lungs, the greater the amount of carbon dioxide released via the pulmonary surfaces. However, most of the additional CO_2 attendant on an increased metabolism is largely eliminated by the cutaneous and branchial surfaces. *Amphiuma* and *Siren* exhibit the lowest metabolic rates observed of any amphibians measured at comparable temperatures.

B. PULMONARY BRANCHIAL AND CUTANEOUS GAS EXCHANGE (PBC)

The separate and simultaneous contribution of the gills and skin of *Necturus* and *Siren* in their aquatic respiration, and that of the lungs in aerial respiration was determined in the PBC respirometer (Guimond and Hutchison, 1972, 1973a). The results of these experiments are given in Table II and summarized in Figure 3.

1. *Necturus*

The gills of *Necturus* oscillated continually in the course of the experiments and, under those conditions, played a prominent role accounting for over 60% of the total $\dot{V}_{O_2}$ and $\dot{V}_{CO_2}$ at 25°C. The cutaneous surfaces exhibited a decreased contribution with increased temperature. For example, $\dot{V}_{CO_2}$ was 43·31% of the total at 5°C, but dropped to 26·86% at 25°C. The pulmonary gas exchange seldom exceeded 10% of the total at any of the experimental temperatures. Shield and Bentley (1973) ligated the gills of *Necturus* at 22°C and observed a 42% reduction in total gas exchange compared to normal animals. These findings support earlier observations based on indirect evidence that the gills of *Necturus* are important in respiration. Other investigators (Kneeland, 1859; Robin, in discussion in Hughes, 1966; Fanelli and Goldstein, 1964), however, have concluded that the gills are insignificant in respiration and are not essential to the survival of *Necturus*. It is reasonable that these authors might consider the gills unessential in gas exchange since, under some circumstances (e.g. low temperature, air-saturated water), this animal can manage without the use of gills. The frequency of gill movements in *Necturus* is quite variable and factors such as animal activity, temperature and aquatic oxygen tension affect the rate of gill oscillations and function. In the AA experiments we observed that in cool, aerated water (5°C, 15°C;

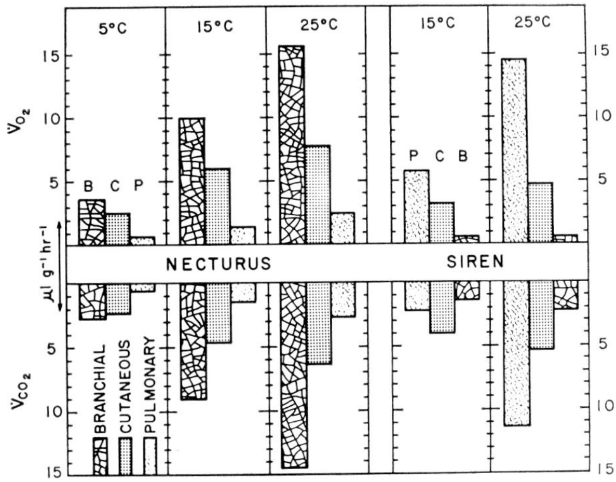

FIG. 3. Summary histogram depicting the relative contribution of the pulmonary branchial and cutaneous gas exchange surfaces for *Necturus* and *Siren* at different temperatures.

P_{O_2} = 130–155 mmHg) where the animal remained undisturbed, the gills seldom waved and were mostly retracted to the side of the head; but as temperatures increased, or the animal became active, the gills always oscillated. It is under the latter circumstances that the gills assume their prominent and indispensable role.

We were able to show that the efficient gills of *Necturus* are capable of reducing the aquatic P_{O_2} from 155 mmHg to approximately 40 mmHg with little difficulty (Guimond and Hutchison, 1972). Lenfant and Johansen (1967) reported that the arterial P_{O_2} of *Necturus* is about 35 mmHg. Thus the gills of *Necturus* efficiently utilize oxygen to low tensions—much lower than ordinarily encountered in nature and which approximate the normal blood $P_{a_{O_2}}$. Since the gills in the PBC experiments were fully expanded and always waved, the gas exchange value for the PBC animals probably represent a figure close to the maximal branchial capacity. Lenfant and Johansen (1967) noted that if *Necturus* is taken out of water at 20°C the gills collapse and this is quickly followed by a sharp rise in $P_{a_{CO_2}}$ and a marked reduction in $P_{a_{O_2}}$ despite frequent pulmonary ventilations and gulping movements; but as soon as the animal is returned to the aquatic medium, the gills begin to function and the blood gas tensions quickly return to normal. Their observations

TABLE II

Pulmonary, branchial, and cutaneous (PBC) respirometry in Necturus *and* Siren. *Oxygen consumption ($\dot{V}_{O_2}$) and carbon dioxide production ($\dot{V}_{CO_2}$) are given in μl g^{-1} hr^{-1}. Figures in parentheses represent percent gas exchange*

Species	Temp. (°C)	$\bar{x}$ Body Wt. (g)	Hrs.	N	$\dot{V}_{O_2}$ Pulmonary $\bar{x}$ ± SE	$\dot{V}_{O_2}$ Branchial $\bar{x}$ ± SE	$\dot{V}_{O_2}$ Cutaneous $\bar{x}$ ± SE	$\dot{V}_{CO_2}$ Pulmonary x ± SE	$\dot{V}_{CO_2}$ Branchial $\bar{x}$ ± SE	$\dot{V}_{CO_2}$ Cutaneous x ± SE
Necturus	5	146	54	18	0·67 ± 0·11 (9·65)	3·72 ± 0·51 (53·60)	2·55 ± 0·34 (36·75)	0·69 ± 0·12 (12·06)	2·61 ± 0·42 (45·63)	2·42 ± 0·30 (42·31)
	15	162	84	28	1·37 ± 0·09 (8·39)	10·00 ± 0·72 (61·24)	4·96 ± 0·31 (30·37)	1·27 ± 0·13 (8·45)	9·01 ± 0·61 (59·95)	4·75 ± 0·26 (31·60)
	25	143	72	24	2·53 ± 0·22 (9·71)	15·69 ± 1·33 (60·21)	7·84 ± 0·48 (30·08)	2·81 ± 0·23 (11·81)	14·53 ± 1·26 (61·33)	6·39 ± 0·37 (26·86)
Siren	15	400	32	8	5·77 ± 0·29 (61·58)	0·50 ± 0·06 (5·34)	3·10 ± 0·17 (33·08)	2·10 ± 0·23 (27·74)	1·39 ± 0·20 (18·36)	4·08 ± 0·32 (53·90)
	25	458	32	8	14·40 ± 0·19 (73·66)	0·49 ± 0·03 (2·51)	4·66 ± 0·10 (28·84)	11·25 ± 0·85 (60·00)	2·12 ± 0·21 (11·31)	5·38 ± 0·64 (28·69)

further demonstrate the importance of the gills in gas exchange while the contribution of the lungs remains unimpressive.

The PBC experiments also showed that the cutaneous surfaces of *Necturus* can participate considerably in gas exchange, especially at lower metabolic rates (Table II). These findings contradict the earlier estimations, based on anatomical studies, which ascribed very little cutaneous function to the overall respiration of this species (Noble, 1925; Baker, 1949). Bishop (1926) best summarized the respiratory habits of *Necturus* when he remarked that "the skin alone may furnish the necessary oxygen in well-aerated water, that the gills are used under ordinary stream and lake conditions, and that the lungs are called into service as a last resort." He emphasized, however, that the lungs remain inefficient; the latter statement was verified by the PBC experiments. When *Necturus* was in the AA respirometer it hardly surfaced for air; however, in the PBC respirometer *Necturus* often gulped air (as frequently as 10–20 times per hour), and in spite of this increase in ventilation rate, the lungs never contributed materially to the total gas exchange in this animal.

2. *Siren*

In other studies, we observed that the gills of *Siren* are much less involved in gas exchange than in *Necturus* and play only a minor role in respiration, and are more effective in CO_2 elimination than in O_2 absorption (Table II; Fig. 3). The branchial $\dot{V}_{O_2}$ never exceeded 5% of the total $\dot{V}_{O_2}$ while the $\dot{V}_{CO_2}$ accounted for from 11 to 18% of the total. A similar phenomenon is observed in some air-breathing fish where the vestigial or reduced gills function very little in oxygen uptake and are more important in elimination of carbon dioxide (Johansen, 1968; Johansen and Lenfant, 1968). These findings are in opposition to the statements of Negus (1965) who observed that practically all gas exchange occurred across the gills of *Siren*, but reinforce the observations of LeConte (1824), Barnes (1826), Vaillant (1863), Cope (1885), Viosca (1924), Czopek (1965), Shield and Bentley (1973), and Bentley and Shield (1973) that the gills of *Siren* are relatively unimportant in gas exchange.

The unusual thickness of the branchial epithelium (Vaillant, 1863; Cope, 1885), the large number of anastomoses or gill by-passes that can shunt blood away from the branchial filaments (Vaillant, 1863; Baker, 1949; Darnell, 1949), the smallness of the gills relative to body size (Fig. 1), the simplicity of the gill structure which accounts for only 2% of the total respiratory capillaries (Czopek, 1965) and the failure of these organs to oscillate periodically, are all factors which account for the relatively minor performance of these organs in respiration. The gills of the PBC *Siren* were always extended and well perfused; and we suggest that the branchial gas exchange values obtained for *Siren* reflect a maximum exchange capacity.

The cutaneous surfaces of *Siren* can contribute to gas exchange (33% of total $\dot{V}_{O_2}$ and 54% of total $\dot{V}_{CO_2}$ at 15°C). However, as in *Necturus*, this participation becomes relatively less important at higher temperatures (Table II, Fig. 3). These data show that as metabolic demands increase, the unmodified cutaneous surfaces are unable to increase materially their respiratory capacity while the specialized and efficient gas exchangers (gills in *Necturus*, lungs in *Siren*) assume a greater prominence. The phenomenon of restrictive integumentary capacity, especially in $\dot{V}_{O_2}$, at higher metabolic rates is also common in other Amphibia. For example, the cutaneous $\dot{V}_{O_2}$ in *Rana pipiens* exhibits less than a two-fold increase from 15°C to 25°C, while the pulmonary contribution almost quadruples (Guimond and Hutchison, 1968).

The development of external gills of the various species of Amphibia is closely related to the functional needs of these organs (Noble, 1931). *Siren*, which is capable of considerable cutaneous respiration and has efficient lungs, has little need for a third respiratory surface and hence has reduced gills. *Necturus*, on the other hand, exhibits pulmonary exchange and the gills are therefore essential to supplement the cutaneous surfaces in respiration and hence their elaborate structure. This phenomenon of functional development of external gills is also found within a given species. For example, *Necturus* kept in cool, well oxygenated water has gills that are atrophied compared to the gills of animals kept in warmer, non-aerated water. The gills of *Siren*, kept underwater for over two weeks, appeared to be larger and more perfused than they were before submergence began (Guimond and Hutchison, 1976).

3. *Pulmonary Capacities*

(a). *Necturus* and *Siren*. The PBC experiments also provided additional information on the pulmonary capacities of *Necturus*, *Siren* and *Amphiuma*. In the course of the PBC experiments the animals were restrained on a hardware cloth platform. Such restraint resulted in increased breathing rates (Guimond and Hutchison, 1972, 1973a).

The increase in pulmonary ventilation was accompanied by a slight rise in pulmonary gas exchange in *Necturus*, but resulted in a marked augmentation in the pulmonary contribution of *Siren*. The lungs of *Siren* accounted for about 74% of the $\dot{V}_{O_2}$ and 60% of the $\dot{V}_{CO_2}$. The latter finding was surprising since amphibians typically remove most of the CO_2 via the skin. Ultsch (1972) also showed that, depending upon size, the pulmonary $\dot{V}_{O_2}$ of *Siren* at 25°C ranged from 42 to 75% of the total $\dot{V}_{O_2}$. The lungs of *Siren* are thus very effective in gas exchange and can actually eliminate more CO_2 aerially than cutaneously. The African lungfish *Protopterus*, which was thought to remove most of the CO_2 via the gills, is likewise capable of eliminating more CO_2 via the pulmonary surfaces with increased pulmonary ventilation (f) (McMahon, 1970). The lungs as respiratory organs are anatomically more efficient than the skin as gas exchangers, since the former have a thinner epithelium and a denser capillary network with individual capillary diameters some 20 to 30% larger than those of the skin (Czopek, 1965). Therefore, with an increase in f and $\dot{Q}$ (perfusion rate) via the lungs, more O_2 and CO_2 can be exchanged at the pulmonary surfaces. The cutaneous surfaces can only increase $\dot{Q}$ and vasodilation (Poczopko, 1959). However, in the absence of an adequate ventilatory process, the increased cutaneous exchange does not compare with that of the efficient lungs, and we observe the consequent reduction in the relative participation of the skin and gills in total respiration of *Amphiuma* and *Siren* at elevated temperatures. Johansen (1963) and Toews, *et al.*, (1971) have shown that the pulmonary circulation of *Amphiuma* increases immediately after each breath. This synchronization of lung inspiration with improved circulation further enhances the efficiency of pulmonary gas exchange. Similar adjustments have also been reported for lungfish (Johansen and Lenfant, 1968) and such adjustments most probably exist in *Siren*.

(b) *Morphology of the lungs.* A gross examination of the lungs of the species investigated shows why an increase in the ventilation rate occasions a rise in the aerial respiration in *Siren*, but has relatively little effect on the aerial gas exchange in *Necturus*. Figure 4 shows the air-dried lungs of the four species

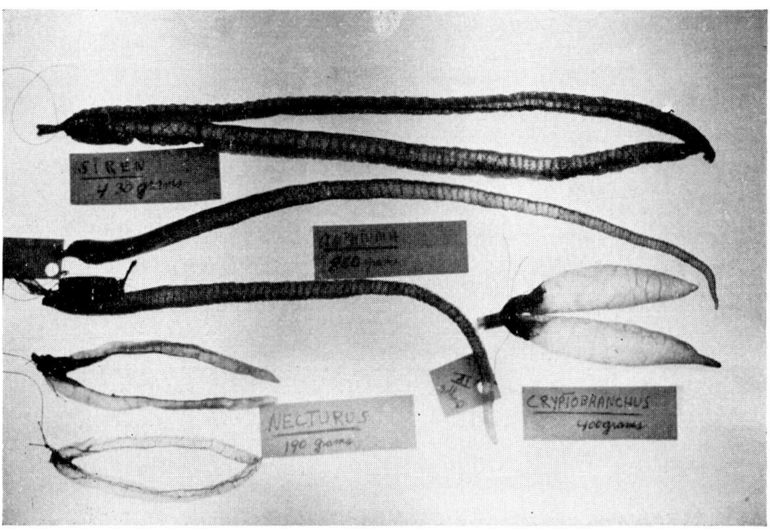

FIG. 4. Preserved lungs of *Amphiuma*, *Cryptobranchus*, *Necturus*, and *Siren*. Note that the lungs of *Necturus* and *Cryptobranchus* are mere transparent and poorly vascularized sacs; the lungs of *Amphiuma* and *Siren*, on the other hand, are practically opaque due to the numerous internal septae and the extensive vascularization of these organs. For purposes of comparison, the lungs of *Siren* in this photograph are 35 centimeters in length.

and illustrates the differences in pulmonary elaboration and vascularization.

The lungs of *Necturus* and *Cryptobranchus* are thin-walled, transparent, sac-like organs, poorly vascularized and devoid of septa. *Amphiuma* and *Siren*, on the other hand, have comparatively elaborate lungs that are long (about $\frac{2}{3}$ of body length), internally honeycombed with numerous septa which increase the pulmonary respiratory surface area by at least a factor of four (Czopek, 1965); these organs are well-vascularized with an extensive capillary network, and the walls of the lung and the septa have large concentrations of smooth muscle (Czopek, 1962a; Stolk, 1962; Willnow, 1964, 1968). From 61 to 65% of the respiratory capillaries are located in the lungs of *Siren* and *Amphiuma* (Czopek, 1965). These findings are in opposition to the views of Orr (1972) who stated that *Siren* is an animal "possessing gills and lacking lungs at maturity".

Although the lungs of *Cryptobranchus* and *Necturus* are not as elaborate as those of *Amphiuma* and *Siren*, they are by no means vestigial. Radiograms (Guimond and Hutchison, 1972, 1973b) show that the lungs of these animals, especially those of *Cryptobranchus*, occupy a considerable volume in the intact animal. Since these organs do not contribute significantly to the total gas exchange, they most probably serve as hydrostatic organs.

(c) *Ventilatory Mechanism in Amphiuma and Siren.* The mechanism of pulmonary ventilation was examined with X-ray cineradiographic techniques (Guimond and Hutchison, 1973a, 1974). The ventilation in both species occurs in a definite sequence. In *Amphiuma*, as an example, we observed that expiration first occurs when the snout extends just above the surface of the water; during this phase the white air-shadow on the cineradiographic film suddenly disappears as the lungs collapse; this process is probably facilitated by the abundant smooth muscles in the lung walls. On inspiration the buccal cavity dilates and draws in air; the nares close and the buccal floor raises and forces the air contained within the buccal cavity into the lungs. Since the buccal volume is considerably smaller than that of the lungs, several buccal fillings are required to inflate the lungs fully. Toews (1971b) also described the consecutive buccal fillings in *Amphiuma*. *Siren* has the same breathing pattern, with the exception that air enters and leaves via the open mouth rather than through the nares as in *Amphiuma*. The cineradiogram showed that expiration results from the apparently active contraction of the lungs themselves and is not associated with buccal movements as suggested for *Amphiuma* by Toews *et al.*, (1971). The method of pulmonary ventilation in *Amphiuma* and *Siren* is similar to that for the South American lungfish, *Lepidosiren paradoxa*, (Bishop and Foxon, 1968), and for the bowfin, *Amia* (Horn and Riggs, 1973).

Toews *et al.*, (1971) observed that lung P_{O_2} for *Amphiuma* seldom exceeded 100 mmHg immediately after each filling. Their data suggest that 53% of the original gases are renewed with each breath. Our observations of cineradiographic films lead us to conclude that the percentage renewal of lung air in *Amphiuma* and *Siren* is greater. Lung ventilation in other amphibians is less effective than in *Amphiuma* and *Siren*. Price and Pierce (1963) found a lung volume of 3·6 ml for 35 g *Rana pipiens*. Guimond and Hutchison (1968) found that at 25°C the pulmonary tidal volume for individuals of this species of comparable weight is 0·35 ml; this represents, therefore, an exchange of approximately 10% of the lung air with each pulmonary ventilation. A 300 g *Rana catesbiana* has a lung volume ranging from 30 to 60 ml (Gans, De Jongh and Farber, 1969); this represents an exchange of approximately 20 to 25% of the lung air with each ventilation. *Amphiuma* and *Siren*, on the other hand, exchange at least 53% with each ventilation.

The lungs of *Amphiuma* and *Siren*, therefore, are not only superior anatomically as gas exchange organs when compared to those of *Cryptobranchus* and *Necturus* (Fig. 4), but they also possess an efficient ventilatory mechanism that assures considerable renewal of pulmonary air with each inspiration. *Siren* utilizes from 16 to 26% of the inspired oxygen (Guimond and Hutchison, 1973a). Toews *et al.* (1971) showed that *Amphiuma* utilizes up to 50% of pulmonary oxygen between breaths. These pulmonary O_2-utilizations are much greater than rates previously reported for other amphibians (1–3%; Whitford, 1973). *Amphiuma* and *Siren*, therefore, seem to possess the most efficient pulmonary apparatus of any amphibian yet investigated.

(d) *Amphiuma and Siren*. On the basis of the similarity in lung structure, effective pulmonary ventilation, and the AA aerial gas exchange values of *Amphiuma* and *Siren*, we suggest that if *Amphiuma* were to increase its ventilatory rate as *Siren* did in the PBC experiments, it would also exhibit the same pronounced increase in pulmonary $\dot{V}_{O_2}$ and $\dot{V}_{CO_2}$ that characterized the PBC *Siren*. Guimond and Hutchison (1974) reported that when AA *Amphiuma* and *Siren* ventilated their lungs about the same number of times they recorded a similar $\dot{V}_{O_2}$; this demonstrates that for animals of comparable size, the pulmonary efficiencies are similar.

Toews *et al.* (1971) suggested that the lungs of *Amphiuma* are probably not important in CO_2 elimination since neither lung nor blood P_{CO_2} levels increase materially during a dive; this may be true at lower metabolic rates (5°–15°C) and where the water is normoxic (e.g. high O_2 and low P_{CO_2}). *Amphiuma* is frequently found in shallow, stagnant, weed-choked and warm bodies of water which are often hypoxic and hypercarbic. Under such conditions, the efficient lungs provide the needed O_2 and can assist in the elimination of CO_2. This is of special significance in hypercarbic waters where the CO_2 gradient between the skin and water is diminished. Even in normoxic waters, Guimond and Hutchison (1974) observed a 2·5× increase in pulmonary $\dot{V}_{CO_2}$ when temperatures rose from 15°C to 25°C.

C. TOTAL AERIAL RESPIRATION (TA)

Notable differences occur in the total metabolism of some of the giant salamanders when kept out of water (TA animals) and compared with other animals maintained in their normal aquatic medium (AA animals; cf. Guimond and Hutchison, 1976). Reductions of 30% and 25% were noted in the metabolism of TA *Necturus* and TA *Cryptobranchus* when compared to their AA counterparts at 25°C (Table III). Bentley and Shield (1973) reported an even greater reduction in the metabolism of *Necturus* (42% drop) when they monitored respiration in and out of water at 22°C. When *Necturus* is exposed to air the delicate gills collapse and eliminate a major portion of the available respiratory surface. Because the lungs hardly participate in respiration, the burden of gas exchange is shifted to the cutaneous surface; but since the skin alone is unable to meet the requirements at higher temperatures (Lenfant and Johansen, 1967; Guimond and Hutchison, 1972; Bentley and Shield, 1973), only a slight rise over the 15°C $\dot{V}_{O_2}$ values is observed in the TA *Necturus* at 25°C (Table III).

Since Jones (1967) showed that a decrease in O_2 availability causes a reduction in the metabolism of diving frogs, it is conceivable that the same factor resulted in the lowered $\dot{V}_{O_2}$ at higher temperatures as a consequence of the reduced respiratory surface area due to the collapsed gills. When *Cryptobranchus* is out of water, its body rests flat on the substratum, which essentially eliminates from one-third to one-half the functional cutaneous surface. Since the lungs are inefficient as respiratory organs, the reduced cutaneous respiratory surface results in a lower $\dot{V}_{O_2}$.

The TA *Amphiuma* and TA *Siren* exhibited a greater $\dot{V}_{O_2}$ at all temperatures than their AA counterparts; but the differences in most instances were insignificant and were directly related to increased activity (routine metabolism as opposed to standard metabolism). Since the lungs are efficient gas exchangers, these animals were able to augment the $\dot{V}_{O_2}$ to accommodate the greater activity. Morgan *et al.* (1967) showed that the metabolism of *Amphiuma* rises precipitously between 20°C and 30°C, a temperature range often associated with the normal habitat (Ultsch, 1973a). In such situations the lungs are essential to assuage the respiratory distress induced by the inability of the integument to provide adequate exchange.

D. EFFECTS OF PROLONGED SUBMERGENCE OR EXPOSURE TO AIR

In view of the intermediate position between air breathing and water breathing held by some of the giant salamanders, we investigated the viability of these forms under conditions of prolonged submersion in air-saturated water or exposure to a moisture saturated aerial environment, for a maximum of two weeks (336 hours), at temperatures of 5°C, 15°C, and 25°C (Guimond

TABLE III

Total oxygen consumption rates ($\dot{V}_{O_2}$) *of* Cryptobranchus *and* Necturus *in aquatic-aerial (AA) respirometers and in air (TA). Values are in* $\mu l\ g^{-1}\ hr^{-1}$

Species	Aerial-aquatic (AA)					Aerial (TA)				
	Temp. (°C)	($\bar{x}$) Body Wt. (g)	Hrs.	N	$\dot{V}_{O_2}$ $\bar{x}$ ± SE	Temp. (°C)	($\bar{x}$) Body Wt. (g)	Hrs.	N	$\dot{V}_{O_2}$ x ± SE
Cryptobranchus	5	418	120	24	8·68 ± 0·69	5	343	95	19	6·41 ± 0·60
	15	511	115	23	20·84 ± 0·99	15	387	110	22	13·53 ± 0·91
	25	423	115	23	30·92 ± 1·03	25	344	100	20	23·58 ± 1·23
Necturus	5	214	40	8	5·46 ± 0·46	5	133	155	31	6·56 ± 0·30
	15	206	40	8	11·18 ± 0·59	15	129	135	27	11·93 ± 0·43
	25	212	40	8	26·52 ± 0·70	25	110	180	36	18·42 ± 0·58

and Hutchison, 1976). *Cryptobranchus* and *Siren* were the most adaptable and survived both forced submersion and air exposure for the full two weeks. *Amphiuma* tolerated all the aerial experiments as did *Necturus* in the submersion experiments. Although *Amphiuma* survived submergence at 5°C and 15°C, all animals perished within two days at 25°C (Table IV). Thus *Amphiuma* is an obligate air breather at elevated temperatures; at lower temperatures, the metabolic requirements are handled satisfactorily by the integumentary exchange, but at higher temperatures the lungs are essential for the accommodation of increased metabolic demands.

Necturus kept in air exhibited a decreased vitality with an increase in temperature: mean survival time of 11·18 days at 5°C, 6·11 days at 15°C, and only 3·48 days at 25°C. These findings are higher than those of the earlier literature (e.g. 3–5 hours; Kneeland, 1859, Bishop, 1926; Harris, 1959a), but are in general agreement with Bentley and Shield (1973), who observed that *Necturus* survived five days out of water at 22°C. At lower temperatures, the cutaneous surfaces of *Necturus* maintain the animal for several days, but at elevated temperatures the skin is unable to handle the increased respiratory demands and the animal succumbs within 3·5 days at 25°C. These animals can survive for several days in air while exhibiting about one-fourth to one-half the normal rate of oxygen consumption (Guimond and Hutchison, 1976; Bentley and Shield, 1973) and *Necturus* can survive for at least nine hours under water in the complete absence of oxygen (Bentley and Shield, 1973). Thus, *Necturus* in air may utilize anaerobic energy sources. In addition to the debilitating effects of oxygen lack, *Necturus* kept in air may also succumb to the effects of an unfavourable increase of CO_2 which is largely removed by the gills.

It is unclear why *Siren* survived for two weeks and *Amphiuma* only two days when both species were submerged at 25°C. It is conceivable that under such circumstances the normally inactive gills of *Siren* may sufficiently increase their function and, along with cutaneous respiration, provide the requisite oxygen demands. *Amphiuma*, on the other hand, can only resort to its limited cutaneous gas exchange. Another possibility, assuming that the gills of *Siren* remain relatively functionless, is that *Siren* has a more effective cutaneous respiration than *Amphiuma*; but this seems unlikely since the epidermis of *Siren* is over twice as thick as that of *Amphiuma* (Czopek, 1965). Finally, perhaps *Siren* exhibits a greater tolerance to induced anoxia and can resort to a more efficient anaerobic respiration than that reported for *Amphiuma* (Rose, Zambernard and Pogany, 1965; and Rose and Zambernard, 1966).

The *Amphiuma* and *Siren* used in the submergence experiments were relatively small (200 g for *Amphiuma*; 300 g for *Siren*), and within these size ranges *Amphiuma* is less tolerant to forced submersion than *Siren* at 25°C. However, as these organisms become larger the ability to utilize cutaneous gas exchange during submergence is further diminished as a result of the less favourable ratio of surface area to weight; consequently, the larger animals must rely to a greater extent on aerial respiration. In organisms which have options for air breathing or water breathing (e.g. *Amphiuma* and *Siren*), size exerts a determinative role in the selection of the most favourable avenue for gas exchange. At 25°C the aerial $\dot{V}_{O_2}$ represented 42% of the total for a 482 g *Siren*, 53% for a 907 g animal and 56% for a 1200 g specimen (Guimond and Hutchison, 1974). Ultsch (1973b, 1974) also found that larger *Siren* obtained a greater percentage of their oxygen from air than small animals (e.g. 1200 g, 75% aerial $\dot{V}_{O_2}$; 90 g, 42% aerial $\dot{V}_{O_2}$). Ultsch (1973b) also found that 90 g animals survived submergence in air-saturated water for at least two weeks, while animals larger than 800 g perished within two days or less at 25°C. In smaller adult forms *Siren* has a greater tolerance of forced submergence than *Amphiuma*, but as the size of the animal increases it also becomes an obligate air breather, especially in individuals weighing more than 800 g.

Necturus and *Siren* can extract oxygen from an aquatic medium with a P_{O_2} of less than 1·5 mmHg (Bentley and Shield, 1973). Lenfant and Johansen (1967) showed that the blood of *Necturus* has a very high affinity for oxygen. These aforementioned factors combined with the ability of *Necturus* to undergo facultative anaerobiosis, enables this animal to sustain brief periods of anoxia.

Amphiuma (Knepton, 1954) and *Siren* (Viosca, 1924; Cagle and Smith, 1939; Carr, 1940; Cockrum, 1941; Freeman, 1958) have been found alive some distances from water in periods of drought, and in some instances burrowed two to three feet in the ground; some of those animals may have been in their burrows for several months. Reno *et al.* (1972), and Gehlbach

TABLE IV

Survival times for Amphiuma *submerged in water and for* Necturus *in air*

Species	Temp. (°C)	$\bar{x}$ Wt (g)	N	Survival Hrs. $\bar{x}$ ± SE	Survival Days
AMPHIUMA	5	279	6	336·00 ± 0·00*	14
(submerged)	15	220	9	336·00 ± 0·00*	14
	25	174	11	48·00 ± 9·39	2·0
NECTURUS	5	127	12	268·25 ± 21·44	11·48
(in air)	15	131	13	146·62 ± 24·69	6·11
	25	124	12	83·67 ± 24·05	3·48

* Experiment ended after 14 days.

et al. (1973) reported that *Siren intermedia* burrows into the bottom of a drying pond and aestivates for over a year. Aestivation or summer dormancy is a response to drought and/or heat and has been described in other amphibians (Mayhew, 1965; Ruibal, Tevis and Roig, 1969; Seymour, 1973a, b). Such a dormancy is well known in the African lungfish, *Protopterus* (Smith, 1930; Thomson, 1969). Like the lungfish, these *Siren* secrete a mucous cocoon, decrease their metabolic rate and heart rate, and the body shrinks as the fat stores are depleted. The similarities of lungfish and *Siren* cocoons and other metabolic adjustments are convergent adaptations to periodic droughts. Rose (1966), Gunther (1968) and Martof (1969) have shown that both *Amphiuma* and *Siren* can undergo prolonged inanition. Some *Amphiuma* survived up to three years without food, while *Siren* has survived 5 years and lost 87% of its original weight before succumbing to starvation.

In addition to an ability to burrow and survive prolonged inanition *Amphiuma*, like *Necturus*, exhibits a capacity for facultative anaerobiosis, and this animal can tolerate anoxia in pure nitrogen for at least six hours (Rose *et al.*, 1965; Rose and Zambernard, 1966). Other investigators (Hack and Helmy, 1964; Helmy and Hack, 1964) have also concluded that these animals can readily undergo anaerobiosis.

E. RELATION OF RESPIRATORY HABITS TO HABITAT

The various respiratory strategies utilized by the giant aquatic salamanders show a remarkable adaptation to the kinds of habitats which these animals ordinarily occupy. For example, *Cryptobranchus* is most frequently found in cool, well oxygenated streams or rivers (Cochran, 1961). However, recent field studies (Hutchison, Engbretson and Turney, 1973; Hillis and Bellis, 1971; Nickerson and Mays, 1973) also reveal that during the warm season water temperatures may vary from 22°C to 27°C. But as Hillis and Bellis (1971) pointed out, nearly all of the 151 hellbenders captured in their study were located at sites of swift, well-aerated and shallow portions of the stream. Under such conditions, the flattened body form of the hellbender with its profusely vascularized skin folds (with epidermal capillaries almost extending to the surface) permits the integument to function as an efficient gas exchange organ.

Necturus has a wider distribution than *Cryptobranchus* and is found in diverse conditions from clear waters of lakes and streams to weed choked bays and coves, and in canals and drainage ditches (Bishop, 1943). In the latter situation the oxygen content of the water can become reduced; under such circumstances the gills increase their intermittent waving motions, and are extremely efficient in extracting O_2 from the water. Also, the ability of *Necturus* to undergo anaerobiosis, to remove O_2 from waters of very low O_2

tensions and to survive prolonged periods at metabolic rates 30 to 40% below normal are additional adaptive features that allow these organisms to maintain themselves during excursions in hypoxic waters.

Amphiuma and *Siren*, on the other hand, are typically found in warmer, shallow and occasionally hypoxic and hypercarbic waters. Under those situations, the large and efficient lungs provide the necessary O_2 and can assist materially in the elimination of CO_2. The low intrinsic metabolic rates, the efficient pulmonary apparatus, the ability to withstand prolonged inanition and to undergo facultative anaerobiosis, along with their capacity to extract O_2 from waters of low oxygen tensions, are all adaptive features which permitted these forms to extend their range and occupy niches that would ordinarily be alien and inhospitable to most amphibians.

The ability of amphibians to obtain suitable O_2 at higher te mperatures possibly determines the optimum temperature for each species and therefore their distribution. Whitford (1964) showed that lunged salamanders with efficient buccal pumping mechanisms (e.g., *Taricha granulosa*, *Ambystoma tigrinum*) are found in warmer areas than those with less efficient buccal mechanisms (e.g., *Desmognathus* and *Gyrinophilus*). *Cryptobranchus* and *Necturus* have little pulmonary ventilation and this has resulted in their general confinement to the cooler waters of the eastern, central and northern areas of the United States. During the warmer months, the *Necturus* from the Great Lakes region become scarce as they apparently migrate to the deeper and cooler waters until the late fall or early winter (Lemberger, personal communication).

Siren and *Amphiuma* on the other hand have an efficient pulmonary mechanism and are found along middle eastern, south-eastern and southern drainage areas of the United States. The temperatures in some of the stagnant waters where *Siren* are found are as high as 29°C (Ultsch, 1972). The amount of dissolved oxygen in these stagnant waters would be considerably less than in waters of lower temperatures. *Necturus* and *Cryptobranchus* would certainly be at a disadvantage in such environments. Due to the increased metabolic rate attendent on the higher temperatures, the gills of *Necturus* would be in perpetual oscillation; this in turn would result in a considerable expenditure of energy and hence, reduce the efficiency of the branchial movements. *Cryptobranchus*, which has no gills would be in even greater difficulty.

The lower preferred temperatures and sluggish and nocturnal habits of amphibians compared to higher vertebrates may be related to the greater respiratory efficiency of the latter groups at higher temperatures. However, the problems of desiccation also play a substantial role in establishing the habits and distribution of amphibians.

Although amphibians have a varied assortment of respiratory surfaces, only one of these gas exchangers is ordinarily elaborated into an efficient respiratory organ, while the others are secondary or supplemental in capacity. *Siren* has elaborate lungs and reduced gills; *Necturus* has simple sac-like lungs, but elaborate and efficient gills; *Amphiuma* has no gills, but well-developed pulmonary surfaces. All three species use cutaneous respiration, but its importance in oxygen uptake is more apparent at lower temperatures or in CO_2 elimination at higher temperatures. *Cryptobranchus* has neither gills nor efficient lungs; however, its flattened body form, excess folds of skin, and extensive cutaneous vascularization confers to the skin of this amphibian a superb respiratory quality.

References

Baker, C. L. (1945). The natural history and morphology of *Amphiuma*. *J. Tenn. Acad. Sci.* **20**, 55–91.

Baker, C. L. (1949). Comparative anatomy of the aortic arches of the urodeles and their relationship to respiration and degree of metamorphosis. *J. Tenn. Acad. Sci.* **24**, 12–40.

Barnes, D. H. (1826). An arrangement of the genera of batracian animals with a description of the more remarkable species including a monograph on the more doubtful reptiles. *Am. J. Sci. Arts* **11**, 268–297.

Bentley, P. J. and Shield, J. W. (1973). Respiration of some urodele and anuran amphibia – II. In air, role of the skin and lungs. *Comp. Biochem. Physiol.* **46A**, 29–38.

Bishop, I. R. and Foxon, G. E. H. (1968). The mechanism of breathing in the South American lungfish, *Lepidosiren paradoxa*: a radiological study. *J. Zool. Lond.* **154**, 263–271.

Bishop, S. C. (1926). Notes on the habits and developments of the mud-puppy *Necturus masculosus* (Raf.). *N.Y. State Mus. Bull.* **268** 60p.

Bishop, S. C. (1941). The salamanders of New York. *N.Y. State Mus. Bull.* 134 p.

Bishop, S. C. (1943). "Handbook of Salamanders". 555 p. Comstock Publ., Ithaca.

Bruner, H. L. (1914). The mechanism of pulmonary respiration in animals with gill clefts. *Morph. Jb.* (Leipzig) **48**, 63–82.

Cagle, F. R. and Smith, P. E. (1939). A winter aggregation of *Siren intermedia* and *Triturus viridescens*. *Copeia*, 1939, 232–233.

Carr, A. F. (1940). A contribution to the herpetology of Florida. *Univ. Fla. Publ. Biol. Sci. Ser.* **3(1)**, 1–118.

Cochran, D. M. (1961). "Living Amphibians of the World." Doubleday, New York, p. 199.

Cockrum, L. (1941). Notes on *Siren intermedia*. *Copeia*, 1941, 265.

Conant, R. (1958). "A Field Guide to Reptiles and Amphibians." Houghton Mifflin and Co., Boston. 366. p.

Cope, E. D. (1885). The retrograde metamorphosis of *Siren*. *Am. Nat.* **19**, 1226–1227.

Czopek, J. (1962a). Smooth muscles in the lungs of some urodeles. *Nature, Lond.* **193**, 728.

Czopek, J. (1962b). Vascularization of respiratory surfaces in caudata. *Copeia*, 1962 (**4**), 576–587.

Czopek, J. (1965). Quantitative studies on the morphology of respiratory surfaces in amphibians. *Acta anat.* **62**, 296–323.

Darnell, R. M. (1949). The aortic arches and associated arteries of caudate Amphibia. *Copeia*, 1949, 18–31.

Dolk, H. E. and Postma, N. (1927). Über die Haut-und die Lungenatmung von *Rana temporaria*. *Z. vergl. Physiol.* **5**, 417–444.

Fanelli, G. M. and Goldstein, L. (1964). Ammonia excretion in the neotenous newt *Necturus maculosus* (Raf.) *Comp. Biochem. Physiol.* **13**, 193–204.

Foxon, G. E. H. (1964). Blood and Respiration, p. 151–209. *In* "Physiology of Amphibia." (Ed. J. A. Moore). Academic Press, New York, London and San Francisco.

Freeman, J. R. (1958). Burrowing in the salamanders *Pseudobranchus striatus* and *Siren lacertina*. *Herpetologica* **14**, 130.

Gage, S. H. and Gage, S. P. (1886). Pharyngeal respiratory movements of adult amphibia under water. *Science* **7**, 395.

Gans, C., DeJongh, H. J. and Farber, J. (1969). Bullfrog (*Rana catesbeiana*) ventilation: how does the frog breathe? *Science, N.Y.* **163**, 123–125.

Gehlbach, F. R., Gordon, R. and Jordan, J. B. (1973). Aestivation of the salamander, *Siren intermedia*. *Am. Midl. Nat.* **89(2)**, 455–463.

Gibbes, L. R. (1853). Description of *Menobranchus punctatus*. *J. Bost. Soc. Nat. Hist.* **6**, 369–373.

Guimond, R. W. and Hutchison, V. H. (1968). The effect of temperature and photoperiod on gas exchange in the leopard frog, *Rana pipiens*. *Comp. Biochem. Physiol.* **27**, 177–195.

Guimond, R. W. and Hutchison, V. H. (1972). Pulmonary branchial and cutaneous gas exchange in the mud puppy, *Necturus maculosus maculosus* (Rafinesque). *Comp. Biochem. Physiol.* **42A**, 367–392.

Guimond, R. W. and Hutchison, V. H. (1973a). Trimodal gas exchange in the large aquatic salamander, *Siren lacertina* (Linnaeus). *Comp. Biochem. Physiol.* **46A**, 249–268.

Guimond, R. W. and Hutchison, V. H. (1973b). Aquatic respiration: an unusual strategy in the hellbender, *Cryptobranchus alleganiensis alleganiensis* (Daudin). *Science, N.Y.* **182**, 1263–1265.

Guimond, R. W. and Hutchison, V. H. (1974). Aerial and aquatic respiration in the congo eel, *Amphiuma means means* (Garden). *Resp. Physiol.* **20**, 147–159.

Guimond, R. W. and Hutchison, V. H. (1976). Effects of prolonged submergence or exposure to an aerial environment on the viability of several species of giant aquatic salamanders. (In manuscript).

Gunther, G. (1968). Further notes on weight changes of starving *Amphiuma means*. *Herpetologica* **24**, 180–181.

Hack, M. H. and Helmy, F. M. (1964). An analysis of melano-proteins from *Amphiuma* liver and from a human liver melanoma. *Proc. Soc. exp. Biol. Med*, **116**, 348–350.

Harris, J. P. (1959a). The natural history of *Necturus*: I. Habits and habitats. *Field and Lab.* **27**, 11–20.

Harris, J. P. (1959b). The natural history of *Necturus*: II. Respiration and phonation. *Field and Lab.* **27**, 71–77.

Helmy, F. M. and Hack, H. (1964). Comparative histochemistry of liver (*Amphiuma* and rat) and the melanin pigment cell system of *Amphiuma* liver. *Acta Histochem.* **18**, 194–212.

Hillis, R. E. and Bellis, E. D. (1971). Some aspects of the ecology of the hellbender, *Cryptobranchus alleganiensis alleganiensis. J. Herptol.* **5**, 121–126.

Holbrook, J. E. (1842). "North American Herpetology." Vol. 5, 87–117. J. Dobson, Philadelphia.

Horn, M. H. and Riggs, C. D. (1973). Effects of temperature and light on the rate of air breathing of the bowfin, *Amia calva. Copeia*, 1973, 653–657.

Hughes, G. M. (1966). Evolution between air and water. *In* "Ciba Foundation Symposium, Development of the lung." (Eds A. V. S. Reuk and R. Porter), pp. 64–84. J. A. Churchill, London.

Hutchison, V. H., Whitford, W. G. and Kohl, M. (1968). Relation of body size and surface area to gas exchange in anurans. *Physiol. Zool.* **41(1)**, 65–85.

Hutchison, V. H., Engbretson, G. and Turney, D. (1973). Thermal acclimation and tolerance in the hellbender *Cryptobranchus alleganiensis. Copeia* 1973, 805–807.

Johansen, K. (1963). Cardiovascular dynamics in the amphibian *Amphiuma tridactylum*, Cuvier. *Acta Physiol. Scand. Suppl.* **217**, 82 p.

Johansen, K. (1968). Air-breathing fishes. *Sci. Am.* **219(4)**, 102–111.

Johansen, K. and Lenfant, C. (1968). Respiration in the African lungfish, *Protopterus aethiopicus*. II. Control of breathing. *J. exp. Biol.* **49**, 453–468.

Jones, D. R. (1967). Oxygen consumption and heart rate of several species of anuran amphibia during submergence. *Comp. Biochem. Physiol.* **20**, 691–707.

Kneeland, S. (1859). On the breathing apparatus of the *Menobranchus. Bost. Soc. Nat. Hist.* **6**, 428–430.

Knepton, J. C., Jr. (1954). A note on the burrowing habits of the salamander, *Amphiuma means means. Copeia*, 1954, 68.

Krogh, A. (1904). On the cutaneous and pulmonary respiration of the frog. *Skand. Archiv. Physiol.* **15**, 328–419.

LeConte, J. (1824). Description of a new species of *Siren* with some observations on animals of a similar nature. *Ann. Syc. Nat. Hist.* **1(1)**, 52–58.

Lenfant, C. and Johansen, K. (1967). Respiratory adaptations in selected amphibians. *Resp. Physiol.* **2**, 247–260.

Lenfant, C., Johansen, K. and Hanson, P. (1970). Bimodal gas exchange and ventilation-perfusion relationship in lower vertebrates. *Fed. Proc.* **29**, 1124–1129.

Luckhardt, A. B. and Carlson, A. J. (1921). Studies on the visceral sensory nervous system. II. Lung automatism and lung reflexes in the salamanders *Necturus* and *axolotls. Am. J. Physiol.* **54**, 122–137.

Martof, B. S. (1969). Prolonged inanition in *Siren lacertina. Copeia* 1969, 285–289.

Mayhew, W. W. (1965). Adaptations of the amphibian, *Scaphiopus couchi*, to desert conditions. *Am. Midl. Nat.* **74**, 95–109.

McMahon, B. R. (1970). The relative efficiency of gaseous exchange across the lungs and gills of an African lungfish, *Protopterus aethiopicus. J. exp. Biol.* **52**, 1–15.

Miller, W. S. (1900). Contributions from the anatomical laboratory of the University of Wisconsin. 1. The lung of *Necturus maculatus*. *Bull. Univ. Wiscon. Sci. Ser.* **2(3)**, 203–215.

Miller, W. S. (1905). The blood and lymph vessels of *Necturus maculatus*. *Am. J. Anat.* **4**, 445–452.

Mitchill, S. (1822). The *Proteus* of the North American lakes. *Am. J. Sci. Arts.* **4**, 181–183.

Mitchill, S. (1824). Observations on several reptiles of North America which seem to belong to the family of *Proteus*. *Am. J. Sci.* **7**, 63–69.

Morgan, L. R., Singh, R. and Fisette, R. (1967). Relationship of oxygen consumption and cytochrome oxidase succinic dehydrogenase activities in the *Amphiuma means*. *Comp. Biochem. Physiol.* **20**, 343–349.

Negus, V. (1965). "The Biology of Respiration." 228 p. E. and S. Livingstone, Ltd., London.

Nickerson, M. A. and Mays, C. E. (1973). The Hellbenders: North American "Giant Salamanders". *Milwaukee Publ. Mus. Publ. Biol. Geol.*, No. 1, 106 p.

Noble, G. K. (1925). Integumentary, pulmonary, and cardiac modifications correlated with increased cutaneous respiration in the amphibia: a solution to the "hairy frog" problem. *J. Morph. Physiol.* **40**, 341–416.

Noble, G. K. (1931). "The Biology of the Amphibia." McGraw-Hill, New York.

Orr, R. T. (1972). "Vertebrate Biology", 3rd Ed., Saunders, Philadelphia.

Poczopko, P. (1959). Changes in blood circulation in *Rana esculenta* L. while diving. *Zool. Polon.* **10(1)**, 29–43.

Price, J. A. and Pierce, J. A. (1963). The physical properties of the frog lung. *Proc. Soc. exp. Biol. Med.* **112**, 538–541.

Reno, H. W., Gehlback, F. R. and Turner, R. A. (1972). Skin and aestivational cocoon of the aquatic amphibian, *Siren intermedia*, Le Conte. *Copeia* 1972, 625–631.

Rose, F. L. (1966). Weight change during starvation in *Amphiuma means*. *Herpetologica* **22**, 312–313.

Rose, F. L. and Zambernard. (1966). Cardiac glycogen depletion in *Amphiuma means* during induced anoxia. *J. Morph.* **120**, 391–396.

Rose, F. L., Zambernard and Pogany, G. S. (1965). Hepatic glycogen depletion in *Amphiuma* during induced anoxia. *Science, N.Y.* **147**, 1467–1468.

Ruibal, R., Tevis, L. and Roig, V. (1969). The terrestrial ecology of the spadefoot toad *Scaphiopus hammondii*. *Copeia* 1969, 571–584.

Salthe, S. N. (1965). Comparative catalytic studies of lactic dehydrogenases in Amphibia: environmental and physiological correlations. *Comp. Biochem. Physiol.* **16**, 393–408.

Seymour, R. S. (1973a). Energy metabolism of dormant spadefoot toads (*Scaphiopus*). *Copeia* 1973, 435–445.

Seymour, R. S. (1973b). Gas exchange in spadefoot toads beneath the ground. *Copeia* 1973, 452–460.

Shield, J. W. and Bentley, P. J. (1973). Respiration of some urodele and anuran Amphibia - I. In water, role of the skin and gills. *Comp. Biochem. Physiol.* **46A,** 17–28.

Smith, B. G. (1907). The life history and habits of *Cryptobranchus alleganiensis*. *Biol. Bull.* **13**, 5–39.

Smith, H. W. (1930). Metabolism of the lungfish, *Protopterus aethiopicus*. *J. Biol. Chem.* **88**, 97–130.

Stolk, A. (1962). Occurrence of smooth muscles and leiomyoma of the lung in the three-toed *Amphiuma*. *Proc. Acad. Wet. Amst.* **65**, 45–48.

Thomson, K. S. (1969). The biology of the lobe-finned fishes. *Biol. Rev.* **44**, 91–154.

Toews, D. P. (1971a). Factors affecting the onset and termination of respiration in the salamander, *Amphiuma tridactylum*. *Can. J. Zool.* **49**, 1231–1237.

Toews, D. P. (1971b). A mechanism for the selective distribution of blood in the Amphibia. *Can. J. Zool.* **49**, 957–959.

Toews, D. P. (1973). Oxygen consumption in *Amphiuma tridactylum*. *Can. J. Zool.* **51**, 664–666.

Toews, D. P., Shelton, G. and Randall, D. J. (1971). Gas tensions in the lungs and major blood vessels of the urodele amphibian, *Amphiuma tridactylum*. *J. exp. Biol.* **55**, 47–61.

Ultsch, G. R. (1972). Gas exchange and metabolism in the Sirenidae (Amphibia, Trachystomata). Ph.D. dissertation, University of Florida, Gainesville.

Ultsch, G. R. (1973a). A theoretical and experimental investigation of the relationship between metabolic rate, body size, and oxygen exchange capacity. *Resp. Physiol.* **18**, 143–160.

Ultsch, G. R. (1973b). The effects of water hyacinths (*Eichhornia crassipes*) on the microenvironment of aquatic communities. *Arch. Hydrobiol.* **4**, 460–473.

Ultsch, G. R. (1974). Gas exchange and metabolism in the Sirenidae (Amphibia, Chordata) – 1. Oxygen consumption of submerged sirenids as a function of body size and respiratory surface area. *Comp. Biochem. Physiol.* **47A**, 485–498.

Vaillant, L. (1863). L'histoire anatomique de la *Sirene lacertine*. *Ann. Sci. Nat.* (*Zool.*) **19**, 295.

Viosca, P. (1924). A terrestrial form of *Siren lacertina*. *Copeia* **136**, 102–104.

Whipple, I. L. (1906). The ypsiloid apparatus of urodeles. *Biol. Bull.* **10**, 255–297.

Whitford, W. G. (1964). Pulmonary and cutaneous gas exchange in salamanders. Ph.D. dissertation, University of Rhode Island, Kingston.

Whitford, W. G. (1973). The effects of temperature on respiration in the Amphibia. *Am. Zool.* **13**, 305–312.

Whitford, W. G. and Hutchison, V. H. (1963). Cutaneous and pulmonary gas exchange in the spotted salamander *Ambystoma maculatum*. *Biol. Bull.* **124**, 344–354.

Whitford, W. G. and Hutchison, V. H. (1967). Body size and metabolic rate in salamanders. *Physiol. Zool.* **40**, 127–133.

Willey, A. (1919). Branchioderma and Branchiotrema. *Trans. Roy. Soc. Can. Ser.* 3, **12(4)**, 95–104.

Willey, A. (1920). Remarks on the respiratory movements of *Necturus* and *Cryptobranchus*. *Proc. Zool. Soc. Lond.* **43**, 649–651.

Willnow, I. (1964). Die Lunge von *Amphiuma means means*. I. Mitteilung: Morphologie. *Zool. Beitr. Berl.* **10**, 29–84.

Willnow, I. (1968). Die Lunge von *Amphiuma means means*. II. Mitteilung: Diskussion. *Zool. Beitr. Berl.* **14**, 359–386.

Gas Transport Characteristics in an Exclusively Skin-breathing Salamander, Desmognathus fuscus (Plethodontidae)

JOHANNES PIIPER, RANDALL N. GATZ*
and EUGENE C. CRAWFORD, JR.†

Abteilung Physiologie, Max-Planck-Institut für experimentelle Medizin, Göttingen, Bundesrepublik Deutschland

In urodelan Amphibia all three modes of respiration occurring in vertebrates —branchial, pulmonary and cutaneous—are developed. Their relative roles in O_2 uptake and CO_2 output vary greatly during ontogeny and from species to species, and largely depend on the external and internal physiological conditions (cf. Guimond and Hutchison, this volume).

Typically, external gills are the respiratory organ of the larval stages, however, cutaneous respiration is always involved. With numerous exceptions a typical adult Urodele possesses lungs but no gills. In many neotenic forms (including all Sirenidae and Proteidae) gills are retained, but lungs are also present. In others (neotenic *Ambystoma mexicanum* and many cave-dwelling plethodontids) only gills are developed, lungs being absent. Finally, in several, prevalently terrestrial, Urodele families (Hynobiidae, Salamandridae, Ambystomatidae), species in which lungs are reduced, to a varied degree, or are totally absent are to be found. These forms have no gills; they are not neotenic, but rather highly evolved and progressive. The total

* Present address: Battelle Memorial Institute, Centre de Recherche de Genève, Bio-Medical Group, 7, Route de Drize, CH-1227 Carouge-Genève, Switzerland.
† Present address: School of Biological Sciences, University of Kentucky, Lexington, Kentucky 40506, U.S.A.

absence of lungs is characteristic of the whole family Plethodontidae, which is the most widespread group of salamanders in North America and the sole group of Urodeles extending its range into tropical regions (northern South America).

Many of the lungless salamanders inhabit cold mountain streams and their banks. This observation may have ecological, physiological and evolutionary significance and deserves detailed study. However, the main reason for our interest in lungless salamanders was that they provided an excellent opportunity to study cutaneous respiration unadulterated by the simultaneous presence of the other modes of respiration, branchial or pulmonary.

Our experimental study was performed on the Common Dusky Salamander (*Desmognathus fuscus*), of 6·1 g mean body mass from the Allegheny mountains of eastern North America.

The experimental studies comprised: (a) Measurement of O_2 consumption in normoxic and hypoxic conditions (Gatz, Crawford and Piiper, 1974b). (b) Determination of blood gas values in mixed cardiac blood, of O_2 dissociation curve, and of blood buffering properties (Gatz, Crawford and Piiper, 1974a). (c) Measurement of the kinetics of elimination of an inert gas, Freon 22, from the body (Gatz, Crawford and Piiper, 1975).

System of quantities and units

The system of quantities and units recommended for use in gas exchange physiology by Piiper, Dejours, Haab and Rahn (1971) has been adopted.

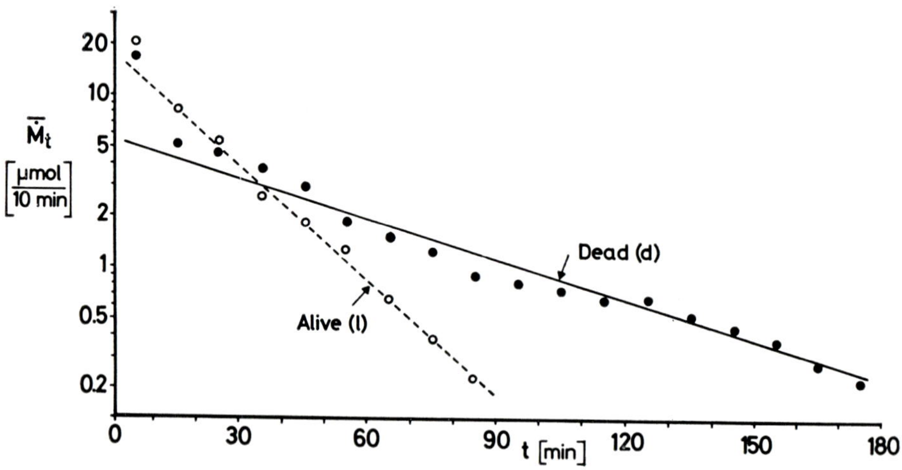

FIG. 1. Elimination of Freon 22 from a specimen of *Desmognathus fuscus*. Ordinate: mean transfer rate during 10 min collecting periods, on logarithmic scale. The rate constant for the slow component, resulting from the slope of the straight lines, is 0·0518/min for the living animal and 0·0174/min for the dead animal. After Gatz *et al.* (1975).

Specifically an amount of a gas is expressed in molar units: mmol (millimole = 10^{-3} mole), μmole (micromole = 10^{-6} mole) or nmole (nanomole = 10^{-9} mole). Conversion to the conventional units like ml STPD is easily performed using the well-known equivalence, 1 mmol = 22·4 ml. Partial pressures of gases are given in torr = mmHg.

I. Circulation

An attempt to estimate the cutaneous blood flow and the cardiac output in *Desmognathus fuscus* from the kinetics of inert gas equilibration has been described in a recent paper (Gatz *et al.*, 1975) and will be summarized in this section.

After equilibrating the animals with 20% chloro-difluoro-methane (Freon 22) in oxygen, the time course of elimination of Freon 22 into atmospheric air was determined in living and dead animals. The efflux of Freon 22 could be described as a sum of two exponentially decaying components (Fig. 1). The slow component, which accounted for the major part of the stored inert gas, was characterized by an average rate constant of 0·05/min in living animals and 0·02/min in dead animals. This difference between the rate constants was attributed to convective transport by blood flow in living animals, since it was reasonable to assume that the diffusing conditions in dead and living animals were identical.

A serial two-compartment model and two variants of a parallel two-compartment model were proposed providing a basis for a quantitative analysis of the data (Fig. 2).

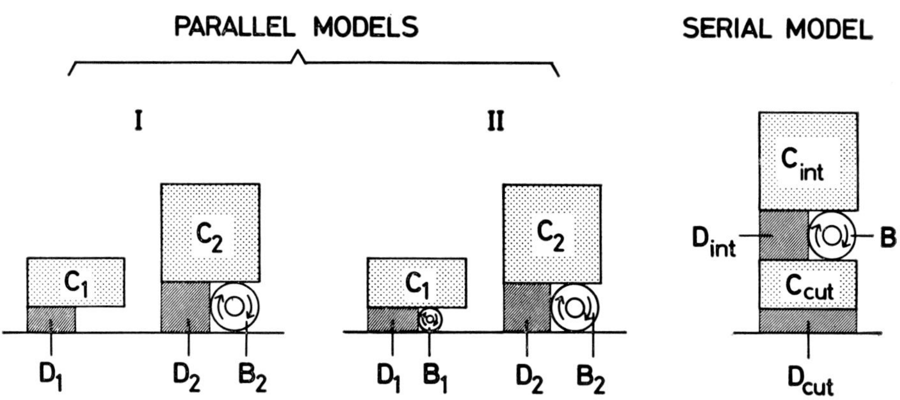

FIG. 2. The models used for analysis of elimination of Freon 22 from the body. C, capacitance of a compartment; D, diffusive conductance of a barrier; B, perfusive conductance. For the dead animal D values are assumed to be the same as in living animal, but B = 0.

The decisive parameters of the models are the following properties of the elements:

(1) Capacitances, C, defined in terms of increment of concentration per increment of partial pressure, are equal to the product of volume and solubility ($= V \cdot \alpha$).

(2) Conductances, both diffusive, D, and perfusive, B, are defined as transfer rate, $\dot{M}$, per partial pressure difference across the element considered, ΔP ($= \dot{M}/\Delta P$).

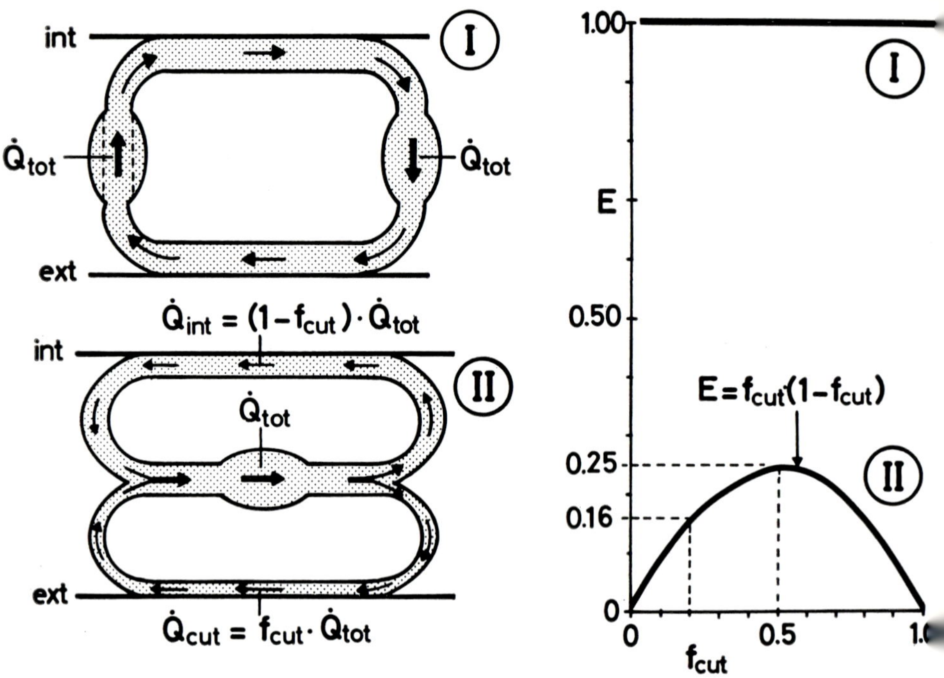

FIG. 3. Gas transfer efficiency of two circulatory models. Model I (typical fishes; mammals, birds) has an efficiency of 100%. In Model II (plethodontid salamanders) the efficiency is much less and depends largely upon the fractional distribution of the cardiac output characterized by the factor f ($= \dot{Q}_{cut}/\dot{Q}_{tot}$). After Gatz *et al.* (1975).

All models yielded similar results for convective conductance due to blood flow, B, in mmol/(min·torr) per g body mass: Parallel Model I, 1·225; Parallel Model II, 1·750; Serial Model, 1·795.

For a circulation system involving no mixing (realized in typical fishes, and in birds and mammals) the convective conductance is given by the product of blood flow, $\dot{Q}$, and the "capacitance coefficient", β, equal to the physical solubility for inert gases, but for CO_2 and O_2 designating the slope of the respective dissociation curves (Piiper *et al.*, 1971):

$$B = \dot{Q} \cdot \beta \quad (1)$$

The anatomical arrangement of the circulatory system in lungless salamanders is quite different and suggests the use of the following simplified model (Fig. 3): the cardiac output is in part directed to the skin, subserving gas exchange with the environment, and in part to the internal organs; the blood returning from both skin and internal organs is completely mixed when reaching the heart.

The convective conductance of such a circulatory system model is given by the following relationship (Gatz *et al.*, 1975):

$$B = f_{cut} \cdot (1 - f_{cut}) \cdot \dot{Q}tot \cdot \beta \quad (2)$$

where f_{cut} is the ratio of cutaneous blood flow to cardiac output ($= \dot{Q}_{cut}/\dot{Q}_{tot}$). This relationship is easily derived by combining mass transport and convective conductance equations for the two parallel circulatory pathways.

Comparison of Eqs (1) and (2) in terms of gas transfer efficiency of the respective system, E, defined as convective conductance per total blood flow and solubility in blood ($= B/(\dot{Q} \cdot \beta)$), reveals the following features.

(1) Efficiency E of the system with complete mixing reaches 25% at most, that of the system without mixing being 100%.

(2) For the system with mixing, E is highest for $f = 0{\cdot}5$, decreasing for both $f > 0{\cdot}5$ and $f < 0{\cdot}5$.

TABLE I

Cardiac output, $\dot{Q}_{tot}$, *and cutaneous blood flow,* $\dot{Q}_{cut}$, *in* Desmognathus fuscus, *calculated from equilibration kinetics of Freon 22 (Gatz* et al., *1975) on the basis of various models and assumptions.* f_{cut}, *assumed ratio* $\dot{Q}_{cut}/\dot{Q}_{tot}$

f_{cut}	Model	$\dot{Q}_{tot}$ $\left[\frac{\mu l}{min \cdot g}\right]$	$\dot{Q}_{cut}$ $\left[\frac{\mu l}{min \cdot g}\right]$
	Parallel I	133	26·6
0·2	Parallel II	190	38·0
	Serial	195	39·0
	Parallel I	85·2	42·6
0·5	Parallel II	122	61·0
	Serial	125	62·5
	"Mean"	**142**	**45·0**

Rather arbitrarily the partitioning factor f was assumed to lie within the range of 0·2 to 0·5. The values of cardiac output, $\dot{Q}_{tot}$, and cutaneous blood flow, $\dot{Q}_{cut}$, calculated from B using alternatively f = 0·2 or f = 0·5 are presented in Table I. The "mean" values, like those in Tables II and III, have been incorporated with the purpose of characterizing the result of the estimations by a single, although very approximate, figure.

II. Gas Transport by Circulation

Application of the circulation model of Fig. 3 (Model II) to analysis of O_2 transport yields the model depicted in Fig. 4. The O_2 content differences between inflowing and outflowing blood, $(C_a - C_m)_{O_2}$ and $(C_m - C_v)_{O_2}$, can be obtained from Fick's principle if the O_2 transfer rate, $\dot{M}_{O_2}$ (meaning both O_2 uptake in skin and O_2 delivery to organs), and the corresponding blood flow values (cutaneous, $\dot{Q}_{cut}$, and internal, $\dot{Q}_{int}$) are known.

The O_2 uptake of *Desmognathus fuscus* in normoxic conditions has been found to equal 32 nmol/(min·g) (Gatz *et al.*, 1974b). Since the O_2 content in mixed (cardiac) blood, $C_{m_{O_2}}$, had been experimentally determined, the $C_{a_{O_2}}$ and $C_{v_{O_2}}$ values could be calculated and the corresponding O_2 pressure and O_2 saturation values were obtained with the aid of the blood O_2 dissociation curve (Gatz *et al.*, 1974a). The values calculated for the various estimates of blood flow are presented in Table II and visualized in the

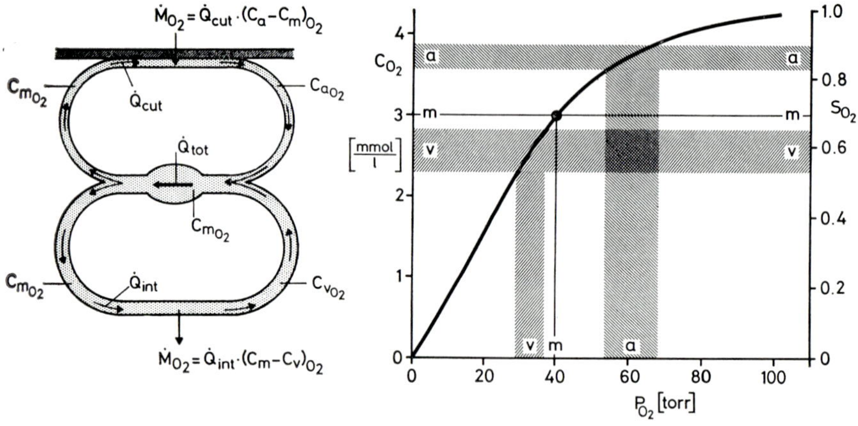

Fig. 4. Left: circulatory model for *Desmognathus fuscus* based on the probable assumption that mixing of venous and arterial blood occurs by the time these bloods reach the heart. The O_2 content differences between inflowing and outflowing blood of both the skin and internal organs can be calculated from Fick's principle knowing the O_2 uptake in the skin and its delivery to internal organs as well as the relevant blood flows, $\dot{Q}_{cut}$ and $\dot{Q}_{int}$. Right: oxygen dissociation curve with experimentally determined mixed cardiac blood value indicated with solid circle. The areas indicated refer to the possible range of values for arterial and venous bloods with respect to O_2 concentration, saturation and partial pressure.

TABLE II

Blood O_2 values in Desmognathus fuscus. *Mixed blood (m): experimental mean values from Gatz* et al., *(1974a) Arterialized (a) and venous (v) blood: calculated from blood flow and O_2 transfer rate (see Text). f_{cut}, assumed ratio $\dot{Q}_{cut}/\dot{Q}_{tot}$*

		C_{O_2} $\left[\frac{mmol}{l}\right]$			S_{O_2} (%)			P_{O_2} (torr)		
f_{cut}	Model	a	m	v	a	m	v	a	m	v
	Parallel I	(4·20)		2·74	(97)		63	(100)		36
0·2	Parallel II	3·88		2·83	89		65	68		37
	Serial	3·86		2·83	89		65	67		37
	Parallel I	3·71		2·29	87		53	63		29
0·5	Parallel II	3·61		2·46	83		57	55		31
	Serial	3·55		2·52	82		58	53		32
	"Mean"	**3·72**		**2·61**	**86**		**60**	**61**		**34**
	Experimental		3·04			0·70			40	

blood O_2 dissociation curve in Fig. 1 (right). The P_{O_2} values range from 53 to 68 torr in arterialized blood, from 32 to 37 torr in venous blood, the value in mixed blood having been experimentally determined at 40 torr. If averaged for all models and for the two assumptions on fractional cutaneous blood flow, the O_2 content difference between arterialized and venous blood is 1·2 mmol/l = 2·7 vol. %, and the O_2 saturation difference, 26%.

For an analogous analysis of CO_2 transport in blood, a CO_2 dissociation curve was constructed (Fig. 6) using the actual mixed blood P_{CO_2} and pH, and the buffer value of true plasma, (13·2 mmol/(l·pH)), according to Gatz *et al.* (1974a). This dissociation curve had the strict meaning of a CO_2 dissociation curve for true plasma. However, since haematocrit and haemoglobin concentration were relatively low, this plot should not be markedly different from a CO_2 dissociation curve for whole blood. Moreover, since the absolute values and the arterio-venous differences of P_{CO_2} were very low, it was considered unnecessary to take into account physically dissolved CO_2 changes.

In order to calculate $(C_v - C_m)_{CO_2}$ and $(C_m - C_a)_{CO_2}$ values, CO_2 production rate was calculated from the O_2 uptake, assuming a respiratory quotient of 0·85, $\dot{M}_{CO_2}$ = 27·2 nmol/(min·g).

The values obtained for the various estimates of cardiac output and cutaneous blood flow are presented in Table III. The P_{CO_2} differences between arterialized, mixed and venous blood are very small, not exceeding 2 torr.

TABLE III

Blood CO_2 values in Desmognathus fuscus. *Mixed blood (m): experimental mean values from Gatz* et al., *(1974a). Arterialized (a) and venous (v) blood: calculated from blood flow and CO_2 transfer rate (see Text). f_{cut}, assumed ratio, $\dot{Q}_{cut}/\dot{Q}_{tot}$.*

f_{cut}	Model	Bicarbonate $\left[\frac{mmol}{l}\right]$ a	m	v	P_{CO_2} (torr) a	m	v
	Parallel I	8·16		9·41	4·7		6·8
0·2	Parallel II	8·44		9·33	5·1		6·6
	Serial	8 45		9·33	5·1		6·6
	Parallel I	8·58		9·79	5·3		7·5
0·5	Parallel II	8·67		9·64	5·5		7·2
	Serial	8·72		9·59	5·6		7·1
	"Mean"	**8·50**		**9·51**	**5·2**		**7·0**
	Experimental		9·15			6·2	

III. DIFFUSIVE VS CONVECTIVE GAS TRANSPORT BETWEEN BODY AND SKIN

In the preceding section it was assumed that all O_2 and CO_2 is transported by blood flow, none by direct diffusion. This important assumption requires confirmation, particularly since the comparison of the exit rate of Freon 22 from dead and living animals (Gatz *et al.*, 1975) has shown that a substantial fraction of this gas is transferred by diffusion.

For simplicity, both diffusive and perfusive transport may be considered as parallel processes occurring independently of each other (Fig. 5). In order to estimate the relative roles of diffusion and perfusion in transport of CO_2 and O_2, diffusive and perfusive conductances must be calculated on the basis of the corresponding values for Freon 22.

For the diffusive conductance or diffusing capacity, D, the following relationship is obtained from Fick's law of diffusion,

$$\mathrm{D} = D\cdot\alpha\cdot\frac{\mathrm{F}}{\mathrm{x}} \tag{3}$$

where, D, α, F and x designate the following properties of the diffusion barrier; D, diffusion coefficient; α, solubility; F, surface area; x, thickness. It is evident that for two gases, X and Y, the following relationship holds:

$$\frac{\mathrm{D_Y}}{\mathrm{D_X}} = \frac{(D\cdot\alpha)_Y}{(D\cdot\alpha)_X} \tag{4}$$

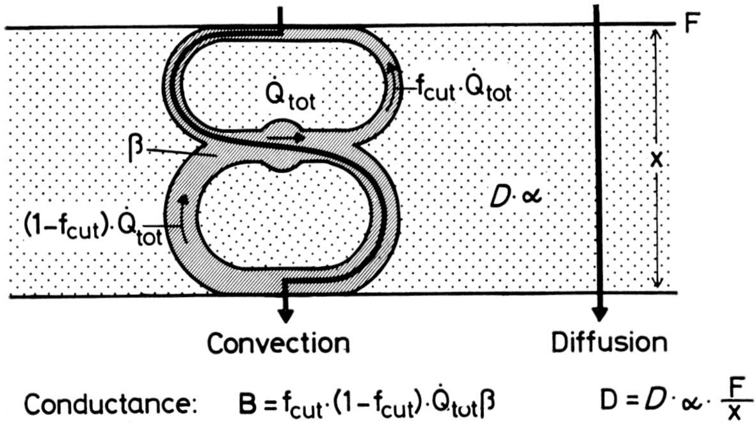

FIG. 5. Diffusion and convection depicted as independent and parallel processes in gas transport in *Desmognathus fuscus*. Quantitative analysis of these two processes is afforded by the equations shown for convective and diffusive conductances.

TABLE IV

Diffusivity and solubility parameters of Freon 22, CO_2 and O_2. X, any parameter; D, diffusion coefficient; α, physical solubility; β, capacitance coefficient of blood. See Text.

Parameter	Value for Freon 22	$\frac{X_{O_2}}{X_{Fr}}$	$\frac{X_{CO_2}}{X_{Fr}}$	$\frac{X_{CO_2}}{X_{O_2}}$
D	$2{\cdot}27 \cdot 10^{-4}\ \frac{cm^2}{min}$ *	1·64	1·39	0·85
α	$57{\cdot}5\ \frac{\mu mol}{l \cdot torr}$	0·032	0·97	30·8†
β_{a-v}	$= \alpha$	0·71	9·8	13·8
β_{a-m}	$= \alpha$	0·63	11·3	18·0
$D \cdot \alpha$	$0{\cdot}13 \cdot 10^{-4}\ \frac{\mu mol}{min \cdot torr \cdot cm}$ *	0·052	1·35	26·2
$\frac{D \cdot \alpha}{\beta_{a-v}}$	= D	0·073	0·138	1·90
$\frac{D \cdot \alpha}{\beta_{a-m}}$	= D	0·083	0·120	1·45

* Calculated, using Graham's law, from O_2 values for connective tissue of frog at 20°C, after Krogh from Altman and Dittmer (1971).

† Ringer solution, 13°C, from Altman and Dittmer (1971).

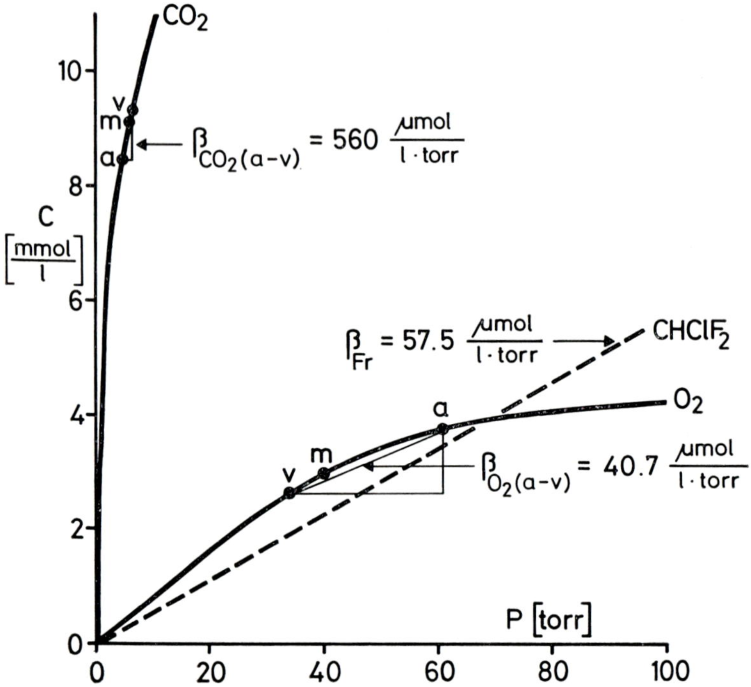

FIG. 6. Blood O_2 and CO_2 dissociation curves for *Desmognathus fuscus* as well as the total body solubility curve for Freon 22 ($CHClF_2$).

The D, α and $(D \cdot \alpha)$ values for Freon 22, CO_2 and O_2, adapted from various sources for use in the present study, are shown in Table IV.

If the convective conductances of two gases, X and Y, are considered, it follows from Eq. (4) that the conductance ratio is:

$$\frac{B_Y}{B_X} = \frac{\beta_Y}{\beta_X} \tag{5}$$

For the inert gas Freon 22 it has been assumed that the solubilities in blood and tissue are equal, i.e. $\beta = \alpha$. For O_2 the β values have been obtained as mean slopes of the line crossing the O_2 dissociation curve at the arterial and venous P_{O_2} values, i.e. at $P_{O_2} = 61$ torr and $P_{O_2} = 34$ torr. This β value is designated as β_{a-v}. In a similar manner, the β_{a-v} values were established for CO_2, using the CO_2 dissociation curve constructed from buffer values for true plasma. The β_{a-v} values thus obtained are presented in Table IV and illustrated in Fig. 6.

The ratio of diffusive conductance to perfusive conductance, D/B, is the decisive parameter for evaluation of the relative role of the two conductances. It follows from Eqs (4) and (5) that D/B is proportional to $(D \cdot \alpha)/\beta_{a-v}$,

$$\frac{(D/B)_Y}{(D/B)_X} = \frac{(D\cdot\alpha/\beta_{a-v})_Y}{(D\cdot\alpha/\beta_{a-v})_X} \tag{6}$$

With the aid of $(D\cdot\alpha)$ and β_{a-v} values the diffusive and perfusive conductances and their ratio were calculated for CO_2 and O_2. The values obtained are presented in Table V.

Since O_2 has a much smaller D than Freon, due to smaller $(D\cdot\alpha)$, but a B of similar magnitude, the contribution of diffusive conductance in comparison to perfusive conductance turns out to be only 3% to 5%, contrasting with 42% to 62% in the case of Freon 22. When CO_2 is compared to Freon 22, the D values are closely similar, due to about equal physical solubility, but the B value of CO_2 is very high because of the high slope of the blood CO_2 dissociation curve at the low P_{CO_2} values. The result is again a relatively minor role of diffusion, relative to perfusion, in CO_2 transport.

The effect of diffusive transport of O_2 and CO_2 could be taken into account in correcting the arterial-mixed and mixed-venous O_2 and CO_2 differences in Tables II and III and in Fig. 4. However, the extent of the corrections appeared not to be large enough to justify their application.

IV. Diffusing Capacity of the Skin

In attempting to analyse the Freon 22 equilibration data two models were assumed (Gatz *et al.*, 1975). In the parallel compartment model no resistance to diffusion was assumed to exist between the cutaneous blood capillaries and the environment. But in the serial compartment model a resistance to diffusion, or finite diffusive conductance, was postulated between the peripheral compartment and environment. When considering gas transport in steady

Table V

Diffusive conductance (diffusing capacity), D, and perfusive conductance, B, in transport of various gases between body and skin in Desmognathus fuscus. *Left hand values have been obtained for the Serial Model, right hand values, for the Parallel Model I*

	Freon 22	O_2	CO_2
D $\left[\frac{\text{nmol}}{\text{min}\cdot\text{torr}\cdot\text{g}}\right]$	0·755–0·765	0·0383–0·0398	1·02–1·033
B $\left[\frac{\text{nmol}}{\text{min}\cdot\text{torr}\cdot\text{g}}\right]$	1·795–1·225	1·28–0·870	17·6–12·0
$\frac{D}{B}$ (−)	0·42–0·625	0·030–0·046	0·058–0·086

state, the compartment volumes are formally suppressed and the resistance to diffusion between the outer compartment and the environment is then between the cutaneous capillaries and the body surface. Thus, the corresponding diffusive conductance, G_{cut}, has the meaning of the diffusing capacity of the external gas exchange organ and is, therefore, comparable to the diffusing capacity of mammalian lungs. It is of interest to calculate the skin D values for O_2 and CO_2 from the value for Freon 22.

The fact that the arterial P_{O_2}, obtained from O_2 uptake and estimated cutaneous blood flow, remained much below the atmospheric P_{O_2} clearly indicates the existence of a considerable resistance to diffusion in the skin. From the O_2 uptake, and P_{O_2} in blood entering the skin ($= P_{m_{O_2}}$) and leaving the skin ($= P_{a_{O_2}}$), a D_{O_2} value can be calculated and compared to the value derived from analysis of the elimination kinetics of Freon 22 according to the serial model. A third method of estimating skin D_{O_2} is based on morphometric data.

A. SKIN D_{O_2} CALCULATED FROM EQUILIBRATION KINETICS OF FREON 22

From the conductance, G_{cut}, equal to 7·09 nmol/(min·torr·g) obtained from the analysis of equilibration kinetics of Freon 22, the diffusing capacity for O_2, D_{O_2}, can be calculated using the conversion factor for $(D \cdot \alpha)$ from Table IV, 0·052. The value obtained is: $D_{O_2} = 0{\cdot}37$ nmol/(min·torr·g) Table VI).

B. SKIN D_{O_2} CALCULATED FROM O_2 UPTAKE AND BLOOD P_{O_2}

D_{O_2} may be calculated from the O_2 uptake, $\dot{M}_{O_2}$, and the mean P_{O_2} difference between environment and cutaneous capillary blood, $\overline{\Delta P}_{O_2}$, according to its definition equation,

$$D_{O_2} = \frac{\dot{M}_{O_2}}{\overline{\Delta P}_{O_2}}$$

The mean environmental-capillary P_{O_2} difference, $\overline{\Delta P}_{O_2}$, can be calculated on the basis of the exponential approach of $P_{m_{O_2}}$ to environmental P_{O_2} ($P_{E_{O_2}} = 152$ torr), but since the extent of the approach is small, simple arithmetic mean of initial and final P_{O_2} differences, i.e. of $(P_E - P_m)_{O_2}$ and $(P_E - P_a)_{O_2}$, is sufficiently accurate.

According to the $P_{a_{O_2}}$ values of Table II the following values are obtained (Table VI):

$$P_{a_{O_2}} = 68 \text{ torr}: \overline{\Delta P}_{O_2} = 98 \text{ torr}, D_{O_2} = 0{\cdot}33 \text{ nmol/(min}\cdot\text{torr}\cdot\text{g)}$$

$$P_{a_{O_2}} = 53 \text{ torr}: \overline{\Delta P}_{O_2} = 106 \text{ torr}, D_{O_2} = 0{\cdot}30 \text{ nmol/(min}\cdot\text{torr}\cdot\text{g)}$$

Considering the relative independence of the D_{O_2} estimates based on Freon elimination kinetics and on O_2 transfer, the agreement of the values appears remarkable.

C. SKIN D_{O_2} CALCULATED FROM MORPHOMETRIC DATA

The calculation is based on Fick's diffusion law in the form of Eq. (1). For $(D \cdot \alpha)_{O_2}$, the value for frog connective tissue from Krogh (1919), $(D \cdot \alpha)_{O_2} = 6{\cdot}75 \cdot 10^{-7}$ μmol/(min·torr·cm), is used. The F and x values have to be estimated from anatomical morphometric studies.

Czopek (1961), who has extensively studied the vasculature of respiratory organs in various amphibians, gives the following values for a specimen of *Desmognathus fuscus* weighing 6·2 g: total surface area of cutaneous and buccopharyngeal capillaries, 3·922 cm^2 per gram body weight; mean thickness of the epidermis, 25·8 μm.

Since the cutaneous capillary net is stated to lie directly under the epidermis, the diffusion distance, x, may be set equal to the thickness of the epidermis, 25·8 μm. For the surface area effective in gas exchange with the environment it seems to be appropriate to refer the O_2 uptake to one half (the outer half) of the total capillary surface area. Thus one obtains $F = 1{\cdot}996$ $cm^2 \cdot g^{-1}$.

Calculation of D_{O_2} using these values yields 0·52 nmol/(min·torr·g). This value is higher than the values derived from physiological measurements (Table VI), but the agreement must be considered rather good in view of the various assumptions and uncertainties involved.

D. D_{CO_2} VALUES

In Table VI D_{CO_2} values have also been included. The values according to he methods A and C are easily obtained using the conversion factors for

TABLE VI

Diffusing capacity of skin for O_2 and CO_2, estimated in various ways

Origin	D $\left[\frac{nmol}{min \cdot torr \cdot g}\right]$ O_2	CO_2
(1) D_{cut} for Freon 22	0·37	9·7
(2) P_m, P_a, $\dot{M}_{O_2}$	0·30–0·33	4·9
(3) Morphometrics, $(D \cdot \alpha)$	0·53	14

$(D \cdot \alpha)$ in Table IV. With method B, using a P_{CO_2} of 5·6 torr (from Table III) and a CO_2 production of 85% of the O_2 uptake, a D_{CO_2} value of 4·9 nmol/(min·torr·g) is obtained, which is half the value computed from Freon equilibration. This disagreement is conspicuous in view of the very good agreement of the corresponding D_{O_2}.

Though there may be several ways of explaining this discrepancy, including the possibility that the $(D \cdot \alpha)_{CO_2}/(D \cdot \alpha)_{O_2}$ ratio of 26·2 (Table IV) may be erroneous, an attractive hypothesis is the following. At the surface of the body, particularly on its lower surface in contact with the ground, there may exist a stagnant layer of gas with slightly increased P_{CO_2} and slightly decreased P_{O_2}. Assuming a $PE_{CO_2} = 2 \cdot 8$ torr corresponding to about 0·4% CO_2, the D_{CO_2} value calculated from Freon 22 equilibration would be reached, whereas a corresponding decrease of ambient P_{O_2} would not exert a significant effect due to the large $\overline{\Delta P}_{O_2}$.

V. Diffusion and Perfusion Limitation in Cutaneous Gas Exchange

Gas transfer between blood and a gaseous medium may be limited by diffusion through the tissue layer between blood and gas and by the blood flow rate (perfusion). Estimation of the quantitative role of gas transfer limitation by diffusion and by perfusion is of basic importance in functional analysis of the performance of a gas exchange organ.

A simple model for study of diffusion and perfusion limitation, along with the equations, is shown in Fig. 7. This approach has been previously applied to analysis of gas exchange of tissue gas pockets (Piiper, 1965). The decisive parameter is the diffusive/perfusive conductance ratio, $D/(\dot{Q} \cdot \beta) = Y$.

An index for diffusion limitation, L_{diff}, is defined as the fractional decrease of transfer rate from its value without diffusion limitation, $\dot{M}_{D=\infty}$, to its actual value $\dot{M}$,

$$L_{diff} = \frac{\dot{M}_{D=\infty} - \dot{M}}{\dot{M}_{D=\infty}}.$$

Accordingly, the index for perfusion limitation, L_{perf}, may be defined as the fractional decrease of transfer rate from the value without perfusion limitation, $\dot{M}_{\dot{Q}\beta=\infty}$, to its actual value, $\dot{M}$,

$$L_{perf} = \frac{\dot{M}_{\dot{Q}\beta=\infty} - \dot{M}.}{\dot{M}_{\dot{Q}\beta=\infty}}$$

The values of L_{diff} and L_{perf} are simple functions of Y. The relationships are formulated and plotted in Fig. 7.

When Y is progressively increased from low values, the transfer is at first purely diffusion-limited. At $Y = 0 \cdot 1$ perfusion limitation begins to be signi-

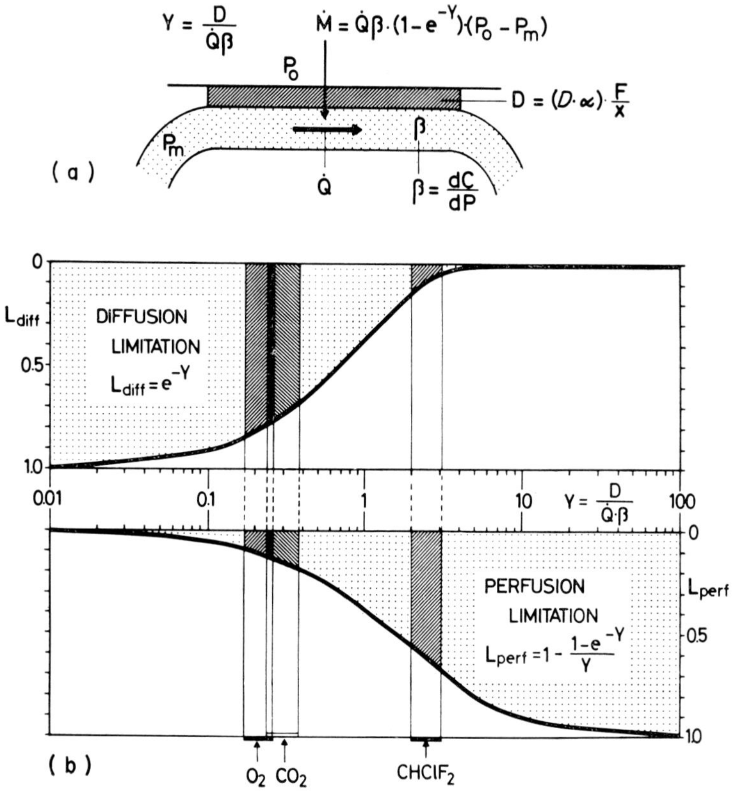

FIG. 7. Diffusion and perfusion limitation. (a) A simple model for analysis. The decisive parameter is the diffusive/perfusive conductance ratio indicated here as $D/(\dot{Q}\beta) = Y$. (b) Plot of diffusion limitation, L_{diff}, and perfusion limitation, L_{perf}, against the diffusive/perfusive conductance ratio, $D/(\dot{Q}\beta) = Y$. The range of Y values estimated for the cutaneous transfer of Freon 22, O_2 and CO_2 in *Desmognathus fuscus* is indicated by the hatched areas. The transfer of Freon 22 ($CHClF_2$) is, for the most part, perfusion-limited, whereas that of O_2 and CO_2 is mostly diffusion-limited.

ficant and diffusion limitation decreases. Above $Y = 3$ the transfer is practically exclusively perfusion-limited.

For Freon 22, $D/(\dot{Q}\beta)$ equal to 2·0 to 3·1 is obtained from the following values:

$$D_{cut} = G_{cut} = 7{\cdot}09 \text{ nmol/(min}\cdot\text{torr}\cdot\text{g)},$$
(Gatz *et al.*, 1975).

$$\beta_{Fr} = 57{\cdot}5 \text{ nmol/(torr}\cdot\text{g)}$$
(from Table IV).

$$\dot{Q}_{cut} = 62{\cdot}5\ \mu\text{l/(min}\cdot\text{g) or } 39{\cdot}0\ \mu\text{l/(min}\cdot\text{g)},$$
depending on the model.

TABLE VII

$D/(\dot{Q}\beta)$ values for cutaneous transfer of Freon 22, O_2 and CO_2 in Desmognathus fuscus

	$f_{cut} = 0{\cdot}5$	$f_{cut} = 0{\cdot}2$
$\left(\frac{D}{\dot{Q}\cdot\beta}\right)_{Fr}$	2·0	3·1
$\left(\frac{D}{\dot{Q}\cdot\beta}\right)_{CO_2}$	0·24	0·37
$\left(\frac{D}{\dot{Q}\cdot\beta}\right)_{O_2}$	0·17	0·26

The $D/(\dot{Q}\beta)$ values for O_2 and CO_2 were calculated on the basis of a comparison of the physical properties of the gases listed in Table IV. As shown in Fig. 7, $D/(\dot{Q}\beta)$ is proportional to the combination of parameters, $(D\cdot\alpha)/\beta$. Using the ratios of $(D\cdot\alpha)/\beta$ values for two gases, the $D/(\dot{Q}\beta)$ values for O_2 and CO_2 presented in Table VII were obtained. These $D/(\dot{Q}\beta)$ are marked in Fig. 7. It is seen that whereas the transfer of Freon 22 between cutaneous blood and environment is prevalently perfusion-limited, diffusion limitation being only slight, the exchange of O_2 and CO_2 is primarily diffusion-limited and much less limited by perfusion. Comparison of O_2 and CO_2 with the relevant properties of Freon 22 in Table IV shows that for O_2 it is mainly the much smaller α that is responsible for the low $D/(\dot{Q}\beta)$ in comparison with Freon 22, but for CO_2 the low $D/(\dot{Q}\beta)$ value is due to the high β_{a-m} value.

The $D/(\dot{Q}\beta)$ values calculated for CO_2 and O_2 are almost identical although the solubility parameters for CO_2 are, by a factor of 20 to 30, higher. But since both α and β are higher, their ratio is not very far from unity.

The very limited extent of diffusion limitation for transfer of Freon 22, as visualized in Fig. 7, helps to understand why, in the analysis of Freon 22 equilibration, use of the parallel model, in which no diffusion limitation was assumed to exist for gas transfer between skin blood and environment, gave results differing little from those obtained on the basis of the serial model in which the presence of such a diffusion resistance was an inherent property of the model (cf. Gatz *et al.*, 1975).

VI. CRITICAL REMARKS AND CONCLUSIONS

The results of the foregoing analysis of gas transport in *Desmognathus fuscus* are subject to severe limitations imposed by unsatisfactory experimental data, ambiguity of the models and a number of required but unwarranted assumptions.

The experimental data have been gathered in different individuals in different conditions. The most significant disparity concerns the environmental temperature. The experiments for determination of O_2 uptake were performed at 13°C, a temperature considered typical for the habitat of this species (Gatz *et al.*, 1974a). Also the blood dissociation curve for O_2 was established at this temperature. The actual mixed blood O_2 and CO_2 values were measured at 13°C in animals acclimatized to 13°C, but somewhat warmed during blood sampling at room temperature. The measurements of Freon 22 equilibration, however, were done at room temperature (22°C) on animals previously stored at lower temperatures. As nothing is known of acute temperature effects or of long term temperature adaptation in *Desmognathus fuscus* corrections could not be applied.

Considering that the circulation rates were probably higher at room temperature than at 13°C, at which temperature O_2 uptake has been measured, the arterio-venous content and partial pressure differences for O_2 and CO_2 might be smaller than calculated in the present analysis.

In calculating the blood P_{O_2} and P_{CO_2} no account was taken of Bohr or Haldane effects, because they have not been established. These effects, if present, would decrease the arteriovenous partial pressure differences.

The physico-chemical properties of tissue and blood regarding diffusivity and solubility of Freon 22, O_2 and CO_2 are in part uncertain for several reasons:

(1) D_{O_2} and α_{O_2} values for frog connective tissue at 20°C were used,

(2) validity of Graham's Law was assumed for calculating D_{CO_2} and D_{Fr} from D_{O_2},

(3) physical solubility and diffusivity of gases were assumed to be the same in all tissues.

Finally, the use of simple compartmental and blood flow models can be expected to lead only to approximations.

These deficiencies suggest that some quantitative corrections may need to be applied to the results and interpretations. However, the general picture of the roles of circulation and diffusion in the transport of O_2 and CO_2 in a lungless plethodontid salamander is considered valid within these limits.

VII. Summary

Combination of experimental data on inert gas equilibration kinetics, on blood gases and on O_2 uptake with anatomical data on the arrangement of the circulatory system and on cutancous vasculature has led to the following picture of respiratory gas transport in the lungless and gill-less terrestrial plethodontid salamander (*Desmognathus fuscus*), of 6·1 g mean body mass.

(1) The cardiac output, estimated at 0·11 to 0·19 ml/min per g body mass, is directed in part (probably less than 50%) to the skin, where O_2 is taken up, and in part to remaining tissues and organs, where O_2 is delivered. In the heart arterialized and venous blood are probably completely mixed. Therefore, three kinds of blood should be distinguished: arterialized, venous and mixed blood. The partial pressure of O_2 in these is estimated at 57–75 torr, 31–37 torr, and 40 torr, respectively.

(2) Transport of O_2 from the skin to the interior of the body and the reverse transport of CO_2 are performed mainly by blood flow, diffusion accounting for less than 8% of the total transfer.

(3) The O_2 diffusing capacity of the skin is estimated at about 0·3 to 0·5 $\text{nmol}\cdot(\text{min}\cdot\text{torr})^{-1}$ per g body mass. The mean partial pressure difference between environmental gas and skin capillary blood is about 100 torr for O_2, 5–6 torr for CO_2.

(4) The exchange of CO_2 and O_2 between skin capillary blood and the environment is prevalently diffusion-limited.

References

Altmann, P. L. and Dittmer, D. S. (Eds) (1971). "Handbook of Respiration and Circulation." Bethesda, Federation of American Societies for Experimental Biology.

Czopek, J. (1961). Vascularization of the respiratory surfaces of some Plethodontidae. *Zool. Pol.* **11**, 131–148.

Gatz, R. N., Crawford, E. C., Jr. and Piiper, J. (1974a). Respiratory properties of the blood of a lungless and gill-less salamander, *Desmognathus fuscus*. *Resp. Physiol.* **20**, 33–41.

Gatz, R. N., Crawford, E. C., Jr. and Piiper, J. (1974b). Metabolic and heart rate response of the plethodontid salamander *Desmognathus fuscus* to hypoxia. *Resp. Physiol.* **20**, 43–49.

Gatz, R. N., Crawford, E. C., Jr. and Piiper, J. (1975). Kinetics of inert gas equilibration in an exclusively skin-breathing salamander, *Desmognathus fuscus Resp. Physiol.* **24**, 15–29.

Krogh, A. (1919). The rate of diffusion of gases through animal tissues, with some remarks on the coefficient of invasion. *J. Physiol. Lond.* **52**, 391–408.

Piiper, J. (1965). Physiological equilibria of gas cavities in the body. *In* "Handbook of Physiology", Sect. 3, Respiration Vol. II (Eds W. O. Fenn, H. Rahn and J. Piiper), pp. 1205–1218. Washington, D.C., American Physiological Society.

Piiper, J., Dejours, P., Haab, P. and Rahn, H. (1971). Concepts and basic quantities in gas exchange physiology. *Resp. Physiol.* **13**, 292–304.

Ventilatory Mechanisms and Problems in some Amphibious Aspiration Breathers (*Chelydra*, *Caiman* – Reptilia)

CARL GANS

Department of Zoology, The University of Michigan, Ann Arbor, Michigan, U.S.A.

I. Introduction

The air-breathing mechanisms of Recent teleosts clearly represent an analogous series of independent adaptive events. Even where homologous structures, such as the branchial chambers, are involved, the scattered appearance of these modifications and the general retention of aquatic respiration by numerous, closely related species suggest that the shift took place more than once (Gans, 1971b). In each case the shift appears to have been driven by the need to utilize gaseous air as a supplementary source of oxygen, and in many cases it seems to have provided protoadaptations that later facilitated temporary emergence from the water.

A parallel but opposite trend appears to have occurred among the reptiles, a group that arose from amphibians after these had completed the transition to a fully terrestrial existence. The diverse attendant specializations apparently permitted them to compete successfully in the aquatic environment and a partially or completely aquatic habitus has developed, apparently independently, in four of the six surviving orders of Recent reptiles. In each case there have been modifications of the recently acquired ventilatory mechanisms, and it seems of interest to take a comparative view of the phenomenon of reverse adaptation to an aquatic situation, in this case of the mechanism of air movement.

The present paper attempts to make such a comparison for one member of each of two groups, turtles and crocodilians. Specifically it is proposed to review breathing in the snapping turtle, *Chelydra serpentina* (Gaunt and Gans, 1969b), and the spectacled caiman, *Caiman crocodilus* (Gans and Clark, 1975), both thoroughly amphibious forms. In doing so, I must take specific exception to the view implicit in some studies that the condition in either or both of these forms represents an intermediate state between fishes and mammals, reflecting either incomplete progress toward an "advanced" condition or a degeneration from some functional (mammalian?) ideal. The use of the single term M. diaphragmaticus for the clearly non-homologous inhalatory muscles of turtles, crocodilians and mammals derives from this view. The discussion should make it amply clear that we are dealing with independently derived groups of species, each primarily adapted to its particular set of niches. Their specializations must be considered from this viewpoint.

The following mechanical analyses were all made on loosely tethered animals by the approaches detailed in Gans (1974) and the original papers. Pulmonary pressures and movements were recorded directly and muscular activities by electromyograms taken with implanted fine wire electrodes. In order to avoid fear-induced responses such as withdrawal bradycardia (Gaunt and Gans, 1969a; also Smith, Allison and Crowder, 1974), many of the experiments were carried out in shielded rooms by remote control and with closed-circuit T.V. monitors. This proved much more necessary for *Caiman* than for *Chelydra*, the latter was generally more irascible and less affected by the presence of an experimenter.

II. Options for Air Movement

Flow of fluid requires a pressure gradient. The magnitude of laminar flow will be directly proportional to the height of the gradient and inversely to the resistance of passage.

To cause gas to move into a chamber one must either raise the gas to a pressure higher than that then pertaining in the chamber or reduce the pressure in the chamber below that of the gas to be moved. The former is the mechanism of forced pumping (characteristic of all air-breathing fishes and Recent amphibians as far as known*), the latter that of aspiration breathing (characteristic of the Recent amniotes—reptiles, birds and mammals).

The forced pumping sequence involves two separate chambers. The gas is first sucked (through open mouth or nostrils) into a primary (buccal) chamber as the volume of the chamber is expanded (floor dropped). The chamber is then closed (mouth shut, nostrils pinched), while the volume is maintained. After closure, muscular activity will constrict the chamber's

*Traumatized force pumpers can apparently survive for some time by use of an inefficient version of aspiration (Bentley and Shield, 1973; Eberle, 1937).

volume and raise its pressure to a supra-atmospheric level. Opening of a path (oesophagus, glottal or laryngeal valve) into the exchange chamber (gut, air sac, branchial diverticulum, lung) permits this to be filled without further muscular effort. Generally exhalation precedes inhalation, thus (a) producing an initial reduction in the volume of the filled space, (b) producing a reduction of the mean "pulmonary" pressure, and (c) reducing mixing of outflow and inflow and thus increasing the oxygen and CO_2 diffusion gradients across the surface of the external gas exchanger. In some fishes the exhaled air is discharged before the start of the inhalatory phase (Rahn, Rahn, Howell, Gans and Tenney, 1971). In frogs the sequence starts with inhalation into the buccal cavity, followed by exhalation from the lung, with the spent gas traversing the buccal cavity but bypassing the inhaled air, and finally passage of this unmixed buccal gas into the lung (Gans, de Jongh and Farber, 1969).

In aspiration breathing the walls of the primary pulmonary chamber are directly deformed increasing or decreasing its volume. This changes the internal pressure, and gas will flow as long as the laryngeal connection is open. In principle the larynx need not serve as a gas-retaining closure as the volume of the pulmonary chamber and hence the pulmonary pressures can be established by direct muscular action. Species finding it advantageous to maintain their pulmonary contents at levels other than atmospheric pressure derive a concurrent advantage from laryngeal closure. This permits the passive maintenance of elevated pressures without the need for continuous muscular activity in the positioning muscles. (Glottal closure, of course, may be maintained elastically.) A closable glott's also prevents accidental entry of food particles or potential macroparasites into the airway.

The two pumping systems show some fundamental differences in their energetics. In aspiration breathing the muscles act directly and the shifting walls are coupled to the pulmonary soft tissues by solid or liquid links. The compliance of the system is good and muscular effort is minimal. Flow may be induced relatively slowly, involving minimum accelerational forces either on the frame or the gaseous medium. Consequently, the pulmonary pressure need depart but slightly from atmospheric.

Forced pumping always requires more energy as the pressure level at the (buccal) pump always has to exceed (or be less than) that in the pulmonary chamber. This imposes requirements on the nature of the valvular closure in the pumping system. The forced pumping system potentially involves control problems in synchronizing the several sets of valves. If the pulmonary chamber's walls are rigid (and its volume fixed) the changes in the pressure of its contents will obviously be maximal. If the chamber is compliant its walls will deform in response to internal pressure. Energy will then have to be supplied indirectly via air which is a highly compressible fluid; this process incorporates a significantly delayed response and a further expenditure of energy.

For a chamber of given coefficient of compliance, the mass of the walls will obviously be the most critical parameter in estimating the energy cost for moving a unit volume of air. This imposes architectural constraints on the location of the lungs and on the integumentary ossifications as well. Not only the mass of the ribs but their sternal connections, as well as the nature of their articulations and muscular connections, will affect the compliance of the system.

It has long been assumed that some of the placoderms used forced ventilation to fill their lungs with air (see Gans, 1971b, for review of the literature). In such species the gas was presumably aspirated into the buccal cavity by a variant of the movements used in shifting water (as well as in suction feeding; Gans, 1969). The same muscles that force water across the gills force the air into the gas-exchange space. This will previously have been emptied by hydrostatic pressure, with the spent gas discharged either via the mouth or across the opercular chamber. Fishes have to move a much greater weight of respiratory fluid per volume of oxygen extracted when pumping water than when pumping air (Dejours, this volume); the muscles involved are consequently capable of rapidly compressing the gaseous contents. This capacity is apt to be greatest when the force-ventilating species is one adapted for suction-feeding as this requires rapid movement of the water. As the potential leakage of gas past the buccal valve is more easily controlled for fast than slow pressurizations, pulse-pumping is the most effective method of forced ventilation. The only differences observed among the various pulse-pumpers concern the path of the air, whether it is aspirated or discharged via the open mouth or through the narial canals.

The transition from pulse-pumping to aspiration breathing probably took place among crossopterygian fishes, i.e. in species that were clearly aquatic (Gans, 1971a,b; Romer, 1972). Apparently the need to overcome the resistance of the water would have added to the breathing work required of the large animals that seem to have operated in warm environments. Such biotopes would only be open to species that could discharge a significant fraction of their CO_2 as a gas. The intrinsic inefficiency of pulse pumping (particularly against the resistance of the water) made the transition advantageous. While pulse pumping patterns are retained in all Recent fishes and, as far as is known, in all Recent amphibians (with the possible exception of caecilians), they have been replaced in reptiles, birds, and mammals. Possibly aspiration breathing was evolved many times. Certainly the morphological basis of ventilation in testudines (Gans and Hughes, 1967; Gaunt and Gans, 1969a; Ireland and Gans, 1973) is non-homologous to that in crocodilians (Gans and Clark, 1975), and does not seem to relate exactly to the conditions in snakes (Rosenberg, 1972), birds (Schmidt-Nielsen, Kanwisher, Lasiewski, Cohn and Breta, 1969) or mammals (Rankin and Dempsey, 1967). [See Wood and Lenfant (1976) for a general discussion of the conditions in reptiles.]

III. The Turtle Pattern

The specializations of turtles are proverbial and there is increasing evidence to support the suggestion that they derive from a line independent from the remainder of Recent reptiles (Parsons, 1959). The major difference from all other vertebrates has been the shift of both sets of limb girdles into the rib cage and the fusion of the latter with integumentary elements and gastralia into an external armour. Multiple physiological adaptations parallel the structural ones. Thus turtles have a tolerance of anoxia greater by an order of magnitude than do other animals (Belkin, 1963, 1968).

Important from the viewpoint of ventilation are the high values of blood bicarbonate (Hughes, 1966; Dessauer, 1970). These permit an effectively higher gradient for the discharge of CO_2, correlated with a more prolonged apnoeic capacity between breaths and the ability to survive with the lung volume reduced by full retraction of the limbs and neck. Especially in aquatic species there may be a variable potential for gas exchange across the pharynx and skin. This is insufficient to supply the demands of active animals near their preferred temperatures, although it does increase the length of the apnoeic periods and permit some gas exchange during submerged hibernation.

The ventilatory mechanism is also unique (Fig. 1). In the terrestrial tortoises (*Testudo*, Gans and Hughes, 1967), the shell serves as a rigid and domed frame. The lungs lie within and attached to the dome, in a space bounded dorsally by bone and ventrally by the visceral mass. In these species the "diaphragmatic" membrane is a non-muscular sheet of connective tissue; consequently the volume available to the lungs reflects (a) the extent to which the girdles and limbs are rotated into the shell, (b) the extent to which the head and neck are withdrawn, and (c) the position of a thin muscular sling (incorporating the transverse muscle) bounding the posterior aperture of the shell.

Ventilation consists of exhalation followed by inhalation; hence flow is diphasic. Pressure changes may be triphasic or even quadruphasic, possibly involving a supplementary rise in pressure (prior to glottal opening) at the beginning of a ventilatory cycle and always involving a return to a resting level after the termination of inhalation. Resting animals usually maintain a slightly positive pressure during apnoea.

In *Testudo graeca* respiration is driven by two sets of muscles, those moving the pectoral girdle (Mm. pectoralis and testocoracoideus) and those shifting the posterior muscular sling (Mm. transversus and obliquus abdominis). Contraction of the pectoral and transverse abdominal muscles shifts inward the muscular walls, reducing the volume of the visceral cavity and compressing the pulmonary contents. If the glottis is open gas will flow out as the pressure rises. In contrast, contraction of the testocoracoid and oblique abdominal muscles will shift the boundaries outward, increasing the volume of the visceral cavity and producing a subatmospheric pressure of the pulmonary contents. As the pressure falls gas will flow into the lung if the glottis is open. The magnitude of muscular activity reflects behavioural conditions and the volume of individual breaths may be enhanced or reduced by pressure changes due to the coincidental positions of limbs and neck (cf. Ireland and Gans, 1973).

IV. The Snapping Turtle

We know that the muscular arrangements associated with the pulmonary system vary extensively among aquatic turtles (Shah, 1962). The lung is but rarely attached to the shell and may be partially or completely loose atop the visceral mass. It may even be freed from diaphragmatic restraint. The diaphragm may be partially (*Chelydra*) or completely invested with muscles, and even the wall of the lung may bear a muscular sleeve (*Lissemys*). Since

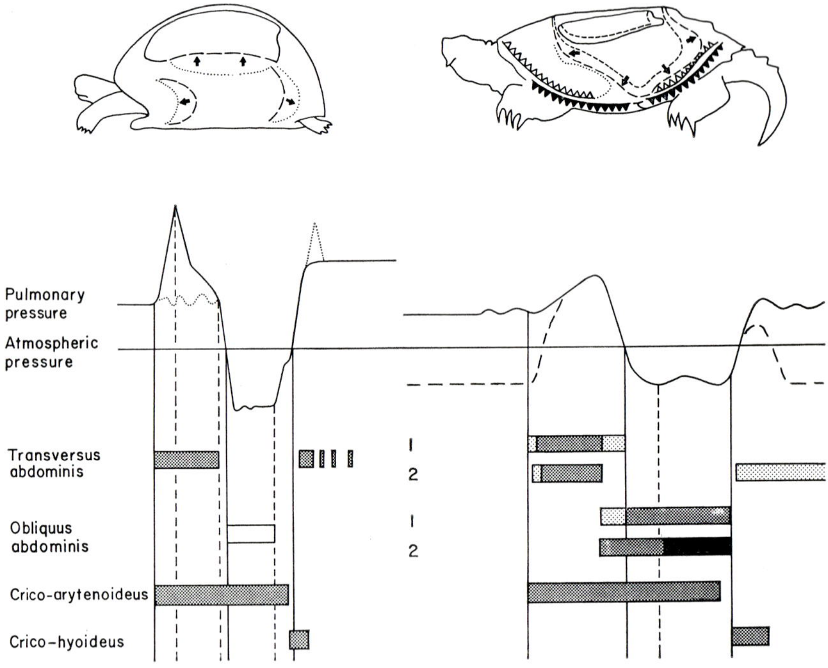

Fig. 1. Diagram comparing the three pressure phases of the ventilatory cycle for *Testudo* (left) and *Chelydra* (right). The broken line in the *Chelydra* pressure record gives the pressure curve when the shell is flooded, the solid line the curve when the animal is on land. Note that *Chelydra* shows more irregularities in muscle activity. These correlate with the amount of water in which the turtle is immersed. (Left side after Gans and Hughes, 1967; right side after Gaunt and Gans, 1969b).

we know nothing about the function of such systems there is no basis for a protracted discussion. Only for the North American snapping turtle, *Chelydra serpentina*, has the ventilatory mechanism been defined by electromyography (Gaunt and Gans, 1969b). It has also been observed that a swimming *Pseudemys* adjusts the volume to which the lungs may inflate, establishing neutral buoyancy by filling the bladder with water and thereby adjusting for the bulk of the visceral mass (Jackson, 1971).

Chelydra is a fairly large turtle. The adults used in the experiments weighed between 500 g and 10 kg. The species has a relatively flat shell that is widely open ventrally, and has large apertures that facilitate free movement of the limbs lateral to the quadruradiate plastron. While this effective predating and scavenging turtle is thoroughly aquatic, it often crosses land between bodies of water. Walking with the shell raised clear of the ground, it covers considerable distances and is markedly more pugnacious on land than in the water.

Like many aquatic turtles *Chelydra* will stay submerged and breathe by extending its neck to the surface. Depending on the depth at which the turtle rests while breathing, its lungs will be subject to 10 cm or more of water pressure which must be overcome in inhalation. Snapping turtles show a marked increase in heart rate as the neck approaches the surface and a further increase during and just after each breath.

The underwater breathing then has analogies to human snorkeling (breathing through a tube) with some complicating factors. As the neck is extended to reach the surface and as it is withdrawn afterwards, the volume of the visceral cavity changes. Retraction also affects the final volume occupied by the lungs, both because of the displacement factor and because every change in the vertical displacement of the gas filled space changes the hydrostatic pressure and hence the relative volume. This should potentially affect the amount of air to be inhaled. As snapping turtles may snorkel while submerged at different depths they will obviously derive a selective advantage from being able to predict, while breathing, the pressure imposed on, or rather the volume occupied by, the lung during apnoea.

The first item noted in experimental analysis is that snapping turtles will, on land, have a slightly negative pulmonary pressure during apnoea. As soon as the level of immersion reaches halfway up the animal's flanks the apnoeic pressure becomes positive and then rises proportional to the depth of immersion. The slightly different structural pattern produces correlated differences in activity from those seen in *Testudo*. Thus exhalation is driven by the diaphragmatic and transverse abdominal muscles and inhalation by the testocoracoid, testoscapular and oblique abdominal muscles. Rotation of the tilted shoulder girdle contributes less to changes in pulmonary volume. The muscular activity also reflects the differences in apnoeic pressure. Each exhalatory muscle (in terms of the strength of the EMG signal) fires most strongly when the animal rests on land and each inhalatory muscle equivalently shows strongest activity when the animal is snorkeling from the deepest immersion. The effect is very clear, and the transition occurs when the water level touches the lateral apertures of the shell (Fig. 2).

Clearly the snapping turtle permits the weight of its viscera to draw air into the lungs when on land. It lets the hydrostatic pressure deflate the lungs, producing passive exhalation when it is snorkeling. When a snapper is disturbed or excited it may fire some normally silent muscles at a variable level. In a resting animal muscular effort is much reduced and concentrated in those muscles for which there is no functional substitute. The shaded envelope in Fig. 3 includes all but three individual breaths (shown as dots) of animals subjected to a slow change of water level; the lines connect the

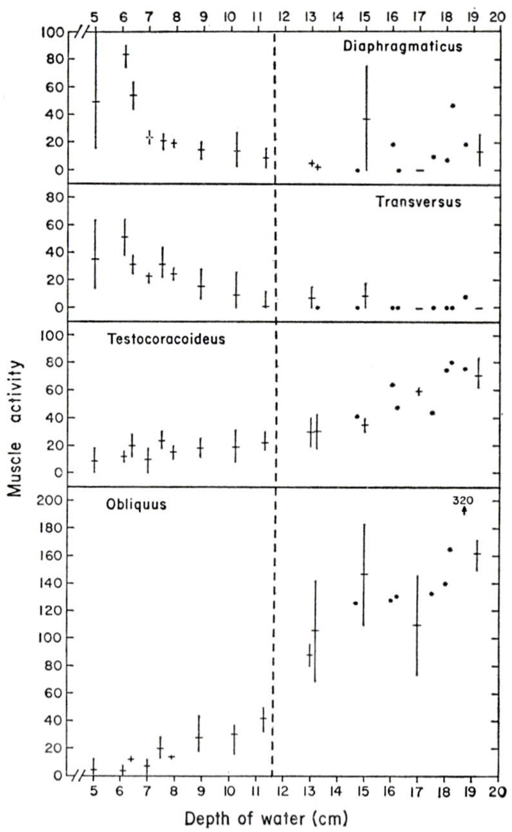

FIG. 2. *Chelydra*. Means and ranges of activity recorded from an electrode in each of the four major respiratory muscles of a single specimen as the depth of water was increased. Units of the Y-axis are arbitrary but equivalent for all electrodes. The dotted line indicates carapacial depth. Note that in excited animals the muscles may fire more than required to empty or fill the lung. (After Gaunt and Gans, 1969b. Courtesy of *Journal of Morphology*.)

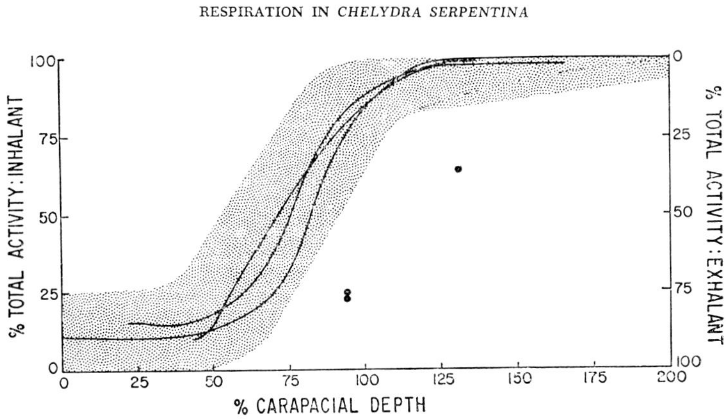

FIG. 3. *Chelydra.* Relative distribution of activity between exhalant and inhalant muscles with change of depth. The solid lines are drawn through the mean values for each depth for each of three turtles. The stippled area encloses all but three (shown as dots) of the approximately 100 individual observations. (After Gaunt and Gans, 1969b. Courtesy of *Journal of Morphology.*)

mean values of electromyographic activity for at least five breaths per depth. Snapping turtles clearly recover some of the extra energy required to overcome visceral drag or hydrostatic pressure during one phase of ventilation by utilizing those forces to replace muscular activity in the opposite phase.

It is as yet difficult to quantify for purposes of comparison EMG signals taken by different electrodes. Consequently it proved impossible to plot the absolute level of activity in all of the inhalatory and exhalatory muscles as a function of depth. However it is suggestive that the sum of the different signals from ventilatory muscles (amplified to approximately equal height peaks) was always least when the open portions of the shell were just flooded; the energy recovery is hence imperfect.

Snapping turtles were also exposed alternately to gradually rising and falling water levels. The data suggest that they do not determine the extent of pulmonary filling from an immediate detection of depth, but rather predict it based on the pressure at the end of the last five breaths. Consequently one sees a hysteresis effect (Fig. 4) with the animal continuously underestimating the depth when the water is rising and overestimating it when it is falling.

The electromyographic analysis provided another set of suggestive data. The wide, slightly buttressed shell apparently deforms significantly even during each breath. This seemingly causes a need for bracing by accessory elements of the frame such as the pelvis. The fan-shaped pelvic protrahent and retrahent muscles consequently fire in concert with the ventilatory muscles, but theirs is a stabilizing rather than a ventilatory action (Fig. 5). They hold

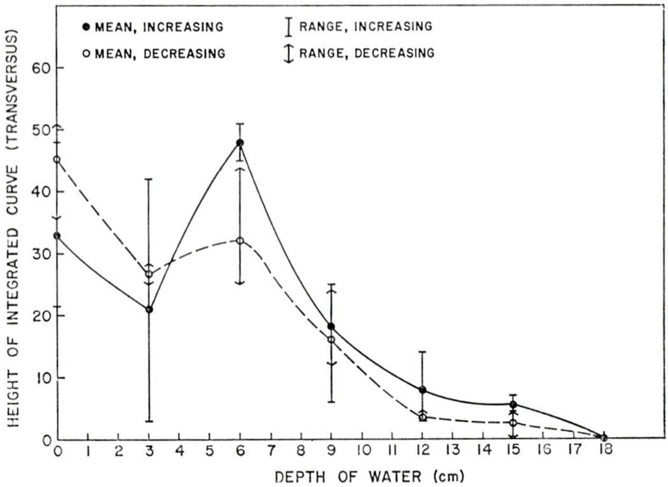

FIG. 4. *Chelydra.* Plots of the height of the integrated record for the *M. transversus abdominis* for an animal that was first submerged (solid line), then exposed (dashed line). The hysteresis observed is typical for all such records of respiratory muscles. The pattern of initial low values (increasing curve) followed by a sudden increase (as the depth of the water becomes noticeable) is common in animals that have been out of water long enough to have adopted a pattern of frequent, shallow ventilations. The carapacial depth of this animal was about 14 cm. (After Gaunt and Gans, 1969b. Courtesy of *Journal of Morphology*).

and maintain the structural patency of the frame against which the forces induced by the ventilatory muscles may act.

V. The Crocodilian Pattern

The 21 closely related species of crocodilians represent the survivors of a once successful group of amphibious predators now universally endangered by destruction of their habitat and greed for their skins. The general arrangement of the crocodilian air passages and lungs differs markedly from that of turtles, but both groups share the characteristic of muscular action against a flexible frame (Fig. 6). The lungs fill the pleural cavity encompassed by the movable rib cage. The space they occupy is posteriorly closed by a membrane, the centre of which is integral with the liver. The straps of the diaphragmatic muscle connect the liver to the pelvis and their contraction increases the volume of the pleural cavity. Activity of the transverse abdominal muscle squeezes the visceral mass and forces the liver anteriorly, decreasing the volume of the pleural cavity. Action of the superficial intercostals folds the ribs posteriorly, slightly reducing the pleural volume, while activity of the deep intercostals rotates the ribs anteriorly and thereby increases the volume.

The trachea terminates anteriorly at the closable glottis, which lies on the

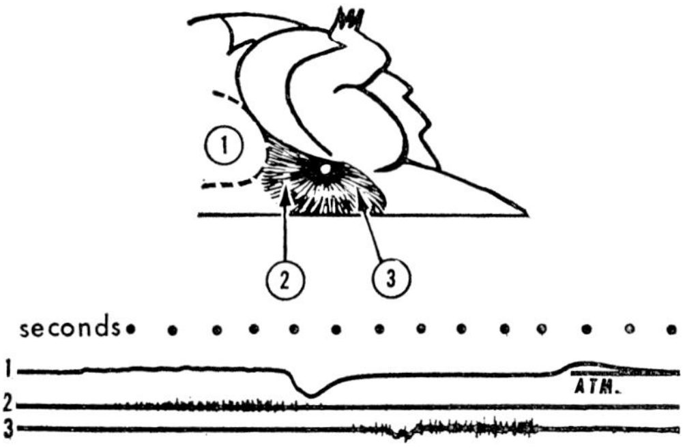

FIG. 5. *Chelydra*. Simultaneous recordings of intrapulmonary pressures (1) and activity in the *Mm. attrahens pelvim* (2) and *retrahens pelvim* (3). This single fan-shaped muscle thus shows differential activity, serving mainly to stabilize the pelvic girdle, in concert with the pulmonary pressure changes. However, these do not facilitate gas flow but instead stabilize the shell.

floor of the posterior pharyngeal cavity where it is supported on a wide hyoid plate (Fig. 7). This space is effectively separated from the anterior part of the mouth by the overlapping pharyngeal folds. The internal nares open at the top of a dome in the roof of the posterior pharyngeal chamber; the elongate narial canals run from here to the dorsal tip of the snout where they end as valved external nostrils. The olfactory chambers branch off this set of tubes and olfactory sampling clearly requires gas flow within the narial canals. This is in good agreement with the report of Huggins, Parsons and Pena (1971), who were able to show electrical activity in the olfactory bulb coincident with EMG signals from the narial dilators.

The floor of the posterior buccal cavity may be raised and lowered by

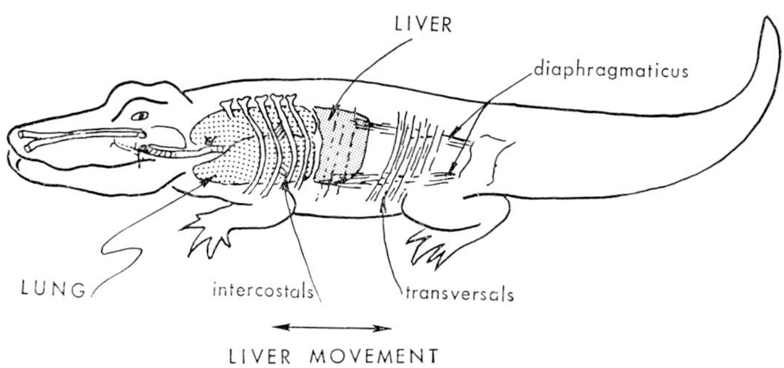

FIG. 6. *Caiman*. Sketch to show the major structures and the muscles tending to influence filling and emptying of the lung. (Modified from Gans, 1970).

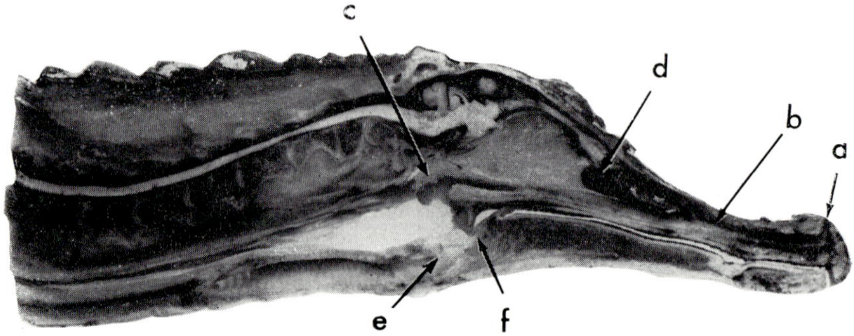

FIG. 7. *Caiman crocodilus*. Photograph of a longitudinal section through the head of a juvenile to show the external naris (a), narial canal (b), internal naris (c), olfactory chamber (d), larynx (e), and buccal division (f).

action of lateral muscles on the wide hyoid plate embedded in its floor. Naifeh, Huggins, Hoff, Hugg and Norton (1970) showed, and Gans and Clark (1975) confirm, that the oscillations of the buccal space are independent of and never associated with the thoracic oscillations resulting from ventilatory cycles. It has generally been assumed that the buccal space stayed patent with the gas crossing it during ventilation. However, this became unlikely when continuous gas analysis (Gans and Clark, 1975) showed no changes in the P_{O_2} and P_{CO_2} of its contents coincident with ventilation.

Cinefluoroscopy of a spectacled caiman (*Caiman crocodilus*), the throat of which had been painted with barium-contrast medium, showed that the chamber is normally collapsed and that the glottis then lies within the cup of the internal nares. This removes the volume of the buccal cavity from the air passage and substantially reduces the respiratory dead space. It is unclear what happens when crocodilians gape while resting on land. The pharyngeal folds visibly flap in and out when an excited *Crocodilus palustris* is breathing with gaping mouth; does air then bypass the nasal passages and thus reduce respiratory dead space and breathing work? Certainly the narial passage, with a volume of less than 1 ml for a 1000 g animal breathing with a tidal volume of 12 ml, accounts for relatively little of the breathing work during rest.

VI. THE CROCODILIAN SUBMERGED

The crocodilian architecture is clearly adaptive to the aquatic aspects of their life. The elongate snout suits an aquatic predator and the elongate narial tubes allow the air intakes (and the eyes) to emerge without disclosing the remainder of the animal. The resultant increase in respiratory dead space then reflects the disadvantageous portion of a selective compromise (Gans, 1974). The flexibility of the system is less than that exhibited by snorkeling turtles as the crocodilian neck cannot be retracted. The position of the head will be determined by the surface of the water; the trunk may float in parallel to this or angle downward to bob when the animal first exhales slowly and then inhales. The caiman's position obviously affects intrapulmonary pressure and the work of breathing.

When a *Caiman crocodilus* rests on land, its ventilatory pressure ordinarily (1) first rises to a superapnoeic level, (2) then drops to the vicinity of atmospheric pressure, (3) then levels off in a "shoulder" of variable extent, then decreases sharply to subatmospheric remaining there for some five seconds, before (4) it starts to rise sharply and smoothly (without overshoot) to the apnoeic pressure (coincident with glottal closure). In contrast, when *Caiman* is submerged its ventilatory pressure will (1) lack the preliminary peak but (2, 3) drop immediately to subatmospheric to remain there for some ten seconds, before (4) rising sharply beyond the apnoeic pressure to which it slowly returns.

The muscular activity exactly reflects and explains the changes of the pulmonary pressure curve (Fig. 8).* The apnoeic pressure is supra-atmospheric

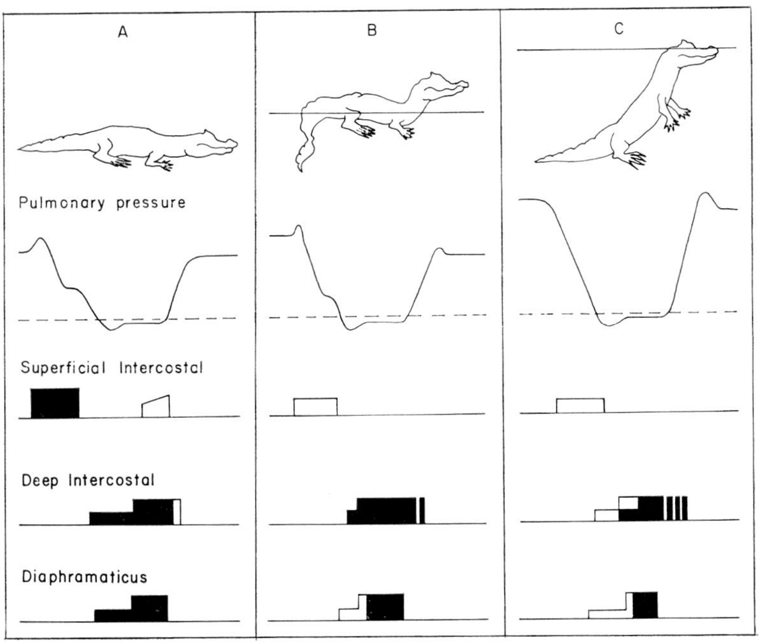

FIG. 8. *Caiman*. Sketches showing the modification of pulmonary pressures and some muscular activities as a function of the depth to which the animal is immersed. (After Gans and Clark, 1975).

* Subsequent studies (Gans and Clark, 1975) indicate that the activities are even more complex than here suggested; the general principles hold, but dorsal and ventral portions of intercostal muscle sheets may act antagonistically.

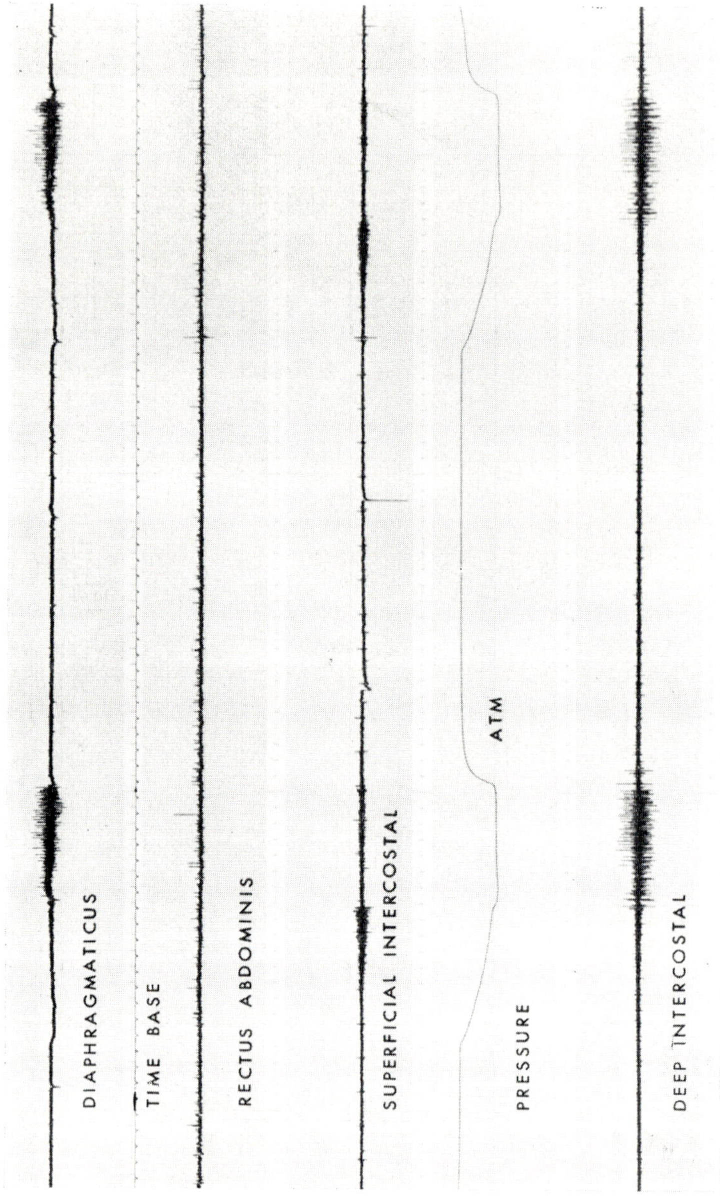

FIG. 9. *Caiman*. Photograph of original record showing the activity of the superficial and deep intercostals during a breath. The simultaneous activity of these two "antagonistic" muscles suggest that they serve a stabilizing rather than a ventilatory function. (After Gans and Clark, 1975).

because the animal's weight on land will be imposed on the sternum and tend to rotate the ribs posteriorly, thus reducing the volume of the pleural cavity. When a submerged animal breathes, hydrostatic pressure aids exhalation. The superficial intercostal and transverse abdominal muscles fire at the start of ventilation. They drive up the initial pressure if the airway is still partially obstructed by incomplete opening of glottal or narial valve and promote outflow as soon as the airway becomes patent. In the terrestrial situation they become inactive by the start of the shoulder; in the aquatic situation they are ordinarily so. These muscles then promote exhalation, and their silence reflects the substitution of hydrostatic pressure for muscular activity. As the magnitude of apnoeic intrapulmonary pressure is a function of the depth of the lung during breathing, the start of exhalation results only in a pressure drop without a coincident or preceding short-term rise.

The shoulder in the descending pressure curve and its disappearance with depth result from the gap between the periods of activity of the exhalatory and inhalatory muscles. When exhalation is driven hydrostatically the initial drop leads immediately to firing of the M. diaphragmaticus continuing the drop to subatmospheric; hence the shoulder in the descending pressure curve disappears.

With flooding of the body, the onset of firing in the M. diaphragmaticus is slower and maximum activity (for each breath) is delayed until the last third of the more prolonged subatmospheric period. The activity of the deep intercostal muscle is delayed even further and it also continues firing through the termination of the subatmospheric phase. No absolute measure of firing magnitude as a function of depth is yet available. Records from some of the superficial intercostals of excited animals breathing on land show that they exhibit a secondary and gradual rise of activity coincident with the peak firing of the deep intercostals during the end of inhalation. This requires a review of the hypothesis that these intercostals are completely antagonistic and serve entirely to power ventilation.

The rigidity of the "collapsed" rib cage and diaphragmatic sheet do not permit the lungs to collapse. Consequently opening of the glottal and nasal valves should permit the pressure of the pulmonary gases to drop to atmospheric, no matter what the depth of immersion. At the start of inhalation the ribs would then be in the most posteriorly rotated position, positioned there by the action of the superficial intercostals, enhanced by hydrostatic pressure when the animal is immersed and by its weight when it rests on land. Contraction of the diaphragmatic muscle pulls against the open end of the costal frame and reduces the pressure within the pleural cavity. Following thereupon, the action of the deep intercostals rotates the ribs anteriorly, indeed causes a rearrangement of their flexible connections, further increasing the volume of the pleural cavity (and the amount of air that may be inhaled). As the end of the inhalatory period approaches, maximum firing assures complete filling. In this context the continuing activity of the deep intercostals, beyond the instant of glottal closure and the simultaneous activity of the superficial intercostal, suggest that these muscles also have a secondary and stabilizing function (Fig. 9). The opposed muscles apparently fix the position of the ribs, maintaining the flexible costal frame rigid in the expanded position; the action of the diaphragmatic muscle may then aspirate. Some intercostal actions are thus analogous to those of the pelvic muscles of *Chelydra*.

VII. Summary

The preceding discussion should have documented the point that the mechanisms of aspiration breathing in turtles and crocodilians do not utilize homologous structures. They are clearly the result of two independent evolutionary paths, as distinct from each other as those leading to the air utilizing schemes in teleost fishes. Each set of breathing mechanisms furthermore represents part of a particular adaptive syndrome selected to cope with the adaptive zone in which these groups survive.

The several animals show different mechanisms of aspirating and discharging air, or reducing respiratory dead space and pumping work, and of utilizing the hydrostatic pressure imposed by the water to compensate for some of the extra work that must be done when breathing under water. In each case we have a group of terrestrial animals that has reinvaded the water. They could then utilize the resources available there, but had to avoid predation and maintain their reproductive potential. The ventilatory adaptations of turtles and crocodilians must be seen in this context. They do not reflect steps in some kind of continuous evolutionary sequence independent of the rest of the organism. In each case the animal has compromised. It has utilized an existing ventilatory pattern and modified it to solve the problem of gas exchange with only intermittent access to free air.

Acknowledgements

In preparing this review I am indebted to many friends. B. Clark, C. O. D. Diefenbach, J. A. Farber, A. S. Gaunt, G. M. Hughes and Fred Pohl helped me to perform these experiments carried out with the support of the U. S. National Science Foundation. B. Clark, P. Dejours, A. S. Gaunt, G. M. Hughes, C. Lenfant, T. S. Parsons and T. Scanlon commented on the manuscript. The responsibility for its final formulation remains my own.

References

Belkin, D. A. (1963). Anoxia: Tolerance in reptiles. *Science, N.Y.* **139**(3554), 492–493.

Belkin, D. A. (1968). Aquatic respiration and underwater survival of two freshwater turtle species. *Resp. Physiol.* **4**, 1–14.

Bentley, P. J. and Shield, J. W. (1973). Ventilation of toad lungs in the absence of the buccopharyngeal pump. *Nature, Lond.* **243**(5409), 538–539.

Dessauer, H. C. (1970). Blood chemistry of reptiles: Physiological and evolutionary aspects. *In* "Biology of the Reptilia". (Eds C. Gans and T. S. Parsons) Vol. 3, pp. 1–72. Academic Press, London, New York and San Francisco.

Eberle, G. (1937). Befall und Tötung von Erdkröten (*Bufo vulgaris*) durch Fliegenbrut. *Natur und Volk* **67(5)**, 250–254.

Gans, C. (1969). Comments on inertial feeding. *Copeia* 1969(4), 855–857.

Gans, C. (1971a). Respiration in early tetrapods – the frog is a red herring. *Evolution* **24(3)**, 740–751.

Gans, C. (1971b). Strategy and sequence in the evolution of the external gas exchangers of ectothermal vertebrates. *Forma et Functio* **3**, 66–104.

Gans, C. (1974). "Biomechanics: Approach to vertebrate biology." x+261 pp. J. P. Lippincott, Philadelphia.

Gans, C. and Clark, B. (1975). Studies on respiration of *Caiman crocodilus* (Crocodilia: Reptilia). (Submitted to *Respiration Physiology*.)

Gans, C. and Gaunt, A. S. (1969). Shell and physiology of turtles. *African Wildlife* **23(3)**, 197–206.

Gans, C. and Hughes, G. M. (1967). The mechanism of lung ventilation in the tortoise *Testudo graeca* Linné. *J. exp. Biol.* **47(1)**, 1–20.

Gans, C., de Jongh, H. J. and Farber, J. (1969). Bullfrog (*Rana catesbeiana*) ventilation: How does the frog breathe? *Science, N.Y.* **163**(3872), 1223–1225.

Gaunt, A. S. and Gans, C. (1969a). Diving bradycardia and withdrawal bradycardia in *Caiman crocodilus*. *Nature, Lond.* **223**, 207–208.

Gaunt, A. S. and Gans, C. (1969b). Mechanics of respiration in the snapping turtle, *Chelydra serpentina* Linné. *J. Morph.* **128(2)**, 195–228.

Huggins, S. E., Parsons, L. C. and Pena, R. V. (1968). Further study of the spontaneous electrical activity of the brain of *Caiman sclerops*: olfactory lobes. *Physiol. Zool.* **41**, 371–383.

Hughes, G. M. (1966). Evolution between air and water. Ciba Foundation Symposium on Development of Lung. (Eds de Reuck and Porter.) pp. 64–80. Churchill Ltd.

Ireland, L. C. and Gans, C. (1973). The adaptive significance of the flexible shell of the tortoise *Malacochersus tornieri*. *Anim. Behav.* **20(4)**, 778–781.

Jackson, D. C. (1971). Mechanical basis for lung volume variability in the turtle *Pseudemys scripta elegans*. *Am. J. Physiol.* **220(3)**, 754–758.

Naifeh, K. H., Huggins, S. E., Hoff, H. E., Hugg, T. W. and Norton, R. E. (1970). Respiratory patterns in crocodilian reptiles. *Resp. Physiol.* **9**, 31–42.

Parsons, T. S. (1959). Nasal anatomy and the phylogeny of reptiles. *Evolution* **13(2)**, 175–187.

Rahn, H., Rahn, K. B., Howell, B. J., Gans, C. and Tenney, S. M. (1971). Air breathing of the garfish (*Lepisosteus osseus*). *Resp. Physiol.* **11**, 285–307.

Rankin, J. and Dempsey, J. A. (1967). Respiratory muscles and the mechanism of breathing. *Am. J. Phys. Med.* **46(1)**, 198–244.

Romer, A. S. (1972). Skin breathing – primary or secondary? *Resp. Physiol.* **14(1–2)**, 183–192.

Rosenberg, H. I. (1973). Functional anatomy of pulmonary ventilation in the garter snake, *Thamnophis elegans*. *J. Morph.* **140(2)**, 171–184.

Schmidt-Nielsen, K., Kanwisher, J., Lasiewski, R. C., Cohn, J. E. and Breta, W. L. (1969). Temperature regulation and respiration in the ostrich. *Condor*, **71**, 341–352.

Shah, R. V. (1962). A comparative study of the respiratory muscles in Chelonia. *Breviora* (**161**), 1–16.

Smith, E. N., Allison, R. D. and Crowder, W. E. (1974). Bradycardia in a free ranging American alligator. *Copeia* 1974(3), 770–772.

Wood, S. C. and Lenfant, C. J. M. (1976). Respiration: Mechanics, control and gas exchange. *In* "Biology of the Reptilia." (Eds C. Gans and W. R. Dawson) Vol. 5. Academic Press, London, New York and San Francisco.

Respiration of Marine Snakes

HAROLD HEATWOLE and ROGER SEYMOUR*

Department of Zoology, University of New England, Armidale, New South Wales, Australia and *Department of Zoology, Monash University, Clayton, Victoria, Australia*

I. Introduction

Most species of marine snakes feed and carry out other vital activities on the sea bottom. The water in which they live, and hence the depth to which they dive, ranges from a few centimetres to about 100 m (Shuntov, 1971; Heatwole and Seymour, 1975a).† They ascend and descend nearly vertically at a rate of about 1 m per 3 sec. Thus a snake diving to 100 m would require 600 sec (10 min) just for its round trip to the surface to breathe. Consequently, long voluntary submergence times are to be expected. Although a variety of factors such as activity and high temperature reduce submergence times, and snakes often make short dives of only a few minutes or a few seconds, they commonly voluntarily submerge for from 5 to 30 min and occasionally remain under water for over an hour (Heatwole, 1975a). The maximum time recorded (in the laboratory) is about 2½ h (M. E. D. Webster, personal communication). Several reports of 6–8 h in the field have not been sufficiently well documented to be accepted without question.

Forcibly submerged snakes often do not survive more than about 5 h.

* This work was supported by National Science Foundation under grants NSF–GA35835, GA34948 and GD34462 to the Scripps Institution of Oceanography for operation of the Alpha Helix Research Program, and by the Australian Research Grant Committees through grants to both authors.

† *Pelamis platurus* is an exception. It is pelagic, primarily inhabiting the surface but diving as a means of thermoregulation and to escape rough surface water.

Survival times have, however, ranged from 21 min to 24 h depending on temperature and amount of dissolved oxygen (Graham, 1974; Dunson and Ehlert, 1971).

Deep dives and prolonged apnoea have special implications for the interrelated problems of respiration, circulation and buoyancy. Graham, Gee and Robison (1975) have discussed buoyancy of *Pelamis platurus*. The present chapter reviews recent work on oxygen consumption, heart rate, and gas transport properties of the blood. Although only a few of the more than 50 known species of marine snakes have been studied from the standpoint of diving physiology, they represent all three of the ophidian families occurring in the sea, i.e., Laticaudidae (sea kraits), Hydrophiidae (sea snakes)* and Acrochordidae (file snakes). Consequently, a broad outline of the physiological capabilities of these animals is beginning to emerge.

II. Oxygen Consumption

The oxygen consumption of sea snakes is similar to that of terrestrial snakes and hence depressed metabolism is not among the sea snake repertoire of diving adaptations. By contrast the marine file snake (*Acrochordus granulatus*) has a relatively low rate of oxygen utilization and its long submergence time may be related in part to its low oxygen needs (Heatwole and Seymour, 1975b). Activity results in elevated oxygen consumption in marine snakes; there is also diel variation in oxygen consumption of sea snakes, more oxygen being used in the day than under equivalent conditions of temperature and activity level at night (Heatwole and Seymour, 1975b).

Because of the relative impermeability to water and ions of sea snake skin (Dunson, personal communication) and its heavily-scaled nature, one would expect cutaneous gaseous exchange of marine snakes to be minimal. However, most species which have been studied to date have considerable oxygen uptake through the skin, in some species accounting for as much as 22% of the total exchange (Table I). *Hydrophis belcheri* was an apparent exception in that it had no measurable cutaneous oxygen uptake when inactive; however, when active it took up 10ml O_2/Kg/h. In sea snakes generally, activity resulted in an increase in cutaneous oxygen uptake of 60–143% of inactive values (Heatwole and Seymour, 1975a). Cutaneous oxygen uptake is not unique to marine, or even to aquatic snakes (Standaert and Johansen, 1974), although most of these have cutaneous uptakes many times greater than the single species of terrestrial snake which has been studied (Table I). Body size and its effect on relative surface area may influence cutaneous oxygen uptake as *Pelamis* had the highest absolute values and the smallest

* A taxonomic assessment at the family level is provided by Burger and Natsuno (1975).

TABLE I

Pulmonary and cutaneous oxygen uptake in inactive snakes

Species	Body weight (g)	Mean $\dot{V}_{O_2}$ (ml) Pulmonary	Cutaneous	Total	Mean Cutaneous Uptake (% of Total)	Authority
SEA SNAKES						
Aipysurus laevis	668	24	7	31	21·6	Heatwole and Seymour 1975b
Aipysurus duboisii	355	39	5	44	10·8	
Lapemis hardwickii	420	99	5	104	5·2	
Hydrophis belcheri	297	45	0	45	0	
Acalyptophis peronii	240	39	5	44	10·8	
Pelamis platurus	7–116	65	18	83	21·7	Graham, 1974
FILE SNAKES						
Acrochordus granulatus	326	13	2	15	13·3	Heatwole and Seymour, 1975b
Acrochordus javanicus	450–2950	8·7	0·7	9·4	7·4	Standaert and Johansen, 1974
TERRESTRIAL SNAKE						
Constrictor constrictor	1250–3400	22·1	0·6	22·7	2·6	Standaert and Johansen, 1974

body size of any species studied (Table I). Within the size range of the other species investigated, there was no consistent relation of the two variables.

The skin may be of even more importance in CO_2 exchange than in O_2 uptake (Standaert and Johansen, 1974; Heatwole and Seymour, 1975b). Graham (1974) found that in *Pelamis* loss of CO_2 into water takes place up to rates of 94% of the standard oxygen consumption.

III. Heart Rate

Heart rate differences between breathing at the surface and apnoea during diving occur widely among air-breathing vertebrate divers, including birds, mammals, reptiles and amphibians (Andersen, 1966). Among the reptiles this occurs in crocodilians, turtles, lizards and freshwater snakes (e.g. see Belkin, 1964; Berkson, 1966; Andersen, 1966; White and Ross, 1966; Bartholomew and Lasiewski, 1965; Johansen, 1959; Murdaugh and Jackson, 1962; Pough, 1973).

In marine snakes there is a marked increase in heart rate associated with breathing, followed by a more gradual decline after the breath. There are several patterns observed depending on level of activity and frequency of breathing (Fig. 1). When inactive snakes breathe at infrequent intervals, heart rate reaches a low, rather stable level during apnoea (designated as the diving heart rate). More frequent breathing in inactive snakes results in failure to achieve the diving heart rate because breathing tachycardia repeatedly occurs before there is adequate time for the diving heart rate to become established (Fig. 1). Oscillations are damped at night relative to the day as a result of a reduction in the level of breathing tachycardia by about 13·8% (Heatwole, 1975b).

Moderate activity, whether caused by spontaneous movement or induced by prodding the snake, results in slight, momentary elevations in heart rate during apnoea which seldom attain values as high as those prevalent at the peak of breathing tachycardia. Vigorous activity makes it difficult to ascertain when breathing occurs but the heart rate becomes elevated and erratic (Fig. 1).

The heart rate increment between inactivity and maximum activity, known as heart rate scope (for activity), is one measure of the ability of the heart to cope with the increased metabolic demands of exercise. Two animals were considered to have undergone maximal activity (vigorous attempts to escape from the tank). It was assumed that their heart rates reflected the combined effect of activity and frequent breathing; consequently the heart rate scope was calculated as the difference between breathing tachycardia during inactivity and heart rate at maximal activity. Heart rate scopes were 13·2 beats min^{-1} (increase of 47% above inactive level) for *Hydrophis elegans* and 5·9 beats min^{-1} (21% increase) for *Aipysurus laevis*. Maximum heart rate scopes for other reptiles are 25–43 beats min^{-1} (12 species of terrestrial lizards), 30 beats min^{-1} (2 species of terrestrial snakes), 40 beats min^{-1} (2 species of turtles) and 25 beats min^{-1} (the tuatara) (Gatten, 1974). It appears that values for sea snakes are lower than those for most reptiles. However, heart rate scope is temperature dependent (Licht, 1965); other temperatures may give higher values for sea snakes, although the present measurements were made at temperatures the snakes normally encounter in nature.

The magnitude of difference between breathing and diving heart rates in marine snakes was considerable. The former averaged from about 17 to 29 beats min^{-1} and the latter from about 5 to 15 beats min^{-1}, giving a mean reduction in heart rate during apnoea of about 39–75% (Table II). Some

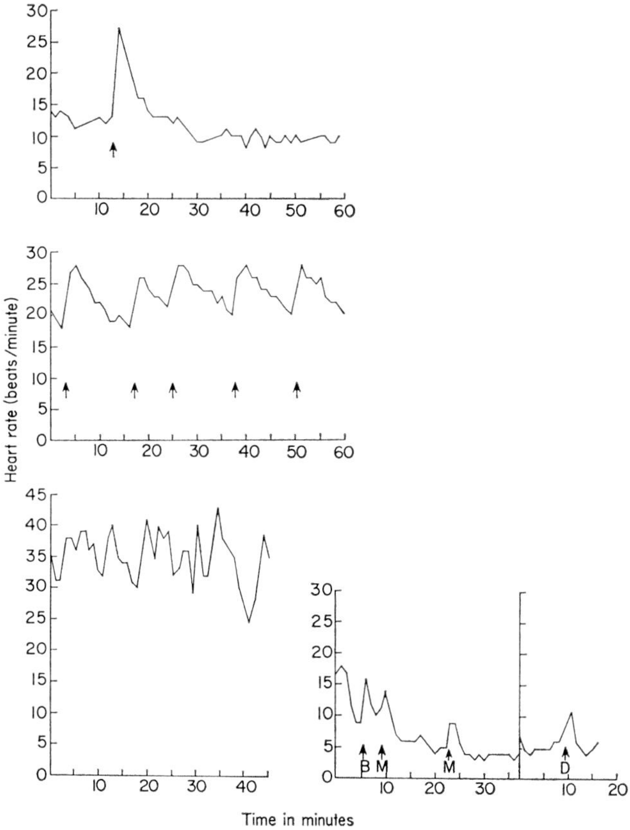

FIG. 1. Heart rates of marine snakes during breathing and apnoea. Top: *Hydrophis elegans*, inactive and breathing infrequently. Centre: *Hydrophis elegans*, inactive and breathing frequently. Bottom left: *Aipysurus laevis*, vigorously active. Bottom right: *Acrochordus granulatus*, occasionally active. Unlabelled arrows and arrows with B indicate breaths, arrows with M indicate spontaneous movements, and arrow with D indicates snake disturbed by prodding.

TABLE II

Characteristics of heart beat in inactive sea snakes and the file snake. N values refer to number of animals used (A), number of minutes of breathing heart rate in sample (B), and number of minutes of diving heart rate in sample (D).

	N	Breathing Heart Rate (BHR) in beats min^{-1}		Diving Heart Rate (DHR) in beats min^{-1}			BHR-DHR		
Species	A, B, D	Mean	Range	Mean	Range	BHR-DHR	in % of BHR	BHR in % of DHR	°C
FILE SNAKE									
Acrochordus granulatus	5, 30, 201	19·2	14–29	4·8	2–9	14·4	75	400	21·5 ± 1·6
SEA KRAIT									
Laticauda colubrina	3, 6, 60	28·8	24–33	9·8	8–12	19·0	66·0	294	21·5 ± 3·0
SEA SNAKES									
Aipysurus laevis	4, 4, 71	21·2	20–24	13·0	10–18	8·2	38·7	163	20·0
Aipysurus duboisii	1, 2, 0	27·0	22–32	—	—	—	—	—	26·0
Lapemis hardwickii	1, 3, 0	16·6	16–18	—	—	—	—	—	19·0
Hydrophis elegans	2, 19, 57	28·1	23–31	14·7	8–22	13·4	47·7	191	22·7 ± 0·3

other diving reptiles have reductions in excess of 95% (see review by Heatwole, 1975b). However, most species with such high values were forcibly submerged, a procedure which increases the level of breathing tachycardia and decreases the diving heart rate, thereby enhancing the proportionate difference between the two (Andersen, 1961; Gaunt and Gans, 1969).

Breathing tachycardia anticipates the actual ventilatory cycle. Even in floating animals which do not undergo any locomotion or other activity prior to breathing, the heart rate speeds up as early as 40 sec prior to initiation of ventilation; immediately after the breathing cycle has ended heart rate again slows down (Fig. 2). A similar pattern occurs in the Javan file snake

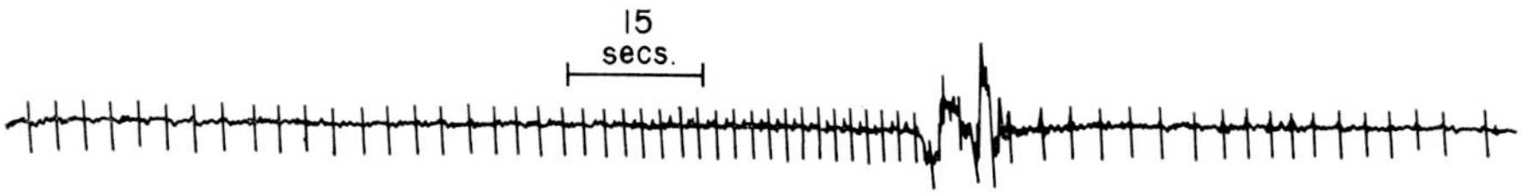

FIG. 2. ECG of a *Hydrophis elegans* during breathing and apnoea. The disturbance to the trace at the right of centre represents a breath.

(Pough, 1973) and in some freshwater turtles (Belkin, 1964). Its significance is probably that the conditions favouring maximum gaseous exchange in the lung are in operation by the time air renewal takes place, a factor of importance to a species which breathes briefly and then immediately undergoes rather extensive apnoea.

Thus far the term "diving bradycardia" has been deliberately avoided in reference to the lowered heart rate during apnoea in marine snakes. In their case it is more accurate to speak of the breathing heart rate as a breathing tachycardia and consider the diving heart rate as the normal base level, not as true diving bradycardia. This is not merely a semantic problem. Typically, diving bradycardia (1) persists even during vigorous underwater activity and (2) is associated with vasoconstriction of peripheral and splanchnic vessels and contraction of an oxygen debt in much of the body, with the consequent conservation of oxygen supplies for the central nervous system and heart (Andersen, 1966). The term diving bradycardia should probably be reserved for lowered heart rate only when these two conditions are met.

Neither condition applies to the heart rate changes of marine snakes which are reported here. There is clearly an activity effect during apnoea (Fig. 1) and as will be shown below, during dives of the length involved in this study sea snake metabolism is almost completely aerobic. Indeed, peripheral vasoconstriction would appear to be disadvantageous to sea snakes as it would reduce blood flow to the skin and hence adversely affect cutaneous gas exchange. Some other diving reptiles show characteristics during submergence similar to those of marine snakes. For example, the sea turtle, *Chelonia mydas*, has diving heart rates which are sensitive to activity and the build up of lactate during dives is slight (Berkson, 1966). By contrast, the alligator has characteristics of true diving bradycardia (Andersen, 1961).

We do not intend to imply that true diving bradycardia *never* occurs in marine snakes. In fact, Webster (personal communication) has data indicating that during extensively prolonged voluntary submergences, a heart rate level lower than the diving heart rate occurs. It is possible that further investigation will show this to be true bradycardia. If so, there may be three heart rate levels in marine snakes, the diving heart rate (base level) from which there are departures upward during breathing (breathing tachycardia) and downward with prolonged submersion (diving bradycardia).

The important point to be made here is that among vertebrates there are two functionally distinct heart rate responses, (1) breathing tachycardia which facilitates equilibration of blood and tissue oxygen and carbon dioxide tensions with those of alveolar gas during brief breathing periods, and (2) diving bradycardia which is associated with establishment of peripheral oxygen debts and the husbanding of oxygen supplies for more vital centres. A terminological distinction between these two phenomena would appear essential. Only the former has been conclusively demonstrated to occur in sea snakes; the latter remains as a possibility.

IV. Respiratory Properties of the Blood

Because of the diving habits of sea snakes, it might be expected that they would show distinctive respiratory adaptations reminiscent of those occurring in diving mammals (Lenfant, 1969). However, preliminary measurements of the respiratory properties of the blood, including haemoglobin concentration, buffering capacity, and plasma bicarbonate, in several species of Indo-Pacific sea snakes showed no significant differences separating them from terrestrial squamates (Seymour and Webster, 1975). Since that initial study, a more rigorous comparison has been made between the banded sea krait, *Laticauda colubrina*, and the black snake, *Pseudechis porphyriacus* (Seymour, in preparation). These similarly sized snakes are in closely related families or subfamilies (depending on one's taxonomic predilections) but differ markedly in life style. The results confirmed the similarity between sea kraits and their more terrestrial relatives (Table III). Specifically, haematocrit, Hb concentration, and O_2 capacity were not significantly different. Blood volume also does not appear to be greater in sea snakes than in terrestrial snakes (W. Dunson, personal communication). These data, together with the observation of a comparatively low O_2 saturation in the arterial blood, indicate that sea snake blood is not adapted for increased O_2 storage.

Many vertebrate divers exhibit intense anaerobic metabolism during forced and voluntary dives so it has been generally assumed that a high blood buffering capacity is advantageous to reduce pH change. The reported values of reptile blood buffering are extremely variable, ranging from −9 to about −300 mmol/pH (Bennett, 1973; Seymour and Webster, 1975). Most reptiles show values between −9 and −40, and sea snakes appear with this class (Seymour and Webster, 1975; Seymour, in preparation). Similarly, the bicarbonate–carbonic acid buffering system, evaluated with calculations of plasma HCO_3^-, is equally represented in sea and land snakes. Several species of both sea and land snakes show relatively low plasma HCO_3^- (between about 7 and 14 mM/l; see Dessauer, 1970, and Seymour and Webster, 1975), but sea kraits and black snakes show relatively high values (Table III). A possible reason for the general failure to find a consistent

TABLE III

Mean respiratory properties of the blood of a sea snake, Laticauda colubrina, *and a terrestrial elapid,* Pseudechis porphyriacus. *N = Number of animals. Data from Seymour (in preparation)*

	L. colubrina			*P. porphyriacus*		
	$\bar{x}$	±S.D.	N	$\bar{x}$	±S.D.	N
Haematocrit (%)	24·3	3·6	7	28·2	6·4	6
Haemoglobin (g%)	6·15	1·23	7	8·61	1·73	6
O_2 Capacity (ml %)	9·3	1·6	7	10·9	1·9	6
Buffering capacity (mmol HCO_3^-/l·pH)	−19·4	10·8	6	−17·6	6·0	5
pH*	7·52	0·05	6	7·68	0·06	10
P_{CO_2} (mmHg)*	31·8	2·3	5	19·0	4·5	9
HCO_3^- (mM/l)	30·8	—	—	26·6	—	—

*Values from various vessels during the resting, air-breathing state

relationship between blood buffering and diving behaviour in reptiles lies in the regular occurrence of anaerobic metabolism in totally terrestrial reptiles which are tolerant to severe acid-base disturbances (Bennett, 1973). In any case, several lines of evidence (discussed below) indicate that natural voluntary diving in sea snakes is accomplished largely aerobically. Moreover, because CO_2 can be readily removed cutaneously, severe acidosis is not normally a danger. Nevertheless, during intense activity, a lag in the O_2 supply system can cause transient anaerobiosis in both sea snakes and terrestrial reptiles.

At Ashmore Reef, O_2 equilibrium curves were prepared with blood from seven species of sea snakes (Seymour and Webster, 1975). Again, the Hb–O_2 affinity in most species was similar to that in terrestrial reptiles, but a pattern emerged which is consistent with relationships discussed for snakes by Johansen and Lenfant (1972). Most sea snakes having a relatively low affinity are characterized by moderate activity when seeking active prey. In these cases, the O_2 transport function of the blood is emphasized at the expense of O_2 storage. On the other hand, the turtle-headed sea snake, *Emydocephalus annulatus*, showed a considerably higher Hb–O_2 affinity than the others; a P_{50} of about 19 mmHg compared to a range of 25 to 40 mmHg in the others. It may be significant that this snake eats only fish eggs (Voris, 1966) and is presumably less reliant on intense activity than are the other species. Hb–O_2 affinity thus appears to be related more closely to activity patterns than to diving, *per se*, in snakes.

In view of the proposal by a number of authors that a high Bohr effect is advantageous for vertebrate divers (Andersen, 1966; Johansen, Lenfant and Grigg, 1966; Lenfant, Johansen, Petersen and Schmidt-Nielsen, 1970), it was surprising to find that a significant Bohr shift could not be demonstrated in sea snakes (Seymour and Webster, 1975). These findings doubtless resulted from the poor resolution of the spectrophotometric technique used to determine O_2 saturation in whole blood so the more precise volumetric technique of Tucker (1967) was later employed for *Laticauda colubrina* and *Pseudechis porphyriacus* blood (Seymour, in preparation). A Bohr shift appears in *Laticauda* but it is quite small compared with the shift in *Pseudechis* (Fig. 3)

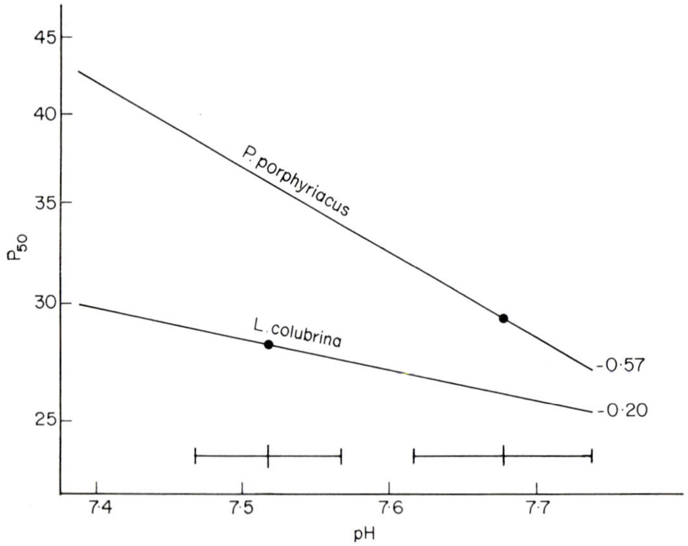

FIG. 3. Relationship between P_{50} and pH in whole blood of the sea krait, *Laticauda colubrina*, and the terrestrial elapid, *Pseudechis porphyriacus*. Means ($\pm$ SD) of arterial pH and Bohr effect ($\triangle \log P_{50}/\triangle pH$) are indicated. Data from Seymour (in preparation).

and other reptiles (Wood and Moberly, 1970; Wood and Johansen, 1974). A large Bohr shift may facilitate O_2 transport at higher P_{O_2} during a dive, but it also may be a liability by inhibiting loading in acidotic snakes following dives involving intense activity.

V. Changes in Level of O_2 and CO_2 in the Lung and Blood During Dives

To supplement the *in vitro* analyses of the blood properties, *in vivo* changes occurring in the aortic blood and the lung were examined during voluntary breath-holding and forced dives of *Lapemis hardwickii* and *Acalyptophis peronii* (Seymour and Webster, 1975). The results of voluntary dives in shallow aquaria are qualitatively similar in these snakes to published data obtained for other reptiles diving spontaneously (Lenfant *et al.*, 1970; Johansen and Lenfant, 1972; Wood and Johansen, 1974). However, a distinguishing feature of the sea snake data is that the O_2 saturation of aortic blood never exceeded 80% despite high lung P_{O_2}.

During discrete episodes of voluntary apnoea, Pa_{O_2} dropped from about 50 to about 30 mmHg which, with the decrease in $P_{A_{O_2}}$ from about 100 to 50 mmHg, indicated that about half of the O_2 available just after a breath, had been used. Breathing appeared to be in response to low P_{O_2} in this case because P_{CO_2} remained quite constant, presumably since cutaneous CO_2 loss can be high (Graham, 1974). Blood pH decreased less than 0·1 units and lactate concentration remained low, never exceeding 15 mg%. It is apparent, therefore, that voluntary diving occurred aerobically under laboratory conditions. That the same usually occurs at sea may be inferred from rates of O_2 uptake from the lung, and field observations of voluntary submergence times. For example, about half of the O_2 reserve in *Acalyptophis peronii* was used during dives averaging 17 min. *Lapemis hardwickii* used about half of its reserve in roughly 8 min. At similar temperatures, *Aipysurus laevis*, *A. duboisii*, and *Emydocephalus annulatus* commonly remain submerged for 5 to 30 min in the field (Heatwole, 1975). Moreover, long recovery times following forced dives involving anaerobic metabolism are inconsistent with field observations of active snakes diving immediately after breathing (Heatwole, 1975). Exceptionally long voluntary submergences of inactive snakes lasting up to 2·5 h are coupled with severe reduction in O_2 reserve but the level of anaerobiosis is not known (Seymour and Webster, 1975). Possible alterations in metabolic rate and cutaneous O_2 uptake should be evaluated during these extended dives.

When voluntary dives were forcibly extended by preventing otherwise unrestrained snakes from reaching the surface for 20 to 30 min, *Lapemis hardwickii* became intensely active. This panic quickly exhausted O_2 reserves and increased anaerobic metabolism as evidenced by blood lactate buildup, sometimes occurring before the dive was terminated. As anaerobiosis is a common phenomenon in well ventilated terrestrial reptiles during intense activity (Bennett, 1973), it is expected that the same should occur in sea snakes. However, the effects of activity may be exacerbated by depletion of O_2 reserves during forced dives. Panic seems characteristic of non-diving animals (Gaunt and Gans, 1969). Indeed, all information we have suggests that sea snakes rely highly on aerobic metabolism. They may become fatigued at relatively low blood lactate levels compared to other reptiles (Seymour and Webster, 1975) and some may die after only about one-half hour in water of low P_{O_2} (Graham, 1974).

VI. Gas Transport and Blood Circulation

Preliminary investigations on two species revealed quite large P_{O_2} differences between the lung and aortic blood which indicate the presence of considerable shunting of venous blood past the lung (Seymour and Webster, 1975). Calculations based on A-a and a-v O_2 differences showed that the

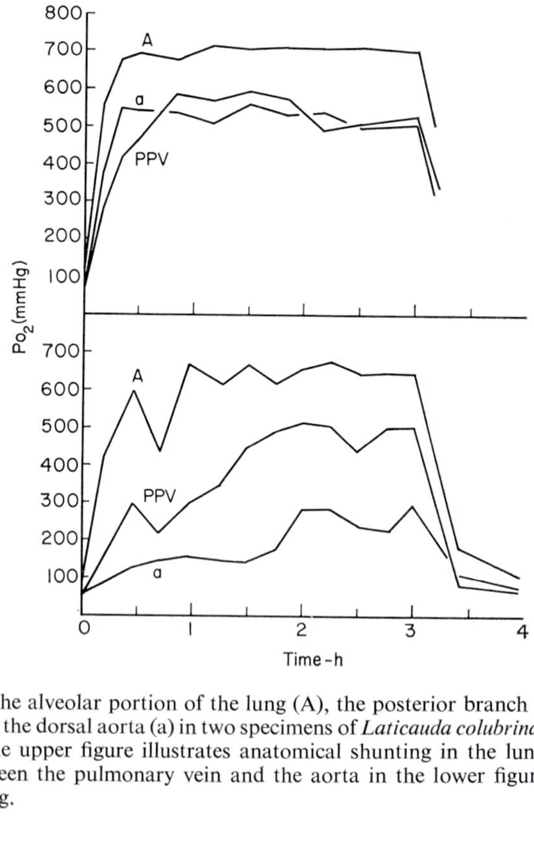

Fig. 4. P_{O_2} in the alveolar portion of the lung (A), the posterior branch of the pulmonary vein (PPV), and the dorsal aorta (a) in two specimens of *Laticauda colubrina* during breathing of pure O_2. The upper figure illustrates anatomical shunting in the lung only, while the difference between the pulmonary vein and the aorta in the lower figure indicates intracardiac shunting.

arterial blood was composed of at least 50% venous blood. Intraventricular shunts commonly develop in forcibly submerged reptiles when there is little or no O_2 remaining in the lung but sea snakes exhibited low Pa_{O_2} even when the lung O_2 content was high. Regardless of where the shunting occurs, the phenomenon may have adaptive value by keeping Pa_{O_2} relatively low when hydrostatic pressure during diving elevates PA_{O_2}. This tends to (1) minimize the danger of losing O_2 through the skin to the sea in deep water and (2) enhance cutaneous O_2 uptake in shallow water. Moreover, since N_2 also can presumably pass transcutaneously, the shunt dilutes the N_2-rich blood, equilibrating in the lung at depth, with relatively N_2-poor blood returning from the periphery. The Pa_{N_2}, therefore, and hence the tissue P_{N_2} may remain substantially lower than that in the lung. This reduces the danger of N_2 bubble formation during decompression after deep or repeated diving (Seymour, 1974).

To determine where the shunt actually occurs, samples from the pulmonary veins and other major blood vessels in *Laticauda colubrina* have been recently obtained via unobstructive catheters. The data will appear in greater detail elsewhere (Seymour, in preparation), but suffice it here to present part of the results which are indicative of the patterns of shunting in this species. When breathing air, the snakes showed quite variable O_2 differences across the lung; differences ranging from a few mmHg to over 90 mmHg appeared, probably representing nonuniformities in the ventilation/perfusion ratio in various regions of the elongate lung. However, anatomical shunts in the lung were indicated by significant differences during breathing of pure O_2. Differences of over 100 mmHg persisted even in the face of CO_2-induced hyperventilation. During the O_2 breathing experiments, aortic P_{O_2} could be either similar or dissimilar to that in the pulmonary vein (Fig. 4). In the latter case, intraventricular shunting must be occurring. It appears, therefore, that the disparity between lung and aortic blood P_{O_2} in sea snakes may be due to both pulmonary and intracardiac shunting. The patterns of shunting are highly variable and their control and physiological correlates remain unclear.

References

Andersen, H. T. (1961). Physiological adjustments to prolonged diving in the American Alligator, *Alligator mississippiensis*. *Acta Physiol. Scand.* **53**, 23–45.

Andersen, H. T. (1966). Physiological adaptations in diving vertebrates. *Physiol. Rev.* **46**, 212–243.

Bartholomew, G. A. and Lasiewski, R. C. (1965). Heating and cooling rates, heart rate and simulated diving in the Galapagos marine iguana. *Comp. Biochem. Physiol.* **16**, 573–583.

Belkin, D. A. (1964). Variations in heart rate during voluntary diving in the turtle *Pseudemys concinna*. *Copeia* (1964), 321–330.

Bennett, A. F. (1972). The effect of activity on oxygen consumption, oxygen debt, and heart rate in the lizards *Varanus gouldii* and *Sauromalus hispidus*. *J. Comp. Physiol.* **79**, 259–280.

Berkson, H. (1966). Physiological adjustments to prolonged diving in the Pacific green turtle (*Chelonia mydas agassizii*). *Comp. Biochem. Physiol.* **18**, 101–119.

Burger, W. L. and Natsuno, T. (1974). A new genus for the Arafura Smooth Seasnake and redefinitions of other seasnake genera. The Snake **6**, 61–75.

Dessauer, H. C. (1970). Blood chemistry of reptiles. *In* "Biology of the Reptilia" (Eds C. Gans and T. S. Parsons), Vol. III, pp. 1–72. Academic Press, New York, London and San Francisco.

Dunson, W. and Ehlert, E. (1971). Effects of temperature, salinity, and surface water flow on distribution of the sea snake *Pelamis*. *Limnol. Oceanog.* **16**, 845–853.

Gatten, R. E., Jr. (1974). Effects of temperature and activity on aerobic and anaerobic metabolism and heart rate in the turtles *Pseudemys scripta* and *Terrapene ornata*. *Comp. Biochem. Physiol.* **48A**, 619–648.

Gaunt, A. S. and Gans, C. (1969). Diving bradycardia and withdrawal bradycardia in *Caiman crocodilus*. *Nature. Lond.* **223**, 207–208.

Graham, J. B. (1974). Aquatic respiration in the sea snake *Pelamis platurus*. *Resp. Physiol.* **21**, 1–7.

Graham, J. B., Gee, J. H. and Robison, F. S. (1975). Hydrostatic and gas exchange functions of the lung of the sea snake *Pelamis platurus*. *Comp. Biochem. Physiol.* (In press).

Heatwole, H. (1975a). Voluntary submergence times in marine snakes. (In preparation).

Heatwole, H. (1975b). Heart rate changes during breathing and apnea in marine snakes. (In preparation).

Heatwole, H. and Seymour, R. (1975a). *In* "Biology of Sea Snakes" (Ed. W. Dunson), University Park Press, Baltimore. (In press).

Heatwole, H. and Seymour, R. S. (1975b). Pulmonary and cutaneous oxygen uptake in sea snakes and a file snake. *Comp. Biochem. Physiol.* **51A,** 399–405.

Johansen, K. (1959). Heart activity during experimental diving of snakes. *Am. J. Physiol.* **197**, 604–606.

Johansen, K. and Lenfant, C. (1972). A comparative approach to the adaptability of O_2Hb affinity. *In* "the 4th Alfred Benson Symposium: O_2 Affinity of Hemoglobin and Red Cell Acid-base Status" (Eds P. Astrup and M. Rorth), Munksgaard, Copenhagen.

Johansen, K., Lenfant, C. and Grigg, G. C. (1966). Respiratory properties of blood and responses to diving of the platypus, *Ornythorhynchus anatinus* (Shaw). *Comp. Biochem. Physiol.* **18**, 597–608.

Lenfant, C. (1969). Physiological properties of blood of marine mammals. *In* "Biology of Marine Mammals" (Ed. H. T. Andersen), pp. 95–115, Academic Press, New York, London and San Francisco.

Lenfant, C., Johansen, K., Petersen, J. A. and Schmidt-Nielsen, K. (1970). Respiration in the freshwater turtle, *Chelys fimbriata*. *Resp. Physiol.* **8**, 261–275.

Licht, P. (1965). Effects of temperature on heart rates of lizards during rest and activity. *Physiol. Zool.* **38**, 129–137.

Murdaugh, H. V. and Jackson, J. E. (1962). Heart rate and blood lactic acid concentration during experimental diving of water snakes. *Am. J. Physiol.* **202**, 1163–1165.

Pough, H. (1973). Heart rate, breathing and voluntary diving of the elephant trunk snake, *Acrochordus javanicus*. *Comp. Biochem. Physiol.* **44A**, 183–189.

Seymour, R. S. (1974). Sea snakes: avoidance of N_2 bubbles following deep dives. *Nature. Lond.* **250**, 489–490.

Seymour, R. S. and Webster, M. E. D. (1975). Gas transport and blood acid-base balance in diving sea snakes. *J. exp. Zool.* **191,** 169–181.

Shuntov, V. P. (1971). *Ékologiya.* **4**, 65–72. Consultants Bureau English translation, 1972.

Standaert, T. and Johansen, K. (1974). Cutaneous gas exchange in snakes. *J. Comp. Physiol.* **89**, 313–320.

Tucker, V. A. (1967). Method for oxygen content and dissociation curves on microliter blood samples. *J. Appl. Physiol.* **23**, 410–414.

Voris, H. K. (1966). Fish eggs as the apparent sole food item for a genus of sea snake, *Emydocephalus* (Krefft). *Ecology*, **47**, 152–154.

White, F. N. and Ross, G. (1966). Circulatory changes during experimental diving in the turtle. *Am. J. Physiol.* **211**, 15–18.

Wood, S. C. and Johansen, K. (1974). Respiratory adaptations to diving in the Nile Monitor Lizard, *Varanus niloticus*. *J. Comp. Physiol.* **89**, 145–158.

Wood, S. C. and Moberly, W. R. (1970). The influence of temperature on the respiratory properties of iguana blood. *Resp. Physiol.* **10**, 20–29.

Epilogue

Comparative study of air-breathing fishes

G. M. HUGHES

This symposium is a landmark in the study of air-breathing fish and it is especially appropriate that it was held in India, at a Zoology Department and in conjunction with an International Congress of Physiology. Indian scientists have studied the structure and functional morphology of their particular air-breathing fishes for many years and in very great detail, but there have been relatively few studies involving physiological techniques. In recent years this has taken place to a greater extent and it is apparent from what was heard during the symposium that a tremendous amount remains to be done. The application of methods and approaches developed during the last 20 years as a result of investigations by scientists with backgrounds in mammalian physiology as well as those based in more classical aspects of zoology has been very fruitful, and it is clear that much more can be done with these fish. It is also especially appropriate that these fish should have been discussed in the wider context to which they belong, namely the group of vertebrate animals which show many diverse adaptations to the different conditions which Dejours has summarized for breathing both in air and in water. It is only by the adoption of such comparative approaches that ideas as to possible mechanisms and knowledge of the availability of suitable techniques can be encouraged. Of special value to those who have had little direct experience of Indian air-breathing fishes is the realization that they are not only of great academic interest, but that their study is also an important economic project to which the Indian Government gives much emphasis. It was extremely valuable to hear about this from someone with the authority of the Deputy Minister of Education.

Already several techniques successfully used with water-breathing fishes have begun to be applied to air-breathing forms. The particular technique which I, myself, first used in comparative studies with fish was to record pressure changes within the buccal and opercular cavities (Hughes, 1960). Figure 1 shows some recordings from two species of Indian air-breathing fishes. The pressure changes show waveforms similar to those known from

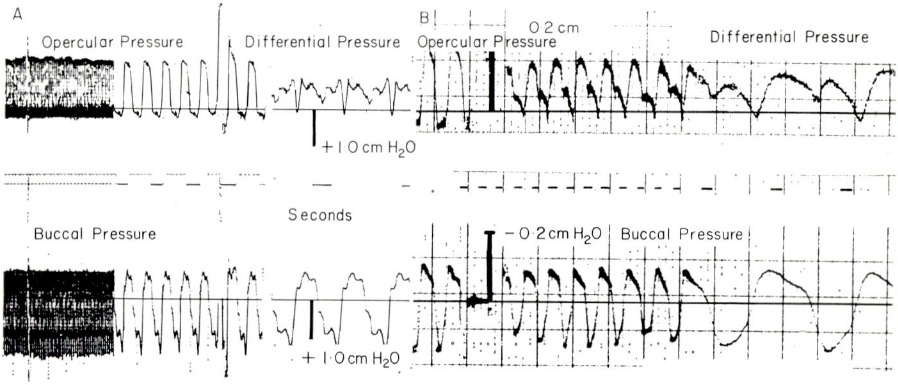

FIG. 1. Pressure recordings from the buccal and opercular cavities of (a) *Saccobranchus fossillis* and (b) *Anabas testudineus*. The cavities were cannulated and pressures recorded using two Sanborn 268 differential transducers. Notice the greater sensitivity in the recordings from *Anabas*. Pressure calibrations are given relative to a zero pressure which is the pressure in the surrounding water. In the recordings of buccal and opercular pressures positive pressures are downward.

other fish. The comparison between *Saccobranchus*, primarily a water-breather, and *Anabas* which depends far more on air breathing, is of interest. In *Saccobranchus* a regular ventilatory rhythm is interrupted by periodic coughs and the amplitude of the pressure changes is far greater than those recorded from *Anabas*. In the differential pressure waveform the buccal and opercular pumps seem to be fairly equally developed and a reversal phase was recorded. In the particular recordings shown in Fig. 1, the opercular pump is the more important and once again there is a reversal in the differential pressure. Under conditions when *Anabas* was hyperventilating, these pressure changes were greater in amplitude and it was frequently found that the buccal pump was the more dominant. It is clear that there must be a whole range of variations in the relative importance of these two pumps among air-breathing fishes. This technique could be applied with some interesting results to some other types of fish in India such as the sucker fish from torrential streams among which there are species in which water both enters and leaves via the opercular openings.

Some preliminary pressure recordings have also been made in *Anabas* following cannulation of the suprabranchial chambers; these give interesting information about the mechanism of ventilation during air breathing. A variety of different patterns have been recorded and these are especially significant in showing changes just before and after individual air-breaths. The actual mechanics of the ventilatory movements needs further elucidation as they are quite complicated as indicated by the cine-photographs shown in Fig. 2. It is clear that a combined study using pressure recordings and cine analysis would be very fruitful. One particular possibility is that water may enter the suprabranchial chambers during the ventilatory sequences, although this has not usually been considered to occur. Some X-ray films have given further indication that this might take place in some other anabantoids (H. M. Peters, personal communication).

Much work needs to be done with these fish and in spite of the considerable morphometric studies summarized by Munshi in the present symposium, it is not yet possible to carry out a sophisticated analysis of these gill systems similar to that described by Scheid and Piiper, for sampling of blood by means of indwelling cannulae has scarcely been possible in Indian species, partly because of their size, but such techniques are developing and soon some of the possibilities discussed by Satchell could be investigated. Studies of this kind would give valuable background techniques for studies on the effects of pollution of these fishes. We know that in India many pollution studies are being directed to both the water-breathing and the air-breathing

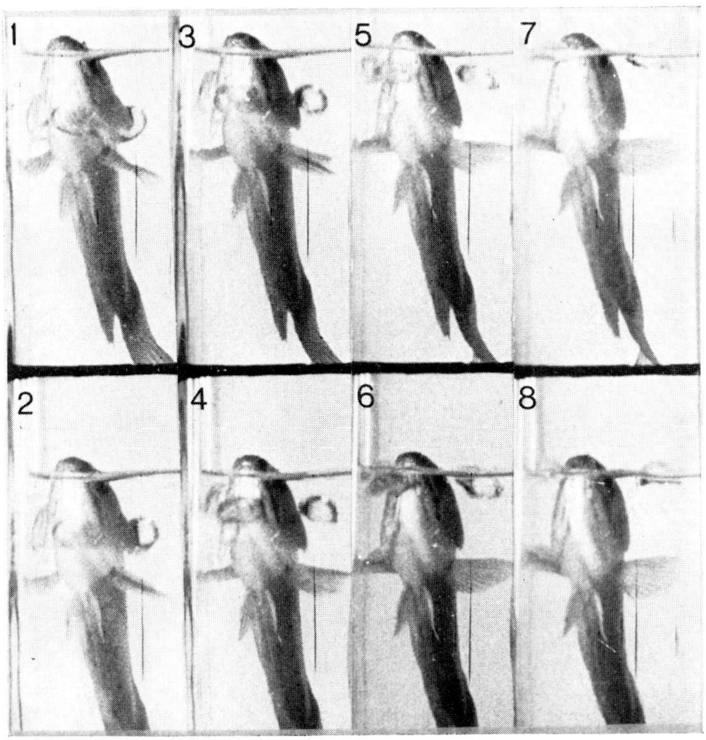

FIG. 2. Frames from a cinefilm (64/sec) taken as *Anabas* takes an air breath from the water surface. Notice the liberation of gas from the opercular slits before the fresh air is taken in at the mouth.

fish. Investigations of gill epithelia and blood characteristics have already given some interesting results (Munshi and Singh, 1971; Mahajan, 1975) which emphasize the value of respiratory parameters in assaying pollution (Hughes, 1976). The value of a morphometric approach has already been indicated in these fish and it is remarkable how changes in relative development of the different respiratory organs can be correlated with changes in life habits of the fish, which in turn are related to the ecology and especially the seasonal characteristics of the environment.

We have all become very much more aware of some of these problems and the paper of Dehadrai and Tripathi indicates how much needs to be done. Clearly the approaches and techniques used by Ultsch and by Guimond and Hutchison, in their work with Amphibia combined with those presented by B. N. Singh, could be of great help in these studies. Temperature changes within the Indian subcontinent are very marked and thus one can see that the aspects surveyed by Rahn and Howell are particularly relevant.

This symposium has also emphasized that air-breathing fish inhabit not only freshwater but also brackish and marine environments. The paper by Graham gives an admirable summary of the wide range of forms which show this air-breathing habit and thus draws attention to problems which need tackling by methods of classical zoology, such as the taxonomy of the mudskippers. Taxonomic problems are in some ways a "nuisance" to those more interested in animal physiology, but recent years have also shown many examples of how important it is to know not only the particular species being investigated but perhaps even the variety. The example quoted by Munshi (p. 76) for two varieties of *Anabas* is very relevant in this context. One variety has a much thicker water-blood barrier than the other; the distinction between thick and narrow-trunked varieties was not appreciated by professional zoologists but discussions with fishermen showed that they had recognized these two varieties for a long time.

There are many adaptations among air-breathing fishes which show functional convergence, but one of the main interests in this field is in the adaptations and homologies of the respiratory systems of lungfish. In this symposium it has not been possible to give them all the attention they deserve, but fortunately this is a branch of the subject which has been documented fairly well in recent years (Johansen, 1975). Yet there are still important areas well known in mammalian respiratory studies which can be applied to the lungs of lower vertebrates with illuminating results. Such topics as the ontogenetic and phylogenetic origins of pulmonary surfactant surveyed by Pattle, and the application of morphometric techniques indicated by Hughes and Weibel are but two examples of the many possibilities. The physiology of these lungs presents stimulating analysis for mammalian physiologists and Wood and Lenfant show how this involves many different systems such as the nervous, muscular and vascular systems. As a modified lung, the swimbladder must rank as one of the most studied organs of the fish body, and Fänge shows how its study continues to illuminate not only basic structural and functional problems but also problems of diving. It is in this context that attention has been given to "reversed" evolution, i.e. animals belonging to groups of predominantly terrestrial vertebrates, which have subsequently adopted an amphibious or aquatic mode of life. The description of ventilation and other adaptations of crocodilians and turtles covered by Gans is an excellent illustration of the value of electromyography, which is another technique that could usefully be applied to Indian fishes. Some recent work on seasnakes described by Heatwole and Seymour gives further evidence of other types of physiological adaptation among secondarily aquatic vertebrates and has led them to discuss important human problems such as "the bends".

We conclude then that the studies described in this volume are of general significance not only because they summarize the present state of work on air-breathing vertebrates, but also because they illustrate important biological principles, such as convergence, homology, analogy, and homeostasis. All of these serve to emphasis the value of an integrated comparative approach to problems of this kind.

References

Hughes, G. M. (1960). A comparative study of gill ventilation in marine teleosts. *J. exp. Biol.* **37**, 28–45.

Hughes, G. M. (1970). Polluted fish respiratory physiology. In "Effects of Pollution on Aquatic Organisms". S.E.B. seminar series. Cambridge University Press (in press).

Johansen, K. (1970). Air-breathing in fishes. *Fish Physiology*, Vol. IV, 361–411. Academic Press.

Mahajan, C. L. (1975). Unpublished observations.

Munshi, J. S. D. and Singh, R. K. (1971). Investigation on the effects of insecticides and other chemical substances on the respiratory epithelium of the predatory and weed fishes. *Indian J. Zootomy* **8**, No. 2. 127–134.

Subject Index

T

U

V

W

X

Y